Lecture Notes in Civil Engineering

Volume 75

Lecture Notes in Civil Engineering (LNCE) publishes the latest developments in Civil Engineering—quickly, informally and in top quality. Though original research reported in proceedings and post-proceedings represents the core of LNCE, edited volumes of exceptionally high quality and interest may also be considered for publication. Volumes published in LNCE embrace all aspects and subfields of, as well as new challenges in, Civil Engineering. Topics in the series include:

- Construction and Structural Mechanics
- Building Materials
- Concrete, Steel and Timber Structures
- Geotechnical Engineering
- Earthquake Engineering
- Coastal Engineering
- Ocean and Offshore Engineering; Ships and Floating Structures
- Hydraulics, Hydrology and Water Resources Engineering
- Environmental Engineering and Sustainability
- Structural Health and Monitoring
- Surveying and Geographical Information Systems
- Indoor Environments
- Transportation and Traffic
- Risk Analysis
- Safety and Security

To submit a proposal or request further information, please contact the appropriate Springer Editor:

- Mr. Pierpaolo Riva at pierpaolo.riva@springer.com (Europe and Americas);
- Ms. Swati Meherishi at swati.meherishi@springer.com (Asia—except China, and Australia, New Zealand);
- Dr. Mengchu Huang at mengchu.huang@springer.com (China).

All books in the series now indexed by Scopus and EI Compendex database!

More information about this series at http://www.springer.com/series/15087

Bibhuti Bhusan Das · Salim Barbhuiya ·
Rishi Gupta · Purnachandra Saha
Editors

Recent Developments in Sustainable Infrastructure

Select Proceedings of ICRDSI 2019

Springer

Editors
Bibhuti Bhusan Das
National Institute of Technology Karnataka
Mangalore, Karnataka, India

Salim Barbhuiya
School of Civil Engineering
Leeds, UK

Rishi Gupta
Department of Civil Engineering
University of Victoria
Viktoria, BC, Canada

Purnachandra Saha
School of Civil Engineering
KIIT Deemed University
Bhubaneswar, India

ISSN 2366-2557 ISSN 2366-2565 (electronic)
Lecture Notes in Civil Engineering
ISBN 978-981-15-4576-4 ISBN 978-981-15-4577-1 (eBook)
https://doi.org/10.1007/978-981-15-4577-1

This Springer imprint is published by the registered company Springer Nature Singapore Pte Ltd.
The registered company address is: 152 Beach Road, #21-01/04 Gateway East, Singapore 189721, Singapore

Preface

Sustainability is the key, and it needs to be practiced. Infrastructure which is a major drive for the growth of nation's economy is dependent on the sustainable practices. Concrete found to be a key material of infrastructure development. Over the years, it is found that an enormous amount of research is being carried out to produce sustainable concrete and construction materials. Even though there is a substantial amount of research being carried out on alternative materials which makes the concrete more green and sustainable, dissemination of knowledge is required to share the concepts and principles. Keeping all these in mind, an International Conference on Recent Developments in Sustainable Infrastructure (ICRDSI-2019) was organised during 11–13 July 2019 at School of Civil Engineering, KIIT Deemed to be University, Bhubaneswar. This book includes selected papers from this conference dealing with several topics, including sustainable materials, computational, analytical and numerical modelling, construction planning and management, corrosion of steel reinforcement and protection measures, life cycle assessment, green construction materials and technologies, smart cities, sustainability and geo-environmental engineering, structural health monitoring system, nanomaterials, and rainwater harvesting analysis and design.

I thank Springer Nature team, co-editors, reviewers, student volunteers, KIIT Management and Staff for their full support and cooperation at all the stages of this event. I do hope that this book will be beneficial to students, researchers and professionals working in the field of Recent Developments in the Sustainable Infrastructure (Materials and Management).

Surathkal, India

Dr. Bibhuti Bhusan Das

Contents

About the Editors

Dr. Bibhuti Bhusan Das is currently serving as Associate Professor at National Institute of Technology Karnataka, Surathkal. Before joining NITK, he has served as Centre Head for National Institute of Construction Management and Research, Goa and Indore Campus. He has worked as a Post-Doctoral Research Associate and Adjunct Faculty in the Department of Civil Engineering at Lawrence Technological University, Southfield, Michigan, USA. His area of research focuses on the "Sustainability of Construction and Building Materials" such as microstructure characterization of materials, non-destructive testing of concrete structures, corrosion of reinforcement and durability studies on concrete and sustainability in construction project management. He is an Associate Member of American Concrete Institute and representing in the International Faculty Network Committee launched by the ACI Education Foundation in 2009.

Dr. Salim Barbhuiya is a Lecturer in Advanced Concrete Technology at the University of Leeds. He joined the University of Leeds in Sep 2018 after working as a Lecturer in Civil Engineering for 5 years at Curtin University Australia. Recognizing his outstanding contribution in advancing Curtin's international reputation, public visibility and diversified resources he was awarded the PVC Award for Internationalization in 2015. Prior to that Dr. Barbhuiya worked as a Post-Doctoral Research Fellow in Queen's University Belfast (2009-10), University of the West of Scotland (2010-11) and Loughborough University (2011-12). He completed his Bachelor's in Civil Engineering from NIT Silchar, India (1999), Master's in Structural Engineering from AIT, Thailand (2005) and obtained his PhD on Material Technology for Sustainable Concretes from Queen's University Belfast (2009). Dr Barbhuiya's main area of research currently is on cement and concrete science with a key emphasis on the use of nanotechnology to characterize and improve their physical and durability properties. In particular, his research work includes studying the mechanisms of concrete deterioration at nanoscale level and relates this to the micro and macro level for engineering applications. His research interests also include improving the engineering and durability properties of geopolymer concrete leading to the development of Standard Specifications.

Dr. Rishi Gupta is an Associate Professor and Director of undergraduate program in the Department of Civil Engineering at the University of Victoria. He received both his Master's and PhD in Civil Engineering (Materials) from the University of British Columbia. His current research is focused on studying the early-age behavior of cement-based composites containing SCMs and fibers. His areas of interest include development of sustainable construction technologies, masonry structures, structural health monitoring, and non-destructive testing. He has more than 15 years of combined academic and industry experience. His industry experience includes working as the Director of Research of Octaform Systems Inc in Vancouver. Dr. Gupta is a Fellow of Engineers Canada and the past chair of the EGBC's Burnaby/New West branch. He is the Chair of the international affairs committee of the Canadian Society of Civil Engineering (CSCE). He is a long standing member of the American Concrete Institute and is also a voting member of several subcommittees of ASTM C 09.

Dr. Purnachandra Saha received his Ph.D. from IIT Bombay and has 10 years of teaching experience. He is currently working as a senior associate professor in the school of civil engineering, KIIT Deemed to be University, Bhubaneswar, Odisha, India. His areas of research are earthquake engineering, structural control, sustainable materials, green building and he has published 82 research articles to date. He has also worked in a multinational company (mainly design of industrial structure before completing his doctoral degree) for about eight years. He is a member of ASCE and The Faculty Advisor of ASCE KIIT University Student Chapter. He is also a Life member of Institute of Engineers (India), Member of National Information Centre of Earthquake Engineering, E-Affiliate member of Earthquake Engineering Research Institute (EERI), and Member of American Concrete Institute (ACI). He is an editorial board member and a reviewer for a number of national and international journals.

Recent Developments in Geotechnical Engineering

Different Techniques Used for Well Foundation Construction Focused on Pneumatic Cassion Technique: A Review

Rudrani Das, Amit Ganguly, and Purnachandra Saha

Abstract Well foundation is a kind of deep foundation that is conventionally allocated beneath the water plane for bridges. Different techniques that are used for the construction of well foundation and well sinking are horizontal drilling by divers, grabbing method, jet air method, jet water method, jack-down technique, vertical cutting edge method, pneumatic caisson technique, etc. The objective of the present study is to review different types of well foundation construction techniques with a case study of pneumatic caisson technique. Pneumatic caisson techniques are mainly used in hilly and rocky areas. Pneumatic caisson is a kind of caisson which is shut at the top and revealed at the bottom when in construction. The water is ceased from the caisson cell or chamber by dint of consolidated air. Pneumatic caisson mainly holds control over the work and preparation of the foundation in the rocky area for the sinking of caisson. Only with an absolute attentive observation, the caisson can be imperiled in a supine position. The bottom of the cell should be closed adequately with concrete as it may have chances to be placed on dry bottom. The excavation takes place, under the very high pressure and worker can go inside for inspection of work, and some kind of health hazard may occur for this method. There are many advantages as well as disadvantages for all the above techniques used for well foundation.

Keywords Well foundation · Air jet · Water jet · Jack down · Pneumatic caisson

R. Das · P. Saha (✉)
School of Civil Engineering, KIIT, Bhubaneshwar, Odhisa, India
e-mail: dr.purnasaha@gmail.com; purnasahafce@kiit.ac.in

R. Das
e-mail: rudranids@gmail.com

A. Ganguly
Simplex Infrastructure Ltd, Kolkata, India
e-mail: amit.ganguly00@yahoo.com

B. B. Das et al. (eds.), *Recent Developments in Sustainable Infrastructure*, Lecture Notes in Civil Engineering 75, https://doi.org/10.1007/978-981-15-4577-1_1

1 Introduction

The wells or as we call it the "caissons" basically dole out a huge section modulus with the modest transverse surface, and hence are competent in taking large vertical and parallel loads even when the unsupported length is immense. They distribute a firm and hefty foundation for overweight loads to deep strata at river or seabed [1]. The existence of wells previously and during 1863 was exceedingly crude. They seemed to have imperiled by people who were completely unskilled of even the elementary principles of putting down the vertical shaft. A lot of methods came into existence since then. In the diving method, the divers went underwater and abdicated the big rocks for making the foundation by the use of drilling machines. In the grabbing method, big pick-up machines like cranes were introduced, and they were used for removing the big rock strata from the ground surface since it was very difficult for humans to go underwater and do the drilling [2]. In the jet air method, the air was bloated from combats through the pipes. As a result, the air barges through the soil mass and the concrete shoal degrading the frictional resistance on the shoal surrounding the boundary [3]. In the jet water method, the hard soil was excavated from the caisson to eliminate the soil around the curb and cutting edge portion [4]. The very next method that was introduced after this was the jack-down method. Soil crutches were alternatively made at the prognosticated locations according to the requirement along the boundary, and the load was capable of being put to use by jacks through fabricated steel girders that are positioned on the peak of staining [3, 5]. In spite of the variable techniques that were used, the well foundation technique still faced many problems especially in the rocky hilly areas where the current of the river was very strong, that is, when the pneumatic caisson technique came into the limelight [6–8]. Lots of literature are available on individual well foundation construction technique. Hence, there is a necessity to organize these data in one place and compare between the techniques. In this context, the main factual reason behind this present study is to review different well foundation construction techniques with a case study of pneumatic caisson technique.

2 Different Techniques Used for Well Foundation

2.1 Jet Air Method

Air jetting is a normal method for minimizing the friction of soil at the outer boundary of the well for enhancing the sinking. Air jetting system was adopted in a very different and effective manner so as to work around the caisson periphery at the same time. It also provided scope for extension and depletion of area in befitting the site fulfillment. When consolidated, air was bloated from combats through the pipes where the air barges through the soil mass and the concrete shoal degrading the frictional resistance on the shoal surrounding the boundary. The fervency of jetting air could be kept on hold by maneuvering single or multiple rows of pipes. Generally, four parallel rows on the inferior half of the caisson are enough to loosen the soil around and above the ambit portion that gives maximum resistance of soil at the time of sinking [3]. The major advantages of jet air method are that it yields smoothly on cohesive soils. Without any adversity, this particular technique can be adopted on all site construction. Cost is minimal, and it does not involve any special effort. The major disadvantages are that air leakage is more and it is not suitable for higher capacities.

2.2 *Jet Water Method*

The jet water technique is adopted for excavating the hard soil from the caisson to eliminate the soil around the curb and cutting edge portion. When digging the soil surrounding, the cutting edge and ambit cannot be eliminated specifically for cohesive soils and hard strata. As a result, the ambit and the cutting edge get freed allowing the well to sink. The liquid jets producing high velocity are adopted in multiple construction methods such as jet grouting, sand draining, and jet piling [9]. The jet water technique yields smoothly on clayey or hard strata. These construction methods are effectively used because they are high in digging efficiency, provide high energy-density rates, and require but a relatively small and simple device for producing high-velocity liquid jet. It is usually in the strata with subterranean water or at the seafloor that high-velocity liquid jets are used for piercing and crushing. The water encountered in such locations rapidly decelerates the liquid jet velocity and minimizes the working adequacy of the jet [3]. The advantages of water jet method are that it is a versatile and cost-effective cutting process. Minimal fixturing is required. It perforates most materials without starting holes. It does not have any heat-affected zone. The major disadvantage of water jet method is that the well may have a chance of tilting. The mechanism enforcing this particular hypothesis has successfully increased the distance traveled by high-velocity liquid jets in water. Even though there was such a high-velocity jet liquid, an impairment appeared in which there were fluctuations in the distance traveled by jet liquid which gave rise to irregular casualties in watertight wall construction.

2.3 *Jack-Down Technique*

Jack-down technique on a very general level means driving the caisson into the bed surface by appertaining the load from the peak of the steining by the help of jacks from soil crutches [10]. Soil crutches were alternatively made at the prognosticated locations according to the requirement along the boundary and the load was capable of being put to use by jacks through fabricated steel girders that are positioned on the peak of steining. The loads that are applicable to the jacks depend on the size, shape, and depth of caisson and the subsoil strata. The jacks are operated discretely or comprehensively, and load on each jack can also be assorted to control the tilts or slope of caisson during sinking. The soil crutches are fundamentally friction piles and are drafted to take reaction of jacks. However, friction is contemplated only in the length of anchors underneath final founding level of caisson. Soil friction circling caisson is decreased by air jetting and soil beneath cutting edge and curb is excavated by water jetting as and when necessary. System is very tidy, swift, and efficacious, and sinks the caisson in true vertical position with administered and guided tilts, shifts, and rotation [3]. The supremacy of jack-down method lies in elementary mechanization of sinking process and can be adopted on projects, especially with higher profundity of foundations and huge size of caisson in all types of strata. With mechanized system and controlled operations of sinking, speed of construction and tilts and shifts of caisson are productively controlled which are the two major factors in construction of caisson foundations. Provisions in Indian codes for steining thickness are mostly based on the supposition that thickness of steining should be adequate for self-sinking of caisson to lower sinking efforts and to avoid extravagant loading during sinking. Therefore, substantial reduction in concrete and thereby in cost can be made by lowering thickness of steining with jack-down method of sinking. The rapid completion of the project due to swift advancement of sinking and construction of foundation shall significantly affect the benefit–cost ratio. Economics of jack-down method is of course a point for study but it is only the initial investment in acquisition of jack-down equipment, while the operational cost is very emblematic. Though the soil crutches are non-recoverable, then overall effect on the total cost of the project by using these mechanized methods after economizing the designs and with greater speed of construction shall be very nominal. Jack-down method is more useful and efficient in controlling the tilts and shifts in sinking of caissons and could be tried on crucial projects [5]. The major disadvantage is that it is a tough process and the labor cost is high.

2.4 *Pneumatic Caisson Technique with a Case Study*

Pneumatic caisson technique is generally done in the hilly and rocky areas. Places like Arunachal Pradesh, Himachal Pradesh, and parts of North Bengal are famous for pneumatic caisson technique [3]. When the depth of the water is very high and there is huge current in the water, the divers or the grabbing methods are not able to abdicate the big rocks for making the foundation, and that time the pneumatic caisson technique was introduced. The pneumatic caisson technique consists of on-ground construction of a reinforced concrete caisson having a working chamber for the workers to work freely as shown in Figs. 1, 2, and 3.

In this case study, the location of the bridge was in Arunachal Pradesh in River Dalai which is connecting Brahmaputra river. That particular location of the bridge pier was totally on rocky area and that is why pneumatic caisson technique was used there.

Plan of the well -

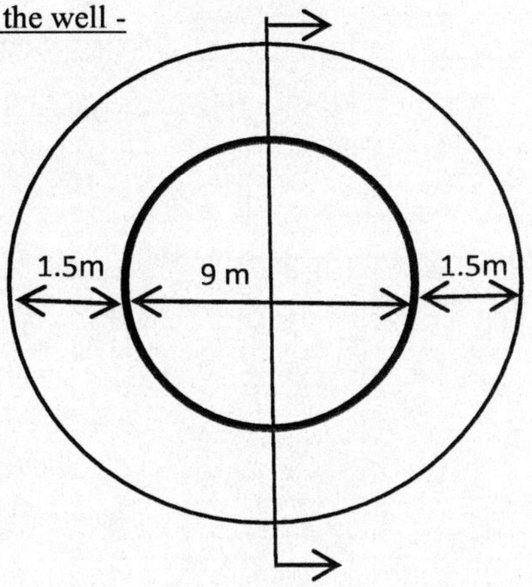

Fig. 1 Plan of the well

Section of the well with airlock and muck chamber -

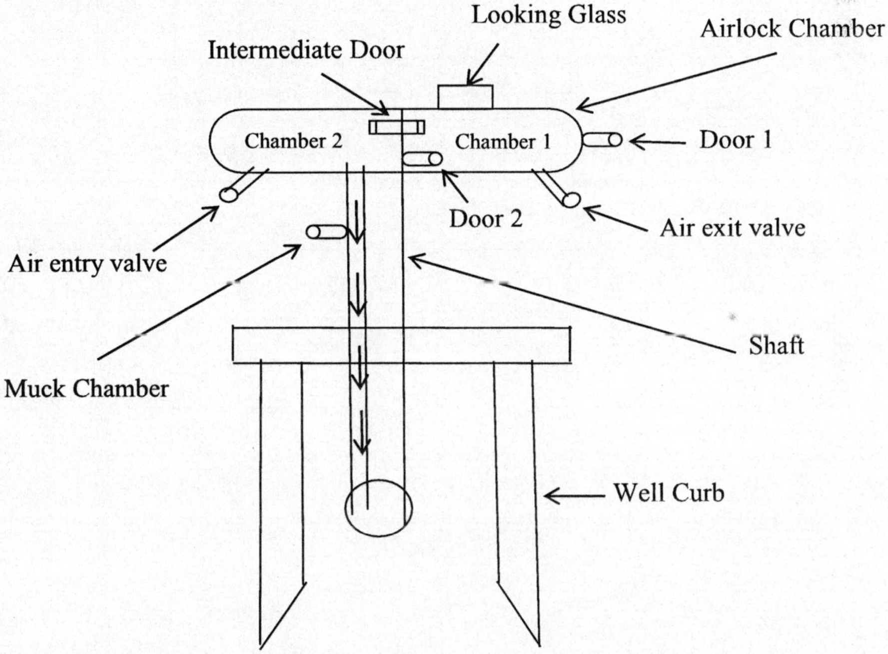

Fig. 2 Section of the well with airlock and muck chamber

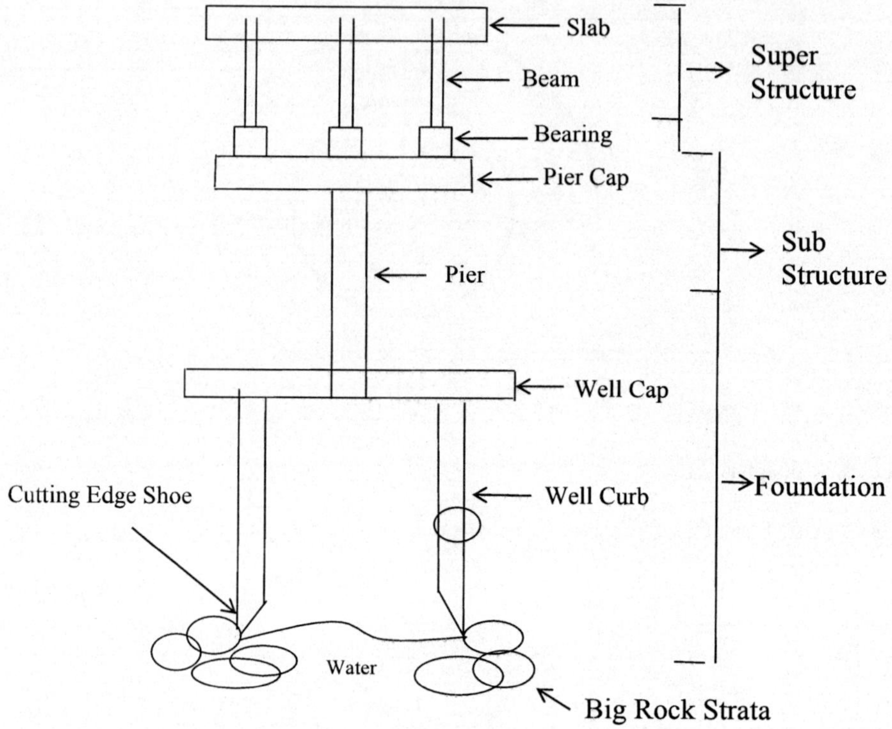

Fig. 3 Schematic diagram of the well foundation

The flowchart of activities that are typically involved in the construction and sinking of pneumatic caisson technique for bridge construction has been illustrated below:

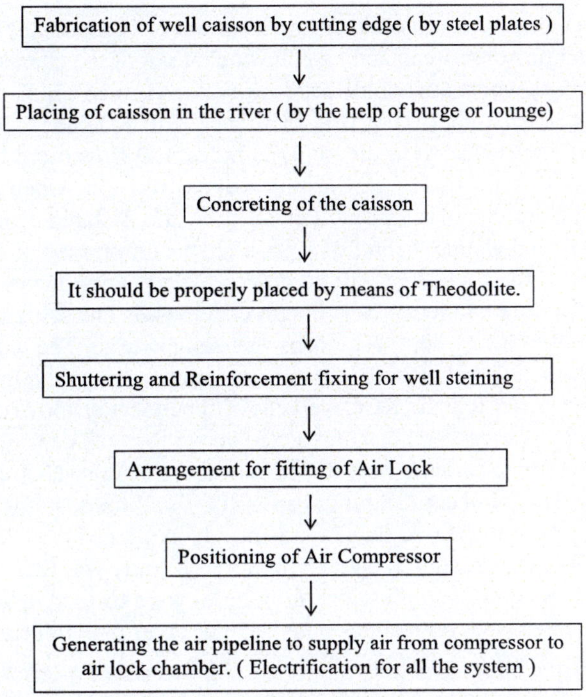

The outer diameter of the well was 12 m and inner diameter was 9 m, and the steining width of the well was 1.5 m as shown in Fig. 1 and through the inside diameter of the well, the working shaft fitted with airlock chamber had been temporarily fitted (Fig. 2).

The workers enter the airlock chamber 1 through door 1 because the air pressure in chamber 1 is normal pressure and is equal to the nature's pressure but in chamber 2, the air pressure is very high. So there is a door 2 situated in between the two chambers which are closed due to the high air pressure. After entering into chamber 1, the workers sit there for sometime. There is also a looking glass on chamber 1 so that the other workers who are outside can help by keeping a watch if any of the workers who are inside fall sick. After sometime, the intermediate valve connecting both the chambers is opened so that the high pressure in chamber 2 is passed to chamber 1 through the valve. The pressure generated from chamber 2 comes in a very slow manner to chamber 1 because the workers won't be able to take the high pressure suddenly. After sometime, the pressures in both the chambers become equal. Only when both the pressures become equal, the door 2 opens. Now, the workers from chamber 1 enter the chamber 2 and go down through a staircase fitted with a temporary working shaft (made of steel). Now, after the workers have come down, they will start the manual excavation by use of jackhammer, spade, shovel, etc.

Again, by the same air pressure method with intermediate door connecting in chamber 2, there is also an arrangement to remove the muck by means of muck bucket and electrical winch. The workers who are working down fill the bucket with muck and pieces of boulders or stones and send the bucket upward to the workers who are standing up so they can abdicate the bucket out of the chamber. Now, there will be a lot of cases where the boulders are really huge in size which can't be broken manually or by jackhammers. For this case, jack drill machine is introduced for creating holes in the rocks. After creating the holes, ED (Electrical Detonator) and gelatin are used for controlled blast. Blasting will only be done when all the workers will come out of the chamber and will stand at a safe distance from the blasting area and also the well chamber should be properly decompressed of air. As a result, the well penetrates the rocks and will go down. After it goes down, again steining with RCC will be done to increase the height of the well. The same method will be followed unless the cutting edge penetrates the design strata. The well will now be in a stable condition. Then bottom plugging and intermediate plugging will be done for more stability. After that, an RCC well cap will be located above the well. Then, an RCC pier and a pier cap will be constructed above the well cap so as to carry out the load from the superstructure (i.e., the bridge slab) through substructure (i.e., the pier, pier cap, and the well cap) and foundation (i.e., well foundation) to the ground.

A clear diagram of the above statement is briefly explained (see Fig. 3).

The main difficulty which was noticed in this case study was that sometimes due to rapid decompression or compression, air bubbles were formed inside the body of the workers which caused tremendous body ache or other fatal injuries. The remedy provided to this particular difficulty was that there was a provision of a medical chamber where sick workers were taken care and by means of air pressure, the air bubbles from the body were removed. Also, regular medical test was carried out for each of the workers. After coming out of the chamber, each of the workmen was provided with war milk. Majorly, for the safety and welfare of workmen, the following precautions should be taken, i.e., accurate control of pressure, sufficient air circulation, slow decompression and duplicate and spare equipments (Teng). Apart from the safety and welfare of workmen, the main advantage that the pneumatic caisson technique hold is allocating dry working conditions that is good for locating concrete. This technique is suitable for heavy bridges. This technique is also well suited for foundations for which other methods might cause settlement of adjacent structures. The major disadvantages of pneumatic caisson technique are that the construction of pneumatic caissons is much expensive than open caissons. During working on the various constructional activities, a proper care has to be taken; otherwise, it may lead to fatal accidents and lastly, the labor cost is very high for this technique.

3 Discussion and Conclusion

In this study, diving method, grabbing method, jet air method, jet water method, jack-down method, and pneumatic caisson technique have been presented. The diving method is the simplest among this but due to shortage of workers and high life risk, this method is not applicable everywhere. The biggest advantage of grabbing method is that the whole process is done mechanically by the use of cranes but even then it had a disadvantage. Sometimes, the crane could not hold the big strata rocks because the rocks were totally wet and stuck to the ground. To prevent this difficulty, air jet method was used. Air jet method is an organized and effectual turnpike for minimizing the friction of soil surrounding caisson. Air jet technique works well in cohesive soil but in this method air leakages were more. This method was not suitable for working in places which had high pressure capacity. Because of this major drawback of air jet method, water jet method was introduced. The jet water technique is adopted for excavating the hard soil from the caisson to eliminate the soil around the curb and cutting edge portion. Even with such a high-velocity liquid jet, an impairment had been noted, i.e., there were fluctuations in the distance traversed by the liquid jet which lead to tilting of the well steining. Soil crutches were alternatively made at the prognosticated locations according to the requirement along the boundary, and the load was capable of being put to use by jacks through fabricated steel girders that were positioned on the peak of steining. Hence, the caisson shifting and tilting and also sinking operations were adequately kept in control in a mechanized way by the jack-down method. This particular method was a very effective method but this particular method was not really suitable for hilly and rocky areas having high depth of water and huge water current. That was when pneumatic caisson technique was introduced. In this method, the water was excluded from the caisson chamber by means of compressed air as a result of which the well became totally dry. The workers were able to work freely because of the dry condition. Because of the airlock chamber, the workers were also able to adapt and work in high pressure capacity. If the workers were not able to break the hard rock strata, ED and gelatin were used for controlled blasting. This technique is suitable for heavy bridges and is also well suited for foundations for which other methods might cause settlement of adjacent structures. The pneumatic caisson technique was a very effective technique for well foundation construction but it had some disadvantages too. The labor cost is very high for this technique, and since the workers have to work in high pressure capacity, sometimes they can have serious health hazards if they don't take proper precautions.

References

1. Hilal (2014) Design and construction of well foundations
2. Collins Engineers, Inc (2010) 123 North Wacker Drive, Suite 300 Chicago, Illinois 60606 (2010). Underwater bridge inspection. U.S. Department of Transportation Federal Highway Administration, FHWA-NHI-10-027
3. Kapila KK, Rastogi SP (1999) Special method of well sinking adopted at new Nizamuddin Bridge on National Highway-24 in New Delhi. In: Colloquium, foundations for major bridges: design and construction; New Delhi in IABSE reports, vol 80, pp 277–284
4. Chowdhury I, Tarafdar R, Ghosh A, Dasgupta SP (2017) Dynamic soil structure interaction of bridge piers supported on well foundation. Soil Dyn Earthq Eng 97:251–265
5. Sheriff RE (2015) Glossary of terms used in well logging. Soc Explor Geophys 35(6), December, 1970, 130.194.20.173
6. Yea GG, Kim TH (2012) Vertical cutting edge forces measured during the sinking of pneumatic caisson. Mar Georesour Geotechnol 30(2):103–121

7. Butler WP (2004) Caisson disease during the construction of the Eads and Brooklyn Bridges: a review. Undersea Hyperb Med 31(4):445–59. PMID 15686275
8. Aichel OG (1916) The caisson as a new element in concrete dam construction; a proposal made in connection with the Columbia River Power Project. Spon & Chamberlain, New York
9. Yahiro T, Yoshida H (1977) High-velocity jet digging method. U.S. Patent Documents, Continuation of Ser. No. 523,647, Nov. 13, 1974
10. Yoshida H, Yamada M, Hayashi A (2017) Wider application of Jack-down construction method. Frontier Service Development Laboratory, Research and Development Center of JR East Group Tokyo Construction Office (Previously at Frontier Service Development Laboratory, 359944834, vol 82, pp 250–298

Strengthening of Soil Subgrade Using Bio-Enzyme

T. Shil, R. Pradhan, Satyajeet Nanda, and Benu Gopal Mohapatra

Abstract Soil stabilization is frequently used in construction. In recent years, bio-enzyme base soil stabilization has been promoted by various enzyme manufacturers. These manufacturers are claiming the high increase in soil shear strength, reduction in soil liquid limits, and swelling properties. In this paper, a review as well as experimental study was carried out on the use of bio-enzyme in soil stabilization. Various studies suggest the use of bio-enzyme and indeed increase the soil shear strength properties up to 35% for clay. The strength gaining of bio-enzyme treated soil is very slow. Few other studies suggest that bio-enzyme not increases enough shear strength and it may not be useable to all kind of clay minerals.

Keywords Bio-enzyme · UCS · Soil stabilization

1 Introduction

In the construction industry, a cost-effective and environmentally friendly construction practice is an important challenge for engineers. Soil stabilization is a method of improving the strength and durability of soils. Soil stabilization has been frequently employed in the construction of roads and airfield pavement. In chemical soil stabilization, chemical additives to the soil alter the geotechnical properties of soil by physically combining with soil particles. Soils stabilized with lime and cement show a large increase in the peak strength and stiffness; however, this strength increment comes with shrinkage and cracks in the stabilized soil and sometimes possesses

T. Shil (✉) · R. Pradhan · S. Nanda · B. G. Mohapatra
School of Civil Engineering, KIIT DU, Bhubaneswar, India
e-mail: ts199408@gmail.com

R. Pradhan
e-mail: ramapradhan68@gmail.com

S. Nanda
e-mail: satyajeet.nandafce@kiit.ac.in

B. G. Mohapatra
e-mail: dean.civil@kiit.ac.in

B. B. Das et al. (eds.), *Recent Developments in Sustainable Infrastructure*, Lecture Notes in Civil Engineering 75, https://doi.org/10.1007/978-981-15-4577-1_2

unfavorable resilience modulus. In recent years, many different kinds of bio-enzyme-based soil stabilizer have been introduced by various private agencies. Conflicting reports on its strength gaining and enhancement of other soil properties have been reported in the literature. In general, the bio-enzyme is noncorrosive, nontoxic, and inflammable liquid which extracted from the vegetable and fruit extract, which can be easily mixed with the water and subsequently with the soil. The manufactures of bio-enzyme claim, soil, stabilized with bio-enzyme improves the strength of soil, increases the life of structures, increases the CBR value, and reduces the consistency limits. Bio-enzyme treated soil produces more stable and permanent structures, which can resist wear and tear, infiltration, and weathering of soil particles. Contrary to the above, studies like Khan and Taha [6] and Rauch et al. (1993) found no significant enhancement in soil properties by stabilizing the soil with the bio-enzyme. This paper reviews the change in soils' index and engineering properties of bio-enzyme treated stabilized soils. The comparison is made between the conventional soil stabilizers (like cement) with the bio-enzyme.

2 Soil Stabilization

Soil stabilization can basically a protein which ways like bring particles more closer to each other, by changing the composition of soil grains, facilitating the soil particles to devolve strong bond with each other. The soil stabilization can be categorized into two parts: Mechanical stabilizer and stabilization by mixing various admixtures. Based on the type of admixtures, the soil stabilization can be named in many different ways as lime stabilization, cement stabilization, fly ash stabilization, rice husk ash stabilization, bituminous stabilization, thermal stabilization, electrical stabilization, stabilization by geo-textile and fabrics, recycled and waste products, and thermal stabilization. The selection of the method depends on the modification of properties sought in shear strength, compatibility, and permeability of the soil and on the depth and thickness of compressible layer, type of structure, allowable, total and differential settlements, available time and equipment, local experience, environmental factors, and their costs. Because of the large variety of soils, no one method is ever successful in more than a limited number of soils. Many of the times, mechanical stabilization does not induce enough strength enhancements; in these scenarios, chemical soil stabilization may be taken into consideration. Chemical stabilization is an effective method for soil stabilization for a wide range of soils; however, it comes with disadvantages like swelling, shrinkage, cracks, and difficulty to mix admixture with the soil. Table 1 shows the disadvantage of various stabilization methods commonly used in practice.

Fine-grained soil with a high percentage of clay content possesses low shear strength and high compressibility. The nature and type of bonding between soil particles and water molecules influence much of the clay behavior. The soil contented a high amount of Smectite group of clay mineral (montmorillonite) possesses small particle size (10A0 to 10 μ), high cation exchange capacity (80 to 100 meq/100 g),

Table 1 Disadvantages of different stabilization methods

Method of stabilization	Disadvantage
Addition or removal of soil particles	In the case of longest roadway, replacement of soil is not adequate
Organic chemical	Eco-friendly, low cost with higher bearing capacity
Lime	Dust problem, slurry application
Cement	Uneconomical as requirement of cement quantity is higher
Membrane stabilization	Costly in nature as well as careful laying needs to be done

large specific surface area (50 to 120 m2/g), and very common in occurrence (Avsar et al. 2009; Fityus and Buzzi 2009; Nowamooz and Masrouri 2008; Huang and Wu 2007; Sabtan 2005; Mitchell 2003). The above characteristics of Smectite group of clay mineral facilitate the clay particles to hold a very high amount of water molecule which subsequently induces a high amount of swelling in the soil. Swelling in soil can produce very detrimental effects on the geotechnical structures like shallow foundation, road subgrade, and railways (Hasan et al. 2016). A significant amount of clay content in soil produces compressibility or volume change. Smectite group of clay mineral shows high compressibility characteristics. High volume change in soil mass can produce large differential settlement which may lead to the extensive damage to the structures (Assadi and Shahaboddin 2009; Avsar et al. 2009; Chen et al. 2007; Ferber et al. 2009; Huang and Wu 2007). Soil stabilization in fine-grained soil must address the issue of volume change, low shear strength, and swelling. Usually, chemical stabilization and preloading with a quick drainage system employed to stabilize the fine-grained soil. Table 2 shows the results of Trivedi et al. [4] and Haralambos [9] where fine-grained soil was stabilized. About 20 to 30% in UCS and 5% in CBR value were increased after the stabilization.

Loose granular soil sometime needs stabilization. Both mechanical and chemical stabilizations are employed for stabilization of loose granular soil. The main aim of this stabilization is to increase the shear strength properties of soil. Table 2 shows the results of Phummiphan et al. [5] where shear strength from the UCS test increased by 128%. The state of stabilizing coarse-grained soil is unsaturated; therefore, studying these materials in terms of unsaturated soil mechanics is now getting momentum (Wheeler and Sivakumar 1995).

Table 2 Percentage change in soil properties with admixture

Author name	Type of soil	Methodology	Dosage (%)	Index properties of treated soil	Strength properties of treated soil
Jyoti S. Trivedi et al. [4]	Fine-grained organic soil (OH)	Using Fly Ash in subgrade stabilization.	10	1.9% (LL), 7.89% (OMC), –2.13% (MDD)	4.87% (CBR)
			20	5.04% (LL), 3.86% (OMC), –2.13% (MDD)	4.93% (CBR)
Phummiphan et al. (2015)	Lateritic Soil	Using high calcium fly ash		–	UCS (after 7 to 90 days of curing) + 128.8%
Haralambos [9]	CL, ML, SM, GP-GM, GC	Portland cement	3	–	USC increased from 21 to 100%, increases more in coarse-grained soil
			7	–	USC increased from 21 to 51%

3 Bio-Enzyme

The enzyme is basically a protein which acts as a catalyst that speeds up a chemical reaction without becoming part of the end product. The bio-enzyme being used in commercial applications in the agriculture industry and was found to have stabilization qualities in certain soil types. Bio-enzyme is nontoxic, nonflammable, noncorrosive liquid enzyme formulation fermented from vegetable extracts that improve the engineering properties of the soil. Enzymes cause the water typically held on the exterior of a clay particle to be shed and in turn blanket the clay particle causing the particle to become neutral in charge, which pulls the particle closer together. This not only reduces the water that can be absorbed in the clay structure but it strengthens the soil and bonds the structures together. Enzymes are environmentally friendly, and organic products that when used properly can provide great soil stabilization qualities for road construction, roadbeds, airstrips, storage yards, access roads, and military applications. The bio-enzyme is applied to the soil in the liquid form.

The surface of clay particles usually possesses negatively charge. The dipolar water molecule attracts toward the negatively charged clay surface and produces water double layer around the clay particles. The bigger the double layer, the more would be the swelling in the soil. The bio-enzyme replaces the absorbed double-layer water with the organic cations; consequently, the thickness of double layer reduced and soil tendency to swelling reduced extensively. Furthermore, the presence of organic cations reduces the ability of clay particle to react with water, which makes the soil particles more stable and intact the mechanical compaction effort applied to the clay particle during the construction.

The effect of bio-enzyme on the clay particle is categorized as follows.

Clay water effect	Enzyme reduces the size of double layer of clayey, thereby reducing swells in clayey in the soil.
Cation exchange	Thickness of double layer is reduced permanently, which reduces the water absorption by clay, which in turn reduces plasticity in clay and increases the molecular structure of clay.
Enzyme stabilization effects	Organic cations have large flat structures approaching the size of clay particles, which then neutralizes the negative charge of clay by covering it.
Other effects	It facilitates the removal of pore water leading to higher densities and reduced void percentages, leading to a realignment of particles, thereby reducing the permeability of the soil.

Table 3: Summarized results of various studies on bio-enzyme treated soil. Various studies and manufactures of bio-enzyme for soil stabilization have claimed the following enhancement and economic benefit:

1. Cuts cost by 10–20%.
2. Higher CBR value/higher road strength: TerraZyme base structures have a much higher CBR value.
3. Lower the maintenance cost by 30–50%.
4. Save construction time by 50%.
5. Pavement thickness is reduced by 5–10% in the phase of trials.
6. Environment-friendly and bio-degradable product.
7. The product has been accredited by IRC.

Table 3 Percentage change in properties of bio-enzyme treated soil

Author name	Type of soil	Dosage	Change in index properties of treated soil in percentage	Change in strength properties of treated soil in percentage
Agarwal and Kaur [8]	Black cotton soil.	0.25 ml/5 kg		20.27% (UCS)
		4 ml/5 kg		6.25% (UCS)
Shankar et al. [1]	Lateritic soil	0.029 ml/kg	– 4.33% (LL), –0.4% (PL)	35.29%(CBR),118.05% (UCS)
		0.050 ml/kg	– 6.67% (LL), –4% (PL)	24% (CBR), 84.71%(UCS)
Saini and Vaishnava [12]	Clayey silt	3.0 m^3/200 ml	– 10.71% (LL), –8.82% (PL), 3.65% (MDD)	3.65% (CBR), 15.87% (CBR)
		1.5 m^3/200 ml	– 21.15% (LL),–4.68% (PL), 15.8% (MDD)	
Nair and Joseph (2016)	CH-MH	200 ml/3 m^3		6.68% (CBR), 36.85% (UCS)
		4 ml/5 kg		6.73% (CBR), 166.58% (UCS)

4 Experimental Investigation

(a) General

In this present study, an attempt was made to present bio-enzyme an eco-soil stabilizer mixed with soil to increase its index and strength properties. The specific objectives of the present study are as below:

- To carry out literature review for understanding soil stabilized with bio-enzyme and in particular to improve the soil subgrade.
- To determine the optimum dosage of bio-enzyme based on index and strength parameters using laboratory experiments.

Soil used in this experiment was collected from Nayagarh, Odisha (20.1654 °N, 85.0233 °E). Soil was dark black in color. Soil was transported to Laboratory of School of Civil Engineering, KIIT University and kept in for drying. Sieve analysis was conducted for the natural soil using wet method of sieving. It is found that the soil has a fine content (more than 60%). The physical properties of the soil tests are conducted as per relevant IS codes as shown in Table 4.

(b) Sample methodology

In this experiment (UCS), water was accurately measured with measuring jar, which is obtained from optimum moisture content of natural soil tested earlier and thoroughly mixed with calculated soil mass and kept the mixed soil in a tightly wrapped polythene bag for 24 h, and then sample specimen was molded. In case of dilution of bio-enzyme, it is mixed with distilled water with a hypodermic needle in a plastic bottle; then required amount of mix was taken out to mix with soil mass; and kept the mixed bio-enzyme mixed soil in a tightly wrapped polythene bag for 24 h, and then sample specimen was molded. For natural soil, initially volume of UCS specimen

Table 4 Summary of index properties of natural soil

Sl.no.	Physical property	Natural soil	IS code
1	Specific gravity	2.58	IS 2720-3-1 (1980)
2	Soil classification	CI	IS 1498 (1970)
3	Consistency limits		
	Liquid limit	35.5%	IS 2720-5 (1985)
	Plastic limit	19%	IS 2720-5 (1985)
	Shrinkage limit	16.5%	IS 2720-6 (1972)
4	Free swell index	21%	IS 2720-40 (1977)
5	Optimum moisture content	15.78%	IS 2720-8 (1983)
6	Maximum dry density	18.10 kN/m^3	IS 2720-8 (1983)
7	Unconfined compressive strength	270 kN/m^2	IS 2720-10 (1991)
8	California bearing ratio (Soaked)	2.48	IS 2720-16 (1987)

Fig. 1 UCS sample preparation

was calculated and according to that mass of soil was calculated; mass of soil then thoroughly mixed with OMC of soil; by using hydraulic pressing tool, soil specimen was then compressed in three consecutive layers and wrapped with plastic wrapper in order to avoid entry and exit of moisture into the soil specimen. After that, immediately soil specimens were kept in curing tank, majorly immersed in for curing of 7 and 28 days (Fig. 1).

(c) Mix proportion

Avijeet Agencies Pvt. Ltd company mix design was adopted in this study for improving engineering properties of soil. Though several trial mixes were conducted referring to different authors' mix designs, it was finalized with company mix designs as it was having different dilution proportions as compared to different authors' mix design. The mix design proportions are given in Table 5.

Table 5 Mix proportion of diluted bio-enzyme mixed with soil

Mix design	Dosage	Dilution ratio	Amount of raw bio-enzyme (ml)	Amount of diluted bio-enzyme (ml)	Added water to soil (ml)
Dosage 1	5 m^3/lt	1:1000	0.109	109.23	OMC-109.23
Dosage 2	7.5 m^3/lt	1:1000	0.073	72.82	OMC-72.82
Dosage 3	10 m^3/lt	1:1000	0.055	54.61	OMC-54.61
Dosage 4	12.5 m^3/lt	1:1000	0.044	43.69	OMC-43.69

Table 6 UCS of natural soil after 7 and 28 days

Type of soil	Curing in days	Unconfined compressive strength (qu), kN/m^2	Undrained shear strength or cohesion (cu), kN/m^2	Strain, %
Natural soil	7	270.89	135.44	1.26
Natural soil	28	278.59	139.29	1.27

Table 7 UCS of treated soil after 7 days of curing

Type of soil	Dosage, m^3/lit	Unconfined compressive strength (qu), kN/m^2	Undrained shear strength or cohesion (cu), kN/m^2	Strain, %
Sample 1	5.0	322.91	161.45	1.47
Sample 2	7.5	338.70	169.35	1.54
Sample 3	10.0	325.33	162.66	1.48
Sample 4	12.5	310.94	155.47	1.42

Table 8 UCS of treated soil after 28 days of curing

Type of soil	Dosage, m^3/lit	Unconfined compressive strength (qu), kN/m2	Undrained shear strength or cohesion (cu), kN/m2	Strain, %
Sample 1	5.0	341.52	170.76	1.58
Sample 2	7.5	361.08	180.54	1.64
Sample 3	10.0	347.73	173.86	1.58
Sample 4	12.5	334.77	167.38	1.52

(d) Results

Unconfined Compressive Strength (UCS) Test- The UCS value with a dosage of 5 m^3/lit, 7.5 m^3/lit, and 10 m^3/lit has been determined and presented in Tables 7 and 8.

Variation of compressive strength of natural soil and bio-enzyme treated soil with different dosages are shown as follows (Figs. 2 and 3).

From Tables 6 and 7, it is found that the UCS is increased up to 7.5 m^3/lit of dosage, and further increase in dosage has reduced the UCS value (Fig. 4).

5 Discussion

Tables 2 and 3 showed the percentage change in soil index and engineering properties when soil stabilized with commonly used admixture and bio-enzyme. From these tables, it can be seen that soil properties change with both commonly used admixture

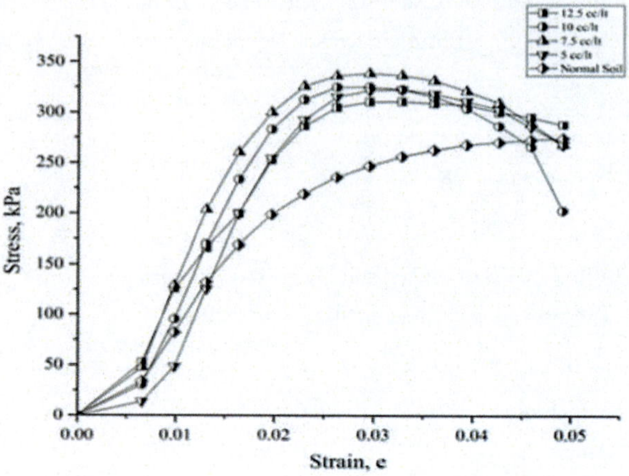

Fig. 2 UCS at 7 days of curing

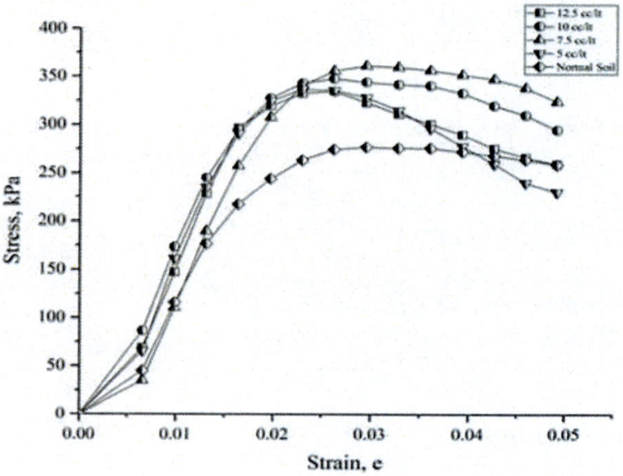

Fig. 3 UCS at 28 days of curing

and bio-enzyme. Saini and Vaishnava [12] observed in bio-enzyme treated soil that CBR value increases after a curing period of 2 weeks. The result exhibited that the dosage of 2 m^3/200 ml was optimum because the consistency limits were reduced and the soaked CBR increased after curing. Amount of clay plays a major role in the variation of consistency limits. Marginal change in MDD exhibited which is 1.486 gm/cm3 to 1.633 gm/cm3. OMC was observed to be decreased from 23.00 to 20.40% but OMC and Liquid Limit, and plastic limit decrease. UCS was observed to increase from 3.53 KN/m2 to 8.86 KN/m2. CBR value increases with increased curing period. Initially, CBR value was 1.19% but with enzyme stabilization CBR value was 5.80% after 4 weeks, which shows an increase of 387% from the original soil. Eujine [15] investigated the use of bio-enzyme in subgrade soil using the optimum dosage of 200 ml/m^3 LL increases in the first 2 weeks and later decreases and enhances the UCS up to 12 times the untreated soil. Dosage suggested in between 210 and 230 ml/m3.

Fig. 4 Comparison of UCS results at 7 and 28 days

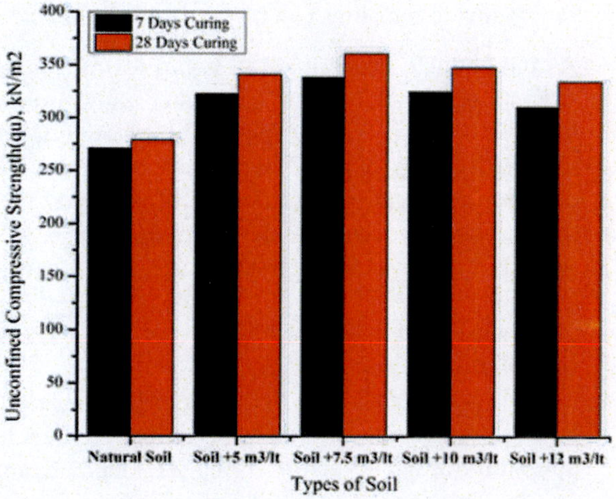

Agarwal and Kaur [8] investigated the effect of bio-enzyme on compressive strength of black cotton soil. The result demonstrated that unconfined compressive strength of soil increases to 200% with a dosage of 1 ml/5 kg. The result also demonstrated that in stabilized soil UCS of soil got higher in the curing period of 7 days. Nair and Joseph (2016) investigated the effects of bio-enzymatic stabilization on clayey sand. The result demonstrated that using a dosage of 400 ml/m3 of TerraZyme Atterberg limits was reduced to 11% as compared to original soil. The result also demonstrated that using an optimum dosage of 400 ml/m3 unconfined compressive strength of soil gets increased to 180 kN/m2, whereas original soil was having 140 kN/m2. The result also demonstrated that the CBR value of stabilized soil to 5% as compared to original soil. Interestingly, Shankar et al. [1] found a large increase in USC using bio-enzyme in coarse-grained soil. Khan and Taha [6] and Rauch et al. (1993) use bio-enzyme in fine-grained soil. This emphasized that bio-enzyme may not react equally with all types of clay, and its cation exchange capacity need to be evaluated for each and every soil before use in practice. In both these studies, no significant change in soil properties has been observed. Compared to the other commonly used chemical stabilizers like cement or lime, the process of strength gain is very slow.

6 Conclusions

In the present study, experimental study has been conducted to investigate the effect of bio-enzyme as a soil-stabilizing agent. Index and shear strength properties were studied at a well-controlled laboratory condition. UCS tests were conducted on large diameter.

Following conclusions can be drawn from the above research:

- As bio-enzyme is available in liquid form, it is easy to mix with the soil. The strength gaining of bio-enzyme treated soil is very slow, and it may need at least 7 to 28 days to observe any appreciable changes in the soil properties. Bio-enzyme may not react with all types of clay; therefore, it is very essential to conduct a detailed investigation before any practical uses.
- It is essential to determine the optimum dosage, while using bio-enzyme as a soil stabilizer, as it was found in this research that at higher dosage the shear strength may decrease.
- The shear strength from UCS test has been increased in the range of 20 to 35% at optimum dosage of 7.5 m3/lit.
- It was found that bio-enzyme indeed brings positive changes to the soil properties; however, extensive laboratory investigation must be conducted to determine the optimum dosage, effect of curing, efficiency of mixing, type of soil, and water quality. The same should be verified in field condition before implementation.

References

1. Shankar AU, Rai HK, Mithanthaya R (2009, July) Bio-Enzyme stabilized lateritic soil as a highway material. Indian Roads Congr J 70(2)
2. Baby M, Gowshik A, Rajeshwar AK, Mohanasundram M (2016) Experimental study of expansive soil stabilized with TerraZyme. Int J Eng Res Technol 5:897–899
3. Brazetti R, Murphy SR (2001, May) Objective performance measurement of actual road sites treated with an organic soil stabilizer. In: First road transportation technology transfer conference in Africa, Dar Es Salaam, Tanzania
4. Trivedi JS, Nair S, Iyyunni C (2013) Optimum utilization of fly ash for stabilization of sub-grade soil using genetic algorithm. Procedia Eng 51:250–258
5. Phummiphan I, Horpibulsuk S, Sukmak P, Chinkulkijniwat A, Arulrajah A, Shen SL (2016) Stabilisation of marginal lateritic soil using high calcium fly ash-based geopolymer. Road Mater Pavement Des 17(4):877–891
6. Khan TA, Taha MR (2015) Effect of three bio enzymes on compaction, consistency limits, and strength characteristics of a sedimentary residual soil. Adv Mater Sci Eng, 2015
7. Li YJ, Li L, Dan HC (2011) Study on application of TerraZyme in road base course of road. Appl Mech Mater 97:1098–1108. Trans Tech Publications
8. Agarwal P, Kaur S (2014) Effect of Bio-Enzyme stabilization on unconfined compressive strength of expansive soil. Int J Res Eng Technol 3(5):30–33
9. Haralambos SI (2009) Compressive strength of soil improved with cement. In: 2009 International foundation congress and equipment expo American society of civil engineers international association of foundation drilling pile driving contractors association
10. Shaka PM, Shaka SM (2016) Laboratory investigation on Black cotton soils and Red soil stabilized using Enzyme. Int Res J Eng Technol 3(6):325–330
11. Panchal S, Khan MM, Sharma A (2017) Stabilization of soil using bio-enzyme. Int J Civil Eng Technol 8(1)
12. Saini V, Vaishnava P (2015) Soil stabilization by using TerraZyme. Int J Adv Eng Technol 8(4):566
13. Zhang X, Zhang XP, Peng HT, Xia Q, Wang J (2013) Relation of microstructure and unconfined compression strength of soil stabilized with TerraZyme. Adv Mater Res 664:760–763. Trans Tech Publications

14. Gayatri SN, Shyla AJ (2016) Effect of bio enzyme and geotextile on strength characteristics of sub grade soil. Int Res J Eng Technol 03:1891–1897
15. Eujine GN, Somervell LT, Chandrakaran S, Sankar N (2014) Enzyme stabilization of high liquid limit clay. Elec J Geotech Eng 19
16. Sen J, Singh JP (2015) Stabilization of black cotton soil using bio-enzyme for a highway material. Int J Innovat Res Sci Eng Technol 4(12):553
17. Stan C, Ciobanu V (2012) Using enzymatic emulsions to reinforce road layers. Bulletin of the Transilvania University of Brasov. Forest Wood Ind Agric Food Eng, Series II, 5(1):109
18. Shirsath HA, Joshi SR, Sharma V (2017) Effect of bio-enzyme (TerraZyme) on the properties of sub grade soil of road. In Proceedings of the international conference on recent trends in civil engineering, science and management (pp 53–58)
19. Sravan MV, Nagaraj HB (2015, June) Preliminary study on use of TerraZyme as a bio stabilizer along with cement and lime in compressed stabilized earth blocks. In First international conference on bio-based building materials, Clermont-Ferrand (pp 1–8)

Analysis of Seepage from a Triangular Furrow with Negligible Freeboard Considering Soil Capillarity Using Inverse Hodograph and Conformal Mapping Technique

Kshyana Prava Samal, G. C. Mishra, and Nayan Sharma

Abstract Furrow irrigation is a conventional method of irrigation. It has lower application efficiency in comparison to drip irrigation. The obvious reasons for lower efficiency are (i) excess seepage losses and (ii) inappropriate furrow spacing. Close spacing of the furrows leads to unwise use of land resources, whereas wide spacing leads to non-availability of irrigation water to crops grown in the central part of the furrow ridge. In furrow irrigation method, generally a freeboard is provided to prevent overtopping of water during irrigation water supply. Overtopping for a small period is not harmful to plants. But a freeboard of considerable height would lead to loss of cropping area. In this study, the seepage analysis is carried out for a furrow which has negligible freeboard or ridge height considering uniform soil capillarity. For such a section, the stream line originating at the water surface level in the furrow moves close to the horizontal ground surface before becoming a parabolic surface. The locus of the phreatic line is found for different side slopes of the furrow section. There is a need to find the seepage loss and the loci of the phreatic lines, and hence appropriate spacing of the furrows was found out. The analysis has been carried out considering the symmetry of the flow domain while solving using inverse hodograph method. In this study, seepage analysis is carried out for a triangular furrow which has negligible freeboard considering uniform soil capillarity. It was found that the seepage loss and the lateral spreading vary quasi-linearly with capillary suction head and due to capillary action, first water moves in horizontal direction and then after moving certain distance follows a parabolic path.

Keywords Furrow irrigation · Seepage loss · Spacing of furrows · Capillarity · Hodograph method

---

K. P. Samal (✉)
School of Civil Engineering, KIIT Deemed to be University, Bhubaneswar, Odisha, India
e-mail: kshyanaprava.samalfce@kiit.ac.in

G. C. Mishra · N. Sharma
Water Resources Department and Management, IIT, Roorkee, Uttarakhand, India
e-mail: govindamishra2007@gmail.com

N. Sharma
e-mail: nayanfwt@iitr.ac.in

B. B. Das et al. (eds.), *Recent Developments in Sustainable Infrastructure*, Lecture Notes in Civil Engineering 75, https://doi.org/10.1007/978-981-15-4577-1_3

1 Introduction

Furrow irrigation is the method of water application in which water moves in the furrows and the crops are grown on ridges. Hence, it is a conventional method of irrigation practiced in tropical region in India to grow vegetable crops, which has lower application efficiency in comparison to drip irrigation [4, 5]. It is observed that about half of the water delivered in a furrow gets lost as seepage and other losses [1]. Generally a freeboard is provided in a furrow to prevent overtopping of water during irrigation water supply. Seepage from a triangular furrow with sufficient freeboard, which does not restrict capillary rise in the furrow ridge, has been analyzed by Samal and Mishra [2]. Overtopping for a small period is not harmful to plants. But a freeboard of considerable height would lead to loss of cropping area. Risenkampf (vide. [3] studied the case of seepage from a very shallow ditch having negligible freeboard considering capillarity. His study reveals that capillarity influences the water to move in lateral direction which otherwise would have gone vertically downward under the action of gravity as the depth of water in the ditch is very small. The capillarity also increases the flow rate.

2 Statement of the Problem

The inverse hodograph, complex potential pertaining to half of the flow domain, and the mapping onto auxiliary t-plane are depicted in Fig. 1a through f.

Mapping of the Inverse Hodograph (dz/dw) Plane onto the Auxiliary *t*-plane

According to the Schwarz–Christoffel transformation, the conformal mapping of the inverse hodograph domain onto the lower half of the auxiliary t-plane, the vertices A, B, P, C, S, D having been mapped onto $-\infty,\ 0, \lambda,\ 1,\ \gamma, \infty$, respectively, on the real axis of the t-plane, is given by

$$\frac{\mathrm{d}z}{\mathrm{d}w} = M_3 \int_0^t \frac{\mathrm{d}\tau}{\tau^{1-\alpha}(\tau-\lambda)^{1-2}(1-\tau)^{1-(1/2-\alpha)}(\tau-\gamma)} + N_3 \tag{1}$$

Parameters $\lambda, \gamma, M_3,$ and N_3 are unknown. The integration constant N_3 is governed by the lower limit of integration. Since for vertex B, $t = 0$, and $\frac{\mathrm{d}z}{\mathrm{d}w} = 0$, the constant $N_3 = 0$. Equation (1) is rewritten as

$$\begin{aligned}\frac{\mathrm{d}z}{\mathrm{d}w} &= M_3 \int_0^t \frac{\{(\tau-\gamma)+(\gamma-\lambda)\}}{\tau^{1-\alpha}(1-\tau)^{(1/2)+\alpha}(\tau-\gamma)}\mathrm{d}\tau \\ &= M_3\left[\int_0^t \tau^{\alpha-1}(1-\tau)^{(\frac{1}{2}-\alpha)-1}\mathrm{d}\tau - \frac{(\gamma-\lambda)}{\gamma}\int_0^t \frac{(1-\tau/\gamma)^{-1}}{\tau^{1-\alpha}(1-\tau)^{(1/2)+\alpha}}\mathrm{d}\tau\right]\end{aligned} \tag{2}$$

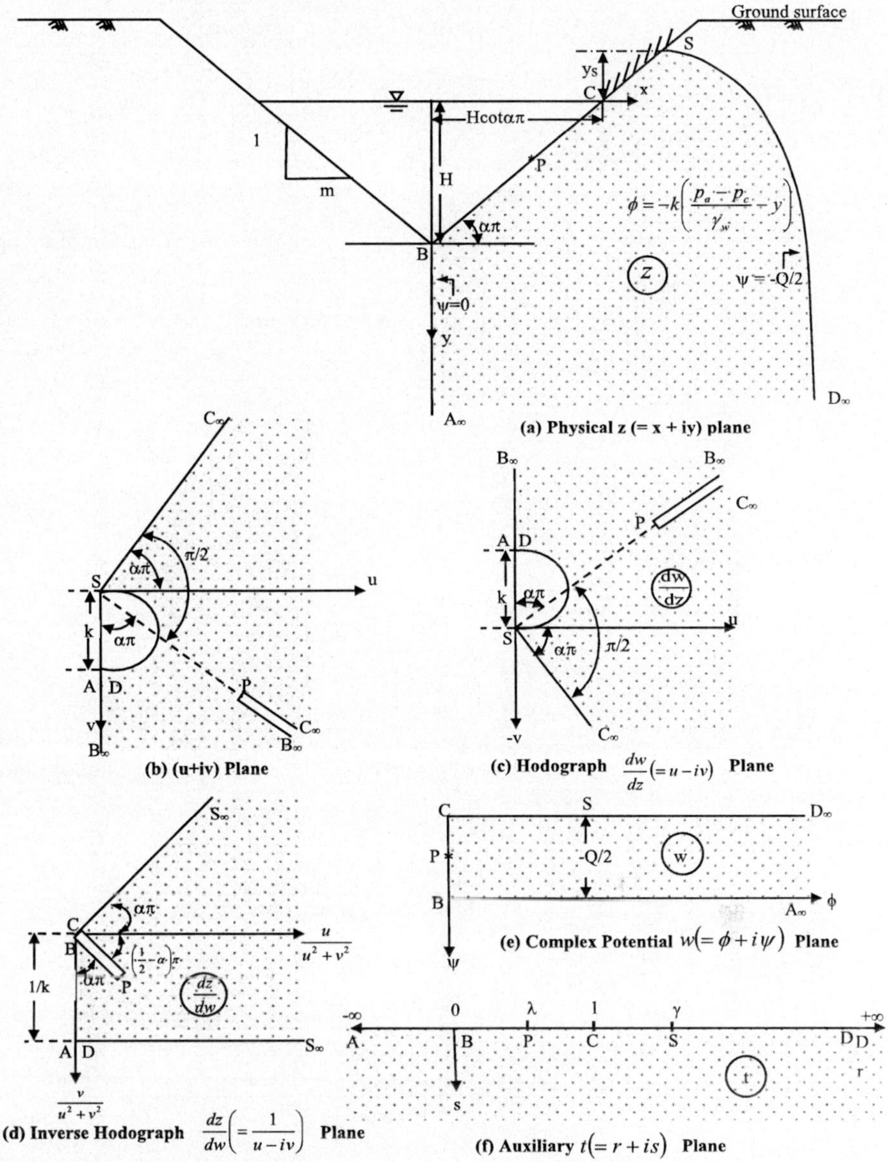

Fig. 1 Steps of conformal mapping

Expanding the term $(1 - \tau/\gamma)^{-1}$ binomially and integrating

$$\frac{dz}{dw} = M_3\left\{\beta_t(\alpha, \frac{1}{2} - \alpha) - \frac{(\gamma - \lambda)}{\gamma}\sum_{n=1}^{\infty}(\gamma)^{1-n}\beta_t(n + \alpha - 1, \frac{1}{2} - \alpha)\right\}; 0 \le t \le 1 \tag{3}$$

For vertex C, $t = 1$, and $\frac{dz}{dw} = 0$. Incorporating this condition in (3) and simplifying

$$\left\{B(\alpha, \frac{1}{2} - \alpha)\right\} - \left\{\frac{(\gamma - \lambda)}{\gamma}I_\gamma\right\} = 0 \tag{4}$$

where

$$I_\gamma = \left\{\sum_{n=1}^{\infty}\gamma^{1-n}B\left\{(n + \alpha - 1), (\frac{1}{2} - \alpha)\right\}\right\}$$

Equation (4) yields

$$\lambda = \gamma\left[1 - \frac{B(\alpha, \frac{1}{2} - \alpha)}{I_\gamma}\right] \tag{5}$$

Evaluation of M_3

Going around vertex S in t-plane, there is a jump equal to $(-i/k)$ in inverse hodograph plane. Putting $\tau - \gamma = re^{i\theta}$; $d\tau = re^{i\theta}\, i\, d\theta$

$$-\frac{i}{k} = M_3 \lim_{r\to 0}\int_0^{\pi}\frac{\{(\gamma + re^{i\theta} - \lambda)\}}{(\gamma + re^{i\theta})^{1-\alpha}(1 - \gamma - re^{i\theta})^{(1/2)+\alpha}(re^{i\theta})}re^{i\theta}i\mathrm{d}\theta$$

$$= M_3\int_0^{\pi}\frac{(\gamma - \lambda)}{\gamma^{1-\alpha}(1 - \gamma)^{(1/2)+\alpha}}i\mathrm{d}\theta = M_3\frac{(\gamma - \lambda)}{\gamma^{1-\alpha}(1 - \gamma)^{(1/2)+\alpha}}i\pi$$

$$= M_3\frac{(\gamma - \lambda)\pi}{\gamma^{1-\alpha}(\gamma - 1)^{(1/2)+\alpha}(-1)^{(1/2)+\alpha}}i$$

Solving for M_3

$$M_3 = \frac{\gamma^{1-\alpha}\{\gamma - 1\}^{\frac{1}{2}+\alpha}}{k\pi(\gamma - \lambda)}\left\{e^{i\pi(\frac{1}{2} - \alpha)}\right\} \tag{6}$$

3 Mapping of the Complex Potential Plane onto *t*-Plane

The Schwarz–Christoffel transformation of the *w*-plane onto *t*-plane is given by

$$\frac{\mathrm{dw}}{\mathrm{d}t} = \frac{M_4}{t^{1/2}(1-t)^{1/2}} \tag{7}$$

Integrating

$$w = 2M_4 \sin^{-1}\sqrt{t} + N_4 \tag{8}$$

At vertex B, $t = 0$, and $w = 0$; hence, $N_4 = 0$. For vertex C, $w = -iQ/2$ and $t = 1$; hence,

$$M_4 = -iQ/(2\pi) \tag{9}$$

Incorporating N_4 and M_4 in (8)

$$w = \left(\frac{-iQ}{\pi}\right)\sin^{-1}\sqrt{t} \tag{10}$$

or

$$\left(\frac{wi\pi}{Q}\right) = \sin^{-1}\sqrt{t} \tag{10a}$$

For $t \geq 1$

$$\frac{wi\pi}{Q} = \frac{\pi}{2} + \frac{1}{i}\ln\left(t \pm \sqrt{t^2-1}\right) \tag{11}$$

At $t = \gamma,\ w = \varphi + i\psi = kh_c - \frac{iQ}{2}$
Incorporating this condition in (11)

$$\left(kh_c - \frac{iQ}{2}\right)\frac{i\pi}{Q} = \frac{\pi}{2} + \frac{1}{i}\ln\left(\gamma \pm \sqrt{\gamma^2-1}\right) \tag{12}$$

Solving for h_c and dividing by H

$$\frac{h_c}{H} = \frac{1}{\pi}\frac{Q}{kH}\left\{-\ln\left(\gamma - \sqrt{\gamma^2-1}\right)\right\} \tag{13}$$

As selection of $\ln\left(\gamma + \sqrt{\gamma^2-1}\right)$ would yield a negative value of h_c, this term is not considered.

4 Relationship Between z- and t-Plane for $0 \le t \le 1$

$$\frac{dz}{dt} = \frac{dz}{dw} \cdot \frac{dw}{dt}$$

For $0 \le t' \le 1$

$$\frac{dz}{dt} = M_3 \left\{ \beta_t\left(\alpha, \frac{1}{2} - \alpha\right) - \frac{(\gamma - \lambda)}{\gamma} \sum_{n=1}^{\infty} (\gamma)^{1-n} \beta_t\left(n + \alpha - 1, \frac{1}{2} - \alpha\right) \right\} \left\{ \left(\frac{-iQ}{2\pi} \right) \frac{1}{t^{1/2}(1-t)^{1/2}} \right\} \tag{14}$$

Integrating

$$\int_0^{t'} dz = M_3 \left(\frac{-iQ}{2\pi} \right) \int_0^{t'} \left\{ \begin{array}{r} \beta_t\left(\alpha, \frac{1}{2} - \alpha\right) \\ - \frac{(\gamma - \lambda)}{\gamma} \sum_{n=1}^{\infty} \gamma^{1-n} \beta_t\left(n + \alpha - 1, \frac{1}{2} - \alpha\right) \end{array} \right\} \frac{dt}{t^{1/2}(1-t)^{1/2}} + z_B \tag{15}$$

or

$$z(t') = -\frac{iM_3 Q}{2\pi} \int_0^{t'} \left\{ \frac{\beta_t\left(\alpha, \frac{1}{2} - \alpha\right) - \frac{(\gamma - \lambda)}{\gamma} \sum_{n=1}^{\infty} \gamma^{1-n} \beta_t\left(n + \alpha - 1, \frac{1}{2} - \alpha\right)}{t^{1/2}(1-t)^{1/2}} \right\} dt + z_B \tag{16}$$

At $t' = 1$, $z = z_c$

$$z_c = \left(-\frac{iM_3 Q}{2\pi} \right) \int_0^{1} \left\{ \frac{\beta_t\left(\alpha, \frac{1}{2} - \alpha\right) - \frac{(\gamma - \lambda)}{\gamma} \sum_{n=1}^{\infty} \gamma^{1-n} \beta_t\left(n + \alpha - 1, \frac{1}{2} - \alpha\right)}{t^{1/2}(1-t)^{1/2}} \right\} dt + z_B \tag{17}$$

$$z_c - z_B = H \cot(\alpha\pi) - iH = \left(-\frac{iQ}{2\pi} \right) \left\{ \frac{\gamma^{1-\alpha} \{\gamma - 1\}^{\frac{1}{2}+\alpha}}{k\pi(\gamma - \lambda)} e^{i\pi\left(\frac{1}{2} - \alpha\right)} \right\} I_1 \tag{18}$$

or

$$H \cot(\alpha\pi) - iH = \left(\frac{Q}{2\pi} \right) \left\{ \frac{\gamma^{1-\alpha} \{\gamma - 1\}^{\frac{1}{2}+\alpha}}{k\pi(\gamma - \lambda)} \right\} I_1 [\cos(\alpha\pi) - i \sin(\alpha\pi)] \tag{19}$$

where

$$I_1 = \int_0^1 \left\{ \frac{\beta_t(\alpha, \frac{1}{2} - \alpha) - \frac{(\gamma-\lambda)}{\gamma} \sum_{n=1}^{\infty} \gamma^{1-n} \beta_t(n + \alpha - 1, \frac{1}{2} - \alpha)}{t^{1/2}(1-t)^{1/2}} \right\} dt$$

I_1 has been evaluated in Appendix 1.
Equating the real parts on either sides in (19)

$$H \cot(\alpha\pi) = \left(\frac{Q}{2\pi}\right) \frac{\gamma^{1-\alpha}\{\gamma - 1\}^{\frac{1}{2}+\alpha}}{k\pi(\gamma - \lambda)} I_9 \cos(\alpha\pi) \tag{20}$$

Solving for $Q/(kH)$

$$\frac{Q}{kH} = \frac{2\pi^2(\gamma - \lambda)}{\sin(\alpha\pi)\left\{\gamma^{1-\alpha}\{\gamma - 1\}^{\frac{1}{2}+\alpha} I_9\right\}} \tag{20a}$$

The same relation is obtained equating imaginary parts on either sides of Eq. (19).

5 Relation Between z- and t-Plane for $1 \le t \le \gamma$

$$\frac{dz}{dw} = M_3 \int_1^t \frac{(\tau - \lambda)}{\tau^{1-\alpha}(1-\tau)^{\frac{1}{2}+\alpha}(\tau - \gamma)} d\tau = \frac{M_3}{(-1)^{\frac{1}{2}+\alpha}(-1)} \int_1^t \frac{(\tau - \lambda)}{\tau^{1-\alpha}(\tau - 1)^{\frac{1}{2}+\alpha}(\gamma - \tau)} d\tau$$

$$\frac{dz}{dt} = \frac{dz}{dw} \cdot \frac{dw}{dt}$$

$$= \left\{ \frac{M_3}{(-1)(-1)^{\frac{1}{2}+\alpha}} \int_1^t \frac{(\tau - \lambda)}{\tau^{1-\alpha}(\tau - 1)^{\frac{1}{2}+\alpha}(\gamma - \tau)} d\tau \right\} \left[\left(\frac{-iQ}{2\pi}\right) \frac{1}{t^{1/2}(1-t)^{1/2}} \right] \tag{21}$$

Integrating

$$z(t') = \frac{iQM_3}{2\pi(-1)^{\frac{1}{2}+\alpha}(-1)^{1/2}} \int_1^{t'} \left[\left\{ \int_1^t \frac{(\tau - \lambda)d\tau}{\tau^{1-\alpha}(\tau - 1)^{\frac{1}{2}+\alpha}(\gamma - \tau)} \right\} \frac{1}{t^{1/2}(t-1)^{1/2}} \right] dt + z_c \tag{22}$$

Incorporating M_3

$$z(t') = \left\{ \frac{\left(e^{i\pi/2}\right) e^{i\pi\left(\frac{1}{2}-\alpha\right)}}{(e^{-i\pi})^{\frac{1}{2}+\alpha}(e^{-i\pi})^{1/2}} \right\} \left\{ \frac{Q}{2\pi} \right\} \left\{ \frac{\gamma^{1-\alpha}\{\gamma-1\}^{\frac{1}{2}+\alpha}}{k\pi(\gamma-\lambda)} \right\}$$

$$\int_1^{t'} \left[\left\{ \int_1^t \frac{(\tau-\lambda)\mathrm{d}\tau}{\tau^{1-\alpha}(\tau-1)^{\frac{1}{2}+\alpha}(\gamma-\tau)} \right\} \frac{1}{t^{1/2}(t-1)^{1/2}} \right] \mathrm{d}t + z_c$$

$$= \left\{ \frac{Q}{2\pi} \right\} \left\{ \frac{\gamma^{1-\alpha}\{\gamma-1\}^{\frac{1}{2}+\alpha}}{k\pi(\gamma-\lambda)} \right\} \int_1^{t'} \left[\left\{ \int_1^t \frac{(\tau-\lambda)\mathrm{d}\tau}{\tau^{1-\alpha}(\tau-1)^{\frac{1}{2}+\alpha}(\gamma-\tau)} \right\} \frac{1}{t^{1/2}(t-1)^{1/2}} \right] \mathrm{d}t + z_c \tag{23}$$

At $t = \gamma$

$$z_S = \left\{ \frac{Q}{2\pi} \right\} \left\{ \frac{\gamma^{1-\alpha}\{\gamma-1\}^{\frac{1}{2}+\alpha}}{k\pi(\gamma-\lambda)} \right\} \int_1^{\gamma} \left[\left\{ \int_1^t \frac{(\tau-\lambda)d\tau}{\tau^{1-\alpha}(\tau-1)^{\frac{1}{2}+\alpha}(\gamma-\tau)} \right\} \frac{1}{t^{1/2}(t-1)^{1/2}} \right] \mathrm{d}t + z_c$$

$$= \left\{ \frac{Q}{2\pi} \right\} \left\{ \frac{\gamma^{1-\alpha}\{\gamma-1\}^{\frac{1}{2}+\alpha}}{k\pi(\gamma-\lambda)} \right\} I_2 + z_c \tag{24}$$

where $I_2 = \int_1^{\gamma} \left[\left\{ \int_1^t \frac{(\tau-\lambda)d\tau}{\tau^{1-\alpha}(\tau-1)^{\frac{1}{2}+\alpha}(\gamma-\tau)} \right\} \frac{1}{t^{1/2}(t-1)^{1/2}} \right] \mathrm{d}t$

I_2 has been evaluated in Appendix 2.

The length $(z_S - z_c)/H$ is given by

$$(z_S - z_c)/H = l/H = \left\{ \frac{Q}{2\pi^2 kH} \right\} \left\{ \frac{\gamma^{1-\alpha}\{\gamma-1\}^{\frac{1}{2}+\alpha}}{(\gamma-\lambda)} \right\} I_2 \tag{25}$$

6 Locus of the Phreatic Line

$$\frac{dz}{dw} = \frac{M_3}{(-1)^{\frac{1}{2}+\alpha}} \int_{\infty}^{t} \frac{(\tau-\lambda)}{\tau^{1-\alpha}(\tau-1)^{\frac{1}{2}+\alpha}(\tau-\gamma)} d\tau + N_4 \tag{26}$$

At point ∞, $\frac{dz}{dw} = \frac{i}{k}$; hence, $N_4 = \frac{i}{k}$.

$$\begin{aligned}\frac{dz}{dw} &= \frac{M_3}{(-1)^{\frac{1}{2}+\alpha}}\int_{\infty}^{t} \frac{\{(\tau-\gamma)+(\gamma-\lambda\}}{\tau^{1-\alpha}(\tau-1)^{\frac{1}{2}+\alpha}(\tau-\gamma)}d\tau + \frac{i}{k}\\ &= \frac{M_3}{(-1)^{\frac{1}{2}+\alpha}}\int_{\infty}^{t} \tau^{\alpha-1}(\tau-1)^{\left(\frac{1}{2}-\alpha\right)-1}d\tau + \frac{M_3}{(-1)^{\frac{1}{2}+\alpha}}\int_{\infty}^{t}\frac{(\gamma-\lambda)}{(\tau)^{1-\alpha}(\tau-1)^{\frac{1}{2}+\alpha}\tau\left(1-\frac{\gamma}{\tau}\right)}d\tau+\frac{i}{k}\\ &= \frac{M_3}{(-1)^{\frac{1}{2}+\alpha}}\int_{\infty}^{t} \tau^{\alpha-1}(\tau-1)^{\left(\frac{1}{2}-\alpha\right)-1}d\tau + \frac{M_3(\gamma-\lambda)}{(-1)^{\frac{1}{2}+\alpha}}\int_{\infty}^{t}\frac{\left(1-\frac{\gamma}{\tau}\right)^{-1}}{(\tau)^{2-\alpha}(\tau-1)^{\frac{1}{2}+\alpha}}d\tau+\frac{i}{k}\end{aligned} \tag{27}$$

Expanding the term $\left(1-\frac{\gamma}{\tau}\right)^{-1}$ binomially

$$\frac{dz}{dw} = \frac{M_3}{(-1)^{\frac{1}{2}+\alpha}}\int_{\infty}^{t} \tau^{\alpha-1}(\tau-1)^{\left(\frac{1}{2}-\alpha\right)-1}d\tau + \frac{M_3(\gamma-\lambda)}{(-1)^{\frac{1}{2}+\alpha}}\int_{\infty}^{t}\sum_{n=1}^{\infty}\frac{\left(\frac{\gamma}{\tau}\right)^{n-1}}{\tau^{2-\alpha}(\tau-1)^{\frac{1}{2}+\alpha}}d\tau+\frac{i}{k} \tag{28}$$

To convert the improper integral (as the lower limit of integration is ∞) a substitution $\tau = \frac{1}{\xi}$ is made. Accordingly, (28) reduces to

$$\begin{aligned}\frac{dz}{dw} &= \frac{M_3}{(-1)^{\frac{1}{2}+\alpha}}\int_{o}^{1/t}\left(\frac{1}{\xi}\right)^{\alpha-1}\left(\frac{1-\xi}{\xi}\right)^{\left(\frac{1}{2}-\alpha\right)-1}\left(\frac{-d\xi}{\xi^2}\right)+\frac{M_3(\gamma-\lambda)}{(-1)^{\frac{1}{2}+\alpha}}\\ &\quad\int_0^{1/t}\frac{\sum_{n=1}^{\infty}(\gamma)^{n-1}\xi^{n-1}}{\left(\frac{1}{\xi}\right)^{2-\alpha}\left(\frac{1-\xi}{\xi}\right)^{\frac{1}{2}+\alpha}}\left(\frac{-d\xi}{\xi^2}\right)+\frac{i}{k}\\ &= \frac{-M_3}{(-1)^{\frac{1}{2}+\alpha}}\int_{o}^{1/t}\xi^{\frac{1}{2}-1}(1-\xi)^{\left(\frac{1}{2}-\alpha\right)-1}d\xi - \frac{M_3(\gamma-\lambda)}{(-1)^{\frac{1}{2}+\alpha}}\\ &\quad\int_0^{1/t}\sum_{n=1}^{\infty}(\gamma)^{n-1}(\xi)^{\left(n+\frac{1}{2}\right)-1}(1-\xi)^{\left(\frac{1}{2}-\alpha\right)-1}d\xi+\frac{i}{k}\\ &= \frac{-M_3}{(-1)^{\frac{1}{2}+\alpha}}\left\{\beta_{\frac{1}{t}}\left(\frac{1}{2},\frac{1}{2}-\alpha\right)+(\gamma-\lambda)\sum_{n=1}^{\infty}(\gamma)^{n-1}\beta_{\frac{1}{t}}\left(n+\frac{1}{2},\frac{1}{2}-\alpha\right)\right\}+\frac{i}{k}\end{aligned} \tag{29}$$

For

$$\gamma < t \le \infty,$$

$$\frac{dw}{dt} = \left(\frac{-iQ}{2\pi}\right)\frac{1}{t^{\frac{1}{2}}(1-t)^{\frac{1}{2}}}$$

$\frac{\mathrm{d}w}{\mathrm{d}t}$ being real and positive.

$$\frac{\mathrm{d}w}{\mathrm{d}t} = \frac{Q}{2\pi} \frac{1}{t^{\frac{1}{2}}(t-1)^{\frac{1}{2}}} \tag{30}$$

and

$$\frac{\mathrm{d}z}{\mathrm{d}t} = \frac{-M_3}{(-1)^{\frac{1}{2}+\alpha}} \left(\frac{Q}{2\pi}\right) \frac{\beta_{\frac{1}{t}}\left(\frac{1}{2}, \frac{1}{2}-\alpha\right) + (\gamma-\lambda) \sum_{n=1}^{\infty} (\gamma)^{n-1} \beta_{\frac{1}{t}}\left(n+\frac{1}{2}, \frac{1}{2}-\alpha\right)}{\sqrt{t(t-1)}} + i\left(\frac{Q}{2\pi k}\right) \frac{\mathrm{d}t}{\sqrt{t(t-1)}} \tag{31}$$

Incorporating M_3

$$M_3 = \frac{\gamma^{1-\alpha}\{\gamma-1\}^{\frac{1}{2}+\alpha}}{k\pi(\gamma-\lambda)} \left\{e^{i\pi\left(\frac{1}{2}-\alpha\right)}\right\}$$

$$\frac{\mathrm{d}z}{\mathrm{d}t} = \left(\frac{-1}{(-1)^{\frac{1}{2}+\alpha}}\right) \left\{\frac{\gamma^{1-\alpha}(1-\gamma)^{\frac{1}{2}+\alpha}}{k\pi(\gamma-\lambda)}\right\} \left\{e^{i\pi\left(\frac{1}{2}-\alpha\right)}\right\} \left(\frac{Q}{2\pi}\right) \frac{\beta_{\frac{1}{t}}\left(\frac{1}{2}, \frac{1}{2}-\alpha\right) + (\gamma-\lambda) \sum_{n=1}^{\infty} (\gamma)^{n-1} \beta_{\frac{1}{t}}\left(n+\frac{1}{2}, \frac{1}{2}-\alpha\right)}{\sqrt{t(t-1)}} + i\left(\frac{Q}{2\pi k}\right) \frac{\mathrm{d}t}{\sqrt{t(t-1)}} \tag{32}$$

$$\frac{\mathrm{d}z}{\mathrm{d}t} = \frac{\mathrm{d}x}{\mathrm{d}t} + i\frac{\mathrm{d}y}{\mathrm{d}t}$$

Considering the real parts in (34)

$$\frac{\mathrm{d}x}{\mathrm{d}t} = \left\{\frac{\gamma^{1-\alpha}(1-\gamma)^{\frac{1}{2}+\alpha}}{k\pi(\gamma-\lambda)}\right\} \left(\frac{Q}{2\pi}\right) \frac{\beta_{\frac{1}{t}}\left(\frac{1}{2}, \frac{1}{2}-\alpha\right) + (\gamma-\lambda) \sum_{k=1}^{\infty} (\gamma)^{n-1} \beta_{\frac{1}{t}}\left(n+\frac{1}{2}, \frac{1}{2}-\alpha\right)}{\sqrt{t(t-1)}} \tag{33}$$

Integrating

$$x(t') = \int_{\infty}^{t'} \left\{\frac{\gamma^{1-\alpha}(1-\gamma)^{\frac{1}{2}+\alpha}}{k\pi(\gamma-\lambda)}\right\} \left(\frac{Q}{2\pi}\right) \frac{\beta_{\frac{1}{t}}\left(\frac{1}{2}, \frac{1}{2}-\alpha\right) + (\gamma-\lambda) \sum_{k=1}^{\infty} (\gamma)^{n-1} \beta_{\frac{1}{t}}\left(n+\frac{1}{2}, \frac{1}{2}-\alpha\right)}{\sqrt{t(t-1)}} \mathrm{d}t + x|_{t=\infty} \tag{34}$$

As $x|_{t=\infty} = \frac{Q}{2k}$,

$$x(t') = \int_{\infty}^{t'} \left\{ \frac{\gamma^{1-\alpha}(1-\gamma)^{\frac{1}{2}+\alpha}}{k\pi(\gamma-\lambda)} \right\} \left(\frac{Q}{2\pi}\right) \frac{\beta_{\frac{1}{t}}\left(\frac{1}{2}, \frac{1}{2}-\alpha\right) + (\gamma-\lambda)\sum_{k=1}^{\infty}(\gamma)^{n-1}\beta_{\frac{1}{t}}\left(n+\frac{1}{2}, \frac{1}{2}-\alpha\right)}{\sqrt{t(t-1)}} \mathrm{d}t + \frac{Q}{2k} \tag{34a}$$

Dividing by H,

$$\frac{x(t')}{H} = \int_{\infty}^{t'} \left\{ \frac{\gamma^{1-\alpha}(1-\gamma)^{\frac{1}{2}+\alpha}}{k\pi(\gamma-\lambda)} \right\} \left(\frac{Q}{2\pi}\right) \frac{\beta_{\frac{1}{t}}\left(\frac{1}{2}, \frac{1}{2}-\alpha\right) + (\gamma-\lambda)\sum_{k=1}^{\infty}(\gamma)^{n-1}\beta_{\frac{1}{t}}\left(n+\frac{1}{2}, \frac{1}{2}-\alpha\right)}{\sqrt{t(t-1)}} \mathrm{d}t + \frac{Q}{2kH}$$

Substituting

$$t = \frac{1}{\xi^2}, \sqrt{t(t-1)} = \sqrt{\frac{1}{\xi^2}\left(\frac{1}{\xi^2}-1\right)} = \sqrt{\frac{1-\xi^2}{\xi^4}}, \quad dt = -\frac{d\xi}{\xi^3}$$

$$\frac{x(t')}{H} = -\left\{ \frac{\gamma^{1-\alpha}(1-\gamma)^{\frac{1}{2}+\alpha} Q}{2\pi^2 kH} \right\} \left[\int_0^{\sqrt{1/t'}} \frac{B_{\xi^2}\left(\frac{1}{2}, \frac{1}{2}-\alpha\right)}{\sqrt{\frac{1}{\xi^2}\left(\frac{1}{\xi^2}-1\right)}} \frac{\mathrm{d}\xi}{\xi^3} + \int_0^{\sqrt{1/t'}} \left\{ (\gamma-\lambda)\sum_{n=1}^{\infty}\gamma^{n-1} \frac{B_{\xi^2}\left(n+\frac{1}{2}, \frac{1}{2}-\alpha\right)}{\sqrt{\frac{1}{\xi^2}\left(\frac{1}{\xi^2}-1\right)}} \frac{\mathrm{d}\xi}{\xi^3} \right\} \right] + \frac{Q}{2kH} \tag{35}$$

$$\frac{x(t')}{H} = -\left\{ \frac{\gamma^{1-\alpha}(1-\gamma)^{\frac{1}{2}+\alpha} Q}{2\pi^2 kH} \right\} \left[\int_0^{\sqrt{1/t'}} \frac{B_{\xi^2}\left(\frac{1}{2}, \frac{1}{2}-\alpha\right)}{\xi\sqrt{1-\xi^2}} \mathrm{d}\xi + \int_0^{\sqrt{1/t'}} \left\{ (\gamma-\lambda)\sum_{n=1}^{\infty}\gamma^{n-1} \frac{B_{\xi^2}\left(n+\frac{1}{2}, \frac{1}{2}-\alpha\right)}{\xi\sqrt{1-\xi^2}} \mathrm{d}\xi \right\} \right] + \frac{Q}{2kH} \tag{36}$$

Expanding in complete beta function in hypergeometric series of second kind to remove the singularity at $\xi = 0$, in (36)

$$\frac{x(t')}{H} = -\left\{ \frac{\gamma^{1-\alpha}(1-\gamma)^{\frac{1}{2}+\alpha} Q}{2\pi^2 kH} \right\} \left[\int_0^{\sqrt{1/t'}} \frac{(\xi^2)^{1/2}_{2} F_1\left(\frac{1}{2}, \frac{1}{2}+\alpha, \frac{3}{2}, \xi^2\right)}{\xi\sqrt{1-\xi^2}} \mathrm{d}\xi + \int_0^{\sqrt{1/t'}} \left\{ \frac{(\gamma-\lambda)\sum_{n=1}^{\infty}\gamma^{n-1}(\xi^2)^{n+\frac{1}{2}}_{2} F_1\left(n+\frac{1}{2}, \frac{1}{2}+\alpha, n+\frac{3}{2}, \xi^2\right)}{\xi\sqrt{1-\xi^2}} \mathrm{d}\xi \right\} \right]$$

$$+ \frac{Q}{2kH}$$

$$\frac{x(t')}{H} = -\left\{\frac{\gamma^{1-\alpha}(1-\gamma)^{\frac{1}{2}+\alpha}Q}{2\pi^2 kH}\right\}\left[\begin{array}{c}\int\limits_0^{\sqrt{1/t'}} \frac{{}_2F_1\left(\frac{1}{2},\frac{1}{2}+\alpha,\frac{3}{2},\xi^2\right)}{\sqrt{1-\xi^2}}\mathrm{d}\xi \\ + \int\limits_0^{\sqrt{1/t'}}\left\{\frac{(\gamma-\lambda)\sum\limits_{n=1}^{\infty}\gamma^{n-1}\xi_2^{2n}F_1\left(n+\frac{1}{2},\frac{1}{2}+\alpha,n+\frac{3}{2},\xi^2\right)}{\sqrt{1-\xi^2}}\mathrm{d}\xi\right\}\end{array}\right] + \frac{Q}{2kH} \tag{37}$$

Further substituting $\frac{1}{2}\sqrt{1/t'}[1+\eta] = \xi, \quad \mathrm{d}\xi = \frac{1}{2}\sqrt{1/t'}\,\mathrm{d}\eta$ in (37)

$$\frac{x(t')}{H} = -\left\{\frac{\gamma^{1-\alpha}(1-\gamma)^{\frac{1}{2}+\alpha}Q}{2\pi^2 kH}\right\}\left[\begin{array}{c}\frac{1}{2}\sqrt{1/t'}\int\limits_{-1}^{1} \frac{{}_2F_1\left(\frac{1}{2},\frac{1}{2}+\alpha,\frac{3}{2},\xi^2\right)}{\sqrt{1-\xi^2}}d\eta \\ + \frac{1}{2}\sqrt{1/t'}\int\limits_{-1}^{1}\left\{\frac{(\gamma-\lambda)\sum\limits_{n=1}^{\infty}\gamma^{n-1}\xi_2^{2n}F_1\left(n+\frac{1}{2},\frac{1}{2}+\alpha,n+\frac{3}{2},\xi^2\right)}{\sqrt{1-\xi^2}}d\eta\right\}\end{array}\right] + \frac{Q}{2kH} \tag{38}$$

Computation of $y(t')/H$

$$\frac{\mathrm{d}y}{\mathrm{d}t} = \frac{Q}{2\pi k}\frac{1}{t^{1/2}(t-1)^{1/2}}$$

$$y(t') = \frac{Q}{2\pi k}\int\limits_{\gamma}^{t'}\frac{dt}{t^{1/2}(t-1)^{1/2}}$$

$$\frac{y(t')}{H} = \left(\frac{Q}{2\pi kH}\right)\int\limits_{\gamma}^{t'}\frac{dt}{(t)^{1/2}(t-1)^{1/2}} \tag{39}$$

Substituting $t = \cosh^2\xi$; $\mathrm{d}t = 2\cosh\xi\ \sinh\xi\ \mathrm{d}\xi$ and incorporating the limit at $t = \gamma$; $\xi = \cosh^{-1}\sqrt{\gamma}$; at $t = t'$, $\xi = \cosh^{-1}\sqrt{t'}$

$$\begin{aligned}\frac{y(t')}{H} &= \left(\frac{Q}{kH}\right)\left[\cosh^{-1}\sqrt{t'} - \cosh^{-1}\sqrt{\gamma}\right] \\ &= \left(\frac{Q}{kH)}\right)\frac{1}{\pi}\left[\ln\left\{\sqrt{t'}+\sqrt{t'-1}\right\} - \ln\left\{\sqrt{\gamma}+\sqrt{\gamma-1}\right\}\right]\end{aligned} \tag{40}$$

7 Results and Discussion

The following steps are followed for computing results:

Step 1: Assuming γ, λ is obtained from (5).
Step 2: Using the values of γ and λ the corresponding Q/kH is obtained from (20a).
Step 3: h_c/H is obtained from (13) using the value of γ and the corresponding Q/kH is obtained from steps 1 and 2.
Step 4: l/H is obtained from (25) using the computed values of γ, λ, and Q/kH.
Step 5: Finally loci of the phreatic line is obtained from (38) and (40).

8 Variation of Seepage Losses, *Q/KH* with Capillary Suction Head, h_c/H, for Different Side Slopes of the Furrow

The variations of $\frac{Q}{kH}$ with $\frac{h_c}{H}$ for different values of m (i.e., $m = 0.5$, 1, 1.5 and 2) are shown in Fig. 2. As seen from the figure, $\frac{Q}{kH}$ varies quasi-linearly with $\frac{h_c}{H}$.

9 Variation of *L/H* with Capillary Suction Head, h_c/H, for Different Side Slopes of the Furrow

Variations of $\frac{l}{H}$ with suction head $\frac{h_c}{H}$ for different values of m (i.e., $m = 0.5$, 1, 1.5, and 2) are presented in Fig. 3. $\frac{l}{H}$ is quasi-linearly proportional to the capillary suction head, $\frac{h_c}{H}$. l/H is more in case of steeper slope, i.e., $m = 0.5$ and less in case of flatter slope, i.e., $m = 2$.

10 Loci of the Phreatic Lines for Different $\frac{h_c}{H}$ and *M*

The loci of phreatic lines are presented in Figs. 4 and 5 for m (i.e., $m = 0.5$ and 1) and suction head, h_c/H. As seen from the figures, with increase in capillary suction head, the phreatic lines widen in lateral direction. A phreatic line joins asymptotically with the vertical at $\frac{x}{H} = 0.5\frac{Q}{kH}$.

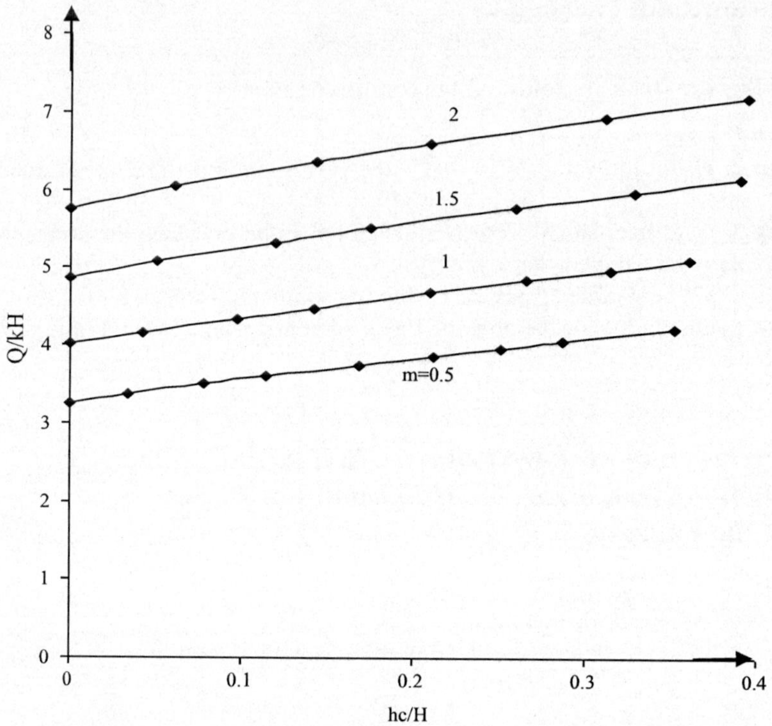

Fig. 2 Variation of seepage losses, Q/kH, with capillary suction head, h_c/H, for different side slopes of triangular furrows

11 Conclusions

1. The seepage $\frac{Q}{kH}$ varies quasi-linearly with capillary suction head $\frac{h_c}{H}$.
2. The lateral spreading, $\frac{l}{H}$, is also quasi-linearly proportional to the capillary suction head $\frac{h_c}{H}$.
3. Due to capillary action, first water moves in horizontal direction and then after moving certain distance follows a parabolic path.

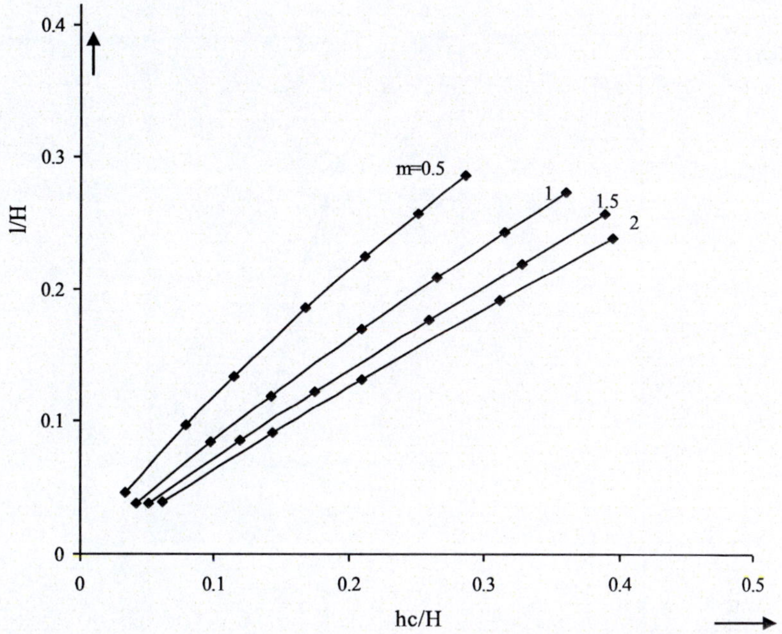

Fig. 3 Variations of l/H, with capillary suction head, h_c/H, for different side slopes of triangular furrows

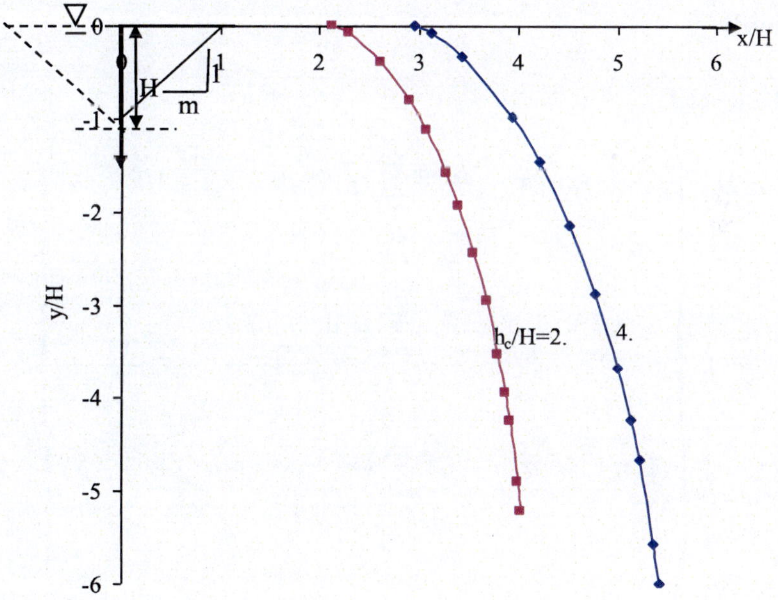

Fig. 4 Loci of phreatic lines for various h_c/H for $m = 1$

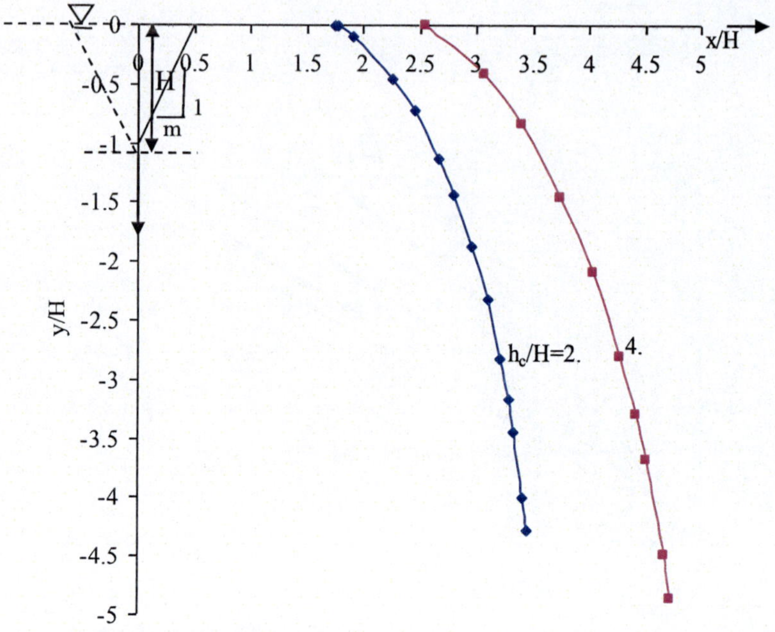

Fig. 5 Loci of phreatic lines for various h_c/H for $m = 0.5$

Appendix 1

Evaluation of I_9

$$I_9 = \int_0^1 \left\{ \frac{\beta_t(\alpha, 1/2-\alpha) - \frac{(\gamma-\lambda)}{\gamma} \sum_{k=1}^{\infty} (\gamma)^{1-n} \beta_t(n+\alpha-1, \frac{1}{2}-\alpha)}{t^{1/2}(1-t)^{1/2}} \right\} \mathrm{d}t$$

$$= \int_0^{1/2} \left\{ \frac{\beta_t(\alpha, 1/2-\alpha) - \frac{(\gamma-\lambda)}{\gamma} \sum_{k=1}^{\infty} (\gamma)^{1-n} \beta_t(n+\alpha-1, \frac{1}{2}-\alpha)}{t^{1/2}(1-t)^{1/2}} \right\} \mathrm{d}t$$

$$+ \int_{1/2}^{1} \left\{ \frac{\beta_t(\alpha, 1/2-\alpha) - \frac{(\gamma-\lambda)}{\gamma} \sum_{k=1}^{\infty} (\gamma)^{1-n} \beta_t(n+\alpha-1, \frac{1}{2}-\alpha)}{t^{1/2}(1-t)^{1/2}} \right\} \mathrm{d}t$$

$$= I_{9a} + I_{9B}$$

In the integrand of $I_{9a,}$ $t = 0$ is a singular point. For removing this singularity, substituting $t = \xi^2$, dt $= 2\xi d\xi$ and accordingly incorporating the limits at $t = 0$, $\xi = 0$; and at $t = 1/2$, $\xi = \frac{1}{\sqrt{2}}$,

$$I_{9a} = 2\int_{o}^{1/\sqrt{2}}\left\{\frac{\beta_{\xi^2}(\alpha,\frac{1}{2}-\alpha)-\frac{(\gamma-\lambda)}{\gamma}\sum_{n=1}^{\infty}(\gamma)^{1-n}\beta_{\xi^2}(n+\alpha-1,\frac{1}{2}-\alpha)}{\sqrt{(1-\xi^2}}\right\}d\xi$$

Further making a substitution $\xi = \frac{1}{2\sqrt{2}}(1+\eta),\ d\xi = \frac{1}{2\sqrt{2}}d\eta$

$$I_{9a} = \frac{1}{\sqrt{2}}\int_{-1}^{-1}\left\{\frac{\beta_{\xi^2}(\alpha,\frac{1}{2}-\alpha)-\frac{(\gamma-\lambda)}{\gamma}\sum_{k=1}^{\infty}(\gamma)^{1-n}\beta_{\xi^2}(n+\alpha-1,\frac{1}{2}-\alpha)}{\sqrt{(1-\xi^2}}\right\}d\eta$$

$$I_{9b} = \int_{1/2}^{1}\left\{\frac{\beta_t(\alpha,1/2-\alpha)-\frac{(\gamma-\lambda)}{\gamma}\sum_{k=1}^{\infty}\gamma^{1-n}\beta_t(n+\alpha-1,1/2-\alpha)}{t^{1/2}(1-t)^{1/2}}\right\}dt$$

In the integrand of I_{9b} $t = 1$ is a singular point. Substituting $1 - t = \xi^2$, dt $= -2\xi$ dξ and changing the limits $I_{9b} =$

$$\left[\int_{o}^{1/\sqrt{2}}\left\{\frac{\beta_{1-\xi^2}(\alpha,\frac{1}{2}-\alpha)-\frac{(\gamma-\lambda)}{\gamma}\sum_{n=1}^{\infty}\gamma^{1-n}\beta_{1-\xi^2}(n+\alpha-1,\frac{1}{2}-\alpha)}{\sqrt{(1-\xi^2)}}\right\}2d\xi\right.$$

Further making a substitution $\xi = \frac{1}{2\sqrt{2}}(1+\eta),\ d\xi = \frac{1}{2\sqrt{2}}d\eta$

$$I_{9b} = \frac{1}{\sqrt{2}}\left[\int_{-1}^{1}\left\{\frac{\beta_{1-\xi^2}(\alpha,\frac{1}{2}-\alpha)-\frac{(\gamma-\lambda)}{\gamma}\sum_{n=1}^{\infty}\gamma^{1-n}\beta_{1-\xi^2}(n+\alpha-1,\frac{1}{2}-\alpha)}{\sqrt{(1-\xi^2)}}\right\}d\eta\right.$$

I_{9b} can be evaluated using Gaussian weights and abscissa.

$$I_9 = I_{9a} + I_{9b}$$

Appendix 2

Evaluation of I_{10}

$$I_{10} = \int_{1}^{\gamma} \left[\int_{1}^{t} \frac{(\tau - \lambda)}{\tau^{1-\alpha}(\tau - 1)^{\frac{1}{2}+\alpha}(\gamma - \tau)} d\tau \right] \frac{\mathrm{dt}}{\mathrm{t}^{1/2}(t-1)^{1/2}}$$

$\tau = 1$ is a singular point. Making a substitution

$$\tau - 1 = \xi^m,\ \tau = 1 + \xi^m,\ \mathrm{d}\tau = \mathrm{m}\xi^{m-1}\,\mathrm{d}\xi$$

$$I_{10} = \int_{1}^{\gamma} \left[\int_{0}^{(t-1)^{1/m}} \frac{(1 - \lambda + \xi^m) m \xi^{m-1}}{(1+\xi^m)^{1-\alpha} \xi^{m(\frac{1}{2}+\alpha)}(\gamma - 1 - \xi^m)} \mathrm{d}\xi \right] \frac{\mathrm{d}t}{t^{1/2}(t-1)^{1/2}}$$

Assuming $m = \frac{2}{1-2\alpha}$

$$I_{10} = m \int_{1}^{\gamma} \left[\int_{0}^{(t-1)^{1/m}} \frac{(1 - \lambda + \xi^m)}{(1+\xi^m)^{1-\alpha}(\gamma - 1 - \xi^m)} \mathrm{d}\xi \right] \frac{\mathrm{d}t}{t^{1/2}(t-1)^{1/2}}$$

For removing the singularity at $t = 1$, substituting $t - 1 = \eta^2$, $\mathrm{t} = 1 + \eta^2$, $\mathrm{dt} = 2\eta\,\mathrm{d}\eta$

$$I_{10} = m \int_{0}^{\sqrt{\gamma-1}} \left[\int_{0}^{\eta^{1-2\alpha}} \frac{(1 - \lambda + \xi^m)}{(1+\xi^m)^{1-\alpha}(\gamma - 1 - \xi^m)} \mathrm{d}\xi \right] \frac{2\,d\eta}{\sqrt{1+\eta^2}}$$

making a substitution

$$\eta = 0.5\sqrt{\gamma - 1}(1 + Y);$$

$$\mathrm{d}\eta = 0.5\sqrt{\gamma - 1}\,\mathrm{d}Y$$

$$I_{10} = m\sqrt{\gamma - 1} \int_{-1}^{1} \left[\int_{0}^{\eta^{1-2\alpha}} \frac{(1 - \lambda + \xi^m)}{(1+\xi^m)^{1-\alpha}(\gamma - 1 - \xi^m)} \mathrm{d}\xi \right] \frac{\mathrm{d}Y}{\sqrt{1+\eta^2}}$$

Assuming

$$\xi = 0.5\eta^{1-2\alpha}(1 + X)$$

$$d\xi = 0.5\eta^{1-2\alpha}\,\mathrm{dX}$$

$$I_{10} = m\sqrt{\gamma - 1}\int_{-1}^{1}\left[\int_{-1}^{1}\frac{(1-\lambda+\xi^m)0.5\eta^{1-2\alpha}}{(1+\xi^m)^{1-\alpha}(\gamma-1-\xi^m)}\mathrm{d}X\right]\frac{\mathrm{d}Y}{\sqrt{1+\eta^2}}$$

$$= m\sqrt{\gamma-1}\sum_{i=1}^{N}W(i)\sum_{j=1}^{N}W(j)F\{\xi(Y(i),X(j)),\eta(Y(i))\}$$

The integration is carried out using 96 Gaussian weights and 96 abscissas, i.e., $N = 96$.

References

1. Carter DL (1985) Controlling erosion and sediment loss on furrow-irrigated land. Reprinted from soil erosion and conservation, soil conservation society of America
2. Samal KP, Mishra GC (2016) Analysis of seepage from a triangular furrow considering soil capillarity using inverse hodograph and conformal mapping technique. ISH J Hydraul Eng. http://dx.doi.org/10.1080/09715010.2016.1213144
3. Harr ME (1962) Groundwater and seepage. McGraw-Hill, New York
4. Tiwari KN, Mal PK, Singh RM, Chattopadhyay A (1998) Response of Okra (Abelmoschus Esculentus (L.) Moench.) to drip irrigation under mulch and non-mulch condition. Agric Water Manage 38:91–102
5. Tiwari KN, Mal PK, Singh RM, Chattopadhyay A (1998) Feasibility of drip irrigation under different soil covers in tomato. J Agric Engrg 35(2):41–49

Bearing Capacity Analysis Based on Optimization of Single Layer Depth of Reinforcement Below Rectangular Footing

Bandita Paikaray, Sarat Kumar Das, Benu Gopal Mohapatra, Sahil Pritam Swain, and Sabyasachi Swain

Abstract This paper presents a study based on laboratory model tests conducted with rectangular footing on unreinforced and geosynthetics reinforced crusher dust. The test was conducted at loose (RD = 29%) and medium dense (RD = 64%) state of crusher dust taking the variation of single layer depth of geosynthetic placed below foundation to find the maximum ultimate bearing capacity. The single layer depth was varied at a depth of 0.2, 0.25, 0.3 and 0.35*B* for loose state of crusher dust and for medium dense state, it was varied with 0.3, 0.4 and 0.5*B* (where *B* is the width of the footing). Also geogrid along with geosynthetic (stagged one above another) was used below footing at the optimum depth to find their combined effect. It has been observed that the inclusion of reinforcement at various depths provided below footing gives higher ultimate bearing capacity compared to unreinforced condition at both relative density states. For loose state, the optimum single layer depth of geosynthetic was found out to be 0.3*B* and for medium dense state, it was found to be 0.4*B*. The percentage variation of ultimate bearing capacity of reinforced crusher dust with respect to unreinforced at loose state (0.3*B*) was 112.56% and for medium dense state (0.4*B*) was 71.15%. Increase in relative density along with addition of reinforcement layer to the crusher dust has great impact on the bearing capacity of footing. At both the reinforced and unreinforced conditions, higher density produced higher result than the loose dense state of the material. Combination of geosynthetic and geogrid used as a single layer below footing at optimum depth for RD 29%

B. Paikaray (✉) · S. P. Swain · S. Swain
School of Civil Engineering, KIIT DU, Bhubaneswar, India
e-mail: bandita.paikarayfce@kiit.ac.in

S. P. Swain
e-mail: sahilpritam12@gmail.com

S. Swain
e-mail: sabyasachiswain009@gmail.com

S. K. Das
I.I.T(ISM), Dhanbad, India
e-mail: saratdas@iitism.ac.in

B. G. Mohapatra
KIIT DU, Bhubaneswar, India
e-mail: bmohapatrafce@kiit.ac.in

B. B. Das et al. (eds.), *Recent Developments in Sustainable Infrastructure*, Lecture Notes in Civil Engineering 75, https://doi.org/10.1007/978-981-15-4577-1_4

and RD 64% resulted in higher ultimate bearing capacity compared to the single reinforcement and the percentage variation with respect to unreinforced condition, it is found out to be 121.85% at RD 29% and 74.46% at RD 64%.

Keywords Model test · Crusher dust · Rectangular footing · Geosynthetic · Relative density · Ultimate bearing capacity

1 Introduction

For the last two decades in Civil Engineering, application of synthetic material such as geogrid, geonet, geomembrane and geotextile has been known as a common technique to increase the ultimate bearing capacity of soils. Reinforcing soil for increasing the load-carrying capacity has been a common practice and used widely since nineteenth century. Many researchers have used reinforcement as synthetics material in different forms like geosynthetic, geogrid and geonet and in many other forms to study its effect on bearing capacity by model testing or by analytical/numerical method. It can be observed that most of the researches have studied on behaviour of strip and circular footing in reinforced case to find the effect of reinforcement on bearing capacity [1–6] but square and rectangular footings are most common in practice [7–12]. All model testing have been conducted on different shapes of footings to find their best effect on bearing capacity. The effect of reinforcement, effect of ring footing, effect of shape of footing and effect of contamination have been studied by different researchers. In above tests, the effect of optimization of reinforcement layer has not been conducted. Also using an industrial waste like crusher dust in huge amount has not been reported except sand. In present study, crusher dust has been taken as foundation material. Present study involves utilisation of crusher dust as a foundation material in bulk. The annual production of quarry dust is roughly about 1.48 billion tonnes (2016 statics). It can be used in road pavements, filling pits and abundant mines and in foundations based upon its suitability and ease of availability in nearby areas. In present study, isolated rectangular (15 cm × 30 cm) footing for model testing on unreinforced and reinforced (geosynthetic and geogrid) condition at both loose and dense states of RD 29% and RD 64% has been taken and the effect of reinforcement depth below footing at both density conditions has been optimised. The objective of study is to determine the optimum depth of placing reinforcement and also the combined effect of geosynthetic and geogrid on bearing capacity at the same optimum depth of placement.

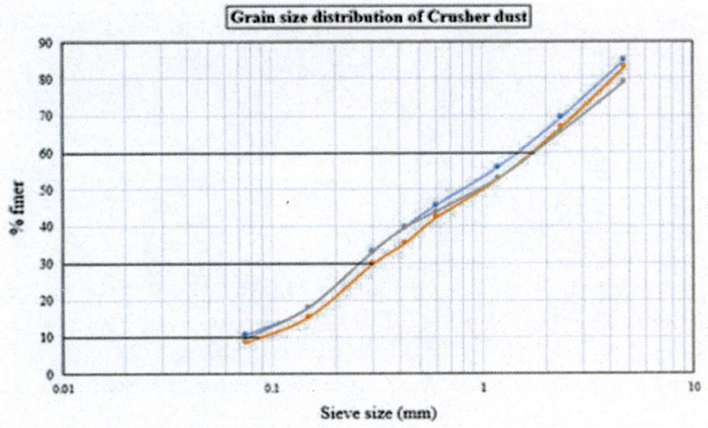

Fig. 1 Grain size distribution of crusher dust

2 Materials and Methods

2.1 Material

2.1.1 Crusher Dust

For the complete experimental work, crusher dust was used in the foundation. It was collected from Khurda, Odisha, India. The gradation of the curve was found to be poorly graded soil showing the property of sand (SP). From the grain size distribution (Fig. 1), the coefficient of uniformity (C_u) and coefficient of curvature (C_c) were found to be 20.6 and 0.519, respectively. The maximum (e_{max}) and minimum (e_{min}) void ratios were found out to be 0.591 and 0.323, respectively, by dynamic compaction test. Sand rain technique was adopted based on the relative density finalised for falling the material from a fixed height from the bottom of the tank. Each time a 10 cm depth of fill was maintained to obtain a uniform density throughout the tank.

2.1.2 Geosynthetics

In the present study, a nonwoven geosynthetic which is black in colour was used as a reinforcement. It improves the ground stabilisation, good in tension and enhances the tensile property of the soil. It also distributes the load across the large area. Similarly, biaxial geogrid was also used as a reinforcement in present study. The tensile strengths of geosynthetic geogrid were found to be 8.4 and 4.3 kN/m, respectively, and the thicknesses of sheets were measured to be 1.15 mm and 0.48 mm, respectively. Geosynthetic was extensively used in the experiment at different depths (as a

Fig. 2 Geosynthetic and geogrid

first layer) to find out its optimum depth of giving maximum bearing capacity. The schematic view of reinforcements is shown in Fig. 2.

2.2 *Methodology*

2.2.1 Footing Model

Footing of rectangular shape with length, width and height of 30 cm, 15 cm and 2.5 cm, respectively, having polished base was used in all the tests. It was used as a surface footing in the present study. The ball bearing was paced at the centre of the groove to transfer the load, Fig. 3.

Fig. 3 Model footing used for model test

2.2.2 Test Programme

Experimental test was conducted on a test tank of size 1.6 m (L) × 1.0 m (B) × 1.2 m (H). The boundary effect was minimised based upon the size of the tank and the test plate. Concentric vertical load was applied through the load cell which was connected to the hydraulic jack from the top and to the attachment rod in the bottom setting to the ball bearing placed over the test plate for transferring the load during test. Two LVDTs were connected to the diagonal side of the plate placed vertically for measuring the settlement.

3 Experimental Procedure

The waste material 'crusher dust' was poured by sand rain technique from a desired height to maintain the relative densities of 29 and 64% inside the tank. Uniformity was maintained by filling with a mass of 269 and 294 kg for loose and medium dense states with a uniform layer of 10 cm each time till the top of the reinforcement layer reached. In case of combination of both geosynthetic and geogrid, one was placed above another after reaching the required depth. The top layer spacing of reinforcement from the surface of the material was taken as 'u'. The load was applied with a seating load of 0.3 kN to make the surface levelled and then load was applied gradually till the material failed in shear, Fig. 4.

Getting the load and settlement data from the experiment till the failure, the load settlement curve was drawn and the ultimate load-carrying capacity was determined.

Fig. 4 Load-bearing test on rectangular footing

4 Result and Discussion

4.1 Effect of First Layer of Reinforcement

Model testing was conducted varying the first layer depth '*u*' of reinforcement below footing and the corresponding bearing pressure versus settlement was noted plotted to graph, Fig. 5, at RD 29%. The depth of first layer reinforcement was varied as 0.2, 0.25, 0.3 and 0.35*B* from the bottom of the base of footing. When compared to the unreinforced condition, it can be clearly seen that the inclusion of geosynthetic reinforcement is showing higher ultimate bearing capacity. As the depth of placement of single layer reinforcement below footing increases, the UBC (ultimate bearing capacity) value also increases, but there is no significant increase in the UBC value for increase in the '*u*' value from 0.3 to 0.35*B*. It can be terminated that the provision of reinforcement layer beyond $u/B = 0.35$ will not contribute to any significant increase in bearing capacity of the footing. For unreinforced condition of crusher dust of RD = 29%, the failure is similar to the punching shear failure with loose state of material without provision of any reinforcement to resist the failure. But when the reinforcement is included the failure surface is showing a local shear failure with little surface heave around the footing (RD 29%).

Similarly, bearing pressure is changing with changing the depth of first layer reinforcement for relative density of 64% which is compared with the unreinforced condition, Fig. 6. The first layer depth of reinforcement was varied as 0.3, 0.4 and 0.5*B*, from the base of the footing.

From the figure, it can be seen that there is a decrease in the UBC value after $u/B = 0.4$ implying the higher ultimate obtained at 0.4*B*. This phenomenon can be visualised, as the *u*/*B* exceeds a particular depth [13], and the major soil failure

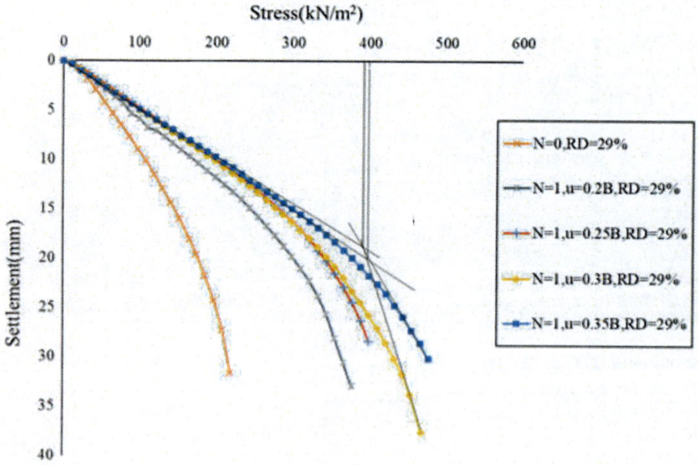

Fig. 5 Bearing pressure versus settlement at RD29%

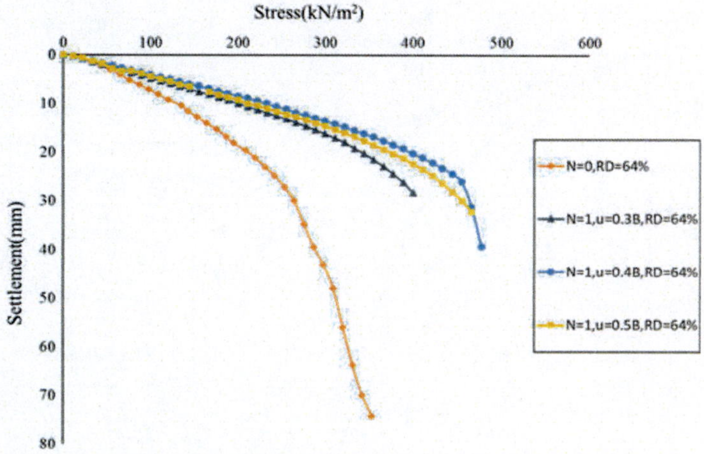

Fig. 6 Bearing pressure versus settlement at RD29%

zone lies above the reinforcement layer and the reinforcement laid below it has no significant contribution towards increase in UBC hence decreasing. It is reaching to the conclusion that the reinforcement layer beyond $u/B = 0.4$ is not contributing any significant increase in bearing capacity of the footing for RD = 64%. In this case, the failure observed is general shear failure with sufficient surface heave around the footing which can be easily visible with naked eye, Fig. 7.

The failure zone depth below footing depends upon the type of footing, shape of footing, relative density and also the type of reinforcement along with the tensile strength of the same.

Fig. 7 Nature of failure surface below footing

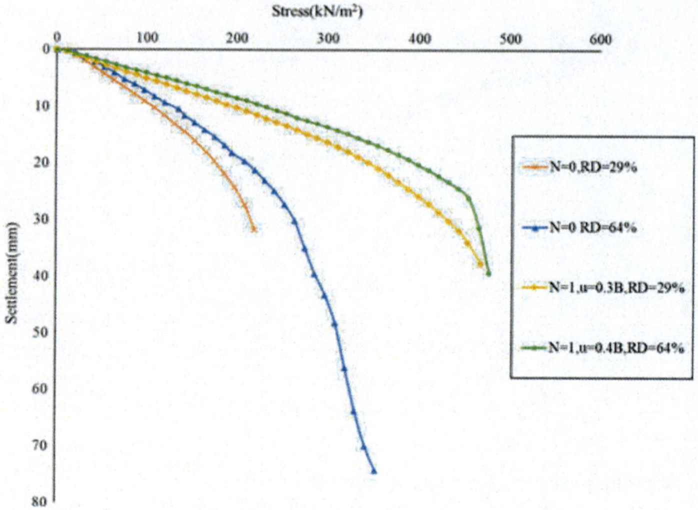

Fig. 8 Density variation at optimum depth

4.2 *Effect of Density on Bearing Capacity of Footing*

The comparison between unreinforced crusher dust at both the density state and with reinforcement at optimum depth is presented in Fig. 8.

The medium dense condition shows a higher UBC as compared to the loose condition. It is more densely packed due to high frictional coefficient, so imparting more bearing capacity. RD 64% shows an increase of 53% in ultimate bearing capacity when compared to the RD 29% in unreinforced condition. From the graph, it can be clearly seen that a great increase in the bearing capacity of the soil after inclusion of reinforcement is obtained. The optimum depths were found out to be $u/B = 0.3$ for loose condition and $u/B = 0.4$ for medium dense condition.

4.3 *Effect of Combination of Reinforcement*

Based on the result obtained at optimum depth, another layer of geogrid reinforcement was added to the single layer reinforcement to find their combined effect. The result obtained at both the density is tabulated below, Table 1. It shows the variation of ultimate bearing capacity (UBC) on crusher dust for two relative densities RD = 29% and RD = 64% and it is compared with the unreinforced condition. For RD = 29%, the reinforcements are provided at $u/B = 0.3$ and for RD = 64% the reinforcements are provided at $u/B = 0.4$ keeping no spacing (staged one above another). It is clear from the table that the medium dense condition is giving higher UBC compared to the loose dense condition by 19.45% at the optimum depth. It is due to the denser

Table 1 Bearing capacity variation with variation of RD and reinforcement

Type	*N*	RD = 29%	RD = 64%	Percentage variation for density (%)
		UBC (kN/m^2)	UBC (kN/m^2)	
–	0	183	278	51.9
GS	1	389	445	14.4
GG + GS	2	406	485	19.45

packing of crusher dust compared to the loose dense condition. The geogrid layer is provided over the geosynthetic layer to provide better confinement of the crusher dust as it has large apertures so as to limit the lateral movement of the soil. But on the same time it has been observed that the increase in UBC due to geogrid is not showing much higher value compared to single layer of geosynthetic at the optimum depth.

5 Conclusions

The bearing capacity of rectangular footing placed on unreinforced and reinforced crusher dust with different relative densities by varying the first layer depth of reinforcement was studied. Based on the experimental studies, following conclusions are drawn:

- Addition of reinforcement giving higher ultimate bearing capacity to rectangular footing when it is compared with the condition without addition of the same.
- As the density increases from low to medium dense, there is an increase in UBC value by 42%.
- For RD = 29%, the optimum depth of first layer of geosynthetic reinforcement at which the UBC value is found to be maximum is 0.3 times width (B) of the footing and for RD = 64%, it is found to be 0.4B. There was an increase of 112.5% in RD = 29% and 71.15% in RD = 64% in UBC when compared to unreinforced condition.
- For reinforced case, RD = 64% is showing higher UBC as compared to RD = 29% by 15%.
- Combination of geosynthetic with geogrid at the optimum depth is increasing the UBC (not appreciable change) by 4.4% at RD 29%, whereas for RD 64% it is increasing by 9%.
- The provision of both geogrid and geosynthetic reinforcement in one layer shows a very little improvement in UBC and settlement when compared to the reinforced condition when reinforcement was placed at optimum depths for both loose and medium dense states.

Bibliography

1. Patra CR, Das BM, Atalar C (2005) Bearing capacity of embedded strip foundation on geogrid-reinforced sand. Geotext Geomembr 23(5):454–462
2. Abtahi SA, Boushehrian AH (2018) Experimental behaviour of circular foundations on oil contaminated sand. Sci Iran, 1–18
3. Basudhar PK, Saha S, Deb K (2007) Circular footings resting on geotextile-reinforced sand bed. Geotext Geomembr 25(6):377–384
4. El Sawwaf M, Nazir A (2012) Behaviour of eccentrically loaded small-scale ring footings resting on reinforced layered soil. J Geotechn Geoenviron Eng 138(2), 376–384
5. Elsaied AE, Saleh NM, Elmashad ME (2015) Behaviour of circular footing resting on laterally confined granular reinforced soil. HBRC J 11(2), 240–245
6. Ismail Ibrahim KMH (2016) Bearing capacity of circular footing resting on granular soil overlying soft clay. HBRC J 12(1):71–77
7. Abu-Farsakh M, Chen Q, Sharma R (2013) An experimental evaluation of the behaviour of footings on geosynthetic-reinforced sand. Soils Found 53(2), 335–348
8. Latha GM, Somwanshi A (2009) Bearing capacity of square footings on geosynthetic reinforced sand. Geotext Geomembr 27(4):281–294
9. Paikaray B, Das S, Mohapatra B (2018) Interference of two shallow square footings on geogrid reinforced crusher dust. In: Sustainable construction and building material, vol 25. Lecture notes in civil engineering, pp 41–60
10. El Sawwaf M, Nazir AK (2010) Behaviour of repeatedly loaded rectangular footings resting on reinforced sand. Alexandria Eng J 49(4):349–356
11. Saha S, Deb K (2017) Bearing capacity of rectangular footings on multilayer geosynthetic-reinforced granular fill over soft soil. Int J Geomech 17(9)
12. Yetimoglu T, Wu JTH, Saglamer A (1994) Bearing capacity of rectangular footings on geogrid reinforced sand. J Geotechn Eng 120(12):2083–2099
13. Guido VA, Chang DK, Sweeney MA (1986) Comparison of geogrid and geotextile reinforced slabs. Can Geotech J 23:435–440.

Decolorization of Congo Red Using Synthesized Titanate Nanotubes (TNTs)

Saismrutiranjan Mohanty, Sanjib Moulick, and Sanjoy Maji

Abstract Congo red (CR, $\lambda_{max} = 497$ nm, $C_{32}H_{22}N_6Na_2O_6S_2$) removal efficiency from aqueous media was evaluated on synthesized titanate nanotubes (TNTs). TNTs were synthesized hydrothermally and characterized by SEM/EDX and XRD. The as-prepared TNTs were applied on Congo red to adsorb and/or photodegrade. Adsorption, in the absence of light, followed by photodegradation using UV irradiation (UV-C, 360 nm) was conducted. The results of the adsorption experiments suggested that a dose of 4 g/L TNTs could adsorb ~86% CR with a shaking speed of 520–540 rpm within 45 min at room temperature and pH ~6.8 ± 0.2 (pH of de-ionized water), when the initial CR concentration was kept at 6 mg/L. Under the same experimental conditions, the UV light was allowed to irradiate the same experimental suspension for another 5 h to achieve the maximum CR removal efficiency ~98%. The adsorptive behavior of TNTs was evaluated using Langmuir isotherm and was found to be 12.04 mg/g. The feasibility of adsorption was understood by D–R isotherm. Adsorption is followed Langmuir and Freundlich isotherms in a close proximity. Positive free energy change value, ΔG^0, suggests the process to be non-spontaneous and R_L (dimensionless equilibrium parameter) value implies the favorability of adsorption process.

Keywords Titanate nanotubes · Congo red · Adsorption/photodegradation · Isotherms

S. Mohanty · S. Maji (✉)
Department of Chemistry, School of Applied Sciences, KIIT Deemed to be University, Bhubaneswar, India
e-mail: maji.sanjoy@gmail.com

S. Mohanty
e-mail: saismruti1994@gmail.com

S. Moulick
School of Civil Engineering, KIIT Deemed to be University, Bhubaneswar, India
e-mail: smoulickfce@kiit.ac.in

B. B. Das et al. (eds.), *Recent Developments in Sustainable Infrastructure*, Lecture Notes in Civil Engineering 75, https://doi.org/10.1007/978-981-15-4577-1_5

1 Introduction

Dyes are basically coloring materials and are responsible for the pollution of environment. About 20% of dyes are lost as effluent from textile industries and in dying processes [1–11]. Different types of industries like food, paper, cosmetics, plastic, and textile utilize dyes for coloring the produced products. Dyes are very hard to degrade due to their synthetic origin and complex aromatic molecular structure. Congo red, a diazo dye, blue in acid solution and red in alkaline solution, is used as a biological stain or as an indicator. It is carcinogenic and needs to be removed from aqueous media for human consumption.

Commonly used treatment techniques include adsorption, flocculation, biological treatment, reverse osmosis, ultra- as well as nano-filtration, and photocatalysis [3]. Commonly used photocatalysts are TiO_2 (anatase, P25), SnO_2, ZnO, etc. The bandgaps for TiO_2, SnO_2, and ZnO are 3.2 eV, 3.6 eV, and 3.37 eV respectively. Among all the photocatalysts, TiO_2 is used more frequently as it transfers electron to molecular oxygen faster than the other available photocatalysts due to its low bandgap. Furthermore, it is readily available, cheap, and eco-friendly in nature. In the present study, synthesized titanate nanotube (TNT) has been employed to remove Congo red (CR) from aqueous media.

2 Materials and Methods

Congo red (1000 mg/L) was prepared using $C_{32}H_{22}N_6Na_2O_6S_2$ (SRL Chemicals) and diluted as per the requirement. Titanium dioxide (TiO_2) powder was procured from Loba chemicals. The concentrations of CR were recorded at 497 nm (λ_{max}).

All apparatus were washed properly with tap water followed by double distilled water (bio-age water purification) followed by dilute acid treatment. Remi (R-24) was used for the centrifugation of the samples. The pH measurements were done using a digital pH meter (Wensar, WPH-10). Sartorius electronic balance, BSA224S-CW, was used for measurement of mass. Shimadzu, UV-1800, was employed to record spectra. For powder XRD analysis, X-ray diffractometer (Shimadzu XRD 6100) was used with a Cu Kα radiation. For SEM/EDX studies, FEI-SEM (Apreo LoVac) was utilized.

TiO_2 powder (anatase, P-25, 25 g) was mixed with 100 mL of 10 M aqueous NaOH solution and taken into a stainless steel hydrothermal unit and placed in a hot air oven for 14 h at 150 °C. The positive charge Na^+ gets encapsulated in between the edge-shared TiO_6 and octahedral layers of $Na_2Ti_3O_7$. In this critical situation, a positive strong generated static attractive force holds Na^+ and TiO_6 units tightly together prohibiting the rolling of these layers into the nanotubes. In hydrothermal treatment over a period of time, Na^+ is gradually un-encapsulated with intercalated H_2O molecules into the interlayer space of TiO_6 sheets.

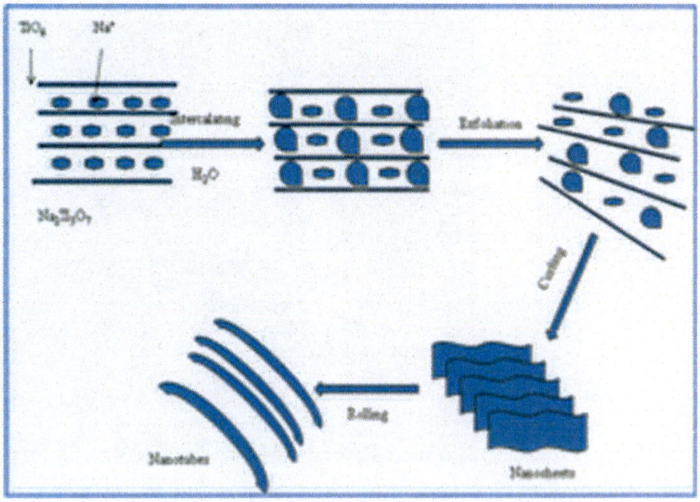

Fig. 1 Schematic presentation of the formation of TNT

Larger size of H_2O molecule than Na^+ helps in enlargement of interlayer distance during Na^+ un-encapsulation, leading to weakening of attractive static interaction force existing between TiO_6 and octahedral layers. Finally, layered $Na_2Ti_3O_7$ particle gradually exfoliates to form numerous sheet-shaped products and finally gets rolled to nanotubes. The nanotube formation mechanism is shown in Fig. 1 [11, 12]. The titanate nanotubes, formed by alkaline hydrothermal treatment, are generally crystalline in nature. However, there are several controversial issues regarding the exact structure of these materials [11, 13]. The produced material was washed properly and repeatedly with double distilled water to attain pH ~ 7 and then dried up in a hot air oven at 100 °C for duration of 6 h to generate TNT in solid powder form. This was reserved in a black container in dry condition for further use.

The prepared TNT was again treated with 0.01 M HCl (3 g TNT in 10 mL, 6 times in an hour), to replace Na^+ by H^+. The suspension was repetitively washed in double distilled water after each time of acid treatment to achieve pH ~ 7 and dried at 100 °C for 6 h in a hot air oven. The resulting material is known as HTNT. Both TNT and HTNT were characterized by SEM/EDX and XRD [13]. It is worth to mention here that in the optimal conditions, efficiency of HTNT is comparable with TNT; hence, the entire study was carried out using TNT only.

The evaluation of phase and structure of the synthesized samples was carried out using an X-ray diffractometer (Shimadzu XRD 6100) with Cu K_α radiation (wavelength $\lambda = 0.15406$ nm) in the scan range $2\theta = 20$–80° with a data acquisition rate of 0.033° per step.

Experimental suspensions (50 mL) with varying concentrations of CR (1–8 mg/L) were taken in a 500 mL beaker (surface area 55.44 cm^2). TNTs of different doses (1–6 g/L) were mixed in the CR suspension. The pH of the solution was fixed at 6.8 ± 0.2. Beaker with suspension was then put on an electrically operated magnetic

Fig. 2 Experimental set up

stirrer with an rpm of 520 ± 20 for around 45 min to achieve the maximum removal of solute on TNT surfaces under dark conditions and around 5 h under UV light (UV-C, 360 nm) for photodegradation in the presence of adequate amount of oxygen at room temperature of 25 ± 2 °C. The distance between the CR suspension level in the beaker to the UV light was kept at 13.2 cm. The experimental setup used in this study was covered with a black cloth to prevent the irradiation of the hazardous UV radiation to the surrounding (Fig. 2).

3 Results and Discussions

During the formation of nanotube structures, wrapping along a definite crystallographic axis results in the widening of peaks along this direction and therefore assigning Miller indices to those peaks are difficult. In addition to that, the titanate nanotubes easily undergo phase transformation during the physical or chemical treatments (viz., heating, acid treatment, etc.) [11, 13]. The X-ray Diffraction (XRD) patterns of TNT sample (Sample 1) and HTNT sample (Sample 2) were measured in order to determine the crystalline phases present in the nanostructures (Fig. 3). It can be noted from the figure that the samples do not contain any rutile or anatase phase of TiO_2. Further, diffraction peaks are present in both the samples except a little change in intensity. The diffraction peaks of the sample can be linked with $A_2Ti_nO_{2n+1}$ crystallographic series with $n = 5$ and A = H, Na [13]. In case of HTNT, no additional peaks in the XRD pattern appeared except a little change in intensity for all the peaks as compared to that of TNT sample. It can be noted from Fig. 3 that the XRD pattern of HTNT sample (Sample-2) matches fairly well with the monoclinic (*C*2/*m*) $H_2Ti_5O_{11}$, H_2O structure (JCPDF-00-044-0131). These results suggest that HTNT sample originates from sodium titanate phase. A comparison of the observed 2θ values (in degree) and interplanar spacing (d_{hkl}) along with the standard values

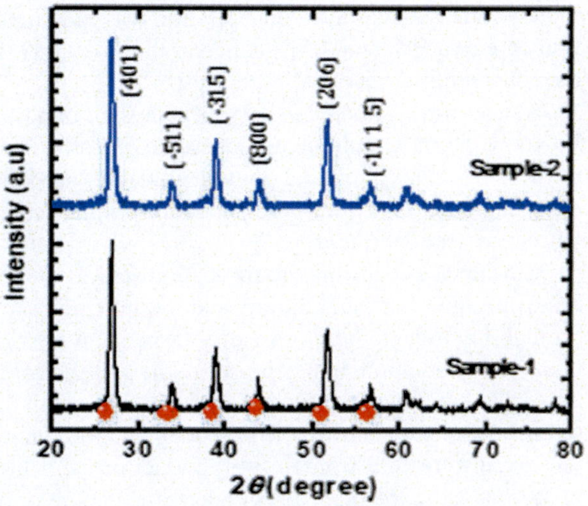

Fig. 3 X-ray diffraction (XRD) pattern of Sample-1 (TNT) and Sample-2 (HTNT). The (red) solid dots represent the peak positions corresponding to monoclinic $H_2Ti_5O_{11}$, H_2O (JCPDF-00-044-0131)

is summarized in Table 1. SEM/EDX study was conducted using FEI-SEM (Apreo LoVac) for both TNT and HTNT. However, with some impurities, the main composition of the synthesized precursors was found to be the titanium, as P-25 is ~99% pure experiments were conducted at different concentrations of CR (1–8 mg/L) with different doses of TNT (1–6 g/L) to determine the optimum conditions for removal of CR from suspension. The optimum removal percentage of CR from synthetic sample was found at the CR concentration of 6 mg/L with a TNT dose of 4 g/L. The maximum adsorption and photodegradation were achieved at 45 min and 5 h, respectively, with maximum removal efficiency of CR at ~98%. In a similar way, the decrease in percentage of removal with respect to increase in concentration for CR beyond 6 mg/L was caused due to the fixed precursor sites with respect to chemical

Table 1 Comparison of the observed 2θ values (in degree) and interplanar spacing (d_{hkl})

Sl. no.	Lattice planes	2θ value (in degree)			Interplanar spacing (d_{hkl}) in Å		
		Standard	Experimentally measured		Standard	Experimentally measured	
			Sample 1 (TNT)	Sample 2 (HTNT)		Sample 1 (TNT)	Sample 2 (HTNT)
1	(401)	26.197	26.87	26.92	3.399	3.315	3.310
2	(−511)	33.640	33.80	33.67	2.662	2.649	2.659
3	(−315)	38.525	38.87	38.96	2.335	2.315	2.310
4	(800)	43.627	43.76	43.87	2.073	2.067	2.062
5	(206)	51.008	51.70	51.50	1.789	1.766	1.773
6	(−11 15)	56.253	56.59	56.60	1.634	1.625	1.625

moiety. The residual dye concentration was measuered by drawing a standard calibration curve (Abs =0.051x concentration + 0.089, $R^2 = 0.9993$) of Congo red in the concentration range of 0–20 mg/L.

Contact time (solute and solvent) plays an important role on removal efficiency. To know the maximum uptake of solute on sorbent surface with time, two types of studies—(i) adsorption and photodegradation and (ii) simultaneous adsorption and photodegradation—were performed at optimum conditions (TNT dose: 4 g/L and CR concentration: 6 mg/L).

In the first case, the maximum adsorption was found to occur within 45 min (~86% removal) under dark conditions and the maximum removal of ~98% was achieved through photodegradation of same suspension within 5 h (Fig. 4) of shaking time under UV irradiation. It may be due to the presence of fresh and active surface sites in the beginning which were exhausted gradually with time causing the suppression of removal efficiency during the photodegradation. Photodegradation of dye is caused due to transfer of electron(s) from conduction band to valence band of semiconductor photocatalyst in the presence of adequate amount of oxygen in the suspension.

In the second case, removal efficiency of CR was high in the beginning and then suppressed (Fig. 4). The obtained results may be due to the presence of fresh TNT photocatalyst surfaces and may be due to the simultaneous occurrence of adsorption and photodegradation of the solute molecules. Slowly decreasing activity of precursors' sites may have suppressed the removal of CR as it reflects in photodegradation also.

Langmuir as well as Freundlich isotherm studies were performed (Fig. 5) in order to evaluate the maximum adsorption capacity of TNT onto Congo red. The linearized forms of Langmuir isotherm (Eq. 1) and Freundlich isotherm (Eq. 2) are presented as follows:

$$(1/q_e) = (1/Q\mathrm{max}) + (1/bQ\mathrm{max}) \times (1/C_e) \tag{1}$$

Fig. 4 Kinetics of adsorption and photodegradation

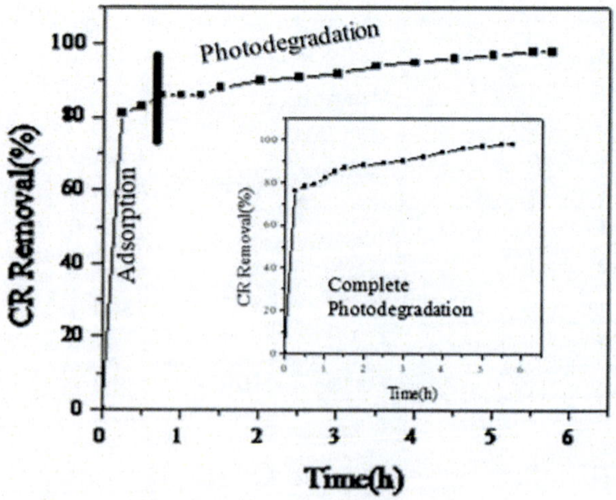

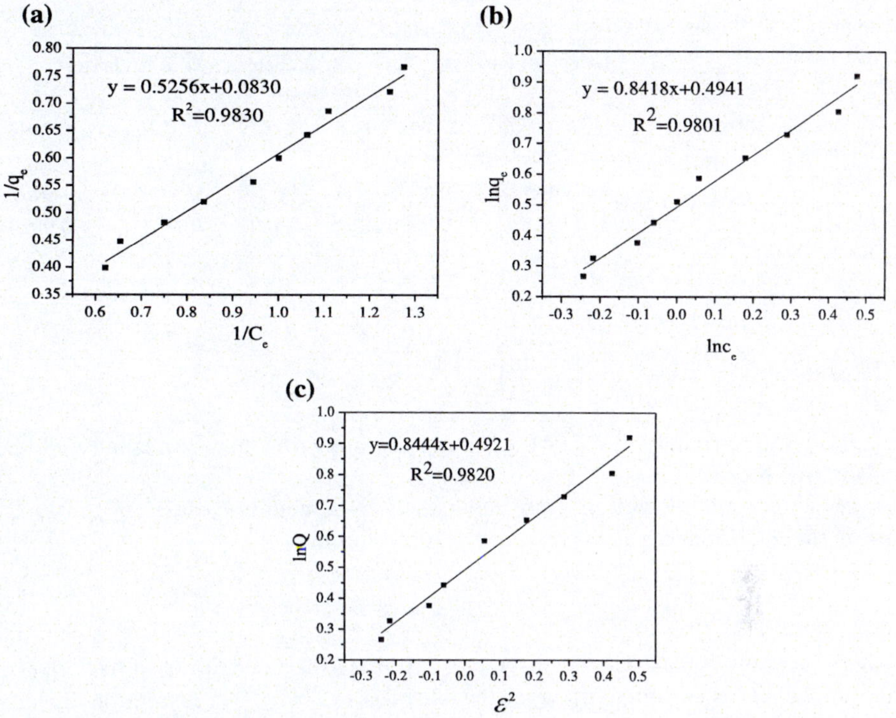

Fig. 5 Isotherms **a** Langmuir, **b** Freundlich, **c** D–R

$$\ln(q_e) = \ln(k_f) + (1/n)\,[\ln(C_e)] \tag{2}$$

where q_e = quantity of CR adsorbed on TNT surface, Q_{max} = maximum adsorption capacity, b = adsorption constant associated with the enthalpy of adsorption, C_e = equilibrium CR concentration,

n and k_f are adsorption intensity and adsorption capacity, respectively.

The linear Langmuir isotherm was obtained as $1/q_e = 0.5256(1/C_e) + 0.0830$ ($R^2 = 0.9830$) with maximum adsorption capacity (Q_{max}) of 12.04 mg/g. The Freundlich isotherm was drawn as $\ln q_e = 0.8418(\ln C_e) + 0.4941$ ($R^2 = 0.9810$). The calculated value of n ($1.1879 > 1$) shows (Table 2) the higher CR adsorption capacity toward TNT. The calculated dimensionless equilibrium parameter using the following equation ($R_L = 0.5135$) suggested the favorable adsorption [14].

$$R_L = \frac{1}{(1 + bC_0)} \tag{3}$$

The Gibbs free energy change (ΔG^0) value was estimated using the following equation:

Table 2 Isotherm constant values

	Langmuir isotherm		Freundlich isotherm		D–R isotherm	
Congo red synthetic sample (6 mg/L) at	Q_{max} (mg/g)	12.04	n	1.1879	E (kJ/mol)	0.7692
Temperature	b (L/mg)	0.1579	k_f	1.6390	Q (mg/g)	1.6357
(298 ± 2 K)	R^2	0.9830	R^2	0.9801	R^2	0.9820
ΔG^0 (kJ/mol)	1.641					

$$\ln\left(\frac{1}{b}\right) = \frac{\Delta G^0}{RT} \tag{4}$$

ΔG^0 was found to be 1.641 kJ/mol suggesting the non-spontaneity of the adsorption process.

Feasibility of adsorption was understood by Dubinin–Radushkevich (D–R) isotherm [14, 15] and was expressed as (Fig. 5) follows:

$$\ln Q = \ln Q_m - k\varepsilon^2 \tag{5}$$

where ε represents the Polanyi constant and expressed as $\varepsilon = RT[\ln(1 + (1/C_e)]$, Q is the amount of CR adsorbed (mg/g) on the unit mass of TNT, Q_m is adsorption capacity (~11.00 mg/g), k is a constant associated to adsorption energy, R = universal gas constant, and T = absolute temperature (K). The mean free energy of adsorption (E =~3.829 kJ/mol) indicates the adsorption of CR onto TNT is of physical adsorption as the obtained result is lower than 8 kJ/mol and was estimated using the following equation [16]:

$$E = -(2k)^{-0.5} \tag{6}$$

Under UV irradiation, the activated CR molecules (dye) were converted into short-lived active transient free radicals species on contacting with TNT surfaces transferring the produced negative charged electron(s) on it. This produced electron(s) were injected into the conduction band of TNT linked through the nanotube sheets. This process could also be conducted by radical dye molecules. The injected electrons were detained by the surface adsorbed O_2 molecules yielding $O_2^{\bullet-}$ and $HO^{\bullet}$ radical. The dye molecules were rapidly mineralized by the radicals. The water quality parameters prior to and after the treatment are presented in Table 3. It can be noted that except slight decrease in dissolved oxygen, other water quality parameters remained approximately same. The probable photodegradation mechanism of the dye in the presence of UV light with TNT as the photocatalyst is presented as follows [11, 17]:

$$\text{Dye} + h\nu \rightarrow \text{Dye}^* \tag{7}$$

Table 3 Water quality pre- and post-treatment

Parameter	Pre-treatment	Post-treatment [CR] = 6 mg/L
pH	8.2	8.4
EC (μs/cm)	288	290
Fe_{tot} (mg/L)	0.67	0.66
Cl^- (mg/L)	230	230
DO (mg/L)	6.1	5.8
TDS (mg/L)	47	47
Tem (°C)	27	28
ORP (mV)	98	102
Alk (mg/L)	2.7	2.8
Hardness as $CaCO_3$ (mg/L)	78	78.2
Salinity (PSU)	0.15	0.2
F^- (mg/L)	5.8	5.8
HCO_3^- (mg/L)	109	109
NO_3^- (mg/L)	22.2	22.2
SO_3^- (mg/L)	4.6	4.6
Na^+ (mg/L)	16.6	16.6
K^+ (mg/L)	3.4	3.4
Ca^{++} (mg/L)	28.3	28.4
Mg^{++} (mg/L)	11.8	11.7

$$\text{Dye}^* + \text{TNT} \rightarrow \text{Dye}^{\cdot} + \text{TNT (e)} \tag{8}$$

$$\text{TNT (e)} + O_2 \rightarrow \text{TNT} + O_2^{\cdot-} \tag{9}$$

$$O_2^{\cdot-} + \text{TNT (e)} + 2H^+ \rightarrow H_2O_2 \tag{10}$$

$$H_2O_2 + \text{TNT (e)} \rightarrow OH^{\cdot} + OH^- \tag{11}$$

$$\text{Dye}^{\cdot} + O_2\left(\text{or } O_2^{\cdot-} \text{or } {}^{\cdot}OH\right) \rightarrow \text{Degraded mineralized products} \tag{12}$$

4 Conclusion

Efficiency of TNT in removal of CR from aqueous media through adsorption and photodegradation at room temperature was studied. Both adsorption and photodegradation (UV-C, 360 nm) contributed to remove CR ~ 98% within 5.75 h at a TNT dose of 4 g/L and CR concentration of 6 mg/L at a pH ~ 6.8 ± 0.2 (pH of de-ionized water). The adsorptive capacity of TNT using Langmuir isotherm was found to be 12.04 mg/g. Adsorption followed Langmuir and Freundlich isotherms in a close proximity. Positive free energy change values and dimensionless equilibrium parameter suggest the process to be non-spontaneous, favoring the adsorption process.

Acknowledgments The authors acknowledge SERB, DST, Government of India, for the financial support (vide contract no. ECR/2016/001315).

References

1. Ratna, Padhi BS (2012) Pollution due to synthetic dyes toxicity & carcinogenicity studies and remediation. Int J Environ Waste Manag 3(3), 940–955
2. Robinson T, Mcmullan G, Marchant R, Nigam P (2001) Remediation of dyes in textile effluent: a critical review on current treatment technologies with a proposed alternative. Biores Technol 77(3):247–255
3. Guillard C, Lachheb H, Houas A, Ksibi M, Elaloui E, Herrmann JM (2003) Influence of chemical structure of dyes of, pH and of inorganic salts on their photocatalytic degradation by TiO_2 comparison of the efficiency of powder and supported TiO_2. J Photochem Photobiol A: Chem 158(1):27–36
4. Crini G (2006) Non-conventional low-cost adsorbents for dye removal: a review. Biores Technol 97(9):1061–1085
5. Han RP, Ding DD, Xu YF, Zou WH, Wang YF, Li YF, Zou LN (2008) Use of rice husk for the adsorption of congo red from aqueous solution in column mode. Biores Technol 99(8):2938–2946
6. Wu YJ, Zhang LJ, Gao CL, Ma JY, Ma XH, Han RP (2009) Adsorption of copper ions and methylene blue in single and binary system on wheat straw. J Chem Eng Data 54(12):3229–3234
7. Gupta VK, Jain R, Varshney S (2007) Removal of reactofix golden yellow 3 RFN from aqueous solution using wheat husk—an agricultural waste. J Hazard Mater 142(1–2):443–448
8. Fujishima A, Rao TN, Tryk DA (2000) Titanium dioxide photocatalysis. J Photochem Photobiol, C 1(1):1–21
9. Tokuhisa H, Hammond PT (2003) Solid-state photovoltaic thin films using TiO_2, organic dyes, and layer-by-layer polyelectrolyte nanocomposites. Adv Func Mater 13(11):831–839
10. Al-Thabaiti SA, Hahn R, Liu N, Kirchgeorg R, So S, Schmuki P, Basahel SN, Bawaked SM (2014) NH_3 treatment of TiO_2 nanotubes: from N-doping to semimetallic conductivity. Chemica Commun 50(59):7960–7963
11. Liu R, Yang W-D, Chueng H-J, Ren B-Q (2014) Preparation and application of titanate nanotubes on dye degradation from aqueous media by UV irradiation. J Spectrosc 2015, Article ID 680183, 1–9
12. Wei M, Konishi Y, Zhou H, Sugihara H, Arakawa H (2005) Formation of nanotubes TiO_2 from layered titanate particles by a soft chemical process. Solid State Commun 133(8):493–497
13. Thorne A, Kruth A, Tunstall D, Irvine JTS, Zhou W (2005) Formation, structure, and stability of titanate nanotubes and their proton conductivity. J Phys Chem B 109(12):5439–5444

14. Singh TS, Pant KK (2004) Equilibrium, kinetics and thermodynamic studies for adsorption of As(III) on activated alumina. Sep Purif Technol 36(2):139–147
15. Kundu S, Gupta AK (2006) Adsorptive removal of As(III) from aqueous solution using iron oxide coated cement (IOCC): evaluation of kinetic, equilibrium and thermodynamic models. Sep Purif Technol 51(2):165–172
16. Mahramanlioglu M, Kizilcikli I, Bicer IO (2002) Adsorption of fluoride from aqueous solution by acid treated spent bleaching earth. J Fluorine Chem 115(1), 41–47
17. Mahlambi MM, Ngila CJ, Mamba BB (2015) Recent developments in environmental photocatalytic degradation of organic pollutants: The case of titanium dioxide nanoparticles—a review. J Nanomater 2015, Article ID 790173, 1–29

A Case Study of Geotechnical Investigations for Soft Lateritic Soil Site in South Goa

Priyanka Bhadke and Purnanand P. Savoikar

Abstract All civil engineering structures require foundation to rest on sufficiently strong soil or rock strata. Lateritic soils have been encountered in abundance in most of the parts of Goa and are considered to be one of the problematic soils. These soils are highly porous, permeable and slightly acidic with low pH values. In the present study, a site for G+3 RCC structure located at the village Cacora Curchorem in South Goa is selected where the construction was in progress for Part A and B of the building which was resting on the rock/hard soil. However, during the excavation work for the part C, a very soft lateritic soil was encountered for the depths exceeding 10 m. The present paper discusses the geotechnical investigations and plate load tests conducted on the site to ascertain the safe bearing capacity of the soil and the suitable ground improvement technique for the ongoing construction work at the site.

Keywords Lateritic soil, plate load tests, south goa · Ground improvement

1 Introduction

Laterites are very important building materials and are normally found in the world in between the tropical and subtropical regions. They are observed in the regions of high rainfall and high temperature and are formed due to weathering action of these agencies. They contain various minerals like iron, aluminium, quarts and others in varying percentages. The term Laterites was coined by Buchanan [4] wherein he referred to them as soft rock which can be cut with knife (*Later* means bricks in Latin). Buchanan [4] used this terminology "laterites" first while dealing with soft rock from Peninsular region of India. According to him, this rock can be cut into bricks which can be air-dried to produce harder material and can be used for construction of buildings. Figure 1 shows a typical lateritic soil profile in Goa.

P. Bhadke (✉) · P. P. Savoikar
Civil Engineering Department, Goa Engineering College, Farmagudi, Goa 403401, India
e-mail: piyubhadke306@gmail.com

P. P. Savoikar
e-mail: psavoikar@gmail.com

B. B. Das et al. (eds.), *Recent Developments in Sustainable Infrastructure*, Lecture Notes in Civil Engineering 75, https://doi.org/10.1007/978-981-15-4577-1_6

Fig. 1 Soil profile at Balli, Goa resting on weathered phyllites [5]

Laterite is developed by intensive and prolonged chemical weathering of the underlying parent rock which is termed as laterization. Laterites and lateritic soils are found in the world in hot and wet tropical areas in countries like Cambodia, Thailand, Kenya, Vietnam, Malawi, India, etc. Laterites exist both in soil and rock forms and are rich sources of iron and aluminium. Figure 2 shows the distribution of different soil types over the world.

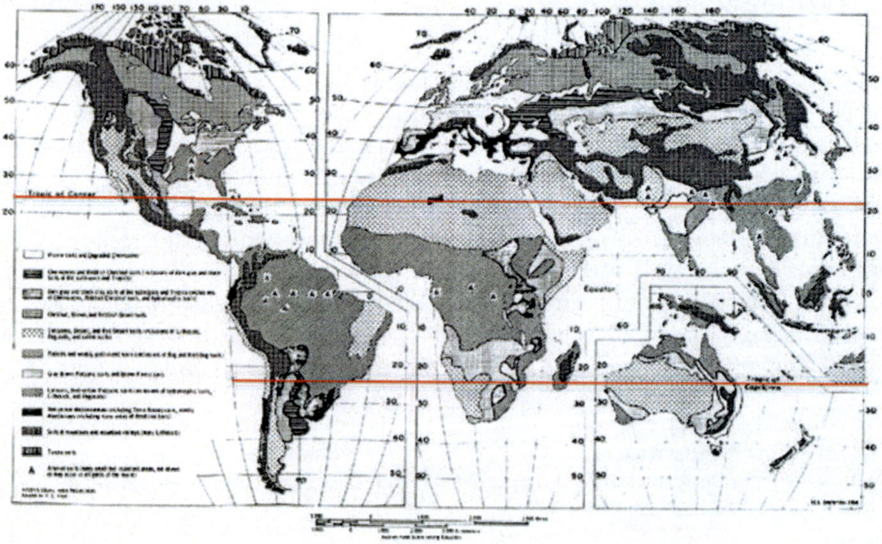

Fig. 2 Distribution of different types of soils all over the world (www.nzdl.org)

2 Laterites in Goa

Laterites are extensively encountered in the Western Coastal region including Maharashtra, Goa, Karnataka and Kerala. Any construction activity like roads, railways or mining requires placement of soil without official support. Also, the rainfall is very heavy along the Western Ghats, thus the stability of lateritic slopes has become a major problem. Some of the districts like Ratnagiri and Sindhudurg in Maharashtra are considered as landslide prone zones. Various geologists in Goa has worked extensively on soils and rocks in Goa. Prominent among them were Fernandes [6], Mascarenhas [8], Kalavampara [7], Widdowson [10] and their works are extensively covered by Mascarenhas and Kalavampara [9]. The extensive research work on geology of Goa is presented by Desai [5] and the other researchers. Figure 3 shows various laterite capped plateaux regions of Goa [10].

Amu et al. [2] reported that the sugarcane straw ash improved the geotechnical properties like moisture content and unconfined compressive strength of the soil sample. Amadi et al. [1] worked on lateritic soils in selected sites in Minna and reported that the lateritic soils have higher values of maximum dry density and California bearing ratio while liquid limit, plasticity index and optimum moisture content is lower. It is suitable for subgrade, sub-base and base material. In the present

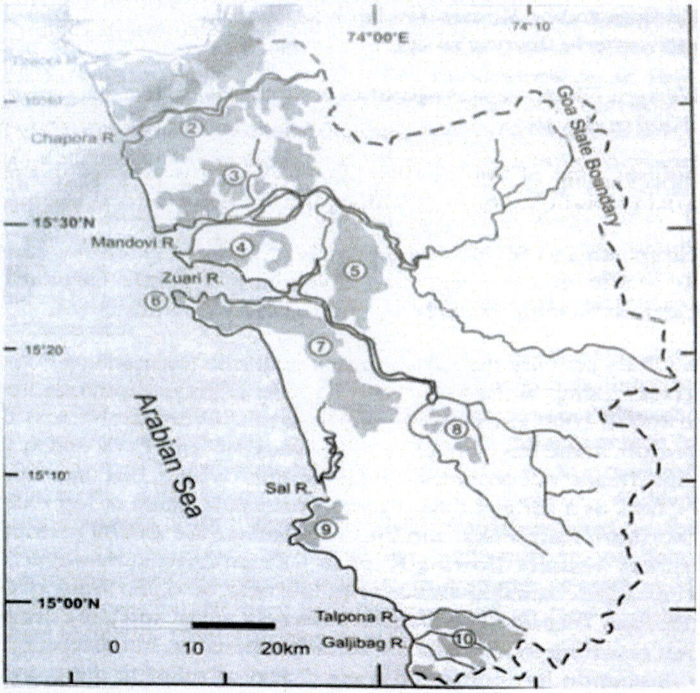

Fig. 3 The various laterite capped plateaux regions of Goa [10]

study, the geotechnical properties of typical lateritic strata at a site located at Cacora village of Quepem Taluka in South Goa are studied.

3 Geotechnical Properties of Lateritic Soil in Cacora Goa

In the present study of lateritic soils, a typical site is selected in Cacora village of Quepem Taluka in South Goa district where G+3 RCC structure was proposed. The schematic layout of the building is shown in Fig. 4. In the Phase I and II, hard lateritic rock was found, while in Phase III-A and B, hard soil was observed. While excavating foundations for the Phase III-C, a very soft reddish-yellowish soil was observed. Hence, it was decided to do boreholes at the site to ascertain the extent of soft soil and design the foundations accordingly. Three locations were chosen for the boreholes and two locations for the plate load tests as shown in Fig. 4a. Borehole depth of 10 m was chosen depending upon the height of the proposed structure. The borehole data indicated the presence of Kaolinite and Gibbsite with less amount of Limonite clay up to 1.5 m depth. Thereafter, Limonite and Gibbsite were available for depth up to 4.5 m. Higher content of Gibbsite and Kaolinite was observed for depth of up to 6 m while Limonitic argillaceous clay/silt was observed from 6 to 10 m depth. Soil colour was observed to be reddish to yellowish with very fine texture. The details of the geotechnical properties of the soil are shown in Table 1.

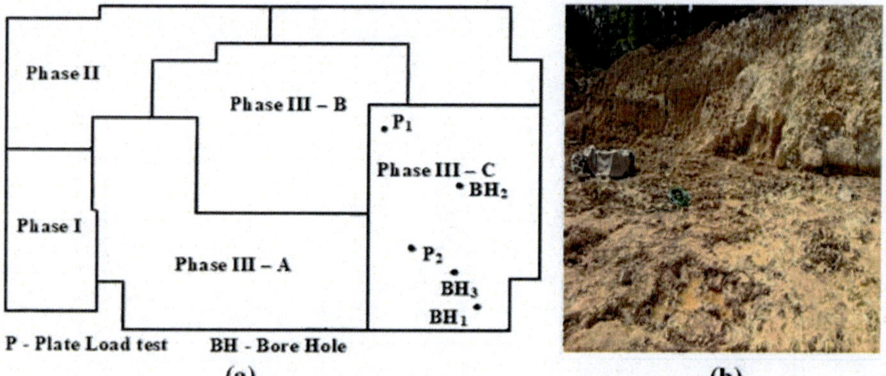

Fig. 4 **a** Building layout with location of boreholes and plate load test and **b** soft lateritic soil at the site

Table 1 Geotechnical properties of soil from all boreholes

Depth (m)	Unit weight (kN/m^3)	NMC (%)	*G*	Liquid limit (%)	Plastic limit (%)	PI (%)	Free swell index	Cohesion *c* (kN/m^2)	Angle of internal friction (°)
0–1.5	14–17	17–35	1.68–1.81	32–38	29–35	3–5	5–6	1–12	15–30
1.5–3.0	15–21	20–36	1.83–1.92	26–36	24–33	2–3	5–6	1–35	10–30
3.0–4.5	16–21	17–38	1.82–1.88	25–34	22–29	1–5	6–6.3	10–20	15–30
4.5–6.0	16–21	24–28	1.65–1.92	26–34	25–30	2–4	5–7	10–20	18–30
6.0–7.5	16–20	26–28	1.46–1.90	20–36	33–NP	–	5.1–6.6	4–10	10–25
7.5–9.0	17–18	30–37	1.55–1.94	24–35	31–NP	–	5–6.9	0–4	20–32

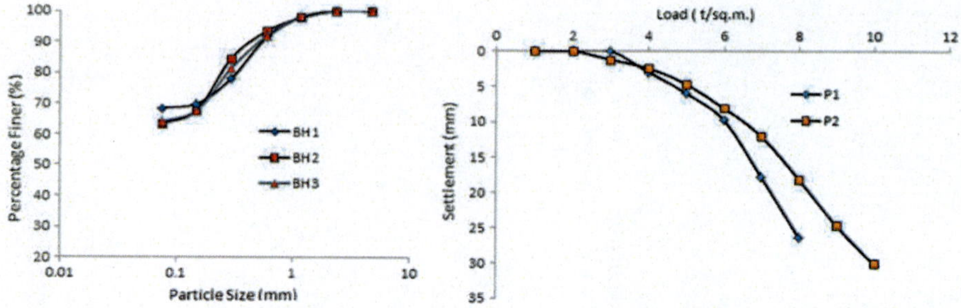

(a) (b)

Fig. 5 **a** Particle size distribution curve and **b** load-settlement curve

3.1 *Grain Size Analysis Tests*

The grain size analysis tests were conducted as per IS:2720 (Part 4)–1985 Methods of Test for Soils Part 4 Grain Size Analysis. The particle size curve for wet sieve analysis is shown below in Fig. 5a. The Uniformity Coefficient (C_u) and Coefficient of Curvature (C_c) values cannot be calculated since D_{60} and D_{30} data is not available from the graph. Based on the results soil can be classified as silty clay.

3.2 *Atterberg Limits and Strength Properties*

Tests for determining the liquid limit and plastic limit of the three soil samples were conducted as per IS:2720 (Part V)–1985 Tests for Soils—Determination of Liquid and Plastic Limit. Specific gravity test of the sample was conducted as per IS:2720 (Part. III) Sect/2-1980 of Fine, Medium & Coarse-Grained Soil. Free swell index tests were conducted on all the three soil samples as per IS:2720 (Part 40)–1977—Methods of test for soils: Determination of free swell index of soils. Direct shear tests were conducted on all the three soil samples as per IS:2720 (Part 13)-1986—Direct shear test. The observed values of various geotechnical properties are tabulated below for each borehole in Table 1.

3.3 *Plate Load Test Results and Safe Bearing Capacity of Soil*

Two plate load tests were conducted at the locations marked in Fig. 4. The full set-up of facilities of loading platform along with loading arrangement, hydraulic jack, pressure gauges and settlement gauges were arranged as per the provisions of IS:1888-1982. A circular plate of 40 cm diameter and 2.5 cm thick was used. The pressure was applied in the increments of 8.75 t/m^2 and average settlement readings were recorded. The final readings of each test were recorded and plotted in Fig. 5b. The ultimate load for the first test was observed to be 27.5 t/m^2 and that for second test was 28.12 t/m^2. The safe bearing capacity was then calculated for a factor of safety of 3.5. The safe bearing capacity at location 1 was 7.86 t/m^2 and at location 2 was 8.04 t/m^2. The settlement at ultimate load was observed to be 9.7 mm (max). The estimated settlement of foundation based on plate size of 400 mm and for raft width of 17.59 m was observed to be 28.70 mm. The settlement of raft computed as per IS:8009-Part 1 (1976) from the assumed values of Young's Modulus of soft soil $= 1 \times 10^4$ kN/m^2, $\mu = 0.4$ [3] and $I = 1.53$ was observed to 23.87 mm which is less than the permissible value of 75 mm for raft.

4 Foundation Recommendations in the Site

The safe bearing capacity from plate load tests at a depth of 0.5 m below the existing level was 75 kN/m^2 while from direct shear tests conducted on soil samples from same site, the lowest value of SBC at same depth is 79.58 kN/m^2 (at Borehole No. 1). For calculation of settlement of foundation raft, the E value of soil assumed is 10000 kN/m^2 (based on $E = 1000c$, National Research Council of Canada (1975) and lowest value of $c = 1$ kN/m^2). Considering the significant depth of 5 m, the settlement of footing of 2.5 m width will be PL/AE $= (98264.4 \times 5)/(930.73 \times 10000)$ $= 52.79$ mm. Since this settlement is high for isolated footings and that the SBC of soil is low, isolated footings are not recommended as they will cover more than 50% of area and also settlements are high. Hence, in the present case, the raft foundation is recommended. It is also recommended to replace the existing soft soil with better quality soil to improve the load-settlement characteristics. A raft foundation was recommended at a depth of 1.5 m below the present ground level at the site. This raft shall be placed on a 1 m thick compacted layer consisting of well-graded and well-compacted murrum. For this, soil may be excavated from a depth of 2.5 m below existing ground level and may be replaced with well-graded murrum compacted in three layers (total thickness of 1 m) and raft shall be constructed on this compacted murrum layer.

5 Problems Associated with Laterite Soil

Lateritic soils are softer when wet and harder when dried. It has low-bearing capacity. The major issues with lateritic soils in Goa are landslides and settlement problems. Landslides occur mainly due to human intervention, rains and also due to soft nature of lateritic soils. Soft lateritic soils are also prone to settlements. Figure 6 shows some of the typical landslides that have occurred in Goa in the recent past.

Fig. 6 Landslides at **a** Khandepar Goa (2018), **b** Mardol Goa (2016), **c** Vasco-da-gama Goa (2014) and **d** Porvorim—Goa (2014)

6 Solutions for Soft Lateritic Soil

Laterite soils have been encountered in abundance in most of the parts of Goa and are considered to be one of the problematic soils. It is highly permeable, slightly acidic with low pH values. Based on type of soil, suitable ground improvement techniques will be suggested. Some of the ground improvement techniques are listed below.

Compaction: Compaction is the oldest mode of ground improvement in which densification of soil is done by expulsion of air from voids using dynamic loading. It is useful for shallow depth and mostly used for improvement of highway foundation.

Mechanical stabilization: When soil of shallow depth is required to improve, this method may be useful. In this method some inert material (aggregate, gravel) is mixed with soil and compacted to required density.

Wooden piles: There may be need of ground treatment to improve the bearing capacity of the soil. Timber piles can be useful for the structure to stand on soft clayey soil with little or negligible distress over the years.

Vibro-replacement: Vibro-replacement with stone columns is a subsoil improvement method. In this method, large-sized column of coarse backfill materials (stones) is installed in the soil by using special depth vibrators. The stone columns are used in the cases of soft and saturated clays to improve the bearing capacity of the soil underlain by sufficiently hard stratum at reasonable depths. Vibro-replacement with stone columns technique can also be used for different types of soils, from soft clays to lose sands, by forming reinforcing elements of low compressibility and high shear strength.

7 Conclusions

Laterites and lateritic soils are observed within the tropics and are formed by intensive and prolonged chemical weathering of the underlying parent rock which is termed as lateritization. Laterites are observed in Peninsular India, Karnataka, Kerala and Goa. Lateritic soils become soft when wet and they become strong after drying and forms a good foundation/building material. In Goa region, the lateritic soil was observed at the chosen site at Cacora Goa which was found to be very soft and having low liquid limit and plasticity index. Since for depth up to 10 m no strong strata was observed, deep foundations like piles or ground improvement techniques like stone columns cannot be adopted. Hence, the raft foundation is recommended after replacing the existing soft soil with better quality soil to improve the load-settlement characteristics.

References

1. Amadi AN, Akande WG, Okunlola IA, Jimoh MO, Francis Deborah G (2015) Assessment of the geotechnical properties of lateritic soils in Minna, north central Nigeria for road design and construction. Am J Min Metall 3(1):15–20
2. Amu OO, Ogunniyi SA, Oladeji OO (2011) Geotechnical properties of lateritic soil stabilized with sugarcane straw ash. Am J Sci Ind Res 2(2):323–331
3. Bowles JE (1996) Foundation analysis and design, 5th edn. The Mc-Graw Hill Companies Inc., New York
4. Buchanan F (1807) A journey from Madras through the countries of Mysore, Canara and Malabar, vol 1–3. East India Company, London, pp 1–1500
5. Desai AK (2018) Geology and mineral resources of Goa, 1st edn. New Delhi Publishers, New Delhi
6. Fernandes OA (2009) Physiography and general geology of Goa. In: Mascarenhas A, Kalavampara G (eds) Natural resources of Goa: a geological perspective. Geological Society of Goa, Miramar Goa, pp 11–24
7. Kalavampara G (2009) Synopsis of mining operations in Goa. In: Mascarenhas A, Kalavampara G (eds) Proceedings of natural resources of Goa: a geological perspective. Geological society of Goa. Miramar Goa, pp 97–118
8. Mascarenhas OA (2009) Evolution of laterite in Goa. In: Proceedings of natural resources of Goa: a geological perspective. In: Mascarenhas A, Kalavampara G (eds) Geological society of Goa. Miramar Goa, pp 35–68
9. Mascarenhas A, Kalavampara G (2009) Natural resources of Goa: a geological perspective. Geological Society of Goa, Miramar Goa
10. Widdowson M (2009) Evolution of laterite in Goa. Proceedings of natural resources of Goa: a geological perspective. In: Mascarenhas A, Kalavampara G (eds) Geological society of Goa. Miramar Goa, 35–68

Geotechnical Properties of Lateritic Rocks in North Goa

Mrudula Ingale and Purnanand P. Savoikar

Abstract Laterite rock is rich in iron and aluminium and is thought to have formed in tropical warm and humid regions like Goa. Generally, all laterites are of rusty-red colouration due to high content of iron oxide. Owing to difficulties in construction of foundations on the lateritic strata, geotechnical investigations of such areas has drawn considerable attention. In the present study, samples of laterite rock were obtained from boreholes at Dona Paula and Porvorim Goa sites. All these sites are closer to the coastal belt in North Goa. Various tests were conducted on these rock samples to study the geotechnical properties of the lateritic rocks in Goa and also the geology of laterites in Goa. These tests include, UCS, unit weight, water absorption test, etc. were also studied. Correlation is obtained about variation of UCS with *L*/*D* ratio for the sites under consideration.

Keywords Geotechnical · Lateritic rock · UCS · *L*/*D* ratio

1 Introduction

The landforms of Goa exhibit a polycyclic landscape that has evolved largely during Cenozoic under humid tropical climatic conditions. The State of Goa is largely categorised as follows:

- The coastal tract
- Sub-Ghat region and
- The high ranges of the Western Ghats.

Laterites are deposits that are more recent. Laterite covers the major portion of Goa and typically occurs as plateau landforms. In Goa, only the laterite and the younger sand, soil and alluvium deposits are of era comparable to the advent of human beings.

M. Ingale (✉) · P. P. Savoikar
Civil Engineering Department, Goa Engineering College, Farmagudi, Goa 403401, India
e-mail: mrudulaingale2@gmail.com

P. P. Savoikar
e-mail: psavoikar@gmail.com

B. B. Das et al. (eds.), *Recent Developments in Sustainable Infrastructure*, Lecture Notes in Civil Engineering 75, https://doi.org/10.1007/978-981-15-4577-1_7

Over the past few million years, the laterite, sand and soil seen today have developed [15]. Laterite, alluvium and sand cover most of the geological formations in Goa, as one expect in the wet tropical climate (Fig. 1).

The study of geotechnical properties of laterites is important because of its uncertain behaviour. Laterite may be considered as a soft rock as compared to other rocks. Geotechnical properties of laterites in Goa are not yet studied by the researchers. The study of characterization and evolution of primary and secondary laterites in north-western Bengal basin was studied by Ghosh and Gucchait [6]. Kasthurba et al. [9] determined the in situ features of laterite such as colour, texture, composition and hardness by field research of laterite in Kerala for construction purposes. Meshram and Randive [10] reported on the geochemical study of laterites of the Jamnagar district. Widdowson [14] studied the evolution of laterites in Goa. The work summarised the incidence of laterites in Goa, distribution and development. Goan laterite development is positioned within the geomorphological and geological (i.e. morphotectonic) development of India's western continental margin.

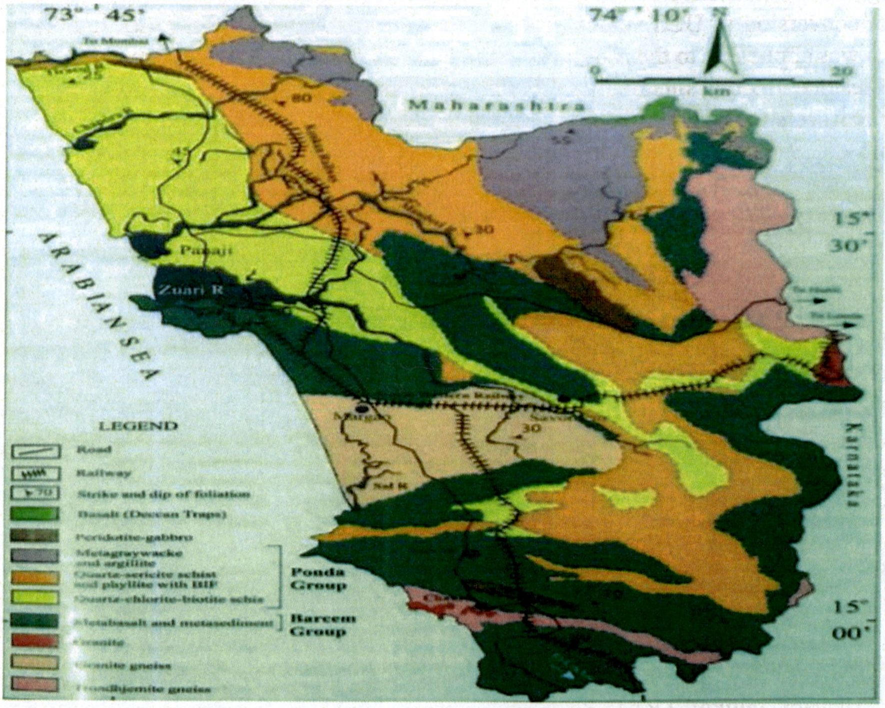

Fig. 1 Geological sketch map of Goa revised after Geological Survey of India, 1996 (modified after Dessai [4])

Turk and Dearman [13] proposed an equation for correlation of UCS test results to *L*/*D* ratio of 2. An equation was given for normalisation of test results to a length-to-diameter ratio of two and 50 mm diameter. Within the range of 1–4 of *L*/*D* ratio error was found to be less than 10%.

Agustawijaya [1] conducted UCS on the number of samples of soft rock considering dry and saturated conditions. From the results, it was observed that the strength reduced remarkably on saturation. One of the causes of strength reduction is weathering. Distinctly weathered rock gives less strength than partially weathered rock. UCS values for the soft rock were found to be lesser than 20 MPa, while for some samples value approached as low as 1 MPa, which is very low for rock material. An equation relating UCS and *L*/*d* ratio was also proposed. It was also reported that *L*/*D* ratio has no significant influence on UCS values. As regards, influence of dry density on UCS, it was reported that UCS increases with dry density.

Ergun and Nilsun [5] investigated the effect of *L*/*D* ration and shape on UCS for the dry cores of seven rocks with *L*/*d* ratios of 1–2.5. For the tested rocks, correlation formula was obtained. UCS values obtained for lesser loading time did not show significant difference as compared to loading time of 8–10 min. Derived equation for conversion of UCS values obtained on different *L*/*d* ratios to standard value of 2.5 was applicable to the nine rock types such as basalt, pink andesite, grey andesite, limestone, marble, siltstone, dubham dolomite, westerly granite and mizuho trachyte which were tested during the study.

2 Occurrence and Distribution of Laterites

In 1807, Buchanan [3] first defined and named in Southern India a laterite formation. He named it from the Latin term Later laterite, meaning bricks. Laterites are iron and aluminium rich residues of chemical decomposition and leaching of parent rocks in situ under warm, humid tropical climate. Alternatively, they could be described as low temperature and pressure alteration products of weathering of rocks under warm, monsoon climate. The process of prolonged chemical weathering of rock is called as lateralisation.

Figure 2a shows distribution of laterites in Goa in the various plateaux as shown below:

Pernem taluka: Harmal—Keri and Mandre—Morji Plateaux
Bardez taluka: Mapusa—Assagao, Saligao, Porvorim, Vernem and Aguada Plateux
Tiswadi taluka: Dona Paula—Taleigao, Bambolim, Kadamba and Merces Plateaux
Ponda taluka: Ponda area Plateaux
Mormugao taluka: Vasco de Gama headland, Dabolim and Mormugao plateaux
Quepem taluka: Quepem area Plateaux
Canacona taluka: Cabo de Rama and Canacona area Plateaux.

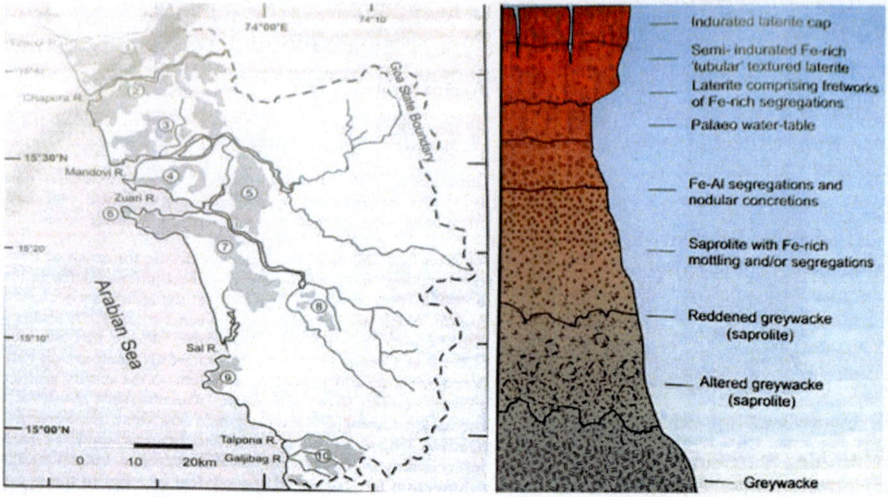

(a) (b)

Fig. 2 **a** Map of distribution of key laterite-capped plateaux of Goa and **b** Idealised profile of an 'in-situ' laterite [14]

3 Geology of Laterites in Goa

3.1 Laterite Profile

A laterite profile is a typical field-section of a laterite (Fig. 2b), from surface to the parent rock. The term 'profile' refers to an 'outline' and not a 'cross-section' of an outcrop. The upper part of the section is generally red, hard and massive, which is referred to as 'lateritic crust'. It consists of a fretwork of oxides and hydroxides of iron and aluminium. The lower part of it is vermiform in nature and grades into a tubular, semi-indurated laterite. Below this there is a prominent thick horizon of variegated colours and is dominated by clay minerals is called as 'lithomarge'. It is generally nodular concretionary at the top and mottled below. The mottled portion consists of kaolinite along with patches and accumulations of iron oxyhydroxides. The saprolite beneath the mottled zone is relatively impermeable and corresponds to 'plasmic zone' of the profile. A significant portion of the saprolite shows bleaching and constitutes the 'pallid zone'.

3.2 *Geotechnical Properties of Laterites in North Goa*

In the present study, four lateritic rock sites were selected from North Goa district which is along or near to the coastal belt of Goa. Undisturbed lateritic rock samples were collected from the boreholes driven for soil investigations at two sites in Donapaula Goa and two sites in Porvorim Goa. All the four sites are located along the coastal belts of Goa (200–1500 m from the coast). Unconfined compression test (IS:9143), bulk density and water absorption test (IS:1124-1974) were conducted on the rock samples collected.

3.3 *Borehole Details*

Boreholes were driven up to the depth of 12.5 m. The rock samples collected from all the four sites were analysed for mineral content. It was observed that for depth up to 6 m, the laterite rock with mineral composition of Limonite ($Fe_2O_3 \bullet nH_2O$) followed by brown/red coloured laterite with high iron content (gibbsite/kaolinite) and very less hematite content (brownish red/yellow in colour) was observed up to 6 m depth. The structure of rock was porous/concretionary. For the depth of 6–12.5 m, the mineral composition varied from Goethite and Limonite (brown/red colour) for depth up to 7.5 m followed by laterite rich in limonite (brownish yellow) with gibbsite/kaolinite was also observed. Goethite/haematite was not present in appreciable quantities. More or less same profile was observed at all the four boreholes. The Rock Quality Designation (RQD) value varied from 7 to 26% for top 6 m and up to 7% for remaining half depth was observed. Table 1 shows the geotechnical properties of rock such as UCS, bulk density and water absorption at the Porvorim sites for different *L*/*D* ratios while Table 2 shows the corresponding properties at Dona Paula sites. The variation of UCS with *L*/*D* ratio is plotted in Fig. 3 while variation of UCS with bulk density is shown in Fig. 4. The all lateritic rocks exhibited the water absorption of 5–10%.

4 Results and Discussions

4.1 *Influence of* L/D *Ratio on UCS*

Figure 3 shows variation of UCS with *L*/*D* ratio for rock specimens with different *L*/*D* ratios. It can be seen that the scatter of UCS values is large indicating that there is no significant influence of *L*/*D* ratio on UCS of rock samples collected from all four sites. Similar conclusions were also drawn by Agustawijaya [1].

Table 1 Geotechnical properties of Porvorim sites 1 and 2

L/*D* ratio	UCS (kN/m^2)	Density (kN/m^3)	Water absorption (%)
1.15	3156	20.3	4.92
1.15	883	19.7	6.08
1.8	2777	20.3	5.58
2.1	3871	21.6	5.93
2.5	1370	17.4	11.26
3	2447	19	6.82
3	1388	21.7	4.59
3.2	3038	20.3	6.32
3.3	3903	21.3	5.21
3.5	2300	20.9	6.62
1.8	5238	20.3	5.36
1.9	295	18.4	6.3
1.9	3143	20.4	7.82
2	1262	17.9	7.77
2.2	4698	20.5	6.88
2.3	1742	10.5	4.99
2.6	2313	18.9	6.65
2.6	5481	20.2	7.61
3.4	2104	20.2	5.82

Table 2 Geotechnical properties of Dona Paula sites 1 and 2

L/*D* ratio	UCS (kN/m^2)	Density (kN/m^3)	Water absorption (%)
1.06	1270	20.7	4.92
1.11	6375	22.8	3.2
1.29	2542	22	4.83
1.57	2919	22.8	4.73
1.6	4193	23.9	5.43
1.66	2644	21.4	3.63
1.11	1235	21.1	3.43
1.2	1335	20.7	2.92
1.26	3599	23.2	4.72
1.6	1890	22.1	6.31
1.73	5779	24	4.29

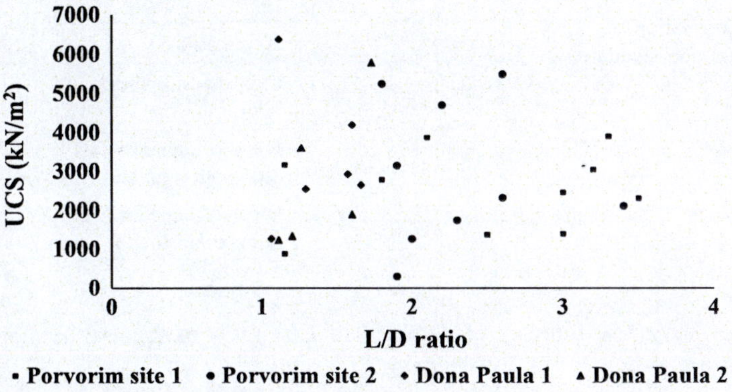

Fig. 3 Influence of *L*/*D* ratio on UCS

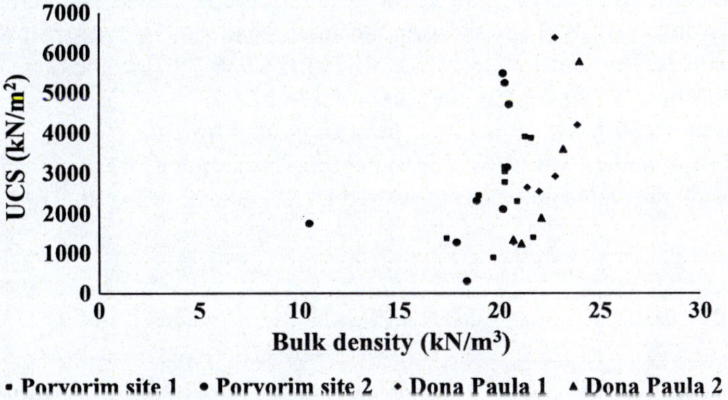

Fig. 4 Influence of bulk density on UCS for all four sites

4.2 Influence of Dry Density on UCS

Figure 4 shows variation of UCS with dry density. It can be seen that as dry density increases, UCS also increases. The values of UCS may be seen clustered around the value of density in the range 16–25 kN/m^3 since for lateritic rock sites, the density varies from 16 to 25 kN/m^3.

4.3 Comments on UCS Values

It is evident from plot of UCS with *L*/*D* ratio that there is wide scatter of values. Also, for the same site and for same *L*/*D* ratio, the value of UCS may be different. In general, *L*/*D* ratio of 2 is assumed to be the standard ratio for testing the rock

Table 3 Correlation equations for non-standard *L*/*D* ratios

Equation	Applicability	References
$\frac{\sigma_c}{\sigma_m} = \frac{1}{0.848+0.304\frac{D}{L}}$	Coal measures rocks in UK	[7, 12]
$\frac{\sigma_c}{\sigma_m} = \frac{1}{0.88+0.24\frac{D}{L}}$	For non-standard *L*/*D* ratios other than 2	[2]
$\frac{\sigma_c}{\sigma_m} = \frac{1}{0.875+0.25\frac{D}{L}}$	For non-standard *L*/*D* ratios other than 2	[11]

specimens. Very few number of samples of *L*/*D* ratio of 2 were available in the present case to comment on influence of *L*/*D* ratio.

Change in *L*/*D* ratio gives different values of UCS which may sometimes be misleading. The factors on which UCS may depend are shape and size of specimen, rate of loading, degree of saturation, etc. which are termed as external factors while defects, porosity, mineralogy, degree of weathering are termed as internal factors [13]. According to ISRM [8] the test specimen shall not be less than 54 mm in diameter and *L*/*D* ratio of 2.5–3, while ASTM (D2938-79) specifies test specimen of diameter not less than 48 mm and *L*/*D* ratio of 2–2.5.

Various co-relation equations have been suggested for correction in UCS values in the case of non-standard *L*/*D* ratios (other than 2) as shown in Table 3, where σ_c is the corrected UCS value and σ_m is measured UCS value for non-standard *L*/*D* ratios.

5 Conclusions

In the present study, water absorption, unconfined compressive strength and unit weight of the rock samples collected from different sites of Goa was obtained. The mineral composition of all rocks indicated the presence of Limonite in higher quantities followed by Gibsite and Kaolinite. Hematite and Goethite were observed in very less quantities. It can be concluded that the unconfined compressive strength of the laterite rock not necessarily increases with the increasing depth, sometimes it reduces for the intermediate layer of soil and again increases. This reduction in UCS values may be due to the porous and concretionary structure of the laterites found at different depths, varying mineral contents and degree of lateritization/alteration. It is required to develop standard correlation equation for UCS for non-standard *L*/*D* ratios as many times UCS is used in determining safe bearing pressures on rocks.

References

1. Agustawijaya DS (2007) The uniaxial compressive strength of soft rock. Civil Engineering Dimension 9(1):9–14
2. ASTM D-2938-79 (1980) Standard method of test for unconfined compressive strength of rock core specimens. Annual Book of ASTM Standards, Part 19, pp 440–443
3. Buchanan F (1807) A journey from Madras through the countries of Mysore, Canara and Malabar. East India Company, London 1–3:1–1500
4. Dessai AK (2018) Geology and mineral resources of Goa, 1st edn. New Delhi Publishers, New Delhi
5. Ergun T, Nilsun H (2009) The effect of length to diameter ratio of test specimens on the uniaxial compressive strength of rock. Bull Eng Geol Environ 68(4):491–497
6. Ghosh S, Guchhait SK (2015) Characterization and evolution of primary and secondary laterites in Northwestern Bengal Basin, West Bengal, India. J Palaeogeogr 4(2):203–230
7. Hobbs DW (1964) Rock compressive strength. Colliery Eng 41:287–292
8. ISRM (1979) Suggested methods for determining the uniaxial compressive strength and deformability of rock materials. Int J Rock Mech Min Sci, Abstr. 16, 135–140
9. Kasthurba AK, Santhanam M, Mathews M (2007) Investigation of laterite stones for building purpose from Malabar region, Kerala state, SW India—Part 1: field studies and profile characterisation. Constr Build Mater 21:73–82
10. Meshram RR, Randive KR (2011) Geochemical study of laterites of the Jamnagar district, Gujarat, India: implications on parent rock, mineralogy and tectonics. J Asian Earth Sci 42:1271–1287
11. Protodyakonov MM (1969) Method of determining the strength of rocks under uniaxial compression. In: Protodyakonov MM, Koifman MI et al (eds) Mechanical properties of rocks. Translated from Russian, Israel Program for Scientific Translations, Jerusalem, pp 1–8
12. Szlavin J (1974) Relationships between some physical properties of rock determined by laboratory test. Int J Rock Mech Min Sci Geomech Abs 11:57–66
13. Turk N, Dearman WR (1986) A correction equation on the influence of length to diameter ratio on the uniaxial compressive strength of rocks. Eng Geol 22:293–300
14. Widdowson M (2009) Evolution of laterite in Goa. In: Mascarenhas A, Kalavampara G (eds) Proceedings of natural resources of Goa: a geological perspective. Geological Society of Goa, Miramar Goa, pp 35–68
15. Widdowson M, Gunnel Y (1999) Lateritization, geomorphology and geodynamics of a passive continental margin: Konkan and Kanara coastal lowlands of Western Penninsula India. In: Thiry M, Simon-Coincon R (eds) Proceedings of paleoweathering, paleosurfaces and related continental deposits, vol 27. International Association of Sedimentologists Special Publication, pp 245–274

Monitoring, Control and Mapping of Landslides in Goa

Mandira Faldesai and Purnanand P. Savoikar

Abstract Natural disasters like earthquakes, landslips, avalanches, floods, cyclones, droughts and volcanic eruptions have been distressing mankind since the commencement of civilization which credit up to 4% of total yearly deaths. Worldwide, landslides are assessed to cause about 1000 deaths annually apart from destructing resources worth millions of dollars. Landslides are one of the many natural disasters causing immense destruction and loss of lives civil structure, communication routes, agricultural and forestland across the earth. Approximately 15% of the land area of India is susceptible to landslide out of which 0.09 million sq. km come under Western Ghats and Konkan hills including Goa. Goa contains hilly mountainous terrain, in combination with sandy coastal plains. Being one of the most beautiful places, it is now facing serious scarcity of land to be used for residential purpose. Construction activity on hilly areas has been exploiting the scenic view which has also led to a number of landslides during rains. Slopes have been cut for many recent developments in the state which can pose a great threat to inhabitants living there due to lack of stability checks of the slopes. Landslide countermeasure is a strong issue hence there is crucial need for enhanced and modern techniques for steadying slopes and to avert landslides which serve the drive for this study. In the present study, various slopes in the state are being examined. Literature have been reviewed on various landslide related topics. In the present work, the stability of landslide-prone sites in Goa is studied. The work presents mapping of various landslides that occurred in Goa which is not mentioned in the Geological survey of India.

Keywords Landslides · Goa · Mapping · Monitoring · GSI

M. Faldesai (✉) · P. P. Savoikar
Civil Engineering Department, Goa Engineering College, Farmagudi, Goa 403401, India
e-mail: Mandira115@gmail.com

P. P. Savoikar
e-mail: psavoikar@gmail.com

B. B. Das et al. (eds.), *Recent Developments in Sustainable Infrastructure*, Lecture Notes in Civil Engineering 75, https://doi.org/10.1007/978-981-15-4577-1_8

1 Introduction

The Indian State Goa lies within the shore region called Konkan in Western India. It is encircled by Karnataka to east and south and Maharashtra to the north, with extended narrow strips of the Arabian Sea starting its Western coast. Goa is located between 14°53′54″N and 15°40′00″N and 73°40′33″E and 74°20′13″E. Two-thirds of Goa is enclosed by laterite having a thickness of 2–25 m. The annual rainfall in Goa is 995 mm, July being the wettest month. Every year during monsoon, landslides occur in the breakable hill slopes of the mountain ranges which causes direct loss to life and property, and indirect loss by blocking transportation routes and communication routes. In order to decrease the risk of mass movements, it is necessary to tackle landslides effectively. Awareness of landslides among people is an important phenomenon. Most of the development is now taking place on hill slopes and in areas which have gradient which increases the risk of landslides. There is a need to preserve delicate ecosystem areas which includes mountain slopes, low-lying area, khajans, private forests, conservation zones. Earlier Goa was considered stable as far as slopes and landslides are concerned but due to unplanned development it has become susceptible to landslides. Slopes have been cut for many recent developments in the state of Goa which can pose a great threat to inhabitants living there due to lack of checking the stability of the slopes hence causing landslides. Landslides are also triggered due to earthquake shock. Earthquake of magnitude 4.8 on Richter scale at Koyana dam sent mild tremors across Goa. The tremors were felt at Porvorim, Old Goa and Margao indicating that the reservoirs can induce earthquake and earthquakes induce landslides.

Many development projects such as extension of roads, highways, bridges construction are been taken care by the Goa government. Recently Goa has observed many landslides, mainly due to heavy rains during monsoons causing damage to life and property. Hence landslide monitoring and control measures are essential.

2 Literature Review

An extensive literature review is done to classify the research work conducted on landslide stability analysis and slope steadiness methods.

Chowdhury et al. [1] conducted studies on the issues which are of critical importance to landslide hazard risk assessment and management and concluded that the landslide research should include rational and comprehensive methods for the assessment of hazard and risk. Yang li and Huang Fu [2] worked on the effect of nonlinearity and dilation on slope stability and reported that the nonlinear failure criteria dilation has important effects on the slope stability. Hence, these parameters should be properly evaluated. Prashant and Mukherjee [3] conducted studies on soil nailing for slope stabilization near railway tracks. Functioning and monitoring of nails were described with numerous parameters to be monitored and proper instrumentation process to be used. Dutta and Sarma [4] carried out landslide susceptibility zoning of a potential landslide-prone area located in Guwahati city in the state of Assam using empirical method. Landslide incidence map was compared to landslide susceptibility map and the collective percentages of the witnessed landslide values versus ranked LSI values were calculated and plotted.

Shiny and Satyam [5] carried out stability analysis of Malbasey landslide during 2011, India–Nepal earthquake. It was reported that the Malbasey slope is highly prone to landslides during earthquakes. Elayaraja and Chandrasekharan [6] investigated landslides at Chinnabikatty at Nilgiris along NH67 in the state of Tamilnadu. It was concluded that the Chinnabikatty is a site of repeated failure due to landslides. Choudhury et al. [7] conducted a field case study of recent landslide disaster site at Pune—India using PLAXIS software and concluded that the heavy rainfall in the area is one of the major factors in movement of soil mass. It was found that the topsoil overlying basalt rock is clay soil of medium plasticity. Transitional failure occurred due to slipping of soil mass from basalt rock. Singh and Som [8] conducted analysis on earthquake caused landslides in India, claiming hundreds of lives in the Himalayan lands of India. Expansion of regression equations showed that in Himalayan atmosphere, landslide may trigger even with unseen earthquakes distressing longer distances having earthquake magnitude more than 8 M.

3 Review of Landslides in Goa

About 0.42 million sq. km or 12.6% of land area, is prone to landslides in India, of which about 1/5th lies in Konkan, Western Ghats including the Nilgiris (Tamil Nadu, Karnataka, Kerala, Maharashtra and Goa). The Geological Survey of India (GSI) [9], is a Government of India Ministry of Mines society, for conducting geological surveys and studies of India. It gives detailed information about the various landslides that have occurred in India. Figure 1 shows some of the typical landslides that have occurred in Goa.

(a) (b) (c)

Fig. 1 Landslides at **a** Ambaji Fatorda [10], **b** Chorla Ghat and **c** Khandepar Goa

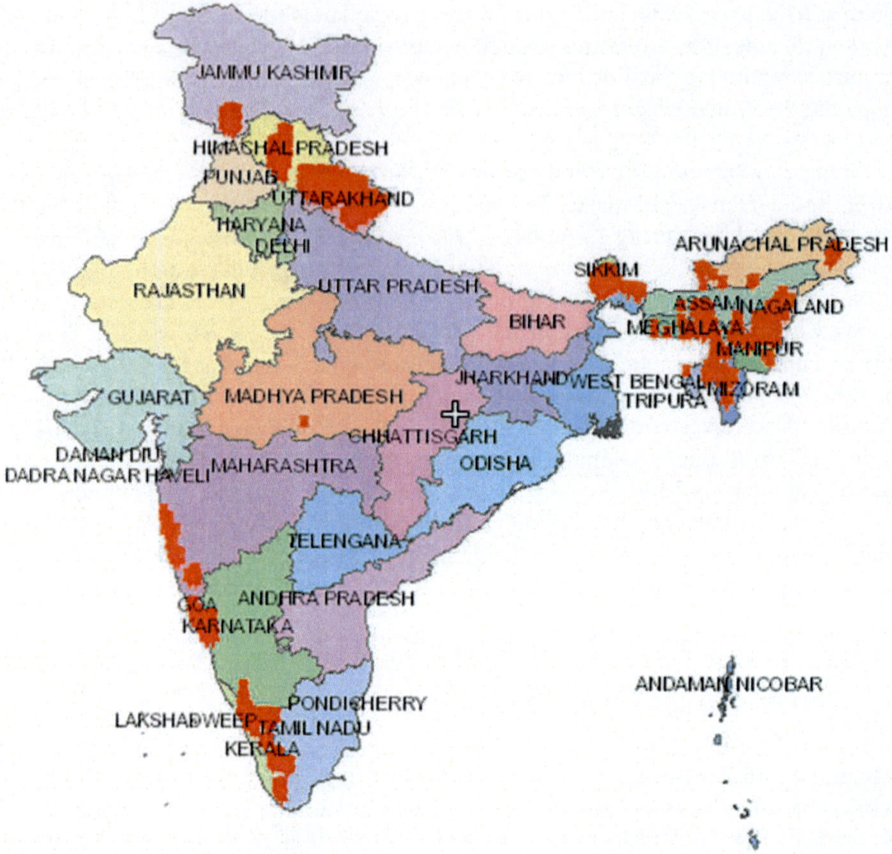

Fig. 2 Inventory map of landslides in India (*source* GSI)

An inventory map of landslides shows the position, three-dimensional extent and type of landslides in a region. 12 landslides which are not mentioned in the landslide inventory map of Goa are studied and mapped using ArcGIS software. Figure 2 shows the inventory map of landslide in India while Fig. 3 shows the inventory map of landslides for Goa.

4 Mapping of the New Landslides in Goa

Damaging effects of landslides relative to life and economy in the world are very harsh. Mapping of landslides that have occurred in the past, helps the risk managers in assessment and reduction of landslide related hazards using the remote sensing data and GIS. Various kinds of maps are used to portray hazard from landslides. These maps can be used to identify potential instability, or probabilities based on factors

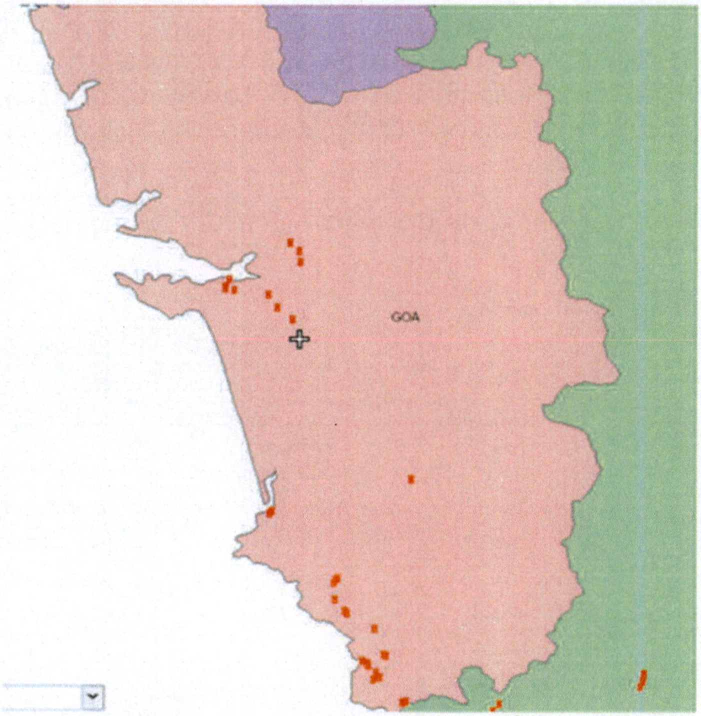

Fig. 3 Inventory map of landslides in Goa (*source* GSI)

such as rainfall, slope angle, soil type and stages of earthquake shaking. Landslide inventory maps show landslide locations, slope features, extent and the cause/type of landslide which are helpful in predicting the hazard for an area. Landslide vulnerability maps describe the relative probability of future land sliding based exclusively on the inherent properties of a site. Landslide hazard maps show the possible areas prone to landslides in a given area. It also shows the chances of itinerant downslope in a given distance.

In the present study, the various landslides that have occurred in last 4–5 years are studied and their geotechnical properties are evaluated. ArcGIS software is used for mapping these landslides that have occurred in Goa in recent years on to the landslide map of Goa. ArcGIS is a geographic information system for operating with maps and managing geographic information. Various landslides which took place recently and which were not mentioned in the landslide inventory map of Goa in the Geological Survey of India site are studied and mapped using ArcGIS software. Twelve landslides were studied and data was collected. Each site was visited and their coordinates were noticed using Google maps. Using the coordinates different landslide locations were marked on the Goa geological map in the ArcGIS software. All the data of the respective landslides were created in Excel sheet and then transferred to ArcGIS software. Hence when snapped on a specific landslide all the data conforming to that will appear in a tabular format. In the present study, the various landslides that have occurred in Goa were at the following places and year: Borim (2018), Khandepar (2018), Chorla (2017), Narvem (2015), Aquem (2017), Arpora (2015), Neura (2016), Ponda (2016), Mardol (2016), Nagarcem (2014), Raia (2014), Ambaji (2014).

Table 1 gives the details of typical landslides in Goa. Figure 4 shows the procedure of marking new landslides on the Goa map using ArcGIS Software (Fig. 5).

Table 1 Details of typical landslides at Goa

Parameters	Ambaji—Goa	Chorla Ghat—Goa	Khandepar—Goa
District	South Goa	North Goa	North Goa
NH SH location	Ambaji, Margao	Chorla ghat, SH31 between Goa–Karnataka	Khandepar
Latitude	15.2954828	15.6325631	15.4201855
Longitude	73.9579137	74.1182475	74.0406736
Activity	Inactive	Active	Stabilized
Geology	In situ soil	In situ soil	In situ soil
Height	13.8 m	18 m	20 m
Inclination of slope	71.60°	16.5	–
Material type	Debris	Debris	Debris
Moment type	Slide	Slide	Slide
Moment rate	Rapid	Rapid	Rapid
Failure mechanism	Shallow translational	Shallow translational	Shallow translational
Geomorphology	Moderately dissected hills	Highly dissected hills	Highly dissected hills
Hydrological condition	Damp	Damp	Damp
Land use/land cover	Sparse vegetation	Good vegetation	Good vegetation
Triggering	Rainfall	Rainfall and anthropogenic	Rainfall and anthropogenic
Geoscientific cause	Cutting of slope	Road cut leading to destabilization and modification of slope	Road cut leading to destabilization and modification of slope
Remedial measure suggested	Soil nail wall, gabion-reinforced wall	Soil nail wall, vegetation, gabion reinforced wall.	Soil nail wall, vegetation, benching

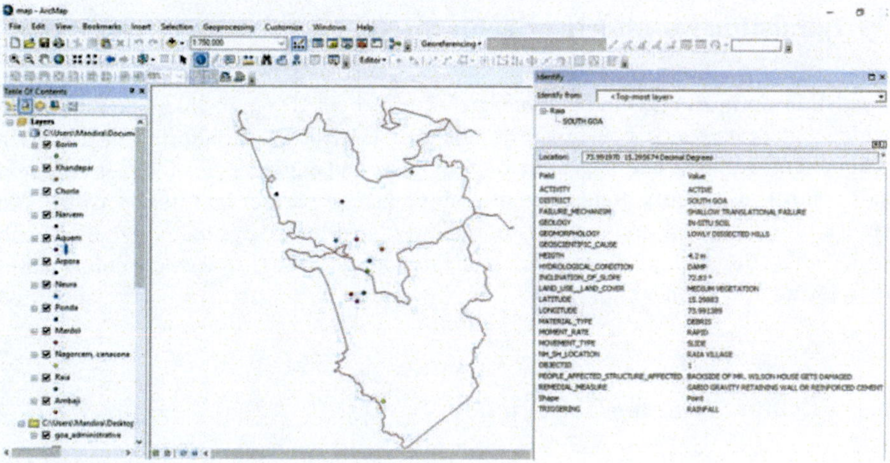

Fig. 4 Marking various landslides using ArcGIS software

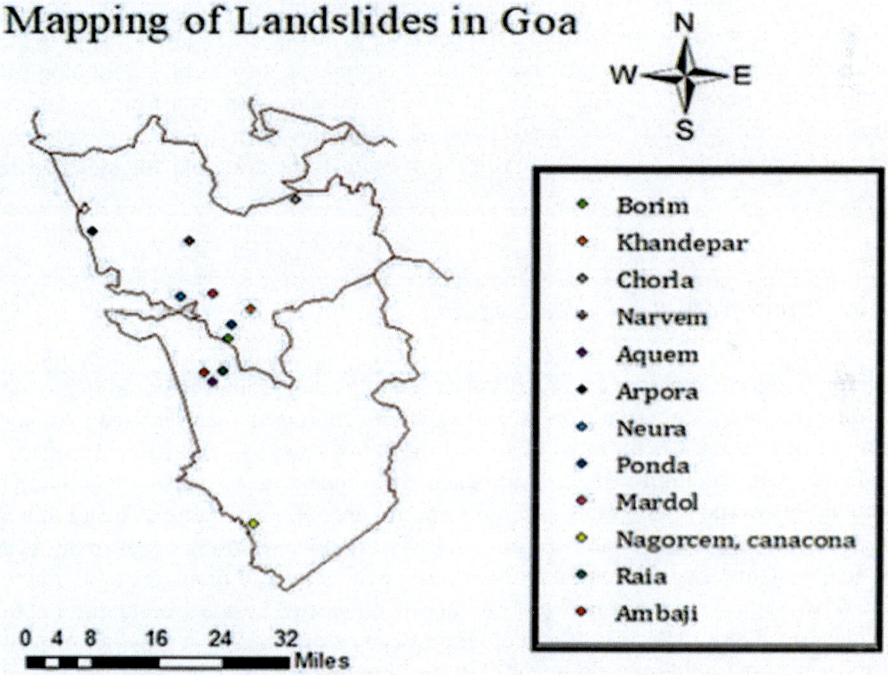

Fig. 5 Latest landslides in Goa mapped on Goa map using ArcGIS Software

5 Landslide Monitoring Techniques

Landslide monitoring is vital to forecast the performance of landslides and estimating which outbursts can trigger great numbers of landslides. Monitoring of landslide parameters is an influential weapon of geologists and engineers for the measurement and control of stability conditions of a slope and to predict its future growth. The choice of instruments and methods of quantity to follow hangs on the types of bend which affect the method of constancy analysis and accordingly the whole deformation monitoring operation.

5.1 Remote Sensing Techniques

Remote sensing utilizes satellite or aerial sensors to gather data about the object or area. It is in the form of the photographs. The inspecting and administering system is the most suitable method of recording present ground situations, and evaluating the results on environment. Satellite images of high decree are used for mapping the landslide inventory with factors like surface morphology, structural and lithological features, land cover, temporal changes, etc. Digital elevation data from space are also of interest for landslide characterization, where the accurateness of the present satellite systems is adequate, though enhancements are desirable for small-scale landslides.

5.2 Photogrammetric Techniques

Photogrammetric methods are broadly used in confirming ground activities in ground settling discoveries in mining areas, and mainland photogrammetry is used in monitoring engineering structures. Diminished time of investigation, rapid 3D coordinates and monitoring of unlimited points are the main compensations of using photograph techniques. Aerial photographs offer an overall viewpoint of a large area and limits of present slides and are outlined on aerial photographs. Surface and near surface drainage channels can be outlined. Important associations in drainage, topography and other natural components which are hardly connected on the ground turn out to be clear on photographs. Current photographs can be diverged with previous ones to inspect the continuing development of slides. Photographs can be reviewed at any time, in any place and by anyone. The Global Positioning System (GPS) can act as substitute for surveying to support geotechnical assessments of sharp slopes by offering 3D coordinate time series of movements at distinct points on the slipping surface.

5.3 Geotechnical Techniques

Geotechnical instruments are used widely in the monitoring of constructions.

Extensometers: It is used for determining stress–strain in tensile tests. Several devices are used to calculate change in distance to determine compaction or upheaval of soil, strain in rocks, and ripping of rock layers around driven tunnels and slope stability. Extensometer processes the axial movement between a number of reference points in the same axis which can be connected within a borehole or on slope.

Inclinometers: Inclinometers are instruments that are fitted in boreholes pierced into the landslide mass. It measures the curving of originally straight boreholes, thus classifying any adjustment in inclination of the borehole casing. It determines slip surfaces and zones of movement and also reveals the depth of the failure planes. This is classically achieved with gravity-operated tilt sensor.

Geophones: These are devices that can measure vibration associated with movement. A geophone operates by converting the ground movement (velocity) into voltage and hence are useful as early warning systems. They can detect landslides on the basis of frequency composition, amplitude and duration of the vibration signal.

Piezometers: Lots of landslides are activated by slope saturation through heavy rainfall. Magnitude of pore water pressures and piezometric levels form a significant part of slope stability analysis. Threshold levels are defined to provide early warning of circumstances that may lead to catastrophic failure. Piezometer evaluates the pore pressure of groundwater within a geological structure and offers an indication of the build-up stresses and strains within the rock mass. Usual types of borehole piezometers are vibrating wire, pneumatic and faucet piezometers.

Tiltmeters: Tiltmeter kinds of various liquid, electrolytic, vibrating wire and pendulum types are used in engineering surveys. They are more sensitive and are useful on extremely gradual moving rotational failures. Tiltmeters can also influence the direction of movement, to delimit the areas of deformation and to determine the mechanism of movement. They also provide warnings of impended slide movement.

Crack meter: A crack meter is a convenient tool in the early detection of distorting mass movements. The device measures the displacement between two points which display symbols of split. Crack meters are also used to inspect movement across surface cracks and joints in concrete, rock and soil. It is a lined displacement transducer with anchors and joints. When the crack opens, the movements will be transformed into electrical signals that can be read with a data logger.

Figures 6 and 7 show some of the instruments used in monitoring the soil mass movement. Presently no much instrumentation has been done to monitor the landslides in Goa. However, some of the above techniques can be effectively used to monitor such soil mass movements.

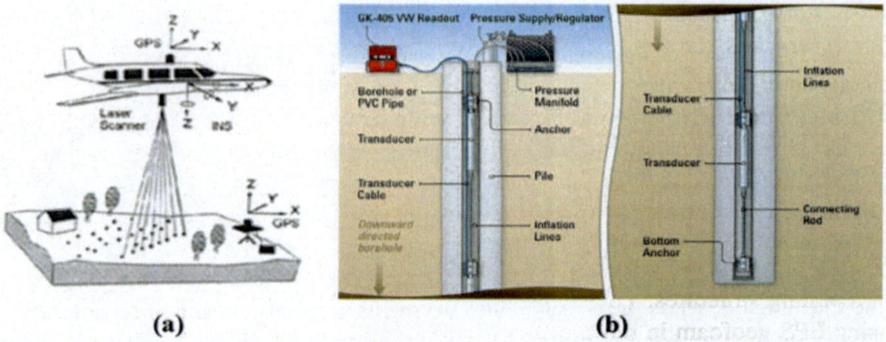

Fig. 6 Landslide monitoring techniques **a** Remote sensing **b** Extensometers (*source* Google)

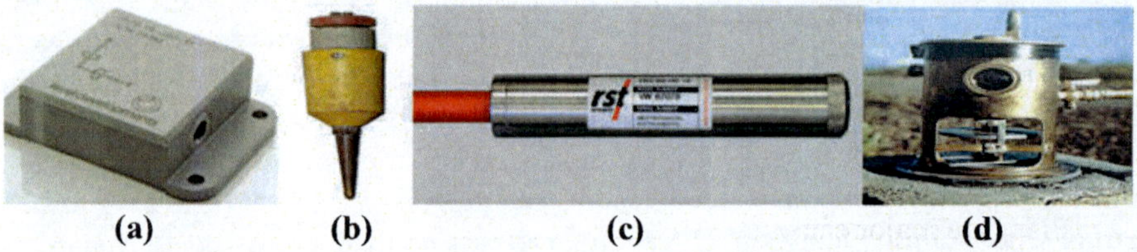

Fig. 7 Geotechnical landslide monitoring techniques **a** Inclinometer sensor, **b** Geophone **c** vibrating wire piezometer and **d** tiltmeter (*source* Google)

6 Control Measures for Landslides

Retaining walls: Retaining walls, such as gravity, cantilever, counterfort, sheet pile walls, are normally used for retaining earth, depending upon type and height of earth to be reserved. Soil type, slope angle, groundwater physiognomies and other specifics are considered before deciding on the proper solution.

Gabion wall: A gabions are steel wire netted boxes which are filled with stones or concrete and are used for retaining the soil and embankments for various civil engineering and military uses. It is a wall made of stacked stone-filled gabions tied together with wire, laid at an angle to the slope or stepped back into the slope for better stability. The life of gabions walls generally depend upon the life of the wire and not on the fillings of the basket. Some gabion producers guarantee a structural consistency of 50 years.

Soil nailing: Soil nailing is a technique used to support soils which are susceptible to collapse. In this technique, soil is reinforced with steel rods which are inserted into pre-drilled holes and then grouted. This technique is generally used where other stabilization techniques cannot be implemented due to space restraint.

Ground anchors: Stabilization of slopes can be done by the applying active forces to the unstable ground. These forces raise the normal stress and resist friction along the creeping surface. Anchors are used to connect the face to each other by a beam frame, which is generally made of reinforced concrete.

Geosynthetic strengthening: Geosynthetics are porous, flexible, man made cloths which act to reinforce and escalate the stability of structures such as earth fills, and thereby permit steeper cut slopes and less grading in hillside terrain. Geotextiles, Geogrids, Geomembranes, geocells, geocomposites are some of the geosynthetic domestic products which are comprehensively used in retaining structures. Tie back retaining wall, reinforced earth wall, retaining wall using EPS geofoam, cantilever retaining wall with relieving shelves are some of the other recent progresses in retaining structures. The lateral pressure coming on the structures is reduced by using EPS geofoam in place of the backfill soil. This will result in lesser pressure on stem and hence the design of stem will be economical. In Goa, reinforced earth walls, gabions, soil nailing and gabions with soil nailing have been commonly used to retain the earth.

7 Conclusions

Landslides are major cause of concern in developing countries. The triggering factor like rainfall, high level of development activities, cutting of vegetation, etc. are the major reasons for landslides. Various slope protection measures like RCC retaining walls, reinforced earth walls, soil nailed walls, gabion walls, tie back wall, etc. need to be implemented considering the various factors like height of the slope, type of the soil, availability of space for construction, cost, long-term durability, etc. New techniques like cantilever retaining walls with relieving platform and controlled yielding techniques using EPS geofoam are also being explored as retaining measures. The present study of recent landslides in Goa has been used in mapping the landslide spots in Goa onto the landslide inventory map of Goa and will be helpful for designers, builders and engineers to assess the landslide susceptibility of the various areas in Goa, before undertaking any construction activity.

References

1. Chowdhury RN, Flentje PN, Koko C (1999) Aspects of recent landslide research at the University of Wollongong. In: Proceedings of 9th international conference and field trip on landslides, pp 1–8
2. Yang X, Huang F (2009) Slope stability analysis considering the joined influences of nonlinearity and dilation. J Cent S Univ Technol 16(2):292–296
3. Prashant A, Mukherjee M (2012) Soil nailing for stabilization of steep slopes near railway tracks. Final report, Research Designs & Standard Organizations, Lucknow
4. Dutta PJ, Sarma S (2013) Landslide susceptibility zoning of the Kalapahar Hill, Guwahati, Assam, India using a GIS based heuristic technique. Int J Remote Sen Geosci 2(2):49–55

5. Shiny N, Satyam N (2015) Assessment of Malbasey landslide (West Sikkim) during 2011 India-Nepal earthquake. In: Proceedings of the 6th international geotechnical symposium on disaster mitigation in special geoenvironmental conditions, Chennai, India
6. Elayaraja S, Chandrasekaran S (2015) Analysis of landslide at Chinnabikatty along Mettupalayam—Ooty National Highway in Nilgiris, India. In: Proceedings of the 6th international geotechnical symposium on disaster mitigation in special geoenvironmental conditions, Chennai, India
7. Choudhury D, Kumar A, Patil M (2015) Theory to reality of failures of slopes including forensic analysis of an Indian Landslide Disaster. In: Proceedings of SLOPE 2015, international conference on landslides and slope stability, Bali, Indonesia, vol 6, pp 1–7
8. Singh H, Som S (2016) Earthquake triggered landslide—Indian Scenario. J Geolog Soc India 87(2):105–112
9. Geological Survey of India, www.gsi.gov.in
10. Hede A (2015) A study of monitoring and control of landslides in the State of Goa. Masters degree thesis, Goa University

State of Art on Load-Carrying Capacity and Settlements of Stone Columns

Manita Das and Ashim Kanti Dey

Abstract Construction of structures over a soft clay deposit is probably the most challenging task for a geotechnical engineer. Structures built over the soft soil experience a huge settlement, which may become the main cause of failure. Thus, for any construction, improvement of the soft clay properties is necessary. Stone columns are frequently used to improve the soft soil nowadays. Stone columns speed up the rate of consolidation of the soft soil and thereby increase the load-carrying capacity and lower the settlement value. A lot of studies have been reported on the behavior of soft clay reinforced with stone columns; many theories have also been developed. An Indian Standard is also available to determine the bearing capacity and the settlement behavior of stone columns. However, the theories are not well accepted because of wide variations of results. The present study includes a review of published theories and a comparison of results based on a typical problem.

Keywords Stone column · Bearing capacity · Settlement · Ground improvement · Soft clay

1 Introduction

Construction of stone columns is a widely used technique for the improvement of soft clayey soil. Basically, the stone columns improve the stiffness of the soft clay, thereby improve the load-carrying capacity of the soft clay and also act as a drain for a speedy drainage of entrapped water, and thereby accelerate the consolidation settlement. Stone columns were first introduced by Moreau et al. in 1830 [1]. They used the stone column of diameter 0.2 m and length 2 m for the foundation of iron-works for the military purpose in Bayonne, France. Moreau declared a huge amount

M. Das (✉)
Civil Engineering Department, ITER, SOA (Deemed to be University),
Bhubaneswar, Odisha, India
e-mail: manitadas@soa.ac.in

A. K. Dey
Civil Engineering Department, NIT, Silchar, India
e-mail: ashim_kanti@yahoo.co.in

B. B. Das et al. (eds.), *Recent Developments in Sustainable Infrastructure*, Lecture Notes in Civil Engineering 75, https://doi.org/10.1007/978-981-15-4577-1_9

of improvement in soft soil with the use of stone columns in form of a reduction in settlement and increase in load-carrying capacity. Steuermann introduced the vibro-compaction technique of stone column construction for the first time in 1939. The use of the vibroflotation technique was introduced in USA and Germany in the 1940s. The depth of treatment of soft soil by using stone column was introduced to 20 m at the end of the 1950s. In 1955, the use of stone column spread to Japan, and later, this technique of ground improvement was widely used in China and other countries. In India, the use of stone columns started in the 1970s.

It is reported that the stone columns are able to support any structure constructed on very soft to firm clayey soils and loose silty sands with fines more than 15% [2, 3]. Stone columns are being applied to support many structures like embankments, raft foundations, liquid storage tanks and other low-rise structures. As per IS 15284-1 (2003), the stone columns work most effectively in clayey soils with undrained shear strength ranging from 7 to 50 kPa [4]. Depending upon the method of installation and the installation equipment, the stone columns can be formed up to a depth of 30 m [5]. However, as per IS 15284-1 (2003), the optimal value of length of a stone column is four times its diameter.

Maheshwari and Khatri discussed the nonlinearity of stone column-reinforced soil [6]. Saroglou et al. also discussed the properties of stone column-reinforced soft soil [7]. Rajesh and Jain (2015) studied the influence of permeability affecting stone column-reinforced ground [8]. As per the available records, the use of stone columns started in 1830. However, the first theory on the determination of the capacity of a stone column and the extent of improvement of properties of soft clay was published in 1983 [2]. Most of the studies were conducted in the laboratory, and a few studies were conducted in the field. The present paper discusses a state of art on determination of load-carrying capacity [4, 9–11] and estimation of settlement [12–14] for structures resting on stone columns. At the end, a comparison of results obtained from different theories is also presented.

2 Installation Techniques

There are two types of installation techniques for construction of stone columns, namely vibroflotation technique and rammed technique [15]. In both the techniques, bore holes are first made in the soft soil and are filled with well-graded stones in layers. The size of stones varies from 25 to 40 mm, and spacing of holes varies from 2 to 3 times the column diameter. In the vibroflotation technique, a vibroflot is used to compact the stone, and in rammed technique, a rammer is used to compact the stone. The insertion and compaction of stones are carried out simultaneously in layers. The column installation effect decreases with the increase in radial distance [16]. During compaction, the diameter of the hole increases by 33.33% of the original diameter [17].

3 Failure Mechanism

Up to 1974, researchers only studied on field experiments. After that the researchers started for laboratory model tests on stone column. In 1974, Hughes and Withers obtained the bulging effect up to a depth four times the column diameter [1]. They also obtained the bulging failure in case of a group column like a unit cell. As per unit cell concept, every single column in a group acts individually, and no interaction effects are considered. Therefore, the capacity of a group column is considered as the summation of the capacity of individual columns. But later it was found that there must be some interaction effect in between two adjacent columns. Therefore, the concept of unit cell was proved partially wrong [18]. In 1995, Hu conducted the study on laboratory model tests of group stone columns and concluded that the general shear failure occurs in case of a group stone column [19]. In 1997, Rao et al. conducted the laboratory model tests in single and group columns and observed the characteristics of spacing between the stone columns [20]. They reported that as the column comes closer to the bulging zone, the bearing capacity of a group column increases due to confinements. In 2002, Bae et al. conducted both laboratory model test and numerical analysis with FEM method and found the general shear failure in a conical shape in case of a group stone column [21]. In a group stone column, the failure mode was obtained as the combination of lateral deflection and bulging failure of stone columns [22]. From the previous studies of the failure of a stone column, we cannot conclude anything with full confirmation because of much confusion related to the failure in case of a group of stone columns. Therefore, elaborated studies are needed to know about the details of failure as per soil conditions and stone column pattern whether single or group pattern and in case of a group column, the different type of arrangements, i.e. triangular, rectangular, etc.

4 Bearing Capacity Determination

When stone columns are loaded, they deform by bulging near the surface. The adjoining soil imparts a confining effect which helps the stone column to take the vertical load. Moreover, a stone column is not a uniform composition of concrete but an assemblage of stone aggregates. Thus, the accepted formula of load-carrying capacity of a pile cannot be applied to a stone column.

There are various methods for the determination of the bearing capacity of a stone column. The methods are briefly discussed below.

As per IS code [4], the load-carrying capacity of the ground treated with stone columns is obtained by adding the contribution of three components:

(a) the resultant capacity obtained from the resistance provided by the surrounding soft soil against the lateral deformation, i.e. bulging under the axial load, Q_1 as expressed in Eq. (1),

(b) capacity resulting from an increase in the resistance provided by the surrounding soft soil because of the surcharge over it, Q_2 as expressed in Eq. (2) and
(c) the bearing support, which the intervening soil provides between the columns, Q_3 as expressed in Eq. (3).

$$Q_1 = \frac{\sigma_v \pi D^2/4}{\text{FOS}} \tag{1}$$

where

FOS factor of safety = 2,
σ_v limiting axial stress $= \sigma_{rL} K_{p_{\text{col}}}$,
σ_{rL} limiting radial stress $= \sigma_{r_0} + 4c_u$,
c_u undrained shear strength (undisturbed) of the soft clay surrounding the column,
σ_{r0} initial effective radial stress,
$K_{p_{\text{col}}}$ coefficient of passive earth pressure $= \tan^2\left(45° + \frac{\varphi_c}{2}\right)$,
D column diameter and
φ_c shearing resistance angle of stone.

$$Q_2 = \frac{K p_{\text{col}} \Delta\sigma_{r_o} A_s}{2} \tag{2}$$

where

$\Delta\sigma_{r_o}$ increase in mean radial stress of the soil and
A_s area of the stone column.

$$Q_3 = q_{\text{safe}} A_g \tag{3}$$

where

A_g Intervening soil area for each column $= 0.866S^2 - \frac{\pi D^2}{4}$ for triangular arrangement and
q_{safe} the safe bearing pressure of soil.

Thus, safe load on each column,

$$Q = Q_1 + Q_2 + Q_3. \tag{4}$$

Afshar and Ghazavi (2012) used an imaginary retaining wall such that it stretches vertically from the stone column edge and presented a theoretical expression as presented in Eq. (5) to find the bearing capacity of the column [10]. The bearing capacity of the column is expressed as

$$q_{\text{ult}} = C_c N_c \overline{q} N_q + \frac{1}{2} W \gamma_c N_c \tag{5}$$

where, $N_c = 2\frac{\cos\frac{\varphi_c}{2}}{\cos\frac{\varphi_s}{2}}\frac{\sqrt{K_{pcc}}}{K_{as}}$, $N_q = \frac{K_{pc}}{K_{as}}\frac{\cos\frac{\varphi_c}{2}}{\cos\frac{\varphi_s}{2}}$ and $N_\gamma = \tan\eta_a\left(\frac{K_{pc}}{K_{as}}\frac{\cos\frac{\varphi_c}{2}}{\cos\frac{\varphi_s}{2}} - \frac{\gamma_s}{\gamma_c}\right)$ where K_{p_c} = lateral passive earth pressure coefficient, K_{as} = lateral active earth pressure coefficient, γ_c = unit weight of the stone material, γ_c = unit weight of soft soil, φ_s and φ_c are the friction angles of soft soil and stone column material, respectively, q = surcharge pressure on passive region surface, η_a = angle between active wedge and horizontal direction and K_{as} = lateral active earth pressure coefficient.

$$K_{as} = \frac{\cos^2\varphi_s}{\cos(\delta_1)\left[1 + \sqrt{\frac{\sin(\varphi_s+\delta_1)\sin(\varphi_s)}{\cos(\delta_1)}}\right]^2}$$

$$K_{pc} = \frac{\cos^2\varphi_c}{\cos(-\delta_2)\left[1 - \sqrt{\frac{\sin(\varphi_c+\delta_2)\sin(\varphi_c)}{\cos(-\delta_2)}}\right]^2}$$

$$K_{pc_c} = K_{pc}\left(1 + \frac{c_w}{c_c}\right)$$

$$\delta_1 = \frac{\varphi_s}{2} \text{ and } \delta_2 = \frac{\varphi_c}{2}$$

where

c_w is the wall–soil interface cohesion whose value varies between $0.3c_c$ for stiff soil and c_c for soft soil. In the absence of experimental data, the recommended c_w value is $0.45c_c$.

W = width of continuous strips for each row of the stone columns $= \frac{A_s}{S}$ where A_s is the horizontal cross-sectional area of the column and S is the center to center distance between two subsequent stone columns.

$$\eta_a = \varphi_s + \tan^{-1}\left(\frac{-\tan\varphi_s + C_1}{C_2}\right)$$

where

$$C_1 = \sqrt{\tan\varphi_S(\tan\varphi_s + \cot\varphi_s)(1 + \tan\delta_1\cot\varphi_s)}$$

and

$$C_2 = 1 + (\tan\delta_1[\tan\varphi_s + \cot\varphi_s])$$

Etezad et al. (2015) investigated the bearing capacity of stone columns (group) in soft soil [11]. They presented an analytical model for the prediction of the bearing capacity of the stone column-reinforced soft soil under rigid raft foundation considering general shear failure mechanism. The authors presented the forces which act on the soft soil section and composite soil section with the failure a mechanism as shown in Figs. 4 and 5. They expressed an equation for the ultimate bearing capacity of the stone column as shown in Eq. (6). They also presented the expression of bearing capacity factors (N_c, N_q and N_γ) as shown in Eqs. (7), (8) and (9).

$$q_{ult} = C_{comp}N_c + qN_q + \frac{1}{2}B\gamma_{comp}N_\gamma \tag{6}$$

where

$$N_\gamma = 2\cos(\psi - \varphi)\left\{\frac{\frac{F}{\gamma_{comp}B^2}\cos\delta\left[(a-1)\cdot(\tan\psi - \cot\theta_1) + \left(\frac{a\cdot e^{\theta_1 \tan\varphi_{comp}}}{\sin\theta_1}\right)\right]}{\frac{1}{3}\cos(\psi - \varphi_{comp}) + \sin(\psi - \varphi_{comp})\left(\frac{1}{3}\tan\psi - \frac{\tan\psi - \cot\theta_1}{2}\right)} - \frac{\frac{1}{12\sin^3\theta_1\left(9\tan^2\varphi_{comp}+1\right)}\left[e^{\theta_1\tan\varphi_{comp}} - (3\tan\varphi_{comp}\sin\theta_1 + \cos\theta_1)\right] - \frac{\tan\psi - \cot\theta_1}{24}}{\frac{1}{3}\cos(\psi - \varphi_{comp}) + \sin(\psi - \varphi_{comp})\left(\frac{1}{3}\tan\psi - \frac{\tan\psi - \cot\theta_1}{2}\right)}\right\} - \frac{\tan\psi}{2}, \tag{7}$$

$$\frac{F}{B^2} = \gamma_c\left\{\frac{\frac{H1^2\sin\theta^*\tan(90-\theta^*)\cos\varphi_c}{3B^2\sin^2\theta_2}\left[-\cos(\theta^*+\theta_2) + \cos\theta^* e^{\theta_2\tan\varphi_c}\right]^2\cdot\left[\frac{1}{2}\cos(\theta^*+\theta_2) + \cos\theta^* e^{\theta_2\tan\varphi_c}\right]}{\cos\delta\left[a\cdot\sin\theta_2 + \sin\theta^*\cos(\theta^*+\theta_2)\right] - \sin\delta\cos\theta^*\cos(\theta^*+\theta_2)} + \frac{\frac{\xi\cdot H1^2\cos^3\theta^*}{3B^2\sin^2\theta_2\left(9\tan^2\varphi_c+1\right)} - \frac{H1^2\cos^2\theta^*\cos^2(\theta^*+\theta_2)}{3B^2\sin\theta_2}}{\cos\delta\left[a\cdot\sin\theta_2 + \sin\theta^*\cos(\theta^*+\theta_2)\right] - \sin\delta\cos\theta^*\cos(\theta^*+\theta_2)}\right\},$$

$$N_q = \cos(\psi - \varphi_{comp})\frac{\frac{A}{B}\cdot\cos\delta\left[\frac{2H_1\cdot m}{B} - (\tan\psi - \cot\theta_1)\right]}{\left[\frac{1}{4}\cos(\psi - \varphi_{comp}) + \frac{1}{2}\sin(\psi - \varphi_{comp})\left(\frac{-\tan\psi}{2} + \cot\theta_1\right)\right]} \tag{8}$$

and

$$N_c = \tan\psi + \cos(\psi - \varphi_{comp})\frac{\frac{D\cdot C_c}{B\cdot C_{comp}}\cdot\cos\delta\left[\frac{2H_1\cdot m}{B} - (\tan\psi - \cot\theta_1)\right] + \frac{1}{4\sin^2\theta_1\tan\varphi_{comp}}\left(e^{2\theta_1\tan\varphi_{comp}} - 1\right) - C_{comp}}{\left[\frac{1}{4}\cos(\psi - \varphi_{comp}) + \frac{1}{2}\sin(\psi - \varphi_{comp})\left(\frac{-\tan\psi}{2} + \cot\theta_1\right)\right]} \tag{9}$$

By using all these equations, the bearing capacity of a stone column can be found out. The researchers also developed some design charts for N_c, N_q and N_γ against the friction angle of stone column materials. The values of these charts are converted into tabular form and presented in Tables 1, 2 and 3.

The angles θ_1, θ_2, θ^*, φ_c, φ_{comp} and δ can be better understood from Figs. 3 and 4 of reference [11].

Table 1 Variation of N_c against friction angle of stone column material (φ_s) for various native soil friction angles (φ_c) (after Afshar and Ghazavi [10])

φ_s	φ_c										
	30	32	34	36	38	40	42	44	46	48	50
0	8.5	9.2	9.99	10.3	11.8	12.48	13.1	14.97	15.18	17.54	20
5	9.85	10.04	11.2	12.4	12.58	13.7	15	16.3	18.05	20.05	22.5
10	11	11.7	12.49	13.1	14	15.85	17.29	18.4	20.5	22.72	25.95
15	12.6	12.75	13.75	14.9	16.4	17.6	19.04	21.92	23.90	26.73	29.95
20	13.98	14.99	15.98	17.35	18.10	20.05	22.1	24.92	27.5	30.03	33.8

Table 2 Variation of N_q versus friction angle of stone column material (φ_s) for various native soil friction angles (φ_c) (after Afshar and Ghazavi [10])

φ_s	φ_c										
	30	32	34	36	38	40	42	44	46	48	50
0	3.05	4.01	4.09	4.55	5	5.54	6.02	6.52	7.1	7.5	7.85
5	4.85	5.0	5.11	5.9	6.5	6.8	7.1	7.55	8.05	8.95	10
10	5.95	6.25	6.8	7.1	7.5	7.82	8.3	9.7	10.04	11.95	12.55
15	7.48	7.55	7.85	8.05	9.85	10.04	11.45	12.5	13.4	15.03	17.2
20	9.02	9.95	10.25	11.9	12.49	13.6	15	16.9	17.95	20.05	22.51

Table 3 Variation of N_{γ} versus friction angle of stone column material (φ_s) for various native soil friction angles (φ_c) for $\left(\frac{\gamma_s}{\gamma_c}\right) = 1.2$. (after Afshar and Ghazavi [10])

φ_s	φ_c										
	30	32	34	36	38	40	42	44	46	48	50
0	4	4.1	5	5.11	6.1	7.45	9.1	10	12.02	14.05	16.95
5	5	5.06	6.11	7.15	8.05	10	11.2	12.98	15.05	18.1	22.5
10	6	7.45	8.3	9.95	10.07	12.55	14.95	17.5	20.03	24.95	30
15	9.05	10	11.8	12.64	15	17	20	24.02	27.7	32.55	40
20	12.4	13.97	15.02	17.48	20	23.12	22.45	32.41	37.5	44.95	55

where $c_{comp} = A_s c_s + (1 - A_s)c_c$,
$\gamma_{comp} = A_s\gamma_s + (1 - A_s)\gamma_c$,
and A_s = replacement ratio

$$= \frac{A_{col}}{s^2}$$

for a square column pattern.

$$\varphi_{comp} = \tan^{-1}[A_s\mu_s \tan\varphi_s + (1 - A_s)\mu_c \tan\varphi_c],$$

where

A_{col} cross section area of the column,
s spacing between the columns,
c_s cohesion of stone column's material,
c_c cohesion of clay,

$$\mu_s = \frac{n}{1 + (n-1)A_s}$$

and

$$\mu_c = \frac{1}{1 + (n-1)A_s},$$

where n = stress ratio

$$= \frac{\text{vertical stress in the granular material}}{\text{vertical stress in cohesive soil.}}$$

The authors presented the following charts for the solution:

(a) N_γ versus shearing resistance angles for the reinforced soil for $\frac{\gamma_{comp}}{\gamma_c} = 1$, $\frac{\gamma_{comp}}{\gamma_c} = 1.2$, $\frac{\gamma_{comp}}{\gamma_c} = 1.4$, $\frac{\gamma_{comp}}{\gamma_c} = 1.6$, $\frac{\gamma_{comp}}{\gamma_c} = 1.8$ and $\frac{\gamma_{comp}}{\gamma_c} = 2$.
(b) N_q versus shearing resistance angles for the reinforced soil.
(c) N_c versus shearing resistance angles for the column-reinforced soil for $c_{comp}/c_c = 0.2$, $c_{comp}/c_c = 0.4$,$c_{comp}/c_c = 0.6$ and $c_{comp}/c_c = 0.8$.

The charts for the determination of N_q were converted into tabular form which is presented in Table 4.

The remaining charts can be obtained from reference [11].

By using these charts, the bearing capacity factors and thereby the bearing capacity of stone columns for different friction angles can be found out.

Table 4 Variation of N_q with the shearing resistance angles for the reinforced ground (after Etezad et al. [11])

ϕ_c	ϕ_{comp}						
	15	20	25	30	35	40	45
0	2.5	2.53	2.6	3.2	4.9	6.4	7.5
6	2.51	2.8	3.5	5	6.1	7.8	10
12	3.5	4.1	5	6.2	7.45	10.13	13.1
18			7.3	8.5	11.3	14.84	17.9
24				12.65	15.5	19.99	24.98
30						30	39.85

5 Settlement Calculation of Stone Columns

Various investigators conducted the theoretical study on settlement of stone columns and developed some solutions [15, 23–26]. As per IS 15284 (part 1) 2003 [4], the settlement of the treated ground can be calculated with the use of reduced stress method. As per IS code, by using the expression presented in Eqs. (10), (11) and (12), the settlement of stone columns can be calculated. This method is based on the replacement ratio (a_s) and the stress concentration factor (n). The stress concentration factor can be estimated as

$$n = \frac{\text{vertical stress in the composite soil, } \sigma_g}{\text{vertical stress in soft soil, } \sigma_s} \tag{10}$$

And, the replacement ratio is the ratio of the area of the stone column to the equivalent area represented by a stone column, which is estimated as

$$a_s = \frac{A_s}{A_s + A_g} \tag{11}$$

where A_s represents the area of the stone column and A_g represents the plan area of the soil for the column.

In case of a stone column, the settlement of stone column mainly occurs due to the consolidation of the soil. IS code suggests the consolidation settlement (S_t) as

$$S_t = m_v \sigma_g H \tag{12}$$

where

m_v modulus of volume decrease of soil,
σ_g vertical stress in surrounding soil, and
H thickness of treated soil.

Castro (2016) conducted an analytical study and obtained the expression for the settlement of stone columns below rigid footings [23]. He considered a horizontal slice at a depth of 'z' of the unit cell as shown in Fig. 1 and investigated the settlement over there.

He found out the settlement of each layer and obtained the total settlement of the stone columns by summing up all these values. To find out the total settlement, he used Eq. (13).

$$S_z = \sum_{1}^{i} \varepsilon_{z,i} \Delta t_i \tag{13}$$

where

Δt_i thickness of ith slice and

$\varepsilon_{z,i}$ vertical strain at ith slice.

For stone column,
oedometric (constrained) modulus,

$$E_{\mathrm{mc}} = E_{\mathrm{c}} \frac{(1 - \vartheta_{\mathrm{c}})}{(1 + \vartheta_{\mathrm{c}})(1 - 2\vartheta_{\mathrm{c}})}$$

For soft soil, oedometric modulus,

$$E_{\mathrm{ms}} = E_{\mathrm{s}} \frac{(1 - \vartheta_{\mathrm{s}})}{(1 + \vartheta_{\mathrm{s}})(1 - 2\vartheta_{\mathrm{s}})}$$

where E_{c} and E_{s} are Young's modulus of stone column materials and soft soils, respectively.

ϑ_{s} and ϑ_{c} are Poisson's ratio of soft soil and stone column, respectively.

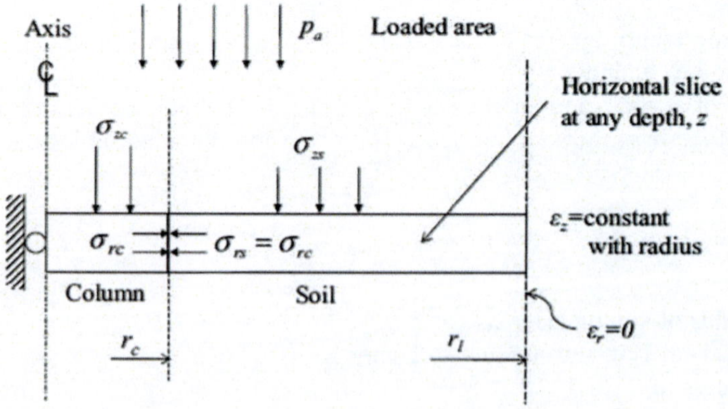

Fig. 1 Analytical model showing the horizontal slice (after Castro [23])

Shear modulus of stone column materials, $G_c = \frac{E_s}{2(1+\vartheta_c)}$.
Shear modulus of soft soil, $G_s = \frac{E_s}{2(1+\vartheta_s)}$.
Lame's constant for stone column materials, $\lambda_c = E_{mc} - 2G_c$.
Lame's constant for soft soil, $\lambda_s = E_{ms} - 2G_s$.
Active earth pressure coefficient
for the stone column, $K_{ac} = \frac{1-\sin\varphi_c}{1+\sin\varphi_c}$
and for dilatancy angle, $K_c = \frac{1-\sin_c}{1+\sin_c}$,
where ψ is the dilatancy angle.
Column constant, $C_E = \frac{3\lambda_c + 2G_c}{1+2K_{ac}K_c + \frac{\lambda_c}{G_c}(1-K_{ac}-K_c+K_{ac}K_c)}$.
For each slice,
area replacement ratio, $a_r = \frac{Nd_c^2}{(D+\frac{Z}{2})^2}$,
where N is the number of stone columns in a group pattern, d_c is the stone column diameter and Z is the depth of the slice considered.

The average vertical stress at each depth,

$$p_a(z) = p_a(0)\frac{D^2}{\left(D+\frac{Z}{2}\right)^2},$$

where $p_a(0)$ is the applied load and D is the diameter of footing.

$$F^* = \frac{\lambda_c - \lambda_s}{a_r(\lambda_c + \lambda_s + G_c + G_s) + (\lambda_c + G_c - G_s)}$$

$$E_{ml}^{e} = a_r E_{mc} + (1-a_r)E_{ms} + F^* a_r(\lambda_s(1-a_r) - \lambda_c(1+a_r))K^*$$

$$= \frac{K_{ac} - \frac{1}{C_E}\lambda_s}{\frac{a_r(\lambda_s+G_s)-G_s}{C_E(1+a_r)} + K_{ac}K_c}$$

$$E_{ml}^{p} = (1-a_r)E_{ms} + \frac{(1-a_r)}{(1+a_r)}a_r\lambda_s K^* + \frac{a_r}{K_{ac}}J^*,$$

where $J^* = \lambda_s + \frac{a_r(\lambda_s+G_s)-G_s}{1+a_r}K^*$.

$$Y = \frac{\left(K_{0s}\gamma_s' - K_{ac}\gamma_c'\right)E_{ml}^{e}}{G_c[2K_{ac} + F^*(1+a_r)] - \lambda_c(1-K_{ac})[1-F^*(1+a_r)]}$$

$$p_a^y = Yz$$

If $p_a \le p_a^y$, then $\varepsilon_z = \frac{p_a}{E_{ml}^{e}}$.
If $p_a > p_a^y$, then $\varepsilon_z = \frac{p_a^y}{E_{ml}^{e}} + \frac{p_a^p}{E_{ml}^{p}}$.

Zahmatkesh and Choobbasti (2010) defined a term called the settlement reduction ratio (SRR) as $\mathrm{SRR} = \frac{\text{settlement of the composite ground}}{\text{settlement of ground without stone column}}$ for evaluating the settlement of soft clay reinforced with stone columns [25]. SRR is also defined as $\mathrm{SRR} = \frac{E_o}{E_{eq}}$, where

E_O Young's modulus of ground without stone column and
E_{eq} Young's modulus of the composite ground.

Zhang et al. (2013) suggested the settlement of stone columns as the summation of the total compression deformation of the stone column (S_p) and the settlement of the underlying unreinforced layers (S_s) as shown in Eq. (14) [26]. Therefore, they divided the unit cell into *N* elements and found out the settlement of each element differently and obtained the total settlement by summing up settlements of all the elements. The total compression deformation of the column (S_p) and the settlement of the underlying unreinforced layers (S_s) can be obtained from Eqs. (15) and (16). Thus, the total settlement becomes

$$S = S_p + S_s \tag{14}$$

where

$$S_\mathrm{p} = \sum_{i=1}^{N} \Delta s_{\mathrm{p},i} \tag{15}$$

and

$$S_\mathrm{s} = \sum_{i=1}^{N_\mathrm{s}} \frac{q_i}{E_{si}} H_i \tag{16}$$

q_i = vertical stress due to the transfer of the applied stress (q) down into the *i*th subjacent unreinforced soil layer,

H_i is the thickness, E_{si} is the compression modulus of the *i*th subjacent unreinforced soil layer,

and N_s represents number of the soil layers.

$$\Delta s_{\mathrm{p},i} = l_i \times \frac{\sigma_{\mathrm{zp},i}}{E_\mathrm{p}} \times \frac{1 - \mu_\mathrm{p} - 2\mu_\mathrm{p}^2}{(1 - \mu_\mathrm{p}) - 2\mu_\mathrm{p} k_i}$$

where μ_p is Poisson's ratio of the column and E_p is Young's modulus of the column.

$$\begin{aligned} \sigma_{\mathrm{zp},i} &= \text{uniform vertical stress} \\ &= \frac{E_\mathrm{p}}{1 - \mu_\mathrm{p} - 2\mu_\mathrm{p}^2} \left[(1 - \mu_\mathrm{p}) - 2\mu_\mathrm{p} k_i \right] \varepsilon_{z,i} \end{aligned}$$

$\varepsilon_{z,i}$ is the vertical strain of the column at the ith slice.

Christian and Carrier (1978) carried out model tests with stone columns and compared the load-settlement behavior with that obtained through commercially available software PLAXIS [27]. In a theoretical study, they used the following set of equations to find out the settlement ratio (SR) which is the ratio of the settlement of the composite soil to the settlement of the soft soil without a stone column.

$$s_{\mathrm{E}} = \mu_0 \mu_1 \frac{qB}{E} \tag{17}$$

$$E_{\mathrm{eq}} = \frac{\sigma}{s} \tag{18}$$

$$\varepsilon = \frac{s}{L}$$

$$\mathrm{SR} = \frac{E_0}{E_{\mathrm{eq}}},$$

where q represents the applied footing load, E represents elastic modulus of the soil and μ_1 and μ_0 are the constant values which depend on the thickness between the footing base and hard strata and the depth of the footing, respectively. σ is the average applied stress, s is the settlement of the footing, ε is the average strain and L is the thickness of the clay bed. E_0 represents Young's modulus of the unreinforced ground, and E_{eq} is the equivalent secant modulus with the assumption the whole soil medium is homogeneous.

Greenwood and Thompson (1984) presented one chart of area ratio versus settlement ratio to find out the settlement of treated soil for a constant undrained strength of soil [15]. The chart is shown in Fig. 2.

ΔH ettlement of untreated soil,

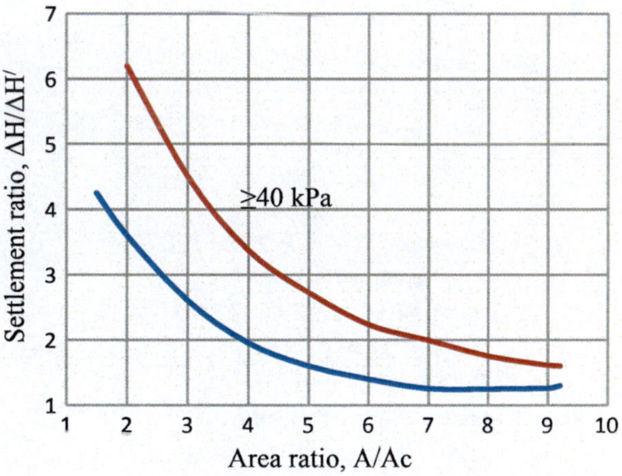

Fig. 2 Approximate settlement reduction for ground reinforced with stone columns (after Greenwood and Thompson [15])

$\Delta H'$ Settlement of stone column treated soil,
A_c Area of stone column and
A Area supported by column.

6 Design Example Problem

Problem statement:
To obtain the bearing capacity and settlement of a footing of 1 m diameter resting on a group of stone columns of diameter 0.5 m and length of 4 m arranged in a triangular pattern with spacing at 2.5 times the diameter of column,
friction angle of the column, $\varphi = 43°$,
undrained cohesion of clay, $c_u = 25$ kPa,
unit weight of stone column material, $\gamma_s = 20\text{kN/cu} \cdot \text{m}$,
unit weight of clay, $\gamma_c = 17\frac{\text{kN}}{\text{cu}} \cdot \text{m}$,
liquid limit, $w_L = 55\%$ and
water content $= 34\%$.
Solution:

A. Bearing Capacity Determination
Bearing capacity of untreated soil as per IS code

$$q_u = c_u N_c s_c d_c i_c + q N_q s_q d_q i_q = 25 \times 5.14 \times 1.3 \times 1 \times 1 + 17 \times 0 = 167.05\,\text{kPa}$$

The bearing capacity of a group of stone columns is obtained by different methods as discussed below:

1. IS 15284 (part 1) 2003 [4]:

(a) Bearing capacity due to bulging of column:

$$Q_1 = \frac{(\sigma_v \pi D^2)/4}{2}$$

$K_0 = 0.6$ for clays.
$\sigma_{v0} = 17$ kPa.
$\sigma_{r0} = 20.4$ kPa.
$\sigma_{rL} = 20.4 + 4 \times 25 = 120.4$ kPa.
$k_{p_{c0l}} = (\tan(45 + 21.5))^2 = 5.29$
$\sigma_v = 120.4 \times 5.29 = 636.916$ kPa.
Yield load $= 636.916 \times \frac{\pi \times 0.5^2}{4} = 125.058$ kN
Safe load on column, $Q_1 = \frac{125.058}{2} = 62.53$ kN

(b) Capacity due to a surcharge:
Increase in mean radial stress, $\Delta\sigma_{r0} = \frac{q_{safe}}{3}(1 + 2K_0)$.
For $\varphi = 43°$,

$N_c = 5.7.$
So, $q_{safe} = \frac{25 \times 5.14}{2.5} = 51.4,$

$$A_s = \frac{\pi \times 0.5^2}{4} = 0.1963\,\text{m}^2,$$

$$\Delta\sigma_{r0}\frac{51.4 \times (1 + 2 \times 0.6)}{3} = 37.7\ \text{kPa and}$$

$$Q_2 = \frac{k_{p_{col}}\Delta\sigma_{r0}A_s}{2}$$

$= \frac{5.29 \times 37.7 \times 0.1936}{2} = 19.3kN.$

(c) Bearing capacity provided by the intervening soil:

$$Q_3 = q_{safe} \cdot A_g$$

$$A_g = 0.866s^2 - (\pi \times d^2)/4.$$

$$= 0.866 \times (1.25)^2 - \frac{\pi}{4} \times 0.5^2 = 1.157\,\text{m}^2$$

$$Q_3 = 51.4 \times 1.157 = 59.47\,\text{kN}$$

Therefore, the total bearing capacity, $Q = Q_1 + Q_2 + Q_3$
$= 62.53 + 19.3 + 59.47 = 141.3\,\text{kN}.$
Therefore, ultimate bearing capacity,

$$Q_u = 282.6\,\text{kN} = \frac{282.6}{0.866 \times (1.25)^2} = 208.85\text{kPa}.$$

2. By Etezad et al. [11]:
   The ultimate bearing capacity of stone column,
   $q_{ult} = C_{comp}N_c + qN_q + \frac{1}{2}B\gamma_{comp}N_\gamma.$
   Replacement ratio,
   $A_s = \frac{\frac{\pi}{4} \times (0.5)^2}{0.866 \times (1.25)^2} = 0.58$ for a triangular column pattern.

$$c_{comp} = 0.58 \times 0 + (1 - 0.58) \times 25 = 10.5\ \text{kN/m}$$

$$\gamma_{comp} = 0.58 \times 20 + (1 - 0.58) \times 17 = 22.1\ \text{kN/m}$$

$$n = \frac{\sigma_s}{\sigma_g} = \frac{636.916}{167.05} = 3.812$$

$$\mu_s = \frac{n}{1 + (n-1)A_s} = \frac{3.812}{1 + (3.812 - 1) \times 0.58} = 1.45$$

$$\mu_c = \frac{1}{1 + (n-1)A_s} = 0.38$$

So,

$$\varphi_{comp} = \tan^{-1}[A_s \mu_s \tan\varphi_s + (1 - A_s)\mu_c \tan\varphi_c]$$
$$= \tan^{-1}\left[0.58 \times 1.45 \times \tan 43^\circ + (1 - 0.58) \times 0.38 \times \tan 0\right] = 38.1^\circ$$

From the design charts, they presented the bearing capacity factors are obtained as shown below:

For $\frac{\gamma_{comp}}{\gamma_c} = 1.3$ and $\varphi_{comp} = 38.1^\circ,\ N_\gamma = 7.05$

For $\frac{C_{comp}}{C_c} = \frac{10.5}{25} = 0.42$ and $\varphi_{comp} = 38.1^\circ,\ N_C = 33$

For $\varphi_{comp} = 38.1^\circ$, $N_q = 5.38$

So, the ultimate bearing capacity,

$$q_u = 10.5 \times 33 + 0 + 0.5 \times 22.1 \times 1 \times 7.05 = 424.4 \text{ kPa.}$$

3. By Afshar and Ghazavi [10]:

The bearing capacity of stone column, $q_{ult} = C_c N_c + \bar{q} N_q + \frac{1}{2} W \gamma_c N_\gamma$.

Here,

$$\delta_1 = \frac{\varphi_s}{2} = \frac{43^\circ}{2} = 21.5^\circ$$

and $\delta_2 = \frac{\varphi_c}{2} = \frac{0^\circ}{2} = 0^\circ$.

$c_w = C_c$ for soft soil $= 25$ kPa.

$K_{pc} = \frac{\cos^2 0^\circ}{\cos(-0^\circ)\left[1 - \sqrt{\frac{\sin 0^\circ \cdot \sin 0^\circ}{\cos 21.5^\circ}}\right]} = 1.$

$$K_{pc_c} = 1(1 + 1) = 2.$$

$$K_{as} = \frac{\cos^2 43^\circ}{\cos(21.5^\circ)\left[1 + \sqrt{\frac{\sin(43+21.5)^\circ \sin(43^\circ)}{\cos(21.5^\circ)}}\right]^2} = 0.1748.$$

So,

$$N_c = 2\frac{\cos\frac{0^\circ}{2}}{\cos\frac{43^\circ}{2}} \frac{\sqrt{2}}{0.1748} = 17.39,$$

$$N_q = \frac{1 \times 1}{0.1748 \times \cos 21.5^\circ} = 6.15,$$

$$C_1 = \sqrt{\tan 43^\circ(\tan 43^\circ + \cot 43^\circ)(1 + \tan 21.5^\circ . \cot 43^\circ)} = 1.6307$$

and $C_2 = 1 + (\tan 21.5°[\tan 43° + \cot 43°]) = 1.8$.
So, $\eta_a = 43° + \tan^{-1}\left(\frac{-\tan 43° + C_1}{1.8}\right) = 64.374°$ and
$N_\gamma = \tan 43^0\left(\frac{1}{0.1748}\frac{cos0°}{cos21.5°} - \frac{20}{17}\right) = 10.366$.
Now, W = width of continuous strips for each row of stone columns $= \frac{A_s}{S}$
$= \frac{0.196}{1.25} = 0.1571$.
So, $q_{\text{ult}} = 25 \times 17.39 + 0 + \frac{1}{2} \times 0.1571 \times 17 \times 10.366 = 448.592$ kPa.

B. Settlement Determination
For untreated soil,
settlement,

$$S = \frac{C_c H}{1 + e_0} \log\left(\frac{\sigma_0 + \Delta\sigma}{\sigma_0}\right)$$
$$= 230.1\text{mm}.$$

For soils treated with stone column, the following methods are applied for calculation:

i. IS 15284, PART-1, 2003 [4]:
The settlement due to consolidation is

$$S_t = m_v \sigma_g H$$

Initial stress, $\sigma_0' = 10 \times 4/2 = 20\,\text{kPa}$ and
σ_v' = change in stress $= \frac{167.05 \times \frac{\pi}{4} \times 1^2}{3 \times \frac{\pi}{4} \times 3^2} = 6.185\,\text{kPa}$.
Compression index, $c_c = 0.009(w_L - 10\%)$

$$= 0.009 \times (55 - 10)$$

$= 0.405$.
So, $e = 0.405 \times \log_{10} \frac{5.28 + 20}{20} = 0.0412$.
Again, water content = 34%,
$e_0 = w_s \times G = 0.34 \times 2.6 = 0.884$ and
$m_v = \frac{0.0412}{(1 + 0.884) \times 5.28} = 4.14 \times 10^{-3}$.
Again, $\sigma_g = \mu_g$,
$n = 3.812$,
$a_s = 0.58$,

$$\mu_g = \frac{1}{1 + (n - 1)A_s} = \frac{1}{1 + (3.812 - 1) \times 0.58} = 0.38 \quad \text{and}$$

Settlement of stone column, $S_t = 4.14 \times 10^{-3} \times 6.185 \times 4 = 0.08744\,\text{m} =$ 102.43 mm..

(ii) By Castro [23]:
Let us consider, Young's modulus for soft soil, $E_s = 25,000\text{kPa}$,

Young's modulus for column materials, $E_c = 55,000\text{kPa}$,
Poisson's ratio for soft soil, $\vartheta_s = 0.35$ and
Poisson's ratio for stone column material, $\vartheta_c = 0.33$.
Since vertical stress in surrounding soil, $\sigma_g = 167.05\text{kPa}$,
applied load, $p_a(0) = \left(167.05 \times \frac{1.157}{3}\right)\,\text{kN} = 64.43\,\text{kN}$.
For stone column,
oedometric (constrained) modulus, $E_{\text{mc}} = \frac{55000\times(1-0.33)}{(1+0.33)(1-2\times0.33)} = 81490.49\,\text{kPa}$.

For soft soil, oedometric modulus, $E_{\text{ms}} = \frac{25000\times(1-0.35)}{(1+0.35)(1-2\times0.35)} = 40123.456\,\text{kPa}$,

where E_c and E_s are Young's modulus of stone column materials and soft soils.

Shear modulus of the column materials, $G_c = \frac{55000}{2(1+0.33)} = 20676.69$.
Shear modulus of soft soil, $G_s = \frac{25000}{2(1+0.35)} = 9259.26$.
Lame's constant for stone column materials,

$$\lambda_c = 81490.49 - 2 \times 20676.69 = 40137.11.$$

Lame's constant for soft soil,

$$\lambda_s = 40123.456 - 2 \times 9259.26 = 21604.94$$

Coefficient of active earth pressure
for stone column, $K_{\text{ac}} = \frac{1-\sin 43^\circ}{1+\sin 43^\circ} = 0.189$
and for dilatancy angle, $K_c = \frac{1-\sin 0^\circ}{1+\sin 0^\circ} = 1$.
Column constant, $C_E = 45335.5$.
The whole soil profile is divided into 4 slices of 1 m thickness.
For the slice whose distance from top of the stone column, $z = 2$ m,
area replacement ratio, $a_r = 0.0625$.
The average vertical stress at each depth, $p_a(z) = 64.43 \times \frac{1^2}{\left(1+\frac{2}{2}\right)^2} = 16.106\,\text{kPa}$.

$$F^* = 0.328.$$

$$E_{ml}^e = 16666.82\,\text{kPa}.$$

$$K^* = 0.76.$$

$$E_{ml}^p = 7874.60\,\text{kPa}.$$

$J^* = 2060.95\,\text{kPa}$.

$$Y = 34.14.$$

Table 5 Settlement value for stone column

Depth in m	p_a in kPa	a_r in %	E^e_{ml} in kPa	E^p_{ml} in kPa	p^y_a in kPa	ε_z in %	S_Z in mm
0–1	64.43	0.25	4835.077	5867.3	0	3.59	35.9
1–2	28.64	0.111	4391.2	4839.76	32.375	1.998	19.98
2–3	16.1	0.0625	86666.82	16666.82	68.28	3.16×10^{-1} %	3.16
3–4	10.3	0.04	46355.51	18178.7	20.36	1.83×10^{-1}	1.83

Yielding load, $p^y_a = 34.14 \times 2 = 68.28$ kPa.
Here, $p_a < p^y_a$.
So, strain at this layer, $\varepsilon_z = 3.16 \times 10^{-3}\%$,
Δt_i = thickness of ith slice = 1 m and $\varepsilon_{z,i}$ = vertical strain at ith slice = $3.16 \times 10^{-3}\%$.
Settlement, $S_z = 3.16 \times 10^{-3} \times 1\,\text{m} = 3.16 \times 10^{-3}\,\text{m} = 3.16$ mm.
Similarly, for other slices, the settlement values are listed in Table 5.
The total footing settlement becomes $S_z = \sum_1^i \varepsilon_{z,i} t_i$.
So, the total settlement is 60.87 mm.

iii By Zhang et al. [26]:
Total settlement,
S = total compression deformation of the stone column + settlement of the underlying unreinforced layers = $s_p + s_s$.
For the first layer, $\sigma_{zp,i} = 64.43$ kPa.

$$\sigma_{zp,i} = 64.43 = \frac{E_p}{1 - \mu_p - 2\mu_p^2}\left[(1 - \mu_p) - 2\mu_p k_i\right]\varepsilon_{z,i}$$

$$= \frac{25000}{1 - 0.35 - 2 \times 0.35^2}[(1 - 0.35) - 2 \times 0.35 \times k_i]\varepsilon_{z,i}$$

$k_i = 0.922$
Now, $\Delta s_{(p,i)} = 1 \times \frac{49.6}{55000} \times \frac{(1-0.33-2\times0.33^2)}{(1-0.33)-2\times0.33\times k_i)} = \frac{0.00173}{(0.67-0.66k_i)} = \frac{0.00173}{0.05752} = 0.00735$.
So, $s_p = 29.5$ mm
Again, $S_s = \sum_{i=1}^{N_s} \frac{q_i}{E_{si}} H_i$
For the first layer, $q_1 = 64.43$ kPa.
So, settlement of the first layer, $s_1 = \frac{64.43}{25000} \times 1 = 2.58 \times 10^{-3}$ m.
For the second layer, $q_2 = q_{2i} + q_2$,
overburden pressure, $q_{2i} = \gamma z_2 = 19 \times 1 = 19$ kPa,
pressure due to external load, $q_2 = \frac{64.43}{(0.5+1\times\tan 30^\circ)^2} = 55.51$ kPa and
$q_2 = 19 + 55.51 = 74.51$ kPa.

Settlement of the second layer, $s_2 = \frac{74.51}{25000} \times 1 = 2.98 \times 10^{-3}$ m.
For the third layer, $q_3 = q_{3i} + q_3$,
overburden pressure, $q_{3i} = \gamma z_3 = 19 \times 2 = 38$ kPa,
pressure due to external load, $q_3 = \frac{64.43}{(0.5+2\times\tan 30°)^2} = 23.53$ kPa and
$q_2 = 38 + 23.53 = 61.53$ kPa.
Settlement of the third layer, $s_3 = \frac{61.53}{25000} \times 1 = 2.46 \times 10^{-3}$ m.
For the fourth layer, $q_4 = q_{4i} + q_4$,
overburden pressure, $q_{4i} = \gamma z_3 = 19 \times 3 = 57$ kPa,
pressure due to external load, $q_4 = \frac{64.43}{(0.5+3\times\tan 30°)^2} = 12.932$ kPa and
$q_4 = 57 + 12.932 = 69.932$ kPa.
Settlement of the fourth layer, $s_4 = \frac{69.932}{25000} \times 1 = 2.8 \times 10^{-3}$ m.

$$s_s = s_1 + s_2 + s_3 + s_4 = \left(2.58 \times 10^{-3} + 2.98 \times 10^{-3} + 2.46 \times 10^{-3} + 2.8 \times 10^{-3}\right) \text{m} = 0.00938 \text{ m} = 9.38 \text{ mm}$$

So, the total settlement becomes $s = s_p + s_s = 29.5 + 10.82 = 40.32$ mm.

(iv) By Greenwood and Thompson [15]:

Area of stone column, $A_c = \frac{\pi}{4} \times 0.5^2 = 0.1963 \text{ m}^2$.

Area supported by column, $A = 0.866 \times (1.25)^2 = 1.353 - \frac{\pi}{4} \times 0.5^2 = 1.157$.

Undrained cohesion of clay, $c_u = 25$ kPa.
$\frac{A}{A_c} = \frac{1.157}{0.1963} = 5.85$.
By using Bowel's chart for $c_u = 25$kPa, $\frac{\Delta H}{\Delta H'} = R = 1.95$
Considering settlement of untreated soil, $\Delta H = 230.1$ mm,
settlement of stone column treated soil, $\Delta H' = \frac{230.1}{1.95} = 118$ mm.
The summary of all the results is presented in Table 6.

Table 6 Summary of all results

Method	Bearing capacity in kPa	Settlement in mm
For untreated soil		
	167.05	230.1
For soil treated with stone column		
IS 15284, PART-1, 2003	208.85	102.43
Etezad et al. [11]	424.4	
Afshar and Ghazavi [10]	448.592	
Castro [23]		60.87
Zhang et al. [26]		40.32
Greenwood and Thompson [15]		118

7 Conclusions

Out of many published research papers, only a few papers have been referred to in this review paper. Selection of the referred papers was based on the availability of bearing capacity and settlement formulae. It is noticed from the present study that there is a wide variation in the results of bearing capacity and settlement suggested by different researchers. The following conclusions are drawn from this review paper:

(i) Bearing capacity of the untreated soil is improved by 1.5–3 times with installation of the stone column in a triangular pattern at a spacing of 2.5 times its diameter.
(ii) The Indian Standard code [4] gives a conservative result on the improvement of bearing capacity, hence a detailed study and upgradation of the codal provision are required.
(iii) The expected consolidation settlement of untreated soil is reduced by 2–4 times with installation of stone columns.
(iv) The most conservative result was obtained by Greenwood and Thompson [15].
(v) The codal provision in IS code [4] suggests a possible reduction of 0.4–0.5 times the expected settlement of untreated column.
(vi) The overall settlement is to be reduced to an effective use of stone column below building foundation.

References

1. Hughes JMO, Withers NJ (1974) Reinforcing of soft cohesive soils with stone columns. Ground Eng 7(3):42–49
2. Barksdale, R. D., Bachus, R. C., (1983) "Design and Construction of Stone Columns Volume II, Appendixes."Federal Highway Administration Office of Engineering and Highway Operations Research and Development, Washington
3. Lo SR, Zhang R, Mak J (2010) Geosynthetic-encased stone columns in soft clay: A numerical study. Geotext Geomembr 28(2010):292–302. https://doi.org/10.1016/j.geotexmem.2009.09.015
4. IS 15284-1 (2003) "Design and construction for ground improvement-Guidelines." Part-1: Stone columns [CED 43: Soil and foundation engineering], ICS 93.020
5. Black, J. A., Sivakumar, V. Madhav, M. R., and Hamill, G. A. (2007). "Reinforced stone columns in weak deposits: laboratoryModel study."Journal of geotechnical and geoenvironmental engineering, ASCE, Vol. 133, No. 9, https://doi.org/10.1061/(asce)1090-0241(2007)133:9(1154)
6. Maheshwari P, Khatri S (2010) Nonlinear response of footings on granular bed-stone column-reinforced poor soil. Int J Geotech Eng 4(4):435–443. https://doi.org/10.3328/IJGE.2010.04.04.435-443
7. Saroglou H, Antoniou A, Pateras S (2009) Ground improvement of clayey soil formations using stone columns: A case study from Greece. Int J Geotech Eng 3(4):493–498. https://doi.org/10.3328/IJGE.2009.03.04.493-498
8. Rajesh S, Jain P (2015) Influence of permeability of soft clay on the efficiency of stone columns and geosynthetic-encased stone columns – a numerical study. Int J Geotech Eng 9(5):483–493. https://doi.org/10.1179/1939787914Y.0000000088

9. Poorooshasb, H. B., Meyerhof, G. G (1996) "Consolidation settlement of rafts supported by stone columns", Journal of Southeast Asian Geotechnical Society, 1996, Vol. 27, No. 2, pp. 84-97, 1996
10. Afshar J. N., Ghazavi, M., (2012) "A simple analytical method for calculation of bearing capacity of stonecolumn." International Journal of Civil Engineering, Vol. 12, No. 1, Transaction B: Geotechnical Engineering, January 2014
11. Etezad, M., Hanna, A. M., Ayadat, T. (2015). "Bearing capacity of a group of stone columns in soft soil."InternationalJournal of Geomechanics,ASCE, ISSN 1532-3641/04014043(15)
12. Ambily., A. P., Gandhi, S.R., (2004) "Experimental and theoretical evaluation of stonecolumn in soft clay."ICGGE
13. Ambily, A. P., Gandhi, S. R., (2007) "Behavior of stone columns based on experimental and fem analysis." Journal of geotechnical and geoenvironmental engineering, ASCE, 2007.133:405-415, DOI:10.1061/(ASCE)1090-0241(2007)133:4(405)
14. Niroumand, H., Kassim, K. A., Yah, C. S., (2011) "Soil improvement by reinforced stone columns based on experimental work." EJGE, Vol. 16 [2011], Bund. L
15. Bowles, J. E.,(1988) "Foundation analysis and design."4th edition,McGraw-Hill, Singapore
16. Guetif et al (2007) Improved soft clay characteristics due to stone column installation. Comput Geotech 34(2007):104–111. https://doi.org/10.1016/j.compgeo.2006.09.008
17. Taube, G., Martin, P.E., P.G. (2002) "Stone Columns for Industrial Fills", Nicholson Construction Company, Cuddy, Pennsylvania
18. Bachus, R. C. and Barksdale, R. D. (1984), "Vertical and lateral behaviour of model stone columns.", Proceedings of the International Conference on In-situ Soil and Rock Reinforcement, Paris, 99–104
19. Hu, W. (1995) "Physical modeling of group behavior of stone column foundations", Ph.D. Thesis, University of Glasgow
20. Rao NS, Reddy KM, Rumar PH (1997) Studies on groups of stone columns in soft clays. Geotechnical Engineering-Bangkok 28(2):165–182
21. Bae, W. S., Bang, W. S., Byung, C. A. (2002) "Behavior of foundation system improved with stone columns", Proc. 12th Int. Offshore and Polar Engineering Conf., Kitakyushu, Japan, 675–678
22. Ghazavi M, Afshar JN (2013) Bearing capacity of geosynthetic encased stone columns. Geotext Geomembr 38(2013):26–36. https://doi.org/10.1016/j.geotexmem.2013.04.003
23. Castro J (2016) An analytical solution for the settlement of stone columns beneath rigid footings. Acta Geotech 11:309–324. https://doi.org/10.1007/s11440-014-0358-4
24. Sexton, B. G., McCabe, B. A., Castro, J., (2014) "Appraising stone column settlement prediction methods using finite element analyses." ActaGeotechnica (2014) 9:993–1011, https://doi.org/10.1007/s11440-013-0260-5
25. Zahmatkesh, A., Choobbasti, A. J., (2010) "Settlement evaluation of soft clay reinforced by stone columns, considering the effect of soil compaction." IJRRAS 3 (2), May 2010
26. Zhang, L., Zhao, M., Shi, C., Zhao, H., (2013) "Settlement calculation of composite foundation reinforced with stone columns."International journal of geomechanics, ASCE, 13(3): 248–256, https://doi.org/10.1061/(asce)gm.1943-5622.0000212
27. Christian JT, Carrier WD (1978) Janbu, Bjerrum and Kjaernsli's chart reinterpreted. Can Geotech J 15:123–128

Performance Study of Single Helix Embedded in Cohesionless Soil Under Pull-Out Load

Neha Nasreen, Ashok Kumar Khan, and Sitam Suvam Jena

Abstract In this investigation, an endeavour is made to understand the pull-out behaviour of single helix anchor embedded in cohesion less deposit. Physical tests were conducted using a laboratory-scale model tests tank of size 800 mm × 800 mm × 800 mm. A total of 18 tests were conducted in laboratory in which helix anchor was installed in loose and medium dense dry sand. A single model helix anchor made of steel was subjected to vertical pull-out load in a test tank with dry sand. Various parameters like diameter of helix were varied as 75 mm, 100 mm and 125 mm, embedment ratio was varied as 2, 4, 6 and relative density of sand was varied as 30 and 60% to understand their influence on the pull-out behaviour. A series of load uplift curves were drawn to understand the pull-out behaviour. It was observed that the pull-out capacity significantly varied with embedment ratio as well as the diameter of helix. But the increase in uplift capacity of helix anchor is less affected by increase in relative density in comparison to diameter of helix as well as embedment ratio. The results of the experimental investigation were validated using a numerical study based on PLAXIS 2D software. When the uplift load was varied with displacement from the PLAXIS examinations demonstrated commonly great concurrence with the physical modelling acquired from the exploratory outcomes.

Keywords Cohesionless deposit · Displacement contours · Embedment ratio · Helix anchor · PLAXIS 2D · Pull-out load · Uplift

N. Nasreen (✉) · S. S. Jena
School of Civil Engineering, KIIT Deemed to be University, Bhubaneswar, Odisha, India
e-mail: nehansrn@gmail.com

S. S. Jena
e-mail: sitam567@gmail.com

A. K. Khan
Department of Civil Engineering, NIT, Jamshedpur, Jharkhand, India
e-mail: akk.nitjsr@gmail.com

B. B. Das et al. (eds.), *Recent Developments in Sustainable Infrastructure*, Lecture Notes in Civil Engineering 75, https://doi.org/10.1007/978-981-15-4577-1_10

1 Introduction

Due to rapid urbanisation, all over the world there is a necessity to construct high rise structures like transmission towers, radar towers, street lamps which are subjected to high lateral forces coming from wind load. Such lateral forces cause an overturning moment on the foundation of the structure which causes instability to the structures. Earlier massive foundations are being used to provide stability to the structure which is quite heavy, expensive and time taking to construct. Now a new construction method is increasing its popularity by the use of ground anchors. Helix anchor is a ground anchor consists of one or more helix blade welded to the central shaft and is used to resists uplift forces coming over the structures. Anchors are driven in the ground by applying a rotating torque on it in which it is screwed into the ground. Helix anchor has a number of advantages over other construction method such as it is easy to instal, can be used immediately after the installation, produces minimum noise and vibrations while installation, and can be installed at any angle. Helix anchors are very much useful for offshore structures like submerged pipelines and submerged tunnels. These are subjected to high buoyant forces coming over it and hence large uplift forces act on such structures. Hence helix anchor is used to resists these high uplift forces. Helix anchor is used as a tieback in retaining wall, used as underpinning for failed foundation. These applications of helix anchor necessitate the development of theories and in depth understanding of it. Despite the increase in its use, its behaviour has not been understood properly over the last 20 years.

In this investigation, the ultimate capacity of the single helix anchor is surveyed subsequent to playing out a progression of laboratory-scale model pull-out tests on helix anchors in sand. The tests have been conducted to determine the effect of the embedment ratio, helix diameter and sand density against the uplift capacity of helix anchor. The research facility tests were led in medium loose and medium dense dry sand and load uplift curves are obtained. The experimental results have been validated using 2D finite element analysis based on PLAXIS 2D. The deformation of sand medium around the helix is observed on PLAXIS 2D. A series of parametric comparative graphs are plotted to understand the contribution of each parameter to the axial pull-out capacity of the helix anchor.

Alexander Mitchell was the first person to use helix anchor in 1836. He used helix anchor to build Maplin lighthouse near Thames River in England. In 1980s more than 100 lighthouses were constructed using helix anchor. Many researchers have worked on helix anchor. Ghally et al. [1] carried tests on single helix anchor in three different states, i.e. dense, medium and loose state in sand. The outcome of the study is the pull-out capacity of helix anchor is a function of diameter of helix, embedment ratio and sand density. He gave two modes of failure, namely, shallow and deep mode of failure. Also gave analytical equation to determine the pull-out capacity of single helix anchor. Ilamparuthi and Shreni [2] studied the response of anchor in two-phase materials under uplift. He studied the effect of suction and upward seepage. Depths of embedment, relative compactness of sand and seepage velocity are the variables. He concluded that suction contributes marginal in sand;

upward seepage reduces the uplift capacity of anchor. He used PLAXIS to validate the experimental results and has shown good agreement. Niroumand and Kassim [3] conducted tests on uplift capacity of square anchor plate in dense sand. He has done experimental tests and validated using PLAXIS 2D using soil hardening model and a relatively good agreement found between the two results. Nazir et al. [4] studied the execution of single vertical helix anchor embedded in dry sand. He concluded that embedment ratio and density of sand affects the uplift capacity but the shaft diameter ratio does not influence the uplift capacity. He gave empirical equations to determine the uplift capacity in loose as well as dense sand. He observed failure mechanism in loose and dense state of sand. Local failure surface was observed for loose state while for dense sand, the failure surface was of truncated cone. Mukherjee and Mittal [5] performed tests on uplift capacity of group of helix anchors in sand. He observed that the pull-out capacity of anchor increases with number of helix anchor, increase in embedment ratio, as well as the number of helix blade in an anchor. He used PLAXIS 3D to validate the results. Demir and Ok [6] present the uplift response of multi helix anchor in cohesive soil. Used PLAXIS 2D to better understand the behaviour. He concluded that the uplift resistance of multi helix anchor depends on embedment ratio and anchor spacing ratio. He demarcates the difference between shallow and deep mode of failure in PLAXIS 2D. Ghosh and Samal [7] performed numerical investigation on the pull-out capacity of single helix anchor embedded in homogeneous soil deposit using finite element analysis ABACUS. They varied the number of helix plates, ratio of spacing between the helix; plate to the diameter of the plate, as well as the depth of upper and lower helix plate. Ghaly and Hanna [8] carried out Ultimate pullout resistance of single vertical anchor. He concluded that embedment ratio and density of sand affects the uplift capacity. The compression and uplift bearing capacities of helical piles in cohesion less soil has been carried out by Abdel et al. [9]. He summarised that the compression and uplift loads are increased with increasing of the embedment ratios D/d and helix diameter d. Ilamparuthi et al. [10] described the two modes of failure develop within the soil mass depending on the anchor embedment ratio shallow anchor behaviour & deep anchor behavior was the outcome of this paper. Nazir et al. [11] analysed that the uplift capacity increases with the increase of embedment ratio and sand density however the shaft diameter ratio is not significantly influent to the uplift capacity. Mittal & Mukharjee [12] observed that the ultimate pullout capacity of helical screw anchor increases with increase in the embedment depth, number of anchors & number of screw blades in the anchor.

2 Materials and Methodology

2.1 Experimental Study

The experimental study involves a series of laboratory-scale pull-out tests on model helix anchors embedded in dry cohesionless deposit. In this study, the behaviour of single screw helix anchor under the influence of vertical uplift load was studied. The testing programme was conducted on three different model helix anchors having three different helix diameters, the helix anchors were embedded in sand in three different embedment ratios. Pull-out tests were conducted in two different sand densities.

2.1.1 Test Materials

The following materials were used in the study:

(a) Cohesionless deposit: River sand from Kharkai River was used as cohesionless deposit. The sand was collected from river banks and air-dried under sunlight. The dried sand was sieved through 4.75 mm sieve to remove any gravel-sized pebbles (Tables 1 and 2).
(b) Model helix anchor: Model helix anchors used in the study made up of mild steel. Helix bearing plates were fabricated using 3 mm thick mild steel plate. Circular steel plate equal to the diameter of the helix bearing plate was cut and formed into a helix. The helix was welded to solid steel shaft with uniform diameter of 25 mm. The shaft was sharpened on the bottom end to form the pilot point. A thread arrangement was done on the upper end of the shaft to which a square pile cap was attached (Fig. 1).

Table 1 Index properties of sand

Property	Value
Specific gravity, G_s (at 27 °C)	2.67
Effective size, D_{10}	0.40 mm
Effective size, D_{30}	0.58 mm
Effective size, D_{60}	0.90 mm
Coefficient of uniformity, C_u	2.25
Coefficient of uniformity, C_c	0.93
Maximum dry density, $\gamma_{d\text{-max}}$ (kN/m^3)	16.56
Minimum dry density, $\gamma_{d\text{-min}}$ (kN/m^3)	14.16
ϕ (loose) (°)	31
ϕ (medium dense) (°)	33

Table 2 Engineering properties of sand

Property	Medium loose sand	Medium dense sand
Height of pour (cm)	60	100
Dry density 'γ_d' (kN/m^3)	14.94	15.48
Relative density (%)	30	60
Friction angle ϕ (°)	31	33
Cohesion '*c*' (kN/m^2)	0	0
Elastic modulus '*E*' (kN/m^2)	8500	9000
Poisson's ration 'μ'	0.3	0.3

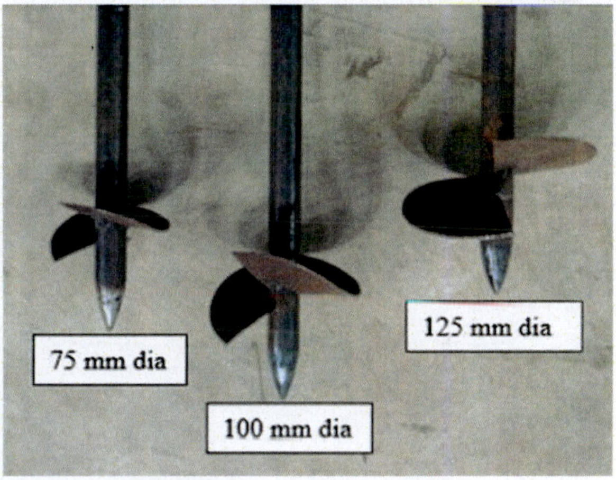

Fig. 1 Helix anchors

2.1.2 Test Variables

The diameter of the helix bearing plate (*D*), embedment of the helix bearing plate (*B*) and density of the fill (γ) are varied in this investigation. 18 different tests were carried out by varying the above parameters. The variable parameters used to depict the tests were expressed in a non-dimensional form called embedment ratio, *B*/*D*.

The nomenclature of the test number in test series is defined *X*–*Y*–*Z*, where *X* is diameter of the helix in mm, *Y* is the embedment ratio, and *Z* is the relative density of the sand. For example, test number 100-4-60 represents the pull-out test of 100 mm diameter helix anchor embedded in an embedment ratio 4 in the sand with relative density of 60%.

2.1.3 Test Setup

Test bed-loading frame consists of testing tank and loading system. Test tank was made of wood of the following dimensions: 0.80 m × 0.80 min plan and 0.80 m high. The inner surface of the tank was lined with smooth polyethylene sheet to reduce friction between sand and wooden walls. The loading frame consists of two heavy steel columns and a horizontal crosshead that supports two frictionless pulleys. The test bed of known density was prepared by rainfall technique from a predetermined height. Depending on the diameter of helix and the embedment ratio, the position of bearing plate as well as the position of pile cap was marked on the inner face of the tank. The sand was then rained into the tank from the prescribed height. Sand was rained into the tank in layers of 20 cm thick. The uniformity of placement density was checked at every layer by a dynamic penetrometer. After topping the tank off to the ideal stature, the sand surface was levelled with a ruler attached to a spirit level. The dial gauges pointers were rested on the pile cap. Care was taken to maintain perfect verticality of the dial gauge pointer. A normal uplift load was applied in increments of 0.005 kN to the model helix. The uplift was measured by averaging two dial gauge readings (Fig. 2).

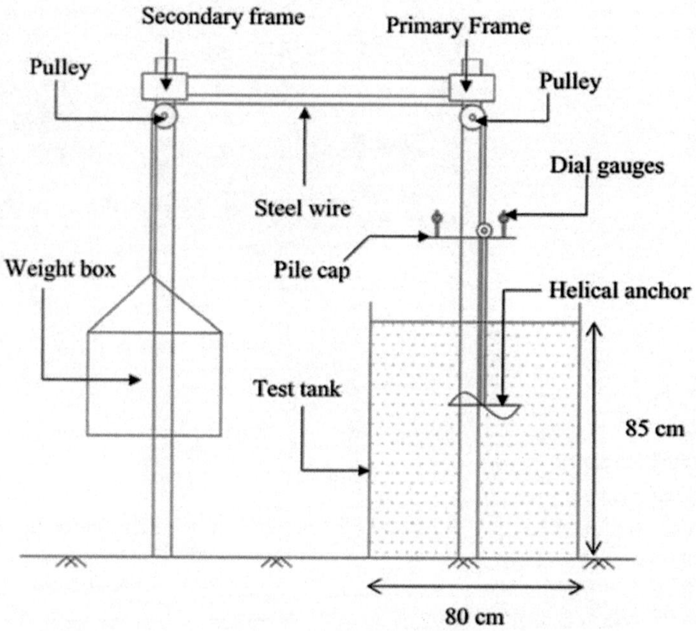

Fig. 2 Test setup

3 Numerical Analysis

Finite element analysis is a robust mathematical tool that is used to resolve many engineering problems. In finite element method, continuous media is divided into number of finite elements. PLAXIS is finite element method based software which is used to solve many geotechnical engineering problems. In this study, PLAXIS has been used to validate the experimental results. Axisymmetric model has been used for the creation of the model. Soil has been modelled using Mohr-Coulomb model in which five input parameters are required, namely, Elastic modulus (E) and Poisson's ratio (μ) for soil elasticity; Angle of internal friction (ϕ) and cohesion (c) for soil plasticity, and (ψ) as an angle of dilatancy. The elastic modulus, angle of internal friction has been calculated by performing triaxial test of sand. The bearing plate of the helix anchor is modelled as plate element. Hence, the pitch of the helix bearing plate is ignored and the helix bearing plate is modelled as circular plate element. PLAXIS divides the entire media into finite elements. Elements can be either 6 noded or 15 noded. 15 noded triangular elements have been used in comparison to 6 noded to provide greater accuracy in the calculation of stress and deformations of every point. The material properties and model parameters for the soil clusters are entered in material data sets tab. Fully automatic mesh generation is possible through PLAXIS. Mesh densities of five different types are available ranging from very coarse, coarse, medium, fine and very fine. In this study, medium mesh has been used which provides good results with better accuracy. The soil structure interface properties are input by using stiffness reduction factor, i.e. R_i given as 0.67 in this study.

4 Results and Discussions

The uplift capacity of helix anchor has been determined experimentally and the experimental results have been validated using numerical approach PLAXIS. Various graphs were plotted to determine the effects of diameter of helix, embedment ratio of helix as well as the density of helix. Pull-out load versus uplift curves has been plotted.

Figures 3 and 4 show the variation of pull-out load with uplift of a helix anchor by varying embedment ratio by keeping the diameter of helix as well as the relative density of sand as constant. It can be inferred from the above graphs that as the embedment ratio of anchor is increased from 2 to 4 and then to 6 pull-out loads of the helix anchor are increasing. This is due to the fact that overburden stress increases above the helix. The deeper the embedment of the helix the higher will be the overburden stress thus it requires a higher uplift load to pull-out the anchor.

Figures 5 and 6 show the variation of pull-out load with uplift of a helix anchor in which diameter of helix anchor has been varied by keeping the embedment ratio as well as the relative density of sand as constant. The graph shown above is for

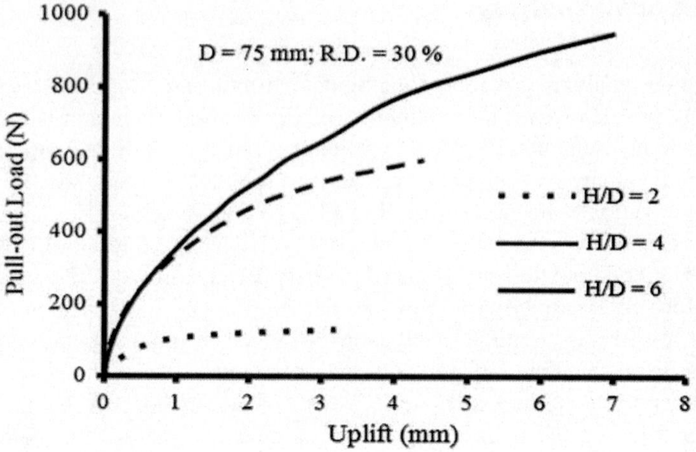

Fig. 3 Load versus uplift curve with varying embedment ratio

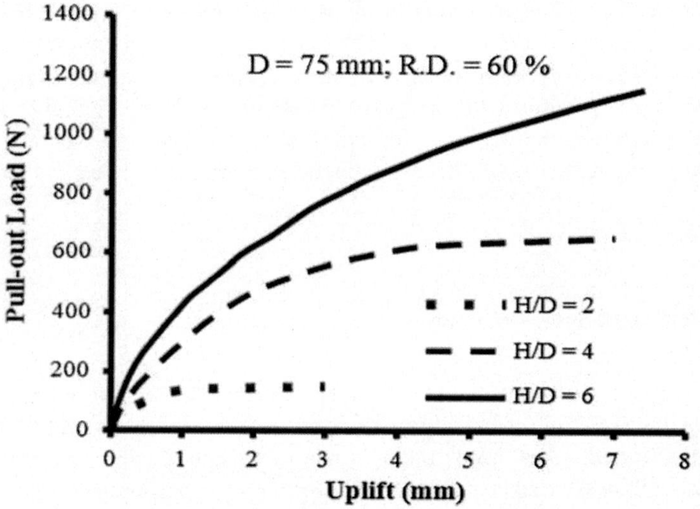

Fig. 4 Load versus uplift curve with varying embedment ratio

embedment ratio of 2 similarly graphs have been plotted for embedment ratio of 4 and 6. It can be concluded that as the diameter of helix plate is increasing the ultimate uplift load is increasing, which is due to the increase in the soil mass above the helix bearing plate which ultimately increases the pull-out load of the anchor. The variation of pull-out load with density of sand has been indicated in Fig. 7 by keeping the embedment ratio as well as the diameter of helix as constant. The above figures are for diameter of 75 mm similar graphs has been plotted for diameter of 100 mm as well as 125 mm. From the graphs, we can conclude as the density is increased

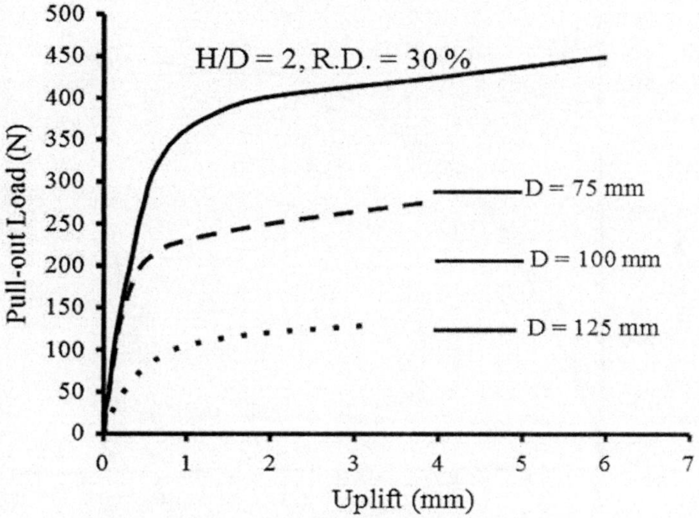

Fig. 5 Load versus uplift curve with varying diameter of helix

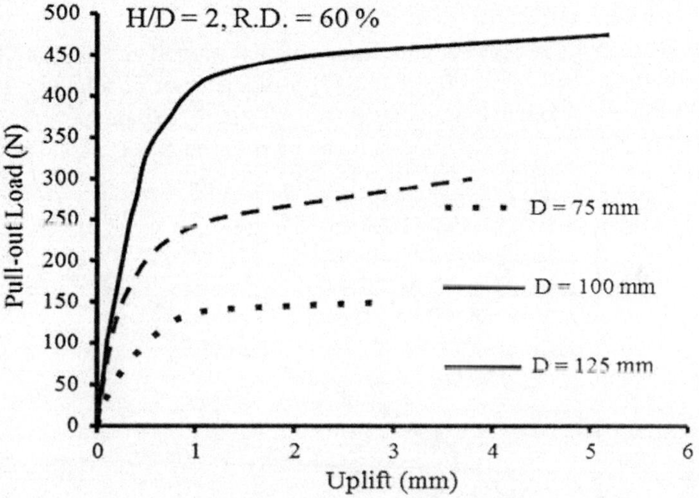

Fig. 6 Load versus uplift curve with varying diameter of helix

from 30% to medium dense state 60% the pull-out capacity of helix anchor tends to increase. However, the increase in uplift capacity ranges from 5 to 32% for the figure mentioned above. This is due to the fact that the friction angle of sand will increase with an increment in sand density. A higher friction angle will produce higher frictional resistances which will then increase its uplift capacity.

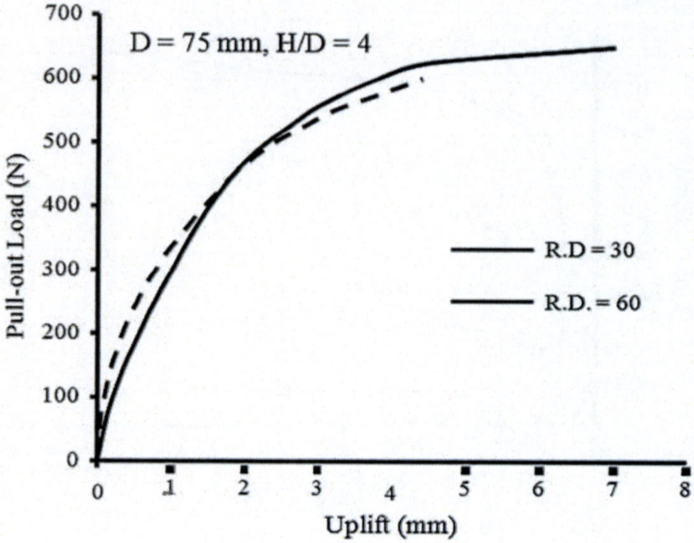

Fig. 7 Pull-out load versus uplift curves with varying density

The comparison curves between the experimental as well as the PLAXIS results have been plotted as shown in Figs. 8 and 9 for diameter of 75 mm. similar graphs have been plotted for other diameters. From the graph, it can be concluded that the PLAXIS is showing a fairly good agreement with the experimental results. The

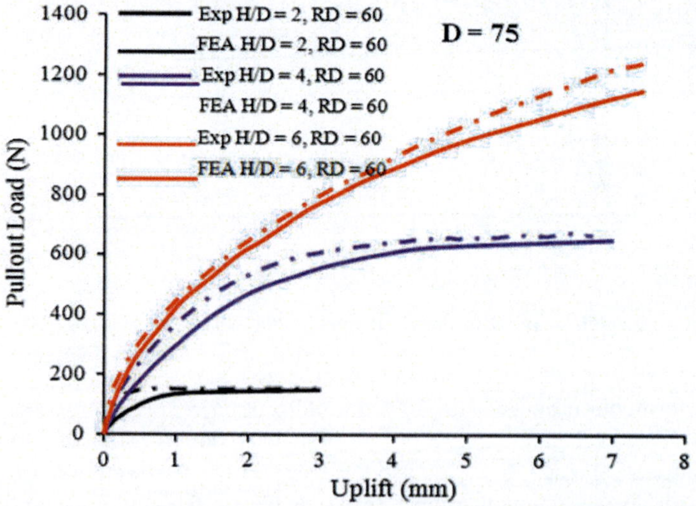

Fig. 8 Pull-out load uplift curves variation for experimental & Finite element for D = 75 mm & RD = 60%

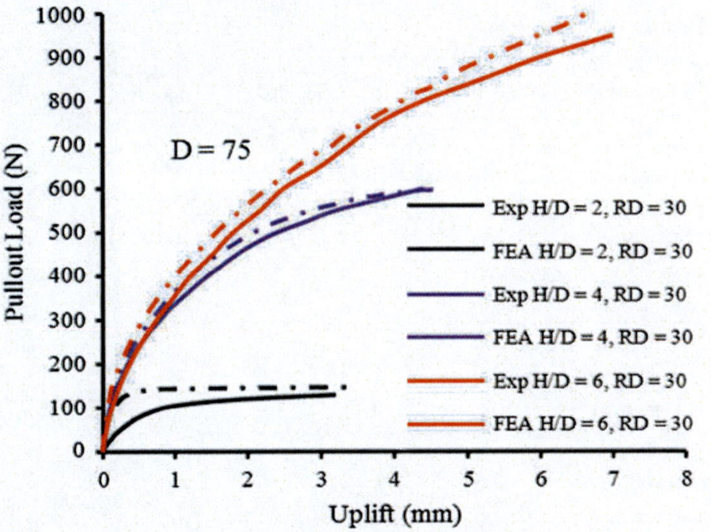

Fig. 9 Pull-out load uplift curves variation for experimental & Finite element for D = 75 mm & RD = 30%

values of pull-out capacity of the helix anchors obtained from the laboratory have been tabulated in Table 3. The pull-out load of helix anchors has been calculated by double tangent method. To validate the results acquired from the lab tests, comparison of its results was done with PLAXIS2D results. PLAXIS results have also been shown in the same table.

Table 3 shows the comparison between the results, it is clear that the percentage difference between experimental results and PLAXIS result is not more than20%. In PLAXIS, we get output as load uplift curve, displacement contours illustrating the failure mode of anchor, stress contours and many others. Load uplift curve as well as the displacement contours have been analysed in PLAXIS. Figures 10, 11, and 12 below show the displacement contours of helix anchor of diameter 75 mm with different embedment ratios likewise it has been analysed for another diameter of helix anchor also. From the figure, we can say that for an anchor at embedment ratio of 2 and 4 the displacement contours are reaching the ground surface and for an embedment ratio of 6, it is within the soil medium and not affecting the surface of soil (Table 4).

5 Conclusions

An attempt has been made to comprehend the behaviour of model helix anchor in cohesionless deposit. Different parameters have been studied to understand the influence on pull-out capacity of helix anchor. Experimental investigations were

Table 3 Summary of test variables

Test name	Dia of helix (mm)	Depth of helix (mm)	Embedment ratio '*B*/*D*'	Height of pour (cm)	Relative density (%)
75-2-30	75	150	2	60	30
75-4-30	75	300	4	60	30
75-6-30	75	450	6	60	30
75-2-60	75	150	2	100	60
75-4-60	75	300	4	100	60
75-6-60	75	450	6	100	60
100-2-30	100	200	2	60	30
100-4-30	100	400	4	60	30
100-6-30	100	600	6	60	30
100-2-60	100	200	2	100	60
100-4-60	100	400	4	100	60
100-6-60	100	600	6	100	60
125-2-30	125	250	2	60	30
125-4-30	125	500	4	60	30
125-6-30	125	750	6	60	30
125-2-60	125	250	2	100	60
125-4-60	125	500	4	100	60
125-6-60	125	750	6	100	60

carried out on model helix anchor in dry cohesionless deposit and numerical analysis has been done to better understand. The following conclusions can be drawn

(1) The embedment ratio plays an important role on the uplift capacity of helix anchor. Uplift capacity of helix anchor increases with an increase in embedment ratio.
(2) Pull-out capacity of helix anchor increases with increase in diameter of helix for same relative density as well as the embedment ratio.
(3) Uplift capacity of helix anchor in loose sand will always be lower than in dense sand. Hence it increases with increase in sand density.
(4) Numerical analysis has been studied using PLAXIS and it shows a fairly good agreement with experimental results.
(5) Displacement contours have been Plotted using PLAXIS. From the contours, it can be concluded that the embedment of 2 and 4 shows a shallow mode of failure in which the displacement is extending up to the ground while for an embedment of 6 contours lies within the soil media indicating deep mode of failure.

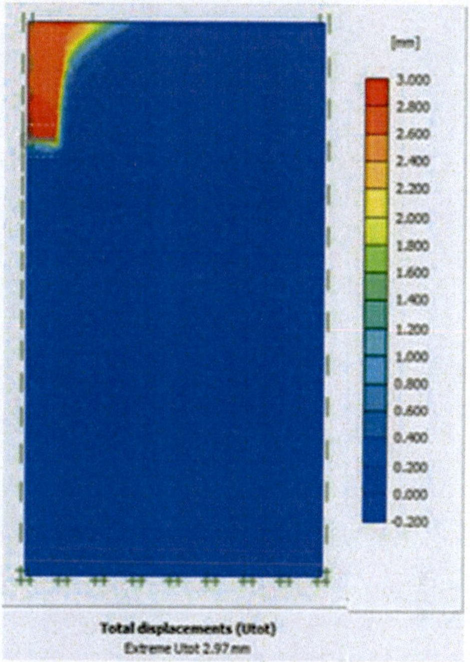

Fig. 10 Displacement contours of 75-2-30

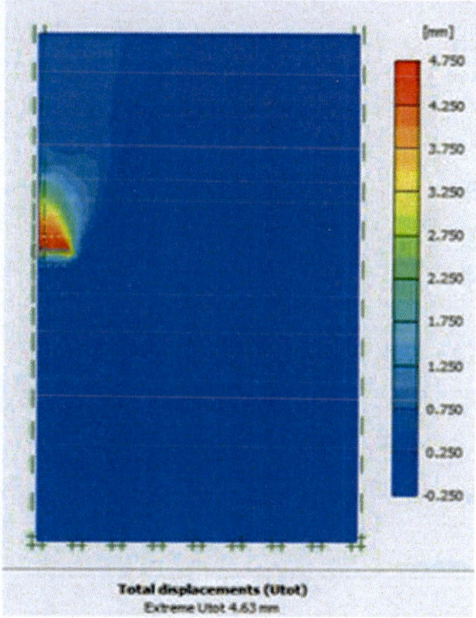

Fig. 11 Displacement contours of 75-4-30

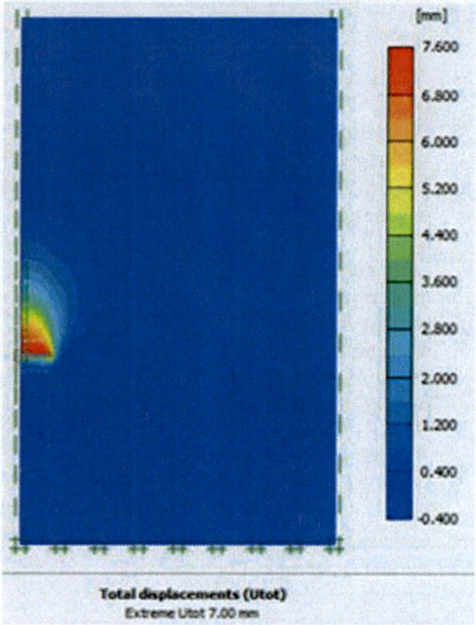

Fig. 12 Displacement contours of 75-6-30

Table 4 Comparison between experimental and PLAXIS

Test data	Experimental		Finite element analysis		% Difference
	Qu (N)	Uplift (mm)	Qu (N)	Uplift (mm)	
75-2-30	110.5	0.68	132.5	0.28	19.91
75-2-60	140	0.76	152.5	0.3	8.92
75-4-30	478	1.65	510	1.54	6.69
75-4-60	600	2.3	590	1.76	−1.66
75-6-30	750	3.2	790	3.2	5.33
75-6-60	820	2.4	860	2.8	4.87
100-2-30	235	0.55	274	0.5	16.59
100-2-60	254	0.64	287	0.36	12.99
100-4-30	1040	3.5	1080	4	3.84
100-4-60	1100	2.65	1096	3	−0.36
100-6-30	1530	6.2	1550	5.6	1.30
100-6-60	1590	5.3	1858	7.58	16.85
125-2-30	385	0.79	461	0.48	19.74
125-2-60	435	0.68	485	0.45	11.49
125-4-30	1630	3.75	1760	4.45	7.97
125-4-60	1870	4.1	2060	4.4	10.16
125-6-30	2540	6.95	2880	7.7	13.38
125-6-60	3350	7.676	3380	7.52	0.89

References

1. Ghally A et al (1991) Ultimate pull-out resistance of single helical anchor. Can Geotech J:661–667
2. Ilamparuthi K, Shreni V (2009) Response of anchor in two-phase material under uplift. In: Indian geotechnical conference, GEOtrendz, pp 618–622
3. Niroumand H, Kassim KA (2010) Analytical and numerical studies of horizontal anchor plates in cohesion less soils. Electron J Geotech Eng (EJGE) 15:281–292
4. Nazir R, Chuan HS, Niroumand H, Kassim KA (2013) Performance of single vertical helical anchor embedded in dry sand. Measurement 49:42–51
5. Mukherjee S, Mittal S (2015) Vertical uplift capacity of a group of equallyspaced helical screw anchors in sand. In: International conference on structural engineering and construction and management, pp 25–31
6. Demir A, Ok B (2015) Uplift response of multi-plate helical anchors in cohesive soil. Geomech Eng 8(4):615–630
7. Ghosh P, Samal S (2016) Ultimate pullout capacity of isolated helical anchor using finite element analysis, IGC 2016 15–17 December 2016, IIT Madras, Chennai, India
8. Ghaly A, Hanna A (1994) Ultimate pullout resistance of single vertical anchor. Can Geotech J:661–667
9. Abdel-Rahim HH, Taha YK, El din El sharif Mohamed W (2013) The compression and uplift bearing capacities of helical piles in cohesionless soil:2055–2064
10. Ilamparuthi K, Dickin EA, Muthukrisnaiah K (2002) Experimental investigation of the uplift behaviour of circular plate anchors embedded in sand. Can Geotech J 39:648–664
11. Nazir R, Chuan HS, Niroumand H, Kassim KA (2014) Performance of single vertical helical anchor embedded in dry sand. Measurement 49:42–51
12. Mittal S, Mukherjee S (2013) Vertical uplift capacity of a group of helical screw anchors in sand. Indian Geotech J 43(3):238–250

Soft Clay Stabilization with Steel Slag

Ankit Kumar, Sumon Saha, and Rana Chattaraj

Abstract Steel slag is a waste material generated from steel industry. Millions of tonnes of steel slag are generated in each year throughout the world. Substantial portions of steel slag are accumulated each year due to un-utilization. The accumulation of steel slag not only occupies a huge area of land, but it also pollutes the surface water as well as groundwater. Bulk utilization of steel slag is the need of the hour. Soil stabilization is an area where bulk utilization of steel slag is possible. In this paper, an attempt has been made to utilize the steel slag in improving the geotechnical strength characteristics of soft clayey soil. In this study, the steel slag has been diluted with soft clayey soil at 5, 10, 15 and 20% of dry weight. Sieve analysis, Atterberg's limits, free swell index (SI), specific gravity, compaction, unconfined compressive strength and triaxial test were conducted on the virgin soil as well as on blended soil. UCS tests were performed at 0, 7, 14 and 28 days of curing. The strength characteristic of the blended soil was found to be improved with the increase in percentage of steel slag and curing period.

Keywords Fly ash · Steel slag · Soil stabilization · Atterberg's limit · Compaction · UCS and triaxial test

1 Introduction

In India expansive soils spread about 0.8×10^3 million km^2 regions which are roughly 20% of surface area. The swelled soil may apply enough strain to split sideways, carports, storm cellar floors, pipelines and even establishments; making it harmful for structures if not sufficiently treated. Hence to improve the expensive soil steel

A. Kumar · S. Saha · R. Chattaraj (✉)
KIIT Deemed to be University, Bhubaneswar, India
e-mail: rana.chattarajfce@kiit.ac.in

A. Kumar
e-mail: arnabdebnath00@gmail.com

S. Saha
e-mail: sumonsaha.4211@gmail.com

B. B. Das et al. (eds.), *Recent Developments in Sustainable Infrastructure*, Lecture Notes in Civil Engineering 75, https://doi.org/10.1007/978-981-15-4577-1_11

slag is used as additives. The worldwide steel creation was 1360 million tonnes as per the report of World Steel Association (WSA-2017) producing around 170 to 250 million tonnes of steel slag annually as a waste. India is the third-most elevated maker of rough steel in Asia, as per the WSA report 2014. Huge number of steel making plants produces around 29 million metric tonne of waste material every year. Disposal of steel slag to the natural environment is becoming more and more difficult for reasons like confined assets, ecological impacts, costs and so on. Because of the cementitious property of steel slag by dissecting, its mineral creation, physical and chemical properties, a thought of utilizing steel slag fines to improve the bearing capacity of weak ground has been put forward.

Shalabi [1] performed experimental analysis to find out the effect of varying percentage of steel slag on clayey soil. The tests showed that there is a continuous decrease from 50 to 30% in the liquid limit (LL) with increase in basic oxygen steel slag (BOS) content. There is a decrease in plastic limit (PL) from 21 to 11% with increase in slag content up to 30%. Biradar [2] investigated the effect of F grade fly ash and steel slag to stabilize clay soils. Various tests were conducted on the modified soil by mixing fly ash and steel slag at different percentages, i.e. 0, 5, 10, 15 and 20% by the weight of the soil. It is reported that if clayey soil stabilized by steel slag then this amended soil attains higher CBR value than fly ash. Akinwumi [3] suggested that the optimum % of steel slag was found to be 8%, addition of which increased its unsoaked CBR values by 40% and its UCS by 66.7% while the LL, PL and plasticity index reduced by 6.3, 4.0 and 2.3%. Zha et al. [4] studied how the expansive soil would behave when it is stabilized with fly ash. The test results showed that the fly ash and lime-fly ash treatment in expansive soil decreased the swell potential and shrinkage of the soils. Free swelling and swell pressure are reduced with different types fly ash content in soil. Free swelling and swell pressure of fly ash stabilized soil also depend on days of curing. Ashango and Patra [5] studied that steel slag, rice husk ash and quick lime can be used as a additives to stabilize the expansive soil. The optimum mix was found to be 65% expansive soil+20% steel slag+5% lime+10% rice husk ash based on UCS value. Poh et al. [6] studied the effects of BOS on stabilization of English China clay (ECC) and Mercia mudstone (MM). The stabilization of ECC and MM is carried out by the use of three BOS slag, namely, BOS A, B and C collected from three several steel manufacturing sites in the United Kingdom. Two different activators that are quicklime and sodium metasilicate were also used in this investigation along with mixture of BOS slag. Peter et al. [7] conducted lab experiment for enhancement of subgrade behaviour of expansive soil with coir waste. They have found that dry density is constantly decreasing with increase in coir pith and coir fibre content. However, addition of 0.5% coir pith did not have any influence on dry density but decreases optimum moisture content (OMC). The OMC of amended soil suddenly reduced at 0.2% of coir fibre content but after that the value of OMC increases with coir fibre content in amended soil. Dang et al. [8] used agricultural waste bagasse ash for remediation of expansive soils. The test results showed that the free swell ratio decreases as % of bagasse increases. Durotoye et al. [9] studied the change in the engineering properties of expansive soil by the use of varying percentage of common salt. The results show that swelling parameters such as PL, LL, SL, specific gravity, OMC, free swell index drastically decreases with increase in % of NaCl. However, strength parameters such as MDD, unsoaked and soaked CBR, compressive strength increases with increase in NaCl content.

2 Materials

Soil used in the research work has been collected from Cuttack district near Chaudwar. A total of 150 kg of clayey soil was collected from a depth of 20ft during the excavation of soil for the purpose of construction of a water tank. The soil was packed in 10 gunny bags. Some amounts of soil were also packed in polythene to avoid moisture loss of the soil so that we could further determine its moisture content. The physical properties of untreated clayey soil have been shown in Table 1. Steel slag was collected from the dump yard of Rourkela steel plant where the slags were disposed after the production of steel material. A total of 45 kg of steel slag were collected and packed in 3 gunny bags which was then transported to the geotechnical lab of Civil engineering Department, KIIT University. It was then oven-dried as it was exposed to natural weather for about two months. The samples which were collected had sizes ranging from 50 to 100 mm which was then crushed to particle size below 425 μ in the material testing lab. The physical properties of the soil are given in Table 1.

In this study, various types of test were carried out such as sieve analysis, hydrometer analysis, pycnometer test, swelling index test, standard proctor test, UCS, triaxial test. Sieve and hydrometer analysis are carried out to evaluate the grain size of the particle. Grain size analysis was done in the lab as indicated by [10]. The particle size distribution curve is represented in Fig. 1. In this research, liquid limit and plastic limit for untreated and treated clay with increasing %steel slag are determined. The liquid limit and plastic limit were resolved according to methods indicated in [11]. Liquid limit was obtained using Cassagrande's apparatus. [12] is followed to evaluate specific gravity of amended and non-amended soil. In this research, pycnometer

Table 1 Physical properties of untreated soil

Silt+Clay (%)	52
Sand (%)	48
Liquid limit (%)	47.2
Plastic limit (%)	15.45
Shrinkage limit (%)	8.65
Plasticity index (%)	31.75
G	2.47
SI (%)	47
Soil type	CI

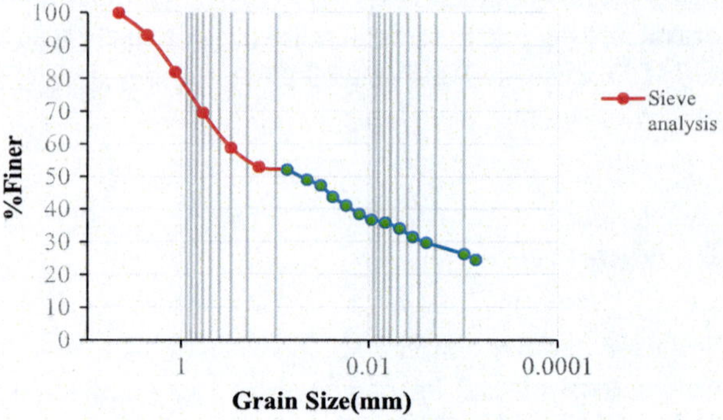

Fig. 1 Grain size distribution curve

test has been conducted three times for each sample and the average had been taken. Specific gravity of untreated soil as well as steel slag has been found out individually as well as specific gravity of mixture of untreated soil + increasing % of steel slag was also determined. Free swell index is done as per [13]. On submergence in water, free SI is the percentage enhance in volume of the soil. The standard proctor tests were conducted on untreated clayey soil and on clayey soil mixed with basic oxygen steel slag (BOS) according to Indian standards [14]. The test was conducted by placing sample in a metallic mould having an inner diameter of 100 mm and a height of 120 mm. placing of each sample was done in three layers. 2.5 kg of load was subjected to each layer and 25 blows of tamping were provided to each layer. Variation in MDD and OMC is observed by varying the amount of steel slag in the clayey soil. The unconfined compressive strength test is conducted to measure the strength of a material. It is a type of triaxial test in which the confining stress is zero. This test is applicable for cohesive soil. UCS tests were performed as per [15]. These tests were conducted on 7.5 cm × 15 cm size specimen samples. UCS tests were conducted for non-amended and amended soil samples with addition of 0, 5, 10, 15 and 20% BOS at their OMC and MDD values obtained from compaction tests. Samples were cured for duration of 7, 14 and 28 days for UCS test. Triaxial compression test is conducted as per [16]. In this test, the sample is subjected by constant confining stress (σ_3) whereas deviator stress (σ_3) is increased till the soil sample fails in shear. In this research, the test is repeated with identical samples using three different values of confining pressure, namely, 100, 150 and 200 kPa.

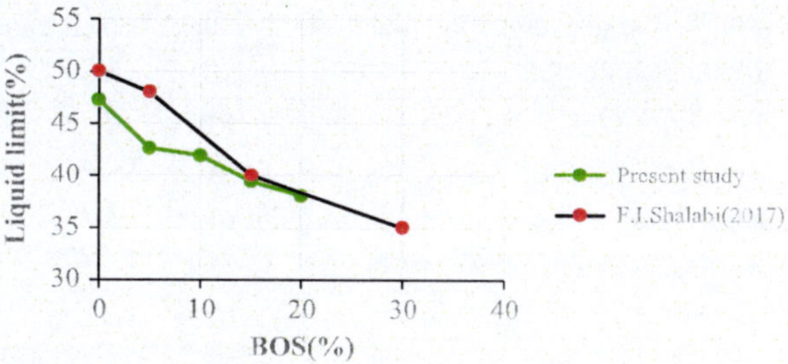

Fig. 2 Variation of LL of treated soil with BOS content

3 Results and Discussion

3.1 Grain Size Analysis

In this research, the soil is classified based on Indian Standard Soil Classification System. Sieve analysis and hydrometer analysis were carried out for grain size analysis. A graph is plotted showing % finer versus grain size curve as in Fig. 1. From the figure, it is known that 48 percent of particles are retained over 75 μ sieve and 52% of particles pass through 75 μ sieve. By the use of A-line chart and A-line equation: $I_p = 0.73(W_{L\%}{-}20)$, the soil is classified as intermediate plastic clay (CI). From the grain size distribution curve, the amount of clay is found to be 22% and silt to be 30%.

3.2 Atterberg Limit

With 0, 5, 10, 15 and 20% steel slag additive, liquid limit (LL) and plastic limit (PL) were conducted on each sample. The LL signifies compressibility. Higher the LL higher will be the compressibility. Figure 2 represents that the LL has decreased with increase in %BOS from 47.2% for 0% BOS additive to 38% for 20% BOS additive. The plasticity index (PI) also decreases continuously with increase in %BOS from 31.75% for 0% steel slag to 19.88%.

3.3 Specific Gravity

The specific gravity was found out for the untreated soil sample and with soil samples treated with different % of steel slag. Average value of specific gravity is taken by testing three times on each sample. Specific gravity of the steel slag, i.e. 3.02 is found to be higher than the specific gravity of untreated soil, i.e. 2.45. The variation of specific gravity is shown in Table 2.

Table 2 Physical properties and plasticity index of amended and non-amended soil

Mixture proportion	OMC %	MDD gm/cc	Specific gravity	PI
S	1.57	20.4	2.47	31.75
S+5 BOS	1.592	20	2.54	26.04
S+10 BOS	1.604	19.7	2.57	24.3
S+15 BOS	1.683	19.4	2.65	20.98
S+20 BOS	1.726	18.2	2.72	19.88

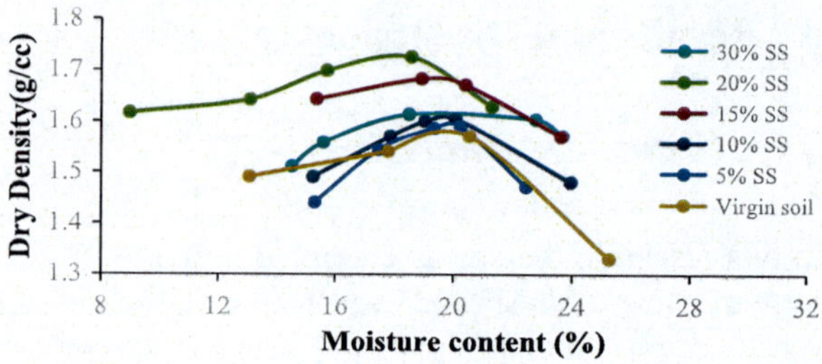

Fig. 3 variation moisture content with dry density of treated soil

3.4 *Compaction Characteristics*

Table 2 represents the variation of OMC and MDD on the basis of BOS content in soil. It has been shown that OMC reduced linearly with percentage of BOS content in modified soil. In Fig. 3, it is found that the dry density of mixture reaches maximum at 20% of marble dust after that dry density was observed to be reduced.

3.5 *Unconfined Compressive Strength (UCS)*

Figure 3 shows the changes of UCS with BOS content at 0, 7, 14 and 28 days of curing. It is shown that UCS value enhances up to 15% addition of BOS in amended soil. Addition of further BOS resulting decreases of UCS value. With increase in no of curing days, the UCS value is found to be increased. The increase in UCS up to 15% addition of BOS may be attributed to the fact of cation exchange mechanism between steel slag and soil particles (Fig. 4).

3.6 *Triaxial Compression Test*

The relationship of axial stress difference and BOS with varying curing days at confining pressures of 100, 150 and 200 kPa, respectively, is represented by Figs. 5, 6 and 7. It shows that with increase in confining pressure the axial strength increases

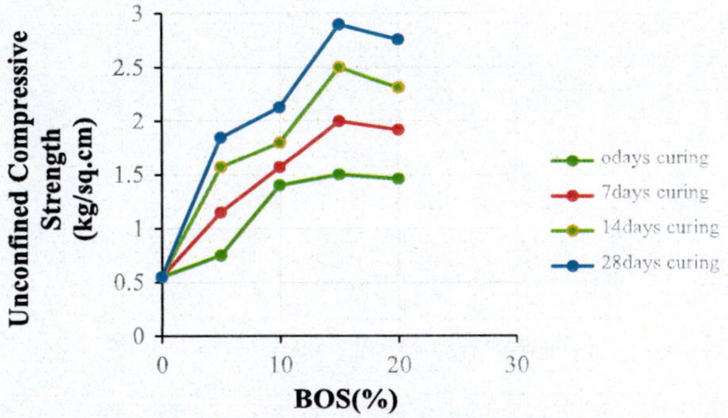

Fig. 4 Relationship between UCS and %BOS at varying curing days

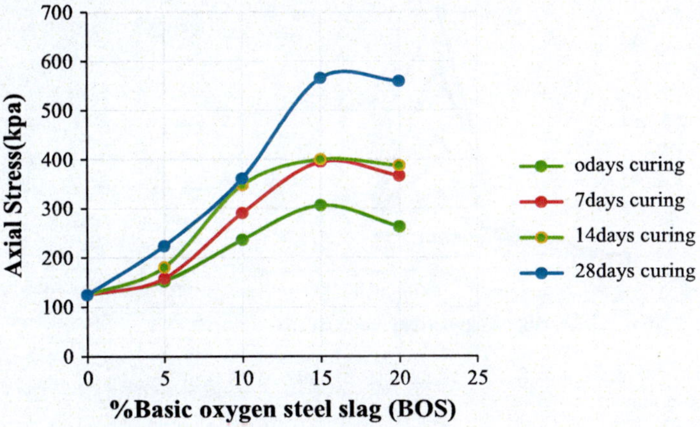

Fig. 5 Relationship between maximum stress difference and %BOS at confining pressure of 100 kPa

Fig. 6 Relationship between maximum stress difference and %BOS at confining pressure of 150 kPa

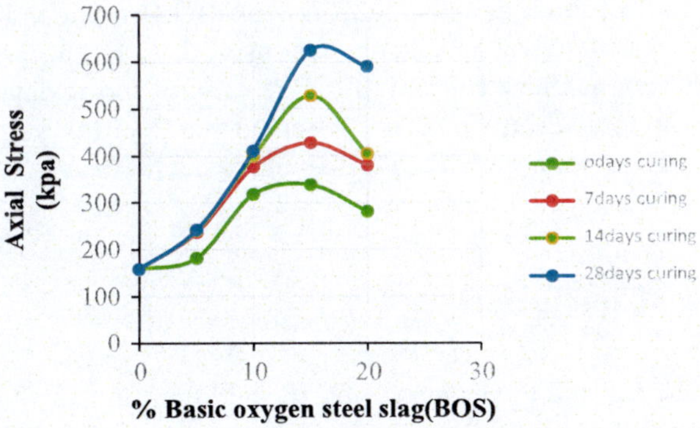

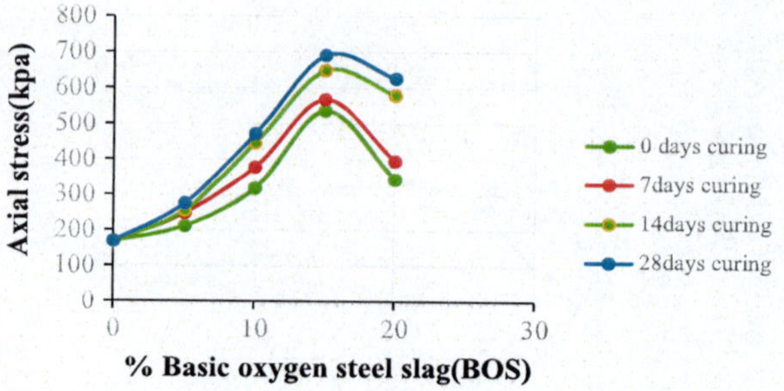

Fig. 7 Relationship between maximum stress difference and %BOS at confining pressure of 200 kPa

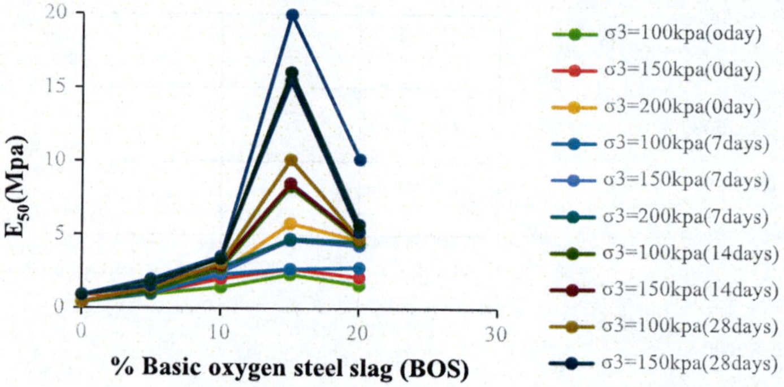

Fig. 8 Relationship between Secant modulus(E_{50}) and %BOS at different confining pressures and different curing days

which shows hydration reaction is getting accelerated with higher confining pressure. With increase in %BOS, the maximum stress difference is also increasing up to 15% BOS addition and decreases at 20% BOS addition. There is increase in maximum stress difference with increase in curing days. Figure 7 shows the relationship between Secant modulus E_{50} and %BOS at varying curing days. E_{50} is a measure of stiffness of the soil. From Fig. 7, it is observed that E_{50} value increases with increase in amount of BOS up to 15%, then it is found to decrease at 20% BOS (Fig. 8).

4 Conclusion

The LL and PI decrease with the addition of steel slag in virgin soil. The specific gravity of black cotton soil increases from 2.47 to 2.72 with percentage increase in steel slag. SI decreases from 47% for virgin soil to 11.11% for 20% steel slag addition (4 times decrease). MDD increases from 1.57 to 1.72 g/cc with steel slag addition up to 20% and the OMC decreases from 20.4 to 18.2%. The optimum % of steel slag was obtained to be 15% based on both UCS tests and triaxial compression tests. The UCS value is found to be maximum at 15% steel slag addition and the secant modulus is also larger for 15% steel slag addition which indicates higher stiffness.

References

1. Shalabi FI, Asi IM, Qasrawi HY (2017) Effect of by-product steel slag on the engineering properties of clay soils. J King Saud Univ Eng Sci 29(4):394–399
2. Biradar KB, Kumar UA, Satyanarayana DP (2014) Influence of steel slag and fly ash on strength properties of clayey soil: A comparative study. Int J Eng Trends Technol (IJETT) 14(2)
3. Akinwumi I (2014) Soil modification by the application of steel slag. Periodica Polytech Civil Eng 58(4):371–377
4. Zha F, Liu S, Du Y, Cui K (2008) Behavior of expansive soils stabilized with fly ash. Nat Hazards 47(3):509–523
5. Ashango AA, Patra NR (2016) Behavior of expansive soil treated with steel slag, rice husk ash, and lime. J Mater Civ Eng 28(7):06016008
6. Poh HY, Ghataora GS, Ghazireh N (2006) Soil stabilization using basic oxygen steel slag fines. J Mater Civ Eng 18(2):229–240
7. Peter L, Jayasree PK, Balan K, Raj SA (2016) Laboratory investigation in the improvement of subgrade characteristics of expansive soil stabilised with coir waste. Trans Res Procedia 17:558–566
8. Dang LC, Hasan H, Fatahi B, Jones R, Khabbaz H (2016) Enhancing the engineering properties of expansive soil using bagasse ash and hydrated lime. Int J Geomate 11(25):2447–2454
9. Durotoye TO, Akinmusuru JO, Ogbiye SA, Bamigboye G (2016) Effect of common salt on the engineering properties of expansive soil. Int J Eng Technol 6(7):233–241
10. IS-2720-(Part-IV) (1985) Grain size analysis
11. IS-2720-(Part-V) (1985) Determination of liquid and plastic limit
12. IS-2720-(Part III) (1980), Determination of specific gravity
13. IS-2720-(Part VI) (1972) Determination of shrinkage factors
14. IS-2720-(Part VII) (1980) Determination of compaction parameters
15. IS-2720-(Part X) (1991) Determination of Unconfined compressive strength
16. IS-2720-(Part XI) (1983) Determination of shear strength parameters of soil from unconsolidated undrained triaxial compression test without measurement of pore water pressure
17. World steel production report on crude steel production 2014

Review on Study of Soil Stabilization by Application of Microorganisms

Sourya Snigdha Mohapatra and P. K. Pradhan

Abstract In recent years when there is a scarcity of desired land sites to satisfy the rapid development of infrastructures in metro cities, the engineers are facing great challenge of improving the properties of the soil which are going to carry the transferred load from the superstructure, ex: different types of buildings, bridges and roadways. The soil improvement techniques nowadays are emerging by the implementation of various significant ways to enhance the mechanical engineering properties of different soil types, like different types of expansive soils and minerally active soils. The soil treatment methods with chemical and cement grout are used increasingly by the engineers. The chemicals and cement used as a soil stabilizing material affect the natural environmental properties of the soil, hence, hindering the groundwater flow. To overcome their ill effect on soil and underground water, more sustainable method is the need of the time. Hence, engineers now are using the most readily available microorganisms and biochemical processes which are naturally active in subgrade soils to enhance the mechanical parameters of earth in survivable and cost-effective way. The main purpose of this kind of implementation is to improve both physical and engineering parameters of soil to use and reuse the soil when needed. The most important uses bio-clogging and bio-cementation are introduced. Bio-clogging is the process which produces the void clogging materials through biological activity which affects the pore spaces and permeability of soil. Bio-cementation, the process in which particle linking materials are generated because of the biological methods in situ for which the resisting capacity of the earth increases significantly. Bacteria which can grow in absence of oxygen and microaerophilic are the most eligible bacteria for the bioremediation of the soil.

Keywords Microorganisms · Bio-cementation · Bio-clogging · Mechanical parameters of soil

S. S. Mohapatra (✉) · P. K. Pradhan
Department of Civil Engineering, VSSUT, Burla 768018, Odisha, India
e-mail: sourya.snigdha130@gmail.com

P. K. Pradhan
e-mail: pkpradhan1@yahoo.co.in

B. B. Das et al. (eds.), *Recent Developments in Sustainable Infrastructure*, Lecture Notes in Civil Engineering 75, https://doi.org/10.1007/978-981-15-4577-1_12

1 Introduction

Nowadays, construction of any structure on such soils which do not offer favourable characteristics is a greater challenge for the engineers. Availability of weak soil or geotechnical properties is making the civil engineers to avoid such type of sites. Generally, stabilization of the soil is a method of improving the in situ soil quality and enhancing the soil strength. Materials like lime, cement, fibres and geosynthetics can be used for the soil stabilization. Few techniques adopted for stabilization harm the soil environment, make them toxic and cause underground water contamination. Microbial soil remediation has come into focus for strengthening the soil properties with an effective environmental impact. The process of microbial application to the soil is more eco-friendly in comparison to other conventional treatment methods. There are two approaches in microbial geotechnology: (bio-clogging and bio-cementation) [2, 4]. Bio-clogging is the process in which void clogging materials are produced due to biological activity so that the void ratio and permeability of soil can be decreased. Bio-binding is the microbial process in which soil particles are allowed to bind with each other with effect of microbial activity present in situ. Every process implemented results in changing different geotechnical properties of soil [4, 9]. This paper has concentrated on different microbes, the process of their growth and mechanism, their different geotechnical applications to improve the soil properties and briefly explains the geotechnical survivability using bioremediation for soil remediation.

2 Soil Earth with Effect of Microbes

Naturally, present soil includes some microorganisms which are precisely adoptable for different environmental effects, as they exist for over 3.5 billion years. However, the microbes which are present in soil may not be sufficient for the effective improvement of the soil properties. Hence, it is in need to add more amount of microbes to the soil for the stabilization so that it will make the eco-friendly as well as for avoiding the use of toxic chemicals on soil. The table given below shows different microorganisms followed by the environmental researchers to study their éffect on different soil properties.

The anaerobic bacteria *Sporosarcina pasteurii* (*S. pasteurii*) [7, 10] causes Calcite precipitation through microbial activity in porous soil to enhance urea hydrolysis through the enzyme urease. The surface of the soil particles as well as void spaces is clogged by the calcium carbonate deposition precipitated by the microbes which in turn reduces the porosity and permeability of the soil [14, 15].

3 Mechanism of Microbes to Geotechnical Engineering

The theory of microbial geotechnology comprises two methods: bio-clogging and bio-cementation. Bio-clogging helps in producing void filling materials which reduce the volume of voids in the soil making the soil denser (Tables 1, 2 and 3).

Bio-cementation helps in enhancing the potency and hardness properties of the soil by implementing microorganisms and binding matters into the soil [4].

In the process of bio-clogging, impervious layer of algae and Ciano bacterial biomass is formed [1]. Bacteria with and without oxygen, oligotrophic microaerophilic bacteria and bacteria inducing nitrification cause the generation of

Table 1 Different microorganisms followed by the environmental researchers for studying their effect on different soil properties [2]

Serial no.	Microorganism	Biological process	Influence of the microorganisms
1	*Bacillus pasteurii*	Cementation process	Increment in axial strength, high elasticity, pore space decreases to 90%
2	*Sporosarcina pasteurii*	Urease activity	Pore space decreases by 90%, strength significantly increases
3	Urease enzyme-driven mineralization	Enzyme to hydrolyze urea	Increment in unconfined compression strength and reduction in permeability
4	*Bacillus pasteurii*		The performance is influenced by the effective size of the sand
5	*Sporosarcina pasteurii*	$CaCo_3$ precipitation	Visible improvements in unconfined compressive strength and stiffness, and reduction in permeability of the treated sands
6	Tryptic Soy Broth was used for the cultivation of fastidious and non-fastidious microorganisms	Use of Tryptic Soy Broth	Decrease in plasticity, improvement in shrinkage and swelling characteristics
7	*Bacillus pasteurii*, *Sporosarcina pasteurii*, *Spoloacto bacilus*, Clostridium and Desulfotomaculum	Urea hydrolysis	Shear strength improvement, unconfined compressive strength, improvement in stiffness and liquefaction of the soil

Table 2 Microbial processes involved in bio-clogging [2]

Microorganisms grouped Physiologically	Activities of bio-clogging	Essential conditions for bio-clogging	Useful geotechnical applications
Algae and cyanobacteria	Impermeable layer of biomass Formation	UV light penetration and supply of nutrient	Reduction of water infiltration into slopes and control of seepage
Slime producing bacteria in presence and absence of oxygen	Production of slime in soil	Presence of oxygen and medium with ratio of C:N > 20	Provide cover for soil erosion control and slope protection
Oligotrophic microaerophilic bacteria	Slime production in soil	Low concentration oxygen and medium with low concentration of carbon source	Reduce drain channel erosion and control seepage
Bacteria involved in nitrification	Slime production in soil	Ammonium and oxygen present in soil	Reduction in erosion of canal channel
Bacteria reducing sulphate content.	Undissolved sulphides of metals production	Oxygen free conditions: soil including sulphate and carbon sources	Grout curtains are formed to minimize the mitigation of heavy particle minerals and living dirt
Bacterium increasing ammonification	Due to increased pH and release of CO_2 Insoluble carbonates of metals formed in soil	NH_4 and soluble mineral salt present in soil	Prevention of screeching of soil dams
Bacteria reducing iron in soil and water	Insoluble ferrous and ferric salts generation and precipitation in soil	Oxygen free ailments changes to presence of oxygen conditions.	Prevention of liquefaction of soil made dams and dykes

slime in earth. Bacteria which reduces sulphate causes generation of insoluble sulphides of metals and bacteria with effect of ammonification causes formation of insoluble carbonates of metals. Bacteria those help in iron reduction causes formation of ferrous solution and precipitation of insoluble ferrous and ferric salts and hydroxides in soil.

In the process of bio-cementation, sulphate-reducing bacteria causes production of binding of the soil particles with sulphides metals and helps in binding the particles with carbonates metals produced during hydrolysis of urea.

Table 3 Microbial processes that can potentially lead to bio-cementation [3, 13]

Microorganisms grouped physiologically	Activities of bio-cementation	Essential conditions for bio-cementation	Useful soil applications
Bacteria involved in sulphate reduction	Undissolved sulphides of metals production	Oxygen free conditions: sulphate and carbon sources present in soil	Increases the stability of slopes in canals and dams
Bacteria involved in ammonification	Increased pH and released CO_2 Produces insoluble carbonates of metals in soil	NH_4 and soluble mineral salt	Increases liquefaction-potential of sand, enhances stability of retaining walls and also increases load-carrying capacity of foundations
Bacteria involved in iron reduction	Generation and precipitation of ferrous and insoluble ferrous and ferric salts	Oxygen free ailment changes to oxygen presence ailments; presence of ferric minerals	Earth densification, resistance towards liquefaction and increase in bearing capacity

4 Soil Engineering Properties

Engineering parameters of low strength soil may be very much poor and cause erosion. Due to the increase in infrastructure the soil needs to be stabilized within the time. The traditional method of synthetic injecting techniques is of big budget and need much more injection holes for treating large number of volumes. The potential of the MICP by providing urea and calcium in different sandy soil samples has been evaluated by many researchers and it has been observed that consolidation of sand by *B. pasteurii* reduces the pore space up-to 50% and hydraulic conductivity up-to 90% in the places where the binding/cementation takes place.

Heightening in load-carrying capacity and hardness of cylindrical samples of sand upon microbial calcium carbonates has been supplied. Urea treated MICP has implemented for suppressing dust, reducing flow of water in granular media, improving and enhancing soil properties, stabilizing slopes and strengthening liquefiable soils. The implementation of biological treatment to coarse soil and expansive soil resulting increase the soil strength, reduce the hydraulic conductivity, void space and swelling properties [3, 11].

5 Microbial Geotechnology in Application

The biological treatment of soil is used widely in soil remediation in several ways. Bacteria can be injected into the soil in many processes. Also in some cases, the bacteria can be mixed properly with the soil and placed in the required area. Some of the fields where improvement is seriously needed are

i. Soil Reinforcement/stabilization for enhancing the Stability of slopes and retaining walls.
ii. Increase in the load-carrying capacity of footings.
iii. Reducing liquefaction of the expansive soil.
iv. Treatment of paved surface.
v. Prevention of attrition and failure of slopes.
vi. Binding of the micro and nanoparticles on open surfaces to reduce dirt levels in soil.
vii. Increase in parrying the offshore structures to erode the sediments within or beneath gravity foundations and pipelines.
viii. Stabilization of contamination from soil by hog-tieing.
ix. Control in corrosion in sea and river banks.

6 Implementation on Field

Till now some of the field methods have implemented during which microorganisms have been thoroughly utilized for both increasing the resistance and hardness of soil and to decrease permeability of soil [1, 9, 14]. For the effective implementation of the bacteria, the microbial suspension is prepared in the laboratory in sufficient amount and then injected to the weak soil on field as in traditional method. Grouting methods can be implemented for the process of bio-clogging and bio-cementation. The process adopted for the implementation and details of the specific field project influences the cost of microbial treatment on soil.

7 Conclusions

1. Overall, different microorganisms improve and enhance the mechanical properties of the soil in situ.
2. Microorganisms producing calcium carbonate increase the bearing capacity, strength and stiffness of the soil as well as reduce the porosity and permeability of soil with improving susceptibility towards liquefaction [11, 16].
3. The MICP process is very cost-effective and eco-friendly method for soil treatment.

4. MICP treated soil offers resistance towards liquefaction as well as prevents damage due to seismic loading [15].
5. Facultative anaerobic and microaerophilic bacteria are the most suitable microbes for the treatment of needy soil in most convenient environmental conditions [14].
6. The performance of the microbial stabilized soil in the field is needed to be observed.

References

1. Dejong Jason T, Fritzges Michael B, Nusslein K (2006) Microbially induced cementation to control sand response to undrained shear. J Geotech Geoenviron Eng 132(11):1381–1392
2. Chou C-W (2007) Bioimprovement of geotechnical properties of sandy soils. M.Sc. thesis, University of Maryland
3. Ahmed A, Hussain I (2008) Enhancing the stability of fine grained soil using biological approach. Electron J Geotech Eng 1:1–11
4. Volodymyrand I, Jian C (2008) Applications of microorganisms to geotechnical engineering for bioclogging and biocementation of soil in situ. Rev Environ Sci Biotechnol 7(2):139–153
5. Kavazanjian E, Karatas I Jr. (2008) Microbiological improvement of the physical properties of soil. In 6th international conference on case histories in geotechnical engineering, Arilngton, VA, pp 1–10
6. Mortensen BM, Haber MJ, Dejong JT, Caslake LF, Nelson DC (2011) Effects of environmental factors on microbial induced calcium carbonate precipitation. J Appl Microbiol 338–349
7. Ng W-S, Lee M-L, Hii S-L (2012) An overview of the factors affecting microbial- induced calcite precipitation and its potential application in soil improvement. Int Schol Sci Res Innov 6(2):683–689
8. Yasuhara H, Neupane D, Hayashi K (2012) Experiments and predictions of physical properties of sand cemented by enzymically–induced carbonate precipitation. Soil Found 52(3):539–549
9. Wei-Soon NG, Min-Lee L, Tan CK, Hii SL (2013) Improvements in engineering properties of soils through microbial—induced calcite precipitation. KSCE J Civil Eng 17(4):718–728
10. Pusadkar SS, Chavan R (2014) Effect of effective size of sand on behavior of microorganism reinforced sand. In International conference on Recent Trends in Engineering and Technology, Mount Zion College of Engineering, Pudukottai, Tamilnadu, pp 127–132
11. Pusadkar SS (2014) Swelling and shrinkage characteristics of biologically treated black cotton soil. International conference on 'Geo-environmental issues and sustainable urban development' (GEN- 2014), MNNIT, Allahabad
12. Shahrokhi R, Zomorodian SM, Naizi A, O'Kelly BC (2014) Improving sand with microbial—induced carbonate precipitation. Ground Improve Proc Inst Civil Eng 1–14
13. Ukken ET, Krishnan A (2015) Microbial-induced calcite precipitation for soil improvement. Int J Adv Res Trends Eng Technol (IJARTET) 2
14. Umar M, Kassim KA, Chiet KTP (2016) Biological process of soil improvement in civil engineering: a review. J Rock Mech Geotech Eng 1–8
15. AnbuPeriasamy Chang-Ho Kang, Shin Yu-Jin, So Jae-Seong (2016) Formations of calcium carbonate minerals by bacteria and its multiple applications. Spring Plus Spring Open J 5-250:1–26
16. Chang I, Im J, Cho G (2016) Introduction of microbial biopolymers in soil treatment for future environmentally-friendly and sustainable geotechnical engineering. Sustainability 8(251):1–23

Parametric Study on Foundation Retrofitting Using Micro-piles

Vishal Singh and Benu Gopal Mohapatra

Abstract Improvement in techniques of building construction has led to the incorporation of new ideas to ease the lifestyle of the new generation at minimal cost. Thus, this paper aims to determine the effect of additional floors added to the existing strucutre, which further leads to the increase in load on the building's foundation. The effect of additional superimposed load on the settlement characteristics of foundation modified through micro-pile-based retrofitting is studied. Initially, a three-storey building with isolated footing is considered for the analysis and a few additional storeys are incorporated to the existing structure which increases the superimposed load on the foundation. Commercially available FE softwares such as STAAD Pro. and PLAXIS-3D were used for the determination of load on foundation of the building and then modeling of footing, respectively. Four layers of soils containing clay, lean clay crust, soft lean clay, and soft fat clay are considered in the first model, whereas soil layers of soft clay and hard clay are taken for the second model. Diameters and the number of piles are different in different footings (designed as per superimposed load) of the building model. Results of the study revealed that reduction in settlement by nearly about 34, 58, and 66% are obtained for addition of fourth, fifth, and sixth-floor loading, respectively for the first model, and for the second model, the settlement is reduced by nearly about 30, 50, 62, and 67% for the addition of third, fourth, fifth, and sixth-floor loading with micro-piles when compared to the respective model without micro-piles.

Keywords Micro-pile, Foundation retrofitting · PLAXIS-3D, Settlement, Additional storeys

---

V. Singh
PhD. Research scholar, Department of Civil Engineering, Birla Institute of Technology and Science, Pilani 333031, India

B. G. Mohapatra (✉)
Professor & Director, School of Civil Engineering, KIIT University, Bhubaneswar 751024, India
e-mail: bmohapatrafce@kiit.ac.in

B. B. Das et al. (eds.), *Recent Developments in Sustainable Infrastructure*, Lecture Notes in Civil Engineering 75, https://doi.org/10.1007/978-981-15-4577-1_13

1 Introduction

Increase in population has led to the infrastructure needs, and thus upgradation in building structures is found necessary while it has now become a challenge in the civil engineering works. Upgrading the building foundation to withstand the additional load due to the addition of few storeys to the existing building is a difficult and sensitive work. In most of the cases the buildings are not structurally safe to receive the additional loads; therefore some method of strengthening is mandatory that can enhance the strength of the foundation to bear the additional load which is getting added to the building with increase in number of storeys. Foundation of the building must be designed as per the soil available in the region which can be categorised as loose, medium and dense, Bowles [1]. The method which is found very useful in this case is "micro-pile under-pinning method". A micro-pile is a small diameter (<300 mm), drilled, and grouted pile that is typically reinforced, Sivakumar [2]. The objective of underpinning work using micro-piles is to create new foundations such that the existing load or additional load to the building can be wholly or partially transferred to the soil without harmful movements, Thorburn and Littlejhon [3].

2 Literature Review

Under-pinning of historic/old structures has been done by some researchers like Dahiru et al. [4], Hon and Demesak [5], Gazzarrini et al. [6] and Majdi and Giffen [7] using micro-piles method. In this context, Dahiru et al. [4] studied the main cause of failure of structure in Nigeria, and it was due to foundation failure. The failure of foundation was predicated due to many reasons and in different regions, like poor workmanship, cheap material, lethargic supervision, lack of co-ordination and testing of soil, and so on. About 81.3% people agreed for the under-pinning operation for the building but 18.7% did not agree. Later on, physical observation and result of interviews show that the buildings are in good condition.

Hon and Demesak [5] discussed about the construction methods of foundation of Lehigh Valley Hospital New Medical Building. Here, due to the construction of the basement, it was to affect the nearby one-storey General Service building and three-storey Anderson building's foundation. Again, like the construction mechanism as explained by Gazzarrini et al. [6], the nearby building's foundations were under-pinned using micro-piles and jet-grouting technology. The foundation of new hospital building was below the existing foundation of nearby buildings. Thus, the nearby buildings were needed to be under-pinned. Majdi and Giffen [7] have discussed about the under-pinning operation held in the New York City of USA in the columns of the hotel Commodore/Grand Hyatt. The Grand Central Terminal Station is close to Grand Hyatt and three subway lines (4/5/6) are used to pass beneath the hotel. The columns which were heavily loaded were retrofitted using under-pinning method. In a study by Davie and Senapathy [8], a 3000-ton stucture with high capacity mini-piles were used to overcome the settlement of mat foundation of a power plant where heavy equipments need to be placed. Hooda and Mittal [9] have investigated the micro-piles experimentally to check its ability to withstand the horizontal pressure with three types of piles, positive batter piles, negative batter piles and vertical piles. A case study by Robinson [10] describes about the use of micropiles while underpinning a building in Vail, Colorado during a renovation project where drilling of micropiles in the old foundation work was done.

3 Present Study

The present study aims to investigate the influence of various parameters on the performance enhancement of the structure retrofitted by micro-piles. The scheme of study includes:

I. Use of micro-pile under-pinning method in the isolated foundation of the structure modeled.
II. Conducting the parametric study identifying the variable parameters which governs the under-pinning of foundation.
III. Identifying the influence of the micro-pile diameter and numbers of micro-piles under each footing of the building model.
IV. Identifying the effect on settlement of the footing with and without the inclusion of micro-piles.

For the study, a modeling has been done in STAAD Pro. consisting of two bays and three storeys building as a frame structure of beam column and slab. For loading consideration gravity loading for dead load, live load on slab, masonry load on beams (both external and internal), and parapet wall load at the top floor is taken. The loading combinations are tired as much as possible but mainly 1.5(D.L + L.L), 1.2(D.L + L.L ± E.L), 1.5(D.L + L.L), 0.9(D.L ± E.L), 1.2(D.L + L.L + W.L − X), 1.2(D.L + L.L + W.L − Y) and designed as per IS 875 (part-1), IS 875 (part-2), IS 875 (part-3), IS 1893 (part-1):2016, IS 456-2000, SP-16. Again, loading on each column is calculated and their values are given in Table 1 for additional loads also, which is to be applied to the model of foundation in PLAXIS-3D. According to the soil layers given as per Tables 4 and 9, the bearing capacity in Case-I is 135 kN/m^2 and in Case-II it is 227 kN/m^2.

The loading calculated using STAAD is given in Table 1 for different storeys. Figures 1 and 2 represent the two models of soil used in the parametric study. The data for footing model is given in Table 2, where unit weight of footing is 25 kN/m^3 and Young's modulus is 28 × 10^6 kN/m. Similarly, the data of the soil layers are given in Table 3. Table 4 consists of the data for the embedded piles properties, where unit weight of micro-piles is 25 kN/m^3 and Young's modulus is 28 × 10^6 kN/m, and Figs. 3 and 4 represent the elevation and loading frame of the building model, respectively.

Table 1 Surface load calculations of columns on respected footing applied during analysis

Floors added to building	Ultimate central column load (kN) (W_u)	Ultimate side middle column load (kN) (W_u)	Ultimate side column load (kN) (W_u)	Applied surface load on central column footing (kN/m^2)	Applied surface load on side middle column footing (kN/m^2)	Applied surface load on side column footing (kN/m^2)
Second	1258.60	877.72	579.13	7866.25	9752.48	6434.82
Third	1606.11	1121.61	740.77	10038.18	12462.33	8230.86
Fourth	1953.66	1365.49	902.42	12210.37	15172.11	10026.92
Fifth	2301.21	1609.37	1064.07	14382.56	17881.88	11823.00
Sixth	2648.76	1853.25	1225.71	16554.75	20591.66	13619.00

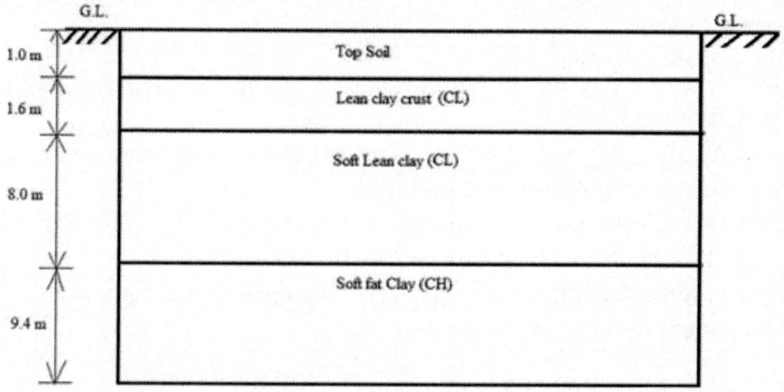

Fig. 1 Four layers of soil for parametric study [Model-I]

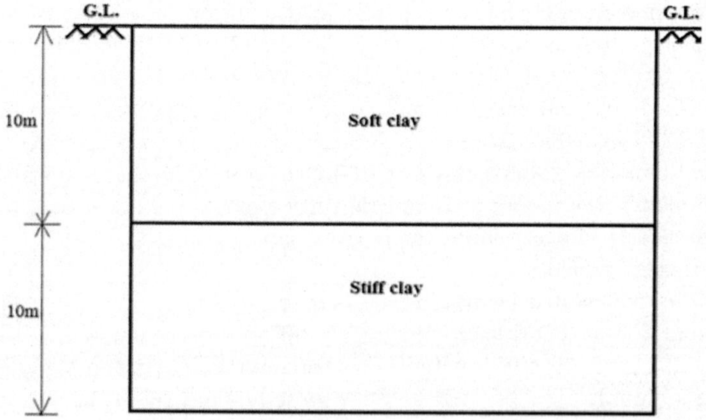

Fig. 2 Two layers of soil for parametric study [Model-II]

Table 2 Footing (pile cap) properties adopted in 3D finite-element modeling

Parameter	Model	Footing (G + 2) central column	Footing (G + 2) side-middle column	Footing (G + 2) side column	Unit
Material type	N.A.	Plate, isotropic	Plate, isotropic	Plate, isotropic	N.A.
Depth (d)	Soil Model-I	0.410	0.360	0.30	m
	Soil Model-II	0.46	0.40	0.30	
Poisson's ratio (ν)	Soil Model-I	0.15	0.15	0.15	N.A.
	Soil Model-II	0.15	0.15	0.15	
Sizes	Model-I	3.0 × 3.0	2.5 × 2.5	2.1 × 2.1	m^2
	Model-II	2.1 × 2.1	1.8 × 1.8	1.5 × 1.5	

3.1 Parametric Study

In these studies two building models were prepared and they were loaded in software for analyzing the settlement of footing with the addition of each storey load as per Table 1. Here, not just pile under footing is loaded but diameter and spacing of pile are also maintained in each footing (by keeping the pile length of 16 m as fixed in design) according to the load coming over footing from respective columns. The pile length, diameter, reinforcement, and so on in a micro-pile under-pinning depend on various parameters like the soil properties in which it is going to place, the load coming from the super-structure, and maximum loaded part of the structure and the footing. Hence, a parametric study is equally important to understand this behavior under several possible conditions. This will help the practicing engineers to save the construction time and money with safe vertical enhancement of the building after retrofitting its foundation. The following parameters are pointed out based on the literature review and considered to be the most appropriate in this present parametric study (Figs. 5, 6, 7, and 8). The Hardening type and Mohr-column type soil models represented in Table 3 are considered as per PLAXIS-3D manual by Brinkgreve [11].

- Soil profile and its parameter.
- Loading due to increase in building storey.
- Diameter of pile.

Micro-piles have diameter lying between 75 and 300 mm, Sivakumar [2]; thus, this pile can be placed in existing footing without disturbing the reinforcement of the footing while drilling as per the spacing of the footing reinforcement.

3.2 Soil Profile and Its Parameters

Soil profile and the parameters like cohesion (c), angle of internal friction (φ), depth of each soil layer (h), Young's modulus (E), and so on are very important parameters

Table 3 Soil parameters adopted in finite-element modeling [Model-I and Model-II]

Parameter	Clay	Lean clay crust	Soft lean clay	Soft fat clay	Soft clay	Hard clay	Unit
General	Model-I				Model-II		
Material model	Hardening soil	Mohr-coulomb	Mohr-coulomb	Mohr-coulomb	Hardening soil	Hardening soil	N.A.
Drainage type	Undrained-A	Drained	Drained	Drained	Undrained-A	Undrained-A	N.A.
Unit weight above phreatic level (γ_{unsat})	16.0	17	17	17	17.0	19.0	kN/m^3
Unit weight below phreatic level (γ_{sat})	16.0	19.10	17.60	17.10	17.0	19.0	kN/m^3
Young's modulus (E')	N.A.	15 × 10^3	20 × 10^3	25 × 10^3	12.5 × 10^3	35 × 10^3	kN/m^2
Secant stiffness E_{50}^{ref})	2 × 10^3	–	–	–	12.5 × 10^3	35 × 10^3	kN/m^2
Cohesion	25	10	(7–9)	10	20	30	kN/m^2
Friction angle	30	15	12	9	24	24	°
Poisson's ratio	0.2	0.4	0.4	0.4	0.3	0.3	N.A.

Table 4 Embedded pile properties adopted in finite-element modeling

Parameter	Model	Embedded pile central footing	Embedded pile side middle footing	Embedded pile side footing	Unit
Diameter (dia.)	Model-I	0.17	0.2	0.25	m
	Model-II	0.12	0.14	0.25	
Skin friction ($T_{top, max}$)	Model-I	245	288	360	kN/m
	Model-II	1087	1264	2260	
Skin friction ($T_{bottom, max}$)	Model-I	150	175	218	kN/m
	Model-II	471.5	549	983	
Resistance force (F_{max})	Model-I	2.04	2.82	4.40	kN
	Model-II	4.37	6.0	19.0	
Length of pile (L)	Model-I	16	16	16	m
	Model-II	16	16	16	

Fig. 3 3D model of building

which govern the strength and bearing capacity of soil. Parameters are given in Table 3 for different cases.

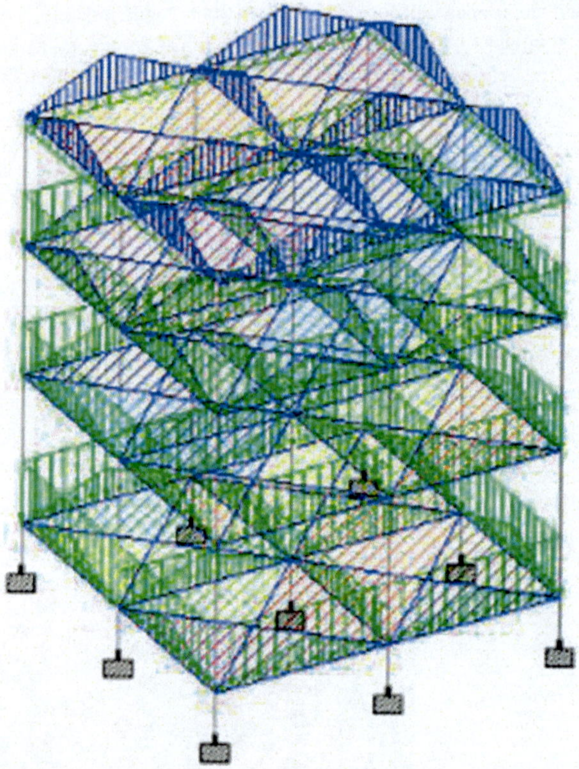

Fig. 4 Loading on $G + 2$ building

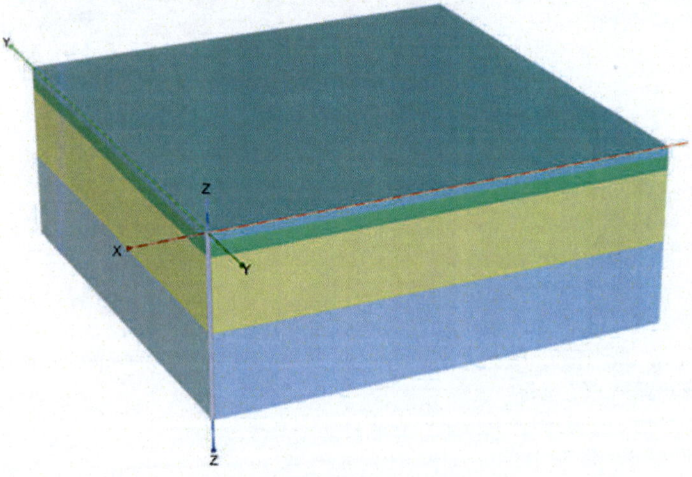

Fig. 5 Four layers of soil as per the parametric study consideration

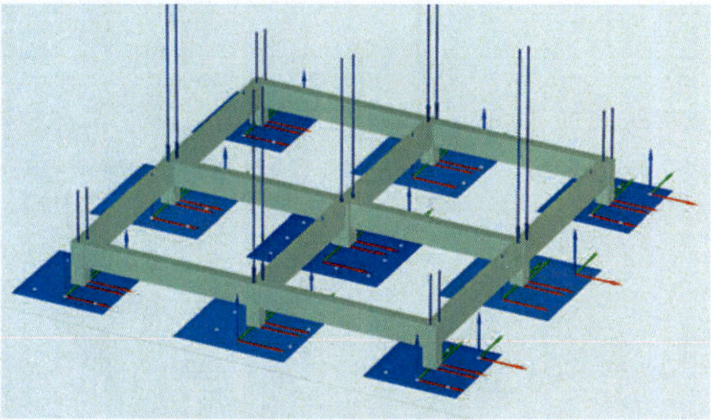

Fig. 6 Model of $G + 2$ building's footing without micro-pile and surface loading

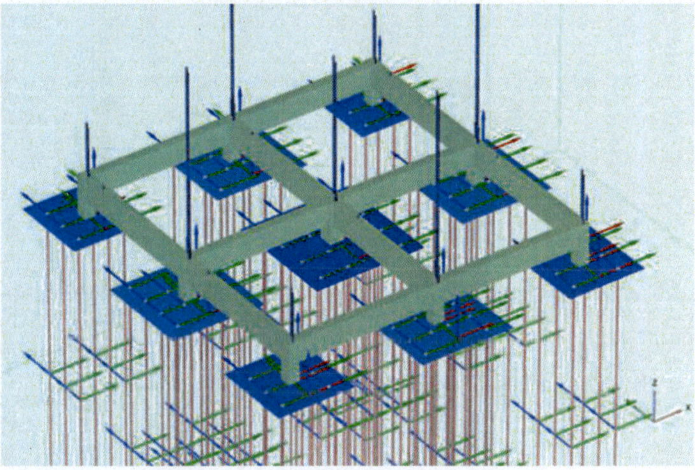

Fig. 7 Model of $G + 2$ building's footing with micro-pile and surface loading

3.3 Loading Due to Addition of Storeys in Building

To study the effect of loading on footing without under-pinning and with under-pinning due to addition of few extra storeys to the existing structure, few additional storeys are added and the behavior of the footing is observed for both without and with micro-pile under-pinning. Loading considered as per Table 1.

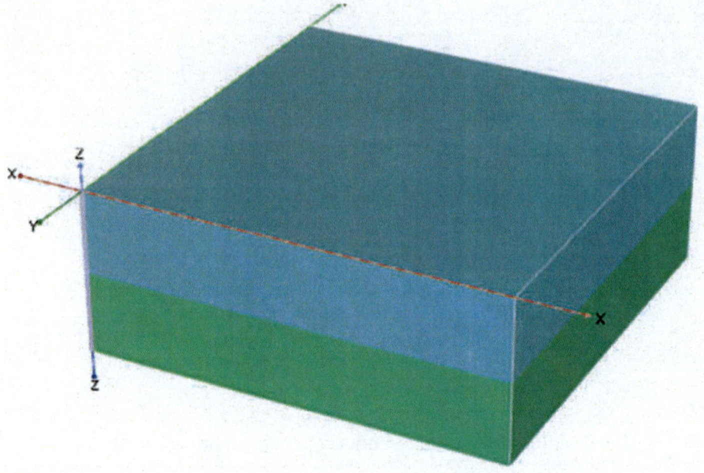

Fig. 8 Two layers of soil as per the parametric study consideration

3.4 Diameter of Micro-pile

As the footing of existing building is supposed to be subjected to loading due to additional storey and the footings are under-pinned using micro-pile, this micro-pile under footing helps in resisting the settlement of the building and thus the change in diameter of the pile affects the settlement and bending moment and shear forces in the pile too. Different sizes of pile diameter have been selected for the analysis based on the spacing available in the footing reinforcement of the existing three-storey building. The use of pile diameter also varies to accumulate that number of micro-piles obtained in design in the footing of existing structure.

4 Results and Discussion

With the application of load on all the footings of the designed three-storey building as surface load to the columns, the settlements are recorded on building for the study of without using micro-pile. Figure 9 represents the deformed mesh and the footing settlement at second floor loading without micro-pile. The settlement of building with every additional load due to additional stories is calculated and their values are given in Table 5. Figure 10 represents the deformed mesh of settlement of the footing model in PLAXIS-3D using micro-pile. The values of settlement are given in Table 5, and the table shows that how the building settlement can be minimized by using micro-piles to the existing foundation. It has been found that the settlement of the building can be resisted with the inclusion of micro-piles in the existing footing but the diameter to be used must be maintained with spacing while placing (Table 6).

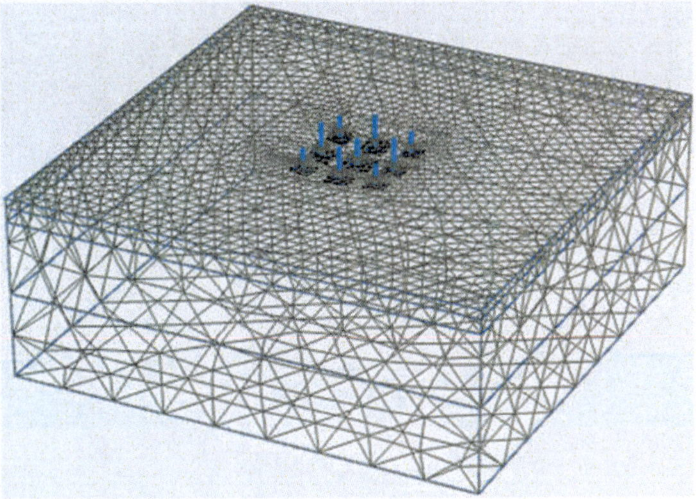

Fig. 9 Deformed mesh settlement of building's footing

Table 5 Applied surface load and settlement of Model-I and Model-II

Building with successive floors	Applied surface load on central column footing (kN/m^2)	Applied surface load on side middle column footing (kN/m^2)	Applied surface load on side column footing (kN/m^2)	Overall settlement footing of building without micro-pile (mm)	Overall settlement of building with micro-pile (mm)	Overall settlement footing of building without micro-pile (mm)	Overall settlement of building with micro-pile (mm)
				Model-I		Model-II	
Second	7866.25	9752.48	6434.82	26.66	N.A.	36.28	N.A.
Third	10038.18	12462.33	8230.86	37.65	N.A.	57.10	40.16
Fourth	12210.37	15172.11	10026.92	51.75	33.93	86.29	42.92
Fifth	14382.56	17881.88	11823	70.21	29.27	122.1	45.97
Sixth	16554.75	20591.66	13619	93.14	31.42	165.2	53.90

5 Conclusion

In this present study, a three-storey building with isolated footing has been considered. The performance evaluation of this building under various loads has been carried out. The effect of additional successive storey to the existing building has been studied in terms of settlement of the foundation by keeping in mind the punching shear failure of column and micro-pile both in the existing footing. Thus, the building model has been loaded to up to sixth and seventh storey load maximum in Model-I and Model-II, respectively, such that the punching failure could be avoided. It was observed that the addition of successive storey increases the settlement of

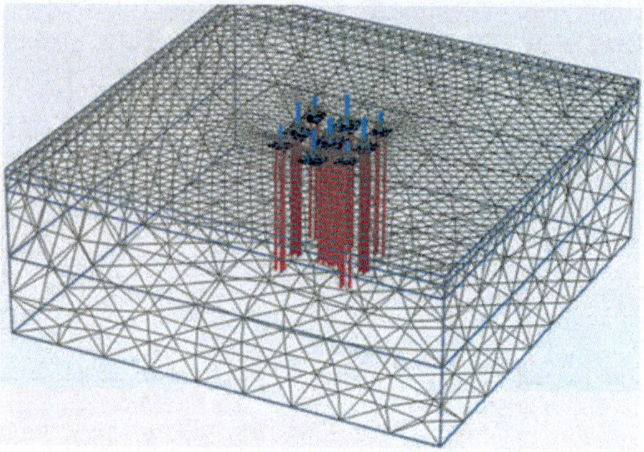

Fig. 10 Deformed mesh settlement of three-storey building's footing with micro-pile

Table 6 Number of micro-piles applied under each footing

Building with successive floors	No. of piles provided in central column	No. of piles provided in side middle column	No. of piles provided in side column	No. of piles provided in central column	No. of piles provided in side middle column	No. of piles provided in side column
	Model-I			Model-II		
Second	N.A.	N.A.	N.A.	N.A.	N.A.	N.A.
Third	N.A.	N.A.	N.A.	4	3	2
Fourth	9	5	3	6	4	2
Fifth	13	8	4	9	6	2
Sixth	17	10	6	12	7	4

Note Allowable settlement is kept 40 mm in case of isolated footing in plastic clay. Load calculated for additional floors are given in Table 5. The existing building load up to addition of second floor is considered where there is no requirement of under-pinning but as soon as the fourth floor load is considered in case of Model-I settlement is beyond the permissible limit and thus micro-piles are required. It has been observed that for fifth and sixth floors loading the required numbers of pile to support the structure are coming as a fraction number in design. But in practical design the number of piles is taken as a next higher whole number. This consideration reduces the settlement in fifth and sixth floors loading consideration compared to fourth floor loading. (It can be observed in Fig. 11.) From Table 7 it is observed that there is a settlement reduction due to the introduction of micro-pile under each footing because the settlement of isolated footing from second to third floors loading is within the permissible limit of 40 mm as per the allowable settlement criteria, thus there is no requirement of providing micro-pile to the foundation for retrofitting works. However, from addition of fourth to sixth floors, loadings have been made to settle the footings more than 40 mm and thus these were retrofitted using micro-piles and reduction in settlement is observed. But in case of Model-II, the micro-piles are required with the addition of third floor load, so it may be due to variation in sizes of footing as compared to Model-I. Here, the settlement varies smoothly without any sudden hike in the graph, which can be observed from Fig. 12

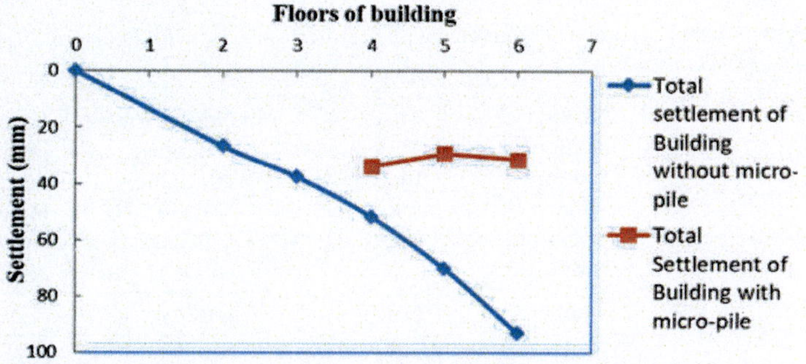

Fig. 11 Additional storeys versus settlement [Model-I]

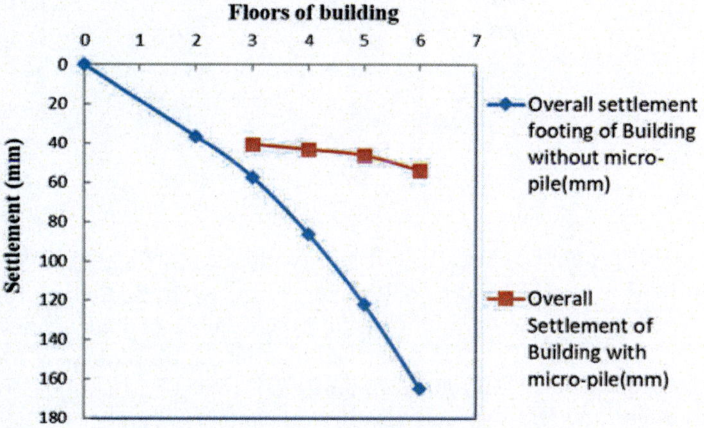

Fig. 12 Additional storeys versus settlement [Model-II]

the building and sometimes it exceeded the permissible limit. To reduce the settlement below the permissible limit (40 mm considered in this study) and to make the foundation safe, the idea of underpinning the foundation with micro-pile has been introduced. With and without the use of micro-pile, an extensive parametric study on building models with addition of successive storey has been carried out. Based on the parametric study, the following conclusion can be drawn:

i. For the soil profile of Case-I: Top soil (with $\varphi = 30°$ and $c = 25$ kN/m^2), lean clay crust (with $\varphi = 15°$ and $c = 10$ kN/m^2), soft lean clay (with $\varphi = 12°$ and $c = 7$ kN/m^2), and soft fat clay (with $\varphi = 9°$ and $c = 10$ kN/m^2). Settlement for the isolated footings was observed by changing the number of floors stepwise from second to sixth floor. The results showed that maximum settlement for the isolated footing exceeded 40 mm from fourth floor onwards. Using the micro-pile under-pinning method for such case the maximum settlement for the isolated footing was reduced by approximately 34% for fourth floor, 58% for fifth floor, and 66% for sixth floor.

Table 7 Percentage reduction in settlement (without micro-pile vs. with micro-pile)

1	2	3	4	5	6
Additional floors	Model	Max. settlement of footing without micro-pile (mm)	Max. settlement of footing with micro-pile (mm)	Net settlement of footing after piling: col^n•(4) – G + 2 settlement col^n•(3) (mm)	% Reduction in settlement due to micro-pile col^n•(3) and col^n•(4)
Second	I	26.66	N.A.	N.A.	N.A.
Third		37.65	N.A.	N.A.	N.A.
Fourth		51.75	33.93	7.27	34.43
Fifth		70.21	29.27	2.61	58.31
Sixth		93.14	31.42	4.76	66.26
Second	II	36.28	N.A.	N.A.	N.A.
Third		57.10	40.16	3.88	29.66
Fourth		86.29	42.92	6.64	50.26
Fifth		122.1	45.97	9.69	62.35
Sixth		165.2	53.90	17.62	67.37

ii. For the soil profile of Case-II: Soft soil (with $\varphi = 30°$ and $C = 25$ kN/m^2), stiff soil (with $\varphi = 15°$ and $C = 10$ kN/m^2) in Case-II. Settlement for the isolated footings was observed by changing the number of floors stepwise from second to sixth floor. The results showed that maximum settlement for the isolated footing exceeded 40 mm from third floor onwards. Using the micro-pile under-pinning method for such case the maximum settlement for the isolated footing was reduced by approximately 30% for third floor, 50% for fourth floor, and 62% for fifth floor and 67% for sixth floor.

Acknowledgments This work has been done by the first author during his M.Tech program at KIIT deemed to be University, Bhubaneswar, Odisha and the author also likes to thank Dr. (Prof.) Nishant Roy, BITS Pilani, Rajasthan campus, for his kind help and support.

References

1. Bowles JE (1997) Foundation Analysis and Design. The McGraw-Hill Companies
2. Sivakumar Babu GL (2012) NPTEL course on ground improvement using micro-pile. Indian Institute of science
3. Thorburn S and Littlejhon GS (1993) Underpinning and retention. Spinger-Science + Business Media, B.V
4. Dahiru D, Salau S, Usman J (2014) A study of underpinning methods used in the construction industry in Nigeria. Int J Eng Sci (IJES) 1:5–13
5. Hon RK, Demcsak MR (2010) Design and construction of an underpinning and earth-retaining system for Lehigh Valley hospital building. In: The earth retention conference, pp 342–351

6. Gazzarrini P, Kokan M, Jungaro S, Hunt D (2016) Roger Arena, South Tower, Vancouver, BC, Canada: jet grouting for the structural underpinning of the existing Georgia viaduct foundations and temporary excavation support. Geotech and Struct. Eng congr 960–970
7. Majdi RY, Giffen R (2017) Underpinning historic structures at grand central station. Struct Congr 604–613, New York
8. Davie JR, Senapathy H (2015) Underpinning a 3000-Ton structure with high-capacity mini-piles. Deep Found ASCE 6:647–654.
9. Hooda MTD, Mittal A (2017) Study of micro-piles subjected to lateral loading and oblique pull "review", GSP 136 innovations in grouting and soil improvement. Int H Adv Res Sci Eng 6:369–374
10. Robison MS (2010) Excavation support and micro-pile underpinning in Vail, Colorado. Geo-Trends, 112–118
11. Brinkgreve RBJ, Engin E, Swolfs WM (2013) PLAXIS-3D Manual

Stabilization of Clayey Soil with Marble Dust

Arnab Debnath, Sumon Saha, and Rana Chattaraj

Abstract In this study the effectiveness of marble dust to improve the mechanical properties of clayey soil has been studied. Marble has high lime content, which acts as the main factor for soil stabilization. 25–30% marble dust (MD) is generated during marble quarries and stones are cut as a block. In this study marble dust was added in the range of 0–30% to the virgin soil by weight. Various tests like Atterberg's limits, compaction test, and strength tests were performed on untreated soil as well as on treated soil. The results obtained in case of treated soil were compared with the results obtained from the untreated soil. Unconfined compressive strength (UCS) tests were performed at different curing intervals. It was found that the soil strength increased with the increase in marble powder content and curing periods.

Keywords Marble dust · Clayey soil · Soil stabilization · Compaction · Curing and UCS

1 Introduction

Owing to the presence of montmorillonite mineral in clayey type of soil, it swells with addition of water and shrinks with evaporation of water. If any structure is constructed on this type of soil, a lot of damage may be induced to the structure due to the swelling and shrinkage of the underneath soil. This is why these types of soils need to be stabilized. There are several techniques to improve the clayey soil such as by adding different types of additives like marble dust, fly ash, rice husk, cement, and so on. In this study, MD has been used to stabilize the soil. The MD is generated when marble blocks are cut into small pieces. Marble dust is stored in sedimentation

A. Debnath · S. Saha · R. Chattaraj (✉)
KIIT Deemed to be University, Bhubaneswar, India
e-mail: rana.chattarajfce@kiit.ac.in

A. Debnath
e-mail: arnabdebnath00@gmail.com

S. Saha
e-mail: sumonsaha.4211@gmail.com

B. B. Das et al. (eds.), *Recent Developments in Sustainable Infrastructure*, Lecture Notes in Civil Engineering 75, https://doi.org/10.1007/978-981-15-4577-1_14

pond. Different types of industries such as construction, cement, ceramic industry, and paint industry use very little amount of marble dust. The objective of this study is to utilize the waste of MD for solving different environmental problems, such as leaching, dusting and take high disposal area.

Hebhoub et al. [1] indicate that the amount of marble waste generated by extraction process can be up to 80% of the total volume of stone extracted. Okagbue and Onyeobi [2] evaluated the geotechnical properties of red tropical soils treated with marble dust. It is reported that plasticity is reduced and compressive strength increased with the addition of marble dust. Bharti et al. [3] noticed that the optimum moisture content (OMC) of the soil is reduced due to lime, which reacts with moisture. Brooks et al. [4] suggested that maximum dry density (MDD) of soil is decreased with limestone dust content due to cation exchange reaction that reduces the weight volume ratio. The CBR value of existing soil increased with limestone content due to cation exchange, flocculation agglomeration reaction, and cementitious gel formation. Sreekumar and Mary Rebekah Sharmila [5] performed an experiment where different percentages of marble dust were added (3, 6, 9, 12, and 15%) at different curing periods of 3, 7, and 14 days to the soil sample to determine the basic properties of treated soil such as Atterberg limits, compaction characteristics, and strength characteristics. The CBR value of the untreated and treated soil at 0 and 9% marble dust was recorded to be 5.19 and 8.83%, respectively. Addition of marble dust beyond 9% results in decrease in strength. Saygili [6] suggested that the swelling potential of clayey soil can be improved with the addition of marble dust. Marble dust stabilized clay soil gain more than 48% compressive strength from the untreated clay soil. Sabat [7] evaluated that the expensive soil can be improved by adding ceramic dust at different proportion. As ceramic dust cohesion lessens, the cohesion of the treated soil is reduced and angle of internal friction improved. The unconfined compressive strength of soil increased with the addition of marble to the soil. Li et al. [8] reported that marble can be used to replace either cement or paste. The strength and microstructure concrete can be improved if marble dust is used as a paste in cement. Khatic [9] reported that marble dust and rice husk ash can be used to make brick that have more strength than burn clay brick. Okagbue and Onyeobi [2] suggested that the treated expensive soil gains the maximum compressive strength but after that strength started to decrease as excess amount of marble dust cannot react with soil. Saboya Jr et al. [10] suggested that the expensive clay soil can be used to make ceramic brick by adding marble powder. Zorluer and Demirbas [11] suggested that the mixture of marble dust and fly ash can also be used to stabilize the base materials. In that case only fly ash is not sufficient to improve the base materials that is why the author suggested the mixture of MD and fly ash tobe used.

The aim of this study is to improve the strength of clayey soil by treating with MD. To evaluate the performance of the treated soil, different types of tests such as liquid limit, standard proctor, and UCS tests are carried out.

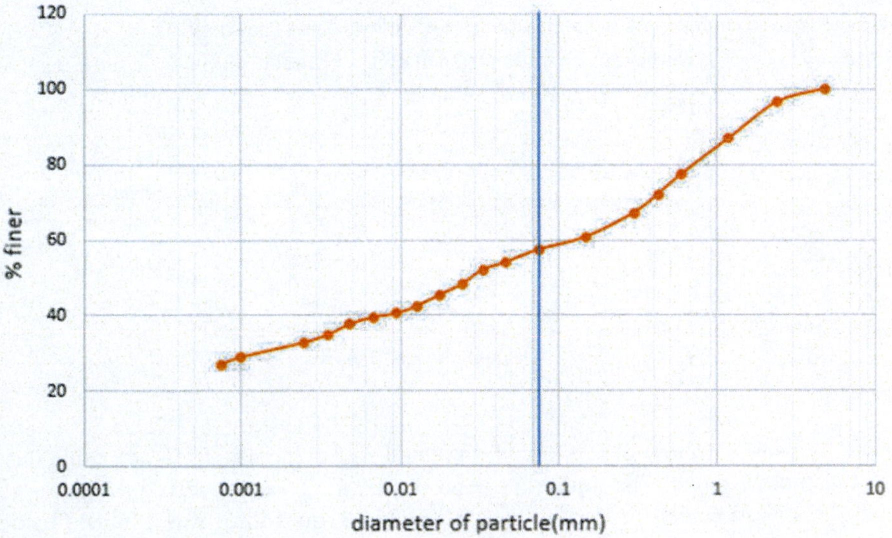

Fig. 1 Grading curve of soil

2 Materials

The clay soil was collected from Cuttack area for this study. 38% clay and 56% silt lie in this soil by weight. Figure 1 presents the particle size distribution curve. The specific gravity and natural moisture content of this soil is 2.49 and 40.07%, respectively.

MD was collected from marble processing factory in Bhubaneswar. In geological terms, marble is basically a metamorphic rock which is formed when limestone undergoes very high thermal and pressure energy, and this process is known as metamorphosis. When marble quarries and stones are cut as blocks, 25–30% of the total marble becomes unusable waste in the form of marble powder. Marble has very high lime (CaO) content, which acts as the main factor for soil stabilization. Quick lime (CaO) is the main constituent of marble dust and it is the main component that helps in the stabilization of the soil.

3 Tests Procedures

Identification of particle was evaluated according to IS-2720 part IV. Particle size distribution was evaluated by dry sieve and hydrometer analysis. First, the soil is dried, then sieve analysis has been conducted. In this research pycnometer test has been conducted three times for each sample and the average had been taken. Specific gravity of untreated soil as well as marble dust has been found out individually as

Table 1 Physical properties and plasticity index of untreated and treated soil

Mixture proportion	Optimum moisture content (%)	Maximum dry density (gm/cc)	Specific gravity	Plastic index (PI)
S	20.9	1.56	2.439	31.75
S + 5MD	20.16	1.644	2.484	25.95
S + 10MD	19.42	1.652	2.515	23.91
S + 15MD	18.35	1.669	2.547	22.14
S + 20MD	17.38	1.682	2.51	21.94
S + 25MD	16.55	1.685	2.527	20.25
S + 30MD	16.12	1.71	2.54	19.25

well as specific gravity of mixture of untreated soil + increasing % of marble dust was also determined. The liquid limit and plastic limit were resolved according to the methods indicated in IS-2720 part V (1985). Liquid limit was obtained using Cassagrande's apparatus. Standard proctor test had been conducted according to IS-2720 part VII for determination of optimum moisture content and maximum dry density of soil with different proportion marble dust. First, the mold was filled with the soil sample in three different layers. Each layer is compacted by 25 numbers of blows. The MDD and OMC of the soil with different percentage of marble dust content are shown in Table 1.

3.1 UCS

UCS test is used to determine the compressive strength of the sample. Unconfined compressive strength test had been done according to IS-2720 part I. A strain flow at the rate of 1% per min was applied during the test. Strength of the sample was evaluated at different curing periods. At least three samples were prepared at each composition of soil and marble dust.

4 Experimental Design

The value of maximum compressive strength was determined by response surface methodology (RSM) using Design-Expert software 11. Systematic error can be minimized by using response surface methodology. RSM minimizes the number of tests. The relationship between response and independent variable are evaluated by RSM. The combined effects of independent variables like percentage of MD content and curing interval (days) can be evaluated by experiment based on RSM. In that experiment range of marble dust content and curing interval lie between 0 and 25% and 0 and 30 days, respectively.

5 Results and Discussion

5.1 *Specific Gravity (G)*

The specific gravity of the virgin soil and MD have been evaluated to be 2.439 and 2.837, respectively, which indicated that marble dust is heavier than the soil particles. The values of specific gravity of treated soils at different percentage of marble dust have been listed in Table 1. Specific gravity of the treated soil increased with marble dust content as the marble dust was replacing the soil, which is lighter compared to its own weight.

5.2 *OMC and MDD*

Standard proctor test on virgin soil shows the OMC and MDD value as 20.9% and 1.56 g/cc. It is seen that the OMC of the treated soil decreases with the increase in marble content but the corresponding MDD increases. The graph of MDD versus marble dust content in amended soil is shown in Fig. 2.

5.3 *Atterberg Limit*

The value of PI of untreated soil as well as treated soil is shown in Table 1. If marble content is increased in the treated soil, plastic index decreases. As marble dust is cohesionless, the value of liquid limit of treated soil decreases.

5.4 *Unconfined Compressive Strength*

The soil with minimum of 10–13% marble dust content gained maximum compressive strength at 28 days. The maximum compressive strength of amended soil was found to be 2.8 kg/cm^2, which was obtained at 10% marble dust content by weight and after 28 days. The compressive strength of treated soil started decreasing after 20% of marble dust content. A mathematical model was developed to determine the compressive strength (σ) as a function of marble dust content (MD) and curing period

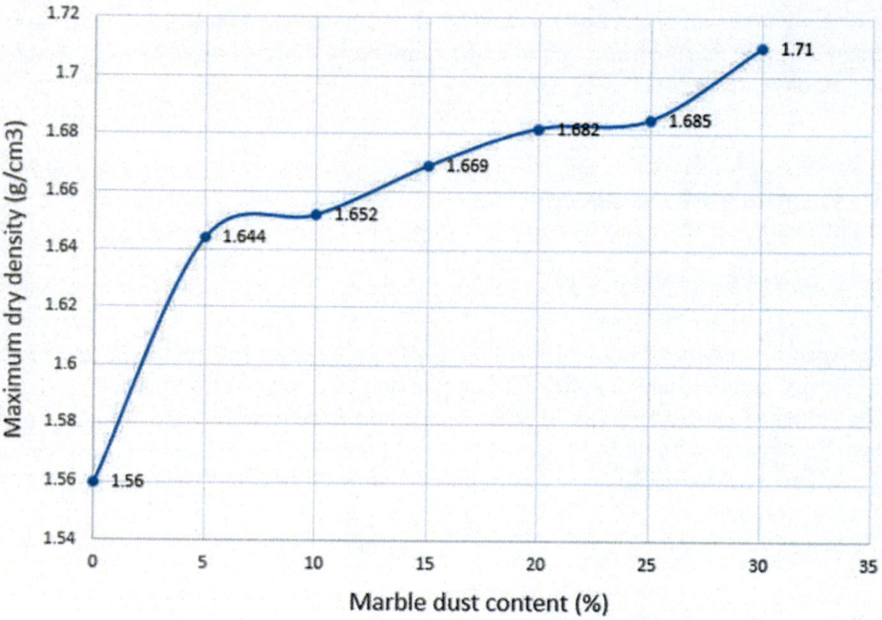

Fig. 2 Variation in maximum dry density (MDD) with marble dust content

(CP). A best fit surface (Fig. 3) was generated from the factorial analysis. Based on the mathematical model, an empirical equation has been proposed to estimate the strength of the treated soil at various curing period and marble dust. The proposed equation is valid for curing period and percentage of marble dust in the range of 0–30 days and 0–25%, respectively.

$$\sigma = 0.747 + 0.0385 \times CP + 0.241 \times MD - 7.5 \times 10^{-4} CP \times MD - 4.337 \times 10^{-4} \times CP^2 - 9.135 \times 10^{-3} \times MD^2$$

where σ in kg/cm^2, CP in days and MD in g/cc. Figure 4 presents that after 15 days, the value of UCS of treated soils increases slightly. The graph of UCS with different percentage of marble dust content at 0, 5, 10, 15, 20, and 28 days is shown in Fig. 5.

6 Conclusion

Grain size distribution curve and plasticity index of the virgin soil indicates that it was a fine-grained soil with medium plasticity. It is seen that specific gravity of MD is more than non-amended soil. It is observed that addition of marble dust decreases plasticity index substantially; this may be due to the non-cohesive nature of the marble dust. In case of standard proctor test, it is seen that MDD of the treated soil

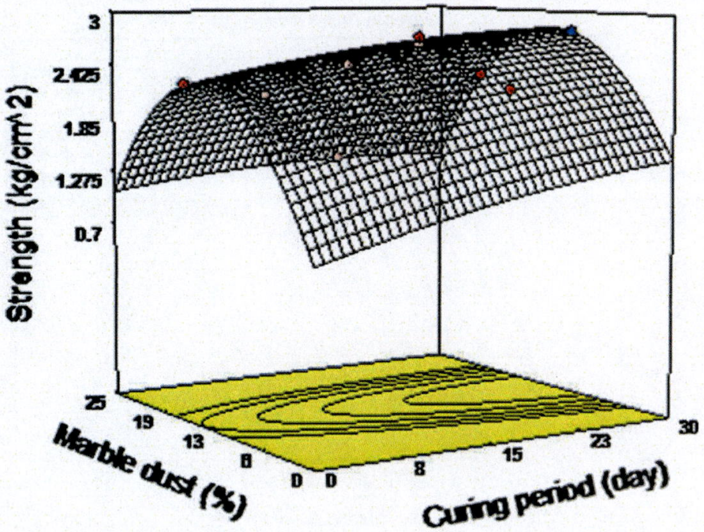

Fig. 3 Predicted surface for relationship among strength, marble dust and curing periods

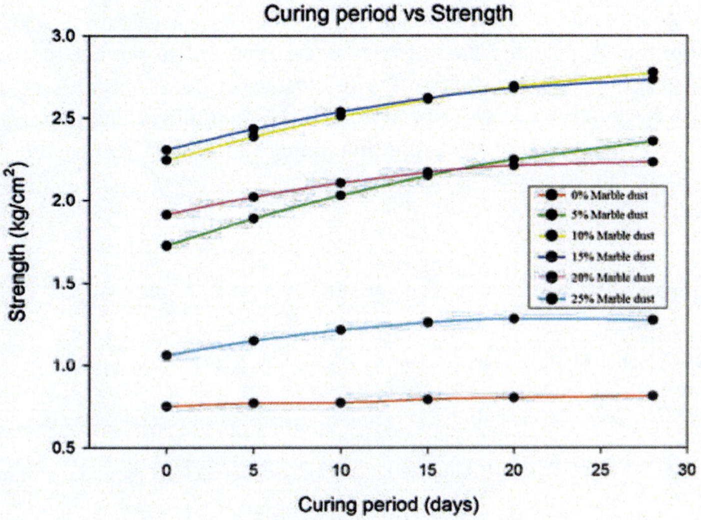

Fig. 4 Variation in strength with curing period for treated soils

increases continuously with the increase in marble dust content whereas the OMC decreases. To investigate the unconfined compressive strength, a mathematical model was developed for the design of the experimental program by using a software package called Design Expert-7. More than threefold increase in unconfined compressive strength was observed due to the addition of marble dust. It is observed that the rate

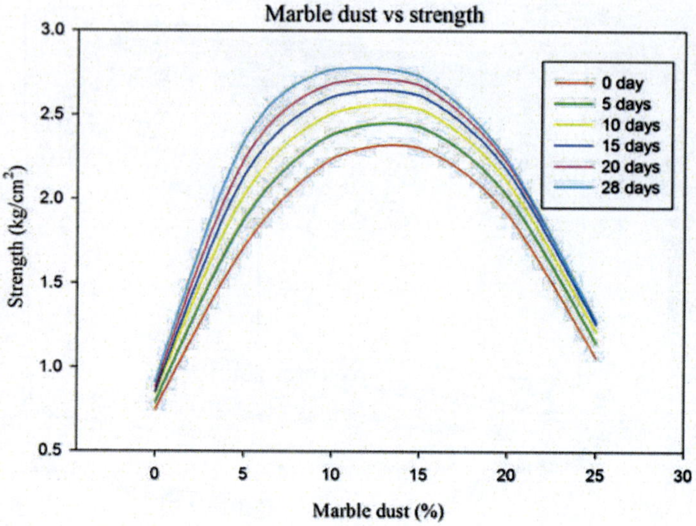

Fig. 5 Variation in strength with percentage of marble dust content for treated soils

of strength gain is maximum at the early stages of curing. Based on the experimental investigation, an empirical equation has been suggested to predict the strength of the treated soil at different combination of curing period and marble dust content. This empirical equation can be used to predict the unconfined compressive strength of MD treated clayey soil at the preliminary stage of any project by the practicing engineer where lab tests' data are not available.

References

1. Hebhoub H, Aoun H, Belachia M, Houari H, Ghorbel E (2011) Use of waste marble aggregates in concrete. Constr Build Mater 25(3):1167–1171
2. Okagbue CO, Onyeobi TUS (1999) Potential of marble dust to stabilise red tropical soils for road construction. Eng Geol 53(3–4):371–380
3. Bharti Jy P, Khan W, Kumar P, Gupta KD (2017) Combined effect of addition of marble dust and fly ash on expansive soil. J Mater Civil Eng
4. Brooks R, Adeoye FF, Takkalapelli KV (2010) Geotechnical properties of problem soils stabilized with fly ash and limestone dust in Philadelphia. J Mater Civ Eng 23(5):711–716
5. Babu S, Mary S (2017) Soil stabilization using marble dust. Int J Civil Eng Technol (IJCIET) 8(4):1706–1713
6. Saygili A (2015) Use of waste marble dust for stabilization of clayey soil. Mater Sci 21(4):601–606
7. Sabat AK (2012) A study on some geotechnical properties of lime stabilised expansive soil–quarry dust mixes. Int J Emerg Trends Eng Dev 1(2):42–49
8. Li LG, Huang ZH, Tan YP, Kwan AKH, Chen HY (2019) Recycling of marble dust as paste replacement for improving strength, microstructure and eco-friendliness of mortar. J Clean Prod 210:55–65

9. Khatik TD, Saklecha PP, Shende NP (2017) An experimental investigation on use of rice husk ash and marble dust as smart building material in brick. J Emerg Technol Innov Res 4(07):22–26
10. Saboya F Jr, Xavier GC, Alexandre J (2007) The use of the powder marble by-product to enhance the properties of brick ceramic. Constr Build Mater 21(10):1950–1960
11. Zorluer I, Demirbas A (2013) Use of marble dust and fly ash in stabilization of base material. Sci Eng Compos Mater 20(1):47–55

Recent Developments in Structural Engineering

Seismic Performance of Polynomial Friction Pendulum Isolator (PFPI) on Benchmark Cable-Stayed Bridge

Pratik Sen and Purnachandra Saha

Abstract The performance of polynomial friction pendulum isolator (PFPI) applied to the benchmark cable-stayed bridge is explored. Seismic performance of the PFPI is evaluated with the basic friction pendulum system (FPS) based on the evaluation criteria stated in the phase I benchmark problem. The surface curvature of the PFPI is varied using a polynomial function to alleviate the drawbacks of FPS whose surface is spherical. The energy-absorbing capacity and the horizontal flexibility during friction are the two important parameters developed in the PFPI system. A parametric study has been done to determine the optimum value of the PFPI key parameters. The numerical outcome designated that the PFPI system is seismically superior to that of the conventional FPS isolated system because of its variable curvature corresponding to the ground motion excitation. It is also noticed that a maximum deck displacement along longitudinal direction has been significantly reduced by PFPI than FPS.

Keywords Seismic isolation · PFPI · FPS · Benchmark cable-stayed bridge

1 Introduction

Cable-stayed bridges (CSB) have turned out to be prevalent all through the world and are likewise significant lifeline structures. Engaging style, enhanced stiffness over the substructure, and full and effective usage of structural materials are the major advantages of CSB. But this structure is vulnerable to seismic force because of its low damping and flexibility. For straight comparison between the diverse procedures for control over a particular kind of structure, different researchers have created benchmark problems. This helps the researchers to evaluate the different sensors, devices, and algorithms designed through this benchmark problem. The Bill Emerson Memorial Bridge built in Cape Girardeau, Missouri, USA is considered

P. Sen · P. Saha (✉)
School of Civil Engineering, Kalinga Institute of Industrial Technology, Deemed to be University, Bhubaneswar, Odisha, India
e-mail: dr.purnasaha@gmail.com; purnasahafce@kiit.ac.in

P. Sen
e-mail: sen.pratik709@gmail.com

B. B. Das et al. (eds.), *Recent Developments in Sustainable Infrastructure*, Lecture Notes in Civil Engineering 75, https://doi.org/10.1007/978-981-15-4577-1_15

as Benchmark problem on the CSB which has been originated by Dyke et al. [1]. This benchmark problem sets out a few evaluation criteria based on the performance of proposed control systems over the uncontrolled bridge. A huge research on the employment of frictional isolators has been conducted, and the performance is found to be very indifferent toward serious varieties in the frequency obtained due to excitation of base. This component is the most significant bit of the friction type isolators when contrasted with the elastomeric bearings. Advantage of friction-based isolators is that it guarantees the highest transmissibility of acceleration equivalent to the extreme constraining force of friction [2, 3].

The passive control devices which do not require any external power for its operation have been explored by different researchers in order to mitigate the response of seismic force on civil engineering structures effectively. The isolators which develop resisting forces by means of sliding friction are one of them. The friction pendulum system (FPS), among the friction base isolators, is found to be most popular since it uses gravity action as a force restoring mechanism by and very easy to install. Re-centering mechanism and sliding mechanism are the characteristics of the FPS. The spherical sliding surface of FPS enables its time period of vibration to be stable. Several researchers performed numerous studies on the behavior of FPS like Tsopelas et al. [4] performed experimental research on seismic control of bridge using sliding isolation. Tsai [5] formulated the finite element model of the study of FPS to achieve seismic isolation. The seismic analysis of a continuous bridge with the help of sliding isolation using systematic approach is performed by Wang et al. [6]. Jangid [7] explored the optimal friction coefficient of FPS for the least acceleration response of a base-isolated structure under different types of earthquake ground motions, and significant sliding displacement in the FPS for lower values of friction coefficient is observed. The performance of FPS system for the seismic response control of the CSB has been investigated by Soneji and Jangid [8]. For the benchmark CSB, Saha and Jangid [9, 10] performed the comparative studies on the isolation systems and figure out that huge device displacement for near-field earthquake (Gebze 1999 earthquake). Benicia-Martinez Bridge, California and American River Bridge, California, etc. are some of the practical structures where FPS had been installed for the seismic protection. However, from extant literature on FPS, certain disadvantages of it have been detected owing to spherical sliding surface which requires a precise intensity of ground excitation. Saha and Jangid [9] investigated the performance of different types of elastomeric as well as sliding friction applied in Benchmark Cable-Stayed Bridge and observed that sliding friction isolator like FPS performs better than the elastomeric isolators. Saha and Jangid [10] state that hybrid seismic control strategies consisting of isolators and VFD in Benchmark Cable-Stayed Bridge and observed better performance of sliding friction isolators. Saha and Jangid [11] studied different types of sliding isolation systems applied to benchmark cable-stayed bridge. VCFPS and VFPS are more robust than other sliding isolators [12]. Though FPS has been relatively independent of the ground motion frequency contents, the highest excitation intensity has a powerful impact on the design of FPS. Generally, FPS may be ineffective other than the designed particular intensity of excitation. PFPI [13] which has variable curvature of the sliding surface may eliminate drawbacks

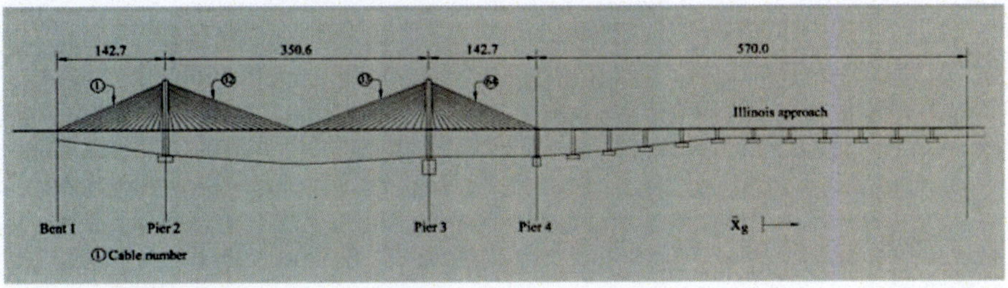

Fig. 1 The benchmark cable-stayed bridge [1]

regarding FPS system. The success of PFPI in other structures leads us to study its performance for phase I benchmark cable-stayed bridge. Saha et al. [14] investigated the performance of PFPI on benchmark highway bridge. The deck displacement of bridge is reduced substantially by using PFPI than conventional FPS. Saha et al. [15] studied the hybrid system using PFPI with FV spring damper applied to benchmark highway bridge significantly reduced the bridge responses.

The objectives of this present research paper study are (i) to examine seismic performance of PFPI for phase I benchmark cable-stayed bridge and (ii) to conduct a parametric study to determine the variational effect of important isolator parameters on the responses of the CSB.

2 Benchmark Cable-Stayed Bridge

The benchmark problem on CSB is based on Bill Emerson Memorial Bridge crossing the Mississippi River near Cape Girardeau, Missouri (Fig. 1). The location of the bridge is under the seismic zone of New Madrid and serves as the primary river crossing. Significant seismic forces have been assumed in this bridge design.

3 Evaluation Criteria

To enable direct comparison between control systems, a set of evaluation criteria was developed in the benchmark problem considering the following earthquake [1]: (i) El Centro, 1940; (ii) Mexico City, September 1985; and (iii) Gebze, Turkey, 1999.

4 The Uncontrolled Responses of Cable-Stayed Bridge

The CSB responses subjected to an earthquake in the longitudinal direction of the deck are (i) stay cable forces, (ii) behavior of towers, and (iii) deck displacement, which should be within the range of 0.2–0.7 times of the failure tension [1]. The two configurations of the bridge assumed for without control analysis are (a) the deck is fixed longitudinally to the piers and (b) the deck is free and supported only by the stays. Restricted displacements of the deck have been shown in configuration (a) but exhibit higher base shear and unacceptable cable tension. In configuration (b) shear and moment are 45.6% and 58.7% of those of configuration (a), respectively, but exhibit huge deck displacement [1].

5 Friction Pendulum System (FPS)

The FPS isolator comprises an articulated slider which moves on a stainless steel spherical surface (Fig. 2). Restoring force for the system has been developed by the movements of the slider over the spherical surface. The natural time period of vibration (T_b) of a rigid mass supported on FPS is dependent on the concave surface radius of curvature (R_c) and is determined from the pendulum equation:

$$T_b = 2\pi\sqrt{\frac{r_c}{g}} \tag{1}$$

where 'g' is denoted as the acceleration due to gravity. The resisting force (f) of the FPS is given by Tsai [5]:

$$f = k_b x_b + F_x \tag{2}$$

where k_b is the stiffness of the isolator; x_b and F_x denote the isolator displacement and the frictional force, respectively.

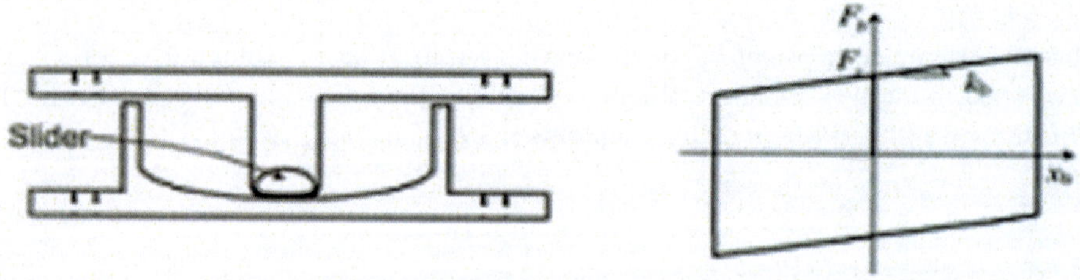

Fig. 2 Schematic and ideal force–deformation diagram of FPS

6 Polynomial Friction Pendulum Isolator (PFPI)

PFPI is like as FPS isolator, but unlike FPS's spherical surface, the geometry of the sliding surface in PFPI isolator is communicated as a polynomial function *y*(*x*). The horizontal displacement of the slider is measured along the x-axis, and the y-axis of the PFPI isolator is symmetrical of the sliding surface. Thus, the PFPI has a polynomial sliding surface with variable curvature [13].

The horizontal shear force of the PFPI isolator is for $y'(x) \ll 1$:

$$U(x) = Wy'(x) + \mu W Z_x \tag{3}$$

where "W is the vertical load, μ is the friction coefficient, and $y'(x)$ is the second-order derivative of the polynomial equation."

Secant stiffness of this isolator:

$$K_r(x) = \frac{u_r(x)}{x} = \frac{Wy'(x)}{x} \tag{4}$$

The instantaneous isolation time period $T(x)$

$$T(x) = 2\pi\sqrt{\frac{x}{gy'(x)}} \tag{5}$$

Fifth-order polynomial equation can be chosen for defining PFPI

$$y'(x) = \frac{u_r(x)}{W} = ax^5 + cx^3 + ex \tag{6}$$

$$y''(x) = \frac{k_r(x)}{W} = 5ax^4 + 3cx^2 + e \tag{7}$$

The mechanical property of PFPI depends on the three polynomial coefficients *a*, *c*, and *e*.

K_1 and K_2 be the isolator secant stiffness at two different isolator displacements D_1 and D_2.

Normalized initial stiffness (K_0) at $x = 0$, i.e., $y''(0) = K_0$.

Isolator secant stiffness (K_1) at $x = D_1$, i.e., $y''(D_1) = K_1$.

Isolator secant stiffness (K_2) at $x = D_2$, i.e., $y''(D_2) = K_2$.

Solution of this simultaneous equation leads to

$$a = \frac{-k_0 + k_1}{-5(D_1^2)} \tag{8}$$

$$c = \frac{2(-k_0 + k_1)}{3(D_1^2)} \tag{9}$$

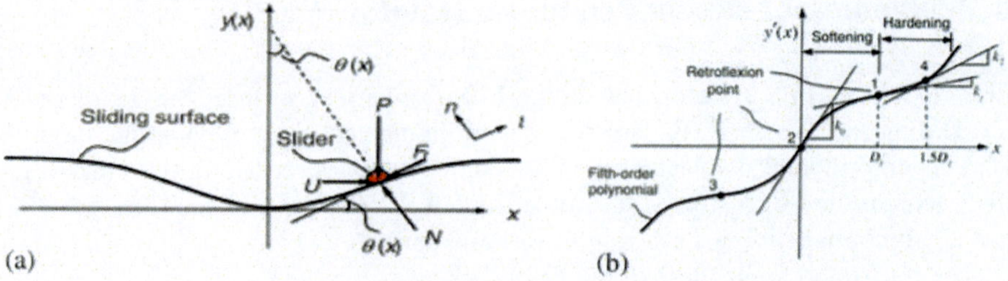

Fig. 3 **a** Force component of the slider of PFPI, **b** Normalized restoring stiffness [13]

$$e = k_0 \tag{10}$$

Convert the polynomial coefficients a, c, and e into the K_0, K_1, and D_1. Isolator stiffness,

$$K_j = \frac{4\pi^2}{T_j^2} \quad (\text{for } \; j = 0, 1, 2) \tag{11}$$

The three parameters to be determined in the PFPI design are k_0, k_1, and D_1 (Fig. 3).

7 Governing Equations of Motion

In the structural system, the dynamic equation exposed to seismic loads when there is a solitary part in the excitation or when the excitation is consistently connected at all backings of the structure

$$M\ddot{U} + C\dot{U} + KU = -M\Gamma\ddot{x}g + \wedge f \tag{12}$$

"For multiple supports",

$$M\ddot{U} + C\dot{U} + KU = \wedge f - MR_s + M_g\ddot{U}_g - CR_s + C_g\dot{U}_g \tag{13}$$

"where $\ddot{U}$ (m/s^2) is the second time derivative of the displacement response vector U (m); M, C, and K are the mass, damping, and stiffness matrices of the structure; f (N) is the vector of control force inputs; $\ddot{x}g$ (m/s^2) is the ground acceleration; Γ is a vector of zeros and ones, relating the ground acceleration to the bridge degrees of freedom (DOF); and $\wedge$ is a vector relating the force(s) produced by the control device(s) to the bridge DOFs [1]". C_g, M_g, and K_g matrices are the damping, mass matrices, and stiffness, respectively. "$R_s = -K^{-1}$ kg is the pseudo-static influence vector which describes the influence of support displacements on the structural displacements [1]".

8 Numerical Study

A number of numerical simulations have been carried out in MATLAB (1997) and SIMULINK (1997) for determining the evaluation criteria as has been mentioned in the CSB benchmark problem. The natural time period of the bridge in the configuration (a) is 3.45 s and that of configuration (b) is 6.18 s. Time history analysis has been executed for the three specified earthquake ground motions [1] to acquire the seismic responses of the CSB. The isolators are placed in the various locations of the CSB as mentioned in the benchmark problem.

The parameters influencing the PFPI isolator performance are "(i) normalized initial stiffness (k_0), (ii) normalized stiffness at retroflection point 1 (k_1), (iii) isolator displacement at retroflection point 1 (D_1), and (iv) coefficient of friction (μ)".

For getting the optimal value of D_1, a parametric study has been executed out considering $k_0 = 7.00$ N/m, $k_1 = 0.00$ and $\mu = 0.05$; the peak responses obtained from the CSB corresponding to the different values of first displacement (D_1) are shown in Fig. 4.

It is examined from Fig. 4 that variation of D_1 of PFPI has a significant influence on peak responses of benchmark CSB. On the peak base shear response (J_1), peak shear at deck (J_2), peak base moment (J_3), peak moments at deck (J_4), and peak deck displacement (J_6) are almost independent of first displacement (D_1) of the PFPI for El Centro and Mexico earthquakes. On the other hand, J_1, J_2, J_3, and J_4 are decreasing significantly with the increase in the D_1 of the PFPI for Gebze earthquake. But J_6 is increasing significantly with the increase in D_1. Therefore, D_1 equal to 0.25 m has been chosen which holds good for all the considered earthquakes. As the Gebze earthquake is a near-field earthquake with very high displacement spectra, values of specified responses increase with increase in D_1.

The variation in peak responses of the CSB for various values of normalized initial stiffness (k_0) and normalized stiffness at retroflection point 1 (k_1) of the PFPI isolator considering D_1 and coefficient of friction (μ) are equal to 0.25 m and 0.05, respectively, have been performed. k_0 and k_1 of the PFPI isolator have a considerable influence on the peak responses of the CSB, and it depends on the type of ground motion as observed from Figs. 5 and 6. Therefore, the parameters with normalized initial stiffness (k_0) = 21.5 N/m and normalized stiffness at reflection point 1 (k_1) = 0 have been chosen as an optimum for all earthquakes, i.e., El Centro (1940), Mexico (1985), and Gebze (1999).

To determine the norm seismic response and variation of peak of the benchmark bridge, a study has been executed by varying coefficient of friction (μ) within a range of 0.05–0.08 keeping $k_0 = 21.5$ N/m, $k_1 = 0$, and $D_1 = 0.25$ m. It can be observed from Fig. 7 that with an increment in frictional coefficient for El Centro earthquake, peak shear at the deck, peak base shear, peak base moment, peak moments at the deck,

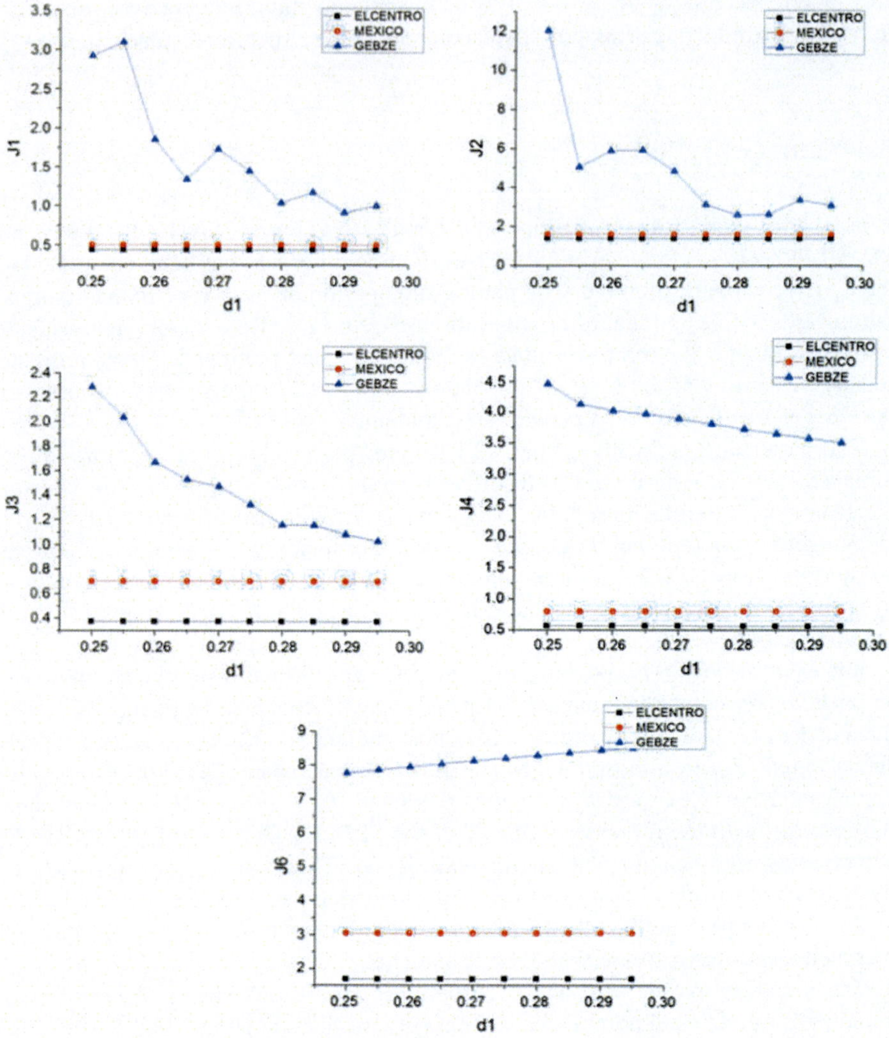

Fig. 4 Effect of D_1 of the PFPI on J_1, J_2, J_3, J_4, and J_6

and peak deck displacement are nearly constant. On the other hand, increasing the friction coefficient for Mexico and Gebze earthquake, peak base shear, peak shear at the deck, peak base moment, peak moments at the deck, and peak deck displacement are decreasing. The study shows that the higher the value of the frictional coefficient (μ), the more beneficial it is for the structure but for the current study, frictional coefficient (μ) has been taken as 0.05, as is recommended for conventional FPS (Fig. 8).

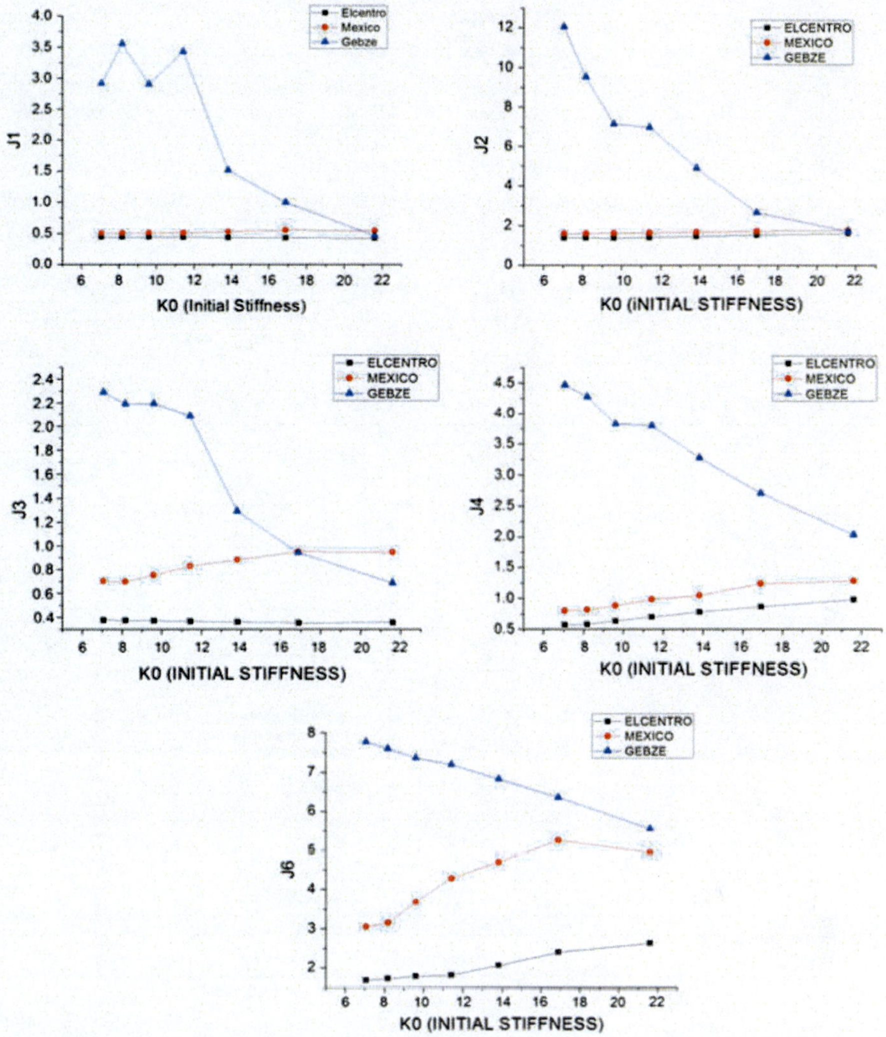

Fig. 5 Effect of k_0 of the PFPI on J_1, J_2, J_3, J_4, and J_6

Time history analysis of the CSB has been performed considering the chosen values of FPS and PFPI, and evaluation criteria are tabulated in Table 1. From the table, it can be observed that PFPI reduces the maximum base shear in case of Gebzee earthquake. The maximum deck shear for El Centro earthquake has been reduced which is advantageous for the normal design of bridge. On the use of PFPI, the maximum base moment for Mexico city earthquake seems to have been reduced significantly which shows a good superiority of PFPI. The maximum deck moment

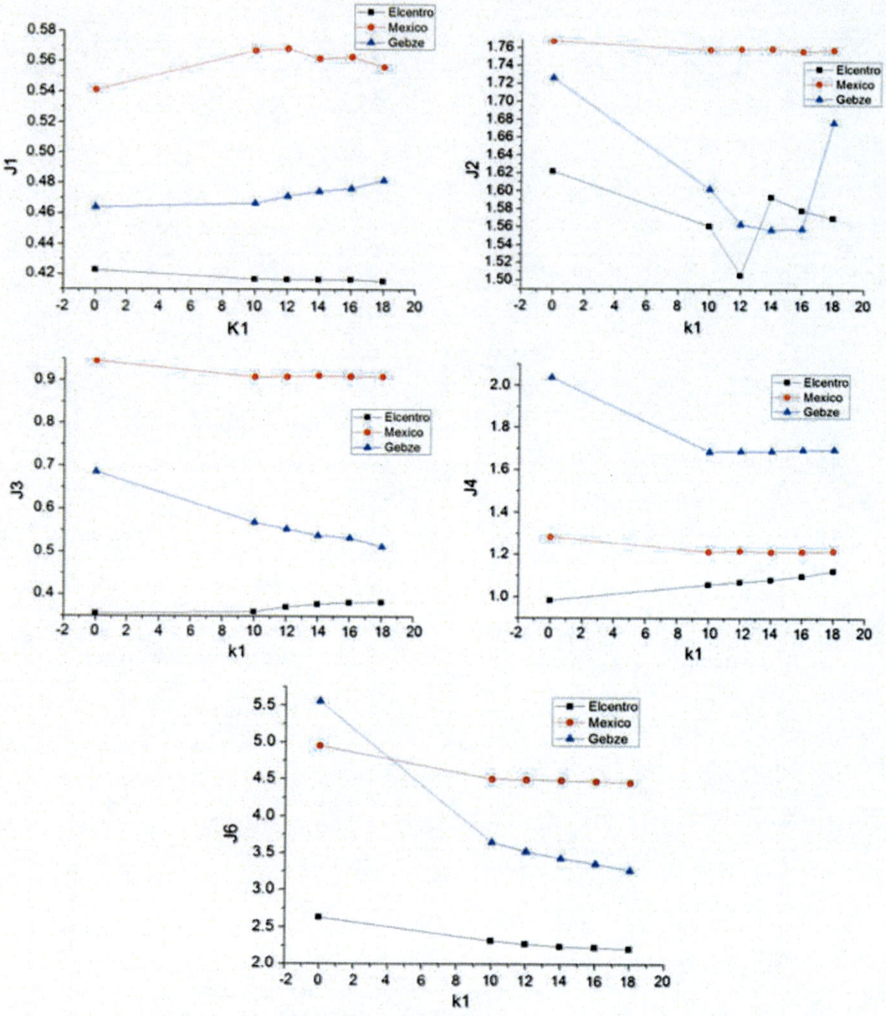

Fig. 6 Effect of k_1 of the PFPI on J_1, J_2, J_3, J_4, and J_6

and maximum cable deviation show a considerable reduction for El Centro and Mexico city earthquakes when using PFPI as an isolator. By clinical examination of Table 1, it is found that PFPI significantly reduced the displacement of deck for Gebze earthquake which is a major concern of the FPS system. The polynomial curved surface of the PFPI makes it possible.

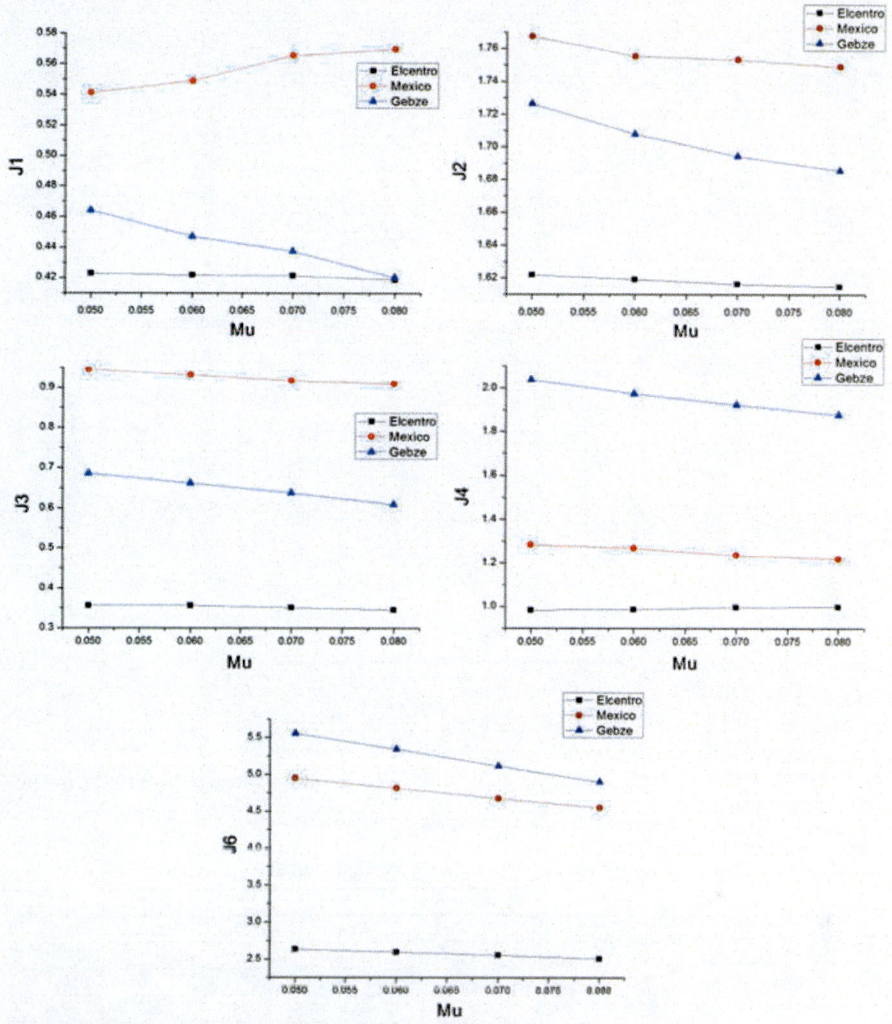

Fig. 7 Effect of μ of the PFPI on J_1, J_2, J_3, J_4, and J_6

9 Conclusions

In the present study, an endeavor has been made to exhibit benchmark CSB using polynomial friction pendulum isolator (PFPI). Phase 1 benchmark CSB is considered for this investigation. The execution of the CSB under FPS and PFPI is considered using the hysteretic model, and the outcomes are classified as evaluation criteria referenced in the benchmark CSB to determine the better one. Totally, 24 number of isolators are put in this evaluation model. Parametric investigations have been done by shifting the significant parameters of each isolator to determine the optimal value.

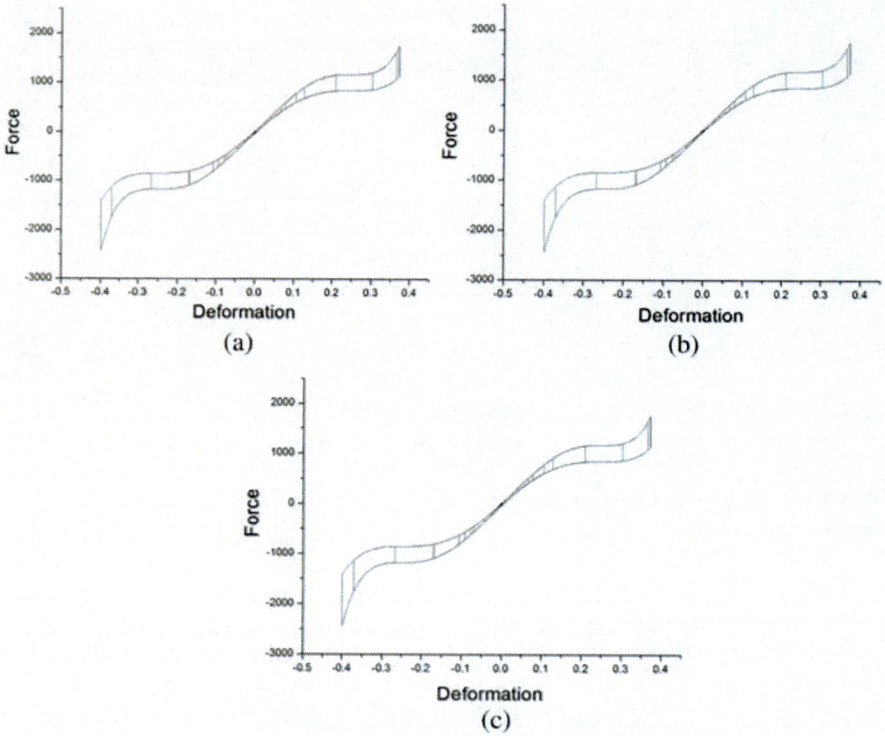

Fig. 8 Force–deformation behavior of PFPI for **a** El Centro (1940), **b** Mexico City (1985), **c** Gebze (1999) earthquakes

Table 1 Maximum evaluation criteria for El Centro, Mexico, and Gebzee earthquakes

		FPS			PFPI		
		Elcentro	Mexico	Gebze	Elcentro	Mexico	Gebze
j_1	Max Base Shear	0.44649	0.54214	0.379	0.42268	0.54122	0.46398
j_2	Max Deck Shear	1.3724	1.8164	2.2646	1.6219	1.7674	1.7265
j_3	Max Base Moment	0.3812	0.87586	0.84252	0.35628	0.9453	0.6856
j_4	Max Deck Moment	0.88393	1.1429	3.6218	0.98384	1.2837	2.038
j_5	Max Cable Deviation	0.25482	0.085254	0.31797	0.30484	0.14394	0.28607
j_6	Max Deck Displacement	3.1981	6.4359	10.538	2.6333	4.9524	5.5542
j_7	Norm Base Shear	0.26138	0.42896	0.42388	0.28142	0.53377	0.55539
j_8	Norm Deck Shear	1.7834	1.7901	4.2105	1.5938	1.6745	2.2827
j_9	Norm Base Moment	0.53377	0.80714	1.595	0.40245	0.77354	1.0298
j_{10}	Norm Deck Moment	2.2264	2.6173	7.5295	1.2838	2.2371	3.6131
j_{11}	Norm Cable Deviation	0.03751	0.013962	0.044568	0.05019	0.024456	0.047981
j_{12}	Max Control Force	0.00010	6.82E-05	0.00020159	0.0023082	0.0017248	0.0054532
j_{13}	Max Device Stroke	2.0998	3.2411	5.7776	1.729	2.494	3.1175
j_{14}	Norm Ductility	0.00019	0.00022	0.0012613	0.0060293	0.00583	0.019875
j_{15}	Total Power	3.10E-05	3.00E-05	0.00011895	0.00095696	0.0007766	0.0018743
j_{16}	Number Device	24	24	24	24	24	24

Based upon the exploration carried out on the CSB for the seismic response control, the conclusions deduced are as follows:

i. Regardless of the structure being flexible, introducing PFPI decreases the critical seismic reaction of the benchmark CSB.
ii. The decrease in the base shear reaction of the towers is accomplished around 42–68% for every one of the seismic ground motions and the types of the indicated isolator.
iii. The decrease of seismic reactions relies upon the types of ground motion and kinds of isolator.
iv. PFPI significantly reduced the deck displacement in the case of Gebze earthquake which has been a major concern or the FPS system. The polynomial curved surface of the PFPI makes it possible.
v. Contrasting the estimations of assessment criteria displayed, it tends to be concluded that PFPI is increasingly vigorous and the performances of this isolator are superior to customary FPS.

References

1. Dyke SJ, Caicedo JM, Turan G, Bergman LA, Hague S (2003) Phase I benchmark control problem for seismic response of cable-stayed bridges. J Struct Eng ASCE 129:857–872
2. Mostaghel N, Tanbakuchi J (1983) Response of sliding structures to earthquake support motion. Earthquake Eng Struct Dynam 11:729–748
3. Mostaghel N, Khodaverdian M (1987) Dynamics of resilient-friction base isolator (R-FBI). Earthquake Eng Struct Dynam 15:379–390
4. Tsopelas P, Constantinou C, Okamoto S, Fujji S, Ozaki D (1996) Experimental study of bridge seismic sliding isolation system. Eng Struct 18:301–310
5. Tsai CS (1997) Finite element formulations for friction-pendulum seismic isolation bearings. Int J Numer Meth Eng 40:29–49
6. Wang Y-P, Chung L-L, Liao W-H (1998) Seismic response analysis of bridges isolated with friction pendulum bearings. Earthquake Eng Struct Dynam 27:1069–1093
7. Jangid RS (2005) Computational numerical models for seismic response of structures isolated by sliding systems. Struct Control Health Monitor 12:117–137
8. Soneji BB, Jangid RS (2006) Effectiveness of seismic isolation for the cable-stayed bridge. Int J Struct Stab Dyn 6:77–96
9. Saha P, Jangid RS (2008) Comparative performance of isolation systems for benchmark cable-stayed bridge. Int J Appl Sci Eng 6:111–139
10. Saha P, Jangid RS (2009) Seismic control of benchmark cable-stayed bridge using passive hybrid systems'. IES J Part A: Civil Struct Eng 2:1–16
11. Saha P, Jangid RS (2016) Seismic control of benchmark cable-stayed bridge using variable sliding isolators. J Struct Eng 43:69–90
12. Saha P (2015) Seismic control of benchmark cable-stayed bridge using variable friction pendulum isolator. Adv Struct Eng 1271–1282
13. Lu LY, Lee TY, Juang SY, Yeh SW (2013) Polynomial friction pendulum isolators (PFPIs) for building floor isolation: an experimental and theoretical study. Eng Struct 56:970–982
14. Saha A, Saha P, Patro SK (2017) Polynomial friction pendulum isolators (PFPIs) for seismic performance control of benchmark highway bridge. Earthquake Eng Eng Vib 16(4):827–840

15. Saha A, Saha P, Patro SK (2018) Seismic protection of the benchmark highway bridge with passive hybrid control system. Earthquakes Struct 15(3):227–241

Limit State Design and Factor of Safety: An Overview

B. Jena

Abstract An attempt is made to look into various aspects of limit state design of steelworks in buildings and the relevance of Indian Standards (IS: 800-1984 [14]) to such a design. Provisions of British Code (BS: 5950-1985 [7]) are kept in view while reviewing the other international standards. Various design provisions in IS: 800-84 are examined and discussed. The anatomy of global factor of safety consisting of partial safety factors imposed on characteristic loads and characteristic strengths are critically examined with statistical backgrounds (Borges and Gastanheta 1971 [5]) and compared with various international recommendations. The so-called characteristic imposed and wind loads stipulated in Indian Standards (IS: 875-1987 [15]) have been compared with those proposed in British Code (CP-3-1972/BS: 6399-1984 [10]) and the American Code (ANSI: A58.1-1982 [3]). The observed discrepancies in the loading standards, affecting the overall factor of safety of the structure, have been highlighted.

1 Introduction

The aim of structural design is to produce a safe and economical structure fulfilling its required purpose. Theoretical knowledge of structural analysis must be combined with the knowledge of design principles to give a safe design. Over the years, various methods of analysis and design have been developed to ensure the following:

(1) Under worse loading the structure is safe.
(2) Under normal working conditions, the deformation and cracking are within limits and the structure is durable.

Despite difficulties in assessing the precise loading and the variations in strength of material, these conditions are to be met.

There are basically three design methods to achieve safe and workable structure:

(1) Elastic design based on permissible stress method,

B. Jena (✉)
School of Civil Engineering, KIIT Deemed to be University, Bhubaneswar, India
e-mail: bjenafce@kiit.ac.in

B. B. Das et al. (eds.), *Recent Developments in Sustainable Infrastructure*, Lecture Notes in Civil Engineering 75, https://doi.org/10.1007/978-981-15-4577-1_16

(2) Plastic design/load factor method, and
(3) Limit state designs.

The elastic design, based on permissible stress method, is the traditional method where a factor of safety is imposed on the yield stress of steel. The structures are analyzed by elastic theory, and the sections are sized so that the permissible stresses are not exceeded. The method has proved to be uneconomical.

The plastic theory developed takes into account the behavior of the material beyond the yield point and then finding the load that causes the structure to collapse. Then, the working load is obtained by dividing the collapse load by a suitable load factor.

Finally, the Limit State Design has been developed to take into account all the conditions that make the structure unfit for use. It takes into account the variability of loads and material strength. It imposes partial safety factors on loads and material strength. The magnitudes of these factors are varied both for ultimate and working loads in order to achieve maximum economy. For concrete structures, the Limit State Design has been adopted by various standards [1, 8, 13]. Similar revisions have also been made for the design of steel structures in British Standards [7] as well as in American Standards [2]. The revision of IS: 800-84 incorporating Limit State Design is therefore overdue. Some of the salient features of Limit State Design in BS code are discussed below.

2 Provisions of Limit State Design in BS: 5950-85

The BS: 449-69 has now been revised to BS: 5950-85 incorporating the provisions of Limit State Design. The limit states considered are (i) ultimate limit states and (ii) serviceability limit states.

The ultimate limit states includes (a) yielding, rupture, buckling, and transformation into mechanism, (b) stability against overturning and sway, and (c) fracture and fatigue.

Serviceability limit states include mainly deflection under working loads and the durability of the structure.

Partial safety factors "γ_m" and "γ_f" are imposed on characteristic strength and characteristic loads, respectively, to get the design strength and the design loads as below.

Design Strength = Characteristic Strength/γ_m and Design Load = γ_{f*} Characteristic Load, where γ_m is taken as 1.0 for steel and $\gamma_f = 1.4$ for dead load and 1.6 for imposed loads. When wind load is included, $\gamma_f = 1.2$.

The characteristic strength for steel is the minimum yield strength, and the characteristic loads are the nominal working loads specified in CP: 3-72.

Perry–Robertson's formula [6, 16] has been extended for the design of compression members and members under bending. Under combined bending and axial loads, linear interaction relationships have been adopted. Sectional capacity is determined using plastic modulus as well as elastic modulus depending on the type and class of sections [7].

For rigid frames, elastic analysis based on sub-frames has been adopted and the effective length factors are based on the work of Professor Wood [18].

Plastic designs based on mechanism conditions have also been recommended for rigid frames provided (i) the frame is effectively braced against side sway, (i) the load is predominantly static, and (iii) the stress–strain diagram has a plateau at least six times the yield strain.

Under service conditions, deflections are checked due to unfactored loads.

3 Provisions in IS: 800-84

The design provisions are based on traditional elastic permissible stress method with a stress factor varying from 1.5 to 1.67. It is very much in line with BS: 449-69. Design provisions for columns under axial compression in IS: 800-62 adopted secant formula [11, 17] and when revised in 1984, it changed over to Rankine–Merchant formula [12] without adequate explanation. Rankine–Merchant formula has also been extended for the design of members under bending compression. Linear interaction relationships have been adopted for design of members under combined loading.

For rigid frames, elastic analysis based on limited frame method has been adopted. The effective length factors are those proposed by Wood [18].

Plastic method of analysis and design is proposed in line with Beedle [4]. A load factor 1.7 has been proposed for dead and imposed loads. The working loads are those given in IS: 875-64/87. These loads are in line with BS code CP-3-72 (BS: 6399-84).

Both permissible stress and the plastic design methods are inadequate to predict the true factor of safety of a structure. These design methods are now being replaced by Limit State Design, which is based on partial safety factors imposed on characteristic strength and characteristic loads.

The characteristic values and the partial safety factors are discussed below against their statistical backgrounds [5].

4 Characteristic Strength

The strengths of material upon which the design is based are those strengths below which results are unlikely to fall. The strength distribution is assumed to be approximately normal as in Fig. 1.

The characteristic strength is then defined as that value below which it is unlikely more than 5% of the results will fall:

$$R_{ch} = (m_r - 1.64 s_r) = m_r(1 - 1.64 v_r),$$

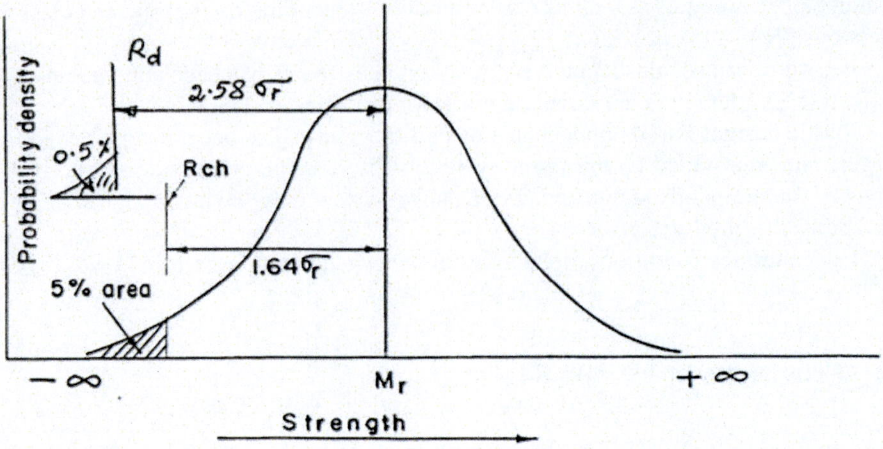

Fig. 1 Normal distribution of strength with critical values

where v_r is the coefficient of variation of resistance.

The above relationship reflects the method and control of manufacture, quality of constituents, and the nature of the material.

5 Partial Factor of Safety for Material 'γ_m'

If the design strength of the material is defined as that strength corresponding to 0.5% fractile [5] as shown in Fig. 1, then γ_m = characteristic strength/design strength = $R_{0.05}/R_{0.005} = (1 - 1.64v_r)/(1 - 2.58v_r)$. Considering v_r for concrete 0.20 and v_r for steel 0.10, we get $\gamma_{mc} = 1.39$ as against 1.50 in IS: 456-2000 and 1.40 in BS: 8110-85 and $\gamma_{ms} = 1.13$ as against 1.15 in IS: 456-2000 and BS: 8110-85, BS: 5950-85 takes $\gamma_{ms} = 1.0$.

6 Characteristic Loads

If the load distribution is assumed to be approximately normal as in Fig. 2, the characteristic load is given by

$$F_{ch} = (m_f + 1.64s_f) = m_f(1 + 1.64v_f),$$

where v_f is the coefficient of variation of load. It means that the accepted probability of 5% of the loads being larger than F_{ch}.

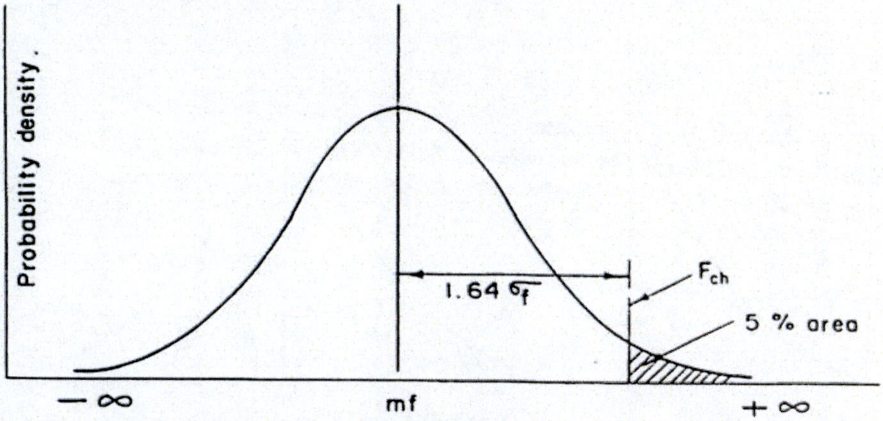

Fig. 2 Normal distribution of load with its characteristic value

7 Partial Factor of Safety for Loads 'γ_f'

The distributions of load and the resistance are as shown in Fig. 3a. Failure takes place when the load is greater than the resistance, i.e., $F > R$ or when $D = (R - F)$ is negative. Now $m_d = (m_r - m_f)$, $\sigma_d = \sqrt{(\sigma_r^2 + \sigma_f^2)}$ and $v_d = \sigma_d/m_d$. Failure will take place when $(m_d - \beta\sigma_d) = 0$ as shown in Fig. 3b. The load enhancement factor is then given by γ_f = minimum resistance/characteristic load $= m_r(1 - 2.58v_r)/m_f(1 + 1.64v_f) = \mu(1 - 2.58v_r)/(1 + 1.64v_f)$. For probability of failure of 10^{-5} and for $v_r = 0.15$, $v_f = 0.20$, we get $\beta = 4.27$, $\mu = 3.25$, and $\gamma_f = 1.50$.

The value of $\gamma_f = 1.50$ has gone into the IS: 475-2000 for dead and imposed loads for the design of concrete structures.

However, after a detailed review of the literature [5], it is observed that dead load is deterministic in character and can well be represented by a normal distribution curve, whereas live load and wind load can be better represented by Extreme Type-I. Assuming the probability of failure of 10^{-5} and the coefficient of variation of resistance 15%, the design factors of safety for dead, live, and wind loads are indicated in Table 1 for various coefficients of variations of loads.

Further, when wind load is in combination with dead and live load, a probability factor 0.75 is multiplied to get the combination effect. From Table 1, the following combinations result.

(i) Dead + Live
$$U = 1.53D_k + 1.64L_k$$
(ii) Dead + Wind
$$U = 0.9D_k + 0.75 * 1.64L_k$$
(iii) Dead + Live + Wind
$$U = 0.75(1.53D_k + 1.64L_k + 1.64W_k),$$

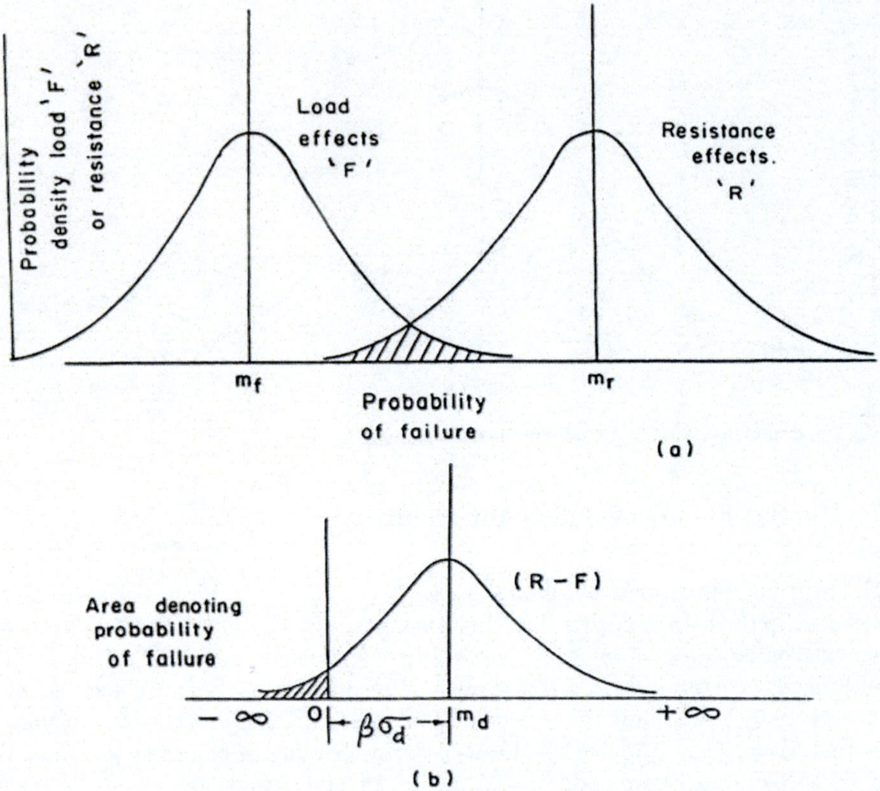

Fig. 3 Distribution of load and resistance at probability of failure

Table 1 Design factors of safety

Loading	Distribution of loads	v_f (%)	Distribution of resistance	v_r (%)	Load factor γ_f
Dead load	Normal	10	Normal	15	1.53
Live load	Extreme Type 1	20	Normal	15	1.64
Wind load	Extreme Type 1	20	Normal	15	1.64

where D_k, L_k, and W_k refer to characteristic loads defined as that load above which is not more than 5% of results fall. Not much information is available on characteristic loads as defined above. In the present limit state designs, specified loads are those design loads used in the working stress method and designated as service loads, and these loads are quite different from the characteristic loads defined above. As a matter of comparison, the load factors adopted by various national and international codes are compared in Table 2.

Table 2 Comparison of design load factors for ultimate limit state

(A) *Dead and live loads*	
(1) U = 1.4D + 1.7L	ACI: 318-77, AISC
(2) U = 1.4D + 1.6L	BS: 8110-85, BS: 5950–85
(3) U = 1.35D + 1.5L	CEB-FIP-78
(4) U = 1.5D + 1.8L	AS: 1480-74, AS: 1280-75
(5) U = 1.5D + 1.5L	IS: 456-2000
(B) *Dead and wind loads*	
(1) U = 0.90D + 1.3W	ACI: 318-77, AISC
(2) U = 0.90D + 1.4W	BS: 8110-85, BS: 5950-85
(3) U = 1.0D + 1.5W	CEB-FIP-78
(4) U = 0.90D + 1.5W	AS: 1480-78, AS: 1280-75
(5) U = 0.90D + 1.5W	IS: 456-2000
(C) *Dead, live and wind loads*	
(1) U = 1.0D + 1.3L + 1.3L	ACI: 318-77, AISC
(2) U = 1.2(D + L + W)	BS: 8110-85, BS: 5950-85
(3) U = 1.2(D + L + W)	CEB-FIP-78
(4) U = 1.25D + 1.25L + 1.5W	AS: 1480-74, AS: 1250-75
(5) U = 1.2(D + L + W)	IS: 456-2000

8 Loadings

8.1 Live Load

A detailed review of loading standards for live loads specified for various buildings in ANSI-72 and CP-3 is shown in Table 3. It is observed that except items like 17, 2a, and 13a, the recommendations of ANSI are largely on the higher side. The recommendations of IS: 875 are very much in line with CP-3.

8.2 Wind Load

As per American Code [3], the design wind pressure is given by

$$P = C_p q_f = 0.00256\, C_p \left(K G_f \right) V_{30}^2,$$

where P = total pressure in lb/sqft, q_f = effective velocity pressure, C_p = pressure coefficient,

K_z = velocity pressure coefficient depending on the exposure and height h,

G_f = gust factor depending on the type of exposure and dynamic characteristics of the structure, and

V_{30} = basic wind speed in miles/hour (map speed).

On the other hand, the British code [10] suggests that pressure exerted on any part of the structure is given by

Table 3 Comparison of live loads on buildings

Use		ANSI: 72 A 58.1 (kN/m^2)	CP: 3-Chp.V, (kN/m^2)	ANSI/CP-3
1	Armories and drill rooms and halls	7.2	5.0	1.44
2	Assembly halls or places of assembly, public halls, theater, cinemas			
	(a) Fixed sittings (b) Without fixed sitting	2.9 4.8	4.0 5.0	0.73 0.96
3	Balconies	2.9	1.5	1.93
4	Corridors, hallways, passageways, aisles, public place, footbridge between buildings	4.8	4.0	1.20
5	Dance halls and ballrooms	4.8	5.0	0.96
6	Dining rooms and restaurants	4.8	2.0	2.40
7	Garages (light vehicle)	2.4	2.5	0.96
8	Gymnasiums (main floor, balconies)	4.8	5.0	0.96
9	Hospitals			
	(a) Wards, private rooms, bedrooms (b) Operating rooms, laboratories	1.9 2.9	2.0 2.0	0.95 1.45
10	Hotels and motels			
	(a) Guest rooms, bedrooms (b) Public rooms, vestibules	1.9 4.8	2.0 5.0	0.95 0.96
11	Kitchen	–	3.0	–
12	Libraries			
	(a) Reading room (b) Stacking room	2.9 7.2	2.5 6.5	1.16 1.11
13	Manufacturing units, factories, etc.			
	(a) Light (b) Heavy	6.0 12.0	7.5 10.0	0.80 1.20
14	Office buildings			

(continued)

Table 3 (continued)

Use		ANSI: 72 A 58.1 (kN/m²)	CP: 3-Chp.V, (kN/m²)	ANSI/CP-3
	(a) Offices (b) Filing and storage space (c) Offices with computing and data processing	2.4 6.0 –	2.5 5.0 3.5	0.96 1.20 –
15	Residential: private apartments, boarding houses, guest houses, hostels, lodgings houses, etc.	1.9	1.5	1.27
16	Reviewing stand (grandstand)	4.8	5.0	0.96
17	Schools: classrooms	1.9	3.0	0.64
18	Stairs and exit ways	4.8	3.0	1.60
19	Storage warehouses			
	(a) Light (b) Heavy	6.0 12.0	5.0 9.6	1.20 1.25
20	Stores: (departmental store)			
	(a) First floor (b) Upper floor	4.8 3.6	4.0 –	1.20 –
21	Stage floor	7.2	5.0	1.44

$$F = C_p q = K C_p (S_1 S_2 S_3 V_{10})^2,$$

where $K = 0.613$ in SI units (N/mm^2 and m/s) $= 0.0612$ in metric units (Kgf/m^2 and m/s) $= 0.00256$ in imperial units (lb/ft^2 and miles/hour),

C_p = pressure coefficient, q = dynamic pressure, S_1 = topography factor, S_2 = factor takes into account the ground roughness, and S_3 = statistical constant and is taken as unit for a probability level of 0.63 and period of exposure of 50 years.

The effective velocity pressure q_f given by ANSI and the dynamic pressure q given by CP-3 for an exposure "open country with no obstruction" are compared in Fig. 4 for various wind speeds and different heights of structure. It is observed that the predictions of ANSI are almost 1.75 times of those proposed in CP: 3/IS: 875.

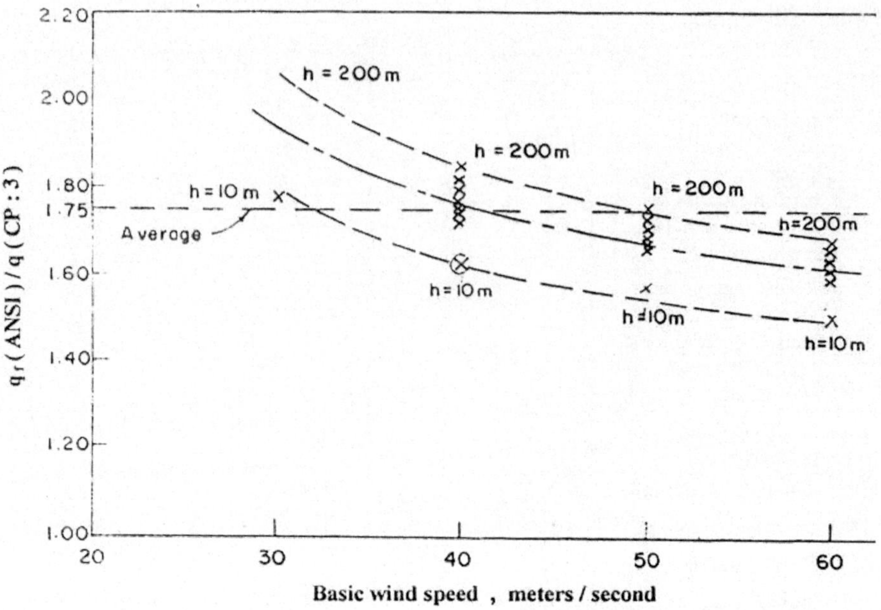

Fig. 4 Comparison of wind loads specified in ANSI and CP:3

9 Conclusion

(i) The revision of IS: 800-84, incorporating the Limit State Design, is overdue. Any change or modification of the existing design tools should be well commented while implementing the Limit State Design.

(ii) The dead load being more deterministic in character compared to imposed loads, the load factors should be different for dead and imposed loads. A single load factor 1.50 imposed on dead and imposed loads as in IS: 456-2000 and a load factor 1.70 imposed in IS: 800-84 for plastic analysis and design are inconsistent with the nature of variability of the dead and imposed loads.

(iii) The nominal imposed and wind loads specified in British code, CP-3 and American code, and ANSI–82, being used as characteristic loads, differ widely, thereby significantly altering the global factor of safety of structure. Clearly, more research is necessary for the areas of characteristic loads to justify any Limit State Design.

References

1. ACI-318 (1995) Building code requirements for structural concrete. American Concrete Institute, MI
2. AISC (1970) Manual of steel construction. American Institute of Steel Construction, NY
3. ANSI A58.1 (1982) Minimum design of loads for buildings and other structure. American National Standard Institute Inc., NY
4. Beedle LS (1958) Plastic design of steel frames. Wiley, NY
5. Borges JF, Gastanheta M (1971) Structural safety. Laboratorio Nacional De Engenharia Civil, Lisbon
6. BS: 449 (1969) Specification for the use of structural steel in building. British Standard Institution, London
7. BS: 5950 (1985) Structural use of steel work in building. British Standard Institution, London
8. BS: 8110 (1985) Structural use of concrete. British Standard Institution, London
9. CEB-FIP (1978) Model code for concrete structure, 3rd ed. C Comite Eura International du Beton
10. CP: 3 (1972) Basic data for the design of buildings. British Standard Institution, London
11. Galambos TV (1968) Structural members and frames. Printice Hall, New Jersey
12. Horne MR, Merchant W (1965) The stability of frames. Pergamon Press, Oxford
13. IS: 456 (2000) Indian standard code of practice for plain and reinforced concrete. Bureau of Indian Standards, New Delhi
14. IS: 800 (1984) Indian standard code of practice for general construction in steel. Bureau of Indian Standards, New Delhi
15. IS: 875 (1987) Indian standard code of practice for design loads for buildings and structures. Bureau of Indian Standards, New Delhi
16. MacGinley TJ, Ang TC (1988) Structural steel work-design to limit state theory. Butterworth, London
17. Timoshenko SP, Gere JM (1961) Theory of elastic stability. Mc Graw-Hill, NY
18. Wood RH (1974) Effective lengths of columns in multi-storey buildings. Struct Eng 52, July, August, September, London

Yield Behaviour of Three Edge Simply Supported Two-Way Slab Under Concentrated Loading

Sushree Sangeeta Panda, Subham Ghosh, and Bhagabat Jena

Abstract Failure of slab takes place forming a specific mechanism with flat facets between linear plastic hinges at which relative rotation occurs, called yield-line ring. The present study has made an attempt to study the behaviour of two-way RCC slab of size 1 m × 1 m of thickness 0.1 m at failure, simply supported on three sides and fourth side being free, under a centrally applied concentrated load. In total three-slab specimens having compressive strength of concrete, percentage of reinforcement and grade of reinforcement being kept constant are tested as an explorative test. Formation of yield lines is observed and collapse load has been determined. Johansen's yield-line theory has been adopted to predict the upper-bound collapse load of the slabs. The theoretical prediction is compared with the experimental values. Certain discrepancy between theoretical and experimental values is observed. The load deflection relationship has been observed till the collapse of the slab.

Keywords Johansen's yield line · Two-way slab · Tensile membrane action · Compressive membrane action · Upper bound

1 Introduction

A reinforced concrete slab is a common structural element of modern buildings. In general, we come across two types of slabs. One-way slab with length-to-breadth ratio is greater than or equal to two, whereas in two-way slabs it is less than two. Various design techniques as well as analytical methods have been adopted to analyse the slab and its behaviour under different boundary conditions and loading. Yield-line method is one such technique in which the ultimate load-carrying capacity of

S. S. Panda (✉) · S. Ghosh · B. Jena
School of Civil Engineering, KIIT Deemed to be University, Bhubaneswar, Odisha, India
e-mail: sushree.pandafce@kiit.ac.in

S. Ghosh
e-mail: subhamghoshcese@gmail.com

B. Jena
e-mail: bjenafce@kiit.ac.in

B. B. Das et al. (eds.), *Recent Developments in Sustainable Infrastructure*, Lecture Notes in Civil Engineering 75, https://doi.org/10.1007/978-981-15-4577-1_17

slab is determined based on applying limit state method using rigid-plastic analysis provided with different boundary conditions, shape, size and reinforcement ratio. The method provides an upper-bound solution of collapse load of the slabs. The method was first proposed by Ingerslev [1] and later developed by Johansen [2]. According to Johansen's Yield Line Theory [2], under the action of load, the slab undergoes some small deflection, and further increase in the load leads to failure of the slab under large deflection. The slab attains a specific pattern due to the formation of discrete plastic hinges called yield line. When the slab reaches its plastic limit it causes flat facets due to the formation of linear plastic hinges called yield lines and relative-rotation takes place-at-the yielding lines.

2 A Brief Literature Review

Burgess [3] provided a new analytical approach to tensile membrane action of reinforced concrete slabs at large deflections. Gong et al. [4] observed that equally reinforced slabs fail forming a fan-like mechanism satisfying yield-line theory under loads. In case of unequally reinforced slab called orthotropic, it has the probability to fail under negative yield-line pattern. According to Ouyang and Suaris [5], in conventional yield-line theory by analytically approximating the strain softening of the concrete and by plotting the load deflection curve, beyond the peak load can be predicted. Beyond the peak, the load decreases suddenly and if the reinforcement provided is ductile and has a good bond with the concrete then the load will increase again due to the formation of tensile membrane action in the slab section. Thavalingam et al. [6] found that by using rigid-plastic theory, the load-carrying capacity of the slab can be estimated by yield-line method. Thakkar and Pandey [7] in their work focused on failure analysis of an aluminium alloy plate based on progressive failure that is similar to yield-line patterns observed in concrete slabs. Zhang and Dong [8] conducted tests on specimens at larger displacements having boundary condition of single edge-clamed and another single edge-supported simply. Moment-carrying capacity of the slabs was found to be enhanced by 26.6% by implementing yield-line theory until 1/15th of the span length reached the maximum vertical displacement. According to Park [9], in slabs the reinforcement acts as a plastic membrane and the developed tensile membrane action in the reinforcement zone of the slabs can achieve ultimate flexural strength. Lim et al. [10] mentioned that during progressive collapse analysis, tensile membrane action has very less influence on the boundary condition, whereas the catenary action plays a vital role in the boundary condition. Muspratt [11] the paper represents the design of slabs using elastic design methods that do not conclude the yielding of the reinforcement complete graphical design using distribution of bending moment contours using lower bond theorem and later on yield-line theory was used to determine the ultimate moment-carrying capacity.

The present work is an exploratory experimental work taken to realise yield-line pattern developed for two-way square slab simply supported on three sides, free on fourth side and subjected to concentrated loading at its centroid. From literature, for such loading type and support condition, theoretically presumed yield pattern as per Johansen's yield line [2] has yet not experimentally been observed. Experimental formation of yield lines is observed and collapse load has been determined. Johansen's yield-line theory [2] has been adopted to predict the upper-bound collapse load of the slabs. The theoretical prediction is compared with the experimental values.

3 Experimental Investigation

Three two-way RCC slabs of dimensions 1 m × 1 m and depth of 100 mm were prepared having constant concrete compressive strength, percentage and grade of reinforcement. The bottom and side clear cover provided for reinforcement was 15 mm. Reinforcement of 8 mm diameter bars of grade Fe 500 were provided at a spacing of 200 mm c/c both way perpendicular to each other. The slab specimens confirm to (IS-456, 2000) design as under reinforcement. The slabs were made torsionally restrained at the corners over a distance Span/5 = 200 mm by providing corner reinforcement. Figure 1 shows the reinforcement cage prepared for casting the slab. All test slabs were subjected to same manner of moist curing and testing after casting. Testing was done by applying concentrated load at the centroid of the slab and noting the corresponding deflection induced till failure. The test slabs were simply supported on three edges as shown in Fig. 2.

Fig. 1 Reinforcement cage for slab

Fig. 2 Test setup frame

4 Result and Discussion

The compressive strength of concrete used in casting the slabs is found to an average of 33.7 MPa (Table 1). The applied load till failure and its corresponding deflection at the centre of the test specimens are given in (Table 2) and its plotting is given in Fig. 3. Figure 3 shows more or less linear response till the yield of reinforcement and beyond yielding, it takes a non-linear path. Ultimate failure loads taken by the specimens are 60, 57 and 60 kN, the average being 59 kN whereas the corresponding deflections are found to be 16.02, 12.92 and 14.5 mm, the average being 14.48 mm.

The crack pattern at failure of the slabs is shown in Fig. 4. It is observed that the cracks is initiated from the centre and travels towards the centre of the free edge. Further with increase in load, gradually the crack travels towards the corners opposite to the free edge forming flat facets at failure. The nature of yield lines formed experimentally matches more or less to Fig. 5 as the theoretical rigid-plastic Johansen's yield line [2], however with some discrepancy that is obtained in the experiment. The discrepancy may be attributed to experimental error. For Fig. 5, for a square slab, the theoretical collapse load is given by Pillai and Menon [12].

Table 1 Concrete compressive strength, f_{cu}, of test specimen

Specimen no.	f_{cu} (MPa)	Average f_{cu} (MPa)
1	32.3	33.7
2	34.47	
3	34.2	

Table 2 Test result of slab specimen

Sl no.	1	2	3	4	5	6	7	8	9	10	11
Slab 1											
Jack load (kN)	0	7	16	24	30	35	40	45	50	55	60
Deflection (mm)	0	0.75	1.31	1.99	2.44	3.06	5.77	7.92	9.52	12.22	16.02
Slcb 2											
Jack load (kN)	0	9	16	25	30	34	41	45	50	57	Nil
Deflection (mm)	0	0.55	0.95	1.55	2.26	3.01	4.76	5.91	8.11	12.92	Nil
Slab 3											
Jack load (kN)	0	9	19	26	35	43	49	55	60	65	Nil
Deflection (mm)	0	0.5	0.75	1.1	3.4	5.1	7.35	9.5	14.5	Nil	Nil

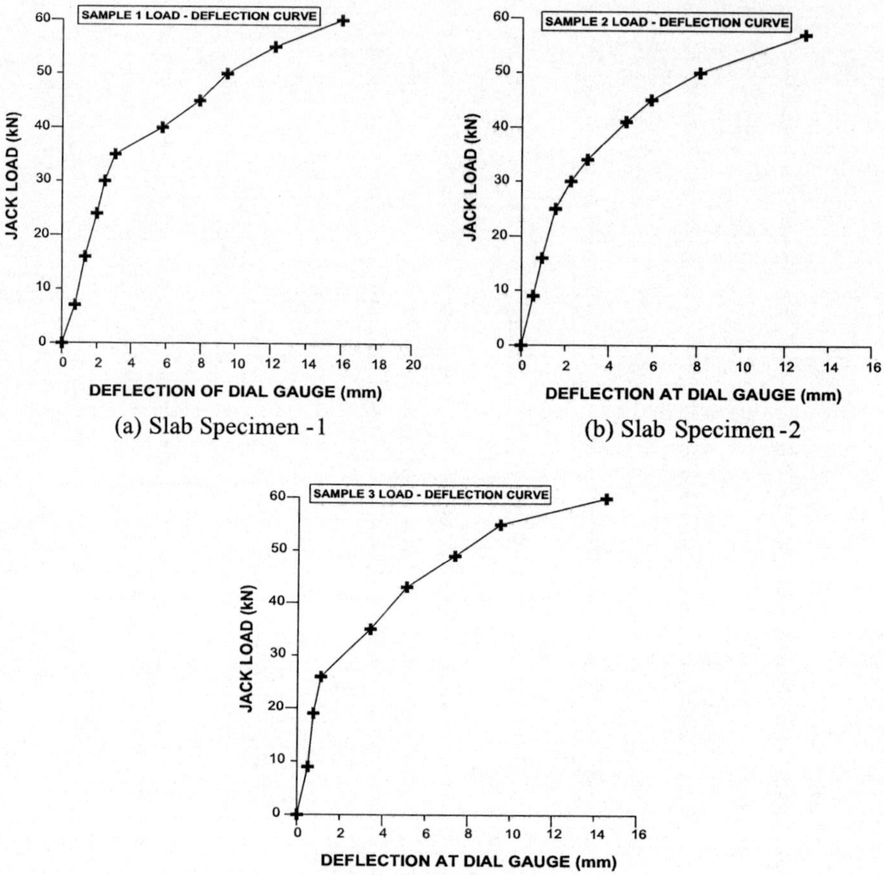

(a) Slab Specimen -1 (b) Slab Specimen -2 (c) Slab Specimen -3

Fig. 3 Load versus centroidal deflection curve of test slabs (**a**, **b**, **c**)

$$w_u = 14.14 \times m/l^2 \tag{1}$$

where w_u = collapse load per unit area,

l = size of side of square slab = 1 m for the test slab,

m = ultimate moment per unit length for isotropic slab is calculated for the test specimen (area of steel, $A_{st} = 452.16$ mm^2, Fe 500 grade reinforcement being 8 mm diameter at 200 mm c/c, effective depth 75 mm, $f_{cu} = 33.7$ MPa) as 4.777 kN m per m length. Then, from Eq. (1), w_u is found to be 67 kN/m^2. The experimental collapse concentrated load has been obtained as an average 59 kN, which can be approximately assumed to be 59 kN/m^2 which is a little less than theoretical value of 67 kN/m^2 from Eq. (1) satisfying it as an theoretical upper-bound collapse load.

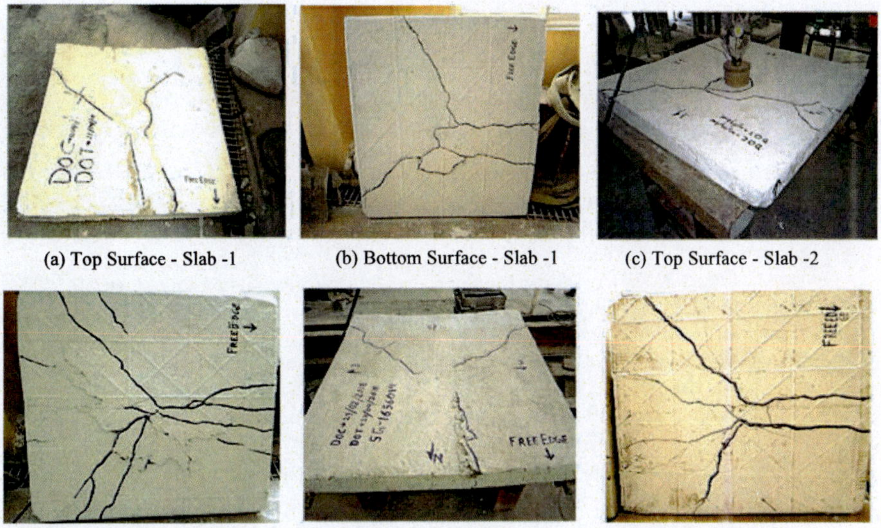

(a) Top Surface - Slab -1 (b) Bottom Surface - Slab -1 (c) Top Surface - Slab -2

(d) Bottom Surface - Slab -2 (e) Top Surface - Slab -3 (f) Bottom Surface - Slab -3

Fig. 4 Failure crack pattern of test slabs (**a**, **c**, **e**—top surface; **b**, **d**, **f**—bottom surface)

Fig. 5 Theoretical yield lines of square slab simply supported on three sides

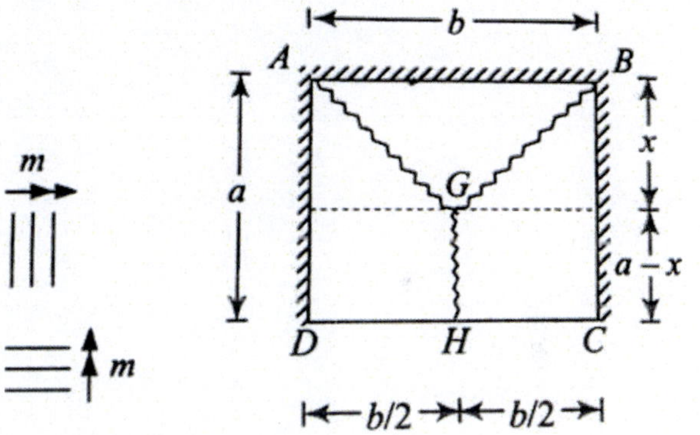

5 Conclusions

From the present exploratory experimental work carried out on two-way square slab simply supported on its three sides, it may be concluded as

- Load deflection is somewhat linear till yield point and non-linear beyond.
- The ultimate collapse load for the test slabs is found to be an average 59 kN whereas theoretical yield-line collapse load is 67 kN approximately satisfying upper-bound solution
- Experimental failure pattern more or less matches with theoretical positive yield-line pattern.

References

1. Ingerslev A (1923) The strength of rectangular slabs. J Inst Struct Engrs 1:3–14
2. Johansen KW (1943) Brudlinieteorier. Copenhagen; [English translation "Yield-line theory". Cement and Concrete Association, London, 1962]
3. Burgess I (2017) Yield-line plasticity and tensile membrane action in lightly-reinforced rectangular concrete slabs. Elsevier, p 195
4. Gong J, Zhang Y, Han S (2011) Ultimate bearing capacity of reinforced concrete slab carrying concentrated load. J Eng Mech 877–886
5. Ouyang C, Suaris W (2015) RC rectangular slabs with edge restraints. J Eng Mech 2146–2165
6. Thavalingam AJ, McKeown JJ, Sloan D (1998) A computerised method for rigid-plastic yield-line analysis of slabs. Comput Struct 601–612
7. Thakkar BK, Pandey PC (2006) Damage-coupled progressive failure and yield line analyses of metal plates. J Eng Mech 487–497
8. Zhang DS, Dong YL (2011) Experimental behavior of one-way concrete slabs at large displacements. Appl Mech Mater 1035–1039
9. Park (2016) Ultimate strength of rectangular concrete slabs under short-term uniform loading with edges restrained against lateral movement. ICE Publishing, pp 125–150
10. Lim P, et al (2017) Investigations of tensile membrane action in beam-slab systems under progressive collapse subject to different loading configurations and boundary conditions. Elsevier, pp 520–536
11. Muspratt MA (1978) Elastic analysis of slabs. Build Environ 13 (1):51–59
12. Pillai SU, Menon D Reinforced concrete design, 3rd ed. McGraw-Hill

Torsional Buckling Analysis of a Bar Member

Lovely Sabat and Chinmay Kumar Kundu

Abstract During buckling of column, it is assumed that the column would buckle as the cross section bends in the plane of symmetry. But in some problems of buckling failures of column, it would be either due to twisting or due to combined effect of bending and twisting. Such a combined effect of bending and twisting in a structure is known as torsional buckling. In the present work, a thin-walled bar of cross section $(b \times t)$ with the length '*l*' is studied by applying uniform axial compression. The differential equation for the deflection curve and the differential equation for torsional buckling are presented. The expressions for total moment, torque and torque per unit length are derived and finally the expressions for the critical stresses and critical load for torsional buckling failure are derived. A numerical example is solved. The critical stress and critical load are calculated.

Keywords Buckling · Torsional buckling · Critical stress · Critical load

1 Introduction

In general cases of buckling of columns, it is assumed that the column would buckle as the cross section bends in the plane of symmetry. But in some problems of buckling failures in a column it would be either due to twisting or due to combined effect of bending and twisting. Such a combined effect of bending and twisting on a structure is known as torsional buckling [1, 2]. Single symmetric cross-section members with or unsymmetric cross-section members may experience combined twisting and translation movement about the shear centre. Such type of failure occurs if the section has very low torsional rigidity, like in case of a bar member of thin-walled open section (e.g. angle sections, tee-shaped members, etc. made of thin-walled members). These types of members can undergo torsional buckling before flexural buckling. In case of columns with central loading, there are three different buckling loads out of which

L. Sabat (✉) · C. K. Kundu
School of Civil Engineering, KIIT Deemed to be University, Bhubaneswar, India
e-mail: lovelysabt2000@gmail.com

C. K. Kundu
e-mail: chinmay.kundufce@kiit.ac.in

B. B. Das et al. (eds.), *Recent Developments in Sustainable Infrastructure*, Lecture Notes in Civil Engineering 75, https://doi.org/10.1007/978-981-15-4577-1_18

one load will correspond to combined torsional + flexural mode in sections with two symmetric axes [3].

Flexural buckling load about the weak axis always has the lowest value, and thus we usually ignore the torsion buckling load in case of double symmetric sections. In unsymmetrical sections, buckling will always occur in torsional flexural mode irrespective of the shape and dimensions. But non-symmetric sections are rarely used [4].

2 Formulation of Torsional Buckling

Considering a bar subjected to uniform axial compression, due to which torsional buckling may occur. The idea of torsional buckling is the axis of the bar remaining straight and each flange buckles by rotating about z-axis.

To find the compressive force due to which torsional buckling occurs, we have to consider the deflection of the flange during buckling. This whole analysis can be explained by considering a usual example of buckling of a strut with pin ends on both sides and it is subjected to centrally applied force '*P*'.

Initially, the strut was straight, and then let us assume that the '*P*' force reaches its maximum value 'P_{cr}' so that the strut can have a slightly deflected form of equilibrium.

The bending stresses will develop due to the deflection and it is added with the initial uniformly distributed compressive force and the initial compressive stresses will act on slightly rotated cross section (Figs. 1, 2 and 3).

Taking a section '$m - n$' on the deflected shape of the strut with the length of the section '*dz*'. Considering the deflected shape to be straight, the differential equation for the deflection curve will be from the differential equation for bending of beam columns,

$$EI\frac{d4y}{dx4} + P\frac{d2y}{dx2} = q \tag{1}$$

where

EI = flexural rigidity,

q = lateral load.

In this case, $q = 0$, so Eq. (1) will be

$$EI\frac{d4y}{dx4} + P\frac{d2y}{dx2} = 0. \tag{2}$$

The deflection $y = v$ and the deflection is about z-axis, so Eq. (2) will be

$$EI_Z\frac{d4v}{dz4} = -P\frac{d2v}{dz2} \tag{3}$$

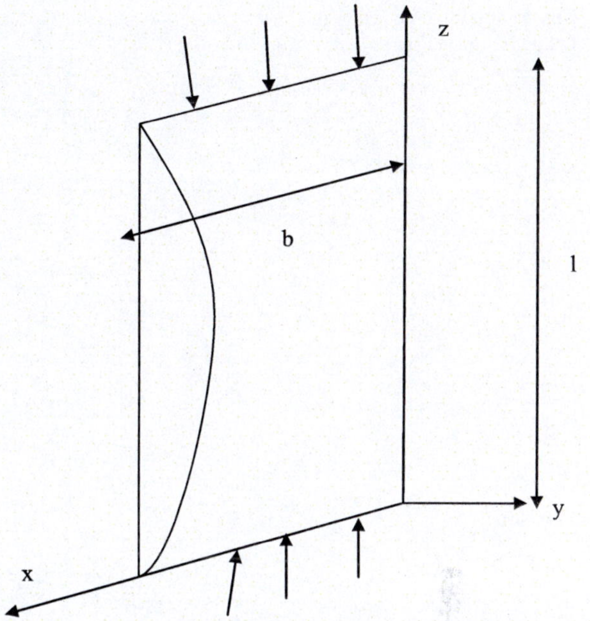

Fig. 1 Torsional buckling of flanges

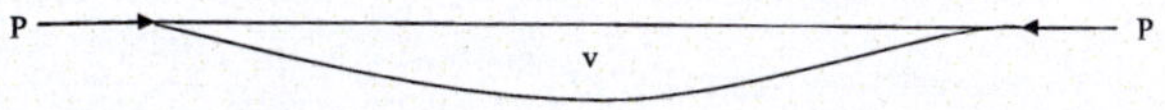

Fig. 2 Deflection of the flange

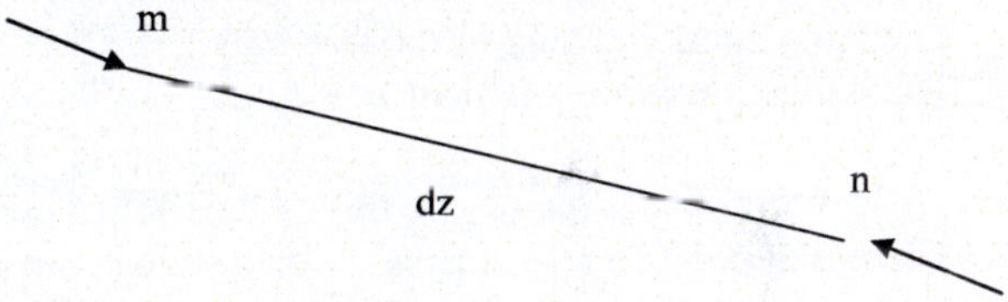

Fig. 3 A section $m - n$ from the deflected shape

Now, to find the bending stresses and the deflection curve, we assume that the strut is loaded by an imaginary lateral load, i.e. due to the compressive force, a lateral load is generated which results in the deflection of the flange. The intensity of this imaginary lateral load is

$$q = -P\frac{d2v}{dz2} \tag{4}$$

Considering the torsional buckling of the bar, at the critical loading condition, the buckled form of equilibrium is retained past the compressive stresses which acts on the cross section of the longitudinal fibres (Fig. 4).

Taking an element $m - n$ as a thin strip,

The thickness of the strip $= t$,

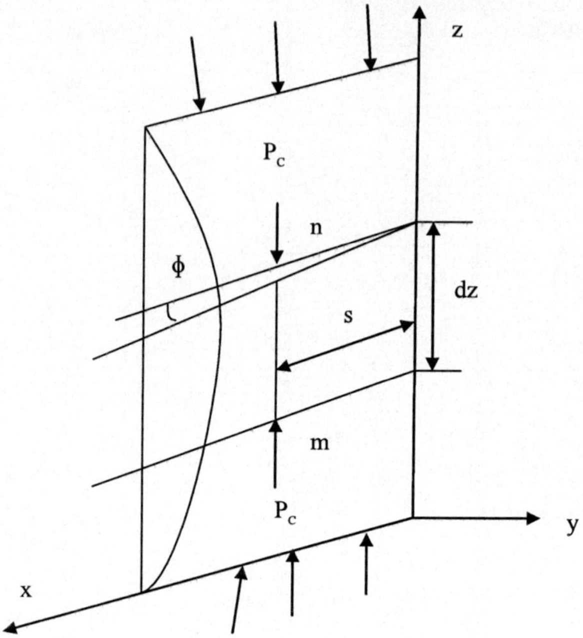

Fig. 4 Torsional buckling of the bar with section $m - n$

The length of the strip = dz,
The small twist angle of the cross section = Φ,
The distance between the strip and the z-axis = s^2,
The cross-sectional area of the thin strip $(A) = t \times ds$,
The deflection in the element in y-direction due to torsional buckling $(v) = s \times \Phi$,
The compressive force (P_c) acting at the rotated ends $m - n = \sigma \times tds$,
The initial compressive stresses at the rotated ends of $m - n = \sigma = \frac{Pc}{A}$,
Now, $P_c = P$, so substituting the value of P_c in Eq. (4), we get

$$q = -(\sigma tds)\frac{d2v}{dz2} \tag{5}$$

$$q = -(\sigma tsds)\frac{d2\Phi}{dz2} \tag{6}$$

Taking moment about z-axis of 'q' acting on the element,

$$M = (qdz \times s) \tag{7}$$

$$M = -(s^2 tds)\frac{d2\Phi}{dz2}dz \tag{8}$$

Now, the torque acting on the element between the two sequential cross section (T) will be

$$T = \int -(\sigma s t d s) \frac{d2\Phi}{dz2} dz \times s \tag{9}$$

$$T = -\sigma \frac{d2\Phi}{dz2} dz \int t s^2 ds \tag{10}$$

Now,

$$\int t s^2 ds = I_0 \tag{11}$$

So,

$$T = -\sigma \frac{d2\Phi}{dz2} dz \times I_0 \tag{12}$$

where I_0 = polar moment of inertia of the cross section about the shear centre which is same as the centroid. The torque/unit length (m_z) will be

$$m_z = -\sigma \frac{d2\Phi}{dz2} \times I_0 \tag{13}$$

This value of m_z can be used for any shape of cross section when the shear centre and centroid coincides.

Now, for the differential equation for torsional buckling, we can use the differential equation for non-uniform torsion of thin-walled open section, i.e.

$$M_t = C \frac{d\Phi}{dz} - C_1 \frac{d3\Phi}{dz3} \tag{14}$$

Differentiating with respect to dz,

$$\frac{dMt}{dz} = C \frac{d2\Phi}{dz2} - C_1 \frac{d4\Phi}{dz4} \tag{15}$$

where Φ = angle of twist,
C = torsional rigidity,
C_1 = warping rigidity,
J = torsion constant and
M_t = torque.
M_t and m_z act along the element of the twisted bar,

$$m_z = \frac{dMt}{dz} \tag{16}$$

So, substituting Eq. (16) in Eq. (15), we get

$$-m_z = C\frac{d2\Phi}{dz2} - C_1\frac{d4\Phi}{dz4} \tag{17}$$

$$m_z = C_1\frac{d4\Phi}{dz4} - C\frac{d2\Phi}{dz2} \tag{18}$$

Now, substituting Eq. (13) in Eq. (18), we get

$$-\sigma\frac{d2\Phi}{dz2} \times I_0 = C_1\frac{d4\Phi}{dz4} - C\frac{d2\Phi}{dz2} \tag{19}$$

$$C_1\frac{d4\Phi}{dz4} - (C - \sigma I_0)\frac{d2\Phi}{dz2} = 0 \tag{20}$$

From Eq. (20), we can calculate the value of

1. The critical value of initial compressive stresses.
2. The value of the critical load (P_{cr}).

For this case, the warping rigidity is zero, i.e. $C_1 = 0$.
So, Eq. (20) reduces to

$$-(C - \sigma I_0)\frac{d2\Phi}{dz2} = 0 \tag{21}$$

$$C - \sigma I_0 = 0 \tag{22}$$

$$\sigma_{cr} = \frac{C}{I_0} \tag{23}$$

where $C = GJ$,
$G =$ shear modulus,
$I_0 = \frac{1}{3}tb^3$ and
$J = \frac{1}{3}bt^3$.
Therefore,

$$\sigma_{cr} = \frac{Gt^2}{b^2} \tag{24}$$

and

$$P_{cr} = \frac{\sigma_{cr}}{A} \tag{25}$$

where $A =$ area of the cross section $= b \times t$.

Equation (25) shows that the initial compressive stresses (σ_{cr}) do not depend on the length of the bar or member because here we have neglected the flange resistance offered because of the bending in the direction perpendicular to the flanges [5].

3 Example

Considering a bar of length 3 m, with thickness of 20 mm and the width of the member as 40 mm. Value of E is 200 GPa, μ is 0.3. The value of yield stress is 250 MPa and the FOS is 2.5. Thus, the critical buckling load and stress are calculated as follows (Fig. 5):

$$\text{Length of the member } (l) = 3\,\text{m}$$
$$\text{Thickness of the member } (t) = 20\,\text{mm}$$
$$\text{Width of member} = 40\,\text{mm}$$
$$E = 200\,\text{GPa}$$
$$\mu = 0.3$$
$$\lambda_y = 250\,\text{Mpa}$$
$$\text{FOS} = 2.5$$

Fig. 5 Bar with dimensions

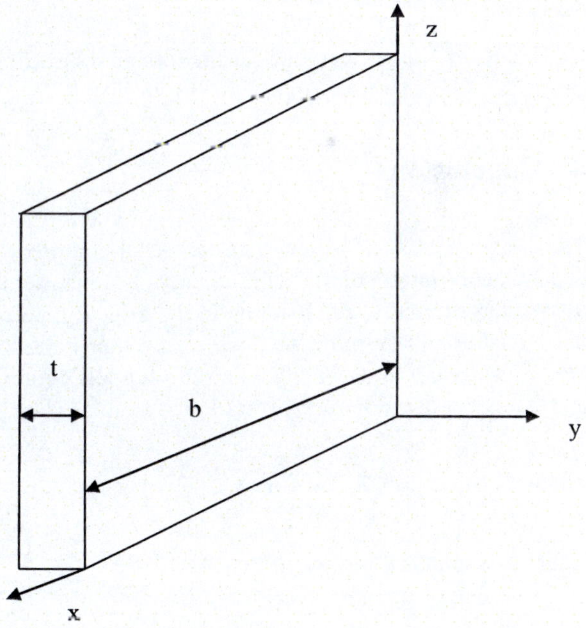

Now,

$$\begin{aligned} G &= \frac{E}{2(1+\mu)} \\ &= \frac{200}{2(1+0.3)} \\ &= 76.92\,\text{GPa} \end{aligned}$$

Then, the initial compressive stress (σ_{cr}),

$$\begin{aligned} \sigma_{cr} &= \frac{Gt^2}{b^2} \\ &= \frac{76.92 \times 10^3 \times 20^2}{40^2} \\ &= 19230\,\text{N/mm}^2 \end{aligned}$$

Then, the critical load (P_{cr}),

$$\begin{aligned} P_{cr} &= \frac{\sigma_{cr}}{A} \\ &= \frac{19230}{40 \times 20} \\ &= 24.037\,\text{N}. \end{aligned}$$

Therefore, the critical load for the torsional buckling to occur (P_{cr}) = 24.037 N.

4 Conclusions

The above expressions show that the initial compressive stress (σ_{cr}) values do not depend on the length of the bar or member because here we have neglected the flange resistances offered to the bending in the direction perpendicular to the flanges. But if we wish to obtain more accurate result, then each flange has to be considered to be uniformly compressed plate and the support conditions being simply supported on three sides and completely fixed on one side. Then, the result obtained will be

$$\sigma_{cr} = \left(0.456 + \frac{b^2}{l^2}\right) \times \frac{\pi^2}{6(1-\mu)} \times \frac{Gt^2}{b^2} \tag{26}$$

where b = width of flange,
l = length of flange,
μ = Poisson's ratio,
t = thickness of flange,

$G =$ shear modulus.

Buckling can occur in a thin-walled open section by the action of bending or by assemblage of bending and twisting and out of the two ways of action, whichever is critical that will depend on the shape and dimensions of the cross section. Hence, we should consider the torsional flexural buckling in designing such members. This can be done by calculating an equivalent slenderness ratio and using the same column strength curve as for flexural buckling.

Bibliography

1. Timoshenko SP, Gere JM (2010) Theory of elastic stability, 2nd ed. McGraw Hill Education, India
2. Vlasov VZ (1961) Thin walled elastic beams, 2nd ed. Israel Program for Scientific Translation
3. Kollar LP, Pluzsik A (2012) Bendingand torsion of composite beams (torsional-warping shear deformation theory). J Reinf Plastic Compos 31(7):441–480
4. Bhavikatti SS (2000) Strength of material. Vikash Publishing House Pvt. Ltd.
5. Gere JM (2004) Mechanics of materials, 6th ed.

Design and Analysis of Transmission Tower Under Wind Loading

Swabarna Roy and Chinmay Kumar Kundu

Abstract Transmission towers play an important role in power sector and in human life. Structural design of lattice transmission tower is a crucial topic among the civil engineers. Various researches have been done on steel tower response to wind loading. Wind loads are randomly applied dynamic loads which are a major concern in case of transmission towers which are at high altitude from the ground level. In this paper, a 33 kV double circuit transmission tower with a height of 21 m has been modeled. The tower is designed having a square base width of 6 m situated in wind zone—IV having a basic wind speed of 50 m/s. The model has been analyzed manually as per the IS code. The design of the steel structure is based on the Indian Standard Code IS 800-2007 under limit state design which is the revised version of the code. The design wind force on each component of the tower has been calculated according to IS 875 (Part 3): 1987. In this paper, the analysis method used for calculation of axial forces is considered to be linear in behavior with two-dimensional approaches and the structure is considered to be determinate by ignoring the horizontal members. The results obtained have been further validated using STAAD-Pro v8i software in order to compute the accuracy of calculation. Temperature load is an important parameter in the design of transmission tower which has been considered. The sag tension for different ambient temperatures has been determined under different sag scenarios of the conductors.

Keywords Transmission towers · Static analysis · Wind load · STAAD-Pro

1 Introduction

In structural engineering aspect, most wind damage to structures occurs due to strong winds which are more prevalent in structures of higher altitude. Wind is not a steady phenomenon due to natural turbulence and gustiness present in it. But when averaged

S. Roy (✉) · C. K. Kundu
School of Civil Engineering, KIIT University, Bhubaneswar, India
e-mail: swabarnaroy0210@gmail.com

C. K. Kundu
e-mail: chinmay.kundufce@kiit.ac.in

B. B. Das et al. (eds.), *Recent Developments in Sustainable Infrastructure*, Lecture Notes in Civil Engineering 75, https://doi.org/10.1007/978-981-15-4577-1_19

over a long time duration produces a static load on the structures. A lattice transmission tower is a steel structure constituting of one or multiple triangular units with straight members having their ends connected at nodes or joints. This tall structure under this time-dependent load will experience dynamic oscillations because of the fluctuating component and gustiness of wind.

Static analysis can be done to determine the failure occurring due to wind load [1]. Wind load analysis for design is done by using Indian Standard Code of Practice for design load (other than earthquake) IS 875 (Part 3): 1987. The analysis and design of structures to resist wind-induced loads are based on different wind zones [2]. Probabilistic analysis has been done by many researchers showing that the wind load is high on the conductors than on the towers [3, 4]. In this paper, the main objective is to study the effect of wind load on transmission tower. For this, a transmission tower of 21 m which includes the ground clearance (h1), maximum sag of the lower most conductors wire (h2), vertical spacing between the conductor wires (h3), and vertical distance of the ground wire from the upper most conductor wire (h4) has been modeled [5, 6]. It has a square base width of 6 m and top width 1.5 m. It is located in wind zone 4 having basic wind speed of 50 m/s. In this paper, static analysis is done to evaluate the forces acting on the member elements. The along wind force acting at each nodal joint along with the horizontal and vertical reactions at the support is found out. The sag tension calculation for conductor at different ambient temperatures has been determined under different sag scenarios of the conductors using the parabolic equation. Now, the tower is modeled using STAAD [7, 8]. The lateral forces which occur because of wind loading acting on each panel joint are obtained. Further, the calculated results were validated with the STAAD-Pro obtained results in this paper.

2 Geometry of Transmission Tower

A transmission tower model of 21 m having a square base dimension 6 × 6 m and top dimension 1.5 × 1.5 m located in open terrain is modeled as shown in Fig. 1. The tower has right panels having column size ISA 130 × 130 × 10 mm, horizontal member having section size ISA 80 × 80 × 6 mm, diagonals, and other secondaries are of size ISA 65 × 65 × 6 mm. Tower members are designed as angle sections.

3 Analysis of Transmission Tower

3.1 Tower Specification

Transmission tower type: 33 kV double circuit

Bracing pattern: Pratt system

Conductor configuration: Vertical configuration

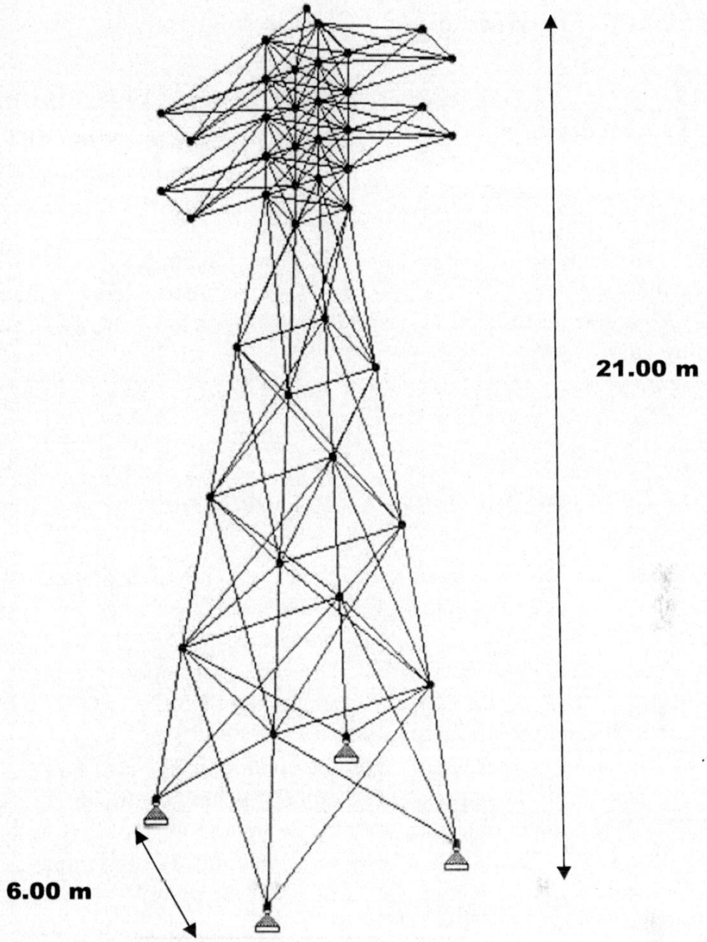

Fig. 1 Model of 21-m-high lattice tower

Cross Arm: Pointed
Conductor material: ACSR
Minimum ground clearance of conductor: 0.7460 m
Ultimate tensile strength of conductor: 6734.0 kg
Factor of safety: 4
Modulus of elasticity final (E_1): 80000.0 kg/m^2
Modulus of elasticity initial (E_2): 0.4675 kg/m^2
Maximum temperature of conductor = 75 °C
Everyday temperature = 32 °C
Minimum temperature = 0 °C.

3.2 Calculation of Wind Load

Wind load analysis on tower has been carried out considering IS code IS 875 (Part 3): 1987. The value of design wind speed has been considered for design wind pressure calculated [8, 9].

Basic wind speed $V_b = 50$ m/s

Wind zone—4, Terrain category—3

Reference wind speed $V_r = V_b/K_0 = 50/1.375 = 36.36$ m/s

Design wind speed $V_d = V_r * K_1 * K_2 = 41.24$ m/s (where K_1 is risk factor having value 1.07, K_2 is terrain roughness coefficient having value 1.06, K_0 is conversion factor having value 1.375)

Design wind pressure $P_d = 0.6 \times V_d^2 = 1021$ N/m^2.

3.3 Wind Load and Sag Tension for Conductor

The sag tension for the conductor is calculated using Indian Standard Codes of Practice for use of structural steel in overhead transmission line tower. As per the specification in the Indian Standard Code of Practice IS. 5613: Part 2: Sec 1. The maximum temperature is 75 °C for ASCR, everyday temperature is 32 °C with design wind pressure of 0, 75, 100%, minimum temperature is 0 °C with design wind pressure of 0 and 36% and design wind pressure of 0 and 36% [10, 11].

For 33 kV, the wind pressure applied on the conductor $F_{wc} = \frac{2}{3} P_d \cdot C_{dc} \cdot D \cdot G_c = 1.25$ kg/m, where P_d is design wind pressure, C_{dc} is drag coefficient for conductor which is 1, D is the diameter of conductor taken as 0.0181 m, and G_c is gust response factor taken here as 2.6. The total wind pressures on conductor are found as follows:

For 100% wind, $W_{100\%} = 1.25 \times 1.00 \times 2.6 = 3.25$ kg/m,

For 36% wind, $W_{36\%} = 0.36 \times 1.25 \times 1.00 \times 2.6 = 1.17$ kg/m, and

For 75% wind, $W_{75\%} = 0.75 \times 1.25 \times 1.00 \times 2.6 = 2.44$ kg/m.

The sag tension is calculated using the parabolic equation as

$$T_2^2\left\{T_2 - \left[T_1 - \frac{W^2 \times q_1^2 \times L^2 \times E_c \times A}{24 \times T_1^2}\right]\right\} = \frac{W^2 \times q_1^2 \times L^2 \times E_c \times A}{24} \tag{1}$$

Dividing the above equation by A and using the notation "f" for stress and $\delta = \frac{w}{A}$, we get

$$f_2^2\left\{f_2 - \left[f_1 - \frac{\delta^2 \times q_1^2 \times L^2 \times E_c \times A}{24 \times f_1^2}\right]\right\} = \frac{\delta^2 \times q_1^2 \times L^2 \times E_c \times A}{24} \tag{2}$$

Now the temperature factor is given by ($t \times \alpha \times E_c$) where $t = t_2 - t_1$. Therefore, the values of temperature factor for different temperatures, i.e., 5 °C, 75 °C, 32 °C, and 50 °C are found to be 38.448, 61.232, 0, and 25.632, respectively [12].

The tension factor is calculated for different conditions with varying wind pressure applied on the conductor [13, 14]. Firstly, considering still condition, i.e., no wind pressure is applied, the tension factor "Z" is given by

$$Z = \frac{\delta^2 \times q_1^2 \times L^2 \times E_c}{24} \quad (3)$$

For this, the value of loading factor (q_1) is 1, we get the value of tension factor as 3568312.33. At 36% wind pressure on conductor (loading factor q_2 is 2.068), it is 15260330.14. At 100% wind pressure on conductor (loading factor q_3 is 5.127), the value of tension factor is 93797118.2. The sag tensions considering different cases are found out and tabulated in Table 1.

Table 1 Sag tension table for different cases

Case no.	Description of cases	Tension at mid span of conductor (kN)	Deflected sag (S_D) (m)	Vertical sag (S_V) (m)
I	Conductor temperature of 5 °C and 100% wind pressure applied	1631.27 kg = 16.31 kN	1.62	1.45
II	Conductor temperature of 5 °C and 36% wind pressure applied	1624.34 kg = 16.24 kN	0.66	0.65
III	Conductor temperature of 75 °C and no wind pressure applied	1587.4 kg = 15.87 kN	0.32	0.00
IV	Conductor temperature of 32 °C and 100% wind pressure applied	1683.49 kg = 16.83 kN	1.57	1.40
V	Conductor temperature of 5 °C and no wind pressure applied	1743.93 kg = 17.43 kN	0.29	0.00

Table 2 Wind load acting at each panel joint

Panel joint	Height (m)	Wind load (kN)
B	6	5.6535
C	12	3.7965
D	14	1.9785
E	16	3.021
F	18	3.3496
G	20	3.3496
H	21	4.3306

3.4 *Wind Load on Tower Panels*

The lateral force due to wind loading on each of the panel joint is determined by the product of the wind intensity and the area exposed to windward direction. Only normal operating condition and no broken condition of the wires are considered. Taking column size ISA 130 × 130 × 10 single angle, horizontal member having section size ISA 80 × 80 × 6 single angle, diagonals, and other secondaries of size ISA 65 × 65 × 6 single angle, the wind load at each panel joint is calculated as shown in Table 2.

3.5 *Stresses in the Members of the Tower*

The stresses in the different members are determined by using approximate method. For this, the horizontal and secondary members of the tower are ignored, hence converting the tower to a determinate plane frame. Horizontal reaction at the base of tower is found to be 12.7465 kN, max., bending moment is calculated as 405.699 kNm, and vertical reaction at the support is found to be 73.7636 kN. The member forces for each vertical and diagonal member due to wind load at each joint are calculated manually and shown in Table 3.

Table 3 Member forces in various members

Members	Forces (kN)
F_{AB}	68.254
F_{BC}	44.6147
F_{CD}	18.104
F_{DE}	54.741
F_{EF}	30.064
F_{FG}	30.064
F_{HG}	3.115

3.6 Deflection

The horizontal and secondary members of the structure have been ignored, thus reducing it to a plane frame. In order to determine the deflection at any point, the externally applied loads are removed and a unit load is applied at the particular joint in the vertical direction. Figure 2 shows the unit load is applied at joint B, for finding the vertical and horizontal deflections. The deflections both horizontal and vertical at different panel joints are shown in Table 4.

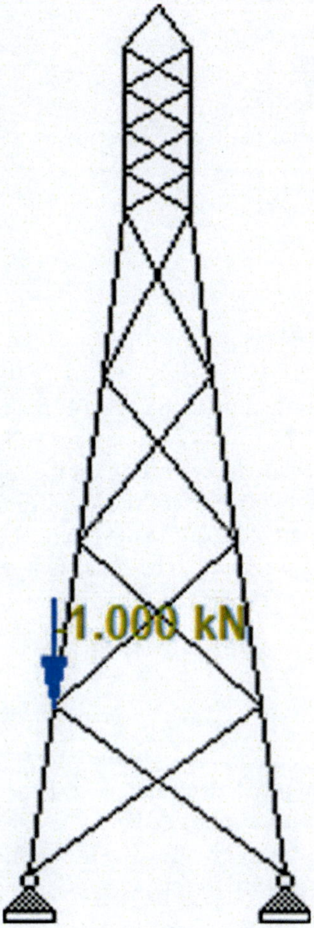

Fig. 2 Unit load is applied at joint B to find the vertical and horizontal deflections

Table 4 Deflections both horizontal and vertical at different panel joints

Panel joint	Maximum deflection	
	Horizontal (mm)	Vertical (mm)
B	0.60	−0.3907
B′	0.60	0.3907
C	1.0508	−0.508
C′	1.0508	0.508
D	2.1016	−1.012
D′	2.1016	1.012
E	2.127	−1.254
E′	2.127	1.254
F	4.1524	−1.464
F′	4.1524	1.464
G	6.1778	−1.686
G′	6.1778	1.686
H	9.6728	0

4 Results and Discussion

The analysis has been performed using STAAD-Pro and the steps for the analysis have been shown in Fig. 3. The results obtained are validated with the STAAD-Pro v8i. The compressive and tensile forces acting on the leg member and the bracings of the tower obtained from the software are shown in Table 5. The stresses in various leg members and bracings members of the tower are shown in Table 6. The results obtained by the STAAD-Pro are found to be consistent with the manually calculated results.

5 Conclusion

Steel transmission tower is highly indeterminant structure and manual calculation of it is very complex and software results needed to be carried out in order to validate the results. The tower is found to be effective in resisting the wind load which is acting directly on the tower. Based on above study, the following conclusions are drawn:

1. The wind loading is predominant in tall towers playing an important role in its analysis and design.
2. The member in the vertical plane predominantly receives maximum number of loads as compared to the horizontal and inclined members.

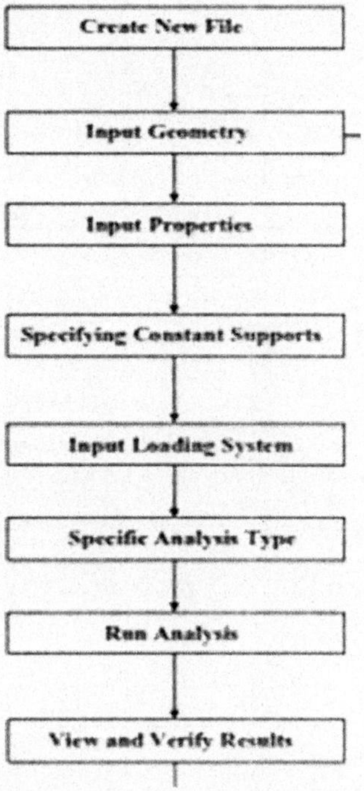

Fig. 3 Steps of analysis using STAAD-Pro

3. The vertical members of the tower remain in tension or carry the tensile force. The horizontal and the diagonal bracings of the lattice steel tower undergo compression under the action of wind load on the tower.
4. Apart from wind load, the temperature load is an important parameter in the design of transmission tower. Sagging of the conductor plays an important role at different temperature variants.

Table 5 Compression and tensile force acting on the tower obtained from STAAD-Pro

Leg member					Bracings				
Beam	L/C	Node	Tension (kN)	Compression (kN)	Beam	L/C	Node	Tension (kN)	Compression (kN)
15	1	17	49.197	0.077	69	1	1, 23	5.064	0.156
14	1	16	43.68	0.087	84	1	2, 17	8.427	0.187
13	1	15	36.819	0.162	83	1	17, 22	2.553	0.124
1	1	4	31.903	0.136	70	1	16, 23	7.048	0.124
31	1	29	20.73	1.194	71	1	16, 21	3.952	0.092
30	1	28	1.864	0.747	82	1	15, 22	4.363	0.092

Table 6 Stresses in the various members

Leg member			Bracings		
Beam	L/C	Axial N/mm^2	Beam	L/C	Axial N/mm^2
15	1	1.904593	69	1	0.2377
14	1	1.40668	84	1	−0.07022
13	1	0.994886	83	1	0.072043
1	1	0.436852	70	1	0.37932
31	1	0.39903	71	1	0.1971
30	1	0.297757	82	1	−0.110226
29	1	0.163288	81	1	0.35821
5	1	0.062018	72	1	0.66559

References

1. Lindley D, Willis DJ (1974) Wind loads on transmission line towers. In: Fifth Australasian conference on hydraulics and fluid mechanics, pp 224–234
2. Panwar S, Kaushik Y, Singh A, Sharma N (2016) Structural analysis and design of steel transmission tower in wind zones II and IV: a comparative study. Int J Eng Technol Manag Appl Sci 4(5):168–177
3. Alam MJ, Santhakumar AR (1994) Probabilistic wind loadings on transmission line structures in India. Eng Struct 16(3):181–189
4. Albermani F, Kitipornchai S (2003) Numerical simulation of structural behaviour of transmission towers. In: Conference proceeding, University of Queensland, pp 1–16
5. Chay MT, Albermani F, Wilson R (2006) Numerical and analytical simulation of downburst wind loads. Eng Struct 28:240–254
6. Chen X, Li C, Xu J (2015) Failure investigation on a coastal wind farm damaged by super typhoon: a forensic engineering study. J Wind Eng Ind Aerodyn 147:132–142
7. Mahesh N, Ranga Rao V (2017) Design and estimation of electric steel tower. Int J Civil Eng Technol 8(1):646–652
8. Code of practice for design loads (other than earthquake) for buildings and structures (IS 875, Part 1-1987, Reaffirmed 1997 Second Revision), 1989. Bureau of Indian Standards, New Delhi
9. Code of practice for design loads (other than earthquake) for buildings and structures (IS 875, Part 3-1987, Reaffirmed 1997 Second Revision), 1989. Bureau of Indian Standards, New Delhi
10. Oliver SE, Moriarty WW, Holmes JD (2000) A risk model for design of transmission line systems against thunderstorm downburst winds. Eng Struct 20:1173–1179
11. Zhang W, Zhu J, Liu H, Niu H (2015) Probabilistic capacity assessment of lattice transmission towers under strong wind. Front Built Environ 1:1–12
12. Preeti C, Jagan Mohan K (2013) Analysis of transmission towers with different configurations. Jordan J Civil Eng 7(4):450–460
13. Sonowal DB, Bharali JD et al. (2015) Analysis and design of 220 kV transmission line tower: a conventional method of analysis and Indian code-based design. J Mech Civil Eng 40–49
14. Hossain SA, Samanta B (2018) Structural optimization and design of electric transmission tower. Int J Innov Res Sci Eng Technol 7(6):7116–7123

Structural Optimization of Microwave Antenna Tower Subjected to Wind Load

Swabarna Roy and Chinmay Kumar Kundu

Abstract The microwave antenna towers are important in the telecom industry as they are used to broadcast microwave transmission from one location to another. These open-latticed steel towers are highly indeterminate structure. In this paper, a 50 m microwave antenna tower mounted with a hollow hemispherical dome of 2 m diameter is considered. It is subjected to a multiple combination of wind, dead loads, and the self-weight of microwave antenna. The tower is designed optimally for least weight. It is having a base width of 6 m and is located in Bhubaneswar City having a basic wind speed of 50 m/s. The various sections chosen are angle, pipe, and tube section with tower having X-type bracing. Weight is considered as an objective parameter which is function of independent variables (base width, number of panel, and panel height ratio). The design wind force on each panel of the tower has been calculated according to IS 875 (Part 3): 1987. It has been seen that with angle section X bracing there is a reduction in weight of about 14% than the weight before optimization. The results show that angle sections are best suited for the design of microwave antenna as compared to pipe and tube sections.

Keywords Microwave antenna tower · Optimization · Wind load

1 Introduction

In every developing country, there is a high requirement of electricity consumption. The transmission towers are being used to fulfill the need of transmitting electric power to various regions. While in communication industry, tall towers are required for transmitting information signals through antennae. These are being used by different agencies such as radio, telecommunication, television, and defense for communicating signals from one place to the other. This tall structure is generally made of standard steel of angle, pipe, or tube section having a square base. These are of

S. Roy (✉) · C. K. Kundu
School of Civil Engineering, KIIT University, Bhubaneswar, India
e-mail: swabarnaroy0210@gmail.com

C. K. Kundu
e-mail: chinmay.kundufce@kiit.ac.in

B. B. Das et al. (eds.), *Recent Developments in Sustainable Infrastructure*, Lecture Notes in Civil Engineering 75, https://doi.org/10.1007/978-981-15-4577-1_20

lattice type consisting of leg, primary, secondary bracings, and carrying a microwave antenna attached to it. Various researchers have contributed on analysis and design of tower using different configurations and bracing patterns. Static and dynamic analysis has been done on transmission tower under different load conditions and K-type bracing compared with X-type bracing which showed lesser deflection and least weight of tower [1]. Analysis is done on transmission line tower and reduction in weight is obtained by angle section of thickness ranging from 5 to 6 mm [2]. Various analytical studies have been done by many researchers on the transmission tower using optimization technique [3]. Also, the X-type bracing pattern with angle section shows maximum weight reduction after optimization [4]. Non-linear structural analysis has been done by many researchers on guyed antenna tower and showed maximum weight reduction [5, 6].

Structural design optimization is a mathematical approach that concerns in finding the maxima and minima function subject to some constraints [7, 8]. This involves various optimization techniques to find the best possible design in terms of weight, reliability, and thus the overall cost. In this paper, a 50 m microwave antenna tower mounted with a hollow hemispherical dome of 2 m diameter is considered. It is subjected to a multiple combination of wind, dead loads, and the self-weight of microwave antenna. The model is created in FEAST software. The tower is designed having a square base width of 6 m and located in Bhubaneswar City which lies in wind zone—IV, having a basic wind speed of 50 m/s. The various sections chosen for the design of the members are angle, pipe, and tube section each having X-type bracing system. The design wind force on each panel of the tower has been calculated according to IS 875 (Part 3): 1987 [9, 10]. The calculated wind loads are assigned at each node from top to bottom of the tower and analysis is done in STAAD-Pro v8i software.

In this paper, the optimum design of the antenna tower subjected to wind load and self-weight is considered as an objective parameter which is function of independent variables (base width, number of panel, and panel height ratio). The tower with angle section and X-type bracing system shows a maximum reduction in weight as compared to pipe and tube sections. Hence, it is designed optimally for least weight.

2 Modeling of Tower

An attempt has been made in analysis and modeling of microwave antenna tower using STAAD-Pro and finite-element-based FEAST software. The modeling of 50 m microwave antenna tower is done as a steel structure. For the analysis, the loads considered here are the dead load of the tower, self-weight of a hollow hemispherical dome of 2 m diameter mounted on top, and the wind load. Three section properties considered for design are angle section, pipe section, and tube section. A comparative study has been made to find the most effective section which gives the least weight. The total view of the tower in both 3D and front is shown in Fig. 1.

Fig. 1 View of the microwave antenna, (**a**) 3D view, (**b**) front view

The X-type of bracing system is kept constant and the total weight of the tower is calculated. The bracing pattern of the lower two panels is different than the rest of the panels and is shown in Fig. 2.

3 Load Calculation on Microwave Antenna Tower

Tower of 50 m height and having 20 number of panels is considered for analysis. The tower is having base width of 6 m and top width of 2 m. Calculation of wind load on the tower by static method is done as per IS 875 (Part 3): 1987 [10]. It is located in Bhubaneswar City having a basic wind speed of 50 m/s. The design wind

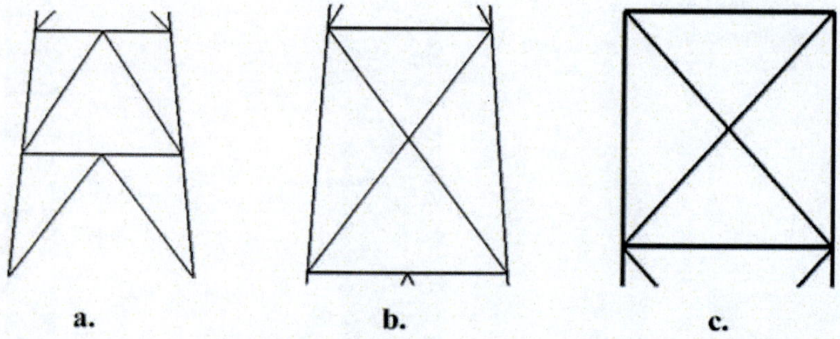

Fig. 2 Configuration of the X bracing at, **a** lower two panel, **b** third panel, **c** topmost panel

speed is obtained from various parameters which include the risk factor coefficient (k1) which is 1.06, terrain, height, and structure size factor (k2) for category 3 and class B is assumed and taken as 1.09, and topography factor (k3) is assumed as 1.2. The design wind pressure at 50 m above is found to be 2.883 kN/m^2. A hollow hemispherical dome of 2 m diameter having a self-weight of 10 kN is considered in the present study. It is assumed to be mounted at a height of 50 m, i.e., at the top panel of the tower. The wind load acting on the hemispherical dish on the front face and rear face is found from bowl wind coefficients (c_f) as per the code. The values of c_f are 1.4 and 0.4 for front face and rear face, respectively. The solidity ratio for different panel, terrain factor, and design wind pressure at various heights of the tower are calculated and shown in Table 1.

The wind forces are computed and analysis is completed by applying the forces at each node [11, 12]. The structure is reduced to a plane frame and the displacement is found in post-processing mode in FEAST. The maximum displacement at the top 50 m level due to wind load is found out to be 53.66 mm. Figure 3 shows the maximum and minimum displacements of the tower after the application of the wind

Table 1 Design wind pressure at various heights of the tower and solidity ratio for different panels

Panel no. from top	Height (m)	Terrain factor coeff. k^2	Design wind pressure Pz (N/m^2)	Solidity ratio	Overall force coeff. c_f	Pz. c_f (N/m^2)
1–5	10	1.075	3332.2	0.245	3.075	10246.5
6–10	20	1.045	3148.8	0.245	3.075	9682.7
11–15	30	1.005	2912.4	0.245	3.075	8955.6
16	34	0.964	2679.6	0.204	3.28	8789
17	38	0.926	2472.5	0.165	3.475	8592.05
18	42	0.88	2232.9	0.165	3.475	7759.6
19	46	0.832	1996.03	0.134	3.63	7245.6
20	50	0.808	1882.5	0.101	3.795	7144.2

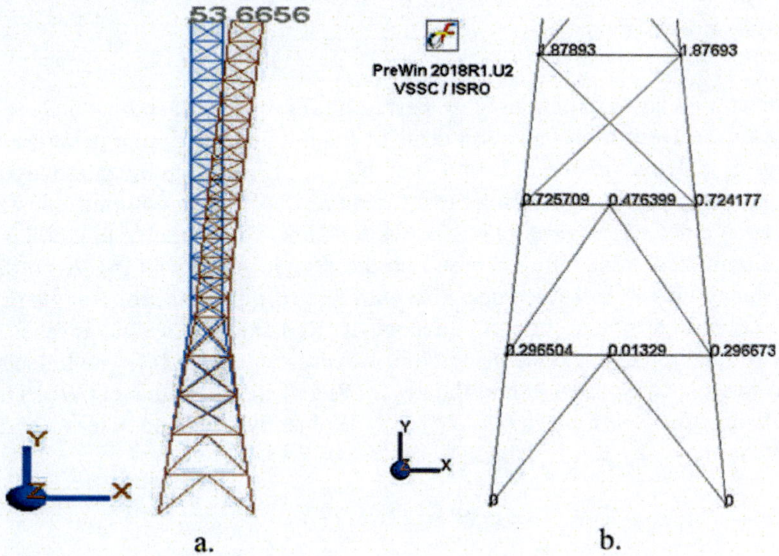

Fig. 3 Displacement of the node of the tower, **(a)** maximum, **(b)** minimum

load. The wind load acting on the front face of the microwave antenna is found to be 12.68 kN acting on the two top node of the tower. Displacements at each node in X-direction and Y-direction are found out and shown in Table 2.

Table 2 Displacements at each node in X- and Y-directions

Node no.	Displacement in X-direction (mm)	Displacement in Y direction (mm)	Node no.	Displacement in X-direction (mm)	Displacement in Y-direction (mm)
1	53.6349	1.81749	20	25.2726	1.6492
2	53.6339	−1.81164	22	22.0222	1.58672
4	50.003	−1.81069	24	18.9083	1.51069
6	50.0034	1.81619	26	15.9593	1.41983
8	46.3755	1.81201	20	25.2726	1.6492
10	42.7581	1.80353	22	22.0222	1.58672
12	39.1611	1.78938	24	18.9083	1.51069
14	35.5973	1.7682	26	15.9593	1.41983
16	32.0821	1.73865	28	13.2061	1.31293
18	28.6336	1.69942	30	10.6821	1.18879

4 Tower Optimization

Optimization is an art of obtaining the best results or the optimal output under a given circumstance. Design optimization deals in minimizing the design parameters like cross-sectional area, material, height, weight, etc. [13, 14]. Optimization algorithm executes iteratively by comparing the different results until an optimum satisfactory result is obtained under a desired objective function. The objective function is to be either minimized or maximized based on the desired objective [15]. According to the problem, the tower is to be designed with three different sections, angle section, pipe section, and tube section simultaneously. Reduction in weight is taken as an objective function and is done using Hooke–Jeeves algorithm [6]. The independent design variables are base width of the tower (*B*) and number of panel (*NP*). The two variable function is written as $f(B, NP)$ and the objective function in terms of design variable is shown by Eq. 1 given by

$$W = f(B, NP) \tag{1}$$

Hooke–Jeeves algorithm is a numerical method for finding the minimum or optimal solution from a multivariable function. It is used for minimizing a real-valued function $f(x)$ for $x \varepsilon R^n$. The process of solving is basically divided into two parts, exploratory steps and pattern search. The advantage of using this method is that the process moves from one point to the next point in the direction of the vector until the function value stops. Then it repeats the process to find a better point. The various steps involved in optimization process are shown in Fig. 4.

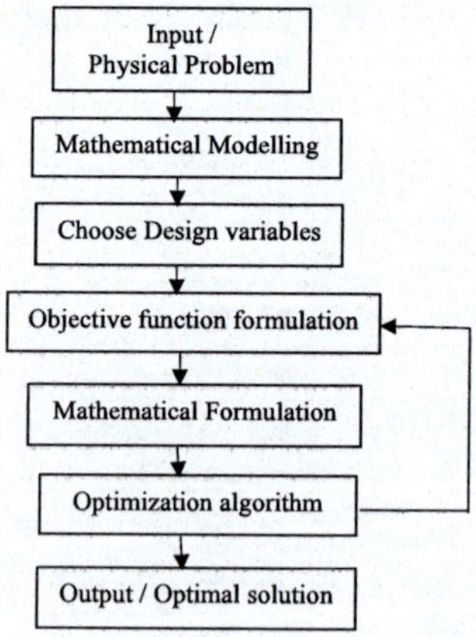

Fig. 4 Steps in optimization process

5 Result and Discussions

Optimization algorithm follows a number of steps or iterations in finding out the optimal value. The geometric parameters such as the base width of the tower (*B*) and the number of panel (*NP*) are treated effectively as design variables. Objective function was finding out the least weight (*W*) among the three cross-sectional shapes. X-type bracing pattern was used for the diagonal members which mostly carries the compressive force. A total of five iterations have been done in the numerical process.

Optimization process for tower while having angle section gives an optimal value of 60.47 kN in the third iteration for specific values of the design variables. The values obtained for each number of iterations are shown in Table 3. Tower with pipe section gives minimum weight of 51.2 kN in the fourth iteration as shown in Table 4. In case of tube section, the tower gives minimum weight in the second iteration. The weight is found to be 76.76 kN which is more than that of angle and pipe section. Table 5 shows the iteration results for tube section.

The optimization process gives the value of minimal weight of the tower for each cross-sectional shape. A comparative study has been made to find out the most effective section. A graph has been plotted for number of iterations performed in the process and the weight of the tower as shown in Fig. 5.

The results of the optimization process for microwave antenna tower show a maximum reduction in weight for different sections. The tower with pipe section has an optimal weight smaller than that of angle and tube section. The optimal weight is about 44% smaller than the other two sections. The tower with angle section gives the maximum reduction in weight after optimization. It has been seen that with angle section X bracing there is a reduction in weight of about 14% than the weight before optimization. While in case of pipe section and tube section the percentage reduction

Table 3 Iteration results of optimization process for tower with angle section

Design variables	Iteration number				
	1	2	3	4	5
Base width *B* (m)	6	5.7	5.4	5	4.7
Number of panel *NP*	20	21	22	23	24
Weight *W* (kN)	65.22	62.76	60.47 (optimal)	60.65	61.2

Table 4 Iteration results of optimization process for tower with pipe section

Design variables	Iteration number				
	1	2	3	4	5
Base width *B* (m)	6	5.7	5.4	5	4.7
Number of panel *NP*	20	21	22	23	24
Weight *W* (kN)	65.02	51.84	51.42	51.2 (optimal)	52.93

Table 5 Iteration results of optimization process for tower with tube section

Design variables	Iteration number				
	1	2	3	4	5
Base width B (m)	6	5.7	5.4	5	4.7
Number of panel NP	20	21	22	23	24
Weight W (kN)	79.8	76.76 (optimal)	76.8	80.04	81.6

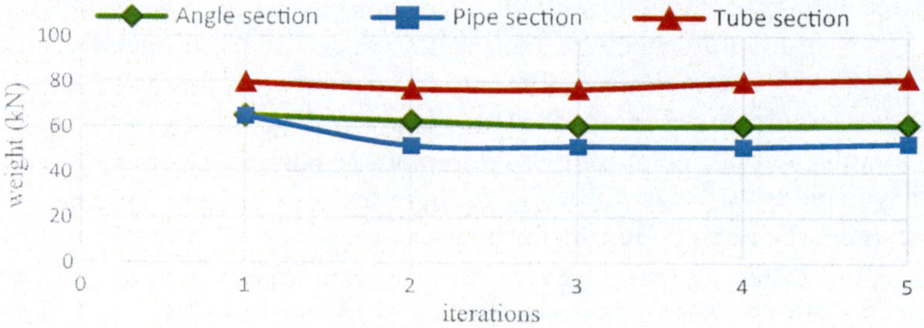

Fig. 5 Variation of weight using angle, pipe, and tube section

Table 6 Percentage weight reduction after optimization process

Section type	Weight before optimization	Weight after optimization	% Reduction in weight
Angle section	69.89	60.47	13.48
Pipe section	56.83	51.2	9.91
Tube section	84.11	80.04	4.84

in weight is about 10% and 5%, respectively. The percentage reduction in weight after optimization using three different sections is shown in Table 6.

6 Conclusion

In this study, an attempt is made to design and analyze a microwave antenna tower using STAAD-Pro and finite element software FEAST. In the analysis, wind load is taken as the main load and wind forces are computed by applying the forces at each node. The results are tabulated under wind load. Optimization is done for weight reduction considering three different sections for direct comparison. Based on the investigations performed, the following conclusions are drawn:

1. The nodal displacement for each node due to wind load is found out and the value of maximum displacement obtained is found to be 53.66 mm.

2. The tower with pipe section has an optimal weight smaller about 44% than that of angle section and tube section.
3. The pipe section and tube section show the percentage reduction in weight by about 10% and 5%, respectively.
4. It has been seen that with angle section X bracing there is a reduction in weight of about 14% than the weight before optimization.
5. Therefore, results obtained show that tower can optimally be designed using angle section.

References

1. Karthik S, Sowjanya GV (2015) Static and dynamic analysis of transmission line tower under seismic loads. Int J Eng Res Technol 4(08):29–33
2. Sakthivel T, Sanjeevi R (2015) Analysis and design of multi circuit transmission line tower. Int J Emerg Technol Comput Sci Electron 13(1):56–61
3. Pahwa S, Tiwari V, Jatwa H (2014) Analytical study of transmission tower subjected to wind and seismic loads optimization technique. Int J Emerg Technol Adv Eng 4(9):375–383
4. Gopi SP (2014) Analysis and design of transmission line tower. Int J Mod Eng Res 4(1):116–138
5. Green WH (1985) Minimum weight sizing of guyed antenna tower. J Struct Eng 111(10):2121–2136
6. Galeb AC, Khayoon AM (2013) Optimum design of transmission towers subjected to wind and earthquake loading. Jordan J Civil Eng 7(1):70–92
7. Hadimani P, Kulkarni S (2017) Static and dynamic analysis of transmission line tower. Int J Emerg Res Manag Technol 6(2):147–154
8. Hossain SA, Samanta B (2018) Structural optimization and design of electric transmission tower. Int J Innov Res Sci Eng Technol 7(6):7116–7123
9. Code of practice for design loads (other than earthquake) for buildings and structures (IS 875, Part 1-1987, Reaffirmed 1997 Second Revision), 1989. Bureau of Indian Standards, New Delhi
10. Code of practice for design loads (other than earthquake) for buildings and structures: Wind loads (IS 875, Part 3-1987, Reaffirmed 1997 Second Revision), 1989. Bureau of Indian Standards, New Delhi
11. Preeti C, Jagan KM (2013) Analysis of transmission towers with different configurations. Jordan J Civil Eng 7(4):450–460
12. Siddesha H (2010) Wind analysis of microwave antenna towers. Int J Appl Eng Res 1(2):574–584
13. Srivastava PK, Simant RG (2014) Structural weight optimization of a bracket using Ansys. Int J Res Eng Technol 3(10):96–99
14. Yong Zhi X, An Le M (2014) Analysis and design of integrated optimization of 1.5 MW wind turbine tower. Adv Mater Res 971(973):958–961
15. Raghavendra T (2012) Computer aided analysis and structural optimization of transmission line tower. Int J Adv Eng Technol 3(3):44–50

Seismic Analysis of a Concrete Gravity Dam Using ABAQUS

Sourav Sarkhel, Jyotiprakash Padhi, and Anil Kumar Dash

Abstract The safety against seismic loads of the dams existing at present and the design of new earthquake-resistant dams is under rigorous research. The designs of the earthquake-resistant dams along with the evaluation for safety of the existing dams depend upon the reliable analytical procedure that can determine the stresses and deformations induced by the earthquake. Therefore, this study is carried out to perform the modal analysis, study the influence of hydrostatic forces, and also to determine the effect of dynamic loads on the dam with rigid and flexible foundation by using ABAQUS. In this study, modal, static, and dynamic analysis of Koyna dam located in Maharastra, India was performed by using ABAQUS. Dynamic analysis of the dam is done under two conditions: dam with a rigid base and dam with a flexible base. North–South Component of the El Centro data was used for the dynamic analysis of the Koyna Dam. The crest of the dam undergoes a displacement of 0.016 m in case of flexible foundation which is higher compared to the displacement of the crest without the dam with rigid foundation, which is 0.0107 m. Major principal stress and shear stress value is found to be higher in case of flexible foundation as compared to rigid foundation of the dam. In case of rigid foundation, maximum displacement is found within the crest of the dam for all the time steps due to dynamic load, whereas Maximum displacement observed within the center part of the dam for the time steps of 0.6 and 0.8 s and at the base of the dam for the 1 s time step due to dynamic load in case of flexible foundation. The major principal stress of the dam is found to be lower (1.04E+09 N/m^2) in case of flexible foundation as compared to the rigid foundation (1.55E+09 N/m^2). The shear stress of the dam is found to be higher in case of flexible foundation (6.95E+09 N/m^2) as compared to the rigid (5.28E+08 N/m^2) foundation.

S. Sarkhel
Dream Institute of Technology, Kolkata, West Bengal, India
e-mail: souravsarkhel.ofc@gmail.com

J. Padhi (✉) · A. K. Dash
KIIT Deemed to be University, Bhubaneswar, Odisha, India
e-mail: jyotiprakash.padhifce@kiit.ac.in

A. K. Dash
e-mail: anil.dashfce@kiit.ac.in

B. B. Das et al. (eds.), *Recent Developments in Sustainable Infrastructure*, Lecture Notes in Civil Engineering 75, https://doi.org/10.1007/978-981-15-4577-1_21

Keywords Soil–structure interaction · Hydrostatic · Dynamic · ABAQUS

1 Introduction

A dam is an obstruction constructed across a stream or river to restrain water and increases the level of the water behind the barrier allowing the formation of a reservoir used for the electricity generation, irrigation, flood control, and water supply, etc. The failure of these structures may result in the property loss along with damage to the environment. The probability of failure increases due to any seismic activity, which may develop cracks within the structure. Therefore it is necessary to check the stability of the dams against all the forces that the structure has to resist such as the hydrodynamic forces and earthquakes. The first proposed method for the study of rigid concrete gravity dams under seismic activity is by Westergaard [1]. Similar type of analysis was also done by Chopra [2] by considering the upstream face to be inclined to get a better representation of the actual dam. The safety against seismic loads of the dams existing at present and the design of new earthquake-resistant dams is under rigorous research in the past few decades [3, 4]. Lot of researches has been done to determine the behavior of the dam under the seismic load [5, 6]. Later on the finite element method (FEM) was introduced because of its advantages over the established method, since FEM produces more accurate results [7]. The effect of the hydrodynamic forces on dams is determined by the assumption that the water is incompressible [8–10]. But later on, it is found that compressibility of the water has an important role in the seismic analysis [2]. Chopra and Chakrabarti [11], Fenves and Chopra [12, 13], and Hall and Chopra [14] studied the finite element analysis using frequency domain. Later on, Sharan [15] and Sommerfeld [16] developed the simplest boundary conditions while Maity and Bhattacharya [17] and Tsai and Lee [18] developed more complex boundary conditions. The response of a dam is largely affected by considering the type of soil in which the dam is constructed [19]. Pal [20] became the first to perform nonlinear analysis on the Koyna dam. He neglected the effect of the reservoir and assumed the foundation to be rigid. Khosravi and Heydari [21] analyzed the dam under four different conditions by assuming the reservoir to be empty and full along with the foundation to be flexible and rigid. Burman et al. [6] studied the effect of the flexibility and nonlinearity of the foundation on the seismic response of a dam by using SAP 2000. Reddy et al. [22] carried similar studies to analyze the effect of the flexible foundation. In the present paper, a dam structure is analyzed neglecting the effect due to the reservoir water but considering the effect of soil condition.

Table 1 Material properties of the dam and foundation

Material	Young's modulus (N/m^2)	Density (Kg/m^3)	Poisson's ratio
Concrete	3.15×10^{10}	2430	0.235
Foundation soil	6.8923×10^{10}	981	0.3333

2 Methodology

2.1 *Description of Dam*

In this study, modal, static and dynamic analysis of Koyna dam located in Maharastra, India was performed by using ABAQUS for the rigid and flexible foundation. Koyna dam has a height of 103 m with a crest width of 14.8 m, length of 807.2 m, and 70 m wide at its base. The capacity of the dam is 2797.4 Mm^3. The material properties of the dam as well as the soil utilized for the analysis are summarized in Table 1. In order to model the soil–structure interaction, a soil domain of depth 310 m is considered. The North–South Component of the El Centro earthquake data used for the evaluation of the results in the present study. The magnitude of the earthquake was 6.9 in Richter scale and had an extreme intensity of X in Mercalli's intensity scale.

2.2 *Static and Dynamic Analysis in ABAQUS*

In the analysis of the structure using ABAQUS, the geometric model is prepared first according to the dimension of the Koyna dam used in this study. The geometric model of the Koyna Dam is presented in Fig. 1. After making the model, the material properties are provided (Table 1). The structures are then assembled from the assembly module and the required type of analysis is chosen from the step module. The mesh is generated by selecting the type of mesh properties and suitable element type. In this study, four-node bilinear plane strain type element is chosen and the shape of the elements is considered to be quadrilateral. In order to carry out the modal and hydrostatic analysis of only the dam structure, the base of the dam is considered to be rigid or flexible. For the dynamic analysis, Displacement/Rotation type of boundary condition is applied. The rotation and the translational movement in the vertical direction are considered to be zero, since the dam is analyzed only for the horizontal acceleration of the earthquake. For the dam–foundation interaction model, the dam is tied to the modeled soil domain and the Displacement/Rotation boundary condition is applied to the soil domain. The dam–foundation interaction is created among the soil and the dam and the edges of the soil domain except on the edge in which the interaction takes place are made fixed. The surface to surface type

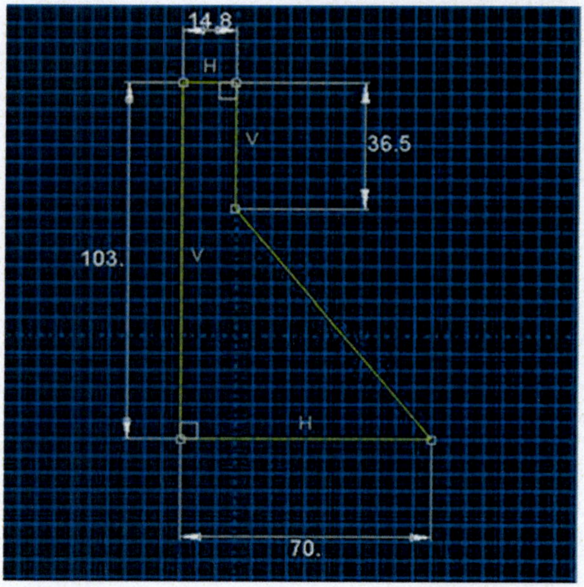

Fig. 1 Geometric model of dam

of interaction between the soil and the dam is created for the interaction problem of the soil and the structure. The sliding is restricted to finite limit and the tangential behavior of the contact is kept to be rough.

3 Results and Discussions

3.1 Modal Analysis of Dam with Rigid Foundation Using ABAQUS

The modal analysis is carried out and the first ten natural frequencies along with the mode shapes are obtained by using ABAQUS. The dam mesh is generated using four-node bilinear plane strain quadrilateral elements. The total number of elements generated is 85 with 105 numbers of nodes. The first ten frequencies corresponding to the mode numbers 1, 2, 3, 4, 5, 6, 7, 8, 9, and 10 are 2.9161, 7.8182, 11.9132, 15.8829, 24.1343, 25.2690, 33.7364, 37.5544, 40.8527, 40.8527, and 41.1362 Hz, respectively.

The first 10 mode shapes are obtained through the modal analysis of the dam with empty reservoir and rigid foundation. Mode shapes 1 and 2 of the empty dam with rigid base from ABAQUS are shown in Fig. 2. The maximum displacement is observed at the crest for the mode shapes 1–6. As the frequency increases, the

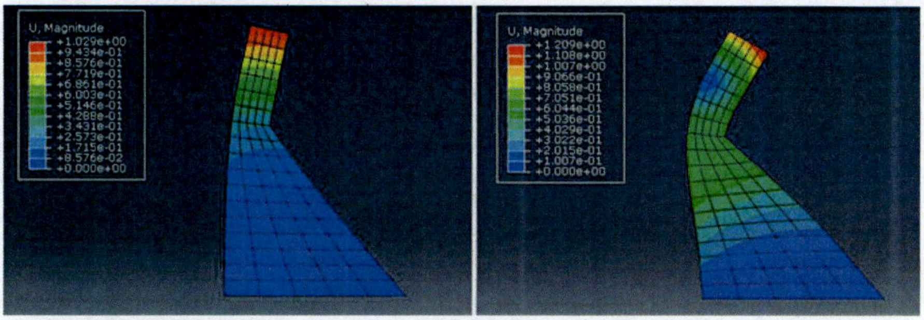

Fig. 2 Mode shape 1 and 2 of empty dam with rigid base from ABAQUS

Table 2 Comparison of the first four natural frequencies obtained from ABAQUS with literature without considering the interaction

Mode number	Frequency	
	Sarkar et al. [23]	ABAQUS
1	3.002	2.9161
2	7.953	7.953
3	10.848	11.9132
4	15.640	15.8829

maximum displacement gradually shifts from crest to the base of the dam. Uniform displacement is observed for lower frequencies, whereas with increase in natural frequency nonuniform displacement was found throughout the dam. Maximum displacement was found towards the toe of the dam in the mode shapes 8 and 10.

3.2 *Comparison of First Four Natural Frequencies of Dam with Rigid Foundation*

Table 2 compares the results of first four natural frequencies obtained from ABAQUS with the results from the reported literature [23]. It can be observed that good conformity has been achieved between the results of the present study with the reported literature.

3.3 *Modal Analysis of Dam with Flexible Foundation Using ABAQUS*

The modal analysis of the dam with empty reservoir and flexible foundation is carried out and the first ten natural frequencies obtained are recorded. The first ten frequencies corresponding to the mode numbers 1, 2, 3, 4, 5, 6, 7, 8, 9, and 10 are 2.8693, 7.1284,

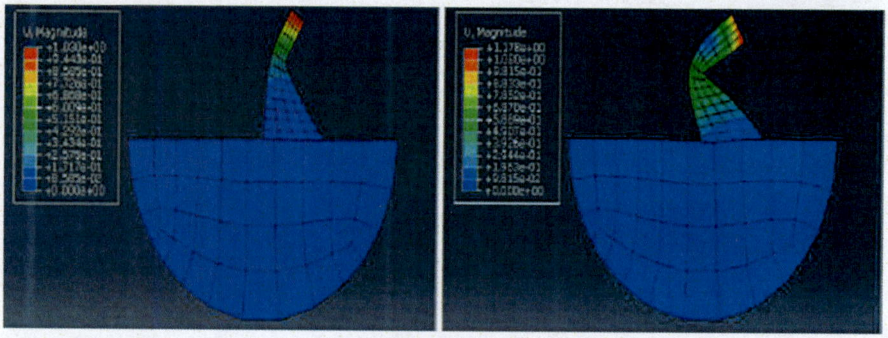

Fig. 3 Mode shape 1 and 2 of dam with flexible foundation

9.9294, 13.9756, 20.8041, 21.7313, 29.6948, 33.9417, 36.3918, and 37.6962 Hz, respectively.

The first ten mode shapes corresponding to the natural frequencies are obtained by the modal analysis of the dam with empty reservoir and flexible foundation (Fig. 3). The deflections in each of the mode shapes are different from each other. The deflection observed from mode shape four to six varies throughout the dam with less deflection of the soil. As the natural frequency increases, larger deflection of the dam along with the foundation soil is noticed. The mode shapes obtained from the modal analysis for the flexible foundation were compared with that of the rigid foundation. Mode shapes of flexible base were found to be more distorted as compared to rigid base. In case of rigid foundation, the displacement is uniform from mode shapes 1–6 whereas in case of flexible foundation except mode 1, all other mode shapes have nonuniform displacement.

The first natural frequency obtained through the analysis of ABAQUS and the reported literature (Chopra 1980) is compared. The natural frequency estimated from ABAQUS analysis is 2.8415 Hz which matches well with the natural frequency of the reported literature, i.e., 2.9325 Hz.

3.4 Hydrostatic Analysis of a Dam with Rigid Foundation Using ABAQUS

Hydrostatic analysis is performed and the results of different stresses and deformations are obtained by using ABAQUS. The type of element and number of modes are similar to modal analysis. The maximum and minimum total deformation of the dam was found to be 0.107 m and 0.0 m, respectively. The contour map of total deformation due to hydrostatic pressure is presented in Fig. 4.

The equivalent (von Mises), major principal stress and shear stress of the dam due to the effect of the hydrostatic pressure is given in Table 3 and contour maps of shear and major principal stresses are shown in Fig. 5. Major principal stress varied from

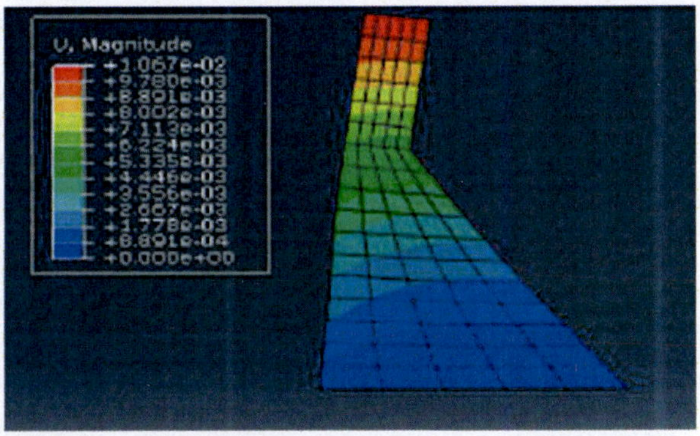

Fig. 4 Contour map of total deformation due to hydrostatic pressure

Table 3 Equivalent, major principal and shear stress value on the dam due to hydrostatic pressure

Parameters	Equivalent (von-Mises) stress (N/m^2)	Major principal stress (N/m^2)	Shear stress (N/m^2)
Minimum	4.838E+02	−1.651E+0.4	−9.510E+02
Maximum	3.033E+06	2.566E+06	1.334E+06

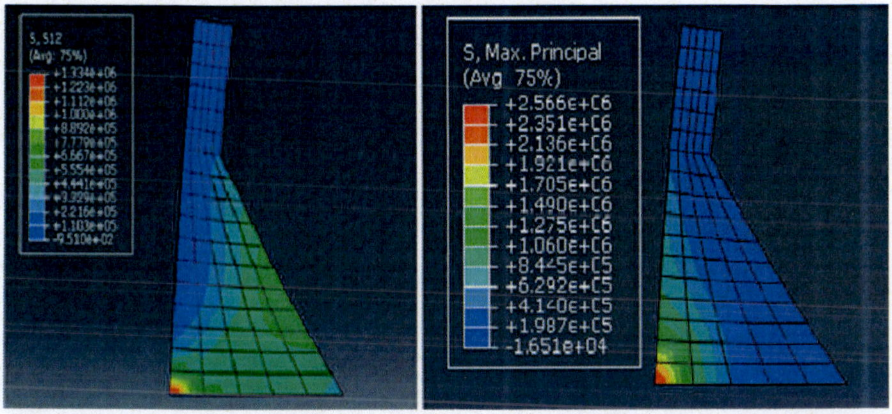

Fig. 5 Contour map of shear stress and major principal stress due to hydrostatic pressure

−1.651E+0.4 to 2.566E+06 N/m^2 whereas shear stress varied from −9.510E+02 to 1.334E+06 N/m^2. It can be observed from Fig. 5 that, maximum shear and major principal stress is detected in the heel of the dam whereas minimum stress is found within the crest in both cases. The variation of both shear stress and major principal stress within the entire body of the dam is similar.

3.5 Hydrostatic Analysis of a Dam with Flexible Foundation

The hydrostatic analysis of the dam with flexible foundation for displacement is presented in Fig. 6. The crest of the dam undergoes a displacement of 0.016 m in case of flexible foundation which is higher compared to the displacement of the crest with rigid foundation, which is 0.0107 m. Maximum displacement is detected at the crest of the dam. Major principal stress and shear stress obtained from the hydrostatic analysis of a dam with flexible foundation is shown in Fig. 7. The maximum value of major principal stress and shear stress is 2.782E+06 N/m^2 and 1.538E+06 N/m^2 respectively. It can be also detected from Fig. 7 that, the maximum major principal stress as well as shear stress is observed at the heel of the dam. Minimum stress value is observed within the crest of the dam which is similar to the dam with rigid foundation.

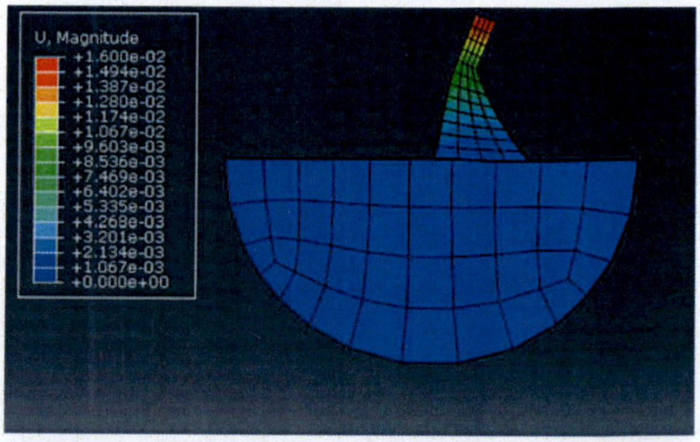

Fig. 6 Contour map of total displacement of the dam with flexible foundation due to hydrostatic pressure

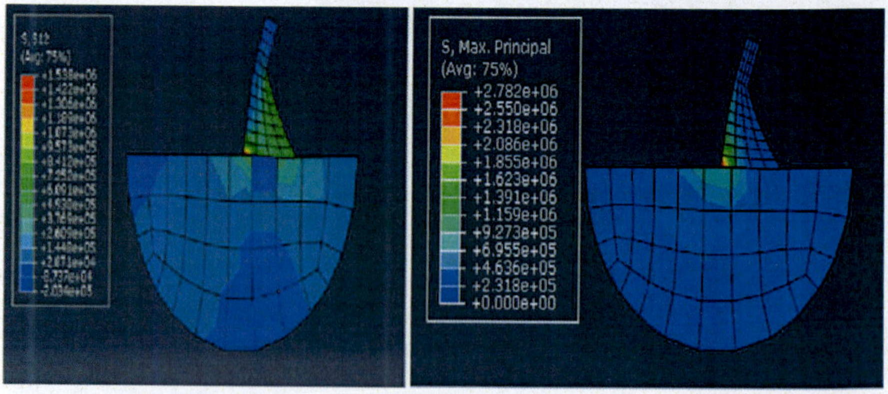

Fig. 7 Contour map of shear stress and major principal stress of dam with flexible foundation under hydrostatic force

Table 4 Maximum values of displacement, von Mises stress, major principal stress and shear stress at time step 1 s

Case no.	Foundation condition	Maximum displacement (m)	Major principle stress (N/m^2)	Shear stress
1	Rigid	7.026E−02	3.143E+0.6	1.660E+07
2	Flexible	6.412E−02	9.669E+0.6	4.656E+06

3.6 *Dynamic Analysis*

The dynamic analysis of the modeled dam is analyzed by using El Centro earthquake data and different responses of the dam such as the displacement, von Mises stress, major principal stress and the Shear stress are found out using ABAQUS. Dynamic analysis of the dam is done under two conditions: dam with a rigid base and dam with a flexible base. The rigidity of the dam is obtained by fixing the base. Table 4 represents the data of various responses of the dam for rigid and flexible condition at time step 1 s. Higher values of displacement and stresses are found in case of rigid foundation as compared to flexible foundation of dam (Table 4).

The total displacement of the dam with rigid and flexible foundation at 1.0 s is presented in Fig. 8, due to the dynamic load. The maximum displacement value is observed at the crest for rigid as well as flexible foundation for the time steps of 0.2 and 0.4 s. In case of rigid foundation, maximum displacement is found within the crest of the dam for all the time steps. For flexible foundation, maximum displacement is observed in the center part of the dam for the time steps of 0.6 and 0.8 s and at the base of the dam for 1 s time step.

The major principal stress developed within the body of the dam due to dynamic load is computed by using the earthquake data of El-Centro with an equal time spacing of 0.2 s. The contour map of major principal stress of rigid and flexible foundation for time step of 1.0 s is presented in Fig. 9. The pattern of major principal stress is changing for each time step for both rigid as well as flexible foundation. Contour map of the major principal stress for rigid foundation is different from flexible foundation.

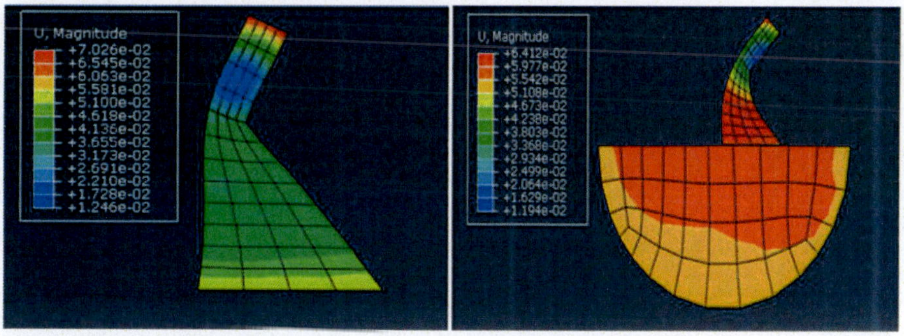

Fig. 8 Contour map of displacement of a dam with rigid and flexible foundation at 1 s

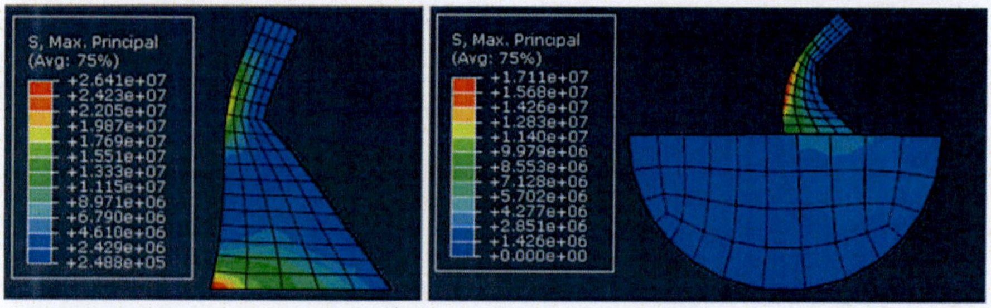

Fig. 9 Contour map of major principal stress of a dam with rigid and flexible foundation at 1 s

Major principal stress observed at the heel of the dam in case of rigid foundation whereas within the center part of the dam for flexible foundation of 1 s time step.

Maximum shear stress is found within a small area of the dam for 0.2 s time step whereas the stress was observed within a large portion of the dam in case of 0.4 s time step for rigid foundation. The shear stress distribution within the body of the dam at 0.2 s time step is entirely different at 0.4 s time step for flexible foundation. Maximum shear stress is found at the base of the dam with rigid foundation for 0.8 s time step whereas at the heel in case of flexible foundation for 0.6 s time step.

4 Conclusion

In this present study, an effort has been made to determine the response of a concrete gravity dam with rigid and flexible foundation by using ABAQUS. The results obtained from the modal analysis are compared with the reported literature to check the accuracy of the work. Finally, the interaction is modeled and dynamic analysis is done using ABAQUS. As the frequency increases, the maximum displacement gradually shifts from crest to the base of the dam. Uniform displacement is observed for lower frequencies whereas with increase in natural frequency nonuniform displacement was found throughout the dam. Due to the hydrostatic effect, the crest of the dam undergoes a displacement of 0.016 m in case of flexible foundation which is higher compared to the displacement of the crest of the dam with rigid foundation, which is 0.0107 m. Major principal stress and shear stress value is found to be higher in case of flexible foundation as compared to rigid foundation of the dam. The major principal stress of the dam is found to be lower (1.04E+09 N/m^2), in case of flexible foundation, as compared to the rigid foundation (1.55E+09 N/m^2). The shear stress of the dam is found to be higher in case of flexible foundation (6.95E+09 N/m^2) as compared to the rigid (5.28E+08 N/m^2) foundation.

References

1. Westergaard HM (1933) Water pressure on dams during earthquake transactions. ASCE 98:418–433
2. Chopra AK (1968) Earthquake behavior of reservoir-dam systems. J Eng Mech Div ASCE 94:1475–1500
3. Chopra AK, Chakrabarti P, Gupta S (1980) Earthquake response of concrete gravity dams including hydrodynamic and foundation interaction effects, Report No. UCB/EERC-80/01, Earthquake Engineering Research Center, University of California, Berkeley, California
4. Calayir Y, Dumanoğlu AA, Bayraktar A (1996) Earthquake analysis of gravity dam-reservoir systems using the Eulerian and Lagrangian approaches. Comput Struct 59 (5):877–890
5. Burman A, Maity D, Sreedeep S (2010) Iterative analysis of concrete gravity dam-nonlinear foundation interaction. Int J Eng Sci Technol 2(4)
6. Burman A, Reddy B, Maity D (2008) Seismic analysis of concrete gravity dams considering foundation flexibility and nonlinearity. In: 12th International conference on computer methods and advances in geomechanics (IACMAG), 1–6 October 2008, Goa, India
7. Mandal KK, Maity D (2016) Transient response of concrete gravity dam considering dam-reservoir-foundation interaction. J Earthquake Eng. https://doi.org/10.1080/13632469.2016.1217804
8. Zangar CN, Haefeli RJ (1952) Electric analog indicates effects of horizontal earthquake shock on dams. Civ Eng 22(4):54–55
9. Zeidan BA (2013) Hydrodynamic analysis of concrete gravity dams subjected to ground motion
10. Zeikiewicz OC, Bettes P (1978) Fluid structure dynamic interaction and wave forces. An introduction to numerical treatment. Int J Numer Meth Eng 13:1–16
11. Chopra AK, Chakrabarti P (1981) Earthquake analysis of concrete gravity dams including dam-water-foundation rock interaction. Earthquake Eng Struct Dynam 9:363–383
12. Fenves G, Chopra AK (1984) Earthquake analysis of concrete gravity dams including reservoir bottom absorption and dam-water-foundation rock interaction. Earthquake Eng Struct Dynam 12:663–680
13. Fenves G, Chopra AK (1985) Effects of reservoir bottom absorption and dam-water-foundation rock interaction on frequency response functions for concrete gravity dams. Earthquake Eng Struct Dynam 13:13–31
14. Hall JF, Chopra AK (1982) Hydrodynamic effects in the dynamic response of concrete gravity dams. Earthquake Eng Struct Dynam 10:333–345
15. Sharan SK (1987) Time domain analysis of infinite fluid vibration. Int J of Num Methods in Engg 24:945–958
16. Sommerfeld A (1949) Partial differential equations in Physics. Academic Press, New York
17. Maity D, Bhattacharyya SK (1999) Time-domain analysis of infinite reservoir by finite element method using a novel far-boundary condition. Finite Elem Anal Des 32:85–96
18. Tsai CS, Lee GC (1991) Time-domain analyses of dam-reservoir system. II: substructure method. J Eng Mech 117(9):2007–2026
19. Bayraktar A, Hancer E, Akkose M (2005) Influence of base-rock characteristics on the stochastic dynamic response of dam-reservoir foundation systems. Elsevier
20. Pal N (1976) Seismic cracking of concrete dams. J Struct Div ASCE 102:1827–1844
21. Khosravi S, Heydari MM (2013) Design and modal analysis of gravity dams by Ansys parametric design language. Walailak J
22. Reddy BV, Burman A, Maity D (2008) Seismic response of concrete gravity dams considering foundation flexibility. Indian Geotech J 38:187–203
23. Sarkar R, Paul DK, Stempniewski L (2007) Influence of reservoir and foundation on the nonlinear dynamic response of concrete gravity dams. J Earthq Technol 44(2):377–389

Seismic Analysis of RC Framed Tall Structures with Plan Irregularity

Saurav Kar and Tapas Sadhu

Abstract For the last few decades, architects, designers and structural engineers have focused on various advanced architectural aspects of reinforced concrete and steel structures. Regular plan buildings are quite commonly existent in India, while irregular plan buildings are in practice. These asymmetrical plan buildings, which are constructed in seismic prone areas, are most vulnerable and prone to subsequent damage during earthquakes. Seismic excitations generate the most destructive forces on structures. In accordance with newly revised Indian Code for Earthquake Provisons i.e. IS 1893:2016. In this present paper, high rise RC buildings have been modelled using a software package like STAAD.Pro V8i. Various models of irregular plans (e.g. U-, L-, T- and X-shaped) each of 25 storeyed (75 m each) have been taken into account. All the building models are located in high intensity seismic zone IV. As per IS 1893:2016, equivalent static analysis and dynamic response spectrum method have been conducted on all the building models. Even though various irregularities (such as torsional irregularity, stiffness irregularity) affect the seismic behaviour, this paper primarily concentrates on the effect of geometric irregularity on various RC structures. The paper also highlights both on which irregular shape of building plan gives the severe most response towards ground motion and also the accuracy of dynamic analysis conducted on high rise RC structures compared to static analysis.

Keywords Asymmetrical structures · Torsional irregularity · Plan irregularity · Oscillation · Floor displacement

S. Kar (✉) · T. Sadhu
Department of Civil Engineering, Heritage Institute of Technology, an Autonomous Institute affiliated to Maulana Abul Kalam Azad University of Technology (MAKAUT formerly known as West Bengal University of Technology), Kolkata 700107, West Bengal, India
e-mail: sourav.kar@heritageit.edu

T. Sadhu
e-mail: tapas.sadhu@heritageit.edu

B. B. Das et al. (eds.), *Recent Developments in Sustainable Infrastructure*, Lecture Notes in Civil Engineering 75, https://doi.org/10.1007/978-981-15-4577-1_22

1 Introduction

Earthquake excitations pose vulnerable effects on the existing structures. There is a prime concern for design of buildings for seismic vulnerability as there is an involvement of socioeconomic loss. Seismic behaviour on structures is quite different from wind loading patterns, which results in elastic deformations of structures.

Perfect regularity in structures is considered to be an idealization which rarely occurs. Major national and international codes suggest that irregularity in plan and elevations are different perspectives but in actual they are the combination of these two. The differential study between irregular and regular buildings is itself a paradigm shift. As far as plan irregularity is concerned, past earthquakes suggest that this type of asymmetric behaviour is due to nonuniform mass, stiffness and strength distribution throughout the structure, resulting in severe floor rotations (torsional response) and floor displacements. In the past few decades, numerous developments have been described [1]. Overall geometry and configurations are considered to be most critical during the seismic performance of the structure. The seismic oscillatory motions which are transmitted to the foundation primarily depends on factors such as (a) geometric plan configurations, (b) plan aspect ratio and (c) slenderness ratio of the building.

1.1 Asymmetric Building Plan

It is observed that seismic performance of buildings basically depends on the fundamental configurations of convexity and concavity of lenses. Buildings with concave-shaped plans have load paths directly transferring the seismic forces to its base, while the latter ones create an indirect load paths generating stress concentrations at the location of bending. Concave type buildings are more preferable compared to convex types. Coupling of torsion effects in asymmetrical building structures are extensively investigated by simple one-storey model easily while multi-storeyed building models are quite difficult to analyse and are more complex. Inelastic behaviour of structures is of high interest because of its ability of structures predominantly depends on the resistance offered by the ductile members and energy dissipation.

Numerous researches have been conducted for the past decade to inspect the seismic behaviour of unsymmetrical structures. A comparison of static and dynamic seismic analysis on a multistorey building is conducted. Linear seismic analysis is conducted on the building by static method (Seismic Coefficient Method) and dynamic method (Response Spectrum Method) using STAAD-Pro as per the IS-1893-2002-Part-1 [2]. Similarly, another dynamic analysis is conducted on a 30-storeyed building located in zones II and III as per codal provision [3]. Previous works focussed on the performance of in-plane bending of irregular structures and the pivotal parameters on which the seismic response depends [4–6]. Furthermore,

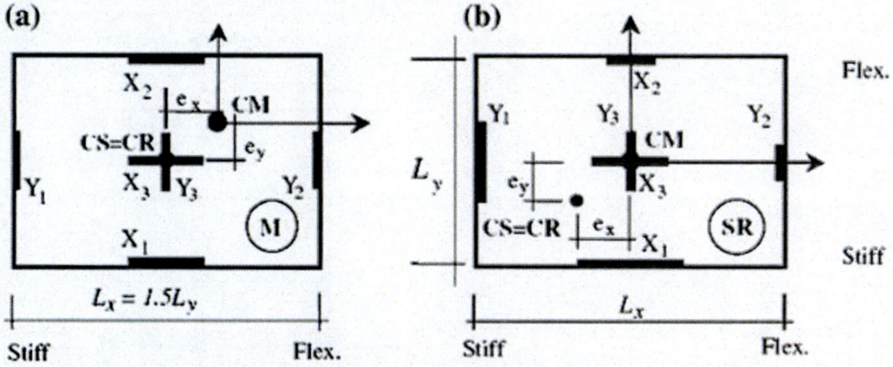

Fig. 1 Mass eccentric model and stiffness eccentric model [8]

some interesting research shows the plastic coupling on torsional response in comparison to elastic deformations [7, 8]. Major conclusions of parametric study primarily performed on torsionally stiffened mass eccentric systems as shown in Fig. 1.

2 Selection and Design of Building Type

In this particular study, as earlier mentioned, variously shaped Reinforced Concrete tall structures have been taken into consideration for study. Irregular shapes like U-, T- X- and L-shaped RC structures (plan) have been taken into account.

All these buildings are residential buildings which are located in zone IV, hard soil. All the buildings are 25-storeyed each. M25 concrete grade is used with yield strength of steel taken as 415 N/mm^2. Thickness of infill walls has been taken as 0.25 m, clear height of infill wall as 2.45 m. Burnt clay brick walls are considered with a unit weight of 20 kN/m^3 and opening in walls are taken as 20%. Dead load acting on beams transferred from infill walls is taken as 10 kN/m^2. Imposed load on the structures is taken as 3 kN/m$^{2.}$ The basic building model is shown in Fig. 2. As far as structural elements are concerned, all the beams sections are taken as 300 mm × 400 mm, columns as 400 mm × 400 mm and slab thickness as 150 mm thick. A typical elevation of the five-storeyed building structure is shown in Fig. 3.

3 Various Seismic Analyses Adopted

In this present study, two types of seismic analyses have been conducted on all the building models. Initially, equivalent static analysis in accordance with IS 1893:2002 and 2016 have been conducted and then response spectra method has been adopted.

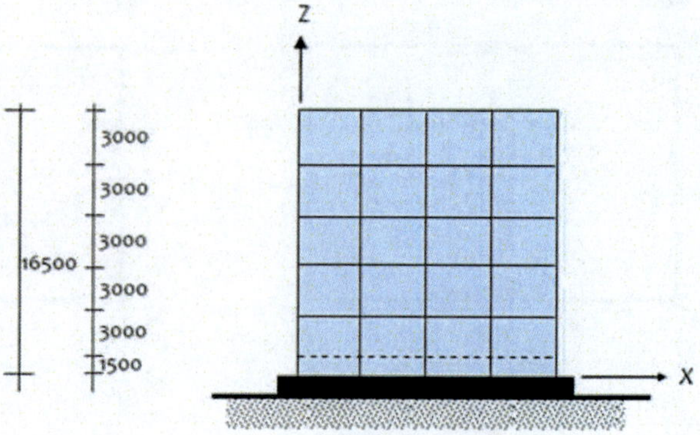

Fig. 2 Typical elevation of five-storeyed RC structure considered in this study

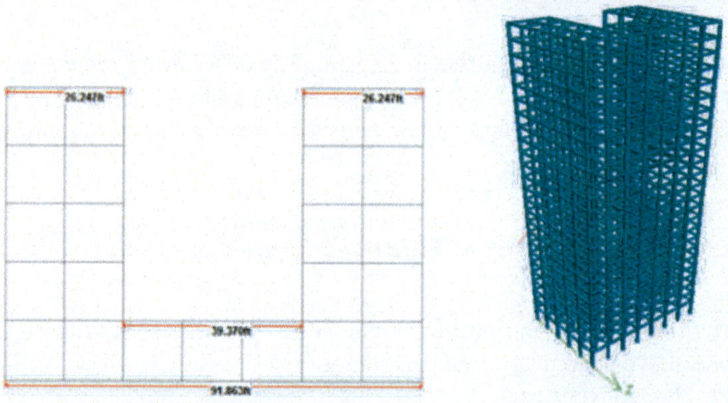

Fig. 3 Geometric configurations of U-shaped structure

The fundamental equation for single degree-of-freedom structure is based on the Eq. (1)

$$m\ddot{x}(t) + c\dot{x}(t) = kx(t) = -m\ddot{x}_g(t). \tag{1}$$

3.1 Equivalent Static Analysis Adopted (IS 1893, 2002, IS 1893, 2016)

For equivalent static analysis, based on the natural time period of the structure will be calculated as per Eq. (2)

$$T_a = \frac{0.09h}{\sqrt{d}}, \tag{2}$$

where T_a = Natural time period of the structure and d = *Base dimension of the building at the plinth level.*

Similarly, the design horizontal seismic coefficient may be calculated as per Eq. (3)

$$A_h = \frac{ZIS_a}{2Rg}, \tag{3}$$

where Z = seismic zone factor, I = importance factor, S_a/g = average response acceleration coefficient and R = response reduction factor.

The total design lateral force or design seismic base shear (V_b) along any principal direction will be calculated by the Eq. (4) and the distribution of lateral forces along all the floors can be calculated by Eq. (5).

$$V_b = A_h W, \tag{4}$$

where W = total seismic weight of the building.

$$Q_i = \frac{W_i h_i^2}{\sum_{j=1}^{n} W_j h_j^2}, \tag{5}$$

where Q_i = design lateral forces, W_i = Seismic weight at ith floor, h_i = height of the ith floor measured from the base, n = number of storeys.

It is known to all that according to IS 1893:2016, response spectra method should be conducted for irregular buildings based on modal analysis.

3.2 *Dynamic Analysis Adopted Based on Response Spectrum (IS 1893, 2002, IS 1893, 2016)*

The equation of MDOF system is given by the following:

$$[m]\{\ddot{x}(t)\} + [c]\{\dot{x}(t)\} + [k]x(t) = -[m]\{r\}\ddot{x}_g(t). \tag{6}$$

Modal mass M_k of mode k can be calculated as per Eq. (7)

$$M_k = \frac{\left[\sum_{i=1}^{n} W_i \phi_{ik}\right]^2}{g \sum_{i=1}^{n} W_i (\phi_{ik})^2}, \tag{7}$$

where ϕ_{ik} = mode shape coefficient at floor i of mode

Similarly, modal participation factor needs to be calculated as per Eq. (8) and design lateral force as per Eq. (9) and storey shear force as per Eq. (10)

$$P_k = \frac{\sum_{i=1}^{n} W_i \phi_{ik}}{\sum_{i=1}^{n} W_i (\phi_{ik})^2}, \tag{8}$$

$$Q_{ik=A_k \phi_{ik} P_k}; \tag{9}$$

$$V_{ik} = \sum_{j=(i+1)}^{n} Q_{ik} \tag{10}$$

where A_k = design horizontal acceleration spectrum using natural time period of oscillation T_k of mode k obtained from dynamic analysis.

4 Seismic Behaviour of Variously Shaped RC Structures

4.1 Various Irregular Plan Structures

For all the irregular plan structures, respective plan and elevations are given below. The plan and elevation U-shaped structure are shown in Fig. 3. Similarly, the plan and elevations for X-shaped, L-shaped and T-shaped structures are shown in Figs. 4, 5 and 6, respectively.

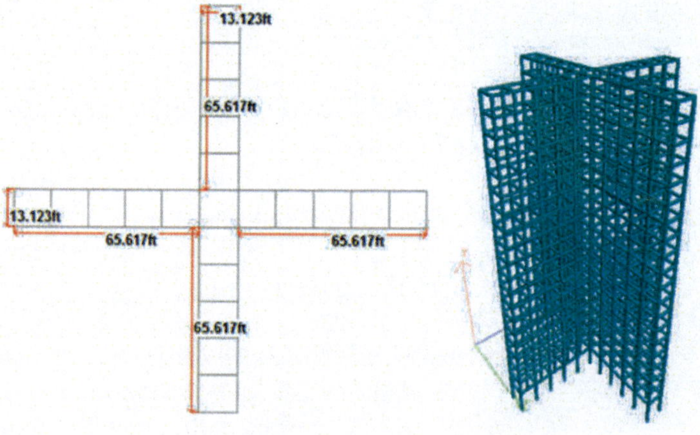

Fig. 4 Geometric configurations of X-shaped structure

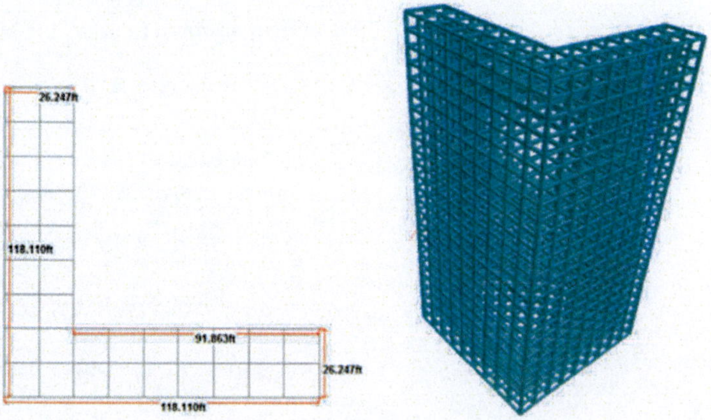

Fig. 5 Geometric configurations of L-shaped structure

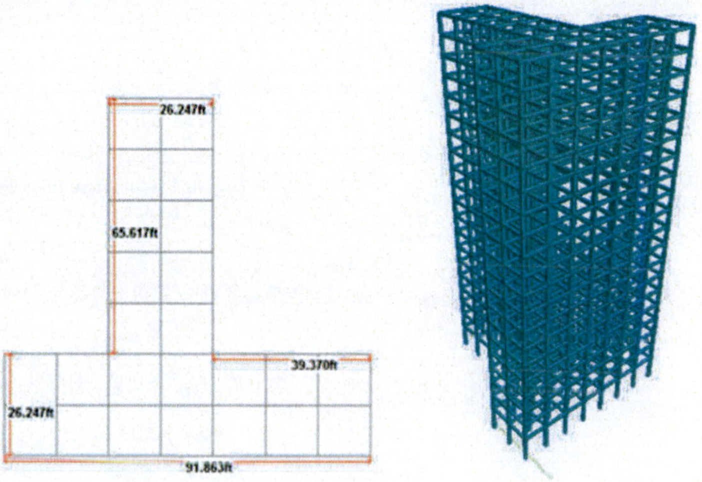

Fig. 6 Geometric configurations of T-shaped structure

4.2 Results Based on Equivalent Static Analysis

The variation of maximum floor or storey displacement versus floor height is shown in Fig. 7 for U-shaped structure, Fig. 8 for C-shaped structures and subsequently Figs. 9 and 10 for L- and T-shaped structures. All the graphs shown for various unsymmetrical plan structures have been represented based on the worst condition effects considered along X-direction and Z- direction.

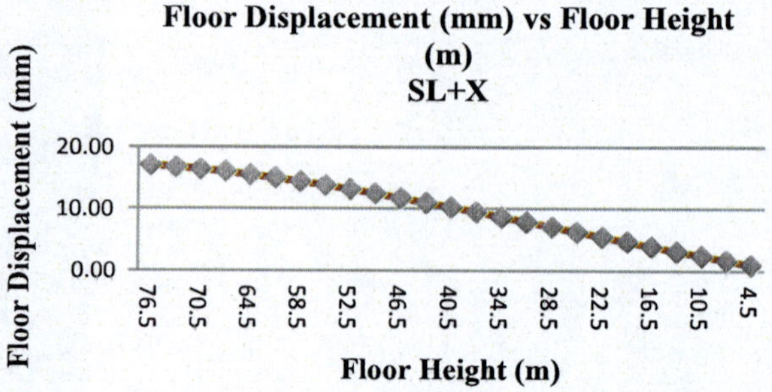

Fig. 7 Max floor displacement versus floor height (for U-shaped structure)

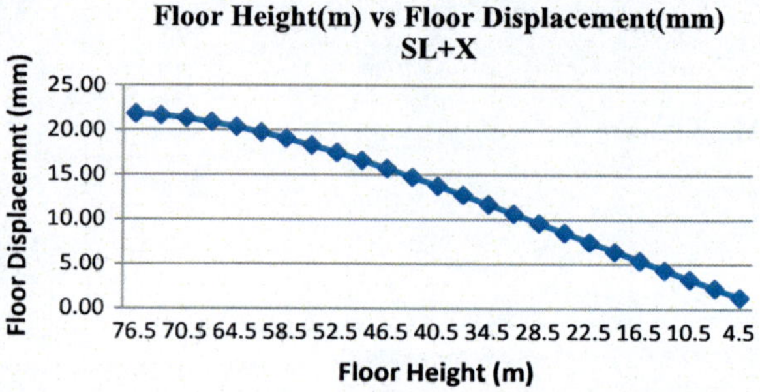

Fig. 8 Max floor displacement versus floor height (for X-shaped structure)

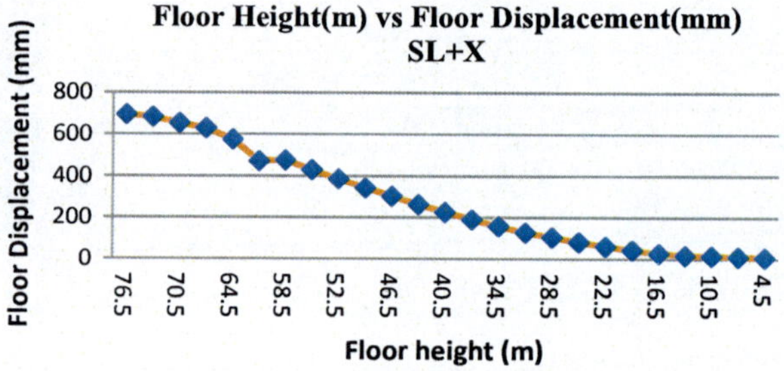

Fig. 9 Max floor displacement versus floor height (for L-shaped structure)

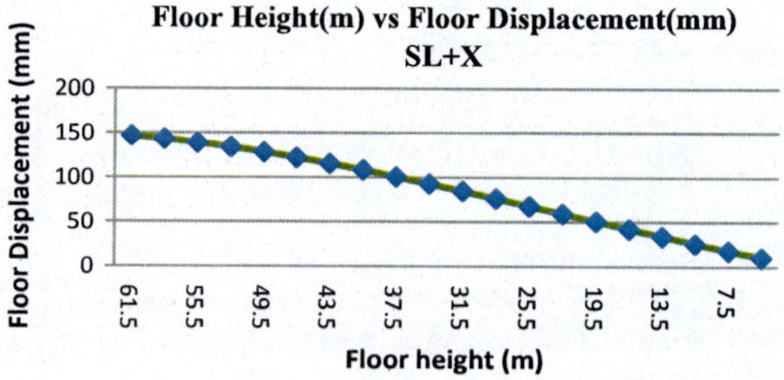

Fig. 10 Max floor displacement versus floor height (for T-shaped structure)

4.3 Results Based on Dynamic Analysis

4.3.1 Modes of Oscillations

The first six modes include undesirable oscillations like diagonal, translation, torsion, opening–closing and dog-tail wagging for complex shapes as shown in Figs. 11, 12, 13 and 14.

Results show that structures with plan complexity with long end projected portion and interior corners are observed to undergo translational and torsional behaviour

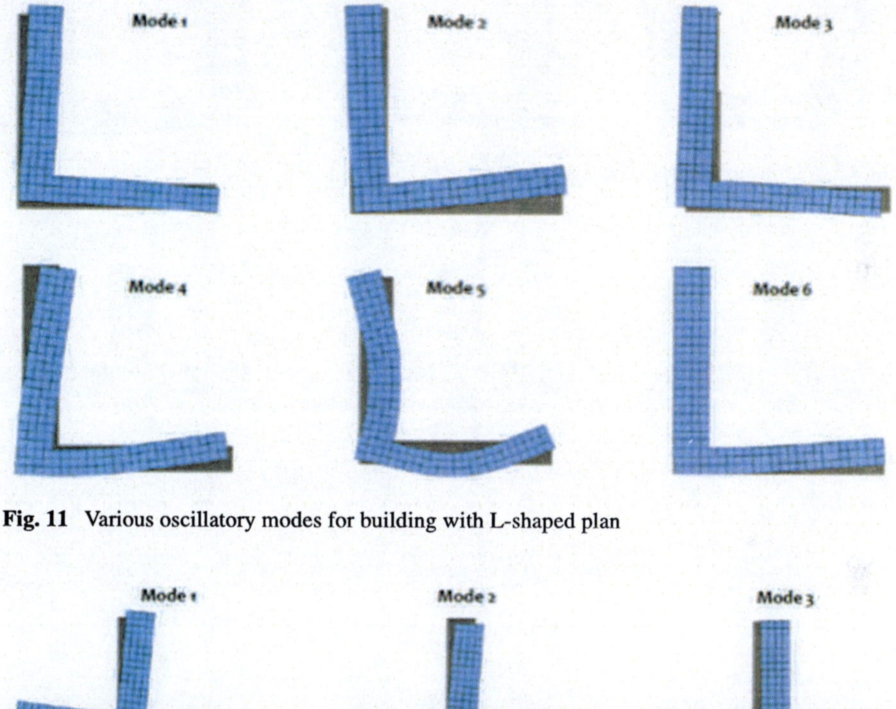

Fig. 11 Various oscillatory modes for building with L-shaped plan

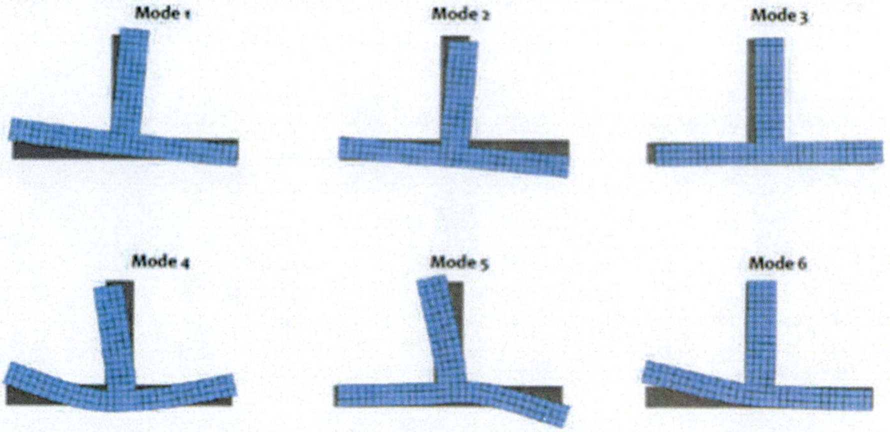

Fig. 12 Various oscillatory modes for building with T-shaped plan

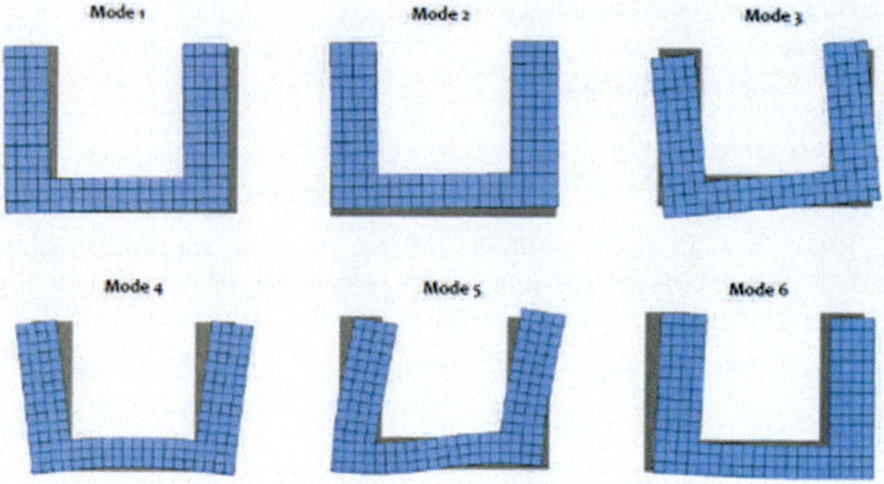

Fig. 13 Various oscillatory modes for building with U-shaped plan

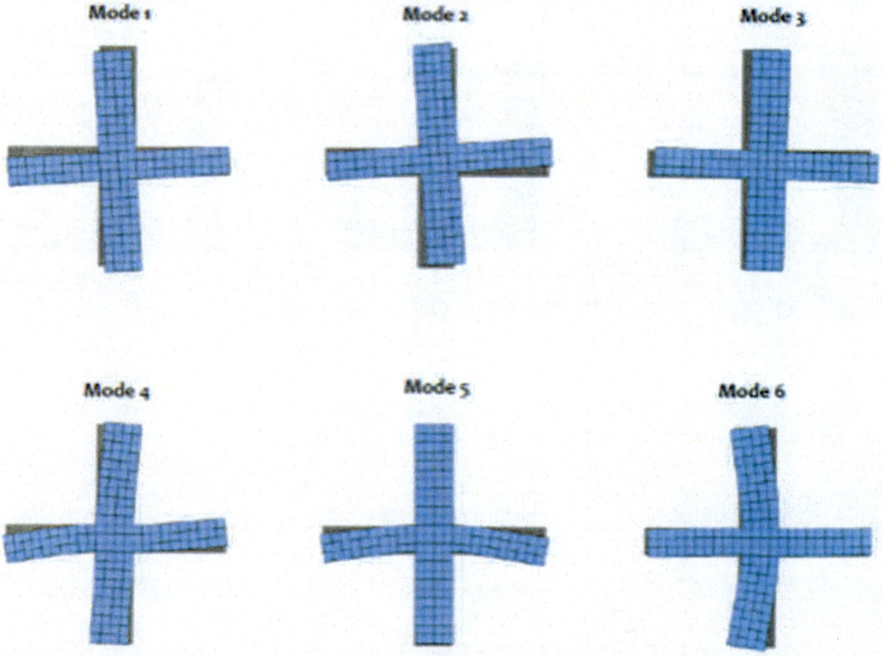

Fig. 14 Various oscillatory modes for building with X-shaped plan

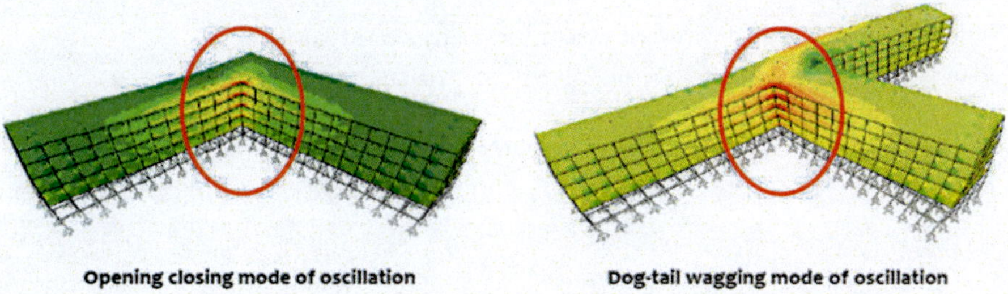

Fig. 15 High-intensity stress concentration in L-shaped and T-shaped plan buildings

of oscillation. Dog-tail wagging mode of oscillation is exhibited by T-shaped plan in which a long slender projection oscillates while the remaining portion remains stationary. On the other hand, re-entrant corners in L-shaped and T-shaped structures posses a high-intensity damages and subsequently increases the vulnerability of the structure as shown in Fig. 15.

5 Results and Discussions

5.1 Comparative Results Among All the Irregularly Shaped Structures

The maximum displacements of building in different storeys along both +X and + Z direction for all four types of irregular plan structures have been compared and shown in figures. Also, the natural time periods and the base shear values of all types of structures (U, X, L, inverted-T) are given below in the Table 1.

From the results obtained, it is observed that the values of base shear decrease with increase in storey height with maximum value at the base and minimum value at the top floor in both X and Z directions for a fixed number of storey as expected. It is also observed that floor displacements go on increasing with the increase in storey height with maximum value at the top. It is also significant enough to mention that the maximum rooftop displacements (Δ_{max}) for U-shaped and X-shaped structures are well within the range of 16–22 mm, but for L-shaped and T-shaped structures the rooftop displacements (Δ_{max}) reached higher values of about 600–630 mm. In case of L-shaped and T-shaped RC structures, chances of vulnerability is very high. The reason behind such abnormality can be predicted due to plan properties in those storeys where the location of centre of mass is changed in +X and +Z directions. It is clear that higher values for maximum displacement (Δ_{max}) of the storeys in both X and Z direction irrespective of the plan shape is obtained. Table 1 also shows that among all the four irregularly shaped plan building structures, L-shaped and T-shaped plan structures imparts more damage and severity to the structural stability.

Table 1 Comparative study between various Irregularly shaped structures

Plan shape	Base shear (kN)	Natural time period	Max storey displacement (mm)
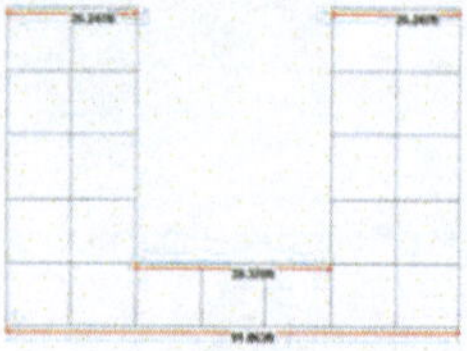	V_b (+X) = 692.97 kN V_b (+Z) = 585.36 kN	T_x = 1.300 s T_z = 1.539 s	Δ_x = 16.99 mm Δ_z = 13.755 mm
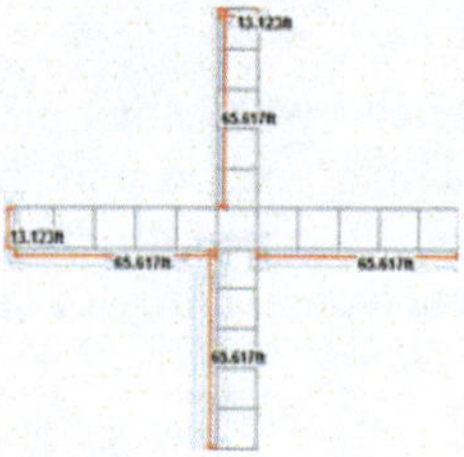	V_b (+X) = 610.32 kN V_b (+Z) = 515.54 kN	T_x = 1.310 s T_z = 1.539 s	Δ_z = 20.11 mm Δ_x = 21.83 mm
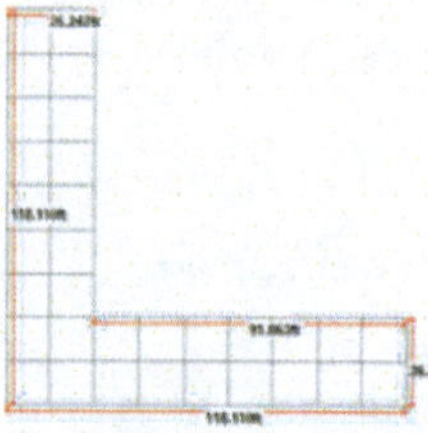	V_b (+X) = 793.412 kN V_b (+X) = 793.412 kN	T_x = 1.911 s T_z = 1.914 s	Δ_x = 694.4 mm Δ_z = 639.89 mm
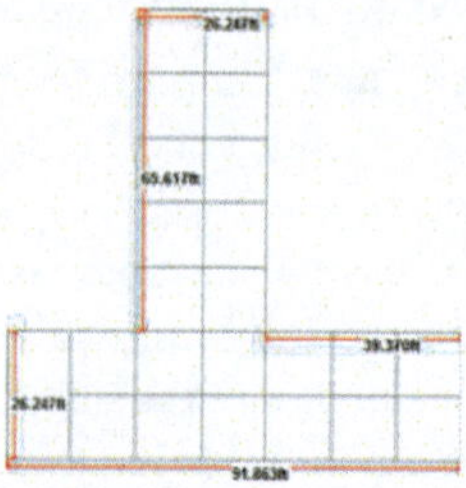	V_b (+X) = 1103.83 kN V_b (+X) = 1101.82 kN	T_x = 1.30 s T_z = 1.432 s	Δ_x = 123.99 mm Δ_z = 118.23 mm

Even the peculiar mode of oscillation and higher stress concentration in L-shaped and T-shaped plan structures, they are most vulnerable to seismic excitations.

6 Conclusions

The current study on seismic behaviour of various irregular plan structures is conducted, some of the major observations have been pointed out. Symmetric R.C.C buildings like the U, X and inverted-T shaped ones generated lesser magnitudes of base shear in both +X and +Z directions when compared to unsymmetrical L-shaped R.C.C building and L-shaped building generated at least 15% higher magnitude of base shear in both +X and +Z directions due to its irregularity.

Base shear values were a cumulative addition of the lateral loads experienced at each storey level with minimum base shear at the top storey and maximum base shear at the bottom storey.

Lateral loads magnitudes increased at a steady rate as one moved higher from a lower storey to upper storey, reaching maximum at the topmost floor level. Irregularly shaped buildings undergo more deformation and hence regular shape building must be preferred. From the analytical studies conducted, it can be stated that even if unsymmetrical plan buildings are constructed, then provision of reinforcement details is a big concern. Results show that there are maximum probabilities for generation of high intensity stresses at interior junction and corners. Well confined reinforcement should be provided in accordance with Indian code for ductile reinforcement for earthquake resistant structures. Well graded steel like Fe 500 and Fe 550 D should be considered as the minimum permissible grade of steel reinforcement to be used for construction purposes.

References

1. Rutenberg A (2002) EAEE task group (TG) 8: behaviour and irregular and complex structures-progress since 1998. In: Proceedings of the 12th European conference on earthquake engineering
2. Gottala A et al (2015) Comparative study of static and dynamic seismic analysis of a multistoried building. Int J Sci Technol Eng 2(01)
3. Sharma M et al (2014) Dynamic analysis of multi-storeyed regular building. IOSR J Mech Civil Eng (IOSR-JMCE) 11(1):37–42
4. Paulay T (1997) Seismic torsional effects on ductile structural wall systems. J Earthquake Eng 1(4):721–745
5. Fajfar P (2000) A nonlinear analysis method for performance-based seismic design. Earthquake Spectra 16(3):573–592
6. Bosco M et al (2002) Effects of in elevation irregularity on the elastic seismic response of in-plan asymmetric buildings. In: Proceedings of the third European workshop on the seismic behaviour of irregular and complex structures
7. Peruš I, Fajfar P (2002) On inelastic seismic response of asymmetric single-storey structures under bi-axial Excitation. In: Proceedings of the third European workshop on the seismic behaviour of irregular and complex structures
8. Peruš I, Fajfar P (2005) On the inelastic torsional response of single-storey structures under bi-axial excitation. Earthquake Eng Struct Dyn 34:931–941

Behaviour of RC Deep Beams Strengthened with Externally Bonded GFRP Fabrics: An Experimental Study

A. Kumari and A. N. Nayak

Abstract This paper aims to study on shear behaviour of reinforced concrete (RC) deep beams strengthened with externally bonded GFRP (Glass fibre reinforced polymer) fabrics. Six numbers of RC deep beams were constructed. Three beams were left as it is to count as the reference beams and the rest three beams were retrofitted with GFRP sheets in various patterns. Two of these beams were wrapped with single layer and one beam was wrapped with double layers of GFRP and tested under single-point loading. The load at initial crack, peak load and contribution due to GFRP were recorded. A key discussion has been made on the enhancement of shear capacity of strengthened beams with respect to the un-strengthened beams. A comparison has also been made between the obtained experimental strength and predicted shear strength evaluated from available design guidelines.

Keywords Cracking · GFRP · Reinforced concrete deep beams · Shear failure · Strengthening

1 Introduction

A deep beam mainly defined as a beam with depth comparable to span. These beams have a wide application in high-rise building, transfer floor, offshore, foundation and pile cap, etc. Because of its geometrical configuration deep beam mostly failed under shear rather than flexure. Thus shear strengthening of deep beam is of a challenging task. In the recent years, fibre reinforced polymer (FRP) composite has been considered as an important material for retrofitting of RC structures because of its advantageous characteristics, such as high tensile strength, lightweight and corrosion resistance [1, 2]. Numerous scientists and industry professionals across the world has been examined the viability of FRP as a substitute for reinforcement bars [1–4]. Several research works have already focused on retrofitting of regular RC beams using FRP (CFRP/GFRP/AFRP) composites in order to study their effectiveness [5–10].

A. Kumari (✉) · A. N. Nayak
Department of Civil Engineering, VSSUT Burla, Sambalpur, Odisha 768018, India
e-mail: archanakumaripadhi@gmail.com

B. B. Das et al. (eds.), *Recent Developments in Sustainable Infrastructure*, Lecture Notes in Civil Engineering 75, https://doi.org/10.1007/978-981-15-4577-1_23

Though, extensive research works have been documented on behaviour and performance of RC deep beams [11–16] only few works have been reported [17–20] on FRP strengthened RC deep beams. Zhang et al. [17] investigated the enhancement in shear capacity of the externally bonded CFRP strengthened deep beams. Islam et al. [18] carried out an experimental study on deep beam strengthened with CFRP sheets and has been proved a strength enhancement of about 40%. Lee et al. [19] studied on behaviour of T-section deep beams externally bonded with CFRP fabrics and showed a remarkable strength gain attributed to CFRP strengthening. Javed et al. [20] reported that, the development of first shear crack has been delayed due to presence of CFRP sheets and thereby increased the shear capacity.

From the past reviews it is observed that only few works have been conducted on shear strengthening of deep beams using externally bonded CFRP composite. In contrast to CFRP, GFRP is cheap, easily available and hence can also be used for externally strengthening of RC deep beams. This paper aims to explore the effectiveness of GFRP sheets for strengthening of RC deep beams.

2 Experimental Investigation

In the current research, six deep beams were casted; three deep beams were tested without GFRP. The rest three beams were tested after strengthening with externally bonded GFRP sheets in different layers. All the above six specimens were tested under single-point loading. Casting of deep beams, material property of GFRP, strengthening of GRFP, test setup, testing and observation, results and discussion are presented in the next subheads.

2.1 Casting of Specimen

Six rectangular deep beams of size, 1000 mm × 500 mm × 100 mm each, were constructed using M_{30} grade concrete. At the same time, three sets of concrete cubes of standard size, i.e. 150 × 150 × 150 mm were also prepared in order to obtain the desired compressive strength. The prepared beams and cubes were allowed for curing in water for the required age. The reinforcement detailing of the deep beam is shown in Fig. 1.

2.2 Testing of GFRP

The GFRP coupons were made by binding the GFRP sheets with the binder and shown in Fig. 2. The binding materials used were Epoxy (CY230) and Hardener (HY951). The above were mixed in the ratio (9:1) and used to bind the FRP. The

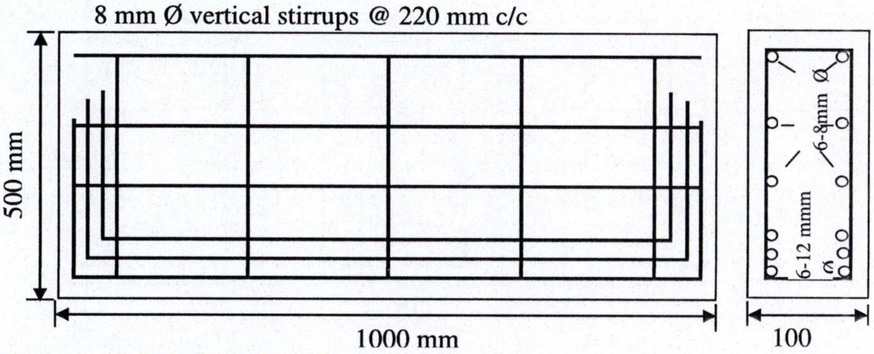

Fig. 1 Reinforcement detailing of deep beam

Fig. 2 Dimensions and photograph of GFRP coupon

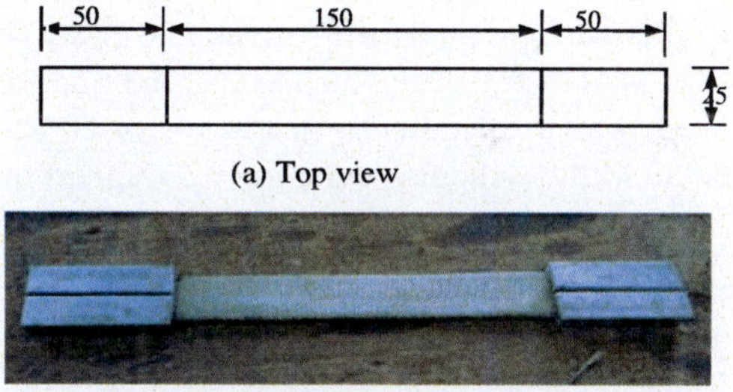

(a) Top view

(b) Coupons prepared from GFRP sheets for testing

All dimensions are in mm

GFRP coupons were tested after being cured properly and various parameters like Young's modulus, both stress and strain at 0.2% yield, load at break, and load at peak were obtained. The average values of the parameters, i.e. load at peak and break, strain and stress at 0.2% yield, and Young's modulus were found to be 7.280 kN, 6.442, 1.231%, 101.40 Mpa and 9825 Mpa, respectively.

2.3 *Wrapping of GFRP*

In the current investigation, 0.275 mm thick GFRP sheet was used for the external shear strengthening of deep beams. Initially, the binder was applied to the prepared concrete surface in the ratio of Epoxy to Hardener is 9:1. Then, bonding of GFRP fabrics was done by putting one layer upon another layer followed by the binder coating. Then the excess binder was squeezed out using a roller to expel the air void in between FRP and the concrete surface. Thereafter, the beams were left for air curing for a period of one week before testing. Three deep beams were kept as it

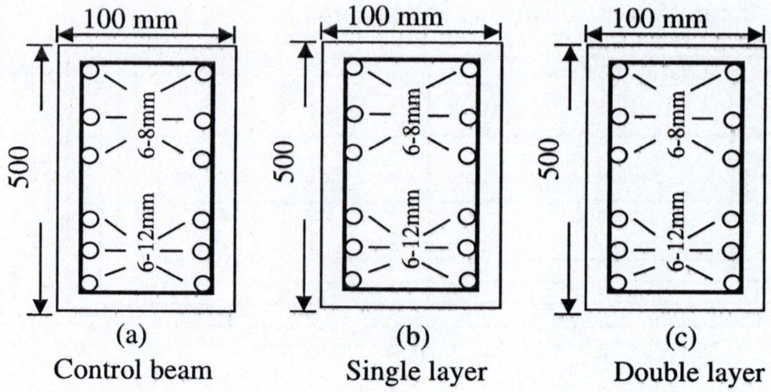

Fig. 3 Cross section of the GFRP wrapped deep beam

is without GFRP and were designated as RCD1, RCD2 and RCD3. Other two deep beams were strengthened with one layer of GFRP fabrics throughout the beam length in U-pattern and designated as RCDF1 and RCDF2, respectively. The remaining one beam was wrapped with two layers of GFRP fabrics and designated as RCDF3. A detail of the wrapping scheme is presented in Fig. 3.

2.4 Test Setup and Instrumentation

The beams were tested in the 100T capacity universal testing machine (UTM) under one point loading. The dial gauge readings at three positions were recorded in order to check the deflections against the load increase as shown in Fig. 4. The loading arrangement is schematically presented in Fig. 4. The final test setup is presented in Fig. 5.

Fig. 4 Schematic diagram of the test set

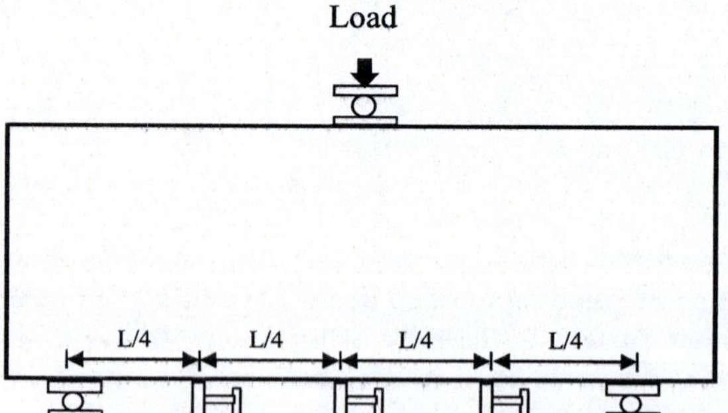

Fig. 5 Complete test set up in UTM

Fig. 6 Crushing of concrete at left support at 370 kN

The beam RCD1 (control beam) was placed directly on the roller supports keeping the effective span of 780 mm. A first hairpin shear crack was observed at the shear span of RCD1 under a load of 310 kN. Further with an increase in the loading, the crack propagated towards the support and loading point and the concrete at left support was crushed corresponding to a load of 370 kN (Fig. 6). Subsequently, the beam failed due to bearing failure at supports at a load of 378.5 kN. In case of beam RCD2 (control beam) to avoid crushing of concrete, bearing plates of size 50 mm × 150 mm were provided at two supports and under the loading point. First diagonal crack was seen at the shear zone at a load of 280 kN. Later, upon increasing the load, bulging of concrete at the loading zone was found at 400 kN. After reaching an ultimate load of 464.5 kN the load drop was observed. The load came to 420 kN and the beam failed completely due to bearing failure at the loading point (Fig. 7). It is found that the failure load of RCD2 is higher than that of RCD1 due to provision of bearing plates at supports for eliminating bearing failure. In order to avoid bearing failure at loading point of RCD3 (control beam), a bearing plate of 200 mm × 100 mm was provided under the loading point. First diagonal crack appeared at the shear span under a load of 450 kN (Fig. 8). Later, the crack directed towards the support as well as loading point. As a result, concrete was crushed under the loading point at 520

Fig. 7 Bearing failure of RCD2 at 464.65 kN

Fig. 8 Diagonal crack in left shear of RCD3 at 450 kN

kN load. With further increasing the load, RCD3 failed due to bearing failure at the loading point under the load of 555.55 kN. It is to be noted that the failure load of RCD3 was higher than that of RCD2 due to provision of a wider bearing plate at the loading point. RCDF1 was wrapped by 1 layer GFRP. As like the first control beam RCD1, this beam was also tested without bearing plate under the supports. As the load increased, the concrete at the right support started crushing inside the GFRP sheets at a load of 380 kN. Gradually the load increased and the beam failed due to bearing failure at the right support due to crushing of concrete. The failure load of RCDF1 was 397.8 kN (Fig. 9). In RCDF2 (1 layer GFRP) bearing plates were provided at both supports and load point as in RCD3. Moreover, the thickness of the bearing plate under the loading point was increased to double. As the load increased, the GFRP sheets started delaminating from the beam surface followed by tearing of GFRP. At a load of 560 kN, the beam failed in shear followed by both the GFRP delaminations and tearing of GFRP. After removing the GFRP fabrics from the concrete surface, the shear crack was seen at the shear span as shown in Fig. 10. It is also observed that, by providing thicker bearing plate at the loading and support point, the bearing failure can be avoided. In the case of RCDF3, same test setup as in RCDF2 was made. As the load increased, the GFRP sheets started delaminating

Fig. 9 Bearing failure of RCDF1 at 397.8 kN

Fig. 10 Shear crack in left shear span of RCDF2

Fig. 11 Spalling of concrete inside RCDF3

in an inclined manner at a load of 460 kN at the shear span. Later, the concrete was observed to be bulge near the loding point under a load of 700 kN. As a result, the GFRP sheets got delaminated completely at that region. The ultimate load taken by the beam was 710.65 kN. The failure of the beam was due to spalling of concrete (Fig. 11) followed by shear failure.

3 Test Results and Discussions

The obtained test data, i.e. compressive strength, cracking/delamination load and ultimate strength, are presented in Table 1. It is observed that though RCD1, RCD2 and RCD3 are identical with respect to dimension and steel reinforcement detailing but the collapse load for each one of them is not the same. The failure load of RCD3 is higher than that of RCD1 and RCD2. Likewise, the failure load of RCD2 is higher than that of RCD1. It is due to the reason that, bearing plates were not introduced at supports and loading point for RCD1. For RCD2 bearing plates were introduced at supports only. For RCD3 the bearing plates at supports as well as

Table 1 Experimental results obtained from the test

Beam	28 days average characteristic cube strength (N/mm^2)	Cracking/delamination load (kN)	Ultimate load (kN)
RCD3	40.29	310.00	555.55
RCDF2	39.38	560.00	584.1
RCDF3	40.17	460.00	710.65

under the loading point were provided for eliminating the bearing failure at supports or loading point. Hence RCD3 is taken as control beam. In case of RCDF1, though one layer of GFRP was wrapped around the three sides of RC deep beam, but no bearing plates were introduced at supports as well as loading points due to which failure occurred at a lower load of 397.8 kN, followed by crushing of concrete at right support (bearing failure). The failure load of this beam is also lower than that of the beams without GFRP (RCD2 and RCD3). However, the beam RCDF2 wrapped with one layer of GFRP and supported with bearing plates exhibited higher load than that of RCDF1. Hence, it is worth mentioning that bearing area at the loading and support point is playing an important role in failure pattern and shear strength of deep beams. Moreover, the beam wrapped with two layers of GFRP sheets and supported with bearing plates at both the loading and support points exhibited highest ultimate load of 710 kN.

3.1 Load Versus Deflection

The load versus displacement curves for RCD3, RCDF2 and RCDF3 are shown in Fig. 12. From Fig. 12, it is observed that maximum displacements are 9.2 mm, 12.1 mm and 10.6 mm for beams RCD3, RCDF2 and RCDF3, respectively, indicating not only enhancement of shear strength, but also enhancement of ductility.

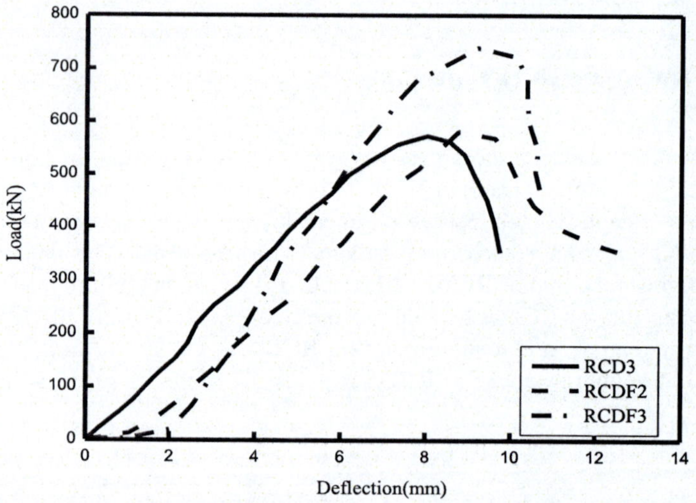

Fig. 12 Load versus deflection graph at mid-span

4 Comparison of Measured Load to Calculated Load from the Source Available in the Literature

4.1 ACI 318-99

The contribution due to concrete can be evaluated by using Eq. (1)

$$V_c = \begin{cases} \left(3.5 - 2.5\dfrac{M_u}{V_u d}\right) \times \left(0.15\sqrt{f_c} + 120\rho_w \dfrac{V_u d}{7 \times M_u}\right) b_w d \\ \text{less than } 0.5\sqrt{f_c} b_w d \text{ (MPa)} \\ \text{Eqs. (11)–(29), ACI 318-99 Code} \end{cases} \tag{1}$$

The contribution due to shear reinforcement can be evaluated using Eq. (2)

$$V_s = \left[\frac{A_v}{s}\left(\frac{1 + \frac{l_n}{d}}{12}\right) + \frac{A_{vh}}{s_2}\left(\frac{11 - \frac{l_n}{d}}{12}\right)\right] f_y$$
$$\text{(Eqs. (11)–(30), ACI 318-99 Code)} \tag{2}$$

4.2 ACI 440.2R-02 Code

The FRP contribution to shear strength was calculated using Eq. (3) of ACI 440.2R-02 code.

$$V_f = \frac{A_{fv} f_{fe} (\sin\alpha + \cos\alpha) d_f}{s_f} \tag{3}$$

The comparison of the measured ultimate load with calculated ultimate load is given in Table 2.

Table 2 Comparison of measured load to calculated load from the source available in the literature

Beam	ACI 318-99		ACI 440	Total shear force	Total calculated load	Experimental load	P_{EXP}/P_{CAL}
	V_C	V_S	V_f	V	P_{CAL}	P_{EXP}	
RCD3	114.39	116.85	0	231.24	462.48	555.55	1.20
RCDF2	114.39	116.85	77.08	308.32	616.64	584.1	0.94
RCDF3	114.39	116.85	88.09	319.33	638.66	710.65	1.11

5 Conclusion

From the experimental investigation, the following conclusions are drawn for deep beams externally wrapped with GFRP fabrics.

1. The shear capacity of RC deep beam can be upgraded using externally bonded GFRP sheets.
2. The ultimate strength of strengthened beams increases as compared to control beam.
3. Initial shear cracks/tearing of GFRP/de-bonding of GFRP got delayed in case of retrofitted beams.
4. The ultimate load of RCDF3 retrofitted with two-layer u-wraps GFRP was found to be higher as compared to RCDF1 and RCDF2 (retrofitted with 1 layer). The shear strength enhancement is proportional to increase in the GFRP layer. The findings may be verified with further testing of more number of deep beams retrofitted with increased layer of GFRP fabrics.
5. Bearing failure in deep beams is mainly dependent on the area of bearing plate provided at supports as well as at load point.
6. The predicted shear strength from the source available in the literature is conservative with the test results.

References

1. Khalifa A, Nanni A (2002) Rehabilitation of rectangular simply supported RC beams with shear deficiencies using CFRP composites. Constr Build Mater 16(3):135–146
2. Zhang Z, Hsu T (2005) Shear strengthening of reinforced concrete beams using carbon-fibre-reinforced polymer laminates. J Compos Constr 9(2):158–169
3. Mosallam AS, Banerjee S (2007) Shear enhancement of reinforced concrete beams strengthened with FRP composite laminates. Compos Part B 38:781–793
4. Sen T, Reddy HNJ, Shubhalakshmi BS (2012) Shear strength study of RC beams retrofitted using vinyl ester bonded GFRP and epoxy bonded GFRP. Civil Environ Res 2(2):23–33
5. Chaallal O, Nollet MJ, Perraton D (1998) Shear strengthening of RC beams by externally bonded side CFRP strips. J Compos Constr 2(2):111–113
6. Khalifa A, Nanni A (2000) Improving shear capacity of existing RC T-section beams using CFRP composites. Cement Concr Compos 22(2):165–174
7. Li A, Assih J, Delmas Y (2001) Shear strengthening of RC beams with externally bonded CFRP sheets. J Struct Eng 127(4):374–380
8. Adhikary BB, Mutsuyoshi H (2004) Behaviour of concrete beams strengthened in shear with carbon-fibre sheets. J Compos Constr 8(3):258–264
9. Abbas JL, Abd SM (2012) Shear capacity and deflection response of RC beams strengthened in shear with U-shaped CFRP wraps. Eur J Sci Res 82(2):265–282
10. Priyadharshini E, Veerakumar R, Selvamani P, Kaveri S (2016) An experimental study on strengthening of reinforced concrete beam using glass fiber reinforced polymer composites. Indian J Sci Technol 9(2)
11. Zhang N, Tan KH (2007) Size effect in RC deep beams: experimental investigation and STM verification. Eng Struct 29(12):3241–3254

12. Yang KH, Ashour AF (2008) Load capacity of reinforced concrete continuous deep beams. J Struct Eng 134(6):919–929
13. Mohammadhassani M, Jumaat MZ, Ashour A, Jameel M (2011) Failure modes and serviceability of high strength self compacting concrete deep beams. Eng Fail Anal 18(8):2272–2281
14. Kim HS, Lee MS, Shin YS (2011) Structural behaviours of deep RC beams under combined axial and bending force. Procedia Eng 14:2212–2218
15. Suresh GS, Kulkarni S (2016) Experimental study on behaviour of RC deep beams. Int Res J Eng Technol (IRJET)
16. Liu J, Mihaylov BI (2016) A comparative study of models for shear strength of reinforced concrete deep beams. Eng Struct 112:81–89
17. Zhang Z, Hsu CT, Moren J (2004) Shear strengthening of reinforced concrete deep beams using carbon fiber reinforced polymer laminates. J Compos Constr 8(5):403–414
18. Islam MR, Mansur MA, Maalej M (2005) Shear strengthening of RC deep beams using externally bonded FRP systems. Cement Concr Compos 27(3):413–420
19. Lee HK, Cheong SH, Ha SK, Lee CG (2011) Behavior and performance of RC T-section deep beams externally strengthened in shear with CFRP sheets. Compos Struct 93(2):911–922
20. Javed MA, Irfan M, Khalid S, Chen Y, Ahmed S (2016) An experimental study on the shear strengthening of reinforced concrete deep beams with carbon fiber reinforced polymers. KSCE J Civil Eng 20(7):2802–2810

Seismic Behaviour of a Typical Rail Bridge Using North-East India Specific Synthetic Ground Motions Under Multi-support Excitation

Aakash Kumar and Nirmalendu Debnath

Abstract Seismic excitations are time varying and spatially varying waveforms depending on the seismotectonics of a region. Analysis of a long structure like bridges, pipelines, dams using uniform seismic excitation is not a realistic approach as the seismic excitation does not remain stationary at every support location. Thus, multi-support seismic excitation is performed to obtain seismic responses of a structure under a realistic situation. In this paper, a replica of a five-span railway bridge (5 spans @ 26.24 m) located in the seismically active north-east zone of India has been studied. A finite element modelling of the total bridge has been created in the SAP2000 software. Synthetic ground motions of moment magnitude (Mw) 8.7 with wave velocity of 1200 m/s have been generated as per the geotechnical and geological conditions of this region considering Oldham and Dauki faults by Stochastic Model Simulation (SMSIM). The synthetic acceleration response has been used for performing nonlinear time-history analysis for both uniform and multi-support seismic excitation study. Nonlinear time-history analysis has been performed to obtain the dynamic response of the total structure due to seismic excitation. The study concludes that the absolute maximum displacement, shear force, bending moment at the total deck and the piers increase in case of multi-support seismic excitation than uniform seismic excitation. This study shows that the codal provisions should incorporate the effect of multi-support excitation and site-specific ground motion details in the design of important structures to keep them operational in extreme hazardous conditions.

Keywords Rail bridge · Multi-support excitation · Stochastic model simulation · Synthetic ground motion · Nonlinear time-history analysis

A. Kumar (✉) · N. Debnath
National Institute of Technology, Silchar, India
e-mail: akash.mishra2312@gmail.com

N. Debnath
e-mail: nirmalendu.nits.civil@gmail.com

B. B. Das et al. (eds.), *Recent Developments in Sustainable Infrastructure*, Lecture Notes in Civil Engineering 75, https://doi.org/10.1007/978-981-15-4577-1_24

1 Introduction

The purpose of bridges is to overcome any physical obstruction without disturbing the region underneath it. The design and protection of bridges are one of the most important aspects of research and development. In recent years, the span length of the bridges is increasing. Thus, as a solution to the shortcoming of consistent seismic excitation method the adoption of multi-support seismic excitation implements the different ground motion input, i.e. seismic displacements, velocities, accelerations to different bridge foundations. Multi-support excitation was first used in the analysis of the Golden Gate bridge [2]. However, this increase in number of span and the span length results in decrease in the structural stiffness of the multi-span bridges. The spatial variation of the seismic excitation is prominently affected by the multi-support excitation. The effect of multi-support excitation is prominent in large-span structures with multiple number of supports. The effect of excitation is different for different supports because uniform ground motion for all support does not represent the actual scenario. Thus, time-history analysis must be performed considering effect of time delay of ground motion considering geotechnical and geological parameters that may vary with the site location. Synthetic ground motion data are used in the analysis of bridges. The synthetic ground motions of specified magnitude are generated from the original ground motion data along with the implementation of source parameters and site parameters [4]. The method for generating the site-specific ground motion is called synthesizing stochastic process. The site location, geological conditions and geotechnical specifications influence the ground motion. The main objective of this paper is to study the response of a concrete rail bridge of the north-east India under multi-support excitation by generating seismic ground motion from the observed data of a past earthquake.

2 Study Area Details

The study area is mainly concerned to a five-span railway bridge located in Mizoram, India. The region is highly seismically active and is surrounded by various faults. The Eastern Himalayan structures, the Mishmi Massif, the Indo-Myanmar arc, the Brahmaputra Valley and the Shillong plateau comprise the major tectonic background in north-east India. The various faults in the vicinity of the structure are Dauki fault, Kopili fault, Mat fault, Sylhet fault, thrust zone are shown in Fig. 1. The bridge that has been studied in this paper is connecting Kawnpui and Mualkhang Station in Bhairabi–Sairang section of north frontier railway in Mizoram by a new broad-gauge line.

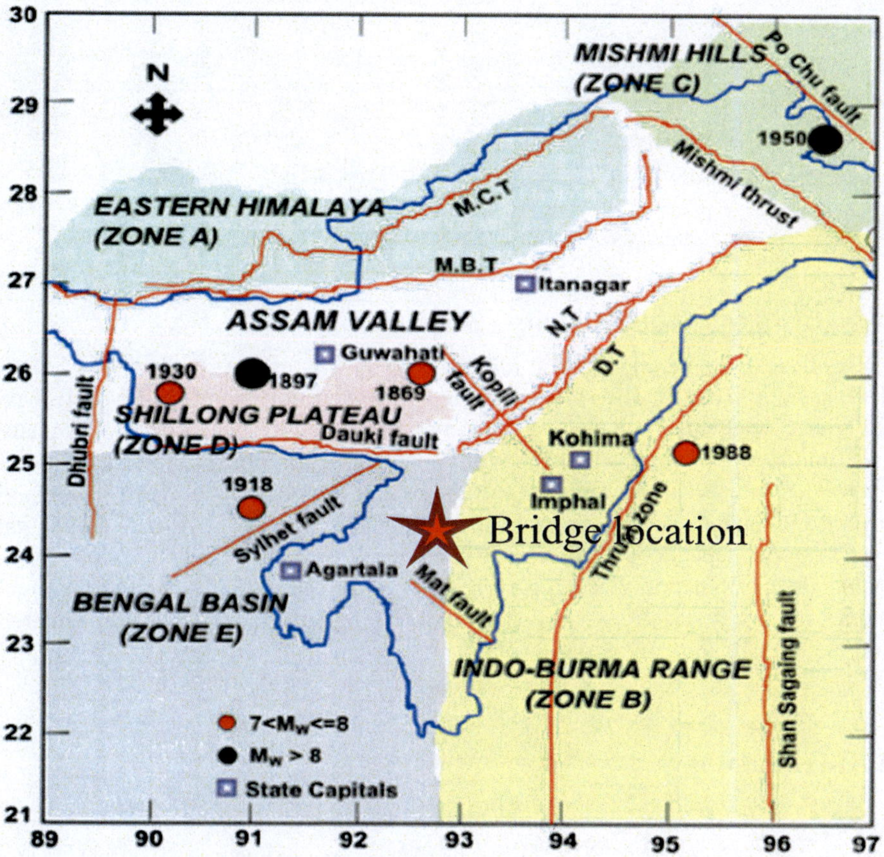

Fig. 1 Map showing different seismic faults surrounding the state of Mizoram and location of the concerned concrete rail bridge

3 Data for Study

3.1 Bridge Details

This paper is related to a concrete beam rail bridge and the schematic diagram of the structure is shown in Fig. 2. The figure is representing the structure on the ground level as the seismic motion for this study is applied at the ground surface level below the piers. The span configuration is 5 spans of 26.2 m each, foundation type is pile foundation below both abutment and piers. Circular tapered shaped pier of diameter 4.0 m and depth 12.5 m is considered. Elastomeric bearing with steel composite girder comprises the superstructure. M40 grade of concrete is used in the design considering seismic zone V.

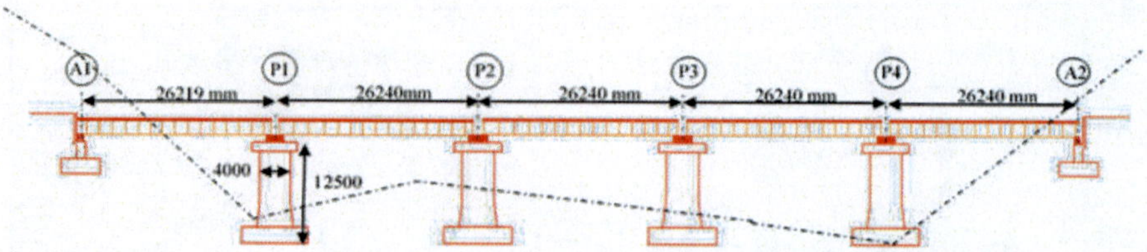

Fig. 2 Schematic diagram of the five-span bridge (5 @ 24 m) located in Mizoram

3.2 Ground Motion Details

The synthetic ground motion [1] is generated by stochastic method SMSIM (Stochastic model simulation) which is a deterministic procedure where a random phase spectrum is modified for a specified duration of time, specific range of earthquake magnitude and also considering the distance of the site from the source. This method is followed mainly in the cases where the earthquake data are not available. The total system works depending on the source parameters, path parameters and site parameters. The combination of all the parameters gives the resultant synthetic ground motion. Four past seismic events in the north-east region of India has been considered and their details are provided in Table 1.

4 Methodology

4.1 Multi-support Excitation Process

The theory [2] behind this method assumes that the displacement vector of a bridge is divided into two parts the superstructure and the substructure. The superstructure components are denoted as t superscript and the substructure notations are denoted with g superscript. Consider N and Ng degrees of freedom for the superstructure and substructure, respectively. The total displacement in this case is sum of relative displacement of the structure with respect to support and quasi-static displacement. The notations *m*, *C*, *K* represent the mass, damping coefficient and stiffness, respectively. The multiple support excitation equation of motion is expressed in Eq. 1. The total structural displacement x^t is a combination of the dynamic displacement x as well as the displacement generated at the supports.

$$\begin{bmatrix} m & mg \\ m_g^T & m_{gg} \end{bmatrix} \begin{Bmatrix} \ddot{x}^t \\ \ddot{x}_g \end{Bmatrix} + \begin{bmatrix} C & C_g \\ C_g^T & C_{gg} \end{bmatrix} \begin{Bmatrix} \dot{x}^t \\ \dot{x}_g \end{Bmatrix} + \begin{bmatrix} K & K_g \\ K_g^T & K_{gg} \end{bmatrix} \begin{Bmatrix} x^t \\ x_g \end{Bmatrix} = \begin{Bmatrix} 0 \\ P_g \end{Bmatrix} \tag{1}$$

Table 1 Parameters used for ground motion simulation [4]

Parameter	1897 Shillong	1950 Assam	1934 Nepal–Bihar	1988 Manipur
Strike	292° N	333.5° N	100° N	284° N
Dip	40° ESE	57.5° SW	30° S	45° E
Focal depth (km)	35	35	20	100
Source/hypocenter (Lat, Lon)	26° N, 91° E	28.38° N, 96.68° E	26.60° N, 86.80° E	26.19° N, 94.89° E
Earthquake magnitude (Mw)	8.1	8.7	8.1	7.2
Maximum magnitude (Mw)	8.7	8.8	8.4	8.2
Fault length (km)	330	600	312	200
Fault width (km)	150	100	80	80
Stress (bar)	159	66	275	83
Shear wave velocity (km/s)	3.25			
Crustal density (g/cc)	2.7			
Kappa	0.05			
Quality factor	342f0.72			
Geometrical spreading	1/R, R < 100 km and 1/R0.5, R > 100 km			
Damping	5			

4.2 Generation of Synthetic Ground Motion

The synthetic ground motion [5] is generated using the basic Eq. 2. The amplitude spectrum $A(r_{ij}, f_k)$ of the ith event that has been recorded at jth station for the kth frequency is represented as a combination of three main functions depending on the site $SI_j(f_k)$, path propagation $P(r_{ij}, f_k)$ and spectral function $SO_i(f_k)$. The parameters are calculated depending on fault configuration that comprises strike, dip, length and width of fault, hypocentre location, maximum earthquake magnitude, stress drop, shear wave velocity, crustal density, site amplification, damping and shear wave quality factor.

$$A(r_{ij}, f_k) = SO_i(f_k)xSI_j(f_k)xP(r_{ij}, f_k). \tag{2}$$

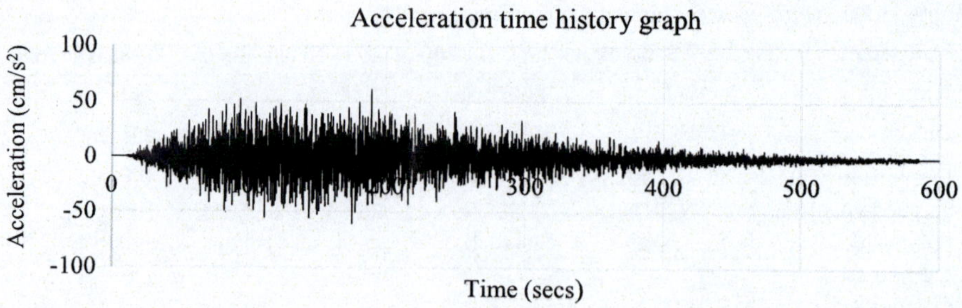

Fig. 3 Synthetic ground motion data simulated for Shillong 1897 earthquake

5 Analysis Procedure

5.1 Synthetic Ground Motion Generation

The synthetic ground motion data has been generated by using stochastic model simulation (SMSIM) from the data mentioned in Table 1 for all the four seismic. The peak ground acceleration of all the four synthetic earthquakes has been found and it is observed that Shillong 1897 data gives maximum PGA of 0.06 g. Thus, this data has been used for the analysis of the structure. The generated motion for 1897 Shillong is shown in Fig. 3.

5.2 Bridge Modelling

The bridge has been modelled in SAP2000. Modal analysis has been performed to check if the structural orientation in the modelling is alright. The extrude view of the bridge model is shown in Fig. 4. The fundamental time period of the structure

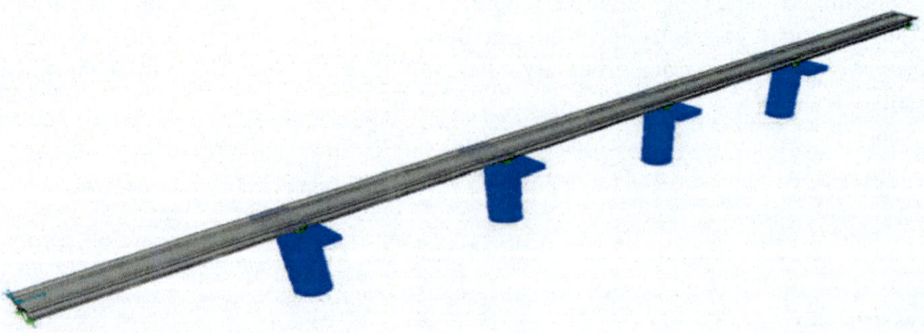

Fig. 4 Extrude view of the bridge modelled in SAP2000

obtained is 1.6 s which is within the permissible range. The different mode shape patterns clarify that the deck of the bridge is highly affected mainly in the lower modes of 1–5 and the higher modes 6–10 majorly affects the piers of the bridge.

5.3 *Application of Uniform and Multi-support Excitation*

The acceleration time-history data has been used for the uniform excitation method where it is considered that all the supports are excited with similar type of excitation at the same time. This acceleration time-history data has been integrated twice by Newmark's β method to obtain the displacement time-history graph for the multi-support excitation. The displacement time-history data has been applied to each support of the bridge pier in a definite load pattern such that time lag of 0.025 s obtained from Eq. 3 has been used. Shear wave velocity of 1200 m/s has been considered till now for class A site.

$$\text{Time lag} = \frac{\text{Bridge span}}{\text{Wave velocity}}. \tag{3}$$

6 Results and Discussions

The multi-support excitation gives higher values with respect to the uniform seismic excitation. The variations have been quantified in terms of percentage. The support moment exceeds by 68%, intermediate span moment exceeds by 138%, shear force exceeds by 218% at supports and 78% at spans, axial force exceeds by 127% at supports and 84% at span and the longitudinal displacement exceeds by 78.5% for the span. The multi-support excitation shows variation in piers also. About 110% of difference from the results of uniform excitation has been observed in case of axial forces and the maximum lateral displacement of each pier. The variations are clearly observed in Figs. 5, 6, 7, 8 and 9 and Table 2. The various references can be followed for the validation of these results. Yoon and Nazmy [6] discussed a four-span concrete bridge where variation of around 7 times in the results of uniform and multi-support excitation has been observed. Again, Hai et al. [3] says a multi-span portal bridge has been analysed and about 5–8 times variation in uniform and multi-support excitation has been observed in case of shear force and bending moments of the structure.

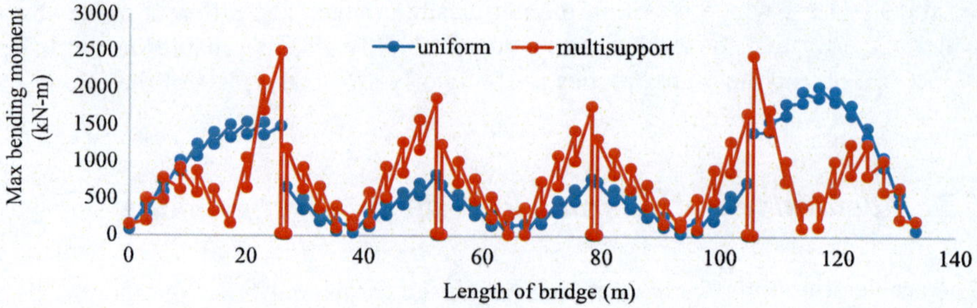

Fig. 5 Variation of bending moment along total span

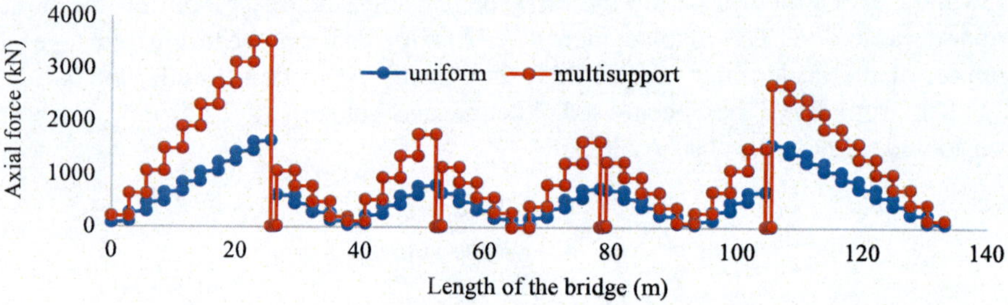

Fig. 6 Variation of axial force along total span

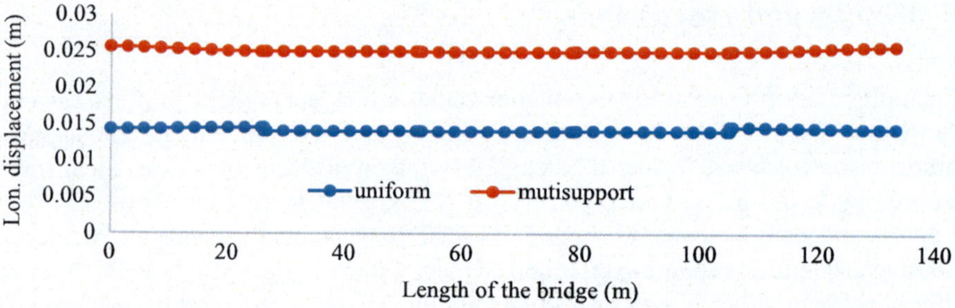

Fig. 7 Variation of longitudinal displacement of total bridge along span

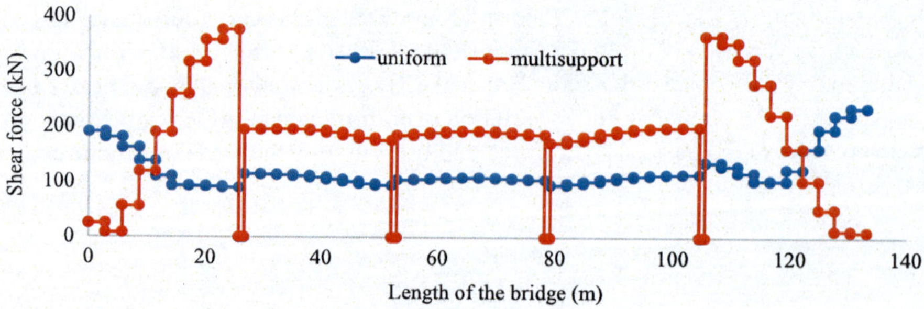

Fig. 8 Variation of shear force along total span

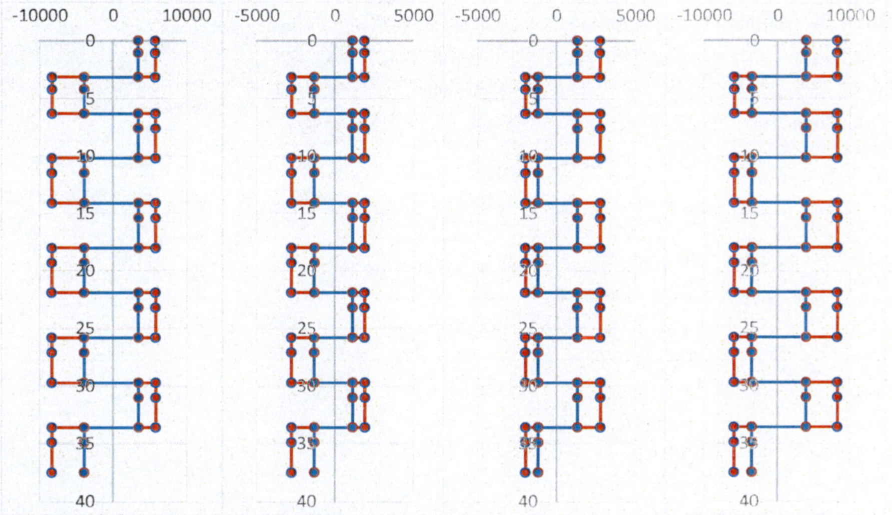

Fig. 9 Variation of axial force (kN) generated in each pier due to multi support and uniform seismic excitation with respect to depth of pier

7 Conclusion

The effect of the multi-support excitation must be incorporated in the design by laws so that the design proceedings are modified that will cause more efficient design of new structures and retrofitting if existing structure for preventing damage. The site location is the guiding criteria for this analysis. If the shear wave velocity of the site location increases, the time lag decreases such that the variation produced by the multi-support excitation will be minutely different from the uniform excitation process. If the analysis is performed for various velocities then the pattern of the change of the results can be depicted. The distance of the site location from the fault zone is also a guiding criterion as the variation of the peak ground acceleration comes into play. The synthetic ground motion changes with the change in the distance of the fault location from the site.

Table 2 Summary of variation in results due to uniform and multi-support excitation methods

Parameter	Position	Uniform excitation	Multi-support excitation	Percentage change (%)
Max. bending moment (kN m)	At supports	1487	2500	68
	Mid of end span	2033	540	−74
	Mid of inter-span	160	382	138
Maxi. shear force (kN)	At supports	118	376	218
	Mid of end span	140	190	36
	Mid of inter-span	108	193	78
Max. axial force (kN)	At supports	1550	3529	127
	Mid of end span	1035	1900	84
	Mid of inter-span	120	530	342
Max. axial force in piers (kN)	Pier 1	3937	8330	112
	Pier 2	1295	2755	113
	Pier 3	1330	2782	109
	Pier 4	3964	8337	110
Max. long displacement (mm)	Span	14	25	78.5
Max. lateral displacement (mm)	Piers	16	35	118

References

1. Boore D (1983) Stochastic simulation of high-frequency ground motions based on seismological models of the radiated spectra. Bull Seismol Soc Am 73(6):1865–1894
2. Chopra Anil K (1995) Dynamics of strucutres. Prentice Hall, New Jersey
3. Hai Z, Li J, Jun-nan L (2013) Nonlinear analysis of pounding between decks of multi-span bridge subjected to multi-support and multi-dimensional earthquake excitation. Springer 20:2546–2554
4. Nath S, Raj A, Thingbaijam K, Kumar A (2009) ground motion synthesis and seismic scenario in Guwahati city—a stochastic approach. Seismol Res Lett 80(2)
5. Nath S, Raj A, Sharma J, Thingbaijam K, Kumar A, Nandy D, Bansal B (2008) Site amplification, Qs, and source parameterization in Guwahati region from seismic and geotechnical analysis. Seismol Res Lett 79(4)
6. Yoon C, Nazmy A (2001) Response factors for bridges subjected to uniform and multiple support excitations. KSCE J Civil Eng 5(2)

Seismic Control of Soft Storey Buildings Using LRB Isolation System

Soumitri Soumyarani Choudhury and Sanjaya Kumar Patro

Abstract Sustainable development not only deals with the protection of environment but also with the development that can be sustained for generations. Large part of the world is subjected to earthquake forces, which demands that structural engineers have to design the buildings so that it can withstand the consequence of these earthquakes. In order to mitigate this problem, seismic base isolation technique is used in recent years. In seismic base isolation, laminated rubber bearing (LRB) is the most extensively used technology, because of its fruitfulness and reliability. The principal purpose of seismic isolation is to lessen the seismic demand on structure. Isolation works on the principle of increasing structural time period. Hence decreasing the vulnerability of earthquake demand on the structure. In urban environment, for the purpose to provide parking facility at the ground storey of building, the concept soft storey has taken its place. However, the buildings with soft storey are most vulnerable to earthquake. In Bhuj earthquake (2001), most of the buildings damaged were of soft storey type. This investigation is to find out the efficacy of laminated rubber bearing isolation system in managing the reciprocating action of the soft storey frame model for the installation of isolation system at the top and at the bottom of soft storey column when subjected to El Centro, 1940 ground motion by using state-space method and a comparison is made between structures with and without infill wall. The response behavior in terms of absolute acceleration, inter storey drift, and storey shear have been investigated, for example, building models. With the use of isolation system, storey shear for all storey level is reduced. When LRB is placed at bottom of soft storey column base shear is reduced by 31.91% and absolute top floor acceleration is reduced by 42.23% that of soft storey building with fixed base.

S. S. Choudhury
Department of Civil Engineering, PMEC, Brahmapur, Odisha, India
e-mail: pmecbam@gmail.com; soumitri.s@gmail.com

S. K. Patro (✉)
Department of Civil Engineering, Veer Surendra Sai University of Technology, Burla, Odisha, India
e-mail: head_civil@vssut.ac.in; litusanjay@yahoo.com

B. B. Das et al. (eds.), *Recent Developments in Sustainable Infrastructure*, Lecture Notes in Civil Engineering 75, https://doi.org/10.1007/978-981-15-4577-1_25

Keywords Earthquake-resistant building · Laminated rubber bearing · Soft storey building · Base isolation

1 Introduction

In practical world, most of the real structures have open ground space for parking and other commercial uses. These structures have to be modeled as soft storey building to know their dynamic response during earthquake motion. In soft storey buildings at the ground storey level, there is sudden decrease in values of mass and lateral stiffness and this results in higher inter storey drift. Many researchers investigated on soft storey buildings with different isolation systems. However, the attempt of analyzing the effect of isolation system by varying its location on soft storey building is few. This report tries to make an effort in the analysis of a soft storey building using laminated rubber bearing isolation system by varying its location position on building. The principal purpose of seismic isolation is to lessen the seismic demand on structure. Isolation is an efficient mean to reduce seismic demand on structures. Isolation works on the principle of increasing structural time period. Hence decreasing the vulnerability of earthquake demand on the structure. During earthquake, for isolated structure the whole structure is given a displacement which reduces the inter storey drift in isolated structure but for fixed base structures during the earthquake motion, inertia tends to retain the structure in its initial position and this results in a large inter storey drift and ultimately leading to failure of the structure.

Laminated rubber bearings are composed of horizontal layers of rubber (natural or synthetic) bonded with steel by vulcanization. For higher vertical loads steel plates supports loads in vertical direction and in horizontal direction, they do not restrict movement of rubber layers. This makes rubber bearing more flexible in horizontal direction and rigid in vertical direction. Sliding base isolator is easy to manufacture and lengthen the time period of vibration of structure. But there are certain major limitations, such as (a) it is prone to wear and tear due to friction, (b) after a long period of inactivity it can settle in the dish, (c) when a small earthquake occurs it can lead to a situation where the pendulum get stuck halfway up the bowl. Rubber bearings are having beneficial to use as they do not break even they are pushed past their design capacity. However, it is having a certain limitation because of its large base displacements and excessive structural vibrations, which results due to flexibility of rubber.

The seismic response, for example, soft storey frame building with laminated rubber bearing when subjected to El Centro 1940 ground motion has been considered to evaluate the efficacy of laminated rubber bearings. State-space approach has been adopted to solve the equation of motion of the idealized example system [1].

2 Mathematical Formulation and Example Structure

The equation of motion of an idealized example system with laminated rubber bearing and its solution is discussed here. The mathematical formulation is developed with commonly used assumptions for isolated structures. The formulation is developed based on the following assumptions:

1. Only horizontal ground motion is acted upon the structure and consequence of vertical motion due to rising of the structure is neglected.
2. The base is assumed to have point of contact with rubber bearing.
3. The overturning effect due to uneven distribution of normal force at rubber bearing is neglected.
4. The response behavior is elastic.

The equation of motion of the example system as shown in Fig. 1 is given by

$$M_0\ddot{x}_0 + C_0\dot{x}_0 + K_0x_0 = -M_0r_0(\ddot{x}_b + \ddot{x}_g) \quad (1)$$

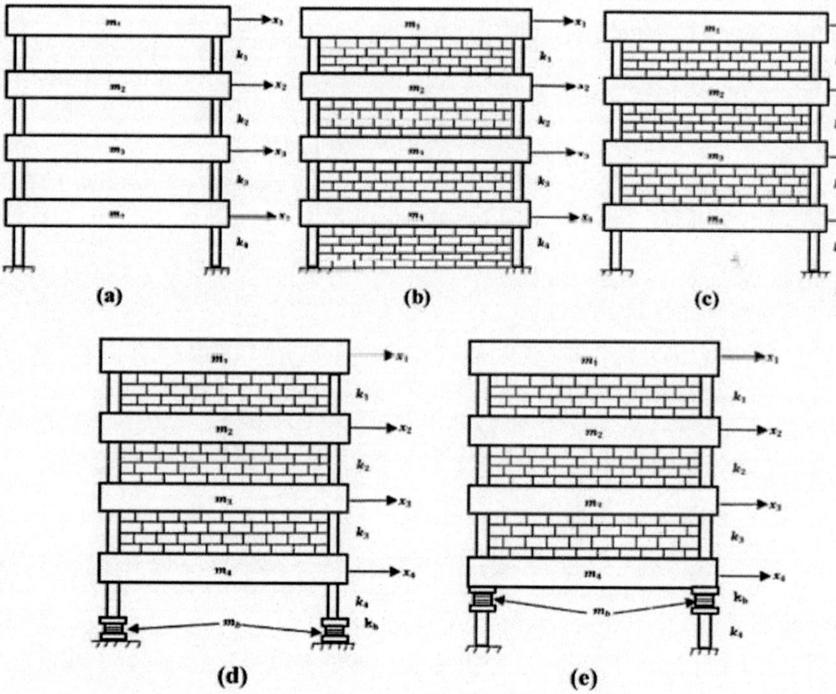

Fig. 1 Model of four-storey building with **a** fixed base model without infill wall, **b** fixed base model with infill wall, **c** soft storey model with fixed base, **d** soft storey model supported on LRB at bottom, **e** soft storey model supported on LRB at top

$$\ddot{x}_b + h(x_b, \dot{x}_b) + \sum_{i=1}^{n} a_i \ddot{x}_i = -\ddot{x}_g \tag{2}$$

with the function $h(x_b, \dot{x}_b)$ depending on the base isolation system used.

For laminated rubber bearing system this function can be defined as

$$h(x_b, \dot{x}_b) = 2\xi_b \omega_b \dot{x}_b + \omega_b^2 x_b \tag{3}$$

where x_b is the relative displacement between the base of the structure and the ground, x_i is the displacement of the ith floor relative to the base, $\ddot{x}_g$ is the horizontal ground acceleration, M_0, C_0, and K_0 represents mass, damping, and stiffness matrices of the structure, respectively, r_0 is a influence vector. ω_b is the natural circular frequency of bearing, and its effective damping ratio ξ_b are defined as

$$2\xi_b \omega_b = \frac{c_b}{m + m_b}, \quad \text{and } \omega_b^2 = \frac{k_b}{m + m_b} \tag{4}$$

where k_b and c_b are the horizontal stiffness and damping of the bearing. The mass ratio a_i is defined as the ratio of the mass of the ith floor to the total mass of the structure.

The mass matrix [M] is given by

$$[M] = \begin{bmatrix} m_1 & 0 & \cdots & 0 \\ 0 & m_2 & \cdots & 0 \\ \vdots & \vdots & \ddots & \vdots \\ 0 & 0 & \cdots & m_b \end{bmatrix}$$

where m_b is the mass of base isolator.

The stiffness matrix is given by

$$[K] = \begin{bmatrix} k_1 & -k_1 & 0 & \cdots & 0 \\ -k_1 & k_1 + k_2 & -k_2 & \cdots & 0 \\ 0 & -k_2 & k_2 + k_3 & \cdots & 0 \\ \vdots & \vdots & \vdots & \ddots & \vdots \\ 0 & 0 & 0 & \cdots & -k_n \\ 0 & 0 & -k_n & \cdots & k_n + k_b \end{bmatrix}.$$

The damping matrix taken is Rayleigh damping. It is one type of proportional damping where the damping matrix is a linear combination of the mass and stiffness matrix, i.e.,

$$[C] = \alpha[M] + \beta[K]$$

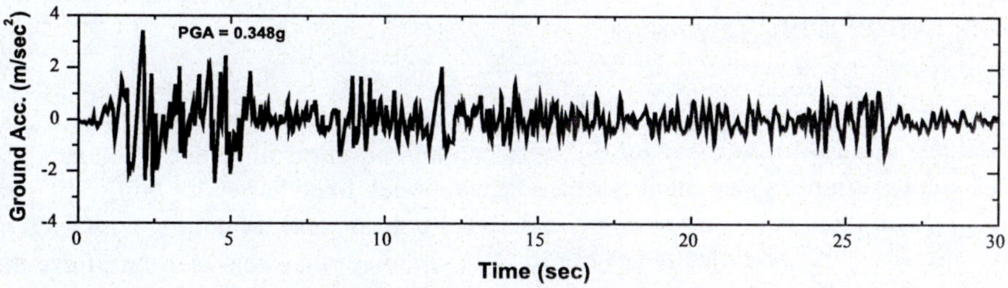

Fig. 2 Acceleration time history of El Centro 1940 ground motion

where α and β are Rayleigh damping coefficient.

Solution procedure: A wide range of dynamic problems uses state-space method for calculating the response of the system. In this paper state-space response using direct integration is used for the analysis of the example structure. In this method response of the system using both the velocity and displacement as independent variables is analyzed, and these variables are called states. The objective of the numerical analysis using the integration method is to integrate the forcing function. In this study, to represent the forcing function delta forcing function method is used.

3 Earthqauke Ground Record

The main characteristic of ground motion includes peak ground accelerations (PGA), duration of strong motion, and its frequency content. Many studies use the NS component of El 1940 acceleration time history which is shown in Fig. 2. Hence the effect of 1940 El Centro ground motion (PGA = 0.348 g) on the response behavior of example system is studied in this paper.

4 Response Evaluation of the Example Building

The four-storey reinforced concrete frame building with, (i) fixed base without infill wall (bare frame), (ii) fixed base with infill wall (fully infill wall), (iii) soft storey with fixed base, (iv) soft storey with LRB at bottom, and (v) soft storey with LRB at top, has been considered in evaluating the seismic performance of laminated rubber bearing isolator. The floor mass for bare frame case is taken as 350.2 kg, for infill 5353 kg and mass of LRB is taken as 4035.6 kg (0.75 times of LRB mass is equals to mass of above floor), respectively. The time period of LRB is taken as 2 s and accordingly the stiffness of LRB is calculated. The damping of the structure and LRB is taken as 5% and 10%, respectively. The stiffness of the floors without infill is taken as 573,600 N/m and stiffness of the floor with infill is taken as 117,861,912 N/m [2].

4.1 Maximum Responses

The seismic responses such as inter storey drift, absolute acceleration, floor displacement, and storey shear for different example systems have been evaluated. A different example system includes bare frame model, fixed base with infill wall, soft storey with fixed base, soft storey with LRB at bottom, and soft storey with LRB at top. Since the prime objective of the LRB base isolation system is to minimize the peak responses, the study of peak responses allows one to evaluate the effectiveness of laminated rubber bearing system.

Figure 3 shows the relative displacement of different storeys with all five building models. From the figure, it can be seen that displacement of fixed base with infill wall is very low compared to bare frame model. This is due to the increase in stiffness of fully infill wall model. When soft storey model is considered, its displacement is much larger than the fully infill wall model because of sudden decrease of stiffness at soft storey level. Considering laminated rubber bearing as isolation system displacement with respect to ground floor for all floor is reduced considerably. However, for soft storey building model with LRB at top displacement with respect to ground floor is increased. This is due to the fact that, as LRB is placed at top of soft storey column, displacement of LRB is more compared to ground floor (soft storey level) due to flexibility of rubber bearing. This increases displacement of all floors above LRB isolation system.

The inter storey drift of all building models is shown in Fig. 4. From the figure, it can be seen that inter storey drift of bare frame model is higher for all floor levels compared to other building models except ground floor. For fully infill wall model, it is lesser for all floor levels because of its increased stiffness due to infill wall. In

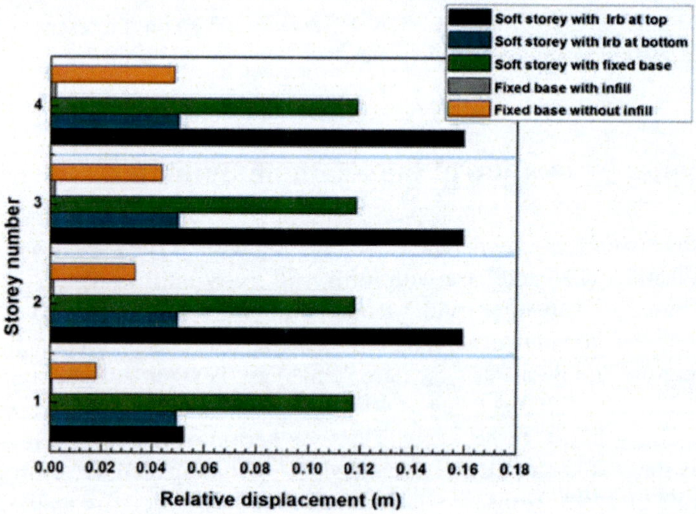

Fig. 3 Relative displacement of different floor for different building models

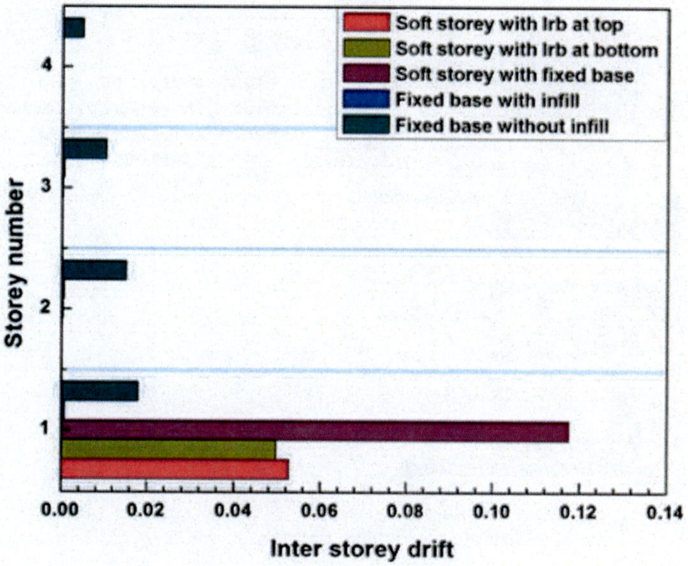

Fig. 4 Inter storey drift of different floor for different building models

case of soft storey building model, inter storey drift for ground floor is higher than all other models and it decreases toward increase in floor level. The first floor inter storey drift of soft storey building with laminated rubber bearing system placed at bottom is reduced by 57.61% that of soft storey building with fixed base, and for soft storey building with LRB at top of soft storey column, it is 55.14%. Considering infill wall inter storey drift for all floor levels is decreased because of isolation system.

The peak absolute acceleration value of all floor levels for all building models is shown in Fig. 5. It can be seen that absolute acceleration of bare frame model is higher for all floor levels excluding ground floor. For ground floor, absolute acceleration of soft storey with fixed base is greater. Considering isolation system absolute acceleration for all floor level is reduced. However, there is a small difference in value of acceleration between soft storey with LRB at top and soft storey with LRB at bottom. Absolute top floor acceleration of soft storey structure building with laminated rubber bearing with isolation system at bottom of soft storey column if reduced by 42.23% that of soft storey building with fixed base, and for soft storey building with LRB at top of soft storey column, it is 7.85%.

The storey shear for all floor levels for all building models is shown in Fig. 6. From the figure, it can be seen that storey shear for fully infill wall model is higher than bare frame model because of increase in mass of fully infill wall model. For soft storey with fixed base, storey shear is higher compared to all building models because of irregularity in stiffness along height of building model. With the use of isolation system, storey shear for all storey level is reduced. Base shear of soft storey building with laminated rubber bearing at bottom of soft storey column reduced by

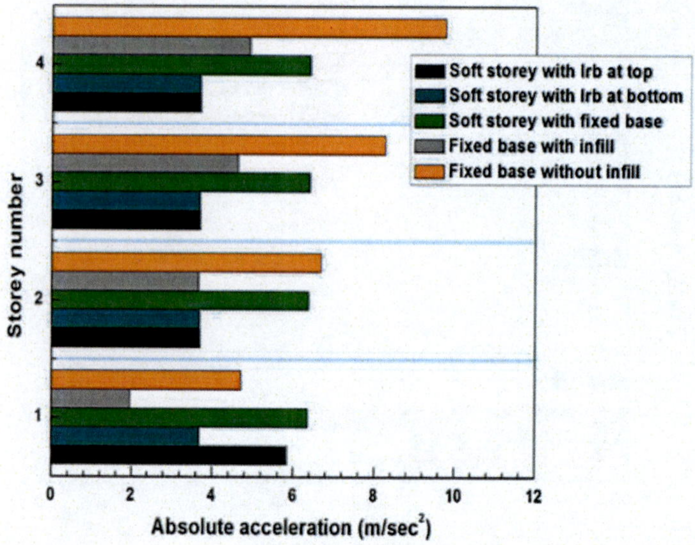

Fig. 5 Absolute acceleration of different floor for different building models

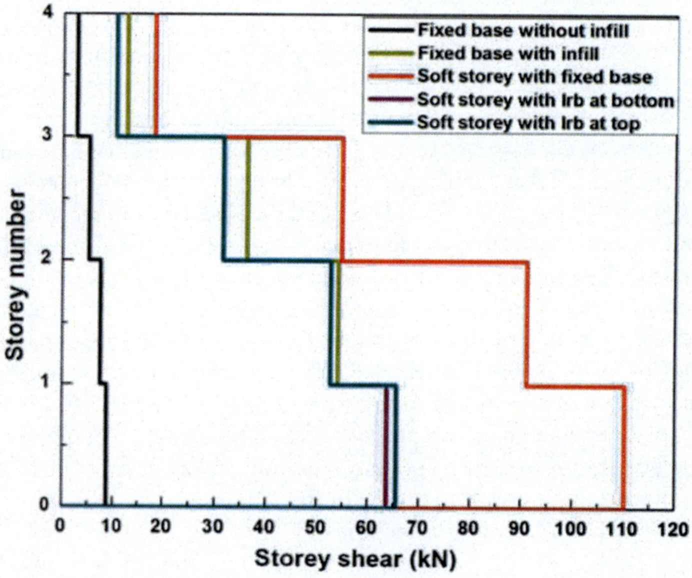

Fig. 6 Storey shear of different floor for different building models

31.91% that of soft storey building with fixed base. This shows the effectiveness of isolation system.

5 Conclusion

In this study research has been carried out on bare frame building, building with fully infill wall, soft storey building with fixed base as well as laminated rubber bearing isolation system. Time history analysis using El Centro ground motion is carried out for all building models. The influence of infill wall and soft storey on seismic response of the building has been investigated. The effect of placement position of laminated rubber bearing on soft storey building is studied as well. Based on the investigations in this paper, it is concluded that

1. The response parameters obtained from the time history analysis which is presented in the form of graph indicates that the reinforced concrete soft storey building with fixed base has undesirable seismic performance due to their insufficient stiffness and drift capacity.
2. The use of laminated rubber bearing in soft storey building shows excellent seismic behavior, in terms of absolute acceleration, base shear and controlled inter storey drift considerably that of soft storey building with fixed base.
3. The position of location of LRB isolator on soft storey building has little impact on seismic response of soft storey buildings.

References

1. Chopra AK (1996) Dynamics of structure, theory and applications to earthquake engineering. Prentice Hall 07632
2. Yang Y-B, Lee T-Y, Tsai I-C (1990) Response of multi-degree-of-freedom structures with sliding supports. Earthquake Eng Struct Dyn 19:739–752
3. IS: 1893 (2002) Indian standard criteria for earthquake resistant design of structures. In: Part 1 General provisions of buildings (fifth revision). New Delhi, Bureau of Indian Standards
4. Chopra AK, Clough DP, Clough RW (1973) Earthquake resistance of buildings with a 'soft' first storey. Earthquake Eng Struct Dyn 1:347–355
5. Hart GC, Wong K (1969) Structural dynamics for structural engineers. Wiley
6. Saiful I, Mohammed J, Mohd Zamin J (2006) Seismic isolation buildings to be a practical reality: behavior of structure and installation technique. J Eng Technol Res 3(4):99–117, ISSN 2006-9790

Comparison of Effectiveness of TLCBD Over LCVA in Vibration Control of Structure Under Non-stationary Earthquake

Shulanki Pal, Bijan Kumar Roy, and Satyabrata Choudhury

Abstract Liquid column vibration absorber (LCVA) is very much useful to mitigate the responses generated by the dynamic forces due to its cost-effectiveness dynamic vibration absorbing property. LCVA is usually a U-shaped tube having different cross-sectional areas in the vertical and horizontal column, normally filled up with water. The basic working principle of LCVA is to dissipate the vibration energy by gravitational force and the viscous force generated by the water present in the column. The damping force generation by the liquid movement through the orifice is very less in case of LCVA. To overcome this aspect, a new system is introduced which is called tuned liquid column ball damper (TLCBD) where a metal ball is installed in the horizontal part of the tube instead of orifice present in the LCVA. In the present study, the optimum performance of the response reduction of the structure is compared for both these systems. The numerical study has been performed by considering a single degree of freedom (SDOF) structure with an LCVA and TLCBD attached on the top of it separately under non-stationary random earthquake. The stochastic structural optimisation has been performed for both the cases and their results are compared. The results are confirmed the disparity between the response reduction of these two damper systems and improved response reduction is achieved for TLCBD system over LCVA system.

Keywords Stochastic structural dynamics · Tuned liquid column ball damper · Liquid column vibration absorber · Optimisation · Response

S. Pal (✉) · B. K. Roy · S. Choudhury
Civil Engineering Department, NIT Silchar, Silchar, Assam, India
e-mail: shulanki@gmail.com

B. K. Roy
e-mail: bijan.roy@rediffmail.com

S. Choudhury
e-mail: scnitsilchar@gmail.com

B. B. Das et al. (eds.), *Recent Developments in Sustainable Infrastructure*, Lecture Notes in Civil Engineering 75, https://doi.org/10.1007/978-981-15-4577-1_26

1 Introduction

Rapid urbanisation and hike in growth rate of population have directed the structures to be more flexible and less in damping properties due to their incremental height. These kinds of vibration sensitive structures are quite susceptive to the dynamic forces like wind and earthquake. To minimise the response generated due to this kind of vibrational forces tuned liquid column damper (TLCD) was introduced by Sakai et al. [1]. TLCD falls under the passive damping devices as there is no need of external source of energy to activate its response mitigation property. The past researchers have studied the response reduction of structures using TLCD and characteristic of different parameters affecting the response under different kind of loadings like sinusoidal loading, seismic loading [2], wind loading [3] and wave motion [4] is notably worthwhile. The optimal damping property for different orifice opening condition of a hybrid TLCD was investigated by Hauron et al. [5]. Further, a lot of modification has been initiated in the TLCD to improve its response mitigation property. LCVA is one of those modifications where the ratio in between the horizontal and the vertical part of the tube retaining the water has been improved [6]. Hitchcock et al. [6, 7] experimented on a rectangular-based unidirectional and bidirectional LCVA to evaluate the characteristics and the response mitigation efficiency considering the structure subjected to sinusoidal loading. In their work, it was concluded that the damping property of the LCVA depends upon the area ratio, length ratio and vertical column height of it. For LCVA limited studies have been performed considering seismic excitation [8]. Different studies have been performed using time history and white noise input. Konar and Ghosh [9] performed a case study considering both the sloshing and oscillating liquid motion of the liquid to evaluate the proper performance of LCVA over other types of liquid column dampers. Chang and Hsu [10] performed a comparative study of LCVA with various other dampers to ensure the better performance of LCVA over others. In Sydney, Samali et al. [11] conducted an experimental study on a 76-storey building subjected to wind load to compare the effectiveness of LCVA over conventional type tuned mass damper. The functional behaviour of LCVA primarily depends upon the three types of forces, first one is gravitational force generated by the liquid mass present, second one is viscous interactive force produced due to liquid interaction with the rigid container and the transitional force developed by the liquid movement within columns. By introducing orifice in the horizontal column part of the LCVA, the efficiency of it can be increased as it introduces better damping property due to head loss [10]. Al-Saif et al. [12] have introduced a ball instead of the orifice in TLCD to modify the damping property of the damper as the ball acts as a rolling orifice. In that study, the structure was subjected to the harmonic loading representing the wind force. In Gur et al. [13] performed a comparative study between the conventional TLCD and TLCBD attached on a SDOF structure subjected to random seismic vibration and verified the efficiency with the effectiveness due to real earthquake ground motion. Normally the gradient-based optimisation procedure is used to evaluate optimised values of the response and other parameters which are less suitable for the nonlinear problems. To overcome this particle swarm optimisation (PSO) can be used to find out the appropriate solution for this kind of nonlinear problem [14, 15].

In this study, the optimised RMS response is compared for LCVA and TLCBD separately. An SDOF structure has been considered with the dampers attached on it independently and the response has been evaluated for different parameters based on properties of damper, structure and excitation. The PSO procedure has been applied for the optimisation process. Non-stationary stochastic earthquake has been considered to compare the effectiveness.

2 Stochastic Response Analysis of LCVA-Structure System and LCVA with Ball-Structure System

The primary structure is considered as SDOF system with the dampers attached on the top of it independently. Figure 1a represents the structure-LCVA system and Fig. 1b shows the schematic diagram of structure-TLCBD system. Here, h denotes the height of the water present in the column. L_e, ρ, B represent the overall length, density of the liquid present and horizontal span of the damper, respectively.

The equation of motion for an LCVA, LCVA-structure system under earthquake motion can be expressed as

$$\ddot{y} + \frac{2C_1}{L_{ee}}\dot{y} + \frac{2g}{L_{ee}}y + p\frac{L_e}{L_{ee}} = -p\frac{L_e}{L_{ee}}\ddot{x}_g \tag{1}$$

$$(1+\mu)\ddot{x} + 2\xi_0\omega_0\dot{x} + \omega_0^2 x + \mu p\frac{L_e}{L_{em}}\ddot{y} = -(1+\mu)\ddot{x}_g \tag{2}$$

where p, L_{ee}, L_{em} denotes the length ratio, total span of an equivalent liquid column and length of an equivalent uniform liquid column of the damper.

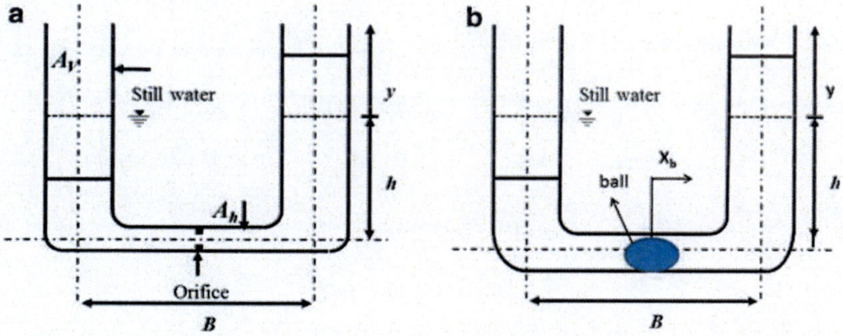

Fig. 1 **a** Schematic diagram of LCVA, **b** schematic diagram of TLCBD

$$L_{ee} = Br + (L_e - B) \text{ and } L_{em} = (B/r) + 2h.$$

Here r is the area ratio and C_1 represents the equivalent linearized damping coefficient and can be expressed as $\sigma_{\dot{y}}\xi r^2/\sqrt{2\pi}$. $\sigma_{\dot{y}}$, ξ is standard deviation of liquid velocity and head loss coefficient, respectively.

$\mu = m_l/m_0$, is the mass ratio, $\gamma = \omega_l/\omega_0$, is the tuning ratio of damper and ω_l can be expressed as $\sqrt{2g/L_e}$. m_l and m_0 are the mass of the liquid and mass of the structure, respectively. ξ_0, ω_0 are the damping ratio and natural frequency of the structure.

The equation of motion for the ball present at the horizontal portion of the TLCBD, the liquid present in the column, structure on which TLCBD is attached can be expressed, respectively,

$$\left(m_b + \frac{J_b}{R_b^2}\right)\ddot{x}_b + d_{eq}\dot{x}_b = \left(\frac{J_b}{R_b^2}\right)(\ddot{x} + \ddot{x}_g) + d_{eq}\dot{x}_l + \left(\frac{2m_l g R_{bt}^2}{L_e}\right)y \quad (3)$$

$$m_l\ddot{y} + (2m_l\xi_l\omega_l)\dot{y} + \left(\frac{2m_l g}{L_e}\right)y = -pm_l(\ddot{x} + \ddot{x}_g) \quad (4)$$

$$\left(m_0 + m_l + \frac{J_b}{R_b^2}\right)\ddot{x} + 2\xi_0\omega_0 m_0\dot{x} + m_0\omega_0^2 x = -\left(m_0 + m_l + \frac{J_b}{R_b^2}\right)\ddot{x}_g - pm_l\ddot{y} + \left(\frac{J_b}{R_b^2}\right)\ddot{x}_b \quad (5)$$

In this equation m_b denotes the mass of the ball and J_b denotes the mass moment of inertia of the ball (=$2m_b R_b^2/5$). d_{eq} is the viscous damping for the motion of the ball due to the liquid present on the tube (= $6\pi\nu R_b$). ν represents the kinematic viscosity of the fluid. ξ_l, ω_l represents the damping in terms of head loss coefficient and the natural frequency of the liquid present in TLCBD. Radius of the ball is denoted by R_b and R_{bt} represents the ratio between the diameter of ball and tube.

The equation of motion for an LCVA-structure or a TLCBD-structure subjected to a seismic excitation can be expressed in a compact form as

$$[M]\{\ddot{u}\} + [C]\{\dot{u}\} + [K]\{u\} = -[M]\{r\}\ddot{x}_g. \quad (6)$$

For LCVA system the mass, stiffness and damping matrix can be denoted as

$$M = \begin{bmatrix} 1 & p\frac{L_e}{L_{ee}} \\ \mu p\frac{L_e}{L_{em}} & 1+\mu \end{bmatrix}, K = \begin{bmatrix} \frac{2g}{L_{ee}} & 0 \\ 0 & \omega_0^2 \end{bmatrix}, C = \begin{bmatrix} \frac{2C_1}{L_{ee}} & 0 \\ 0 & 2\xi_0\omega_0 \end{bmatrix}. \quad (7)$$

For TLCBD system the mass, stiffness and damping matrix can be denoted as

$$M = \begin{bmatrix} 1 & 0 & -\overline{p} \\ 0 & 1 & p\frac{L_e}{L_{ee}} \\ -\overline{\mu} & \mu p\frac{L_e}{L_{em}} & (1+\mu+\overline{\mu}) \end{bmatrix}, K = \begin{bmatrix} 0 & -\overline{k} & 0 \\ 0 & \frac{2g}{L_e} & 0 \\ 0 & 0 & \omega_0^2 \end{bmatrix}, C = \begin{bmatrix} \overline{c} & -\overline{c} & 0 \\ 0 & 2\gamma\xi_l\omega_0 & 0 \\ 0 & 0 & 2\xi_0\omega_0 \end{bmatrix}. \quad (8)$$

Here $\overline{p} = \frac{2}{7}, \overline{\mu} = 8\rho_b R_b R_{bt}^2/15\rho_l L_e$, $\bar{c} = 45\nu/14\rho_b R_b^2$, $\bar{k} = 15\rho_l g/14\rho_b R_b$. ρ_b, ρ_l are the density of ball and the liquid. A nonlinear stochastic structural ground motion at the base of the primary structure has been considered to assess the response of the structure–damper system. A stochastic excitation considering wide-ranging practical situations, a stationary ground motion well known as Kanai–Tajimi model [17, 18] has been used for that. By filtering through soil a white noise excitation acting at the rock bed gives the above.

The filter produces a coloured noise output with Kanai–Tajimi power spectral density (PSD). The equations after filtered condition are stated as

$$\ddot{x}_f(t) + 2\xi_g\omega_g\dot{x}_f(t) + \omega_g^2 x_f(t) = -\ddot{w}(t)(\varphi(t))^2 \tag{9}$$

$$\ddot{x}_g(t) = -2\xi_g\omega_g\dot{x}_f(t) - \omega_g^2 x_f(t) \tag{10}$$

where $\ddot{w}(t)(\varphi(t))^2$ is the white noise intensity at the rock bed with power spectral density $S_0 = 2\xi_g\sigma_{\ddot{z}}^2/\pi\left(1 + 4\xi_g^2\right)\omega_g$ and $\varphi(t)$ is the modulation function proposed by Jennings et al. [19], can be expressed as (Fig. 2)

$$\varphi(t) = \begin{matrix} \left(\frac{t}{t_1}\right)^2 & t < t_1 \\ 1 & t_1 \le t \le t_2 \\ e^{-\beta(t-t_2)} & t > t_2 \end{matrix} \tag{11}$$

Here t_1 and t_2 is assumed as 2 s and 12 s, respectively. β is considered as 0.4 (s^{-1}).

ξ_g, ω_g are the damping ratio and the frequency of the soil, respectively. x_f, $\dot{x}_f$, $\ddot{x}_f$ represents the displacement, velocity and acceleration of the ground. The stochastic structural excitation has been introduced in the formulation by substituting $\ddot{x}_g$ from

Fig. 2 Jenninng's modulation function

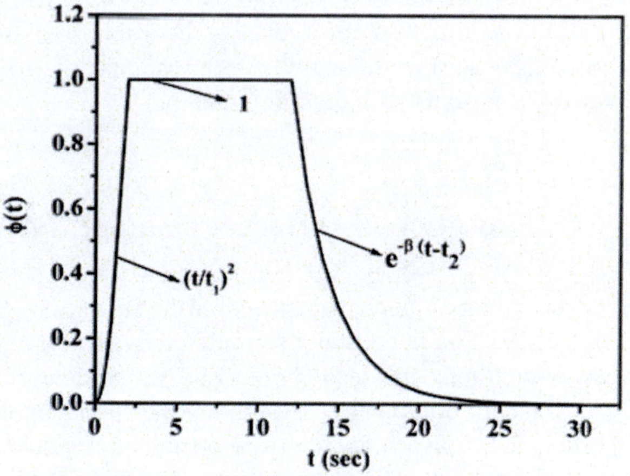

Eq. (10) and combining it with the dynamic equation of structure-damper system, Eq. (6). This dynamic equation can be converted to stead space equation as shown below and responses can be determined.

$$\frac{d\{Y\}}{dt} = [A]\{Y\} + \{w\}. \tag{12}$$

To introduce the state variables, augment state vector is used and can be expressed as $\{Y\} = \left[y, x, x_f, \dot{y}, \dot{x}, \dot{x}_f\right]^T$ for LCVA-structure system and $\{Y\} = \left[y, x_b, x, x_f, \dot{y}, \dot{x}_b, \dot{x}, \dot{x}_f\right]^T$ for TLCBD-structure system.

Here, $[A]$ is an augmented matrix and can be denoted as

$$[A] = \begin{bmatrix} \{0\}_{(n,n)} & \{0\}_{(n,1)} & [I]_{(n,n)} & \{0\}_{(n,1)} \\ \{0\}_{(1,n)} & 0 & \{0\}_{(1,n)} & 1 \\ [M]^{-1}_{(n,n)}[K]_{(n,n)} & \omega_g^2\{r\}_{(n,1)} & [M]^{-1}_{(n,n)}[C]_{(n,n)} & 2\xi_g\omega_g\{r\}_{(n,1)} \\ \{0\}_{(1,n)} & -\omega_g^2 & \{0\}_{(1,n)} & -2\xi_g\omega_g \end{bmatrix} \tag{13}$$

where $\{w\} = \left[\{0\}_{(1,n)} - \ddot{w}\right]^T$.

Here n is the number of response variables. For LCVA $n = 3$ and for TLCBD $n = 4$.

The covariance of the responses evaluation is preferable rather than the direct responses for stochastic analysis. The response covariance matrix can be obtained considering the stochastic structural process to be Markovian [16]

$$\frac{d[C_{YY}]}{dt} = [A][C_{YY}]^T + [C_{YY}][A]^T + [S_{ww}]. \tag{14}$$

The matrix $[S_{ww}]$ comprises rock bed white noise excitation terms and can be expressed as, having all the terms as zero excluding the last diagonal term represented as $2\pi S_0$. At each iteration, the damping and the stiffness matrices terms are required to be updated until the convergence of the response statistics as these are the functions of the response. From the covariance response matrix, the RMSD can be obtained by using Runge-Kutta integration method.

3 Optimisation of Different Damper Systems

The conventional stochastic structural optimization (SSO) problem under random seismic vibration is normally formulated by searching a suitable set of design variables, to minimise the desired objective function over a possible admissible domain, Ω. A standard nonlinear programming problem can be generated by modifying the optimization problem for the system subject to stochastic load. For known properties of the structure and the damper like mass ratio, damping ratio and frequency, the design variables can be achieved. Normally the mean square responses (displacement, acceleration, stress, etc.) the total life-cycle cost or the failure probability of the structure is considered as the objective function in the SSO. In the present analysis, the objective function is the RMS displacement of the SDOF system. The SSO problem so defined leads to a standard nonlinear programming problem [20].

The tuning ratio (γ_{LCVA}) and the head loss coefficient (ξ) are the deterministic optimisation parameters for the LCVA-structure system. Whereas, the tuning ratio (γ_{TLCBD}) and the ball tube diameter ratio (R_{bt}) is the deterministic optimisation parameters for the TLCBD-structure system. As the RMSD of the structure depends upon these corresponding parameters, so to evaluate the response these parameters are needed to be determined. So the design vectors can be described as $\bar{b} = (\gamma_{LCVA}, \xi)$ and $\bar{b} = (\gamma_{TLCBD}, R_{bt})$. To search a suitable set of design variables over a possible admissible domain, Ω, the conventional SSO problem under random seismic vibration is formulated.

$$\text{Find } \bar{b} = (\gamma_{LCVA}, \xi) \text{to minimise } f = \sigma_x \tag{15}$$

$$\text{And Find } \bar{b} = (\gamma_{TLCBD}, R_{bt}) \text{ to minimise } f = \sigma_x. \tag{16}$$

Here σ_x is the RMSD response of the primary structure which can be evaluated by taking root mean square of the covariance matrix.

$$\sigma_x = \sqrt{CYY(2, 2)} \text{ for LCVA} \tag{17}$$

$$\sigma_x = \sqrt{CYY(3, 3)} \text{ for TLCBD.} \tag{18}$$

3.1 PSO Algorithm

To solve this type of nonlinear unconstrained optimisation problems, the optimization algorithm for PSO in MATLAB has been generated. It is a process to simulate simplified model using swarming theory. Individual improvement technique is used for population competition and cooperation. In this process, each solution is called a particle. A group of particle search together in parallel way to explore the best position. By keeping the memory of all the position of each particle, PSO combines both the local and the global search. Self-experience and neighbouring experience is used for finding out the best solution. In this study, the investigation of RMSD for structure using both the dampers has been performed by solving the above matrices given Eqs. 17 and 18 using PSO.

4 Numerical Study

To determine the effectiveness of the LCVA attached on the top of the structure, a primary structure has been modelled as an SDOF system subjected to stochastic earthquake excitation. Comparison of the response of the structure for an LCVA attached on it, with a ball and without a ball in the horizontal part of the tube is the main emphasis of this numerical study. For the present numerical study, the following values are assumed unless mentioned otherwise: Time period of the structure, $t_0 = 1.8$ s, Mass ratio, $\mu = 3\%$, Damping ratio of structure, $\xi_0 = 2\%$, Frequency of the base filter, $\omega_g = 5\pi$ rad/s, Duration of the earthquake, $t = 20$ s, Damping of the soil, $\xi_g = 0.6$, Length ratio of LCVA, $p = 0.75$, Area ratio, $r = 1.5$, Peak ground acceleration, PGA $= 0.3\,\text{g} = 3\sigma_{\ddot{z}}$. For this study, the earthquake considered has been assumed to have an energy content concentrated at the range of 1–4 Hz. The population size considered for PSO is 100 and the number of iteration is 1000.

The RMSD and the associated design variables have been plotted varying with different LCVA parameters like mass ratio and length ratio, structural properties and seismic excitation. In Figs. 3, 4, 5 and 6 the variation of the optimum tuning ratio, optimum head loss coefficient of LCVA, optimum ball tube diameter ratio for the TLCBD and corresponding RMSD has been plotted for the different mass ratio considered. With the increase in the mass ratio the tuning ratio decreases for both these dampers. For LCVA the rate of decrement is more than that of the TLCBD, whereas the optimum head loss coefficient for the LCVA increases with the increase in the mass ratio. The reduction of the ball tube diameter ratio can be noticed with the increment in the mass ratio. Generally, the mass ratio of liquid dampers is considered as 0.5% to 4% of the primary structure. For the tall buildings the mass of the higher storeys are more and addition of more mass in those storeys can affect the other parameters of the structure in a wrong way. From Fig. 6 it is visible that with the increase in the mass ratio up to a certain limit the TLCBD performs better and the

Fig. 3 The optimum tuning ratio with varying mass ratio

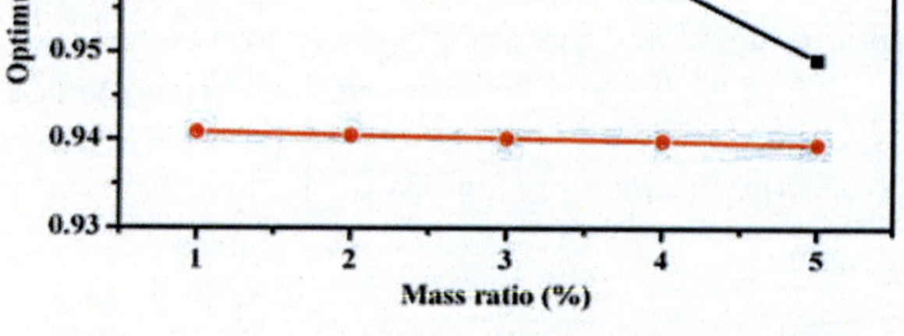

Fig. 4 The optimum head loss coefficient with varying mass ratio

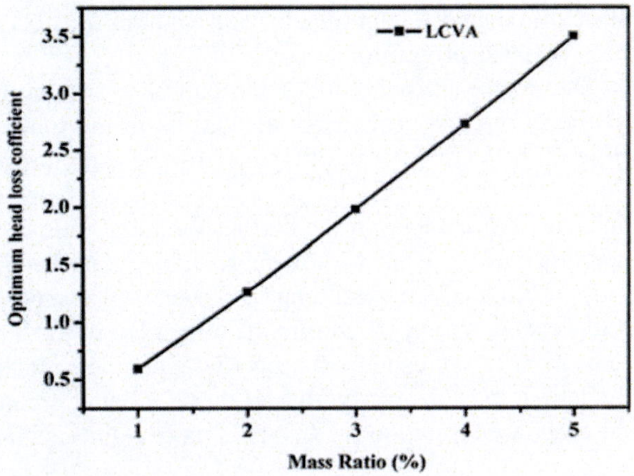

Fig. 5 The optimum ball tube diameter ratio with varying mass ratio

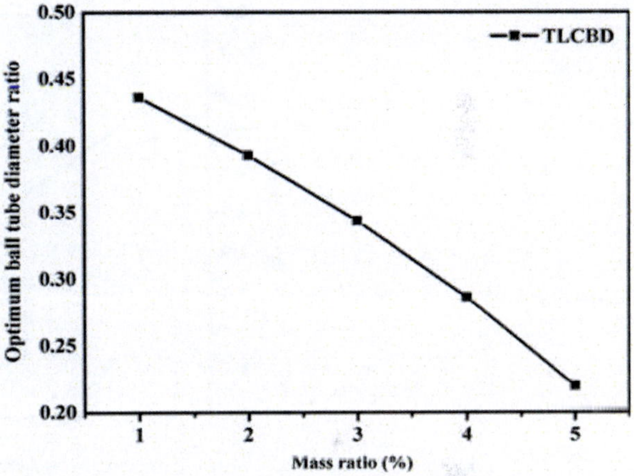

Fig. 6 The RMSD with varying mass ratio

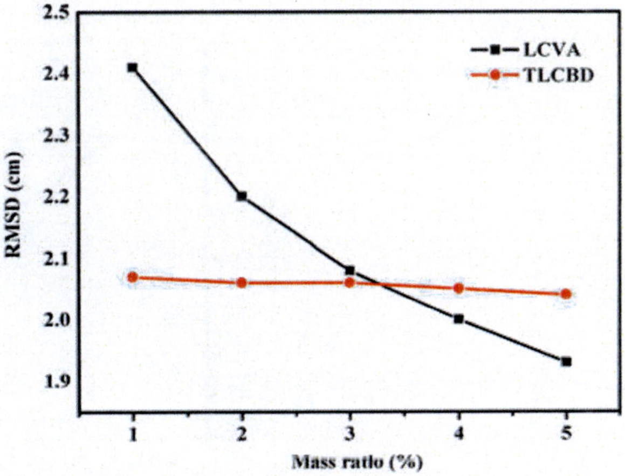

structural displacement remains constant but for LCVA it decreases for higher mass ratio in a less proportion.

Due to the introduction of the ball as a moving orifice, the damping property of the LCVA gets enhanced and results in more effective response mitigation. In Figs. 7, 8, 9 and 10, the response of the structure has been plotted considering the various parameters for an SDOF structure. In Fig. 7 RMSD of the structure has been plotted against the various damping ratios of the structure which shows a decremented response with increase in the damping ratio. As the structural damping ratio increases the maximum part of the excitation gets taken care by structures own high-level damping. Therefore the utilisation of the dampers gets decrease resulting to be more close response for any of the dampers attached on it. With the increase in the length ratio damper with ball shows a better result than LCVA. But the ratio of the response reduction for LCVA is more rigorous than TLCBD. The RMSD of the

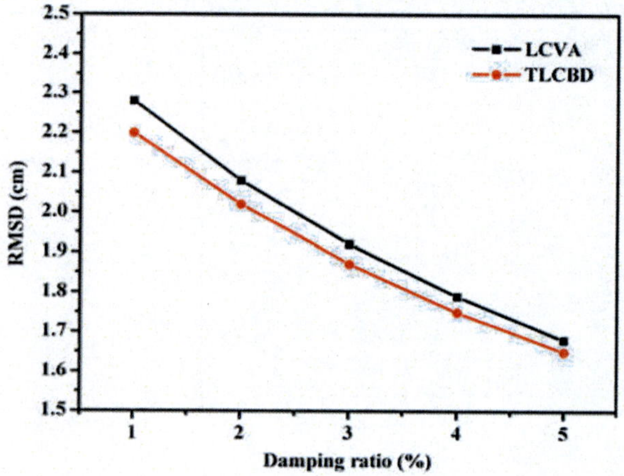

Fig. 7 The RMSD with varying damping ratio of structure

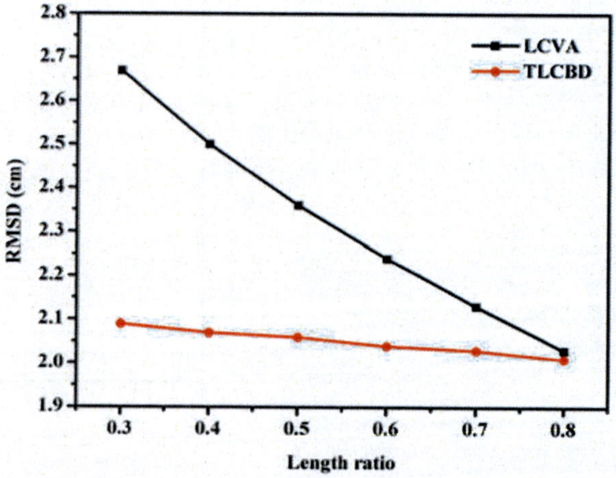

Fig. 8 The RMSD with varying length ratio

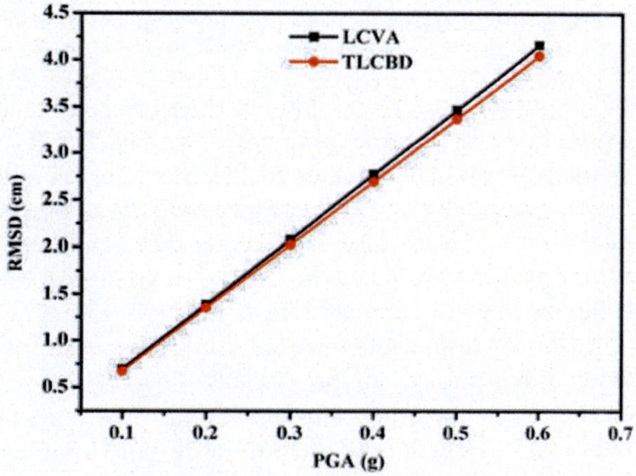

Fig. 9 The RMSD with varying PGA

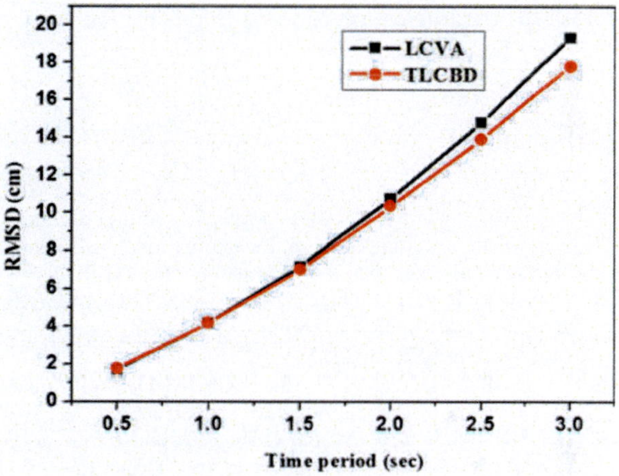

Fig. 10 The RMSD with varying time period of structure

structure has been plotted against different length ratios for both dampers in Fig. 8. For both PGA and different time periods of the structure, the responses have been plotted as shown in Figs. 9 and 10. In these figures, it is clearly noticeable that for the lower cases the LCVA and the TLCBD show almost same result as there is no such effect of the ball movement due to lesser displacement but with the increase in the both the PGA and the time period the TLCBD shows better results than normal conventional type LCVA. From all the graphs shown below, it can be concluded that the TLCBD performs better than the LCVA because of its enhanced damping capacity due to the installation of the ball in place of normal conventional orifices.

5 Conclusions

In the present study, the optimum performance of the response reduction of the structure is compared for both the LCVA and TLCBD systems subjected under non-stationary earthquake. A numerical study using the stochastic structural optimisation has been considered to evaluate the efficiency of dampers subjected to random earthquake. For this PSO is used to evaluate the response parameters considering different values of responsible factors. The optimum tuning ratio, optimum head loss coefficient and the optimum ball tube diameter ratio follow the conventional manner. For TLCBD the response reduction is more than that of LCVA for all the cases considered. For mass ratio, the TLCBD shows better result for the suitable mass ratio considered without harming the various other structural parameters. For damping ratio and the length ratio, the rate of response mitigation is more for TLCBD than the conventional type LCVA with an orifice. From the results, it is clearly visible that the TLCBD shows improved performance in the efficiency than that of the LCVA for all the parameters considered in this regard.

References

1. Sakai F, Takeda S, Tamaki T (1989) Tuned liquid column damper-new type device for suppression of building vibration. In: International proceedings of the international conference on high rise buildings, Nanjing, China, pp 926–931
2. Debbarma R, Chakraborty S, Ghosh SK (2010) Optimum design of tuned liquid column dampers under stochastic earthquake load considering uncertain bounded system parameters. Int J Mech Sci 52(10):1385–1393
3. Wu JC, Shih MH, Lin Y, Yi, Shen YC (2005) Design guidelines for tuned liquid column damper for structures responding to wind. Eng Struct 27(13):1893–1905
4. Lee HH, Wong SH, Lee RS (2006) Response mitigation on the offshore floating platform system with tuned liquid column damper. Ocean Eng 33:1118–1142
5. Haroun MA, Pires JA, Won AYJ (1996) Suppression of environmental induced vibrations in tall buildings by hybrid liquid column dampers. Struct Design Tall Build 5:45–54
6. Hitchcock PA, Kwok KCS, Watkins RD (1997) Characteristics of liquid column vibration absorbers (LCVA) I. Eng Struct 19(2):126–134
7. Hitchcock PA, Kwok KCS, Watkins RD (1997) Characteristics of liquid column vibration absorbers (LCVA) II. Eng Struct 19(2):135–144
8. Wu JC, Chang CH, Lin Y (2009) Optimal designs for non-uniform tuned liquid column dampers in horizontal motion. J Sound Vib 326(1–2):104–122
9. Konar T, Ghosh A (2012) Bimodal vibration control of seismically excited structures by the liquid column vibration absorber. J Vib Control 14(3):385–394
10. Chang CC, Hsu CT (1998) Control performance of liquid column vibration absorbers. Eng Struct 20(7):5811–5816
11. Samali B, Mayol E, Kwok KCS, Mack A, Hitchcock P (2004) Vibration control of the wind-excited 76-story benchmark building by liquid column vibration absorbers. J Eng Mech 130(4):478–485
12. Al-Saif KA, Aldakkan KA, Foda MA (2011) Modified liquid column damper for vibration control of structures. Int J Mech Sci 53:505–512
13. Gur S, Roy K, Mishra SK (2015) Tuned liquid column ball damper for seismic vibration control. Struct Control Health Monitor 22:1325–1342

14. Marano GC (2008) Multiobjective optimization criteria for linear structures subject to random vibrations. Open Civil Eng J 2:75–87
15. Leung AYT, Zhang H, Cheng CC, Lee YY (2008) Particle swarm optimization of TMD by non-stationary base excitation during earthquake. Earthquake Eng Struct Dynamics 37:1223–1246
16. Lutes LD, Sarkani S (1997) Stochastic analysis of structural and mechanical vibrations. Prentice Hall, Upper Saddle River, New Jersey, USA
17. Kanai K (1957) Semi-empirical formula for the seismic characteristics of the ground. Bulletin of Earthquake Research Institute, University of Tokyo 35:309–325
18. Tajimi HA (1960) Statistical method of determining the maximum response of a building during earthquake. In: International proceedings of 2nd world conference on earthquake engineering, pp 781–797
19. Jennings PC, Housner GW, Twai C (1969) Simulated earthquake motions for design purpose. In: Proceedings of 4th world conference on earthquake engineering, Santiago, A-1, pp 145–160.
20. Nigam NC (1972) Structural optimization in random vibration environment. AIAA J 10(4):551–553

Performance-Based Design of Dual System

Durga Mibang and Satyabrata Choudhury

Abstract In the present study, reinforced concrete dual system building have been designed by unified performance-based design (UPBD) method for various target performance objectives. In such buildings, lateral load is mainly taken by shear wall and gravity load is taken by frame. Proper engineering procedure is required for apportioning and designing shear wall and frame members. The designed building are subjected to static and dynamic non-linear analyses. The shear wall is modeled as a layered shell element, whereas the beams and columns are designed as frame elements using ETABS-2016 software. In this study, frame-shear wall building of height 12 storeys has been considered as sample building. The building has been designed with expected strength and demand level of EC-8 spectrum at 0.45 g. The target performance objective is considered as Life safety (LS) performance level with 2% drift. Various performance parameters like performance point, design drift, equivalent damping, and base shear have been found out. It has been observed that the target performance objectives have been achieved.

Keywords Performance-based design · UPBD method · RC dual system · Multi-layer shell element · Plastic hinge

1 Introduction

The dual systems are in common use for multi-storied reinforced concrete buildings. In the past result, it has been found that the dual system buildings enhanced the structural performance of the building during earthquake. In the present study, the frame dual-wall building of 12 storeys is considered where the dual-wall carries 70% of lateral load and frame carries 30% of lateral load. The pre-decided design drift, target performance level and hazard level of earthquake, the selected building

D. Mibang (✉) · S. Choudhury
Department of Civil Engineering, National Institute of Technology Silchar, Silchar 788010, Assam, India
e-mail: durgamibang28@gmail.com

S. Choudhury
e-mail: scnitsilchar@gmail.com

B. B. Das et al. (eds.), *Recent Developments in Sustainable Infrastructure*, Lecture Notes in Civil Engineering 75, https://doi.org/10.1007/978-981-15-4577-1_27

is evaluated using UPBD method and various parameters like equivalent damping, time period, and base shear are calculated according to (Choudhury and Singh). The shear wall is modeled as a layered shell element. The benefits of providing layered shell shear wall are the ability to model prolonged, interconnect, and dual frame. In modeling, the beam which is connected to shear wall is modeled to some extend inside the shear wall shell elements as mentioned in (Fahjan et al.). The shear wall with two layers could be modeled with different techniques (multi-layer shell and Mid-Pier frame). The characteristics of plastic hinges of the shear wall are defined using ASCE 41-13 recommendation or fiber-based Hinge property (Fahjan et al.). Check is done to ensure that the length of shear wall is capable to take the assigned lateral load. The building is designed using expected strength as per ASCE 41-13. The column size is decided on trial basis by keeping the percentage of steel 3–4% and to ensure the capacity design criteria as per 13920:2016. The static non-linear analysis is used to find out the performance level of the building. Design spectrum is considered is EC-8 for type B soil at hazard level of 0.45 g. The target performance objectives considered are LS with 2% drift.

2 Design Method

2.1 Direct Displacement-Based Design (DDBD) of Frame-Wall Structures

The direct displacement-based design of dual system structure is given (Sullivan et al.). In this method, MDOF system has to be converted into an equivalent SDOF a system. The inflection height of the building is established by assigning strength proportion to the member. The height of inflection (h_{inf}) of building is achieved when the moment of wall is zero. Equivalent single degree of freedom (ES_{DOF}) properties are determined by using the following relations:

Design displacement,

$$\Delta d = \frac{\sum_{i=1}^{n} mi\,\Delta i^2}{\sum_{i=1}^{n} mi\,\Delta i} \tag{1}$$

Effective mass,

$$m_e = \frac{\sum_{i=1}^{n} mi\,\Delta i}{\Delta d} \tag{2}$$

Effective height,

$$h_e = \frac{\sum_{i=1}^{n} m_i\,\Delta i h_i}{\sum_{i=1}^{n} m_i\,\Delta i} \tag{3}$$

where h_i = storey height, m_e = effective mass of the structure, h_e = effective height, m_i = seismic mass of i-th floor. Frame and wall ductility demand are determined from Eq. (6) and (7).

The yield displacement profile of wall is given by,

$$\Delta_{i,y} = \frac{\phi_{y,w} h_i h_{\mathrm{inf}}}{2} - \frac{\phi_{y,w} h_{\mathrm{inf}}^2}{6}, \text{ when } h_i \geq h_{\mathrm{inf}} \tag{4}$$

$$\Delta_{i,y} = \frac{\phi_{y,w} h_i^2}{2} - \frac{\phi_{y,w} h_i^3}{6 h_{\mathrm{inf}}}, \text{ when } h_i \leq h_{\mathrm{inf}} \tag{5}$$

Wall ductility demand,

$$\mu_w = \frac{\Delta d}{\Delta_{he,y}} \tag{6}$$

Frame ductility,

$$\mu_{i,f} = \left(\frac{\Delta_i - \Delta_{i-1}}{h_i - h_{i-1}}\right)\frac{1}{\theta_y F} \tag{7}$$

where

$$\theta_{y,F} = \frac{l_b \varepsilon_y}{2 h_b} \tag{8}$$

The equivalent viscous damping for frame and walls is obtained by adding the elastic and hysteretic component together as follows:

$$\xi_{SDoF} = \left(\frac{M_{\mathrm{wall}} \xi_{\mathrm{wall}} + \mathrm{MOT}_{\mathrm{frame}} \xi_{\mathrm{frame}}}{M_{\mathrm{wall}} + \mathrm{MOT}_{\mathrm{frame}}}\right) \tag{9}$$

$$\xi_w = \frac{95}{1.3\pi}(1 - \mu_w^{-0.5} - 0.1.\varphi.\mu_w)\left(1 + \frac{1}{(T_{e_{\mathrm{trial}}} + 0.85)^4}\right) \tag{10}$$

$$\xi_f = \frac{120}{1.3\pi}(1 - \mu_w^{-0.5} - 0.1.\varphi.\mu_f)\left(1 + \frac{1}{(T_{e_{\mathrm{trial}}} + 0.85)^4}\right) \tag{11}$$

where $T_{e_{\mathrm{trial}}} = \frac{N}{6}\sqrt{\mu_{\mathrm{sys}}}$, ξ_{SDoF} is equivalent viscous damping of system, $\mathrm{M_{wall}}$ is wall moment, MOT is frame overturning moment, ξ_{wall} is equivalent damping of wall, ξ_{frame} is equivalent damping of frame. The effective time period of the structure can be carried out from Fig. 1.

The effective stiffness (K_e) is determined by,

$$k_e = \frac{4\pi^2 m_e}{T_e^2} \tag{12}$$

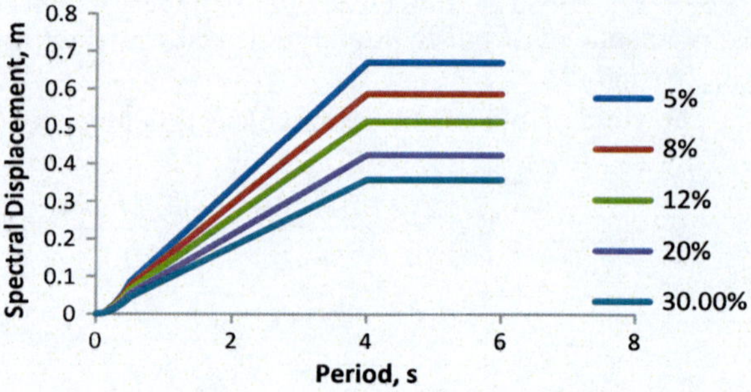

Fig. 1 Displacement spectra for 0.45 g corresponding to EC-8

Here, T_e is obtained from displacement spectra plot corresponding to EC-8.

The base shear (V_b) is distributed throughout the height of the structure using the following expression:

$$V_b = k_e \Delta d \tag{13}$$

where the F_i is the portion of base shear applied at the i-th floor.

$$F_I = \frac{m_i \Delta i}{\sum_{i=1}^{n} m_i \Delta i} \tag{14}$$

2.2 *Unified Performance-Based Design (UPBD) Method for Frame-Wall Buildings*

A DDBD procedure for frame building was proposed by petting and priestly (Pettinga and Priestley). At later stage, this procedure was proceed by Sullivan et al. for frame-shear wall structure and has further by Choudhury and Singh and is presented along with this method. In UPBD method, the beam size and length of wall are known.

$$h_b = \frac{0.5\varepsilon_y l_b}{\theta_d - \theta_{pb}} \tag{15}$$

Depth of the beam can be found from the Eq. (15). The width of beam as kept half to two-third of beam depth.

And also for shear wall, the length of wall (L_W) is obtained as

$$L_w = \frac{\varepsilon_y h_{\text{inf}}}{\theta_d - \theta_{pw}} \tag{16}$$

where L_W is horizontal length of wall, ε_y is yield strain of rebar, h_{inf} is height of inflection, θ_d is design drift and θ_{pW} is plastic rotation in wall. The thickness of wall is obtained from base shear consideration and number of walls used in a particular direction. The inflection height h_{inf} of the dual system building is found by the moments carried by shear wall or frame. The design of the structure is done with expected (mean) strengths of materials, as per FEMA-356 provisions. The load combinations are as below:

$$DL + LL$$
$$DL + LL \pm F_x$$
$$DL + LL \pm F_y$$

where *DL*, *LL*, F_X, F_Y corresponds to dead load, live load, earthquake load in *x* and *y* direction, respectively.

3 Modeling

In the present study, 12-storey reinforced concrete frame-shear wall building is considered as shown in Fig. 2, which has been designed by UPBD (Choudhury 2008) method for target performance objective of 2% drift with life safety (LS) performance level. Analysis and design have been done using ETABS-2016 software. Shear wall has been modeled as multi-layered shell elements. The building is designed with M30 concrete and Fe415 grade steel with expected strength as per FEMA-356. Height of each floor is kept same over the entire height. 3–4% of steel should be maintained for column along with capacity design. A building considered is designed for life safety (LS) performance level. The capacity design of strong-columns and weak-beam was

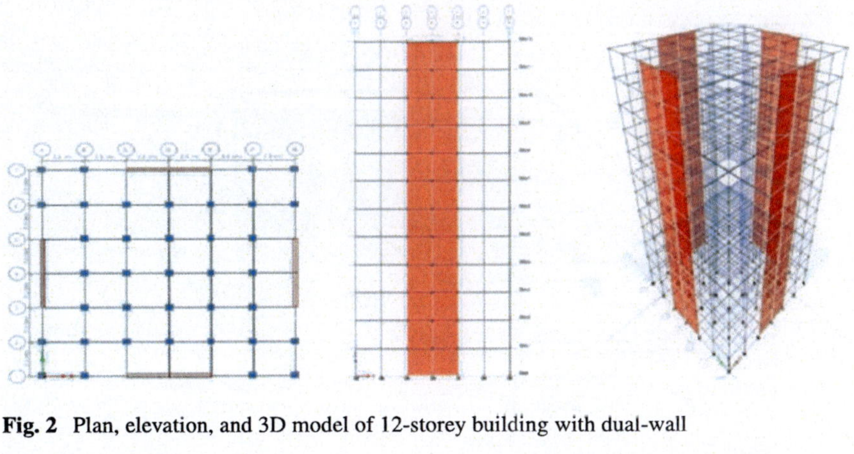

Fig. 2 Plan, elevation, and 3D model of 12-storey building with dual-wall

Table 1 Building and design consideration

Building name	Column sizes		Beam size depth width (mm)	Shear wall thickness (mm)	Target performance
Dual-wall building	Inner column	Outer column			
		750 750			
		700 700			
12 LS	650 650	650 650	950 450	250	LS

satisfied as per IS Code 13920:2016. The design consideration of the building is shown in Table 1 and Fig. 3 shows the inflection height of the building considered and multiple layered shell element has been shown in Fig. 4.

Capacity design is the best approach toward ductile design criteria. In capacity design, the certain structural members are allowed to fail in some intended way,

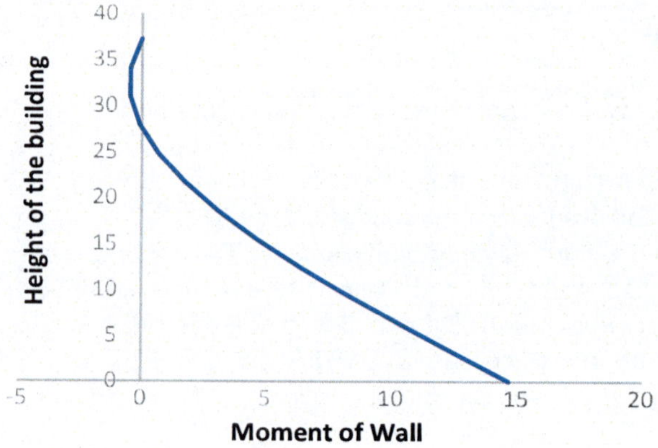

Fig. 3 Inflection height of the considered building

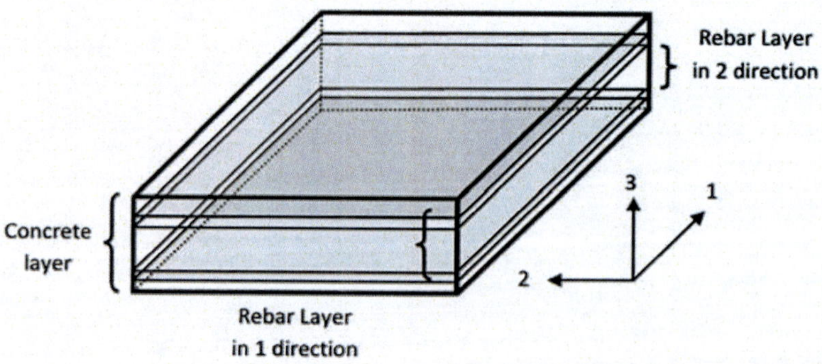

Fig. 4 Multi-layered shell element [7]

whereas some other members are to remain undamaged. This concept applies to beam-column joint where the column is of higher capacity so that the beam fails earlier to make the failure a ductile one. IS code 13920:2016 suggests that the capacity design has to be carried out for ductile and on ductile members. According to IS code mentioned the design should be such that

$$\sum M_{\text{COLUMN}} \geq 1.4 \sum M_{\text{BEAM}}$$

The column sizes were taken by trial keeping the steel 3–4% of the cross-sectional area of column.

4 Analysis

Non-linear analysis is the procedure to obtain earthquake demands on the building. The analysis includes a capacity spectrum method that intersection the capacity curve and reduced demand curve to determine the performance point. From Fig. 5 it can be seen that the performance of the structural member, i.e., beam and column which reached to the target performance level and can be recognized by the formation

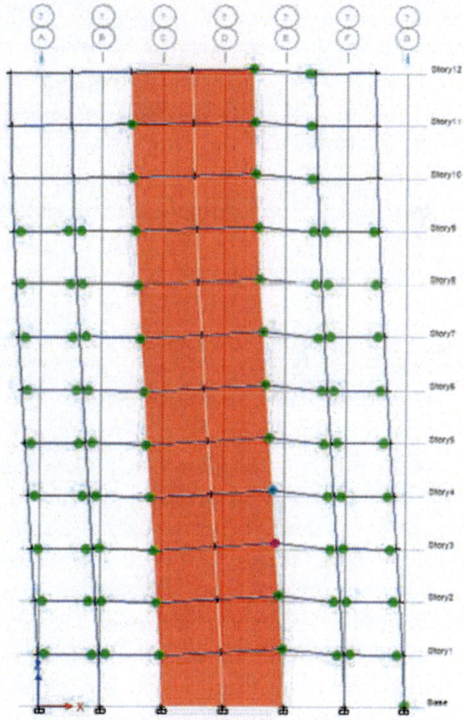

Fig. 5 Formation of plastic hinges

of plastic rotation as displayed in Fig. 5. The non-linear behavior of layered shell element for dual system modeling can be rectified by examining by the stresses in concrete and reinforcement layers which are shown in Fig. 7. Also, since, shell element does not show any hinge formation as its a lumped plasticity element. The behavior of the shell element is understood from the stress diagrams as shown in Fig. 7.

5 Results and Discussion

Non-linear analysis or push over analysis is the procedure to obtaining earthquake demands on building. In Fig. 2 plan, elevation and 3D modeling of the structure considered is shown. For performance-based design of structures, flexural RC members of shear wall in terms of plastic rotations are mentioned in ASCE 41-13. The inflection height for frame-shear wall building is shown in Fig. 3.

(Otani) had put forward a common procedure for analysis of layered shell element shear wall in the past, i.e., non-linear finite element but the design needs some further investigations and to resolve this problem, (Fahjan et al.) has further proposed the rigid-beam which is connected to shear wall and properly placed inside the multi-layer shell elements. These shell elements give a proper bond between rigid beams and it also gives proper connection with other elements, i.e., beams and columns. The beam (rigid) elements which are provided in between the shell elements are generally of highly elastic materials may be 10 times the modulus of elasticity. Modeling of a shell element section is shown in Fig. 4.

The main work of the rigid beams is to minimize the mesh sensitivity effect caused by the inadequacy of the drilling degree of freedom formulation (Fahjan et al.). By using the method (fahjan) has found the moment and other internal forces of shear wall as well as connecting beams. In shell element, the steel bar is provided either one layer or two layers. The main motive of the rigid beams is to reduce the mesh sensitivity formulation by fahjan et al.; by using the method, fahjan has found that the moment and other internal forces shear wall as well as connecting beams give better results. In shell element, the steel bars are provided with either one layer or more and should satisfy the code 13920:2016 criteria.

The performance level attained by the building and the performance point (PP) is shown in Fig. 5, which will indicate its attained performance against its design criterion. Figure 6 also shows the pushover curves carried out for frame-wall building and satisfied the target performance considered. The hinge locations in the building design for several performance stages are shown in Fig. 5. It has been observed from Fig. 5 that the performance of elements, i.e., beam and column has reached the needed performance level, which were recognized by the plastic hinge rotations.

From Fig. 7, it can be observed that its maximum stress is present at the bottom storey of shear wall. To inspect the performance of the shear walls, hinge formation is produce in terms of plastic rotations that could be checked in ACI 40, and ASCE 41-13 acceptance criteria recommendations.

Fig. 6 Performance level for dual system building

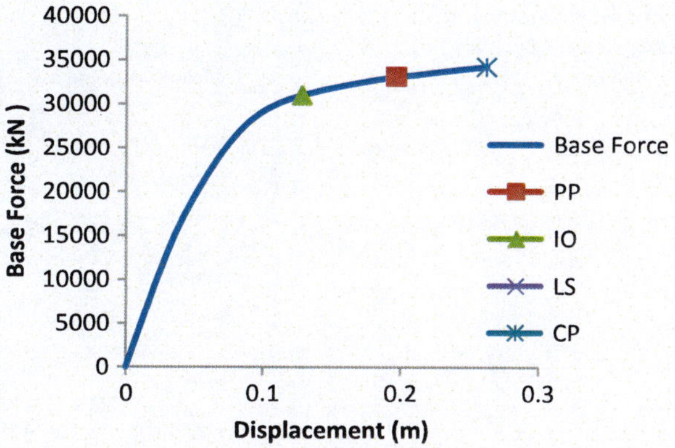

Fig. 7 Stress diagrams of multi-layer shell element

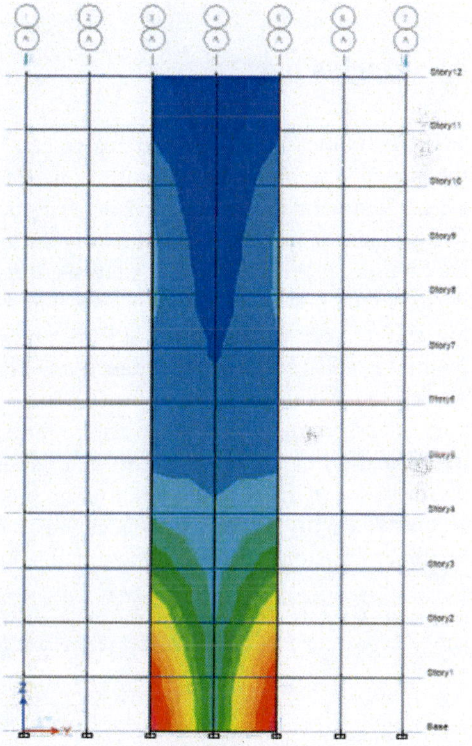

Five earthquakes data are considered to find out the non-linear time history analysis (NLTHA). The inter-storey displacement histories can be found out by non-linear time history analysis. The inter-storey drift diagrams for typical buildings considered have been shown in Fig. 8 and it satisfies the target drift.

Fig. 8 Inter-storey drift diagram for building considered

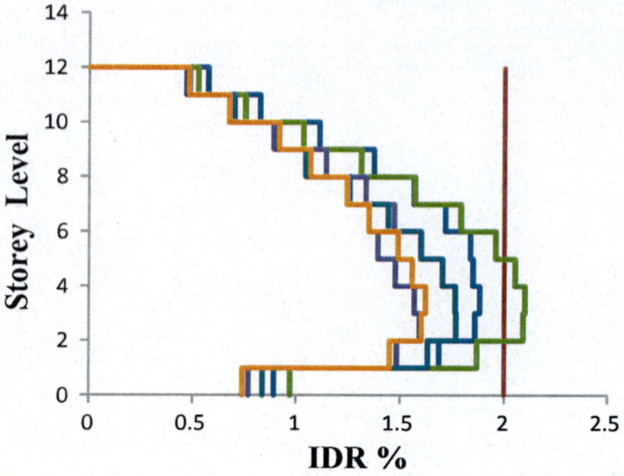

6 Conclusions

The UPBD method of seismic design was proposed only for IO buildings without the presence of infill, the design is done for dual frame building that is designed for LS performance level (PL) for 2%. Using UPBD method has attained desired PL in terms of plastic hinges, thus validating the use of this method for the design of frame-wall building. With push over analysis, it has been found that the building designed is performed better and it can be seen that it has also achieved its target performance level. Here performance point (PP) lies between IO and LS and It indicates that the method is valid for LS performance level for 12-storey building considered. In this paper, we have also discussed the RC shear wall modeled. The outcome of the work done clearly indicates that the shell element provides indirect way to understand the formation of hinges in shear wall which had to be understood. To identify the performance of shear walls, fiber hinge models produce plastic rotations that could be checked with ASCE 41-13 acceptance criteria recommendations. Here we can see that inter-storey drift of structure considered has also satisfied the target drift.

Acknowledgments This work was supported by "TEQIP III" of NIT Silchar, and the authors gratefully acknowledge Prof. Sivaji Bandyopadhyay (Director, NIT Silchar) for providing the financial support.

Bibliography

1. Sullivan TJ, Calvi GM, Priestley MJN (2003) The limitations and performances of different displacement based design methods. J Earthq Eng 7(1;2003):201–241
2. Sullivan TJ, Priestley MJN, Calvi GM (2006) Direct displacement-based design of frame-wall structures. J Earthq Eng 10:91–124

3. Pettinga JD, Priestley MJN (2005) Dynamic behaviour of reinforced concrete frames designed with direct displacement-based design. J Earthq Eng 9(2):309–330
4. Kappos AJ, Panagopoulos G (2004) Performance-based seismic design of 3d RC buildings using inelastic static and dynamic analysis procedures. ISET J Earthq Technol (paper no. 444), 41(1):141–158
5. Choudhury S, Singh SM (2013) A unified approach to performance-based design of RC frame buildings. J Inst Eng India Ser 73–82
6. Mayengbam SS, Choudhury S (2011) An economic comparison of direct displacement based design with is-1893 response spectrum method for RC frame buildings. Int J Civ Struct Eng 2. https://doi.org/10.6088/ijcser.00202010114
7. Fahjan YM, Kubin J, Tan MT (2011) Non linear analysis methods for reinforced concrete buildings with shear walls
8. Hiremath GS, Hussain S (2014) Effect of change in shear wall location with uniform and varying thickness in high rise building. Int J Sci Res (IJSR) 3(10)
9. Paulay T (2002) The displacement capacity of reinforced concrete coupled walls. Elsevier Science Ltd., Engineering Structures 1165–1175
10. ASCE/SEI 41–13 American Society of Civil Engineers Seismic Evaluation and Retrofit of Existing Buildings (2014) Manufactured in the United States of America
11. Euro code- 8: Seismic Design of Buildings (2011) European Commission Joint Research Centre, Europe
12. IS456:2000 (Indian standard) Plain and reinforced concrete code of practice (Fourth Revision) Tenth Reprint APRIL 2007, Bureau of Indian Standards, New Delhi, India
13. IS: 875: Part3 (Indian Standard) Wind Loads on Buildings and Structures, New Delhi, India
14. IS 1893(PART-1):2016 (Indian Standard), Criteria for earthquake resistant design of structure, general provisions and buildings, sixth revision, Bureau of Indian Standard, New Delhi, India
15. IS 13920:2016 (Indian Standard) ductile design and detailing of reinforced concrete structure subjected to seismic force- code of practice(first revision),Bureau Of Indian Standards, New Delhi, India

Study on Optimum Location of Outrigger for High-Rise Building

B. Venkat Rao, T. Manasa Lakshmi, Mallika Alapati, and G. K. Viswanath

Abstract Due to rapid population growth and constraint on land availability, construction sector spread its wings in vertical dimension. As high-rise buildings are stretching toward sky, challenges with structural behavior govern the choice and design of structural system. Lateral load resisting systems popularly in use are moment resisting frames, load bearing structural walls, dual systems, tube system, and moment-resisting frames with outrigger. The intent of the work in this paper is to investigate the performance of outrigger, location optimization, and their efficiency when used in multiple numbers placed at various heights (2/3rd, 1/2th, 1/3rd times the height of building). Models of 30 storey buildings with outrigger and belt truss systems are analyzed for earthquake and wind loads and the lateral drift responses are compared to find the optimum location of outrigger and belt truss systems. The results are interpreted and found that outrigger system can effectively reduce the lateral drift of the building and optimum location of outriggers is found to be at mid-height of building considered along with cap truss.

Keywords Outrigger-belt truss system · Optimization · Tall building · Outrigger location · Lateral drift

1 Introduction

As there is expansion of cities and value of land continues to high, buildings are rising tall and slim. More investment on land and the need to protect farming area has contributed to navigate residential buildings toward higher levels. To make possible for slimmer structures, various structural systems for high-rise building have been

B. Venkat Rao (✉) · T. Manasa Lakshmi
Civil Engineering, V.R. Siddhartha Engineering College, Vijayawada, Andhra Pradesh, India
e-mail: bvrvrsec@gmail.com

M. Alapati
Civil Engineering, VNR VignanaJyothi Institute of Engineering and Technology, Hyderabad, Telangana, India

G. K. Viswanath
JNTU, Hyderabad, Telangana, India

B. B. Das et al. (eds.), *Recent Developments in Sustainable Infrastructure*, Lecture Notes in Civil Engineering 75, https://doi.org/10.1007/978-981-15-4577-1_28

widely used. The lateral load resisting systems in common in high-rise structures are framed system, structural walls, dual systems, outriggers, and tube systems. Among them frames with braced-outriggers proved to be one of the efficient procedures to increase the lateral stiffness of tall building, reducing the storey drifts.

The outrigger member is a cantilever-shape horizontal member and is connected to the outer column and inner core of the structure. Under the action of lateral loads, core undergoes rotation and the outrigger connecting core and the outer column induces tension and compression alternatively in the same members. Besides the outer columns, it is common to locate other outer columns to serve in restricting the rotation of outriggers. Introduction of a deep spandrel girder at the levels of the outrigger around the structure shows the control of tilting of outrigger. Some examples of outrigger system used are US Bank Center (Milwaukee), New York Times Tower (New York), China World Tower (Beijing, China), Taipei 101 (Taiwan), Water Front Place (Brisbane), and Exchange Square (Hong Kong).

2 Literature Review

Parametric study of tall building structures braced with outriggers is carried out by some authors [1] considering the flexibility of the outriggers. Mathematical expressions for optimum location of the outriggers, moments in the core and the storey drift were developed. Some researchers, Kamath, and others [2] studied the behavior and response of the frames with outriggers for high-rise structures for static and dynamic loads. In their study, the 3D structural models with flexural rigidity ratio range of (0.25–2.0) EI_0/EI optimum outrigger location are obtained from the range of 0.975–0.40 (Hs/H), from the top of the building. Studies on different number for steel frame outrigger-belt truss systems were conducted on structures with 20, 30, 40, and 50 storeys to propose the optimal number and location of the outrigger. Babaei [3] studied optimal solutions in the design domain for storey drift and weight of the structure. The degree of stiffening and the level of drift control on the number of the outrigger trusse are studied by Gerasimidis et al. [4].

Location optimization of outrigger for high-rise RCC structure was investigated by Krunal Z Mistry and others. In their study tall building with ten different arrangements of outriggers was considered. Best outrigger location and the performance efficiency of each outrigger were arrived at when pair of outriggers were used [5]. Hoenderkamp presented a preliminary design for outrigger braced high-rise shear walls subjected to lateral forces. One of the outrigger locations is fixed, while the second one location is investigated for optimum position [7]. Shivacharan and others [10] studied the optimum position of outriggers for tall structures with vertical irregularity. Authors concluded that there is 12.78 and 11.5% reduction of lateral displacement and storey drift [10]. Some researchers analyzed tall buildings and found the optimum location of outriggers system for lateral loads and concluded that the topmost deflection was at the top storey when flat slab with core was engaged and is 625.7 mm, which was reduced up to 411.18 mm when first outrigger was provided at mid-height of structure and 335.15 mm when provided with second outrigger at 3/4th height of the structure [11]. Alex and Otto conducted the study on structures braced with multiple outriggers based on the continuum approach. Authors concluded that continuum analysis yields good results for any number of outriggers. Design curves were established by some researchers for estimating the structural responses for any kind of structural configuration [13]. However, in the present work, authors tried to arrive at the optimum location of the braced outrigger in a two-outrigger system with one fixed outrigger location (at top storey).

3 Numerical Study

In the present study, moment resisting building frames with belt trusses in multiple numbers and at different locations are considered for analysis. Linear static and response spectrum analysis are performed. Analysis is done using ETABS and as per the latest codal provisions. Locations of outriggers are taken according to CTBUH guidelines. The selected topology is with two outriggers, one is fixed at top storey, and the location of the second outrigger is optimized for least lateral displacements and storey drifts.

Figure 1 shows the plan area and dimensions of the models as 6 m as bay length with 5 × 5 grid and with central core as opening. Figs. 2 and 3 show the position of outrigger with central core, one outrigger fixed at top storey, and other at mid-height along with top storey, respectively. The height of each storey is 3.5 m. Moment Resisting Frames (MRF) are considered as base structural systems.

Structural steel with yield strength of 250 N/mm^2 is used for all columns, beams, and belt trusses. A value of 2×10^5 N/mm^2 is considered as Young's modulus. A dead weight of 5 kN/m^2 and imposed load of 3 kN/m^2 were considered. Based on IS 1893:2016 earthquake loads are calculated and using the ETABS software options. The seismic zone is considered as Vijayawada city. Basic wind speed is considered as 50 m/sec and the wind pressure is determined according to IS 875:1987 part-3.

4 Results and Discussions

Table 1 displays the results of earthquake and wind analysis of maximum displacement for various outrigger positions.

It is observed from Table 1, for the case of cap truss +1/2nd case the displacements are less showing maximum reduction of 39.44 and 36.68% for seismic and wind effects, respectively. This is because of increase in the lateral stiffness of the structure.

Table 2 shows the result of maximum storey drift for different position of outriggers.

It is observed from Table 2, for the case of cap truss +1/2nd case the displacements are less showing maximum reduction of 29.46 and 33.90% for seismic and wind

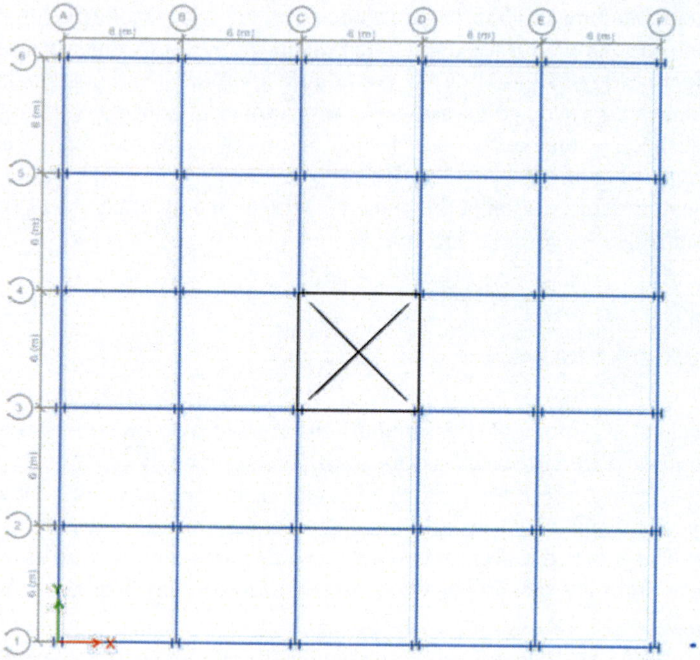

Fig. 1 Plan of the model

effects, respectively. From the above results, the maximum reduction of displacement as well as storey drift is seen when the outrigger is placed at middle of the storey height along the cap truss.

From Fig. 4, maximum percentage reduction in displacement is 36.10% which is at when the outrigger is placed at 0.5 times the height of the building along with cap truss.

Figure 5 shows the reduction of storey drift of 29.46% when the outrigger is placed at a height of 0.5 times height of the building along with cap truss.

5 Conclusions

As discussed in Sect. 4, the following conclusions are drawn

The structural system with two outriggers considered, yielded better structural responses. The results show maximum reduction of 39.44% in lateral displacements when the outriggers are placed at top storey and at 0.5 times the height of the building. Maximum reduction of 33.96% is found in the case of storey drift reduction for

Fig. 2 Outrigger at top storey

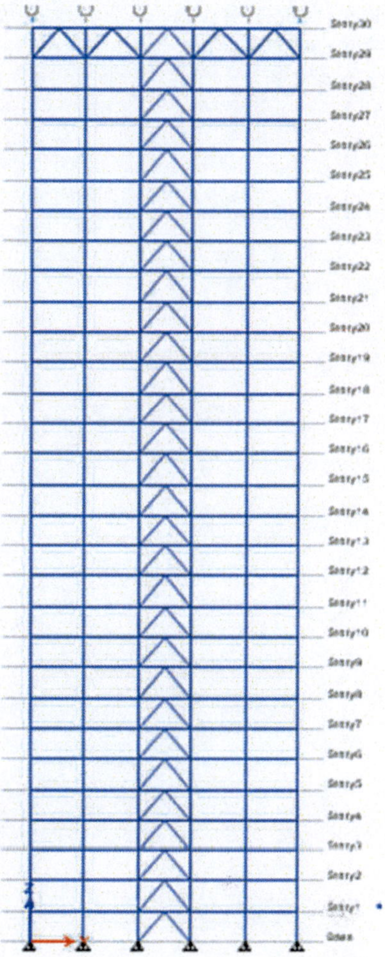

the same case. However, from the present study, it can be concluded that the outrigger structural system proves to be an efficient way to reduce inter-storey drifts and controls top displacements when compared with the frame with only central core. Inverted "V" shape bracing shows good results in all cases considered.

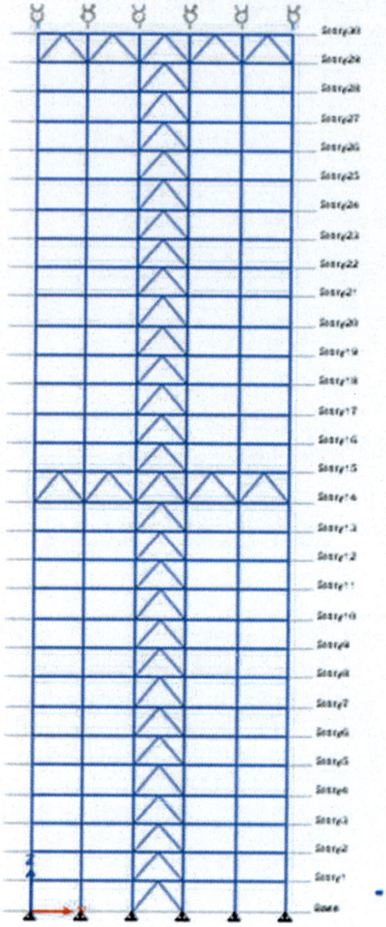

Fig. 3 Outrigger at top storey and mid-height

Table 1 Maximum displacements (mm) at top storey

Model	Earthquake analysis	Reduction (%)	Wind analysis	Reduction (%)
Core only	261.950	–	3156.220	–
Cap truss only	229.198	12.50	2840.365	10
Cap truss +2/3rd	171.436	34.55	2214.700	29.83
Cap truss +1/2nd	158.616	39.44	1998.268	36.68
Cap truss +1/3rd	167.377	36.10	2011.012	36.28

Table 2 Maximum storey drifts (mm)

Model	Earthquake analysis	Reduction (%)	Wind analysis	Reduction (%)
Core only	0.003282	–	0.040688	–
Cap truss only	0.003162	3.65	0.039927	1.87
Cap truss +2/3rd	0.002748	16.27	0.036287	10.81
Cap truss +1/2th	0.002315	29.46	0.026892	33.90
Cap truss +1/3rd	0.002400	26.87	0.027405	32.64

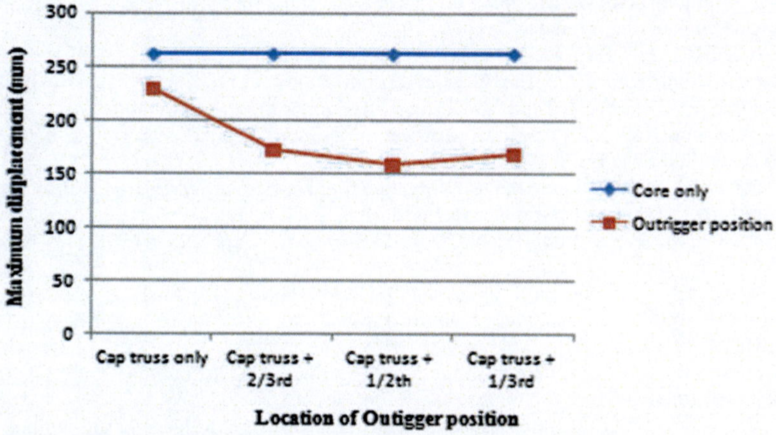

Fig. 4 Maximum displacement of response spectrum analysis

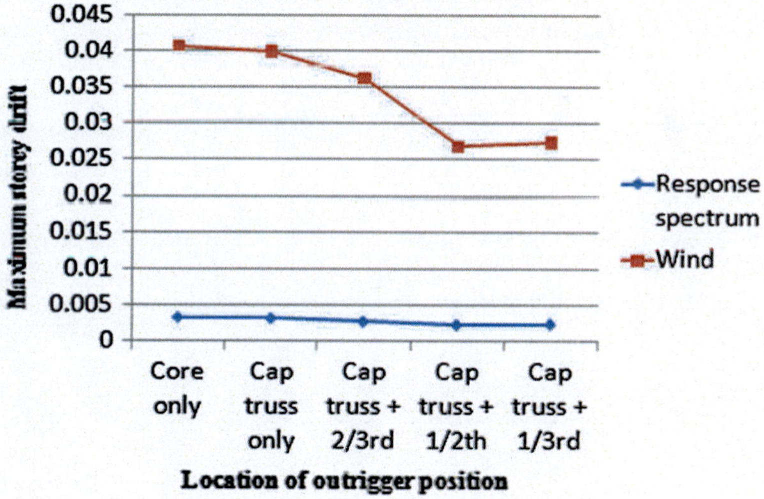

Fig. 5 Maximum storey drifts

References

1. Smith B (1981) Parameter study of outrigger braced tall building structures. J Struct Div 107(10):2001–2014
2. Kamath K, Divya N, Rao AU (2012) A study on static and dynamic behaviour of outrigger structural system for tall buildings. Bonfring Int J Ind Eng Manage Sci 2(4):15–20
3. Babaei M (2017) Multi-objective optimal number and location for steel outrigger-belt truss system. J Eng Sci Technol 12(10):2599–2612
4. Gerasimidis S, Efthyymiou E, Baniotopoulos CC (2009) Optimum outrigger locations of high-rise steel buildings for wind loading. In: Proceedings of the 5th European & African conference on wind engineering. Florence, Italy, pp 19–23
5. Mistry KZ, Dhyani DJ (2015) Optimum outrigger location in outrigger structural system for high rise building. Int J Adv Eng Res Dev 2:2348–6406
6. Choi HS, Ho G, Joseph L, Math N, Outrigger design for high-rise buildings
7. Hoenderkamp JCD (2007) Second outrigger at optimum location on high-rise shear wall. J Struct Des Tall Spec Build 17:619–634
8. Taranath BS (1974) Optimum belt truss location for high-rise structures. Eng J AISC 11:18–21
9. Taranath BS (2011) Structural analysis & design of tall buildings: steel and composite construction. CRC Press, Print ISBN:978-1-4398-5089-3, e Book ISBN:978-1-4398-5090-9, 2012
10. Shivacharan K, Chandrakala S, Narayana G, Karthik NM (2015) Analysis of outrigger system for tall vertical irregularities structures subjected to lateral loads. Int J Res Eng Technol 04(05):84–88
11. Sukhdeve SB (2016) Optimum position of outrigger in G+ 40 RC building. Int J Sci, Technol Eng 2(10):1051–1055
12. Shankar Nair R (1998) belt trusses and basements as virtual; outriggers for tall buildings. Am J Steel Constr 35:140–146
13. Coull A, Lau WHO (1989) Multi outrigger braced structures. J Struct Eng ASCE 115(7)

Seismic Control of Benchmark Highway Bridge Using Fiber-Reinforced Elastomeric Isolator

Jyoti Ranjan Barik and Purnachandra Saha

Abstract In this study, the seismic response of a benchmark highway bridge is scrutinized with a passive control system named Fiber-Reinforced Elastomeric Isolator (FREI). Bearing with fiber reinforcement and elastomeric damping material is called fiber-reinforced elastomeric isolator bearing. Advancement of fiber materials is striking nowadays whose stiffnesses are practically identical to steel. So as per efficient and innovative point of view, this isolator is supposed for seismic analysis. The Phase I problem of the benchmark highway bridge is considered for this study. The ground acceleration is applied at all supports of the bridge. A comparative study on the basis of displacement amplitudes has been encountered under six real ground motions. Evaluation criteria are also established under different boundary conditions which highlight that substantial reduction of seismic response can be achieved by using higher displacement amplitude values of FREI. It is observed that peak responses of base shear and displacement at mid-span are effectively reduced by employing certain range of displacement amplitude values.

Keywords Benchmark highway bridge · Seismic response · Displacement amplitude · Fiber-Reinforced elastomeric isolator

1 Introduction

A highway bridge is imminent whenever an obstruction arises and overcomes the barrier by adding roads in the highway system. Prominently, it should be workable before an earthquake as well as after an earthquake. Different control strategies have been introduced to acquire good seismic resistance. Base isolation plays a crucial role in mitigation of earthquake effects as it increases the horizontal flexibility and

J. R. Barik
Department of Civil Engineering, Nalanda Institute of Technology, Bhubaneswar, Odisha, India
e-mail: jyotiranjanbarik0@gmail.com

P. Saha (✉)
School of Civil Engineering, Kalinga Institute of Industrial Technology, Deemed to Be University, Bhubaneswar, Odisha, India
e-mail: dr.purnasaha@gmail.com

B. B. Das et al. (eds.), *Recent Developments in Sustainable Infrastructure*, Lecture Notes in Civil Engineering 75, https://doi.org/10.1007/978-981-15-4577-1_29

shifts the fundamental period of the structure in the lower spectral acceleration of the structure which ultimately causes reduction of the earthquake-induced force into the structure. However, the relative displacement between ground and structure has been increasing due to enhancement of horizontal flexibility of the system. Researchers have enumerated different control devices and algorithms. Control devices consist of passive, active, hybrid and semi-active strategies for the decrement seismic reaction of the highway bridge.

Recently, research is ongoing in the fields related to the development of new devices and depending on their performance, obsolete technology is being replaced by the new devices. Seismic isolation codes have been growing continually. The presence of this new standard for designing earthquake protective structure, the designer has to take into account these types of new earthquake resistance approach for the safety of the structures.

Fiber-reinforced elastomeric isolator (FREI) is a revolutionary control device that is adopted nowadays in many structures as it gives substantial outputs as compared to the uncontrolled structures. FREI can be a better substitute of HDRB as well as SREI in terms of utilization of fiber polymer rather than steel layers. Somehow FREI can be malleable as it is economical and less sophisticated than other elastomeric isolators. For this benchmark study is 91/5 highway overcrossing structure is considered which is situated in southern California.

2 Fiber-Reinforced Elastomeric Isolator

Bearing with fiber reinforcement and elastomeric damping material sandwiched together is called fiber-reinforced elastomeric isolator bearing. Advancement of fiber materials is striking nowadays whose stiffnesses are practically identical to steel. Fundamentally seismic gadgets can be made utilizing elastic layers sandwiched with thin bidirectional fiber texture layers. Already if there should arise an occurrence of fiber layers, steel plates are utilized to restrict the protruding in the elastomer layers of the isolator. The framework containing such layers, which shows high vertical stiffness to get to the weight of the structure is known as a steel-reinforced elastomeric isolator. Fiber is extraordinary compared to other substitutes of steel and is beneficial while making lesser weight isolators.

2.1 Classification of FREI

Routinely base confinement frameworks are intended to be isotropic and base isolators are constantly thought to be symmetric. The isolators being made as of now are either round or square fit as a fiddle. Fiber-reinforced isolators can be effortlessly sliced to required configuration which isn't conceivable with steel fortified isolators. FREI has been grouped by two viewpoints as specified in Fig. 1 underneath.

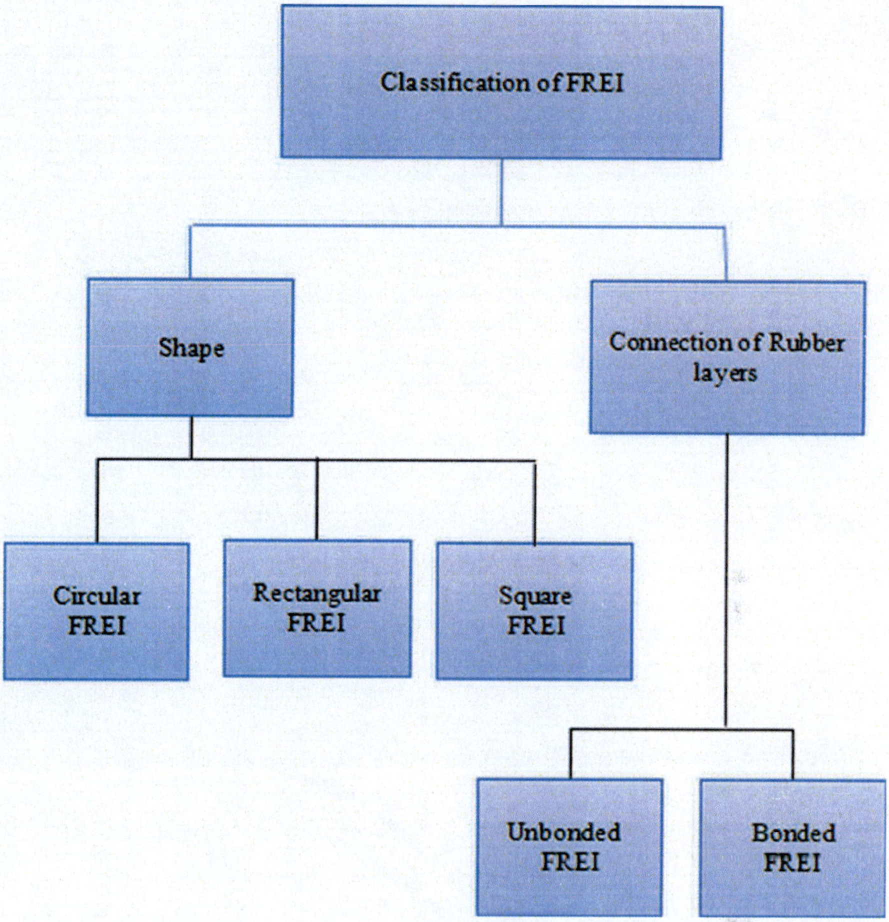

Fig. 1 Classification of FREI. *Source* Das et al. [3]

2.1.1 Bonded FREI

Bonded FREIs have inflexible steel end plates situated at the extreme end surfaces of the isolator as appeared in Fig. 2a, which are utilized to attach it to the superstructure and the substructure. B-FREIs are invaluable over U-FREIs as there is no doubt of slip along these lines no danger of perpetual removals. B-FREIs are proficient to oppose surprising tensile forces. In a laterally deformed B-FREIs, the fiber support layers experience twisting distortion at their finishes because of the internal moments and shear that create in the isolator. Figure 3a synthesizes the shear resistance phenomena of B-FREIs under the action of ground motion.

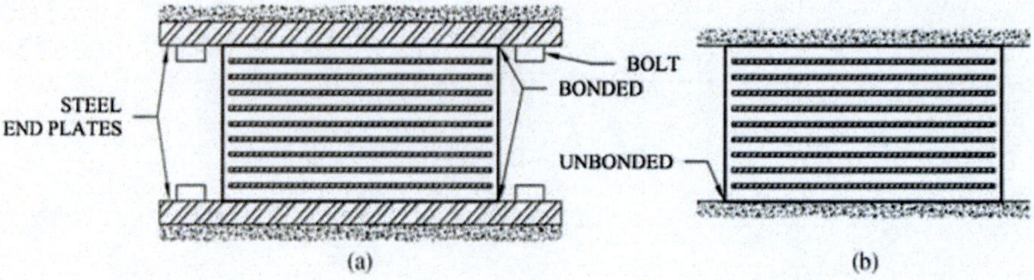

Fig. 2 Bonded and unbonded FREIs. *Source* Osgooei et al. [13]

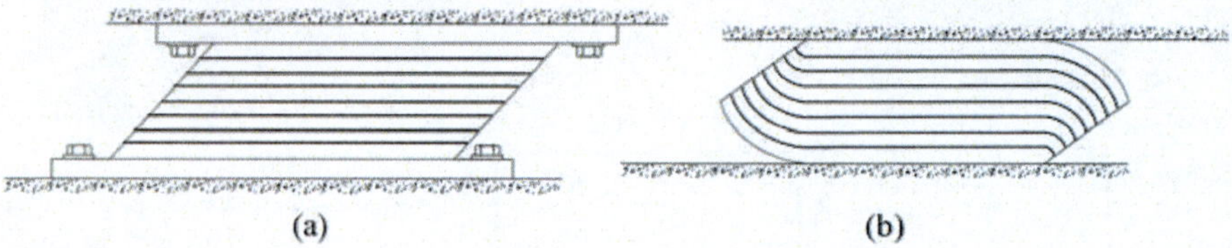

Fig. 3 Lateral deformation phenomena of FREIs. *Source* Osgooei et al. [13]

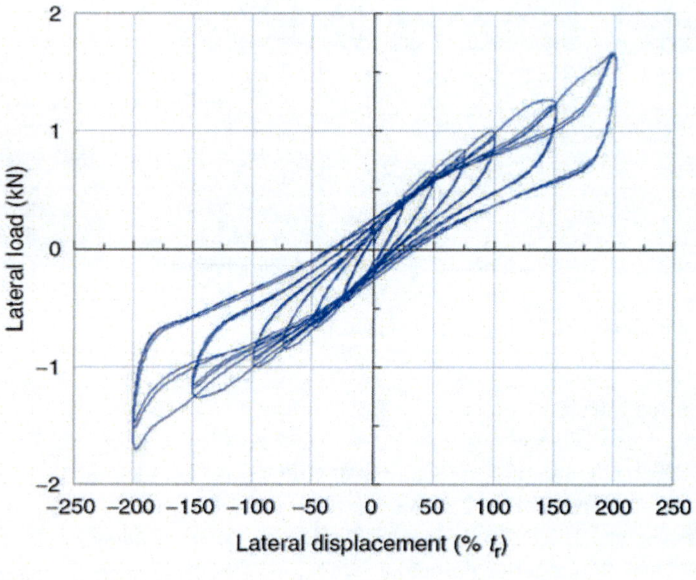

Fig. 4 Force-deformation loop of FREI. *Source* Nezhad et al. [11]

2.1.2 Unbonded FREI

At the point when FREIs are introduced without affixing to the upper and lower bolsters, at that point it is named as unbonded FREIs. Shear forces are passed on from the isolator to the superstructure and substructure through friction along the help surfaces. Because of the nonappearance of the thick end plates and adaptability of fiber support layers, unbonded fiber fortified elastomeric isolator displays a one of a kind rollover distortion under shear as appeared in Fig. 2b. This rollover lifts the seismic productivity by diminishing the successful parallel stiffness of the isolators. There are a few downsides of unbonded application as correlation with bonded isolators. U-FREIs are lacking to oppose the ductile powers and wasteful in circumstances like expansive vertical ground increasing velocities. Figure 3b highlights the possible shear deformation of U-FREIs.

2.1.3 Models of FREI

FREI has several models as listed in Fig. 5. In the present study, Backbone Curve model of fifth-order polynomial is taken into consideration which comprises the simulation of nonlinear elastic spring with a dashpot. The control force of this model is characterized by

$$f_b = f_{sb} + f_{db} \tag{1}$$

where f_{sb} is the force due to stiffness and f_{db} is the damping force. Force f_{sb} is a fifth-order polynomial of displacement 'u'. Hence,

$$f_{sb} = a_1.u + a_2.u^3 + a_3.u^5 \tag{2}$$

$$f_{db} = C\dot{u} \tag{3}$$

where a_1, a_2, a_3 are constant parameters that are obtained from the curve fitting methods on experimental lateral load-displacement hysteresis loops. The displacement velocity components are denoted as u and $\dot{u}$, respectively. 'C' is the damping coefficient which relies on the effective stiffness (k_{eff}) and effective damping ratio (β_{eff}) and defined as (Fig. 4)

$$C = 2\,\beta_{eff}\sqrt{(mk_{eff})} \tag{4}$$

3 Benchmark Highway Bridge

For this benchmark study is 91/5 highway overcrossing structure is considered which is situated in southern California as shown in Fig. 7. The bridge is vulnerable because two main seismic faults are near the location of the bridge. It is a box girder bridge that has two continuous clamps. Figure 6 highlights the elevation view and the top view.

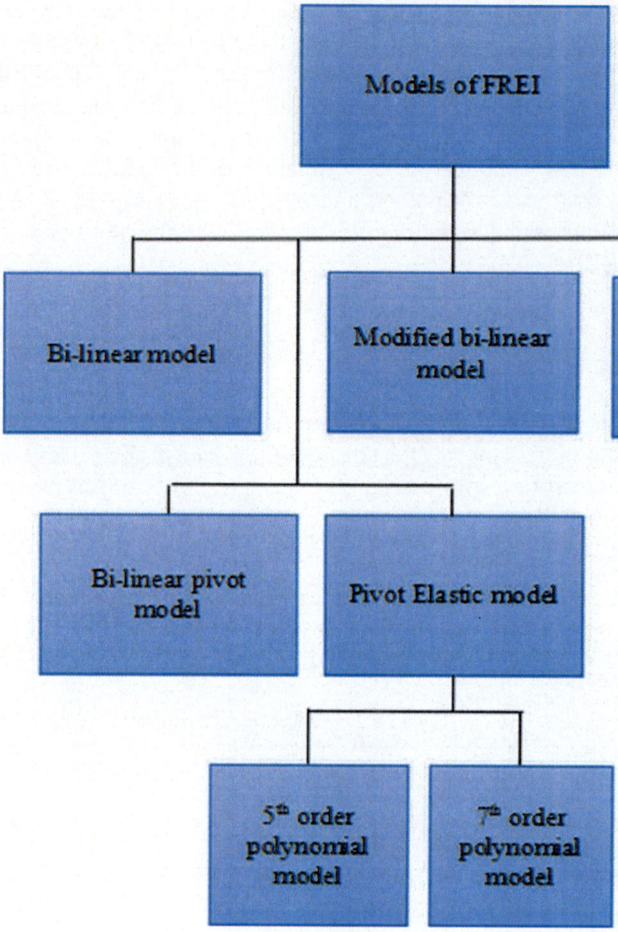

Fig. 5 Classification of FREI models. *Source* Osgooei et al. [13]

Each span is 58.5 m long. It is a four-lane highway bridge with two pillars distorted at 33°. The width of the deck along East-span and along West span is 12.95 m or 15 m. Deck cross section consists of three cells. A prestressed outrigger supports the deck. The length of the outrigger is 31.4 m and height is 6.9 m. The height of the columns is nearly 6.9 m. The mass of the deck and the whole bridge are 3,278,404 kg and 4,237,544 kg, respectively. Deck end and the abutment are isolated with eight fluid dampers and four elastomeric pads are used to minimize the seismic reactions of the bridge as illustrated in Fig. 8.

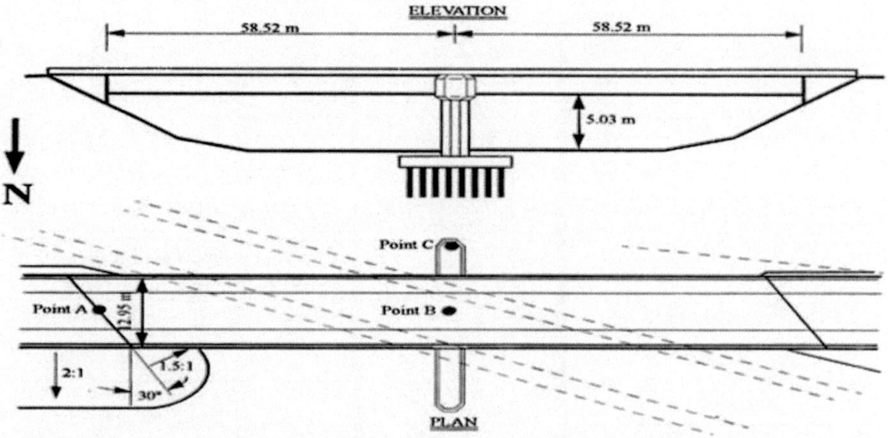

Fig. 6 Side and top view of the benchmark highway bridge. *Source* Agrawal et al. [1]

Fig. 7 91/5 Highway Overcrossing. *Source* Agrawal et al. [1]

3.1 Evaluation Criteria

The structural performance of a system under a seismic excitation is idealized by evaluation criteria. A cluster of 21 evaluation criteria emphasize the efficiency of various seismic devices and algorithms that have been established. These criteria are classified into three categories. 1. Peak responses, 2. Normed responses and 3. Control strategies.

Six earthquakes are selected for the analysis, which are shown in Table 1.

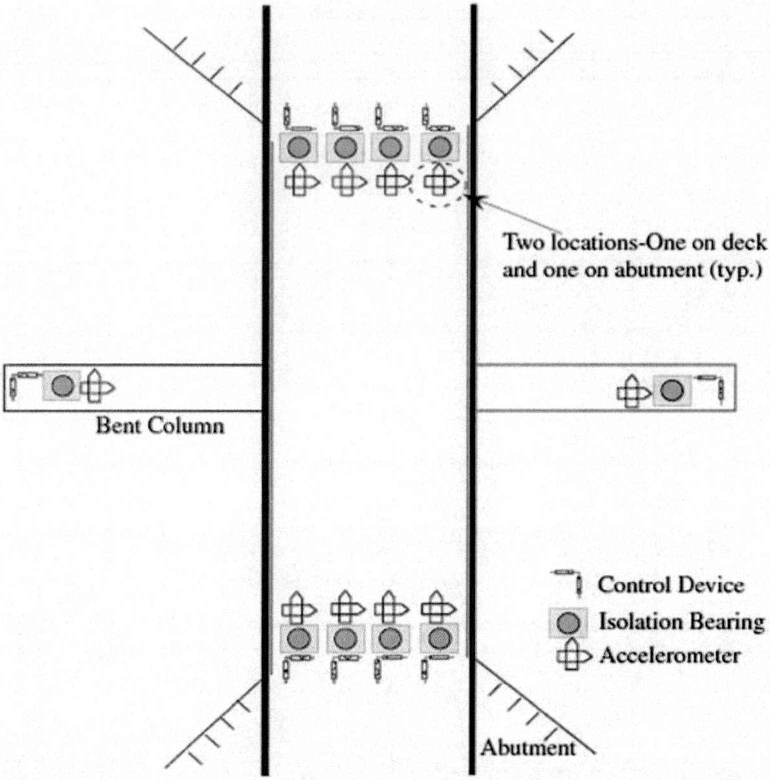

Fig. 8 Control device location on the bridge. *Source* Agrawal et al. [1]

Table 1 List of earthquakes

No.	Earthquakes	Year
1	North palm spring earthquake	1986
2	Chi-Chi earthquake (TUC084 component)	1999
3	Imperial Valley earthquake (El Centro component)	1940
4	Northridge earthquake (Rinaldi component)	1984
5	Duzce earthquake (Bolu component), Turkey	1999
6	Kobe earthquake (Nishi Akashi component)	1995

4 Numerical Study

To inculcate the effectiveness of FREI for the benchmark highway bridge a numerical simulation is done in MATLAB for specified six earthquakes. In this present investigation, several force-deformation behaviors are observed on the basis of selected ground excitation for different displacement amplitudes. The performance of FREI is dependent on its two parameters; damping coefficient and displacement amplitude and velocity component.

It is noticed that while increasing the displacement amplitude values, we are getting the analogous contour of force-displacement diagram essentially at an amplitude of 1.5. Moreover, at displacement amplitude of 2.00, the shape of loop-changed significantly. So it can be concluded that the optimum value of amplitude ratio for this study is 1.5 and the different curvature patterns are observed in Fig. 9.

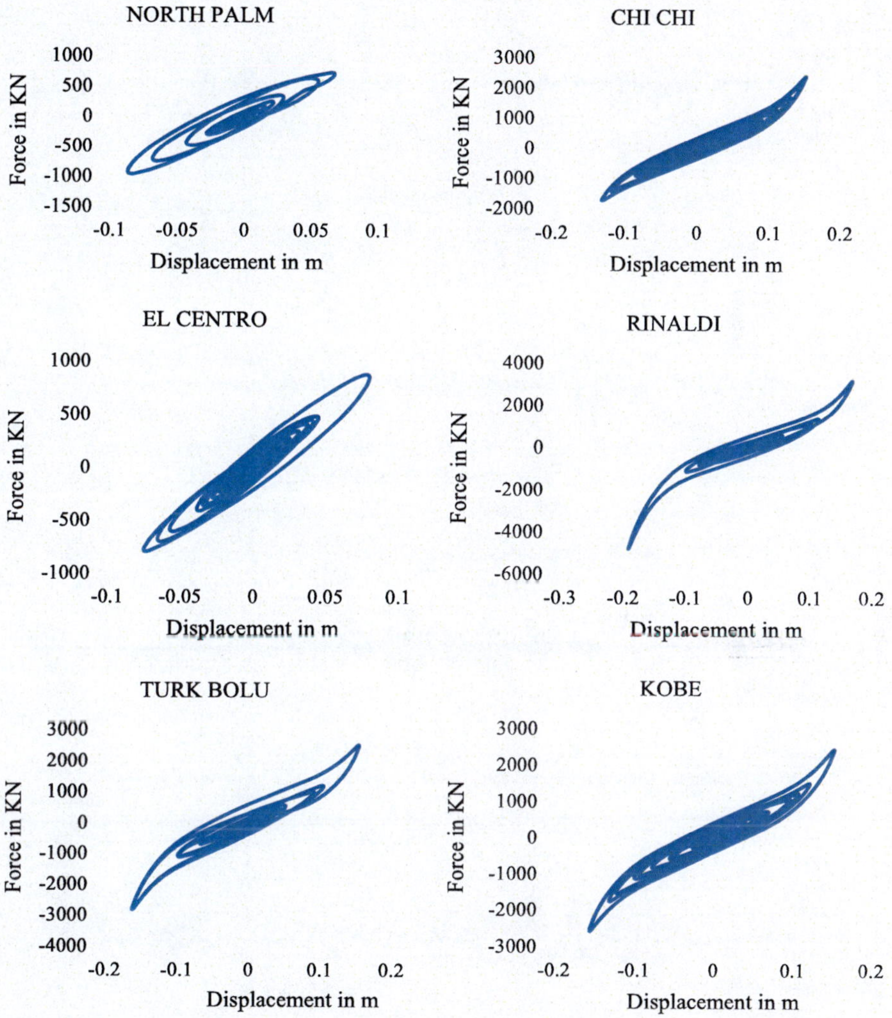

Fig. 9 Force-deformation behavior under different earthquakes of displacement amplitude 1.50

Base shear values of the different earthquakes of different displacement amplitudes are obtained from Fig. 10. Numerically, it is observed that the increment of displacement amplitude values for a particular earthquake, more or less instigated the decrement in the base shear value of the controlled structure accordingly and the variations corresponding to different displacement amplitudes are more or less negligible.

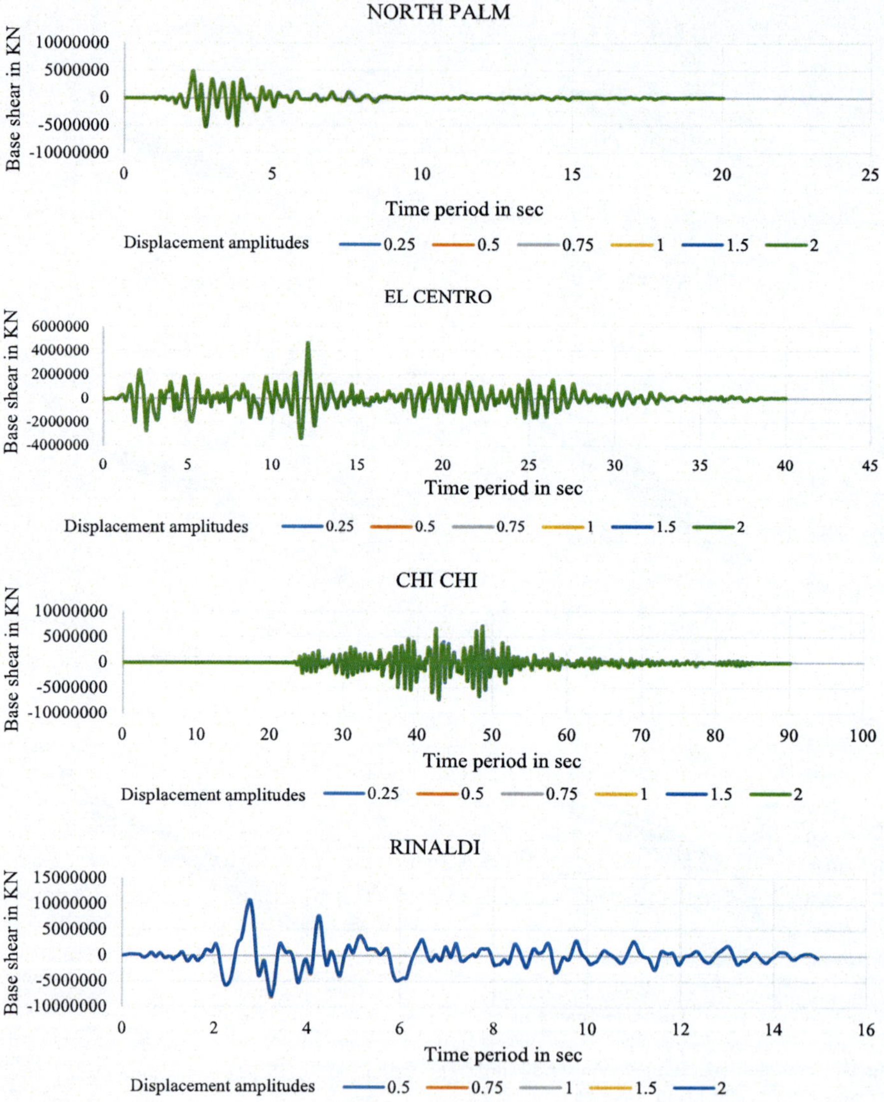

Fig. 10 Comparison of base shear values of different displacement amplitudes of different earthquakes

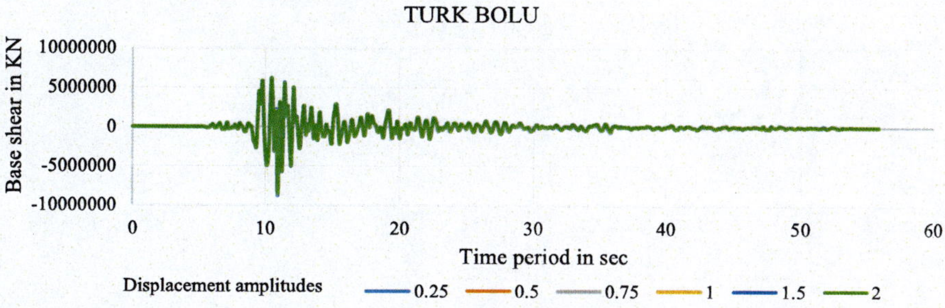

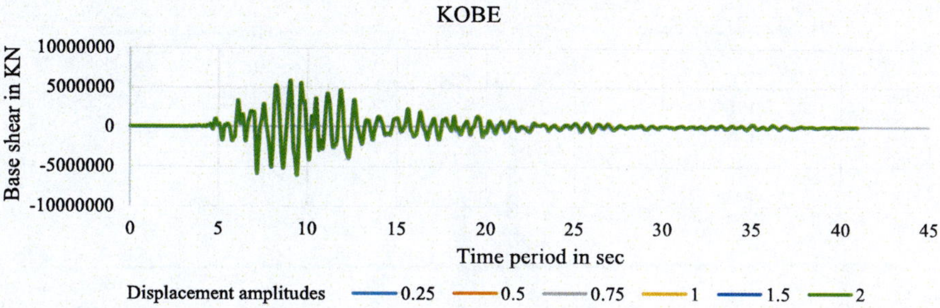

Fig. 10 (continued)

Figure 11 represents the variation of control force values according to the displacement amplitudes of different earthquakes. Fundamentally, it is noticed that for each earthquake, as the displacement amplitude values are increased, there is a marginal decrease in device control force. The control force value attained a maximum at an amplitude of 0.25.

Figure 12 illustrates the comparison of evaluation criteria of different displacement amplitudes on the basis of earthquakes and Fig. 13 enables the comparison of evaluation criteria of different earthquakes on the basis of displacement amplitudes. Peak base shear and overturning moment values are found to be minimum for North palm earthquake for all displacement amplitude values whereas peak displacement at mid-span responses is minimum for Turk Bolu earthquake.

5 Conclusion

Seismic response of highway benchmark bridge with fiber-reinforced elastomeric isolator is scrutinized in this study. To furnish seismic isolation, devices are placed in both the abutments. Total 16 numbers of isolators are implanted in this evaluation model, 8 isolators in each abutment.

The following conclusions can be drawn from the study:

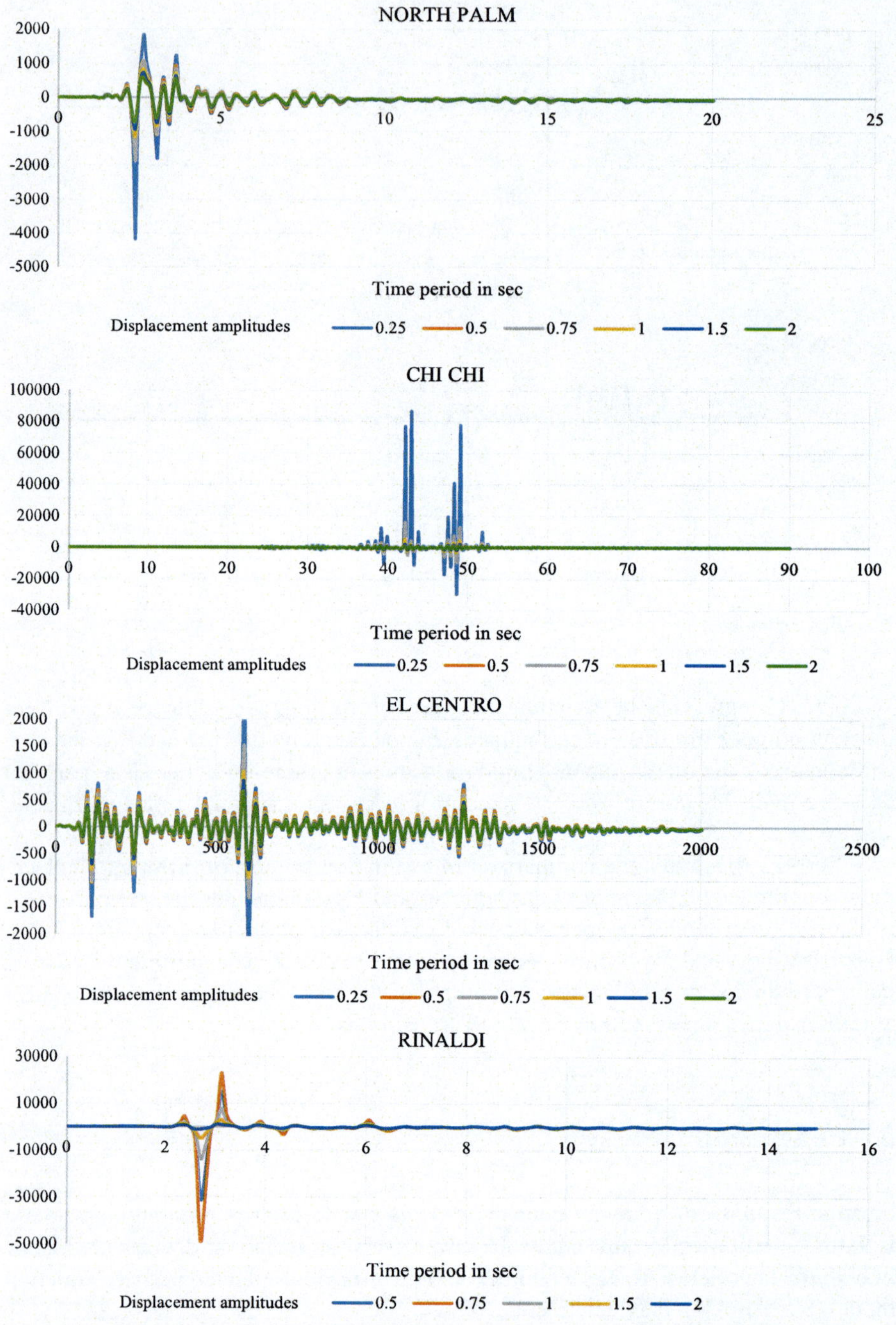

Fig. 11 Comparison of control forces of different displacement amplitudes of different earthquakes

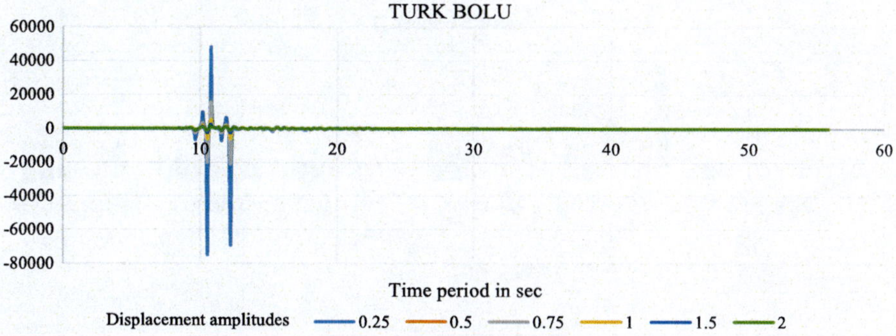

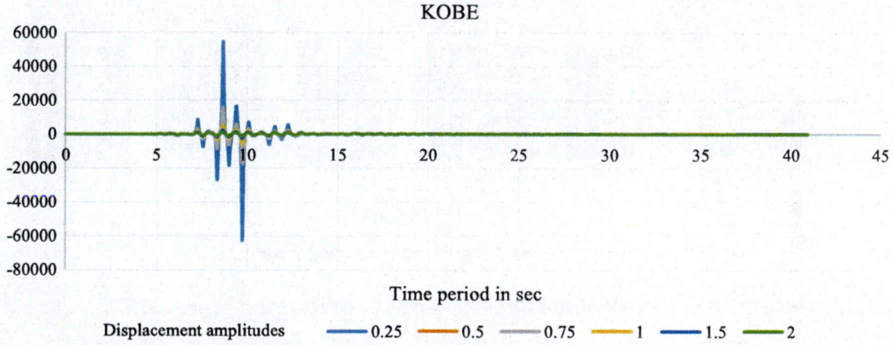

Fig. 11 (continued)

(1) Peak base shear and peak overturning responses are adequately shortened by employing intermediate displacement amplitude of 1.00, whereas peak acceleration at mid-span response factor can be effectively reduced at displacement amplitude of 0.25.
(2) Peak displacement at mid-span response can be effectively reduced by employing higher displacement amplitudes of 1.50 and 2.00.
(3) Among all the investigations, it is witnessed that FREI system of displacement amplitude 1.50 gave optimum performance than other trials.

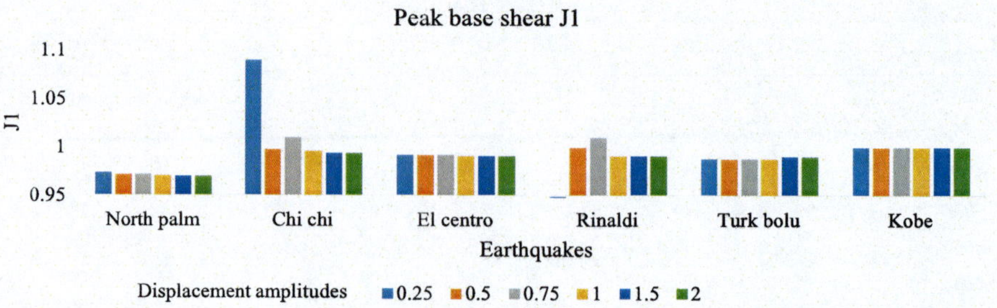

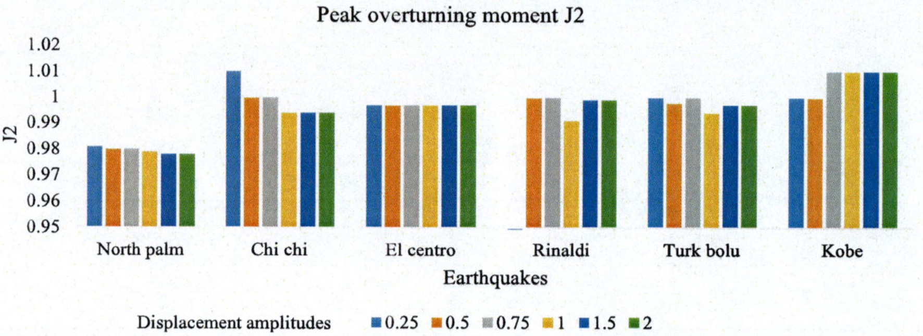

Fig. 12 Comparison of evaluation criteria on the basis of earthquakes

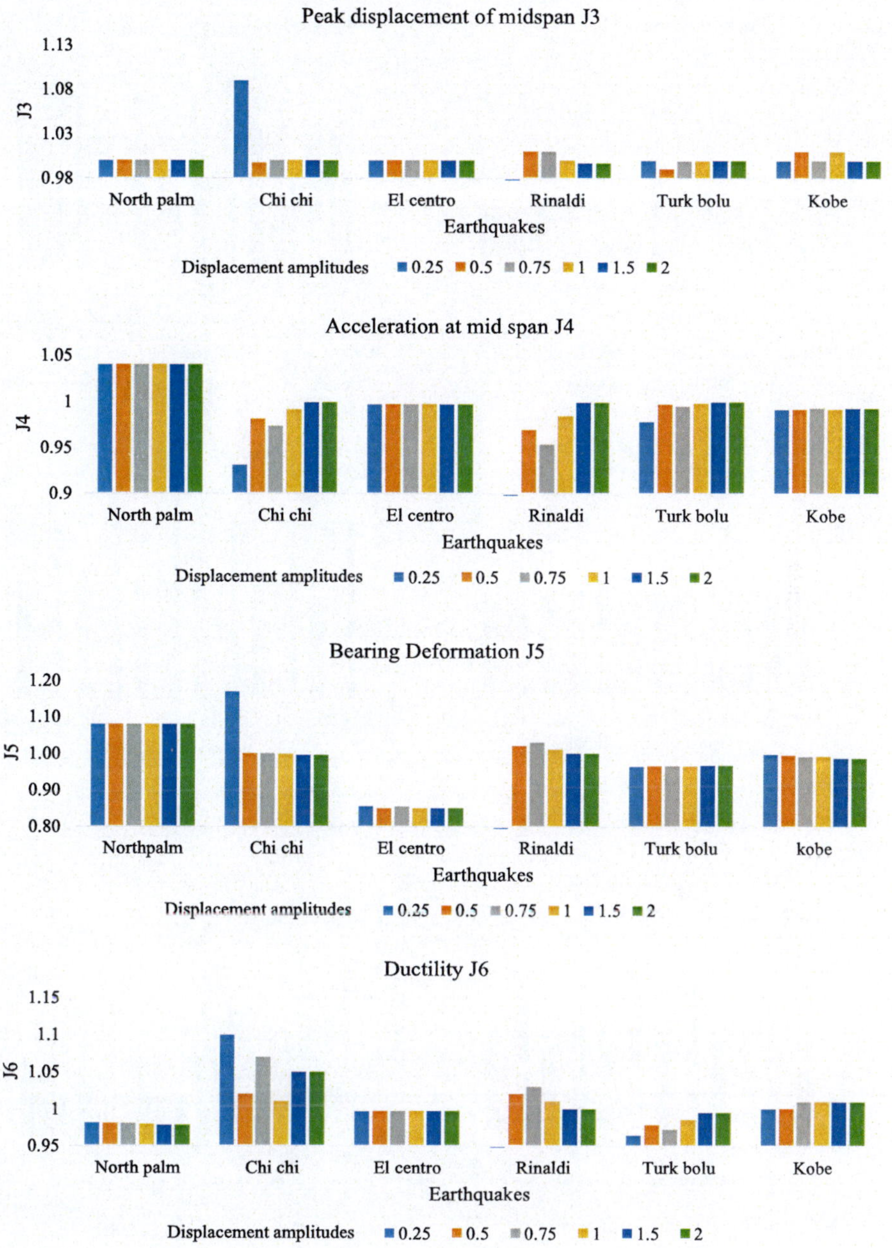

Fig. 12 (continued)

Fig. 13 Comparison of evaluation criteria on the basis of displacement amplitudes

Bibliography

1. Agrawal AK, Tan P, Nagarajaiah S, Zhang J (2009) Benchmark structural control problem for a seismically excited highway bridge—part I: phase i problem definition. J Struct Control Health Monitor 16(5):509–529
2. Al-Anany Y, Tait M (2015) A numerical study on the compressive and rotational behavior of fiber reinforced elastomeric isolators (FREI). Compos Struct 133:1249–1266
3. Das A, Dutta A, Deb S (2012) Modeling of Fiber-Reinforced Elastomeric Base Isolators. WCEE
4. Kelly J (1999) Analysis of fiber reinforced elastomeric isolators. J Struct Eng 2(1):19–34
5. Madhekar SN (2014) "Seismic performance of benchmark highway bridge installed with passive control devices. Struct Eng Convent, IIT Delhi
6. Madhekar SN, Jangid RS (2009) Variable dampers for earthquake protection of benchmark highway bridges. Smart Mater Struct 18
7. Madhekar SN, Jangid RS (2011) Seismic performance of benchmark highway bridge installed with piezoelectric friction dampers. The IES J Part A: Civil Struct Eng 4(4):191–212
8. Madhekar SN, Jangid RS (2012) Use of pseudo-negative stiffness dampers for reducing the seismic response of bridges: a benchmark study. Bull Earthquake Eng 10:1561–1583
9. Narasimhan S (2009) Robust direct adaptive controller for the nonlinear highway bridge benchmark. Struct Control and Health Monit 16:599–612
10. Nezhad H, Tait M, Drysdale R (2011) Bonded versus unbonded strip fiber reinforced elastomeric isolators: Finite element analysis. Compos Struct 93:850–859
11. Nezhad H, Tait M, Drysdale R (2008) Lateral response evaluation of fiber-reinforced neoprene seismic isolators utilized in an unbonded application. ASCE 134:101627
12. Osgooei P, Tait M, Konstantinidis D (2016) Seismic isolation of a shear wall structure using rectangular fiber-reinforced elastomeric isolators. J Struct Eng, ASCE, 04015116(1–10)
13. Osgooei P, Tait M, Konstantinidis D (2017) Non-iterative computational model for fiber-reinforced elastomeric isolators. Eng Struct 137:245–255
14. Panchal VR, Jangid RS (2008) Seismic isolation of bridge using variable curvature friction pendulum system. 14th World Conference on Earthquake Engineering, Beijing, China
15. Saha A, Saha P, Patro SK (2015) Seismic response control of benchmark highway bridge using non-linear FV spring damper. The IES J Part A: Civil Struct Eng 8(4):240–250
16. Saha P (2012) Seismic control of benchmark cable-stayed bridge using variable friction pendulum systems. 15th World Conference on Earthquake Engineering, Lisboa
17. Tan P, Agrawal AK (2009) Benchmark structural control problem for a seismically excited highway bridge part II: sample control designs. Struct Contr Health Monitor 16:530–548

Optimal Design of Hysteretic Nonlinear Energy Sink for Suppression of Limit Cycle Oscillations of a Flapping Airfoil

Omkar Ghosh and Sourav Das

Abstract The aim of this study is to address the suppression of limit cycle oscillations (LCOs) of a flapping airfoil using hysteretic nonlinear energy sink (HNES). A flexible flapping wing with span-wise bending and twisting degrees of freedom incorporated through translation and torsional springs with nonlinear stiffness in heave and pitch is considered in this study. HNES is comprised of a mass connected with primary structure by a purely hysteretic and a linear elastic spring connected in parallel. The nonlinear hysteresis behaviour is characterized by Bouc–Wen model. The performance of HNES with negative stiffness is investigated. Numerical results presented in this paper clearly elucidated the enhanced performance of the proposed HNES with negative stiffness compared with conventional NES and HNES with positive stiffness. An optimization is carried out to find the optimal design parameters of HNES in a deterministic approach by considering uncertainties in initial conditions with the objective of minimizing peak and root mean square response of LCOs. The theoretical study in this paper clearly demonstrated the concept and efficiency of the proposed passive control strategy which can be easily adopted for existing and up-coming wings.

Keywords Passive control · Nonlinear energy sink · Negative stiffness · Bouc–wen model · Optimization

1 Introduction

Passive vibration absorb device has gained attention among researchers because of the simplicity in installation and low maintenance cost. Several passive vibration absorbers consisting of linear mass-spring-dashpot system has been proposed [1]

O. Ghosh (✉)
School of Civil Engineering, KIIT Deemed to be University, Bhubaneswar, Odisha, India
e-mail: sg211352@gmail.com

S. Das
School of Engineering, The University of British Columbia, Okanagan Campus, Kelowna, BC, Canada
e-mail: sourav.das.bony@gmail.com

B. B. Das et al. (eds.), *Recent Developments in Sustainable Infrastructure*, Lecture Notes in Civil Engineering 75, https://doi.org/10.1007/978-981-15-4577-1_30

and widely used in civil, mechanical and aerospace engineering. Natural frequency of the linear vibration absorber is designed close to natural frequency of the main structure so that the energy is partly transferred to oscillator from primary structure by resonance mechanism and dissipated by damping of the oscillator. Tuned mass damper (TMD) is one of the well-established vibration absorbers used to mitigate vibration due to dynamic external excitation for flexible structures. Several investigations have been carried out to investigate the effectiveness of linear TMD to control vibrations of the linear structures [2]. Because this kind of passive vibration absorber has narrow bandwidth, there is a possibility of detuning by degradation of stiffness of the primary structure. Due to development of technology, nonlinear vibration absorbers are used as passive device whose bandwidth is much broader than linear absorber and able to show nonlinearity of restoring force of the oscillator to reduce the dynamic force-induced vibration.

Nonlinear energy sink (NES) has grown popular among the researchers due to wider bandwidth of the absorber. An NES consists of a mass attached with nonlinear stiffness, such that energy transfer takes place passively from vibrating structure to NES [3]. However, unlike TMD, due to broader frequency range, NES can absorb the energy due to external excitation and less prone to detune which may occur due to degradation of stiffness property. Various types of NES have been proposed. The main difference between proposed NES is the arrangement of the nonlinear restoring force.

Limit cycle oscillations (LCO) are the sustained oscillations, occurred due to the interaction between aerodynamic and structural forces in the presence of the nonlinearities. For airfoil LCO, the main aspects are the sub-critical and super-critical oscillations because super-critical oscillation affects the fatigue damage whereas due to discontinuity behaviour of sub-critical LCO, the airfoil may experience the unexpected dynamic behaviour. Therefore, to ensure safety during its operation, control and health monitoring are the essential components to protect them against structural damages induced by vibration. Several researchers studied the LCO control of airfoil using the passive devices in the latter past. Lee et al. [4, 5] have proposed NES to control the flutter instability.

With this view, the present study aims to address the performance of a novel kind of vibration absorber, hysteretic nonlinear energy sink (HNES) which is the combination of a hysteretic spring and a linear elastic spring, connected in parallel. In this study, the following issues are addressed:

- A reduced-order airfoil model is proposed to be used to study the performance of HNES.
- Demonstrate the performance of this passive control strategy with negative stiffness compared with cubic NES and HNES with positive stiffness.
- An optimization is carried out to study the performance envelope of the proposed control strategy using Monte Carlo simulation.

2 Dynamics of Two Degrees of Freedom Airfoil

In this study, a two degrees of freedom airfoil model is used for the investigation. The degrees of freedom are considered as pitch and plunge to describe the airfoil motions which are defined by the heave h and the angle of attack α, respectively, as shown in Fig. 1. The geometrical properties of airfoil are described by the chord length c, the semi-chord length b and the lifting surface S. Elastic, gravity and aerodynamic centre of the airfoil are represented by EC, GC and AC, respectively, as shown in Fig. 1. The aerodynamic centre is the point where pitching moment coefficient of the airfoil does not vary with angle of attack. The mass and inertia moment of the airfoil is denoted by M and I_α, respectively. The static moment of the airfoil is represented by S_α ($=Mx_{CG}$), where x_{CG} is the distance between elastic and gravity centre. The linear stiffness and damping in pitch and plunge modes are represented by k_α, c_α and k_h, c_h, respectively. Due to geometric or localized nonlinearity, the aero-elastic system has shown nonlinear behaviour. To introduce nonlinearity in aero-elastic system, cubic stiffness is considered in plunge and pitch directions, which are represented by k_{h_3} and k_{α_3}, respectively. The governing equations of motion for the two degrees of freedom airfoil model can be formulated as

$$M\ddot{h} + S_\alpha\ddot{\alpha} + c_h\dot{h} + k_h h + k_{h_3}h^3 + L_w = 0 \tag{1}$$

$$I_\alpha\ddot{\alpha} + S_\alpha\ddot{h} + c_\alpha\dot{\alpha} + k_\alpha\alpha + k_{\alpha_3}\alpha^3 - M_w = 0 \tag{2}$$

where L_w and M_w represent the lift force and aerodynamic moment, respectively, occurred due to wind velocity U. In this study, wind flow is assumed to be quasi-steady flow, where the aerodynamic forces and moments depend only on the instantaneous motion of the body surface. Here, the prior history of the flow motion or the influence

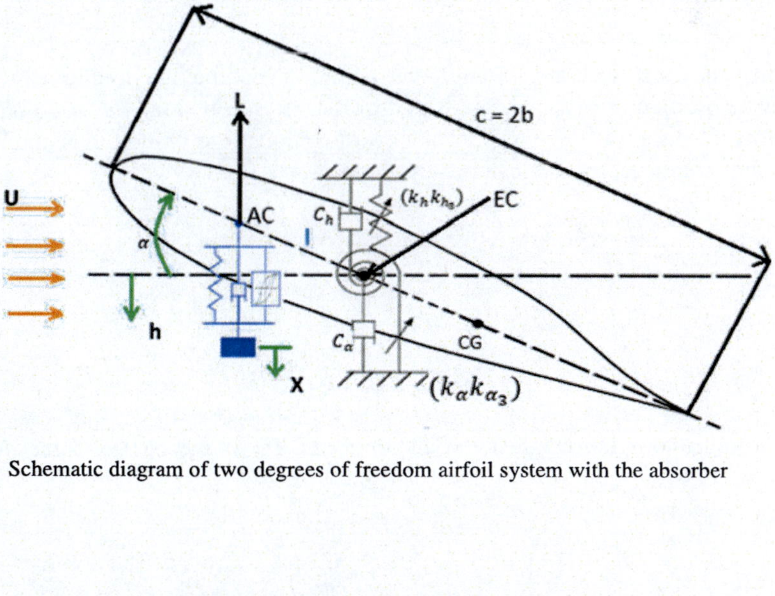

Fig. 1 Schematic diagram of two degrees of freedom airfoil system with the absorber

of the wake is neglected. The lift force and aerodynamic moment acting on the airfoil can be expressed as

$$L_w = \frac{1}{2}\rho S U^2 \frac{\partial C_L}{\partial \alpha}\left(\alpha + \frac{\dot{h}}{U}\right) \tag{3}$$

$$M_w = L_w e = \frac{1}{2}\rho S U^2 e \frac{\partial C_L}{\partial \alpha}\left(\alpha + \frac{\dot{h}}{U}\right) \tag{4}$$

where ρ is the air density, C_L is the lift coefficient and e is the distance between aerodynamic and elastic centre of the airfoil. To reduce the number of system parameters of the airfoil, the dimensionless quantities are introduced which are

$$y = \frac{h}{b};\ r_\alpha = \sqrt{\frac{I_\alpha}{Mb^2}};\ x_\alpha = \frac{S_\alpha}{Mb};\ \tau = \omega_\alpha t;\ \omega_\alpha = \sqrt{\frac{k_\alpha}{I_\alpha}};\ \omega_h = \sqrt{\frac{k_h}{M}};\ \Omega = \sqrt{\frac{\omega_h}{\omega_\alpha}} \tag{5}$$

$$\xi_h = \frac{k_{h_3} b^2}{M\omega_\alpha^2};\ \xi_\alpha = \frac{k_{\alpha_3}}{Mb^2\omega_\alpha^2};\ \zeta_h = \frac{c_h}{M\omega_\alpha};\ \zeta_\alpha = \frac{c_\alpha}{Mb^2\omega_\alpha} \tag{6}$$

The dimensionless aerodynamic parameters are

$$\bar{U} = \frac{U}{b\omega_\alpha};\ \beta = \frac{L_w b}{M};\ \nu = \frac{M_w}{M} \tag{7}$$

After substituting the non-dimensional quantities into the Eqs. (1) and (2), yields

$$\begin{bmatrix} 1 & x_\alpha \\ x_\alpha & r_\alpha^2 \end{bmatrix}\begin{bmatrix} \ddot{y} \\ \ddot{\alpha} \end{bmatrix} + \begin{bmatrix} \zeta_h + \beta\bar{U} & 0 \\ -\nu\bar{U} & \zeta_\alpha \end{bmatrix}\begin{bmatrix} \dot{y} \\ \dot{\alpha} \end{bmatrix} + \begin{bmatrix} \Omega^2 & \beta\bar{U}^2 \\ 0 & r_\alpha^2 - \nu\bar{U}^2 \end{bmatrix}\begin{bmatrix} y \\ \alpha \end{bmatrix} + \begin{bmatrix} \xi_h y^3 \\ \xi_\alpha \alpha^3 \end{bmatrix} = \begin{bmatrix} 0 \\ 0 \end{bmatrix} \tag{8}$$

In this study, the two kinds of nonlinear vibration absorber are used to reduce the limit cycle oscillation of the airfoil. The absorber is placed along the mid-chord of the airfoil at a distance '*l*' from EC. The governing equations of motion of combined airfoil and vibration absorber can be written as

$$M\ddot{h} + S_\alpha\ddot{\alpha} + c_h\dot{h} + k_h h + k_{h_3} h^3 + L_w - F_C = 0 \tag{9}$$

$$I_\alpha\ddot{\alpha} + S_\alpha\ddot{h} + c_\alpha\dot{\alpha} + k_\alpha\alpha + k_{\alpha_3}\alpha^3 - M_w + F_C l = 0 \tag{10}$$

$$m\ddot{x} + F_C = 0 \tag{11}$$

where m represents the mass of the absorber and F_C is the control force by the absorber.

3 Nonlinear Energy Sink

In recent decades, the nonlinear energy sink (NES) has grown popular among the researchers due to wider frequency bandwidth of the absorber. An NES consists of a mass attached with nonlinear stiffness and viscous damper, such that energy transfers from vibrating structure to NES and dissipate through the viscous damper [6–9]. However, unlike tuned mass damper, due to broader frequency range, NES can absorb more energy due to external excitation and less prone to detune which may occur due to degradation of stiffness properties of the primary structure. In this study, an NES of mass m consisting of a spring of linear stiffness k and cubic stiffness k_3, a viscous damping c is considered. The control force by NES can be written as

$$F_{\text{NES}} = c(\dot{x} + l\dot{\alpha} - \dot{h}) + k(x + l\alpha - h) + k_3(x + l\alpha - h)^3 \tag{12}$$

The dimensionless NES parameters are considered as follows

$$\tilde{x} = \frac{x}{b}; \in = \frac{m}{M}; \omega = \sqrt{\frac{k}{m}}; \gamma = \frac{\omega^2}{\omega_\alpha^2}; \lambda = \frac{l}{b}; \zeta = \frac{c}{m\omega_\alpha}; \xi = \frac{k_3 b^2}{m\omega_\alpha^2} \tag{13}$$

After substituting Eqs. (12) and (13) into Eqs. (9), (10) and (11) yields,

$$\bar{M}\ddot{Q} + \bar{C}\dot{Q} + KQ + \bar{F}_c(t) = 0 \tag{14}$$

where

$$Q = \begin{bmatrix} y \\ \alpha \\ \tilde{x} \end{bmatrix}; \bar{M} = \begin{bmatrix} 1 & x_\alpha & 0 \\ x_\alpha & r_\alpha^2 & 0 \\ 0 & 0 & 1 \end{bmatrix}; \bar{C} = \begin{bmatrix} \zeta_h + \in\zeta + \beta\bar{U} & -\in\zeta\lambda & -\in\zeta \\ -\nu\bar{U} - \in\zeta\lambda & \zeta_\alpha + \in\zeta\lambda^2 & \in\zeta\lambda \\ -\zeta & \zeta\lambda & \zeta \end{bmatrix};$$
$$\bar{K} = \begin{bmatrix} \Omega^2 + \in\gamma & \beta\bar{U}^2 - \in\gamma\lambda & -\in\gamma \\ -\in\gamma\lambda & r_\alpha^2 - \nu\bar{U}^2 + \in\gamma\lambda^2 & \in\gamma\lambda \\ -\gamma & \gamma\lambda & \gamma \end{bmatrix}; \bar{F}_c = \begin{bmatrix} \xi_h y^3 - \in\xi(\tilde{x} + \lambda\alpha - y)^3 \\ \xi_\alpha\alpha^3 + \in\lambda\xi(\tilde{x} + \lambda\alpha - y)^3 \\ \xi(\tilde{x} + \lambda\alpha - y)^3 \end{bmatrix} \tag{15}$$

4 Hysteretic Nonlinear Energy Sink

HNES is the combination of a purely hysteretic spring and a linear elastic spring, which are connected in parallel with primary structure, proposed by Tsiatas et al. [10]. The force offered by HNES described by Bouc–Wen model [11, 12], which is expressed as follows

$$F_c(t) = F_c^{el}(t) + F_c^h(t) = akx(t) + (1-a)kDz(t) \tag{16}$$

where $x(t)$ is the displacement time history for HNES, k is the initial stiffness, a is the ratio of post- to pre -yield stiffness, D is the yield displacement. In Eq. (16), the superscript '*el*' and '*h*' denote the elastic and hysteretic component of force offered by HNES. The dimensionless hysteretic variable $z(t)$ is written as

$$\dot{z} = D^{-1}\left[A - \{\beta.\text{sgn}(z\dot{x}) + \gamma\}|z|^n\right]\dot{x} \tag{17}$$

In above equation, sgn(.) represents the signum function. The dimensionless exponential parameter n describes the transition from pre-yield to post-yield response. The constants A, β and γ control the shape and size of the hysteresis loop. In this study, for two degrees of freedom airfoil system, $x(t)$ in Eqs. (16) and (17) is replaced by $(x + \text{l}\alpha - \text{h})$ and the governing equation of motion Eq. (14) can be rearranged as follows

$$\bar{F}_c = \left[\xi_h y^3 - \in \gamma\bar{F} \;\; \xi_\alpha\alpha^3 + \in \lambda\gamma\bar{F} \;\; \gamma\bar{F}\right]^T \tag{18}$$

where $\bar{F} = a(\tilde{x} + \lambda\alpha - y) + (1-a)Dz$ and other system matrices are same as Eq. (15). The state vectors of the system are given as

$$Z(t) = \left[y \;\alpha\; \tilde{x}\; \dot{y}\; \dot{\alpha}\; \dot{\tilde{x}}\right]^T \tag{19}$$

The governing equations of motion in state-space form can be expressed as

$$\dot{Z}(t) = AZ(t) + E\bar{F}_c(t) \tag{20}$$

where the system matrices are given by

$$A = \begin{bmatrix} 0 & I \\ -\bar{M}^{-1}\bar{K} & -\bar{M}^{-1}\bar{C} \end{bmatrix};\; E = -\begin{bmatrix} 0 \\ \bar{M}^{-1}I \end{bmatrix} \tag{21}$$

5 Realizing Negative Stiffness in HNES

Molyneaux et al. [13] proposed the idea of anti-spring or negative stiffness spring for the reduction of oscillation of a vibrating system. Platus et al. [14] extended this concept for vibration isolation system. Negative stiffness property is achieved by using pre-compressed springs. Winterflood et al. [15], Virgin et al. [16], Liu et al. [17] Antoniadis et al. [18] have proposed the implementation of negative stiffness concept for nonlinear isolation system.

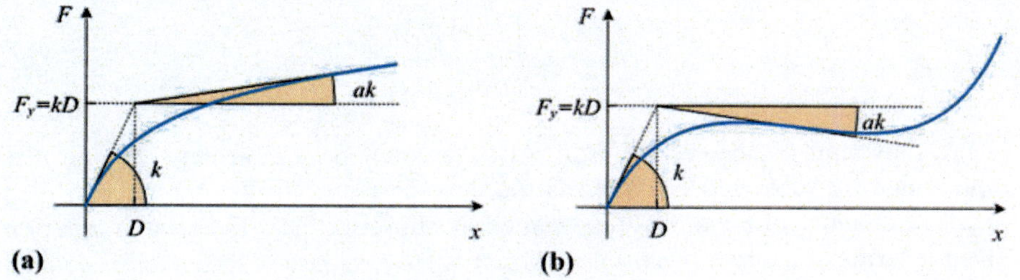

Fig. 2 Typical force-displacement behaviour of HNES with (**a**) positive (**b**) negative stiffness [10]

Tsiatas et al. [10] have proposed a simple approach to obtain the negative stiffness by considering negative values of parameter 'a' in Bouc–Wen model in Eq. (16) which leads to true softening behaviour, as shown in Fig. 2. In this study, the main objective is to investigate the performance of negative stiffness-based HNES over the positive stiffness-based HNES and cubic NES.

6 Flutter Velocity

Flutter is a kind of structural failure of aero-elastic structure which is often lead by dynamic instability. Due to interaction between aerodynamic and structural forces in the presence of structural nonlinearities, the airfoil will exhibit the limit cycle oscillations (LCO). The LCO that occurs for flow velocity larger than flutter velocity is known as super-critical LCO and for flow velocity smaller than flutter velocity is known as sub-critical LCO. In super critical region, airfoil may lead to structural fatigue problems whereas in sub-critical region, discontinuity is observed in LCO and that may influence the sudden change in dynamic behaviour of airfoil. In this context, it is necessary to find the flutter point of the airfoil. The linear flutter velocity is calculated by considering nonlinear term of governing equation as zero and eigenanalysis is performed for matrix 'A' in Eq. (21). The eigenvalues may be real or complex ($z = p_R + i.\omega$) in nature. The airfoil is stable when real part (p_R) of 'z' is less than zero, whereas the airfoil encounters flutter when p_R is more than zero. Using this condition, the flutter velocity is calculated.

7 Performance Function

In order to measure the performance of the controller, performance indices are required to quantify the reduction of the response of the system in terms of maximum response quantities. The first performance evaluation index (J_1) is the maximum displacement ratio which is defined as

$$J_1 = \max\left[\underset{i \in \eta}{t} \frac{|y_{p,c}|}{|y_{p,uc}|} \right] \tag{22}$$

where subscript 'p' denotes the peak. Also, the numerator indicates the controller represented by subscript 'c', whereas the denominator indicates the uncontrolled response quantity of the airfoil represented by subscript 'uc'. The second criterion given in terms of the root mean square (rms) of the response of the airfoil is defined as

$$J_2 = \frac{|y_{\text{rms},c}|}{|y_{\text{rms},uc}|};\ y_{rms} = \sqrt{\frac{1}{T_{\text{end}}} \int_0^{T_{end}} y^2(t)dt} \tag{23}$$

where T_{end} is the total time duration. To realize the effectiveness of proposed HNES which can able to reduce the limit cycle oscillation of airfoil to the lowest level, numerical optimization is performed. In this study, the minimization objective function is considered as

$$J = \min[J_1 + J_2]$$
$$s.t. \begin{cases} \gamma_{\text{min}} \le \gamma \le \gamma_{\text{max}} \\ \zeta_{\text{min}} \le \zeta \le \zeta_{\text{max}} \end{cases} \tag{24}$$

The optimization formulation is considered as constraint optimization in which random variable γ is the stiffness and ζ is the damping of HNES, respectively. The random variables γ and ζ are bounded with the lower bounds γ_{min} and ζ_{min} and upper bounds γ_{max} and ζ_{max}, respectively.

8 Numerical Results

In this section, numerical analysis of proposed control strategy to reduce the limit cycle oscillation of the airfoil is presented. The mass and the coefficients of damping and stiffness of the airfoil are tabulated in Table 1. The fourth-order Runge–Kutta method is adopted here to solve the differential equations of motion for the airfoil. In this study, linear flutter velocity for airfoil is calculated for steady and quasi-steady wind flow when it is uncoupled with NES. From Fig. 3, it is observed that flutter velocity $\bar{U}_F^{(QS)}$ for quasi-steady state becomes 0.87 whereas for steady state it

Table 1 Dimensionless aero-elastic parameters [19]

x_α	r_α	β	ν	Ω	ζ_α	ζ_h
0.2	0.5	0.2	0.08	0.5	0.01	0.01

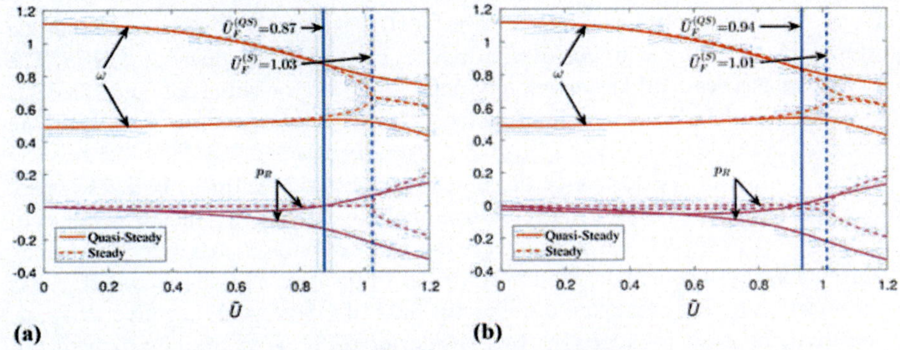

Fig. 3 Frequency and damping of aero-elastic modes for steady and quasi-steady state aerodynamics; (**a**) without aerodynamic damping; (**b**) with aerodynamic damping

($\bar{U}_F^{(S)}$) becomes 1.03, when aerodynamic damping is not considered into the model. The flutter velocities for steady and quasi-steady state become 1.01 and 0.94, when aerodynamic damping is taken into the model. All these values are normalized as discussed in Eqs. (5), (6), and (7), (13).

For this present study, the mass ratio, i.e. mass of the absorber to mass of the airfoil is taken as 5%. It is evident that larger mass of absorber will able to reduce the limit cycle oscillation of airfoil. But, in reality, it is not feasible to provide a larger mass into the airfoil. The absorber should be a negligible mass of the airfoil. Here, the vibration absorber is placed at the leading edge of the profile, i.e. $\lambda = 1$. The first objective is to investigate the effectiveness of HNES with positive stiffness compared to cubic NES to reduce LCO of the airfoil. The normalized stiffness and damping coefficients for the NES and HNES are taken as 0.06 and 0.11. The parameters in Bouc–Wen model are taken as: $a = 0.5$, $d = 0.01$, $n = 1$, $\beta = 0.5$ and $\gamma = 0.5$. Figure 4a is depicted the displacement time histories for uncontrolled, with NES and HNES in heave direction. It is seen that the controlled displacement by NES is approximately

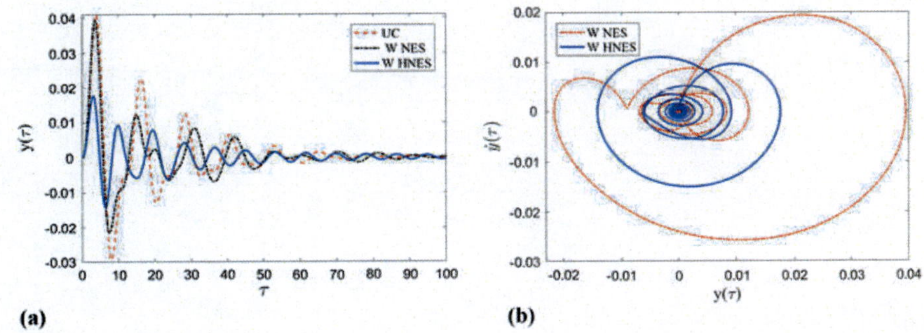

Fig. 4 Sub-critical LCO for quasi-steady wind with initial conditions [0 0 0 0.1]; (**a**) heave displacement time histories; (**b**) heave displacement and heave velocity phase diagram

0.04 whereas the controlled displacement by HNES is 0.02. It is observed that the performance of HNES with positive stiffness is more effective than cubic NES. The Heave displacement and heave velocity phase diagram for with NES and HNES is shown in Fig. 4b. Also the angle of attack LCO and corresponding phase diagram are shown in Fig. 5.

In this study, the performance of HNES with negative stiffness is investigated. The negative stiffness approach is employed here by considering negative values of parameter 'a' in Eq. (16) which depicts the true softening behaviour. The force–displacement hysteresis diagram for HNES with positive and negative stiffness is shown in Fig. 6. To investigate the effectiveness of HNES with negative stiffness, a sensitivity analysis is carried out by considering different values of parameter 'a' ranging from (−0.5) to (0.5). The objective function is considered to quantify the effectiveness of HNES, which is mentioned in Eq. (24). To achieve more reduction, the considered performance index should be minimized. From Fig. 7, it is seen that the performance of HNES is more effective by considering negative stiffness compared to positive stiffness. From Fig. 7, it is concluded that the optimum value of 'a' is

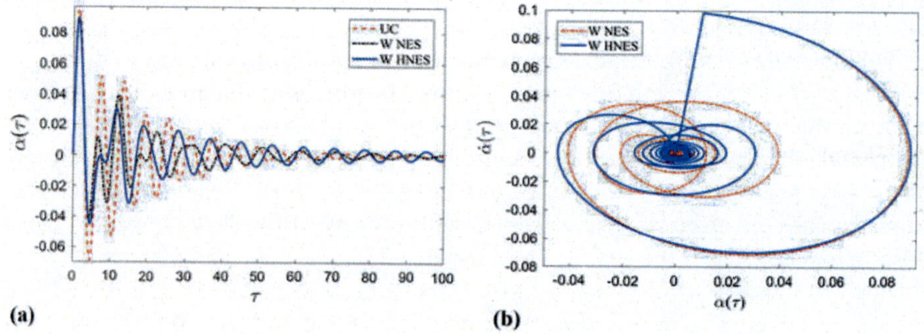

Fig. 5 Sub-critical LCO for quasi-steady wind with initial conditions [0 0 0 0.1]; (**a**) angle of attack displacement time histories; (**b**) pitch displacement and velocity phase diagram

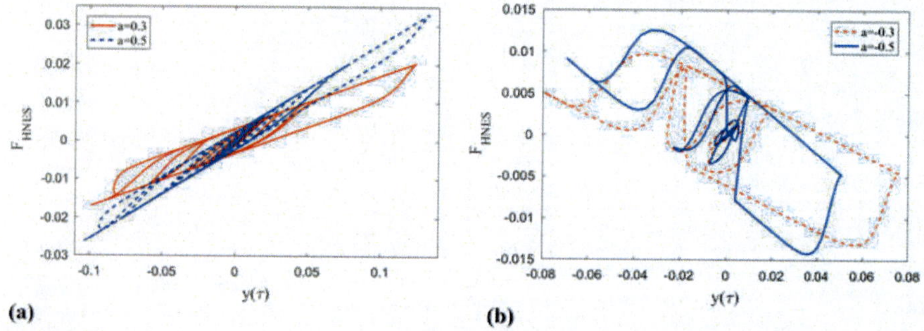

Fig. 6 Force–displacement behaviour of HNES for (**a**) positive stiffness ($a > 0$) (**b**) negative stiffness ($a < 0$)

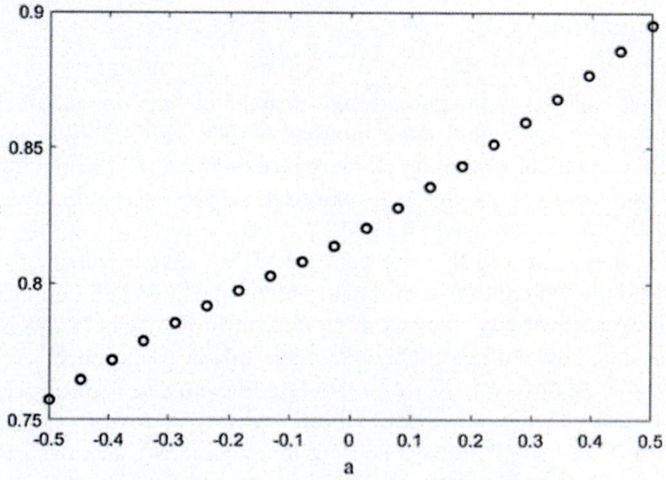

Fig. 7 Performance of HNES for various values of Bouc–Wen parameter '*a*'

(−0.5). As a designer's point of view, it is very much needed to find the optimal values of the absorber. To do that, Monte Carlo simulation-based optimization is carried out for optimal parameters of HNES. For the optimization, the design vectors with lower and upper bounds are considered as $\gamma = [0.01, 60]$ and $\zeta = [0, 0.5]$ of the HNES with negative stiffness, respectively. The different initial conditions are taken into this study to consider the uncertainty in operating conditions. The performance envelope of the HNES for different values of initial conditions and design parameters of HNES is depicted in Fig. 8.

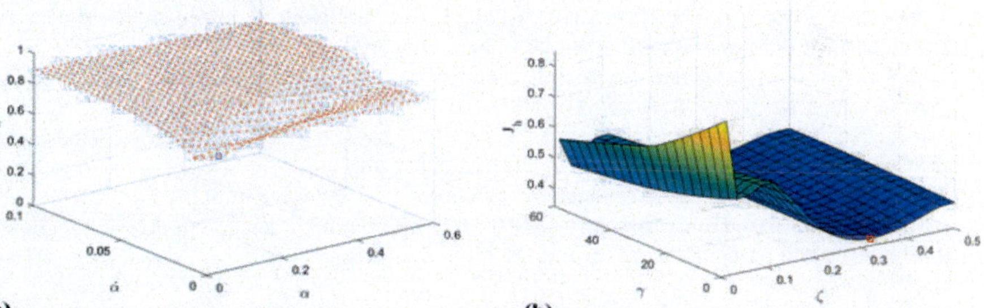

(a) **(b)**

Fig. 8 **(a)** Performance of HNES with varying initial conditions; **(b)** optimal values of stiffness (γ) and damping values (ζ) of HNES

9 Conclusions

The dynamic analysis of benchmark two degrees of freedom airfoil is carried out in this study. The heave and pitch motions for the airfoil are considered for the analysis. Mathematical modelling of the reduced order airfoil model illustrated. The hysteretic nonlinear energy sink is proposed to suppress the limit cycle oscillations of the airfoil. The performance of HNES has been compared with the conventional cubic NES. It is observed that the peak and RMS displacement of the airfoil in the heave and pitch directions significantly reduced using HNES compared to cubic NES. Moreover, sensitivity analysis shows that performance of HNES is better when considering negative stiffness. The advantage of the proposed control strategy is that it can able to reduce LCOs of airfoil significantly in both directions with no significant mass added to the airfoil. HNES is a passive device whose stiffness and damping can't be changed after installing into airfoil. So, as a designer's point of view, optimal design parameters of the device are required considering uncertainty in initial conditions and design vectors. To do that, Monte Carlo simulation is carried out. From the numerical results, it can be elucidated that the promising response reduction is achieved considering optimal parameters (stiffness and damping) of HNES with negative stiffness.

References

1. Casciati S, Chassiakos AG, Masri SF (2014) Toward a paradigm for civil structural control. Smart Struct Syst 14(5):981–1004
2. Den Hartog JP (1956) Mechanical vibrations. McGraw-Hill Book Company, New York, pp 122–169
3. Gendelman Oleg V (2001) Transition of energy to a nonlinear localized mode in a highly asymmetric system of two oscillators. Nonlinear Dyn 25(1–3):237–253
4. Lee Y et al. (2007) Suppression aeroelastic instability using broadband passive targeted energy transfers, part 1: Theory. AIAA J 45(3):693–711
5. Lee YS et al. (2007) Suppressing aeroelastic instability using broadband passive targeted energy transfers, part 2: experiments. AIAA J 45(10):2391–2400
6. Vakakis AF, Rand RH (2003) Nonlinear dynamics of a system of coupled oscillators with essential stiffness nonlinearities. In: ASME 2003 international design engineering technical conferences and computers and information in engineering conference. American society of mechanical engineers
7. McFarland DM, Bergman LA, Vakakis AF (2005) Experimental study of non-linear energy pumping occurring at a single fast frequency. Int J Non-Linear Mech 40(6):891–899
8. Nguyen TA, Pernot S (2012) Design criteria for optimally tuned nonlinear energy sinks —part 1: transient regime. Nonlinear Dyn 69(1–2):1–19
9. Gendelman OV (2011) Targeted energy transfer in systems with external and self-excitation. Proc Inst Mech Eng, Part C: J Mech Eng Sci 225(9):2007–2043
10. Tsiatas George C, Charalampakis Aristotelis E (2018) A new hysteretic nonlinear energy sink (HNES). Commun Nonlinear Sci Numer Simul 60:1–11
11. Bouc R (1967) Forced vibrations of mechanical systems with hysteresis. In: Proceedings of the fourth conference on nonlinear oscillations, Prague, 1967

12. Wen YX (1976) Method for random vibration of hysteretic systems. J Eng Mech Div 102(2):249–263
13. Sapountzakis EJ et al (2017) KDamper concept in seismic isolation of bridges with flexible piers. Eng Struct 153:525–539
14. Platus DL (1992) Negative-stiffness-mechanism vibration isolation systems. In vibration control in microelectronics, optics, and metrology, vol 1619. International Society for Optics and Photonics
15. Winterflood J, Blair DG, Slagmolen B (2002) High performance vibration isolation using springs in Euler column buckling mode. Phys Lett A 300(2–3):122–130
16. Virgin LN, Santillan ST, Plaut RH (2008) Vibration isolation using extreme geometric nonlinearity. J Sound Vib 315(3):721–731
17. Liu Xingtian, Huang Xiuchang, Hua Hongxing (2013) On the characteristics of a quasi-zero stiffness isolator using Euler buckled beam as negative stiffness corrector. J Sound Vib 332(14):3359–3376
18. Antoniadis I et al (2015) Hyper-damping properties of a stiff and stable linear oscillator with a negative stiffness element. J Sound Vib 346:37–52
19. Dowell Earl H (1999) A modern course in aeroelasticity. Meccanica 34(2):140–141

Remaining Life Assessment (RLA) Study and Retrofitting Old and Dilapidated Buildings

T. Sarat Kumar Shroff and Kalpataru Rout

Abstract The building and the structures are basically considered for a particular design life and functionality to meet its purpose. However, due to various factors like design deficiency in the conceptual stage, poor quality of material and methodology in construction stage, lack of maintenance, exposure to aggressive environment, unexpected over loading, external impact, etc., the structures have reduced its strength in due course of time and sometimes causing mishaps endangering life of occupants, surrounding habitats including major financial losses. Periodical examination of health condition with proper diagnosis is thus very important step for ascertaining residual life of the building to enable to take corrective action for improvement of its strength and restoration process. Application of proper methodology by conducting semi-destructive/non-destructive tests with scientific instruments for investigation of structural members as per standard operation and codal provision to be carried out and the interpretation of results to be made to find out residual strength of various members of the building. The analysis will suggest appropriate measures to be undertaken for repair and retrofitting of the members so that the buildings and structures perform better and achieve the prolonged life of building. In this case, two industrial buildings, one from M/s J K Paper Mills, Rayagada, Odisha, and another from M/s Neelachal Ispat Nigam Limited (NINL), Jajpur, Odisha, have been selected as examples. The building selected from J K Paper Mills, Rayagada, Odisha, is an old RCC-framed structure of 24-m height, constructed almost 40 years back. After investigation, it was decided to abandon it and dismantle up to certain height for its safety and stability. In the second case, it is the Ammonium Sulphate Plant (ASP) building of Coke Oven and by-product plant of NINL which was constructed only 10 years back but having signs of distress such as spalling, disintegration and deterioration of concrete, exposure of corroded rebar, spillage of chemicals, dampness in RC members and in-filled walls, etc., because of sulphate and ammonia attack. After investigation, the affected members were identified and restoration measures were undertaken for structural deficient members. After restoration, the tests were

T. S. K. Shroff (✉) · K. Rout
NeelachalIspat Nigam Limited, Kalinganagar Industrial Complex, Jajpur, Odisha, India
e-mail: tskshroff@gmail.com

K. Rout
e-mail: krout1161@gmail.com

B. B. Das et al. (eds.), *Recent Developments in Sustainable Infrastructure*, Lecture Notes in Civil Engineering 75, https://doi.org/10.1007/978-981-15-4577-1_31

re-conducted to check the improvement in the strength of members. This paper will be helpful to all the owner/occupier and engineers looking after the old and dilapidated buildings and industrial buildings having exposure to corrosive environment like steel, power and chemical plants, and housing societies requiring structural audit as per various regulatory and legal provisions of safety and stability.

Keywords Old and dilapidated buildings · Corroded structures · Structural evaluation and rehabilitation

1 Introduction

A building or structure during its service life when questionable of its remaining life to meet its functional requirement and structural stability need to undergo its condition assessment for suitable repair and upgrading work. The option of demolishing an existing structure and rebuilding is a lengthy process and is often not practicable because of space vis-à-vis huge cost involvement, whereas the time and cost of refurbishing the existing one is too less in general in comparison to a new structure. Also when a building or structure is required to be upgraded to withstand the level of increasing the loading criteria, strengthening of its elements is necessarily carried out based on the report of condition assessment. Condition assessment is also a prerequisite for the suitable insurance coverage for remaining service life of the building. Figure 1 shows indicative deterioration of the structure with reference to its designed performance.

First building was an old building of M/s J K Paper Mills, Rayagada, Odisha, constructed in the year 1962. After serving a life of more than 50 years, the building was abandoned for use. The building is an RCC portal-framed structure of 21-M height where various processing equipments were installed. Condition monitoring was carried out and found that few of the structures are unsafe; accordingly the building was dismantled from 21 to 12 M to decrease its load on distressed members for its stability and safety.

Fig. 1 Indicative performance over time

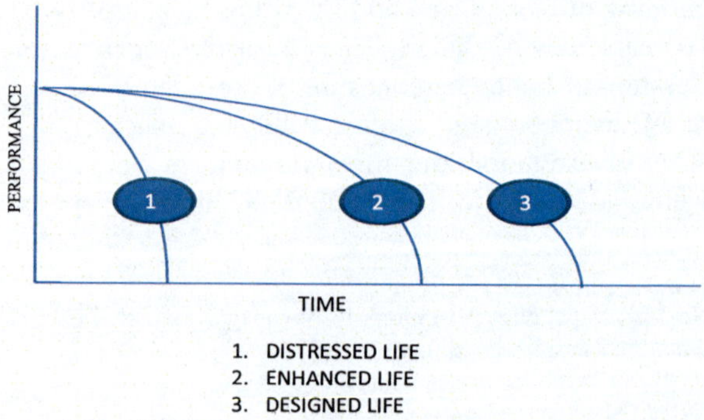

The second building which is explained herein is the 'Ammonium Sulphate Plant' building located in the premises of Coke Oven and by-product plant of M/s Neelachal Ispat Nigam Limited (NINL) which was reported to be distressed and its stability is in doubtful. Various distress features such as severe spalling, disintegration and deterioration of concrete, exposure of corroded reinforcement steel, spillage of chemicals, dampness in RC members and in-filled walls, etc., were observed in the building, therefore called for a detail investigation for overall condition survey of the building and diagnosis of the causes of distress and suggest appropriate remedial measures to increase its life on long-run sustainable basis. The building was constructed in the year 2005–06, an RCC-framed structure with in filled walls and since then in service. The building consists of ammonium sulphate manufacturing unit and a godown. The manufacturing unit comprises of G + 3 upper floors with light weight roofing, whereas godown consists of structural steel roofing trusses supported on RC columns.

2 Methodology

Condition assessment and evaluation is generally carried out in two levels, i.e., preliminary and detailed. Adequate first-hand information is obtained to assess the physical condition of the building at the preliminary investigation level. This includes a walkover survey and collection of basic and design engineering details, construction and maintenance data. Based on the observation made during preliminary survey, detailed investigation is carried out by selecting suitable non-destructive and semi-destructive tests. Figures 2 and 3 innumerate the distressed sample components of the building at various locations and its various tests conducted for diagnosis.

2.1 Physical Inspection

Detail observations made consequence to the field inspection of the building at different levels are mentioned below:

Ground Floor (Ceiling at 4.32-m level)—Severe spalling and deterioration of concrete, snapping of re-bars and exposure of corroded re-bars in RC columns, beams, ceiling slabs and staircase at various locations. Corrosion in structural steel members supporting equipment.

First floor (Ceiling at 8.21-m level)—Severe deterioration of cover concrete, snapping of re-bars and exposure of corroded re-bars in RC columns, beams and ceiling slabs at various locations. Separation cracks between RC members and in-filled walls. Corrosion of structural steel members supporting pipes, hand rails and cable tray at most locations.

Second floor (Ceiling at 14.65-m level)—Same as above with cracks and exposure of corroded re-bars in RC columns, beams and ceiling slabs at a few locations.

Dampness and damp patches in in-filled walls

Deterioration of concrete and exposure of corroded re-bars in RCC column

Ammonium Sulphate Plant at NINL

Separation cracks between RC members and in-filled wall(Typical views)

Ultrasonic Pulse Velocity test on RCC members

Rebound Hammer test on RCC slab

Cover meter studies on RC column

Carbonation test

Fig. 2 Overview NINL ammonium sulphate plant dilapidated building, structures and testing

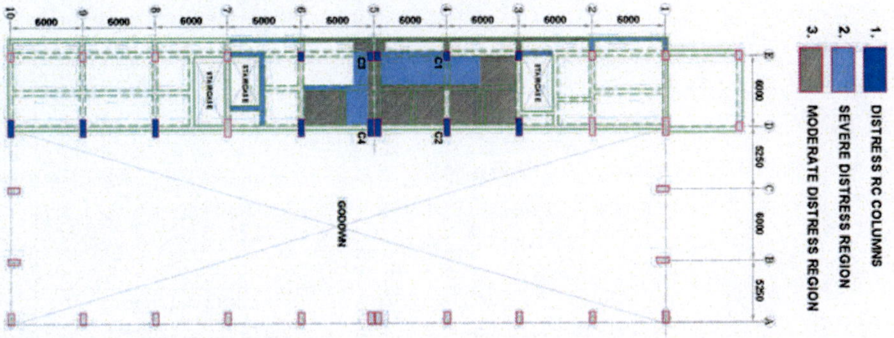

Fig. 3 Layout of ASP building

Dampness and damp patches in in-filled walls at a few locations. Separation cracks between RC members and in-filled walls. Erosion/deterioration of bricks at bottom regions of in-filled walls at a few locations. Corrosion of structural steel members supporting pipes, equipments, staircase and handrails at most locations.

Third floor (Ceiling at 16.90-m level): Same as above.

Godown—Cracks, deterioration of concrete and corrosion of re-bars in RC columns at a few locations. Corrosion stains and peeling of paint in roof trusses, wind bracings and purlins at most of the locations. Erosion/deterioration of bricks at bottom regions of in-filled walls, debonding of plaster in in-filled walls. Deterioration of flagging concrete/foundation protection around the building.

3 Evaluation Tests and Analysis

Considering the observations made as mentioned above, following in situ and laboratory tests are considered to be necessary and carried out for satisfying the overall objective:

Dimensional measurements of structural members—The physical dimension measurement were taken at site, in order to measure and record the observations made as mentioned the same for the various members.

Tests on soil to verify the characteristics and chemical concentration/absorption—Disturbed soil samples were collected for laboratory test selected at random. From the results, it is inferred that the sulphate and pH value of soil within the permissible limits at unaffected regions. However, at affected region the sulphate and pH values are not within the permissible limits. Refer Table 1.

Table 1 Results of soil sample

Sample ID	Region	Grain size distribution				Silt and clay	Atteberg limits (%)			pH	Chloride (%)	Sulphate (%)	
		Gravel	Sand				Liquid limit	Plastic limit	Plasticity limit			SO_3	SO_4
			Coarse	Med	Fine								
1	Unaffected region	17	30	18	4	31	50	26	24	6.64	0.004	0.039	0.047
2	Affected region	12	36	8	1	43	38	26	12	3.29	0.012	1.18	1.41

Remarks The type of soil collected is silty sand/silty sand with clay
Permissible Limits as per IS 456-2000: pH −6 to 9 and Sulphate (SO_3) < 0.2% by mass

Semi-destructive test to estimate strength of existing RC members—In order to accomplish this core test was conducted. The core samples were taken from the RC columns and slabs selected at random for laboratory experiment. The representative cores were tested for its compressive strength as per procedure of IS 516-1959 (latest 2013).

The outcome of the tests in Table 2 indicates that the actual strength of in situ concrete in tested cores is 11.7 N/sq.mm to 36.8 N/sq.mm as against required designed strength of 20 N/sq.mm.

Following are the test results:

RC Columns—Compressive strength 12.1–26.1 N/sq.mm
RC Slabs—Compressive strength 11.7–36.8 N/sq.mm

From above test results, it is evident that the in situ compressive strength of concrete in the tested cores extracted from RCC columns and slabs are varying with respect to designed grade of concrete M20.

Non-destructive tests to determine the quality of in situ concrete in RCC members:

a. **Ultrasonic Pulse Velocity (UPV) Test**—UPV measurement has done on RCC members at random at all approachable regions of the plant building. The test was conducted using 'PUNDITLAB' (Portable Ultrasonic Non-destructive Digital Indicating Tester) equipment from M/s. Proceq, Switzerland, as per the procedure of IS:13311-(Part-I)-1992-(Reaff 2013). The UPV test results in Table 3 implies the strength of concrete in the tested RCC members in unaffected regions falls under the category of 'Medium to Good Concrete' according to given Table 2 of respective std 13311 (Part-I)-1992 (Reaff 2013). However, low-pulse velocity readings were also obtained at a few locations indicating the presence of cracks in interior concrete or debonding of concrete.
b. **Rebound Hammer Test on slabs**—The examination was performed on the RCC slabs at arbitrarily to assess the surface hardness/quality and strength of in situ concrete. The test was conducted using Schmidt Rebound Hammer from M/s. Proceq, Switzerland, as per the guidelines in Indian Standards IS: 13311-(Part-II)-1992-(Reaffirmed in 2013).

   The results given in Table 4 indicate that the strength of in situ concrete in tested region falls in the range of 20–24 N/sq.mm at unaffected regions, whereas at affected regions it is less than 15 N/sq.mm.

Cover Meter Studies—It was performed on various RCC members to find out the thickness of cover concrete, disposition and probable diameter of peripheral embedded re-bars. It is performed by Profometer-5 from M/s. Proceq, Switzerland, as per the guidelines furnished by the manufacturer's manual. The results of the test given in Table 5, revealed cover concrete provided to the re-bars is adequate in most of the tested RC members.

Half-Cell Potential Difference Measurement test (HCPDMT)—This was carried out on various RC members using Copper–Copper Sulphate Half-Cell to

Table 2 Concrete core test results

Sl No	Member and grid identification	Core Length (*l*) (mm) core	Dia (d) (mm)	Core wt. (Kg.)	Failure load (kN)	Core comp. strength (N/sq.mm)	l/d ratio	Correction factor for (l/d) ratio	Revised comp strength (N/sq.mm)	Equiv. cube comp. strength (N/sq.mm)	Type of failure
1	2	3	4	5	6	7	8	9	10	11	12
	0.15-m Lvl										
1	Col-D3	100	73	0.949	57.54	14.84	1.370	0.932	13.83	17.3	Typical compressive failure
2	Col-D4	95	73	0.917	52.14	13.45	1.301	0.924	12.43	15.5	
3	Col-E6	110	73	1.053	85.80	22.13	1.507	0.946	20.95	26.2	
	4.302 Lvl										
4	Col-E4	105	73	1.080	49.58	12.79	1.438	0.939	12.01	15.0	
5	Col-E6	119	73	1.103	39.00	10.06	1.630	0.960	9.65	12.1	
6	Slab-(D-E/4-5)	79	73	0.721	126.82	32.71	1.082	0.901	29.46	36.8	
7	Slab-(D-E/2-3)	90	73	0.950	123.60	31.88	1.233	0.917	29.23	36.5	
	8.210-m Lvl										
8	Col-E6	95	73	0.904	87.52	22.57	1.301	0.924	20.86	26.1	
9	Col-E4	105	73	0.956	40.00	10.32	1.438	0.939	9.69	12.1	
10	Slab-(D-E/2-3)	90	73	0.940	80.00	20.63	1.233	0.917	18.92	23.6	
	14.650-m Lvl										
11	Slab-(D-E/3-4)	83	73	0.722	40.00	10.32	1.137	0.906	9.35	11.7	
12	Slab-(D-E/6-7)	78	73	0.622	67.78	17.48	1.068	0.899	15.72	19.6	
	Crane Foundation (Outside the building)										
13	RC footing	90	73	0.828	49.00	12.64	1.233	0.917	11.59	14.5	

Table 3 Results of ultrasonic pulse velocity test

Sl .	Identification	Grid location	Avg pulse velocity(Km/sec)	Comments
1	2	3	4	5
	0-m lvl–4.32-m lvl			
1	RC columns	A-4	3.4	Quality grading of in situ Concrete. Below 3: Doubtful, 3.1–3.5: Medium, 3.6– 4.5:Good, Above4.5: Excellent as per IS: 13311 (P-1) 1992 Table 2.
2		A-7	3.6	
3		C-10	3.4	
4		D-01	3.7	
5		D-3	3.6	
6		D-4	3.6	
7		D-6	3.8	
8		D-8	3.5	
	0-m lvl–4.32-m lvl			
9	RC columns	E-6	4.3	
10		E-7	3.9	
11	RC beams	D01-E01	3.9	
12		D5-D6	3.8	
13		E6-E7	4.2	
14		B/wD6-D7E6-E7	3.5	
	4.32-m lvl–8.21-m lvl			
15	RC columns	D-4	4.0	
16		D-7	3.4	
17		E-4	3.5	
18		E-6	3.5	
19	RC beams	B/w D3–D4 E3–E4	4.0	
20		D6–E6	3.9	
21		B/w D6–D7 E6–E7	3.7	
22		E4–E5	3.4	
	8.21-m lvl–14.65-m lvl			
23	RC columns	D-4	3.3	
24		D-6	3.2	
25		D-7	3.2	
26	RC beams	D4–E4	3.5	
27		D5–E5	3.4	

Table 4 Rebound hammer test reports

Srl. No.	Identification	Test location	Average rebound number	Comp strength range (N/sq.mm)
1	2	3	4	5
	At 4.32-m lvl			
1	D1–D2 E1–E2	Location – 1	36	22–26
2		Location – 2	34	18–22
3		Location – 3	34	18–22
4	D6–D7 E6–E7	Location – 1	30	14–18
5		Location – 2	32	18–22
6		Location – 3	36	22–26
	At 8.21mt lvl			
7	D3–D4 E3–E4	Location – 1	36	22–26
8		Location – 2	32	18–22
9		Location – 3	32	18–22
10	D4–D5 E4–E5	Location – 1	34	18–22
11		Location – 2	34	18–22
12		Location – 3	36	22–26
	At 8.21mt lvl			
13	D5–D6 E5–D6	Location – 1	32	18–22
14		Location – 2	34	18–22
15		Location – 3	36	22–26
	At 14.65mt lvl			
16	D4–D5 E4–E5	Location – 1	32	18–22
17		Location – 2	32	18–22
18		Location – 3	34	18–22
19	D5–D6 E5–E6	Location – 1	36	22–26
20		Location – 2	32	18–22
21		Location – 3	34	18–22

find out the proneness of corrosion in reinforcement steel. The test was conducted using CANIN equipment from M/s. Proceq, Switzerland as per the standard procedure. The results are tabulated in Table 6. Repercussion indicate the reinforcement steel in the tested RCC members are in 'Uncertainty to High probability of corrosion'. (Moderate to advanced stage of corrosion-as ASTM C876-91, reaff 1999)) resulting in depletion of diameter of embedded re-bars.

Table 5 Results of cover meter studies

Sl. No.	Member Location	Grid marked	Range of concrete cover (mm)
1	2	3	4
	0-m lvl–4.32-m lvl		
1	RC columns	A-3	45–50
2		A-8	50–60
3		D-3	50–55
4		D-6	45–55
5		E-5	40–50
6		E-7	45–50
7	RC beams	D5–D6	40–45
8		D7–D8	40–50
9	RC slabs	D1–D2 E1–E2	15–25
10		D6–D7 E6–E7	20–30
	4.32-m lvl–8.21-m lvl		
11	RC columns	E-4	45–55
12		E-6	45–50
13		D3	40–45
14		D6	45–50
15	RC beams	B/w D5–D6 E5–E6	30–40
16		D5–E5	35–40
17		E4–E5	30–45
18	RC slabs	D3–D4 E3–E4	20–25
19		D4–D5 E4–E5	25–30
20	RC columns	E-3	45–50
21		E-5	50–55
22		D-4	45–50
23		D-6	40–45
24	RC beams	D4–E4	30–35
25		B/w D3–D4 E3–E4	35–40
26	RC slabs	D4–D5 E4–E5	20–25
27		D5–D6 E5–E6	20–30

Table 6 Readings of HCPDMT

Sl. no.	Floor/member identification	Grid identification	Half-cell measurements	Probability of corrosion
	0-m lvl–4.32-m lvl			
1	RC columns	D6	−365 to −390	High
2		D4	−385 to −410	
3	RC beams	D3–D4	−390 to −415	
4		D5–E5	−410 to −430	
5	RC slab	D5–D6 E5–E6	−400 to −410	
	4.32 m lvl–8.21-m lvl			
6	RC column	E4	−380 to −395	High
7	RC beam	D4–E4	−370 to −390	
8	RC slab	D4–D5 E4–E5	−390 to −410	
	8.21- tlvl–14.65-m lvl			
9	RC column	E4	−250 to −350	Uncertainty of corrosion
10	RC beam	D5–E5	−320 to −370	
11	RC slab	D4–D5 E4–E5	−310 to −360	

Test for determination of Carbonation content of RC members—The test undertaken on RCC members using phenolphthalein indicator in dilute methyl alcohol to evaluate the extent of carbonation content in existing concrete. The results of the test are tabulated in Table 7. Outcome of analysis reveals that the cover concrete in most of the tested RC members has carbonated up to reinforcement steel and the cover concrete is failing to protect the reinforcement steel from corrosion.

Chemical composition on concrete samples was carried out for determination of chloride content, sulphate and pH. Representative specimen from the RCC members was taken to laboratory for analysis:

a. **Chloride**: Experiment performed on concrete to determine the chloride content in the concrete. The presence of chloride beyond its permissible limit in the concrete around the embedded reinforcement will result in corrosion. The quantity of chlorides in concrete is determined by chemical analysis and is expressed in terms of chlorides by weight of concrete. The results of the test are tabulated in Table 8. The content of chloride in RC members is falling in the extent of 0.07–1.34 kg/Cu.m as against the permissible limit of 0.6 kg/m cube as per standards.
b. **Determination of Sulphate content on concrete samples**: The presence of sulphates in concrete beyond its permissible limit will result the reaction of calcium present in cement with sulphates, which causes deterioration. The

Table 7 Contents of carbonation

S/No.	Location	Identification mark	Extent of carbonation (mm)	Stipulation
	0-m lvl–4.32-m lvl			Carbonation should not reach up to the reinforcement steel during its life time
1	RC columns	D6	Cover carbonated	
2		D4	Cover carbonated	
3	RC beams	D3–D4	Cover carbonated	
4		D5–E5	Cover carbonated	
5	RC slab	D5–D6 E5–E6	Cover carbonated	
	4.32-m lvl–8.21-m lvl			
6	RC column	E4	Cover carbonated	
7	RC beam	D4–E4	Cover carbonated	
8	RC slab	D4–D5 E4–E5	Cover carbonated	
	8.21-m lvl–14.65-m lvl			
9	RC column	E4	Cover carbonated	
10	RC beam	D5–E5	Cover carbonated	
11	RC slab	D4–D5 E4–E5	Cover carbonated	

procedure for evaluating concentration of sulphates in concrete is generally done by chemical analysis. It is expressed as percentage of sulphates by weight of cement content in concrete. The values are tabulated in Table 8 appended. The sulphate content (SO3) on RC member, indicate that it is in the cover region of RCC and found to be in the range 15.82%–44.25% as against the permissible limit of 4% as per standards.

c. **Evaluation of pH in concrete**: The pH value in fresh concrete is generally in the range of 12–14. But due to carbonation, the pH value will be reduced significantly. Besides, when the pH value falls below about 10, the alkalinity of the concrete will be inadequate for protection of re-bars against corrosion. The findings are appended in Table 8.
The pH of concrete in the tested RC members is in the range of 6.4–8.26 which is lower than the acceptable safe pH value.

Test on water samples—The water samples (washed water stagnated on floor) were collected and tested for chlorides, sulphate and pH value as per the guidelines in IS: 456-2000 (Reaff in 2005) and IS: 3025-1983 (part-24 and 32). The test results are appended in Table 9.

Table 8 Chemical analysis report on hardened concrete

Sl. no.	Sample identification	Chloride content (Kg/Cum)	Sulphate content SO_3 per cent by mass of cement in concrete mix	pH value	Requirements
1	2	3	4	5	6
1	Beam-D3–D4 (4.32-m Lvl)	0.14	43.81	7.30	IS:456-2000 Cl: 8.2.5.2, mentions limits of acid-soluble chloride content in RC containing embedded metal should not exceed 0.6 kg/Cu.m and the Clause: 8.2.5.3 mentions the total water-soluble sulphate content of the concrete mix expressed as SO_3 should not exceed 4% by mass of the cement in the mix pH value preferable <10 as per studies
2	Col-E6 (0mt–4.32-m Lvl)	0.12	43.15	6.40	
3	Col-E1 (0mt–4.32-m Lvl)	1.34	15.82	8.26	
4	Beam-B/w D–E/3-4 (8.21-m Lvl)	0.19	42.71	7.62	
5	Col-E1 (4.32mt–8.21-m Lvl)	1.09	41.22	7.74	
6	Mortars Sample-D6–E6 (0mt–4.32-m Lvl)	0.07	43.22	6.97	
7	Mortars Sample-D9–D10 (0mt–4.32-m Lvl)	0.34	44.25	7.65	
8	Mortars Sample-D6–E6 (0mt–4.32-m Lvl)	0.31	42.37	7.10	

Table 9 Results of water sample

Sl. no.	Particulars	Results	Stipulation of IS 456-2000
1	Chlorides as Cl	70.91 mg/l	500 mg/l max. for RCC
2	Sulphate as SO_3	1630.90 mg/l	400 mg/l max.
3	pH value	3.23	Shall not be less than 6

3.1 Theoretical Analysis and Design Verification

In addition to the above investigative tests, a detailed theoretical structural analysis was carried out for the existing structure. The building was analysed using STAAD PRO V8i software. The structure was modelled for ground, three upper floors and lightweight roofing. The building was analysed as per the existing structural systems and data collected at site. Based on the physical observation, non-destructive test results, the grade of steel and concrete considered are as follows:

Grade of concrete—RC Columns-M15, RC Beams-M20, RC Slabs-M20
Grade of steel—Fe-415
Loadings—Following are the loadings considered in the structural analysis.

Dead load as per IS: 875 (Part I)-1987.

(i) Self-weight of slab(150-mm thick)—3.75 kN/m22
(ii) Loading due to Floor Finishes(Other Area)—3.0 kN/m^2
(iii) Loading due to Floor Finishes(Machinery Area)—1.50 kN/m^2
(iv) Loading due to Roof Finishes—2.00 kN/m^2
(v) Machinery Load (Assumed medium duty)—7.00 kN/m^2
(vi) Truss load—1.5 kN/m^2
(vii) Wall loads

(a) $0.4 \times (4.32 - 0.6) \times 20 - 29.76$ KN/m
(b) $0.23 \times (4.32 - 0.6) \times 20 - 17.11$ KN/m
(c) $0.4 \times (3.9 - 0.6) \times 20 - 26.4$ KN/m
(d) $0.23 \times (3.9 - 0.6) \times 20 - 15.18$ KN/m
(e) $0.4 \times (6.4 - 0.6) \times 20 - 46.4$ KN/m
(f) $0.23 \times (6.4 - 0.6) \times 20 - 26.68$ KN/m

Live load as per IS: 875 (Part-II)-1987

(i) Live load around machinery area—2.0 kN/m^{2}i); Live load on floors—5.0 kN/m^2
ii) Live load on roof (accessible)—1.5 KN/m^2
iii) Live load on roof (inaccessible)—0.75 KN/m^2

Earthquake Load: (Ref: IS 1893:2002)

Zone factor 0.16 (Zone III) Importance factor 1.75
Response reduction factor 3 Damping ratio 'DM' 5%

Wind Load: (Ref: IS: 875 (Part 1)-1987) (Reaff 1997)

Basic wind speed 50 m/sec, K1 factor 1.0 K2 factor 1.09 K3 factor 1.0
Design wind speed (Vz) = 54.65 m/sec
Wind pressure = 0.6* Vz2 = 0.6 *472
Wind pressure = 1.792 kN/sq.m

Load cases considered are:
1: Earthquake Load (EQX) 2: Earthquake Load (EQZ)
3: Dead Load (DL) 4: Live Load (LL)

Load combinations:

5: 1.5 (DL + LL)	18: 1.2 (DL + LL + WX)
6: 1.2 (DL + LL + EQX)	19: 1.2 (DL + LL + WZ)
7: 1.2 (DL + LL + EQZ)	20: 1.2 (DL + LL – WX)
8: 1.2 (DL + LL – EQX)	21: 1.2 (DL + LL W)
9: 1.2 (DL + LL – EQZ)	22: 1.5 (DL + WX)
10: 1.5 (DL + EQX)	23: 1.5 (DL + WZ)
11. 1.5 (DL + EQZ)	24: 1.5 (DL – WX)
12: 1.5 (DL – EQX)	25: 1.5 (DL -WZ)
13: 1.5 (DL – EQZ)	26: (0.9 DL + 1.5 WX)
14: (0.9 DL + 1.5 EQX)	27: (0.9 DL + 1.5 WZ)
15: (0.9 DL + 1.5 EQZ)	28: (0.9 DL – 1.5 WX)
16: (0.9 DL – 1.5 EQX)	29: (0.9 DL – 1.5 WZ)
17: (0.9 DL – 1.5 EQZ)	

The intensities from the critical combination of load case are appraised for the design check as per the guidelines in IS: 456-2000 IS: 875 (Parts I and II)-1987.

4 Isometric View and Discussions

Discussions on Structural Analysis and Design Check for Columns, Beams and Slabs were performed based on the critical forces. The outcome of proof check indicates the existing sections of columns, beams and slabs and area of steel is adequate for the present level of loading. However considering the reduction in diameter of re-bars due to corrosion the sectional dimension of RCC columns, beams and slabs are inadequate, hence calls for restoration (Fig. 4).

5 Inferences

1. The results of soil test infer that the sulphate and pH value of soils within the permissible limits at unaffected regions. However, at affected region the sulphate and pH values are not within the permissible limits, resulting aggressive environment.

Fig. 4 Isometric view of the building

2. It is evident from semi-destructive tests, the in situ compressive strength of concrete in the tested cores extracted from RCC columns and slabs are varying with respect to designed grade of concrete indicating deterioration of concrete on account of chemical environment.
3. The non-destructive test implies the quality and hardiness of concrete in the tested members are found to be satisfactory at unaffected regions. However, at affected region, the quality of concrete is deteriorated and not satisfactory.
4. The half-cell potential measurement test indicates the reinforcement in most of the tested members are in the range of 'Uncertainty to High probability of corrosion' (moderate to advanced stage of corrosion) resulting in depletion of diameter of re-bars.
5. The cover of RCC members has been fully carbonated up to the reinforcement, so concrete has failed to protect the reinforcement steel from corrosion.
6. Findings of chemical analysis on concrete samples states that: (i) chloride content of concrete is more than the permissible limit, (ii) value of pH of concrete is lower than the acceptable pH value, (iii) sulphate content in cover concrete region is more than the permissible limit.
7. The observed distress in RCC members is mainly due to corrosion of reinforcement. Corrosion of rebar is due to carbonation of cover, continuous exposure to the harmful chemical environment and spillage of chemicals over the RC members and improper maintenance over a period of time.
8. Erosion, dampness and deterioration of plaster in infilled walls are due to continuous leakage of chemicals/water on the walls and inadequate maintenance over a period of time.
9. Water stagnation observed over the floor slab and roof slab is mainly due to inadequate slope provided over slab and improper drainage system provided in the building.
10. Separations cracks are essentially due to differential thermal movement of two different materials. (i.e. b/w columns/beams and masonry walls).

11. Examining and verification of the design analysis, it is found that original sections and reinforcement provided in the RC members are adequate to resist the present level of loadings. However, considering chemical analysis results, deterioration in concrete corrosion in reinforcement and extent of distress, strengthening measures are recommended for distressed RC members.

6 Conclusion

The observed distress/deterioration in RC members of 'ammonium sulphate manufacturing unit' is extensive and severe in nature. Further, these distresses (i.e. b/w grid D-E/2-7) in the building calls for immediate restoration measures to ensure safety of the building temporarily for few years. However, considering the present level of distress in RC Beams and ceiling slabs(i.e. b/w grid D-E/2-7), it is advisable to remove and recast distress RC beams and ceiling slabs with all necessary protective measures to ensure a durable structure, otherwise go for some advanced method of repair of concrete.

7 Future Work

Restoration methodology to be used for old, distressed and dilapidated buildings.

Bibliography

1. ASTM C876-91, reaff 1999
2. Bureau of Indian standards, "IS: 456-2000 (Reaffirmed in 2005)"
3. Bureau of Indian standards, "IS: 516-1959 (Reaffirmed in 2013)"
4. Bureau of Indian standards, "IS: 13311-(Part-I)-1992-(Reaffirmed in 2013)"
5. Bureau of Indian standards, "IS: 3025-1983(part-24 & 32)"
6. Bureau of Indian standards, "IS: 875 (Part I)-1987"
7. Report on Evaluation of existing "Ammonium sulphate Plant" building at the premises of NeelachalIspat Nigam Limited, Duburi, Orissa, Feb 2015 by M/s Civil-aid Technoclinicpvt ltd, Bangaluru

History of Finite Element Method: A Review

Lovely Sabat and Chinmay Kumar Kundu

Abstract Finite element method (FEM) is a numerical technique to obtain an approximate solution for problems involving elliptical partial differential equation by dividing the domain into n no of parts of smaller size and applying the boundary conditions to them. In more advance way, it is the analysis of any structure by dividing the whole body into number of small elements and applying the constraints and loads on them and finding the unknowns as per our analysis and the whole analysis are done by the software itself. Initially, the FEM concept was used for solving mathematical formulations in easier way but the development of the FEM concepts and various FEM software, like NASTRAN, ANSYS, etc., made its applications to reach into fields of statistical analysis of structures, linear and non-linear analysis, heat transfer problems and also in bio engineering, nuclear engineering, metallurgical and much more, and this is possible due to the advance types of element in the FEM software. So in the research, the origin and the history of FEM is been studied and the contribution of various researchers has been shown which gives a very clear development idea of FEM and applications of FEM in daily engineering applications from past date till today. The idea of discretization is very old, the mathematical papers on FEA by Schelbach and Courant show the same approach. Earlier, before 1922, also Courant used the finite element ideas in Dirchlet's principle. The FEM we use today involves the contribution of many researchers, namely Turner, Clough, Martin and Topp, Argyris, Bubuska, Aziz, Irons, Melosh and many more, like M. J. Turner at being perfected the direct stiffness method, clough coined the term 'Finite Elements', contribution of B. M. Irons towards FEA was the introduction of shape functions, patch test, text books by Huges and Bathe, Zienkiewicz laid the foundation for further advancement of FEM, and thus the period of 1962–1972 is known as the golden age of FEM and so on.

Keywords Finite element method · Finite element analysis · Finite elements

L. Sabat (✉) · C. K. Kundu
School of Civil Engineering, KIIT Deemed to be University, Bhubaneswar, India
e-mail: lovelysabat2000@gmail.com

C. K. Kundu
e-mail: chinmay.kundufce@kiit.ac.in

B. B. Das et al. (eds.), *Recent Developments in Sustainable Infrastructure*, Lecture Notes in Civil Engineering 75, https://doi.org/10.1007/978-981-15-4577-1_32

1 Introduction

Finite element method (FEM) is a numerical technique used to obtain an approximate solution for the problems involving elliptical partial differential equation with boundary conditions by dividing the whole domain into number of finite elements as per the problem. FEM is used to simulate a given problem by adopting the technique of numerical analysis. For computers to solve these PDEs, which can be considered as one of the mostly used numerical technique has been developed known as finite element analysis uses various types of elements like beam element, plate and shell element, solid element, etc. Mathematics is the necessary technique to understand any problem comprehensively and quantify stresses, strains, displacements of any physical phenomena of a structure and these are described by using partial differential equations. With various softwares of FEM like NASTRAN, ANSYS, FEAST which can be written in any computer language like C, C++, the analysis of any type of engineering problem are possible easily in less time. This technique is adopted by engineers to decrease the number of physical examples, experiments and modify the components in the design procedure for better products in a quicker way.

2 Origin of Finite Element Method

The concept of geometrical division can be seen from the Archimedes time, who for computing the area of a unusual shape, divided the whole shape into equal triangles and quadrilaterals whose areas can be easily worked out and the addition of all areas is the total area of the unusual shape. The whole generalized idea was formulated after much research and it is the result of contribution of many researchers [1, 2]. From the studies, it can be said the name of finite element was coined before few years, but this concept is been adopted from a long time back like to find the circumference of a circle the mathematicians approximated the perimeter of a polygon and calculated the circumference. The FEM technique can be used to analyse three major boundary value problems, namely equilibrium problems, eigen value problems and transient problems. The applications of FEM is not only limited to the aircraft structures analysis, civil engineering structures and mechanical designs but its applications has into the field of heat conduction, geo-mechanics, nuclear engineering, biomedical engineering, electrical machines and electromagnetic, hydrodynamics and water resource engineering and much more [3, 4].

3 History of Finite Element Method

Mathematically, most of the field problems are represented by using differential equations or by an integral expression and such representation may be utilized to formulate the finite element formulations [5, 6]. These formulations are like ready-to-use formulas and are contained in the general FEA programs.

In 1851, Schellbach also implemented the concepts of FEA for deriving the differential equation of the surface of an area enclosed by a closed curve in space. He divided the whole surface into right-angled triangles and, after discretizing the area, he wrote a finite difference expression for the whole area. Figure 1 represents the discretization of a whole body into finite elements.

At this time, Schellbach did not propose any other application or generalization of this idea. But now finite element analysis is regarded as the most adopted way to avoid calculations involving differential equations as in FEA the differential equations are substituted by a set of approximate polynomial equations.

In 1906, researchers noted that the framework having number of bars in a regular pattern which acts as an isotropic elastic body where the concepts of finite element analysis can be applied to solve such type of problems, so this framework method can be treated as the precursor to FEA approach [7].

In 1941, the application of FEA to the problems of plane elasticity and plate bending was seen and reported. The FEM we use today for analysis is likely to have an origin from a German-American mathematician Richard Courant's work publish in his 1943 paper, in which he analysed the torsional capacity of a hollow shaft by discretizing the whole cross section into equal triangles and after discretization, a suitable stress function 'ϕ' is chosen and from the nodal values these stress function is linearly interpolated over each triangle. Some of the researchers consider this work of Courant's to be the source of the present FEM technique.

But the practical application of FEA method bloomed up when the aeronautical engineers further developed the method, apparently without knowing the Courant's work on variational and minimization techniques and the statements for the solution of physical problems. Engineers working in the boeing and aeronautical industry applied the FEM technique for the modelling of the wing skin and swept-back wing [8, 9]. These works and implementations of FEA made a remarkable progress in FEM during the mid-1950s but not many papers were published due to company policies and, during this time, the conventional analysis methods, which were solved on small computers, were replaced by the FEMs.

Fig. 1 Triangular elements

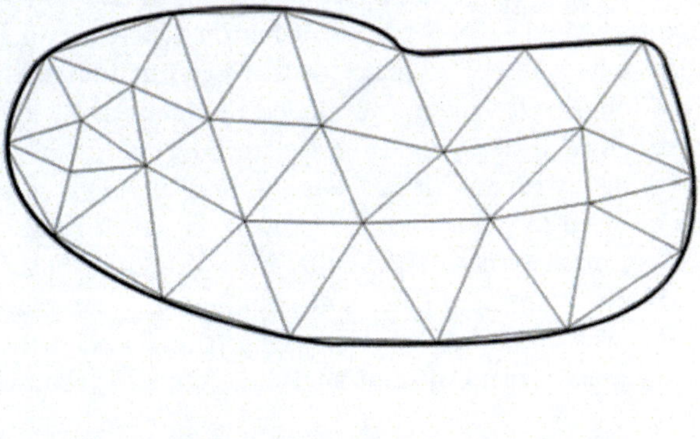

During the mid-1950s, engineers were working on processing approximate methods for calculation of stresses induced in aircraft wings. During the period of 1950–1962, all the aerospace companies were working with the usual force method, while M. J. Turner generalized and perfected the direct stiffness method [10]. The direct stiffness method fit for the automatic computer analysis of unusual structures and also for statically indeterminate structures. The forces, stiffness and the displacements are related in form of equation. The member stiffness relation can be represented in the general form,

$$Q^m = K^m q^m + Q^{om} \tag{1}$$

where m = Number of members.

Q^m = Vector of members characteristic forces.
q^m = Vector of members characteristic displacements.
K^m = member stiffness matrix.
Q^{om} = Vector of members characteristic forces when $q^m = 0$.

Like in case of a simple bar element, each node has 1 degree of freedom, so the equation seem like

$$\begin{Bmatrix} Fx1 \\ Fx2 \end{Bmatrix} = \begin{bmatrix} K11 & K12 \\ K21 & K22 \end{bmatrix} \times \begin{Bmatrix} u1 \\ u2 \end{Bmatrix}. \tag{2}$$

where Fx_1 and Fx_2 are the forces.

K_{11} is the stiffness coefficient.

u_1 and u_2 are the unknown displacements.

During 1952–1953, Turner developed the first continuum based on finite element. Four academicians who were largely responsible for the development of the technology and its applications beyond aerospace industry to various engineering problems during 1950s and 1960s were J. H. Argyris, R. W. Clough, H. C. Martin and O. C. Zienkiewicz.

In 1956, the major names were Turner, Clough, Martin and Topp in the USA who devised a three-noded triangular element to model the wing skin, similar type of work was carried out by Taig in England and by Argyris and Kelsey in Germany and presented several papers of matrix procedure that contained the finite element ideas which can be adopted for solving structural analysis problems [11]. The work by Clough and Wilson introduced the analysis of linear and non-linear portion of a structure at the same time by forming matrix of stiffness coefficients and then merging it with non-linear elements and solving it by stepwise method, so there will be no need to modify all the properties which results in easier analysis. From the papers of Argyris, it shows that his work contributed towars the analysis of thick and thin anisotropic plate and complex shell structures by adopting natural mode FEM. The combined work of Argyris and Kacianauskas shows their contribution to the development of semi-analytical finite element which makes approximation of

3-D fields of stresses developed and displacements occurred in a problem and also showed its applications in analysing the thin-walled beam sections and also showed the use of isoparametric, sub-parametric or super parametric elements [12].

Argyris alone with Tenek contributed towards the FEM development by working on problems of non-linear steady-state heat transfer and also thermoelastic problems with non-linear geometry [13, 14].

Argyris, Clough and Martin learned FEM from Turner. Argyris was a force method expert and was a consultant to boeing in 1950s. He used the conclusions from Turner's group and used the material into his influential 1954 work. Argyris takes the credit of being the first in constructing a displacement assumed continuum element. Clough and martin joined the Turner's group in 1952 and 1953. Clough further developed the method in 1960 and worked to propel this concept into civil engineering applications. Working with Argyris, they named the first generation which extends from 1950 to 1962 [15].

In 1960s, papers by Argyris and Turner show that the FEMwas transformed from a physically based procedure with limited mathematical foundation into its present day method utilizing the variational principles and its applications in the aeroelastic problems and even beyond that [16]. Using the variational principle, the stiffness matrix of a structural element can be obtained,

So, strain energy equation is given by

$$U = \frac{1}{2} \times \iiint_{\Omega}^{\Omega} [[[B]\{d\}]^T [D]]\{B\}\{d\}d\Omega \tag{3}$$

Then applying the variational principle force can be stated as,

$$\{F\} = \frac{\partial U}{\partial\{d\}} = \iiint [B]^T [D][B]d\Omega\{d\}. \tag{4}$$

Now, the strain and the displacement displacement can be related as,

$$\{\varepsilon\} = [\mathrm{B}] \times \{\mathrm{d}\} \tag{5}$$

The constituent relationship matrix can be stated as,

$$\{\sigma\} = [\mathrm{D}] \times \{\varepsilon\} \tag{6}$$

where $[D]$ = Constituent relationship matrix

$[B]$ = Strain displacement matrix

$\{d\}$ = displacement vector

So, from the relation $\{F\} = [K] \times \{d\}$, the elemental stiffness matrix will be,

$$[K] = \iiint [B]^T [D][B]d\Omega \tag{7}$$

Clough coined the name 'Finite Element' in 1960 and thereafter many elements were developed for stress analysis. In 1963, FEA method was widely accepted and used in academia, especially when it was accepted as a form of Rayleigh–Ritz method which is a technique of classical approximation and once the mathematical formulas of the method is constituted, new elements were developed for various problems and this method gained much popularity and started growing exponentially.

In 1964, Zinkiewicz and expert in FDMs worked with Clough to develop FEM further and he wrote the first text book on the subject. The period of 1962–1972 is known as the golden age of FEM. The applications of FEM to non-structural problems had started in 1965 and the first book devoted to FEM appears in 1967. In 1965, Courant has applied finite element analysis concepts to solve the axis-symmetric solids for the heat conduction and transfer analysis and radial and axial displacements for stress analysis [16]. It was founded that the finite element formulation of axis-symmetric problem can be written as,

$$u\,(r, z) = \sum_{i=1}^{n} N_i(r, z)u. \tag{8}$$

$$w\,(r, z) = \sum_{i=1}^{n} N_i(r, z)w. \tag{9}$$

where u_i and w_i are the radial and axial displacements, respectively, at the nodes. N_i is the interpolation function.

Later the finite element concepts were applied to induction motor problems for magnetostatic analysis for finding the nodal unknowns as magnetic potential and also in the field related to seepage flow. But Courant did not mention Schellbach's work in his paper but he suggested a wide generalization which can be applicable to many fields. Zienkiewicz and Cheung presented a broader interpretation of FEM and its applications to general problems and the book by Przemieniecki presented the FEM applied to the problems of stress analysis [17].

During this period, many developments were invented:

1. Melosh worked on conforming elements with Rayleigh–Ritz method based on minimum potential energy principle.
2. Turner worked on ideas of inter-element compatibility for error bounding and convergence.
3. Papers about heat conduction and seepage flow using FEA appeared in 1970s.
4. Computer programs of FEA emerged in late 1960s and early 1970s.
5. During this period, the element approximate continuum models were developed.
6. The displacement formulations dominated the method of analysis, invention of isoparametric formulations, numerical integration, fitted natural coordinates, shape functions, patch test concepts were developed by Irons and co-workers.

Table 1 Development of various softwares

Sl no.	Year	Developments
1	1959	"DAC-1" was developed by General motors and IBM to assist the car designing
2	1960	William Fetter coined the term "Computer Graphic" for his human factors cockpit drawings
3	1965	NASA developed "NASTRAN" as a structural analysis solver tool
4	1970	John Swanson developed ANSYS
4	1977	The first professional FEM P-version code "FIESTA" was initiated by Alberto Peano from ISME
5	1982	The first industrial implementation of FEA "PROBE" was developed by Barna Szabo and Kent Myers for research and aerospace applications
6	1987	RASNA Corp. Developed "MECHANICA"
7	2001	A. Duster proved that the P-version of FEM is the most efficient for plasticity
8	2006	ASME guide was released for verification and validation in computational solid mechanics
9	2008	The actual standard for development of models and simulations were released by NASA
10	2012	"Simulation Governance" a technical requirement in computational solid mechanics software was introduced by BarnaAzabo and Ricardo Actis
11	2013	'SimScale" the 1st cloud-based 3D simulation platform was developed

4 Software Development of Finite Element Method

In late 1960s and early 1970s, the general purpose computer programs for FEA was developed and also the finite element software was incorporated with computer graphics which made FEA more attractive to be used for design purpose as in earlier times, FEA was mainly used to verify the already made or studied structures that had failed. Table 1 shows the development of softwares of FEM over the period of time.

5 Basic Procedure and Applications of FEM

The application of FEM has implemented into static structural problems, steady-state thermal, hydrodynamic problems, magnetostatic analysis, eigen values problems, fluid flow problems, etc. In solving of any types of problem, there are few basic steps that are followed in every analysis of FEM.

Finite element analysis can use more than one material within a single structure such as isotropic, orthotropic and anisotropic. The equilibrium equations for various cases are

1. Linear static:

$$Ku = F \tag{10}$$

2. Linear dynamic:

$$Mu'(t) + Cu'(t) + Ku(t) = F(t) \tag{11}$$

3. Non-linear static:

$$Ku + F_{NL} = F \tag{12}$$

4. Non-linear dynamics:

$$Mu'(t) + Cu'(t) + Ku(t) + F_{NL}(t) = F(t) \tag{13}$$

where $M =$ mass of the structure
$C =$ damping of the structure
$K =$ stiffness of the structure
$F =$ force of the structure
$U =$ displacement of the structure.
In general, the FEM we use today for analysis involves steps of:

1. Discretization of the whole continuum into subdivisions known as finite elements and these elements are interconnected by the nodes.
2. At first, the element type is chosen like beam element, plate element, shell element, solid element and according to it the degree of freedom is considered at each node.
3. Identifying the variables (displacements, stress, temperature, pressure, etc.) on the nodal points.
4. As we do not know the variation of the variable field inside the whole body, it is assumed that this variation is approximated by using a function known as approximating function (ϕ).

$$\{u\} = 1 + x + y + x^2 + xy + y^2 \tag{14}$$

A displacement function is chosen that shows the displacement variation inside the element and then it is approximated in form of a linear function.

$$u = N_1u_1 + N_2u_2 + \ldots \tag{15}$$

$$v = N_1 v_1 + N_2 v_2 + \ldots \quad (16)$$

5. Formation of the elemental and global stiffness matrix for the elements by considering the number and the degree of freedom of the nodal points.

   The total number of degrees of freedom = number Of nodes × DOF of one node

   The size of elemental stiffness matrix = total number of DOF, which is always a square matrix.

   Then the global stiffness matrix is formed by combining all the small elemental stiffness matrices.

$$K_e = t_e A_e B^T D B \quad (17)$$

6. Formation of the elemental and global load matrix for the elements by considering the problem.

$$\{F_e\} = \begin{bmatrix} F1 \\ F2 \\ F3 \end{bmatrix}. \quad (18)$$

   Then the global load matrix is formed by assembling the elemental load matrices.

7. Then by incorporating the boundary conditions to the problem the stiffness matrix are reduced and by applying the simultaneous equation finally the variables and stress resultants are calculated.

$$[K] = \{\delta\} \times \{F\} \quad (19)$$

   The stress in the member is calculated by,

$$\{\sigma\} = [D] \times [B] \times [\delta] \quad (20)$$

   The reactions at the supports are calculated as,

$$\{R\} = [K] \times \{\delta\} - \{F\} \quad (21)$$

6 Conclusions

By 1995, Mackerle estimated that about 3800 journal papers about FEA were being published annually and there are about 380 books, 56,000 papers and 400 conferences proceedings and 310 general purpose FE computer programs. The FEM is been used from more than 50 years in analysis of structures and now it is the basic requirement for any type of engineering analysis in all industrial sectors. The leading software CAD and CAE software covers FEM and CFD applications due to which it has been much easier to analyse composite structures, perform multiphysics modelling, fractures and cracks and lot more.

References

1. Cook RD, Malkus DS, Plesha ME (1989) Concepts and applications of finite element analysis, 3rd edn. Wiley, New Work
2. Zinkiewicz OC (1989) The finite element method. McGraw-Hill, London
3. Bathe KJ (1996) Finite element procedures. Prentice Hall, Englewood Cliffs, NJ
4. Rao SS (2011) The finite element method in engineering, 5th edn. Elsevier, USA
5. Chandrupatla TR, Belegundu AD (2004) Introduction to finite elements in engineering, 3rd edn. Prentice-Hall of India, New Delhi
6. Williamson F (1980) Richard Courant and the finite element method: a future look. Academic Press, Inc., pp 369–378
7. Cen S, Li C, Rajendran S, Hu Z (2016) Advances in finite element method. 2016
8. Clough RW (1990) Finite elements in analysis and design 7(2):89–101
9. Clough RW (1980) The finite element method after twenty five years: a personal view. Comput Struct 12(4):361–375
10. Clough RW, Wilson EL (1979) Dynamic analysis of large structural systems with local nonlinearities. Comput Methods Appl Mech Eng 17–18(1):107–129
11. Argyris J, Tenek L, Olofsson L (1997) TRIC: a simple but sophisticated 3-node triangular element based on a 6 rigid-body and 12 straining modes for fast computational simulations of arbitrary isotropic and laminated composite shells. Comput Methods Appl Mech Eng 145(1–2):11–85
12. Argyris JH, Johnsen ThL, Mlejnek HP (1978) On the natural factor in nonlinear analysis. Comput Methods Appl Mech Eng 15(3):365–388
13. Zienkiewicz OC (1996) Origins, milestones and directions of the finite element method—a personal view. In Handbook of numerical analysis, vol. 4, pp. 3–67
14. Melosh R, Utku S, Islam M, Slama M (1984) An emulator for minimizing computer resources for finite element analysis. Comput Struct 18(4):567–574
15. Turner MJ, Clough RW, Martin HC, Topp LJ (1956) Stiffness and deflection analysis of complex structures. J Aeronaut Sci 23(9):805–823
16. Reddy JN (1996) An introduction to the finite element method. McGraw-Hill, London
17. Reddy JN (2004) An introduction to nonlinear finite element analysis. Oxford University Press
18. Argyris J, Tenek L (1996) Combined steady-state nonlinear heat transfer/thermal post buckling computations in unstiffened and stiffened laminated composite plates and shells. Comput Methods Appl Mech Eng 138:131–185
19. Zienkiewicz OC, Taylor RL, Zhu JZ (2005) The patch test, reduced integration, and non-conforming elements. In: The finite element method set, vol. 1, pp. 329–355

Yield Behaviour of Two-Way Reinforced Concrete Flyash Brick Slab

Sushree Sangeeta Panda and Bhagabat Jena

Abstract The present exploratory experimental work has been attempted to study the yield behaviour of two-way square reinforced flyash brick slab (RBC) compared with that of RCC slab. The RBC slab has been prepared using flyash bricks partially replacing concrete in RCC slab. The bricks are bonded using concrete of same compressive strength as that of control RCC slab. Reinforcement is provided embedded within the concrete bond between the brick units. Thus, it reduces the weight of slab and achieves economy. All the slab specimens with edges discontinuous and simply supported are tested under concentrated load applied at its centroid. Load-carrying capacity at onset of crack and till failure, Load–deflection characteristics, crack pattern, and failure mechanism of the slabs have been studied.

Keywords Yield line · Two-Way slab · Reinforced brick slab · RCC slab

1 Introduction

Concrete has been used as a predominant construction material. Due to environmental concern, the need of the hour is to reduce consumption of cement. These days, flyash bricks have almost replaced the conventional bricks for an eco-friendly sustainable construction practices. In a concrete slab, as concrete is structurally meant for resisting the flexural compressive stress, near and below the neutral axis of slab, concrete strength is not fully utilized but occupies good amount of volume in the slab that can be replaced by cheaper construction material like flyash bricks. Thus, it can reduce the mass and cost of construction. The present work aims at studying the yield behaviour of two-way square reinforced flyash brick slab under concentrated loading.

S. S. Panda (✉) · B. Jena
School of Civil Engineering, KIIT Deemed to be University, Bhubaneswar, Odisha, India
e-mail: sushree.pandafce@kiit.ac.in

B. Jena
e-mail: bjenafce@kiit.ac.in

B. B. Das et al. (eds.), *Recent Developments in Sustainable Infrastructure*, Lecture Notes in Civil Engineering 75, https://doi.org/10.1007/978-981-15-4577-1_33

2 Literature Review

Relevant literature review pertaining to the present work has been appreciated and briefed below [1] introduced semi-fabricated two-layered slab system connected by continuous truss-type shear connector. The first layer was a precast ferro-cement layer consisting of wire mesh and steel reinforcement. It initially acted as a formwork and resisted the tensile stresses in the slab. The second layer consisted of bricks and mortar. Continuous shear connectors were provided to connect the two layers and between the shear connectors, brick units are accommodated. The slab specimens were tested flexurally under two-line loads simply supported at four corners. They studied load–deflection, crack pattern, strain distribution and failure loads of their test slabs. They found that experimental ultimate failure load was comparable to design load as per BS8110. The cracking load was observed to be about 30% of ultimate failure load and the crack pattern was found similar to that observed in RCC one-way slab.

Chahar and Patel [2] experimentally studied reinforced slabs replacing concrete by normal bricks. They observed the load–deflection behavior of reinforced brick slab and compared with that of RCC slab under 4-point loading and two sides simply supported. Large deflection was observed for reinforced brick slabs as compared to RCC slab with little less failure load found compared to RCC slabs. However, they found that the cracking strength of reinforced brick slabs was more than that of RCC slabs with nonlinear load displacement after first crack.

Rabbani [3] studied the performance of masonry slab with no reinforcement, of various dimensions and thickness comparing with RCC slab. Bricks were laid in different pattern: simple bond, English garden bond, herring bone bond. The testing was of 2-point line loading with two sides simply supported. Combined failure both in joint and brick module was found with Herring bone bond masonry slab capable of resisting maximum load.

Based on literature, the current exploratory work has been an attempt to revisit experimentally the behavior of reinforced flyash brick slab under loading. However, in literature, experimental study of two-way brick slab simply supported on four sides and subjected to concentrated load has not been done. Similarly, in literature, flyash brick is not used in making slab for conducting test. At present time, flyash bricks have almost replaced conventional bricks in construction; the present work has adopted flyash bricks in making reinforced brick slab and comparing its behavior with RCC slab. Typically, the load-carrying capacity at onset of crack and till failure, load–deflection characteristics, crack pattern, and failure mechanism of the slabs have been studied.

3 Experimental Investigation

As an exploratory test, three numbers of reinforced flyash brick slabs (RBC) are prepared of size 1.22 × 1.22 m and of thickness 70 mm. One control RCC slab is prepared of the same dimension as that of RBC slab. The compressive strength of concrete, grade of reinforcement, and percentage of reinforcement used for all the test slabs are kept constant. Figure 1 shows the layout diagram for casting the reinforcement brick concrete slabs, made of 8 mm diameter Fe500 grade reinforcement cage as in Fig. 2 with placement of flyash bricks. Figure 3 shows arrangement of RBC slab specimen prior to casting and "Fig. 4" shows the RBC slab after casting in concrete. Finally, top and bottom surfaces of RBC slab are plastered as in Fig. 5 with cement mortar of nominal mix proportion 1:4. The flyash brick size is 25 × 12 × 7 cm and number of bricks accommodated in RBC square slab are 5 numbers length wise in a row with 3 numbers of rows as in Fig. 1. All the slab specimens are subjected to same manner of curing and testing after 28 days of casting. Figure 6 is the test frame with rollers meant for supporting slab for testing under concentrated load acting at the centroid of the slab as in Fig. 7 of which corresponding centroidal deflection is measured as in Fig. 8.

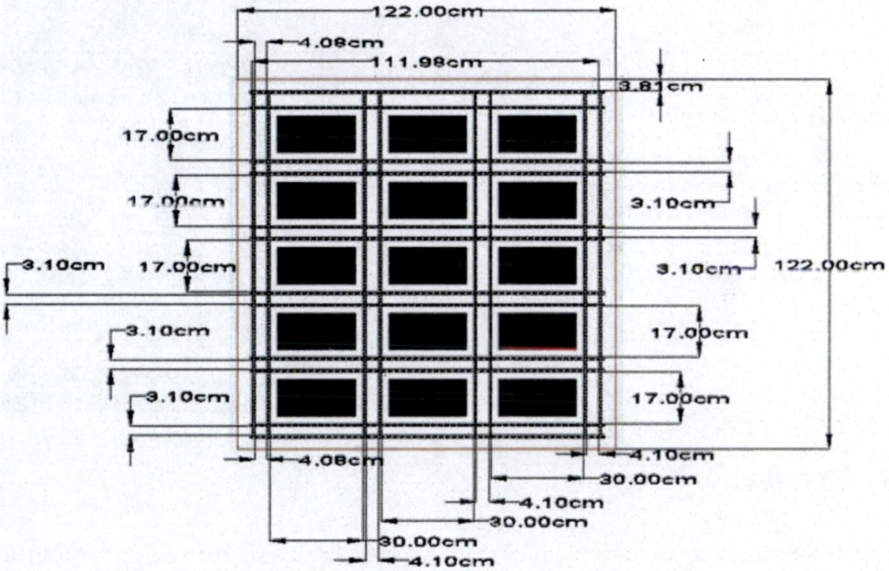

Fig. 1 Layout diagram for casting of RBC slab

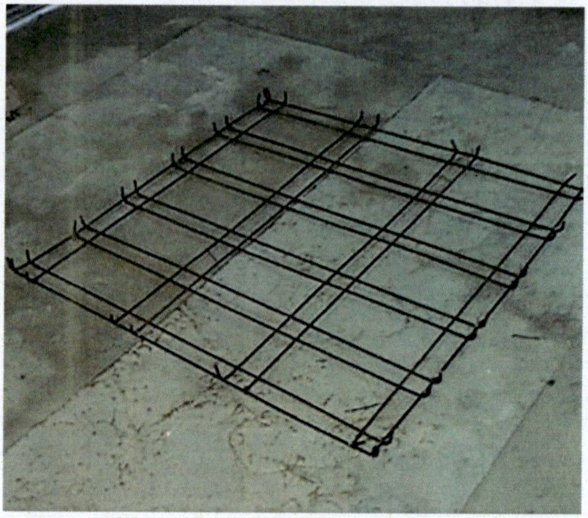

Fig. 2 Reinforcement cage

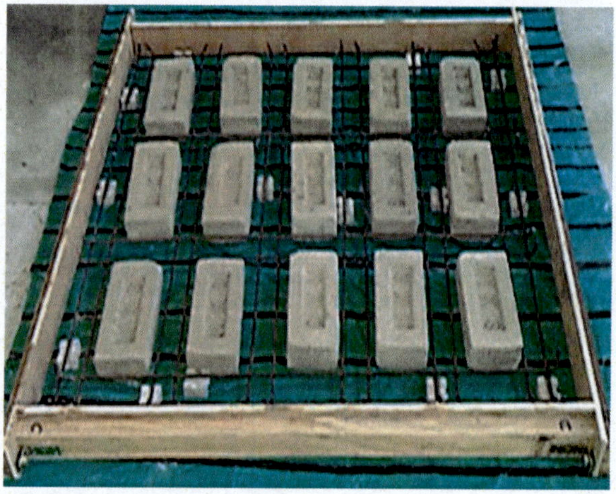

Fig. 3 Arrangement of RBC slab prior to casting

4 Result and Discussion

The average compressive strength of concrete used in casting the control RCC slab and the reinforced brick slab is found at the time of testing to be 38 MPa.

Table 1 shows the test results of the slab specimens. In Table 1, highlighted yellow and green marks indicate load and deflection of the test slabs corresponding to first crack and ultimate failure, respectively. The cracking load for the RBC slab is found to be an average of 29.4 kN, almost half the value of that of RCC slab which is found to be 57.57 kN. This may be attributed to cracks forming due to gradual bond failure around the brick units. The ultimate failure load of RBC slabs is observed to be an

Fig. 4 RBC slab after casting

Fig. 5 RBC slab after surface plaster

average of 97.7 kN and that of RCC slab is 138.57 kN which is about 30% more than that of RBC slab in an average. In case of RBC Slab-1, the bearing plate used is 150 × 150 mm for transferring applied load to the test slab and the failure of the slab has occurred due to punching Fig. 9a. Hence, the bearing plate size is increased to 300 × 300 mm during testing of rest of the slab specimen. It has resulted into increased failure loads for the rest of the specimens (Table 1) than that of RBC Slab-1. From

Fig. 6 Test frame with rollers surface plaster

Fig. 7 Test setup with slab

Fig. 9, the failure pattern of the test slabs is observed to be a combination of yield line pattern assumed as per [4] along with bond failure around the brick units.

From Fig. 10, it is observed that except for the RBC Slab-1 which has failed due to punching, corresponding to load, the centroidal deflection for RBC Slab-2 and RBC Slab-3 is found to be almost same. Figure 10 also reveals that for a constant load P_0, deflection for RCC slab, δ_{RCC} is lesser than that for RBC slab, δ_{RBC}. This indicates that modulus of elasticity for reinforced flyash brick slab is found to be less than that of normal RCC slab.

For concentrated load, P acting on a simply supported rectangular plate, according to Navier's method (Timoshenko and Woinowsky-Krieger), centroidal deflection of the plate, δ is given as,

$$\delta = \alpha a^2 P/D \quad (1)$$

Fig. 8 Under slab dial gauge

Table 1 Test result of load versus deflection

RBC Slab-1		RBC Slab-2		RBC Slab-3		RCC Slab	
Load	Deflection	Load	Deflection	Load	Deflection	Load	Deflection
kN	mm	kN	mm	kN	mm	kN	mm
0	0	0	0	0	0	0	0
7.42	1.7	8.28	0.63	7.14	0.71	3.705	0.37
13.14	3.1	14	1.01			14.82	0.93
18	4.74	18.85	1.46	18.571	1.76	27.93	1.45
24.57	6.4	24	2.11	25.14	2.58	39.045	2.03
30	7.55	30	3.21	29.71	3.55	45.88	2.65
34.57	8.53	40.28	4.65	33.714	4.34	52.155	3.26
39.42	9.52	52	5.96	47.714	5.84	57.57	3.98
45.71	10.83	62.57	8.25	57.71	7.04	68.97	5.63
50.85	11.97	73.42	9.62	63.42	8.7	79.8	7.2
55.42	13.02	84.85	11.18	70	9.48	90.06	9.08
60.28	14.52	94.85	12.44	78.571	11.4	101.46	10.7
		104.85	13.79	84.285	12.17	111.72	12.97
		115.71	16.47	94.57	13.86	121.695	15.45
		123.42	17.93	99.42	15.45	130.53	20.5
		125.14	18.9	107.42	17.2	135.375	25.08
		125.42	18.93			136.23	25.9
						137.085	28
						138.57	28.9

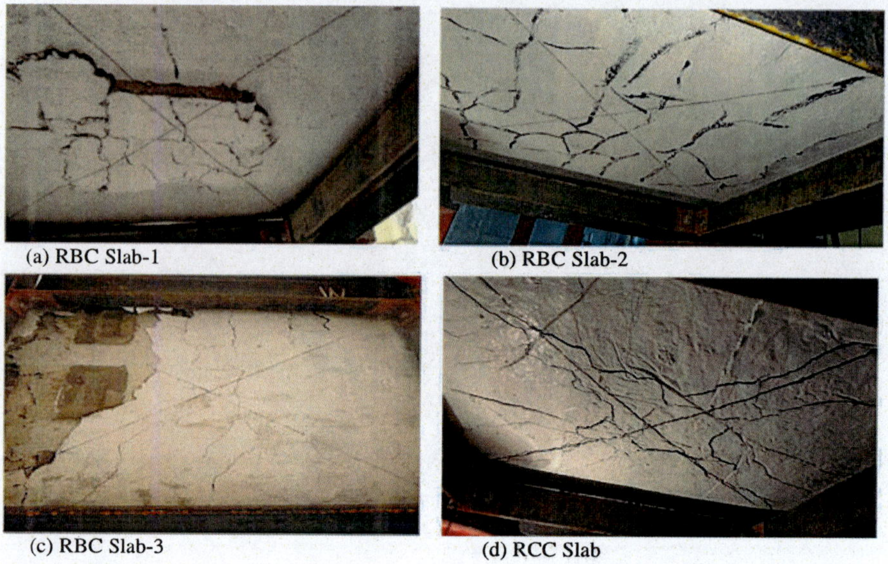

Fig. 9 Bottom surface failure crack pattern of RBC slabs (**a**, **b**, **c**), RCC slab (**d**)

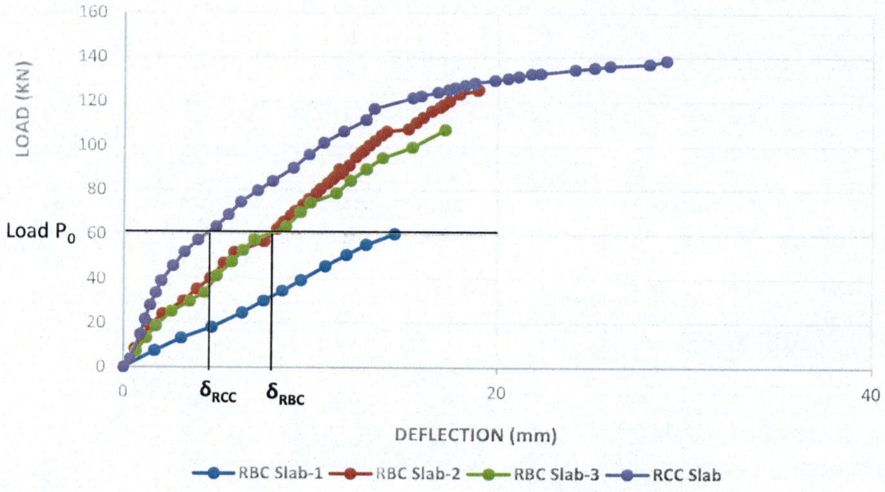

Fig. 10 Load vesus deflection curves of test slabs

where $D = \left(Eh^3\right) / 12\left(1 - \vartheta^2\right)$

E = modulus of elasticity of plate element

h, ϑ = overall depth and Poisson's ratio of plate element resp.

$\alpha = 0.0116$ for b/a = 1

Table 2 Modulus of elasticity of test slabs using Navier's method

Specimen	P/δ in kN/mm	E in kN/mm^2	Average E in kN/mm^2
RBC Slab-1	4.285	1.7	3.875
RBC Slab-2	10	3.97	
RBC Slab-3	15	5.955	
RBC Slab	17.5	6.95	6.95

b, a = support to support distance of two-way slab in Y and X directions, respectively

Assuming $\vartheta = 0.15$ for both RCC and RBC slab material and h = 70 mm for the test slabs, from Eq. (1),

$$E = 0.397P/\delta \;\text{ in kN/mm}^2 \tag{2}$$

Experimental P/δ is obtained by taking initial tangent to load vs deflection curve of test slabs. Using experimental data for P/δ and using Eq. (2), E value is obtained (Table 2). E value obtained for RCC slab using (IS:456, 2000) is found to be 30.8 kN/mm^2 (=$5000\sqrt{fcu}$, and $fcu = 38MPa$) which is much greater value than that for RCC slab by Navier's method which is an elastic analysis. Hence, it requires extensive experimental data to derive something conclusive about the modulus of elasticity of RBC slab.

5 Conclusions

As an exploratory test, three numbers of two-way reinforced flyash brick slab was tested and compared with a control RCC slab under concentrated loading applied at the centroid of the slab. The following conclusions can be made in the experimental investigation,

- The first cracking load for reinforced flyash brick (RBC) slab was found to be almost half of an RCC slab, whereas the ultimate load capacity of the former was about 30% of the normal RCC slab.
- Modulus of elasticity for reinforced flyash brick slab is found to be lesser than that of normal RCC slab as the central deflection is found more than that of RCC slab.
- The failure pattern was found to be a combination of yield line pattern of two-way slab along with bond failure around the brick units.

References

1. Thanoon WA et al (2010) Structural behaviour of ferrocement–brick composite floor slab panel. Constr Build Mater 24(11):2224–2230
2. Chahar AS, Patel RD (2015) August) Studies on RCC and Brick Composite Slabs. IJSTE 2(2):130–134
3. Rabbani ML (2017) A study on low cost roof (Masonary Slab). AJER 6(3):32–36
4. Johansen KW (1962) Brudlinieteorier. Copenhagen; 1943. [English translation "Yield-Line Theory"]. Cement and Concrete Association, London
5. Timoshenko SP, Woinowsky-Krieger S, Theory of plates and shells. Tata McGraw-Hill edition

Recent Developments in Concrete and Building Materials

Mechanical Properties of Fly Ash and Ferrochrome Ash-Based Geopolymer Concrete Using Recycled Aggregate

Srishti Saha, Tribikram Mohanty, and Purnachandra Saha

Abstract Recycled aggregate (RCA) is a term which is used to describe crushed concrete or asphalt from construction debris, which is also used to make other building projects. This method of construction is generally used for making road base or other different infrastructure-based projects. The source of RCA is nothing but the local construction and demolition waste in KIIT Deemed to be University, Bhubaneswar, Odisha. This paper suggests the experimental results and analysis to study the behaviour and feasibility of the use of RCA in concrete form as a replacement of natural coarse aggregate. In this research work, fly ash and ferrochrome ash-based geopolymer concrete are used, and 13 M concentration of caustic soda (NaOH) and sodium silicate (Na_2SiO_3) alkali activators are utilized to synthesize the fly ash and ferrochrome-based geopolymer concrete. Alkali activators to fly ash ratio are constant for all series. Experimental result shows that the strength of geopolymer concrete decrease with an increase in RCA contents up to 50%. Though recycled aggregates, fly ash and ferrochrome ash are waste materials used for making geopolymer concrete, experimental results revealed that up to 20% of replacement of recycled aggregate can achieve the strength and it can be the recommended percentage of RCA in concrete industry.

Keywords Recycled aggregate · Geopolymer · Fly ash · Ferrochrome dust · Mechanical properties

S. Saha · T. Mohanty (✉) · P. Saha
School of Civil Engineering, KIIT Bhubaneswar, Bhubaneswar, Odisha, India
e-mail: tmohantyfce@kiit.ac.in

S. Saha
e-mail: srishti.civil94@gmail.com

P. Saha
e-mail: dr.purnasaha@gmail.com

B. B. Das et al. (eds.), *Recent Developments in Sustainable Infrastructure*, Lecture Notes in Civil Engineering 75, https://doi.org/10.1007/978-981-15-4577-1_34

1 Introduction

The sustainable construction concept was introduced due to the growing concern about the future of our planet because construction industry is a huge consumer of natural resources and simultaneously, waste producer. Conservation, rehabilitation and strengthening of construction industry may be termed as sustainable development [1]. Concrete is one of the most widely used construction materials in the world, mainly due to its favourable features such as durability, versatility, satisfactory compressive strength, cost-effectiveness and availability [2]. Malhotra et al. [3] observed that by using conventional concrete, it was not eco-friendly and causes some negative concerns, for example, the reduction of natural resources, high energy consumption and disposal issues. Lamond et al. [4] reported that in concrete industry coarse and fine aggregates consist of about 60–75% of the total volume of concrete, it can reduce natural aggregates which will have significant impacts in the environment. Davidovits [5] reported that geopolymer concrete is produced by the total replacement of cement. It consists of geopolymeric filler and alkaline liquid. Alkaline solution reacts with Al and Si in a geologically originated material or materials like ferrochrome ash and fly ash for the production of binders. The chemical reactions occurring during this process are known as geopolymerization; hence, the binders produced are known as geopolymers. Duxson et al. [6] observed that that geopolymer concrete produces 5–6 times less energy and 80% less CO_2 emissions than cement. The cement production generated CO_2, which pollutes the atmosphere. The industry produces wastes like fly ash, and slag which is without a doubt dumped on the planet occupies larger areas. The water comes from the chemical industries and discharged through the ground; it contaminates more underwater. Geopolymer concrete does not use any cement; the manufacturing of cement shall be reduced and for this reason the pollutants of atmosphere by means of the emission of CO_2 shall also be minimized. Various types of recycling methods are developed for C & D wastes in the last two decades, and it has been explored and well developed. For instance, recycled concrete aggregate (RCA) has been used to replace natural aggregate (NCA) after being treated [7, 8]. Ceramic wastes obtained from broken roof tiles, blocks, bricks, electrical insulators, etc. are used as aggregates in concrete [9]. Nowadays, prestressed precast concrete is very popular. Since high grade of concrete is used, aggregate from these demolition wastes would have high strength [10]. Several researchers investigated the performance of these wastes as recycled aggregate separately; hence, there is a need to compile and analyse the performance of recycling wastes used in fly ash and ferrochrome-based geopolymer concrete in a single platform. The objective of this paper is to study the effects of different proportions of RCA on mechanical behaviour of geopolymer concrete and compared to that of 0% recycled aggregate.

2 Materials

Source materials used for manufacturing geopolymer concrete using recycled aggregate in this study are fly ash and ferrochrome ash. Ferrochrome ash is the dust which is obtained from gas cleaning process of the industrial manufacturing units of ferrochrome, and it was taken from Balasore Alloys Ltd, Odisha. Fly ash is collected from Sai Ram Fly ash Brick, Bhubaneswar, Odisha. Due to the adding up of those waste materials, the water demand is increased and the workability is reduced. Similarly, cement was replaced by fly ash and ferrochrome ash caused low workability. Maintaining the w/c proportion constant for all mixes, a naphthalene-based superplasticizer is added in order to improve the workability. 'Table 1' gives the chemical composition of source material. The recycled aggregate (RCA) in this research work is taken from old structures, waste sample, which can be dumped outside the laboratory from the Region of KIIT Deemed to be University. The recycled aggregate had been cured properly with a maximum aggregate size of 40 mm. Coarse natural aggregate has maximum size of 10 mm and 20 mm, which are utilized in the experimental work. 'Table 2' denotes the physical properties of C&D waste and natural coarse aggregate (Fig. 1).

Table 1 Chemical composition of industrial waste

Sl no.	Constituent	Ferrochrome ash (FA)	Fly ash
1	SiO_2	27.5	55
2	Al_2O_3	24.7	26
3	CaO	9.06	18.10
4	Fe_2O_3	4.02	19.48
5	MgO	22.5	3.30
6	SO_3	1.92	1.5
7	Na_2O	1.3	–
8	K_2O	0.46	1.79
9	P_2O_5	0.313	–
10	TiO_2	2.196	1.02

Table 2 Properties of RCA & NCA

Property	Recycled coarse aggregate	Natural coarse aggregate
Specific gravity	2.48	2.77
Bulk density (kg/m^3)	1.409	1.781
Loose bulk density (kg/m^3)	1.24	1.33
Water absorption (%)	4.47	0.66

(a) Ferrochrome ash (b) Fly ash (c) C&D waste (d) Recycled aggregate

Fig. 1 **a–d** Waste materials which are used in this research work

3 Methodology

In this study, to make geopolymer concrete by using recycled aggregate, sodium-based alkaline activators are worn. As the attention of caustic soda solution rises in terms of molarity (M), the concrete becomes fragile with rising compressive strength. As well as achieving the desired workability, extra water is necessary to inferior the concentration of the NaOH solution. The caustic soda concentration is kept at 13 M, but the concentration of the Na_2SiO_3 solution was 16.37% Na_2O_3, 34.35% SiO_2 and 49.72% H_2O as alkaline solutions. NaOH pellet is dissolved in H_2O and geared up with sodium silicate solution. Under the mix design condition, sodium hydroxide quantity was 89.75 kg/m^3 at 13 M, and sodium silicate was 224.3 kg/m^3 at 13.3 M. Clear drinking water was used to make the final solution. A mixing of solution Na_2SiO_3 and NaOH, sodium hydroxide of 97% in pure flaking form and sodium silicate were used for the preparation of activator. Chemical compositions of activator were $Na_2O = 77.75\%$, $SiO_2 = 77.75\%$ and water 33.765% by mass. Solution molarity was maintained at 13 M during the entire work. The solution was prepared for at least 24 h before being used to prepare geopolymer concrete.

4 Preparation of Specimen

In this research, six types of mixtures are taken into consideration. 'Table 3' gives the idea of mix proportion and detailed analysis of experimental results of all six types. In the first series which is known as control series, where 100% of natural coarse aggregate geopolymer with fly ash and ferrochrome ash is utilized, it was denoted as GFF0% mixture. Other types of mixtures which contain 10%, 20%, 30%, 40% and 50% (% by wt) of recycled aggregate are denoted as GFF10%, GFF20%, GFF30%, GFF40% and GFF50%, respectively. The ratio of activator to fly ash ratio is kept unchanged for all concrete mixes. The colour of fly ash and ferrochrome ash-based geopolymer concrete is light black. The total quantity of water holds the most significant role in the mix. If mixing of concrete is carried out for a long time, the content bleeds out when the water is high during the mixing period, and the aggregate

Table 3 Mix proportion of all six types of geopolymer concrete mixes

Mix	Fly ash (kg/m³)	Ferrochrome ash (kg/m³)	Recycled aggregate (kg/m³)	Natural coarse aggregate (kg/m³)	Fine aggregate (kg/m³)	NaOH (kg/m³)	Na_2SiO_3 (kg/m³)	Superplasticizer (kg/m³)
GFF0% RCA	189.0 (50%)	189.0 (50%)	0%	1350.88	727.40	27.29	27.29	6.05
GFF10% RCA	189.0 (50%)	189.0 (50%)	135.08	1080.704	727.40	27.29	27.29	6.05
GFF20% RCA	189.0 (50%)	189.0 (50%)	270.176	1051.176	727.40	27.29	27.29	6.05
GFF30% RCA	189.0 (50%)	189.0 (50%)	405.264	945.616	727.40	27.29	27.29	6.05
GFF40% RCA	189.0 (50%)	189.0 (50%)	540.352	810.528	727.40	27.29	27.29	6.05
GFF50% RCA	189.0 (50%)	189.0 (50%)	675.44	675.44	727.40	27.29	27.29	6.05

segregates to give a paste. For compression test, split tensile strength and flexural strength test concrete cubes were having side of 150 mm; the cylindrical specimens were having a diameter of 0.15 m and height of 0.3 m, and prisms were having breadth and height of 100 mm and length of 500 mm as prepared. First RCA and fine aggregate were mixed in a 200 L volume laboratory inclined mixer for about 4–5 min; then fly ash and ferrochrome ash are added according to the specified amount, and mixing was continued for about 3 min. After dry blending, the alkaline solution was added to the dry material along with the superplasticizer and again mixed for another 5 min. After completion of mixing, mixed concrete was poured in a mould. All specimens were cast properly and horizontally in three layers. Each layer was tamped properly and for proper placing of concrete used needle vibrator. The test piece was allowed to stand for 3 days in a mould at room temperature to cure. All specimens were placed in a hot air curing chamber to cure the cubes, cylinders and beams at temperatures between 60 and 70 °C, for up to 24 h. The specimens are then demoulded and cured and kept in open air in the laboratory till the date of testing.

5 Results and Discussion

Different percentages of reused aggregate have direct effect on mechanical properties of geopolymer concrete which was obtained at 7, 28, and 56 days as it was shown in Figs. 2, 3, and 4.

5.1 Effect of RCA on Geopolymer Concrete on Compressive Strength

The effect of geopolymer concrete on compressive strength by using different percentages of recycled aggregate is shown in 'Fig. 2'. The suitabilities of replacement

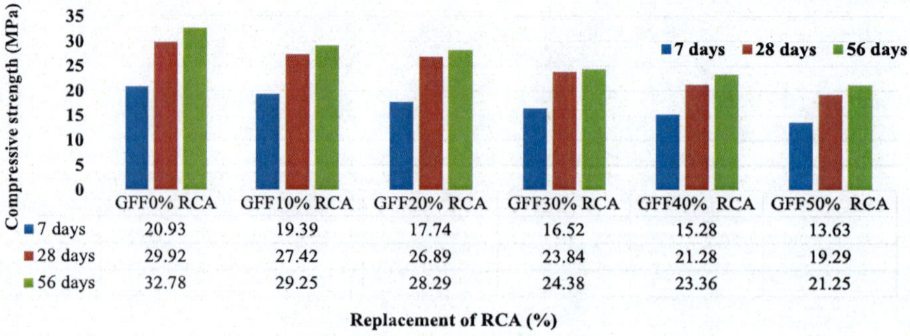

Fig. 2 Compressive strength of RCA at different curing ages

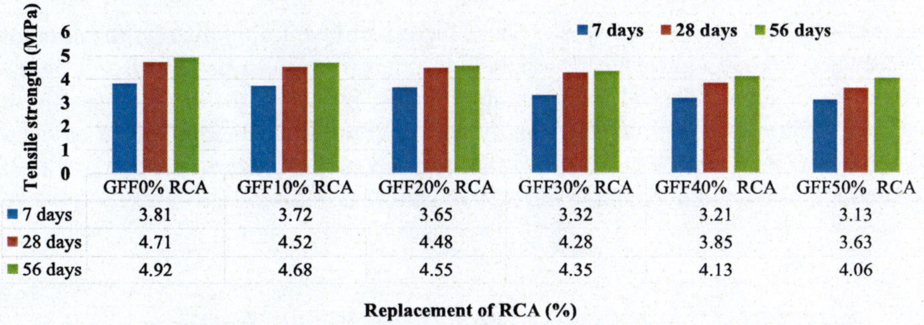

Fig. 3 Tensile strength at different ages of various percentages of RCA

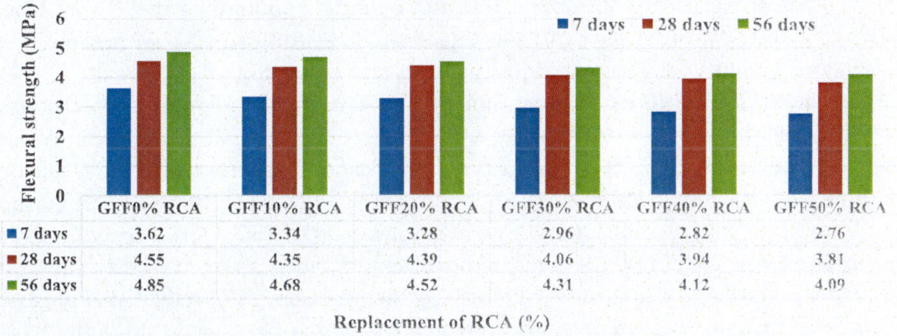

Fig. 4 Flexural strength at different ages of various percentages of RCA

of reused aggregate by natural coarse aggregate on geopolymer concrete were studied. The compressive strength of 150 mm cubes containing 0, 10, 20, 30, 40, and 50% of recycled aggregate, on the part substitution of natural coarse aggregate, was examined. The strength of fly ash and ferrochrome ash-based geopolymer concrete containing 0% recycled aggregate was increased by about 8.99% from 7 to 28 days. At 28–56 days, strength was gradually increased by about 3%.

Experimental results show that by increasing the recycled aggregate contents 10–50%, respectively, the compressive strengths decrease by 1.54%, 3.19%, 4.41%, 5.41% and 7.30% at the age of 7 days as compared to 0% recycled aggregate. Similarly, at 28 days and 56 days, curing the strength gradually decreases to 2.5%, 3.54%, 6.08%, 8.64% and 10.63%, respectively.

It was also noticeable to observe that the highest decrease in strength from 7 to 28 days is nearly 8% if the concrete contains 50% RCA. Test results revealed that strength slightly increases from 28 to 56 days. The source of recycled coarse aggregate was normally compacted concrete structures; for that reason, old mortars always adhered to the RCA. As a result, recycled aggregate causes microcracks throughout the crushing and grinding process, and also RCA are more porous than that of NCA. Moreover, the interfacial transition zone (ITZ) among the old mortar and aggregate is also porous and weak. These factors unfavourably distress the strength of recycled coarse aggregate. Porosity and cracks are detrimental effects on the properties of RCA. Though recycled aggregates, fly ash and ferrochrome ash are waste materials used for making geopolymer concrete, experimental results revealed that up to 20% of replacement of recycled aggregate can achieve strength and it can be the recommended percentage of RCA in concrete industry.

5.2 *Effect of RCA of Geopolymer Concrete on Tensile Strength*

The tensile strengths of 150 mm × 300 mm cylinders containing 0%, 10%, 20%, 30%, 40% and 50% of reused aggregate on the part substitution of NCA are examined at 7 days, 28 days and 56 days, respectively, as shown in 'Fig. 3'.

The suitability of RCA as replacement of NCA in geopolymer concrete was investigated. It was observed that fly ash and ferrochrome ash-based geopolymer concrete containing 50% recycled aggregate gives a maximum decrease in tensile strength which is about 50% at curing age from 7 to 28 days. Similarly, at curing age from 28 to 56 days, the strength was decreased up to 43%. The effect of longer curing age from 28 to 56 days gives higher strength and is about 10–28%. Strength of concrete gradually decreases by increasing the percentages of recycled aggregate up to 50%, by using recycled aggregate from 10 to 50% of enhanced strengths 3.72, 3.65, 3.32, 3.21 and 3.13 MPa at 7 days, 4.71, 4.52, 4.48, 4.28, 3.85 and 3.63 MPa at 28 days and 4.92, 4.68, 4.55, 4.35, 4.13 and 4.06 MPa at curing age of 56 days.

5.3 *Effect of RCA of Geopolymer Concrete on Flexural Strength*

The flexural strengths of 500 mm × 100 mm × 100 mm beams containing 0%, 10%, 20%, 30%, 40% and 50% of reused aggregate, on the part substitution of natural coarse aggregate, are shown in 'Fig. 4'. It was observed that fly ash and ferrochrome ash-based geopolymer concrete containing 0% recycled aggregate maximum increase in flexural strength is about 30%.

Experimental result shows that GFF with 50% RAC has minimum flexural strength of 2.76, 3.81 and 4.09 MPa at curing age of 7 days, 28 days and 56 days, respectively, while other samples are GFF with 10–40% RAC having strengths of 3.34 MPa, 3.28 MPa, 2.96 MPa, 2.82 MPa and 2.76 MPa, respectively, at curing age of 7 days and 4.35, 4.39, 4.06, 3.94 and 3.85 MPa at 28 days curing. Longer curing age gives higher strength; however, from 28 to 56 days, curing strength increases by about 18–30%.

6 Conclusion

A series of experiments were conducted sequentially for the determination of mechanical properties, whereas obtained in the present work, the following conclusions can be made.

- By the addition of RCA along with fly ash, ferrochrome ash gives inferior mechanical properties in comparison to fly ash and ferrochrome ash-based NCA. However, the addition of RCA has marginal effects above properties over control concrete. So all mineral admixtures and recycled aggregate considered in this study are useful in the preparation of sustainable concrete.
- Experimental results show that fly ash and ferrochrome ash-based geopolymer concrete containing 50% recycled aggregate maximum decrease in tensile strength is about 50% at curing age from 7 to 28 days. Similarly, at curing age from 28 to 56 days, the strength was decreased up to 43%.
- It was noticeable to see that concrete containing 50% RCA and maximum strength decreases nearly 8% at curing age from 7 to 28 days. From 7 to 56 days, strength would be decreased by about 8.5% for 0% RCA. Test results revealed that strength slightly increases from 28 to 56 days. Though recycled aggregates, fly ash and ferrochrome ash are waste materials used for making geopolymer concrete, experimental results revealed that up to 20% of replacement of reused aggregate can achieve strength. Generally, in the case of RCA, old adhered mortars are present and interfacial bond between old mortar and aggregate is porous and weak. With the presence of old adhered mortars in recycled aggregate, more porous and also the interfacial bond between the old mortar and the old aggregate is also porous and weak. These factors unfavourably affect the strength of RCA.
- Partial replacement of recycling waste aggregate by natural coarse aggregates up to 20% has resulted in a slight decrease of strength in compression, split and flexural, and higher than that and it decreases considerably.
  Test results show that tensile strength and flexural strength of RCA progressively decrease by increasing the percentage of RCA up to 50%. The research also depicts that RCA is a partial replacement of NCA up to 20%, which is to be considered in the sustainable concrete.
- The chemical composition of fly ash, as well as ferrochrome ash, is compensating each other, because fly ash was rich in silicon dioxide (SiO_2), but the percentages of ferric oxide (Fe_2O_3) and aluminium oxide (Al_2O_3) were more for ferrochrome ash; thus their combination in 1:1 ratio has been established as a good alternative binding material for geopolymer concrete by using recycled aggregate.

References

1. Chen HJ, Yen T, Chen KH (2003) Use of building rubbles as recycled aggregates. Cem Concr Res 33:125–132
2. Yang J, Du Q, Bao Y (2011) Concrete with recycled concrete aggregate and crushed clay bricks. Constr Build Mater 25:1935–1945
3. Malhotra VM, Mehta PK (2002) High-performance, high-volume fly ash concrete: materials, mixture proportioning, properties, construction practice, and case histories
4. Lamond JF, Pielert JH (2006) Significance of tests and properties of concrete and concrete-making materials. ASTM International, West Conshohocken, PA, USA
5. Davidovits, J (1991) Geopolymers: inorganic polymeric new materials. J Therm Anal 37:1633–1656
6. Duxson P, Provis JL, Lukey GC, Van Deventer JS (2007) The role of inorganic polymer technology in the development of green concrete. Cem Concr Res 37:1590–1597
7. Poon CS, Shui ZH, Lam L (2004) Effect of microstructure of ITZ on compressive strength of concrete prepared with recycled aggregates. Constr Build Mater 18:461–468
8. Halicka A, Ogrodnik P (2013) Using ceramic sanitary ware waste as concrete aggregate. Constr Build Mater 48:295–305
9. Gonçalves P, De BJ (2010) Recycled aggregate concrete (RAC)—comparative analysis of existing specifications. Mag Concr Res 62:339–346
10. Gomes M, Brito DJ (2009) Structural concrete with incorporation of coarse recycled concrete and ceramic aggregates: durability performance. Mater Struct 42:663–675

Individual and Combined Effect of Nano- and Microsilica on Cement-Based Product

Wubshet Gebru, Ashoke Kumar Rath, and Dillip Kumar Bera

Abstract In the cement concrete industry, everybody is concerned about high-strength and high-performance behavior of concrete. These properties were improved using micro- and nano-sized pozzolanic mineral admixtures as stated by various authors. In this paper, author has emphasized mostly on using micro- and nanosilica materials with cement. Literature survey revealed that the most suited mineral admixtures for getting highest behaviors are micro- and nanosilica. Nanotechnology is nowadays a buzzword in everyone's tips. Nanotechnology in concrete means the addition of nano-sized materials in the range of 1 and 100 nm to improve the behavior of cement and concrete. Micro- and nano-sized silica particles improve the porous behavior of concrete and also produce more calcium silicate hydrate resulting from chemical reaction of nanoparticles and calcium hydroxide. On the other hand, it makes concrete more compact in microstructure point of view, which improves density, strength, and also durability. This paper also discusses the synergic behavior of nanosilica and microsilica on the property of concrete in comparison with traditionally produced concrete.

Keywords Nanosilica · Microsilica · Nanotechnology · Nanomaterials in concrete · High strength and high performance

W. Gebru · A. K. Rath · D. K. Bera (✉)
School of Civil Engineering, Kiit University, Bhubaneswar, India
e-mail: dberafce@kiit.ac.in

W. Gebru
e-mail: 1754013@kiit.ac.in

A. K. Rath
e-mail: akrath1947@gmail.com

B. B. Das et al. (eds.), *Recent Developments in Sustainable Infrastructure*, Lecture Notes in Civil Engineering 75, https://doi.org/10.1007/978-981-15-4577-1_35

1 Introduction

Currently, concrete is the most abundantly used material in the world. Cement is a basic raw material for making concrete; its production consumes more amount of energy and generates high CO_2 emission to the atmosphere. To protect the environment from pollution and to improve the mechanical and durability property of concrete, researchers and scientists adapted nanotechnology.

Nanotechnology is using constituent material in the nanoscale either through top-down by milling bulk materials into small size or through bottom-up by combining atoms and molecules to form bulk materials, which is called molecular nanotechnology. From this concept, concrete technology has started using constituent materials in nanoscale to produce strong and durable concrete by partially or fully replacement of cement. Carbon nanotubes, silica fume, aluminum oxide, zinc oxide, carbon nanofibers, and micro- and nanosilica particles are the most recently used materials in the construction industry. However, in this review paper, we would discuss specifically influence of micro- and nanosilica particles on the property of concrete.

2 Nanotechnology in Construction Industry

Nanotechnology is one of the recently fasted growing technologies, which has a key role in the different fields of study including civil engineering in the construction sectors.

In the building sector starting from the early civilization period, cement was the most utilized constituent material for the fabrication of cementitious products. As per previously done research, approximately 1.6 tons of raw materials would be required for each ton of cement. As a result in the last two and half decades, researchers have introduced blending materials in the nanoscale for the construction industry. Nowadays, the most alarming problems in the construction industry are atmospheric pollution due to carbon dioxide emission from cement industry, secondly, the more amount of energy consumption for the production of cement, and thirdly the lack of non-renewable component materials for manufacturing process of cement. To reduce this problem, partial substitution of cement by some other supplemental raw materials in nanoscale and microlevel would be the promising techniques.

3 Nanosilica

Among nanomaterials, nanosilica is widely used in the construction industry. It has high pozzolanic property and surface area which help to react with different materials actively. It is known that cement product is composed of nanoscale pore, capillary pore, and some other microscale porous structures. Nanoscale pore would get filled up by nanosize materials to fill the pore and produce denser concrete. Addition of nanosilica to the concrete mix would produce additional calcium silicate hydrate form, where nanosilica interacts with calcium hydroxide which is liberated from the hydration process. To some extent, this extra calcium silicate hydrate form would improve strength and also would decrease the permeability of concrete by filling the pore part of the hardened concrete paste.

3.1 Mechanical Property

3.1.1 Compressive Strength

The mechanical property of high-strength self-compacting concrete was investigated where cement was replaced with nanosilica (NS) in the range of (1, 2, 3, 4, 5%) [2, 3]. Figure 1 shows that 4% addition of nanoparticles was the ideal replacement for binder material of cement. At the early age of reaction process, compressive strength was increased more rapidly as a result of more amount of calcium hydroxide present in the hydration process which reacts quickly with nanosilica and produced additional C-S-H gel to form more denser concrete.

Improvement for microstructures, durability properties, and mechanical performances of ultra-lightweight cement composites (ULCC) by using colloidal nanosilica (CNS) was studied by [4]. The test result indicated that 2% of CNS gives optimum strength in 28 days with both 0.34 and 0.24 w/c ratios, due to compaction of weak zone of a concrete which is found at the contact place of cement paste and aggregate along good pore filing and pozzolanic activity.

Combined effect of various nanosilica particles of size (Fifteen and eighty nanometer) on mechanical property of concrete was stated [5]. In this case, 0.5,

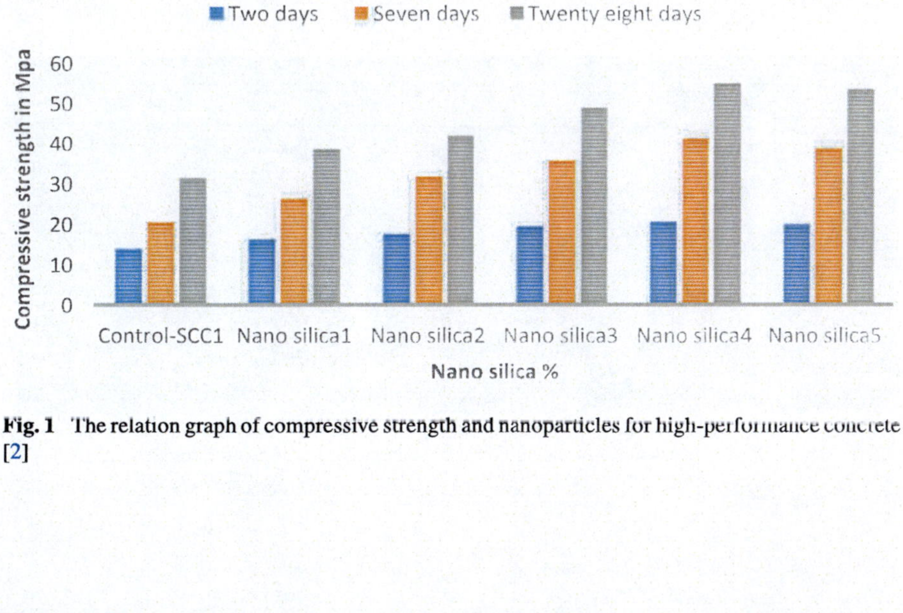

Fig. 1 The relation graph of compressive strength and nanoparticles for high-performance concrete [2]

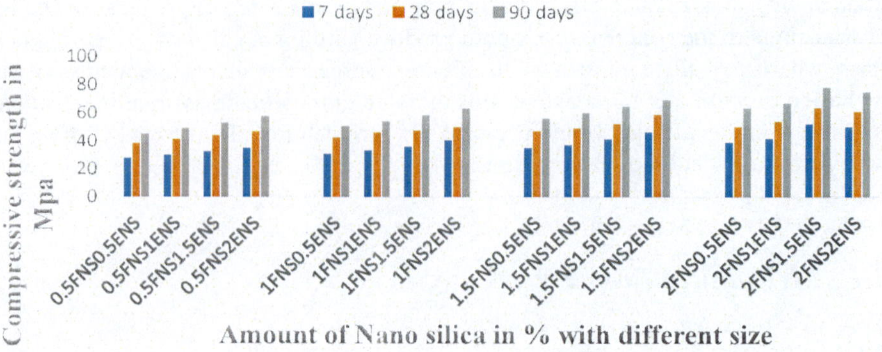

Fig. 2 Compressive strength with different sizes of nanosilica [5]

1.0, 1.5, and 2.0 wt% of NS were used as replacement of cement. From Fig. 2, we can notice that compressive strength was increased till 2% of 15 nm and 1.5% of 80 nm and start decreasing this limit. In general, by curing a concrete in a lime solution, we would achieve high-strength concrete, due to the availability of more amount of calcium hydroxide, which helps nanosilica to react with it and would form extra calcium silicate hydrate. In Figure 2 ENS denotes 80 nm nanosilica and FNS denotes 15 nm nanosilica.

3.1.2 Split Tensile Strength

The consequence of nanoparticles with the addition of different fibers together on the property of cement concrete was investigated [1], where steel fibers in the range of 0.2, 0.3, and 0.5%, polypropylene (0.1, 0.15, and 0.2%), and glass fiber (0.15, 0.2, and 0.3%) in terms of volume replacement. Figure 3 represents tensile strength

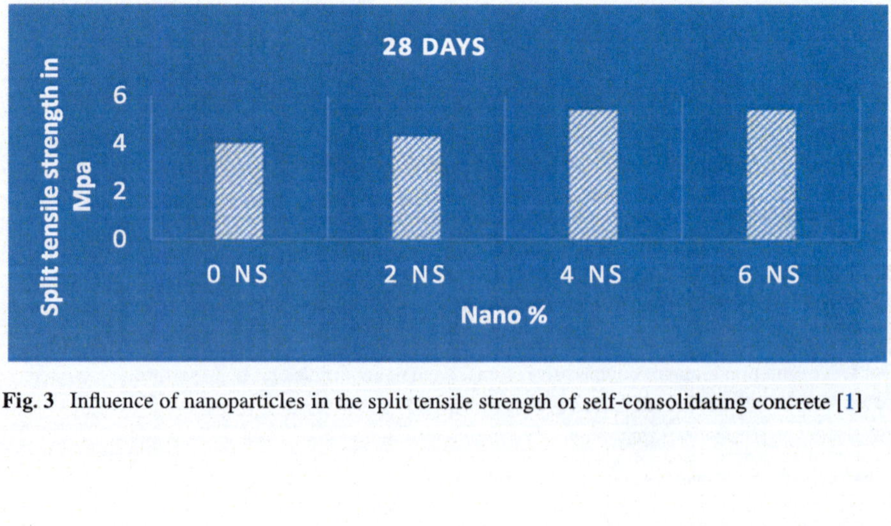

Fig. 3 Influence of nanoparticles in the split tensile strength of self-consolidating concrete [1]

increased up to 4% of NS with the combination of different fibers. The reason for increment of strength was due to the addition of nanosilica which gives more space for mortar and fibers to interact with each other and form a strong concrete product.

The consequence of nanoparticles on the property of concrete and mortar on which 40% of fly ash was replaced with cement constantly was studied [8]. Tensile strength was increased with increasing of nanosilica particles and with the age of concrete. Specifically in the case of 0.5 and 2% nanosilica replacement in 28 days, there would be 10 and 23% strength change, respectively, due to increasing of secondary C-S-H gel which was formed from the reaction of nanosilica particles with calcium hydroxide.

The behavior of high-performance concrete was studied [7] where cement was constantly replaced by 40% copper slag and partially by colloidal nanosilica (0.5 to 3% with 0.5 variation). Due to strong bonding between paste and aggregate, every addition of nanosilica gives more strength than control specimen. However, apart from 2% replacement of nanosilica strength decreased, due to availability of more nanosilica than free lime contents. As a result, extra nano-colloidal silica would act as a filler instead of acting part of hydration process.

3.1.3 Flexural Strength

The consequence of nanoparticles with the addition of different fibers together on the property of cement concrete was investigated [1], where steel fibers in the range of 0.2, 0.3, and 0.5%, polypropylene (0.1, 0.15, and 0.2%), and glass fiber (0.15, 0.2, and 0.3%) in terms of volume replacement. Figure 4 shows that strength increased up to 4% NS with 0.3% steel, 0.2% polypropylene, and 0.2% glass fibers, respectively. In this study, the effect of nano addition was in terms of its pozzolanic and filler effect on the concrete.

The consequence of fly ash and nanoparticles on the property of concrete and mortar was stated [8], where 40% fly ash was constantly added and 0.5, 1, 1.5,

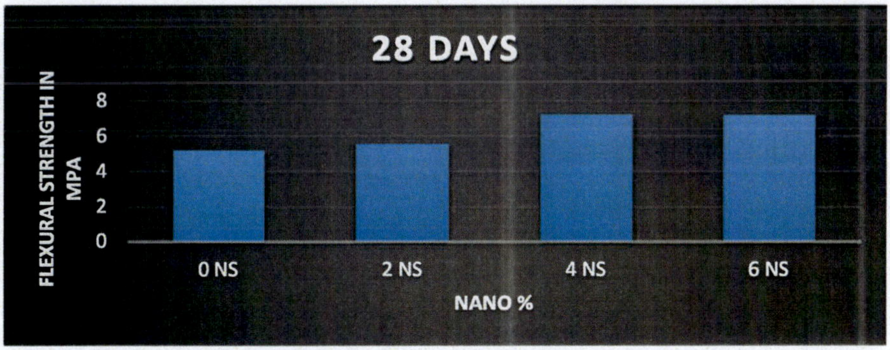

Fig. 4 Effect of nanosilica on flexural strength of concrete [1]

2, and 3% of NS by a limited substitution of cement powder at 0.23, 0.25, and 3.0 water/binder ratio. Result shows that 2% was optimum replacement for better strength caused by the development of additional calcium silicate hydrate gel which is resulted from the combination reaction of nanoparticles and lime liberated from the main hydration process.

3.2 *Durability Property*

3.2.1 Water Absorption

Individual and combined outcome of nanoparticle size of alumina, silica, and titanium on the property of self-compacting concrete which contains fly ash constantly in 25% by weight of cement was investigated [6]. Due to its finer particle size, 3% of NS helps to fill up the voids of the microstructure and formed more compacted and denser concrete with less amount of water absorption.

Mechanical and durability property of self-compacting concrete by limited substitution of cement powder with nanosilica (1, 2, 3, 4, 5%) was analyzed [2]. Figure 5 shows that water absorption decreased up to the addition of 4% nanosilica and beyond that started increasing. In general, amount of replacement nanosilica particles, and curing period and age would result in lowering permeability of the concrete.

The consequence of nanosilica particles on the properties of concrete where copper was constantly substituted as a fine aggregate at a level of 40% has been discussed [7], where 1, 2, 3, 4, and 5% of NS are used as a replacement of cement and give decreased water absorption in the range of 0–2% nano-insertion in terms of weight and water absorption started increasing for 3% addition at 28 days curing.

The consequence of fly ash and nanoparticles on the property of concrete and mortar was stated [8] where 40% fly ash was constantly added and 0.5, 1, 1.5, 2, and 3% of NS by a limited substitution of cement powder at 0.23, 0.25, and 3.0 water/binder ratio. Conventional specimens and specimens with 0.5% NS addition show more absorption due to less hydration process and more pore availability.

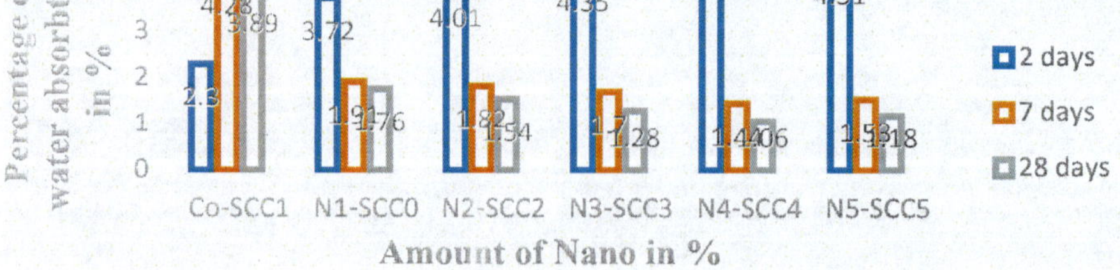

Fig. 5 Percentage of water absorption due to the addition of different amounts of NS [2]

However, later on, 2% NS addition shows less water absorption due to less capillary pore and denser concrete.

3.2.2 Rapid Chloride Permeability Test (RCPT)

The impact of nanosilica particles is on the self-consolidating concrete with a combination of different fibers' addition [1] with 2, 4, and 6% of NS replacement. Nanosilica with the combination of fiber shows a better resistance for chloride penetration into the hardened concrete production of calcium silicate hydrate gel by reacting with calcium hydroxide.

The consequence of nanosilica particles on the properties of concrete where copper was constantly substituted as a fine aggregate at a level of 40% has been discussed [7]. Figure 6 shows that chloride penetration into the hardened concrete decreased up to 2% replacement of colloidal nanosilica, due to spherical shape of nano and copper slag particles which help to give denser microstructure of the concrete.

4 Microsilica

Microsilica is a by-product of the silicon and ferrosilicon fusion production with an average size of 150 nm. It is one of the non-crystalline types of materials which is more finer and approximately 100 times finer than the cement particles. As per previously published research paper, they have stated that approximately 100,000 of silica particles available for each of the cement particles within a cement matrix, when 8% of silica fume was a substituent of cement, used as a partial replacement of cement, which help to protect bleeding and improve weak transition zone of normal concrete. It has a high pore filling effect, and as a result, it can improve the permeability of concrete.

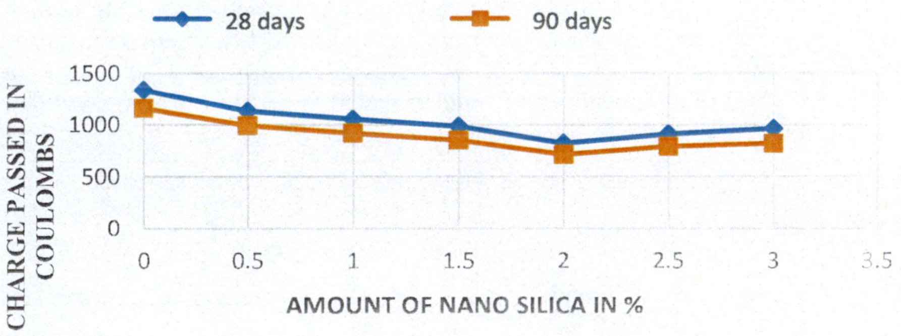

Fig. 6 Chloride penetration test result in terms of charge passed in coulombs [7]

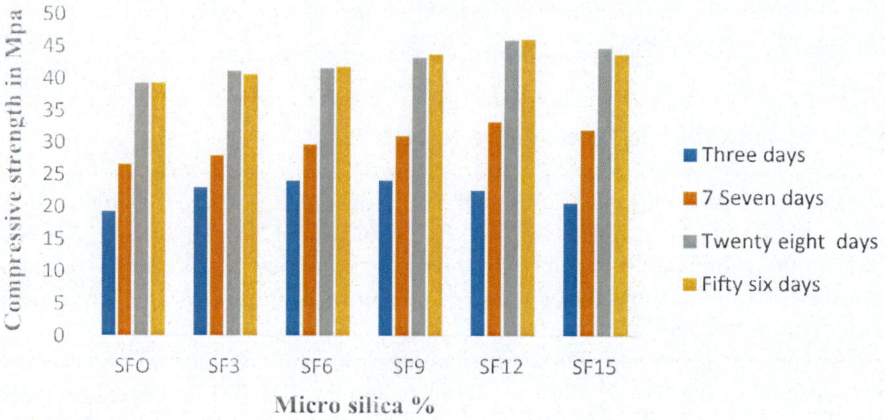

Fig. 7 Compressive strength test result of silica fume in place of cement [11]

4.1 *Mechanical Property*

4.1.1 Compressive Strength

The influence of microsilica particles on the acoustic and mechanical property of cementitious paste by replacing cement [9] with a range of 1, 2, 3, 4, 5, 7.5, and 10% by weight of binder has been investigated due to the formation of additional C-S-H gel by pozzolanic reaction, and its pore filling ability of 5% is optimum to form more denser concrete.

The influence of microsilica on the property of M40 grade of cement concrete by replacing cement was studied [11]. Figure 7 shows that due to the fast reaction between silica fume and fly ash and making of calcium silicate hydrate gel by the inclusion of microsilica and filling ability of 12% MS would produce more packing and denser hardened concrete.

The influence of microsilica on the conventional concrete by replacing cement with 6, 7, and 8% for M25 grade of concrete was studied [12], where 0.7% superplasticizer, water/cement ratio of 0.40, and workability range of 75–100 mm were used. By the help of pozzolanic activity and its gap filing ability, 8% of MS addition was the optimum replacement of cement.

4.1.2 Split Tensile Strength

To overcome the carbon dioxide emission from the production of cement [11], replacement of cement by microsilica in the range of 5, 10, and 15% has been studied, specifically for self-compacting concrete. Addition of microsilica gives better strength up to level of 10% replacement due to extra C-S-H gel formation by pozzolanic reaction. However, beyond 10% replacement of microsilica, strength

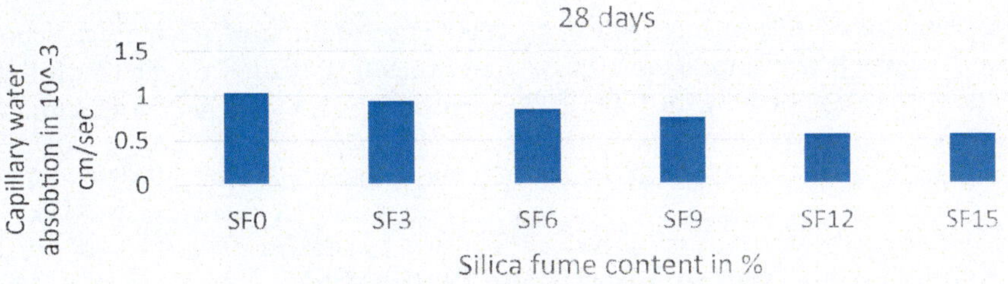

Fig. 8 Capillary water absorption test at different percentages of SF [11]

decreased due to the formation of weak interfacial zone between cement paste and aggregate.

4.1.3 Flexural Strength

The influence of microsilica (MS) on the property of M40 grade of cement concrete where cement was substituted with MS in the range of 0, 3, 6, 9, 12, and 15% has been studied [11]. Figure 8 shows that flexural strength increase with increasing microsilica addition and the maximum strength of 6.5 Mpa was recorded in 15% microsilica addition as a replacement of cement by weight of cement.

4.2 Durability Property

4.2.1 Capillary Absorption Test

Improvement of fiber-reinforced concrete and cement concrete by adding microsilica and titanium oxide was investigated [10], where silica fume in 4.5, 9.5, and 14.5% and ultra-fine TiO2 with 0.5% constantly by weight of cement was used in place of cement. Addition of 9.5% MS was optimum replacement and shows better reduction compared to other specimens.

Microsilica (MS) on the property of M40 grade of concrete where cement was substituted in the limited range of 0% MS, 3% MS, 6% MS, 9% MS, 12% MS, and 15% MS has been discussed [11]. Result shows in Fig. 8 that result was decreased when microsilica increased and 12% was the optimum replacement of cement.

4.2.2 Rapid Chloride Permeability Test (RCPT)

The influence of nano- and microsilica insertion as a limited substitution of slag which is used in alkali-activated slag concrete was discussed [13] with 0, 0.5, 1, 3, and 5%

of nanosilica and 10% of microsilica, respectively. Addition of microsilica separately showed a better chloride resistance than nanosilica, but when both nanosilica and microsilica combined with 1 and 10%, respectively, it would decrease chloride resistivity.

5 Synergistic Effect of Micro- and Nanosilica

5.1 Mechanical Property

5.1.1 Compressive Strength

The combined effect of nanosilica and microsilica on cementitious product has been stated [14], where 3, 5, 7, 10, 15% MS and 1, 2, 3% NS for mortar were mixed separately, and 8%, 10% MS and 2% NS together for concrete mix was applied as a limited substitution of cement powder, where result indicates that 8% of MS and 2% NS give the best result. It has been shown in Fig. 9 that flexural strength value increases maximum after 9% and attains constant at 12% and 15% substitution of MS.

Synergistic effects of MS and NS on the strength and microstructure of mortar of high-performance concrete (HPC) were studied [2] at constant workability. Figure 9 shows that due to its pore filling capacity, 10% MS plus 2% NS would be the optimum addition of both particles to produce strong and denser concrete as shown in Fig. 10.

The strength of M-60 design of cement concrete was studied [15] with the replacement of nano- and microsilica. Result showed that 4–5% of NS show decreased strength within 28 days, due to high heat of hydration. As a result of workability, it was less and harsh concrete would be formed. In general, it is concluded that optimum strength was attained by 1–3% NS range and 9–11% MS, respectively.

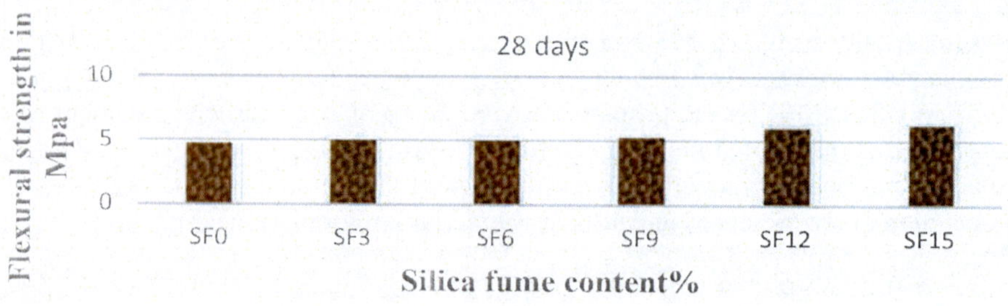

Fig. 9 Flexural strength results at different replacement percentages of MS [11]

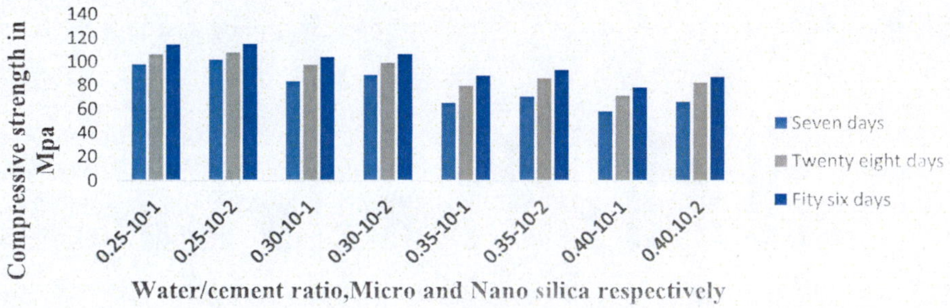

Fig. 10 Mortar strength result with the addition of micro and nano [2]

5.1.2 Split Tensile Strength

The combined effect of micro- with nanosilica particles on the mechanical property of ultrahigh-strength cement concrete was studied [18] with 10% constant amount of silica fume addition and 0.5, 1, 2, and 3% of nanosilica. Result showed that tensile strength increased with increasing nanoparticles up to 2%, with and without the addition of silica fume. However, the combined effect of both silica particles showed a better strength with the range of 2 and 10%, respectively.

5.2 *Durability Property*

5.2.1 Water Absorption

Mechanical property of ultrahigh-performance concrete has been studied [18] with nanosilica (NS) and 10% constant amount of silica fume (SF), where 0.5, 1, 2, and 3% of NS are at same w/b ratio. Results showed that at 2% NS + 10% SF replacement of cement, the concrete water absorption was shown reduced value for binder content of 400 and 500, respectively.

Durability, microstructural, and some other mechanical properties of high-strength self-compacting concrete with the addition of nanosilica (NS) and microsilica (MS) particles have been studied [17]. Result shows in Fig. 11 that a specimen with 10% MS and 2% NS at a binder content of 500 kg per meter cube was achieved a better result.

5.2.2 Rapid Chloride Permeability Test (RCPT)

To find out the integrated influence of micro- and nanosilica particles on the durability of cement concrete, the author [16] proposed a series of mortar mixes with constant workability. Figure 12 shows that the addition of both nanosilica and microsilica

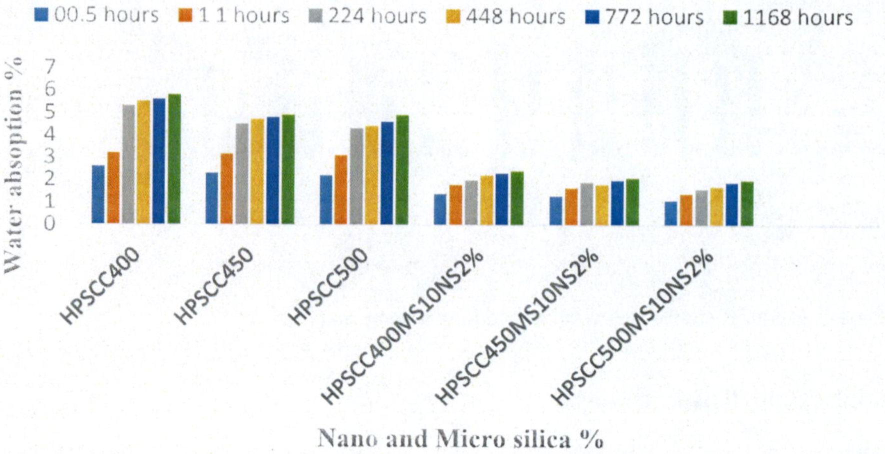

Fig. 11 Water absorption result [17]

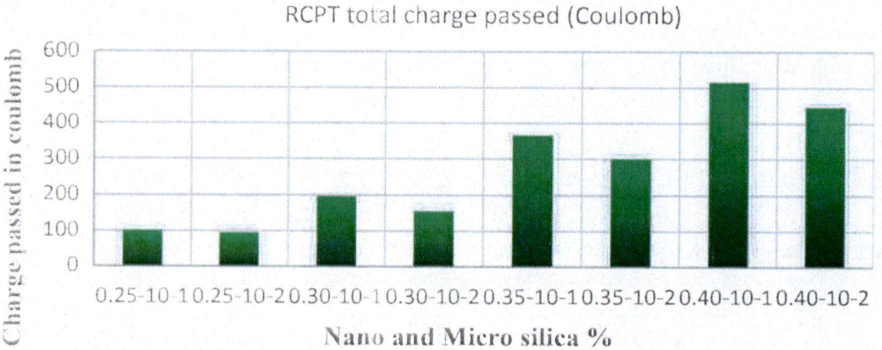

Fig. 12 RCPT result [16]

would give better improvement in chloride resistance than those attained by the addition of microsilica or nanosilica alone at a lower w/c ratio.

The impact of nano- and microsilica on cementitious product was discussed [14] with 3, 5, 7, 10, 15% MS and 1, 2, 3% NS replacement of cement for mortar mix separately, and 8%, 10% MS with 2% NS together for concrete mix. Figure 7 shows that specimens with the addition of nano- and microsilica with less current flow within a given time interval as a result of 2 and 10% together was the optimum replacement.

6 Summary

Nanosilica during the early period of reaction whereas microsilica during later age of hydration process shows a remarkable influence on mechanical property of the concrete, mainly by its highly reactive nature and its pore filing ability of silica particles. However, due to the dispersion problem and use of excess amount, nanosilica more than the presence of calcium hydroxide would lead to decrease of strength.

Combining nanosilica and microsilica particles gives a better improvement of strength of concrete in the early period by reducing the dormant period and also later age strength. At the same time by filling pores of the microstructure of the concrete, chloride resistance, water absorption, and some other chemical attacks would be improved.

Even though so many papers have been published, however, none of them specifies the optimum quantity of nano- and microsilica replacement to improve the property of concrete. Besides, also some other health-related problems of nano- and microparticles product should be studied for future long-term application of nanotechnology in the construction industry.

References

1. Beigi MH, Berenjian J, Omran OL, Nik AS, Nikbin IM (2013) An experimental survey on combined effects of fibers and Nanosilica on the mechanical, rheological, and durability properties of self-compacting concrete. Mater Des 50:1019–1021
2. Nazari A, Riahi S (2011) The effects of SiO2 Nanoparticles on physical and mechanical properties of high strength compacting concrete. Compos Part B 42:570–578
3. Jo BW, Kim CH, Tae GH, Park JB (2007) Characteristics of cement mortar with Nano-SiO2 particles. Constr Build Mater 21(1351–1355):1029
4. Du H (2019) Properties of ultra-lightweight cement composites with Nano-silica. Constr Build Mater 199:696–704
5. Nazerigivia A, Najigivib A (2019) Study on mechanical properties of ternary blended concrete containing two different sizes of Nano-SiO2. Compos Part B 167:20–24
6. Mohseni E, Miyandehi BM, Yang J, Yazdi MA (2015) Single and combined effects of nano-SiO2, nano-Al2O3 and nano-TiO2 on the mechanical, rheological and durability properties of self-compacting mortar containing fly ash. Constr Build Mater 84:331–340
7. Chithra S, Senthil Kumar SRR, Chinnaraju K (2016) The effect of Colloidal Nano-silica on workability, mechanical and durability properties of high performance concrete with copper slag as partial fine aggregate. Constr Build Mater 113:794–804
8. Palla R, Karade SR, Mishra G, Sharma U, Singh LP (2017) High strength sustainable concrete using silica Nanoparticles. Constr Build Mater 138:285–295
9. Shebl SS, Seddeq HS, Aglan HA (2011) Effect of micro-silica loading on the mechanical and acoustic properties of cement pastes. Constr Build Mater 25:3903–3908
10. Karthikeyan B, Dhinakaran G (2018) Influence of ultrafine TiO2 and silica fume on performance of unreinforced and fiber reinforced concrete. Constr Build Mater 161:570–576
11. Nandhini K, Ponmalar V (2018) Passing ability, water and chloride penetration of self compacting concrete using micro silicaas a cementitious replacement material. Revista Română de Mater/Roman J Mater 48(3):362–368
12. Kumar V, Imam A, Srivastava V, Kushwaha AY (2017, November) Effect of micro silica on the properties of hardened concrete. Int J Eng Res Develop 13(11):08–12

13. Behfarnia K, Rostami M (2017) Effects of micro and Nanoparticles of SiO2 on the permeability of alkali activated slag concrete. Constr Build Mater 131:205–213
14. Aladdin MS, Metwally AAE, Omar HK (2018) Synergistic influence of micro-nano silica mixture on durability performance of cementious materials. Constr Build Mater 164:579–588
15. Singh A, Dhillon RS (2017, June) Experimental investigation on effect of microsilica and nanosilica on compressive strength of high strength concrete. Int J Eng Technol Sci Res IJETSR 4(6)
16. Li LG, Zhu J, Huang ZH, Kwan AKH, Li LJ (2017) Combined effects of micro-silica and Nano-silica on durability of mortar. Constr Build Mater 157:337–347
17. Jalal M, Mansouri E, Sharifipour M, Pouladkhan AR (2012) Mechanical, rheological, durability and microstructural properties of high performance self-compacting concrete containing SiO2 micro and Nanoparticles. Mater Des 34:389–400
18. Gesoglu M, Güneyisi E, Asaad DS, Muhyaddin GF (2016) Properties of low binder ultra-high performance cementitious composites:comparison of nanosilica and microsilica. Constr Build Mater 102:706–713

Self-healing Conventional Concrete Using Bacteria

Smruti Pal, Ipsita Mohanty, and Ipsita Panda

Abstract Concrete is most widely used as an epochal building material all over the world. However, the repercussions of its deterioration are inevitable ever since early stage of its service life. It is followed by the fixture and repair works, which is very labor intensive and also involve a lot of capital. This becomes uneconomic when the design period of the concrete is considered. Thus, many conventional surface treatments are applied, and also various alternative techniques are taken for the improvement of its mechanical properties. Self-healing of the affected concrete can not only be of great importance but also it uses the environment-friendly techniques. The present study pertains to the use of bacterial-induced self-fixing concrete which has a significant effect on its mechanical properties as well as crack healing capacity of conventional concrete. Two different species of genus *Bacillus* are incorporated to the concrete which repairs the crack by healing the microcracks occurred in the transition zone as well as enhance the mechanical properties of the concrete. Bacterial concentrations are varied and mixed with the concrete mix to observe the change in mechanical properties, and results show a remarkable increase in the strength of the bacterial concrete. Results indicated that bacteria showed an optimum result when mixed with the concrete in the concentration of 10^8 cfu/ml. The present work illustrated that the production of self-healing concrete by both the species has a capacity to improve the strength properties of conventional concrete.

Keywords Concrete · Bacillus · Mechanical properties · Self-healing · Microcracks · Surface treatment

S. Pal (✉) · I. Mohanty · I. Panda
School of Civil Engineering, KIIT, Bhubaneswar, Odisha, India
e-mail: smruti.pal1509@gmail.com

I. Mohanty
e-mail: ipsita.mohantyfce@kiit.ac.in

I. Panda
e-mail: ipsita.pandafce@kiit.ac.in

B. B. Das et al. (eds.), *Recent Developments in Sustainable Infrastructure*, Lecture Notes in Civil Engineering 75, https://doi.org/10.1007/978-981-15-4577-1_36

1 Introduction

Construction materials are subjected to several environmental and biological factors. Mechanisms' degradation of concrete depends on the potentially aggressive substances via physical, chemical, and environmental factors, in which actions can lead to causing damage to concrete structure. Although there are various types of organic and inorganic products to protect the concrete from damages to make concrete more durable and serviceable, it is still a matter of research to find a more compatible and environmentally friendly method for defending the cement uses and deterioration of concrete structures. Self-healing concrete using bacteria carries the potential to increase durability of concrete structure and reduce the propagation of microcracks to the inner side of concrete structure and have ability to seal the cracks (thereby, reduction of consumption of cement) through bioremediation properties of bacteria (microbial activity of bacteria to seal the cracks up to certain width) which is economical, as the repair and maintenance cost is reduced. One of the prime uses of biomineralization of bacteria is the injecting absorptive media to concrete which goes penchant toward bioremediation.

Biomineralization of bacteria has been adopted in several engineering applications in recent researches. It is referred to as biologically induced mineral precipitation (metabolic process of bacteria to produce sealants), in which bacterial cells allow chemical precipitation of mineral phases by self, creating a microenvironment with conditions that allow to perform MICP. Action of crack healing process leads to $CaCO_3$ precipitation through the metabolic process of bacteria (Chahal 2012; and Li 2011; Siddique and Chahal 2011). Several strategies have been proposed to improve cementitious material's durability and the resiliency of concrete structures through autonomic healing systems. It has been proved from previous researches that bacteria-mediated $CaCO_3$ is compatible with concrete and environmentally friendly among other chemical admixtures and polymers.

Concrete is a highly alkaline material, so it is necessary that the bacteria which have to be added in concrete that should have enough potential of surviving in alkaline environment. MICP (Microbiologically induced calcium carbonate precipitation) of bacteria can help to seal the microcracks in or on the surface of concrete, accelerate/retard the setting process as based on its nutrient given to bacteria, and also can act as binder materials in concrete which lead to enhance various concrete properties [Hassan et al. 2016] referred the self-healing systems that may be broadly sorted into several categories like chemical encapsulation, encapsulation of bacteria, the addition of mineral admixtures or chemicals, and intrinsic self-healing with self-controlled tight crack width [4] investigate the durability improvement of concrete using bacteria in sulfate environment by considering two different bacterial species in various concentrations. [Chandni et al. 2016] identified the calcium carbonate precipitation within the mortar; bacterium may be used for the development of bacterial mortar which indicates that modified mortar can improve the strength of the mortar sample.

Biological environment has been applied on a concrete surface for protection and nutrition of the bacteria, resulting in the reduction of surface permeability of the concrete and providing a biological environment conducive for improving strength and the bacterial activities on the concrete [Farzaneh et al. 2016]. Durability improvement of concrete samples in different environments considering different bacterial strains of different concentrations was investigated [Farzaneh et al. 2015, and Ghosh et al. 2005].

The concept of microcapsule (spore form of bacteria) used as healing agent which was encapsulated and added in the concrete lead to increase in strength and various permeation properties of concrete. To study the effects of parameters, namely, temperature, agitation rate, and pH on the shell thickness, size, and morphology of the microcapsules, various tests were performed. The effect of microencapsulated self-healing materials was shown by experimentally evaluating modulus of elasticity of concrete in the laboratory [Hasssan et al. 2014].

Various bacterial types and species are used in self-healing technology as healing agents including different nutrient mediums and environmental exposures to conduct durability, permeation, elastic modulus, mass, and volume variations tests, which gave positive results for adopting this new emerging technology [all references given below].

2 Research Significance

Various chemical and physical methods have been examined by researchers in recent years to provide an environment in which bacteria used for biological treatment of concrete prolong survival and improve bio-deposition processes on/in concrete. This experimental study is focused to study the effects of alkaliphilic bacteria used in concrete matrix and evaluation of mechanical properties like compressive strength, split tensile strength, and flexural strength of standard concrete (without bacteria) versus self-healing concrete, by incorporating two different bacterial strains with their various concentrations into concrete. Additionally, surface treatment is provided to the controlled crack standard concrete cubes, to show that treatment has proved beneficial in developing the bio-deposition process in concrete technology which can be visualized by photography.

3 Materials and Methods

3.1 Properties of Material

Standard concrete mixtures of M30 grade were designed according to IS:456-2000 and IS:10262-1989. Fineness modulus and bulk specific gravity of the fine aggregate

Table 1 xxxx

Particulars	Properties
Mix proportion (C:FA:CA)	(1:2.55:3.5)
Ordinary Portland cement (RAMCO)	400 kg per m^3
FA (sand 4.75)	960 kg per m^3
CA (20 mm/10 mm)	1480 kg per m^3
Water	180 kg per m^3

were 2.88 and 2.65, and specific gravity of the coarse aggregate was 2.71, respectively. Properties of these materials are given in Table 1.

3.2 *Culture Technique of Bacteria*

Bacillus megaterium and *bacillus subtilis*, both isolated commercially available in KIIT Biotechnology laboratory, were used in this study. Further, this bacterial strain was tested and isolated by using an enrichment culture technique. For the enrichment of bacterial samples, by using the serial dilution technique, bacteria were set out to Petri plate (nutrient agar plate) and by the total plate count method on nutrient agar plates, one colony of each bacterial strain was selected separately, and then it was inoculated in 5 ml of nutrient broth by maintaining pH 12.0 and incubated at 37 °C for 12 to 14 h under shaking conditions (130 rpm) up to specified growth.

When 5 ml of bacterial suspension was prepared, then it took for the cfu count test. Observations are taken to found how much time is required to grow in the definite concentrations that we need in this research work. After that, required amount of liquid bacterial suspension was prepared in a definite concentration and designed amount.

In this paper, three different concentrations (10^6, 10^8, and 10^{10} cfu/mL) of two different types of bacteria (*Bacillus megaterium* and *Bacillus subtilis*) were prepared in a designed required amount separately for each concentration and type.

3.3 *Mechanism of Incorporation of the Healing Agents into Concrete*

Summary of literature reports that bacteria as healing agent can be added in concrete mainly in two ways: (1) Direct raw culture application (liquid suspension, each bacterial cell has single-bonded cell) and (2) encapsulation (capsulated spores of bacteria or solid form of bacteria, each bacterial cell has double-cell layer).

The self-healing technology by using encapsulated bacteria has the potential to provide higher degree of self-healing properties compared with direct application technology. Considering crack width generally less than or equal to 1 mm, spore form of bacteria shows fast healing reaction to cracking in concrete matrix. But the only disadvantage is the preparation of these capsules that are difficult to produce either in a small amount or in huge, and also this method is not cost effective; on the other hand, though these spore of bacteria can survive in very harsh environment, they could not break their cells easily till they are not facing any major cracking or deterioration problems.

But in case of bacteria in the form of liquid suspension they can easily produce in laboratory in a huge amount as per daily basis requirement and their survival in concrete is longer, and also they start their metabolic process to heal the cracks, when their cells get break by any additional load or microcracks.

Therefore, in this finding we used liquid suspension of bacteria to cast the self-healing concrete.

3.4 Hydration Kinetics of Bacterial Solution in the Concrete

In concrete, the use of bacteria can either accelerated or retarded the setting time of concrete depending on the calcium source supplied in the form of calcium lactate, calcium nitrate, calcium glutamate, and calcium formate as nutrient of bacteria.

The use of calcium lactate as nutrient of bacteria retards the setting time of concrete, calcium formate, and calcium nitrate accelerate the setting time of concrete [4].

In this present work, calcium nitrate is used as nutrient of bacteria.

3.5 Mix Design of Concrete and Testing

In this study, M30 grade design concrete mix is prepared to achieve target compressive strength of 38.65 N/mm^2 at 28 days according to BIS:10262:1989. The mix proportion ratio of concrete used as water (0.4): cement (1.0): fine aggregate as sand (2.5): coarse aggregate as gravel (3.5) as described in Table 1. First, the standard concrete specimens were casted and taken out for normal water curing for 7 and 28 days.

Two bacterial suspensions, each having concentrations of 10^6, 10^8, and 10^{10} cfu/mL, were added to dry mix of concrete each separately to make bacterial concrete of M30 grade. At one time, only one concentration of bacterial suspension of specified species used for casting of bacterial concrete for preventing contamination of other bacteria, after several trials in concrete mix proportions to cast the concrete specimens with and without bacteria in different variations.

In the case of preparation of bacterial concrete specimens, liquid suspension of bacteria was added by replacing some quantity of the designed water for that casting

Table 2 Detail mix design of materials used in concrete with mix identity for 1 m^3 volume

Mix identity	Cement (Kg)	FA (Kg)	CA (Kg)	Water (Kg)	Bacterial solution (Kg)	Cell concentration (Cells/ml)	Grade of concrete
MBC0	400	960	1439	160	0	0	30
BM6	400	960	1439	144	16	106	30
BM8	400	960	1439	144	16	108	30
BM10	400	960	1439	144	16	1010	30
SBC6	400	960	1439	162	18	106	40
SBC8	400	960	1439	162	18	108	40
SBC10	400	960	1439	162	18	1010	40

for maintaining the water-to-cement ratio. Detail mix design of materials used in concrete with mix identity for 1 m^3 volume is given in Table 2.

As per IS:516-1956 (for cube and prism) and IS:5816-1999 (for cylinder), cubes for compression test, cylinder for split tensile test, and prisms for flexural test are crushed after a curing period of 7 days and 28 days.

4 Specimen Details

(i) Concrete cubes of size 150 mm conforming to clause 2.1(a) of IS:10086-1982 were cast for compressive strength test,
(ii) $10 \times 10 \times 50$ mm size of prisms conforming to clause 2.1(a) of IS:10086-1982 was cast for flexural strength test,
(iii) 150×300 mm size of cylinder conforming to clause 2.1(a) of IS:10086-1982 was cast for split tensile strength test.

After casting, the specimens were allowed to dry in iron molds for first 24 h at room temperature. After 24 h, those samples were demoulded and placed in the curing chamber for curing room temperature. The specimens were tested after 7 and 28 days of curing period in triplicate. After 7 and 28 days of curing, the specimens were carried out for mechanical properties tests.

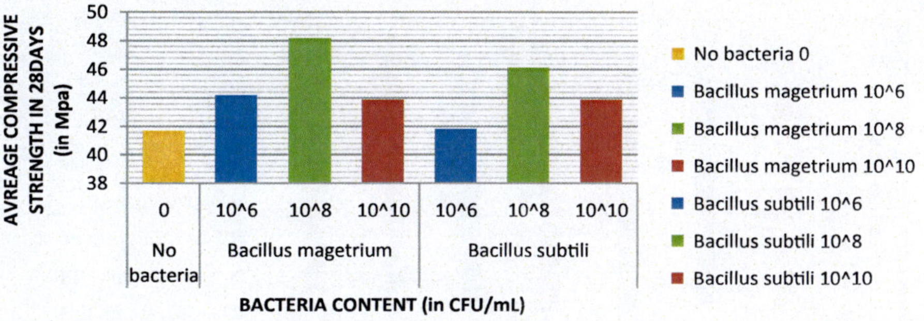

Fig. 1 xxx

5 Results & Discussion

5.1 Compressive Strength

The average compressive strength of standard concrete without bacteria is 38.66 N/mm^2 and 41.675 N/mm^2 after 7 and 28 days of curing period correspondingly, while compared to standard concrete, compressive strength of concrete is higher for every time when different concentrations of bacteria incorporated with concrete mixture to make self-healing concrete with different bacteria and their different cell concentrations. In maximum cases, concrete gains 95% of target strength at the age of 7 days. Maximum 3.079 and 2.974% of compressive strength increase for 7 days for self-healing concrete with *bacillus megaterium* and *bacillus subtilis* of 10^8 cfu/mL, respectively, compared with normal concrete. Maximum 15.615 and 10.63% of compressive strength increase for 28 days for self-healing concrete with *bacillus megaterium* and *bacillus subtilis* of 10^8 cfu/mL, respectively, compared with normal concrete. Both of those bacteria used in the work with different concentrations were increased in compressive strength than standard concrete at 28 days. But concrete with 10^8 cfu/mL gives more strength as compared to standard and other different concentrations of both bacteria, and also shows that *bacillus megaterium* of 10^8 cfu/mL concentration gave more compressive strength results among all. Figure 1 shows the values of compressive strength of concrete at 28 days.

5.2 Flexural Strength

The flexural strength of standard concrete without bacteria is 6.56 N/mm^2 after 28 days of curing period, while compared to standard concrete, flexural strength of bacterial concrete is higher for every time when different concentrations of bacteria incorporated with concrete mixture to make self-healing concrete with different bacterial species. In maximum cases, concrete gains more than 100% of target strength

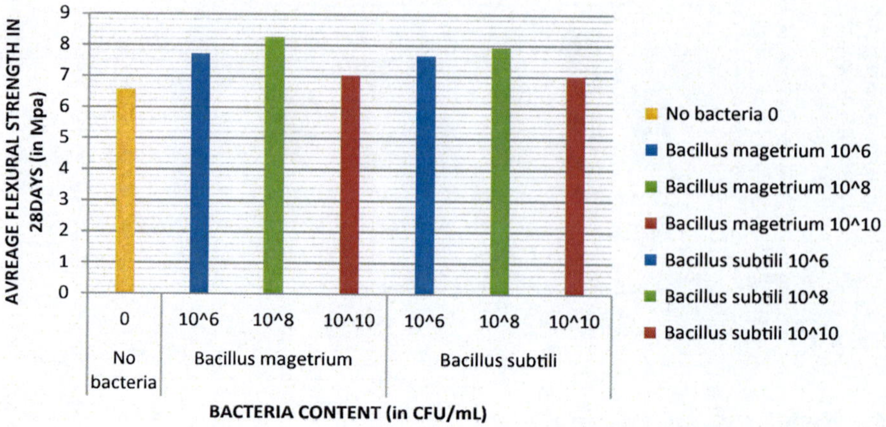

Fig. 2 xxx

at the age of 28 days. Maximum 26 and 21.12% of flexural strength increase for 28 days for self-healing concrete with ***bacillus megaterium*** and ***bacillus subtilis*** of 10^8 cfu/mL, respectively, compared with normal concrete. Figure 2 shows the values of flexural strength of concrete at 28 days. Both of those bacteria used in this work with different concentrations have the potential to increase in flexural strength than standard concrete at 28 days. But concrete with 10^8 cfu/mL gives more strength as compared to standard and other different concentrations of both bacteria. It also shows that *bacillus megaterium* of 10^8 cfu/mL concentration gives more flexural strength results among all other variations. Figure 2 shows the values of flexural strength of concrete at 28 days.

5.3 Split Tensile Strength

The split tensile strength of standard concrete without bacteria is found to be 4.22 N/mm^2 after 28 days of curing period, while compared to standard concrete, split tensile strength of bacterial concrete is higher for every time when different concentrations of bacteria were incorporated with concrete mixture to make self-healing concrete with different bacteria and their different cell concentrations. In maximum cases, concrete gains more than 100% of target strength at the age of 28 days. Maximum 16.33 and 13.64% of split tensile strength increase for 28 days for self-healing concrete with *bacillus megaterium* and *bacillus subtilis* of 10^8 cfu/mL, respectively, compared with normal concrete. Figure 2 shows the values of split tensile strength of concrete at 28 days. Both of those bacteria used in this work with different concentrations have the potential to increase in split tensile strength than standard concrete at 28 days. But concrete with 10^8 cfu/mL gives more strength as compared to standard and other different concentrations of both bacteria. It also shows that *bacillus*

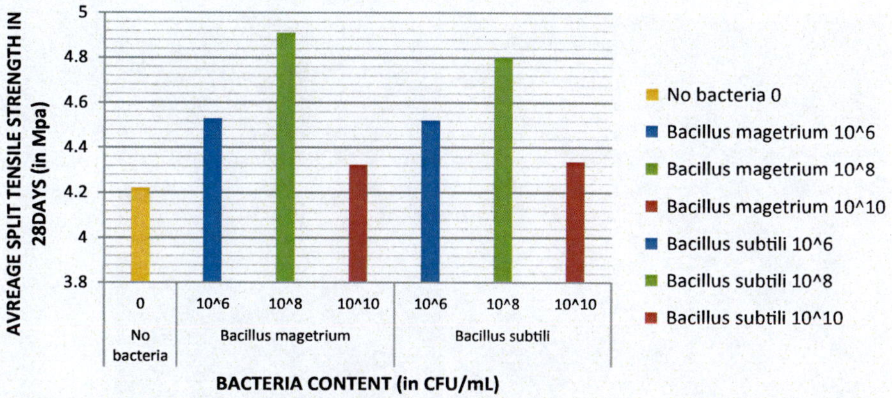

Fig. 3 xxx

megaterium of 10^8 cfu/mL concentration gives more split tensile strength results among all variations. Figure 3 shows the values of split tensile strength of concrete at 28 days.

5.4 *Visualize Crack Healing*

Normal concrete cubes of 150 mm size casted and cured in water tank for 7 days and then taken out for control loading. The loads were applied on the cubes till the crack appears on the surface of concrete cubes. Then, the cracked cubes are taken out for surface treatment. Bacterial liquid suspension of the same concentration of *bacillus subtilis* and *bacillus megaterium* (10^8 cfu/mL) were taken for giving surface treatment by injecting the bacterial suspension on those cracked cubes for conforming the metabolic process of bacteria to heal the crack. It was observed that the cracks took 3 to 4 weeks for healing which can be seen by naked eye or photos. Figure 4 shows the image of cracked cues and treated crack cubes.

6 Conclusion

Present study provides the insight for a new self-healing technological method based on different bacteria with different cell concentrations used for improving the strength of structural concrete. It is clear that MICP of bacteria has the potential to enhance the resistance of concrete toward the process of degradation. The consequences of this research provide a significant contribution toward the understanding of the effects of bacteria on the performance of structural concrete. By observing the results of mechanical properties of concrete matrix, it is concluded that using 10^8 cfu/mL

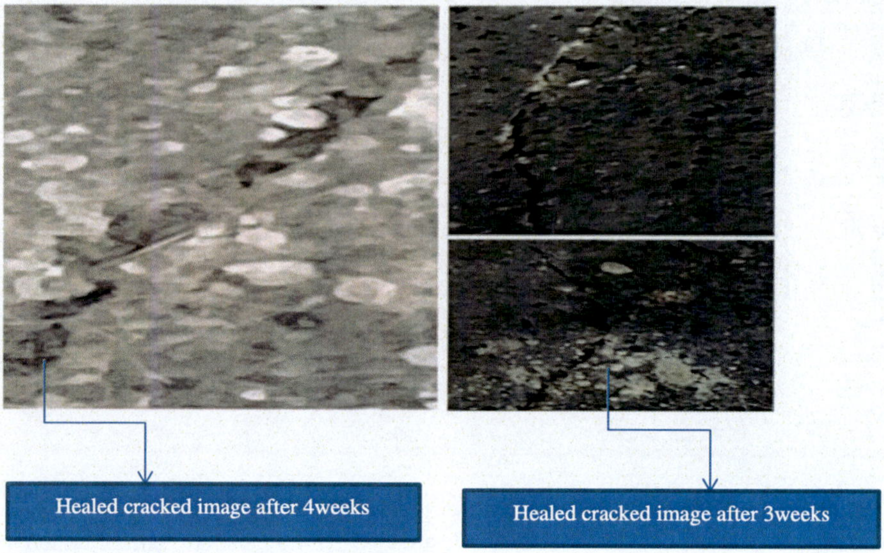

Fig. 4 xxx

concentration of *bacillus megaterium* bacteria gives higher performance among all variations taken in research.

Bibliography

1. Wang JY, Snoeck D, Van Vlierberghe S, Verstraete W, De Belie N (2014) Application of hydrogel encapsulated carbonate precipitating bacteria for approaching a realistic self-healing in concrete. Constr Build Mater 68:110–119
2. Tziviloglou E, Wiktor V, Jonkers HM, Schlangen E (2016) Bacteria based self healing concrete to increase liquid tightness of crack. Constr Build Mater 122:118–125
3. De Muynck W, Cox K, De Belie N, Verstraete W (2008) Bacteria carbonate precipitation as an alternative surface treatment for concrete. Constr Build Mater 22:875–885
4. De Muynck W, Debrouwer D, De Belie N, Verstraete W (2008) Bacterial carbonate precipitation improves the durability of cementitious materials. Cem Concr Res 38:1005–1014
5. Khaliq W, Ehsan MB (2016) Crack healing in concrete using various bio influenced self-healing techniques. Constr Build Mater 102:349–357
6. Kumari C, Das B, Jayabalan R, Davis R, Sarkar P, Effect of Nonureolytic Bacteria on Engineering Properties of Cement mortar. https://doi.org/10.1061/(asce)mt.19435533.0001828
7. Siddique R, Chahal NK (2011) Effect of ureolytic bacteria on concrete properties. Constr Build Mater 25:3791–3801
8. Milla J, Hassan MM, Rupnow T, Evaluation of self healing concrete with microencapsulated calcium nitrate. https://doi.org/10.1061/(asce)mt.19435533.0002072
9. Luo M, Qian CX, Li RY (2015) Factors affecting crack repairing capacity of bacteria based self healing concrete. Constr Build Mater 87:1–7

10. Siddique R, Nanda V, Kunal, Kadri EH, Khan MI, Singh M, Rajor A (2016) Influence of bacteria on compressive strength and permeation properties of concrete made with cement baghouse filter dust. Constr Build Mater 106:461–469
11. Chahal N, Siddique R, Rajor A (2012) Influence of bacteria on the compressive strength, water absorption and rapid chloride permeability of fly ash concrete. Constr Build Mater 28:351–356
12. Chahal N, Siddique R, Rajor A (2012) Influence of bacteria on the compressive strength, water absorption and rapid chloride permeability of concrete incorporating silica fume. Constr Build Mater 37:645–651
13. Luo M, Qian C-x (2016) Influence of Bacteria-based self healing agents on cementitious materials hydration kinetics and compressive strength. Constr Build Mater 121:659–663
14. Siad H, Lachemi M, Sahmaran M, Khandaker M, Hossain A, Mechanical, physical, and self healing behaviors of engineered cementitious composites with glass powder. https://doi.org/10.1061/(asce)mt.19435533.0001864
15. Kim HK, Park SJ, Han JI, Lee HK (2013) Microbially mediated calcium carbonate precipitation on normal and lightweight concrete. Constr Build Mater 38:1073–1082
16. Andalib R, Majid MZA, Hussin MW, Ponraj M, Keyvanfar A, Mirza J, Lee H (2016) Optimum concentration of Bacillus megaterium for strengthening structural concrete. Constr Build Mater 118:180–193
17. Siddique R, Singh K, Kunal, Singh M, Corinaldesi V, Rajor A (2016) Properties of bacteria rice husk ash concrete. Constr Build Mater 121:112–119
18. Wiktor V, Jonkers HM (2011) Quantification of crack-healing in novel bacteria based self-healing concrete. Cem Concr Compos 33:763–770
19. Nosouhian F, Mostofinejad D, Reducing permeability of concrete by bacterial mediation on surface using treatment gel
20. Chen H, Qian C, Huang H (2016) Self healing cementitious materials based on bacteria and nutrients immobilized respectively. Constr Build Mater 126:297–303
21. Xu J, Wang X (2018) Self healing of concrete cracks by use of bacteria containing low alkali cementitious material. Constr Build Mater 167:1–14
22. Wang JY, Soens H, Verstraete W, De Belie N (2014) Self healing concrete by use of microencapsulated bacterial spores. Cem Concr Res 56:139–152
23. Achal V, Mukherjee A, Sudhakara Reddy M, Microbial Concrete way to enhance the durability of building structures. https://doi.org/10.1061/(asce)mt.19435533.0000159
24. Nosouhian F, Mostofinejad D, Hasheminejad H, Concrete durability improvement in a sulfate environment using bacteria. https://doi.org/10.1061/(asce)mt.19435533.0001337
25. Arce GA, Hassan MM, Mohammad LN, Rupnow T, Characterization of self healing processes induced by Calcium Nitrate Microcapsules in cement mortar. https://doi.org/10.1061/(asce)mt.19435533.0001717
26. Gilford J, Hassan MM, Rupnow T, Barbato M, Okeil A, Asadi S, Dicyclopentadiene and sodium silicate microencapsulation for self healing of concrete. https://doi.org/10.1061/(asce)mt.19435533.0000892
27. Xu J, Yao W, Jiang Z, Non-ureolytic bacterial carbonate precipitation as a surface treatment strategy on cementitious materials
28. Van Tittelboom K, De Belie N, De Muynck W, Verstraete W (2010) Use of bacteria to repair cracks in concrete. Cem Concr Res 40:157–166

Use of Autoclaved Fly-Ash Aggregates in Concrete Mixture

Biswaroop Ghosh and Ashoke Kumar Rath

Abstract In all developing countries, coal is used in a large scale. Fly ash, which comes out as a by-product of ignition of coal, is a very dangerous material to handle and it makes a scowl to our mother nature, along with using it as a low-land dumping, fly-ash bricks and replacement of fine aggregate. It also can be used as a replacement for regular aggregate. The grade of fly ash is determined by automated spectroscopy. The fly ash was agglomerated with cement using different proportions. Then, the pellets are produced by using Disc Pelletiser. These pellets got hardened by autoclave technique. The produced aggregates were tested by different aggregate testing procedures. This artificial aggregate will be lightweight having specific gravity less than that of 12.5 mm gravel (3.1) though it has high impact and crushing strength than that of normal aggregates. At the same time, it will also solve the problem of wasting natural resources such as hills.

Keywords Autoclaved technique · Fly-ash pellets

1 Introduction

Coal is a very largely used material for industries like thermal power plant and iron industry. There, a large quantity of coal burns every day. The fly ash got collected around 800 million tonne in our world every year, and it becomes a threat to our society. To solve this problem, fly ash is nowadays used for dumping in low-land, fly-ash bricks and as a partial replacement of sand in concrete. However, fly ash may also be used in lightweight artificial aggregate production. As fly ash has a good pozzolanic property, it mixed with cement (used as a binder) can produce a solid aggregate. In concept of making Green Cities, this aggregate can be used in making buildings, road and other sectors also where aggregates are required [1].

B. Ghosh (✉) · A. K. Rath
School of Civil Engineering, KIIT DU, Bhubaneswar, India
e-mail: biswaroopghosh1996@gmail.com

A. K. Rath
e-mail: akrathfce@kiit.ac.in

B. B. Das et al. (eds.), *Recent Developments in Sustainable Infrastructure*, Lecture Notes in Civil Engineering 75, https://doi.org/10.1007/978-981-15-4577-1_37

The reuse of fly ash in making coarse aggregate replacement is very essential now as it is environment-friendly and economical [2]. The fly ash and cement with a proportion of 10:1, 5:1, 5:2, 5:3, 5:4 and 5:5 are mixed and passed in disc pelletiser to make green pellets. The green pellets are passed through autoclave to make solid surface. Though the aggregates are formed and are circular in shape, these have more pozzolanic properties than normal coarse aggregates and it also can be formed in all sizes such as 4 mm – 40 mm. Due to higher initial cost, these are not widely used in the world. However, there is also a process to make the aggregates: sintering and cold-bonded. Cold-bonded aggregates are produced using minimum energy for curing than sintering [3].

2 Literature Review

N. U. Kockal and T. Ozturan experimental results showed that sintered or cold-bonded lightweight fly-ash aggregates are giving lightweight concretes, having similar characteristics with the normal aggregate concretes. The use of lightweight aggregates in replacement of normal aggregates in concrete production lowers down the strength and stiffness as this aggregate is highly porous but the self-weight of concrete was much lower. However, in respect of concrete sintered, fly-ash aggregate is more permeable than normal aggregate but cold-bonded aggregate gives less permeability than normal one [4].

Manikandan and Ramamurthy suggested that sintering technique aggregates give more strength in respect of cold-bonded aggregate [5].

Priyadarshiny, Ganesh and Santhi have noticed that with the increase in curing period cold-bonded aggregates give more strength than that of other aggregates making techniques [6].

In conventional concrete, weights of aggregates are the main factors according to IS 9142-1979 to show if the aggregate is lightweight or not. The density of concrete using 12.5 mm regular coarse aggregate is of 2100–2800 kg/m^3. This high self-weight of the concrete using regular coarse aggregate may cause the use of high structural members such as reinforcements which causes high economic value to structure. As the aggregate is lightweight in nature, the self-weight of concrete comes up to 1500–1800 kg/m^3 which is much lower than normal aggregate concrete. Thus, we can conclude that it can reduce the dead load than the original one. Hence, economy may be achieved in the design of supporting structural elements. This paper is reviewed on the suitability of using autoclaved fly-ash aggregate strength properties.

3 Fly-Ash Aggregates

Fly-ash aggregates are round-shaped lightweight aggregates and can replace coarse aggregate. It can be produced by (1) sintered, (2) cold-bonded and (3) autoclaved

Fig. 1 Green pellets

technique. In this paper, I have discussed the autoclave technique to make the fly-ash aggregate. This aggregate can be used in all the places as a replacement of normal aggregate (Fig. 1).

4 Materials

4.1 Cement

Portland Pozzolana Cement (P.P.C.) is used in this project according to IS: 12269-1987 [7].

4.2 Fly Ash

Fly ash produced from the burning of sub-bituminous coal is of Class C grade. It has good pozzolanic character. Class C fly ash when mixed with water became harden and over time its strength gets increased. As Class C fly ash has self-cementing property, it gives more strength to the aggregate. It has high calcium, alkali and sulphate which are generally higher in Class F fly ashes.

4.3 Water

Regular water, which was available in lab, was used for making fly-ash pellets, potassium (Table 1).

Table 1 Chemical analysis of cement and fly ash

	Calcium Oxide	Silicon Oxide	Aluminium Oxide	Ferric Oxide	Sodium Oxide	Potassium Oxide	Magnesium Oxide	Density $\frac{\text{Gram}}{\text{CubicCentimeter}}$
Cement	61.45	18.35	2.52	2.12	0.62	0.5	1.54	3.03
Fly ash	1.56	52.75	30.75	8.68	1.38	2.12	1.94	2.78

5 Production Technique of Autoclaved Fly-Ash Pellets:-

5.1 Process of Making Green Pellets

The grain size of the pellets can be produced from 4 mm to 40 mm size by means of agglomeration process as per IS 383-1970 [8]. The process is used to making green fly-ash pellets called as palletisation process. Different pelletisers such as disc type and drum type can be used in making green pellets. In this paper, we used disc-type pelletiser in order to produce green pellets. The efficiency of the production of aggregate depends on various factors, i.e. speed of revolution of pelletiser disc, moisture content, angle of pelletiser disc and duration of pelletisation. The specifications of the disc pelletiser which is used here are as follows [9]:

- Diameter of the disc—0.57 m.
- Depth of the disc—0.25 m.
- Horizontal angle of the disc—45°.
- The speed of rotation of the disc—55 rpm (Fig. 2).

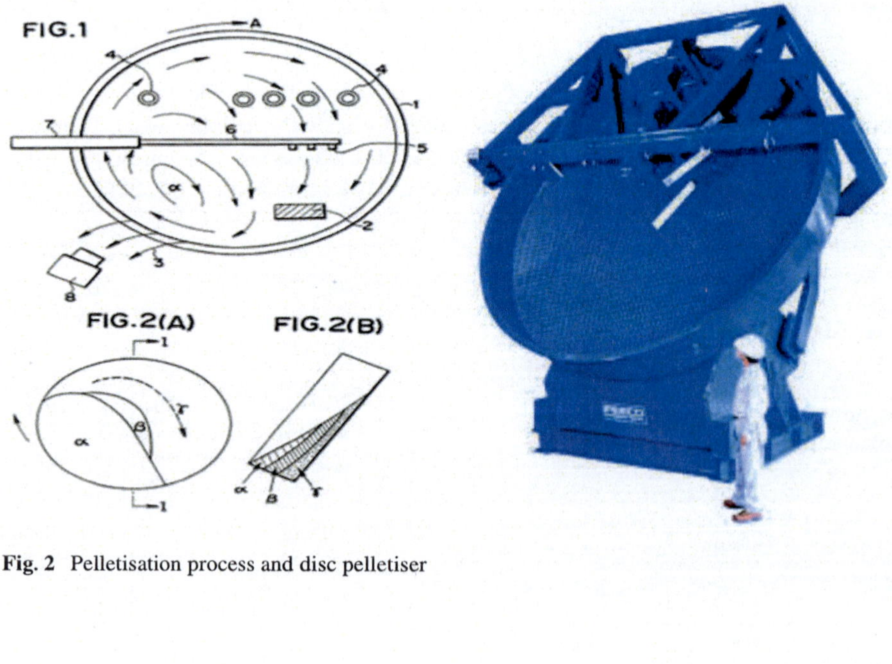

Fig. 2 Pelletisation process and disc pelletiser

5.2 *Hardening of Pellets*

Green pellets are passed through autoclave having 120–150° and 15–20 psi pressure for two to three hours. The final product coming from autoclave can be used in the field as it is stronger than green pellets (Fig. 3).

6 Comparison of Physical Properties of Aggregate

The fly-ash aggregates are porous materials, so for hardening the pellets we have to use cement as binder material. The hardening of the pellets can be done by following the process: cold bonding, sintering and autoclaving. Green pellets can be hardened by sintering, 28 days water curing and autoclaving. Among all these three processes, autoclaving is the most effective process by which strength of the pellets gets increased in a short time span. So, to increase the strength of the pellets after removing them from the pelletiser, we have to pore it into autoclave machine with 120–150° and 15–20 psi pressure for 2–3 h. Study the properties of fly-ash pellets [10].

In this experiment, aggregates of size 10–12.5 mm were used for checking strength of the autoclaved fly-ash aggregates in comparison with regular coarse aggregate. According to IS 2386 part 3 and part 4, different tests were organised to check the strength of the artificial fly-ash aggregate [11] (Table 2 and Figs. 4, 5, and 6).

Fig. 3 Autoclave machine and autoclaved fly-ash pellet

Table 2 Experiments on normal aggregate and autoclaved fly-ash pellets

	UNITS	Normal aggregate (12.5 mm)	Fly-ash: Cement (10:1)	Fly-ash: Cement(5:1)	Fly-ash:Cement (5:2)	Fly-ash:Cement (5:3)	Fly-ash :Cement (5:4)	Fly-Ash:Cement (5:5)	Allowable limit	Reference
Water absorption	%	2.38	10.12	9.79	8.55	8.21	7.87	7.53		IS 2386 Part-III
Crushing value	%	35.65	28.68	30.79	33.06	35.33	38.7	38.78	<45%	IS 2386 Part-IV
Impact value	%	21.720	31.330	29.640	27.870	26.530	24.760	22.810	<45%	IS 2386 Part-IV
Abrasion value	%	3.65	5.12	4.45	4.23	4.18	3.95	3.87	<50%	IS 2386 Part-IV
Bulk density	kg/m^3	1750	1100	1050	980	920	870	830		IS 2386 Part-III
Specific gravity		2.64	1.78	1.8	1.86	1.9	2.01	2.2		IS 2386 Part-III
Size	mm	4.75–20	4.75–20	4.75–20	4.75–20	4.75–20	4.75–20	4.75–20		
Shape		Angular	Circular	Circular	Circular	Circular	Circular	Circular		

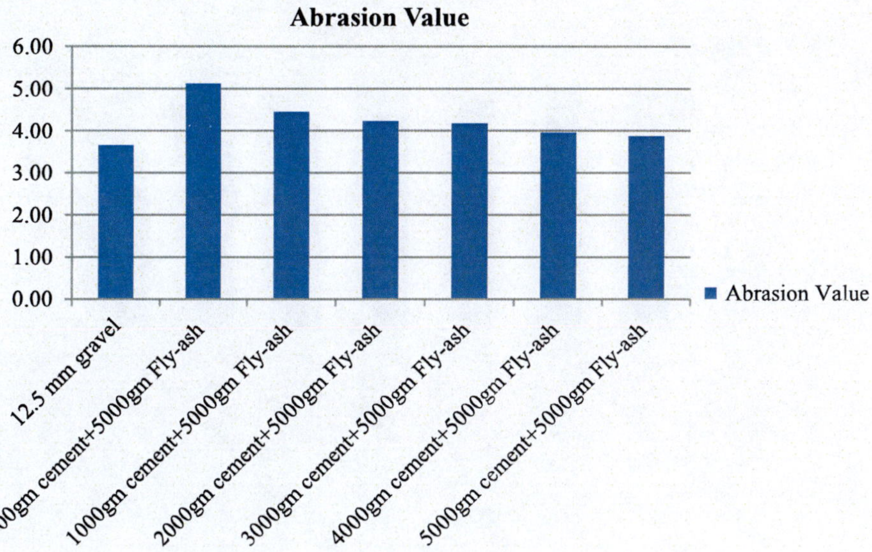

Fig. 4 Abrasion value test of different autoclaved fly-ash pellets

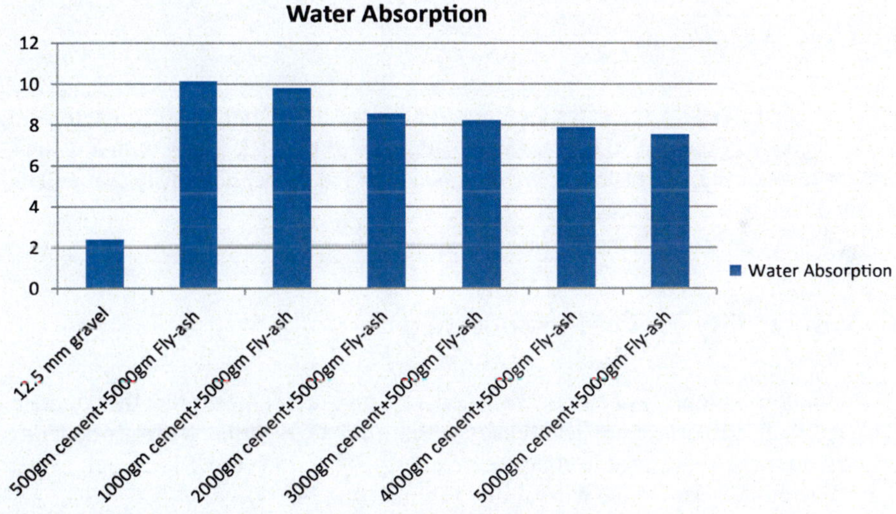

Fig. 5 Water absorption test of different autoclaved fly-ash pellets

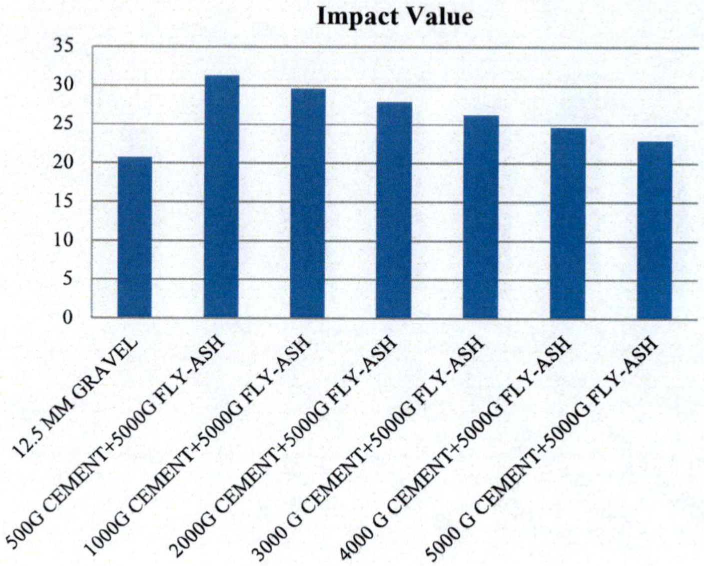

Fig. 6 Impact test of different autoclaved fly-ash pellets

7 Uses in Concrete

The produced fly-ash aggregates are used in M-10 to M-25 and M-40 concrete, and mix design was done. In addition, we use admixtures of 0.6% along with a water–cement ratio of 0.5. Extra 1% of water was added for absorption of fly-ash pellets produced by autoclaved technique.

8 Workability Test of the Specimens

Workability is a property of raw or fresh concrete mixture. It means that the concrete can be placed and can be compacted easily without any segregation. It mainly depends on the amount of water used in concrete mix designs (ACI 1990b). Slump height test is used here to determine the workability (Table 3 and Fig. 7).

9 Comparison of Compressive Strength in Concrete Using Fly-Ash Pellets and Normal Aggregate

See Table 4.

Table 3 Results of slump test

Concrete mixture	Workability (Slump height in mm)				
	M-10	M-15	M-20	M-25	M-40
Conventional concrete	45	47.07	70	81.23	110
Fly-ash pellets concrete (10:1)	46.98	50.66	75	81.98	115
Fly-ash pellets concrete (5:1)	47.23	54.24	75.6	82.58	120
Fly-ash pellets concrete (5:2)	47.05	57.82	78	83.28	128
Fly-ash pellets concrete (5:3)	47.98	61.41	79.23	83.95	135
Fly-ash pellets concrete (5:4)	48.01	64.99	80.32	84.63	145
Fly-ash pellets concrete (5:1)	48.23	68.57	80.52	85.30	150

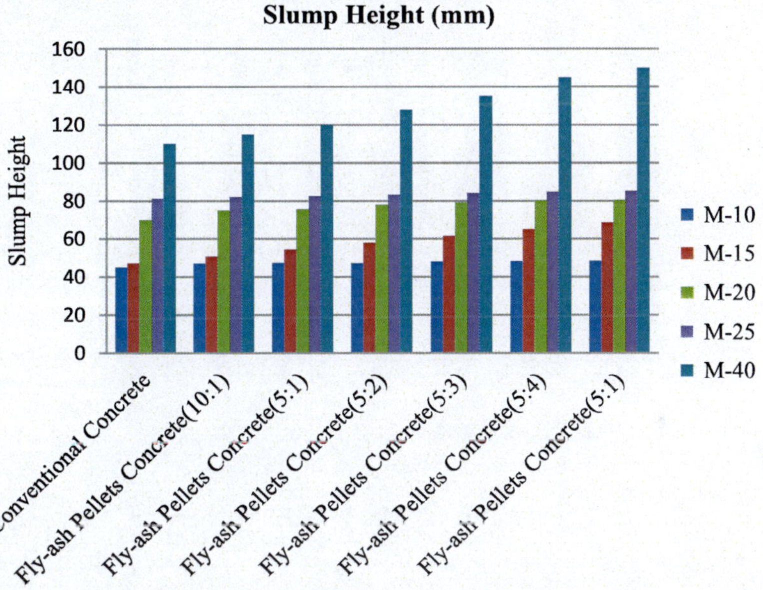

Fig. 7 Slump height test of different autoclaved fly-ash pellets using different concrete mixes

9.1 Cost Efficiency Test in Respect of (0.15 m × 0.15 m × 0.15 m) Cube

The market rate of cement, sand and coarse aggregate are discussed (Table 5 and Fig. 8).

In the above comparison, it is clearly shown that the cost of autoclaved fly-ash aggregate is significantly lower than that of normal aggregate which we are having. The cost of fly-ash pellets is around 30% lower than that of normal aggregate. As

Table 4 The compressive strength of cube and cylinder

Sl. no	Specimen	Days	Grade	Conventional concrete (mpa)	Concrete using fly-ash aggregate					
					Fly-ash: Cement (10:1)	Fly-ash: Cement (5:1)	Fly-ash:Cement (5:2)	Fly-ash: Cement (5:3)	Fly-ash :Cement (5:4)	Fly-ash:Cement (5:5)
1	Cube	28	M-10	16.23	17.45 mpa	17.98 mpa	18.23 mpa	18.75 mpa	19.01 mpa	19.78 mpa
			M-15	21.08	22.45 mpa	22.96 mpa	23.18 mpa	23.78 mpa	24.76 mpa	25.12 mpa
			M-20	27.26	28.13 mpa	28.54 mpa	29.65 mpa	29.75 mpa	29.98 mpa	30.23 mpa
			M-25	31.76	32.25 mpa	32.98 mpa	33.13 mpa	33.53 mpa	33.98 mpa	34.43 mpa
			M-40	62.96	62.98 mpa	63.45 mpa	63.66 mpa	63.78 mpa	63.96 mpa	64.08 mpa
2	Cylinder	28	M-10	15.78	16.13 mpa	16.53 mpa	17.02 mpa	17.15 mpa	17.52 mpa	18.56 mpa
			M-15	20.78	21.34 mpa	22.27 mpa	22.78 mpa	23.01 mpa	23.57 mpa	24.78 mpa
			M-20	26.6	27.01 mpa	27.46 mpa	27.75 mpa	28.31 mpa	28.76 mpa	29.21 mpa
			M-25	31.45	32.05 mpa	32.68 mpa	33.08 mpa	33.23 mpa	33.78 mpa	34.03 mpa
			M-40	61.52	62.24 mpa	62.53 mpa	62.89 mpa	62.97 mpa	63.02 mpa	63.15 mpa

Table 5 Table shows costing difference of conventional aggregate and fly-ash pellets

		Conventional concrete	Fly-ash pellets concrete					
		12 mm Gravel	10:1	5:1	5:2	5:3	5:4	1:1
S. no	Material	Rate (Rs. per kg)	Rate (Rs. per kg)	Rate (Rs. per kg)	Rate (Rs. per kg)	Rate (Rs. per kg)	Rate (Rs. per kg)	Rate (Rs. per kg)
1	Cement	17	17	17	17	17	17	17
2	Fine aggregate	5	6	6	6	6	6	6
3	Coarse aggregate	32	15	18	20	22	24	25

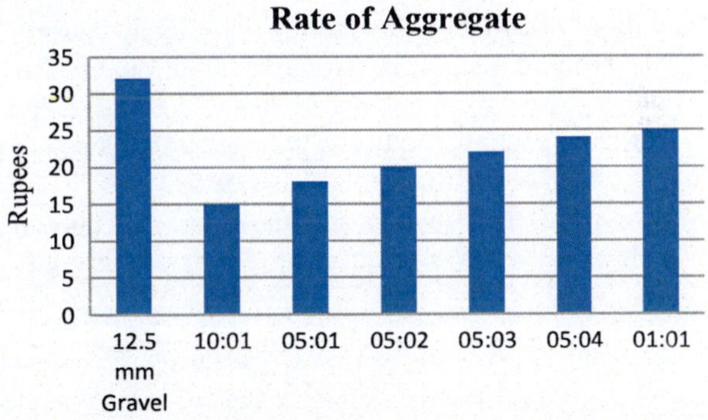

Fig. 8 Cost of different Autoclaved Fly-ash pellets

the aggregate is lightweight and cost efficient so we can say it can give us a better solution in order to replace our normal aggregate. Thus, this aggregate can be used in larger scale construction purpose.

10 Conclusions of the Paper

According to the above experiments and tests performed, some conclusions can be drawn. They are discussed below:

- According to regular coarse aggregate, autoclaved fly-ash pellets give good result as per physical tests done on those aggregates according to IS-2386 part 3 and part 4.
- The new aggregates made with fly ash are giving the great results in crushing value test and abrasion test, by which we can conclude that the fly-ash aggregates produced by autoclave technique has strength bearing capacity similar to that of normal coarse aggregate in the concrete mix.

- The fly-ash aggregates are light in weight than that of normal aggregates. We can conclude it from the variation shown in specific gravity test performed on these aggregates.
- The water absorption is higher than that of normal aggregate.
- In concrete mixture also, this aggregate shows a great strength in different grades of concrete.
- The cost of Autoclaved fly-ash pellets is much lower than that of regular aggregates.
- The source of the regular coarse aggregate is mainly rocks of hills. So with the increasing rate of construction all over the world, the uses of coarse aggregate are increased. The source of regular coarse aggregate is mainly the rocks of hilly areas. In order to get the coarse aggregate the destruction of hilly areas are growing day by day. This causes geological as well as environmental imbalance. For this, earthfalls in hilly region are getting a common issue day by day all over the world. In other way, we can also say that fly ash is the industrial waste having a lot of chemical components, which may cause a lot of physical hazards if they are exposed in open air without precaution. Fly ash of 800 million tonne is coming out in a year from different parts of the world. But only less than 30% is used in different sectors. So this aggregate can replace the coarse aggregate and can help to solve the environmental hazards as well.
- Among all three techniques, sintering, cold bonding and autoclaving, autoclaving is the most efficient technique to make the fly-ash pellets as a replacement.

References

1. Wu H, Sun P (2007) New building materials from fly ash-based lightweight inorganic polymer. Constr Build Mater 21(1):211–217
2. Chauhan SL, Bondre RA (2015) Partial replacement of sand with quarry dust in concrete. Int J Sci Res Publ 5(7):1–4
3. Graytee A, Sanjayan J, Pilot study on the development of synthetic lightweight aggregates. CRC for Low carbon living (p 22). http://www.lowcarbonlivingcrc.com.au/sites/all/files/publications_file_attachments/rp1020_synthetic_aggregate_report.pdf
4. Kockal NU, Ozturan T (2011) Durability of lightweight concretes with lightweight fly ash aggregates. Constr Build Mater 25(3):1430–1438. https://doi.org/10.1016/j.conbuildmat.2010.09.022
5. Manikandan R, Ramamurthy K (2008) Effect of curing method on characteristics of cold bonded fly ash aggregates. Cem Concr Compos 30(9):848– 853. https://doi.org/10.1016/j.cemconcomp.2008.06.006
6. Priyadharshiny P, Ganesh GM, Santhi AS (2011) Experimental study on cold bonded fly ash aggregates. Int J Civ Struct Eng 2(2):493–501
7. 53 grade Ordinary Portland Cement-specification, IS: 12269-1987, 1987
8. Specifications for coarse and fine aggregates, IS: 383-1970, 1970
9. Harikrishnan KI, Ramamurthy K (2006) Influence of pelletization process on the properties of fly ash aggregates. Waste Manag 26(8):846–852. https://doi.org/10.1016/j.wasman.2005.10.012

10. Bijen JM (1986) Manufacturing processes of artificial lightweight aggregates from fly ash. Int J Cem Compos Lightweight Concr 8(3):191–199. https://doi.org/10.1016/0262-5075(86)90040-0
11. Manikandan R, Ramamurthy K (2007) Influence of fineness of fly ash on the aggregate pelletization process. Cem Concr Compos 29(6):456–464. https://doi.org/10.1016/j.cemconcomp.2007.01.002

Optimization of Percentage of AR Glass Fibre Addition to Fly-Ash-Based Self-consolidating Concrete

Asheena Sunny and Nitin Gusain

Abstract Self-consolidating concrete (SCC) is usually referred as flowable concrete which can consolidate without having segregation of aggregates. It uses chemical and mineral admixtures to achieve its properties. The present experimental study investigates the influence of glass fibres with varying proportions (i.e. 0, 1, 1.5, 2 and 2.5% by weight of cement) on rheological and mechanical properties of fly-ash-based self-consolidating concrete. The experiments were conducted by considering water–powder ratio of 0.3656. The guidelines of EFNARC were followed for the mix designing of SCC. Fresh properties were determined by L-Box, Slump Flow and V-Funnel tests and hardened properties were assessed by 28th day compressive, flexural and split tensile strength tests.

Keywords Glass fibre · Fly ash · Superplasticizer · Self-consolidating concrete

1 Introduction

Self-consolidating concrete has a tendency to possess compaction by its self-weight and uniformly fill the entire form work even though there is an existence of congestion in reinforcement. SCC was developed in 1980s by Okamura in Japan due to shortage of labours. Better durability, environment-friendly, lower noise pollution, enhanced aesthetics and cut back in time needed for placement of concrete are the benefits of wide usage of SCC in construction industries. The key characteristics are attained with the aid of chemical admixtures such as superplasticizer and mineral admixtures (like fly ash, rice husk ash, metakaolin, etc.). Unique properties of self-consolidating concrete make it significant in constructability, economic and engineering advantages.

Self-consolidating concrete contains a large amount of powder (particle size < 0.125 mm) which is required to maintain sufficient rheology of concrete which

A. Sunny (✉) · N. Gusain
School of Civil Engineering, KIIT-DU, Bhubaneswar, India
e-mail: Asheena.sunnyfce@kiit.ac.in

N. Gusain
e-mail: nitingusain786@gmail.com

B. B. Das et al. (eds.), *Recent Developments in Sustainable Infrastructure*, Lecture Notes in Civil Engineering 75, https://doi.org/10.1007/978-981-15-4577-1_38

includes segregation resistance, filling and passing abilities of fresh concrete. As the requirement in quantity of cement increases, it causes increased heat of hydration and production cost. From the literature study, it was revealed that addition of fly ash can make improvement in slump flow of SCC by reducing the cost [1]. Also, fly ash utilization in concrete can minimize the hydration heat of cement resulting in reduced cracking [2].

Conventional plain concrete has innumerous micro-cracks within it, and under the application of external load cracks propagate due to weak in tension. However, insertion of fibres into the concrete matrix can enhance the flexural and tensile strengths of concrete [3].

The fibres often gettable are steel, glass, polypropylene, coir, etc. Steel fibres improve ductility, toughness and flexural strength. Increased weight and corrosion damage of steel fibres are its drawbacks, and hence alkali resistant glass fibre, which is recently developed overcomes the defects and can be effectively used in concrete.

The principle aim of the current research work focuses on workability and mechanical properties of fly-ash-based SCC with varying dosages of glass fibre (0, 1, 1.5, 2 and 2.5%) by the weight of cement.

2 Ingredients for the Production of SCC

2.1 Cement

53 grade OPC cement was used and the properties obtained (Table 1) are within the requirement range specified by IS 269–2015.

2.2 Mineral and Chemical Admixtures

Fly ash (Class F) with a specific gravity 2.275 procured from Lafarge RMC plant, Bhubaneswar was used as the mineral admixture.

Superplasticizers are chemical admixtures which improves workability with decline in water content. Superplasticizer used in the study was Cryso Fluid Optima K913 with specific gravity 1.13(+–0.02).

2.3 Aggregates

The fine aggregates (FA) and coarse aggregates (CA) are acquired from structural laboratory, KIIT-DU, Bhubaneswar and confirmed with the specifications provided by IS: 383 (1983). Coarse aggregates used in the current study are of size 10 mm

Table 1 Test results for ordinary Portland cement 53 grade

Test	Values obtained	Requirement range
Setting time—Initial Setting time—Final	165 min 240 min	30 min (minimum) 600 min (maximum)
Fineness	295 m^2/kg	225 m^2/kg (minimum)
Soundness by Lechatilier expansion	1.0 mm	10 mm (maximum)
Soundness by autoclave	0.074%	0.8% (maximum)
Specific gravity	3.15	–
Compressive strength		
72 + 1 h	38 N/mm^2	27 N/mm^2 (minimum)
168 + 2 h	48 N/mm^2	37 N/mm^2 (minimum)
672 + 4 h	60 N/mm^2	53 N/mm^2 (minimum)

Table 2 Experimental observations of coarse and fine aggregates

Particulars	Coarse aggregates(CA)	Fine aggregates(FA)
Fineness modulus	6.18	2.46
Specific gravity	2.68	2.63
Water absorption (percentage)	0.4	0.85

(retaining), combination of 12.5 and 16 mm (passing). The experimental results are stated in Table 2.

2.4 Glass Fibres

AR (alkali resistant) glass fibres of length 12 mm are taken for the experimental study. The property index of glass fibres used is specified in Table 3.

Table 3 AR glass fibre properties

Density	Tensile strength	Elastic modulus	Elongation at break	Water absorption
2.53 g/cm^3	1950–2050 MPa	43–50 GPa	7–9%	<0.1%

Fig. 1 Mould after filling the SCC

3 SCC Mix Design

Many test trials were conducted for regulating dosages of superplasticizer and percentage of fly ash replacement. The guidelines of EFNARC were followed for the mix designing of SCC. The self-compactibility was achieved at 15% replacement of cement with fly ash by keeping superplasticizer dosage of 1.44% of cementitious materials and water–powder ratio of 0.3656. In the present study, coarse and fine aggregate quantities were taken as 897 and 847 kg/m^3, respectively.

4 Casting of Moulds

Fly ash and cement were thoroughly blended in separate containers and then the cementations powder, fine and coarse aggregates poured in concrete mixer as per calculated quantity and dry mixing was done till uniform mix was ensured. Finally, superplasticizer was mixed with water (as per calculation) and added with dry mixture in a mixer. After the proper mixing, slump flow test, V-funnel and L-box test were conducted immediately. And then SCC mix was placed in cube, cylinder and prism moulds. Figure 1 shows the cube mould after filling SCC.

5 Tests Conducted on Fresh SCC

The slump flow, V-funnel and L-box test are the three experimental programmes conducted for obtaining rheological properties of SCC.

Table 4 EFNARC guidelines for fresh properties of concrete

Tests	Minimum–maximum range
Slump flow	650–800 mm
L-box	0.8–1.0
V-funnel	6–12 s

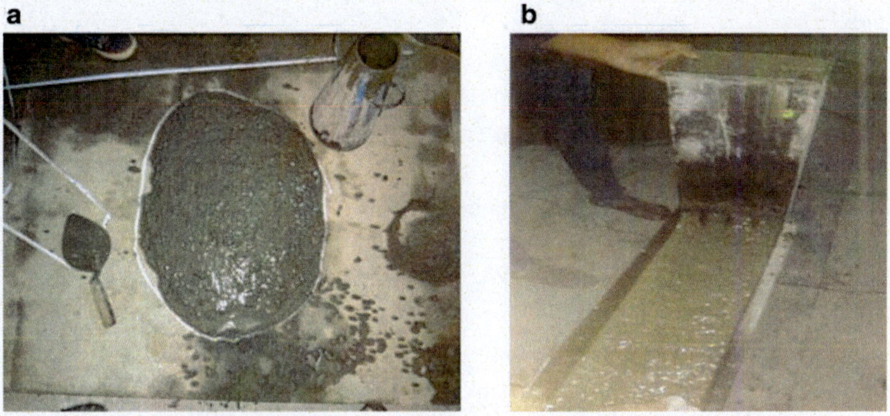

Fig. 2 **a** Slump cone test **b** L-box test

5.1 Slump Flow

Test requires a truncated cone-shaped mould having height of 300 mm and internal dimensions of 200 mm and 100 mm at base and top, respectively. Place a 700 mm square base plate on flat ground and align cone at centre. Pour concrete with the use of trowel and strike out extra concrete without tamping. Lift the cone in upright direction and let the concrete freely flow. Minimum and maximum diameters flowed by the SCC mix were measured and compared with EFNARC guidelines (Table 4). The flow diameter is a direct indication of filling capacity of fresh SCC. Figure 2a shows the flow diameter obtained after the test.

5.2 L-Box

This experiment assesses the passing and filling potential of SCC. Vertical part of L-box is filled with the concrete immediately after the mixing. There is a sliding gate between vertical and horizontal portions, keep it locked until the vertical portion is filled completely. Then uplift the gate and permit SCC to pass through horizontal surface. Tamping rod is used to measure the heights. The ratio of heights is calculated and compared with EFNARC guidelines (Table 4). Test figure is illustrated in Fig. 2b.

5.3 *V-Funnel*

This experiment programme is conducted to calculate flow time, i.e. flowing capability to ensure workability of SCC. The test requires a V-shaped funnel. Then the funnel is filled with concrete and made to flow under gravity. Time required for the flow was calculated. Time calculated should satisfy EFNARC guidelines given in Table 4.

6 Tests Conducted for Mechanical Properties of SCC

SCC was tested for modulus of rupture, compressive and split tensile strength according to IS 516(1959).

7 Test Result and Discussion

7.1 *Rheological Properties*

Slump flow, V-funnel and L-box test were performed in the laboratory on workable SCC. The results obtained are within the prescribed limits of EFNARC guidelines. The workability properties of SCC are shown in Table 5.

Slump flow of fly-ash-based SCC was found to be 708 mm and it was decreased with addition of glass fibre. The peak in L-box ratio was observed in SCC without glass fibres (0.985). It was found that with the addition of glass fibres, the L-box ratio was decreased and V-funnel time was increased.

It was noticed that there was no obstruction found because of addition of glass fibres to the SCC and all the mixes tested have adequate workability. Addition of fibres decreases the rheological property even though the test results were well within the range specified by EFNARC.

Table 5 Test result of fresh SCC

Mix designation	Fly ash %	Glass fibre %	Slump flow value (millimetre)	V-funnel value (second)	L-box value
GFRSCC0	15	0	708	9.1	0.985
GFRSCC1	15	1	695	9.8	0.980
GFRSCC2	15	1.5	693	10.1	0.978
GFRSCC3	15	2.0	690	10.2	0.921
GFRSCC4	15	2.5	685	10.5	0.892

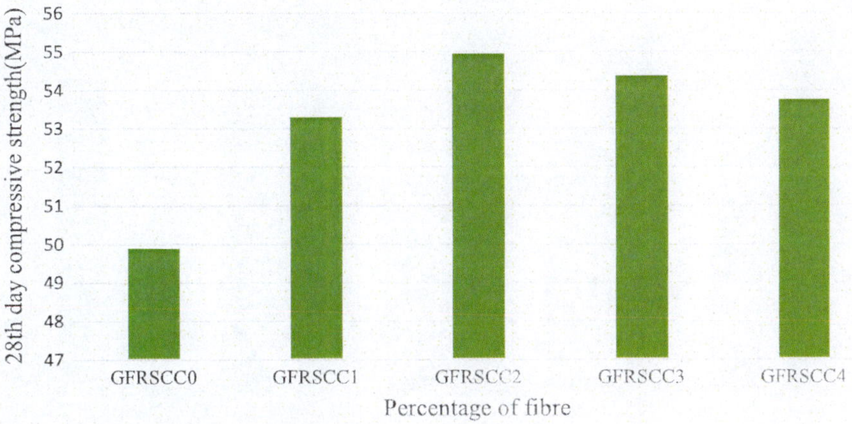

Fig. 3 28th day compressive strength results

7.2 *Hard Properties of SCC*

Hardened strength properties specifically flexural, splitting tensile and compressive strength of SCC with varying dosages of glass fibres were tested.

7.2.1 Compressive Strength

28th day test results are represented in Fig. 3. It was found that incorporation of glass fibres with 1, 1.5, 2 and 2.5% improves compressive strength by 6.8, 10.5, 9 and 7.76%, respectively.

Bridging action of fibres was observed during the time of appearance of tensile cracks. From the experiment, maximum compressive strength was obtained with the addition of 1.5% glass fibre content.

7.2.2 Split Tensile Strength

The 28th day test results of different SCC mixtures are represented in Fig. 4. It was observed that inclusion of 1, 1.5, 2 and 2.5% of glass fibre increases splitting tensile strength by 11.7, 23.5, 29.4 and 33.3%, respectively.

The Split tensile test results of SCC notably increase with incorporation of glass fibre. It has been observed that the noticeable improvement was obtained up to the addition of 1.5% of glass fibre.

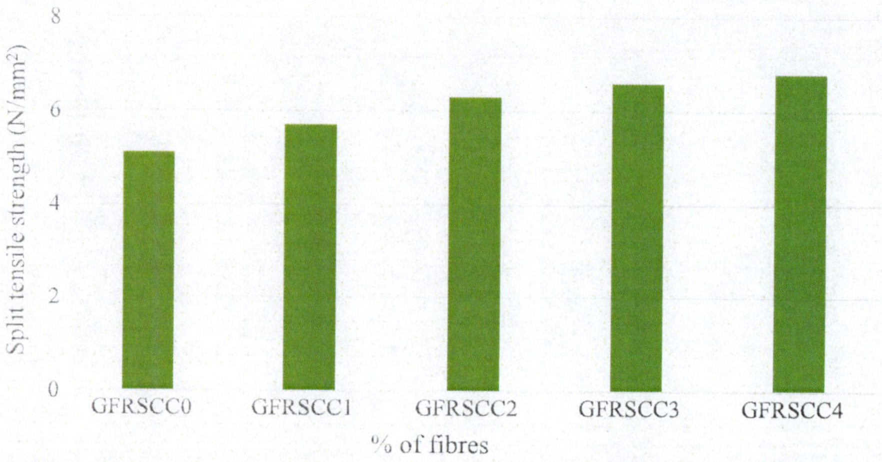

Fig. 4 Splitting tensile strength results

7.2.3 Flexural Strength

The 28th day result self-consolidating concrete with varying percentage of fibres is represented in Fig. 5. It was noted that incorporation of 1, 1.5, 2 and 2.5% of glass fibre increases the modulus of rupture by 22.8, 27.5, 24.6 and 20.22%, respectively.

The flexural strength of SCC was enhanced with incorporation of glass fibre but there is a decline in strength beyond 1.5% of fibres. Thus, 1.5% of glass fibre gives the peak flexural strength of SCC.

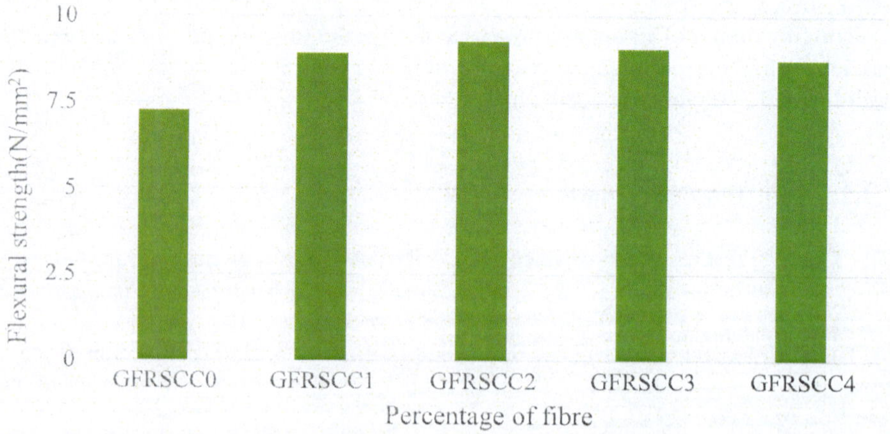

Fig. 5 Flexural strength results

8 Conclusions

Based on the obtained test results, salient features are concluded as follows:

(1) The self-compactibility was achieved at 15% replacement of cement with fly ash by keeping superplasticizer dosage of 1.44% of cementitious material and water–powder ratio of 0.3656.
(2) The workability of SCC with glass fibre reduces with increase in fibre content. However, test results obtained for entire mixes are well within the range recommended by EFNARC.
(3) Incorporation of fibres to SCC improves the 28th day compressive strength by 6.8 to 10.5% and the peak strength was achieved with 1.5% of glass fibre content.
(4) Inclusion of fibres to SCC enhances the split tensile strength by 11.7 to 33.3%, even though noticeable improvement was obtained up to 1.5% of glass fibres.
(5) Flexural strength of SCC was increased from 20.2 to 27.5% by inclusion of glass fibres. However, peak value was noticed at 1.5% glass fibres.
(6) Most enhanced hardened properties followed by inclusion of glass fibre were splitting tensile thereafter flexural and compressive strength.
(7) Therefore, it is concluded that optimum properties are obtained with 1.5% of glass fibre in SCC.

Bibliography

1. Yahia A,Tanimura M,Shimabukuro A,Shimoyama Y (1999) Effect of rheological parameters on self compactibility of concrete containing various mineral admixtures. In: Proc of the forst RILEM international symposium on self-compacting concrete pp 523–535
2. Kurita M, Nomura T (1998) Highly-flowable steel fiber-reinforced concrete containing fly ash. In: Malhotra VM (Ed) ACI. SP 178, pp 159–175
3. Mehta PK, PJM (2006) Monteiro concrete microstructure, properties and materials. Third edition, Tata McGraw-Hill Publishing Co Ltd, New Delhi p 478
4. Ahmad S, Umar A (2018) Rheological and mechanical properties of self-compacting concrete with glass and polyvinyl alcohol fibres. J Build Eng 17:65–74
5. Suresh Babu T, SeshagiriRao MV, Seshu R (2008) Mechanical properties and stress-strain behaviour of SCC with and without glass fiber. Asian J Civil Eng 9(5):457–472
6. Bhubaneshwari P, Murali R (2013) Strength characteristics of glass fiber on bottom ash based concrete. Int J Sci Environ Technol 2(1):90–102
7. Chandraumouli K, SrinivasRao P, Paneerselvam N, SeshadriSekhar T, Sravana P (2010) Strength properties of glass fiber concrete. ARPN J Eng App Sci 5(4):ISSN 1819-6608
8. Rama Mohan Rao P, Rao HS, Sekar SK (2010) Effect of glass fiber on flyash based concrete. Int J Civil Struct Eng 1(3):ISSN 0976-4399
9. Gornale A, Quadri SI, Quadri SM, Ali SMA, Hussaini SS (2012) Strength aspects of glass fiber reinforced concrete. Int J Sci Eng Res 3(7):ISSN 2229-5518
10. Seethapathi M, Senthil Kumar SRR, Chinnaraju K (2016) Study on fly ash based SCC with aggregate replacement. Asian J Res Soc Sci Human 6(9):254–271
11. ShahanaSheril PT (2013) SCC using fly ash and glass fiber. Int J Eng Res Technol 2(9)

12. Nair AP, Asim M (2014) Effect of nano silica on strength and durability properties of GFRSCC. Int J Sci Eng Res 5(7)
13. Deepak Raj A, MerginBenize M, Esther Daisy J, Sri Nikhil M (2014) Experimental methods on GFRSCC. IOSR J Mech Civil Eng 11(2):19–23
14. Ahmad S, Umar A, Masood A (2017) Properties of normal concrete, self-compacting concrete and glass fibre-reinforced self-compacting concrete: an experimental study. In: Proc of the forst RILEM International symposium on plasticity and impact mechanics, procedia engineering 173:807–813
15. Güneyisi E, Atewi YR, Hasan MF (2019) Fresh and rheological properties of glass fiber reinforced self-compacting concrete with nanosilica and fly ash blended. Constr Build Mater 211:349–362

Performance of Concrete with Marble Dust as Supplementary Material: A Review

Srishti Saha, Tribikram Mohanty, and Purnachandra Saha

Abstract Waste marble powder is an inert material obtained from the sawing, shaping and polishing of marble. Those marble dusts are hazardous to the health and atmosphere if it is not disposed properly. The main objective of the paper is to use those powders in construction work as a supplementary material and to observe the performance in concrete. Presently, in this paper, physical, chemical, mechanical properties such as workability, compressive strength, tensile strength and durability such as water absorption by immersion, carbonation properties of concrete with marble dust are studied. The review indicates that in concrete industry marble dust can be used as replacement of sand and it generally improves the mechanical properties of concrete with lower water–cement ratio. In hydration process, marble dust does not play any role but it shows filler effect in concrete and also it shows better performance. Hence, marble dust can be used as an appropriate fine aggregate material for sustainable concrete.

Keywords Marble dust · Fine aggregates · Physical and chemical properties · Strength · Durability properties

S. Saha · T. Mohanty (✉) · P. Saha
School of Civil Engineering, KIIT, Bhubaneswar, Odisha, India
e-mail: tmohantyfce@kiit.ac.in

S. Saha
e-mail: srishti.civil94@gmail.com

P. Saha
e-mail: dr.purnasaha@gmail.com

B. B. Das et al. (eds.), *Recent Developments in Sustainable Infrastructure*, Lecture Notes in Civil Engineering 75, https://doi.org/10.1007/978-981-15-4577-1_39

1 Introduction

The natural mineral resources produced in India are widely used in construction purposes. Among those minerals very high qualities of marbles are used extensively for building purpose in India [1, 2]. Pure limestone is converted to metamorphic rock after chemical reactions which is known as marble. During the processing operations of marble, huge amount of waste is generated which again causes environmental problems. The waste materials are dumped as landfill which not only reduces the fertility and porosity of the soil but also increases the alkalinity of the soil [3]. Moreover, marble powder is extensively used as a mineral for the production of concrete also which can be used as a direct replacement of cement because of its filter effect [4]. The impact of expansion of marble powder as substitution of cement or fine aggregates at substitution proportions of 15, 10, 7.5 and 5% is given [5]. Nowadays, marble powder is very popular. Since marble powder used as replacement of cement or saturated fine aggregates [6], the objective is to review the mechanical and durability properties of concrete using marble dust.

2 Physical and Chemical Properties of Marble Dust

Waste marble powder is an inert material acquired from the sawing and procession operations of marble stone.The physical properties of marble dust rely upon both adhered mortar quality and the amount of followed mortar [5, 7]. Marble waste contains large amounts of calcium oxide (Cao) within the percentage of up to 28.63%, and little bits of ferric oxide (Fe_2O_3) and silicon dioxide (SiO_2) [8]. Research on marble waste demonstrates that the surface area and percentage of fines are high, Which can help in the replacement of cement within certain limit. It was observed that within the hydration process this material could not help to form the calcium silicate gel [9, 10]. A few scientists contemplated that the real density, loose bulk density, specific gravity and fineness of waste marble are accordingly 2.68 g/cm^3, 1.67 g/cm^3, 2.67 and 350 m^2/kg, respectively. It was also observed by the thermal analysis of marble dust that $Caco_3$ is the important organic component. Marble powder shows an endothermic reaction, this generally occurs at temp 740–800 degree centigrade. Within the range of this temperature, there is a sudden loss in weight because of decomposition of calcite ions [11, 12]. Scanning electron magnifying lens with energy-dispersive X-ray spectroscopy expansion was performed on marble waste indicating essentially Cao is the fundamental constituent of the powder. In this way, the physical and chemical behaviours of waste marble are essential parts of the investigation (Fig. 1).

(a) Marble waste pile **(b) Marble waste dumped on the site** **(c) Coarse aggregate from marble waste**

Fig. 1 **a**, **b** and **c** shows different forms of marbles

3 Property of Fresh Concrete

3.1 Workability

Workability, in general, is measured by slump test. It can be described as a consistency measure [13]. The water absorption and w/c ratio affect the workability of marble waste deeply. Depending upon the shape and size of the fine aggregate, grading of aggregate also affects workability.

As compared to round aggregate, annular shape aggregate has higher contact area and this behaviour also shows the granite waste, basalt concrete. As the contact area is high, the absorption of water is also high, and therefore the water–cement ratio will become higher in order to obtain similar workability [14, 15]. It was observed that fine aggregate as a marble sand has lowest water absorption. So, it can be said that with the addition of fine aggregate (marble sand) workability of marble would be improved [16].

It was found that with the addition of marble sand to the concrete it causes cohesiveness due to the behaviour of higher surface area, and as a result workability decreases [14, 16]. With the addition of water to the conventional concrete, saturated marble aggregate is produced and similar workability is found as compared to conventional concrete aggregate, and also with the presence of extra water marble particles, the behaviour of high shape index and smooth surface is shown [17, 18].

It was observed that angular and elongated shape of aggregates shows higher slump value, as well as w/c ratio was an efficient property to achieve the workability. Marble waste as a basalt and granite sand concrete, river sand has directly affected by this property [15, 16].

It was ascertained that due to the addition of water content, accurate proportion of fine and coarse aggregates workability of marble waste should be improved [19]. It was also certified that with the addition of silica fume and marble waste, workability of concrete should be reduced. To overcome this effect, superplasticizers can be used. Using water-reducing agent as a superplasticizer, workability ability of marble waste can be solved from an economic point of view [20, 21].

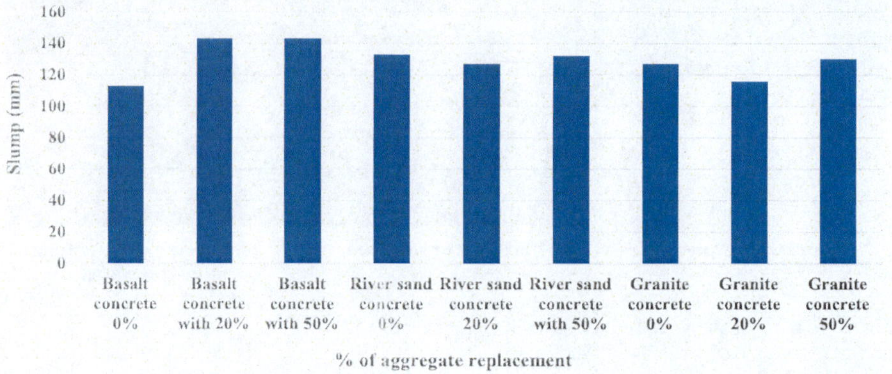

Fig. 2 Workability of marble waste for different percentages of aggregate replacement [14, 15, 22]

Figure 2 shows the workability of marble waste for different percentages of aggregate replacement. It was observed from the figure that basalt concrete and river sand concrete produced better slump when replacement is 50%.

4 Mechanical Properties

4.1 Compressive Strength

Strength durability and structural performance of marble dust are affected by compressive strength of aggregates, therefore becoming one of the most effective properties of hardening of concrete. Properties of marble dust rely on few factors such as use of superplasticizer, mineral dust content, water absorption and curing conditions [14, 22, 23]. Shape and sizes of aggregate affect the compressive performance of marble wastes. As compared to smaller aggregate, bigger size aggregate contains higher amount of adhered mortar. As the amount of adhered mortar is high, the compactness is also high, therefore to achieve better quality of cement matrix [22, 23].

It was seen that at curing age of 7 and 28 days, strength of basalt concrete reduces by the expansion of marble dust. With the increase of water–cement proportion level, strength of a marble dust decreases [23, 24]. With the inclusion of additional water during hydration process, workability increases, yet prompts lesser porosity and a resulting failure of compressive strength [22].

When surface moisture is high, the saturated aggregate shows little decrease in strength as the interfacial bond among the cement paste and reused aggregates is weak. It was found that with the expansion of additional water the reused aggregates immerse, which, therefore, demonstrates the lower compressive strength [14, 22]. With the inclusion of marble dust, the basalt concrete and granite concrete show lower compressive strength for different curing days at 7, 28 and 56 days. It was

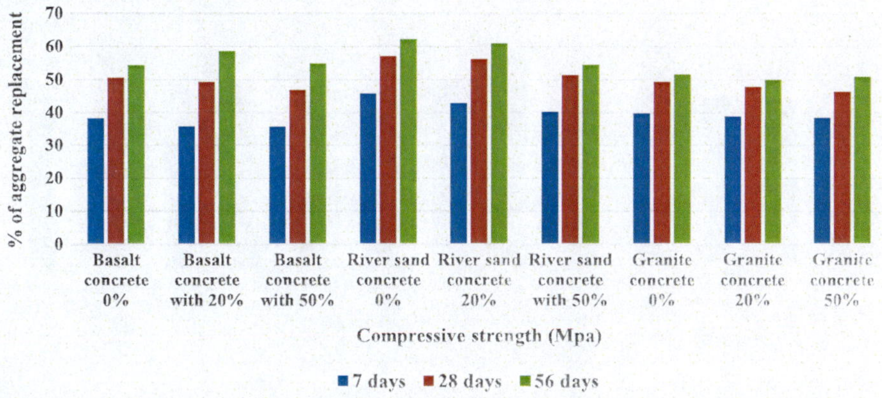

Fig. 3 The compressive strength of marble waste for different percentages of aggregate replacement [22, 23, 25]

seen that the water absorption value of river sand and marble sand is almost the same [24]. Rather than the different types of marble aggregates, basalt concrete has a higher compressive strength. As compared to basalt and sandstone, river sand has a higher strength [14, 22].

Figure 3 represents the compressive strength of marble waste for different percentages of concrete replacement. It was noticed that with increase in the percentages of replacement, strength gradually reduces.

5 Split Tensile Strength

For determining the performance of concrete under tensile stress, the split tensile test is one of the most popular methods for evaluation of tensile strength. Water–cement ratio, chemical mixing method, shape and sizes of coarse aggregates are some factors on which tensile strength of marble dust depends [26, 27]. With the expansion in substitution ratio of various sorts of marble dust, the tensile strength has been observed to be decreasing. In case of compressive strength test, this similar behaviour also occurs. When saturated coarse aggregates were replaced in river sand concrete and granite concrete the split tensile tends to decrease. As compared to granite concrete and river sand concrete, the basalt concrete has a higher tensile strength [26–29].

It was seen that the basalt particles have lamellar geometry shape, which makes weaker zones and prompts a tensile rupture [29]. This condition may have been enhanced because of the substandard productive intermolecular bond among the cement and basalt particles. Chemical response among basalt and concrete glue brought about a decrease of bond quality. Tensile bond strength among basalt and

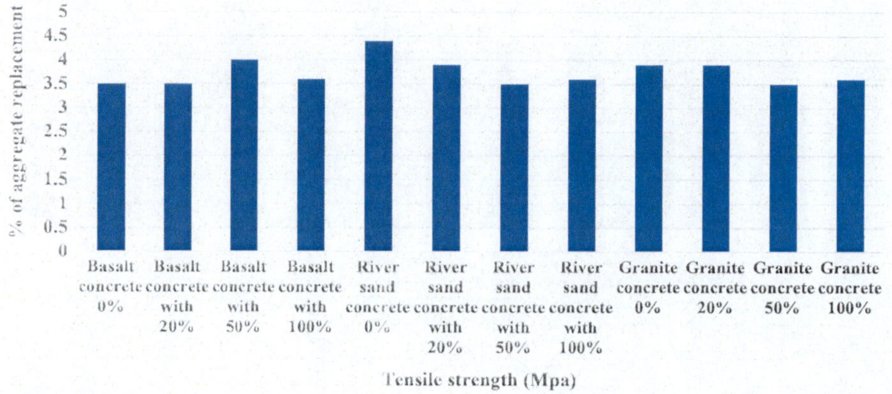

Fig. 4 The splitting tensile strength of marble waste for different percentages of aggregate replacement [22, 26, 28]

concrete glue was lesser when contrasted with limestone and quartzite; however, cleavage strength was higher for limestone than that of basalt concrete [14, 28, 29].

Carbonate content, permeable structure and curing age are few factors on which tensile strength depends. The essential reason behind the low concrete glue bond quality is the manufactured breakdown of feldspars due to their major interaction with the hydrating bond to make soil particles, which swell on retaining water [24, 28]. On the other hand, the substance breakdown of feldspars and other mineral grains on the basalt surface in contact with the solid paste may lessen the surface disagreeableness and weakens the mechanical interlocking effect between the stone surface and the hydration, achieving a weaker bond [14, 25, 26].

Figure 4 represents the tensile strength of marble waste for different percentages of aggregate replacement. It was noticed that strength gradually reduces by increasing the substitution proportion.

6 Durability Properties

6.1 Water Absorption

Water absorption is one of the essential factors which impacts the durability properties of marble aggregates. When there is the replacement of marble sand by basalt and granite sand concrete, it can be observed that basalt and granite sand concrete has a lower absorption of 3.1 and 17.8%, respectively. It was observed that for replacement up to 100% the water absorption of river sand increases by 11.2%. But water absorption of river sand decreases when replacement ratio is 20 and 50%. This deviation occurs due to the behaviours of marble waste [19, 30]. It was observed that nominal aggregate shows better durability properties when it absorbs more water. Air content

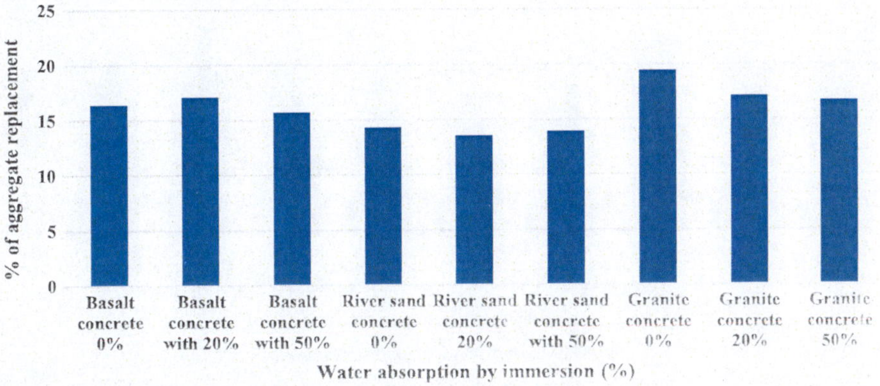

Fig. 5 Water absorption of different types of marble dust [19, 30]

and pore structures are most important properties which affect the water absorption of marble dust [31].

The air content in the blend is vital for water assimilation by immersion accordingly to override extents of over 50% with marble sand, and the concentration of blend is imperilled by the aggregate's harshness. Accordingly, better in aggregate pressing and high air maintenance [19]. The retained air creates an open permeable system, and marble sand demonstrates bring down water retention and by immersion concrete with higher water assimilation is produced [30, 31].

Figure 5 shows the water absorption of marble waste of different percentages of aggregate replacement at 28 days. It was observed from the figure that marble waste as granite concrete produced better water absorption when replacement is 50%.

7 Carbonation Depth

Carbonation is another factor influencing the auxiliary execution of cement. Porosity, water absorption, shape and sizes of aggregate affect the depth of carbonation. It was seen that by incorporation of fine aggregates in granite aggregate, carbonation resistance increases. The carbonation resistance of river sand decreases inside the expansion of fine aggregates [32, 33].

With the expansion of coarse marble aggregates, carbonation resistance of river sand likewise diminishes as it relies upon the porosity. This component is an indirect approach to control the water ingestion by immersion test. Another feature that agrees with this causing is the effect of aggregate measure on the strength of concrete exhibited to carbonation, since the physical and geometrical characteristics of aggregates unequivocally impact the squeezing of the mix and porous structure [31, 33-35].

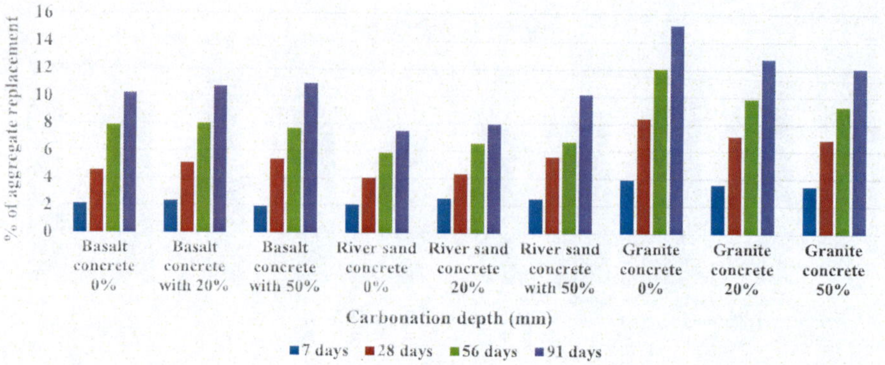

Fig. 6 The carbonation depth of marble waste of different percentages of aggregate replacement [30, 32]

Figure 6 shows the carbonation depth of marble waste of different percentages of aggregate replacement at 7 days, 28 days, 56 days and 91 days. It was observed that with the increase of curing days, carbonation depth of marble waste dust also increases. Marble waste as granite concrete produced better carbonation depth when replacement is 50%.

8 Conclusion

Performance of concrete with marble dust as supplementary materials are analysed in this study. It is observed from the study that

1. Performance of various types of marble dust such as basalt concrete, river sand and granite concrete which is compiled in this study is suitable substitution of fine aggregates or cements. However, marble debris needs suitable treatment before its use.
2. When replacement is 50%, basalt and river sand concrete produce better slump value.
3. The mechanical properties of marble dust, for example, compressive strength and tensile strength, affect the replacement of fine aggregates. Strength of marble dust specimens reduces with the increase in percentage of fine aggregates.
4. The durability properties such as water absorption and carbonation depth affect the performance according to the substitution of fine aggregates.

References

1. Bilgin N, Yeprem HA, Arslan S, Bilgin A, Gunay E, Marsoglu M (2012) Use of waste marble powder in brick industry. Constr Build Mater 29:449–457
2. Pathan VG, Pathan MG (2014) Feasibility and need of use of waste marble powder in concrete production. J. Mech. Civil Eng, 23–26
3. Aruntas HY, Gürü M, Day M, Tekin I (2010) Utilization of waste marble dust as an additive in cement production. Mater Des 31:4039–4042
4. Sarkar R, Das SK, Mandal PK, Maiti HS (2006) Phase and microstructure evolution during hydrothermal solidification of clay-quartz mixture with marble dustsource of reactive lime. J Eur Ceram Soc 26:297–304
5. Ergün A (2011) Effects of the usage of diatomite and waste marble powder as partial replacement of cement on the mechanical properties of concrete. Constr Build Mater 25:806–812
6. Aliabdo A, AbdElmoaty M, Auda EM (2014) Re-use of waste marble dust in the production of cement and concrete. Constr Build Mater 50:28–41
7. Hebhoub H, Aoun H, Belachia M, Houari H, Ghorbel E (2011) "Use of waste marble aggregates in concrete. Constr Build Mater 25:1167–1171
8. Shirulea PA, Rahman A,Gupta RD (2012) Partial replacement of cement with marble dust powder. Int J Adv Eng Res
9. Vardhan K, Goyal S, Siddique R, Singh M (2015) Mechanical properties and microstructural analysis of cement mortar incorporating marble powder aspartial replacement of cement. Constr Build Mater 96:615–621
10. Ran A, Kalla P, Csetenyi LJ (2015) Sustainable use of marble slurry in concrete. J Clean Product 94(2015):304–311
11. Rai B,Naushad K, Kr A, Rushad T (2011) Influence of marble powder/granules in concrete mix. Int J Civ Struct Eng 1:827–834
12. Belaidi ASE, Azzouz L, Kadri E, Kenai S (2012) Effect of natural pozzolana and marble powder on the properties of self-compacting concrete. Constr Build Mater 31:251–257
13. Aruntas HY, Guru M, Dayi M, Tekin I (2010) Utilization of waste marble dust as an additive in cement production. Mater. Des 31:4039–4042
14. Marras G, Careddu N, Internicola C, Siotto G (2010) Recovery and reuse of marble powder by-product. Global stone congress
15. Corinaldesi V, Moriconi G, Taurun RM (2010) Characterization of marble powder for his use in mortar and concrete. Constr Build Mater 24:113–117
16. Karaşahin M, Terzi S (2007) Evaluation of marble waste dust in the mixture of asphaltic concrete. Constr Build Mater 21:616–620
17. Deshmukh JV, Varhade R, Tandel K, Gadekar M (2015) Green cement for sustainable concrete using marble dust. Int J Appl Innov Eng Manage 4:144–150
18. Ergun A (2011) Effects of the usage of diatomite and waste marble powder as partial replacement of cement on the mechanical properties of concrete. Constr Build Mater 25:806–812
19. Binici H, Kaplan H (2007) Yilmaz S Influence of marble and limestone dusts as additives on some mechanical properties of concrete. Sci Res Essay 2:372–379
20. OzturanT ÇeçenC (1997) Effect of coarse aggregate type on mechanical properties of concretes with different strengths. Cem Concr Res 27:165–170
21. Shelke V, Pawde P, Shrivastava R (2012) Effect of marble powder with and without silica fume on mechanical properties of concrete. J Mech Civ Eng (IOSRJMCE) 1:40–45
22. Silva D, Gameiro F, Brito DJ (2014) Mechanical properties of structural concrete containing fine aggregates from waste generated by the marble quarrying industry. J Mater Civ Eng
23. Poon CS, Shui ZH, Lam L, Fok H, Kou SC (2004) Influence of moisture states of natural and recycled aggregates on the slump and compressive strength of concrete. Cem Concr Res 34:31–36

24. Martins P, Brito DJ, Rosa A, Pedro D (2013) Mechanical performance of concrete with incorporation of coarse waste from the marble industry. Mater Res J
25. Coutinho A (1988) Production and properties of concrete. National Laboratory of Civil Engineering, Lisbon, Portugal (in Portuguese)
26. Evangelista L, Brito DJ (2007) Mechanical behaviour of concrete made with fine recycled concrete aggregates. Cem Concr Compos 29:397–401
27. Tasong WA, Lynsdale CJ, Cripps JC (1999) Aggregate cement paste interface. Part I. Influence of aggregate geochemistry. Cem Concr Res 29:1019–1025
28. Odler I, Zurz A (1987) Structure and bond strength of cement aggregate interfaces. MRS Proc 114:21–27
29. Basheer L, Kropp J, Cleland D (2011) Assessment of the durability of concrete from its permeation properties: a review. Constr Build Mater 15:93–103
30. Metha P, Monteiro P (1994) Concrete: structure, properties and materials. Technology dissemination project
31. Amorim P, Brito DJ, Evangelista L (2012) Concrete made with coarse concrete aggregate: influence of curing on durability. ACI Mater Jvol 109:195–204
32. Andre A, Brito DJ, Rosa A, Pedro D (2013) Durability performance of concrete incorporating coarse aggregates from the marble industry
33. Binici H, Shah T, Aksogan O, Kaplan H Durability of concrete made with granite and marble as recycle aggregates. J Mater Proc Technol 208:299–308

Mechanical Properties of Self-compacting Concrete Made of Glass Fibre

Debarshree, Biswajit Jena, Kaliprasanna Sethy, Asish Kumar Pani, and Kirti Kanta Sahoo

Abstract Self-compacting concrete (SCC) which is an extremely floatable, non-segregated concrete can reach easily at the congested formwork and covers the brace without any vibration. This present research work is focused on comparison of the mechanical properties of glass-fibre-reinforced SCC of grade M30. The mechanical properties of SSC like compressive strength, flexural strength, split tensile strength with different ages at 7 and 28 days are evaluated. During the experimental work, the workability is measured by slump flow test, T50 flow test, L-box test and V-funnel test. In this current experiment, long chopped glass fibres of size 12 mm are used to reinforce SCC. Replacement percentages such as 0.05, 0.10, 0.15, 0.20, 0.25 and 0.30% are adopted throughout the research programme. It is observed that at 0.20% replacement, mechanical properties show the better results than control mix and other replacement percentage. Load deflection curve of SCC beam reinforced with glass fibre shows better ductility.

Keywords Compressive strength · Flexural strength · Split tensile strength · Load deflection curve · Sorptivity

Debarshree · A. K. Pani · K. K. Sahoo (✉)
KIIT Deemed to be University, Bhubaneswar, India
e-mail: kirtikanta.sahoofce@kiit.ac.in

Debarshree
e-mail: debashree.civil@gmail.com

A. K. Pani
e-mail: apanifce@kiit.ac.in

B. Jena
DRIEMS, Cuttack, India
e-mail: biswa.tech88@gmail.com

K. Sethy
GCEK, Bhawanipatana, Odisha, India
e-mail: kaliprasanna87@gmail.com

B. B. Das et al. (eds.), *Recent Developments in Sustainable Infrastructure*, Lecture Notes in Civil Engineering 75, https://doi.org/10.1007/978-981-15-4577-1_40

1 Introduction

Self-compacting concrete (SCC) is a highly fluid and stable concrete that can flow consistently under its own weight, pass between the bars and fill in the formwork without the need of compaction. The self-compact ability is achieved by adding a superplasticizer to the mixture and by increasing the amount of fine materials. From different types of studies, it has been shown that fibre reinforcement materials are proved to be efficient as compared to other materials. These types of fibre component are used to improve cracking and fracture toughness of the structure. The local structure has deboning, pulling out and sliding of the fibres which provide the bridging action. The fibres reduce the starting of macrocracking and avoid opening as well as growth of cracks in the structure. Due to these types of components, high demand of energy is required for remarkable crack propagation. Use of low volumetric fibre demand does not affect the elastic performance of the structure.

The glass fibres are two types: (1) continues, (2) discontinues or chopped fibres. Principal choices are below cost, huge capacity, easy and safely functioning and rapid and orderly diffusion expedite comparable mixes [ShahanaSheril, P.T, 2013].

In the current work, SCC having 12 mm chopped glass fibre added in various proportions like 0.05, 0.1, 0.15, 0.20, 0.25 and 0.39%, respectively, have been used.

2 Experimental Programme

2.1 Materials

In this project, Portland slag cement is used according to IS 455:1989. The physical properties are referred from IS 12089:1987 presented in Table 1. In this experiment, 10 mm size of coarse aggregates are used. The required physical properties are referred from IS: 383-1970 (Table 2). Zone III sand is used according to IS: 383-1970 and the properties are in Table 2. Alkali-resistant glass fibre (ARGF) of Young's modulus of 72 GPA and 12 mm long is utilized and the physical properties are given in Table 3. Advantages of this type of glass fibres are little price, better strength, easy and safe behaviour, and then quick and unchanging dispersion simplifying same mixes which in word produce durable concrete. In this research, the SIKA VISCOCRETE 2004 NS superplasticizer is used.

2.2 Mixture Proportion

The experiment of the ordinary concrete and GFRC mix concrete, M30 grade of concrete mix for SCC using subsequent EFNARC code 2005 is prepared. Table 4 shows the mix proportion for both mix. Sika Viscocrete 2004 NS was used to improve

Table 1 Physical properties of Portland slag cement

Sl. no	Particulars		Test results	Requirement of IS:12089:1987
1	Fineness obtained (in M^2/Kg)		389	225 (min)
2	Setting time (minutes)			
	Initial setting time		98	30 (min)
	Final setting time		198	600 (max)
3	Soundness			
	Lechatilier expansion (mm)	1.00		10 (max)
	Autoclave (%)	0.12		0.8 (max)
4	Compressive strength (MPa)			
	72 + 1 h	27.60		16.00
	168 + 2 h	39.50		22.00
	672 + 4 h	56.7		23.00
5	Consistency	31.5%		
6	Fineness	100%		
7	Specific gravity	3.00		

Table 2 Physical properties of coarse aggregate

Sl. no	Coarse aggregate	Fine aggregate
Specific gravity	2.53	2.67
Fineness modulus	6.2	2.28
Water absorption	0.4 (%)	0.85 (%)

Table 3 Physical properties of glass fibre

Type of fibres	CT2024
Length of the fibre (mm)	12
Density of fibre (g/cm3)	2.53
Elastic modulus of fibre (GPa)	43–50
Tensile strength of fibre (MPa)	1950–2050
Elongation at break (%)	7–9
Water absorption	<0.1

Table 4 Approve mix proportion of SCC

Cement in (kg/m^3)	Water in (kg/m^3)	Fine aggregate in (kg/m^3)	Coarse aggregate in (kg/m^3)	Superplasticizer in (kg/m^3)
585.42	234.168	910.889	575.424	2.342
1	0.4	1.56	0.98	0.004

Table 5 Result of fresh concrete

Test	Time (SEC)	Measurement
T500 test	3.8	697 mm
V-funnel test	8	8 s
L-box test	6	7.25/7.4 = 0.98

the workability. To satisfy SCC, the workability was measured by T500 test, L-box test and V-funnel test. Glass fibre percentages of 0.05, 0.10, 0.15, 0.20, 0.25 and 0.30% are added to concrete to make composite concrete.

2.3 *Testing of Fresh Concrete*

T500 and L-box test were conducted to measure the degree of workability. If the concrete can flow above 500 to 700 mm then the slump test value satisfies SCC. Like flow value, in L-box test, concrete can flow in between 8 s. To control the flow ability of self-compacting concrete, V-funnel test is analysed. The test result of workability is shown in Table 5.

2.4 *Preparation of Test Specimens*

Cube of size 150 × 150 × 150 mm, cylinder of size 150 × 300 mm and prisms of size 100 × 100 × 500 mm were taken for conducting mechanical properties. After demoulding, all the samples are cured for 28 days in normal tap water.

2.5 *Sorptivity Test*

Sorptivity test measure for capillary force which utilizes the outlet structure produces liquids to be peaked into the body of the structure. Here, the capillary rice is calculated in concrete cube by putting the cube in water to a depth of 2 to 5 mm deep. The relation among absorption and sorptivity is

$$K = \frac{W}{A\sqrt{t}}$$

where K = sorptivity,
W/A = water absorption per unit area cumulatively,
t = time elapsed.

Table 6 Characterization of mixes

Classification	Fibre content (%)	Description
PSC	0.0	Plain SCC
GFC–0.5	0.05	0.05% GFRSCC
GFC–1	0.1	0.1% GFRSCC
GFC–1.5	0.15	0.15% GFRSCC
GFC–2	0.2	0.2% GFRSCC
GFC–2.5	0.25	0.25% GFRSCC
GFC–3	0.3	0.3% GFRSCC

This was accompanied in laboratory. The time interval selected is 30 min, 1, 2, 6, 24 and 48 h and then remove the sample from the water and weigh it. Weigh the cube till the weight increased, when the weight stays constant and the cube is not gaining the weight at that time, then stop the check.

3 Result and Discussion

3.1 Workability

Different mix proportions are made by adding glass fibre at different percentages. The name of the mixes is described in Table 6. All the mixes are tested through slump flow, T_{50} flow, L-box and V-funnel to satisfy the SCC criteria. Respective values are presented in Table 7. Figures 1, 2, 3 and 4 show the graphical presentation of workability in different forms.

4 Mechanical Properties

After successive curing, all the samples at 7 days and 28 days are tested to quantify the difference between SCC and fibre SCC. Table 8 shows all the values at different days, respectively.

In control mix, 7 days value was 34.25 MPa and the FRSCC value was 36.39 MPa. There is an increase in between them around 5.88% which shows no significant improvement. The peak value of FRSCC shows 0.20% of glass fibre. At 28 days, control mix shows 40.78 MPa and the FRSCC shows 48.95 MPa. Here the glass fibre shows 16.69% of increase at 0.20% of glass fibre. Deviation in the compressive strength in both types of SCC is presented in Figs. 5 and 6.

The percentage improvement of flexural strength for glass fibre over control mix is 5.91% when 0.20% of glass fibre is added in concrete. In control mix, it shows

Table 7 Test result of workability

Sample	Slump flow test (600–750) mm	T_{50} flow test (2–5) sec	L-box test (H_2/H_1) 0.8–1.0	V-funnel test (6–12) sec	Remarks
PSC	697	3.8	0.98	7	Result fulfilled
GFC–0.5	690	6.0	0.96	6	High viscosity
GFC–1	680	4.2	0.92	7.2	Result fulfilled
GFC–1.5	655	5.0	0.88	7.9	Result fulfilled
GFC–2	640	3.9	0.85	8.3	Result fulfilled
GFC–2.5	625	4.3	0.82	8.5	Result fulfilled
GFC–3	520	6.4	0.65	12	High viscosity blockage

Fig. 1 Slump test

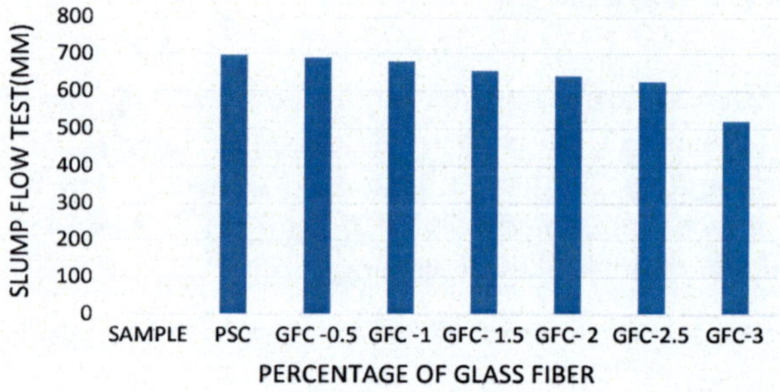

Fig. 2 T_{50} flow test

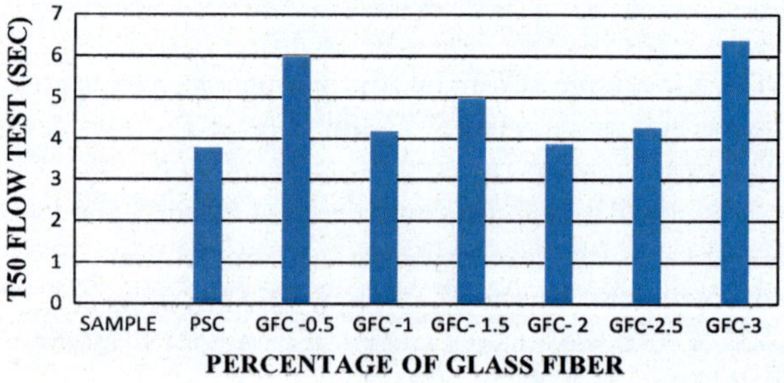

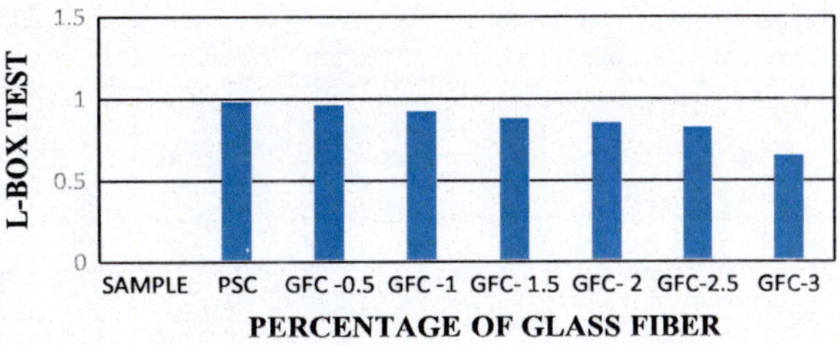

Fig. 3 L-box test

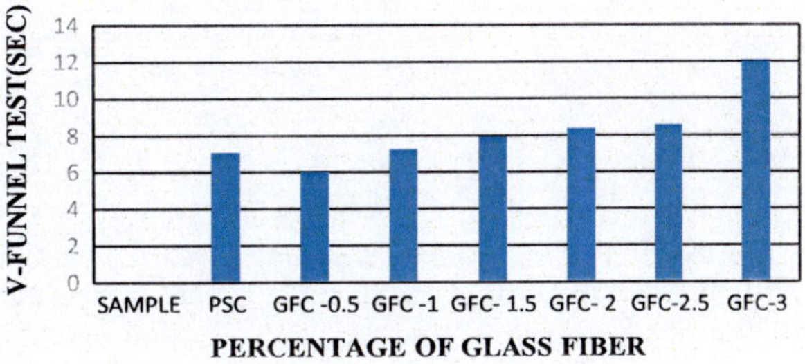

Fig. 4 V-funnel test

Table 8 Test results at 7 and 28 days

Mixes	Compressive strength test at 7 days in (MPa)	Compressive strength test at 28 days in (MPa)	Flexural strength test at 28 days in (MPa)	Split tensile strength test at 28 days in (MPa)
PSC	34.25	40.78	9.06	4.15
GFC–0.5	27.18	36.67	7.6	3.9
GFC–1	25.27	35.29	7.46	3.72
GFC 1.5	28.33	44.96	8.33	4.26
GFC–2	36.39	48.95	9.63	4.35
GFC–2.5	25.17	36.41	8.27	3.57
GFC–3	20.67	32.21	7.8	3.2

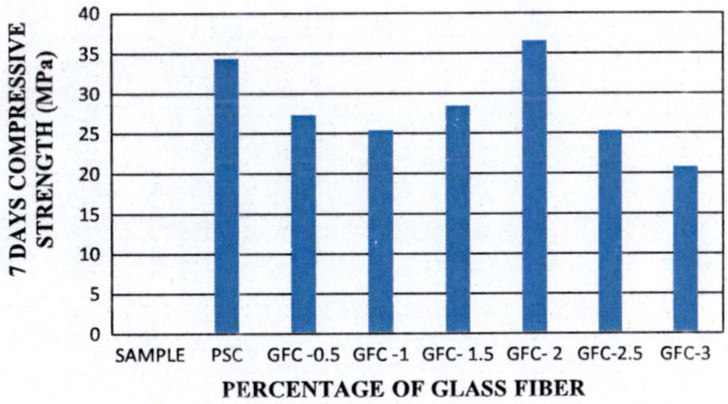

Fig. 5 Deviation of 7 days compressive strength

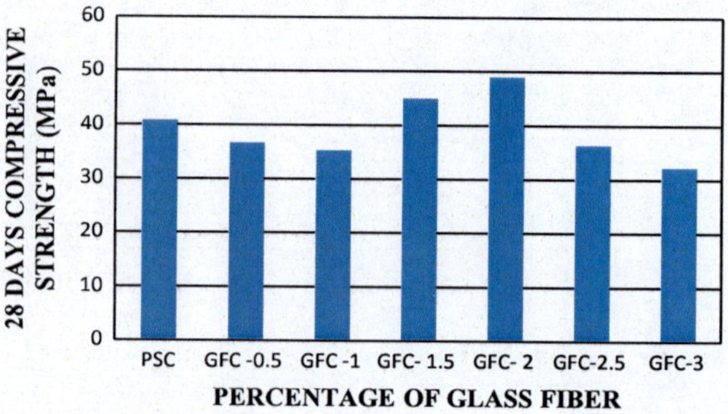

Fig. 6 Deviation of 28 days compressive strength

9.06 MPa and in 0.20% of glass fibre, it shows 9.63 MPa. Figure 7 shows the deviation in the flexural strength for PSC and different fibre SCCs.

Experimental investigation shows control mix having 4.15 MPa and 0.20% of glass fibre shows 4.35 MPa. The glass fibre mixed SCC shows 4.59% of increase at 0.20% of glass fibre. Figure 8 describes about the deviation of split tensile strength.

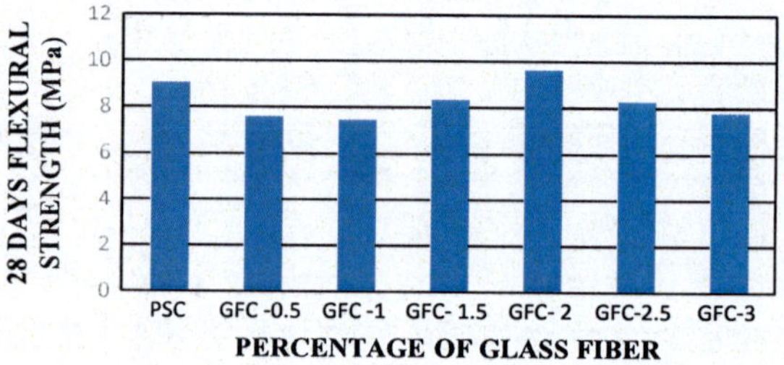

Fig. 7 Deviation of 28 days flexural strength

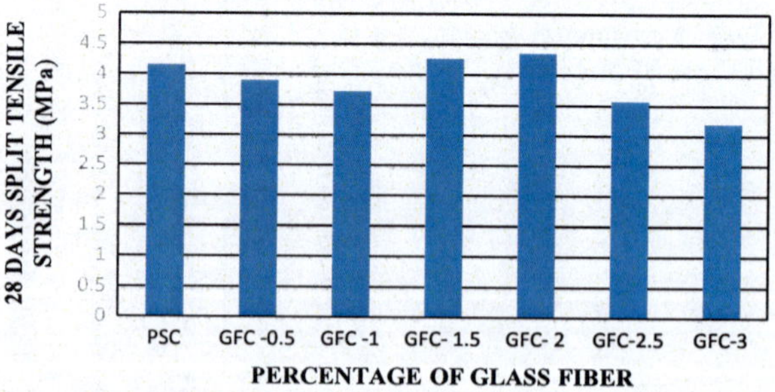

Fig. 8 Deviation of 28 days split tensile strength

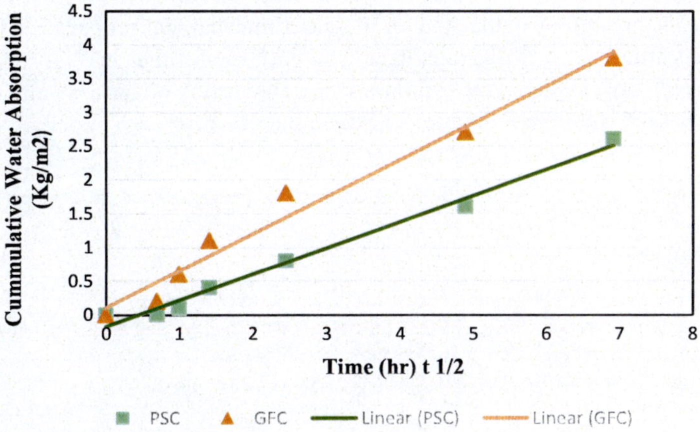

Fig. 9 Sorptivity at different time intervals

Table 9 Sorptivity test results

Sample	Initial dry weight in (gm)	Weight in (gm)					
		At 30 min	At 1 h	At 2 h	At 6 h	At 24 h	At 48 h
PSC	2685	2685	2686	2688	2689	2693	2695
GFC	2450	2452	2454	2455	2457	2459	2461

5 Sorptivity

Figure 9 shows the capillary absorption of water at different time intervals. Here the water absorption through capillary in GFC sample is higher than PSC samples, which indicates the fibres absorbed the water for which the weight of the sample is higher than PSC sample. Figure 9 and Table 9 shows the trend of capillary action for both concretes.

6 Conclusion

1. As glass fibre absorbed more water, it shows the value of slump.
2. Addition of glass fibre in self-compacted concrete enhances the mechanical properties.
3. 0.20% of glass fibre was recognized as optimum doses to increase all the mechanical properties of SSC.
4. At 0.20% of glass fibre, it was observed that compressive strength increased by 5.88% (7 days), and 16.69% (28 days), flexure strength increased by 5.91% (28 days), split tensile strength increased by 4.59% (days), respectively.

5. The GFR concrete is in the state of increased mechanical properties with higher quantity fraction. In fresh state, it showed good performance.
6. In case of sorptivity test, the capillary water absorption of GFC is more than PSC because GFC observed more water due to the glass fibre.

Bibliography

1. Boukendakdji O, Kenai S, Kadri EH, Rouis F (2009) Effect of slag on the rheology of fresh self-compacted concrete. Constr Build Mater 23:2593–2598
2. Chihuahua Jiang, Fan Ke, Fei Wu, Chen Da (2014) Experimental study on the mechanical properties and microstructure of chopped basalt fibre reinforced concrete. Mater Des 58:187–193
3. Collepardi M, Borsoi A, Collepardi S, Croce EN, Passuelo A (2006) Influence of viscosity modifying admixture on the composition of SCC. In: Supplementary volume of eighth CANMET/ACI international conference on super-plasticizers and other chemical admixtures in concrete, Sorrento, Italy, October 29–November 1, pp 253–261
4. Cunha VMCF, Barros JAO, Sena-Cruz JM (2011) An integrated approach for modeling the tensile behavior of steel fibre reinforced self-compacting concrete. Cem Concr Res 41:64–76
5. EI-Dieb AS, Taha MMR (2012) Flow characteristics and acceptance criteria of fibre-reinforced self-compacted concrete (FR-SCC). Constr Build Mater 27:585–596
6. Enfedaque A, Alberti MG, Galvez JC (2014) On the mechanical properties and fracture behaviour of polyefin fibre reinforced self-compacting concrete. Constr Build Mater 55:274–288
7. Hossain KMA, Lachemi M (2004) Self-consolidating concrete incorporating new viscosity modifying admixtures. Cem Concr Res 34:185–193
8. Vijayanand M, Angelescu N, Muthu KU, Puttappa CG, Sundarsana Rao H (2010) The scientific Bulletin of VALAHIA University. Materials and Mechanics–Nr. 5 (year 8)
9. Mallesh M, Shwetha GC, Reena K, Madhukaran (2015) Experimental studies on M30 grade self-compacting concrete. Int J Sci Eng Technol Res (IJSETR) 4(9)
10. Min D, Gaopeiwei, FengNaiqui (2000) The influence of SP and superfine mineral powder on the flexibility. Strength and Durability of HPC, Cem Concr Res 31:703–706
11. Murthy KN, Narasimha Rao AV, Ramana Reddy IV, Vijaya sekhar Reddy M (2012) Mix design procedure for self-compacting concrete. IOSR J Eng (IOSRJEN) 2(9). e-ISSN: 2250-3021, p-ISSN: 2278-8719
12. Abdulhadi M (2012) A compressive study of basalt and polypropylene fibres Reinforced concrete on compressive and tensile behavior. Int J Eng Trends and Technol (IJETT) 9(6)
13. Mailvaganam NP, May (2001) How chemical admixtures produce their effects in concrete. Indian Concr J, 331–334
14. Ouchi M, Okamura H (1997) Effect of superplastisizer on fresh concrete. J Transp Board, 37–40
15. Parra C, Valcuende M, Gomez F (2011) Splitting tensile strength and modulus of elasticity of self-compacting concrete. Constr Build Mater 25:201–207
16. Pereira de Oliveira LA, Castro Gomes JP, Bernardo LFA, Ramos MMM (2013) Evaluation of dry mortar ratio as mix design parameter for steel fibre reinforced self-compacting concrete. Constr Build Mater 40:642–649
17. Raghuprasad PS (2004) Comparative study on different types of blended cement with different grade O.P.0 concrete—an experimental approach. In: ICACC -2004, Proceeding of international conference on advances in concrete and construction, Hyderabad (vol. II, 16–18) December, pp 637–646

18. Ramezanianpour AM, Esmaeili K, Ghahari SA, Ramezanianpour AA (2014) Influence of initial steam curing and different types of mineral additives on mechanical and durability properties of self-compacting concrete. Constr Build Mater 73:187–194
19. ShahanaSheril PT (2013) Self-compacting concrete using fly ash and glass fiber. Int J Eng Res Technol (IJERT) 2(9):2278–0181
20. Shi C, Wu Z, Lv K, Wu L (2015) A review on mixture design methods for self-compacting concrete. Constr Build Mater 84:387–398
21. Thanh L, Muller M, Siewert K, Ludwig H (2015) The mix design for self-compacting high performance concrete containing various mineral admixtures. Mater Des 72:51–62

Study on Mechanical Properties of Steel Fibre Concrete

Asish Kumar Pani and Kirti Kanta Sahoo

Abstract Nowadays, in most of the important constructions, high-strength concrete is very much essential. Achievement of high-strength concrete alone with cement is difficult rather than addition of admixtures. Utilization of fibres in concrete, effective development of flexural rigidity, imperviousness to splitting, resistance to porousness and thawing are incredible. Fibre has strengthening character which is useful for producing high-strength concrete. Therefore, fibre-reinforced concrete (FRC) is accepted world wide rapidly. In this present research, concrete grade of M60 is prepared incorporated with steel fibre. The percentage of steel fibre is maintained from 0.5 to 2% of the weight of binding material. Mechanical properties like compressive strength, flexural strength and split tensile strength are checked to quantify the solo effect of steel fibre on the concrete. It has been reported that all three mechanical properties are increased in comparison to control mix for a steel percentage of 2. Throughout the experimental work ordinary Portland cement is used.

Keywords Steel fibre · Compressive strength · Flexural strength · Split tensile strength

1 Introduction

In the modern construction industry, Ordinary Portland Cement (OPC) is nowadays obsolete. Achievement of high-strength concrete with OPC is not possible. So different admixtures like silica fume, fly ash, nanomaterials, etc. are required to enhance the strength of the concrete. Steel fibre is one material which shows compatibility with concrete. So after long observation, Fibre-Reinforced Concrete (FRC) plays a key role in the construction industry. In the FRC, fibres are distributed arbitrarily which helps in proper blending and enhancing mechanical properties. Due to compatible nature, FRC is called as cement-based composite structures. Most important

A. K. Pani · K. K. Sahoo (✉)
KIIT Deemed to be University, Bhubaneswar, India
e-mail: kirtikanta.sahoofce@kiit.ac.in

A. K. Pani
e-mail: apanifce@kiit.ac.in

B. B. Das et al. (eds.), *Recent Developments in Sustainable Infrastructure*, Lecture Notes in Civil Engineering 75, https://doi.org/10.1007/978-981-15-4577-1_41

Table 1 Property of cement

Sl. No.	Property	Experimental values	Standard value
1	Fineness (%)	1.72	≤10
2	Specific gravity	3.13	–
3	Consistency (%)	33	–
4	Setting time (Initial) min.	88	>30
5	Setting time (Final) min.	176	<600
Compressive strength (N/mm^2)			
1	3 days	31.76	>27
2	7 days	48.47	>37
3	28 days	56.98	>53

favourable parameters of FRC are flexural rigidity, impenetrability to spitting, and superb porousness and ice resistance. Such properties of FRC make it structural trust goodness. Many fibres are used in the concrete such as steel, glass, carbon, steel and natural. In this present research work, steel fibre is taken into consideration for enhancement of concrete properties. M60 grade of concrete is prepared along with different doses of steel fibre such as 0.5, 1, 1.5 and 2%. At 28 days, the mechanical properties are examined to account the effect of steel fibre.

2 Materials

2.1 Binder

Ordinary Portland Cement (OPC) is used as binder in the current research work. OPC having grade 53 made by RAMCO confirming to IS 4031-1988 is used for the whole experimental work. The physical and mechanical properties of cement are presented in Table 1.

2.2 Fine Aggregate

Sand as fine aggregate having fineness modulus 2.50 to 3.50 and silt content not greater than 4% is taken in the present work. The fine aggregate in this present work is a mix of coarse sand as well as fine sand.

2.3 *Coarse Aggregate*

Hard, dense, durable and clean coarse aggregates are used. It is ensured that it is free from disintegrated pieces and other vegetable matter. The used aggregates are roughly cubical in shape and there is a very less amount of flaky pieces.

2.4 *Steel Fibre*

The steel fibre used is flat crimped stainless steel fibre. The physical characteristics are presented in Table 2.

3 Experimental Methodology

M60 grade of concrete is prepared according to IS 456-2000 and IS 10262-2009, respectively. After design mix, **1:1.82:2.74** is taken as mix proportion. Moulds of size 150 × 150 × 150 mm for compressive strength, 100 × 100 × 500 mm for flexural strength and 150 × 300 mm for split tensile strength are used. Different percentages of steel fibres such as 0.5, 1, 1.5 and 2 are mixed in the concrete. Observation of mechanical properties is recorded after successive 28 days of curing. Curing is done in the normal tap water and kept in the open water tank. The temperature and humidity are about 30 °C and 60%. Mechanical properties are conducted as per IS 516-1959.

4 Results and Discussion

After completion of 28 days curing, specimens are tested as per Bureau of Indian Standards to give the mechanical properties of the concrete.

Table 2 Property of steel fibre

No.	Properties	Description
1	Length	35–45 mm
2	Density	7900 kg/m^3
3	Modulus of elasticity	2.1×10^5 N/mm^2
4	Alkali resistant	Good
5	Acid resistant	Poor
6	Heat resistant	Good
7	Elongation	5–35%

Table 3 Compressive strength of control mix

Compressive strength (MPa)	Average
68.5	68.02
66.6	
69.1	

Table 4 Compressive strength steel fibre concrete

Steel fibre %	Sample 1	Sample 2	Sample 3	Average strength (MPa)
0.5	71.70	71	72.2	71.6
1	72.6	72.1	73.1	72.6
1.5	73.42	73.8	73	73.4
2	75.59	74.4	75.9	75.3

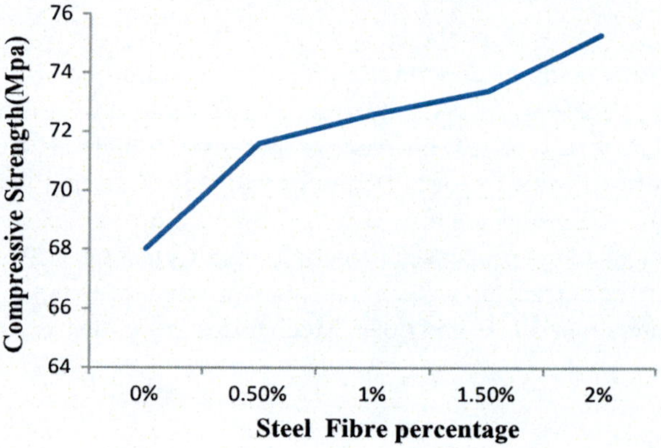

Fig. 1 Relation between compressive strength and steel fibre

4.1 Compressive Strength Test

Tables 3 and 4 show different compressive strength values for different mixes of concrete. In Fig. 1, the copulation between compressive strength and doses of fibres is shown. It is observed that at 2% steel fibres compressive strength shows significant increment. But up to 1% of addition of steel fibre, increment in compressive strength is not so much significant.

4.2 Flexural Strength Test

Tables 5 and 6 show the flexural strength value for both control and steel fibre concrete. Like compressive strength, same trend is observed. At 2% of steel fibre, it

Table 5 Flexural strength of control mix

Flexural strength (MPa)	Average
7.61	7.46
7.20	
7.58	

Table 6 Flexural strength of steel fibre concrete

Steel fibre %	Sample 1	Sample2	Sample 3	Average strength (MPa)
0.5	7.9	7.6	8.1	7.87
1	8.12	8.4	8.62	8.38
1.5	9.1	8.9	8.42	8.81
2	8.5	9.5	9.12	9.04

Fig. 2 Relation between flexural strength and steel fibre

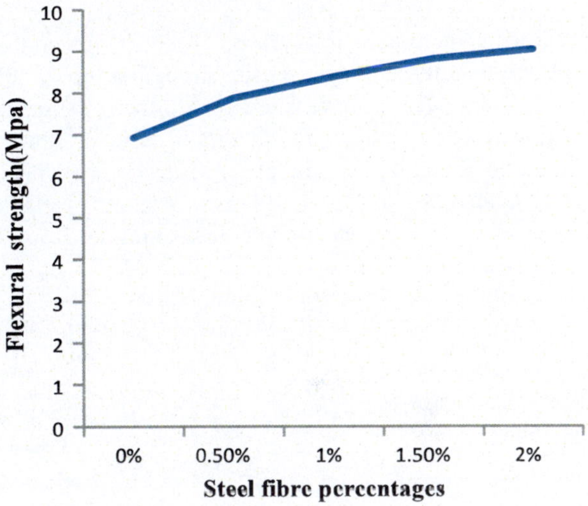

is noticed that maximum flexural strength is achieved. But at 0.5% of steel fibre, it is noticed that increment percentage is not significant.

Figure 2 shows the trend between flexural strength and doses of steel fibre.

4.3 *Split Tensile Strength Test*

Tables 7 and 8 show the tensile splitting strength value for both control and steel fibre concrete. Like compressive strength and flexural strength, same trend is observed. At 2% of steel fibre, it is noticed that maximum split tensile strength is achieved. But most important factor is that at minimum level of addition of steel fibre (0.5%)

Table 7 Split tensile strength of control mix

Split tensile strength (MPa)	Average split tensile strength (MPa)
5.1	4.57
4.11	
4.66	

Table 8 Split tensile strength of steel fibre concrete

Steel fibre %	Sample 1	Sample 2	Sample 3	Average strength (MPa)
0.5	7.92	7.5	7.8	7.4
1	7.1	7.94	7.9	7.64
1.5	8.1	7.7	7.9	7.9
2	8.2	8	7.95	8.05

significant increment is achieved. Also, it was noticed that there is no so much evidential increase in split tensile strength after 1% addition of steel fibre. Figure 3 shows the trend of split tensile strength for both concretes.

It was noticed that after addition of steel fibre, mechanical properties were increased gradually. For compressive strength, 10.7% increase; for flexural strength, 21.17% increase and for split tensile strength, 76% increase were achieved. It was observed that for the steel percentages ranging from 0.5 to 2 there was a gradatory increase in every strength test. The mechanical properties are minimum for 0.5% and maximum for 2% steel fibre volume.

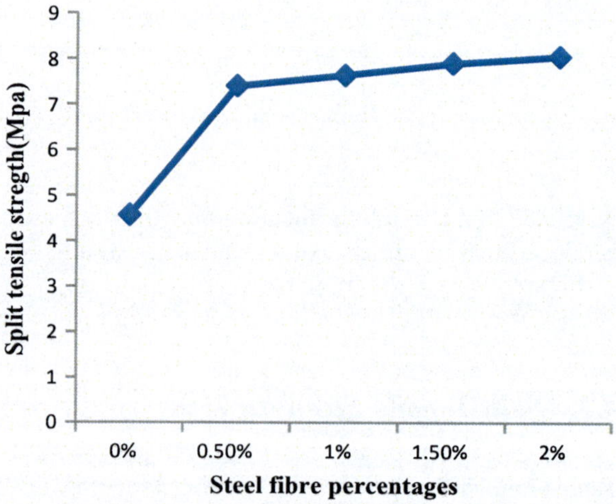

Fig. 3 Relation between split tensile strength and steel fibre

5 Conclusion

The succeeding conclusions were taken from the research.

1. It is found that mechanical properties are gaining its highest value at 2% steel.
2. It is noticed that compressive strength increases up to 10.7%.
3. It is noticed that flexural strength increases up to 21.17%
4. It is found that increment about 76% was observed in case of split tensile strength.
5. Steel fibres can be utilized in all construction types due to water-repelling properties.

Bibliography

1. Gunneswara Rao TD, Rama Seshu D (2003) Torsion of steel fibre reinforced concrete members. Department of Civil Engineering, National Institute of Technology, Warangal
2. Bhikshma V, RavandeKishor, NitturkarKalidas (2005) Mechanical properties of fibre reinforced high strength concrete. Recent advances in concrete and construction tech 6–8
3. Shende AM1, Pande AM2 (2011) Experimental study and prediction of tensile strength for steel fiber reinforced concrete. Int J Civil Struc Eng 1(4)
4. Balouch a SU, Forth a JP, Granju b J-L (2009) Surface corrosion of steel fibre reinforced concrete. Schoolof Civil Engineering, University of Leeds, UK
5. Shweta P, Kavilkar R (2014) Study of flexural strength in steel fibre reinforced concrete. Int J Recent Develop Eng Technol 2(5)
6. Bencardino F, Rizzuti L, Spadea G, Swamy RN (2008) Stress-strain behaviour of steel fiber-reinforced concrete in compression. J Mater Civil Eng
7. Soroushian P, Bayasi (1991) Fibre-type effect on the performance of steel fibre reinforced concrete. ACI Mater J
8. Soulioti D, Barkoula NM, Paipetis A, Matikas TE, Shiotani T, Aggelis DG (2009) Acoustic emission behaviour of steel fibre reinforced concrete under bending. Department of Materials Science and Engineering, University of Ioannina
9. Neves RD, Fernandes de Almeida JCO (2005) Compressive behaviour of steel fibre reinforced concrete. Structural Concrete
10. Deluce JR, Vecchio FJ (2013) Cracking behaviour of steel fiber-reinforced concrete members containing conventional reinforcement. ACI Struct J 10(3)
11. Al-Sakiny ZH (2002) Engineering properties of ultra high strength fibre reinforced concrete. M.Sc. Thesis, University of Technology, Baghdad (2002)
12. IS-456: 2000 (1978) Code of practice for plain and reinforced concrete (Fourth revision), Bureau of Indian Standards, New Delhi
13. Achilleos C, Hadjimitsis D, Neocleous K, Pilakoutas K, Neophytou PO, Kallis S (2011) Proportioning of steel fibre reinforced concrete mixes for pavement construction and their impact on environment and cost. Department of Civil Engineering and Geomatics, Cyprus University Of Technology, Archbishop Kyprianos
14. Baran E, Akis T, Yesilmen S (2011) Pull-out behavior of prestressing fibres in steel fiber reinforced concrete. Department of Civil Engineering, Atilim University, Incek-Ankara, Turkey
15. Gunneswara Rao TD (2001) Effect of inclusion of steel fibres on the behaviour of reinforced concrete beams under different torsional states. Ph.D. thesis, Kakatiya University, India

Sustainable Infrastructures (Materials and Management)—High-Strength Nanoconcrete with the Replacement of Nanoflyash

Abhijeet Prasad Dash and Kirtikanta Sahoo

Abstract Using nanolevel adjustments in the development of advanced materials is the new trend now, and this nanotechnology can help in enhancing the properties of construction materials. The nanomodification of materials has wide areas of applications in changing and improvising the strength and durability of concrete. An effort was taken to explore the new areas of concrete using nanoflyash. Class-F flyash was crushed to nanoform in the high-energy planetary ball mill to produce nanoflyash. A high-strength concrete of M70 grade was prepared with normally available materials (NCC), and then the cement was replaced with 10, 15 and 20% of bulk flyash to produce flyash high-strength concrete (FAC). Then 2, 4, 6 and 8% of nanoflyash was applied to the FAC to get even stronger concrete or to compensate the strength loss due to addition of flyash to this high-strength concrete. The compressive strength and workability of this nanoconcrete were figured out and there was increase in strength in the range of 10 to 24% for various percentages of addition of nanoflyash to flyash high-strength concrete (FAC) after the comparison with NCC and FACs.

Keywords Nanoflyash · M70 · Nanoconcrete · Compressive strength · Workability

1 Introduction

Nanotechnology has changed our perception and will be continuously changing our insight, prospects and abilities to control the materials in a deeper level. These growth and improvements will certainly affect the construction world and civil engineering materials. Nanobinding materials or nanomodified engineering materials with a nanosized cementitious component or other nanolevel particles may be the next revolutionary evolution. Flyash replacement to the cement is also in the surge in order to develop durable, green and cost-effective concrete but with some compromise in

---

A. P. Dash · K. Sahoo (✉)
KIIT Deemed to be University, Bhubaneswar, Odisha, India
e-mail: kirtikanta.sahoofce@kiit.ac.in

A. P. Dash
e-mail: avgtdash@gmail.com

B. B. Das et al. (eds.), *Recent Developments in Sustainable Infrastructure*, Lecture Notes in Civil Engineering 75, https://doi.org/10.1007/978-981-15-4577-1_42

its strength. An experimental effort was made to create nanoflyash and then made concrete by putting nanoflyash to replace the cement. Class-F flyash was crushed to nanolevel in the planetary ball mill to produce nanoflyash. The surface of the flyash was made to rough and more reactive from smooth, glassy and inert. A high-strength concrete of M70 grade was casted, and then 10, 15 and 20% of cement was replaced with bulk flyash. Then the 10% bulk flyash in concrete was again replaced with 2, 4, 6 and 8% nanoflyash. The workability in terms of slump value and compressive strength of concrete with nanoflyash were figured out and there was an increase in the compressive strength of concrete. The workability of nanoflyash concrete found to be significantly more than that of NCC and some of FACs.

2 Experimental

2.1 Materials

A by-product of coal-fired power generating plants flyash or flue ash generally arrested in electrostatic precipitators. Design experts often specify strict necessities, such as quantities of Class-F flyash, slag, low-alkali cement, the use of a non-reactive aggregates, etc.; to avoid alkali–silicate reaction (ASR)-related distress in structures. Using class-F flyash and high-energy ball milling, we have tried to achieve the particles at nanolevel called nanoflyash or nFlyash. Ordinary Portland cement 53 grade with specific gravity 3.14 was procured from local dealer. Naphthalene-based superplasticizer Fosroc Conplast SP-430 was used.

2.2 Methodology

2.2.1 Preparation of Nanoflyash

The details of flyash sample used for the preparation of nanoflyash were given in Table 1. Flyash sample was then oven dried in 100 °C for 3 h to make it free from moistures present in it. Then it has been sieved for 15 min using IS sieve 75 micron and the sieved sample was taken for dry ball milling. A high-energy planetary ball mill (Model: Retsch, PM 100, Germany) was being used for the grinding process. In a stainless steel chamber, tungsten carbide and zirconia balls of 10 mm diameter sizes were being used for achieving nanosized particles. The milling duration was 10 h. The

Table 1 Chemical composition of flyash, weight in %

SiO_2	Al_2O_3	Fe_2O_3	TiO_2	CaO	MgO	Na_2O	K_2O
60.31	28.5	5.4	1.87	0.93	0.63	0.57	0.80

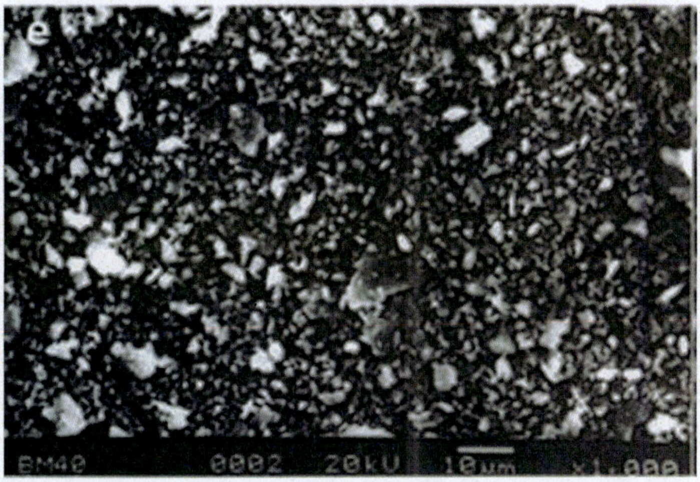

Fig. 1 SEM image after 8 h of dry ball milled flyash

rotation speed of the planet carrier was 300 rpm. The BPR (ball-to-powder weight ratio) of 10:1 was loaded to the mill. It was a dry milling to avoid agglomeration; no Toluene was being used as the medium with an anionic surface active. The ball mill was to be stopped at every hour for 15–25 min to avoid agglomeration and then rubbing the surface, and then again loaded for another 1 h. Then the powdered sample was collected in every 2 h of milling. Figure 1 represents SEM image of sample and Fig. 2 shows the XRD pattern of the respective sample.

In the period of only 10 h of ball milling, the size of flyash reduced to nanolevel from microlevel. The crystallite size of flyash drastically reduced and the percentage of crystallinity was reduced to 38 from 63% and it became more amorphous.

2.2.2 Preparation of High-Strength Concrete M70 Grade

Ordinary Portland cement 53 grade with specific gravity 3.14 was procured from local dealer. Naphthalene-based superplasticizer 'Fosroc Conplast SP-430' was used. Sand zone III and coarse aggregate 10 mm down were used. Trials have been taken to get to the appropriate proportion and mix. The first trial mix did look under sanded and higher quantity of admixture making it set slower. Strength result was not at all good at just about 62.76 MPa. It is necessary to increase the water–cement ratio and reduction of admixture percentage. The second trial mix looked cohesive but still there was need of the reduction in admixture. The second trial had given us some improved but not sufficient result of 66.03 MPa. So we have taken the third trial which proved to be appropriate mix as well as giving the good result in the strength of 77 MPa. From various trials taken, trial-3 has been taken for its good result which is provide in Tables 2 and 3 respevctively.

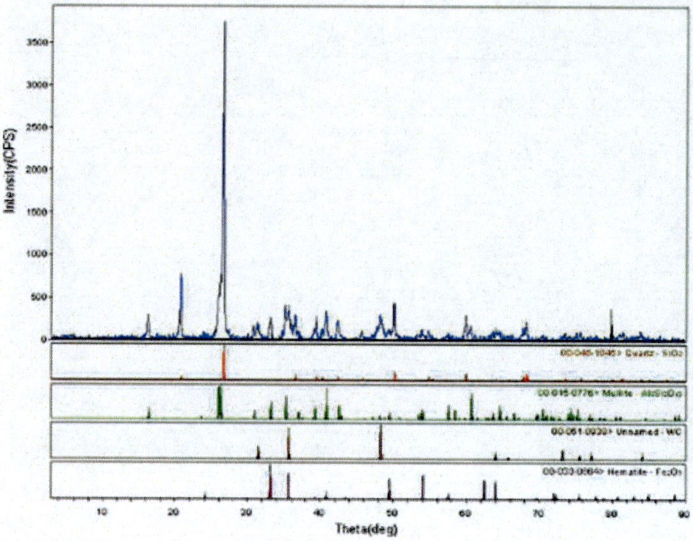

Fig. 2 XRD patterns of 10-h dry ball milled flyash

Table 2 28 days strength of normal cement concrete (NCC)

Cube no.	Slump value in mm	Avg. 28 days compressive strength in MPa
1	75	76.97 ≈ 77
2	75	
3	75	

Table 3 28 days strength of flyash concrete (FAC) at 10, 15, 20% replacement

Mix designation	Slump value in mm	Avg. 28 days compressive strength in MPa
FAC10	50	72.60
FAC15	70	63.85
FAC20	80	55.65

2.2.3 Preparation of Nanoflyash Concrete

The strength in higher percentages of flyash replacement in high-strength concrete (NCC) showed us that the strength is decreasing partly as compared to the NCC.

Table 4 Strength of nanoconcrete with different percentages of nanoflyash

Mix designation (%)	Slump value in mm	Avg. 28 days compressive strength in MPa
10nF2	60	72.87
10nF4	60	78.61
10nF6	70	85.17
10nF8	80	76.97

This opened the way to apply the flyash at nanolevel at certain replacement to the cement and flyash as well. The nanoflyash was applied at different percentages of 2, 4, 6 and 8% with bulk flyash in 10% FAC. Meaning of 10nF2% is 10% flyash with 2% nanoflyash replacement which is provide in the Table 4.

3 Results and Discussion

The replacement of flyash to the high-strength concrete as shown in the Graph 1 indicates a decrement in the strength of concrete gradually from a level of 77 MPa to a level of 55 MPa of about 30% and to compensate this 30% loss in strength it encourages the nanoflyash replacement to the cement. Graph 2 indicates increase in workability (as slump value) as there is increase in the percentage of flyash. There is 40–130% increase in workability with addition of up to 20% of flyash. The variation of the 28 days compressive strength of nanoconcrete with different percentage nanoflyash replacement of cement is shown in Graph 3, which shows that there is a healthy percentage of increase in strength. For 10% flyash replacement to the cement, the strength of M70 grade high-strength concrete is lowered by 6–8%

Graph 1 28 days compressive strength

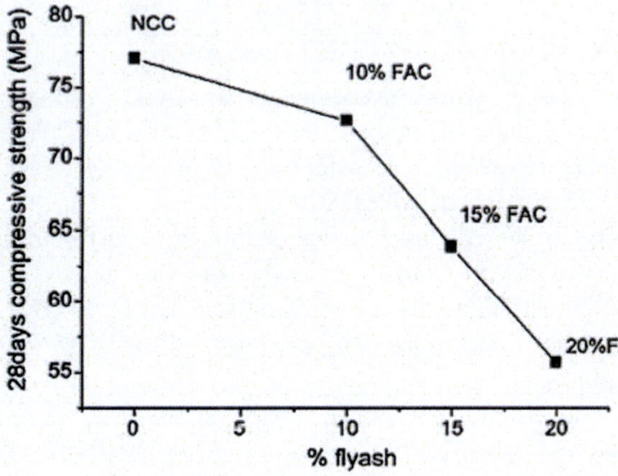

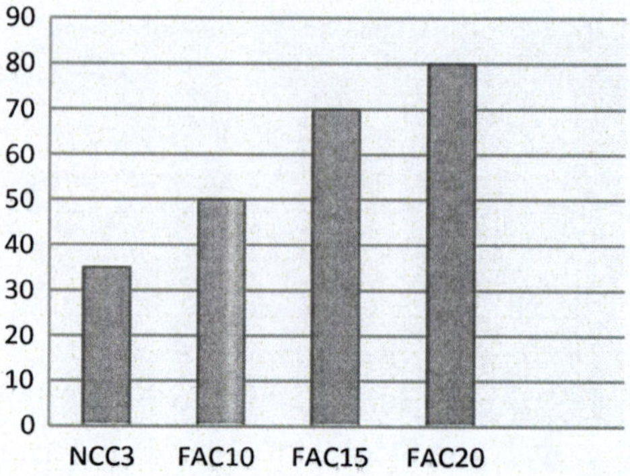

Graph 2 Workability (slump in mm)

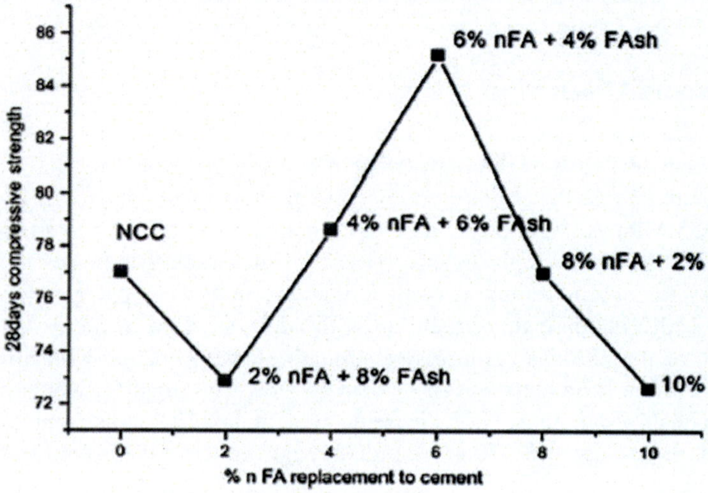

Graph 3 28 days compressive strength of nanoflyash concrete (nFAC)

up to 72 MPa, which has been compensated by replacing 2, 4, 6 and 8% nanoflyash. The percentage of increase in strength shown in Graph 3 indicates an increase of 10% of strength up to 85 MPa from 77 MPa of NCC and an increase of 18% strength from 72.60 MPa of 10% FAC.

The strength of concrete was decreasing with the addition of flyash and the addition of nanoflyash to concrete increases the workability, where some strength gain was also there with addition of nanoflyash. Up to 40–100% increase in workability of nanoflyash concrete as compared to NCC of M70 grade and 10% FAC is shown in Graph 4.

From Graph 5, it can be observed that the strength as well as workability is procured in the mix with 6% nanoflyash + 4% flyash. The strength gain is up to 10%

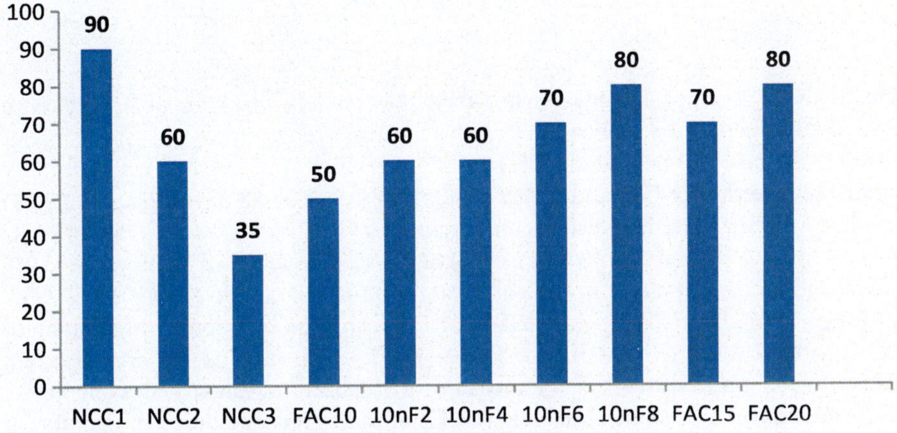

Graph 4 Workability (slump in mm) of all trial NCCs, FACs and nFACs

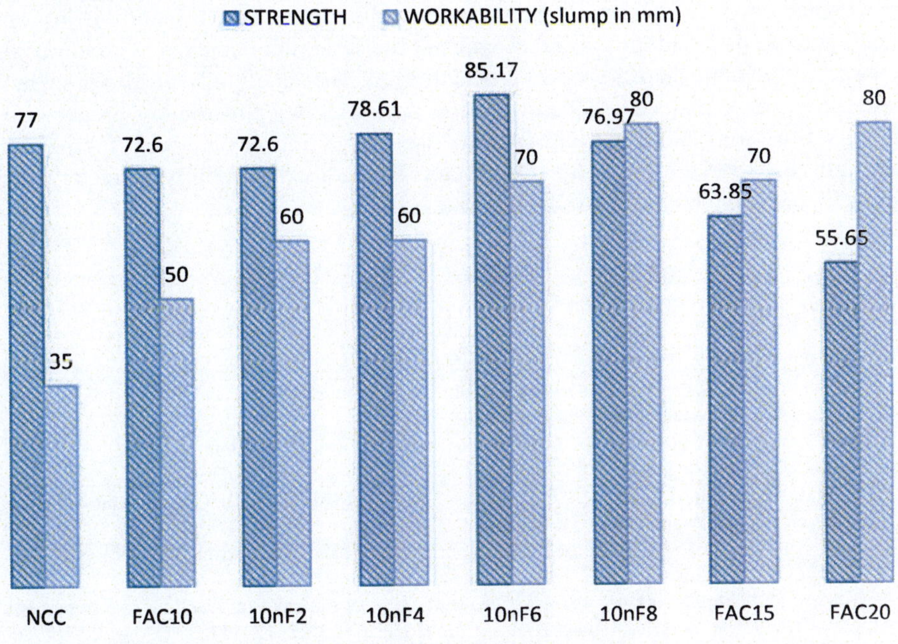

Graph 5 Strength and workability relationships of various mixes

as compared to NCC and there is more than 100% gain in the value of workability compared to NCC.

4 Conclusion

Due to the addition of nanolevel materials to the concrete which are obviously more effective and active, the nanoconcrete with nanoflyash replacement was giving higher result than that of high-strength concrete of M70 which was prepared with normally available materials (NCC) and flyash concrete (FAC). The 28 days strength of M70 grade concrete with nanoflyash was found to be 03 to 10% higher than that of NCC and 09 to 18% higher than that of 10% flyash-based concrete of same grade (10% FAC) and up to 50% more in case of 20% flyash-based concrete (20% FAC). The 28 days strength of (6% nanoflyash + 4% flyash) combination in high-strength concrete were found to be giving the highest result for all variable proportions taken. There is also up to 40 to 100% increase in workability of nanoflyash concrete as compared to normal cement concrete (NCC) of M70 grade. Addition of nanoflyash to high-strength flyash concrete leads to increase in both strength and workability as well. When the percentage of addition of nanoflyash (here 8%) increases, the strength graph decreases, making the cube weight lesser which is leading us to a conclusion of more structural expansion and volume as well. This study is just an attempt to prepare nanoflyash and to improve the strength of concrete with minimal cost and efforts, but there are lots of tasks left to do. The nanoflyash can also prepared with some other easier methods in coming days as today's method of preparation is very time-consuming and the quantity we are getting is very less. Taking the strength characteristics, there are other factors like study of durability characteristics of nanoconcrete which can also be the next leap.

References

1. Ge G, Gao Z (August 2008) Application of Nano technology and Nano materials in construction. Adv Int Constr Edu Res Prac, 235–238
2. Li G (June 2004) Properties of high volume flyash concrete incorporating nano-SiO_2. Cem Concr Compos, 1043–1049
3. Raki L, Beaudoin J, Alizadeh R, Makar J, Sato T (March 2010) Cement and concrete nanoscience and nanotechnology. Materials, 924–930
4. Li H, Xiao HG, Yuan J, Ou J (March 2003) Microstructure of cement mortar with nano-particles. Compos Part B: Eng 35
5. Gaitero JJ, Campillo I, Guerrero A (2008) Reduction of the calcium leaching rate of cement paste by addition of silica nano-particles. Cem Concr Res 38:1112–1118
6. Mann S (November 2006) Nanotechnology and construction. Nano forum, First edition, 5–9
7. Swamy, RN, Ali ARS, Theodorakopoulos DD (1983) Early strength flyash concrete for structural applications. ACIJ, 414–423
8. Balaguru P, Chong K (2006) Nanotechnology andconcrete research opportunities. In: Proceedings of ACI Session, pp 19–28
9. ACI Committee 363R (2004) State-of-the-Art report on high strength concrete. ACI Manual of Concrete Practice, Part –I
10. Thomas K, Satpathy SK, Manna I, Chakraborty KK, Nando GB (August 2007) Preparation and characterization of Nano-structured materials from flyash. Nano Res Lett, 397–404

11. Arjayan PM, Ebbesen TW (1996) Nanometre-size tubes of carbon. Rep Progr Phy 60(1997):1025–1062
12. Shah SP, Konsta-Gdouto MS, Metaxa ZS, Mondol P (2009) Nanoscale modification of cementitious materials. Nanotechnol Constr, 125–130

Light Weight Concrete Using Light Weight Expanded Clay Aggregate and Dry Saw Dust

Honey Mishra and Ankit Jena

Abstract The dynamic nature of concrete to be shaped into an assortment of structures along with its ability to withstand tremendous compressive loads has made it the most widely used construction material of this era. Light weight concrete is characterized with a lower density than normal concrete viz *two-third of normal concrete.* Hence, it allows a greater degree of freedom while considering slender sections at some places leading to creation of increased workable space. In this paper, the properties (physical and mechanical) of LWC using LECA and DSD have been discussed. Going through literature review it was noticed that the features of LECA and DSD include lightweight, thermal insulation, superior acoustic insulation, earthquake resistance, and higher compressive strength. Specimen cubes of dimension $150 \times 150 \times 150$ mm were tested for various proportions of LECA and DSD. With the replacement combination of 7.5% LECA as coarse aggregate and 10% DSD as fine aggregate (by weight), it has been observed that the desired compressive strength was achieved with a weight reduction of approximately 100 kg/m^3. However, further successive increment in percentage of LECA and DSD shows a decreasing trend in compressive strength. Thus, to choose between LWC and NC, a careful analysis should be done keeping in mind the higher cost and skill required to achieve the perfect proportioning of the above-said materials.

Keywords Light weight expanded clay aggregate (LECA) · Dry saw dust (DSD) · Light weight concrete (LWC) · Light weight concrete aggregate (LWCA) · Normal concrete (NC) · Compressive strength

1 Introduction

Majority of the research emphasize on higher performance of concrete, which means a cost-efficient material is needed which fulfills the performance criteria along with

H. Mishra (✉) · A. Jena
Department of Civil Engineering, GIFT, Bhubaneswar, India
e-mail: honey.mishra240@gmail.com

A. Jena
e-mail: jena.ankit@gmail.com

B. B. Das et al. (eds.), *Recent Developments in Sustainable Infrastructure*, Lecture Notes in Civil Engineering 75, https://doi.org/10.1007/978-981-15-4577-1_43

durability. LWC may be categorized as the concrete that contains expanded agents, which results in an increase in the volume of the mixture while providing other characteristics such as lowering the dead load. The advantage of light weight concrete includes maintaining large voids so as not to form laitance layer or cement films when placed on walls. Light weight concrete is preferred over normal concrete for its lower overall cost, faster building rates, lower transportation and handling costs, and low density. Light weight concrete has considerably lower strength than normal concrete made up of coarse aggregates. Its significance is emphasized where light weight gets preference over strength of the concrete. These are favorable for elevated structures, curved members where the dead weight of the concrete plays a vital role, and hence the ratio of strength to weight is crucial than the strength. Another efficiency of light weight concrete is its low thermal conductivity, which can be attributed to its high void percentage. As the strength is fundamentally influenced by the coarse aggregates, the workability depends mostly on the fine aggregates, and hence concrete is sometimes made with conventional fine aggregates and coarse aggregates which are lightweight. Normally, the LWCA is immersed in water for 24 h prior to its use.

2 Experimental Program

Materials Used

2.1 LECA

LECA is round compressed coarse aggregate-like structure which is manufactured by burning natural clay at 1200 °C. It is hard and is full of voids which are interconnected to form a honeycomb-like structure within the aggregates. The aggregate size varies from 0 to 30 mm and depending on the final use these aggregates are processed and graded. Lighter weight, durability, and heat resistance are the chief characteristics of LECA. These aggregates absorb approximately 5–10% water when immersed for 1 h and 50–100% absorption occurs for longer immersion periods (Table 1 and Fig. 1).

Table 1 Physical properties of LECA

Sl no.	Property	Value
1	Specific gravity	0.56
2	Water absorption	18%
3	Impact value	49.68

Fig. 1 Figure showing LECA

2.2 *Dry Saw Dust*

Saw dust is a by-product of wood and timber industry. It consists of powdered residue from various soft woods and hard woods after sawing and related activities (Fig. 2).

From Table 2, the specific gravity of dry saw dust (DSD) 0.37 places it in the category of lighter weight aggregates, while the specific gravity of sand and coarse aggregates is categorized as common rock group whose specific gravity ranges from 2.62 to 3.

Fig. 2 Figure showing dry saw dust

Table 2 Physical properties of dry saw dust

Sl no.	Property	Value
1	Loose bulk density Mg/m^3	0.08
2	Compacted bulk density Mg/m^3	0.12
3	Water absorption (%)	20%

Table 3 Physical properties of cement

Sl no.	Property	Value
1	Normal consistency	30%
2	Initial setting time	30 min
3	Final setting time	8 h
4	Specific gravity	3.15
5	Fineness of cement	3%

Table 4 Physical properties of sand

Sl no.	Property	Value
1	Specific gravity	2.62
2	Water absorption	1%
3	Bulking of sand	11.9%
4	Fineness of sand	2.57

2.3 Cement

In this experiment, OPC 53 grade cement confirming to IS 12269-1987 is used (Table 3).

2.4 Fine Aggregate

Sands which confirm to zone II and passing through 4.75 mm IS sieve are taken (Table 4).

2.5 Coarse Aggregate

The granites which are locally available and retained on 20 and 10 mm IS sieves are used (Table 5).

3 Methodology

In this study, M35 concrete is used, the concrete mix design for standard conventional concrete is prepared by referring IS 10262:2009 (Table 6).

The mix design is done using the lighter weight aggregates such as LECA and DSD. These are partially substituted for conventional coarse aggregate and fine

Table 5 Physical properties of coarse aggregates

Sl no.	Property	Value
1	Specific gravity	2.67
2	Water absorption	0.5%
3	Fineness modulus	3.44
4	Crushing value	12.40%
5	Impact value	12.50%
6	Abrasion value	14.5%
7	Flakiness index	35.85%
8	Elongation index	45.45%

Table 6 Mix proportion

Material	Quantity
Cement	400 kg/m^3
Water	172 lit/m^3
Fine aggregates	642 kg/m^3
Coarse aggregate	1165 kg/m^3
Water–cement ratio	0.43

aggregate with various percentages varying from 0 to 10% in various combinations (Table 7).

The specimens were exposed to standard curing which was carried out at room temperature for 28 days. After curing was done, the specimens were then subjected to compressive strength testing machine to determine the values of corresponding strengths.

Table 7 Replacement combinations

Mix	Percentage of LECA	Percentage of saw dust
M0	0	0
M1	5	5
M2	5	7.5
M3	5	10
M4	7.5	5
M5	7.5	7.5
M6	7.5	10
M7	10	7.5

Table 8 Fineness of cement

Sieve size	Sample taken (g)	Passing of sample (g)	Percentage passing (%)
90 micron	100	90	90

Table 9 Consistency of cement

Sl no.	Weight of sample (g)	Percentage of water added (%)	Penetration value (mm)
1	300	26	32
2	300	28	20
3	300	30	7

Standard consistency of cement = 30

4 Results and Discussions

4.1 *Cement Fineness*

The fineness of cement is determined by dry sieving process, confirming to IS: 4031 (Part 1)—1996 (Table 8).

4.2 *Cement Consistency*

The consistency of cement is when the Vicat's plunger penetrates to a point of 5–7 mm from the bottom of the Vicat's mold when it is tested (Table 9).

4.3 *Slump Test*

Workability of concrete is determined by doing slump cone test either at laboratory or at construction site. This test is done to check the homogeneity as well as quality of the concrete (Table 10).

It can be seen from the above table that with the increase in the percentage of DSD for a particular percentage of LECA, the slump value reduces which indicates that addition of DSD reduces the workability, whereas with increase in percentage of LECA the workability increases. Values of 60, 70, and 115 mm were obtained for 5, 7.5, and 10% replacement of LECA, but were reduced to 57 and 55 mm in M2 and M3. The slump values got reduced for 7.5 and 10% replacement of saw dust to 67 and 63 mm for mixes M5 and M6. The reason for reduction in slump values can

Table 10 Slump value

Mix	% of LECA	Percentage of saw dust	Slump value (mm)
M0	0	0	45
M1	5	5	60
M2	5	7.5	57
M3	5	10	55
M4	7.5	5	70
M5	7.5	7.5	67
M6	7.5	10	63
M7	10	7.5	115

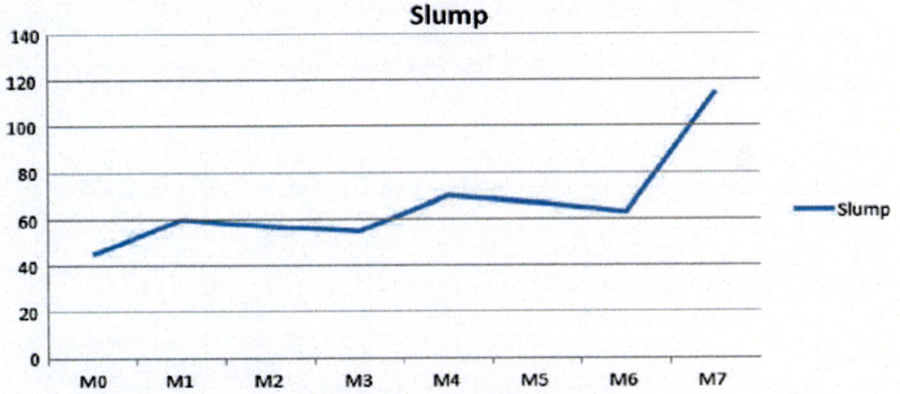

Fig. 3 Graph showing slump values for different mix proportions

be attributed to the fact that the increase in amount of DSD resulted in absorption of water from the mix thereby reducing the workability (Fig. 3).

4.4 Compressive Test of Concrete Cube

Compressive testing machine is used to find the compressive strength of cube and cylindrical specimens. For each set of tests, average of three specimens is taken into consideration.

The control mix M0 exhibited 30.32 and 40.44 N/mm^2 strength at 7 days and 28 days of curing. But as the fraction of LECA and DSD was increased, the strength started to decrease (Fig. 4).

Form Tables 11 and 12 of compressive strength and weight value, it can be seen that with the simultaneous increase of percentage of LECA and DSD, the compressive strength and weight at 7 days and 28 days goes on decreasing. It can be inferred

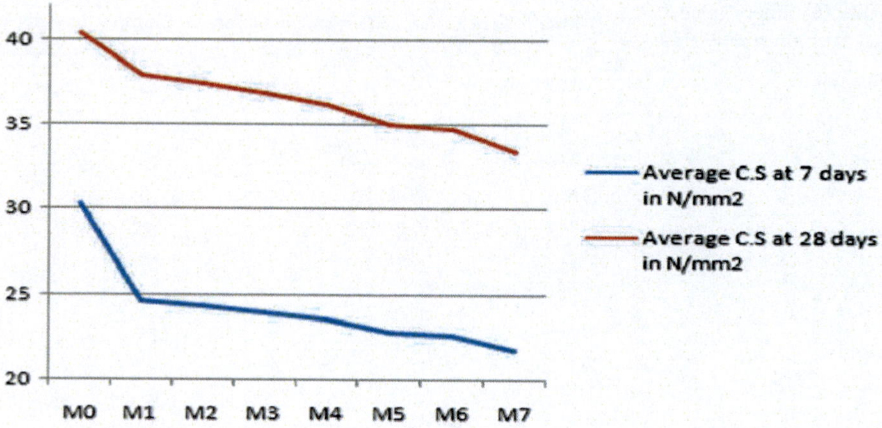

Fig. 4 Graph showing compressive strength for 7 days and 28 days of conventional and light weight concrete (LWA)

Table 11 Compressive strength value

Mix	Percentage of LECA	Percentage of saw dust	Average C.S at 7 days (N/mm^2)	Average C.S at 28 days in (N/mm^2)
M0	0	0	30.32	40.44
M1	5	5	24.62	37.88
M2	5	7.5	24.32	37.42
M3	5	10	23.96	36.87
M4	7.5	5	23.54	36.23
M5	7.5	7.5	22.75	35.03
M6	7.5	10	22.58	34.75
M7	10	7.5	21.74	33.46

Table 12 Weight value

Mix	LECA (%)	Percentage of sawdust (%)	Average weight at 7 days (kg)	Average weight at 28 days (kg)
M0	0	0	8.58	8.75
M1	5	5	6.97	7.21
M2	5	7.50	6.76	7.06
M3	5	10	6.53	6.84
M4	7.50	5	6.42	6.63
M5	7.50	7.50	6.24	6.42
M6	7.50	10	6.12	6.34
M7	10	7.50	5.87	6.28

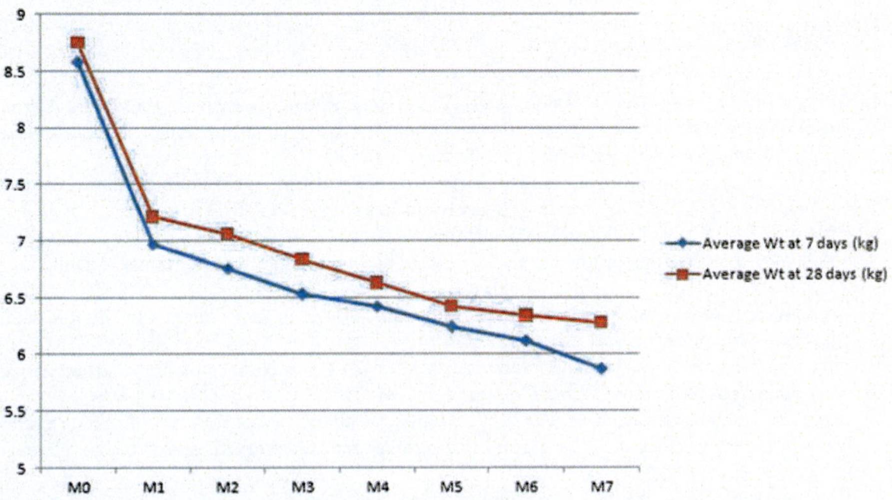

Fig. 5 Graph showing variation of weight for 7 days and 28 days of conventional and light weight concrete (LWC)

that the reduction in strength is a result of decrease in density. Also the presence of saw dust elicits lignocellulolytic decomposition which results in overall reduction of strength of the concrete. Hence, the saw dust should be tested for proportional presence of these materials and some pre-treatment of the saw dust may be done to reduce the effect (Fig. 5).

5 Conclusion

It can be observed from the graph of slump value that for M6 mix the slump value is 63 mm, this concrete is generally used in all concrete constructions with light reinforcement. Hence, it can be a feasible replacement of conventional concrete in construction of residential buildings. When compressive strengths of conventional concrete and light weight concrete are compared, it can be noticed that the strength of light weight concrete for both 7 days and 28 days is close to that of conventional concrete. This indicates that with LECA and saw dust substituted partially, satisfactory strength of concrete can be achieved with the optimum mix of M6. From the graph of average weight of light weight concrete and conventional concrete, it can be inferred that the criteria for light weight is also achieved.

Bibliography

1. Minapu LK, Ratnam MKMV, Rangaraju U (2014) Experimental study on Light Weight Aggregate Concrete with Pumice stone, silica fume and fly ash as a partial replacement of coarse aggregate. Int J Innovative Res Sci, Eng Technol 3:18130–18138 ISSN 2319-8753
2. Uglyanitsa AV, Gilyazidinova NV, Zhikharev AA, Kargin AA (2015) Study of reinforcement corrosion in expanded clay concrete. Hous Build Res Centre 11:307–310
3. Payam S, Lee JC, Mahmudc HM, Mohammad AN (2018) A comparison study of the fresh and hardened properties of normal weight and light weight aggregate concretes. J Build Eng 15:252–260
4. Corinaldesi V, Moriconi G (2015) Use of synthetic fibres in self compacting light weight aggregate Concretes. J Build Eng 4:247–254
5. Suraneni P, Fu T, Azad VJ, Isgor OB, Weiss J (2018) Pozzolanicity of finely ground light weight aggregates. Cement Concr Compos 1(5):214–218
6. Sergey AM, Anna Yu Z, Galina SS (2015) Production technology of waterproof porous aggregates based on alkali silicate and non-bloating clay for concrete of general usage. Cement Concr Compos 111:540–544
7. Boudaghpour S, Hashemi S, A study on light expended clay aggregate (leca) in a Geotechnical view and its application in green house and greenroof cultivation
8. Shetty MS, Concrete Technology Theory and Practice. S. Chand Publications, Fifth Revised Edition 2005, New Delhi
9. Parhizkar T, Najimi M, Pourkhorshidi AR (2012) Application of pumice aggregate in structural light weight concrete. Asian J Civil Eng (Build Hous) 13(1):43–54
10. Sivalinga Rao N, Radha Ratna Kumari Y, Bhaskar Desai V, Swami BLP (2013) Fibre reinforced light weight aggregate (natural pumice stone) concrete. Int J Sci Eng Res 4(5), ISSN 2229–5518
11. Banthia N, Trottier J (1994) Concrete reinforced deformed steel fibbers, part 1: Bond-slip mechanisms. ACI Mater J 91(5):435–446
12. Compione G, Mindess S, Zingone G (1999) Compressive stress-strain behavior of normal and high- strength carbone-fiber concrete reinforced with steel spirals. ACI Mater J 96(1):27–34
13. Balaguru P, Ramakrishnan V (1987) "Properties of light weight fiber reinforced concrete", fiber reinforced concrete-properties and applications, SP105. Detroit, Michigan, American Concrete Institute, pp 305–322
14. Sheth A, Goel A, Vekatram Pai BH (2014) Properties of concrete on replacement of coarse aggregates & cementitious materials with styfoam & rice hush ash respectively. Am J Eng Res (AJER) 03:268–271
15. Bhaskar Desai V, Sathyam A (2014) Some studies on strength properties light weight cinder aggregate concrete. Int J Sci Res Publ 4(2), ISSN 2250-3153
16. Hossain KMA (2013) Properties of volcanic pumice based cement & light weight concrete. Cement Concr Res 34(2004):283–291
17. Shafigh P, Hassanpour MM, Razavi SV, Kobraei M (2011) An investigation of flexural behaviour of reinforced light concrete beams. Int J Phys Sci 6(10):2414–2421, ISSN 1992-1950
18. Yu QL, Spiesz P, Brouwers HJH (2013) Development of cement based light weight composites. Elsevier 44:17–29

A Study on Properties of Concrete Using Silica Fume and Brick Aggregate

Srishti Saha and Joyanta Pal

Abstract An extensive investigation was carried out for recycling of demolished brick aggregate concrete as coarse aggregate. This paper explains behaviour of brick aggregate concrete with addition of SF while partial replacement of OPC. Natural coarse aggregate was replaced by brick aggregate and OPC was replaced by SF in various substitutions, i.e. 5, 6, 7, 8, 9, 10, 12 and 14% (by weight), respectively. The objective of the paper is to improve strength and also make durable concrete by using brick aggregate and SF. An experimental investigation has been carried out for concrete samples having silica fume at a constant slump. Effect of brick aggregate with silica fume on workability, compressive strength, dynamic Young's modulus of elasticity and UPV was investigated. It was found that strength of concrete increases with an increase in SF contents up to 10%, and beyond this limit the strength will be gradually decreasing. It was concluded that the use of SF up to 14% of brick aggregate concrete can be effectively utilized in performance of concrete.

Keywords Brick aggregate · Silica fume · Compressive strength · Dynamic young's modulus of elasticity · Ultrasonic pulse velocity

1 Introduction

Concrete is used extensively for construction purpose because of its structural stability, strength, durability, easy availability of its constituent materials, formability to any shape, etc. While producing the cement huge amount of carbon dioxide is emitted which is responsible to create imbalance in the atmosphere and as a result of that global warming occurs [1–3]. Akhtaruzzaman et al. [4] observed that when natural coarse aggregate was replaced by well-burned brick and crushed bricks, the crushed brick exhibits higher strength as compared to well-burned brick. J Dang et al. [5]

S. Saha (✉)
School of Civil Engineering, KIIT Deemed to Be University, Bhubaneswar, Odisha, India
e-mail: srishti.civil94@gmail.com

J. Pal
Civil Engineering Department, National Institute of Technology, Agartala, Tripura, India
e-mail: joyantanita@yahoo.com

B. B. Das et al. (eds.), *Recent Developments in Sustainable Infrastructure*, Lecture Notes in Civil Engineering 75, https://doi.org/10.1007/978-981-15-4577-1_44

reported that when 50% replacement of crushed concrete brick was done the compressive strength for 7 days and 28 days reduces by 14% and 20%, respectively. Debieb et al. [6] also show that due to the degree of substitution, strength decreases from 20 to 30% when CA and FA were replaced in concrete by coarse and fine crushed bricks. Khatib [7] and Poon et al. [8] also suggested that for replacement of fine aggregate by crushed brick as fine aggregate, it requires similar characteristics. Fly ash, GGBFS, RHA and SF are some industrial by-products which can be used in concrete industry as a partial replacement of cement [9, 10]. Silica fume is known as a by-product which is being made with the combination of silicon metal and ferrosilicon alloy. The usage of SF in the industry has been provided as waste production in concrete technology [11, 12]. Tasdemir et al. [13] reported that SF has to be used in concrete as a compulsory product. This conclusion was made because of drastic improvement in the characteristics of cement-paste aggregate which is nothing but the weakest zone of concrete matrix. Mitchell et al. [14] demonstrated the pattern of XRD for SF also changed a bit till 7 days after mixing with calcium hydroxide solution which was saturated and extensive C-S-H formation occured after 91 days. Larbi et al. [15] also show the consumption percentage of SF was only 75% after continuous hydration of 90 days. The drastic improvements happen in the properties of concrete since silica fume has become most valuable by-product and it was used in industry in an extensive manner between the pozzolanic materials in concrete and cement technologies. The replacement of SF percentage to obtain the strength of concrete for a duration of 28 days ranges from 10 to 20 percent. The objective is to study the performance of concrete used in brick aggregate and silica fume.

2 Materials and Methodology

2.1 Cement, Brick Aggregate, Water, Superplasticizer

The cementitious materials used have been OPC, 43 Grade as per IS: 8112-1989 [16]. Ordinary sand has maximum size of 4.5 mm according to the Indian code IS 383-1970 [17]. According to the IS 456: 2000 [18], clean water can be utilized which was free from impurities and chemical compounds. Consequently, Master Rheobuild 1125-high range retarding superplasticizer, which was basically sulphonated naphthalene polymer based superplasticizer was used. Keeping the water–cement ratio constant for all mixes, a naphthalene-based superplasticizer is added in order to maintain the workability.

Fig. 1 Brick aggregate

2.2 *Brick Aggregate*

The bricks were crushed by jaw crusher which produces flaky and elongated bricks. The production of breaking of bricks was done by hammer which produces angular aggregate and bricks having optimum size of 10 mm are used in this research purpose and waste bricks were collected from NIT Agartala, Barjala, Jirania. The mechanical and physical qualities were tested according to the IS standard 2386-1963. The specific gravity of the aggregate was found 1.95 after rigorous testing (Fig. 1).

2.3 *Silica Fume*

Silica fume as shown in Fig. 2 is one type of mineral admixture, popularly well known as micro-silica; it is a consequence of the by-product formation of metallic silicon or iron–silicon alloys. SF can be utilized in a wide range of procedures for enhancement of the characteristics of concrete. A dry powder of SF is readily available in the market; moreover, silica fume is a very improvident pozzolan when it is worn in concrete owing to its finer particle size, larger available surface area and

Fig. 2 Silica fume

Table 1 Chemical composition of OPC and SF (% by weight)

Sl. no.	CConstituent	OPC	Silica fume
1	Silicon oxide (SiO_2)	22.03	96
2	Aluminium oxide (Al_2O_3)	3.03	0.1
3	Iron oxide (Fe_2O_3)	4.45	0.7
4	Calcium oxide (CaO)	66.02	0.09
5	Magnesium oxide (MgO)	0.77	0.15
6	Sulphur trioxide (SO_3)	3.97	–
7	Sodium oxide (Na_2O)	1.09	0.05
8	Potassium oxide (K_2O)	1.2	0.39
9	Loss of ignition	0.88	0.7

higher composition of silicon dioxide. Silica fume for this study has been collected from Elcom Company, Navi Mumbai, Maharashtra, India. Table 1 shows chemical composition of OPC and SF.

2.4 *Preparation of Concrete Samples*

A nominal mix proportion of 1:2.04:2.32, 1 component cement, 2.04 element fine aggregate, 2.32 element brick aggregate with w/c proportion of 0.5 is utilized in this work. OPC was replaced by SF at different percentages, i.e. 5%, 6%, 7%, 8%, 9%, 10%, 12% and 14%, respectively. Details of mix proportion for brick aggregate concrete containing different percentages of SF are given in Table 2. Based on the mix proportion, cubes of 150 mm are cast for the purpose of compressive strength, ultrasonic pulse velocity test and dynamic modulus of elasticity. Moulds were kept ready and the inner side was properly oiled. Concrete was poured inside the moulds. Every specimen be cast up to three layers and packed down on a vibrating table until the elimination of air bubbles on or after the concrete. All specimens should be enclosed by means of plastic sheets and held in reserve for 24 h at room warmth. Subsequent to 24 h the specimen was opened and kept in curing chamber with sprinkling water for 28 days of casting uninterrupted.

2.5 *Test Procedure*

For obtaining the workability of concrete, slump test was done as per IS: 7320-1974 and IS: 1199-1959 [19, 20]. In this work, slump value was taken as 50 $\pm$ 75 mm. According to IS: 516-1959, compressive strength test of the samples was prepared [21]. The UPV test is the most frequently used practice among all the non-destructive tests of concrete samples which were performed as per IS 13311 (1992) part- [22].

Table 2 Mix design proportions of brick aggregate using different percentages of silica fume

Sl no	Mix id	Mix ratio	Replacement of silica fume (%)	Cement (Kg/m^3)	Silica fume (Kg/m^3)	Brick aggregate (Kg/m^3)	Natural aggregate (Kg/m^3)	Fine aggregate (Kg/m^3)	Water (Kg/m^3)	Superplasticizer (Kg/m^3)
1	NA0SF	20	0	342.78	0	0	699.1	795.87	171.4	3.42
2	BA0SF	20	0	342.78	0	699.1	0	795.87	171.4	3.42
3	BA5SF	20	5	325.64	17.139	699.1	0	795.87	171.4	3.42
4	BA6SF	20	6	322.22	20.567	699.1	0	795.87	171.4	3.42
5	BA7SF	20	7	318.78	24.00	699.1	0	795.87	171.4	3.42
6	BA8SF	20	8	315.36	27.42	699.1	0	795.87	171.4	3.42
7	BA9SF	20	9	311.93	30.85	699.1	0	795.87	171.4	3.42
8	BA10SF	20	10	308.50	37.28	699.1	0	795.87	171.4	3.42
9	BA12SF	20	12	301.65	41.13	699.1	0	795.87	171.4	3.42
10	BA14SF	20	14	294.79	47.99	699.1	0	795.87	171.4	3.42

3 Experimental Result and Discussion

3.1 *Fresh Property*

Workability of concrete using brick aggregate and OPC blended with silica fume in various percentages on replacement of cement was studied by means of slump tests. An experimental study was carried out of brick aggregate concrete samples having silica fume at a constant slump. Workability of concrete made with brick aggregates and blended silica fume in various percentages was monitored by various trials through slump tests. Plasticizers are added by keeping a constant slump value 50–75 mm. and slump standards for different mixes are revealed in Fig. 3.

3.1.1 Effect of Brick Aggregate on Workability of Concrete by Using Different Percentages of Silica Fume

The measurement of workability was done after 15–20 min of mixing with respect to slump. Initially, slump value was 110 mm by using natural aggregate with 0% silica fume, and it was observed that when coarse natural aggregate was replaced by brick aggregate the slump value decreases from 110 mm to 70 mm. When we use brick aggregate with silica fume (5% cementitious material), then it was found that the slump value decreases from 70 mm to 63 mm. As amount of silica fume increases by 6, 7, 8, 9, 10, 12 and 14% of cementitious material, it was observed that slump value decreases up to 45 mm. Workability of waste clay brick depends on the water absorption, porosity and pore size of aggregate and also increase of frictional resistance contributed by the effective surface area of silica fume is the main reason behind the reduction of slump value of waste aggregate comparison to normal concrete.

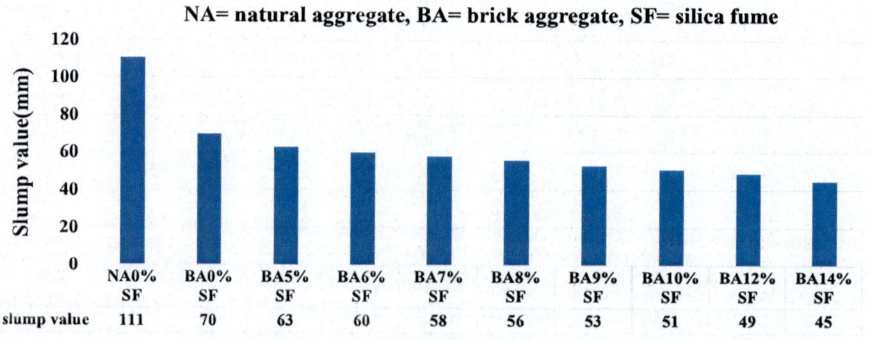

Fig. 3 Workability of concrete by using brick aggregate with different percentages of silica fume

4 Mechanical Properties

The effects of brick aggregate and SF on different properties of concrete are reported below.

4.1 *Compressive Strength of Concrete*

Determining the strength of concrete compressive strength is an important characteristic which was an essential criteria in structural design and detailing prospect. Figure 4 represents the strength development of concrete cubes at curing age of 28 days for different proportions.

4.1.1 Effect of Brick Aggregate on Compressive Strength of Concrete by Using Different Percentages of SF

The suitability of brick aggregate and SF of concrete was investigated. The strength of 0.15 m cubes containing NA 0% silica fume is having strength 27.4 MPa at 28 days. Experimental result shows that the strength of brick aggregate containing 0% silica fume was 12.71 MPa. But when 100% replacement of crushed concrete brick was done, the compressive strength for 28 days reduces by 20%, respectively. This indifferent behaviour may be due to the higher absorption capacity and good control of brick aggregate grading to a larger content. Due to this property, recycled clay brick aggregate can absorb large quantity of water which may obstruct the hydration of cement at early age. It was found that strength of brick aggregate gradually increases with the addition of SF content up to 10% to a maximum value of 17.79 MPa,

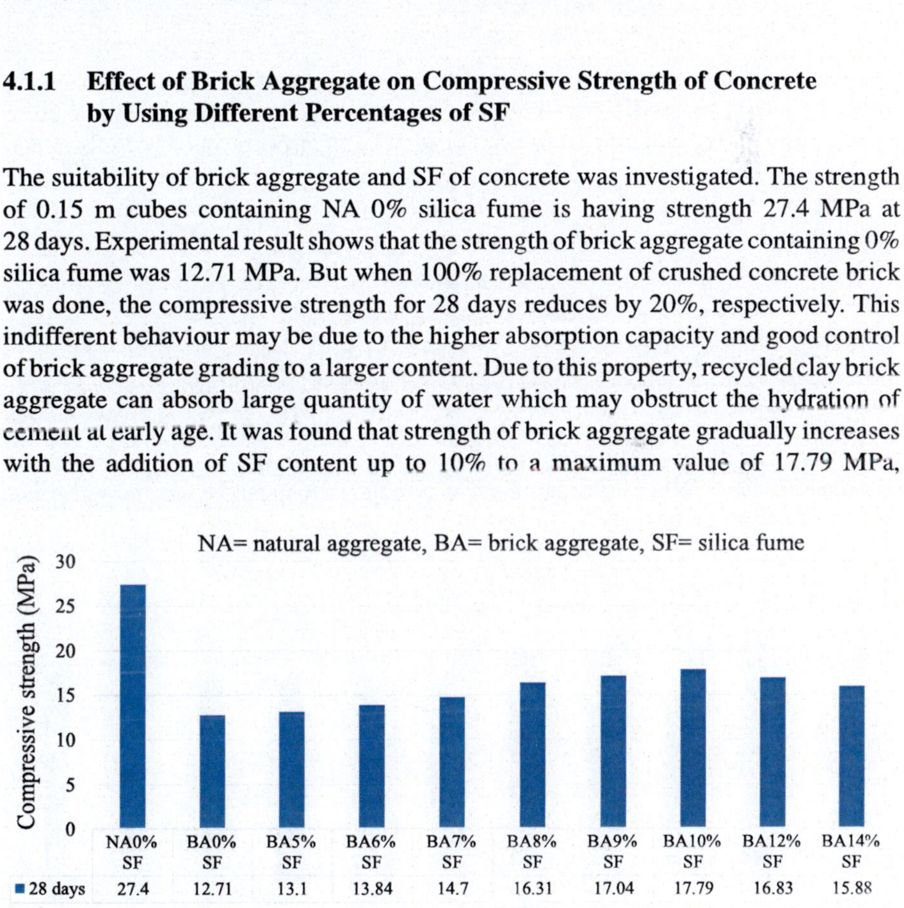

Fig. 4 Compressive strength of concrete by using brick aggregate with different percentages of silica fume

which is 40% more than the initial value. It can also be found that by increasing the SF contents by 12% and 14%, respectively, the compressive strength decreases to 16.83 MPa and 15.88 MPa, respectively, at the age of 28 days, which was 10 and 19% less than BF 0% SF. By incorporating brick aggregate with 10% SF, strength increases about 20% at curing age of 28 days.

The strength development in brick aggregate mixes which contains SF is possible because of the presence of more amount of silica which releases hydrogen after chemical reaction, and with calcium hydroxide extra calcium silicate hydrate gel is made, which is directly responsible for the development of mechanical property of concrete.

4.2 Ultrasonic Pulse Velocity Test

For determining the quality of concrete, the most important non-destructive testing method is a UPV method. It measures the time duration taken by the ultrasonic pulse to pass through the concrete to be examined. The time of travel of pulse between initial on set and the reception is measured electronically. The average speed of wave propagation is measured by dividing the path length among the transducer by the time of travel. Results of UPV test are shown in Fig. 5.

4.2.1 Effect of Brick Aggregate on UPV Test of Concrete by Using Different Percentages of Silica Fume

Inclusion of brick aggregate decreases the UPV for every mix compared to NA0% silica fume. At 28 days of curing, the UPV of NCA with 0% silica fume mix was recorded 5184 m/s. When brick aggregate was replaced to natural concrete aggregate,

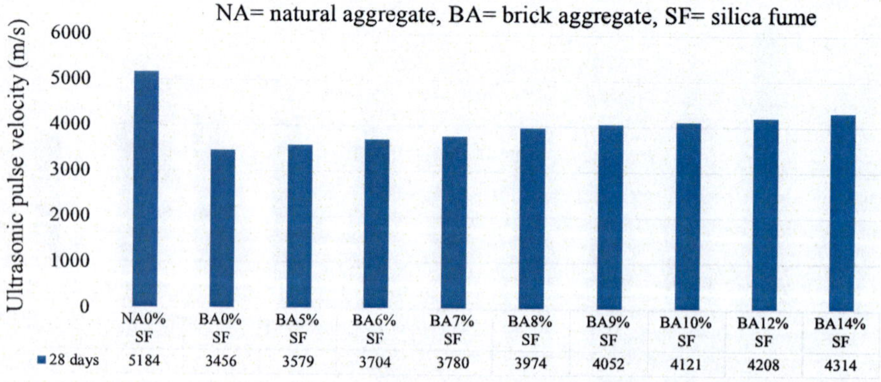

Fig. 5 UPV value of concrete by using brick aggregate various percentages of silica fume

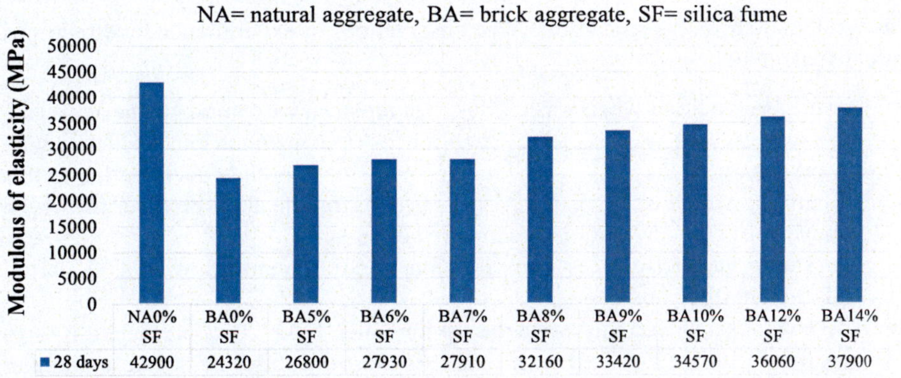

Fig. 6 Modulus of elasticity of concrete by using brick aggregate with different percentages of silica fume

15% less UPV value was recorded as compared to that of natural aggregate. Adhered mortar and low adhesiveness are some property on which strength depends. The strength gain of concrete containing brick aggregate is less as compared to natural aggregate because an industrial brick waste old adhered mortar is present and bonding between the mortar and aggregate is weak and porous.

With the addition of brick aggregate with silica fume content of 5 to 14%, the increase in UPV value by concrete mixes was observed 2.48, 3.24, 5.18, 5.96, 6.65, 7.52 and 8.58%, respectively, and the probable reason for increase in UPV value may be the increase in density of concrete due to the presence of silica fume which is finer material. Experimental values revealed that UPV values for brick aggregate mixes exceeded 3400 m/s. Therefore, each mix can be stratified under superior quality concrete in accordance to IS 13311 (part 1)-1992. It is established that brick aggregate can be more durable up to 14% replacement of SF.

4.3 *Dynamic Young's Modulus of Elasticity*

The measured elastic modulus of natural coarse aggregate and brick aggregate concrete containing different percentages of SF is also found at 28 days curing age and it was shown in Fig. 6.

5 Conclusion

A series of experiments were conducted sequentially for the determination of mechanical properties where natural coarse aggregate has been replaced by brick aggregate and cement has been replaced partially with industrial waste material as mineral admixture like silica fume. The observations of experimental investigations are discussed below:

- It was observed that when waste brick was replaced by natural coarse aggregate 20% less strength was recorded. This indifferent behaviour may be, when they are pre-saturated they act as self-curing agents with in the concrete, due to this property by which brick aggregate can absorb large quantity of water for which it may obstruct the hydration of cement at early age.
- The slump value decreases with addition of SF content up to 14% as partial replacement of cement.
- The compressive strength increases till the silica fume content reaches 10% to maximum value of 17.79 MPa which is 40% more than brick aggregate containing 0% silica fume.
- Maximum gain in strength was observed in brick aggregate containing 14% silica fume.
- Inclusion of SF content from 0 to 14% ultrasonic pulse velocity (UPV) value got increased by 25%.
- Further, by increasing the SF content 0 to 14%, modulus of elasticity value got increased from 24320 MPa to 37900 MPa which makes concrete more elastic.
- The properties like compressive strength, UPV value, and modulus of elasticity of brick aggregate are improved with the addition of silica fume, because of the presence of more amount of silica which releases hydrogen after chemical reaction, with calcium hydrooxide that makes extra calcium silicate hydrate gel.

References

1. Rashad AM, Hosam El-Din H. Seleem, Shaheen AF (2014) Effect of silica fume and slag on compressive strength and abrasion resistance of HVFA concrete. Int J Concr Struct Mater 8:69–81
2. Pradhan D, Dutta D (2013) Influence of silica fume on normal concrete. Int J Eng Res Applic 3:79–82
3. Dilip Kumar Singh R, Sil A (2012) Effect of partial replacement of cement by silica fume on hardened concrete. Int J Emerg Technol Adv Eng 2:472–475
4. Akhtaruzzaman A, Hasnat A (1983) Properties of concrete using crushed brick as aggregate. ACI Con Int Des Constr 5:58–63
5. Dang J, Zhao J, Hu W, Du Z, Gao D (2018) Properties of mortar with waste clay bricks as fine aggregate. Constr Build Mater 166:898–907
6. Debieb F, Kenai S (2008) The use of coarse and fine crushed bricks as aggregate in concrete. Constr Build Mater 22:886–893
7. Khatib JM Properties of concrete incorporating fine recycled aggregate. Cem Concr Res 35:63–769
8. Poon CS, Chan D (2007) The use of recycled aggregate in concrete in Hong Kong. Res Conserv Recyl 50:293–305
9. Ajileye Victor F (2012) Investigations on microsilica (Silica Fume) as partial cement replacement in concrete. Global J Res Eng 12:17–23

10. Raveendran KG, Rameshkumar V et al (2015) Performance of silica fume on strength and durability of concrete. International Journal of Innovative Research in Science, Engineering and Technology 4:10162–10166
11. Seshasayi LVA, Ramaseshu D, Shankaraiah R (2001) Effect of cement Replacements by Fly ash and silica fume on compressive strength of concrete. Int Concr Abs Port 199:581–594
12. Singh P, Mohd, Khan A, Kumar A (2016) The effect of concrete by partial replacement of cement by silica fume. Int Res J Eng Res Technol 3:118–121
13. Darshita T, Anoop P (2014) Study of strength and workability of different grades of concrete by partial replacement of fine aggregate by crushed brick and recycled glass powder. Int J Sci Res 3:141–145
14. Mitchell DRG, Hinczak I, Day RA (1998) Interaction of silica fume with calcium hydroxide solutions and hydrated cement pastes. Cem Concr Res 28:1571–1584
15. Larbi JJ, Fraay ALA, Bijen JM (1990) The chemistry of the pore fluid of silica fume—blended cement systems. Cem Concr Res 20:506–516
16. IS: 8112-1989, specification for 43 grade Ordinary Portland cement. Bureau of Indian Standards, New Delhi, India
17. IS: 383-1970, specifications for coarse and fine aggregates from natural sources for concrete, Bureau of Indian Standards, New Delhi, India
18. IS: 456-2000, Plain and Reinforced concrete—code of practice, Bureau of Indian Standards, New Delhi, India
19. IS: 7320-1974, Specifications for concrete slump test apparatus, Bureau of Indian Standards, New Delhi, India
20. IS: 1199-1959, Methods of sampling and analysis of concrete, Bureau of Indian Standards, New Delhi, India
21. IS: 516-1959, Methods of tests for strength of concrete, Bureau of Indian Standards, New Delhi, India
22. IS: 13311-1-1992, Methods of Non-destructive testing of concrete, part 1: ultrasonic pulse velocity. Bureau of Indian Standards, New Delhi, India

Comparison of Performance of Admixtures on Workability

Alima Fernandes and K. G. Guptha

Abstract Conventional concrete differs from the modern concrete with the use of fifth additive known as admixtures. These admixtures change the desired properties in an unimaginable manner. In the present study, effect of low, medium and high end admixtures is considered to evaluate their performance with varying dosages. Marsh cone studies followed by setting time were studied following IS procedures. This study finds the range of dosages of admixtures for better performance and taking workability into consideration. Water–cement ratios ranging from 0.2 to 0.6 and dosage varying from 0.3 to 1.5% by weight of cement including a control were considered in the study. Results are comparable with their performances for suitable recommendations for the field studies.

Keywords Admixture · Dosage · Workability · Performance

1 Introduction

Admixtures are the fifth additives used in concrete other than cement, sand, aggregate and water. These chemicals improve different properties of concrete based on the type of admixture used in the concrete. Accelerators, retarders, air entraining, water reducing, plasticizers and super-plasticizers are some types of admixtures. Some admixtures may change one or more properties of concrete.

Super-plasticizers are classified into four families, namely, lignosulphonates, sulphonated melamine formaldehyde, sulphonated naphthalene formaldehyde and polycarboxylic ether. Studies have shown that negative electrostatic dispersion occurs between the cement particles in polycarboxylic ether polymers containing long lateral chain, and hence these admixtures greatly improve cement dispersion, reducing water demand to high extent compared to naphthalene-based admixtures.

A. Fernandes (✉) · K. G. Guptha
GEC, Farmagudi, Ponda, Goa, India
e-mail: alimaferns2204@gmail.com

K. G. Guptha
e-mail: kgg@gec.ac.in

B. B. Das et al. (eds.), *Recent Developments in Sustainable Infrastructure*, Lecture Notes in Civil Engineering 75, https://doi.org/10.1007/978-981-15-4577-1_45

The present study was done by using six different chemicals classified under three types, namely, low end, mid end and high end admixtures. Conplast SP 430 ES2 and Conplast SP 600 are low end admixtures based on sulphonated naphthalene and organic polymers, respectively. Auramix 201 and Auramix 300 plus are mid end admixtures while Auramix 402 and Auramix 450 are high end admixtures all based on polycarboxylic ethers.

Workability of concrete is defined as the flowability of concrete. With the use of super-plasticizers, flowable concrete can be produced by keeping water–cement (w/c) ratio as low as 0.3. The factors affecting workability are mainly water–cement ratio, type of admixtures and dosage of admixture subjected to compatibility of cement.

Generally, marsh cone studies are carried out to optimize the dosage to fix the workability of mixes. The present study aims at determining the minimum required water–cement ratio for the type of admixture being selected for it to perform. Study reveals clear differentiation of water–cement ratio to that of flow value corresponding to the cement used. These studies were carried out under the control environment as referred in the relevant codes. Other physical and chemical data of the admixtures were pre-assumed as per data sheets supplied by respective firm.

The results obtained were checked for their consistent repeatability of results. All tests were carried out as per relevant IS codes.

2 Materials

Laboratory experiments were carried out on cement admixtures selected and Marsh cone test was carried out for interpretation of results. Detailed description of each of the materials used is listed as follows.

2.1 Cement

Cement is the binder material in the concrete. It is chemically active constituent and its reactivity is effective upon the addition of water. All tests were carried out on OPC 43 grade Ultratech cement having produced in 38 week. It was reconfirmed that it is following the stipulations as per IS 8112: 2013 (43 grade). Care was taken to remove any lumps if found due to surface moisture by sieving through IS test sieve having aperture size 90 microns.

2.2 Admixtures

Admixtures are selected from Fosroc constructive solutions in such a way that the selected combination comprising of low, medium and high end to evaluate their

Table 1 Properties of admixtures

Properties	Admixtures					
	Conplast SP 430 ES2	**Conplast SP 600**	**Auramix 201**	**Auramix 300 plus**	**Auramix 402**	**Auramix 450**
Appearance	Brown liquid	Brown liquid	Light brown liquid	Light-yellow-coloured liquid	Light-yellow-coloured liquid	Light-yellow-coloured liquid
Specific gravity	1.238 to 1.240	1.175 to 1.240	-	-	-	-
Chloride content	Nil to BS 5075	Nil as per IS:456 and BS:5075	Nil to IS:456	Nil to BS:5075	Nil as per BS:5075 part I	Nil to IS:456
Normal dosage range	0.6–1.5 litres per 100 kg of cementitious material	0.5%–1% by weight of cement	0.3%–1.5% by weight of total cement or binder content	0.3–2.0 litres per 100 kg of cementitious material	0.3–1.5 per 100 kg of cementitious material	0.5–3.0 litres per 100 kg of cementitious material
Alkali content	Typically less than 72.0g Na_2O equivalent per litre of admixture	-	-	Typically less than 1.5g Na_2O equivalent per litre of admixture	-	-
Volumetric mass	-	-	1.05 ± 0.02 kg/litre @ 20°C	1.08 ± 0.02 kg/litre @ 20°C	-	1.09 to 1.11 @ 20°C

performances. Table 1 shows certain important test results extracted from data sheet which are in good agreement with the test carried out at the laboratory as per IS 9103.

2.3 Water

Tap water supplied by the municipality was used for all analysis in the present study. It was also noted that the water used was satisfying the requirements as per IS 456:2000.

2.4 Marsh Cone

Marsh cone, a conical brass vessel held on a wooden stand with top diameter 155 mm covered with 1.5 mm IS test sieve, cone height 290 mm having 10 mm aperture diameter satisfying the requirements, was used in all the analysis.

3 Methodology

The methodology adopted for the experimental study included water–cement ratio varying from 0.2 to 0.6 and admixture dosage varying from 0.3 to 1.5% by weight of cement. Flow chart showing the proposed methodology of the analysis is depicted in Fig. 1 for super-plasticizers, ultra-plasticizers and hyper-plasticizers.

Typical test procedure adopted had the following steps:

1. Cement paste was prepared with water–cement ratio varying from 0.2 to 0.6.
2. Care was taken to see that there are no lumps during the entire process of mechanical mixing.
3. Prepared sample was used to find flow time with marsh cone having 10 mm diameter.
4. Corresponding flow time was measured.
5. Experiment was repeated for consistent reproducible values.
6. Tests were also conducted for dosage varying from 0.3 to 1.5% by weight of cement for repeatability.
7. Similar procedure was adopted for mid end and high end admixtures.
8. All readings are recorded for reference.

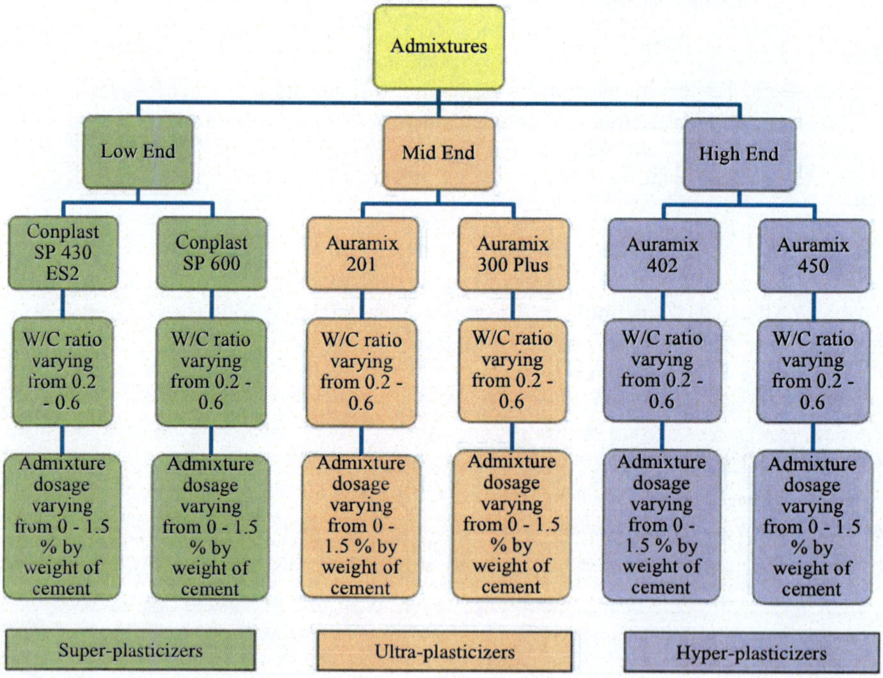

Fig. 1 Flow chart of methodology proposed

4 Results and Discussions

Based on the experiments conducted for parameters chosen, it is depicted in Fig. 1 and results are summarized and discussed as follows.

General observations are drawn based on the data collected from the experiments. Behaviour of admixtures based on water–cement ratios and corresponding flow values is represented in figures in the subsequent articles. General comparison of behaviour of admixtures with that of control is discussed as follows.

4.1 Controls

Referring to Table 2 and Fig. 1, it was observed that the general trend was linear and with the increasing water–cement ratio, flow values also improved. It appears only if minimum water–cement ratio of 0.5 and above the flow values tends to improve. Further increase in water–cement ratio improved flow values with a phenomenal increase of 50% in comparison with a water–cement ratio of 0.5 over 0.6. IS 456: 2000 does not permit for any exposure condition a water–cement ratio of 0.6. Beyond this ratio the strength drastically reduces and hence code does not permit. In the present study, water–cement ratios are limited to only 0.6 and not beyond.

Referring to Table 3 and Fig. 2, it is observed that up to a water–cement ratio of

Table 2 Flow time for sample without admixture

Admixture dosage (In % by weight of cement)	W/C ratio	Flow time (in secs)
0	0.2	-
	0.3	-
	0.4	-
	0.5	24.55
	0.6	14.15

Table 3 Flow time for 0.3% admixture dosage

Admixture dosage (In % by weight of cement)	W/C Ratio	Flow time (In Secs)					
		Conplast SP 430 ES2	Conplast SP 600	Auramix 201	Auramix 300 Plus	Auramix 402	Auramix 450
0.3	0.2	-	-	-	-	-	-
	0.3	-	-	-	-	-	-
	0.4	277.23	149.35	33.95	85.87	29.84	40.95
	0.5	23.26	23.93	15.92	24.41	14.64	14.23
	0.6	13.67	13.53	13.31	13.39	10.70	10.24

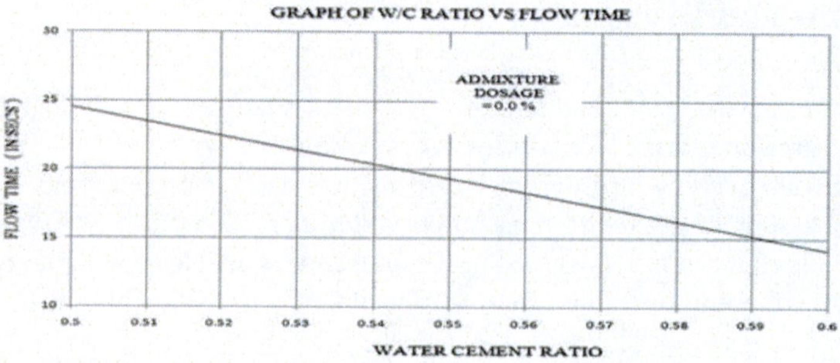

Fig. 2 Behaviour of flows: w/c ratio versus flow time for 0.3% admixture dosage

0.4 none of the admixtures responds to the flow time. With the increase in the water–cement ratio by 0.1 from 0.4, flow values improved to a record range of 80–90%. This was observed for low end admixtures. With further increase in water–cement ratio from 0.5 to 0.6, only marginal improvement was observed. Hence, low end admixtures are active only with higher water–cement ratios. In case of mid end admixtures and high end admixtures with the improvement in water–cement ratios, flow values have improved. However, the improvement beyond 0.5 water–cement ratio did not show any appreciable improvement. From the figure, it is also observed that, what low end admixtures able to achieve the flow time at higher water–cement ratios the mid end and high end admixtures are able to achieve similar targets with water–cement ratio as low as 0.4. With a specific reference to higher mid end and higher high end, admixtures did not perform the way in which lower end in the respective category behaved. However, marginal lower values of flow were recorded in all the trials executed. Hence, it can be concluded that Auramix 300 plus and Auramix 450 have under performed in comparison to Auramix 201 and Auramix 402. From Fig. 2, the optimum water–cement ratio for getting better flow values for all the admixtures containing 0.3% admixture dosage by weight of cement used is 0.49.

Referring to Table 4 and Fig. 3, it is observed that high end admixtures respond to lower water–cement ratios of 0.3 at this dosage. However, other admixtures will respond to a water–cement ratio of 0.4 only. It was also observed that the trend of the curves was similar to each other. Further, with the increase in the water–cement

Table 4 Flow time for 0.6% admixture dosage

Admixture dosage (In % by weight of cement)	W/C Ratio	Flow time (In Secs)					
		Conplast SP 430 ES2	Conplast SP 600	Auramix 201	Auramix 300 Plus	Auramix 402	Auramix 450
0.6	0.2	-	-	-	-	-	-
	0.3	-	-	-	-	364.84	925.1
	0.4	125.23	85.51	23.24	55.52	23.23	27.94
	0.5	22.47	23.34	12.41	20.41	12.66	10.83
	0.6	13.12	12.93	10.35	12.66	9.13	9.54

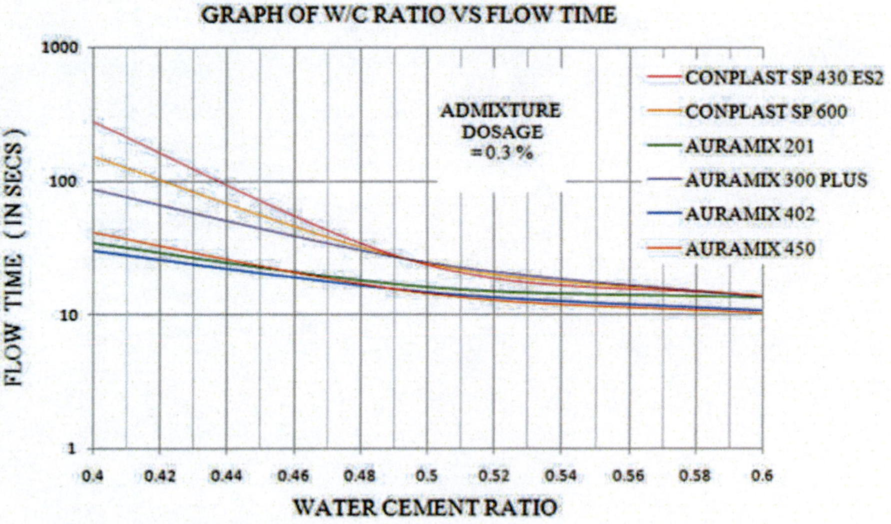

Fig. 3 Behaviour of flows: w/c ratio versus flow time for 0.6% admixture dosage

ratio by 0.1 from 0.3, flow values improved for a record range of 90–95% in case of high end admixtures. However, water–cement ratios beyond 0.5 showed marginal improvement in flow value for these admixtures. In case of low end admixtures with the increase in the water–cement ratio by 0.1 from 0.4, flow values improved for a range of 70–80% and further increase in the water–cement ratio showed marginal improvement, while mid end admixtures showed marginal improvements in flow values for water–cement ratios beyond 0.4.

From Fig. 3, the optimum water–cement ratio for better flow values for high end admixtures is 0.42 and for mid and low end admixtures is 0.51 for 0.6% admixture dosage by weight of cement used.

Referring to Table 5 and Fig. 4, it is observed that high end admixtures and Auramix 201 showed drastical improvement in the flow value for a record range of 90–97% by increasing water–cement ratio by 0.1 from 0.3. However, only marginal improvements are observed for a water–cement ratio beyond 0.4 for these admixtures.

Table 5 Flow time for 0.9% admixture dosage

Admixture dosage (In % by weight of cement)	W/C Ratio	Flow time (In Secs)					
		Conplast SP 430 ES2	Conplast SP 600	Auramix 201	Auramix 300 Plus	Auramix 402	Auramix 450
0.9	0.2	-	-	-	-	-	-
	0.3	-	-	812.29	-	199.57	333.64
	0.4	101.09	77.28	19.65	45.29	16.63	19.32
	0.5	21.84	22.91	11.62	18.86	10.70	9.79
	0.6	12.75	12.43	10.20	12.05	8.81	8.81

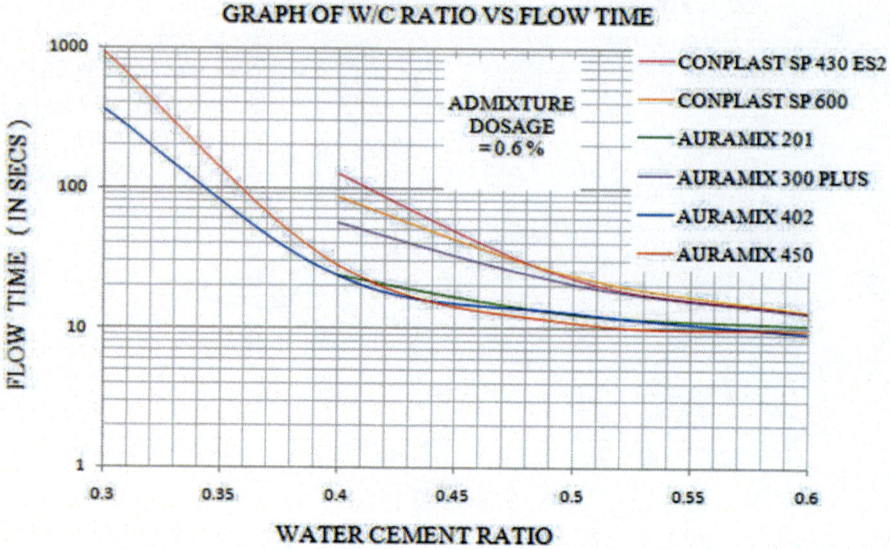

Fig. 4 Behaviour of flows: w/c ratio versus flow time for 0.9% admixture dosage

About 70% improvement is achieved by low end admixtures for a water–cement ratio ranging from 0.4 to 0.5. Auramix 300 plus did not show any appreciable improvement for lower as well as higher water–cement ratios in comparison to Auramix 201.

The optimum water–cement ratio resulting better flow for high end admixtures and Auramix 201 is 0.42 while that of low end admixtures and Auramix 300 plus is 0.5 for 0.9% dosage by weight of cement used, respectively.

Referring to Table 6 and Fig. 5, it is observed that increase in water–cement ratio by 0.1 from 0.3 drastically improved flow value for a range of 90–95% and further increase in the water–cement ratio showed marginal improvements for high end admixtures. Auramix 201 also shows similar improvements in the flow value. However, flow values recorded by Auramix 450 for water–cement ratios beyond 0.3 are similar in comparison to Auramix 402 concluding that no improvement is achieved by adding 1.2% dosage of Auramix 450 over Auramix 402. Similarly, Conplast SP 600 also shows no improvement in comparison to Conplast SP 430 ES2

Table 6 Flow time for 1.2% admixture dosage

Admixture dosage (In % by weight of cement)	W/C ratio	Flow time (in secs)					
		Conplast SP 430 ES2	Conplast SP 600	Auramix 201	Auramix 300 Plus	Auramix 402	Auramix 450
1.2	0.2	-	-	-	-	-	-
	0.3	-	-	420.45	-	160.96	155.5
	0.4	90.08	52.97	16.99	40.93	14.37	15.08
	0.5	21.67	21.28	10.64	18.06	10.00	9.60
	0.6	12.60	12.20	9.93	11.71	8.81	8.78

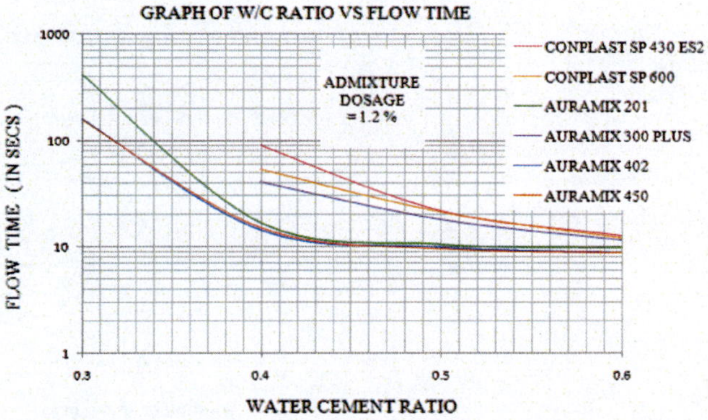

Fig. 5 Behaviour of flows: w/c ratio versus flow time for 1.2% admixture dosage

Table 7 Flow time for 1.5% admixture dosage

Admixture dosage (In % by weight of cement)	W/C Ratio	Flow time (In Secs)					
		Conplast SP 430 ES2	Conplast SP 600	Auramix 201	Auramix 300 Plus	Auramix 402	Auramix 450
1.5	0.2	-	-	-	-	-	-
	0.3	-	-	272.6	-	142.02	130.98
	0.4	60.74	50.86	16.51	35.96	13.01	12.79
	0.5	18.66	20.36	10.01	17.95	9.60	9.21
	0.6	10.84	10.38	9.79	10.08	8.78	8.63

beyond 0.5 water–cement ratios. However, the trends in the flow remain same for all the admixtures.

From Fig. 5, the optimum water–cement ratio for high end and Auramix 201 is 0.42 and that of low end and Auramix 300 plus is 0.5 for 1.2% dosage by weight of cement used.

Referring to Tables 6 and 7, it is observed that no appreciable improvement is obtained by adding 1.5% dosage compared to 1.2% dosage for all the admixtures. Hence, it can be concluded that optimum dispersion of cement particles is reached beyond 1.2% admixture dosage irrespective of the type of admixture used. From Fig. 6, the optimum water–cement ratio for high end and Auramix 201 is 0.41 and that of low end and Auramix 300 plus is 0.5 for 1.5% dosage by weight of cement used.

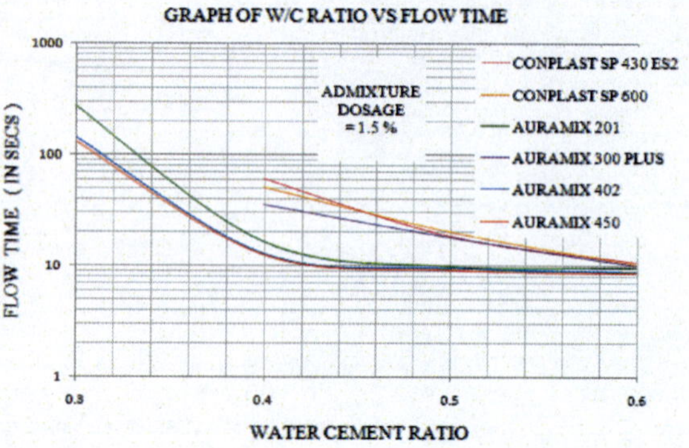

Fig. 6 Behaviour of flows: w/c ratio versus flow time for 1.5% admixture dosage

5 Conclusions

Conclusions drawn from the experimental study are comparable and are listed as follows:

1. Flow value linearly improves for higher water–cement ratios requiring at least 0.5 water–cement ratio when no admixture is used.
2. In case of 0.3% admixture dosage, minimum water–cement ratio required drops down by 0.1 from 0.5. The improvement in low end admixtures ranges from 80 to 90% when water–cement ratio is increased by 0.1 from 0.4 while the other admixtures show marginal improvement.
3. In case of 0.6% admixture dosage, high end admixtures are active with as low as 0.3 water–cement ratio. However, none of the other admixture responds to the flow time corresponding to 0.3 water–cement ratio. Improvement ranging from 70 to 80% in mid end admixtures and a record range of 90–95% in high end admixtures is achieved for optimal water–cement ratios.
4. In case of 0.9% admixture dosage, high end admixtures and Auramix 201 result 90–97% improvement in the flow value from 0.3 to 0.4 water–cement ratio. Appreciable improvement in the flow is achieved by low end admixtures at this dosage.
5. In case of 1.2 and 1.5% admixture dosages, the samples achieve similar flow time concluding optimal dispersion has occurred in the cement particles.
6. It can also be concluded that low end admixtures are active with higher water–cement ratios, whereas high end admixtures are highly active at low water–cement ratios.
7. The reduction in the flow value is parabolic in case of admixtures used and similar trend is achieved irrespective of type and dosage used unlike the linear reduction in the flow value as compared to controls.

Bibliography

1. Shah DS, Shah MP, Jayeshkumar Pitroda J (2014) Chemical admixtures: a role in modern concrete materials and technologies. National conference on: Trends and challenges of civil engineering in today's transforming world
2. Jadhav Durgesh (2016) Compatibility of chemical admixture with cement: Marsh cone test. Int J Adv Mech Civil Eng 3(3):99–103
3. Shah SNR, Aslam M, Shah SA, Oad R (2014) Behaviour of normal concrete using superplasticizer under different curing regime. Pak J Eng Appl Sci 15:87–94
4. Nayak NV, Jain AK (2018) Concrete sustainability. Published by Ane Books Pvt Ltd
5. IS 456: 2000. Plain and reinforced concrete- code of practice (Fourth Revision)
6. IS 9103: 1999 (Reaffirmed 2004), Concrete Admixtures—Specification (First Revision)
7. IS 8112: 2013. Ordinary Portland Cement, 43 Grade—Specification (Second Revision)

Study of Heat of Hydration in Mass Concrete

K. G. Guptha, T. Mohan, Guptha Tejas, Khan Bahadur, Chari Sanmit, Peixoto Cosyma, and Raut Manasi

Abstract Heat of hydration generated by cementitious materials is one of the major setbacks in case of mass concreting. Controlling heat and managing with ambient temperature is a herculean task while working with them. The differential potential if not maintained leads to cracks. Hence, it is essential to control this heat especially in the core. As per IS 456-2000 and ACI specifications, the temperature in the core shall not exceed 70 °C. However, there is no technical or literature support available on this matter barring experience. This study focuses on tracking of after effects of release of heat of hydration to a greater extent in mass concreting. Study also focuses on relationships between various ingredients of concrete and their contributions of heat. Understanding the heat of hydration in core is carried out using a special device called SMARTROCK2—a wireless sensor placed on rebars before the concrete pour. This measures the temperature development at regular intervals. Delivering high-accuracy data that can be accessed from any device through Android or iOS app. Study compiles heat of hydration data on the prototype.

Keywords Smartrock · Pile cap · Heat of hydration

1 Introduction

Cement is a fine grounded material used for binding the matrix of cement concrete. It is hydrophilic material in its fresh form. This hydrophilic property ensures binding and further as it becomes hard and makes concrete a homogenous material in counteracting loads. Cement has different properties depending upon its makeup

G. Tejas (✉) · K. Bahadur · C. Sanmit · P. Cosyma · R. Manasi
GEC, Farmagudi, Goa, India
e-mail: sajetg@gmail.com

K. G. Guptha · K. Bahadur · C. Sanmit · P. Cosyma · R. Manasi
Civil Engineering Department, GEC, Farmagudi, Goa, India
e-mail: kgg@gec.ac.in

T. Mohan · K. Bahadur · C. Sanmit · P. Cosyma · R. Manasi
Goa Site, DBL, Farmagudi, Goa, India
e-mail: tmohan@dbl.co.in

B. B. Das et al. (eds.), *Recent Developments in Sustainable Infrastructure*, Lecture Notes in Civil Engineering 75, https://doi.org/10.1007/978-981-15-4577-1_46

in view of its composition and fineness. Good quality raw materials will make the manufacturing process of cement simple and consistent.

When water is introduced, hydration products are formulated. The principal hydrated product is calcium silicate hydrate (C–S–H) in addition to calcium hydroxide. This paste of cement is termed as ettringite. The properties of the cement can be altered by proportioning four compounds mainly C_3S, C_2S, C_3A, and C_4AF. Its setting time from flash setting can be altered by introducing gypsum. It is a well-known fact that the reaction of cement involves the following:

$$2C_3S + 11H \rightarrow C_3S2H8\ (C - S - H) + 3CH\ (\text{calcium hydroxide}) \quad (1)$$

$$2C_2S + 9H \rightarrow C_3S2H8\ (C - S - H) + CH\ (\text{calcium hydroxide}) \quad (2)$$

As stated earlier, the cement reaction with water is an exothermic reaction, and hence it liberates heat. This heat is termed as heat of hydration. The amount of heat produced is primarily related to the proportion of C_3S present in it. Controlling of heat is an important factor for cement to achieve the strength. It is reported that cement generally produces 89–90 cal/g in 7 days and 90 to 100 cal/g in 28 days [15]. Further, this depends on the fineness of the cement. This heat of hydration if not controlled will introduce cracks in the material. For thin members, this may not be that prominent but matters in mass concrete.

Mass concrete is defined by the American Concrete Institute as "any volume of concrete in which a combination of dimensions of the member being cast, the boundary conditions, the characteristics of the concrete mixture, and the ambient conditions can lead to undesirable thermal stresses, cracking, deleterious chemical reactions, or reduction in the long-term strength as a result of elevated concrete temperature due to heat from hydration." In other words, mass concrete can defined as "any volume of concrete with dimensions large enough, requires measures to be taken to cope with generation of heat from hydration of the cement and attend volume change to minimize cracking." This has been historically associated with large structures such as dams, bridge foundations, and other large volume placements. The study deals with compiling data regularly collected from the Smartrock2 for the period of 28 days from the day of casting and analyzing for attainment of strength and maturity with time. All the studies are conducted on a prototype pile cap of a major project in Goa having 26.3 × 27 × 4 m.

2 Components of the System

Following material components were used in the study.

2.1 Pile Cap

System of load transfer on to a group of piles is through a pile cap. It is a thick concrete mat that rests on concrete or timber piles driven into soft or unstable ground to provide a suitable stable foundation. It usually forms part of the foundation of a multi-story building, bridge structure, or support base for heavy equipment mobilization. The structure under consideration has two pile caps in the marine section each comprising three parts, viz., behavior of pile cap in emitting heat with very large size with cooling system, smaller size mass concrete with similar mix design as that of prototype without cooling system, and smaller size mass concrete without GGBS and cooling system. The present study focuses on part 1 of pile cap (Fig. 1). Details of the pile cap are presented in Table 1. Mix proportions used per cubic meter are shown in Table 2 and corresponding quantities used are listed in Table 3.

Fig. 1 Pile cap

Table 1 General specifications of pile cap

Sr. no	Parameter	Measurement
1.	Dimensions	26.3 × 27 × 4 m
2.	Grade of concrete	M_{55}
3.	GI pipes for cooling system in pile cap	32NB (nominal bore)
4	Date of casting	August 12, 2018
5	Minimum temperature	27.90 °C
6	Maximum temperature	30.80 °C

Table 2 Mix proportions (SSD condition) per cubic meter

Sr. no	Material	Quantity in kg
1	Cement	225
2	GGBS	275
3	Coarse aggregate 20 mm	661
4	Coarse aggregate 10 mm	520
5	Crushed sand	730
6	Water	145
7	Admixture (0.50% by weight of cement) high end	2.50
8	Corrosion inhibitor	3.0
	Total mass	2561.5

Table 3 Quantity of materials in the pile cap

Sr. no	Material	Quantity in kg	Remarks
1	Cement	6,39,000	Aggregates are in SSD condition
2	GGBS	7,81,000	
3	Coarse aggregate 20 mm	18,77,836	
4	Coarse aggregate 10 mm	14,75,437	
5	Crushed sand	20,72,774	
6	Water	4,11,800	
7	Admixture(0.5% by weight of cement)	7,100	
8	Corrosion inhibitor	8,520	
9	Total mass	72,73,467	

Fig. 2 Cooling pipes system of Part 1

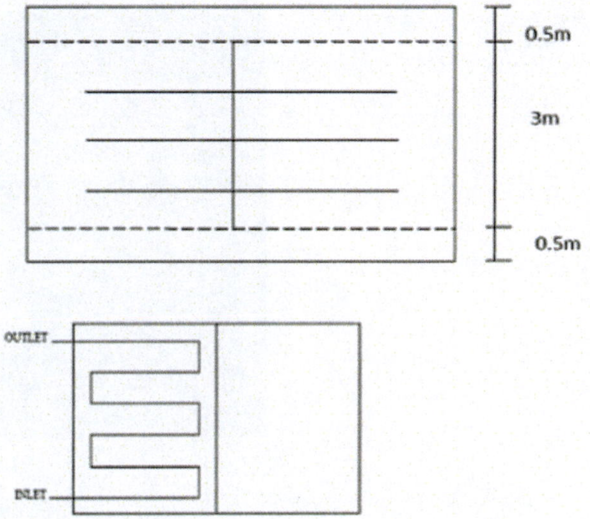

Fig. 3 Layout of cooling system

2.2 *Cooling Pipe System*

The pile cap needs an additional system of cooling to control heat of hydration. This was planned based on the theoretical calculation of likely temperature rise during the pour of concrete. The system has four layers with eight inlets and outlets. Typical system is represented in Fig. 3 as layout. The prototype is also shown in Fig. 2 such that the proposed piping system is embedded into form work and tied as a sacrificing cooling system. This is an interconnected pipe network system whose design depends upon material type, diameter, spacing between pipes, and the temperature of the inlet water and the flow rate of the water.

Four layers of GI pipe 32NB have been used in a grid of 1 m x 1 m horizontally and at a distance of 0.75 m vertically. Temperature at the inlet was maintained as 30 °C (to avoid cracking in concrete due to sudden temperature change due to introduction of cold water) while the final temperature was recorded as 58 °C. Inlet water temperature of 30 °C was maintained for 4 days by recycling the same water using ice in cooling tower. 5HP motors were used to maintain the velocity and pressure in the system.

2.3 *Smartrock2*

It is mobile app-based wireless sensor for monitoring the temperature of concrete from fresh to hardened stage. It estimates strength of concrete based on the concept of maturity. It has a long battery life of 3–4 months. The device gives a real-time data display, photos, and blueprint of temperature and strength. It allows easy data sharing between devices with the facility of generating reports in PDF.

Fig. 4 SmartRock2 sensor G-285

Smartrock2 sensor G-285 has been used in the pile cap. This measures temperature development at regular intervals to deliver high-accuracy data and enable contractors to make appropriate decision regarding temperature control during the curing stages. The sensors were placed on the rebars before the concrete pour and it commences its work of generation of data from the moment the concrete is poured till completion of the project. It is totally free from wires and can be accessed from a smartphone device or tablet through Android or iOS app unlike many other sensors in the market.

The sensor is added to a section like floor, column, wall, footings, etc., as shown in Fig. 4. These are scanned using Bluetooth in the smartphone. Sensors are tagged 15 min prior to their activation. The insulation tape is wrapped around the bars to avoid the electrostatic current from the electrical equipments used at the sites before their installation. The two activation wires are twisted firmly and tightly around the reinforced bars to ensure secured installation of the sensor. In order to maximize the wireless range, the distance is restricted to 1" (2.5 cm) from the concrete surface.

The information obtained from Smartrock2 is used for optimization of curing conditions, quality control in the field, estimation of strength (ASTM C1074), formwork removal timing, post-tensioning, opening road to traffic, and monitoring temperature gradients on mass concrete.

2.4 *Retrieving Data*

Smartrock2 device as stated earlier is a wireless system that can transmit data up to 8 meters (26 feet) when embedded within 5 cm from the concrete surface. It shares real-time temperature, maturity, and strength data and graphs to a smartphone. Several factors will affect the operating range of the sensors to the type of smartphone used. Curing state of the concrete, physical installation location, depth and surrounding obstacles to name a few. The user only needs to enter the pouring time and select the appropriate concrete mix proportion with the smartphone. Once the pouring date and time are set and the application is connected to the sensor, data will be transmitted from the sensor. The real-time temperature value will be transmitted to the smartphone and displayed on the screen.

3 Results

Data retrieved from the sensor is focussing mainly on the core temperature. This is depending on the following:

a. effectiveness of the grid of the cooling system;
b. inlet water temperature;
c. velocity of flow in the pipe;
d. maturity, strength gain as the time elapses by the process of hydration, and heat dissipation
e. within and around the core;
f. effectiveness of external cooling system;
g. ambient temperature and humidity present;
h. ingredients of the mix.

4 Conclusions

The retrieved data is graphically represented in Fig. 5a, b, c, and d. The relations are analyzed and concluded as follows:

(i) The relation between strength gain and temperature is not a smooth parabolic curve rather a stepped curve. This indicates that ingredients responsible for gain in strength are released in batches but not continuously. This is confirmed from Fig. 5. The temperature drops down and remains more or less uniform for a given instantaneous release of the ingredients. Once the strength is gained, further release is observed. Step width reduces with gain in strength from step to step.
(ii) Figure 5b exhibits maturity of concrete with time. Delayed or slow maturity was seen initially but rate of gain improves after 48 h of pour of concrete. Exponential rise in maturity is seen only after a week's time after the pour of concrete.
(iii) Behavior of gain of strength of concrete at the core is different from that of the surface. Core shows a trend similar to that of strength gain as shown in Fig. 5a. This effect is reconfirmed and appears in good agreement with Fig. 5c.
(iv) Rise in the temperature at the core improved continuously and reached to a peak value of 76.05 °C at the end of 52 h. of pouring of concrete. With the better

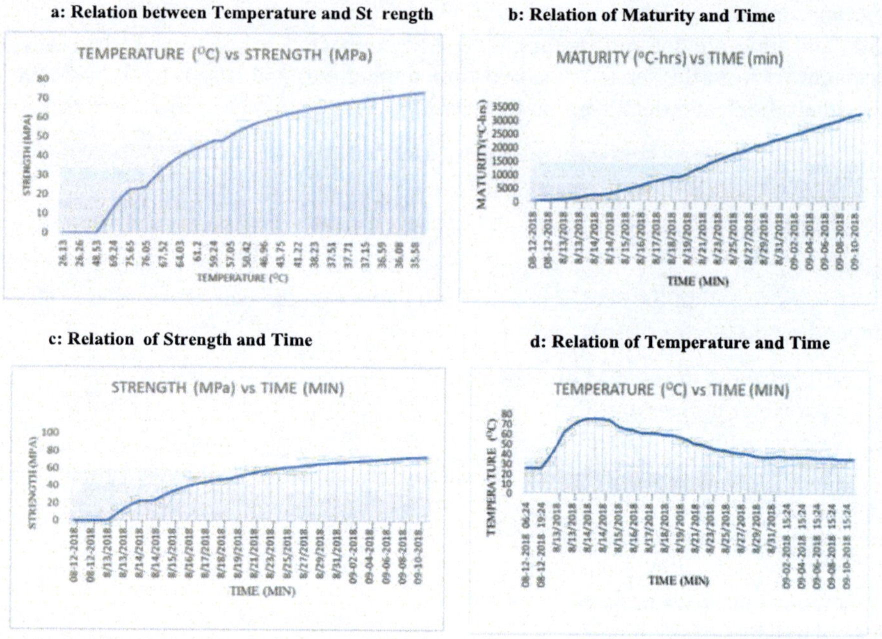

Fig. 5 Relationships among the ingredients

judgement and theoretical calculations, provision of cooling system is very well justified. In absence of this arrangement of cooling system, the structure would have shown cracks, which makes the structure week.

Bibliography

1. Shell LM (2015) Monitoring temperatures in concrete construction using IR thermometer. www.concreteinternational.com
2. Wang K, Ge Z et al. (2016) Report From centre for transportation research and education Lowa State University. www.Ctre.Iastate.Edu
3. Reddi SA LTD:concrete making material presentation
4. Dawood AO, Article on hydration of cement
5. BalaKrishna B, Abdul Awal ASM et al. (2013) Influence of high volume fly ash in controlling heat of hydration of concrete. Int J Eng Res Appl (IJERA) 3(2)
6. Dambhare MG, Shikla N et al. Causes and controlling factors of heat of hydration during mass concrete
7. Zhu Z, Chen W, Xie Z, Zhang G (2018) The Open Constr Build Technol J. www.benthamopen.com/TOBCTJ
8. Slaiai K, Presentation on heat of hydration in mass concrete
9. Kim SG (2010) Effect of heat generation from cement hydration on mass concrete placement
10. Gajda J, Vangeem M (2002) Controlling temperatures in mass concrete. J Concr Int
11. Concrete technology today vol 18/ Number 2 (July 1997). www.portcement.org

12. Abbas ZH, Majdi HS (2017) Contents taken from case studies in construction materials. www.elsevier.com/locate/cscm. Accessed 12 Jan 2017
13. Smart Rock2 sensor details by PTSG(Post Tension Services(Gujarat) LLP)
14. IS:7861(Part1) 1975: Code of practice for extreme weather concreting recommended practices for hot weather concreting
15. Shetty MS (2005) Concrete technology

٢

Study on Specific Compressive Strength of Concrete with Fly Ash Cenosphere

S. K. Patel and A. N. Nayak

Abstract The properties of concrete, such as compressive strength and density, are the key properties that control the size and self-weight of any structure. Hence, concrete with high strength-to-density ratio (specific strength) can reduce the vulnerability of a structure against earthquake and also can reduce the operational energy consumption. Therefore, this investigation is focused on the study of the variations in specific compressive strength (SCS) of concrete incorporating fly ash cenosphere (FAC) as an alternative of natural fine aggregate (NFA). By doing so, the pollution of soil, water and air due to disposal of FAC can also be solved. To fulfil the above objectives, concrete mixes are prepared by replacing various fractions of NFA with FAC. The compressive strength and oven dried density of these concrete mixes are evaluated. Thereafter, the SCS of all these concrete mixes are calculated. The outcomes of the above study reveal that both density and compressive strength of FAC-based concrete are less than those of normal concrete. The decrease in density is due to the lightweight nature of the FAC and reduction in strength is due to the inferior bonding between FAC and cement paste. It is also found that the rate of reduction in compressive strength is higher than that of density. As a consequence, the SCS also gradually decreases with increase in the replacement of NFA with FAC. Therefore, it needs further investigation to enhance the SCS of concrete by increasing the compressive strength with minimum variation of density.

Keywords Fly ash cenosphere · Compressive strength · Oven dried density, Specific compressive strength

S. K. Patel (✉) · A. N. Nayak
Department of Civil Engineering, Veer Surendra Sai University of Technology, Burla 768018, Odisha, India
e-mail: sudeepatel@gmail.com

A. N. Nayak
e-mail: nayakan1964@gmail.com

B. B. Das et al. (eds.), *Recent Developments in Sustainable Infrastructure*, Lecture Notes in Civil Engineering 75, https://doi.org/10.1007/978-981-15-4577-1_47

1 Introduction

Concrete is considered as the most preferred construction material over the world as compared to other materials due to the easy availability of its ingredients. The ingredients required for preparing concrete are cement, natural coarse aggregate (NCA), natural fine aggregate (NFA) and water. Out of the above ingredients, in a concrete mix, the NCA and NFA together occupy about 70% of its total volume. Again, out of this aggregate volume, the NFA used is approximately 30%. Hence, to fulfil this demand of NFA, the construction industry extracts sand from the natural sources which causes the environmental disorders like riverbanks' erosion and sliding, alteration of riverbed, lowering of groundwater, erosion of engineering structures, etc. [1, 2]. Hence, an appropriate substitute of this NFA can save the environment from these natural disorders caused by the construction industry. In this regard, the hollow spherical fly ash cenosphere (FAC), whose diameter varies within 5 to 500 μm with shell thickness 5–10% of diameter [3], can be considered as a sustainable alternative. The FAC is a by-product obtained from the burning of coal in thermal power plants. Being a waste product, these FACs along with fly ash are dumped into the ash ponds. In India itself, only 67.13% of the total produced fly ash (196.44 million tonnes) could be utilised in the year 2017–18 and rest were left into the ash pond [4]. In these ash ponds, the FAC floats on the surface of water due to its low density (400–800 kg/m^3) and gets carried away by air. Further, if it goes into the respiratory track of humans and animals through the air, it causes cardiovascular and respiratory illness [5]. Along with this, the fly ash and FAC also contaminate the soil, water and disturb the ecological balance of the environment.

On the other hand, the lightweight nature of the FAC as compared to NFA can be helpful for reducing the density of concrete. The concrete with reduced density can decrease the size of the elements of a structure, thereby the self-weight of the total structure. This makes the structure less vulnerable against earthquake. Further, the high specific strength; low cost of construction; superior thermal, acoustic and fire resistance property of lightweight concrete (LWC) [6–8] are also suitable for structures like long span bridges, non-structural elements of high rise buildings and offshore structures, etc. [9–12].

As of now, so many investigations have already been carried out on the lightweight cement mortar containing FAC as one of its ingredients [13–15]. The above investigations show that the lightweight cement mortar prepared with FAC as fine aggregate has superior properties like low thermal expansion, heat insulation, sound absorption, fire resistance, etc. [13–15]. But at the same time very few investigations are available in the literature, which concerned with the properties of lightweight concrete prepared by considering FAC as one of its ingredients [16–18]. It is found from these investigations that due to the increase in FAC content in concrete, both compressive strength and density of concrete decrease.

Out of the above, only one investigation [16] has studied the variations of specific compressive strength (SCS) property of concrete by using FAC as an alternative of NFA. As the properties of concrete ingredients along with FAC vary from place to place, the outcome of this investigation cannot be generalised. Hence, in this research work, it is intended to investigate the SCS property of the concrete prepared by using FAC as a replacement of NFA in various percentages. On successful replacement of NFA with FAC in concrete, the environmental problem caused due to sand mining and disposal of FAC can be solved. Further, it will also be helpful for producing a sustainable LWC by using a hazardous industrial waste.

2 Experimental Details

2.1 Materials

Cement of OPC 43 grade satisfying the requirement of IS 8112 [19] is used as binder. The chemical composition of OPC is presented in Table 1 which reveals that that the calcium oxide (CaO) and silicon dioxide (SiO_2) are the major constituents of OPC. Locally available 20 mm and downgraded granite aggregates and river sand are used as NCA and NFA. The grading curve of NCA (Fig. 1) confirms the requirement of IS 383 [20] and NFA shown in (Fig. 2) confirms the requirement of zone-II sand as per

Table 1 Chemical composition OPC and FAC (%)

Composition	OPC	FAC
SiO_2	23.25	56.24
Al_2O_3	5.743	33.4
Fe_2O_3	4.337	2.82
CaO	59.7	1.46
MgO	0.547	0.7
SO_3	3.754	0.24
Na_2O	0.339	0.02
K_2O	1.329	1.64

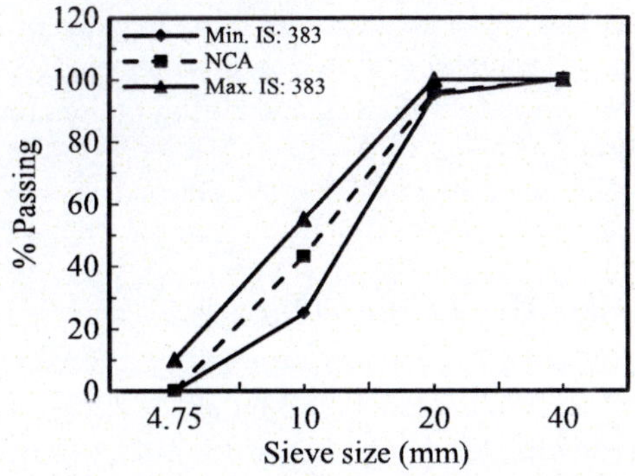

Fig. 1 Gradation of NCA

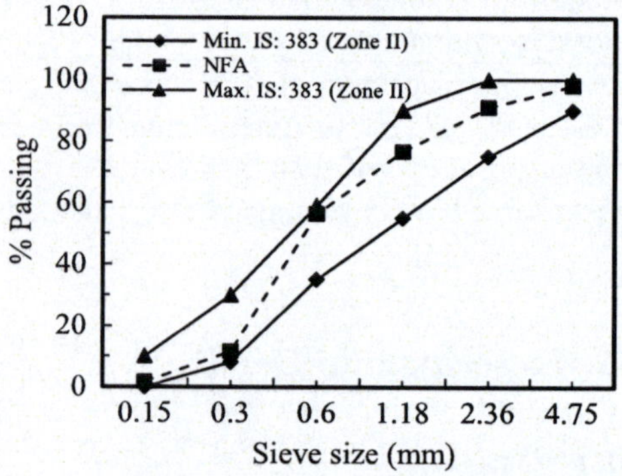

Fig. 2 Gradation of NFA

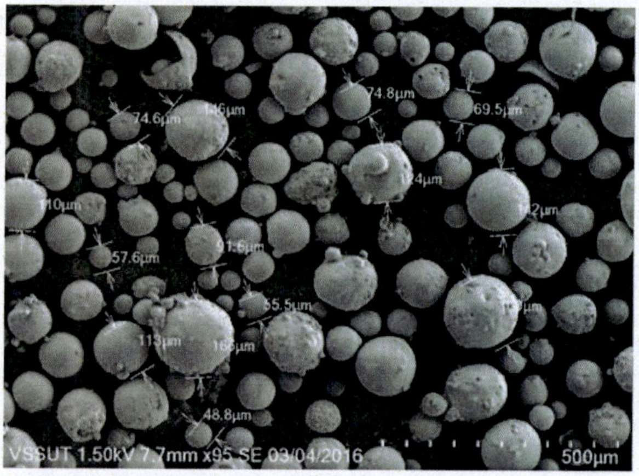

Fig. 3 SEM image of FAC

IS 383 [20]. FAC used is collected from the ash pond of Vedanta Aluminium Limited, Jharsuguda. The SEM image (Fig. 3) and gradation curve of FAC (Fig. 4) confirm the hollow spherical shape of FAC with particle size around 150 μm. Again, the chemical composition and XRD image of FAC shown in Table 1 and Fig. 5 confirm the SiO_2 as a major constituent of FAC in the form of quartz and mullite.

The physical properties of aggregates presented in Table 2 reveal the lightweight nature and higher water absorption capacity of FAC as compared to NFA and NCA. Water used for mixing and curing of concrete satisfies the requirement of IS: 10500 [21].

2.2 *Mix Proportions*

In order to prepare the samples for the investigation, a control mix was designed by following the guidelines of IS: 10262 [22]. The ingredients used are cement, NFA

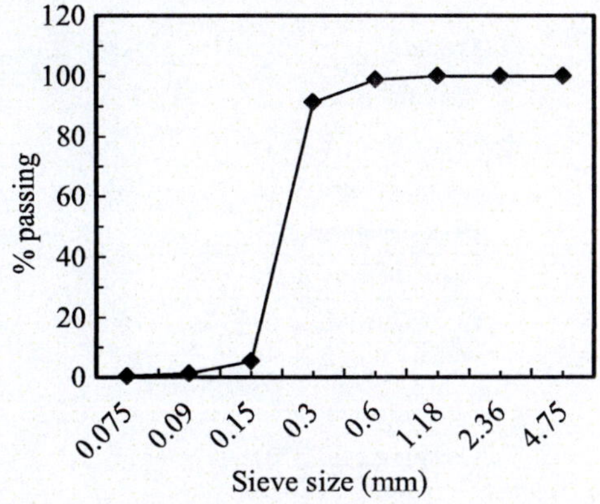

Fig. 4 Gradation of FAC

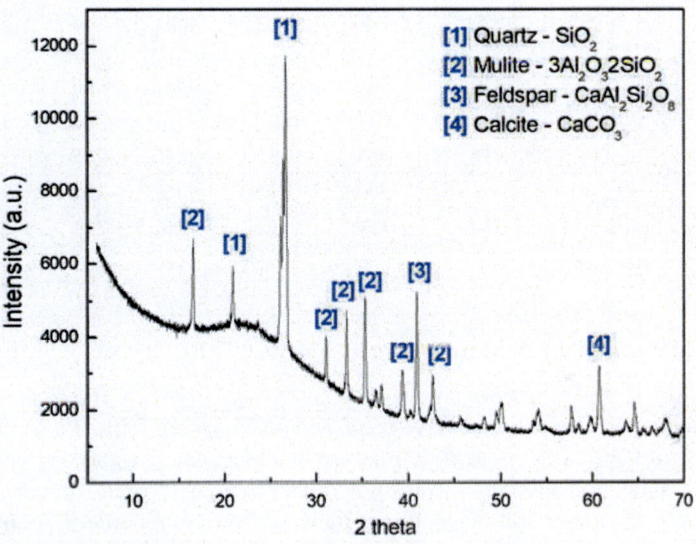

Fig. 5 XRD analysis of FAC

and NCA with water–cement ratio 0.5. In this mix design, the NCA and NFA are considered to be in saturated surface dry condition. Beyond this, five more concrete mixes were prepared by replacing 20, 40, 60, 80 and 100% of NFA by FAC. In all these mixes, an additional amount of water required to saturate the replaced amount of FAC in concrete was added to maintain the slump within 80 ± 5 mm. For calculating the additional water, the difference of water absorption between FAC and NFA is taken into consideration. The details of mix proportion for all the concrete mixes are given in Table 3.

Table 2 Physical properties of aggregates

Property	NFA	FAC	NCA
Specific gravity	2.58	0.76	2.8
Water absorption (%)	0.7	14	0.23
Free moisture (%)	Nil	Nil	0.02
Bulk density (kg/m^3)			
Loose density	1488	454	1518
Compacted density	1600	483	1666
Impact value (%)	–	–	22.43
Abrasion value (%)	–	–	18.47
Crushing value (%)	–	–	19.41

Table 3 Details of mix proportion per m^3 of concrete

Designation	Replacement % of NFA	Cement (kg)	NFA (kg)	FAC (kg)	NCA (kg)	Free water (kg)	Additional water (kg)
C0	0	410	646	–	1162	205	9
C20	20	410	517	32.9	1162	205	13
C40	40	410	388	65.7	1162	205	18
C60	60	410	259	98.6	1162	205	22
C80	80	410	130	131.5	1162	205	27
C100	100	410	0	164.31	1162	205	31

2.3 *Preparation of Specimen and Curing*

The required amount of cement NFA, FAC and NCA as per Table 3 was dry mixed for 2 min with the help of a concrete mixture in a constant rate and thereafter again continued mixing for another 2 min by adding water. After that the concrete was removed from the mixer and the slump of the concrete was recorded. Then the cube specimens were casted by filling up the respective mould in required number of layers as per standards [23, 24]. The specimens were taken out of the moulds after 24 h and immersed in water and maintained at a temperature of 27 ± 2 °C for curing.

2.4 *Testing of Specimens*

The density and compressive strength tests of concrete were conducted as per ASTM C 642 [23] and IS: 516 [24]. After 28 days, the 100 mm cube samples were removed from curing. Then the saturated, submersed and oven dried densities were recorded. The oven dried densities were calculated based on the procedure of ASTM C 642 [23].

Similarly, the 150 mm cubes in surface dried condition were tested for compression after 28 days of curing by applying the load through a compressive testing machine of capacity 2000 KN at 14 $N/mm^2/min$ [24]. At last, the ratio between compressive strength and respective oven dried density of each concrete mix was found out to get the SCS results. Each result presented in this investigation is the average of three such samples tested as per the above details.

3 Result and Discussion

3.1 Density of Concrete

The oven dried density of the control concrete (C0) was found to be 2421.76 kg/m^3. For all other concrete mixes, i.e. C20, C40, C60, C80 and C100 are found to be 3.8, 7.7, 11.8, 16 and 20.1% less than the density of C0. The oven dry densities of all concrete mixes considered in this investigation are presented in Fig. 6. It is inferred from Fig. 6 that the oven dry density continuously reduces with increase in FAC content of concrete. The lower density of FAC as compared to the NFA is liable for this type of reduction in density. A similar type of decreasing trend is also observed in the investigations conducted by McBride et al. [16] and Satpathy et al. [18]. The maximum decrease in oven dried densities of concrete due to 100% replacement of NFA with FAC in both of these investigations is found to be 22 and 24.6% less than their control specimen. This marginally higher reduction of oven dried density of concrete mix as compared to the present investigation is due to the use of higher amount of NFA in the control mix, i.e. 681 and 690 kg by McBride et al. [16] and Satpathy et al. [18], respectively. The lowest density of C100 (1933.73 kg/m^3) in this investigation is marginally higher than the density requirement of structural LWC (1920.00 kg/m^3) as per ACI 213 [6].

Fig. 6 Oven dry density of concrete

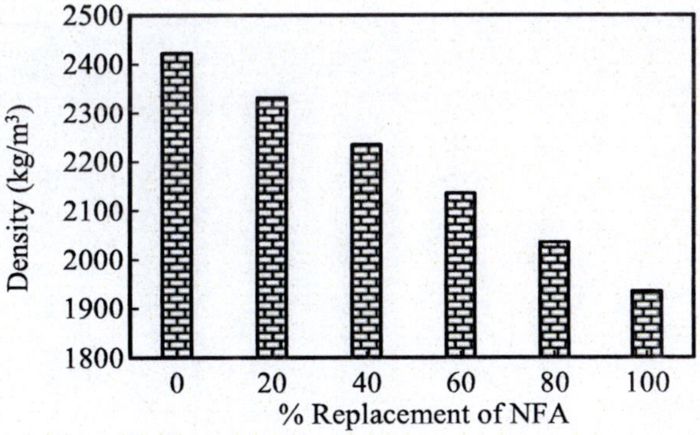

3.2 Compressive Strength of Concrete

The 28 days compressive strength of control concrete (C0) is found to be 37.25 MPa and that of the concrete mixes C20, C40, C60, C80 and C100 are found to be 8.3, 20.1, 42.6, 39.1 and 45.1% less than that of C0. The compressive strengths of all these concrete mixes are presented in Fig. 7. This figure reveals that the compressive strength reduces with the increase in FAC content from 0 to 60%, and thereafter marginally increases at 80% replacement level and reduces again at 100% replacement level. This reduction in compressive strength due to the increase in FAC content in concrete is due to the weaker interfacial bond strength between FAC and cement paste. Further, the marginal enhancement in strength at 80% replacement level may be due to the optimum packing of FAC within the concrete mix.

Similar kind of variation in 28 days compressive strength is also reported in the investigations conducted by McBride et al. [16] and Satpathy et al. [18] as shown in Fig. 8. The highest reductions in strength of concrete due to 100% replacement of

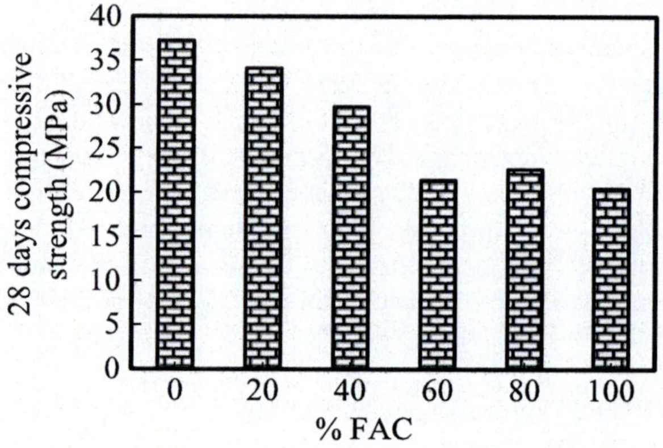

Fig. 7 Compressive strength of concrete

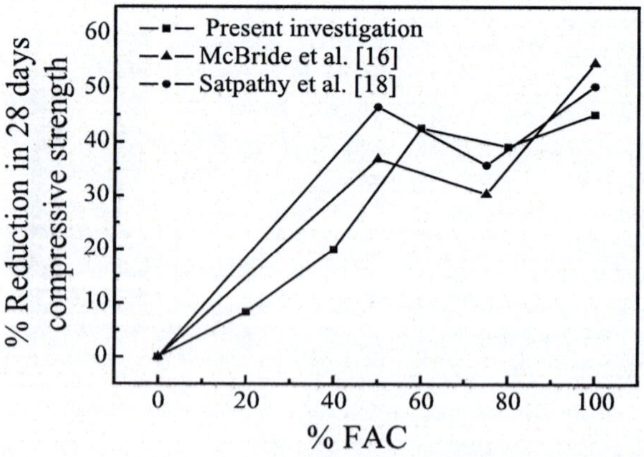

Fig. 8 Reduction in strength of concrete

NFA with FAC in both of these investigations are found to be 54.67 and 50.39% less than the strength of control specimens. This difference in percentage reduction of compressive strength between McBride et al. [16], Satpathy et al. [18] and present investigation is due to the variation in material property, mix design and casting environment. McBride et al. [16] used 389 kg of cement with 681 kg of NFA, Satpathy et al. [18] used 400 kg of cement with 690 kg of NFA, whereas in the present investigation, the quantity of cement used is 410 kg with 646 kg of NFA for producing 1 m^3 of concrete. The compressive strength of concrete prepared by using 100% FAC in place of NFA in this investigation (20.44 MPa) is higher than the strength requirement of structural LWC (17 MPa) as per ACI 213 [6].

3.3 *Specific Compressive Strength*

The SCS of control concrete is 15.4 Nm/kg, which reduces by 4.8, 13.4, 34.9, 27.6 and 31.3% for C20, C40, C60, C80 and C100, respectively. The SCS of all the above concrete mixes considered in this investigation is presented in Fig. 9. It is inferred from this figure that the variation of the SCS of concrete due to increase in FAC percentage is almost similar to the variation of compressive but the rate of reduction is less than that of the compressive strength (Fig. 10).

Similarly, out of the three variations shown in Fig. 10, the rate of reduction of compressive strength is higher than that of the dry density due to which the rate of reduction of specific strength lies within the remaining two.

The correlation between SCS and replacement percentage of FAC (FAC%) in concrete is shown in Fig. 11. The formula obtained from the regression analysis with R^2 value 0.81 is as shown below:

$$SCS = 0.0542 \times (FAC\%) + 15.227 \quad (1)$$

Fig. 9 Specific compressive strength of concrete

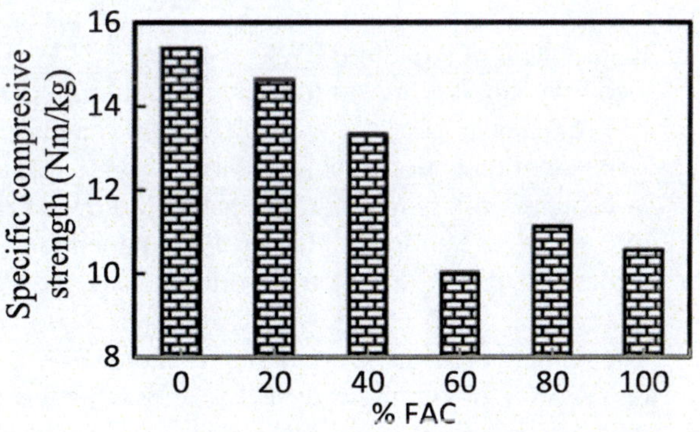

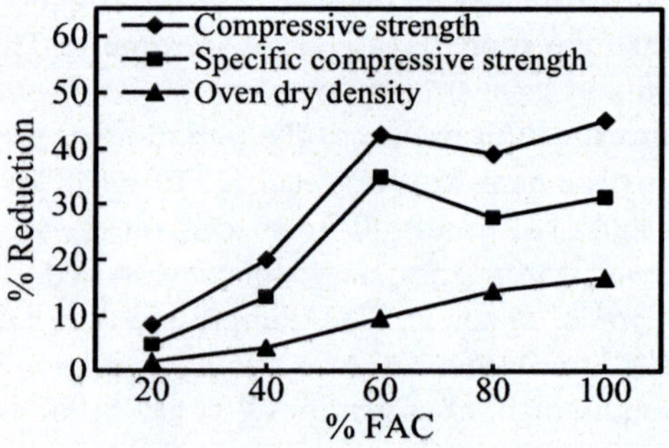

Fig. 10 Percent reduction in strength and density of concrete

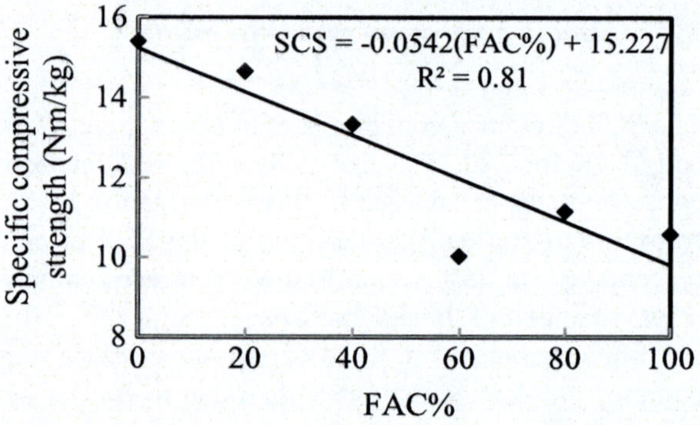

Fig. 11 Relation between SCS and FAC%

4 Conclusion

In this investigation, the oven dried density and compressive strength of different concrete mixes prepared with various fraction of FAC were found out. Further, the SCS of these concrete mixes was calculated, analysed and following conclusion are drawn:

1. Concrete prepared by using 100% FAC in place of NFA satisfies the strength requirement of structural LWC as per ACI 213 [6] and the density of this mix is also very much nearer to the density requirement of structural LWC as per ACI 213 [6]. Hence, concrete with 100% FAC may be considered as structural LWC.
2. The rate of reduction in compressive strength due to the increase in FAC content in concrete mix is higher than that of the oven dried density and less than that of compressive strength. Hence, the obtained rate of reduction for SCS is found to be higher than that of oven dried density and less than that of compressive strength.
3. The derived relation between the SCS and FAC percentage shown in Eq. 1 can be used for predicting the SCS of any concrete mix by using the percentage replacement of NFA with FAC.

This investigation may further be extended to study all the physical and mechanical properties along with the microstructural and durability properties of concrete prepared by using various percentages of FAC in place of NFA. Moreover, additional supplementary cementitious materials may also be used to enhance the above-said properties of concrete.

References

1. Kondolf GM (1994) Geomorphic and environmental effects of instream gravel mining. Landscape Urban Plann 28(2–3):225–243
2. Padmalal D, Maya K, Sreebha S, Sreeja R (2008) Environmental effects of river sand mining: a case from the river catchments of Vembanad lake, Southwest coast of India. Environ Geol 54(4):879–889
3. Vassilev SV, Menendez R, Diaz-Somoano M, Martinez-Tarazona MR (2004) Phase-mineral and chemical composition of coal fly ashes as a basis for their multicomponent utilization. 2. Characterization of ceramic cenosphere and salt concentrates. Fuel 83(4-5):585–603
4. Central Electricity Authority, New Delhi, Report on fly ash generation at coal/lignite based thermal power stations and its utilization in the country for the year 2017–18, December 2018
5. Fenelonov VB, Mel'gunov MS, Parmon VN (2010) The properties of cenospheres and the mechanism of their formation during high-temperature coal combustion at thermal power plans. KONA Powder Part J 28:189–208
6. ACI Committee 213 (2003) Guide for structural lightweight-aggregate concrete. American Concrete Institute
7. Zongjin L (2011) Advanced concrete technology. Wiley
8. ACI Committee 216 (1997) Standard method for determining fire resistance of concrete and masonry construction assemblies. American Concrete Institute
9. Chandra S, Berntsson L (2002) Lightweight aggregate concrete, science, technology, and applications
10. Thomas M, Bremner T (2012) Performance of lightweight aggregate concrete containing slag after 25 years in a harsh marine environment. Cem Concr Res 42(2):358–364
11. Bogas JA, Gomes A, Pereira MFC (2012) Self-compacting lightweight concrete produced with expanded clay aggregate. Constr Build Mater 35:1013–1022
12. Sales A, De Souza FR, Dos Santos WN, Zimer AM, Almeida FDCR (2010) Lightweight composite concrete produced with water treatment sludge and sawdust: thermal properties and potential application. Constr Build Mater 24(12):2446–2453
13. Blanco F, García P, Mateos P, Ayala J (2000) Characteristics and properties of lightweight concrete manufactured with cenospheres. Cement Concr Res 30(11):1715–1722
14. Tiwari V, Shukla A, Bose A (2004) Acoustic properties of cenosphere reinforced cement and asphalt concrete. Appl Acoust 65(3):263–275
15. Barbare N, Shukla A, Bose A (2003) Uptake and loss of water in a cenosphere–concrete composite material. Cem Concr Res 33(10):1681–1686
16. McBride SP, Shukla A, Bose A (2002) Processing and characterization of a lightweight concrete using cenospheres. J Mater Sci 37(19):4217–4225
17. Chia KS, Liu X, Liew JYR, Zhang MH (2014) Experimental study on creep and shrinkage of high-performance ultra-lightweight cement composite of 60 MPa. Struct Eng Mech 50(5):635–652
18. Satpathy HP, Patel SK, Nayak AN (2019) Development of sustainable lightweight concrete using fly ash cenosphere and sintered fly ash aggregate. Constr Build Mater 202:636–655
19. IS: 8112 (1989) Indian standard specification 43 grade ordinary portland cement specification. Bureau of Indian Standards

20. IS: 383 (1970) Indian standard specification for coarse and fine aggregate from natural sources. Bureau of Indian Standards
21. IS: 10500 (2012) Indian Standard drinking water specifications. Bureau of Indian Standards
22. IS-10262 (2009) Indian standard code for concrete mix proportioning guidelines (first revision). Bureau of Indian Standards
23. ASTM C. 642–13 (2013) Standard test method for density, absorption, and voids in hardened concrete
24. IS: 516 (1959) Indian standard methods of tests for strength of concrete. Bureau of Indian Standards

Sustainability of Copper Slag as a Building Material Under Elevated Temperature

Binaya Patnaik, Chandrasekhar Bhojaraju, and T. Seshadri Sekhar

Abstract River sand is the standard practice of fine aggregate used for the concrete production. In this present era of rapid urbanisation, to meet the increasing demand for natural sand by the construction industry, massive scale depletion of the river bed is being carried out which is causing a considerable negative impact on our environment. Hence, it is highly imperative to find sustainable fine aggregates to meet the global demand without disturbing our ecosystem. Copper slag is one such sustainable material which has a promising future to be used as an alternative to river sand. This paper presents a study on finding the optimum dosage of copper slag (CS) for partial replacement of sand in concrete. Further, as part of durability study, the impact of elevated temperature of 200, 400 and 600 °C for 12 h exposure period on strength characteristics of copper slag blended concrete has been presented and been compared with that of normal concrete. The results indicate that copper slag concrete has excellent resistance to weight and strength loss at an elevated temperature of 200, 400 °C compared to normal concrete; however, at 600 °C copper slag concrete shows similar trends like normal concrete. In the present experimental study, M20 & M30 concrete grades were used.

Keywords Copper slag · Compressive strength · Elevated temperature · Weight · UPV

B. Patnaik (✉)
Civil Engineering Department, Gambella University, Gambella, Ethiopia
e-mail: binaya7708@gmail.com

C. Bhojaraju
Construction Engineering Department, University of Quebec, École de Technologie Supérieure (Ets Montreal), Montreal, Canada
e-mail: chandra.sekhar@etsmtl.ca

T. S. Sekhar
NICMAR-CISC, Hyderabad, India
e-mail: ss.tirumala@gmail.com

B. B. Das et al. (eds.), *Recent Developments in Sustainable Infrastructure*, Lecture Notes in Civil Engineering 75, https://doi.org/10.1007/978-981-15-4577-1_48

1 Introduction

Copper slag (CS) is a waste product, which is generated during the copper extraction from its ore [1]. According to ICSG (International Copper Study Group), worldwide copper production is valued to be 20 million tonnes in 2017 [2]. For producing every tonne of copper about 2.2 ton of CS is generated [1], which creates a waste disposal issue for the copper manufacturing industries. Huge piles of CS can be seen around the industries, which is a major concern as it also creates environmental pollution. CS has been used as a landfilling material, and also the preliminary strength studies have indicated that CS can be used as a partial replacement of sand in preparation of concrete. The durability of the concrete is a significant worry these days due to several types of deteriorations. Concrete is expected to perform satisfactorily in the working environment during its designed life span. The concrete constituent materials, water–cement ratio and exposed environment, etc. majorly affects the durability of concrete. Curing, cover to reinforcement and compaction, etc. are also few factors which affect the durability of concrete. This research paper presents the preliminary experimental investigation results for finding the optimum percentage of CS to be used in concrete as a replacement of sand. Further, the impact of elevated temperature on CS concrete (with an optimum percentage of copper slag) has been presented and been compared with that of normal concrete.

2 Literature Review

Balendran et al. [3] considered mainly durability and strength parameters of HPC having pulverised fuel ash, metakaolin and silica fume as pozzolans and exposed to elevated temperature. He observed that silica fume inclusion causes reduced performance of concrete concerning spalling and strength at higher temperatures. The conclusion drawn was the addition of metakaolin and pulverised fuel ash improves the residual strength and durability. Bo et al. [4] investigated the impact of high temperature varied from 100 to 900 °C on the mechanical properties of high strength concrete (HSC). It was concluded that the elastic modulus decreases quickly at elevated temperature. Chakrabarti et al. [5] carried out an experimental investigation on concrete exposed to elevated temperature. The conclusion made was, the strength loss and lost strength recovery when exposed to a higher temperature are mainly due to the dehydration of concrete. Apart from that he also found that at less than 100 °C, the compressive strength increases due to the hydration of un-hydrated concrete. At higher temperature, the compressive strength is severely affected which also depends on the extent of exposure. It was seen that with rehydration up to 80% of the original strength is regained. Srinivasa Rao et al. [6] did an experimental investigation to study the impact of high temperatures (50–250 °C) on various strength properties of HSC made with ordinary Portland cement and Portland pozzolana cement at different ages. It was observed that Portland pozzolana cement concrete performs superior to ordinary Portland cement concrete by holding more residual compressive strength. Poon et al. [7] did experiments to investigate the effect of high temperature on ordinary concrete and high strength pozzolanic concrete. He noticed that the pozzolanic concrete had better performance when compared to normal concrete. Above 400 °C, both concrete loses their strength quickly. Both types of concrete had high deformation and spalling when exposed to high temperature.

3 Material Properties

Portland cement **Ultratech-53 grade (Hyderabad, India)** was used for this experimental investigation as classified by the BIS Specification IS:12269-1987. Table 1 presents its chemical composition. Angular crushed granite metal with a maximum size 20 mm was used for coarse aggregate. The fineness modulus (FM), specific gravity and water absorption were 7.1, 2.637 and 1.1%, respectively, for the quartz sand. Quartz sand having significant size fractions 1.2, 0.6, 0.3, 0.15 each corresponding to 11.5, 18.7, 26.9 and 30.2, respectively, and 4 mm aggregates was used as filler materials in the present investigation. The fineness modulus, specific gravity and water absorption were used 2.43, 2.601 and 1.2%, respectively, for the quartz sand. The compacted state bulk density was found to be 1550 kg/m^3 and in loose state 1414 kg/m^3. Apart from this, air-cooled, black glassy and irregular copper slag having fineness modulus 3.3, specific gravity 3.47 and water absorption 0.24% was used in the present investigation. The compacted state bulk density was found to be 2024 kg/m^3 and in loose state 1898 kg/m^3. Their chemical properties are presented in Table 2 as per the product data sheet. The combined grading of aggregates has been presented in Figs. 1 and 2 presents the CS grading.

Table 1 Chemical composition of cement

S. no.	Composition	Contribution (in %)
1	Alkali oxides	0.6
2	Sulphuric anhydride	1.93
3	Magnesia	1.12
4	Insoluble residue	1.14
5	Alumina iron ratio	1.18
6	Lime Saturation Factor (LSF)	0.82
7	Loss of ignition	1.39

Table 2 Chemical composition of copper slag

S. no.	Composition	Contribution (in %)
1	Silica—SiO_2	33.52
2	Iron—Fe_2O_3	55.8
3	Aluminium—Al_2O_3	3.8
4	Calcium—CaO	3.14
5	Magnesium—MgO	0.72
6	Sodium—Na_2O	0.4
7	Potassium—K_2O	0.76
8	Titanium—TiO_2	0.5
9	Copper—Cu	0.99

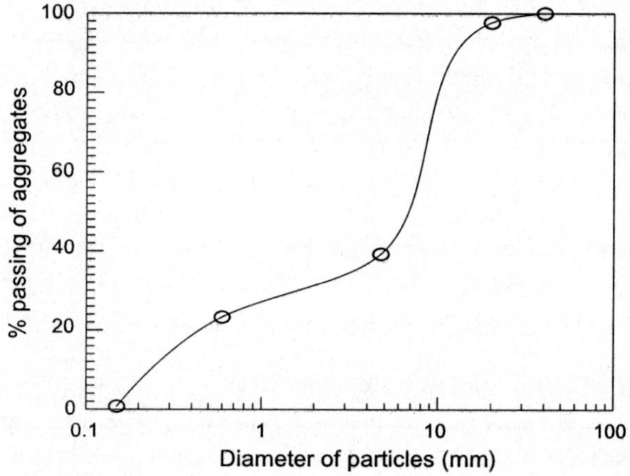

Fig. 1 Combined grading of aggregates

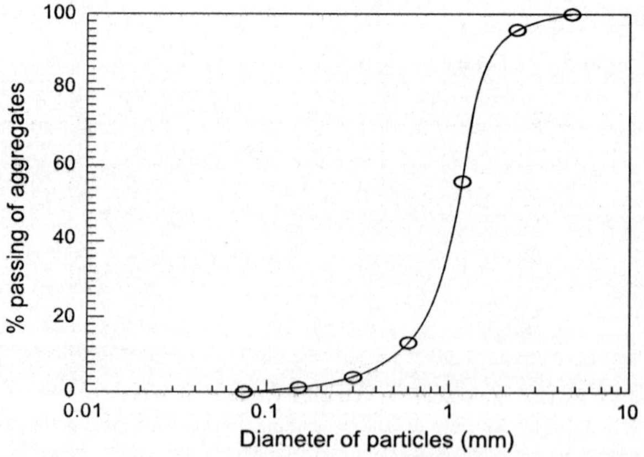

Fig. 2 Grading of copper slag

Table 3 Mix ratios and quantities of constituent materials of concrete

Grade	Cement (kg/m^3)	Fine aggregate (kg/m^3)	Coarse aggregate (kg/m^3)	Water/cement ratio	Water (kg/m^3)	Mix ratios
M-30	350	703.6	1164	0.5	175	1:2.01:3.326
M-20	320	712	1178	0.55	176	1:2.225:3.68

4 Mix Design and Different Mixes

M30 and M20 grades of concrete were designed as per IS 10262–2009. For predicting the optimum dosage of CS as a partial replacement of sand in concrete, six different mixes (CS0, CS10, CS20, CS30, CS40 and CS50) were considered by partially replacing sand from 0 to 50% with CS. After identifying the optimum dosage of CS based on the compressive strength test, test specimens with an optimum percentage of CS were used to investigate the effect of high temperature and been compared with that of normal concrete. The mix proportions are presented in Table 3.

5 Casting and Testing Procedure

The essential parameters for preparing a decent concrete such as appropriate mixing, compaction and proper curing were taken care during the concrete test sample casting process. For mixing, the concrete Pan mixture available in the laboratory was used and each batch was mixed for about 3–4 min. After a day of casting the test samples were demoulded and were adequately cured with potable water for 28, 90 and 180 days. For studying the effect of elevated temperature, the concrete specimens were taken out after 28 days of curing and surface dried by a dry cloth followed by putting in an electric oven for 12 h at specified temperatures (200, 400 and 600 °C). The test specimens were taken out of the electric oven after 12 h, and the loss of weight and strength were measured.

6 Discussion of Results

6.1 Effect of CS in Concrete as Sand

The effect of using CS by partially replacing sand from 0 to 50% on compressive strength at various ages have been presented in Fig. 3. The compressive strength test setup has been shown in Fig. 4.

It can be observed that the compressive strength is increasing from OPCC (0% replacement) to 40% replacement of CS. A further rise in replacement of sand by

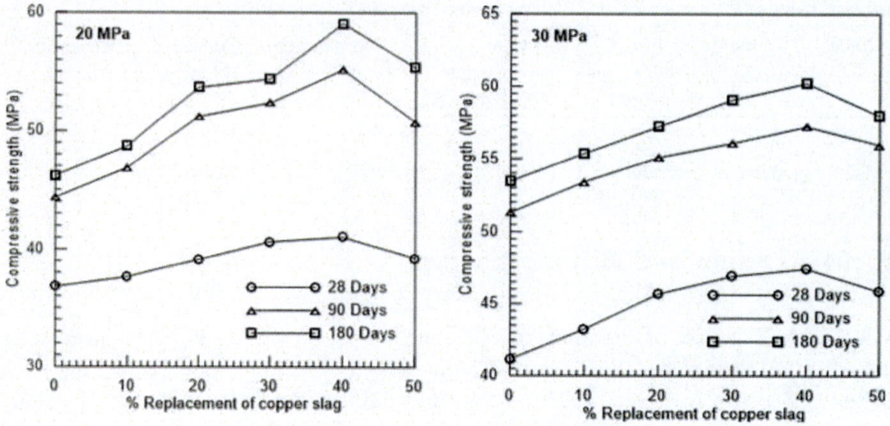

Fig. 3 Variation of strength with percentage replacement of CS

Fig. 4 Compressive strength test setup in CTM

CS affects the compressive strength. Similar pattern was observed for concretes of grades M30 and M20. A similar trend of strength variation can be seen at different ages of testing. The test results exhibited that the workability of concrete increases significantly with increasing the CS proportion in the concrete mix because of the coarser particles, glassy surface and low water absorption of CS, thereby the strength properties are also improved. Beyond 40 percentage, the free water content increases in the concrete which results decreasing in the compressive strength. Based on this, it has been established that the optimum dosage of CS to be 40% as partial replacement of sand in concrete.

6.2 Effect of Elevated Temperature on CS Concrete

The influence on weight, UPV and compressive strength of different mixes of M30 and M20 grade controlled concrete and CS concrete exposed to high temperatures of 200, 400 and 600 °C for 12 h followed by dry cooling to room temperature are presented in Tables 4 and 5. The concrete test specimens subjected to elevated temperature have been shown in Fig. 5. The graphical representation of impact on weight, UPV and compressive strength on copper slag concrete and normal concrete after being exposed to high temperatures have been presented in Figs. 6, 7 and 8, respectively. A copper slag concrete specimen after being subjected to high temperature of 600 °C for 12 h has been shown in Fig. 9. It is clearly depicted from the Tables 4 and 5 that, when OPCC and CSC concrete are subjected to high temperature of 200, 400 and 600 °C for 12 h followed by dry cooling to room temperature, the loss in weight is increasing with rise in the temperature. It can be observed that the weight loss for CS concrete is less compared to the OPCC. Similar pattern can be witnessed for concrete grades M30 and M20.

It can be observed that, when OPCC and CS concrete are exposed to high temperatures of 200, 400 and 600 °C for 12 h followed by dry cooling to room temperature, with increase in the temperature the UPV loss is increasing. It can also be observed that at 200 and 400 °C exposure, the loss in UPV for CS concrete is lesser compared to OPCC; however, at 600 °C the percentage of loss of UPV of CSC is higher than that of OPCC. Similar pattern can be seen for both concrete grades (M30 and M20).

When OPCC and CS concrete are exposed to higher temperatures 200, 400 and 600 °C for 12 h followed by dry cooling to room temperature, the percentage of loss in compressive strength is increasing with the increase in temperature. It can also be observed that at 200 and 400 °C exposure the loss in compressive strength in CS concrete is comparatively lesser than that of OPCC while at 600 °C the percentage of loss of compressive strength of CSC is higher than that of OPCC. Similar pattern can be seen for both concrete grades (M30 and M20).

Higher percentage of loss in UPV and compressive strength for CSC at 600 °C compared to OPCC is due to the higher thermal expansion of CSC compared to OPCC at high temperature. Therefore, thermal cracks are developed on the surface of CSC due to higher thermal strain resulting in higher loss of UPV and compressive

Table 4 Variation in wt., UPV and comp. st. of M20 grade concrete at elevated temperature

Mix	Temp (°C)	Initial wt. (kg)	Final wt. (kg)	% of loss in wt.	UPV-initial (km/s)	UPV-final (km/s)	% of loss in UPV	Initial comp. strength	Final comp. strength	% of loss in comp. st.
M20-CS-SFRC-0-0	200	2.16	2.07	3.99	4.26	2.61	38.73	36.8	32.82	10.82
	400	2.44	2.33	4.47	4.26	1.61	62.12	36.8	28.72	21.96
	600	2.31	2.19	5.14	4.28	0.49	88.63	36.8	23.52	36.09
M20-CS-SFRC-40-0	200	2.26	2.18	3.63	4.46	2.82	36.8	41.05	37.29	9.16
	400	2.26	2.17	4.11	4.45	1.79	59.83	41.05	33.18	19.17
	600	2.27	2.16	4.71	4.46	0.39	91.21	41.05	25.28	38.42

Table 5 Variation in wt., UPV and comp. st. of M30 grade concrete at elevated temperature

Mix	Temp (°C)	Initial wt. (kg)	Final wt. (kg)	% of loss in wt.	UPV-initial (km/s)	UPV-final (km/s)	% of loss in UPV	Initial comp. strength	Final comp. strength	% of loss in comp. st.
M30-CS-SFRC-0-0	200	2.19	2.11	3.56	4.43	2.73	38.33	41.16	36.99	10.13
	400	2.33	2.24	4.07	4.43	1.84	58.57	41.16	32.87	20.14
	600	2.37	2.27	4.55	4.33	0.56	86.97	41.16	26.92	34.6
M30-CS-SFRC-40-0	200	2.43	2.35	3.25	4.48	2.97	33.82	47.41	43.25	8.77
	400	2.49	2.39	3.86	4.48	1.89	57.73	47.41	38.73	18.31
	600	2.43	2.33	4.15	4.48	0.54	87.92	47.41	29.77	37.21

Fig. 5 Concrete specimens subjected to elevated temperature

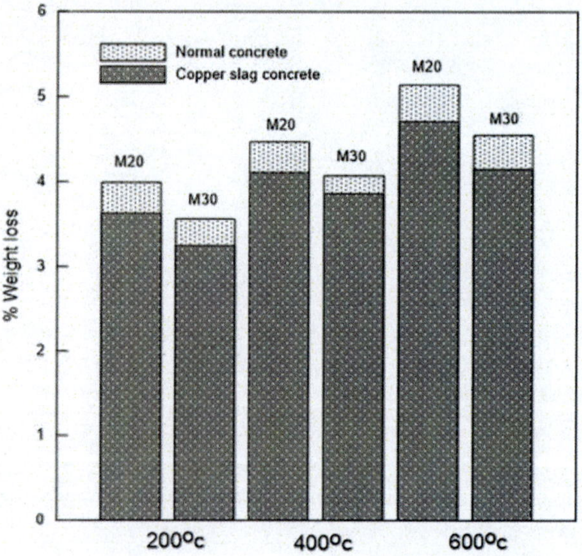

Fig. 6 Effect of elevated temperature on weight of concrete

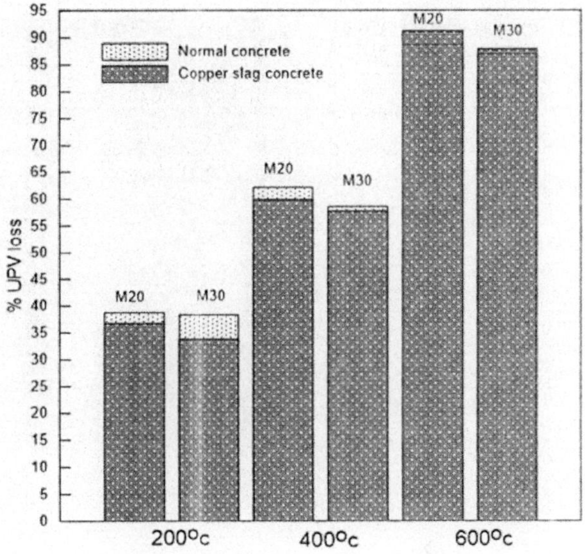

Fig. 7 Effect of elevated temperature on UPV of concrete

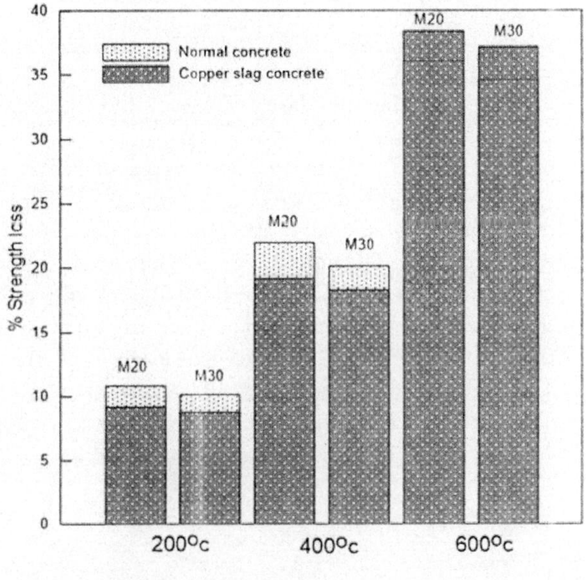

Fig. 8 Effect of elevated temperature on strength of concrete

strength compared to OPCC. Sever spalling and deformation are found in copper slag concrete when exposed to a temperature higher than 600 °C which is mainly attributed to the high pore pressure generated by the internal moisture in the highly impermeable and dense copper slag concrete.

Fig. 9 Concrete specimens after exposed to 600 °C temperature

7 Conclusion

- At elevated temperature, copper slag concrete has better resistance to weight loss as compared to controlled concrete.
- CS concrete has excellent resistance to strength loss at an elevated temperature of 200 and 400 °C compared to normal concrete. However, at 600 °C temperature copper slag concrete shows similar trend like normal concrete due to the low transition temperature of copper slag.
- Outside 400 °C, both normal and copper slag concrete lose their strength quickly.
- Use of the optimum percentage of copper slag (40%) gives the best results to enhance compressive strength at room temperature because of its dense microstructure. The highly impermeable dense microstructure of copper slag concrete at elevated temperature resists the internal moisture to escape causing high pore pressure. This results in the development of microcracks fast followed by the occurrence of spalling and deterioration of strength.
- The higher percentage of loss in compressive strength at 600 °C temperature is also attributed to the high thermal expansion of copper slag.

References

1. Bipra G, Jana RK, Prenchand M (2003) Characterstics and utilization of copper slag—review. Resour Conserv Recycl 39(4):299–313
2. The World Copper Factbook 2018. International Copper Study Group
3. Balendran RV, Rana TM, Magsood T, Tang WC (2002) Strength and durability performance of HPC incorporating pozzolans at elevated temperatures. Struct Survey 20(4):123–128
4. Bo W, Xiao-Ping S, Li H, Yuan J (2002) Effect of high temperature on residual mechanical properties of confined and unconfined high strength concrete. ACI-Mater J 99(4):399–407

5. Chakrabarti SC, Sharma KN, Mittal A (1994) Residual strength in concrete after exposure to elevated temperature. Indian Concr J (December Issue) 713–717
6. Srinivasa Rao K, Potha Raju M, Raju PSN (2006) Effect of elevated temperature on compressive strength on HSC made with OPC and PPC. Indian Concr J (August Issue) 43–48
7. Poon C-S, Azhar S, Anson M, Wong Y-h (2002) Performance of metakaolin concrete at elevated temperatures. Cement Concr Compos 25:83–89

Properties of High Density Concrete with Mill Scale Waste as Aggregates for the Production of Washing Machine Ballasts

Vishnu S. Gavandi and K. G. Guptha

Abstract One of the notable environmental issues dealt by the society is the generation and management of solid waste. Steel industry is one of the major contributors of solid waste generation all over the world. There have been many debates to minimize the production of such wastes, but apart from this, enough prominence should be given to utilize the same. One such waste generated in the steel industry is mill scale waste. Mill scale is formed on the surface of steel plates or sheets during its production in rolling mills. The specific production of mill scale constitutes to about 3 to 4% of the steel produced in the industry. Globally, millions of tonnes of mill scale are generated on annual basis. To some extent, it can be reused in steel industries and cement plants but very rarely research is conducted to explore the possibility of its use as aggregates in concrete. This study concentrates on the utilization of mill scale waste for the production of ballasts used in washing machines. Ballasts act as counterweights in washing machines and require a standard weight. Thus, high density concrete is used to produce ballasts occupying lesser volume for the given weight. Mill scale acts as a high density aggregate and is used as a complete replacement of fine and coarse aggregates. The studies carried out show that a density of 3100 kg/m^3 is achieved for the designed mix concurrently meeting the strength requirements of the ballasts as per the industrial requirements.

Keywords High density concrete · Ballast · Aggregates · Mill scale waste · Solid waste management

1 Introduction

Waste management has been a persistent problem dealt by environmentalists all over the world for decades. These wastes cause considerable amount of damage to the ecosystem, if not correctly monitored and effective steps are taken to reduce

V. S. Gavandi (✉) · K. G. Guptha
Department of Civil Engineering, Goa College of Engineering, Farmagudi, Goa, India
e-mail: vishnugavandi01@gmail.com

K. G. Guptha
e-mail: kgg@gec.ac.in

B. B. Das et al. (eds.), *Recent Developments in Sustainable Infrastructure*, Lecture Notes in Civil Engineering 75, https://doi.org/10.1007/978-981-15-4577-1_49

its production to minimum possible extent. Over the years, there have been many studies conducted to understand waste generation and its uses as a recyclable and reusable material. These wastes can be classified under different categories. One such form of waste is the solid industrial waste. One of the major contributors of solid industrial waste is the steel industry which produces tonnes of waste annually. Wastes produced in steel industry can sometimes also be hazardous in nature. Although the solid wastes produced can be reused in the industry itself, large amounts of it are still abundantly leftover in the process.

Concrete is a widely used material worldwide. "Today, second only to water, concrete is the most consumed material, with three tonnes per year used for every person in the world" [1]. It has numerous applications and depending on its uses it constitutes different materials like cement, water and aggregates like sand and construction metal. Generally, these constituents are naturally available materials which are mined for its use in concrete. As it is an established fact that concrete is very frequently used in large quantities, the natural aggregates are extracted on a large scale. This causes depletion of valuable natural resources disturbing ecological balance. For this reason, researches have conducted tests to incorporate wastes as a partial or total replacement for natural aggregates. This paper concentrates on utilization of mill scale waste generated in the steel industry in concrete to develop a high density mix and its application in the production of ballasts used in washing machines which acts as a counterweight to stabilize the machine when in use.

1.1 Mill Scale Waste

"Mill scale is formed on the outer surfaces of plates, sheets or profiles when they are being produced by rolling red or hot iron steel billets in rolling mills" [2]. These contain both iron in its elemental form and three types of iron oxides, namely, wustite (FeO), hematite (Fe_2O_3) and magnetite (Fe_3O_4), forming the major constituents of mill scale with iron content up to about 70% along with non-ferrous metals and alkaline compounds [3]. Globally millions of tons of mill scale waste are produced in iron and steel industries. The specific production of mill scale constitutes to about 3 to 4% of the steel produced in the industry. Generally, it can be reused in the industry itself and also in cement manufacturing plants; however, mill scale which is not reusable, ends up in landfills [4]. The chemical properties of mill scale waste are given in Table 1 [5].

The mill scale waste sample procured is shown in Fig. 1.

Table 1 Chemical properties of mill scale waste

Element	Fe (total)	Fe^{+2}	Fe^{+3}	FeO	S	P	Mn	Si
Weight (%)	69.33	49	12.5	7.83	0.33	0.22	0.51	0.9

Fig. 1 Mill scale waste used in concrete

1.2 Washing Machine Ballasts

During the wash cycle of the washing machines, it experiences movement due to the spinning drum. In order to balance this spinning a heavy block of concrete is placed which acts as a counterweight to provide inertia to the equally heavy electric motor which drives a very heavy gearbox which in turn drives the inner tub [6]. Since these motor components are very heavy and the forces exerted by them are quite high, the counterweight is required to be quite heavy itself. The weight of these ballasts depends on the type and capacity of the washing machines. As per the typical requirement of weight for the ballasts, if standard concrete is used to cast the same, the volume of the ballasts works out to be quite high. This causes the washing machine to have a larger body, increasing its production cost. Thus, high density concrete is used to produce ballasts occupying lesser volume for the given weight. For achieving higher density of concrete, mill scale waste serves as a high density aggregate which is a perfect replacement of the natural aggregates. "The density of heavyweight concrete is based on the specific gravity of the aggregates and the properties of the other components of concrete" [7]. Concrete with density more than 2600 kg/m^3 is generally termed as high density or heavy weight concrete. As discussed, mill scale waste will be utilized in the design mix to achieve a high density concrete, completely replacing natural aggregates.

2 Materials and Methodology

Along with cement, ground granulated blast-furnace slag (GGBS) was used as a binding material. Different types of admixtures were tried and tested to achieve the desired flow and strength. Fibres are also used to avoid cracks and to maintain workability. The ballasts were required to have a density of 3100 kg/m^3 $\pm$ 1%.

Also, early strength gain was required in order to facilitate faster production and dispatching the same to be installed in the washing machine. Thus a strength of 35 Mpa was desired at the end of 7 days of curing.

2.1 *Cement*

In the design mix, 43 grade ordinary Portland cement (OPC) confirming to IS 8112: 2013 was used. The specific gravity of cement as worked out in lab as per IS: 2720 (Part III/sec I)—1980 was determined as 3.14.

2.2 *Ground Granulated Blast-Furnace Slag (GGBS)*

GGBS was used as a binding material as well as to maintain the amount of fines in the concrete matrix. It served as a partial replacement for cement and was used for durability requirements. Also a proper proportioning of GGBS and cement ensured the accuracy required in maintaining the density to 3100 kg/m^3. Specific gravity of GGBS was determined to be 2.85.

2.3 *Water*

Clean potable water free of any impurities was used both for casting as well as curing the cubes.

2.4 *Mill Scale Waste*

Mill scale was procured from a local firm in Goa. Maximum size was found to be 10 mm. Sieve analysis was conducted on the sample as per IS: 2720 (Part 4)—1985 passing through a set of sieves arranged in descending order starting from 4.75 mm sieve. A dry sample of 2000 grams was used for the analysis. The cumulative percentage passing through each sieve is given in Table 2. Three trials were conducted on the same sample for accuracy. Figure 2 shows the particle size distribution for mill scale waste.

The fineness modulus, effective size in mm (D_{10}), uniformity coefficient (D_{60}/D_{10}), coefficient of curvature ($D_{30}^2/D_{60} \times D_{10}$) and average size of particles in mm (D_{50}) calculated from the particle size distribution curve are as shown in Table 3 [2].

Table 2 Sieve analysis of mill scale waste

Sieve sizes	Cumulative percentage passing		
	Trial 1	Trial 2	Trial 3
4.75	88.10	85.70	85.60
2.36	81.55	80.55	79.20
1.18	56.80	55.85	53.40
0.60	42.25	42.20	40.65
0.30	21.15	19.80	26.50
0.15	7.45	6.90	13.00
0.075	3.40	3.50	8.65
Pan	0.65	1.00	5.85

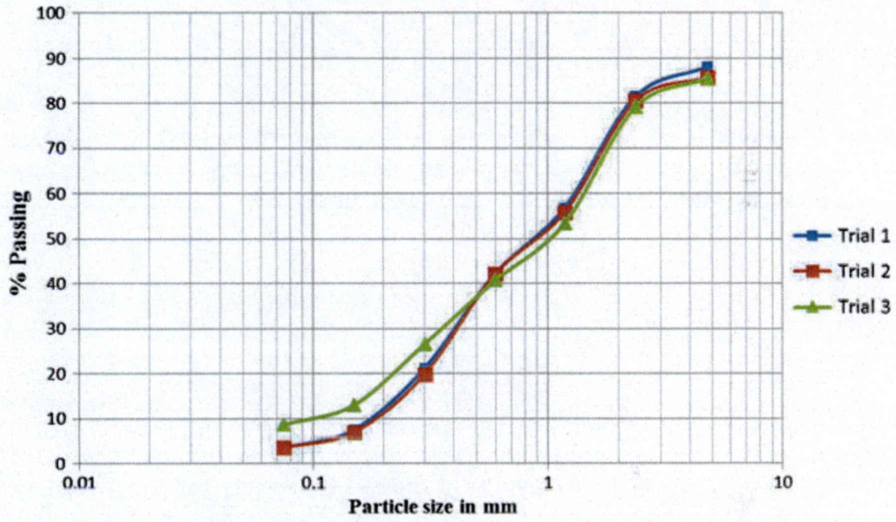

Fig. 2 Particle size distribution for mill scale waste

Table 3 Characteristic properties of mill scale waste

Property	Value
Fineness modulus	4.97
Effective size	0.21
Uniformity coefficient	7.00
Coefficient of curvature	1.48
Average size of particles	1.02

Table 4 Technical data of the fibres

Property	Value
Colour	Orange
Equivalent diameter (mm)	0.91 ± 5%
Density (kg/m^3)	1000
Length (mm)	55 ± 5%
Shape	Wavy

Specific gravity of the sample was determined to be 6.20. Dry loose bulk density (DLBD) was found to be 2865 kg/m^3 as per IS: 2386 (Part III)—1963.

2.5 *Polypropylene Fibres*

Mapei Mapefibre IT 39 NV type of polypropylene fibres were used in the design mix. This type of special fibres helps in maintaining workability of concrete. They also prevent any shrinkage cracks that may appear after drying. It has a form ratio of 60 and has a wavy shape. Technical specifications are given in Table 4.

2.6 *Chemical Admixtures*

Due to high cement content, water–cement ratio also increases and gives rise to shrinkage cracks. It plays an important role in strength gain and final density of the mix. Since an accuracy of only 1% is allowed for density variations, various chemical admixtures are added to control the water–cement ratio and for strength enhancements. Different types of admixtures used in different trials of casting are briefly discussed.

2.6.1 Dynamon SP 508

It is used for a high level of workability and high compressive strength of concrete at an early age. This chemical reduces the water requirement while maintaining the workability and flow.

2.6.2 MYK Remicrete PC 30

It is used for similar purpose as Dynamon SP 508. It is generally used for high-performance concrete and provides the necessary flowability and early gain of strength.

2.6.3 MYK Bond Latex

Known for its waterproofing and bonding applications, this chemical is used for forming a better bond within the constituting materials of concrete forming a compact mix and helps in early setting of concrete.

2.6.4 Mapei Mapefluid IF 328

It is used to form a durable and high-performance mix. Mapei Mapefluid IF 328 is used as a blister reducing agent to form a smooth and well-finished concrete. It also accelerates the hydration of cement.

2.6.5 Epko KP 200

EPKO KP 200 is a corrosion inhibiting chemical working on a bipolar inhibition mechanism. Since ballasts are likely to experience damp conditions, this admixture is used to prevent excessive corrosion to it.

2.7 Methodology

In order to work out the proportions of concrete, a preliminary mix was first designed to achieve a density of 3100 kg/m^3. Keeping density as first preference, various trials were casted with varying constituents to achieve the required density. Similarly keeping density in check different variations in proportioning was tried to achieve the desired strength. In order to maintain density, wet mix was filled in a DLBD measuring vessel with a volume of 3 litres. By mass volume relation, corrections are done to achieve the required wet density. The industry requirement of 3100 kg/m^3 is for the dry set concrete; hence, during casting wet density which is slightly lower than dry density, in this case, is kept lower in relation to the output required. Based on experimentations, wet density works out to be about 3 to 5% less than the dry density.

The process in sequence involves working out proportions, mixing, density corrections, casting, demoulding, curing and testing.

3 Results and Discussions

Various trials were casted and tested for its density and compressive strength. Cubes of 10 $\times$ 10 $\times$ 10 cm were used for casting. Fibres used throughout the trials are considered at 1 to 5 kg/m^3 of the design mix. Generally, the extent of fibres used is made not to exceed 2 kg/m^3 to avoid balling.

Trial 1 was designed for a cement content of 780 kg/m^3. Dynamon SP 508 was used at the rate of 0.8% of cement and Mapefluid IF 328 and EPKO KP 200 were both used at the rate of 0.4% of cement. The density achieved was higher but strength was almost attained. Hence to reduce the density, mill scale content was reduced in the next trial which apart from reducing the density produced slight decrease in strength too. Although in the first two trials, strength and density were almost achieved, the desired flow required to maintain the workability to fill the moulds of the ballasts was not attained. Hence in trial 3, water–cement ratio was increased without changes in the dosage of admixtures. This produced a mix with good density but an ordinary strength gain at the end of 7 days. In trial 4, to maintain workability without increasing water–cement ratio, Remicrete PC 30 was used at a dosage of 0.5% and reducing Dynamon to 0.4%. MYK Bond Latex was also introduced to try and increase the bond of the constituting materials. This mix produced excellent strength at 7 days curing but the density was increased by about 5%. Moving ahead with trial 5, cementitious content was reduced and Dynamon was increased to 1% replacing Remicrete PC 30 and Bond Latex. In this trial, both the desired dry density and strength requirements were attained. Different proportions of different trials are given in Table 5. The results for various trials are shown in Table 6.

Trial 5 was selected as the final design mix and its proportions in kg per one bag of cement are given in Table 7.

Through the experiments, it is clearly seen that water–cement ratio plays a major role in order to secure the proper balance between density and early strength of the mix. Too high of a water–cement ratio for workability requirements causes drastic changes in both density and strength. Also the proportioning of GGBS in the mix was of prime importance, as it was noted that slight changes in amount of GGBS varied the density of the mix. Since early strength was a prerequisite for the research, GGBS content was kept as low as possible as it is known for long-term strength gain. Thus the strength of the mix is expected to increase moderately beyond 7 days of

Table 5 Proportioning of various constituents of concrete

Sr. no.	Material	Mass (Kg/m^3)				
		Trial 1	Trial 2	Trial 3	Trial 4	Trial 5
1	Cement	780	898	898	1081	1047
2	GGBS	129.2	175.8	175.8	143.75	83.33
3	Mill scale waste	2480	1980	1980	2212.5	2141
4	Water	280	361.2	386.5	312.5	333
5	Mapefibre IT 39 NV	1.44	1.44	1.44	1.88	1.88
6	Dynamon SP 508	6.24	6.86	6.86	4.88	10.44
7	MYK Remicrete PC 30	–	–	–	6.00	–
8	Mapefluid IF 328	3.12	3.43	3.43	4.88	4.22
9	EPKO KP 200	3.12	3.43	3.43	4.88	4.22
10	MYK Bond Latex	–	–	–	4.88	–

Table 6 Density and compression strength of concrete

Trial no.	Dry density (Kg/m^3)	Compressive strength (Mpa)		
		1 day	3 day	7 day
Trial 1	3370	13.00	31.88	34.50
Trial 2	3059	18.02	30.16	32.88
Trial 3	3062	10.97	21.16	24.60
Trial 4	3276	4.77	37.38	47.35
Trial 5	3095	8.18	32.02	36.35

Table 7 Proposed constituents of concrete per one bag of cement in Kg

Cement	GGBS	Mill scale waste	Water	Mapefibre	Dynamon SP 508	Mapefluid IF 328	EPKO KP 200
50	3.98	102	15.92	0.09	0.50	0.20	0.20

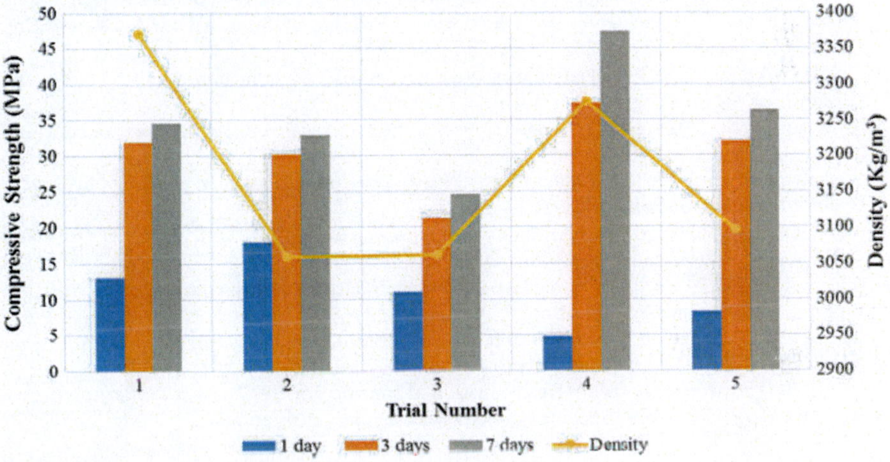

Fig. 3 Variation of strength and density for various trials

curing. The variations of compressive strength and density for different trials are graphically represented in Fig. 3.

4 Conclusions

Mill scale waste serves as a complete replacement for natural aggregates and hence makes it an eco-friendly design. This helps in solid waste management by utilizing mill scale waste and concurrently reduces the depletion of natural aggregates. A density of 3100 kg/m^3 $\pm$ 1% is achieved with mill scale waste as heavy density aggregates which is a prerequisite for the ballasts. The necessary workability required for compaction is achieved with Dynamon SP 508 and Mapefluid IF 328, which also contributes to the early gain of strength. The slump achieved in laboratory was interpreted as a collapsed case which is ideal for the workability criteria. It has been observed that the average of 7 days compressive strength obtained is 36.35 Mpa, which meets the strength requirements for faster production of the ballasts. Also, no corrosion marks were observed in the cubes despite high iron content in mill scale waste.

References

1. Gagg CR (2014) Cement and concrete as an engineering material: an historic appraisal and case study analysis. Eng Fail Anal 40:114–140
2. Akhinesh K et al. (2015) Study of the compressive strength of concrete with various proportions of steel mill scale as fine aggregate. IOSR J Mech Civil Eng (IOSR-JMCE) 104–109 (e-ISSN: 2278-1684, p-ISSN: 2320–334X)
3. Martín MI et al. (2012) Production of sponge iron powder by reduction of rolling mill scale. Ironmaking Steelmaking 39(3):155–162
4. Singhal A et al. (2015) Effect of sand replacement by mill scale on the properties of concrete. Int J Eng Technol Sci Res 2:60–67
5. Gaballah NM et al (2013) Production of iron from mill scale industrial waste via hydrogen. Open J Inorg Non-metallic Mater 3:23–28
6. Nice K (2000) How washing machines work. *HowStuffWorks*, HowStuffWorks, 31 Oct 2000, web. home.howstuffworks.com/washer2.htm
7. Ouda AS (2015) Development of high-performance heavy density concrete using different aggregates for gamma-ray shielding. Prog Nucl Energy 79:48–55
8. IS: 8112—2013: Indian Standard Ordinary Portland cement, 43 grade—specification
9. IS: 2720 (Part 4)—1985: Grain size analysis
10. IS: 2720 (Part III/sec I)—1980: Determination of specific gravity
11. IS: 2386 (Part III)—1963: Specific gravity, density, voids, absorption and bulking

Effect of Bacillus Cohnii Bacteria on the Properties of Concrete

S. Jena, B. Basa, and K. C. Panda

Abstract Concrete plays an important role in this era of rapid urbanization. But concrete is quite prone to crack formation, which affects its durability. If immediate precaution is not taken then cracks tend to expand further and require costly repair. Due to the cracks, an easy path is developed in the structure, through which water, oxygen and carbon dioxide, etc. from air penetrates into the concrete which leads to the decrement in durability of concrete. To overcome this disadvantage, various crack healing techniques have come into limelight; one of them is self-healing bacterial concrete. This paper is aimed at finding out the influence of *Bacillus Cohnii* bacterium on the properties of concrete. Bacterial cell count, i.e. colony-forming unit (CFU) of 10^5 and 10^{10} cells/ml were selected for the present work. Then 10^3 and 10^8 cells/cm^3 bacterial concentrations have been added to concrete for 10^5 and 10^{10} cells/ml, respectively. Specimens were tested after different intervals of curing period. When compared to control mix (without the addition of bacteria), it is noticed that concrete with *Bacillus Cohnii* bacteria shows increment in compressive, flexural and split tensile strength in all curing periods for both cell concentrations 10^3 and 10^8 cells/cm^3. The highest strength is achieved when 10^3 cells/cm^3 bacterial concentration have been added to concrete for 10^5 cells/ml.

Keywords *Bacillus cohnii* · Compressive strength · Split tensile strength · Flexural strength

1 Introduction

After knowing the fact that crystal formation is quite a typical behaviour in bacterial species, it has been utilized vastly in various fields, i.e. oil industries, civil engineering, geological engineering, etc [1]. A few examples of the applications are: plugging

S. Jena (✉) · B. Basa
Institute of Technical Education and Research, SOA (Deemed to be University), Bhubaneswar, India
e-mail: sjena4203@gmail.com

K. C. Panda
Government College of Engineering, Kalahandi, Bhawanipatna, India

B. B. Das et al. (eds.), *Recent Developments in Sustainable Infrastructure*, Lecture Notes in Civil Engineering 75, https://doi.org/10.1007/978-981-15-4577-1_50

of rock system for oil recovery enhancement and protection of ornamental stones [2–8]. Bacteria species could come in use in these applications as these applications require the use of calcium carbonate precipitate which are available in bacteria [9, 10]. Usage of bacterial concrete solves one of the most important vulnerabilities of concrete which is crack formation [11–15]. Crack formation not only decreases the life span of concrete, but also affects reinforcement of concrete as it results in corrosion. Through the cracks, water, oxygen and chloride enter into the concrete, which results in chemical reaction causing a shortening of concrete activity life. Usage of microbial concrete is also economic because, as the cracks remain unhealed for more days, they require more and more money to heal which is not in the case of bacterial concrete, since the healing process of cracks begin from the very moment they are formed.

The saviour of infrastructure industries is the cutting-edge technology of self-healing concrete. Self-healing concrete uses a completely new and unconventional method of dealing with the problem of crack formation [16]. This concrete can heal itself. Conventional concrete, to some extent is self-healing, since it can block the formation of further cracks in concrete by the method of hydration of un-hydrated microparticles. It can also be done by including some external agents in the concrete which can autogenously quicken the process of healing of concrete [17]. Since self-healing concrete is the newest and most promising solution to many concrete related problems, we need to use economic and environment-friendly method to achieve this. Usage of microbial concrete is hence taken into consideration. Till today, three most important bacterial metabolism processes have been found to be very useful for calcium carbonate precipitation. First one is hydrolysis of urea using enzymes [18–21]. The alternate mechanism is the oxidation of organic carbon [22–25]. The third pathway is the denitrification process under anoxic condition [26]. Out of the three mechanisms, the hydrolysis of urea is the most effective and the easiest one to perform.

The objective of this work is to observe the effects of bacteria and bacterial calcite precipitation on the various properties of concrete. For this purpose, *Bacillus cohnii* bacterium was chosen and its effect on compressive, flexure and split tensile strength were observed.

2 Experimental Details

2.1 *Materials*

For this work Ordinary Portland Cement (OPC), Natural Fine Aggregate (NFA), Natural Coarse Aggregate (NCA), *Bacillus cohnii* bacterium and potable water were taken. OPC-43 grade was utilized which is grey in colour and acquired in fine powdered form. NFA available in zone II was utilized for the current study. NCA supplied

Table 1 Physical characteristics of NFA and NCA

Characteristics	Results (IS: 383-1970) [27]	
	NFA	NCA
Fineness modulus	2.73	6.92
Specific gravity	2.69	2.78
Water absorption	0.85	0.24

Table 2 Characteristics of *Bacillus cohnii*

NCMR accession no.	MCC 2819
Taxonomic designation	*Bacillus* Cohnii
Strain Designation	LAP217
Source of isolation	Lonar Lake water sample
Location	Village: Lonar, Dist.: Buldhana, State: Maharashtra, India
Medium name and no.	34c (Alkaline Nutrient Agar)
Growth conditions (pH/Temp. °C)	10/28–30 °C
Incubation (days/h)	24–48 h
Sub culturing period (days)	1 month
Reference	Int J Syst Bacteriol (1980) 30:225

from Khurda, Odisha was used which is having size in between 10 and 20 mm. Different properties of fine aggregate such as specific gravity, water absorption and bulk density results are shown in Table 1. Bacterial samples were ordered from MCC, Pune which was in a freeze-dried condition. The detail description of pure culture for *Bacillus cohnii* is given in Table 2.

2.2 Mix Proportion

M30 grade of concrete was outlined according to standard specification IS: 10262-2009 [28]. The mix proportion was 1: 1.491: 2.69. Two kinds of concrete mixes were prepared, first mix is concrete without bacteria, second mix is concrete added with *Bacillus cohnii* bacterium. Bacterial cell count, i.e. colony-forming unit (CFU) of 10^5 cells/ml and 10^{10} cells/ml were selected for the present work and the bacterial cell count were added with concrete by referring to Jonkers et al. [23]. Two sets of concrete with bacteria were prepared, i.e. Bacterial cell concentration was added in concrete as 10^3 and 10^8 cells/cm^3 for 10^5 and 10^{10} cell/ml, respectively. Tables 3 and 4 show the mix identity and mix calculation of test sample, respectively.

Table 3 Mix identity of test sample

Mix identity	Bacterial cell count	Bacterial cell concentration in concrete ($Cell/cm^3$)
MSC0	0	0
MSC5	10^5 cells/ml	10^3 cells/cm^3
MSC10	10^{10} cells/ml	10^8 cells/cm^3

Table 4 Mix quantity per m^3 of concrete

Mix identity	MSC0	MSC5	MSC10
Cement (Kg)	442.85	442.85	442.85
CA (Kg)	1191	1191	1191
NFA (Kg)	660	660	660
Water required (Kg)	186	186	186
Bacteria (Kg)	0	10	10
Water added (Kg)	186	176	176

2.3 *Bacterial Culture*

For this experimental work, bacterial sample of *Bacillus cohnii* was taken and maintained in agar Petri plate. After that to grow the bacteria, a readymade Nutrient Hi Veg broth (Yeast Extract of 2.0 g, Beef Extract of 1.0 g, 5.0 g of Peptone, NaCl of 5.0 g, Agar of 15.0 g) was used. It was grown at 37 °C in a shaker incubator. To calculate the cell concentration with the help of spectrophotometer, Optical density test was carried out. Bacterial culture concentration of 10^5 cells/ml and 10^{10} cells/ml were maintained in the samples.

2.4 *Casting and Testing of Specimen*

OPC with NCA, NFA and bacteria were weighed and put in the concrete mixer and it was altogether mixed in dry condition until the point when the mixture becomes homogeneous. Then the required measure of water for each mix was included. Immediately after mixing for deciding workability of fresh concrete, slump test was done. Prior to casting of specimens in steel moulds, they were vibrated with the assistance of table vibrator. Then concrete specimen was casted and remoulded in the following 24 h. From that point, the specimens were permitted to cure in potable water for a time period of 7, 14 and 28 days.

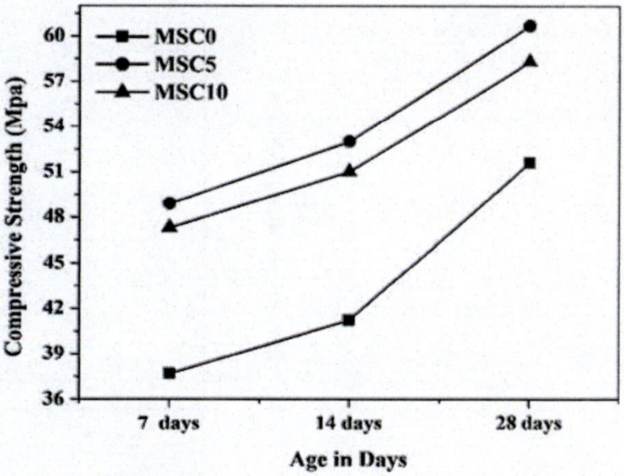

Fig. 1 Comparison of compressive strength of control concrete and bacterial concrete

3 Hardened Concrete Test Results

3.1 Compressive Strength

The compressive strength of specimen is tested after 7, 14, 28 days. Figure 1 shows the comparison of compressive strength between control concrete and bacterial concrete, i.e. concrete added with *Bacillus cohnii.*

It is observed that the compressive strength of concrete mix with *Bacillus cohnii* with cell concentration 10^3 cells/cm^3, increases up to 29.81, 28.54 and 17.61% at 7, 14 and 28 days, respectively, in comparison to concrete without bacteria. While, in concrete mixture having *Bacillus cohnii* cell concentration 10^8 cells/cm^3, the compressive strength increases up to 25.59, 23.69, 12.98% at 7, 14 and 28 days, respectively, in comparison to concrete without bacteria. The measured compressive strength of concrete mixes containing bacteria in different concentrations gives higher value in comparison to control specimen, i.e. concrete without bacteria. The increase in early strength is more in comparison to 28 days strength.

3.2 Split Tensile Strength

A test is conducted to measure split tensile strength of concrete specimen after 7, 14, 28 days. Figure 2 shows the comparison of split tensile strength between control concrete and bacterial concrete, i.e. concrete added with *Bacillus cohnii.*

It is noticed that the split tensile strength of concrete mix having *Bacillus cohnii* cell concentration 10^3 cells/cm^3, increases up to 26.13, 25.13 and 24.39% at 7, 14 and 28 days, respectively, in comparison to concrete without bacteria. While in concrete mix with *Bacillus cohnii* cell concentration 10^8 cells/cm^3, the split tensile

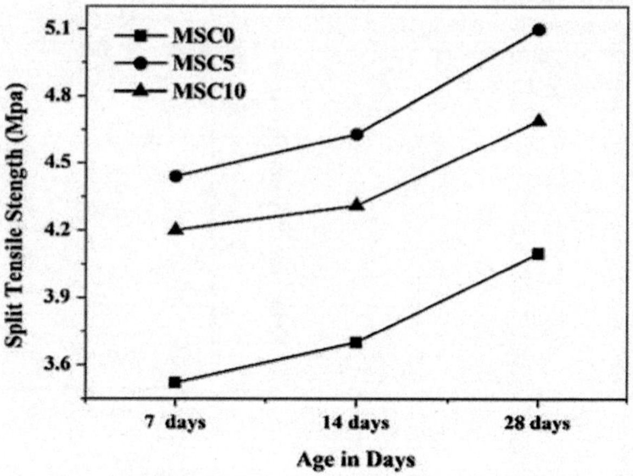

Fig. 2 Comparison of split tensile strength of control concrete and bacterial concrete

strength increases up to 19.31, 16.48, 14.39% at 7, 14 and 28 days, respectively, in comparison to concrete without bacteria. The highest percentage change is observed after 7 days curing period, i.e. 27.42% in case of cell concentration 10^3 cells/cm^3. The measured split tensile strength of almost all concrete mixes with bacteria in different concentrations gives comparatively higher value than control specimen, i.e. concrete without bacteria.

3.3 *Flexural Strength*

There is a test conducted to measure the flexural strength of specimen after 7, 14, and 28 days. Figure 3 shows the comparison of flexural strength between control

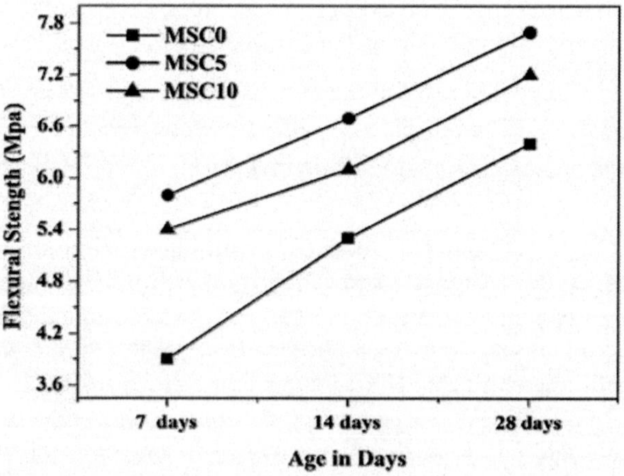

Fig. 3 Comparison of flexural strength of control concrete and bacterial concrete

concrete and bacterial concrete, i.e. concrete added with *Bacillus cohnii.*

The flexural strength of concrete mix containing *Bacillus cohnii* with cell concentration 10^3 cells/cm^3, increases up to 48.71, 26.41 and 20.31% at an interval of 7, 14 and 28 days, respectively, in comparison to concrete without bacteria. While on the contrary, *Bacillus cohnii* cell concentration 10^8 cells/cm^3, the flexural strength increases up to 38.46, 15.09, 12.5% at 7, 14 and 28 days, respectively, in comparison to concrete without bacteria. The measured flexural strength of concrete mixes containing bacteria in different concentrations gives comparatively better value than control specimen, i.e. concrete without bacteria.

4 Microscopical Study

Figure 4 shows the SEM of control concrete and Fig. 5a, b shows SEM of bacterial concrete.

Rod-shaped bacteria of different sizes are observed in Fig. 5a and precipitation of calcite on the surface of concrete is observed in Fig. 5b. A comparison of the control and bacterial concrete specimens after a span of 28 days of curing has shown that bacterial concrete is more compact and denser due to $CaCO_3$ precipitation by bacteria and has more compressive strength than normal concrete.

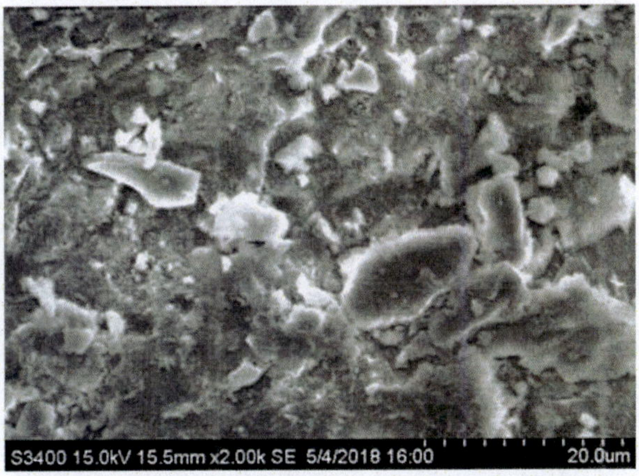

Fig. 4 Microscopical observation of control concrete

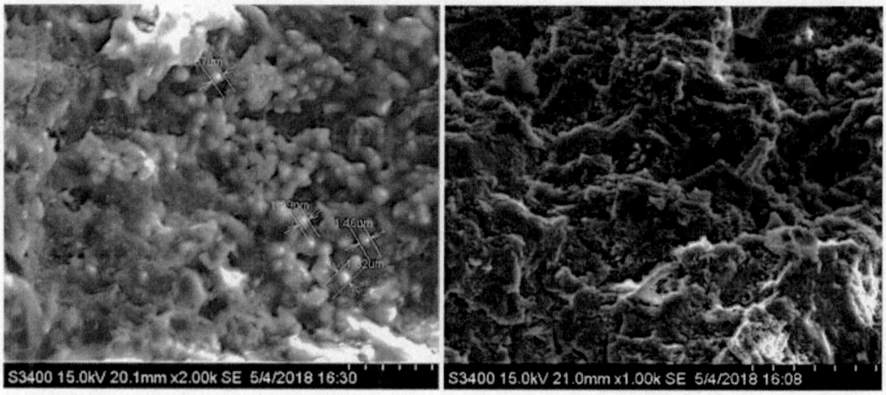

(a) (b)

Fig. 5 Microscopical observation of bacterial concrete

5 Conclusions

The above-shown results lead to the following conclusions.

- In contrast to control mix, concrete having *Bacillus cohnii* bacteria shows increment in compressive, flexural strength and split tensile strength in all curing period for both cell concentration 10^3 and 10^8 cells/cm^3.
- At 28 days curing period, concrete with cell concentration 10^3 cells/cm^3 gives highest compressive strength, i.e. 60.7 Mpa and with cell concentration 10^8 cells/cm^3 gives 58.31 Mpa compressive strength which is lowest.
- Concrete with cell concentration 10^3 cells/cm^3 gives highest split strength, i.e. 5.1 Mpa and with cell concentration 10^8 cells/cm^3 gives 4.69 Mpa split tensile strength which is the lowest.
- Concrete with cell concentration 10^3 cells/cm^3 gives highest flexural strength, i.e. 7.7 Mpa and with cell concentration 10^8 cells/cm^3 gives 7.2 Mpa flexural strength which is the lowest.
- The highest strength is achieved when cell concentration of 10^3 cells/cm^3 have been added to concrete for 10^5 cells/ml.
- Strength increases with addition of bacteria up to certain cell concentration but after that level of cell concentration strength of the structure decreases.
- From SEM it is confirmed that *Bacillus cohnii* bacterium successfully precipitates calcite. Due to the deposition of calcite, pores of the concrete are getting plugged, which is the main reason for increase in strength.

References

1. De Muynck W, De Belie N, Verstraete W (2010) Microbial carbonate precipitation in construction materials: a review. Ecol Eng 36(2):118–136
2. Tiano P, Biagiotti L, Mastromei G (1999) Bacterial biomediated calcite precipitation for monumental stones conservation: methods of evaluation. J Microbiol Meth 36(1–2):139–145
3. Gollapudi UK, Knutson CL, Bang SS, Islam MR (1995) A new method for controlling leaching through permeable channels. Chemosphere 30(4):695–705
4. Finnerty WR, Singer ME (1983) Microbial enhancement of oil recovery. Nat Biotechnol 1:47–54
5. Castanier S, Le Metayer-Levrel G, Perthuisot JP (1999) Ca-carbonates precipitation and limestone genesis - the microbiogeologist point of view. Sediment Geol 126(1–4):9–23
6. De Muynck W, Verbeken K, De Belie N, Verstraete W (2010) Influence of urea and calcium dosage on the effectiveness of bacterially induced carbonate precipitation on limestone. Ecol Eng 36(2):99–111
7. Mitchell JK, Santamarina JC (2005) Biological considerations in geotechnical engineering. J Geotech Geoenviron Eng 131(10):1222–1233
8. MacLeod FA, Lappin-Scott HM, Costerton JW (1988) Plugging of a model rock system by using starved bacteria. Appl Environ Microbiol 54(6):1365–1367
9. Rodriguez-Navarro C, Rodriguez-Gallego M, Chekroun KB, Gonzalez-Muñoz MT (2003) Conservation of ornamental stone by myxococcus xanthus-induced carbonate biomineralization. Appl Environ Microbiol 69(4):2182–2193
10. Hammes F, Verstraete W (2002) Key roles of pH and calcium metabolism in microbial carbonate precipitation. Rev Environ Sci Biotech 1(1):3–7
11. Bachmeier KL, Williams AE, Warmington JR, Bang SS (2002) Urease activity in microbiologically-induced calcite precipitation. J Biotechnol 93(2):171–181
12. Achal V, Mukherjee A (2015) A review of microbial precipitation for sustainable construction. Constr Build Mater 93:1224–1235
13. Siddique R, Chahal NK (2011) Effect of ureolytic bacteria on concrete properties. Constr Build Mater 25(10):3791–3801
14. Wang JY, Ersan YC, Boon N, Belie N De (2016) Application of microorganisms in concrete: a promising sustainable strategy to improve concrete durability. Appl Microbiol Biot 100(7):2993–3007
15. Bang SS, Lippert JJ, Yerra U, Mulukutla S, Ramakrishnan V (2010) Microbial calcite, a bio-based smart nanomaterial in concrete remediation. Int J Smart Nano Mater 1(1):28–39
16. Wu M, Johannesson B, Geiker M (2012) A review: self-healing in cementitious materials and engineered cementitious composite as a self-healing material. Constr Build Mater 28(1):571–583
17. Huang HL, Ye G, Qian CX, Schlangen E (2016) Self-healing in cementitious materials: materials, methods and service conditions. Mater Des 92:499–511
18. Day JL, Ramakrishnan V, Bang SS (2003) Microbiologically induced sealant for concrete crack remediation. In: American society of civil engineers 16th engineering mechanics conference, Seattle American
19. Wang J, Dewanckele J, Cnudde V, Van Vlierberghe S, Verstraete W, De Belie N (2014) X-ray computed tomography proof of bacterial-based self-healing in concrete. Cem Concr Compos 53(7):289–304
20. Bang SS, Galinat JK, Ramakrishnan V (2001) Calcite precipitation induced by polyurethane-immobilized Bacillus pasteurii. Enzyme Microb Technol 28(4–5):404–409
21. Ramachandran SK, Ramakrishnan V, Bang SS (2001) Remediation of concrete using micro-organisms. ACI Mater J 98(1):3–9
22. Jonkers HM (2007) Self-Healing Concrete: A Biological Approach. In: Van der Zwaag S (ed) Self-healing materials: an alternative approach to 20 centuries of material science. Springer Inc., The Netherlands, pp 195–204

23. Jonkers HM, Thijssen A, Muyzer G, Copuroglu O, Schlangen E (2010) Application of bacteria as self-healing agent for the development of sustainable concrete. Ecol Eng 36(2):230–235
24. Luo M, Qian CX, Li RY (2015) Factors affecting crack repairing capacity of bacteria-based self-healing concrete. Constr Build Mater 87:1–7
25. Luo M, Qian CX (2016) Influences of bacteria-based self-healing agents on cementitious materials hydration kinetics and compressive strength. Constr Build Mater 121:659–663
26. Ersan YC, Verbruggen H, De Graeve I, Verstraete W, De Belie N, Boon N (2016) Nitrate reducing CaCO3 precipitating bacteria survive in mortar and inhibit steel corrosion. Cem Concr Res 83:19–30
27. IS: 383-1970, Indian standard specification for coarse and fine aggregates from natural sources for concrete, (second revision). Bureau of Indian Standards, New Delhi
28. IS: 10262 (1982), Recommended guidelines for concrete mix design. Bureau of Indian Standards, New Delhi, India

Effect of PVC Dust on the Performance of Cement Concrete—A Sustainable Approach

M. Manjunatha, K. Vijaya Bhaskar Raju, and P. V. Sivapullaiah

Abstract In a current scenario, about 2.7 billion tonnes of cement is produced every year, by this lot of CO_2 is emitted to the environment, to overcome the environmental problems cementatious material (PVC DUST) is used as an alternate material in concrete. While manufacturing PVC pipes, lot of PVC dust is generated. PVC dust is a fine powder obtained while manufacturing PVC pipes, this waste will create landfilling and environmental problems; due to the contamination chlorinated chemicals in PVC dust the workers working in PVC industries are suffering from diseases like blood cancer, lung cancer and brain tumour, if it is exposed to environment, the people who stay in the surrounding they also get caused by these diseases. Chemical composition of PVC dust is having silica and other compounds so that effectively this is used in concrete as a mineral admixture to enhance the mechanical properties of cement concrete to overcome the above-mentioned serious diseases and environmental problems. Overall the main importance of proposed research is to overcome the environmental pollution problems by using PVC dust in cement concrete, to overcome the landfilling and disposal problems of PVC dust. As a sustainable approach, use of waste PVC dust in concrete will give optimum results. This paper reports on the investigation and implementation of PVC dust as a replacement of cement in concrete mixes for all types of constructions. In concrete, with a view to reduce the use of cement content in mortar and concrete without affecting their performance. For this it must be assured that concrete prepared with PVC dust has good mechanical strength. The characteristic properties of concrete were studied with controlled mix. Concrete mixes with 0 to 25% replacement of PVC dust and other waste materials in concrete. As a sustainable approach, this thesis will overcome environmental pollution, landfilling and disposal problems and economical viability.

M. Manjunatha (✉)
Research Scholar—GST, GITAM University, Bengaluru, India
e-mail: madikeaimanjunath@gmail.com

K. Vijaya Bhaskar Raju
Research Supervisor/Director—GST, GITAM University, Bengaluru, India
e-mail: kvbraju@hotmail.com

P. V. Sivapullaiah
Pro-Vice Chancellor, GITAM University, Bengaluru, India
e-mail: sivapullaiah@gmail.com

B. B. Das et al. (eds.), *Recent Developments in Sustainable Infrastructure*, Lecture Notes in Civil Engineering 75, https://doi.org/10.1007/978-981-15-4577-1_52

Keywords Concrete mix durability PVC (Poly vinyl chloride) dust · Strength and workability

1 Introduction

Cement is the most commonly available and used as construction material all over the world is just behind the water as planets most consumed resource. Cement causes more environmental pollution in all the stages of process in form gas, dust, noise, vibration, etc. During the manufacture of cement, CO_2 is emitted to the environment; about 900 kg of CO_2 is emitted to environment for 1000 kg of Portland cement produced and it counts around 10% of global manmade CO_2 emission to environment. Due to this, environment gets polluted day by day as cement is consumed as a building material and the cost of the cement is also in increasing order; to overcome this an alternate solution of replacement of cement is required as a sustainable approach to protect environmental pollution. In this research, we used PVC dust as a partial replacement of cement up to 20% in concrete.

PVC dust is a waste material obtained from PVC industries in form of dust accumulated in boiler and floor; cutting powder dust PVC consumes about 40% of total chlorine produced, approximately about 16 million tonnes of chlorine per year worldwide. Maximum use of chlorine will cause environmental pollution. At numerous points of PVC life cycle, very large quantities of hazardous organochlorine byproducts are accidentally formed and this formation is hazardous to human health and environment. The chemical mixture formed in the synthesis of manufacture of PVC products contains extremely hazardous and long-lived pollutants such as chlorinated dioxins, chlorinated furnas, polychlorinated biphenyls, hexachlorobenzene and octachlorostyrene, in addition it consists of other chemicals also. Because of their persistence, the PVC dust wastes are highly bioaccumulative and toxic to public health and environment. PVC byproducts and wastes generated contain highly toxic dioxins, phthalates, metal stabilizers, particularly led, cadmium and organotins which are hazardous to public health and environment. If it is exposed to environment in the form of land dumping or open dumping, people will suffer from diseases like lung cancer, brain cancer, brain tumour, etc. A small amount of PVC dust is recycled and also used for the manufacture of slippers, shoes, etc. To overcome the impact of PVC dust in environment and health burdening, it is a best way to use as a small amount of replacement in cement concrete. In this paper, we replaced cement by PVC dust from 0 to 20% in the dosage of 0, 5, 10, 15, 20% and experimentally designated as P_0, P_5, P_{10}, P_{15} and P_{20}.

2 Materials and Methods

2.1 Materials

2.1.1 Cement

In this research, we used ordinary Portland cement of 53 grade confirming to IS: 12269-1987 as a main binding material. Physical properties of OPC 53 grade are shown in Table 1 and heavy metal presence in both PVC dust and cement is shown in Table 2.

Table 1 Physical properties of cement and PVC dust

Property	Cement	PVC dust	Requirements as per BIS 12269–1987
Colour	Greyish	Light Greyish	–
Particle size	3–100μ	3–100μ	–
Grade	53	–	–
Type	OPC	–	–
Specific gravity	2.90	1.86	–
Initial setting time	30 min	58 min	30 min (min)
Final setting time	420 min	1080 min	600 min (max)
Fineness	235 m^2/kg	290 m^2/kg	225 m^2/kg (min)
Soundness (Le Chatelier)	2.5 mm	7.8 mm	10 mm (max)

Table 2 Heavy metal presence in cement and PVC dust

Parameters	Cement	PVC dust	Test method
pH @ 24.70 C	11.65	8.54	IS 2720:PART-26:1987
Conductivity @ 250 C, μs/cm	8260	2190	IS 14767: 2000
Loss on drying at 105 °C, % (w/w)	0.3	1.5	CEN/TC W1: 2003(E)
Loss on ignition at 550 °C (Dry Basis), % (w/w)	0.9	29.1	CEN/TC W1: 2003(E)
Total chromium as Cr, mg/kg	37.4	26.1	US-EPA 3050B: 1996
Copper as Cu, mg/kg	0.4	1.8	US-EPA 3050B: 1996
Iron as Fe, mg/kg	0.3	16.1	US-EPA 3050B: 1996
Zinc as Zn, mg/kg	7.8	7.5	US-EPA 3050B: 1996
Nickel as Ni, mg/kg	3.6	1.2	US-EPA 3050B: 1996
Lead as Pb, mg/kg	1.6	0.9	US-EPA 3050B: 1996
Cadmium as Cd, mg/kg	0.1	$\leq$0.5	US-EPA 3050B: 1996

Fig. 1 Materials

2.1.2 PVC Dust

Polyvinyl Chloride (PVC) is the most widely used thermoplastic polymer. PVC finds a deserve applications in construction industry. Waste PVC dust is collected from the locally available industries and used as a partial replacement of cement in concrete in dosages of 0, 5, 10, 15 and 20%. Physical properties of PVC dust and heavy metal presence is shown in Tables 1 and 2 (Fig. 1).

2.1.3 Fine Aggregate

Manufactured sand (M-sand) of size less than 4.75 mm is used as fine aggregate in concrete confirming zone II of IS: 383-1970 is used. Table 3 gives the properties of M-sand.

Table 3 Properties of M-sand

Property	Result
Fineness modulus	2.66
Specific gravity	2.635
Bulk density	1698 kg/m^3
Water absorption	1.45%

Table 4 Properties of coarse aggregate

Property	Result
Fineness modulus	6.03
Specific gravity	2.72
Bulk density	1466 kg/m^3
Water absorption	0.35%

2.1.4 Coarse Aggregate

Coarse aggregate is granular material produced from the crushed rock of nominal size 20 mm is used. Table 4 gives the properties of coarse aggregate.

2.1.5 Water

In general, the quality of water used for mixing and curing of concrete should be drinking water. Water–cement ratio in concrete will control many properties in both fresh and harden state of concrete such as workability, compressive strengths, water tightness, durability, drying shrinkage and potential for cracking.

2.2 Methods

2.2.1 Concrete Mix Design

In this experimental investigation, M-30 grade concrete mix design is used as per the guidelines and according to Indian standard code IS:10262-2009. Total cement quantity of 488 kg/m^3, fine aggregate 684 kg/m^3 and coarse aggregate 1120 kg/m^3 along with water–cement ratio 0.40 is used.

2.2.2 Casting and Curing of Concrete Specimens for Investigation

All the ingredients of concrete as per M-30 grade mix design were mixed in concrete mixer. For each % of replacement, totally 9 concrete cubes of size 150*150*150 mm,

9 cylinders of size 150 mm diameter and 300 mm depth and 9 concrete beams of size 100*100*500 mm were casted for experimental investigation. All the ingredients of concrete were mixed in concrete mixers and compacted by using table vibrator. For each percentage of replacement slump test is conducted to study the workability of concrete. Totally 45 concrete cubes, 45 concrete cylinders and 45 concrete beams were prepared and cured in curing tank for 7, 14 and 28 days.

2.2.3 Testing of Specimen

Testing on all the prepared specimen is carried out as per Indian standards. Compression strength test is done on prepared concrete cubes of size 150*150 mm as per IS: 516-1959, split tensile test on prepared concrete cubes of size 150 mm diameter and 300 mm depth as per IS: 5816-1979 and flexural strength test on prepared concrete beams of size 100*100*500 mm as per IS: 516-1959. All the tests done in the laboratory are shown in the Fig. 2.

3 Results and Discussion

3.1 Slump Test

Slump cone test is done to measure the workability of fresh concrete at all the varying replacement levels of 0–20% PVC dust. As the percentage of replacement of PVC dust increases the slump value of concrete decreases. For 0 and 5% replacement stage it is observed that the concrete mix is cohesive and thereafter concrete becomes sticky for 10, 15 and 20% dosage of PVC dust (Fig. 3 and Table 5).

3.2 Compressive Strength

Table 6 shows the compressive strength of all the specimens and Fig. 4 shows compressive strength of the concrete with varying % of PVC dust dosage from 0–20%. From these results, it is clearly observed that on each replacement level of PVC dust, compressive strength of concrete decreases at curing time 7, 14 and 28 days. Controlled concrete mix (P_0) shows higher value of compression strength when compared to concrete prepared with PVC dust. As the percentage of PVC dust increases, the compression strength decreases. To overcome environment pollution and health burdening, as a sustainable approach we can use PVC dust in concrete as a replacement up to 15% dosage, thereafter compression strength of concrete decreases less than concrete mix grade M-30. It clearly shows that PVC dust affects the strength of concrete if the dosage of PVC dust is more than 15%.

Fig. 2 Curing and testing of concrete specimens

Fig. 3 Slump values with various proportions of PVC dust

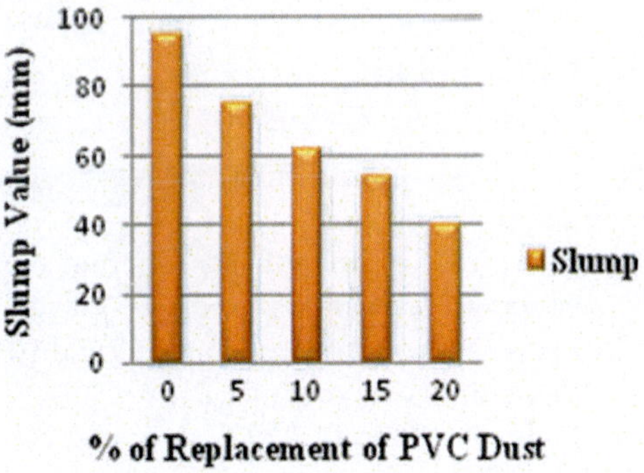

Table 5 Slump test results

Percentage of replacement	Slump test results (mm)
0	95
5	75
10	62
15	54
20	40

3.3 *Split Tensile Strength*

Figure 5 shows tensile strength of concrete specimens with varying percentages of PVC dust. It is clearly observed that tensile strength of the concrete is reduced when PVC is used as partial replacement of cement. Gradually tensile strength is decreasing.

3.4 *Flexural Strength*

Figure 6 gives flexural strength results of concrete specimens with varying percentages of PVC dust. It is clearly observed that flexural strength of the concrete is increased for 5 and 10% dosage of PVC dust when compared to conventional concrete. Thereafter for 15 and 20% dosage flexural strength decreases as per IS code regulations 5, 10, and 15% dosage of PVC dust meets the requirements of compressive, split tensile and Flexural strength. Thereafter it clearly indicates that PVC dust affects the strength parameters of concrete if the dosage of PVC dust is more than 15%.

4 Conclusions

A. Compressive strength of concrete at all replacement levels increases as the curing period increases.
B. Compressive strength of concrete decreases as percentage of PVC dust in concrete increases.
C. Flexural strength of concrete at 5 and 10% of dosage is increased when compared to nominal concrete mix P_0.
D. It clearly shows that PVC dust affects the strength parameters of concrete if the percentage of usage of PVC dust is more than 15%.
E. As per IS code regulations, 5, 10 and 15% dosage of PVC dust meets the requirements of compressive, split tensile and flexural strength.

Table 6 Strength results of partially replaced concrete at 7, 14 and 28 days

Sample designation	Total % of replacement (PVC dust)	Compressive Strength (N/mm^2)			Split Tensile Strength (N/mm^2)			Flexural Strength (N/mm^2)		
		7 Days	14 Days	28 Days	7 Days	14 Days	28 Days	7 Days	14 Days	28 Days
P_0	0	24.44	33.92	37.25	2.94	3.02	3.22	2.90	3.53	4.47
P_5	5	22.07	30.07	35.40	2.65	2.77	3.02	2.97	3.61	4.58
P_{10}	10	21.92	28.29	34.67	2.45	2.56	2.87	3.04	3.77	4.67
P_{15}	15	18.51	26.81	30.22	2.20	2.35	2.65	2.74	3.25	4.22
P_{20}	20	17.47	26.07	28.43	1.56	1.89	2.07	2.21	2.78	3.41

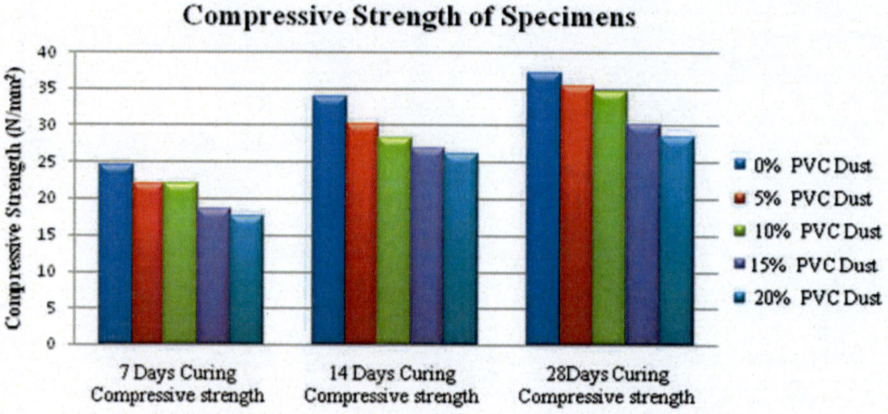

Fig. 4 Compressive strength of concrete with different percentages of PVC dust

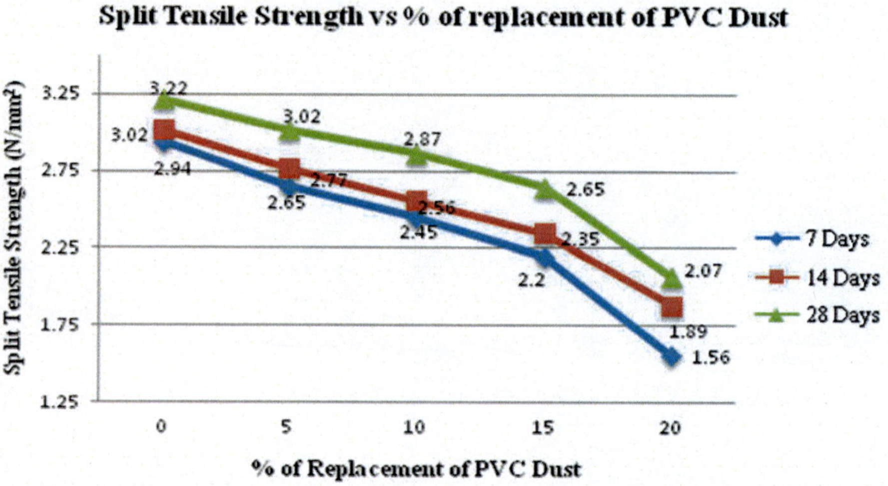

Fig. 5 Split tensile strength of concrete with different percentages of PVC dust

F. Percentage of replacement of PVC dust increases the workability of concrete decreases.
G. To overcome dumping problems of PVC dust it is a better way to use as a replacement of cement in concrete.
H. As a sustainable approach to overcome environmental pollution, health burdening of PVC dust is a good way to use as a replacement of cement in concrete up to 15%.

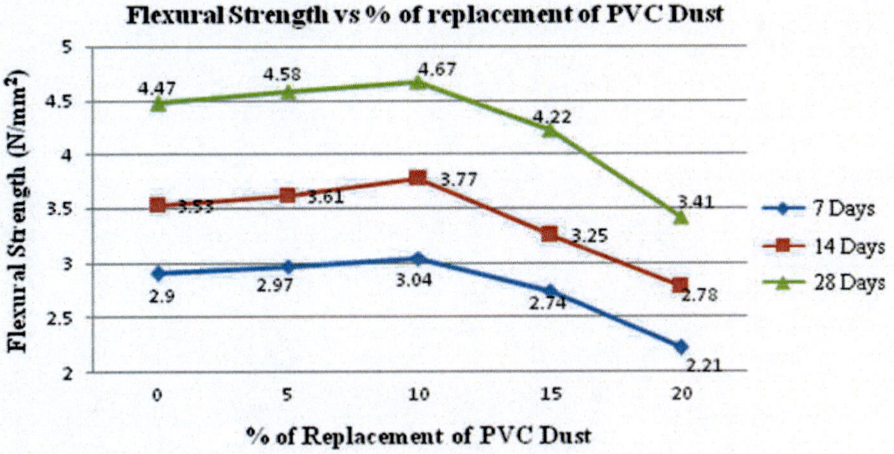

Fig. 6 Flexural strength of concrete with different percentages of PVC dust

Bibliography

1. Surekha, Dr. Chandrashekhar (2015) Experimental investigations on properties of concrete with silica fume, GGBS and PVC dust. Int J Res Appl Sci Eng Technol (IJRASET) 3(I1), June, ISSN: 2321-9653
2. KN Janardhanan, Mohana priya (2018) Experimental study on cement concrete with pvc powder and quarry dust as partial replacement to fine aggregate. Int Res J Eng Technol (IRJET) 05(0), June, ISSN: 2395-0072
3. Nihat ATMACA1, Adem A, Mohammed, Ali Ihsan (2018) Strength and shrinkage properties of self compacting concretes incorporating waste pvc dust. Int J Energy Eng Sci 3–1:47–57
4. IS: 516-1959. Indian Standard Methods of Test for Strength of concrete. Bureau of Indian Standards, New Delhi
5. Nevillie AM (2002) Properties of concrete—fourth and final edition. Pearson Education Limited, Essex

An Overview of Strength and Durability Aspects of Concrete Using PET Fibres

Shruti K. Chodankar and Purnanand P. Savoikar

Abstract As a widely used construction material, concrete plays a very important role in the construction industry. Concrete reinforced with fibrous material is called as fibre reinforced concrete. Fibre reinforced concrete (FRC) is a new structural material that is becoming increasingly important due to rise in concrete structural strength and integrity. Rapid urbanisation and industrialization and overuse of plastics in day to day life all over the world have resulted in large accumulation of plastic waste materials all over. Polyethylene terephthalate (PET) waste is obtained from disposable PET bottles is the major cause of solid disposal. This research is expected to encourage the use of waste plastics in the construction industry to regulate the pollution of the environment and the harm caused to humans. Use of plastic fibres into the concrete as a reinforcing material has attracted widespread attention from all over the world due to its long-term benefits especially in the construction industry. In the current paper, reviewing of the impact of PET fibres on properties of concrete is undertaken.

Keywords Strength · Durability · Polyethylene terephthalate (PET) · Fibre reinforced concrete · Polypropylene (PP)

1 Introduction

Concrete is relatively brittle and weak in tension. Strength and durability aspects of concrete can be altered by adding cementitious material, fibres, etc. to impart strength, durability and improving tensile properties. Concrete is good in compression while it is has low resistance to tension, cracking and impact. By adding fibres to the concrete, these characteristics can be enhanced. During the mixing process, the fibres are spread and mixed randomly into the concrete, improving certain characteristics such as flexibility, tensile strength and durability. In particular, PET fibre reinforced

S. K. Chodankar (✉) · P. P. Savoikar
Department of Civil Engineering, Goa Engineering College, Farmagudi, Goa 403401, India
e-mail: Chodankar.shruti@yahoo.in

P. P. Savoikar
e-mail: psavoikar@gmail.com

B. B. Das et al. (eds.), *Recent Developments in Sustainable Infrastructure*, Lecture Notes in Civil Engineering 75, https://doi.org/10.1007/978-981-15-4577-1_53

concrete with PET fibre exhibits outstanding features in inhibiting crack generation in early stages. Polypropylene, steel, glass, asbestos, jute, coir and nylon are the prevalent kinds of fibres used by scientists.

2 Literature Review

Plastics have become an indispensable component of contemporary life and the quantity of plastic used annually has grown tremendously and has become a severe environmental issue. In order to solve this problem, plastic has been tried in different forms in concrete. A thorough review of the research work on use of waste plastics in concrete is done. Ochi et al. [1] carried out combustion test of the PET fibres and showed that carbon dioxide was the main component of the gas liberated. Trace amounts of carbon monoxide were produced but no toxic gas was identified during the combustion experiment. Pereira de Oliveira and Castro-Gomes [2] investigated the effect of 0–1.5% recycled PET fibres in mortar and carried out compressive strength, flexure test and water absorption tests. Sung et al. [3] carried out experimental study on recycled PET fibres to understand the effect on the concrete and its behaviour. It was seen that the increase in the fibre volume fraction of recycled PET fibres exhibited marginal decrease in compressive strength and elasticity modulus. Test findings showed that free drying shrinkage strain improved reinforced concrete specimens for recycled PET fibres compared to specimens without fibre reinforcement. Bagherzadeh et al. [4] evaluated various strength properties of concrete including water and air absorption and shrinkage cracking. It was observed that addition of PP fibres resulted in control of microcracks. It was found that aspect ratio plays an important role in reinforcing the concrete. Fraternali et al. [5] found that the dosage of 1% showed remarkable improvement in thermal resistance, mechanical properties and ductility of concrete. Foti [6] observed that addition of small amount of fibres resulted in better post cracking performance of concrete. Okan and Cengiz [7] studied the durability aspects of concrete with PP fibres and flyash and reported reduction in drying shrinkage.

Fernando et al. [8] performed an experimental study to understand the behaviour of PET fibres in concrete and reported that the flexural and impact strength can be enhanced by addition of PET fibres. Ramadevi and Manju [9] carried out an investigation and reported that PET fibres reduce concrete weight. Compressive strength enhanced with 2% replacement of the sand with PET bottles fibres while it reduced when replaced up to 4 and 6%. Foti [10] reported that the PET fibres are more advantageous than other types of fibres especially from the economic and ecological point of view. Fraternali et al. [11] showed that there is remarkable improvement in toughness and the alkali resistance of cement mortar. Borg et al. [12] carried out an excellent investigation on waste PET bottles without processing through plastic melting and fibres spinning. The main two aspects were studied. First, fibres were prepared by shredding the plastic bottles and secondly, the compressive strengths were carried out for analysing the fibres and their bond.

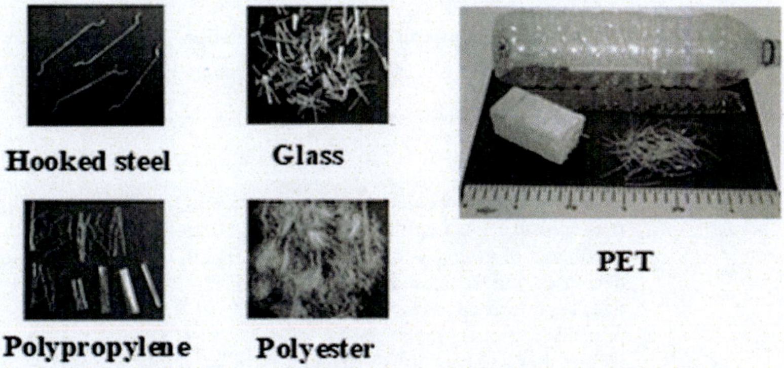

Fig. 1 The various types of fibres used in concrete

Kumaran et al. [13] carried out detailed experimental study using three concrete grades to evaluate the properties of concrete with and without fibres. Different experiments were conducted for compressive strength of cube and cylinder, split tensile strength, flexural strength, water absorption, etc. Tamilarasan and Abinayakeerthana [14] conducted detailed investigations on plastic fibre reinforced concrete with by partially replacing fine aggregates by M-sand with distinct proportions. It was observed that there is enhancement in strength properties due to better bonding with M-sand. The outcome of this research proved that river sand can be totally substituted with produced sand and that the plastic fibres do not adversely affect the mechanical features of concrete.

3 Materials and Methods

Figure 1 shows the commonly used fibres in reinforced concrete. Mainly used fibres are steel, polyethylene (PE), PP and PET fibres. The post consumed PET mineral water bottles have been also used. The length of the fibres was kept in the range of 10–50 mm and the fibre contents were from 0.5 to 6% max. Compressive, tensile and flexural strength, workability, water absorption and flexural strength tests were conducted on the specimens cast. The combustion test of the PET fibres was conducted, carbon dioxide was seen as the main component.

4 Results and Discussions

Review of the literature stated that there is a strong potential for using PET fibres for improved behaviour of concrete. It was reported that recycled PET fibres increases the various strength properties of concrete as listed in Tables 1 and 2. Most of the

Table 1 Influence of utilisation of PET fibres on mechanical properties

Property of concrete	Results	References
Combustion test	Combustion test indicated the release of CO_2 but no toxic gas was detected	Ochi et al. [1]
Water absorption	PET possesses lower wetting tension than PVA. Capillarity coefficient of mortar without fibre decreased with increase in mortar age. There was appreciation in porosity and water absorption values after addition of fly ash and fibre in concrete	Ochi et al. [1], Pereira de Oliveira and Castro-Gomes [2], Kumaran et al. [13], Gu and Ozbakkaloglu [15]
Drying shrinkage	Free drying shrinkage strain was found to increase with use of PET fibres. Marginal change in freeze–thaw resistance of PP fibre concrete. No cracks noticed when long deformed fibres were used in concrete	Sung et al. [3], Okan and Cengiz [7], Borg et al. [12], Gu and Ozbakkaloglu [15]

researchers have studied the different aspect ratio and carried out various tests like compressive strength, split tensile strength, flexural tests. The review of effect of addition of PET fibres in concrete on its various properties is presented in Tables 1 and 2.

5 Conclusions

Based on the present review of research work on utilisation of plastic/PET waste in concrete, it can be concluded that PET/PP fibres have been effectively utilised in concrete. Introduction of PET plastic waste fibres in concrete not only improves the strength of concrete, but at the same time also reduces the environmental concerns caused due to open dumping of PET bottles in surroundings. Compressive strength, split tensile strength and flexure strength were found to be increased. Combustion test indicated the release of carbon dioxide but carbon monoxide was in the form of traces.

Table 2 Influence on utilisation of PET fibres on strength properties

Property of concrete	Results	References
Compressive strength	In most of the cases, slight to significant increase in compressive strength were reported when fibre content was 1–2%. With enhancement in fibre content, the strength was found to decrease. Some authors reported no significant change in mortar compressive strength. In some cases, decrease in strength of concrete was also reported	Ochi et al. [1], Pereira de Oliveira and Castro-Gomes [2], Sung et al. [3], Fraternali et al. [5], Foti [6], Fernando et al. [8], Ramadevi and Manju [9], Borg et al. [12], Kumaran et al. [13], Tamilarasan and Abinayakeerthana [14], Gu and Ozbakkaloglu [15]
Flexural strength	Most of the researchers reported considerable increase in flexural strength when up to 2% PET fibres were used. When shorter length fibres were used, increase in strength was lower. Some researchers reported slight decrease in flexure strength on use of PET fibres	Pereira de Oliveira and Castro-Gomes [2], Sung et al. [3], Bagherzadeh et al. [4], Fernando et al. [8], Borg et al. [12], Kumaran et al. [13], Tamilarasan and Abinayakeerthana [14], Ramadevi and Manju [9], Gu and Ozbakkaloglu [15]
Toughness test	A modest to significant effect on toughness of concrete with PET fibres was reported by most of the researchers. Fibres were reported to act as reinforcement absorbing large amount of energy. In few cases reduction of toughness was observed	Pereira de Oliveira and Castro-Gomes [2], Bagherzadeh et al. [4], Foti [6], Fernando et al. [8]
Pull out test	It was reported that PET fibre has better adhesive strength	Ochi et al. [1]
Bending test	Increase in strength was reported	Ochi et al. [1], Pereira de Oliveira and Castro-Gomes [2]
Split tensile strength	Split tensile strength was found to be increased for fibre content up to 2%. In case of full replacement with M-Sand, tensile strength increased. In some cases, lower split tensile strength was observed	Bagherzadeh et al. [4], Ramadevi and Manju [9], Kumaran et al. [13], Tamilarasan and Abinayakeerthana [14], Gu and Ozbakkaloglu [15]

References

1. Ochi T, Okubo S, Fukui K (2007) Development of recycled PET fibres and its application as concrete-reinforcing fibre. Cem Concr Compos 29:448–455
2. Pereira de Oliveira LA, Castro-Gomes JP (2011) Physical and mechanical behaviour of recycled PET fibre reinforced mortar. Constr Build Mater 25:1712–1717
3. Sung BK, Na Hyun Y, Hyun YK, Jang-Ho JK, Young-Chul S (2010) Material and structural performance evaluation of recycled PET fibres reinforced concrete. Cem Concr Compos

32:232–240
4. Bagherzadeh R, Abdol-Hossein S, Masoud L (2011) Utilizing polypropylene fibres to improve physical and mechanical properties of concrete. Text Res J 82(1):88–96
5. Fraternali F, Vincenzo C, Rosaria C, Gianvittorio R, Luciano F, Loredana I (2011) Experimental study of the thermo-mechanical properties of recycled PET fibres-reinforced concrete. Compos Struct 93:2368–2374
6. Foti D (2011) Preliminary analysis of concrete reinforced with waste bottles PET fibres. Constr Build Mater 25:1906–1915
7. Okan K, Cengiz DA (2011) The durability properties of polypropylene fibres reinforced fly ash concrete. Mater Des 32:1044–1049
8. Fernando P, Oscar RK, Philippe JP, Gleize HR (2012) Mechanical properties of recycled PET fibres in concrete. Mater Res 15(4):679–686
9. Ramadevi K, Manju R (2012) Experimental investigation on the properties of concrete with plastic PET (bottle) fibres as fine aggregates. Int J Emerg Technol Adv Eng 2
10. Foti D (2013) Use of recycled waste pet bottles fibres for the reinforcement of concrete. Compos Struct 96:396–404
11. Fraternali F, Ilenia F, Carmen P, Erminio P, Luciano F (2013) On the use of R-PET strips for the reinforcement of cement mortars. Compos: Part B 46:207–210
12. Borg RP, Owen B, Liberato F (2016) Early age performance and mechanical characteristics of recycled PET fibre reinforced concrete. Constr Build Mater 108:29–47
13. Kumaran M, Nidhi M, Bini PR (2015) Evaluation of strength and durability of waste plastic mix concrete. In: International conference on technological advancements in structures and construction, "TASC-15", pp 10–11
14. Tamilarasan S, Abinayakeerthana R (2016) Experimental studies on waste plastic fibre reinforced concrete using M-sand as fine aggregate. J Mech Civ Eng 2:2456–1479
15. Gu L, Ozbakkaloglu T (2016) Use of recycled plastics in concrete: a critical review. Waste Manag

Strength and Sorptivity of Concrete Using Fly Ash and Silpozz in Marine Environment

T. Jena and K. C. Panda

Abstract This article presents the strength and sorptivity of concrete using fly ash (FA) and silpozz exposed to seawater. Conventional concrete made with 100% cement. First blended concrete series made with 0% FA and 10–30% silpozz replaced with cement and second series made 10% FA and 10–30% silpozz. The studied parameters are compressive strength for 7, 28, 90, 180 and 365 days and flexural strength and split tensile strength for 7, 28 and 90 days of seawater curing (SWC) and normal water curing (NWC) samples. Modulus of elasticity and strength reduction factor (SRF) in bond strength along with slip of 28 days SWC samples are also studied. Water absorption and sorptivity were observed as durability indicator. It reveals from the present investigation that incorporation of silpozz helps to improve the strength and reduces sorptivity of concrete against seawater.

Keywords Compressive strength · Deterioration · Fly ash · Silpozz · Sorptivity

1 Introduction

The growth of concrete technology is completely based on its strength but not only that should be considered but the coarseness degree of various environmental circumstances to which concrete is being exposed for life long as should be considered and have given lot of importance. So, finally both compressive strength and its nature of durability are considered clearly at different design phases and its relationship also is taken into account. While estimating the conditions of present structures, concrete's durability is the most important consideration to be taken into account in designing new structures. Construction of concrete is gradually very cost-effective as its durability has been higher so its becoming one of the most complex things.

T. Jena (✉)
Department of Civil Engineering, ITER, SOA (Deemed to be University), Bhubaneswar 751030, Odisha, India
e-mail: trilochanjena1975@gmail.com

K. C. Panda
Department of Civil Engineering, GCEK, Bhawanipatna, Kalahandi 766002, Odisha, India
e-mail: kishoriit@gmail.com

B. B. Das et al. (eds.), *Recent Developments in Sustainable Infrastructure*, Lecture Notes in Civil Engineering 75, https://doi.org/10.1007/978-981-15-4577-1_54

When one understands the concrete's fundamental basic of durability, all starts finding new methods to increase its service life of all the new and upcoming structures. Concrete's long-time durability potential is the key to knowledge as it consists of uncommon materials such as recycled aggregate and high silica aggregates with mineral admixtures. Although concrete is probably to get damaged up to some extent, ensuring good durability of concrete and minimizing the rate of damage. A few review papers are presented regarding seawater exposure condition such as Sunil [1] experimented for period of one year both in freshwater and seawater curing and concluded that the compressive strength of blended concrete decreases as the exposure condition increases. Menon et al. [2] investigated the influence of high strength concrete incorporated with FA, ground granulated blast furnace slag (GGBS) and silica fume (SF) with superplasticizer under severe exposure seawater tidal zone without serious weakening. Wegian [3] observed that the degradation can be minimized incorporating high volume content of cement in seawater. Anwar and Roushdi [4] found the resistance against environmental deterioration by enhancement of properties of concrete containing OPC, SF and FA in laboratory seawater. Panda and Prusty [5] enhanced the strength properties using silpozz as a pozzolanic material. Jena and Panda [6–9] enhanced the mechanical and durability properties of marine concrete and diffusion of chloride and carbonation is minimized due to incorporation of FA and silpozz. Seyed et al. [10] investigated the compressive strength and water absorption with chloride ion penetration by using 5–25% rice husk ash (RHA) incorporated with 10% micro silica (MS) and found 15% RHA with 10% MS gives optimized results. Kumar et al. [11] investigated the better strength and sorptivity of concrete by utilizing the sugarcane bagasse ash and SF in concrete. Kartini et al. [12] observed the strength and sorptivity of blended concrete incorporating 20–30% rice husk ash (RHA) replaced with OPC.

The main objective is to study the strength properties as well as water absorption and sorptivity of blended concrete incorporated with FA and silpozz with superplasticizer in marine environment.

2 Methodology

2.1 Materials

The text of your paper should be formatted as follows: In this study, OPC 43 grade cement was used. The physical properties of OPC determined as per IS code 8112-1989 [13] such as initial setting time, final setting time, standard consistency, specific gravity and fineness are 165 min, 360 min, 34%, 3.15 and 333 m^2/kg, respectively. Fineness modulus of coarse aggregates 7.0 and specific gravity 2.86 was used. Specific gravity of fine aggregate 2.67 and fineness modulus 3.03 (zone-III) was used. Normal water having pH value 6–8, seawater of Bay of Bengal, Puri beach and CERA HYPERPLAST XR-W40 high-end superplasticizer (SP) was used. The experimental

Table 1 Physical properties of fly ash and silpozz

Physical properties	FA	Silpozz
Specific gravity	2.12	2.3
Bulk density (gm/cc)	1.2	0.23
Specific surface (m^2/g)	33	17
Particle size (Micron)	34	25
Color	Gray	Gray black
Physical state	–	Solid non-hazardous

Table 2 Chemical composition of cementitious materials

Oxides (%)	Cement (OPC)	Silpozz	FA
SiO_2	20.99	88.18	58.13
Al_2O_3	6.05	1.61	31.00
Fe_2O_3	6.01	0.56	4.10
Carbon	–	2.67	–
CaO	62.74	1.59	0.60
MgO	1.33	1.63	0.10
K_2O	0.40	1.67	0.90
Na_2O	0.04	–	0.05
SO_3	1.82	–	0.12
TiO_2	0.025	–	1.63
Others	–	2.09	0.011
Loss on ignition (%)	1.14	0.04	0.29

value of coarse and fine aggregates is confirming to IS code 383-1970 [14]. Silpozz contains amorphous silica which is produced from rice husk burning in a controlled temperature about 700 °C. This ash produced is amorphous in nature. Silpozz is a very good super pozzolan and it is used in place of SF or microsilica. Physical properties of FA and silpozz supplied by the supplier are presented in Tables 1 and 2 which represent chemical composition of cementitious materials.

2.2 *Mix Design and Identity*

The design mix is prepared by confirming to IS code 10262-2009 [15] and targeted for M30. The obtained design mix ratio was (1: 1.44: 2.91) with water to binder ratio 0.43.

The strength properties and durability observations were studied and their percentage decrease in strength was evaluated. Table 3 and Table 4 present details of cementitious materials with SP and details of concrete mix quantity, respectively.

Table 3 Details of cementitious materials with SP

Mix identity	OPC (%)	FA (%)	Silpozz (%)	SP(%)
MC100F0S0	100	0	0	–
M1C90F0S10	90	0	10	0.20
M1C80F0S20	80	0	20	0.29
M1C70F0S30	70	0	30	0.40
M1C80F10S10	80	10	10	0.22
M1C70F10S20	70	10	20	0.33
M1C60F10S30	60	10	30	0.47

Table 4 Details of mix quantity

Mix identity	Cement (kg/m^3)	Fine aggregate (kg/m^3)	Coarse aggregate (kg/m^3)	FA (kg/m^3)	Silpozz (kg/m^3)	Water	SP (kg/m^3)
MC100F0S0	434.32	624.77	1264.97	0	0	186.76	–
M1C90F0S10	390.88	624.77	1264.97	0	43.44	149.40	0.781
M1C80F0S20	347.45	624.77	1264.97	0	86.86	149.40	1.005
M1C70F0S30	304.00	624.77	1264.97	0	130.33	149.40	1.229
M1C80F10S10	347.45	624.77	1264.97	43.44	43.44	149.40	0.781
M1C70F10S20	304.00	624.77	1264.97	43.44	86.86	149.40	1.005
M1C60F10S30	260.60	624.77	1264.97	43.43	130.33	149.40	1.229

3 Results and Discussions

3.1 *Fresh Concrete Properties*

The slump value and compaction factors were measured to know the workability fresh concrete. The minimum slum is 34 mm and maximum slum is 42 mm. The minimum value of compaction factor is 86.20% and maximum 96.20% observed.

3.2 *Hardened Concrete Properties*

The hardened concrete properties were determined as per IS code 516-1959 [16]. The decrease in strength depends upon the time of exposure, quality of materials, concentration of CO_2, diffusion rate of chloride into concrete samples and other harmful constituents present in the seawater.

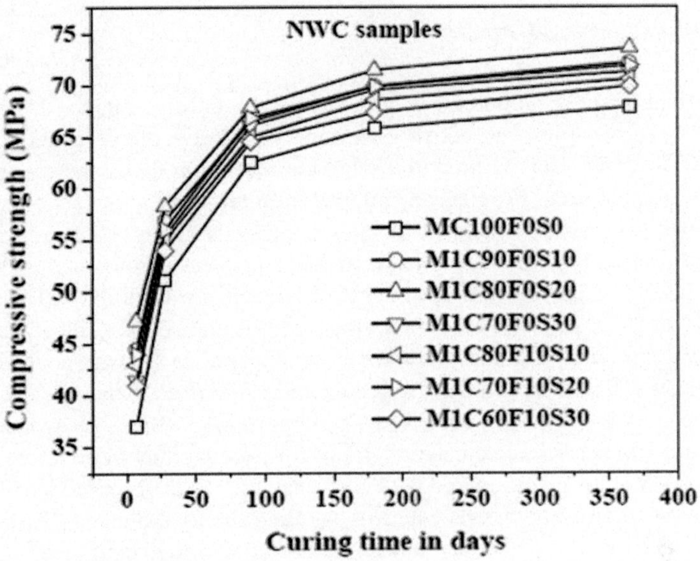

Fig. 1 Compressive strength versus curing time in days (NWC samples)

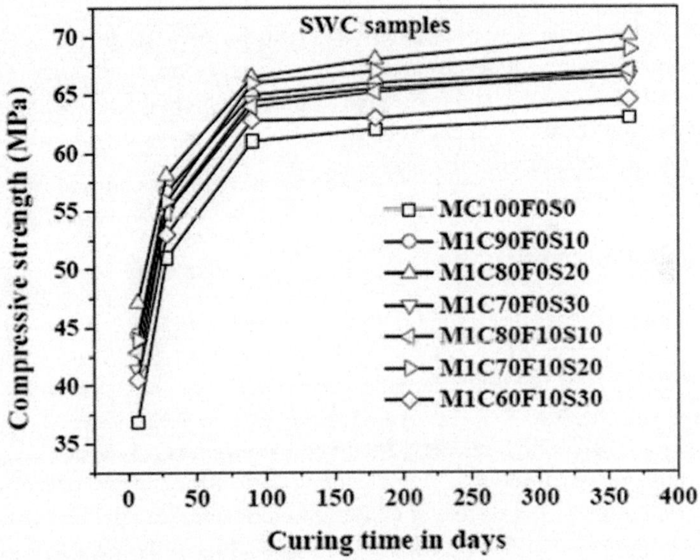

Fig. 2 Compressive strength versus curing time in days (SWC samples)

3.3 Compressive Strength

Figures 1 and 2 show compressive strength versus curing time in days for both NWC and SWC samples, respectively. For one year study, the percentage decrease in compressive strength is limited to 8% but the minimum deterioration percentage is 4.2% for sample up to FA 10% and silpozz 20% replaced with OPC. The control sample shows decrease in strength at all ages of curing which is observed from Fig. 2. The reduce in strength gain in seawater happened due to the formation of hydrochloric acid and $CaCO_3$ when CO_2 present in seawater reacts with $Ca(OH)_2$. The intrusion of chloride and CO_2 restricted incorporating with FA and silpozz, thus a decrease in strength is minimizing. The compressive strength decreases for sample of 0% FA and 20% silpozz is 2% after 3 months of exposure and 4.8% after 12 months as compared to normal concrete. From the above study, concrete works better in seawater blended with FA and silpozz. The decrease in strength for conventional concrete is 7% at 365 days of SWC curing compared with NWC samples. It is observed that the rate of deterioration is more in between 28 and 90 days and slowly reduces to 2% from 90 to 365 days. The silpozz-based concrete without FA is performed better in normal water but the when blended with 10% FA and 20% silpozz performed better in seawater in long-term basis at least for a period of one year which is experimented from this study.

The percentage reduction in strength seems to be very small for one year but the costal structures are vulnerable for long-term exposure conditions. Therefore, it may be recommended that the resistance against concrete deterioration in marine environment is significantly enhanced by using FA and microsilica present in silpozz with proper doses of high-end SP.

3.3.1 Flexural Strength

Flexural strength versus curing time in days is shown in Fig. 3 and Fig. 4 for NWC and SWC, respectively. After comparing between NWC and SWC samples, the percentage decrease in flexural strength is 2% at 90 days and 1.4% at 28 days for 10% FA and 30% silpozz.

The minimum percentage decrease in flexural strength is 0.37% at 7 days, 0.67% at 28 days and 1% at 90 days for FA 10% and 20% silpozz. The rate of decrease in strength increases as the exposure condition elongates. The percentage decrease in flexural strength increases up to 2% at 90 days. The combined effect of FA and silpozz works better up to 30% replacement of OPC and more than that the DF increases.

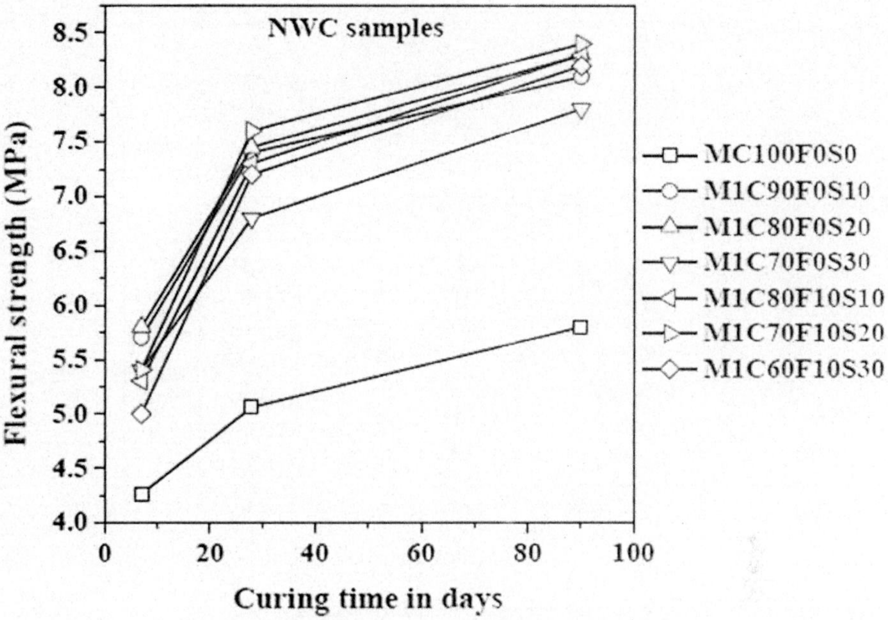

Fig. 3 Flexural strength versus curing time in days (NWC samples)

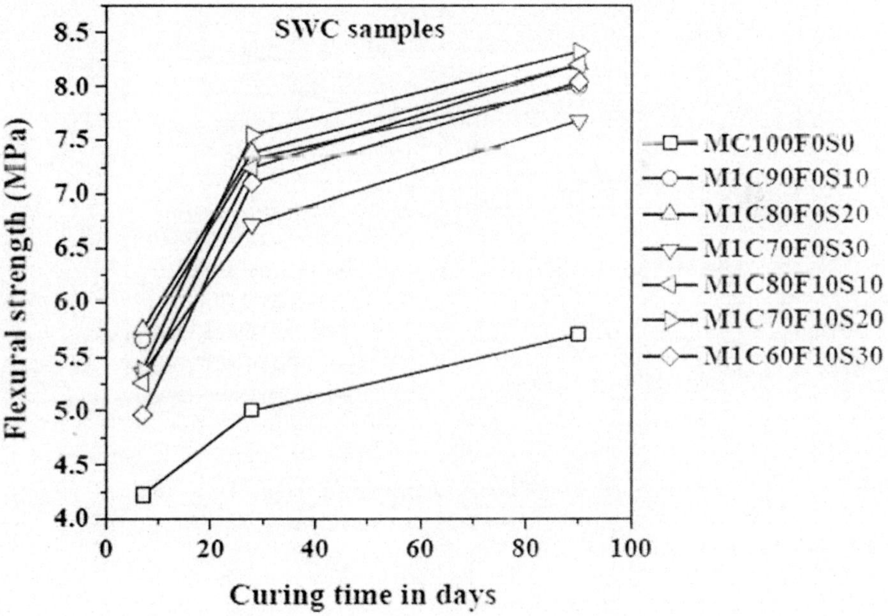

Fig. 4 Flexural strength versus curing time in days (SWC samples)

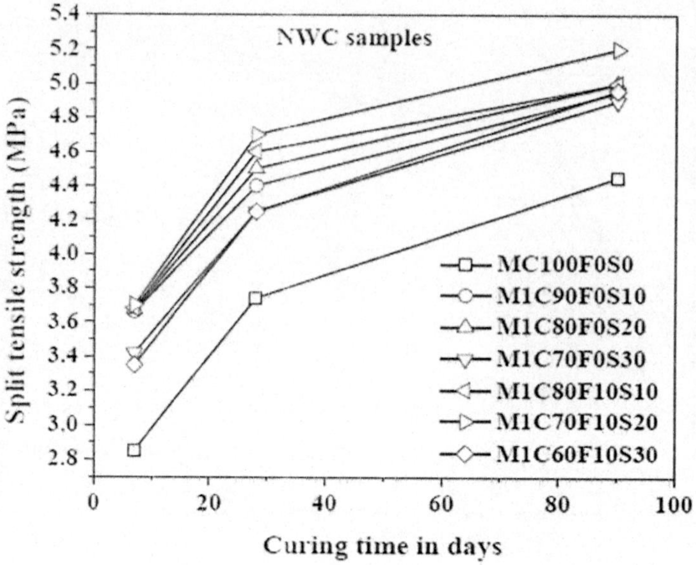

Fig. 5 Split tensile strength versus curing time in days (NWC samples)

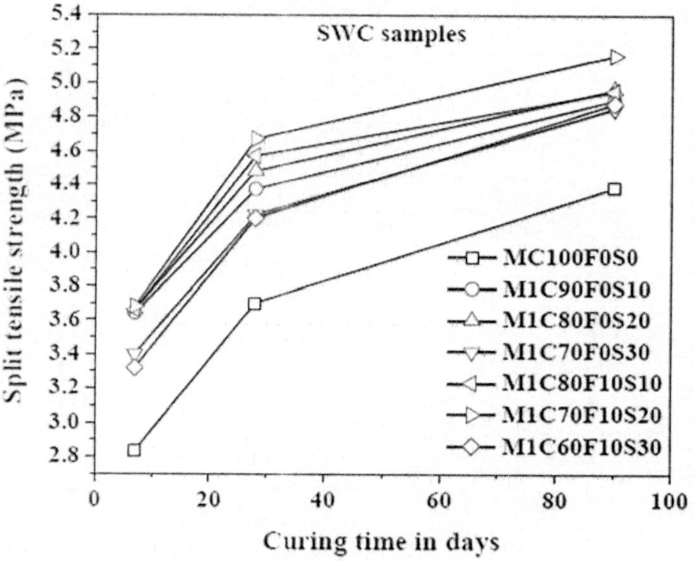

Fig. 6 Split tensile strength versus curing time in days (SWC samples)

3.3.2 Split Tensile Strength

The split tensile strength versus curing time in days is shown in Figs. 5 and 6 for both SWC and NWC samples, respectively. It is observed from the Figs. 5 and 6 that some of the middle samples having almost equal split tensile strength at whole curing period. The sample up to 10% FA and 20% silpozz represents minimum value as compared to the normal concrete at 90 days.

The deterioration of concrete increases as the percentage of replacement of cement increases more than 30%. The addition of SP plays an important role to decrease the deterioration of concrete in seawater.

3.3.3 Modulus of Elasticity

The young's modulus of elasticity is the function of concrete which reveals its elastic nature mainly of aggregate and cement paste in proper relative ratio. It is relatively constant with keeping its stress at a low point because it develops a crack in concrete paste at a level of higher stress. Modulus of elasticity concrete ranges from 30 to 50 GPa. Figure 7 shows the modulus of elasticity (MPa) versus types of mix for 28 days SWC samples from which the mix having 0% FA and 10–20% silpozz gives same value of modulus of elasticity as compared to control sample. The 10% FA and 20% silpozz sample gives the highest value of elasticity as compared to other samples

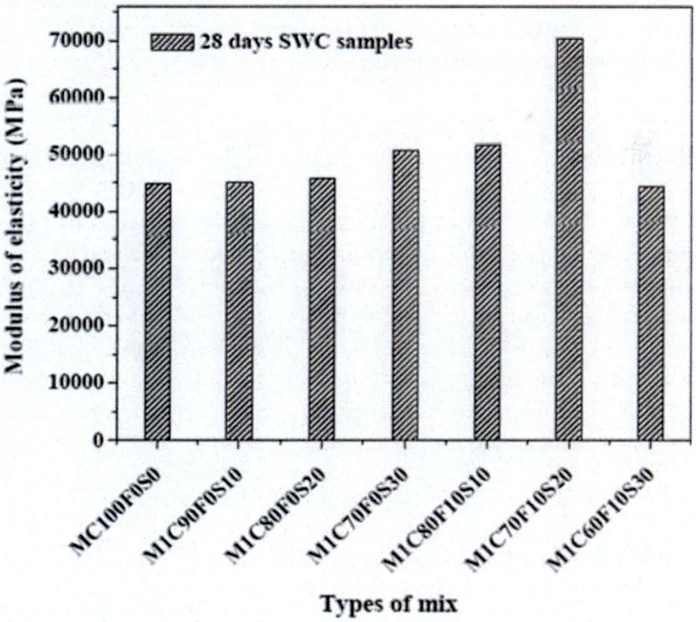

Fig. 7 Modulus of elasticity versus types of mix

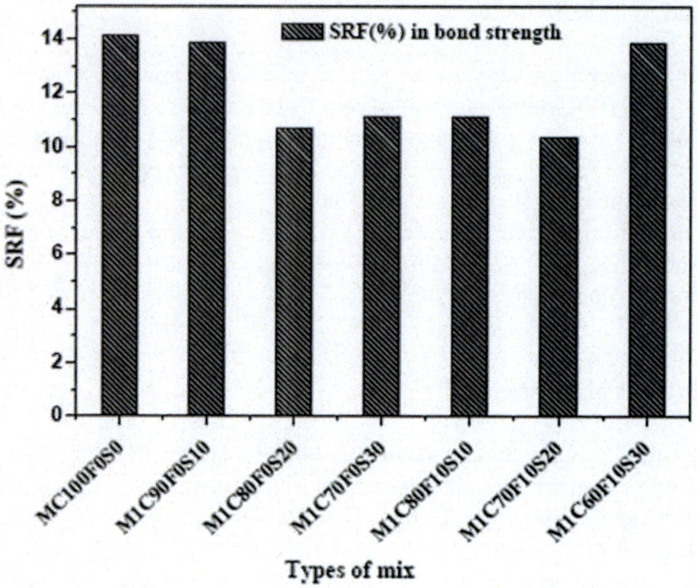

Fig. 8 SRF (%) versus types of mix

including control sample. Silpozz-based samples mixed with 10% FA give higher value of modulus of elasticity than the other samples after 28 days of SWC. There is no change up to 20% silpozz but suddenly the modulus of elasticity increases when 10% FA and 20% silpozz is incorporated.

3.3.4 Pull Out Test

The pull out test is conducted on specimens confirming to IS 2770-1(1967) [17]. The diameter of the tor bar is 12 mm with 60 mm bond length and a total length of 120 mm. The rate of loading was 0.1kN/s on the samples. The SRF is 10.35% which is observed from Fig. 8. The SRF for control specimen is 14.11% at 28 days of curing which shows the highest value. Figures 9 and 10 represent the slip in 'mm' versus types of mix. The maximum slip is 1.40 mm for control sample and minimum slip is 1 mm for silpozz-based sample. The slip is controlled for FA and silpozz-based specimen as compared to the control specimen.

4 Durability Properties

Deterioration of structural components in marine climate is generally associated with the chloride penetration into concrete causing damage. The WSC known as free

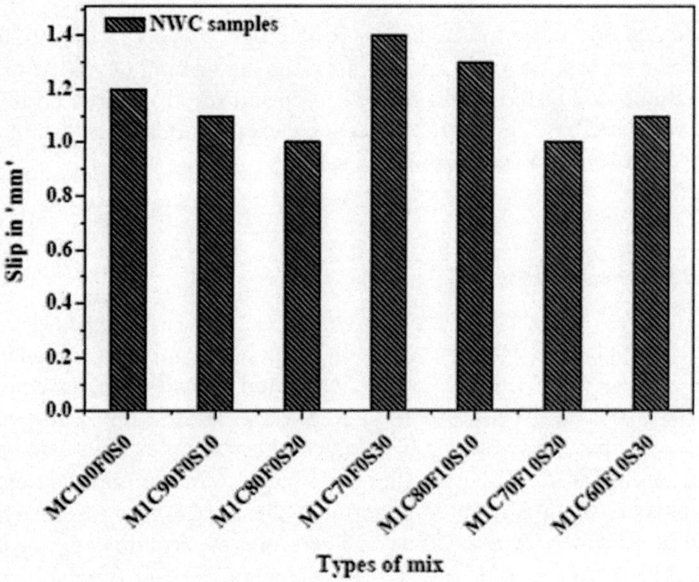

Fig. 9 Slip in 'mm' versus types of mix

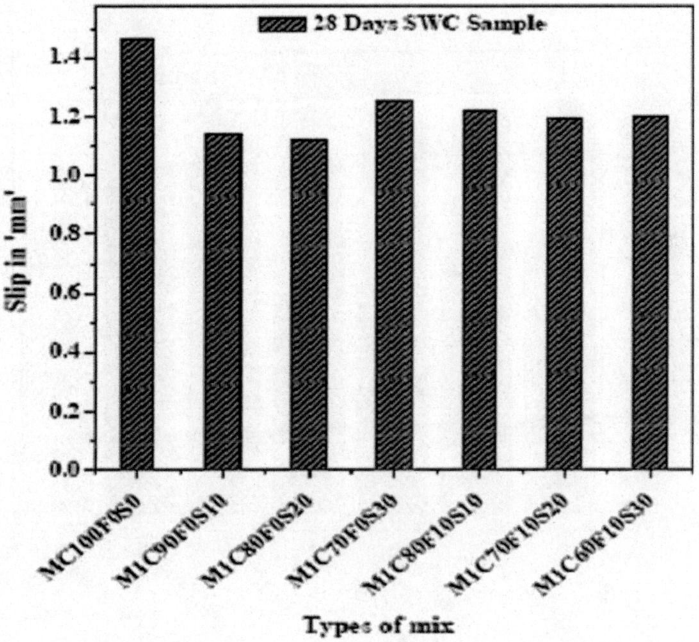

Fig. 10 Slip in 'mm' versus types of mix

chloride and ASC known as total chloride are determined as per IS 14959-2001 (Part-1I) [18]. Water absorption and sorptivity are done confirming to ASTM C 1585–04 [19]. The samples are tested for water absorption and sorptivity after curing 28 days in normal water and SWC precast samples. The test procedures are described below for both water absorption and sorptivity tests.

4.1 Water Absorption

The graphical plot between water absorption and concrete mixes is shown in Fig. 11. It is observed from Fig. 11 that the water absorption capacity for conventional concrete is 2.6% and 2.08% both in NWC and SWC precast samples, respectively. It is also observed that SWC precast blended concrete samples have less absorption capacity as compared to NWC samples. It is clear from the observations that the precast samples have already cured in normal water for 28 days and then they have been cured in seawater for next 28 days. Therefore, absorption capacity decreases for hardened matured stage. When the replacement percentage of silpozz increases, the absorption capacity decreases. When FA is replaced, again absorption capacity increases and after replacing silpozz, it counteracts the absorption capacity and

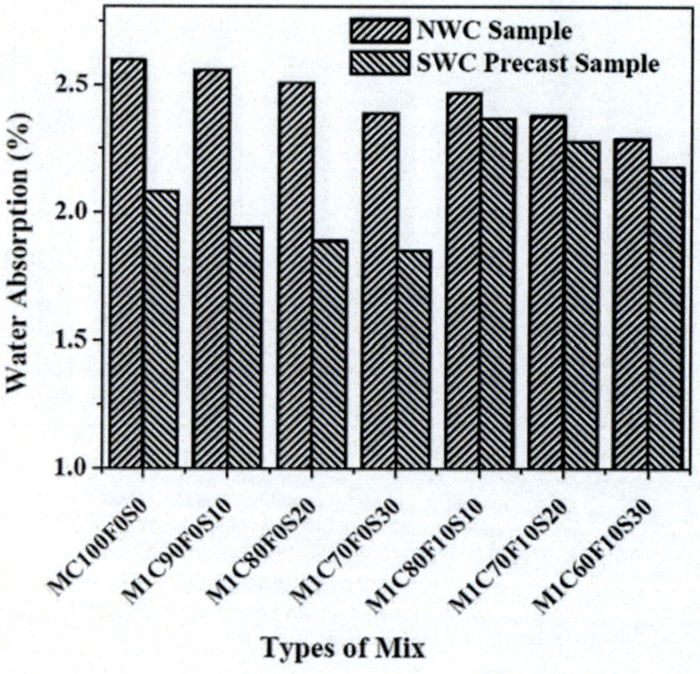

Fig. 11 Water absorption (%) versus types of mix

reduces the rate of absorption. After replacing FA, the absorption capacity increases for both NWC and SWC precast samples. Silpozz-based samples have higher resistivity than the composite samples having FA and silpozz. The absorption capacity depends on the porosity of concrete, particle size of FA and silpozz and their inner structural mechanism of concrete. The dense and compact structured concrete may give less absorption capacity.

4.2 *Sorptivity*

The comparison results of the sorptivity test between NWC and SWC precast samples have been shown in Fig. 12. It is observed from Fig. 12 that the conventional concrete has the highest sorptivity value among all mixes both in NWC and SWC precast samples. The samples M1C70F10S20 and M1C60F10S30 have same sorptivity values for both NWC and SWC. The value of sorptivity decreases by the partial replacement of silpozz and the value increases after partial replacement of FA. The result obtained from the silpozz-based SWC samples is less than the NWC composite samples partially replaced with FA and silpozz. In normal water curing, a significant amount of calcium hydroxide migrates from specimens' pores into surrounding water resulted in less dense matrix hence sorptivity is more but in precast samples the hardened

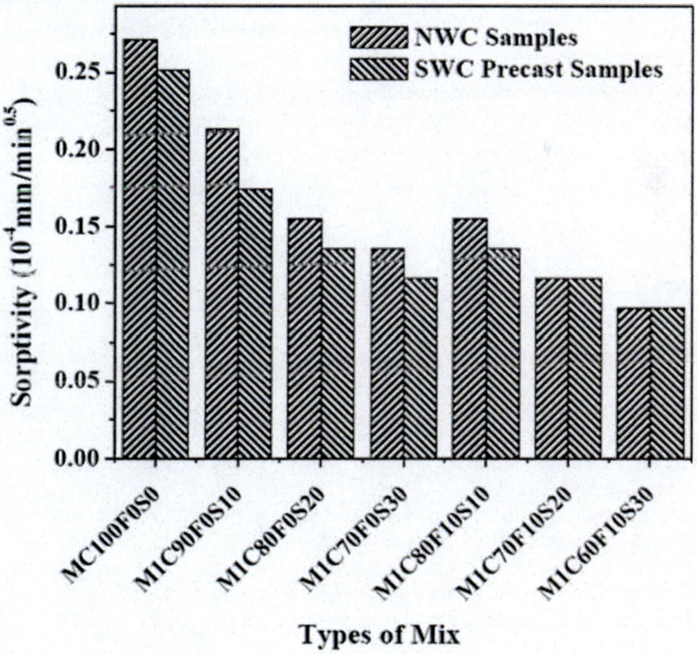

Fig. 12 Sorptivity versus types of mix

stage provides dense matrix. Then after the precast samples were tested which gives less sorptivity. The higher water–cement ratio with 20% silpozz replacement, upon evaporation leaves voids spaces in concrete specimen leads to higher absorption but in lower water–cement ratio, doses of SP result in high impermeable concrete and less absorption capacity.

5 Conclusions

A few concluding remarks are presented:

- The concrete deterioration is not significantly changed which is observed from the study. The SRF is found to be less than 3% and 8% for 10% FA up to 20% silpozz at 90 and 365 days of exposure, respectively.
- Here a different trend is observed in split tensile strength. And it may be due to 90 days test. Silpozz gives high strength at early ages.
- The higher value of young's modulus of elasticity is found to be higher for higher compressive strength samples.
- The slip is observed 1 mm and 1.19 mm for 10% FA with 20% silpozz sample in NWC and SWC, respectively, in 28 days.
- The SWC precast samples have better performance against water absorption. When the replacement percentage of silpozz increases, the absorption capacity decreases. After replacing FA the absorption capacity increases for both in NWC and SWC precast samples.
- The SWC precast samples are showing lower sorptivity value than the NWC samples after 28 days of curing. The conventional concrete has the highest sorptivity value among all mixes both in NWC and SWC precast samples.

References

1. Kumar S (2000) Influence of water quality on the strength of plain and blended cement concretes in marine environments. Cem Concr Res 30:345–350
2. Menon AH, Radin SS, Zain MFM, Trottier JF (2002) Effect of mineral and chemical admixtures on high strength concrete in sea water. Cem Concr Res 32:373–377
3. Wegian MF (2010) Effect of sea water for mixing and curing on structural concrete. IES J Part A Civ Struct Eng 3(4):235–243
4. Anwar M, Roushdi M (2014) Improved concrete properties to resist saline water using environmental by-product. Water Sci J 27:30–38
5. Panda, KC, Prusty, SD (2015) Influence of silpozz and rice husk ash on enhancement of concrete strength. Adv Concr Constr 03(03):203–221
6. Jena T, Panda KC (2015) Effect of fly ash and silpozz on strength and durability properties of concrete in sea water. Indian J Sci Technol 8(29):1–7
7. Jena T, Panda KC (2015) Influence of sea water on strength and durability properties of concrete. Adv Struct Eng 03:1863–1873 Springer India

8. Jena T, Panda KC (2017) Compressive strength and carbonation of sea water cured blended concrete. Int J Civ Eng Technol 8(2):153–162
9. Jena T, Panda KC (2018) Mechanical and durability properties of marine concrete using fly ash and silpozz. Adv Concr Constr 6(1):47–68
10. Seyed AZ, Farshad A, Farzan D, Mojtaba A (2017) Rice husk ash as a partial replacement of cement in high strength concrete containing micro silica: evaluating durability and mechanical properties. Case Stud Constr Mater 7:73–81
11. Kumar ST, Balaji KVGD, Rajasekhar K (2016) Assessment of sorptivity and water absorption of concrete with partial replacement of cement by sugarcane bagasse ash and silica fume. Int J Appl Eng Res 11(8):5747–5752
12. Kartini K, Mahmud HB, Hamidah MH (2010) Absorption and permeability performance of Selangor rice husk ash blended grade 30 Concrete. J Eng Sci Technol 5:1–16
13. IS: 8112-1989. 43 grade OPC specifications (first revision). Bureau of Indian Standards, New Delhi, India
14. IS: 383-1970. Indian Standard Specification for course and fine aggregates from natural sources for concrete (second revision). Bureau of Indian Standards, New Delhi, India
15. IS: 10262-2009. Guide lines for concrete mix design proportioning. Bureau of Indian Standards, New Delhi, India
16. IS: 516-1959. Methods of tests for strength of concrete. Bureau of Indian Standards, New Delhi, India
17. IS 2770-1 (1967) Methods of testing bond in reinforced concrete, part 1: pull-out test, CED 2: cement and concrete. Bureau of Indian Standards, New Delhi, India
18. IS: 14959:2001 (part-II) Method of test determination of water soluble and acid soluble chlorides in mortar and concrete. Bureau of Indian Standards, New Delhi, India
19. ASTM C 1585-04. Standard test method for measurement of rate of absorption of water by hydraulic cement concretes

Effect of FRC Due to Saline Water and Urea Solution

Sagar Sarangi, Anand Kumar Sinha, Birendra Kumar Singh, and Devendra Mohan

Abstract Fibre-reinforced cement mortar is a composite material used to introduce short fibre for enhancing some properties of the mortar matrix. Mortar cubes of 1% hybridization by using poly-propylene (PP), coconut coir (CC) and scrap rubber (TR) fibre were prepared and various properties were compared with a control mix specimen after 28 days. Compressive strength test was conducted for determining the changes in physical properties of the mortar due to the effect of salinity and urination. Study on interfacial bonding between matrix and reinforcement carried out by using SEM analysis. The test result suggested that fibre-reinforced plaster was able to achieve adequate strength if curing water contains salinity of 3.5% (35 g/L) and crop nutrient (NPK-46-0-0) of 40% w/v dissolved form in freshwater. This type of concrete could be used in non-structural components like warehouse flooring, industrial floor, water tank, benches or footpath near sea beaches, etc.

Keywords Durability · Hybrid fibre-reinforced concrete · Compressive strength · Saline water · Urea solution

1 Introduction

Blended cement is an innovative idea for reducing industrial solid waste. By utilizing cementitious material like GGBS, fly ash, Silica fume, alccofine, etc. with the replacement of OPC is one of the emerging technologies for producing different types of binding material. Blended GGBS with Ordinary Portland Cement known as hydraulic cement is preferable for construction of hydraulic structures. As concrete is a brittle material; addition of fibres into the concrete matrix results that concrete becoming somewhat homogeneous and isotopic by virtue of this ductile material [1]. By applying stress, microcrack present in the conventional concrete rapidly

S. Sarangi (✉) · A. K. Sinha · B. K. Singh
Birla Institute of Technology, Mesra, Ranchi, India
e-mail: sagarsarangicit@gmail.com

D. Mohan
IIT (BHU), Varanasi, India

B. B. Das et al. (eds.), *Recent Developments in Sustainable Infrastructure*, Lecture Notes in Civil Engineering 75, https://doi.org/10.1007/978-981-15-4577-1_55

increased and leading the concrete to have low flexural strength and impact resistance. From the ancient period to till date, various experimental works are carried out by introducing a reinforcing agent with cement mortar and plain cement concrete. Hybridization is the process of introducing two or more different types of fibres inside the single matrix. Nowadays hybrid fibre reinforced concrete of both natural fiber and synthetic fibre is possible and the experimental investigation already published by [2, 3].

Growth of crack developed due to plastic shrinkage on green concrete, permeability and carbonation depth in harden concrete can be reduced by introducing limited proportion of PP fibre as a reinforcing agent with concrete matrix reported by [4, 5]. Naturally cultivated coconut coir fibre is one of the natural fibre having impressive tensile properties. By introducing as reinforcing agent can improve the tensile properties of the matrix along with reducing the production cost of conventional one [6].

Rubber (tyres) occupies large surface area while dumping as solid waste. By the reuse of the waste with concrete is able to reduce the surface area and simultaneously it reduces the environmental pollutant. Scrap rubber can be utilized as an alternative building material. From the investigation reports of [7–9] suggested that the rubber fibre-reinforced concrete could be used as seismic resistance material to sustain dynamic.

The scarcity of potable drinking water is a major issue worldwide. Treated seawater might be fulfilling the water requirement by the construction industry for both mixing and curing purposes. Beyond 3.5% salinity causes surface efflorescence and dampness. So it is not recommended for plaster finish work by [10]. However, less than 500 ppm chloride or 1000 ppm SO_3 concentration could provide satisfactory results as compared with portable drinking water reported by [11, 12].

2 Objective

It is required to make concrete somewhat ductile by introducing various reinforcing agents. In corporation of discrete and randomly oriented coir fibre, Polypropylene fibre and chopped scrap rubber (tyres) are used as a reinforcing agent. Cement reinforced plaster can be able to improve it's ductility, in controlling the cracks resistance to environmental effects.

3 Working Methodology

3.1 Materials

Hydraulic blended cement confirm to BIS (IS 455:1989), UltraTech PSC cement was consumed as a binding material in this investigation. Physical and chemical properties of binding material are notified in Table 1. Clean grained silica particle, not more than 4.75 mm was utilized for fine aggregate. Physical properties of sand were tested and conformed by BIS (IS 383:1970) and the results are shown in Table 2. "Fiber guard" white monofilamentous fibrillated Poly Propylene (PP) fiber supplied by "M/S Bajaj

Table 1 Chemical and physical property of cement

Test	Observed value	Req. as per IS 455:1989 (Max)
Chemical test MgO	5.28%	10%
SO_3	1.09%	3%
S	0.65%	1.5%
Loss of ignition	1.15%	5%
Insoluble residue	1.11%	4%
Total chloride content	0.046%	0.10%
K_2O	1.48%	
Na_2O	0.3%	
Physical test		
Specific gravity	3.10	
Fineness	349 m^3/Kg	
Soundness	1 mm	
Standard consistency (%)	32%	
Initial setting time (Min)	50 Not less than 30 min	
Final setting time (Min)	600	

Table 2 Physical properties of fine aggregates

Properties	Value of fine aggregate
FM	5.17
Sp. gravity	2.67
Bulk density grm/cm^3	1.890
Moisture content	8%
Zone	II

Table 3 Properties of fibre

Properties	Value of PP	Value of coir fibre	Value of rubber fibre
Length (mm)	20	20	20
Diameter (mm)	0.1	0.29–0.83	0.2
Tensile strength (MPa)	0.67	170	24.13
Sp. gravity	0.91	0.88	0.96
Water absorption	Nil	10%	Nil
Aspect ratio	200	68	100
Bulk density (Kg/m^3)	689	1824	700

Reinforcement LLP, Nagpur, Maharastra" were used for this study and it's properties are given in Table 3. Natural brown clean coconut coir (CC) fiber collected from the nearby market of BIT Mesra, Ranchi were used. Its physical properties are shown in Table 3. Scrap tyre (TR) pieces collected from nearby resoling centre and chopped into a uniform length of 20 mm and its properties are also shown in Table 3. Saline water having pH = 7.65 and salinity 3.5% collected from Bay of Bengal located at Puri beach, Odisha, India and its chemical component is shown in Table 4. Fertilizer urea (NH_2-CO-NH_2, density 1.32 g/cm^3, molar mass of 60.06 g/mol) solution was prepared with concentration of 40 gm/lt in distilled water at room temperature having pH of the solution 10 for carried out the urination effect of the mortar.

Table 4 Chemical compositions of Saline water

Ion	Formula wt
Cl^-	35.453
Na^+	22.9898
Mg^{2+}	24.312
SO_4^{2-}	96.06
Ca^{2+}	40.08
K^+	39.102
HCO_3^-	61.01
Br^-	79.91
Sr^{2+}	87.62
H_2Br_3	61.83
F^-	19.00

Fig. 1 Sample in curing

3.2 Method

There are 22 numbers of mix proportions of cement and fine aggregate mortar (1:3) were prepared with individual and hybrid mix of different fibers with weight fraction (wf) with respect to cement. Both natural and synthetic fibre were added individually with different in proportions. PP fibre varying from 0.1% to 0.4%, coconut coir (CC) fibre varying from 1% to 10% whereas PBS (Poly Butadiene Styrene) scrap rubber (TR) fibre range of 5% to 17% were mixed as mentioned above. Hybrid mix proportion of H1(PP0.30CC0.50TR0.20), H2(PP0.25CC0.40 TR0.35), H3(PP0.333CC0.334TR0.333) were also prepared. One control mortar sample was also cast for comparative study of the properties among the mix proportions.

Two hundred seven mortar sample specimens were prepared with varying ratio of cement for studying variation in reinforced plaster. From all the mix proportion were subjected to submerge curing at room temperature for the period of 28 days. Three samples with saline water and six samples with potable drinking water initially submerged. After 7 day of curing three samples were shifted from potable drinking water bucket to urea solution bucket and then, allowed for the next 28 days to study physical changes in mortar. The sampling process is shown in Fig. 1.

3.3 Compressive Strength

After removing from curing tank and drying in air compressive strength of mortar cubes was determined using the UTM confirmed to [13]. Average of three cubical samples was calculated and the results are shown in Table 5 and Fig. 2.

Compressive strength (MPa) = Fracture Load/Cross sectional area of specimen (Mpa).

Table 5 Compressive strength of the sample

Name of the mix	Sample id	Drinking water (Mpa)	(S) Sea water (Mpa)	% Changes	(U) urea Sol^n(MPa)	% Changes
Plane (1:3)	S0	33.23	32.72	−1.56	40.27	17.47
PP 0.10%	S1	31.86	30.59	−4.17	34.45	7.53
PP 0.15%	S2	33.44	34.58	3.28	37.10	9.86
PP 0.20%	S3	34.66	36.81	5.84	37.83	8.39
PP 0.25%	S4	32.70	38.86	15.84	30.65	−6.69
PP 0.30%	S5	31.07	37.33	16.76	28.75	−8.09
PP 0.35%	S6	28.82	32.71	11.88	37.47	23.09
PP 0.40%	S7	28.36	32.42	12.53	34.87	18.68
CC 10%	S8	36.02	28.41	−26.76	42.31	14.88
CC 7%	S9	35.27	28.60	−23.33	38.95	9.44
CC 5%	S10	33.73	33.70	−0.11	38.64	12.71
CC 2.5%	S11	33.50	34.97	4.20	37.62	10.94
CC 1%	S12	32.71	36.81	11.14	35.78	8.60
TR 5%	S13	32.70	29.63	−10.38	30.65	−6.69
TR 7%	S14	33.52	34.66	3.28	37.72	11.11
TR 10%	S15	33.97	35.89	5.35	38.74	12.31
TR 12%	S16	34.54	36.28	4.78	40.81	15.36
TR 15%	S17	35.07	36.83	4.79	41.90	16.30
TR 17%	S18	36.02	37.94	5.05	44.24	18.59
PP0.30CC0.50TR0.20	S19	36.81	39.88	7.71	38.86	5.28
PP0.25CC 0.40 TR 0.35	S20	35.78	28.60	−25.10	33.73	−6.08
PP0.333CC0.334TR0.333	S21	36.81	30.65	−20.07	31.87	−15.48

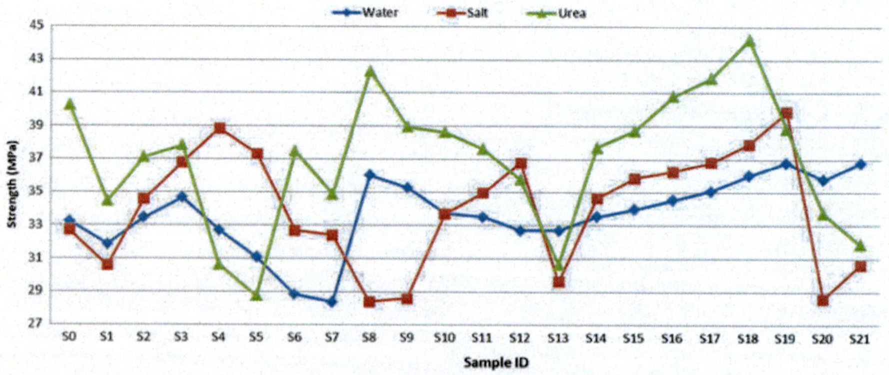

Fig. 2 Graphs of compressive strength of mortar sample

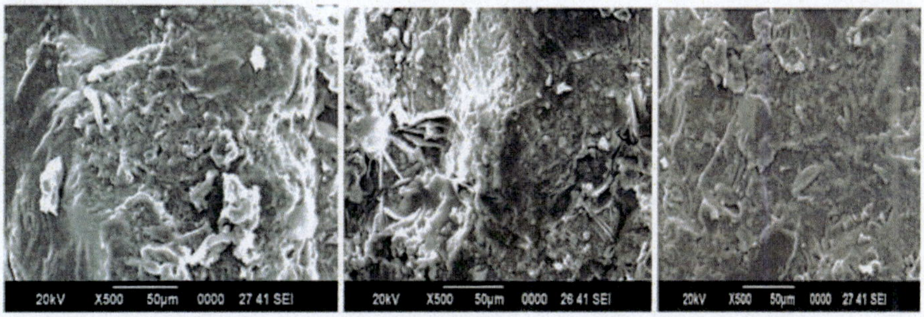

Fig. 3 **a** HFRC control, **b** HFRC submersed with seawater and **c** SEM of HFRC submersed with urea solution of (Sample No-19), Magnification 500

3.4 Morphology Study (SEM Analysis)

The surface morphology of hybrid fibre reinforced plaster mortar sample was carried out according to [13] and the same things are shown in Fig. 3. It shows the porous nature of the untreated hybrid fibre. In the hybrid of PP, CC and TR (untreated) more number of fibre pullout along with microcracks were seen in SEM image and it indicates that micro gaps were present between fibre and the matrix. In the case of treated samples with seawater, due to a chemical reaction between cement and CC fibres, it formed a gel and fills the micro-gap as was evident from the SEM image. Urea, an aggressive Ferro material reacted with matrix which increases growth of crystals in pores of the concrete. But, due to crystallization, it increased the bonding between the minerals present in the composite.

4 Conclusions and Discussion

- Individual fibre-reinforced plaster can withstand both the effect of salinity and urination. It gives good compressive strength as compared with conventional plaster. But, coconut coir fiber reinforced plaster having higher resistant towards the effect of urination as compared to salinity.
- By the addition of PP fibre up to 0.30% of weight fraction of cement is able to resist the effect of salinity by increasing 32.96% whereas due to the effect of urination it reduced 12.26% of compressive strength as compared to conventional.
- From the variation of mixing it seems that PP 0.20% weight fraction is the optimum mixing quantity that improves the durability by resisting both the effect due to salinity and urination.

- Addition of scrap chopped rubber up to 17% by weight of cement is able to save the structure from both the effects.
- By 1% hybridization of a mixture of (S19), PP0.30CC0.50TR 0.20 gives better performance as compared to other mix proportions.
- Strong interfacial bonding between fibre and matrix are found from SEM image due to the effect of salinity whereas weeker in the case of urination effect.
- By finding and adopting an optimum mixing proportion for both natural and synthetic fibres in plain cement plaster are able to reduce the effect due to salinity and urination.

Acknowledgements We acknowledge 'M/S Bajaj Reinforcement LLP Nagpur' for supplying good quality PP fibre, and CIF BIT Mesra, Ranchi for conducting SEM analysis of the samples.

Bibliography

1. Kesavraman S (2017) Studies on metakaolin based banana fibre reinforced concrete. Int J Civil Eng Technol 8(1):532–543
2. Sarangi S, Siddharth (2018) Rubber fiber reinforced concrete-a new building material. Int J Manag Technol Eng 8(5):409–416
3. Siddharth SS (2017) Mechanical & structural characteristic of hybrid FRC. Int J Adv Res 5(2):1425–1430
4. Madhavi TC, Raju LS, Mathur D (2014) Polypropylene fiber reinforced concrete-a review. Ijetae Com 4(4):114–119
5. Saje D et al (2011) Shrinkage of polypropylene fiber-reinforced high-performance concrete. J Mater Civil Eng 23(7):941–952
6. Agrawal AR, Dhase SS, Agrawal KS (2014) Coconut fiber in concrete to enhance its strength and making lightweight concrete. Int J Eng Res Dev 9(8):2278–67
7. Sgobba S, Borsa M et al (2015) Mechanical performance and medium-term degradation of rubberised concrete. Constr Build Mater
8. Sgobba S, Marano GC et al (2010) Use of rubber particles from recycled tires as concrete aggregate for engineering applications. Sec Int Conf Sustain Constr Mater Technol
9. Torgal FP, Shasavandi A, Jalali S (2011) Tyre rubber wastes based concrete: a review. Wastes Sol Treat Opport
10. Abrams Duf (1924) Tests for impure water for rrtirirrg concrete. Am Concr Ins 20:422
11. Building Research Station (1956) Digest No. 90, London, H.M.S.O. Analysis of Water Encountered in Construction. Print
12. Tiwari P, Chandak R, Yadav RK (2014) Effect of salt water on compressive strength of concrete. J Eng Res Appl, ISSN 4(5):38–42. www.ijera.com
13. ASTM C1723-16 (2016) Standard Guide for Examination of Hardened Concrete Using Scanning Electron Microscopy. West Conshohocken
14. ASTM C109/C109M-16a (2016) Standard Test Method for Compressive Strength of Hydraulic Cement Mortars (Using 2-in. or [50-Mm] Cube Specimens). West Conshohocken
15. Bureau of Indian Standards, IS 455:1989. Indian Standard Specification for Portland Slag Cement. Print
16. Bureau of Indian Standards (1970) IS 383:1970. Indian Standard Specification for Coarse and Fine Aggregates from Natural Sources for Concrete. Print
17. Bureau of Indian Standards, IS 455 (1989) Portland Slag Cement - Specification

Effect of Lime on Mechanical Properties of Silica Fume Modified Concrete

Gaurav Udgata, Purnajit Bhowmik, and Silpa P. Das

Abstract The sole objective of the work is aimed to examine the possibility of replacing ordinary Portland cement with condensed silica fume. Silica fume is a waste product from ferro-alloy silicon industry. In this research, the ordinary concrete has been modified by 25% of micro silica fume by replacing Portland cement. Again that same mix has been modified by adding free lime by replacing fine aggregate with different percentages like 3, 5, 7 and 10% by weight with a constant water-cement ratio 0.45 by weight. The cement and silica fume percentage was kept constant and after that the influence of addition of active lime was studied. The cement, sand and aggregate proportion were assumed as 1:1.5:3. Specimen was prepared for various tests like compressive, split-tensile and flexural strength for different curing ages (7, 28 and 56 days). Silica fume modified concrete with 7% of free lime by volume gave the best results.

Keywords Condensed silica fume · Lime · Workability · Compressive strength · Flexural strength and split-tensile strength

1 Introduction

In this construction Era commonly used hazardous Portland cement having sheer danger to ecosystem has led many a researcher to exploit other alternatives like industrial by-products as complementary materials in making of concrete and at the same instance 'also to overcome certain drawbacks found in the normal concrete mix like porosity, low permeability, low durability, low compressive, tensile and flexural strength and less resistance against chemical attack'. To overcome all these

G. Udgata (✉)
Asst. Professor, School of Civil Engineering, KIIT University, Bhubaneswar, India
e-mail: gaurav.udgatafce@kiit.ac.in

P. Bhowmik · S. P. Das
Student, KIIT University, Bhubaneswar, India
e-mail: purnajitbhowmik@gmail.com

S. P. Das
e-mail: pdas.silpa@gmail.com

B. B. Das et al. (eds.), *Recent Developments in Sustainable Infrastructure*, Lecture Notes in Civil Engineering 75, https://doi.org/10.1007/978-981-15-4577-1_56

drawbacks engineers and scientists around the globe are doing many researches for a permanent feasible solution.

The pozzolana (CSF-condensed silica fume) has fascinated the interest of various researchers. Micro silica fume is more or less somewhat a new element in the folks of pozzolans which is introduced to have a reaction with free lime present within the concrete matrix and thereby improving upon the performance. Silica fume has huge concentration of silicon dioxide (85–97%) that contains fine sphere-shaped glassy siliceous particles. Silica fume which is a by-product of silicon-metal production is about 30–100 times finer when compared with cement.

1.1 Silica Fume

Condensed silica fume is highly reactive pozzolanic material as it is extremely fine and highly content with amorphous silicon dioxide. As ordinary Portland cement (OPC) present in concrete starts to have chemical reaction, it liberates calcium hydroxide. To form supplementary binder material, the micro silica fume combines with calcium hydroxide and that supplementary mix is known as hydrate of calcium silicate that is alike to calcium silicate hydrate formed from (OPC). By this supplementary binder, we can change the different features that of conventional concrete (Fig. 1).

Volatized silica fume is the siliceous and aluminous material which is responsible for reaction with CH which decreases the CH content in addition to forming strength contributing cementitious products which are known as 'pozzolanic reaction'. The

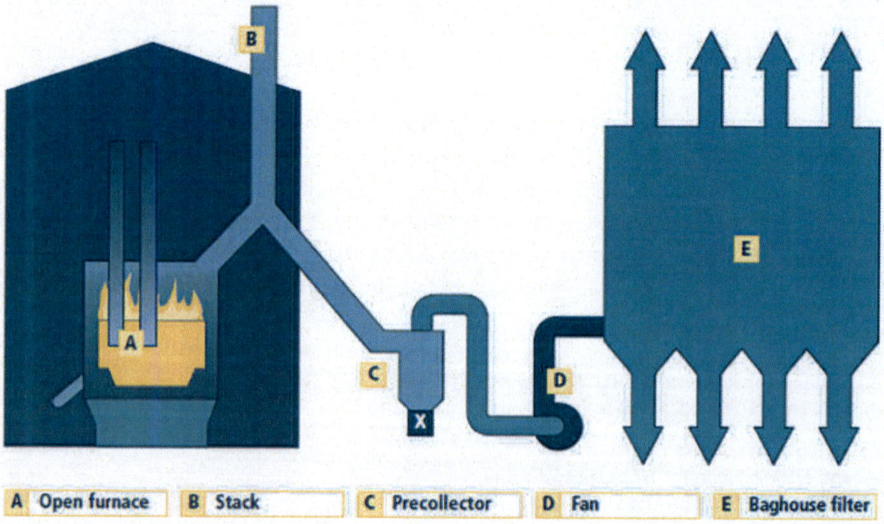

Fig. 1 Manufacture of silicon metal or ferrosilicon alloy

existence of SF in the OPC mixes results significant decrease in the volume of large pores at all ages. In the cement-aggregate bond SF adding up influences the depth of evolution segment in mortars and the degree of the orientation of the Calcium hydroxide (CH) crystals in it.

1.2 Lime

The methods through which lime is transformed to quicklime by the process of heating, subsequently to slaked-lime by means of hydration. It reverts back to CaCO3 through carbonation, this process is known as the 'Lime Cycle'. In food processing industry, to chemically attack plant and animal substances the caustic nature of lime is used. Lime has got many applications in various fields of application but also it has got some limitations like Lime cement usually takes a long duration of time to cure. Lime cement doesn't solidify in water rather stays soft. So, there are certain situations where it can't be utilised (Fig. 2).

2 Literature Review

Safwan A. Khedr et al. (1994) [1] the relative slower growth of strength linked with introducing volatile silica fume in concrete. Pozzolanic action is very slow in nature even though the increase in final strength is notable, it took nearly about 3–7 days for the mix to begin exhibiting the increase. Silica Fume being very fine filler does not demonstrate considerable chemical reaction during the growing stages and because of the associated drop in OPC content. It is practical to anticipate that there would certainly be a plunge in early strength improvement whereas Faseyemi Victor Ajileye et al. (2012) [2] presented that for C30 grade of concrete substitution of cement up to 10% with silica fume leads to a considerable enhancement in its compressive strength. Because of the intense pozzolanic nature of the micro silica and its void

Fig. 2 Lime cycle

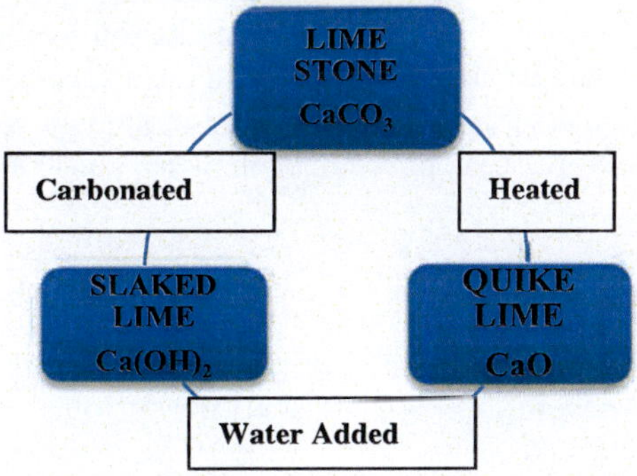

filling ability but interestingly it found there is a decline in its compressive strength when 15% silica fume is mixed as a substitute.

N.K. Amuddhayalli et al. (2012) [3] conclude that pozzolanic materials when added to concrete, additional calcium silicate hydrate is formed. Within the range of 10–15% silica fume replacement level, the optimal 7 and 28 days compressive strength as well flexural strength have been obtained whereas Vikas Srivastava et al. (2010) [4] found that inclusion of silica fume had reduced the workability significantly in all the mixes because they are sticky and cohesive. Thus the addition of silica fume reduces workability. Silica Fume enhanced workability and also increased compressive strength significantly (6–57%) and this increase in compressive strength depends upon its replacement level.

Dilip Kumar Singha Roy et al. (2012) [5] acknowledged that there is a substantial decline in rebound from (35 to 15%) by addition of silica fume that also improved the pumpability of high workability mix that has slump value of above 250 mm and Ashraf Ragab et al. (2015) [6] investigated that amount of hydration increment by utilising limestone cement due to both its chemical effect and physical effect. Replacement of limestone about 5 to 10% associated with outcome of amount of hydration. Whereas limestone fills the voids between the cement elements due to the creation of carbo aluminate segments.

Sri Tudjono et al. (2014) [7] reported that fly ash percentage in concrete can be enhanced by supply of free lime to the concrete. The combined effect of both nano lime and nano fly ash resulted in elevated compressive strength at the very early age as compared to conventional concrete or ordinary concrete. The compressive strength was reached up to 14.81% over controlled specimen by addition of 2.5% of nano fly ash as well as nano lime. Olesia mikhailova et al. (2013) [8] presented that we can use ground dolomite powder as cementitious material which can respond to produce cement with dolomite. Limestones act as a filler able to increase the early strength of conventional concrete. He has been presented two basic reasons for explaining this phenomenon. The first reason was the effect limestone on hydration tricalcium silicate and tricalcium aluminate and second reason was limestone element compose nucleation sites of hydration. The author concludes that less than 5% and more than 25% dolomite limestone by weight decreases the compressive strength after 14 and 28 days of curing age. Beyond 25% of limestone by weight gives the optimal strength. Ashraf Ragab Mohamed, Mona Elsalamawy and Marwa Ragab (2015) [6] investigated that amount of hydration increment by utilising limestone cement due to both its chemical effect and physical effect. Replacement of limestone about 5 to 10% associated with outcome of amount of hydration. Whereas limestone fills the voids between the cement elements due to the creation of carbo aluminate segments.

Table 1 Physical properties of cement used

Parameters	Results	Essential objectives of IS:8112/1987
Fineness (in M^2/Kg)	310	225 (min)
Setting time:		
1. Initial	170 (min)	30 (min)
2. Final	235 (min)	600 (max)
Soundness test:		
1. Le-chatelier expansion (mm)	1.50	10 (max)
2. Auto-clave (%)	0.012	0.8 (max)

3 Materials Used-

3.1 Cement

43 grade OPC conforming to IS: 812/1989 was used in this experimental work. The physical and chemical composition of cement after conducting the test conforms to IS:269/4831 and requirement as per IS: 8112/1989. The physical properties after conducting the test are given in Table 1.

3.2 Aggregate

Coarse aggregate (crushed granite) of size 10 mm (40%) and 20 mm (60%). After conducting the sieve investigation for coarse and fine aggregate the results conforms to the specification of IS:383-1970. The Specific gravity of coarse and fine aggregate was found to be 2.7 and 2.65 along with fineness modulus of 6.2 and 2.47, respectively.

3.3 Silica Fume

Light grey colour silica fume having moisture content less than 3% was used in this study was carried out from a ferrosilicon manufacturing industry named as Krishna Udyog at Kolkata, West Bengal, In India the pozzolanic activity index was in the range of 100–120.

3.4 Lime

Lime is an inorganic material containing calcium in which hydroxides, carbonates and oxides are predominant. This is particularly driven from rocks and minerals, typically from limestone or chalk. As because lime has an adhesive property, it has got a key role in Masonry constructions.

3.5 Water

Filtrated portable water was obtained from laboratory of School of Civil Engineering, KIIT. It was hence used for combining the constituents of concrete. Later the curing process was initiated on it.

4 Experiment Programme

An ordinary mix for normal concrete was prepared and designated as OC0%. Then 25% silica fume by weight was replaced with cement and named as SFMC 25%. Again that same mix has been modified by adding free lime by replacing fine aggregate with different percentages like 3, 5, 7 and 10% by weight with a constant water-cement ratio 0.45 by weight. Various mix of SFMC and ordinary concrete were taken to conduct a compressive analysis on standard moulds of sizes 150 × 150 × 150 mm, BIS moulds having diameter = 150 mm and height = 300 mm was taken to conduct a split-tensile test. The flexural test was conducted on standard BIS specimen of size 100 × 100 × 500 mm. 7,28 and 56 days respectably was taken for curing period (Table 2).

Table 2 Different proportion and water-cement ratio of silica fume modified concrete with lime mix

Mix	Cement (kg)	Silica fume (kg)	Lime (kg)	Fine aggregate (kg)	Coarse aggregate (kg)	Water (kg)	Water-cement ratio
OC0%	41.32	0	0	61.99	123.98	18.59	0.45
SFMC25%	30.99	10.33	0	61.99	123.98	18.59	0.45
SFMCL3%	30.99	10.33	1.85	60.13	123.98	18.59	0.45
0SFMCL5%	30.99	10.33	3.09	58.88	123.98	18.59	0.45
SFMCL7%	30.99	10.33	4.33	57.64	123.98	18.59	0.45
SFMCL10%	30.99	10.33	6.19	55.78	123.98	18.59	0.45

5 Results and Discussion

Compressive-strength test, Flexural-strength test, and Split-tensile strength test were done for the above mixes and the results were checked for 7 days, 28 days and 56 days curing period. The bar graph designating the results for OC and SFMC25% are depicted in Figs. 3, 4 and 5. A comparision of hardened mix results has also been done in Figs. 6, 7 and 8 between normal concrete mix, Silica fume modified mix and also the effect when lime was added as replacement to fine aggregate, respectively.

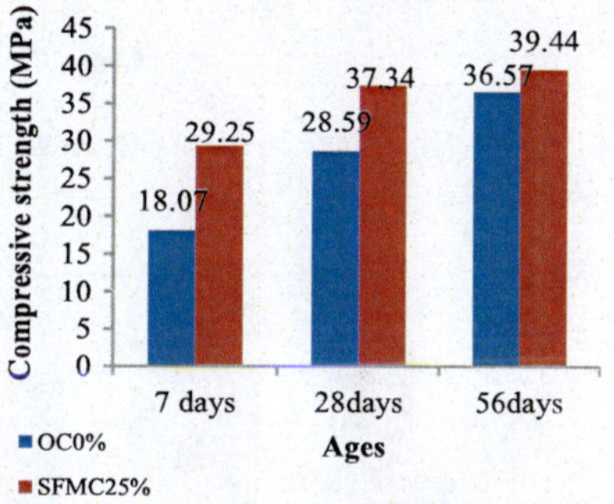

Fig. 3 Compressive-strength Test Results for OC and SFMC25%

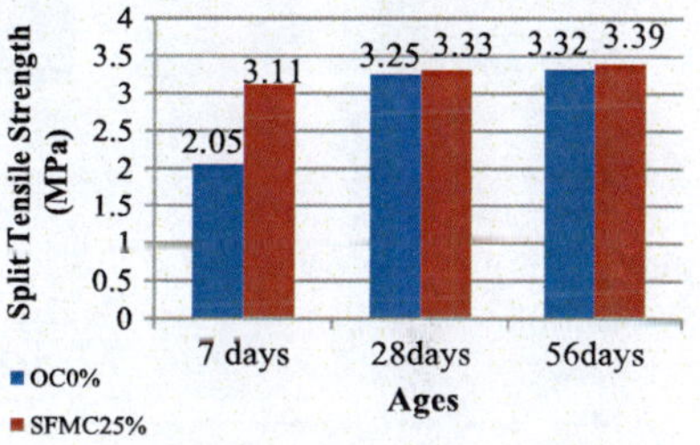

Fig. 4 Split-tensile strength test results for OC and SFMC25%

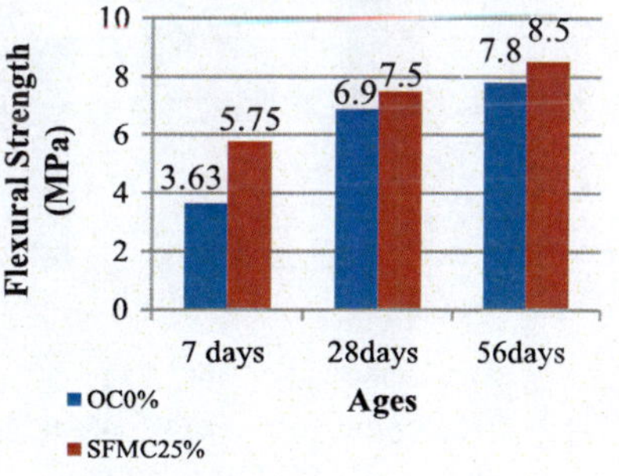

Fig. 5 Flexural-strength test results for OC and SFMC25%

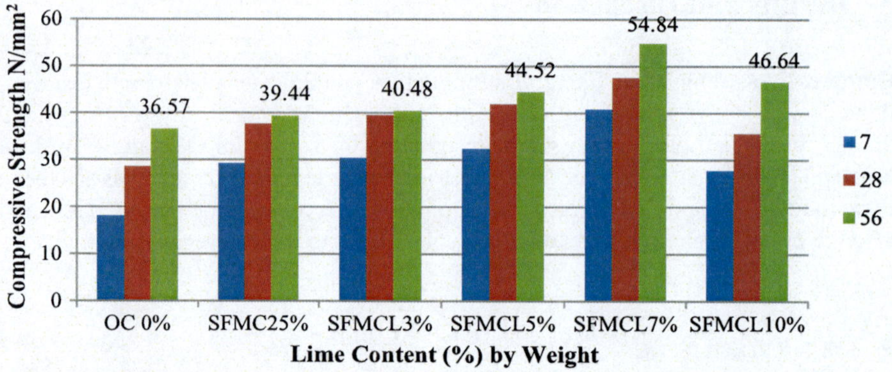

Fig. 6 Relationship between OC, SFMC and SFMCL(increasing percentage) on compressive strength at 7, 28 and 56 days curing period

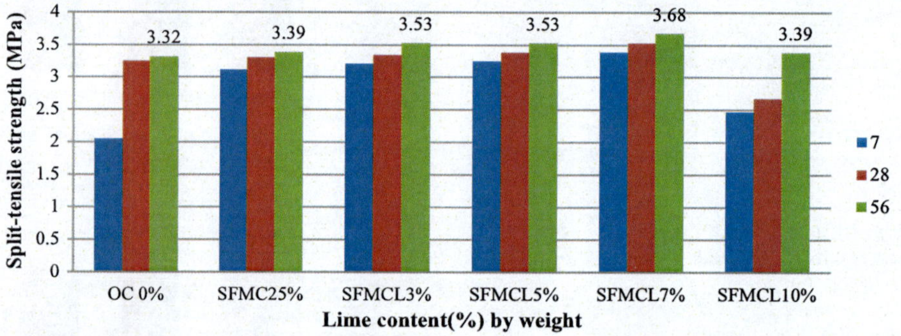

Fig. 7 Relationship between OC, SFMC and SFMCL (increasing percentage) on Split-tensile strength At 7, 28 and 56 days of curing period

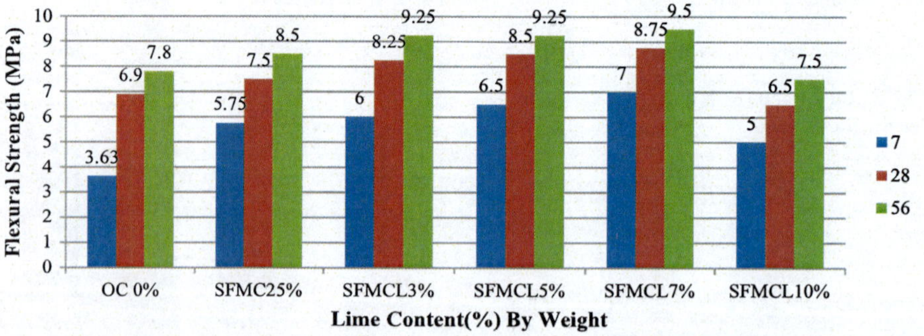

Fig. 8 Relationship between OC, SFMC and SFMCL (increasing percentage) on Flexural strength at 7, 28 and 56 days of curing period

6 Conclusion

(1) Silica fume modified concrete with 25% of cement replacement by weight gave significant test results compared to conventional concrete. There was an increase in compressive strength at 7 days, 28 days and 56 days by 61, 31 and 7%, respectively. Similarly, an increase in split-tensile strength at 7 days, 28 days and 56 days by 51, 2.1 and 2%, respectively, and flexural strength at 7 days, 28 days and 56 days by 58, 8.7 and 9.3%, respectively, was observed.
(2) The pozzolanic nature which leads to the formation of densely packed C-S-H gel results in high compressive strength
(3) Compressive strength, split tensile and flexural strength of concrete was found higher in silica fume concrete produced after replacing fine aggregate up to 7%. There was a decrease in test results beyond 7% replacement
(4) Silica fume modified concrete with 7% fine aggregate replacement was found to achieve a maximum percentage of increase in compressive strength up to 49.95%, increase in split-tensile strength up to 10% and flexural strength up to 21% at the age of 56 days
(5) In high rise building, high-performance concrete with silica fume can be used effectively since high early strength is required.

References

1. Safwan AK, Abou-Zeid MN (1994) Characteristics of silica-fume concrete. ASCE Manag J 6(3), August 1
2. Ajileye FV (2012) Investigation on microsilica (Silica Fume) as patial cement replacement in concrete. Glob J Inc (USA) 12(1), Version 1.0, January
3. Amuddhavalli NK, Mathew J, Effect of silica fume on strength and durability parameters of concrete. Int J Eng Sci Emerg Technol 3(1):28–35
4. Srivastava V, Agarwal VC, Kumar R (2010) Effect of silica fume on mechanical properties of concrete. Youth Edu Res Centre 1(4), ISSN:2278-5213, September
5. Kumar D, Singha Roy AS (2012) Effect of partial replacement of cement by silica fume on hardened concrete. Int J Emerg Technol Adv Eng 2(8), August, ISSN:2250-2459
6. Mohamed AR, Elsalamawy M, Ragab M (2015) Modeling the influence of limestone addition on cement hydration. Alexandria Eng J, 54, I-5, February
7. Tudjono S, Purwanto, Apsari KT (2014) Study the effect of adding nano fliers and nano lime to compressive strength of mortar. Sec Int Conf Sustain Civil Eng Struct Constr Mater Proc Eng 95:426–432
8. Mikhailova O, Yakovlev G, Maeva I, Senkov S (2013) Effect of dolomite lime stone powder on the compressive strength of concrete, 11th Int Conf Modern Build Mater Struct Techn Proc Eng 57:775–780

Polymeric Materials for Repair of Distressed Concrete Structures

Suresh Chandra Pattanaik, Sanjaya Kumar Patro, and Bitanjaya Das

Abstract The concrete repair is not given its due the weightage in civil engineering considering the sustainability issues and environment impact factor that it produces in restoring and conserving the old structures. To build the similar new structure, there is lot of embedded energy involved for demolishing and rebuilding the structure. Unless otherwise the structure is unsafe, the building or structure should be restored and should not be given permission for redevelopment. Before repairing any cracks in concrete structures, it needs to be diagnosed properly. Based on the diagnosis, compatible repair materials need to be selected for the repair. There are different types of repair materials and methodologies available out of which suitable one needs to be selected. The structural audit with non-destructive test (NDT) helps to diagnose the problems. There are different kinds of repair materials and methods available, which need to be specified based on the requirement. In most cases of repair, the failure takes place since the issues of compatibility of repair material are not considered during material selection. Also, during design and execution of a repair project, the planning and surface preparation is not done properly for which repair of same surface continues again and again. The present paper discusses the various types of polymeric materials and methodologies used for repair of distressed concrete structures for a durable concrete repair system.

Keywords Crack repairs · Cementitious grouts · Polymer modified cementitious grouts · Epoxy injection and microconcrete

1 Introduction

The concrete repair is one of the most neglected topics in civil engineering subjects. The structural consultants often show interest for design for any new structure but

---

S. C. Pattanaik · B. Das
School of Civil Engineering, KIIT University, Bhubaneswar 751024, Odisha, India
e-mail: scpattanaik2003@yahoo.co.in

S. K. Patro (✉)
Department of Civil Engineering, VSS University of Technology, Burla 768018, Odisha, India
e-mail: skpatro_ce@vssut.ac.in; litusanjay@yahoo.com

B. B. Das et al. (eds.), *Recent Developments in Sustainable Infrastructure*, Lecture Notes in Civil Engineering 75, https://doi.org/10.1007/978-981-15-4577-1_57

are very much reluctant for addressing the repair and strengthening issues of any old concrete structure. This is due to less exposure to advance repair materials and methodologies in the field of repair and rehabilitation of concrete structures. Even there is shortage of skilled repair contractor to take up the job, only practicing engineers and consultants those who are involved in repair field take the challenges. The construction industry took a boon in India in the year 1950–60. All those buildings and structures are more than 60 years old and have lot of damages and deterioration over the period of time which need repair and maintenance. Based on condition assessment, these structures can be repaired and strengthened. However, the repair industry in India is the unorganized sector of large market size and spread over all the nooks and corners. Thus, the exact figure of repair market is not available. So far, there is no Indian Standard available for repair of concrete structure. But one can refer 'Handbook on Repairs and Rehabilitation of RCC Buildings' published by CPWD for repair guidance. At the same time, one can also refer American Standard ACI-562 [1] or European standard EN 1504 [2] for concrete repair guidance. ACI 562-16 give guidelines to consultants for assessment degree of damages and recommending different suitable strategies for repair and rehabilitation, whereas EN 1504 outlays the brief procedure and characteristic of products to be used for repair, maintain and protect concrete structures. ACI 562 is more exhaustive, whereas EN 1504 is more compressive and concise in ten parts. ACI 562 can be used for practical applications for concrete repair, whereas EN 1504 describes only principles but not application methods in details. To address the issue of repair material requirement, repair material properties and available repair technology, the present paper discusses the detail repair methodology of concrete repair considering Indian practices.

2 Condition Evaluation

The concrete may develop some cracks and distresses during the service life of the structure. The causes of cracks are due to corrosion of reinforcement, bad quality of construction, stress due to overloading, inadequate strength not matching to the design requirement, ageing of the structure and poor maintenance. The crack due to corrosion of reinforcement is most commonly found in every old structure. The corrosion depends upon severity of environment exposure conditions. Before doing any repair to the structure, the condition assessment need to be done and accordingly repair methodology can be established. This also helps to decide protection measures to be given to the structures based on the exposure conditions for achieving a durable service life. Other important feature of condition evaluation is to understand the safety and stability of the structure.

The initial assessment of current condition of structure is to check for cracking, spalling and change of use of structure, if any. The comprehensive condition evaluation should include structural and non-structural defects, original design strength, past history of repair and strengthening and results of different non-destructive tests, partial destructive tests and chemical analysis of concrete samples on selected structural members and locations. The purpose of this condition evaluation should be to establish parameters such as design life, intended use, residual service life and required performance characteristics.

The condition evaluation of concrete structure is being carried out by following tests.

- Concrete quality by Ultrasonic pulse velocity (UPV)
- Concrete surface hardness by Rebound Hammer
- Cover depth measurement by concrete cover meter
- Half-cell potential and rate of corrosion tests to predict the risk of corrosion and quantify the corrosion of reinforcement, respectively
- Electrical resistivity test to check the quality of concrete for protection against corrosion
- Carbonation test to understand the depth of carbonation
- Chemical analysis test for chlorides, sulphate and pH to ascertain the contamination
- Compressive strength of concrete by core test/Cut and Pull out (CAPO) test.

The tests are conducted to determine the extent of deterioration, establish causes of deterioration and identify existing mass transit mechanisms. While doing condition evaluation the basic minimum tests such as rebound hammer, ultrasonic pulse velocity, cover depth, half-cell potential, core test and chemical analysis for pH, chloride and sulphate should be done. The other tests such as electrical resistivity, rate of corrosion and CAPO test should be done at selected important structures or based on the client's requirement. The type of tests, test locations and number of each test should be decided during initial inspection. In general, minimum 10% of structural members should be tested and which can go up to 50% if the structure is old or distressed. However, there is no such standard or specification which describes the number of tests required for condition evaluation. But, higher the number of tests would be better would be condition evaluation of the structure. Analysis of test results should be done considering the original strength and quality of the structures by establishing the co-relation between similar tests to understand the deterioration of structures, calculating the damage rating and evaluating the condition of the structure in a predefined scale. The recommendation and protection measures always form the part of condition assessment.

3 Durability of Concrete Repair

The best repair materials and methods to be specified for the repair project for a long-term durability of repair of concrete structures. The factors of repair material which needs to be considered are compatibility of repair materials with parent concrete and concrete repair bond between old and new surfaces. To achieve a proper bonding between old and new surface, all loose and damage concrete need to removed and surface should be thoroughly rubbed by means of a wire brushing, hand scrubber or a mechanical scariffier depending on area of repair surface. While considering type of bonding agent, one has to consider ambient temperature, moisture, exposure conditions, thickness of repair mortar so that the failure of repair material will never occur at the interface of old and new repair surface. The different types of bonding agent can be considered as styrene butadiene rubber (SBR), polyvinyl acetate (PVA), acrylics and epoxy where SBR and acrylic are mostly used in patch repair of mortars or patch repair of concrete of smaller thickness. Similarly, PVA should not be used for any critical structures since PVA re-emulsified with variation of moisture. Epoxy is most suitable for any kind of structural repair; but locations having high-humid conditions, moisture insensitive epoxy should be used.

The relative importance of repair material properties for a durable repair system can be given in order as drying shrinkage, tensile strength, bond, modulus of elasticity, thermal coefficient of expansion and contraction and compressive strength. The properties and requirement of those repair material should in accordance with repair material as per specification or as given in standard. The most important factor of repair material is drying shrinkage which needs to be even tested of repair material at 1-year period of time and satisfying the drying shrinkage requirement of concrete repair material. The bond and tensile strength of repair material are dependent on each other and taking care of those two properties as per codal provision ensures a durable concrete repair system. In case of repairing old concrete structure, the compressive strength of repair material has got least importance. The compressive strength of concrete repair material should not be more than 20% of the old concrete to match with the compatibility properties or else there may be reversal of stress on old concrete surfaces when using with a very high compressive strength of repair material.

The curing process taken during repair plays a significant role in achieving a durable repair concrete. But the curing process is most neglected one in concrete repair due to the lack of understanding of importance of curing in concrete repair.

4 Polymeric Concrete Repair Methods

There are various polymeric modified concrete repair methods available which can be selected based on the requirement. The polymers can be latex- or resin-based when added to the cement concrete changes the properties of cement concrete by becoming more flexible, reducing the shrinkage, increasing the bond strength of the concrete, having control expansion, improving the tensile strength, becoming less impermeable to moisture and chemical ingress, helping for rapid gain in compressive strength and overall becomes more workable. All these properties help into achieve a durable repair concrete. Again the cracks in concrete may be structural or non-structural types. For non-structural crack, any flexible polymeric crack filling will be helpful to fill the cracks. If the cracks and damages are found in structural members, then they should be strengthened either by injection grouts or strengthened by jacketing with microconcrete.

The fine hairline structural cracks are usually repaired by a low viscous epoxy resin by injecting into the structural members. Wider cracks up to 5 mm width on structural members can also be repaired by injection methods but crack along injection points need to be sealed by epoxy putty so as to avoid any back flow of injection grouts. Cracks on any horizontal surface of the concrete can be repaired by crack filling by gravity if possible depending on locations of the structural members.

4.1 Concrete Repair by Grouting of Cracks

The grout consists of mixture of cement, fine aggregate, water and a polymer such as acrylic, styrene–acrylic, styrene–butadiene rubber or a waterborne epoxy. For easy application, the consistency of such repair material should be such that they can be easily applied by hand for any kind of surfaces. In case of cracks in slabs where it is required to be poured by gravity, accordingly consistency needs to be changed; hence, the constituency varies from still to pourable depending on type of surfaces. This type of polymer-modified repair materials provide better performance than cementitious repair materials with respect to different properties such as bond strength, tensile strength and flexural strength. The cracks are widened in form of 'V' grove and filled with polymeric repair grout (Fig. 1).

The surface is cleaned and saturated to make it suitable for saturated surface dry (SSD) condition before doing any crack repair. Repair materials are mixed by mechanical mixer at site as per required consistency by thoroughly mixing with a drill and paddle mixer.

4.1.1 Crack Filling Up to 5 mm Wide in Concrete

While doing the injecting grouts, PVC nozzle of 10–12 mm or MS packers of 8 mm diameter should be drilled at an interval of 500 mm c/c in columns and beams,

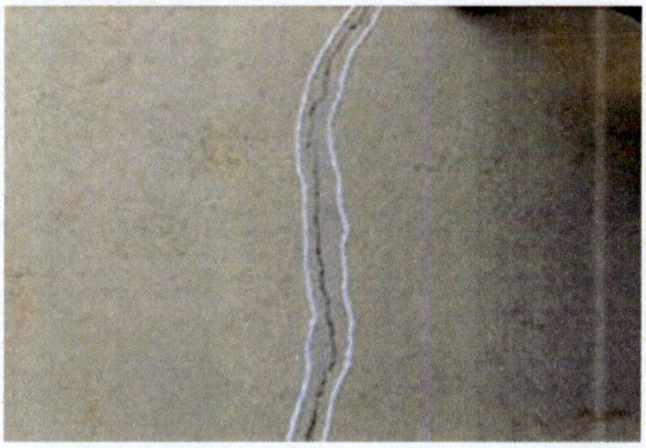

Fig. 1 Crack filling with ready-to-use polymeric crack filling material

and in grid pattern in case of roof slab. The drilled holes should be at least 2–4 mm higher than the size of nozzles/packers. The gap between holes and nozzles/packers need to be filled in with polymer-modified mortar/epoxy putty for setting. For polymer-modified cementitious injection, grouts in 1:3 cement sand mortar mixed with polymer need to be used for sealing of the horizontal cracks. For all vertical cracks, a compressor should be used for injecting the polymer-modified cementitious grouts. The injection should start from lower end and move upward for injecting at other locations. After injection job is completed, PVC nozzles should be cut and sealed with polymer-modified concrete, and surface should be finished with same polymer-modified mortar.

4.1.2 Crack Filling More Than 5 mm in Concrete

In case of wider cracks more than 5 mm, the repair should be carried out with polymer-modified concrete as explained earlier and in addition to that stitching with galvanized steel clamping need to be done.

The clamping should be done with 3–5 mm diameter galvanized steel clamping rod of 150-mm long.

The wire mesh on both the faces of crack needs to be clamped at each 300 mm c/c after the removal of plaster along with cracks for 150-mm wide on each side of the crack. The repair surface along with the clamped mesh needs to be finished with polymer-modified cement mortar.

4.2 Repair of Non-structural Cracks

While selecting repair materials for any non-structural crack repair, a flexible polymeric ready-to use repair material should be selected. The crack should be wider at surface to accommodate the repair material. But, if crack is much wider and appears again and again, in such cases, a flexible material of polyurethane sealant is more suitable [1, 3, 4]. Non-structural cracks are filled with flexible polymeric crack filler material to avoid damp patches inside the rooms. The external crack filling is used for sealing all types of shrinkage cracks and fine map pattern cracks [5].

4.3 Strengthening of Structural Members by Epoxy Injection Grouts

Epoxy resins are injected for hairline cracks to wider cracks up to 5 mm width. A low viscous epoxy should be used for hairline cracks and a medium to gel viscosity epoxy should be used for wider cracks. ASTM C881 'Standard Specification for Epoxy-Resin-Base Bonding Systems for Concrete', [6] should be referred to identify the required criteria for selection of epoxies.

The pressure requirement for different surfaces such as vertical, horizontal or overhead application varies. The highest pressure required for overhead applications for which care should be taken while injecting in overhead applications, so that it does not create any back pressure. The crack size also plays an important role for application with epoxy injection grouts. Wider cracks may create open passage to injected materials for which such wider cracks must be sealed with epoxy putty or polymer-modified mortar between port and port prior to injection application and get fully set. Accessibility to cracks for injection is also important. If only one surface is available for injection accordingly the viscosity of the material should be selected. The epoxy material exhibits very high compressive strength and bond strength for which they are compatible for all types of concrete. In general, the epoxy materials available in the market vary from low viscous to very low viscous resins. One has to select suitable viscous epoxy material as per the specific requirement. Similarly, epoxy can be either moisture-sensitive or -insensitive material for different environment conditions. Under humid conditions, only moisture insensitive epoxy should be used to avoid any failure of materials.

4.4 Strengthening of Structural Members by Jacketing with Microconcrete

Reinforced concrete structural members which are excessively damaged both due to corrosion of reinforcement or due to any other distress can be repaired and strengthened by jacketing with microconcrete. Microconcrete is a ready to use non-shrink polymeric repair material when added with water produces a free-flowing, self-levelling, self-compacting concrete. It is composed of good quality cement, fibres, properly selected aggregates, additives and polymers. The water cement ratio is only 0.14–0.15 to make it a free-flow concrete. Because of rapid hardening, the concrete can attain compressive strength of M30 Grade of concrete within 3 days of casting. The microconcrete can attain M40 to M50 Grade of concrete only at 7 days depending on type of mixtures and polymers in it.

Before doing any microconcrete, the rebar should be cleaned properly by means of wire brushing. The rebar needs to be exposed thoroughly beyond the reinforcement. An acidic rust remover can be used to remove the dust particles from corroded steel. Care should be taken to wash the surface with sufficient water so that no acidic substance remains before the repair material is being placed. After application of few minutes of rust remover, entire rust particles will come out easily after striking the rebar surface. The surface should be cleaned properly with sufficient water jet to remove any contaminant inside the concrete. The rusted steel bar need to be passivated with a zinc reach epoxy coating. An epoxy bond coat needs to be applied on concrete surface before microconcrete. If additional rebars need to be provided then those bars to be embedded in concrete by drilling the holes and fixing the rebars in the drilled holes with polyester resin grouts for anchoring. If rebar needs to be provided as a casing then 8 mm diameter tie bars need to be fixed at 500 mm c/c in original concrete in drilled holes and fixed with polyester resin grouts to act tie bars between old and new reinforcement. All these operations need to be completed before encasing with 100% leak-proof shuttering. In case of pouring in columns, microconcrete can be poured from top and in case of beams the microconcrete should be poured simultaneously at different locations arranged in funnel hopper for pouring microconcrete. Before doing any microconcrete sufficient props and shuttering need to be provided for beams and roof slabs to support the concrete. In case of roof slab, holes need to be drilled at 1.5-m interval and microconcrete should be poured from top simultaneously. Coarse aggregate of 6 mm and down can be replaced with microconcrete and added during mixing but the flow of such mixed concrete material need to be checked and accordingly pouring need to be done at those intervals. During microconcrete, small holes need to be made to release the entrapped gases in microconcrete or else honeycombs will develop at those locations. Shuttering of sides and soffits can be removed after 24 h and 3 days, respectively. Normal water curing for 7 days can be made. The strengthening of a distressed column member by jacketing with microconcrete is shown in Figs. 2 and 3, respectively.

Fig. 2 Additional reinforcements provided for jacketing

Fig. 3 Strengthening of column with microconcrete

5 Conclusion

The following conclusions are drawn:

- The concrete repair industry has lot of potential in terms of market share, sustainability and environment impact factor by restoring and conserving the old structures.
- The selection of suitable compatible repair material and methodology is prime factor for a durable repair system.
- For any non-structural crack repair polymer-modified cementitious material is most suitable.
- The epoxy should be used for strengthening of structural members.
- The structural members having excessive corrosion-related damages and spalling should be repaired by jacketing with microconcrete for structural strengthening.

References

1. ACI 562-13 code requirements for evaluation, repair, and rehabilitation of concrete buildings (ACI 562-13) and commentary
2. EN 1504, European standard, Products and systems for the repair and protection of concrete structures–definitions, requirements, quality control and evaluation of conformity
3. Hand book HB 84-2006: Guide to concrete repair and protection, a joint publication of ACRA, CSIRO and Standards Australia

4. Pattanaik SC (2011) Repair of active cracks of concrete structures with a flexible polyurethane sealant for controlled movement. In Proceedings of the national conference on advances in materials and structures, 'AMAS—2011', Pondicherry
5. ACI 546.3R-14 guide to materials selection for concrete repair
6. ASTM C881 Standard specification for epoxy-resin-base bonding systems for concrete

Thermal Cycles and Its Behavioural Aspects of Glass Fibre Self-compacting Concrete

Seshadri Sekhar Tirumala

Abstract The structural elements when exposed to heat energy because of solar heat in the day time as discussed in Sravana P, Srinivasa Rao P, Seshagiri Rao MV (2006) Effect of thermal cycles on the strength properties of OPC and Fly ash concrete. Indian Concrete J 49–52; Dubal N, Bairagi NK (1996) Effect of thermal cycles on the compressive strength, modulus of rupture and dynamic modulus of concrete. Indian Concrete J 70(8):23–26. 100% elements will undergo decline in strength when exposed. To study the behaviour of thermal cycles on strength characteristics of glass fibre self-compacting concrete mixes (GFSCC) (Dubal N, Bairagi NK 1996 Effect of thermal cycles on the compressive strength, modulus of rupture and dynamic modulus of concrete. Indian Concrete J 70(8):23–26) using AR-glass fibres, the investigation have been carried on various mixes.

Keywords Water–cement ratio · Thermal cycles · SCC · GFSCC · Workability · Glass fibres

1 Preface

Self-compacting concrete (SCC) [3] was initially used in JAPAN for achieving achieve long lasting concrete structures. There was unanswered problem on durability of structures. Preparation of dense concrete and sufficient compaction is required for conventional concrete. In heavily reinforced zone [4], it is difficult to use vibrator and is difficult to have correct bond between steel and concrete. For all these issues, the solution is SCC where the concrete can percolate to all corners of formwork and settle without vibration [5]. SCC concept was used as special concrete that meets the all the requirements of performance and uniformity requirements of ordinary concrete [5] without vibration.

S. S. Tirumala (✉)
National Institute of Construction Management and Research, Hyderabad, India
e-mail: ss.tirumala@gmail.com

B. B. Das et al. (eds.), *Recent Developments in Sustainable Infrastructure*, Lecture Notes in Civil Engineering 75, https://doi.org/10.1007/978-981-15-4577-1_58

2 Research Significance

All the structural elements [6] are open to solar energy all over their life span. The heat will increase and will be very high in mid-noon. At that situation, the heat energy is high than regular. In India, some times it may go up beyond 40 °C; in such places, the surface temperature of the exposed elements such as roof slab, external column, beam and walls may reach somewhere between 80 and 90 °C [1]. The degree of humidity also influences thermal gradient within the elements of structures exposed to solar radiation [6]. Thus, all the structural elements that are exposed to solar radiation are said to experience one thermal cycle in a day, that is, a heating period to a peak value and then subsequent cooling period in a day [2]. All the researchers have worked on ordinary concrete behaviour when subjected to thermal cycles, no researcher have worked on SCC and GFSCC mixes. So to fill the gap, we have carried this experiment [6].

3 Experimental Programme

- To study the behavioural aspects of thermal cycles of GFSCC mixes at different temperatures.

4 Resources Used: [7]

Bonding material: [7] Annular material passing through 12.5 mm, fineness modulus 6.1 and sand with specific gravity 2.55 was used bonding material.

Bonding powder: Powder having fineness 3200 cm^2/gm and specific gravity having 3.1 was used.

Glass fibres: Cem-FIL Anti-Crack HD with filament diameter 14 microns, aspect ratio of 857.1 and modulus of elasticity 72 GPA was used.

Superplasticizer: The yellow colour polycarboxylated ether having pH > 6, density 1.10 $\pm$ 0.01 at 25 °C and chloride ion presence <0.2% was used. A colourless rheodynamic concrete with specific of gravity 1.01 $\pm$ 0.01 at 25 °C, pH 8 $\pm$ 1 used as viscosity-modifying agent.

Fly Ash: Fly ash conforming to type II was used.

5 Experimental Investigation

Various GFSCC cubes, cylinders and beams have been casted with mixes given in Fig. 1. These specimens were exposed to different thermal cycles at temperatures

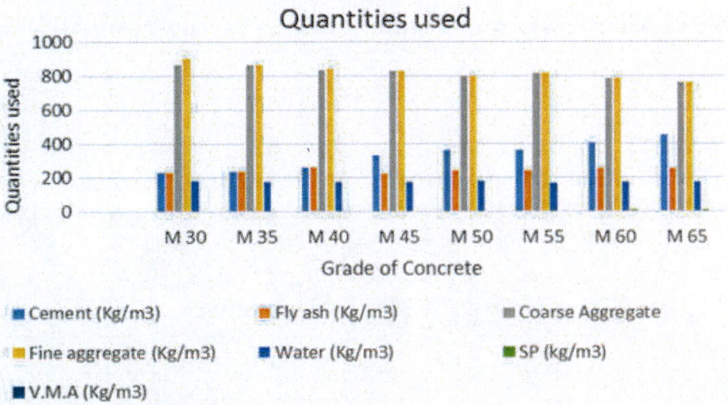

Fig. 1 Materials versus grade of concrete

[7]. The workability behaviour is adjusted by adding admixtures keeping all flowable properties of SCC and GFSCC mixes.

6 Analysis of Results

Behaviour of GFSCC mix at fresh state:It was found GFSCC mixes gave rich mixes. Because of additing glass fibres, the formation of microcracks have been arrested [6].

Behavioural aspects of Thermal Cycles on SCC and GFSCC mix [7]: Figs. 2, 3. 4, 5, 6 and 7 shows 40 60% at 50 °C and 20–30% at 100 °C changes in strength with respect to zero thermal cycles.

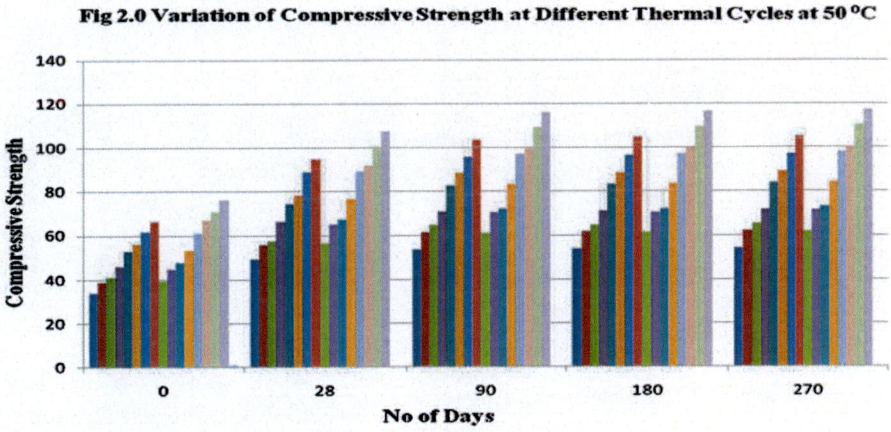

Fig. 2 Strength behaviour in compression of GFSCC mix at 50 °C

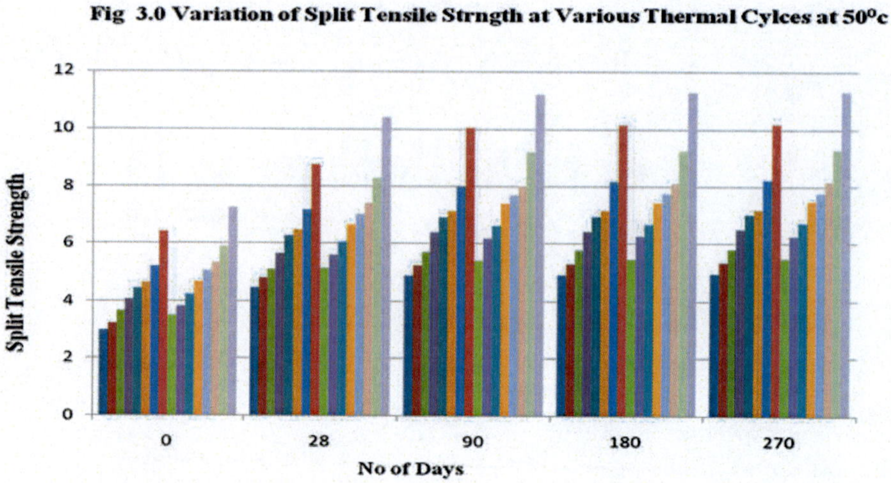

Fig. 3 Strength behaviour in compression of GFSCC mix at 100 °C

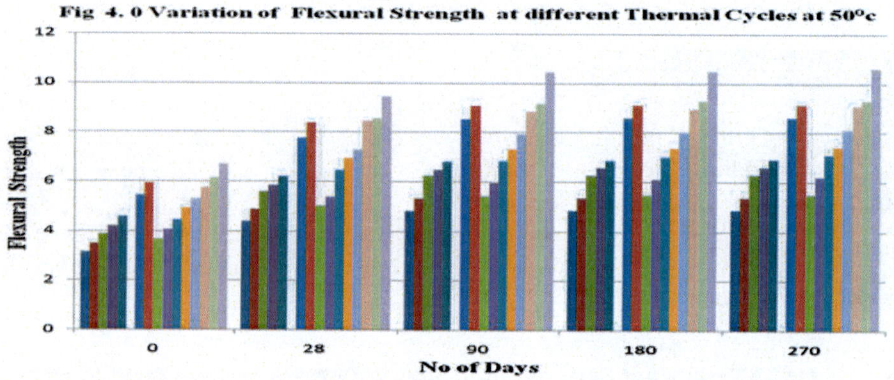

Fig. 4 Strength behaviour in splitting tension of GFSCC at 50 °C

Behavioural aspects of glass fibre on GFSCC mixes [7]: An increment of 15–20% was found as a resultant of glass fibre addition in GFSCC mixes.

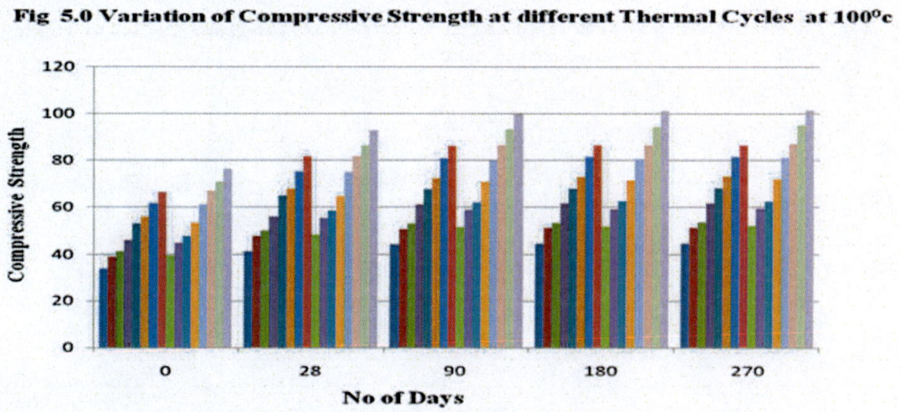

Fig. 5 Strength behaviour in splitting tension of GFSCC at 100 °C

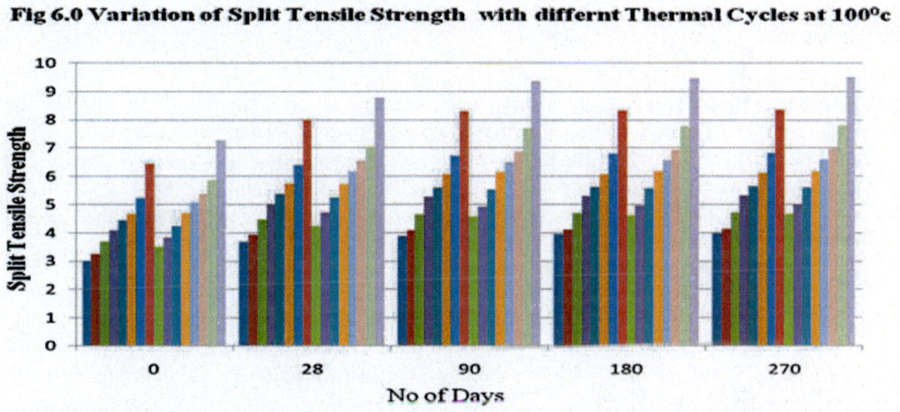

Fig. 6 Strength behaviour in flexure of GFSCC at 50 °C

7 Findings from the Experiment

GFSCC mixes exhibit more strength than SCC to thermal cycles. The percent increment in intense strength of SCC and GFSCC is the result of drying out of calcium silicate–hydroxide gel bond. The Sio_2 present in fly ash forms bond by gelling with lime leading to more strength variation. Glass fibre reduces the formation of microcracks [6]. The E value for fibres is more than concrete [7]. This property gives glass fibres to act as strengthening parameter fresh and hardened state.

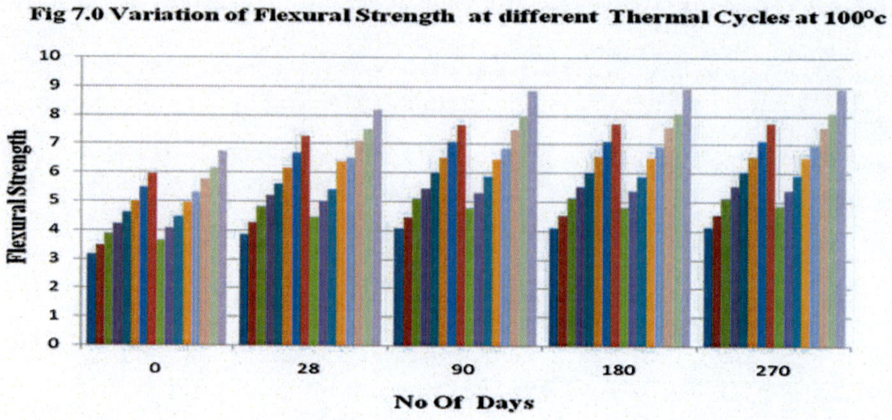

Fig. 7 Strength behaviour in flexure of GFSCC at 100 °C

References

1. Sravana P, Srinivasa Rao P, Seshagiri Rao MV (2006) Effect of thermal cycles on the strength properties of OPC and fly ash concrete. Indian Concrete J 49–52
2. Dubal N, Bairagi NK (1996) Effect of thermal cycles on the compressive strength, modulus of rupture and dynamic modulus of concrete. Indian Concrete J 70(8):23–26
3. Ouchi M, Okamura H (2003) Self-compacting concrete. J Adv Concrete Technol Jpn Concrete Inst 1:5–15
4. Subramanian S, Manu S (2004) Current developments in self compacting concrete. Indian Concrete J 11–22
5. Kumar P, Kaushik SK (2003) Some trends in the use of concrete: Indian Scenario. Indian Concrete J 77(12):1503–1508
6. Srinivasa R, Seshadri Sekhar T (2009) Effect of Thermal cycles on the strength properties of SCC. Indian Concrete J 39–44
7. Seshadri Sekhar T, Srinivasa R (2015) Experimental investigation on the effect of thermal cycles on the strength properties of glass fibre SCC. Indian Concrete J 1–6

Constraints in Journey of Graphene from Laboratory to Life

Sanjukta Sahoo, Trupti Ranjan Mahapatra, and Hemlata Jena

Abstract Graphene extraction through exfoliation was reported in 2001. Researches revealed many attractive properties of graphene, such as excellent strength, good thermal and electrical conductivities, high surface area, its potential to absorb and store harmful gases, etc., for which graphene is being seen as a wonder material. Though multiple benefits with graphene as a material are observed in the research findings, still the adoption of this wonder material is not visible in common public life. Experts opine that new materials take longer time before penetrating into the domain of public use, citing previous instances. However, in the twenty-first-century world, where information travels really fast, it is not unfair to expect greater spread of graphene by this time. The associated factors responsible for the poor availability of this excellent material and the constraints to be overcome for its implementation in common life have been reported in this paper. Firstly, the paper attempts to assess the use scale of graphene in laboratories and in real life. The established graphene institutes and the different research areas are highlighted in this work. Further, the statistical analysis of the research publications in Scopus database is presented to assess the research progress in the field of graphene. The graphene-based products those are commercially available so far also have been reported in this investigation for showing its importance in common life. In the subsequent sections, the constraints in commercialization are presented based on the published literatures and reports.

Keywords Graphene · Commercialization · Up-scaling production

---

S. Sahoo (✉)
KIIT Polytechnic, KIIT, Bhubaneswar, Odisha, India
e-mail: sanjuktasahoo.7@gmail.com

T. R. Mahapatra
Department of Production Engineering, VSSUT, Burla, Odisha, India

H. Jena
School of Mechanical Engineering, KIIT D-U, Bhubaneswar, Odisha, India

B. B. Das et al. (eds.), *Recent Developments in Sustainable Infrastructure*, Lecture Notes in Civil Engineering 75, https://doi.org/10.1007/978-981-15-4577-1_59

1 Introduction

Since its discovery, graphene is being hailed as the next generation material for distinctively attractive properties that present it to be the thinnest, lightest, strongest, best in heat conduction, best in electricity conduction among all the known materials so far. Since this is an allotrope of carbon, the basis of all life forms, it is believed to be an environment-friendly material. Graphene has high surface area, mechanical strength, durability, atomic thickness, nano-sized pores, reactivity toward polar and nonpolar water pollutants imparting high selectivity and water permeability. These properties render a tag of 'wonder material' to graphene. These make graphene the most attractive candidate material to find application in electronics, optical electronics, sensors, biosensors, paints, energy storage device components, construction, and in many other engineering domains. In spite of the lucrative properties, the market penetration of graphene is relatively low. Past experience shows that a new material takes an average of 20 years to make it from laboratory to life. The experience belonged to an age where information speed, accessibility, and research quantities were fairly limited. Considering the present-day scenario, where information technology and research facilities have improved multifold, it is not unfair to expect the lab-to-life journey of graphene to be faster. In the context, the present paper attempts to compile the constraints retarding journey of graphene from related research and business articles.

The paper has broadly two sections. In the first section, existing state of graphene in laboratory and graphene in life has been reported. The state of graphene in laboratory has been discussed analyzing the trend of publications and patents relating to graphene. The patents and market trends have been used to determine the status of graphene in life. The second section compiles the constraints reported by researchers and analysts.

2 Graphene in Lab: The Status

2.1 The Trend of Research Publications

Figure 1 represents the number of graphene-related technical articles indexed in the Scopus database [1]. In Fig. 2, the top ten countries with graphene publications and the number of publications has been furnished. Top ten affiliating institutes and number of publications ascribed to them has been given in Table 1. The table depicts the worldwide affiliates and Indian affiliates.

A close view at Figs. 1 and 2, and Table 1 reveal the increasing interest of scientific community on the graphene material research and supremacy of the China research institutions. The finance available and the policy at institutional levels prevailing across the nations may be attributed as major controlling factors for the wide gap in the publications so far.

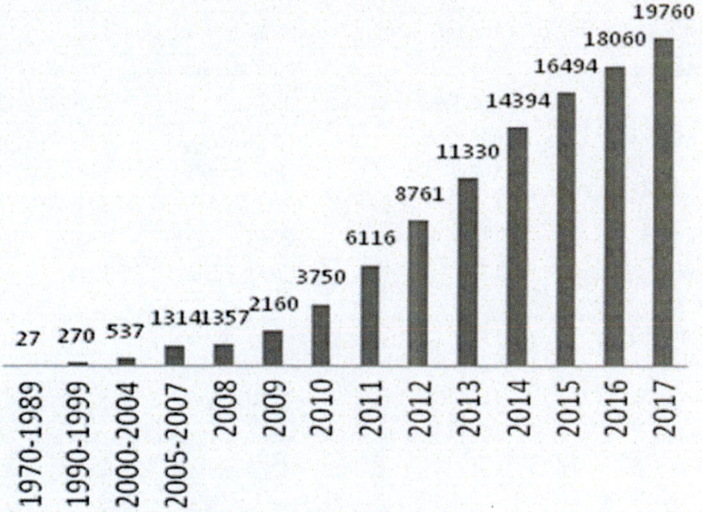

Fig. 1 Number of Scopus indexed publications with keyword "graphene" in titles/abstract (10.10.2018)

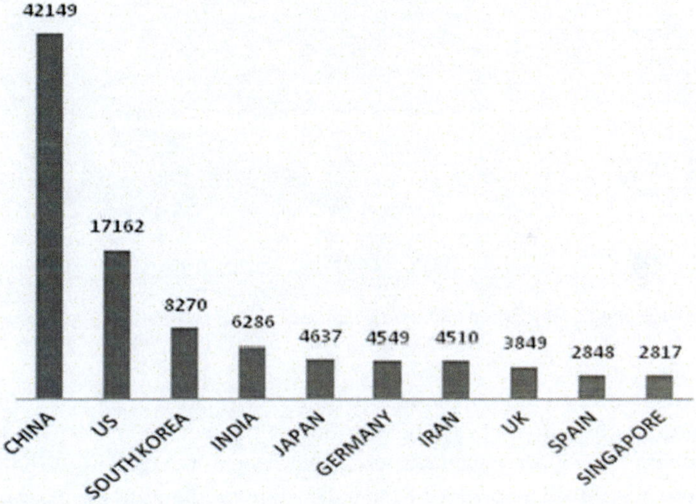

Fig. 2 Country wise 'graphene related publications' as depicted from Scopus (10.10.2018)

2.2 The Trend of the Graphene-Related Patents

Patent landscape report developed by Patseer-Pro published on 20 August 2017 [2] was studied to analyze the data on patents on graphene. Figure 3 displays the number of graphene-related patents granted through 2004–2017 period. The number

Table 1 Affiliation-wise 'graphene-related publications' as depicted from Scopus (10.10.2018)

International status		National status	
Affiliate	Publication number	Affiliate	Publication number
Chinese academy of science	5762	IIT, Kharagpur	53
Ministry of education, China	4945	IISc, Bangalore	52
Tsinghua university	1513	CSIR	48
Nanyang technological university	1378	SRM university	43
University of chinese academy of sciences	1197	Anna university	41
Peking university	1116	VIT, Vellore	38
NUS, Sigapore	1057	IIT, Madras	38
University of science and technology of china	1018	IIT, Delhi	37
Russian academy of sciences	967	CERI	34
Sungkyunkwan university	966	University of Delhi	32

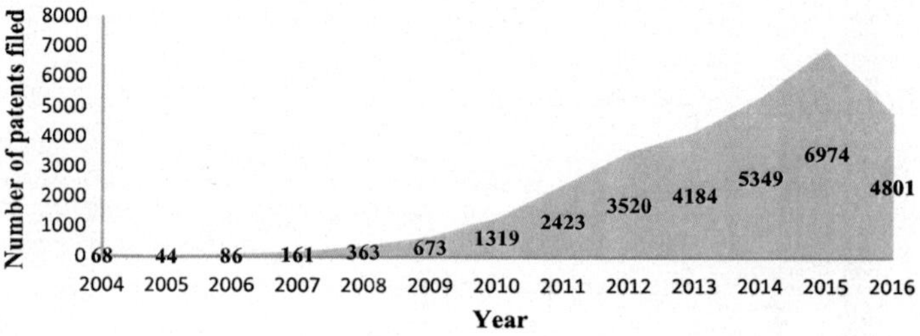

Fig. 3 Year-wise trend of the Patents filed on Graphene-related aspects

of patents has increased more than hundred times during these 13 years. According to the report, number of patents published during this period is 49912 from 30963 patent families. The report presents filtered data from 30000 patents after removing irrelevant records. More than 25% of the patents are filed in graphene being applied to electronics sector.

Table 2 represents the patents on graphene with respect to the application areas. Followed by electronics the composite materials, the sensors, solar cells lead the application area where research has contributed to numerous patents. In the same table, the number of patents on graphene production processes has been represented. The highest number of the patents has been published using Chemical Vapor Deposition (CVD) as the technique for graphene production. Exfoliation is the process, using which Nobel laureates Geim and Novasalev obtained graphene. There are 692 patents published on the technique and subsequent improvements. In each of

Table 2 Number of patents w.r.t. application area, production techniques, and grapheme forms

Application area	Number of patents	Production techniques	Number of patents	Graphene forms	Number of patents
Electronics	7970	CVD	2692	Graphene oxides	8185
Composite Materials	5665	Exfoliation	692	Bilayers	773
Sensors	5228	Epitaxial	653	Quantum dots	773
Solar cells	4965	GO reduction	463	Nano ribbons	371
Optical Electronics	4388	Sonication	440	Monolayer sheets	104
Li-ion batteries	3472	Pyrolysis	228		
Fuel cells	2914				
Aviation	2105				

the method, number of patents is significantly high, which indicates there has been a scope of improving the effectiveness and efficiency of the technologies. Besides pristine graphene, other graphene forms have also attracted significant attention from the research organizations. Most of the patents are on grphene oxide formation followed by quantum dots, bilayers, nanolayers, and monolayer sheets.

2.3 *Graphene in Real Life*

Table 3 furnishes a list of commercial products those use graphene. The list has been drawn from information compiled in the website https://www.graphene-info.com/graphene-products [3].

There is media news which announced several items those are expected to be in market by 2019. Such announcements included Ford's graphene-enhanced cars, Log9's (IIT Roorkey, India) graphene-based filtration products, lightest watch (A collaboration between the University of Manchester, Richard Mille Watches and McLaren Applied Technologies), graphene-enhanced tires (Perpetuus Advanced Materials, UK), graphene-enhanced jackets by Vollebak. The commercialized products so far do not appear to match the progress in research, thus indicating a wide gap between laboratories to life journey of graphene.

Table 3 Commercial products using graphene

The product	Supplier	Remarks
Siren technology security smart packaging	Vorbeck materials, US	Uses graphene-based Ink
Graphene-based flexible touch-panels for mobile devices	Powerbooster Technology, Sanghai	Confirmation essential
Tennis racquets (YouTek Graphene Speed series)	HEAD (American Dutch company dealing with sports equipment), Netherlands	Cost: \$170–\$286
Skis for women		20% more expensive than traditional skis
Quarano bicycle wheels	Vittoria (Wheel producer company)	Better heat dissipation and increased lateral stiffness
Bike frame	Dassi Bikes, UK	Frame weight 750 g
Cycling helmets—Mixino 2014	Catlike, Spain	Light and strong
Cycling shoes-whisper		Light and durable
Fishing rods	Century composites, US	\$469 starting price at Amazon
T-FORCE Gaming line of products: CARDEA ZERO M.2 PCI-E solid-state drive	Team Group	Improved cooling
Graphene oxide (GO) sensor	Biolin Scientific	Enables interaction studies of GO with various analytes
Sensors for real-time detection of molecules	Nano medical Diagnostics, San Diego	Accuracy, portability, cost saving
Earphone diaphragm driver	FIO, China	\$ 24.99 at Amazon
Earphones	Zolo from Ankers	\$99
Graphene-enhanced battery products	Huawei, Vorbeck power strap, Angstron Materials, batteries for electrically driven vehicles (Fisker, Tesla, Graphene Nanochem & Sync, Perpetus Carbon Group & Oxis Energy, gRAFOID, SiNode together with AZ electronic materials, XG Sciences	
Solar cells	ZNShine Solar's G12 evolution era series, yet to come	
Supercapacitors	Skeleton technology, the CRRC, ZapGoCharger, Angstron materials and Sunvault energy	
Graphene ink	Haydale, Vorbeck materials	

2.4 Graphene Journey so Far

The report [4] by the ID Tech-Ex identifies the graphene commercialization period to be consisting of seven phases, the current year belonging to the fifth phase. The first phase as per the report goes back where pioneering firms were established for small-scale production with grants received from various sources. The second phase saw mushroom growth in small volume high-priced sample ideas displaying numerous application fields and caused a significant confusion at the end user. Third phase was a serious and filtering phase, when several companies left the scene and others began to focus on quality, safety, realistic expectations, and consistency. The capacity investment continued whereas the demands were almost inexistent. During the fourth phase, China appeared as a major hub and credible qualification processes advanced. The current phase sees capacity additions, cost reductions and the qualification processes in place. In addition, the revenue growth is also showing acceleration. In future phases 6 and 7, the report forecasts demand acceleration, bulk production, along with the focus on application and profits ahead.

3 Constraints in the Journey from Lab to Life

Several barriers are cited [5] during market entry. The technology challenge is one of the most influencing parameters in market entry of a new product. In case of grapheme, many research articles highlight the technical challenges and the present paper will not reiterate these. In the present paper, the authors have selected a few of the market barriers, where exploration is rather inadequate. Those are: Constraints in production process, Competition from existing products, policies, and economics.

3.1 The Constraints in Graphene Production Process

In a review work, published during 2016 [6, 7], Bhuyan et al. have furnished a division of graphene synthesis methods. The top-down approach consists of mechanical exfoliation, chemical exfoliation, and chemical synthesis methods. The mechanical exfoliation techniques can further be done through adhesive tapes, diamond wedges, shearing fluid flow, etc. Chemical synthesis can be carried out through sonication or through reducing graphene oxide (GO). The bottom-up approaches can be conducted through epitaxial growth, pyrolysis, CVD methods. The graphene synthesis routes vide exfoliation are getting popular recently owing to their capability to produce quality sheets. Besides graphene, GO has also attracted researchers' attention from many forums. The graphites serve as the material input. There are chemical reactions during processing or purification, which depend on chemicals and may result in hazardous wastes. Quality of the output is also a common issue. In Fig. 4, mechanical exfoliation process and GO preparation process [8] have been furnished by flow charts, which would make the constraint identification more clear.

Considering both the processes as displayed in the flow chart, it is evident that the processes have certain commonalities. The graphites serve as the material input. There are chemical reactions during processing or purification, which depend on chemicals and may result in hazardous wastes. Quality of the output is also a common issue. The author will attempt to bring out the constraints [9] on these aspects.

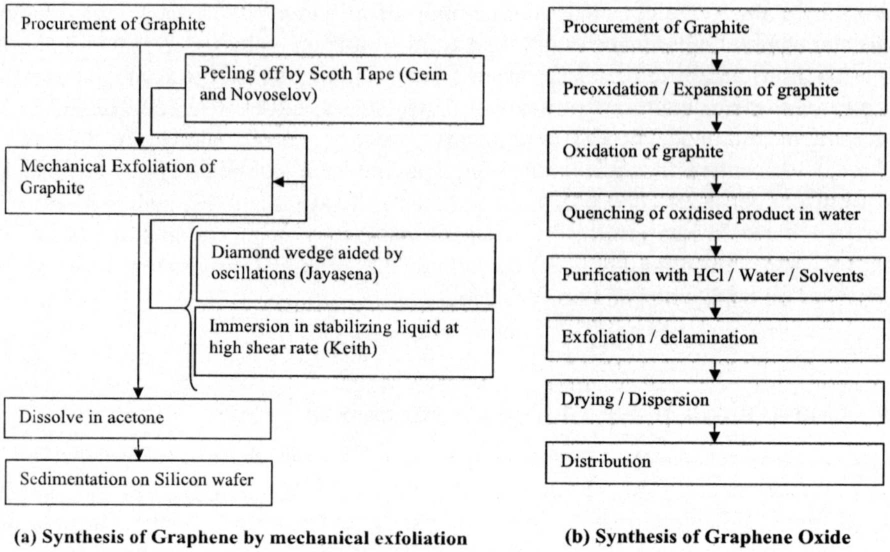

Fig. 4 Flow chart demonstrating production of graphene by mechanical exfoliation (**a**) and production of GO (**b**)

3.1.1 Graphite Sources

Graphites can be naturally mined or synthesized from unstructured carbon. Naturally occurring graphites are of different types. The vein graphite is the purest graphite ore and contains highest carbon content. However, these are produced only at two mines of Sri Lanka in limited quantity. The flake graphites can be procured from China, Canada, Brazil, Australia, and some other countries. China is the largest producer of flake graphite. The price of flake graphite per tonne was in range of 750$–1350$ during 2014 depending on the percentage of carbon, flake size, quantity being procured and delivery location. Price increase is expected posing another constraint to commercial growth. Expanded graphites are the pretreated natural graphites. These have higher specific volume, but because of intercalation and heating requirement, its cost is higher compared to the flake graphites. The synthetic graphites are obtained from unstructured carbon by treatment stages which include heating up to >2500 °C. The synthetic graphites are costly and less crystalline compared to the naturally occurring graphites. The quality of the product is significantly influenced by graphite feed, number of layers required, and defects in raw material.

3.1.2 Reactants

Most of the chemicals used for the purification process are readily available. However, there are certain issues relating to potassium permanganate as this is costly and hazardous. Potassium ferrate can be used as a substitute; however, it is expensive since it is currently not produced in industrial scale. Also, it has another issue of its thermal stability, thus making it unsuitable in case recycling is needed. Toxicity and explosiveness of some intermediate products like nitrous oxides and heptoxide also pose danger to the environment.

3.1.3 Product Quantity and Quality

The quality and quantity of the graphene and the related products are to be viable to meet the customers' demands. The Scotch tape method or graphite wedge methods produce the best quality of graphene; however, these processes are not scalable to be industrially viable. The shear flow methods appear scalable and the industries are yet to adapt. In case of GO, complete oxidation is rarely achieved. For better oxidation efficiency, one can add pre-oxidation set-up, change graphite particle size or increase the reaction times. All of the alternatives add to the cost.

3.1.4 Storage Issues

There are chances of contamination during storage. Also, there are specific packaging requirement, so that the nanosheets do not warp out by the time they reach the customers.

3.2 Competition from Existing Products

In Table 4, a list of products, where graphene is being used has been furnished.

In order to be used in a product, the new material has to displace the existing material and this is a constraint from technical, economical, and managerial issues. A clarification on this aspect has been illustrated here. Graphene has higher electron mobility, higher transparency, better flexibility, and higher mechanical strength compared to the incumbent materials, in the sectors such as touch screens, LEDs, displays, sensors, batteries, fuel cells, solar cells, etc. In case of capacitive touch screens in smartphone, the current used material is the insulator glass, coated with conductive indium tin oxide. However, this material is associated with the issues such as the low transparency of about 90%, not transmissible for UV and IR, slow electron mobility, scarcity of indium, delayed physical vapor deposition, and cracks formed at low load [5]. Graphene, as associated with extremely high transparency, higher electron mobility, fast CVD and higher mechanical strength, can be a better alternative. The other available alternatives for the same purpose besides grapheme, are the silver nanowires, conductive polymers (Clevious material), etc. For graphene to be the substitute of the TIO, and to be used in many others beneficial purposes, it is essential that graphene be available in larger quantity and good quality. Hence,

Table 4 Speciality of graphene

Component	Incumbent material	Issues with the incumbent	Graphene	Other alternatives	References
Capacitive touch screens on smartphones	Insulator glass coated with conductive indium tin oxide	Transparency of about 90%, UV and IR not transmissible, electron mobility, scarcity of indium, delay associated with physical vapor deposition, cracks at low load	Higher transparency, Higher electron mobility, CVD takes less time, higher mechanical strength, more flexible	Silver nanowires, conductive polymers (Clevious material)	[5]

attention and emphasis must be given in the production of good quality graphene in industrial scale for successful applications. It is also imperative that, this must happen before other substitutes take over.

3.3 *Fake Graphenes*

"There can be no quality without quality control." thus opined Peter Boggild [8]. In an article named "War against fake grapheme" published during October 2018, a technical issue of Nature, Boggild expressed the concern over suppliers those passing the graphite flakes as grapheme flakes. In light of the article of Kauling et al. in the Advanced Materials, Boggild restated the findings of Kauling et al. wherein a systematic study of grapheme from 60 suppliers was undertaken. The results were shocking since many of the largely priced grapheme products were found to be consisting of graphite powder mostly. Such unethical practices (known or unknown to producers) result in loss of time and money in part of researchers engaged in grapheme application and development. Besides, it may stand as a factor responsible toward stalling research progress in grapheme technology and its entry from laboratory to life.

4 Concluding Remarks

Graphene is viewed as a wonder material for its lucrative properties. However, there are numerous challenges to be addressed by the time, it reaches main stream of the society. Though there are multiple application areas, the sectors, where large-scale graphene deployment can be achieved are less explored. These sectors can be in construction and manufacturing. Extensive research is essential to understand applicability of graphene. Graphene production technology poses a major technical challenge to the entire globe. The production and the distribution processes are also facing numerous constraints. The researchers are expected to come out with possible solutions so that a material with promise of strength, transparency, conductivity, environmental friendliness can make its journey smoother.

References

1. https://www.scopus.com
2. https://patseer.com/2017/04/patent-landscape-report-on-graphene-by-patseer-pro/
3. https://www.graphene-info.com/graphene-products
4. https://www.graphene-info.com/files/graphene/IDTechEx-graphene-phases-2018.jpg
5. https://www.azonano.com/article.aspx?ArticleID=3942
6. Bhuyan MSA, Uddin MN, Islam MM, Bipasha FA, Hossain SS (2016) Synthesis of graphene. int. Nano Lett 6:65–83
7. Shams SS, Zhang R, Zhu J (2015) Graphene synthesis: a review. Mater Sci Poland 33(3):566–578
8. Boggild P (2018) War against fake grapheme. Nature 562(25):502–503
9. Lowe SE, Zhong YL (2016) Challenges of Industrial scale graphene oxide production. https://doi.org/10.1002/9781119069447.ch13

Simultaneous Optimization of Strength and Acid Resistance in Concrete Composed of Carbonated Fly Ash

Sanjukta Sahoo, Neha Priyadarshini, Navneet Kumar, Bhaabani Sankar Das, and Sabyasachi Dutta

Abstract Fly ash that has captured carbon dioxide is regarded as carbonated fly ash. Use of this as a cement replacement can help in concreting in economically and environmentally beneficial way. For the purpose, the strength and durability aspect need careful investigation. In the current work, three cementitious materials such as fly ash, carbonated fly ash and fine crusher dust have been used as supplementary cementitious materials in concrete for finding strength and durability properties. Three levels of replacement (20, 40, 60%) have been considered to design M30 grade of concrete at water/cement ratio of 0.35. The cube compressive strength of concrete at water-curing age of 7, 28 and 90 days and the acid (5% H_2SO_4) resistance of those 7, 28 and 90 days water-cured concrete at 30 and 90 days acid exposure has been determined and compared with that of control concrete. Multi-response optimization using Taguchi method and grey relational analysis has been used to find process parameters towards simultaneous optimization of strength and durability. The carbonated fly ash concrete with cement replacement of 40% and 90 days of water curing offers simultaneously optimized strength and durability.

Keywords Carbonated fly ash · Concrete · Acid resistance · Grey relational analysis

1 Introduction

Cement being the most important ingredients in concrete, not only plays the vital role in influencing its strength and durability properties but also affects the economical and environmental aspects [1, 2]. As cement manufacturing is highly energy intensive

S. Sahoo (✉)
KIIT Polytechnic, KIIT, Bhubaneswar, Odisha, India
e-mail: sanjuktasahoo.7@gmail.com

N. Priyadarshini
Department of Civil Engineering, VSSUT, Burla, Odisha, India

N. Kumar · B. S. Das · S. Dutta
School of Civil Engineering, KIIT D-U, Bhubaneswar, Odisha, India

B. B. Das et al. (eds.), *Recent Developments in Sustainable Infrastructure*, Lecture Notes in Civil Engineering 75, https://doi.org/10.1007/978-981-15-4577-1_60

and associated with heavy carbon footprint, the alternative of cement with reduced cost and CO_2 emission is always a most demanding searching materials [2, 3]. A number of waste materials with pozzolanic properties have been investigated for their strength and durability enhancing ability as secondary cementitious materials in concrete. Here, the waste management as well as the reduced cost and decreased CO_2 emission have been prioritized to choose a suitable cement substitute that will give equivalent or better strength and durability properties to concrete. Crusher dust is a waste material available in plenty at the rock site near crushing unit. It has been mainly used for replacing sand in mortar or concrete in many research works considering the economical and ecological aspects [4, 5]. Here, the fine crusher dust has been collected to replace cement partially to determine its impact in concrete properties in comparison to that of control concrete. Further, fly ash (FA) is another established supplementary cementitious material, been widely adopted for replacing cement fully or partially in concrete to give significant strength and durability properties [6, 7]. FA when allowed for carbonation reaction, it absorbs CO_2, producing carbonated fly ash (CFA), that is also a waste materials having cementitious properties [8]. Hence, CFA can be considered as a more valuable cement substitute with reduced carbon footprint to produce a safer and green concrete.

In this work, the three cementitious materials such as crusher dust, FA and CFA have been used and the performance of the concrete incorporating CFA (CFA concrete) with various cement replacement levels (20, 40 and 60%) has been investigated for its strength and acid resistance properties and compared with that of the control concrete having no cement replacement and the FA concrete, and crusher dust concrete with the equivalent level of cement replacement. Multi-response optimization using Taguchi method and grey relational analysis also have been furnished to determine the suitability of the materials for producing optimized strength and durability.

2 Experimental Programme

In the present study, the 43 grade (as per IS: 8114-1978) ordinary Portland cement (low-calcium FA (ASTM Class F) was collected from Kaniha, NTPC, Odisha, crusher dust powder was locally collected from Khandagiri site, Bhubaneswar, coarse aggregate (crushed granite, 20 mm), river sand as fine aggregate, polycarboxylate-based superplasticizer and general tap water. CFA, as prepared in the laboratory also has been used as one of the supplementary cementititous materials [8, 9]. Further, a concentrated sulphuric acid (H_2SO_4) of 98% purity and density of 1.84 g/cc is also used in the experimental programme for acid resistance test of concrete. All the properties of the materials used in preparing the concretes for the experimental programme are determined in the laboratory as per the relevant codes of practice. Total ten types of M30 grade concrete such as three CFA concrete (CFC 20, CFC 40 and CFC 60), three FA concrete (FAC 20, FAC 40 and FAC 60), three crusher dust concrete (CDC 20, CDC 40 and CDC 60) and one control concrete (CC) have been prepared for the investigation. A water/binder ratio of 0.35 (constant) with a slump value around a range of 100 mm has been considered for all the mixtures. Concrete cubes of size 150 mm were cast and moulded for both the compressive strength test and acid resistance test in accordance with IS: 516-1959. After 24 h of casting, the cubes were demoulded and kept in water-curing chambers up to the required days of testing. The required cubes of all types concrete for compressive strength test were taken out of the curing chamber at the age of 7, 28 and 90 days and tested in a compressive testing machine of 2000-KN capacity (CTM Digital) at the loading rate of 0.2–0.4 N/mm^2/s. The remaining cubes, those for acid resistance test were taken out of the water-curing chamber after 7, 28 and 90 days, respectively, and kept immersed in sulphuric acid of 5% concentration for the required chemical exposure periods. After 30 and 90 days of acid exposure, the concrete cubes are taken out of the chemical container and tested for their residual compressive strength in the same manner. The respective strength and strength loss of all types concrete were observed and noted down for the analysis.

3 Result and Discussion

See Figs. 1, 2, 3, 4 and 5 and Tables 1, 2 and 3.

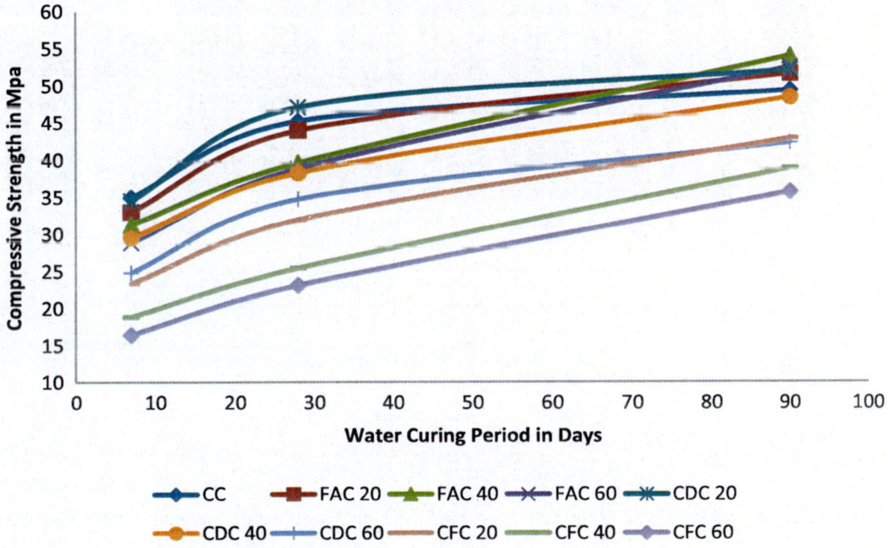

Fig. 1 Compressive strength of concrete specimens

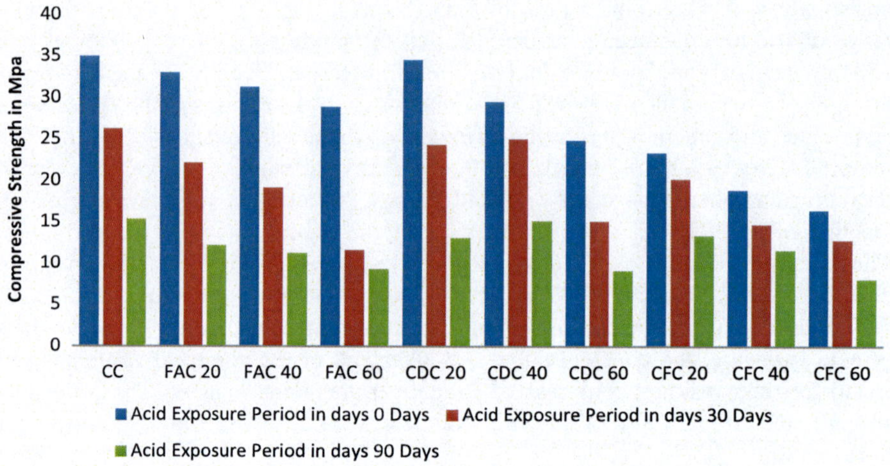

Fig. 2 Compressive strength of 7 days water-cured concrete specimens under acid exposure

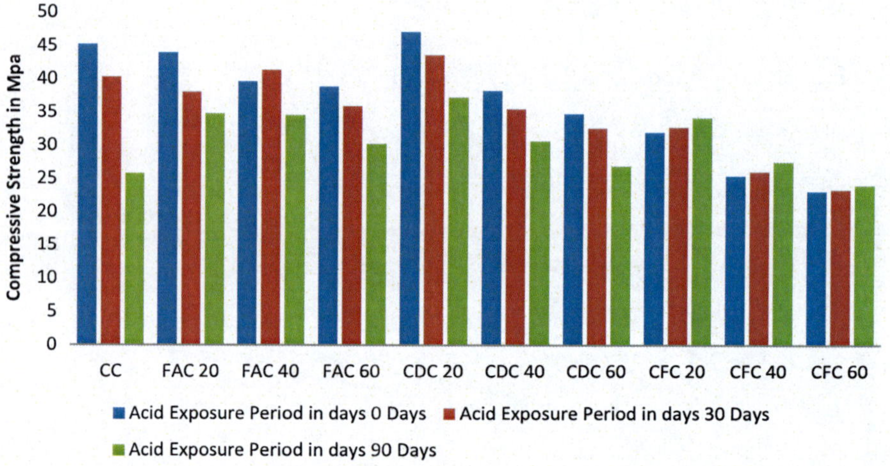

Fig. 3 Compressive strength of 28 days water-cured concrete specimens under acid exposure

4 Discussion of the Results

4.1 Compressive Strength

The compressive strength test result of all the concrete specimens at the water-curing age of 7, 28 and 90 days water curing are displayed in Table 1 and Fig. 1, respectively. The Tables 2, 3 and 4 and the Figs. 2, 3 and 4 display the compressive strength loss value of all the types of concrete at acid exposure period of 0, 30, 90 days after 7,

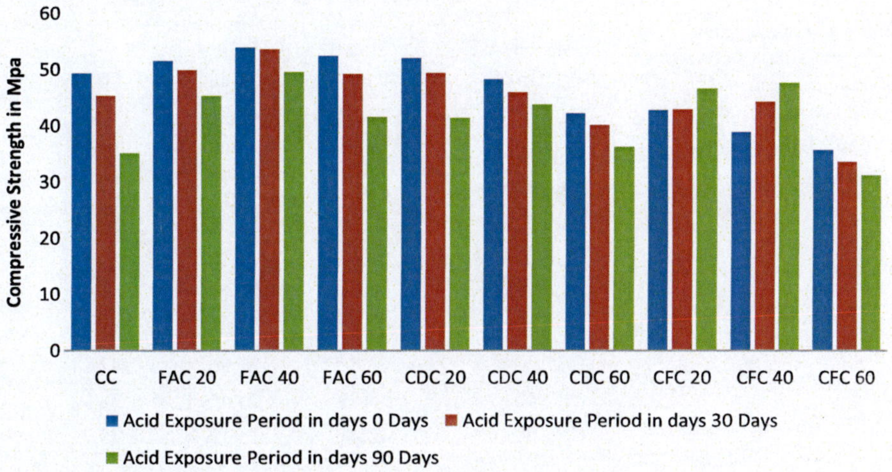

Fig. 4 Compressive strength of 90 days water-cured concrete specimens under acid exposure

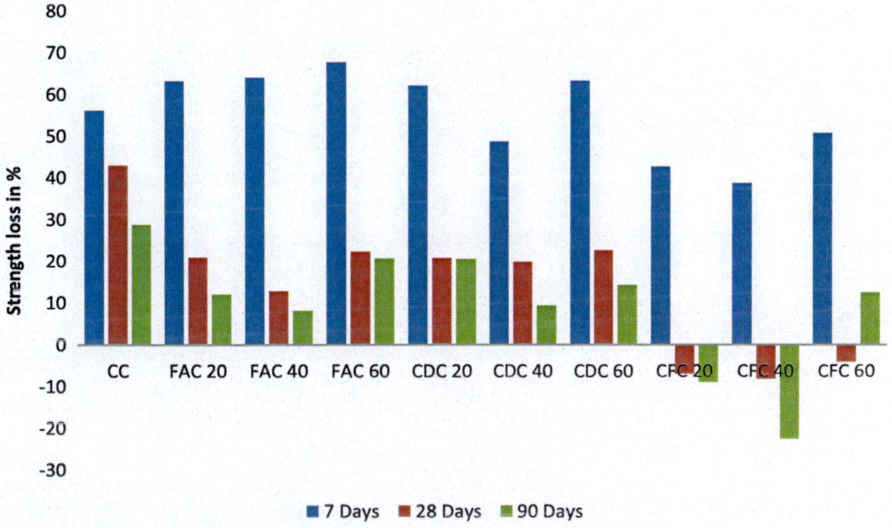

Fig. 5 Strength loss in % in concrete specimens cured for different period under acid exposure of 90 days

28, and 90 days of water-curing ages, respectively. Finally, the Fig. 5 represents the % strength loss rate of all types of concrete after 7, 28 and 90 days of water curing in acid exposure period of 90 days. Out of all the concrete specimens, the CC and CDC20 gained the maximum compressive strength of about 34 MPa within 7 days of water curing followed by FAC20 and FAC40 having a strength gain of about 30 MPa. However, all other concrete, though gained less strength but have overcome

Table 1 Compressive strength (MPa) of concrete specimens at various water-curing periods

Concrete type	Water-Curing period		
	7 days	28 days	90 days
CC	34.99	45.24	49.36
FAC 20	33.02	43.97	51.62
FAC 40	31.29	39.65	54.01
FAC 60	28.89	38.88	52.49
CDC 20	34.57	47.11	52.11
CDC 40	29.51	38.25	48.38
CDC 60	24.82	34.76	42.33
CFC 20	23.33	31.96	42.88
CFC 40	18.83	25.45	38.92
CFC 60	16.41	23.12	35.71

Table 2 Compressive strength (MPa) of concrete specimens after acid exposure

Type of concrete	WCP 7 days		WCP 28 days		WCP 90 days	
	Acid exposure		Acid exposure		Acid exposure	
	30 days	90 days	30 days	90 days	30 days	90 days
CC	26.24	15.33	40.39	25.82	45.39	35.17
FAC 20	22.12	12.15	38.14	34.78	49.97	45.39
FAC 40	19.12	11.25	41.43	34.55	53.65	49.61
FAC 60	11.65	9.31	35.94	30.24	49.24	41.66
CDC 20	24.31	13.09	43.68	37.32	49.45	41.45
CDC 40	25.02	15.14	35.52	30.68	46.03	43.85
CDC 60	15.11	9.11	32.64	26.91	40.15	36.27
CFC 20	20.15	13.37	32.80	34.17	43.03	46.72
CFC 40	14.71	11.54	26.08	27.54	44.37	47.71
CFC 60	12.82	8.08	23.39	24.06	33.62	31.24

the target strength of 7 days water curing. Within the water-curing period of 28 days, CDC20 gained slightly higher strength followed by CC and FAC20. At this age, the strength-gaining rate for CC, CDC20, CDC40 and FAC20 was faster than other types of concrete. However, after 28 days, the strength gaining of CC, CDC20 and CDC60 followed a declined rate. It is worth to be note down that with increased replacement levels; the strength gaining was decreasing for both the crusher dust and FA concrete within 28 days of water curing. However, the FA concrete gained increased strength after 28 days attaining a highest value of 54 MPa at the 90 days water curing. The control concrete performed better strength in the initial period only and after 28 days its strength-gaining rate was declining. The 20% replacement of crusher dust and the 40% replacement of FA displayed the optimum strength gaining in all water-curing

Table 3 Strength loss (%) in concrete specimens after acid exposure of 90 days

Type of concrete	Strength loss %		
	Water-Curing period		
	7 days	28 days	90 days
CC	56.19	42.93	28.75
FAC 20	63.2	20.9	12.07
FAC 40	64.05	12.86	8.15
FAC 60	67.77	22.22	20.63
CDC 20	62.13	20.78	20.46
CDC 40	48.7	19.79	9.36
CDC 60	63.3	22.58	14.32
CFC 20	42.69	−6.91	−8.96
CFC 40	38.71	−8.21	−22.58
CFC 60	50.76	−4.07	12.52

Table 4 Parameter selection for L9 orthogonal array

Parameters	Notation	Level of factors		
		Level 1	Level 2	Level 3
Replacement material	m	FA	Crusher dust	CFA
Curing days	D	7	28	90
Material ratio	r	20	40	60

period. The CFA concrete, though having less compressive strength than other types of concrete, displayed the fast rate strength gaining after 28 days. The increased strength in case of FA-based concretes can be due to the secondary hydration reaction that resulting thickening of CSH gel and thus helps in achieving a higher strength value after 28 days [10, 11].

4.2 *Compressive Strength After Acid Exposure*

For the 7 day's water-cured samples, all the types of concrete suffered a severe strength loss; however, the rate of strength loss for FA concrete and CFA concrete was found very less compared to other types. FAC60 and CFC 40 displayed better acid resistance among all types concrete between 30 days and 90 days of acid exposure. Among all the 28 days water-cured samples, the control concrete suffered the highest strength loss compared to all other concretes between 30 days and 90 days of acid exposure. The FAC60, CDC40 and the CDC60 displayed better acid resistance within 30 days of exposure, whereas the FAC40 and the CFA concrete attained strength gain instead of strength loss in acid exposure. The strength gain may be attributed to the reduced free lime content of the FA and CFA concretes because of secondary hydration reaction. Thus, the reactivity of the concrete have been reduced and hence less affected by the acid ions [12].

For the 90 days water-cured sample, FAC40 has shown the highest strength loss resistance against acid followed by FAC20 and all the crusher dust concrete within 30 days of acid exposure. The CFA concrete up to 40% replacement level displayed the strength gain in acid exposure. The strength loss rate was the highest for the control concrete and the lowest for CDC40 within acid exposure period of 30–90 days. The strength gain also can be attributed to the pore packing and fine lines refinement of the CFA concrete due to deposition of the $CaCO_3$ in the pores that was present in the CFA [13, 14].

5 Multi-response Optimization Using Taguchi Method and Grey Relational Analysis

The responses in the present work are compressive strength of concrete and durability measured in terms of strength loss percentage under acid exposure. The objective of performing multi-response optimization is to determine a proper combination of replacement material, water-curing days and replacement level to maximize the compressive strength and durability simultaneously. The design of experiment prescribed by Taguchi uses a special design of orthogonal arrays to study the entire process parameters' space with less number of experiments. This method seems a practical approach to strategize the parameter selection without going through highly time-consuming and cost-intensive whole scale experiments involving all levels of all parameters. However, the method proves useful with single response. Grey relational analysis coupled with Taguchi's design of experiment proves to be a useful tool in optimizing multiple responses simultaneously [15, 16].

In the experiments, three parameters at three levels are taken to study the concrete responses. The parameters are the replacement materials, water-curing period and the replacement material ratio with respect to the cement in concrete. For three parameters, three level variations L9 orthogonal array is preferred. The parameter selection for the array is presented in Table 4 and the parameters and response as per the L9 Design of experiments are furnished in Table 5.

Grey relational analysis computations can be done in any computing platform. In the present work, MS Excel has been used as the platform for the analysis. The algorithm used here is designed for minimizing both the responses. Since the case involves maximizing the compressive strength and minimizing the strength loss under acid exposure during computation the first responses, those represent compressive strength has been assigned negative signs. Algorithm for the grey relational analysis is presented in schematic Fig. 6.

The notations and their explanation are furnished below.

$x_i \ldots x_n$ are the response variables from experimenst 1 to 9.

Table 5 Parameters and responses

Run	Parameters			Responses	
	m	D	r	Response 1 (compressive strength)	Response 2 (strength loss in %)
	1	1	1	−33.02	63.2
2	1	2	2	−39.65	12.86
3	1	3	3	−52.49	20.63
4	2	1	2	−29.51	48.7
5	2	2	3	−34.76	22.58
6	2	3	1	−52.11	20.46
7	3	1	3	−16.41	50.76
8	3	2	1	−31.96	−6.91
9	3	3	2	−38.92	−22.58

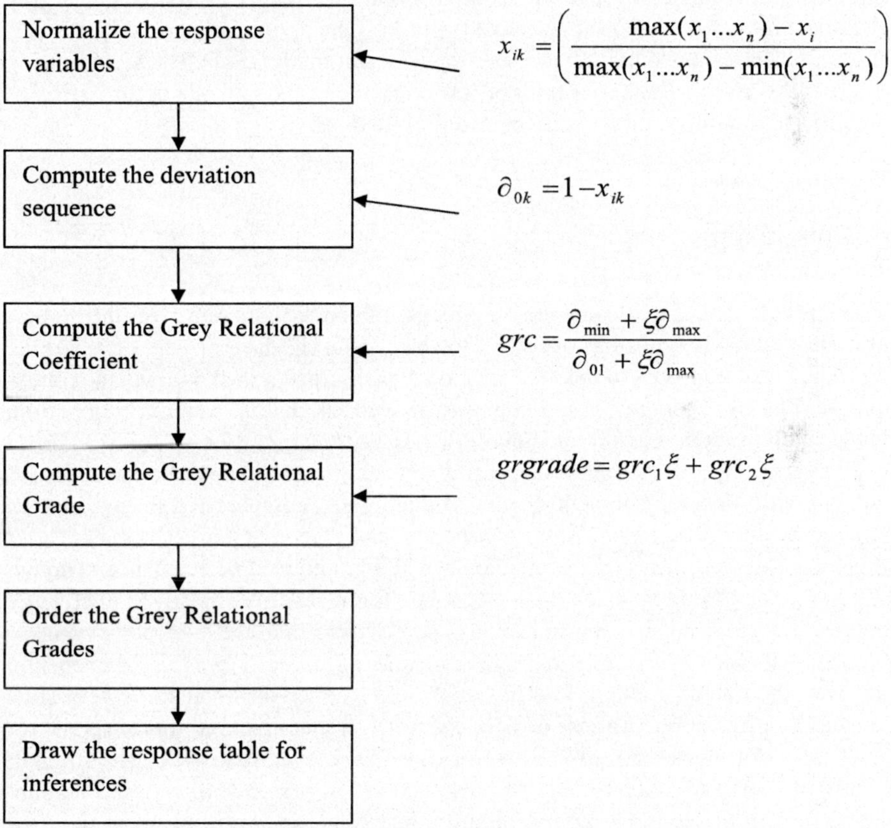

Fig. 6 Algorithm for the grey relational analysis

x_{ik} is the normalized value for the kth response.
∂_{0k} is the deviation sequence for the kth normalized response.
grc is the grey relational coefficient.
ξ is the weightage assigned to the responses.
$grgrade$ is the grey relational grade.

The data pertaining to Table 6 offers the grey relational grades corresponding to the experimental runs. The Rank1 is assigned to the run yielding highest grey relational grade. From the data furnished, it can be inferred that, out of selected nine experiments, the run 9 offers the simultaneous optimization of the strength and durability. The analysis requires the determination of levels for each of the process parameters. Response Table for grey relational grade offers the solution (Table 7).

The mean of grey relational grades for each parameter that is the replacement material, water-curing days and material ratio are furnished in the table. Larger grey relational grade corresponds to the levels for optimization. Row wise highest grey relational grades have been superscripted with '*'. The optimal combination is <m3, D3 an r2>. The above data indicate that replacement material of CFA, water-curing period of 90 days and material ratio of 40% can offer simultaneous optimization of strength and durability under acid exposure of 90 days.

6 Conclusions

The 7 days of water-cured compressive strength of control concrete gives the highest value than all other types of concrete. However, after 28 days of water curing, the 20% replacement of FA and the 40% replacement of crusher dust attained maximum strength. The CFA concrete could not perform well within that water-curing period. However, all the concrete have achieved the minimum target strength for M30 grade concrete at the age of 28 days water curing. At water-curing age of 90 days, the FA concrete with 40% replacement displayed the highest compressive strength value. With increased replacement level, FA concrete have shown better performance after 28 days, whereas crusher dust concrete have shown reduced performance. Though the compressive strength value of CFA concrete found less in comparison to all other concrete, but its strength-gaining rate was much faster than those of other types of concrete. All types of concrete undergone increasing strength loss in acid exposure. However, the rate of strength loss in case of CFA concrete found to be very slow followed by FA concrete. In case of CFA concrete, strength gaining was observed for the 90 days of water-cured samples with replacement level up to 40%. The strength gaining of the CFA concrete may be attributed to the presence of reduced amount of free lime in the concrete after hydration due to the prior utilization of the lime in carbonation process that leads to less reactivity of the concrete materials with the acid ions. Another point may be attributed to the deposition of $CaCO_3$ in case of CFA in the pore space that leads dense packing of the concrete materials thus restricting the acid ion ingress. From the gray relation analysis, it has been found

Table 6 Table showing gray relational analysis

Run	Responses		GRG		Del 0i		grc		grgrade	Order
	RL-I	RL-II								
1	−33.02	63.2	0.460	0.000	0.540	1.000	0.481	0.333	0.407	8
2	−39.65	12.86	0.644	0.718	0.356	0.282	0.584	0.639	0.612	5
3	−52.49	20.63	1.000	0.607	0.000	0.393	1.000	0.560	0.780	2
4	−29.51	48.7	0.363	0.207	0.637	0.793	0.440	0.387	0.413	7
5	−34.76	22.58	0.509	0.579	0.491	0.421	0.504	0.543	0.524	6
6	−52.11	20.46	0.989	0.610	0.011	0.390	0.979	0.562	0.770	3
7	−16.41	50.76	0.000	0.177	1.000	0.823	0.333	0.378	0.356	9
8	−31.96	−6.91	0.431	1.000	0.569	0.000	0.468	1.000	0.734	4
9	−38.92	−22.58	0.624	1.224	0.376	−0.224	0.571	1.808	1.190	1

Table 7 Response table for the grey relational grade

Symbol	Parameters	Grey relational grade			Main effect	Rank
		Level 1	Level 2	Level 3		
m	Replacement material	0.643	0.569	0.760*	0.191	2
D	Curing days	0.392	0.623	0.913*	0.521	1
r	Material ratio	0.637	0.738*	0.553	0.185	3

that the replacement material of CFA, water-curing period of 90 days and material ratio of 40% can offer simultaneous optimization of strength and durability under acid exposure of 90 days.

References

1. Portland Cement Association. Cement and concrete basics.PCA on-line. PCA, Skokie, Illinois, USA. http://www.cement.org/basics/concretebasics_lessononec.asp
2. Kajaste R, Hurme M (2016) Cement industry greenhouse gas emissions—management options and abatement cost. J Clean Prod 112:4041–4052
3. Mohammadi J, South W (2017) Life cycle assessment (LCA) of benchmark concrete products in Australia. Int J Life Cycle Assess 22:1588–1608
4. Kapgate SS, Satone SR (2013) Effect of quarry dust as partial replacement of sand in concrete. Ind Streams Res J 3(5):1–8
5. Hameed MS, Sekar ASS (2009) Properties of green concrete containing quarry rock dust and marble sludge powder as fine aggregate. ARPN J Eng Appl Sci 4(4):83–89
6. Meyer C (2009) The greening of the concrete industry. Cement Concr Compos 31(8):601–605
7. Wu FR, Masuda Y, Nakamura S, Sato S (2006) Influence of mixture proportion and curing condition on contribution of fly ash to strength of concrete. Key Eng Mater 302–303:235–241
8. Sahoo S, Das BB, Mustakim S (2017) Acid, Alkali and chloride resistance of concrete made up of low carbonated fly ash. J Mater Civil Eng ASCE 29(3)
9. Sahoo S (2018) Pore structure characterization in concrete prepared with carbonated fly ash: an experimental and simulation study. In 7th national conference on processing and characterization of materials, NCPCM-2017, IOP Conference Series: Materials Science Engineering, 338
10. Taylor HFW (1997) Cement chemistry, 2nd Edn. Thomas Telford
11. Michael T (2007) Optimizing the use of fly ash in concrete. Portland Cement Association, pp 548–624
12. Ashraf M, Mokal M, Bhattacharya J (2005) Use of fly ash in high performance concrete & self compacting concrete. Fly Ash India, New Delhi
13. Mehta PK (2000) Sulfate attack on concrete: separating myth from reality. Concrete International, Farmington Hills, Michigan, pp 57–61
14. Mehta PK, Palulo JMM Concrete-microstructure, properties and materials. Tata Mcgraw Hill
15. Lin CL (2004) Use of the taguchi method and grey relational analysis to optimize turning operations with multiple performance characteristics. Mater Manuf Process 19(2):209–220
16. Hasani H, Tabatabaei SA, Amiri G (2012) Grey relational analysis to determine the optimum process parameters for open-end spinning yarns. J Eng Fibers Fabr 7(2):81–86

Mechanical and Durability Properties of Fly Ash-Based Geopolymer Concrete

Arani Dutta and Narayan Chandra Moharana

Abstract Portland cement plays a very important part in construction, but the manufacturing process emits almost 5–7% of the CO_2 in the world and it is one of the main causes of global warming. This paper discusses about the alternative material to Portland cement and compares between their strength and durability. The main material in this study is fly ash which is an industrial waste and easily available. In India, every year almost 120 million tons of fly ash is produced in the power plants and fly ash is very rich in silicon and aluminium; that is why it is a very good replacement for cement and this way we can recycle the waste also. Geopolymer is a mixture of fly ash with sodium hydroxide (NaOH) and sodium silicate (Na_2SiO_3). In this paper, geopolymer is made using the combination of M25-grade concrete. The ratio between NaOH and Na_2SiO_3 is kept constant which is 1:2 and the morality of NaOH varied from 10 M to 16 M. In geopolymer concrete, cement is 100% replaced by fly ash. The compressive strength and durability parameter is compared between geopolymer concrete cubes (which is kept at different curing conditions) and concrete cubes in which Portland cement is replaced by 25% of fly ash.

Keywords Global warming · Fly ash · Portland cement · Sodium hydroxide · Sodium silicate

1 Introduction

Portland cement is the principal material used in construction but the manufacturing process is very harmful for environment as it releases 1 tonne of CO_2 per 1 tonne of cement [1]. It is one of the green house gases responsible for global warming and is a great threat to mankind. Nowadays, it is an important issue to develop environmentally friendly and sustainable construction material [2]. The only way to reduce the effect alternative material to Portland cement are blended cement and geopolymers [3]. Geopolymer proves to be the best which completely replaces the Portland cement by aluminosilicate materials like fly ash, blast furnace slag, metakoline, silica

A. Dutta · N. C. Moharana (✉)
Kalinga Institute of Industrial Technology (Deemed to Be University), Patia, Bhubaneswar, India
e-mail: nmoharanafce@kiit.ac.in

B. B. Das et al. (eds.), *Recent Developments in Sustainable Infrastructure*, Lecture Notes in Civil Engineering 75, https://doi.org/10.1007/978-981-15-4577-1_61

fume, etc., which are rich in alumina (Al) and silica (Si) with alkali activators like NaOH, Na2SiO3 KOH, K2SiO3, etc. [4]. On curing at higher temperature, polymerization takes place and products are calcium aluminates silicate hydrate(C–A–S–H) and sodium aluminate silicate hydrate(N–A–S–H) which bind the inert materials to form geopolymer concrete [4–7]. One of the drawback is the heat curing which is not suitable for sight conditions. Normal curing provides lower strength. On other hand, the alkali activator NaOH of higher molar value provides higher strength.

1.1 Effect of Curing on Strength

It was observed that high-temperature cured GPC with rest period about 3 days after casting gains higher strength than GPC cured immediately after casting with low temperature and OPC concrete at 28 days. GPC samples displayed little gain of strength after steam curing was over up to 28 days [8]. As curing temperature increases the strength of the GP mortar increases and optimum temperature is found to be 75 °C. Also the duration of heat curing enhances the strength and optimum heat curing duration is predicted as 2 days. Microstructure of the mortar get weakened at elevated curing temperature and therefore strength get reduced [9]. Compressive strength increases up to 100 °C curing temperature and at 120 °C started decreasing. This loss of strength attributed to the formation of crack due to the loss of moisture at high temperature. Compressive strength of the OPC found to be more than GPC. This is due to heterogeneous microstructure which is due to the disruption of packing of binder at the presence of air void [10]. In comparison to KOH, NaOH activator provides better compressive strength because higher amount of silicate and aluminates monomers are found in case of NaOH. It is attributed that the sodium cations of NaOH provide better geopolymerization due to smaller in size as compared to potassium cation and can easily migrate through the network of moist gel [10] reported that the naphalene-based superplasticizer provides better strength than melamine formaldehyde and polycarboxilate ether-based superplasticizer. Noushini and Castel [11] reported that compressive strength of geoplymer concrete cured at 75 °C for 18 h is found to be about 20 and 15% more than that of heat-cured and normal curing OPC concrete, respectively. Curing at 75 °C for 18 h may be considered as the optimum temperature and duration of curing. Normal curing of GPC at an ambient temperature produces much lower strength and are not suitable for practical purpose. Gunasekara et al. [12] reported that the Geopolymerization continues beyond 90 days up to 365 days and the compressive strength is at per with cement concrete. Compressive strength of GPC and OPC concrete increases with curing period and strength of GPC is higher than that of OPC concrete. Also compressive strength for both the cases are higher for high-temperature curing because the increase in temperature accelerates the geopolymerization in GPC and hydration in OPC concrete [×6]. AdbElaty et al. [2] reported that at higher alkali solution/binder (AS/B) ratio, there is noticeable amount of compressive strength, split tensile strength and modulus of elasticity enhancement and best result is found for 30–40% of NaOH.

Also found (AS/B) plays a rule as that of W/C ratio on OPC concrete. Increase in liquid to binder ratio (L/B) in alkali solution reduces the mechanical properties and vice versa [2]. Chemical activator of multi-compound Na_2SiO_3/NaOH) are found to be most effective for the enhancement of compressive strength [7]. Compressive strength of geopolymer mix is effected by chemical reactivity between the amorphous AlO_3, CaO and Na_2O present in fly ash and alkali activators and this chemical reactivity depends on the median particle size and contents of amorphous SiO_2. Finer is the fly ash particles better is the blending in geopolymer mix due to smaller friction between its components and also increase in exposure area of amorphous SiO_2 to alkali activators. Thus, as a result, there is better geopolymerization [13]. Hadi et al. [13] also reported that fly ash with finer particles has greater coverage of aggregates and forms a dense intertransition zone on aggregate surface which results in higher binding strength. Thus, the optimum value of alkaline to fly ash ratio (AL/F) to achieve optimum vale of compressive strength are different for different fly ash sources depending on particle size and contents of amorphous SiO_2, CaO and Na_2O. Geopolymerization is controlled by chemical equilibrium achieved between amorphous Al_2O_3 and SiO2. Thus, the optimum value of Na_2SiO/NaOH ratio is dominated by characteristic of fly ash. High amount of Si^{4+} Na^+ is liberated from Na_2SiO_3 may congested and inhibits the geopolymerization [13]. Fly ash having higher contents of SiO_2 necessitates higher amount of NaOH to release SiO_2 and other oxides from fly ash to initiate geopolymerization. Larger sized particles in fly ash reduces the surface area exposure of amorphous SiO_2 and Al_2O_3 to alkali activator. On the other hand, unreacted particles exhibits week in geopolymerization and results in strength degradation [13]. It is observed that increase in temperature in hydrothermal curing, enhances the compressive strength. It is attributed to the acceleration of geopolymerization reaction due to the increase in liberation of more reactive alkali at higher temperature [14]. On normal curing at 95% humidity, compressive strength increases with time due to the increase in reactive alkali which accelerates the geopolymerization [14].

1.2 *Workability*

Particle size and shape have dominant influence on the workability of GP mortar. Finer is the fly ash higher is the flow value. On the other hand, if fly ash particles of are spherical and smooth higher is the flow value [9]. Workability of geopolymer concrete increases due to lubricating effect of Na_2SiO_3 on spherical fly ash particles [12]. But high (Na_2SiO_3/NaOH) ratio reduces the workability due to high viscosity of Na_2SiO_3. On the other hand, high-liquid alkaline/fly ash ratio increases the workability. High concentration of NaOH increases the setting time [7]. N-based superplasticizer is most effective in geopolymer matrix activated by NaOH only but PC-based superplasticizer is most effective if the activator is the mixture of Na_2SiO_3 and NaOH. Flow value was tested with slump test and found that the OPC concrete exhibits more slump value than GPC and it is attributed to the viscosity of Na_2SiO_3

GPC increases the cohesiveness. [15]. Polycarboxylate (PC)-based sperplasticizer is found to be more effective to class C fly ash than class F fly ash. It is because negatively charged particles of PC-based superplasticizer get strongly bonded with Ca^{+} ion liberated from Class C fly ash providing negative ion which repulse each other and increases the dispersion capacity. Thus, flow value increases [16]. On the contrary, napthalene-based superplasticizer is more effective for class F fly ash. It is attributed to better chemical stability of the naphalene-based superplasticizer in alkaline environment science the pH value of this superplasticizer ranges from 6 to 9 [16].

1.3 Porosity and Microstructure

Porosity of fly ash-based GPC is found to be more than that of OPC counterpart. It is because Na_2–AlO_3–SiO_2–H_2O gel is less denser than C–S–H gel in OPC concrete. Both heat as well as ambient cured GPC exhibit lower water absorption value than that of OPC concrete. But inappropriate curing condition increases the voids and thus water absorption increases. [11]. In case of fly ash geopolymer concrete, uneven material distribution due to more quantity of coarse particles leads high microporosity [12]. X-Ray tomography test reveals that the voids in Portland pozzolana cement concrete (PPCC) are less than that of GPC but size of the voids in GPC are smaller [15]. Fly ash-based GPC is more porous because the principal reaction product the sodium alumino silicate gel (N–A–S–H) is three-dimensional network product attributes higher porosity. On he other hand, presence of un-reacted fly ash particles also causes high porosity [15, 17]. Sorpitivity coefficient decreases for fly ash-based GPC cured at 75 °C and curing duration 24 h. It is attributed to the formation of denser geopolymer net work which reduces the void and increases the tortuosity. However, increase in heat curing temperature and duration increases the serpitivity coefficient. It is attributed to the elevated temperature which extends the capillary pore net work and thus effects the tortuosity characteristic [11]. GPC specimen displays the high sorptivity value due to large pore structure [17].

1.4 Durability

Geopolymerization continues and corresponding increase in alumino silicate gel fills the cracks and voids. Thus, the durability properties improve with time [12]. Geopolymer concrete exhibits better resistance to aggressive environment laden with acid and sulphate [7]. In some environments, carbonation depth is found to be more in GPC than that in OPC counterpart. In case of fly ash-based GPC, $NaCO_3$ is formed as a primary carbonation product and it would dissolve in water when exposed to weather. Thus, leaching out of carbonation product causes GPC more porous and allow more

CO_2 diffusion through the concrete surface [17]. Chloride penetration and free chloride contents in GPC are found to be more compared to OPC counterpart because of high porosity and thus the GPC concrete are more prone to reinforcement corrosion. In case of OPC concrete, the presence of C_3AF and C–S–H gel binds the chloride contents to form Fidel's salt and thus the rate of chloride penetration diminishes. Also amount of free chloride is found to be less than GPC. High porosity of GPC results in high chloride penetration [17]. Objective of present study to compare the flow value and compressive strength of fly ash-based geopolymer concrete cured at normal temperature and at high temperature in oven, with normal OPC concrete and fly ash concrete with 25% of fly ash replacement. Also comparison of durability parameters such as acid resistance, chloride resistance and sorptivity. Therefore, in this research work, class F fly ash is used as aminosilicate material and mixture of NaOH and Na_2SiO_3 is used as alkaline solution. The ratio between NaOH and Na_2SiO_3 is kept fixed which is 1:2 throughout the experiment but four types of molarities of NaOH is taken 10, 12, 14 and 16 M molar, respectively, to study the effect of molarity on strength and flow value. The change in molarity shows a difference in adhesiveness, flow, strength prominently. Durability and mechanical properties of geopolymer concrete specimen cured in two different curing conditions are compared with that of the OPC concrete and fly ash concrete counterparts. Different curing conditions show a significant difference in strength in geopolymer concrete.

2 Experimental Programme

To fulfil the research objective, experimental programme under taken are discussed in this part. Details about the material properties, mixing, preparation, curing and testing for strength and durability requirements are discussed.

3 Materials

Fly ash used in this study is class F fly ash obtained from NALCO, Angul power plant, Orissa. The physical properties are given as: Specific gravity = 2.9, Water absorption = 15%, Colour—Tan to dark grey. Chemical compositions are given in Table 1 and Fig. 1 for fly ash.

NaOH or caustic soda of make Adity Birla were used in the experiment is procured from Shivam Chemicals, Bhubaneswar. The properties specified by the manufacture

Table 1 Chemical constituents of fly ash

Major constituents	SiO_2	Al_2O_3	Fe_2O_3	TiO_2	CaO	MgO	K_2O	Na_2O
%	59.3	25.86	5.81	1.71	1.07	0.68	1.89	0.07

Fig. 1 Fly ash

Fig. 2 NaOH flakes

are given as chemical formula = NaOH, Melting point = 318 °C, Appearance—White, waxy, opaque crystals, Odour—odourless and Density = 2.13 g/cm^3. Figure 2 shows the flakes of NaOH.

Industrial-grade Na_2SiO_3 was procured from Surendra chemicals, Kolkata. The properties provided by the manufacture are as follows: Chemical formula = Na_2SiO_3, Appearance—White to greenish opaque crystals or liquid, Density = 2.61 g/cm^3 and Melting point = 1,088 °C. Figure 3 shows the liquid Na_2SiO_3.

OPC 53 grade cement from RAMCO Cement Limited was used and physical properties obtained from laboratory test are given as follows: Consistency = 28%, Initial and final setting times = 100 min and 230 min, respectively, Compressive strength after 3, 7 and 28 days are 35.5 Mpa, 47 Mpa and 55.5 Mpa, respectively,

Fig. 3 Liquid Na_2SiO_3

Specific gravity = 3.15. All physical properties satisfy the requirements of IS 12269-1987 [18].

Crushed granite sized 10 mm and 20 mm are used as coarse aggregate. The physical properties are given as Specific Gravity = 2.7, Fineness Modulus = 6.2, Water Absorption = 0.4%. All above physical properties and gradations confirms to the specifications of IS: 383-1970 [19].

Fine aggregates obtained from River Mahanadi conforms to zone III. The physical properties are given as Specific Gravity = 2.65, Fineness Modulus = 2.47, Water Absorption = 0.85% and are conforming to the specification of IS: 383-1970 [19].

Clean potable tap water obtained from the Laboratory of Civil Engineering Department of KIIT, Deemed-to-be-University was used for mixing and curing of concrete.

3.1 Preparation of Concrete Specimens and Details

Three types of cube specimens of size 150 mm × 150 mm × 150 mm [20] were used in the study. Those are (1) normal OPC concrete specimen, (2) fly ash concrete(cement replaced by 25% fly ash and (3) geopolymer concrete specimens (GPC) and detailed in Table 2.

Table 2 Specimen details

Specimen name	Details	Curing conditions
MOP	OPC concrete	Normal water curing
MOF	25% fly ash replacement+OPC	Normal water curing
G10MC	Geopolymer 10 molar	Normal water curing
G10MA	Geopolymer 10 molar	Oven curing
G12MC	Geopolymer 12 molar	Normal water curing
G12MA	Geopolymer 12 molar	Oven curing
G14MC	Geopolymer 14 molar	Normal water curing
G14MA	Geopolymer 14 molar	Oven curing
G16MC	Geopolymer 16 molar	Normal water curing
G16MA	Geopolymer 16 molar	Oven curing

Table 3 Mix proportion for M25 normal OPC concrete cube

Specimen	Cement kg/m^3	Fly ash kg/m^3	Fine aggregate kg/m^3	Coarse aggregate kg/m^3	Water kg/m^3	W/B
MOP	492.5	–	617.23	1105	197	0.4
MOF	369.3	123.2	617.2	1105	197	0.4

Table 4 Ingredients of geopolymer concrete

Specimen	Fly ash kg/m^3	NaOH kg/m^3	Na_2SiO_3 kg/m^3	NaOH:Na_2SiO_3	Fine aggregate kg/m^3	Coarse aggregate kg/m^3	Alkali/binder ratio	W/B
G10M & G10MA	492.5	74	148	1:2	702	1256	0.45	0.4
G12M & G12MA	492.5	74	148	1:2	702	1256	0.45	0.4
G14M & G14MA	492.5	74	148	1:2	702	1256	0.45	0.4
G16M & G16MA	492.5	74	148	1:2	702	1256	0.45	0.4

3.2 *Mix Proportions*

Mix proportion by weight = 1:1.25:2.24 was selected as found from mix design for M 25 using OPC 53 grade cement as per IS:10262-2008. Ingredients for normal concrete and fly ash concrete are given in Table 3, numbers of cube samples were casted 12 for each categories of concrete.

For geopolymer concrete, same the mix proportion as above was used (1:1.25:2.24) by replacing cement by fly ash by 100%. Binder/alkaline ratio was taken 0.45. The ratio of NaOH to Na_2SiO_3 was taken 1:2 which remains constant for all samples. Constituents of different samples are tabulated in Table 4. Numbers of cube samples were casted 27 for each categories of geopolymer concrete.

3.3 *Mixing Procedure*

For normal concrete and fly ash concrete, normal mixing procedure was followed. Cube specimens were casted by placing fresh concrete in three layers and tamping with 25 blows with tamping bar and compaction with vibrating table for 30 s and finally finished with trawling (Figs. 4, 5 and 6).

For geopolymer concrete, samples of alkali solution to total alkali ratio as 1:2 were used. The blending of Na_2SiO_3 and NaOH exerts some amount of heat. To take it down to the surrounding temperature, the arrangement was kept for 24 h before they were blended with fly ash and aggregates. In concrete mixture machine, the

Fig. 4 Preparation of NaOH solution

Fig. 5 Geopolymer concrete cube

Fig. 6 Cubes compaction by vibrating table

Table 5 Workability in term of slump value

Sample	MOP	MOF	G10	G12	G14	G16
Slump in mm	120	125	90	85	70	50

aggregates and fly ash were mixed dry for around 5 min. Then mixing was done with alkali blend and requisite amount of water for 5 min. Cube specimens were casted by placing fresh concrete in three layers and tamping with 25 blows with tamping bar and compaction with vibrating table for 30 s and finally finished with trawling.

3.4 Casting of Cube Specimens

The steel moulds were covered with oil on their inside surface and concrete mix was poured into the moulds in three layers. Each layer was consistently compacted by a tamping bar with 25 numbers of blows and by vibrating table for 1 min. Finally, the top surface was finished utilizing a trowel.

3.5 Curing Process

The ordinary concrete cubes and the fly ash concrete cubes were kept in curing tank up to 7, 28 and 56 days after de-moulding. Half of the geopolymer concrete cubes were placed in the curing tank after de-moulding and another half were kept in the oven for 24 h at 75 °C and after that those were kept in ambient temperature.

4 Testing

4.1 Workability of Fresh Concrete

See Table 5.

4.2 Testing of Hardened Concrete for Compressive Strength

Testing of CTM machine of 2000 KN from AIMIL LTD is available in the Department of Civil Engineering Laboratory of KIIT, Deemed-to-be-University, is shown in Fig. 7. For compressive strength, the specimens were tested for 7, 28 and 56 days

Fig. 7 Cube testing in CTM

and following the test procedure in accordance to IS:516-1959. [21]. Test results in average are summarized in Table 6.

Table 6 Compressive strength of different specimen

Specimen	7 days (N/mm^2)	28 days (N/mm^2)	56 days (N/mm^2)
MOP-Normal OPC	38.157	47.87	53.88
MOF-OPC+25% fly ash replacement	33.87	55.1	59.91
G10MC–G PC10 molar (curing tank)	18.23	20.54	21.52
G10MA–GPC 10 molar(oven curing)	25.29	29.31	35
G12MC–GPC 12 molar(curing tank)	22.2	23.14	24.75
G12MA–GPC 12 molar(oven curing)	45.85	51.2	54.59
G14MC–GPC 14 molar(curing tank)	24.85	26.22	27.98
G14MA–GPC 14 molar(oven curing)	48.4	55.72	56.42
G16MC–GPC 16 molar (curing tank)	30.2	31.29	33.55
G16MA–GPC16 molar(oven curing)	55.62	60.25	62.22

Table 7 Acid attack results

Specimen	Wt. before test	Wt. after test	Wt. Loss in %	Compressive strength after acid attack in N/mm^2	Compressive strength of reference sample N/mm^2	Strength loss (%)
MOF	8.52	8.035	5.69	33.56	64.91	48.0
G10MA	8.358	8.223	1.615	30.617	35.00	12.0
G12MA	8.406	8.376	0.356	48.143	54.59	11.8
G14MA	8.48	8.45	0.354	52.46	58.42	10.2
G16MA	8.473	8.41	0.320	56.85	62.22	08.0

4.3 Test for Durability Parameters

In durability parameters, acid resistance, chloride resistance and sorptivity were tested on the specimens after 28 days of curing drying to constant weight oven at 100 °C.

4.3.1 Acid Attack on Concrete

To study the results of acid resistance and its effect on all categories of specimen, those were immersed in 1% solution on sulphuric acid (H_2SO_4) as described above for 28 days. Cube specimens were taken out one day before test and dried completely up to constant weight by oven at 100 °C. And then the weight was taken and tested for compressive strength including the reference samples which were kept for comparison purpose. Details of test results on weight difference and strength changes are given in Table 7.

4.3.2 Chloride Attack on Concrete

To study the results of chloride attack and its effect on geopolymer concrete specimen and MOF specimen 1% solution on sodium chloride (NaCl) is used. After 28 days of curing and the weight was taken both geopolymer concrete and MOF specimens were immersed entirely into the acid solution and kept for more 28 days. Cube specimens were taken out one day before test and dried completely. And then the weight was taken and tested. Some reference sample was kept for comparison purpose. The weight difference and strength changes are given in Table 8.

Table 8 Chloride attack result

Specimen	Wt. before test	Wt. after test	Wt. Loss/gain %	Compressive strength of reference samples N/mm^2	Compressive strength after chloride attack in N/mm^2	Strength loss (%)
MOF	8.61	8.590	0.232 loss	64.91	63.93	1.00
G10MA	8.39	8.576	2.217 gain	35.00	33.073	5.50
G12MA	8.367	8.566	2.378 gain	54.59	51.52	3.70
G14MA	8.176	8.303	1.553 gain	58.42	56.263	3.69
G16MA	8.337	8.53	1.315 gain	62.22	60.347	3.00

4.3.3 Sorptivity

Geopolymer cubes (16 molar) were cured for 28 days and after that dried in oven at 100 °C. After that the cubes were taken out and three side of the cubes were coated with colour and epoxy to restrict the flow to one side only. The specimens were then immersed in water not more than 5 mm depth. The water absorption were measured for a span of 30 min in an interval provided in the Table 9. Each time the specimen were taken out and excess water was wiped off with the help of a damped cloth. Those were weighted in weight machine.

$I = S.t^{1/2}$, therefore $S = I/t^{1/2}$; Where S = sorptivity in mm, t = elapsed time in mint. I (cumulative infiltration) = $\Delta w/Ad$ Δw = change in weight = W2–W1, where W1 = Oven dry weight of cube in grams, W2 = Weight of cube after 30 min capillary suction of water in grams, A = surface area of the specimen through which water penetrated and d = density of water.

Table 9 Sorptivity results

Time (minute)	Initial weight (kg) (W1)	Final weight (kg) (W2)	Δw = cumulative weight gain (kg)	Density of water (kg/m^3)	Surface area (m^2)	$I = \Delta w/Ad$ (mm)	S
0	8.520	8.520	0.000	997	0.0225	0	0
1	8.520	8.523	0.0035	997	0.0225	0.1778	0.1778
4	8.523	8.526	0.006	997	0.0225	0.2674	0.1337
9	8.526	8.528	0.008	997	0.0225	0.3566	0.11887
16	8.528	8.530	0.010	997	0.0225	0.4457	0.1114
25	8.530	8.531	0.011	997	0.0225	0.4903	0.0895
30	8.531	8.531	0.011	997	0.0225	0.4903	0.0895
Sorptivity = $I/(t)^{0.5}$ = 0.089 $mm/t^{.05}$							

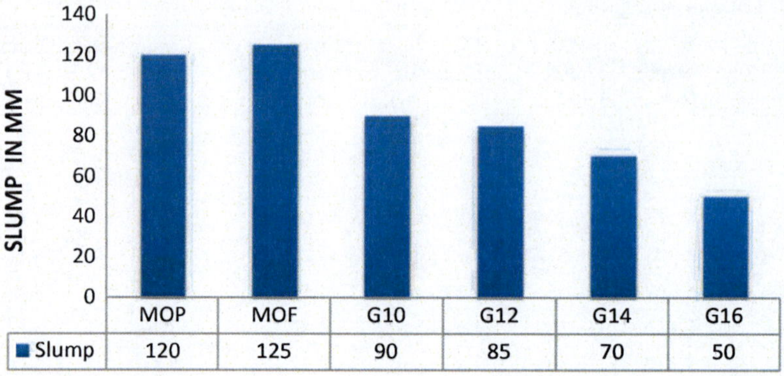

Fig. 8 Variation in workability

5 Result and Discussion

The previous chapter was about the test done and the results. This chapter discuss about the comparison result and the reason.

5.1 *Workability*

Change of workability in term of slump values are represented in the Fig. 8. Slump value of OPC concrete is 120 mm, whereas fly ash concrete displays 125 mm slump. It is due to the spherical particle size of fly ash. On the contrary, the slump value of G10 samples reduces to 90 mm and decreases as molarities of Na_2SiO_3 in GPC increases. It is attributed to the increase in viscosity of Na_2SiO_3 increases the cohesiveness of the GPC mix [15].

5.2 *Compressive Strength Results*

The compressive strength of all mixes at the age of 7, 28 and 56 days are provided in Table 5 and represented in Fig. 9.

OPC concrete and fly ash concrete are found to achieve much better strength up to 56 days. Seven days strength of fly ash concrete is lower than that of OPC concrete due to replacement of 25% fly ash which are not involved in pozzolanic action in early ages. Strength enhancement between 7 days to 56 days in OPC concrete is found to be 41% which is lower than that for fly ash concrete (76%). Fly ash concrete presents higher strength than OPC concrete both at the age of 28 and 56 days and it is due to the additional C–S–H gel produced from pozzolinic reaction [1]. Compressive strength

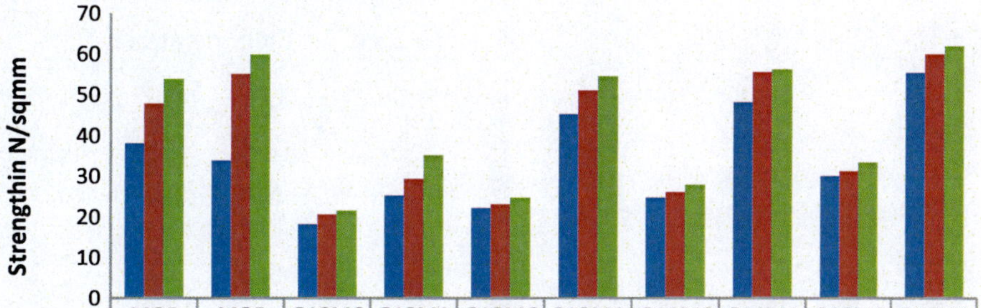

	MOP	MOF	G10MC	G10MA	G12MC	G12MA	G14MC	G14MA	G16MC	G16MA
Series1	38.16	33.87	18.23	25.29	22.2	45.45	24.85	48.4	30.2	55.62
Series2	47.87	55.1	20.54	29.31	23.14	51.2	26.22	55.72	31.29	60.25
Series3	53.88	59.91	21.52	35.2	24.75	54.65	27.98	56.42	33.55	62.22

Fig. 9 Compressive strength comparison

of GPC samples increases with the increase in molarities of NaOH for both the curing conditions. It is observed that the compressive strength of 16 M oven-cured GPC cubes were highest. High-temperature cured GPC specimens show comparatively much more strength as compared to the counterparts which were cured in normal curing tank. Strength enhancement in all GPC sample of both curing conditions between 7 and 56 days are small and about 12%. Strength of temperature-cured GPC with 16 M NaOH is observed to be as per the OPC and fly ash concrete both at the age of 28 and 56 days. Normally cured 16 M GPC samples gains strength up to 30 N/mm^2 which satisfy the requirement of M20 concrete.

The change in compressive strength with the change in molarities of NaOH for both the curing conditions is represented in Fig. 10. The reason for the enhancement

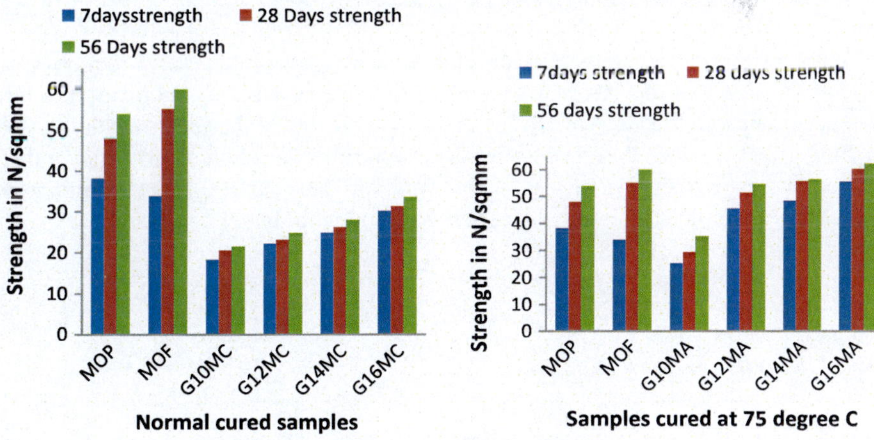

Fig. 10 Variation of compressive strength with change of molarity

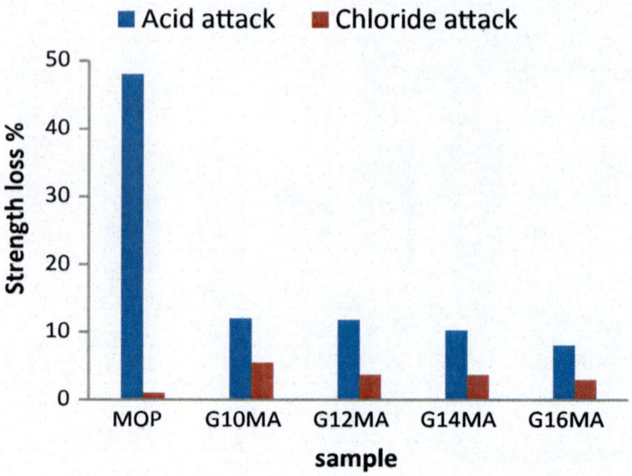

Fig. 11 Strength loss in acid and salt attack

of strength in both the oven-cured and normal cured specimen with increase in the molarity of NaOH on GPC may be attributed to the availability of more alkalis for geopolymerization. Higher strength in temperature-cured GPC sample than the normal cured counterparts may be attributed to the availability of more reactive alkalis at higher temperature.

5.3 *Acid Attack Results*

The test result of acid resistance was given in previous Table 7. The test results shows that the 25% fly ash replacement specimen (MOF) had a drastic fall in strength, at about 48%, whereas the strength loss for 10 M, 12 M, 14 M and 16 M specimen are 12%,11.5% 10.2% and 8%, respectively.

Figure 11 shows the change in strength after acid attack and from the graph it is clearly visible that the G16MA shows greater acid resistivity as compared to the other specimens. In Fig. 12, we can see the weight change of the specimen due to the leaching effect of acid. Weight loss in OPC is 5.0%. The graph clearly shows that the weight loss in geopolymer specimens were very less about 0.35% as compared to the MOF specimen. So it may be concluded that geopolymer concrete poses very good resistance against H_2SO_4 attack.

5.4 *Chloride Attack Results*

Table 7 shows the test results of chloride attack on different specimen. The test result shows that MOF specimen lost only 1% of its strength where as strength loss of geopolymer concrete cube of 10 M, 12 M, 14 M and 16 M was 5.5%, 3.7%,

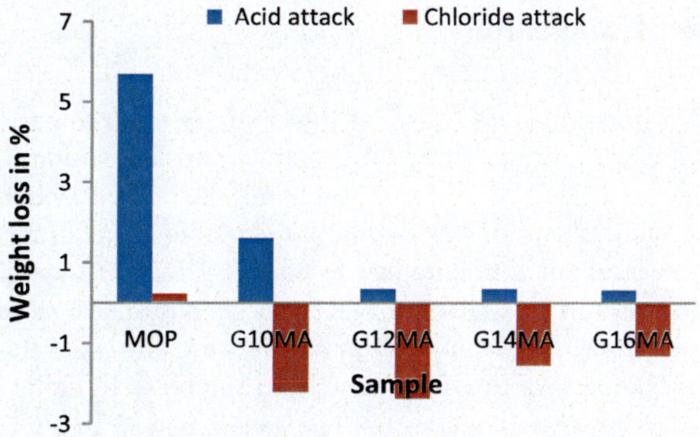

Fig. 12 Change in weight in acid and salt attack

3.69% and 3%, respectively. Figure 12 represents nominal changes in strength in both geopolymer specimens as well as MOF specimen.

Figure 11 presents that there is a loss of weight in MPF sample, whereas there is an increase of weight after the immersion into the NaCl solution. It is due to some amount of salt get deposited into pore spaces the specimens. Percentage gain of weight of GPC samples decreases with an increase in the molarity of NaOH due to pore refinement.

5.5 *Sorptivity Results*

The test results are provided in Table 8 and Fig. 13 represents the variation of soptivity (s)with respect to an increase in time. Sorptivity increases in a rapid manner up to 5 min. It may be due to initial dryness of the specimen and surface porosity. Then the values decreases in irregular manner. It may be attributed to increase in capillary distance towards interior and tortiocity of capillary path.

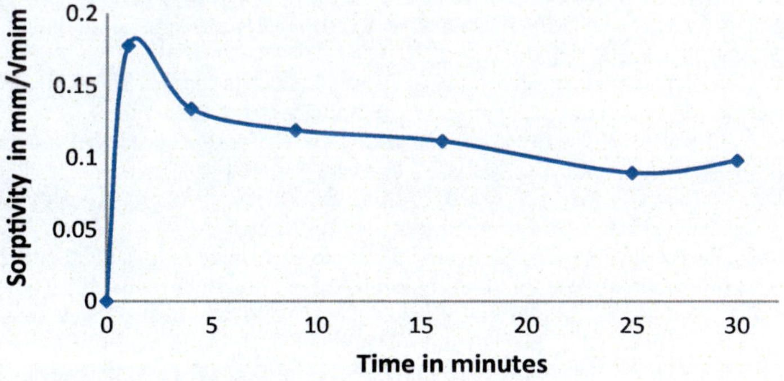

Fig. 13 Variation of sorptivity (s) with respect to time

6 Conclusion

Normal cured GPC displays lower strength in comparison to high-temperature cured GPC and normal cured OPC counterpart and strength increase with the increase in molarities of NaOH. Therefore, it may be recommended for the situations where there is requirement of low strength. Geopolymer concrete shows a very good compressive strength with compare to normal OPC concrete with 25% fly ash replacement. Increase in molarities of NaOH in geopolymer concrete enhances the compressive strength. GPC displays better resistance against acid attack but weak in chloride attack in comparison to OPC counterpart. Increase in molarities of NaOH in geopolymer concrete also enhances the resistance against in acid attack and chloride attack in term of weight and strength loss.

References

1. Nevelle AM (2012) Properties of concrete, text book on concrete tecnology, 4th Edn. Pearsion Education, p 63
2. AdbElaty MAA, Ghazy MF, El Hameed MFA (2017) Optimization of geopolymer concrete by principal component analysis. ACI Mater J Techn Pap 253–264
3. Aslani F (2016) Thermal performance modeling of geopolymer concrete. J Mater Civil Eng 28(1):04015062(1–12)
4. Mehta A et al. (2017) Influence of various parameters on strength and absorption properties of fly ash based geopolymer concrete designed by Taguchi method. Constr Build Mater 150:817–824
5. Albitar M, Mohamed MS, Vasintin P, Drechsler M (2017) Durability evaluation of geopolymer and conventional concretes. Constr Build Mater 136:174–185
6. Singh B, Ishwarya G, Gupta M, Bhattacharyya SK (2015) Geopolymer concrete: a review of some recent developments. Constr Build Mater 85:78–90
7. Ken PW, Ramli M, Ban CC (2015) An overview on the influence of various factors on the properties of geopolymer concrete derived from industrial by products. Constr Build Mater 77:370–395
8. Sarkar PK, Kelly S, Yao Z (2014) Effect of fire exposure on cracking, spalling and residual strength of fly ash geopolymer concrete. Mater Des 63(2014):584–592
9. Chindaprasirt P, Chareerat T, Hatanaka S, Cao T (2011) High strength geopolymer using Fine High calcium Fly ash. J Mater Civ Eng 23(3):264–270
10. Okoye FN, durgaprasad J, Singh NB (2015) Mechanical properties of alkali activated fly ash/Kaolin based geopolymer Concrete. Constr Build Mater 98:685–691
11. Noushini A, Castel A (2016) Effect of heat curing on transport properties of low-calcium fly ash based geo-polymer concrete. Constr Build Mater 112:464–477
12. Gunasekara.C, Law DW, Setunge S (2016) Long term permeation properties of different fly ash geo polymer concretes. Constr Build Mater 124:352–362
13. Hadi MNS, Al-Azzawi M, Yu T Effect of fly ash characteristic and alkaline activator components on compressive strength of fly ash based geo-polymer mortar
14. Allahverdi A, Vafaei M, Maghsoodloorad H (2017) Quality control and assessment of geopolymer cements based on reacted and free alkali. Constr Build Mater 153:274–283
15. Pilehvar S, Cao VD, Szczotok AM, Valentini L, Salvioni D,Magistri M, Pamies R, Kjoniksen AL (2017) Mechanical properties and micro scale changes of geopolymer concrete and Portland cement concrete containing micro-encapsulated phase change. Cement Concrete Res 100:341–349

16. Xie J, Kayali O (2016) Effector super plasticizer on workability enhancement of Class F and Class C fly ash based geopolymers. Constr Build Mater 122:36–46
17. Pasupati K,Berndt M,Sanjayan J, Rajeev P (2017) Durability of low-calcium fly ash based geopolymer concrete culvert in a saline environment. Cement Concrete Res 100:297–310
18. IS 12269-1987, Specification for 53 grade ordinary Portland cement. BIS, New Delhi
19. IS: 383-1970, Specificationfor coarse and fine aggregates from natural sources for concrete, reaff 1993. BIS, New Delhi
20. IS:456-2000 Plain and reinforced concrete code of practice. BIS, New Delhi
21. IS:516-1959 Methods of testing strength of concrete Reaff 1999. BIS, New Delhi

A Review on Mechanical and Microstructure Properties of Reinforced Concrete Exposed to High Temperatures

Sharan Kumar Goudar, Santhosh Kumar Gedela, and B. B. Das

Abstract This paper presents the recent research progress on the response of concrete exposed to fire or high temperatures. The main highlight of this review paper is a compilation of previously reported data regarding the variations in mechanical properties and microstructure properties of concrete when exposed to high temperatures. The concrete structures get deteriorated at the macro- and microscopic levels due to high-temperature exposure. The macro-level damages can be measured with degradation in mechanical properties such as the reduction in compressive strength, weight loss, changes in elastic properties, reduction of bond strength in reinforced concrete, etc. The macro-cracks on the surface of concrete causes spalling which can be observed after exposing the concrete samples to more than 300 °C. The compressive strength of the concrete reduces slightly till 400 °C, and when the temperature increased to 600 °C, there was an exponential reduction in the compressive strength of concrete. Another important parameter is bond strength degradation, which plays a crucial role in durability issues. To understand the deterioration phenomenon and changes in mechanical properties, the changes at the level of the microstructure of concrete need to be understood. Dehydration of products causes deterioration of mechanical properties and weight loss of concrete when exposed to high temperatures. At different temperatures, the microstructure changes and the response of hydration products such as calcium hydroxide (CH), CSH gel, unhydrated cement and capillary water reported by previous researchers are compiled and discussed.

Keywords Microstructure · High temperature · Spalling · Dehydration · Bond strength · Compressive strength

---

S. K. Goudar (✉) · S. K. Gedela · B. B. Das
Department of Civil Engineering, National Institute of Technology Karnataka, Surathkal, Mangalore, India
e-mail: sgcr17@gmail.com

S. K. Gedela
e-mail: santu11112@gmail.com

B. B. Das
e-mail: bibhutibhusan@gmail.com

B. B. Das et al. (eds.), *Recent Developments in Sustainable Infrastructure*, Lecture Notes in Civil Engineering 75, https://doi.org/10.1007/978-981-15-4577-1_62

1 Introduction

Fire, an accidental damage, is a dangerous threat to reinforced concrete structures. It is important to understand the response of reinforced concrete structures when exposed to fire or high temperature. The low thermal conductivity of concrete makes it one of the suitable materials to resist fire or high temperature [1]. However, high temperatures do affect the concrete structures but the response is better compared to other materials [2, 3]. When the concrete structures are exposed to fire or high temperature, there will be reduction in the mechanical properties and that affects the durability. When concrete are subjected to heat, due to non-uniformity of the cement and aggregate elements, changes in mechanical and microstructure properties of concrete are common. Both aggregate and cement elements react to heating in different ways. After cooling down of concrete, some of the reactions are reversible, while others are irreversible which causes a reduction in the strength of concrete [2].

When concrete are exposed to high temperature, until 400 °C, changes in compressive strength of concrete are minimal. During 400–800 °C exposure, the compressive strength of concrete decreases exponentially. The fibers aid to the reduction in compressive strength when exposed to high temperatures. The strength of the concrete also influences the rate of reduction of compressive strength, high strength (HSC) performs better compared to normal strength concrete [4]. There are numerous literatures available on compressive strength reduction of concrete due to the fire exposure. However, the degradation of bond strength due to fire exposure is limited. Few researchers studied only the bond strength degradation by pullout test, but proper microstructure justification for a reduction in bond strength was missing [5–7]. The need of the hour is a systematic study and proper explanation of microstructure changes occurring at the steel–concrete interface and also in bulk concrete.

The pore solution in capillary pores evaporates when the concrete structures are exposed to 100–140 °C and that creates a build-up pressure inside the concrete. This pore pressure is responsible for causing micro-cracks and spalling of concrete [8]. Spalling is a phenomenon where the cover of the reinforced concrete member gets separated from the original structure. The process of cracking is believed to be the same as the spalling phenomenon. Dehydration of the concrete causes the formation of fissures in concrete and causes further cracks [5, 8]. Calcium hydroxide (CH) decomposes at a temperature of 450 °C and converts to calcium oxide (CaO). When calcium oxide comes in contact with moisture after the fire exposure, it expands and causes micro-cracks which ultimately leads to severe cracking and spalling of concrete. The addition of mineral admixtures reduced the formation of CH and ultimately reduces the formation of CaO. When OPC and mineral admixed concretes were exposed to 600 °C temperature, visible cracks were noticed on the surface of OPC concrete, whereas the mineral admixed concrete showed very few micro-cracks on the surface [6, 7, 9, 10]. The mineral admixtures role can be explained in two ways. The first one being, reduction of CH which is consumed during pozzolanic reaction and second is the reduction of capillary pores and an overall improvement in the microstructure of concrete [11]. When concrete exposed to a high temperature of

750 °C, decarbonation occurs in concrete. The calcium carbonate ($CaCO_3$) at 750 °C converts into CaO and CO_2, this process is being termed as decarbonation. Due to the dehydration and decarbonation, CaO content will increase in bulk concrete, which finally expands when comes in contact with moisture and causes severe cracking [12]. The hydrates including CSH and CH will be converted to crystalline-like phase at 1000 °C of fire exposure and the pore system seriously expands which causes spalling of concrete [12]. The effect of high temperature on concretes mechanical and durability properties of concrete are discussed in this review paper.

2 Mechanical Properties of Different Types of Concrete Subject to Elevated Temperature

The main mechanical properties of the concrete structure are compressive strength and bond strength. For normal strength concrete (NSC), compressive strength starts decreasing from 400 °C. The strength of the concrete also influences the rate of reduction of compressive strength, high strength concrete (HSC) performs better compared to normal strength concrete. The fibers in fiber reinforced concrete (FRC) stops the cracks propagation which results in slightly lesser reduction of compressive strength when exposed to high temperatures. The results of different researchers were compiled and a plot of reduction in compressive strength versus exposure temperature was plotted in Fig. 1. The high-temperature exposure reduces the mechanical

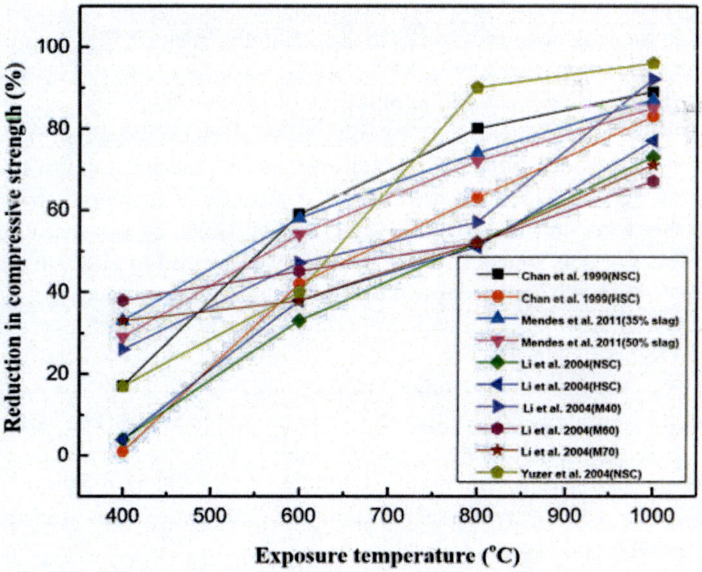

Fig. 1 Effect of high temperature on compressive strength of concrete

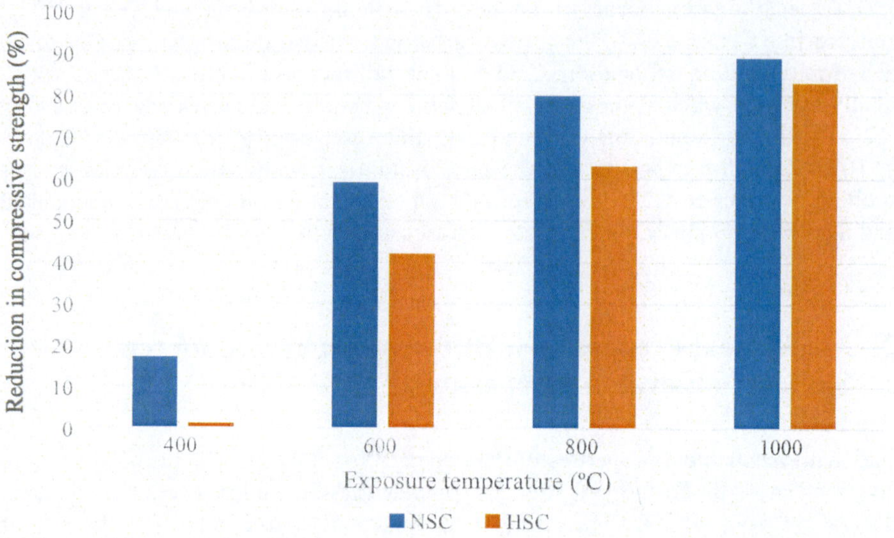

Fig. 2 Reduction in compressive strength of NSC and HSC at different temperatures

properties, reduction in compressive strength increases when exposure temperature increases. Up to 50% reduction in compressive strength was noticed during 400–600 °C exposure. As the exposure temperature increases beyond 600 °C, a severe reduction in compressive strength was observed. The pore solution in capillary pores evaporates and that creates a build-up pressure inside the concrete. This pore pressure is responsible for causing micro-cracks in the concrete structure. These micro-cracks will degrade the compressive strength. It can also be observed from Fig. 1 that HSC and FRC perform better compared to NSC.

Figure 2 shows the comparison of compressive strength reduction of HSC and normal strength concrete (NSC). Marginal strength of concrete was lost till 400 °C, up to 10% for HSC and 15% for NSC. The significant loss in compressive strength was lost in between 400 and 800 °C [1–4, 9, 11]. Some of the researchers also mentioned that the pore density in HSC was less compared to NSC, due to which the pore pressure in HSC concrete will be comparatively more. Because of the high pore pressure, slightly higher reduction of compressive strength was noticed in HSC as compared to NSC [9, 10].

The results of different researchers were compiled and a plot of reduction in bond strength versus exposure temperature was plotted in Fig. 3. The same trend of compressive strength can be noticed for bond strength of reinforced concrete also. As the exposure temperature increases, the percentage reduction of bond strength also increases [13–18]. The presence of calcium hydroxide degrades at a temperature of 450 °C which might cause a reduction in compressive strength. However, a clear explanation for microstructural changes at the steel–concrete interface for causing bond degradation is missing in the literature. The addition of steel fibers in concrete

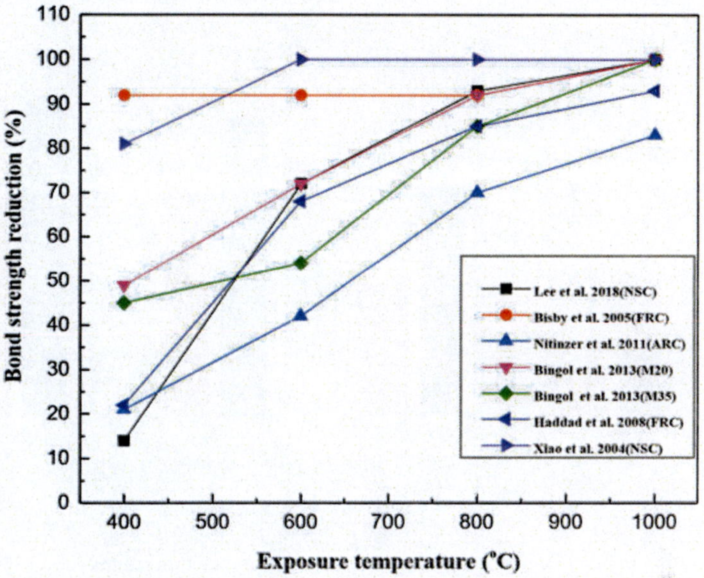

Fig. 3 Effect of high temperature on the bond strength of reinforced concrete

proves to be beneficial in terms of bond strength retention [19, 20]. However, the other fibers such as natural fibers were combustible at low temperatures which exerts expansive pressure which resulted in a slightly higher reduction of bond strength compared to NSC [13, 15]. Overall, the addition of fibers proves beneficial in terms of retaining bond strength as well as compressive strength [13, 16, 17, 20–23]. Figure 4 shows the comparison of reduction of bond strength for NSC and FRC at different

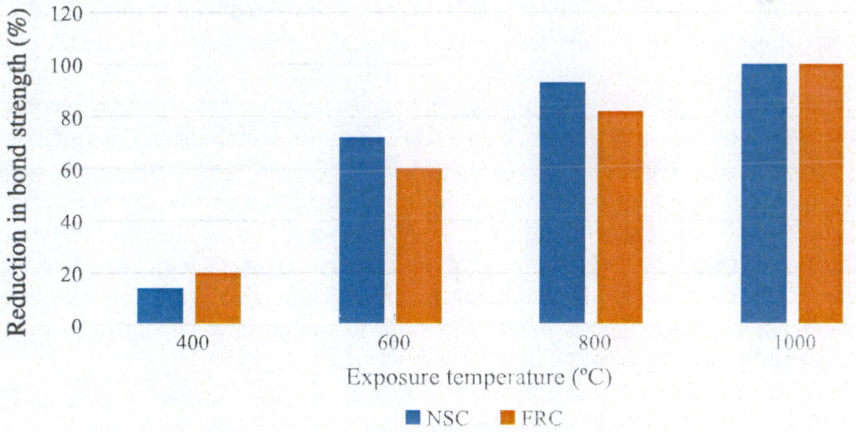

Fig. 4 Reduction in bond strength of NSC and FRC at different exposure temperatures

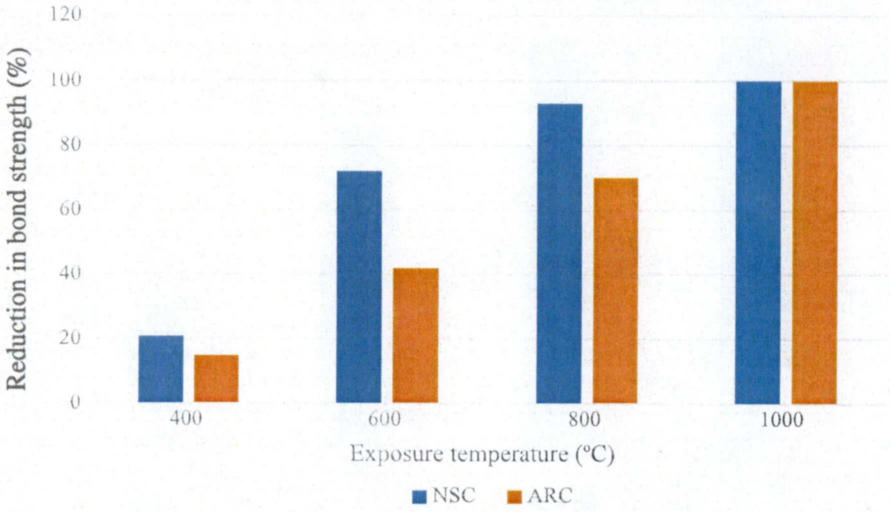

Fig. 5 Reduction in bond strength of NSC and ARC at different temperatures

exposure temperatures. The bond strength of NSC decreased up to 15%, while the bond strength of FRC decreased up to 20% for natural fibers. For the steel fibers, the scenario was quite the opposite. The addition of steel fibers proved to be beneficial in terms of bond strength retention. The spalling effect was reduced due to the addition of steel fibers [5–7].

The aggregate replacement also improved bond strength retention. Netinger et al. [3] reported that the addition of crushed bricks as a replacement of natural aggregates improved the bond strength retention in reinforced concrete. Many researchers [3, 4, 8, 17] reported similar findings and compiled data are shown in Fig. 5.

The addition of mineral admixtures (MA) proved to be beneficial in retaining the compressive strength of concrete when exposed to high temperatures [6, 8, 11, 12, 15, 24]. The mineral admixtures role can be explained in two ways. The first one being, reduction of CH which was consumed during the pozzolanic reaction and the second being reduction of capillary pores and an overall improvement in the microstructure of concrete [11, 19]. The data from several researchers regarding the significance of mineral admixed concrete toward the compressive strength retention are presented in Fig. 6. Mineral admixed concrete performed better when exposed to different temperatures when compared to NSC. It can also be observed that the replacement level of mineral admixtures also plays an important role. The high volume replacement of mineral admixtures had positive effect in terms of compressive strength retention when exposed to high temperatures.

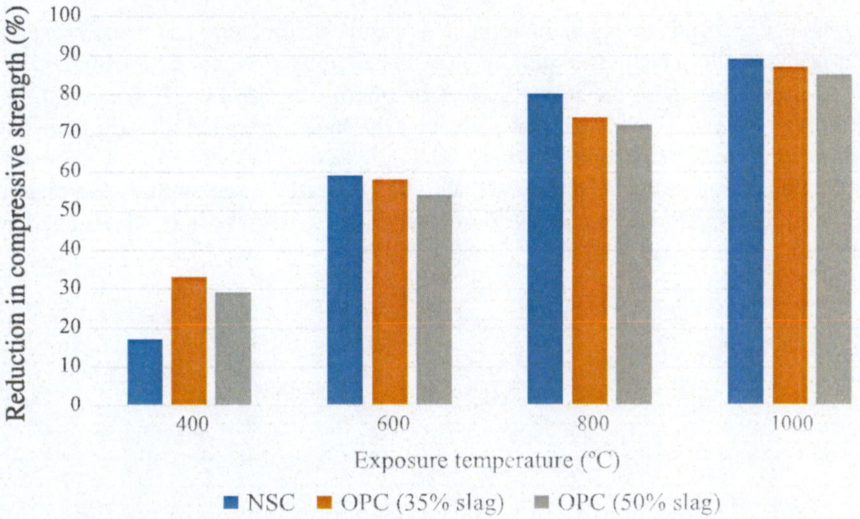

Fig. 6 Reduction in compressive strength of NSC, OPC (35% slag), and OPC (50% slag) at different exposure temperatures

3 Microstructure Properties of the Concrete Subject to Elevated Temperature

The microstructure of the concrete plays a crucial role in determining the mechanical properties. The hydration products undergo physical and chemical changes at high temperature exposure. The pore solution in capillary pores evaporates when the concrete structures are exposed to 100–140 °C and a slight reduction in weight loss was noticed. The evaporated pore water exerts expansive pressure inside the concrete which leads to the formation of micro-cracks [8]. Calcium hydroxide decomposes at a temperature of 450 °C and converts to calcium oxide this phenomenon is generally known as dehydration. After the fire exposure when calcium oxide encounters moisture, calcium hydroxide is reforming, and the phenomenon is generally known as rehydration [11]. The reaction is shown in Eq. 1. Because of the rehydration, newly formed calcium hydroxide expands in volume which ultimately causes severe cracking. The dehydration of calcium hydroxide can be easily recognized from rapid weight loss and consequently analyzed by thermogravimetric analysis [25–27]. The thermogravimetric analysis provides a comparison of water loss of cement paste at different temperatures. The CSH gel also starts to crystallize slowly at 400 °C, which also aid to the cracking process inside the concrete.

Due to the atmospheric carbon dioxide, the concrete undergoes carbonation process, where calcium hydroxide is being converted into calcium carbonate (as shown in Eq. 2). The calcium carbonate at 750 °C converts into calcium oxide and carbon dioxide, this process is being termed as decarbonation (as shown in Eq. 3). Due to the dehydration and decarbonation, calcium oxide content will increase in bulk concrete, which finally expands when encounters the atmospheric moisture and causes severe cracking and spalling [12]. The spalling process generally starts to occur from 200 to 1000 °C depending upon the strength or grade of concrete [6, 11, 12]. The spalling process was faster in low-grade concretes and the high strength concretes showed comparably more resistance to spalling.

The hydrates including CSH and CH will be converted to crystalline-like phase at 1000 °C of fire exposure and the pore system seriously expands which causes spalling of concrete [12].

$$(OH)_2 \leftrightarrow CaO + H_2O \tag{1}$$

$$(OH)_2 + CO_2 \rightarrow CaCO_3 + H_2O \tag{2}$$

$$CaCO_3 \rightarrow CaO + CO_2 \tag{3}$$

4 Conclusions

The high-temperature effects on concrete mechanical and durability properties of concrete were discussed and the following conclusions can be drawn.

- High-temperature exposure of concrete structures results in degradation of mechanical properties. Up to 50% reduction in compressive strength was observed during 400–600 °C exposure and at 800 °C, only 10% of the original compressive strength was retained.
- The high strength concrete and fiber-reinforced concrete performed better compared to normal strength concrete when exposed to high temperatures.
- The bond strength between reinforcing steel and concrete is a sensitive to high temperatures. Significant reduction in bond strength was observed during 400–600 °C exposure and at 800 °C, only 5% of the original bond strength was retained.
- The bond degradation also follows the same trend of compressive strength, as the exposure temperature increases the percentage reduction in strength increases.
- A clear explanation for microstructural changes at the steel–concrete interface for causing the bond degradation is missing in the literature.
- Calcium hydroxide decomposes at a temperature of 450 °C and converts to calcium oxide, which in contact with moisture expands and produces micro-cracks and ultimately spalling of concrete occurs.
- When concrete exposed to a high temperature of 750 °C, decarbonation occurs in concrete. Due to the dehydration and decarbonation, calcium oxide content will increase in bulk concrete, which finally expands when encounters moisture and causes severe cracking and spalling.

- The hydrates including CSH and calcium hydroxide will be converted to crystalline-like phase at 1000 °C of fire exposure and the pore system seriously expands which causes spalling of concrete.

References

1. Chan YN, Peng GF, Anson M (1999) Residual strength and pore structure of high strength concrete and normal strength concrete after exposure to high temperatures. Cem Concr Compos 21(1):23–27
2. Bisby LA, Green MF, Kodur VK (2005) Response to fire of concrete structures that incorporate FRP. Prog Struct Mat Eng 7(3):136–149
3. Netinger I, Kesegic I, Guljas I (2011) The effect of high temperatures on the mechanical properties of concrete made with different types of aggregates. Fire Saf J 46(7):425–430
4. Novak J, Kohoutkova A (2018) Mechanical properties of concrete composites subject to elevated temperature. Fire Saf J 95:66–76
5. Hertz KD, Sørensen LS (2005) Test method for spalling of fire exposed concrete. Fire Saf J 40(5):466–476
6. Ali F, Nadjai A, Silcock G, Abu-Tair A (2004) Outcomes of a major research on fire resistance of concrete columns. Fire Saf J 39(6):433–445
7. Han CG, Hwang YS, Yang SH, Gowripalan N (2005) Performance of spalling resistance of high performance concrete with polypropylene fiber contents and lateral confinement. Cem Concr Res 35(9):1747–1753
8. Fletcher IA, Borg A, Hitchen N, Welch S. Performance of concrete in fire: a review of the state of the art, with a case study of the Windsor tower fire
9. Li M, Qian C, Sun W (2004) Mechanical properties of high-strength concrete after fire. Cem Concr Res 34(6):1001–1005
10. Yüzer N, Aköz F, Öztürk LD (2004) Compressive strength–color change relation in mortars at high temperature. Cem Concr Res 34(10):1803–1807
11. Mendes A, Sanjayan J, Collins F (2008) Phase transformations and mechanical strength of OPC/Slag pastes submitted to high temperatures. Mater Struct 41(2):345
12. Zhang Q, Ye G (2011) Microstructure analysis of heated portland cement paste. Procedia Eng 14:830–836
13. Lee J, Sheesley F, Jing Y, Xi Y, Willam K (2018) The effect of heating and cooling on the bond strength between concrete and steel reinforcement bars with and without epoxy coating. Constr Build Mater 177:230–236
14. Khoury GA (2000) Effect of fire on concrete and concrete structures. Prog Struct Eng Mater 2(4):429–447. Bingöl AF, Gül R (2009) Residual bond strength between steel bars and concrete after elevated temperatures. Fire Saf J 44(6):854–859
15. Haddad RH, Al-Saleh RJ, Al-Akhras NM (2008) Effect of elevated temperature on bond between steel reinforcement and fiber reinforced concrete. Fire Saf J 43(5):334–343
16. Xiao J, König G (2004) Study on concrete at high temperature in China—an overview. Fire Saf J 39(1):89–103
17. Harmathy TZ (1968) Determining the temperature history of concrete constructions following fire exposure. J Proc 65(11):959–964
18. Goudar SK, Das BB, Arya SB (2019) Microstructural study of steel-concrete interface and its influence on bond strength of reinforced concrete. Adv Civ Eng Mater 8(1):171–189
19. Goudar SK, Shivaprasad KN, Das BB (2019) Mechanical properties of fiber-reinforced concrete using coal-bottom ash as replacement of fine aggregate. In: Sustainable construction and building materials 2019. Springer, Singapore, pp 863–872

20. George RM, Das BB, Goudar SK (2019) Durability studies on glass fiber reinforced concrete. In: Sustainable construction and building materials 2019. Springer, Singapore, pp 747–756
21. Goudar SK, Das BB, Arya SB (2019) Combined Effect of marine environment and pH on the impedance of reinforced concrete studied by electrochemical impedance spectroscopy. In: Sustainable construction and building materials 2019. Springer, Singapore, pp 635–649
22. Yadav S, Das BB, Goudar SK (2019) Durability studies of steel fibre reinforced concrete. In: Sustainable construction and building materials 2019. Springer, Singapore, pp 737–745
23. Srikumar R, Das BB, Goudar SK (2019) Durability studies of polypropylene fibre reinforced concrete. In: Sustainable construction and building materials. Springer, Singapore, pp 727–736
24. Nilsen U, Sandberg P, Folliard K (1992) Influence of mineral admixtures in the transition zone in concrete. In: International union of testing and research laboratories for materials and construction. International conference 1992, pp 65–70
25. Harmathy TZ. Fire safety design and concrete
26. Matesová D, Bonen D, Shah SP (2006) Factors affecting the resistance of cementitious materials at high temperatures and medium [0] heating rates. Mater Struct 39(4):455
27. Saito M, Kawamura M (1989) Effect of fly ash and slag on the interfacial zone between cement and aggregate. Spec Publ 114:669–688

Application of Andreassen and Modified Andreassen Model on Cementitious Mixture Design: A Review

K. Snehal and B. B. Das

Abstract Cement is a widely used construction material and its consumption on large-scale causes environmental degradation; thus, more emphasis is being given on industrial by-products as alternative materials to cement for their sustainable usage. It is necessary that varying particle size of supplementary cementitious particles is to be used for filling the voids to form a dense particle-packed concrete. The selection of right combination of material is tedious job by trials involving different replacement materials and the resultant concrete may show unexpected results; thus, a more suitable method is the selection of materials based on optimum packing of particles. To select the optimum size of replacement materials particle packing models are essential, so that a low-cement concrete can be prepared which will be ecological as well as economical with improved density, low porosity and high compressive strength. It is found that there are different models have been developed to achieve optimal packing. However, application of Andreassen and modified Andreassen models for the particle packing of multiple ingredients of cementitious matrix found to be largely being accepted by the researchers. This paper reviews the application of Andreassen and modified Andreassen models for the effective particle packing investigations on cementitious particles. It also reviews the software's employed for designing various cementitious mixtures based on Andreassen and modified Andreassen models.

Keywords Cementitious materials · Particle packing · Andreassen and modified andreassen model · Density · Mix design

K. Snehal (✉) · B. B. Das
Department of Civil Engineering, NITK, Surathkal 575025, India
e-mail: snehalshine090@gmail.com

B. B. Das
e-mail: bibhutibhusan@gmail.com

B. B. Das et al. (eds.), *Recent Developments in Sustainable Infrastructure*, Lecture Notes in Civil Engineering 75, https://doi.org/10.1007/978-981-15-4577-1_63

1 Introduction

The durability and performance of concrete (or any cement-based material) is greatly relied on the level of capillary voids interconnectivity [1, 2]. The appropriate mix proportioning of ingredients with the thorough knowledge of its material characteristic is a key to obtain a sustainable, dense and durable concrete [2]. Concrete is a composite material produced by the mixture of multi-sized (ranging from μm to mm) ingredients such as cement, fine aggregate, coarse aggregate and water. Several researchers have reported that cement production is one of the leading donors to the global warming and global environmental survey states that cement industry produces total CO_2 emission of about 7% over the globe [3]. Considering this, in view researchers have found an alternative by partially replacing cement with industrial by products like fly ash, GGBS, silica fume, etc., as supplementary cementitious materials (SCMs) [3, 4]. At this point of time, there is a need for implementing the theory of particle packing in order to avoid the tiresome task of trial and error technique to optimize the right proportion of ingredients to produce cement-based material from the pool of SCMs [2].

Particle packing is of the interest in research fraternity in the production of concrete, mortar, ceramics, etc. Literature says that dense packing of concrete with aggregate phase as much as possible is one of the effective ways to curtail the use of binder quantity and help in developing an ecological, economical benefited concrete; in addition, it also favours in reducing the level of shrinkage and creep [5, 6]. The traditional method of concrete mix proportioning is a secondary way to measure of particle packing which involves volumetric packing, i.e. by means of an ideal gradation curve of aggregate proportions [2–7]. But, the characteristic of particle packing also depends on the finer particle ingredients of concrete like cement and other cementitious particles. The amount of solid aggregate particles per unit volume expresses the degree of particle packing [8]. The definition of packing density is given as the solid volume of particles to the total volume occupied by the particles and can be mathematically expressed as unit volume minus porosity [2]. The improvement of particle packing density depends on the void filling capacity of larger particles by smaller particles and void spaces between the small particles by further finer particles and so on (typically represented in Fig. 1).

It can be noted from the historical background of particle packing theory that, it is not a newfangled theory but was already in existence during 1890s which was first

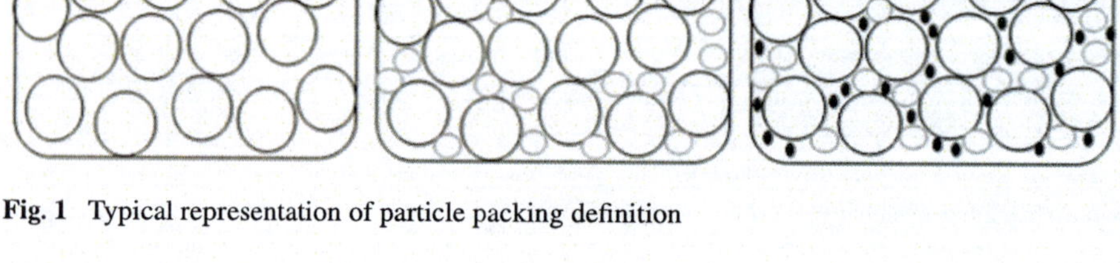

Fig. 1 Typical representation of particle packing definition

suggested by Feret in 1892. He signified that minimal voids in matrix is responsible in achieving maximum strength. Fuller and Thompson in 1907 studied the significance of particle size distribution (PSD) of aggregates and concrete properties based on particle packing theory [9]. The study related to particle packing of aggregates conducted by Suenson in 1911 demonstrated the triangular packing diagrams of aggregates [10]. The basic investigation on packing theory for sphere-shaped particles was started by Furnas in 1931 [11] by assuming that cavity between bigger particles will be filled by smaller particles. It has been noticed that the widespread research was carried out by Powers in 1968 [12] by hosting the importance of void ratio diagram on the basis of particle packing theory and developed a suitable model for concrete and mortar mixtures. However, the attention towards particle size optimization is amplified few decades ago in particularly with the emerging trends in concrete like high strength, high performance, high density, self-compacting concrete, etc. The concept of particle packing is beneficial in contrast to the conventional mix design as it is mainly developed on the basis of physical mechanism. It has been reported by the several researchers that optimization of concrete ingredients can be achieved by means of particle packing model [13, 14]. Particle packing models are based on the concept of theory of particle packing. The applicability of packing theories is chiefly affected by following factors: (a) particle gradation, (b) methods of compaction, (c) shape and size of particles, (d) wall effects (it is the disturbance in packing density due to the presence of isolated coarse particle in the matrix of fine aggregate matrix, Fig. 2a and e loosening effects (it is the disturbance in packing density due to the presence of larger sized small particle which cannot fit in the void spaces of coarse aggregate, Fig. 2b [15].

It is found that there are several particle packing models established by the researchers in order to achieve optimal filling: Furnas [11], Aim [16], Toufar [2–16], Stovall [17], Dewar [18], De Larrard [19, 20], Fuller Thomson [21], Rosin–Rammler [8] and Andreassen models [2–16]. However, application of Andreassen and modified

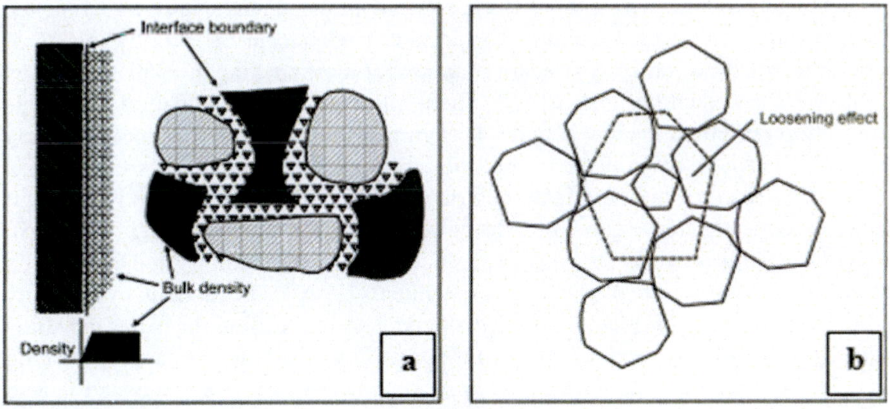

Fig. 2 **a** The wall effect (after Johansen) and **b** The loosening effect (after De Larrard)

Andreassen models for the particle packing of multiple ingredients of cementitious matrix found to be largely being accepted by the researchers and considered as one of the beat known packing model. This paper reviews the application of Andreassen and modified Andreassen models for the effective particle packing investigations on cementitious particles. It also reviews the software's employed for designing various cementitious mixtures based on Andreassen and modified Andreassen models.

2 Categorization of Particle Packing Models

In the past 100 years, numerous models on particle packing theory were suggested by researchers. In general, particle packing can be classified into two entities, i.e. discrete model and continuous model. The broad classification of developed particle packing under these categories with the theoretical consideration is presented in Table 1.

3 Andreassen and Modified Andreassen Models

Fuller and Thomson specified that the properties of concrete can be enriched by adopting concrete mixture with continuous grading and with this concept in 1907 came up with an ideal gradation curve with higher density, which is popularly known as 'Fullers curve' [2, 9, 21]. Fuller curve expression (Eq. 5) with q = 0.5 (factor signifying the modulus of distribution, where curve is modifiable for finer or coarser mix, 2). The model is well known for empirically optimized curve but, the model assumes particle of infinite fineness, i.e. $D_{min} = 0$ [24] with which it may not be able to practically satisfy the ideal distribution of particles.

On the way to improve Fullers equation, Andreasen and Andersen (A&A) carried out a research on particle packing considering the concept of PSD by means of continuous approach and proposed an equation for ideal packing in 1930, i.e. termed as 'Andreassen equation' (Eq. 7, 15). Though the Andreassen equation method is more into theoretical means, it partially signifies the practical concept of particle packing. The Andreasen and Andersen's research and proposed Andreassen equation emphasizes that the void ratio is mainly depends on the distribution modulus 'q'. It has been suggested to use q factor in the between 0.33 and 0.5 as a tuning factor [25], where the q value alters on the particle characteristic, i.e. workability and density and this can be able to achieve by means of experiment [19, 20]. The research by Kennedy et al. made use of 0.45 q factor for asphalt concrete mix design for better densified packing [26] and it is reported by Hummel that higher density of particle packing is achieved by considering q value as 0.4 [27]. However, Andreassen equation also assumed that the fine particles are of infinitely small, i.e. $D_{min} = 0$, which cannot be accept in practical point of view.

Table 1 Classification of developed particle packing under these categories

Particle packing model (PPM)				Theory
Primary classification	Secondary classification	Ternary classification	Quaternary classification	
Discrete model	Binary	Furnas powers [11–12]	–	*Considered for ideal packing of two materials, i.e. fine and coarse particles * d_1 & d_2—diameter; y_1 & y_2—volume fraction and φ_1 & φ_2—packing density of fine and coarse particles, respectively * Two conditions considered: (1) $y_1 \gg y_2$—fine grain dominant $\varphi t = \frac{1}{y2+\frac{y1}{\varphi 1}}$ (2) $y_2 \gg y_1$—coarse grain dominant $\varphi t = \frac{\varphi 2}{1-y1} = \frac{\varphi 2}{y2}$ * Furnas model is valid only if $d_1 << d_2$, if not it also depends on d_1/d_2 ratio * Power proposed an equation to get lesser ratio of void for two-size system i.e. d_{max}/d_{min} where dmax—diameter of maximum size particle; dmin—diameter of minimum size particle
		Aim and Goff [16]	–	* Proposed a simple geometrical model to deal with the wall effect (Fig. 2a) of binary mix and suggested a correction factor in order to avoid excess porosity in the first layer of spherical grains in contact with a plane and smooth wall * $(d_{max}/d_{min})^n$ where n—degree of an 'ideal' curve equation
	Ternary	Toufar model [2–16]	–	Assumption: In this model, spaces between large particles cannot be able to fill by smaller particles only

(continued)

Table 1 (continued)

Particle packing model (PPM)				Theory
Primary classification	Secondary classification	Ternary classification	Quaternary classification	
		Modified Toufar model [17]	–	* The Toufar model is altered by Goltermann et al. [14] and proposed a well-known model 'modified Toufar model'. *In accordance to their work procedure, packing degree of each component is computed and termed as eigen factor (φ_1 and φ_2) * Modified Toufar model fitted better with experimental values compared to the Aim and Toufar models for binary system [22] * Further ternary particle packing also measured and theoretical diagram of packing is produced with respect to modified Toufar model to create the packing density contours *Necessary input to be considered in this model are: density of packing and equivalent diameters of each ingredients
	Multi-modal	De Larrard model [18–20]	LPDM (Linear packing density model)	*This model is proposed by Stovall et al. [17], stating that with the information of size of particle and particle size distribution (PSD) packing density could be calculated * This model performed well in designing the superplasticized cementitious materials
			SSM (Solid suspension model)	* The model proposed with some alterations to LPDM by De Larrard and Sedran and named as SSM * It is a excellent means to validate the high packing density of cementitious materials * An important concept is that this model considered the dissimilarity between real packing density (ϕ) and practical packing density (β) * It is reported that, for a given mix, it could be able to attain the maximum packing density by maintaining the original shape of each particle as well as by positioning one after the other in a mix *This model would be appropriate in case of concentrated suspensions to predict the plastic viscosity

(continued)

Table 1 (continued)

Particle packing model (PPM)				Theory
Primary classification	Secondary classification	Ternary classification	Quaternary classification	
			Modified LPDM	*The LPDM concept was extended with the calculation of eigen packing density and PSD for various blends of all set of size 'i' *This modification is proposed by Glavind et al. and termed as modified LPDM model
			CPM (compressible packing model)	* The model is recommended by De Larrard * This model is independent of LPDM/SSM, rather depends on compaction index 'K' *where packing density is calculated similar to LPDM, while, for different compaction indices the virtual eigen packing density varies
Continuous model	Multi-modal	Fuller Thomson [21]	–	* For obtaining the maximum density, Fuller and Thomson recommended the ideal gradation curves (Fuller's ideal curve) * The Fullers equation that is given as: $CPFT = (d/D)^n 100$ where CPFT = Cumulative percentage finer than; n = 0.45–0.5; d = Particle size; D = larger particle size *Further this expression is modified by Shakhmenko and Birsh [23] for concrete mix proportioning and is represented as: $CPFT = T_n(d_i/d_0)^n$ where n = degree of an ideal curve. T_n = constant, dependent aggregate size (max) and the exponent n

(continued)

Table 1 (continued)

Particle packing model (PPM)				Theory
Primary classification	Secondary classification	Ternary classification	Quaternary classification	
		Andreassen [2–16]	**Modified Andreassen**	*Andreassen proposed an equation for ideal packing considering the particle size distribution via continuous approach. The equation is well known as Andreassen expression, say: $CPFT = (d/D)^q 100$ *Funk and Dinger brought modification to Andreassen equation considering the minimum particle size diameter. This model is popularly known as modified Andreassen model and the equation is given as: $CPFT = \{[(d - d_0)]/[(D - d_0)]\}^q 100$ where CPFT = tcumulative percent finer than, d = particle size, do = particle size distribution of smaller particles, D = the maximum particle size and q = distribution exponent
		Rosin–Rammler [8]	–	* Rosin–Rammler equation considers the characteristic diameters (D') of the various ingredients gradation curve used for concrete *Rosin–Rammler equation is given as: $R(D) = \exp(-D/D')^n$ where R(D) = % Residue fraction; D = diameter, D′ = characteristic diameter, n = 1.04–4 (constant)

*LPDM—Linear packing density model, SSM—Solid suspension model, CPM—compressible packing model

Funk and Dinger specified that in any realistic PSD D_{min} cannot be infinitely small it must have some limited lower size value [28]. Hence, on this baseline Funk and Dinger in 1994 [29] modified the Andreassen's equation considering the minimum particle size (D_{min}) as a finite factor and proposed an equation known as 'modified Andreassen's equation (modified A & A)' (Eq. 8). Figure 3 represents the relative ideal packing curve proposed by Fuller, A & A and and Funk–Dinger for D_{max} = 32 mm and D_{min} = 63 μm [16, 21, 29].

On the other hand, it has been reported that distribution modulus (q) in A & A and modified A & A equations vary depending on the need for workability, i.e. higher the value of q exponent lower is the degree of workability (refer to Table 2). It means that higher the amount of coarse material, larger is the q exponent and if the amount of fine materials are more, then lower is the q value; this has been depicted in Fig. 3 [2]. The amount of fine particles in mixture has the ability to control the water demand and capacity to hold water, in which the exponent q acts as an indicator for volume of finer fraction that could be placed in mixture. This provides a platform for opting the reasonable amount of water and water reducing agents [2]. The selection of distribution modulus should be done carefully so that it should not be too high or too

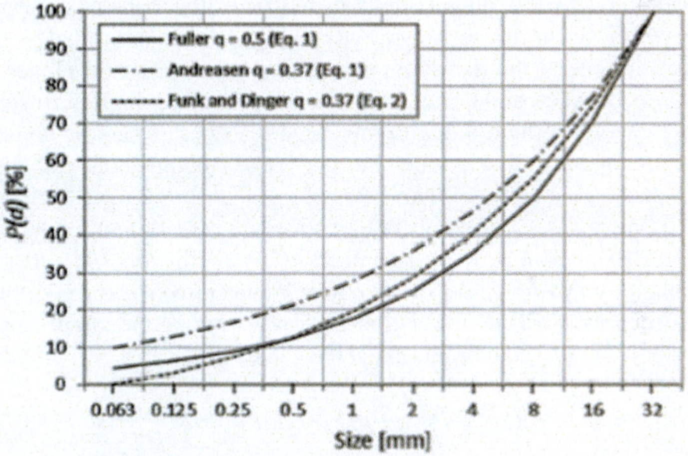

Fig. 3 Idyllic curves proposed by Fuller, Andreasen & Andersen and Funk and Dinger for D_{max} = 32 mm, D_{min} = 63 μm *Source* Fennis and Walraven 2012

Table 2 Relative values of compaction factor and 'q'

Degree of workability	q	Compaction factor
Very Low	0.37	0.8
Low	0.32	0.85
Medium	0.27	0.9
High	0.22	0.95

Source EMMA manual & Table 20 and 22 in SP 23 Handbook

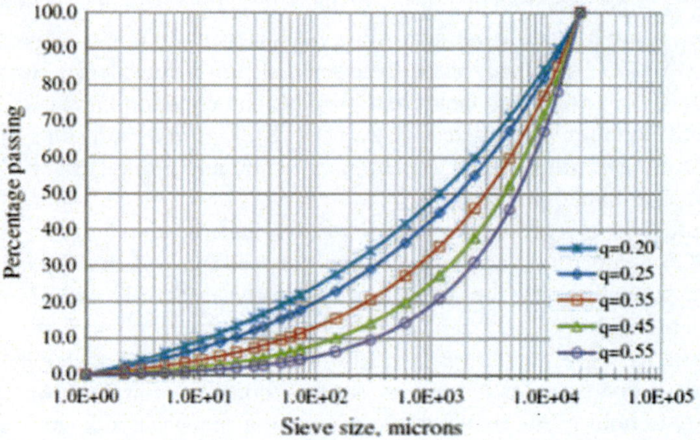

Fig. 4 Ideal modified A & A model curve with q factor ranging from 0.2 to 0.55 *Source* Wang [31]

low in order to avoid the adverse effect on the concrete/mortar mixture. Higher value of q exponent may create segregation effect and lower the value of q exponent may lead to obvious viscosity due to the presence of higher amount of fines. It have been reported that in general the distribution modulus q for recent developed concrete mixtures varies between about 0.21 to 0.37. For producing high performance and conventional concretes the q value in the range 0.25 to 0.3 may be considered by keeping the slump constraint in view. Further for self-compacting concretes, value of q may be considered as lesser than 0.23 [2].

Table 2 Relative values of compaction factor and distribution modulus (q) [30] (referring EMMA manual & Table 20 and 22 in SP 23 Handbook) [2, 10] (Fig. 4).

The evaluation of theoretical particle packing need to consider the summation of the residual squares (RSS) (Eq. 1). In order to accomplish the target curve fitting, the optimization of real (actual) mix must have least RSS value

$$RSS = \sum (Pt - actual - Pt)^2 \tag{1}$$

where Pt-actual: actual PSD curve.

Pt: target particle size distribution curve (A & A model curve).

The gradation with higher particle density and optimum mix ingredients is represented by the 'ideal' curve. But it has been reported that gradation of aggregates alone will not give optimize mix with greater density, the fine particles like cement and other mineral admixtures mixed with water will also play a role in particle packing by accommodating the finer pores [32]. It is easier to optimize the ingredients of mix proportion by adjusting the proportions to the fixed target curve (ideal cure) by means of software. It is made possible with the limited input of parameters, i.e. PSD of the ingredients, and necessary q value can be changed for the required degree of workability (it will be further easier if q value is fixed). A & A and modified A & A models is the well-known, relevant and simple continuous packing model, which considers the continuous distribution of particles and can be adopted for multiple ingredient mixes which is of the need today. However, beyond all, the benefits there is a shortage in consideration of particle characteristics that is shape of the particle which plays a role in particle packing is not considered in the particle packing model that may leads to the overestimate of packing density of mixture. It have been reported that higher packing densities is observed when the model is used in gap-graded mixtures can lead to [3].

The user-friendly software available for particle packing theory based on A & A and modified A & A models is 'LISA' and is presently called as 'EMMA' software.

LISA (*Language-Independent Size distribution Analyser*): It is a software developed by Elkem, Norway. This software was applied to validate the particle packing models for different cementitious mix designs established on the basis of Andreasen PSD model. This software is easy to use because of its simplicity, applicability and accessibility.

Then, now it is called as *EMMA* (*Elkem Materials Mixture Analyser*): It is a user-friendly computerized program developed by Elkem to optimize the concrete and mortar mixes with some modification to the prior software named LISA. This software makes use of PSD of a selected mixture materials (sand, gravel, binders, etc.) by the user and the related to the A & A/modified A & A models. Software enables to compute and display the PSD of mixture composition and predicts the optimum blend of mixture.

EMMA software is also designed for calculating the CO_2 impact of the optimized concrete mixture for particle packing performance. By knowing the CO_2 loading, it enables the user to reduce the carbon footprint through replacing the cement by means of low carbon footprint mineral admixture.

The steps involved in mix proportioning of concrete/mortar with multiple ingredients by EMMA software is represented in a flow chart (Fig. 5).

4 Application of Andreassen and Modified Andreassen Packing Models in Mortar and Concrete Design

The application of particle packing model by means of A & A/modified A & A models approach considering LISA/EMMA software is demonstrated in this section with few researchers' data.

4.1 Cement Mortar

To validate the usage of modified A & A particle packing model to design mortar mixture, a trial mix has been considered and represented in the Table 3 [2]. Where,

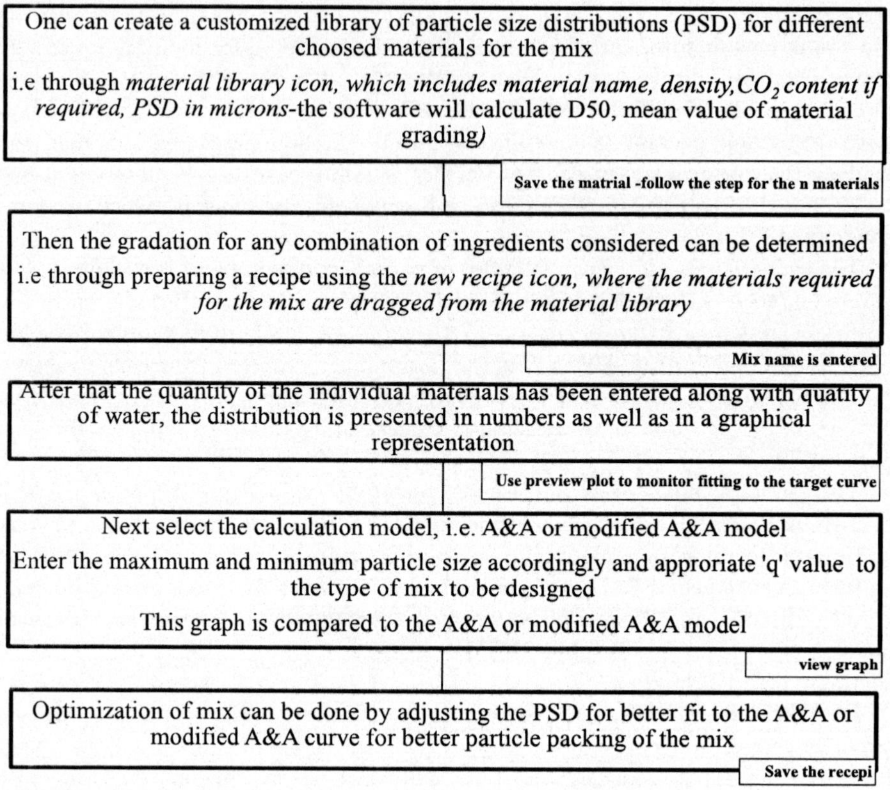

Fig. 5 Flow chart representing the steps involved in mix proportioning of concrete/mortar with multiple ingredients by EMMA software

Table 3 Mix design of mortar mixes with using particle packing model

Ingredients, Kg/m^3	mix A	mix B
Standard sand (G-1)	541	341
Standard sand (G-2)	541	341
Standard sand (G-3)	541	341
Crushed sand (correction)	–	341
Cement, (OPC)	541	541
Quartz powder (inert filler)	–	150
Microsilica	–	60
Water	216	216
SP (PCE)	2	4
w/c	0.4	0.4
w/p	0.4	0.29
Flow spread, percent	60	60

Source Kumar et al. [2]

mix A designates the reference mix designed in accordance to IS 4031-1996. The mix B designates the modified design mix with incorporation of crushed sand, quartz powder and microsilica in order to accommodate the missing regions of particle gradation. The actual gradation curve of the mix is compared with ideal curve of modified A & A model as presented in the Fig. 6. It can be noticed from the figure that mix B fits better to the ideal curve by means of modifying the reference mix with the particle size of missing zones there by representing the better particle packing. It have reported the modified mix (mix B) performed well with respect to 3, 7 and 28 days compressive strength, i.e. approximately 28–30% increase in strength for all the curing ages has been accomplished.

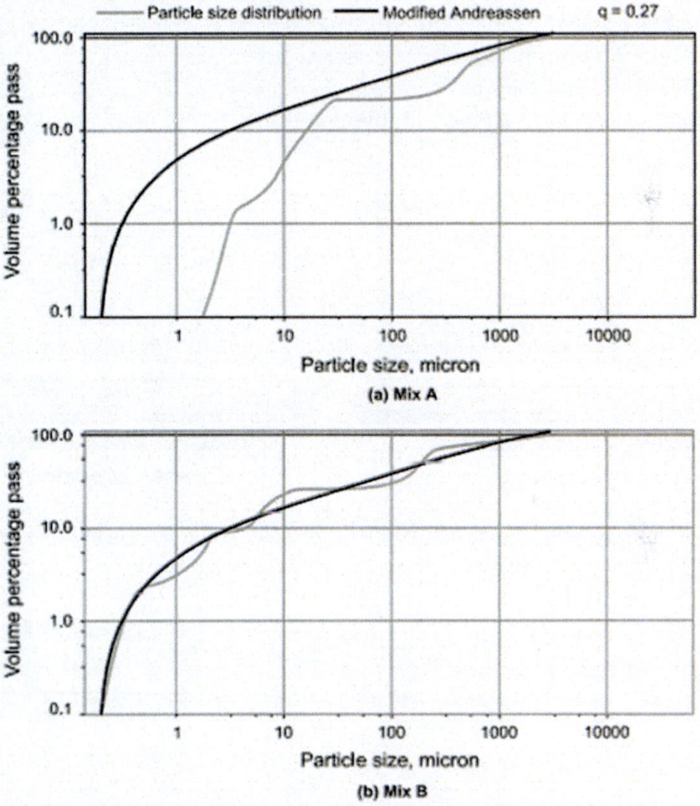

Fig. 6 Andreassen and modified Andreassen packing model application in mortar and concrete *Source* Kumar et al. [2]

4.2 High-Performance and Ultra-High Performance Concrete

The comparative study on optimizing the mix deigns for normal and high-performance concrete conducted by Gopinath et al. [33] made use ACI and BIS mix design methods, as well as EMMA particle packing software (based on A & A/modified A & A models) to modify the mixes. Authors have stated that the designed M30 mixes by ACI and BIS methods were optimized by modifying the mix to fit modified A & A curve by keeping w/c ratio constant. Table 3 presents the comparative mix proportions of modified and unmodified ACI and BIS mixes. Figures 7 and 8 represents the mix design which is designed by means of BIS and ACI methods, respectively and further modified using particle packing model. Authors also investigated on the high-performance concrete containing mineral admixtures like fly ash, silica fume and they have limited their study for optimized mix design by means of particle packing method (Table 5 and Fig. 9). Authors concluded stating

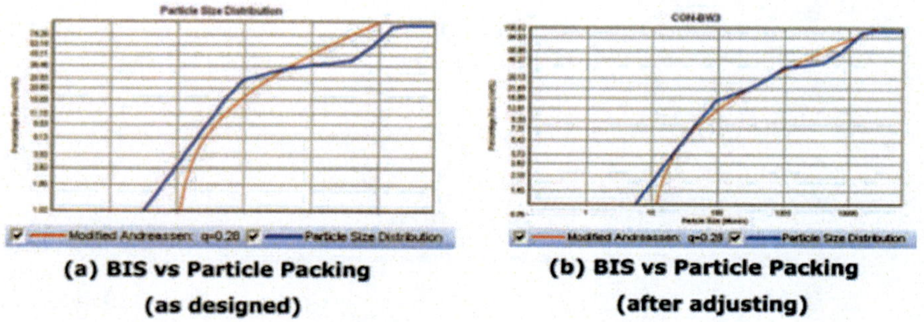

Fig. 7 BIS design mix curve beforehand and afterward optimization using modified A & A particle packing model *Source* Gopinath et al. [33]

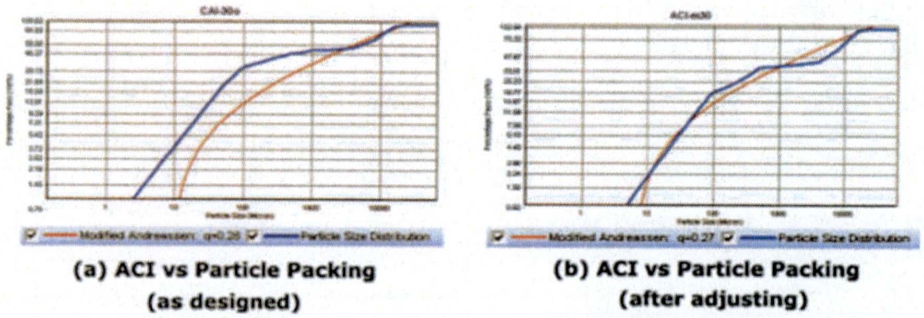

Fig. 8 ACI design mix curve beforehand and afterward optimization using modified A & A particle packing model *Source* Gopinath et al. [33]

Table 4 Mix proportion for normal strength concrete

Mix design	Cement	Fine aggregate	Coarse aggregate	Water	Compressive Strength (MPa)		
	kg/m3				3 day	7 day	28 day
M30-ACI	398	848	888	191	19.46	25.17	40.37
M30-ACI modified	380	460	1300	182.4	17.82	31.97	43.81
M30-BIS	431.67	549.73	1198	194.25	31.13	46.35	53.21
M30-BIS modified	360	660	1199	162	18.33	–	42.07

Source Gopinath et al. [33]

Table 5 Mix quantity for high-performance concrete

Mix design	Quantity (kg/m^3)	Compressive strength (MPa)		
		3 day	7 day	28 day
Cement	420	27.11	42.13	47.53
Fly ash	170			
Silica fume	35			
Quartz sand	500			
Quartz powder	450			
Water	280			

Source Gopinath et al. [33]

that designed mixes modified by the method of particle packing projected closer compressive strength to the design mixes.

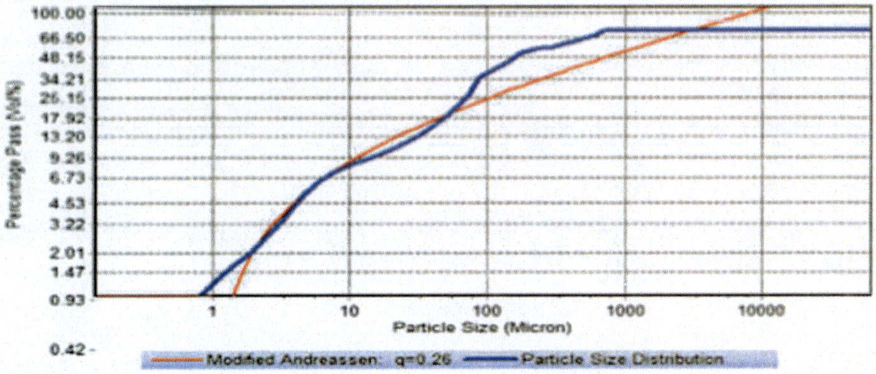

Fig. 9 Mix design curve for high-performance concrete optimized by modified A & A particle packing *Source* Gopinath et al. [33]

The investigation carried out by Li et al. [34] on ultra-high performance concretes (UHPCs) using basalt coarse aggregates (BA) with larger particle size of 16 mm and reported that the particle packing theory has been applied for designing the mix in order to consider optimal proportion of powder content (i.e. cement, microsilica and lime powder).The recipes used by the researchers in their work is presented in the Table 5. Authors have altered the powder content from 900 kg/m^3 to 650 kg/m^3 and basalt aggregates of maximum size from 3 mm to 16 mm, to study the better compactness and strength of UHPC [34]. Modified A & A model with the constant distribution modulus value, i.e. 0.2 has been used in their investigation for optimizing the particle packing among the 12 mixes (UHPC1–UHPC12). Figure 10 represents the target and designed mixes gradation curve of maximum size of basalt aggregate 8 mm and 16 mm, respectively, and Table 6 gives the mix proportion of UHPC. It is clear from the figure that in case of the mixes with maximum aggregate size 8 and 16 mm, UHPC6 and UHPC12 mixes exhibited the much closer curve to the target line (modified A & A curve). However, authors have specified that the maximum strength is obtained for mixed UHPC5 and UHPC11 and reported that for obtaining excellent strength of UHPCs with basalt coarse aggregates, the modulus of distribution q is requested to choose for lower value 0.19 (Table 6).

4.3 *Self-Compacting Concrete*

The study conducted by Brouwers and Radix [35] on the role of distribution of particle size in self-compacting concrete, utilized slag-blended cement, combinations of three sands, gravel and SP. The authors have aimed to reduce quantity of paste (cement, limestone powder and water) as much as possible with certain amount of solid contents (sand and gravel). The three mixes have been considered in their study on the basis of particle packing theory, which contains total powder content of 473 kg/m3–499 kg/m3 (Table 7). Authors have considered Fuller curve, A & A curve and modified A & A curve to validate the mixes, the adopted particle packing model curve has been represented in Fig. 11. It can be observe from the figure that the particle distribution curve of all three mixes fits well with the ideal curve of modified A & A model (with $q = 0.25$) than the other two models. Authors have reported that as sand contributed significantly for PSD, it may not be able to fit perfectly to A & A model, especially in the range of 100 μm–4 mm. Authors have concluded stating that economical SCC mixes can be developed by means of particle packing theory which meets the standards and requirements in fresh state.

4.4 *Geopolymer Concrete*

The experimental investigation carried out by Borges et al. on the fabrication of geopolymer concrete employed Andreassen particle packing method to improve the

Table 6 Mix proportion of UHPC

No	A-b	CEM	mS	LP	S	BA (1–3 mm)	BA (2–5 mm)	BA (5–8 mm)	BA (8–11 mm)	BA (8–16 mm)	Water	SP
		Kg/m³										
UHPC1	3–900	675	45	180	864.5	576.3	0	0	0	0	180	10.8
UHPC2	8–900	675	45	180	588.1	178.6	403.1	313.9	0	0	180	10.8
UHPC3	16–900	675	45	180	304.7	308.3	306.6	247.5	120.9	206.2	180	10.18
UHPC4	8–850	637.5	42.5	170	700.1	100.5	435.7	310.4	0	0	170	10.2
UHPC5	8–800	600	40	160	812.2	22.3	468.4	307	0	0	160	9.6
UHPC6	8–750	562.5	37.5	150	903.6	0	318.8	453.8	0	0	150	9
UHPC7	16–900	675	45	180	242.9	338.9	278.6	251.5	109.9	200.7	207.0	5.4
UHPC8	16–850	637.5	42.5	170	355	259.4	315.4	238.5	118.1	203.1	195.5	5.1
UHP9	16–800	600	40	160	467.2	179.9	352.2	225.3	126.24	205.4	184	4.8
UHPC10	16–750	562.5	37.5	150	579.3	100.4	389	212.2	134.6	207.8	172.5	5.3
UHPC11	16–700	525	35	140	698.7	0.6	445.2	186.9	147.8	209.6	161	4.9
UHPC12	16–650	487.5	32.5	130	782.6	0.0	406.6	221.5	136.2	213.9	149.5	4.6

Source Li et al. [34]

* a (mm) and b (μm)— larger particle size of basalt used and powder content used, respectively, CEM—cement, mS—microsilica, LP—lime powder, S—sand, BA—Basalt aggregate, SP—superplasticizer and water/binder ratio of UHPC1–UHPC6 is 0.2 and UHPC6–UHPC15 is 0.23

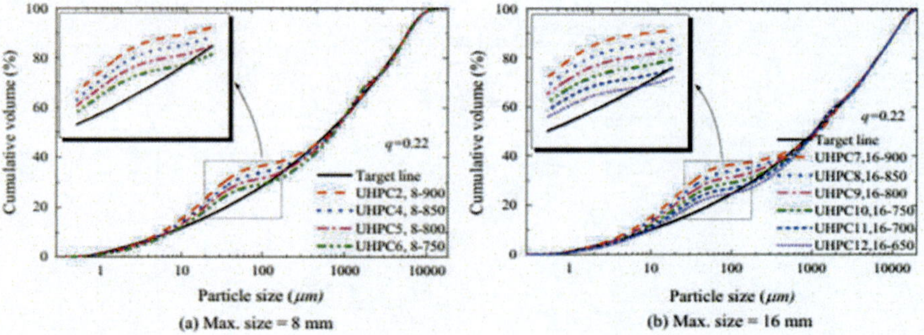

Fig. 10 Gradation of the target and designed curve for UHPCs *Source* Li et al. [34]

Table 7 Mix proportion of self-compacting concrete developed on the basis of packing theory

Material	Mix A [kg/m3]	Mix B [kg/m3]	Mix C [kg/m3]
Cement	310	315	320
Limestone powder	189	164	153
Sand 0–1 mm	–	–	388
Sand 0–2 mm	–	306	–
Sand 0–4 mm	1018	719	628
Gravel 4–16 mm	667	673	687
Water	170	173	174
Superplasticizer	6.0	5.51	5.21
Water/cement ratio	0.55	0.55	0.55
Water/powder ratio	0.34	0.36	0.37
Mass fraction < 250 μm	554	538	687

Source Brouwers and Radix [35]

geopolymer formulations [36]. Authors have considered three q factors, i.e. 0.21, 0.235 and 0.26 with the solution to solid ratio (s/s) of 1.3, 1.4 and 1.5, respectively, in addition two aggregate type, i.e. glass and quartz have been considered in their investigation. Authors have reported that lower value of q factor showed better workability irrespective of the aggregate type. Authors also observed that formulation prepared with 1.3 s/s and q factor 0.26 showed higher viscosity and complexity to compact, while other formulations presented the easy compaction and showed self-compacting effect. The graph plotted by the researchers on the mechanical properties, apparent porosity and density of geopolymers containing quartz and glass aggregate connecting to packing factor q and solution to solid ratio is presented in Figs. 12 and 13. Authors have stated that Andreassen method may be considered as suitable technique to alter the rheological properties for developing various mixes of geopolymers. However, among the considered formulation of geopolymer mixes

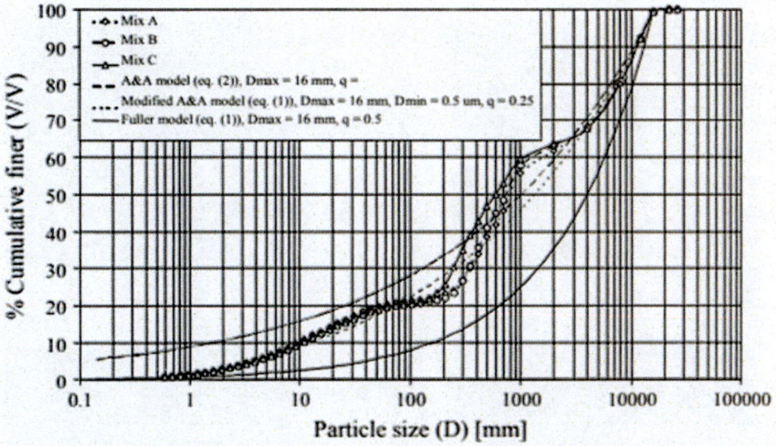

Fig. 11 Particle size distribution of three self-compacting mixes related to ideal curves *Source* Brouwers and Radix [35]

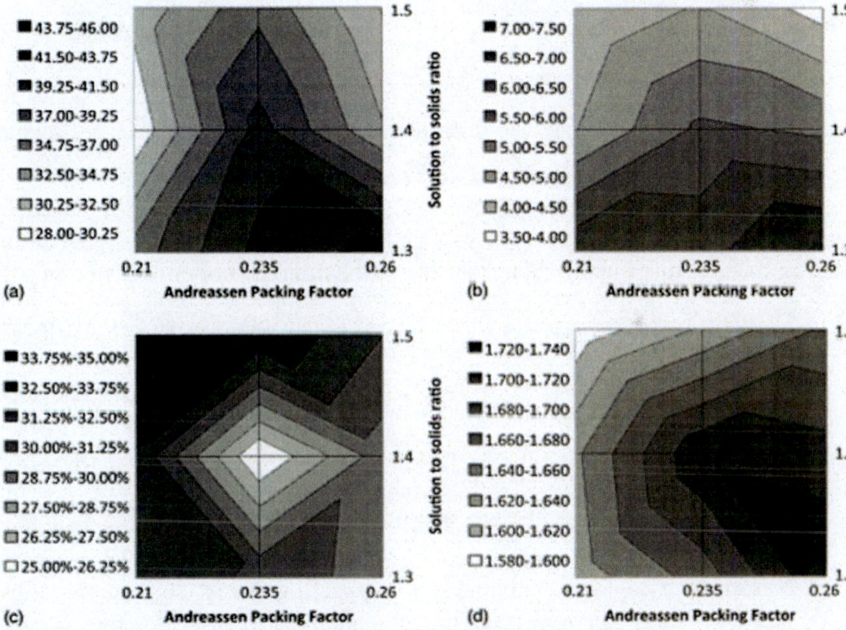

Fig. 12 **a** compressive strength; **b** flexural strength; **c** apparent porosity; **d** apparent density for geopolymers made with quartz aggregate *Source* Borges et al. [36]

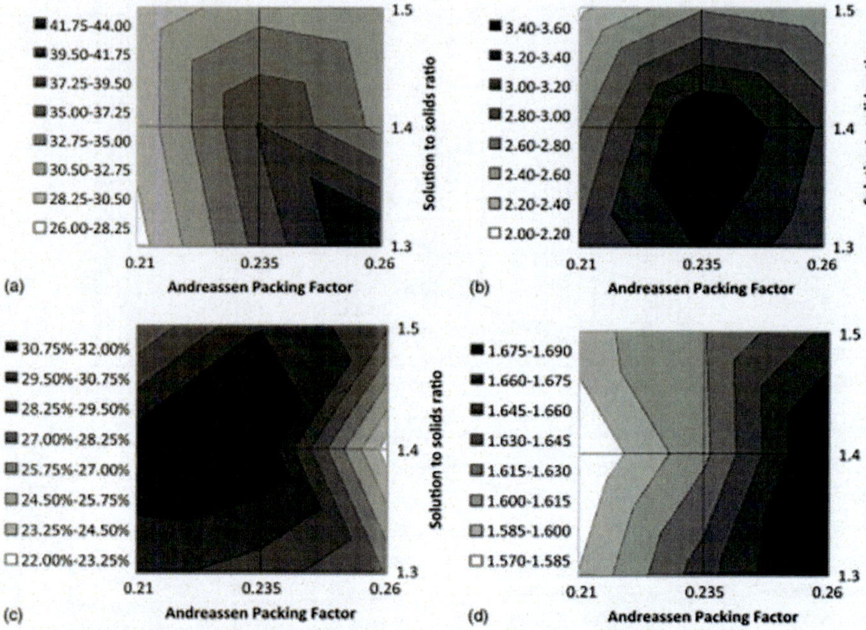

Fig. 13 **a** compressive strength; **b** flexural strength; **c** apparent porosity; **d** apparent density for geopolymers made with glass aggregate *Source* Borges et al. [36]

by the researchers, the mix with s/s of 1.4, q of 0.235 and with quartz aggregate is found to be the best mix in particularly for the fabrication of geopolymer floor tiles.

5 Conclusions

The approach of particle packing theory for obtaining an optimum mix ingredients improvised behaviour of cement composites in addition maintained the adequacy of strength. This review focused on the modified Andreassen model, and this packing model is the best choice for designing the multi-blended mixes and special concrete like high-performance, high-strength and self-compacting mixes. The user-friendly software based on modified Andreassen model made ease in designing the cement composite mixes for optimal packing density.

References

1. Singh PR, Shah ND, Majumdar PK (2018) Effect of density and porosity on the durability of flyash blended concrete. Int Res J Eng Technol (IRJET) 5(2), e-ISSN: 2395-0056

2. Kumar SV, Santhanam M (2003) Particle packing theories and their application in concrete mixture proportioning: a review. Indian Concr J
3. Fennis SAAM, Walgreen JC (2012) Using particle packing technology for sustainable concrete mixture design. Heron
4. Kumar A, Kisku N (2016) Effect of silica fume and fly ash as partial replacement of cement on strength of concrete. Int J Innov Res Sci Eng Technol 5
5. Bleeck TH (2011) Particle packing; an effective approach to optimized design of ultra high strength and self compacting concretes. ADFA J Undergrad Eng Res 4(1)
6. Neville AM (1995) Properties of concrete (4th Edition ed.). Longman House
7. De Larrad (1999) Concrete mixture proportioning—a scientific approach. E & FN Spon, London
8. Mangulkar MN, Jamkar (2013) Review of particle packing theories used for concrete mix proportioning. SS Int J Sci Eng Res 4
9. Fuller WB, Thompson SE (1907) The laws of proportioning concrete. Trans, ASCE 59:67–143
10. Suenson E (1911) Building materials III: stone, pottery, mortar, concrete, artificial stone, glass. Danish
11. Furnas CC (1931) Grading agrregates I-mathematical reltions for beds of broken solids of maximum density. US Bureau of Mines 23(9):1052–1058
12. Powers TC (1968) The properties of fresh concrete. John Wiley & Sons, Inc., New York
13. Rajamane NP, Ranjani IS, Rao BHN, Peter JA, Gopalakrishnan (2004) Use of theory of particle packing for estimation of mean particle sizes of fly ash and silica fume relative to Portland cement. Proc Sec Nat Conf Adv Concr Technol
14. Golterman, Johansen V, Palbol L (1997) Packing of aggregate: an alternative tool to determine the optimal aggregate mix. ACI Mater J, 435–442
15. Mohammed MH, Pusch R, Al-Ansari N, Knutsson SE, Nilsson M, Pourbakhtiar A (2013) Talc-based concrete for sealing borehole optimized by using particle packing theory. J Civil Eng Arch 4(65):440–455
16. Andersen PJ, Johansen V (1991) Particle packing and concrete properties. Material Science of Concrete: II, Skalny J and Mindess S (Edited), The American Ceramic Society, Inc., Westerville, Ohio, 111–147
17. Stovall T, De Larrard F, Buil M (1986) Linear packing density model of grain mixtures. Powder Technol 48:1–12
18. Dewar JD (1999) Computer modeling of concrete mixtures. E and FN Spon, London
19. De Larrard (1997) A new rheometer for soft-to-fluid fresh concrete. ACI Mater J 94(12):237
20. De Larrard F (1998) Why rheology matters? Concr Int 21(8):79
21. Fuller WB, Thompson SE (1907) The laws of proportioning concrete. Am Soc Civil Eng 33:223–298
22. Jones MR, Zheng L, MD (20002) Comparison of particle packing models for proportioning concrete constituents for minimum voids ratio. Newlandater Struct /Mat6riaux et Constr 35:301–309
23. Shakhmenko G, Birsh J (1998) Concrete mix design and optimization. Sec Int Sym Civil Eng Budapest, 1–8
24. Cai W (2011) Effect of particle packing on flow property and strength of concrete mortar. Graduate thesis and dissertations, Lowa state university, 15271
25. Hunger M, Brouwers HJH (2009) Flow analysis of water powder mixtures: application to specific surface area and shape factor. Cement Concr Compos 31:39–59
26. Kennedy T, Huber G, Harrigan E, Cominsky R, Hughes C, Quintus H, Moulthrop J (1994) Superior performing ashphalt pavements (superpave): the product of SHRP asphalt research program. National research council, SI-IRP-A-410
27. Hummel A (1959) Das beton-ABC- Ein Lehrbuch der technology des schwerbetons and des leichtbeton. 11th edn, Wilhelm Ernst and Sohn, Berlin
28. Funk JE, Dinger DR, Funk JEJ (1980) Coal grinding and particle size distribution studies for coal water slurries for high solid content. Final report, Empire state electric energy research corporation, New York

29. Funk JE, Dinger DR (1994) Predictive process control of crowded particulate suspension–applied to ceramic manufacturing. Kluwer academic publishers, Boston
30. www.silicafume.net, Elkem AS
31. Wang X (2014) Proportioning and performance evaluation of self-consolidating concrete. Thesis, Iowa State University, Ames, Iowa
32. Fennis SAAM, Walraven JC (2012) Using particle packing technology for sustainable concrete mixture design. Heron 57(2)
33. Gopinath S, Murthy RA, Ramya D, Iyer NR (2011) Optimised mix design for normal strength and high performance concrete using particle packing method. Arch Civil Eng LVII, 4
34. Li PP, Yu QL, Brouwers HJH (2018) Effect of coarse basalt aggregates on the properties of ultra-high-performance Concrete (UHPC). Constr Build Mater 170:649–659
35. Brouwers HJH, Radix HJ (2005) Self-compacting concrete: the role of the particle size. Radix first international symposium on design, performance and use of self-consolidating concrete SCC
36. Borges PHR, Fonseca LF, Nunes VA, Panzera TH, Martuscelli CC (2016) Andreasen Particle packing method on the development of geopolymer concrete for civil engineering. J Mater Civ Eng 26:692–697

Characterization of Mechanical and Microstructural Properties of FA and GGBS-Based Geopolymer Mortar Cured in Ambient Condition

K. M. Prasanna, Saif Tamboli, and B. B. Das

Abstract Fly ash-based geopolymer mortars require heat curing to achieve its properties, which limits its practical application at ambient conditions. The present study was aimed to accomplish the need for application of fly ash-based geopolymers for practical viability without any heat curing by inclusion of ground-granulated blast furnace slag (GGBS). The results revealed that inclusion of GGBS as a partial replacement to fly ash (FA) in geopolymer mortar, which is cured in ambient curing condition, can be able to achieve required setting time and compressive strength. Amalgamation of GGBS with class FA as binder in geopolymerization lend a hand to attain compressive strength as well as setting time which is analogous to ordinary Portland cement (OPC). Microstructural properties were studied using scanning electron microscopy.

Keywords Geopolymer · Fly ash · GGBS · Curing

1 Introduction

Amongst the worldwide cement industry is leading creators of carbon dioxide (CO_2), generating of about 7% of global CO_2 emissions and its production will be doubled by the year 2020 which is resulting in the emissions CO_2, obviously designates the risk of global warming [1, 2]. Direct CO_2 emission in India in the year 2017 was 588 kg CO_2 per ton of cement and 5% decrease as likened to the baseline of 2010 which is well below the target of 40% reduction to be achieved by 2050 [3]. Second leading producers of cement over the globe next to China is India. Rapid evolution in urbanization the increased the demand in construction sectors where, cement production is anticipated to be more than 600 million tonnes per annum by the end of the year 2025 [4]. Therefore, the need for cement alternatives in the near future is an important matter to be considered.

Thermal power plants utilize coal for the production of electricity and produced fly ash (FA) as a waste product. FA generation and utilization during the year 2017–18

K. M. Prasanna · S. Tamboli · B. B. Das (✉)
Department of Civil Engineering, National Institute of Technology Karnataka, Surathkal, India
e-mail: bibhutibhusan@gmail.com

B. B. Das et al. (eds.), *Recent Developments in Sustainable Infrastructure*, Lecture Notes in Civil Engineering 75, https://doi.org/10.1007/978-981-15-4577-1_64

in India was 196.44 MT and 131.87 MT, respectively, and the percentage utilization stood at 67.13% in the country, which is consumed in the cement, concrete, blocks and tile production, in addition to that few were disposed in landfills and embankments. However, the unutilized FA which is about 32.87% is the matter of concern in environmental point of view. Only 69 out of a total 167 thermal power stations reported a level of utilization of 100% or more as per the report [5]. Survey reports that around the year of 2030, unutilized FA count may go up to 128 MT which will necessitate extra land of 2,300 ha, intensifying the current glitches regarding dumping of FA [6].

Blast furnace slag is produced from steel industry which is generated during the extraction of iron from its ore. The chief portion of GGBS produced in India is mostly used by cement industry, followed by landfills and railway ballast. Typically, GGBS production varies from 300 to 540 kg per tonne for the ore feed containing 60–65% iron. Waste utilization in major steel plants in India is around 100% and more for blast furnace slag yet the concern for disposal remains as the production of iron and steel is increasing upwards every year [7].

Geopolymers were initiated in the era of 1970s by Joseph Davidovits, which is an inorganic polymeric binding material. Geopolymerisation is chemical process where raw alumino-silicate oxides react with highly alkaline environment such as alkali silicate solutions resulting in Si–O–Al bonds, which is an amorphous to semi-crystalline three-dimensional (3D) structures [8]. The exothermic reaction of geopolymerization occurs under atmospheric pressure at temperatures below 100 °C. Actual geopolymer mechanism of setting and hardening is yet to be clarified. However, most of the proposed geopolymerization mechanisms comprise of following stages that progress simultaneously and hence, not possible to differentiate, (1) leaching of Si and Al from the blend aluminosilicate and alkaline aqueous solution, (2) production of oligomers (precursors of geopolymer) entailed of Si–O–Al polymer bonds, (3) polycondensation of oligomer to produce a 3D structure, (4) bonding and encapsulation of the inert materials or ingredients into the geopolymeric 3D framework and (5) finally hardening of the whole system into a polymeric solid 3D structure [9–11].

Lampropoulos and Majidi (2016) investigated on setting time of slag-blended FA geopolymer mortar cured in ambient temperature used potassium hydroxide and potassium silicate as alkaline activator showed a reduction in setting time as the slag content amplified up to 50% with highest compressive strength of 49 MPa [12]. Another study by Saha and Rajasekaran (2017) on the mechanical and microstructural performances of FA-based geopolymer by incorporating GGBS for different concentration of alkali activator and found the setting time decreased considerably with just 10% GGBS replacement and maximum compressive strength for the 50% GGBS replacement level at 16 M NaOH solution [13]. Vidya et al. (2016) investigated on blended FA slag geopolymer mortar using sodium hydroxide solution of 16 M concentration and found the maximum compressive strength at 48 MPa at 28 days for 25% GGBS content and a reduction in strength with further increase of GGBS content [14]. Similarly, Nath (2014) investigated the slag-blended FA geopolymer mortar for slag content up to 30% and a varying alkali binder ratio from 0.35 to 0.45 and a low initial setting time of 41 min is obtained at 30% slag content. Increase in alkali activator content improved the setting time. SEM/EDAX revealed a compact microstructure with EDAX showing an increase in the calcium content with higher GGBS replacement levels [15].

Law and Wordhono (2015) investigated and achieved the highest compressive strength of 62.0 MPa at 28 days for the 50% FA replacement level with GGBS at ambient curing condition [16]. Thakur and Ghosh (2010) investigated the effect of curing temperature and time of curing on FA-based geopolymer mortars and found the optimum curing temperature and time as 85 °C and 48 h, respectively [17]. Ghosh (2017) investigated the effect of increasing alkali content in the mix at 60 °C curing temperature on mechanical properties like compressive strength, bulk density, water absorption, and found that the highest compressive strength is 36 MPa with an optimum alkali content of 8% as well as the SEM results indicated a compact and dense microstructure [18]. Another study by Ismail et al. (2017) studied on the influence of GGBS content and curing temperature on the microstructure using SEM found that a dense microstructure was formed at higher curing temperature of 60 °C as compared to ambient cured mix [19]. It has been reported that when pulverized fuel ash (POFA) was used with GGBS, researchers had found the utmost compressive strength of 92 MPa at 90 days ambient curing for 50% replacement level of POFA by GGBS and the microstructure [20]. Ogawa (2018) investigated the blended FA GGBS on heat-cured geopolymer paste and found a coexistence of C–S–H gel and geopolymeric gel in the microstructure [21]. Whereas, it is reported by Rao and Rao (2015) that setting time reduced with increasing GGBS in the mixes and highest compressive strength of 79 MPa obtained for 16 M activator solution for outdoor curing [22].

Several researches have been carried out on FA-based geopolymer pastes and mortars with elevated curing while it is a known fact that this is not feasible for in situ applications. This study involves investigation on the possibility of using FA-based geopolymers with GGBS incorporation and varying alkali–binder ratios to use the resulting mortar in ambient cured condition. The mechanical properties were reported through compressive strength along with setting times of the geopolymer mortars. Further, microstructural characteristics were investigated using scanning electron microscopy (SEM) to analyse the influence of GGBS incorporation and the variation in alkali–binder ratio on the geopolymer mortar mixtures.

2 Materials

2.1 *Fly Ash*

The FA used in the present study was classified as Class F. The specific gravity and fineness of FA used was to found to be 2.22 and 320 m^2/kg, respectively. FA microscopic image is presented in Fig. 1.

Fig. 1 Microscopic images of FA and GGBS

2.2 GGBS

GGBS is a fine dry powder produced from granular slag. The specific gravity and fineness of GGBS used were found to be 2.80 and 350 m^2/kg, respectively. Microscopic image of GGBS is presented in Fig. 1.

2.3 Alkali Solutions

Commercially available sodium silicate solution (Na_2SiO_3—SS) and sodium hydroxide (NaOH—SH) were used. Laboratory-grade SH was in solid state (flakes/pellets form) and its purity is 97%, which is dissolved in distilled water to prepare 10M solution. Laboratory-grade SS is used in the preparation of alkaline solution along with SH. The pure form of SS is of colourless or white in colour. Commercially available SS in liquid form was used, which was supplied by a local manufacturer. The alkali solution was prepared and kept for 24 h before using in the mix.

2.4 Fine Aggregate

Local available natural river sand, which is clean and dry, passing through IS 480 sieve was used as fine aggregate and it was confirming to zone I as per Indian standard [19]. It has fineness modulus and specific gravity of 2.84 and 2.65, respectively.

Table 1 Mix designations used in the study

FA:GGBS	A/B			
	A series—0.5	B series—0.6	C series—0.7	D series—0.8
100:0	GM1 F100G0 S1	GM1 F100G0 S2	GM1 F100G0 S3	GM1 F100G0 S4
90:10	GM2 F90G10 S1	GM2 F90G10 S2	GM2 F90G10 S3	GM2 F90G10 S4
80:20	GM3 F80G20 S1	GM3 F80G20 S2	GM3 F80G20 S3	GM3 F80G20 S4
70:30	GM4 F70G30 S1	GM4 F70G30 S2	GM4 F70G30 S3	GM4 F70G30 S4
60:40	GM5 F60G40 S1	GM5 F60G40 S2	GM5 F60G40 S3	GM5 F60G40 S4
50:50	GM6 F50G50 S1	GM6 F50G50 S2	GM6 F50G50 S3	GM6 F50G50 S4

3 Methods

In this section, tests performed in the characterization of produced geopolymer mortar along with the mix proportion by varying alkali–binder ratio have been presented.

3.1 Mixture Proportions

In this experimental study, binder to sand (B/S) ratio was 1:3, alkaline to binder (A/B) ratios were taken 0.5, 0.6, 0.7 and 0.8 and variations FA and GGBS were 100%:0%, 90%:10%, 80%:20, 70%:30%, 60%:40% and 50%:50% (by weight). The molar concentration of SH is fixed at 10 M and SS/SH ratio is taken constant has 2.5. The mix designations for the mixes in the present study is present in Table 1 and the complete mixture proportions of various geopolymer mortar samples which are investigated are presented in Table 2.

Mixes with A/B ratios varied in an order of 0.5, 0.6, 0.7 and 0.8 were represented as S1, S2, S3 and S4, respectively. Based on the proportion of varying GGBS and FA in the mixes were designated as 0, 10, 20, 30, 40 and 50, respectively, from 100:0 to 50:50 by weight of addition of GGBS.

3.2 Setting Time

Initial setting time (IST) and final setting time (FST) tests were carried out as per standard procedure followed was according the Indian standard for setting time of concrete (IS 8142:1976) [23].

Table 2 Mix proportions used in the study

Mix designation	Quantity (kg/m^3)					
	Sand	FA	GGBS	Alkaline activator	SS	SH
GM1 F100G0 S1	1105	730	0	365	261	104
GM2 F90G10 S1	1105	657	73	365	261	104
GM3 F80G20 S1	1105	584	146	365	261	104
GM4 F70G30 S1	1105	511	219	365	261	104
GM5 F60G40 S1	1105	438	292	365	261	104
GM6 F50G50 S1	1105	365	365	365	261	104
GM1 F100G0 S2	1032	730	0	438	313	125
GM2 F90G10 S2	1032	657	73	438	313	125
GM3 F80G20 S2	1032	584	146	438	313	125
GM4 F70G30 S2	1032	511	219	438	313	125
GM5 F60G40 S2	1032	438	292	438	313	125
GM6 F50G50 S2	1032	365	365	438	313	125
GM1 F100G0 S3	959	730	0	511	365	146
GM2 F90G10 S3	959	657	73	511	365	146
GM3 F80G20 S3	959	584	146	511	365	146
GM4 F70G30 S3	959	511	219	511	365	146
GM5 F60G40 S3	959	438	292	511	365	146
GM6 F50G50 S3	959	365	365	511	365	146
GM1 F100G0 S4	886	730	0	584	417	167
GM2 F90G10 S4	886	657	73	584	417	167
GM3 F80G20 S4	886	584	146	584	417	167
GM4 F70G30 S4	886	511	219	584	417	167
GM5 F60G40 S4	886	438	292	584	417	167
GM6 F50G50 S4	886	365	365	584	417	167

3.3 Compressive Strength

Compressive Strength of geopolymer mortar was carried out using standard cubes were cast. After demoulding, cube samples were kept at room temperature for required curing ages and these specimens were tested after 3, 7 and 28 days of curing. Compressive strength test was carried as per Indian standard code IS-4031 (part 6), 1988 [24].

3.4 *SEM Studies*

SEM analysis was carried out on 28 days samples of alkali–binder ratio of 0.6 and two samples of ratio 0.8. Samples were vacuumed prior to the test and testing is done for a magnification of 500×, 1000× and 2000×. SEM was performed on broken samples from compressive strength at 28 days of alkali–binder ratio 0.6 cured at an ambient temperature for all GGBS percentages to investigate the general microstructural features. EDAX was performed to understand the nature of reaction products.

4 Results and Discussion

4.1 *Setting Time*

From this experimental investigation, it can be noted that IST and FST of geopolymer mortar have reduced considerably with the increase in GGBS content in the mixes. It is observed from Figs. 2 and 3, that the setting time of mix GM1 F100G0 S4 is maximum of 316 min and 668 min for IST and FST, respectively, were obtained for the alkali–binder ratio 0.8. The IST and FST are minimum for the mix GM6 F50G50 S1 with 22 min and 51 min, respectively, was observed for alkali–binder ratio from 0.5.

It is noted that from figures that IST and FST are increasing with the increase of A/B ratio from 0.5 to 0.8, respectively. With only 10% GGBS incorporation, the initial

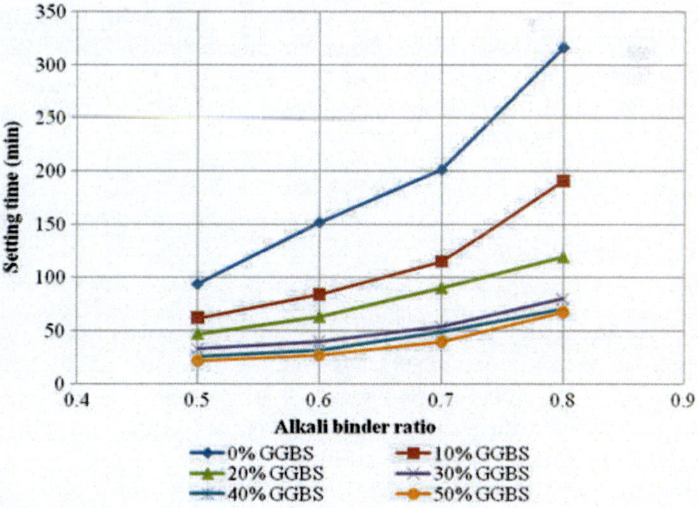

Fig. 2 Initial setting time variation with respect to A/B ratio for different slag percentages

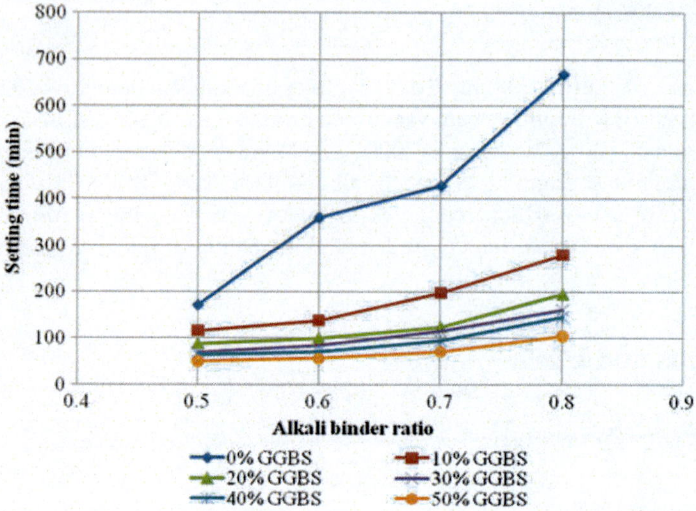

Fig. 3 Final setting time variation with respect to A/B ratio for different slag percentages

setting time was reduced by 40% as compared to FA mixes. Figures 2 and 3 show the variation of IST and FST with alkali–binder ratio for different slag percentages from 0 to 50%. There is a large variation in setting time values of 0% GGBS mixes for all alkali–binder ratios, whereas the setting times of mixes containing GGBS have a relatively linear trend, thus suggesting the possibility of GGBS adding to the stability of the mixes. The setting time values for 30, 40 and 50% GGBS mixes do not vary much for all alkali–binder ratios.

Geopolymer mortar is depending on the development of sodium aluminosilicate hydrate gel (N–A–S–H). However, GGBS is with high-calcium (CaO) content produces geopolymer mortar mixes containing C–S–H gel along with N–A–S–H gel at early duration. Consequently, geopolymer mortar produced with GGBS shows an appreciably lesser time for IST and FST.

4.2 *Compressive Strength*

The compressive strengths were increased with age irrespective of the mix. Highest compressive strength for 3, 7 and 28 days was found for the alkali–binder ratio 0.6. The lowest strengths were reported for alkali–binder ratio 0.8. The compressive strength showed an increasing trend with the increase of GGBS content in the mixes up to 50%. Highest compressive obtained in this study were 34.3 MPa, 41.8 MPa and 56.2 MPa at 3, 7 and 28 days, respectively, for the mixes with alkali–binder ratio 0.6 and 50% GGBS.

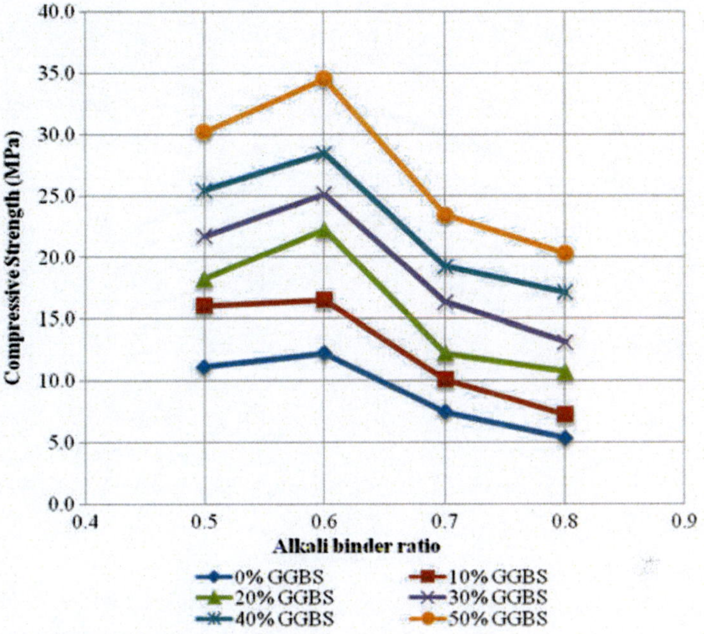

Fig. 4 3 days strength variation for varying A/B ratio

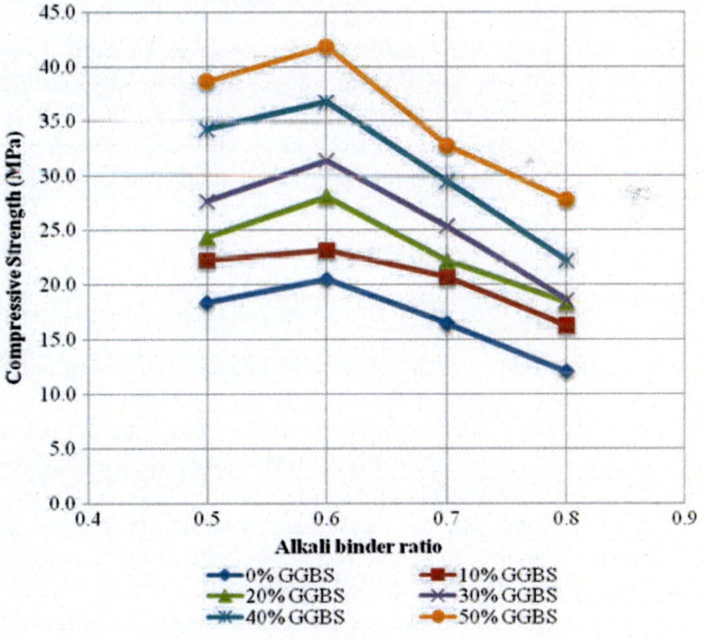

Fig. 5 7 days strength variation for varying A/B ratio

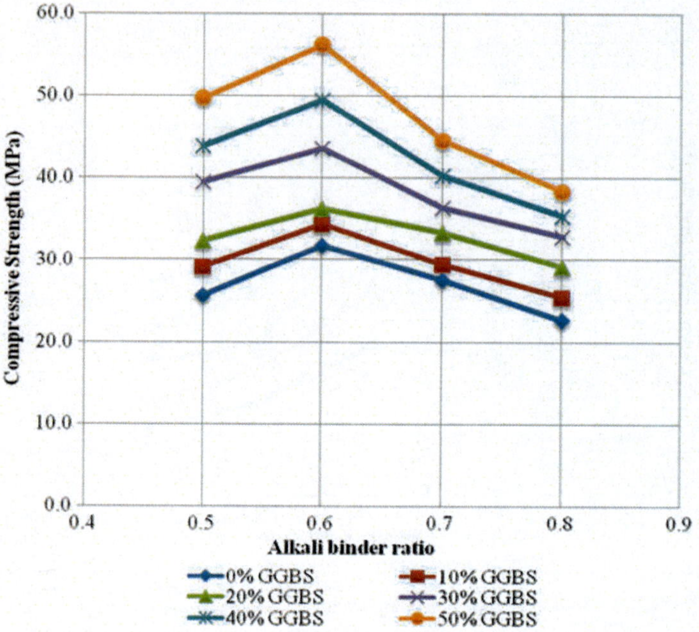

Fig. 6 28 days strength variation for varying A/B ratio

Figures 4, 5 and 6 show the variation of compressive strength with different alkali–binder ratio and different GGBS percentages from 0 to 50%. The 50% mixes in general showed an increase in compressive strengths of 94, 77, 62 and 69% over the mixes containing no GGBS at 28 days. Similarly, the 3-day strength increase was more than 7- and 28-day strength for all mixes of all alkali–binder ratios.

4.3 SEM Studies

The SEM analysis from Figs. 7 and 8 revealed the formation of amorphous phases in the geopolymer mortars. The mix with 0% GGBS content showed more unreacted spherical FA particles than the rest of the mixtures. Few unreacted FA particles were seen in all the mixes. The sample with 10% GGBS showed the gradual development of the geopolymer gel, which covers the crust of the FA particle thereby proving the fact that the geopolymer reaction occurs at the surface of the FA particle. A large number of pores can also be seen in the SEM image of the sample with minor crack widths in the 20% GGBS sample the FA particles seem to be covered with the reaction products and fewer unreacted FA particles are seen as compared with the 10% GGBS sample. The cracks are fewer because of a denser structure of the 20% GGBS sample but the pores are still great in number although with a reduced size. The 30% GGBS sample has lesser pores and a compact structure but the crack width

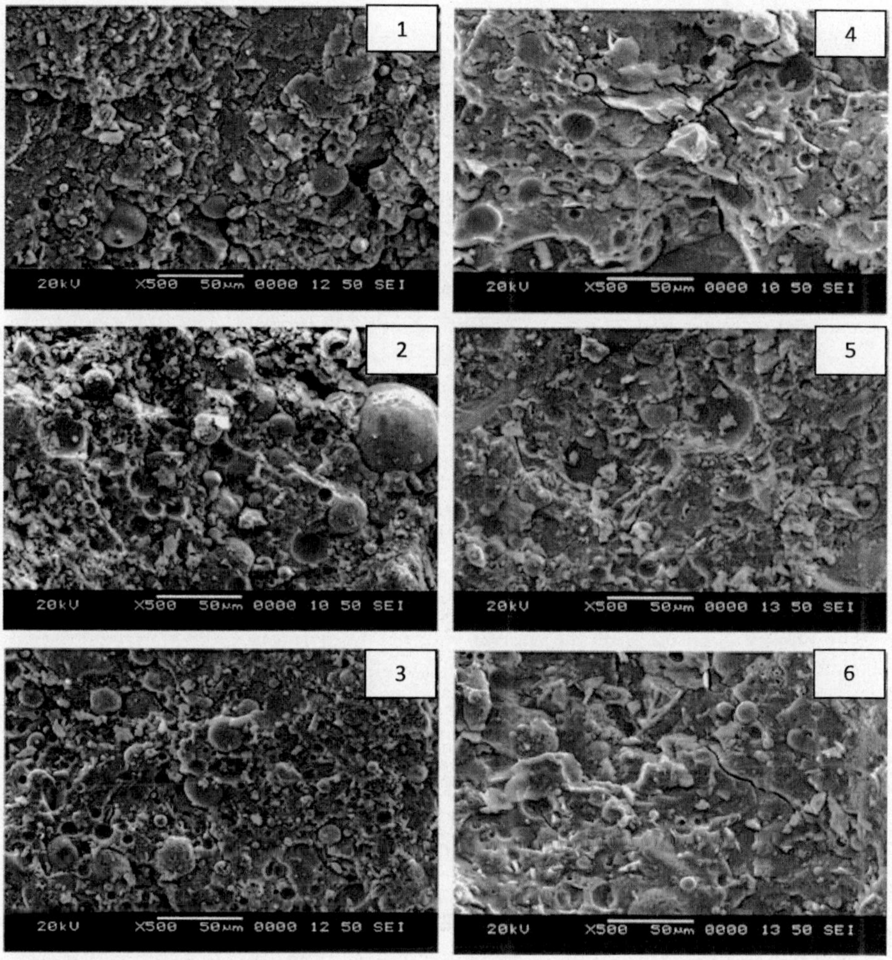

Fig. 7 SEM images of B series mixes (1) GM1 F100G0, (2) GM2 F90G10, (3) GM3 F80G20, (4) GM4 F70G30, (5) GM5 F60G40, (6) GM6 F50G50

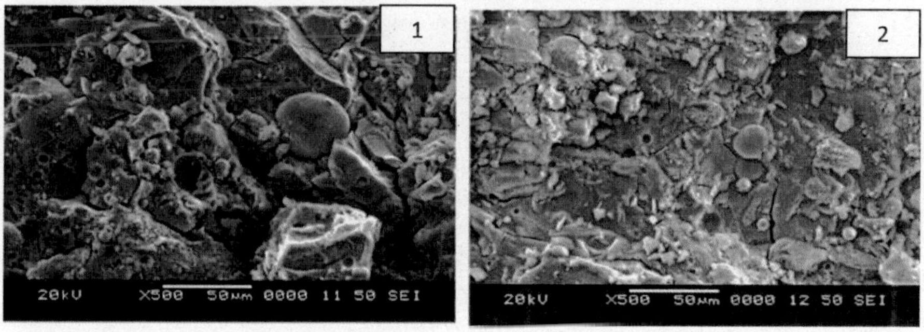

Fig. 8 SEM images of D series mixes (1) GM4 F70G30, (2) GM6 F50G50

has increased considerably over the sample with 20% GGBS samples. In the 40% sample, the cracks have lessened and the structure looks homogenous albeit with a high amount of unreacted GGBS particles. A compact glass-like structure can be seen for the 50% GGBS sample with a few unreacted FA and GGBS particles for which the number of cracks are observed. GM4 F70G30 S4 and GM6 F50G50 S4 showed similar features like B series mixes. For varying alkali–binder ratio, GM6 F50G50 S4 showed more cracks, porous structure and unreacted FA particles too are more as compared with GM6 F50G50 S2.

It was reported that there may be three possible mechanisms for enhanced production of C-A-S-H gel when GGBS was incorporated in the mix. The first mechanism inferred that the presence of surface calcium in GGBS will aid to the increased production of C-A-S-H gel which in turn increases the compressive strength [25, 26]. The calcium presence may cause water deficiency which will increase the alkalinity of mixes through that higher dissolution of existing aluminosilicate [27]. The second mechanism indicates that there is a domination of GGBS in production of C-A-S-H gel in alkali activation process through higher degree of cross-linking and polymerization [28]. The third mechanism inferred that N–A–S–H gel is a secondary product, which reduces the pore volume and enhances the compactness of the gel which increases the compressive strength [29, 30].

The energy spectrums of the mortar samples at different GGBS percentages in the mix for the A/B ratio of 0.6 and the major elements traced in the samples were oxygen, silicon, aluminium, calcium, magnesium, iron and sodium appeared in the sample and same is presented in Figs. 9 and 10. It is observed that as the GGBS content increased in the mixes, the intensity of calcium counts also increased and formed a gel more likely as C–A–S–H gel.

It is found from Table 3 that the elemental composition of the two randomly points in mixes GM1 F100G0 S2 to GM6 F50G50 S2 together with the Ca/Si, Si/Al and Al/Ca ratios. In the GM1 F100G0 S2 mix at both the points, silicon and aluminium are dominant elements; thus, from the Si/Al ratio, it is suggested that the phase present in both the points could be some form of aluminosilicate hydrate (A–S–H). Similar trend is shown by the GM2 F90G10 S2 mix; however, in the GM3 F80G20 S2 mix at point 1, the phase present could be A–S–H and that at point 2 could be calcium silicate hydrate (C–S–H) [31]. Although the points were in the same mix GM3 F80G20 S2, the phases present at the points were clearly different in terms of their elemental composition. GM4 F70G30 S2, GM5 F60G40 S2 and GM6 F50G50 S2 showed calcium silicate hydrate phase as a major binding gel.

From Table 4, for D series mixes, the trend is similar to as shown by B series mixes. On comparing the two sets of data, for the same GGBS content of 30% and 50%, the Al/Ca ratio for D30 and D50 is greater than B30 and B50, respectively.

From Figs. 11 and 12, it can be noticed that average Ca/Si and Si/Al ratios increased when higher amount of FA was replaced with GGBS which showed a positive correlation with the compressive strength of the B series mixes for all percentages of GGBS.

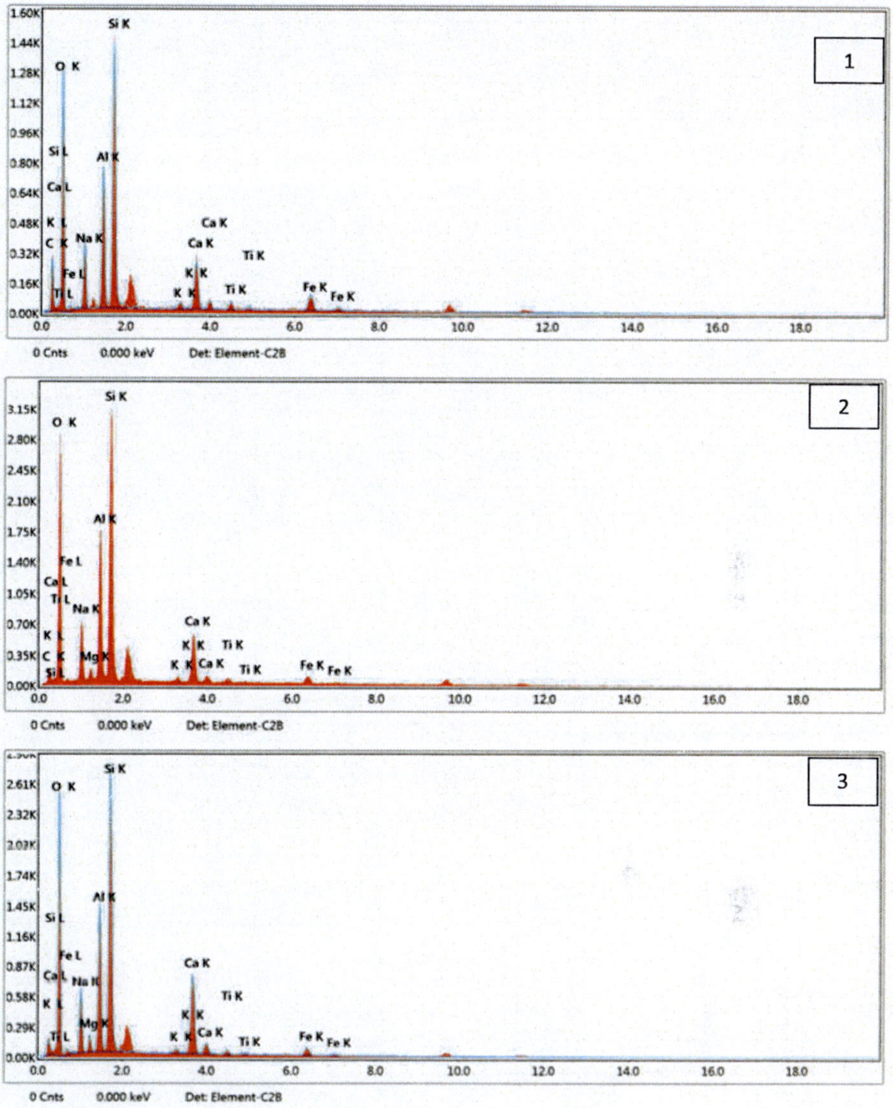

Fig. 9 EDAX spectrums for (1) GM1 F100G0, (2) GM2 F90G10, (3) GM3 F80G20, (4) GM4 F70G30, (5) GM5 F60G40, (6) GM6 F50G50

5 Conclusions

Based on the experimental test results from this study and the following observations can be drawn

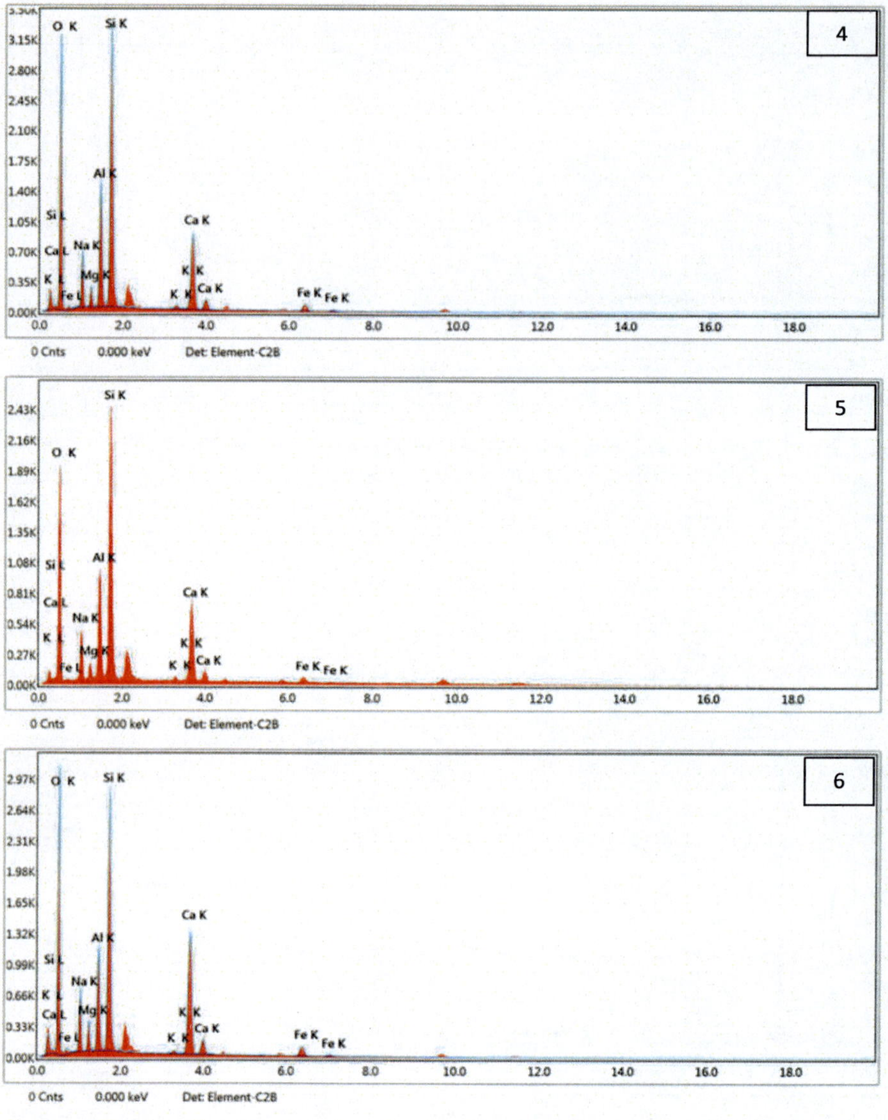

Fig. 9 (continued)

- Increase of GGBS content in the FA-based geopolymer mortar reduces the setting time and it is observed that more than 10% GGBS replacement in the mix reduces the setting time significantly.
- The increase of GGBS content as partial replacement to FA increased the compressive strength irrespective of the alkali–binder ratio at all ages. It is observed that up to 96% increase in compressive strength was attained when 50% FA was replaced by GGBS compared to mix with only FA for alkali–binder ratio 0.5.

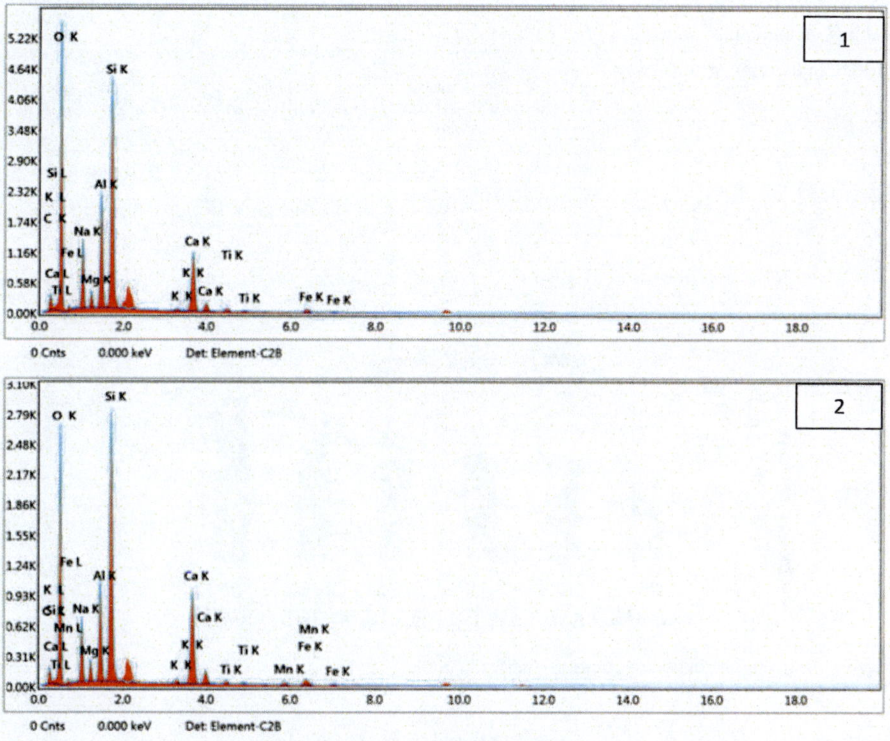

Fig. 10 EDAX spectrums for D series (1) GM4 F70G30 and (2) GM6 F50G50

Table 3 Elemental composition (wt%) in B series mixes

Elemental compositions (wt%)								
B Mixes	Point	Sodium (Na)	Aluminium (Al)	Silicon (Si)	Calcium (Ca)	Si/Al	Ca/Si	Al/Ca
0% GGBS	1	7.3	9.18	18.83	6.4	1.83	0.38	1.43
	2	6.94	13.74	21.08	8.19	1.53	0.38	1.68
10% GGBS	1	7.76	11.84	21.08	7.18	1.78	0.33	1.65
	2	8.28	9.66	17.66	5.77	1.82	0.34	1.67
20% GGBS	1	7.95	11.06	20.92	11.9	1.89	0.47	0.93
	2	7.59	9.71	18.44	8.81	1.87	0.56	1.10
30% GGBS	1	6.4	10.34	18.64	11.33	1.8	0.6	0.91
	2	7.52	9.81	21.35	12.15	2.17	0.59	0.81
40% GGBS	1	7.97	9.88	21.7	14.59	2.19	0.67	0.68
	2	7.27	9.44	22.37	13.11	2.36	0.58	0.72
50% GGBS	1	9.54	8.09	18.32	12.85	2.26	0.7	0.63
	2	7.53	7.38	17.18	16.89	2.32	0.98	0.44

Table 4 Elemental composition (wt%) in D series mixes

Elemental compositions (wt%)								
D Mixes	Point	Sodium (Na)	Aluminium (Al)	Silicon (Si)	Calcium (Ca)	Si/Al	Ca/Si	Al/Ca
30% GGBS	1	8.95	8.98	17.32	9.07	1.92	0.52	0.99
	2	9.02	7.39	16.21	8.22	2.19	0.5	0.89
50% GGBS	1	8.36	8.56	16.96	11.68	1.98	0.68	0.73
	2	7.86	7.17	18.41	13.23	2.56	0.71	0.54

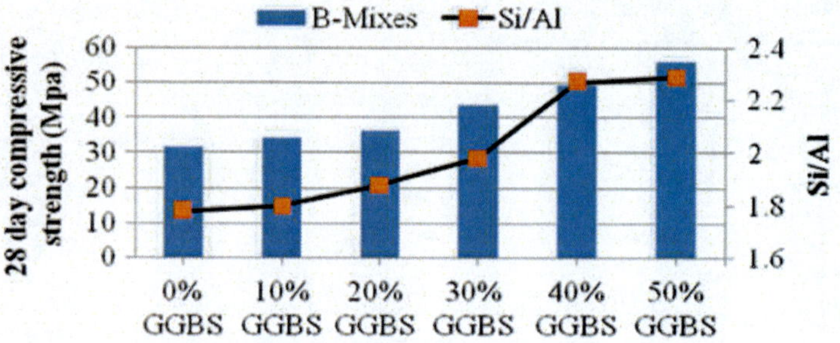

Fig. 11 Correlation between strength and Si/Al ratio

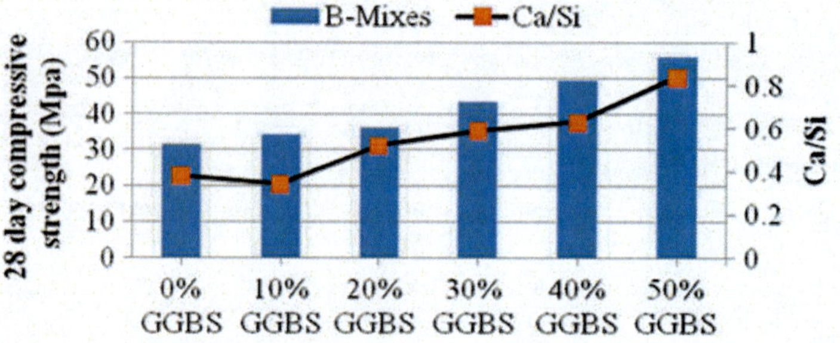

Fig. 12 Correlation between strength and Ca/Si ratio

- Adding up to 50% GGBS content of the total binder achieved strength of mortar up to 56.2 MPa at 28 days of curing. Compressive strength decreased with the increase of alkaline liquid content from 0.6 to 0.8.
- The SEM and EDAX results revealed that FA and GGBS-based geopolymer mortar contains high amount of Ca in turn helps in significant improvement of crosslinking of C–S–H/N–A–S–H chains which gave rise to a denser and compact amorphous gel structures which leads to high degree of polymerization.

References

1. The Cement Sustainability Initiative: Our agenda for action, World Business Council for Sustainable Development, p 20, June 2002
2. Naik T (2008) Sustainability of concrete construction. Pract Period Struct Des Constr 12(2):98–103
3. https://www.wbcsd.org/Sector-Projects/Cement-Sustainability-Initiative/News/Indian-cement-industry-on-track-to-meet-2030-carbon-emissions-intensity-reduction-objectives
4. https://www.ibef.org/industry/cement-india.aspx for cement data
5. http://www.cea.nic.in/reports/others/thermal/tcd/flyash_201718.pdf Fly Ash CEA Report
6. http://flyash2019.missionenergy.org/ for fly ash second part
7. https://ibm.gov.in/writereaddata/files/04242019131825Slag%20I_Steel.pdf for GGBS data
8. Grutzeck MW, Blanco MT, Palomo A (1999) Alkali-activated fly ashes: cement for the future. Cem Concr Res 29(8):1323–1329
9. Van Deventer JGS, Van Jaarsveld JSJ (1999) Effect of the alkali metal activator on the properties of fly ash-based Geopolymer. Ind Eng Chem Res 38(10):3932–3941
10. Wallah SE, Rangan BV, Hardjito D (2002) Study on engineering properties of fly ash based concrete. J Aust Ceram Soc 38(1):44–47
11. Neville AM (2000) Properties of concrete. London
12. Al-Majidi MH, Lampropoulos A, Al Majidi CA (2016) Effect of alkaline activator, water, superplasticizer and slag contents on the compressive strength and workability of slag-fly ash based geopolymer mortar cured under ambient temperature. Int J Civ Environ Struct Constr Arch Eng10(3):308–312
13. Saha S, Rajasekaran C (2017) Enhancement of the properties of fly ash based geopolymer paste by incorporating ground granulated blast furnace slag. Constr Build Mater 146:615–620
14. Sree Vidya M, Vadivel Arun NV (2016) Experimental investigation on blended geopolymer mortar dome using GGBS. Int J Earth Sci Eng 9(3):448–452
15. Sarkar PK, Nath P (2014) Effect of GGBS on setting time, workability and early strength properties of fly ash geopolymer concrete cured in ambient condition. Constr Build Mater: 163–171
16. Law DW, Strano A, Wordhono A (2015) The strength of alkali- activated slag/fly ash mortar blends at ambient temperature. Procedia Eng 125:650–656
17. Thakur RN, Ghosh S (2010) Effect of mix composition on compressive strength and microstructure of fly ash based geopolymer composites. ARPN J Eng Appl Sci 4(4):98–108
18. Ghosh P, Ghosh K (2017) Effect of alkali concentration on mechanical properties, microstructure, zeta potential and electrical conductivity of thermally cured fly-ash blast furnace slag based blended geopolymer composite. Orient J Chem 32(2):704–715
19. Ismail N, Al Hinaii S, El-Hassan H (2017) Effect of GGBS and curing temperature on microstructure characteristics of lightweight geopolymer concrete. In: MATEC web of conferences 120, p 03004
20. Farzadania N, Ali AAA, Demirboga R, Salih MA (2015) Development of high strength alkali activated binder using palm oil fuel ash and GGBS at ambient temperature. Constr Build Mater 93:289–300
21. Saludung A, Ogawa Y (2018) Microstructural and mechanical properties of FA/GGBS based geopolymer
22. Mallikarjuna Rao G, Gunneswara Rao TD (2015) Final setting time and compressive strength of fly ash and GGBS-based geopolymer paste and mortar. Arab J Sci Eng 40:3067–3074
23. IS: 8142 (1976) Indian standard method of test for determining setting time of concrete by penetration resistance
24. IS: 4031 (Part 6) (1988) Method of physical tests for hydraulic cements (Determination of compressive strength of hydraulic cement other than masonry cement), Bureau of Indian Standards, New Delhi
25. Kumar S, Nath SK (2013) Influence of iron making slags on strength and microstructure of fly ash geopolymer. Constr Build Mater 38:924–930

26. Kumar R, Mehrotra S, Kumar S (2010) Influence of granulated blast furnace slag on the reaction, structure and properties of fly ash based geopolymer. J Mater Sci 45(4):607–615
27. Khater H (2011) Effect of calcium on geopolymerization of aluminosilicate wastes. J Mater Civ Eng 24(1):92–101
28. Brough AR, Groves GW, Dobson CM, Richardson IG (1994) The characterization of hardened alkali-activated blast-furnace slag pastes and the nature of the calcium silicate hydrate (C-S-H) phase. Cem Concr Res 24(5):813–829
29. Bernal SA, San Nicolas R, Provis JL, Myers RJ (2013) Generalized structural description of calcium–sodium aluminosilicate hydrate gels: the cross-linked substituted tobermorite model. Langmuir 29(17):5294–5306
30. Liu S, Li Z (2007) Influence of slag as additive on compressive strength of fly ash-based geopolymer. J Mater Civ Eng 19(6):470–474
31. Ismail Idawati et al (2013) Microstructural changes in alkali activated fly ash/slag geopolymers with sulfate exposure. Mater Struct 46:361–373

Fast Setting Steel Fibre Geopolymer Mortar Cured Under Ambient Temperature

K. M. Prasanna, Irambona Theodose, K. N. Shivaprasad, and B. B. Das

Abstract Cement and cementitious materials are being used worldwide as the most popular multipurpose construction materials but the greenhouse gas such as carbon dioxide (CO_2) produced during its manufacturing process creating a huge environmental hazard, thus efforts have been made for alternative binders. Geopolymer binder is new age binder alternative to ordinary Portland cement in infrastructure projects because it is produced from eco-friendly and industrial waste materials. This study was aimed to produce fast setting with ground-granulated blast-furnace slag (GGBS) in fly ash-based geopolymer mortar incorporated with steel fibres cured under ambient temperature. In this research, alkaline to binder ratio was varied from 0.5 to 0.8, crimped steel fibre are varied from 0.5 to 1.5% by total volume of binder and combination of fly ash (FA) and GGBS (100%:0%, 90%:10%, 80%:20%, 70%:30%, 60%:40% and 50%:50%) as binder were used for preparation of fibre geopolymer mortar. The tests conducted include stetting time and flowability of geopolymer mortar, compressive strength and microstructural characterisation of steel fibre geopolymer mortar. The tests for compressive strength were carried out on standard size of mortar samples at curing period of 3, 7 and 28 days. It is noted from the test results that increase in GGBS content setting times were decreased; however, the compressive strength of fly ash-based geopolymer mortar increased. The highest compressive strength at 28 days of curing period was found to be 69.5 MPa, which is obtained with content of 1% of steel fibres and alkaline to binder ratio of 0.6 with 50%:50% binder's proportions. Further, it is observed that the incorporation of steel fibres in plain geopolymer mortar have enhanced the compressive strength and optimum dosage of fibres was found to be 1%.

Keywords Fly ash-based geopolymer · GGBS · Steel fibres · Setting time · Flowability · Compressive strength · SEM

K. M. Prasanna · I. Theodose · K. N. Shivaprasad · B. B. Das (✉)
Department of Civil Engineering, National Institute of Technology Karnataka, Surathkal, India
e-mail: bibhutibhusan@gmail.com

B. B. Das et al. (eds.), *Recent Developments in Sustainable Infrastructure*, Lecture Notes in Civil Engineering 75, https://doi.org/10.1007/978-981-15-4577-1_65

1 Introduction

Nowadays, geopolymeric binders, which are environment-friendly materials, are being used as an alternative binder to ordinary Portland cement (OPC) [1]. It is inorganic material produced by polymerisation of aluminosilicate precursors with alkaline activator solutions [1–3]. The amorphous aluminosilicate materials are commonly used as raw material, to activate the raw materials strong alkali chemicals are needed such as alkali hydroxide and alkali silicate. This is most eco-friendly technology which is an alternative to cementitious materials was found to be geopolymer technology because it could reduce 80% of the CO_2 emissions. Many properties are there to make the geopolymer binder more advantageous than ordinary Portland cement, some of them are: less hydration heat, earlier gaining strength and higher compressive strength, higher chemical resistance, good resistance to acid, good sulphate attack resistance, etc. [4–6].

Researchers reported that higher replacement of the GGBS, with the FA in the production geopolymer binder, accelerates the hydration reaction and led to a faster setting time of the mixture [6–8]. Further, the addition of fibres to cementitious system is to improve the mechanical properties, mainly those that motivate or influence fatigue and tensile stresses. The incorporation of steel fibres in the system considerably improves its properties.

Many researchers reported that effect of different types of fibre on fibre fly ash-based reinforced cementitious systems or geopolymer composites, shown that all types of fibres improved the mechanical properties such as flexural strength, improve the energy absorption and tensile strength, etc., [9–17]. To improve the performance of fibre-based geopolymer composites, main variables that governing are type of fibre and fibre content in the system [9–11, 14]. The workability of fresh geopolymer composite reduced significantly by incorporating fibres, because of higher shear resistance to flow [13]. Hardened properties of the geopolymer composites have improved with increase in percentage of fibre content [7, 14].

In the present study, effect of crimped steel fibres on flowability and setting time of geopolymer mortar incorporating GGBS are measured. Further, the compressive strength behaviour and the microstructural properties of crimped steel fibre geopolymer mortar cured were investigated.

2 Materials

2.1 *Fly Ash*

The FA were analysed for physical properties and test results are presented in Table 1. Based on the chemical composition of FA, it is classified as class F type as per Indian Standards [18].

Table 1 Physical properties of FA and GGBS

Properties	Type of materials	
	FA	GGBS
Specific gravity	2.27	2.85
Fineness (m^2/kg)	315	337

2.2 *Ground-Granulated Blast Slag*

The GGBS used in this work is obtained from Jindal Steel. The physical properties of the GGBS were analysed and it is presented in Table 1.

2.3 *Alkaline Liquid*

In the present study, sodium hydroxide flakes (NaOH) and sodium silicate (Na_2SiO_3) solution were selected as alkaline activator for geopolymerisation. The laboratory grade NaOH is used, was in solid state (flakes/pellets form) and its purity is 97% which is dissolved in distilled water to prepare 10M solution. Laboratory-grade Na_2SiO_3 is used in preparation of alkaline solution along with NaOH. The pure form of sodium silicate solution is of colourless or white in colour. Commercially available Na_2SiO_3 in liquid form was used which is supplied by local manufacturer.

2.4 *Fine Aggregate*

Local available natural river sand, which is clean and dry, passing through IS 480 sieve was used as fine aggregate and it is confirming to zone I as per Indian standards [19]. It has fineness modulus of 2.844 and specific gravity 2.65.

2.5 *Steel Fibre*

Crimped steel fibres of 12.5 mm length, diameter of 0.45 mm and aspect ratio (l/d) of 27.77 were used in this study.

3 Preparation of Mortar

3.1 Mixture Proportions of Test Samples

In this experimental study, binder to sand (B/S) ratio was 1:3; alkaline to binder (Alk/B) ratios were taken 0.4, 0.5, 0.6, 0.7 and 0.8; variation FA and GGBS were 100%:0%, 90%:10%, 80%:20, 70%:30%, 60%:40% and 50%:50% (by weight); and crimped steel fibres content were 0.5%, 1.0% and 1.5% (by weight of binder). The molar concentration of sodium hydroxide (NaOH) is fixed at 10M and Na_2SiO_3/NaOH ratio is taken constant has 2.5. The complete mixture proportions of various geopolymer mortar samples investigated are presented in Table 2.

Table 2 Mix designation for steel fibre geopolymer mortar

Alk/B	Series	B/S	B (%) [FA:GGBS]	Fibre content (%)		
				0.5	1	1.5
0.5	$GM1F_{100}G_{0}$	1:3	100:0	$GM1F_{100}G_{0}S_{1}$	$GM1F_{100}G_{0}S_{2}$	$GM1F_{100}G_{0}S_{3}$
	$GM2F_{90}G_{10}$	1:3	90:10	$GM2F_{90}G_{10}S_{1}$	$GM2F_{90}G_{10}S_{2}$	$GM2F_{90}G_{10}S_{3}$
	$GM3F_{80}G_{20}$	1:3	80:20	$GM3F_{80}G_{20}S_{1}$	$GM3F_{80}G_{20}S_{2}$	$GM3F_{80}G_{20}S_{3}$
	$GM4F_{70}G_{30}$	1:3	70:30	$GM4F_{70}G_{30}S_{1}$	$GM4F_{70}G_{30}S_{2}$	$GM4F_{70}G_{30}S_{3}$
	$GM5F_{60}G_{40}$	1:3	60:40	$GM5F_{60}G_{40}S_{1}$	$GM5F_{60}G_{40}S_{2}$	$GM5F_{60}G_{40}S_{3}$
	$GM6F_{50}G_{50}$	1:3	50:50	$GM6F_{50}G_{50}\ S_{1}$	$GM6F_{50}G_{50}S_{2}$	$GM6F_{50}G_{50}S_{3}$
0.6	$GM7F_{100}G_{0}$	1:3	100:0	$GM7F_{100}G_{0}S_{1}$	$GM7F_{100}G_{0}S_{2}$	$GM7F_{100}G_{0}S_{3}$
	$GM8F_{90}G_{10}$	1:3	90:10	$GM8F_{90}G_{10}S_{1}$	$GM8F_{90}G_{10}S_{2}$	$GM8F_{90}G_{10}S_{3}$
	$GM9F_{80}G_{20}$	1:3	80:20	$GM9F_{80}G_{20}S_{1}$	$GM9F_{80}G_{20}S_{2}$	$GM9F_{80}G_{20}S_{3}$
	$GM10F_{70}G_{30}$	1:3	70:30	$GM10F_{70}G_{30}S_{1}$	$GM10F_{70}G_{30}S_{2}$	$GM10F_{70}G_{30}S_{3}$
	$GM11F_{60}G_{40}$	1:3	60:40	$GM11F_{60}G_{40}S_{1}$	$GM11F_{60}G_{40}S_{2}$	$GM11F_{60}G_{40}S_{3}$
	$GM12F_{50}G_{50}$	1:3	50:50	$GM12F_{50}G_{50}S_{1}$	$GM12F_{50}G_{50}S_{2}$	$GM12F_{50}G_{50}S_{3}$
0.7	$GM13F_{100}G_{0}$	1:3	100:0	$GM13F_{100}G_{0}S_{1}$	$GM13F_{100}G_{0}S_{2}$	$GM13F_{100}G_{0}S_{3}$
	$GM14F_{90}G_{10}$	1:3	90:10	$GM14F_{90}G_{10}S_{1}$	$GM14F_{90}G_{10}S_{2}$	$GM14F_{90}G_{10}S_{3}$
	$GM15F_{80}G_{20}$	1:3	80:20	$GM15F_{80}G_{20}S_{1}$	$GM15F_{80}G_{20}S_{2}$	$GM15F_{80}G_{20}S_{3}$
	$GM16F_{70}G_{30}$	1:3	70:30	$GM16F_{70}G_{30}S_{1}$	$GM16F_{70}G_{30}S_{2}$	$GM16F_{70}G_{30}S_{3}$
	$GM17F_{60}G_{40}$	1:3	60:40	$GM17F_{60}G_{40}S_{1}$	$GM17F_{60}G_{40}S_{2}$	$GM17F_{60}G_{40}S_{3}$
	$GM18F_{50}G_{50}$	1:3	50:50	$GM18F_{50}G_{50}S_{1}$	$GM18F_{50}G_{50}S_{2}$	$GM18F_{50}G_{50}S_{3}$
0.8	$GM19F_{100}G_{0}$	1:3	100:0	$GM19F_{100}G_{0}S_{1}$	$GM19F_{100}G_{0}S_{2}$	$GM19F_{100}G_{0}S_{3}$
	$GM20F_{90}G_{10}$	1:3	90:10	$GM20F_{90}G_{10}S_{1}$	$GM20F_{90}G_{10}S_{2}$	$GM20F_{90}G_{10}S_{3}$
	$GM21F_{80}G_{20}$	1:3	80:20	$GM21F_{80}G_{20}S_{1}$	$GM21F_{80}G_{20}S_{2}$	$GM21F_{80}G_{20}S_{3}$
	$GM22F_{70}G_{30}$	1:3	70:30	$GM22F_{70}G_{30}S_{1}$	$GM22F_{70}G_{30}S_{2}$	$GM22F_{70}G_{30}S_{3}$
	$GM23F_{60}G_{40}$	1:3	60:40	$GM23F_{60}G_{40}S_{1}$	$GM23F_{60}G_{40}S_{2}$	$GM23F_{60}G_{40}S_{3}$
	$GM24F_{50}G_{50}$	1:3	50:50	$GM24F_{50}G_{50}S_{1}$	$GM24F_{50}G_{50}S_{2}$	$GM24F_{50}G_{50}S_{3}$

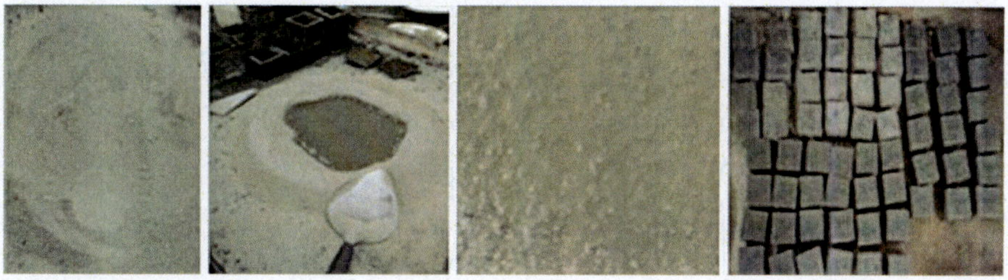

Fig. 1 Sample preparation

3.2 Specimen Preparation and Testing

Geopolymer mortar was prepared by mixing FA and GGBS as binder and sand for 3–4 min manually till the homogeneous dry mixes obtained. The homogeneous dry mixes and alkaline solution of required quantity were mixed properly with hand for around 3 min to ensure homogeneity. Twenty-four series of geopolymer mortar were prepared by varying the steel fibres percentage (0.5–1.5%) as well as different alkaline to binder ratio (0.5–0.8) for mini flow table test. The proportion of FA and GGBS as binder was considered constant as 50%:50%. On the other hand, geopolymer mortar were prepared for setting time of mortar by varying the percentage of GGBS (from 0 to 50% of binder) as well as alkaline to binder ratio (0.5–0.8). Setting time was measured on the geopolymer mortar as per the procedures as indicated by IS: 8142-1976 [20].

Steel fibre geopolymer mortar specimens are prepared for compressive strength test with dimensions of 70.6 × 70.6 × 70.6-mm moulds. After 24 h, these moulds are removed and the specimens are stored in ambient conditions as shown by Fig. 1. Seventy-two sets of geopolymer mortar were prepared by varying the steel fibres content as well as alkaline to binder ratio. Three mortar cubes are tested in each set and average compressive strength are assessed at 3, 7 and 28 days [21]. Further, samples are calculated for scanning electron microscope (SEM) analysis is carried out on the broken pieces of the original sample after the end of compressive tests.

4 Results

4.1 Setting Time

Fly ash-based geopolymer mortar are measured for initial (IST) and final setting time (FST) are evaluated using concrete penetrometer with different alkaline to binder ratio and varying of GGBS content. The experimental test results obtained are presented in Table 3. The results from the table show that setting time largely depend on alkaline solution. As the quantity of alkaline increases, the setting time also increases

Table 3 Setting time test results of geopolymer mortar for Alk/B of 0.5–0.8

Binder composition	IST (min)				FST (min)			
	Alk/B ratio				Alk/B ratio			
	0.5	0.6	0.7	0.8	0.5	0.6	0.7	0.8
$F_{100}G_0$	94	152	201	316	172	361	429	668
$F_{90}G_{10}$	62	84	115	191	116	138	198	280
$F_{80}G_{20}$	47	63	90	119	89	101	125	196
$F_{70}G_{30}$	33	40	54	80	69	84	115	162
$F_{60}G_{40}$	26	32	49	71	64	71	95	147
$F_{50}G_{50}$	22	27	40	68	51	57	72	105

and also as the percentage of GGBS reduced both the IST and FST due to higher content of CaO, which accelerates the hydration reaction in the mixture [7].

IST and FST with respect to alkaline to binder ratio are presented in Figs. 2 and 3. From Fig. 2, it is observed that the comparison of IST of all alkaline to binder ratios with respect to the increase in GGBS content. Based on the variation of alkaline to binder ratio from 0.5 to 0.8, the results show that the reduction in initial setting time is about 142–236%. GGBS replacement in the mixture with FA significantly reduces the IST. The results show that the when GGBS varies from 0 to 50% by weight of binder the initial setting time decreased about 76–82% depending on variation of alkaline to binder ratio.

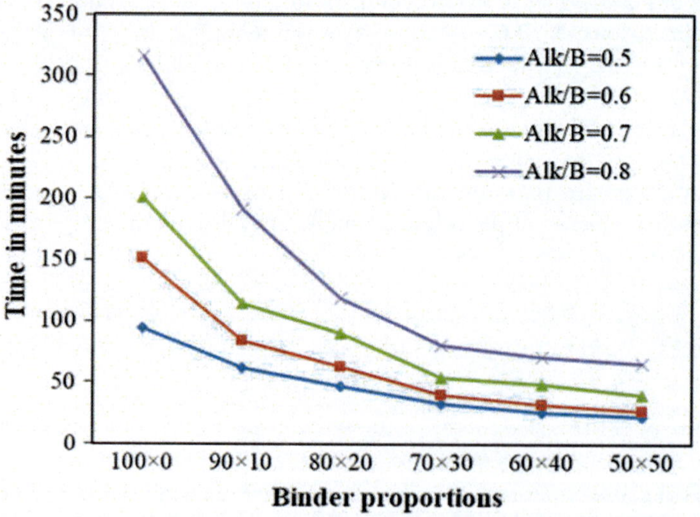

Fig. 2 Comparison of initial setting time for all Alk/B with respect to variation of binder compositions

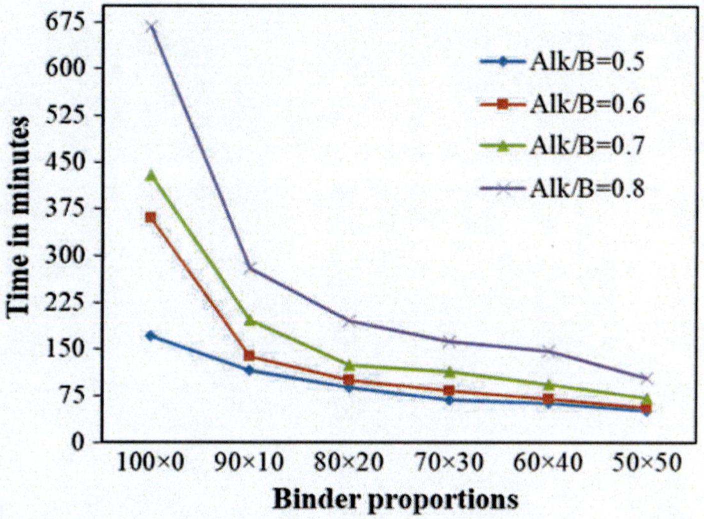

Fig. 3 Comparison of final setting time for all Alk/B with respect to variation of binder compositions

Further from Fig. 3, it is to be noted that the comparison of FST of all alkaline to binder ratios with respect to the increase in GGBS content shows that additions of GGBS in the mixture significantly reduces the FST. The results show that the, when slag varies from 0 to 50% by weight of binder, the FST decreases about 70–84% depending on the variation of alkaline to binder ratio. Based on the variation of alkaline to binder ratio, the figure shows an increase of about 106–288% in FST when alkaline to binder ratio varies from 0.5 to 0.8 with variation of GGBS content.

4.2 *Mini Flow Table Test Results*

Flowability test has been performed to evaluate the impact of incorporating steel fibres in plain geopolymer mortar on the flowing capacity. The average of four diameters measured for flow test for different trial runs are presented in Table 4. Variation of all average flow diameters are presented in Fig. 4. From figure, it is observed that the increase in fibre content decreases the flow diameter in all alkaline to binder ratios. Incorporation of crimped steel fibres in plain geopolymer mortar decreases its flow diameter between 23.56 and 30.82% for alk/b ratio of 0.5, between 24.77 and 33% for alk/b ratio of 0.6, between 24.7 and 32.82% for alk/b ratio of 0.7 depending on the fibre percentage used for each mix. Further, the flow percentage of all alkaline to binder ratios with different variations of steel percentage are presented in Fig. 5. From figure, it is noted that the percentage of flow is increased with an increase in alk/b ratio. It is also clearly seen that incorporation and increase in steel fibres has a greater impact on flowing ability of geopolymer mortar in all mixes.

Table 4 Mini flow table test results

Alk/B ratio	Steel (%)	Average diameter (cm)	Flow (%)
0.5	0	15.83	58.3
	0.5	13.10	31.0
	1	12.55	25.5
	1.5	12.10	21.0
0.6	0	19.21	92.1
	0.5	15.65	59.5
	1	14.90	49.0
	1.5	14.45	44.5
0.7	0	23.90	139.0
	0.5	18.83	88.3
	1	18.23	82.3
	1.5	18.00	80.0
0.8	0	25.00	155.0
	0.5	21.25	112.5
	1	20.58	105.8
	1.5	20.28	102.8

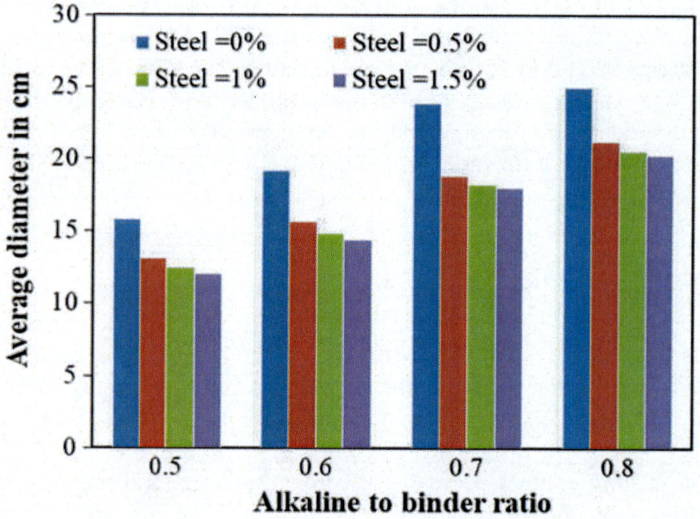

Fig. 4 Average flow diameter with respect to variation of fibres content

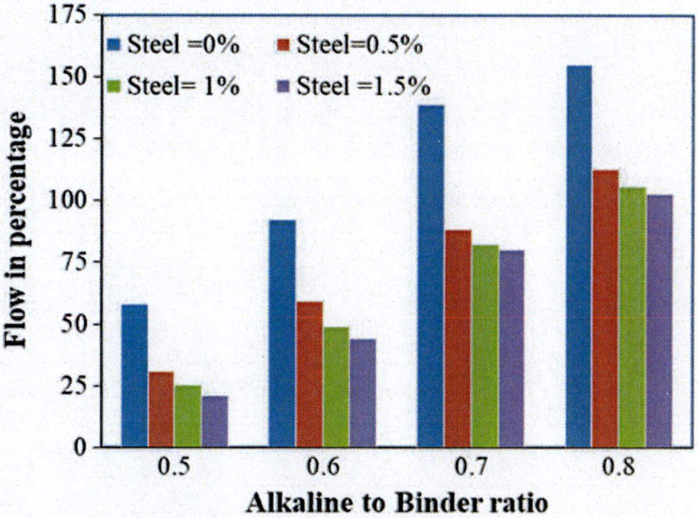

Fig. 5 Flow in percentage with respect to variation of fibres content

4.3 Test Results on Compressive Strength

Compressive strength was done based on IS 4031-1988 by applying a compressive loading on specimens by using a compressive testing machine.

4.3.1 Effect of Binder Proportions and Alkaline to Binder Ratio

Experimental test results of compressive strength on different alk/b ratio and different binder proportions are presented in Table 5. From table, it is observed that increasing GGBS content from 0 to 50%, the strength of geopolymer composite has improved in all mixtures; this is due to the presence in calcium content (CaO) in the mixtures which improved the compressive strength of specimens.

Further, it is observed that alkaline to binder ratio of 0.6 showed greater compressive strength values than others (0.5, 0.7 and 0.8) for all variations of steel fibre content and also noted that compressive strength increased up to 0.6, beyond alk/b ratio of 0.6, the compressive strength values decreased (Fig. 6).

4.3.2 Effect of Crimped Steel Fibres

The variation of compressive strength of geopolymer composites with different percentages of fibre content along with different alkaline to binder ratio are presented in Figs. 7, 8, 9 and 10. It can be observed from figures that the increase in percentage of crimped steel fibres up to 1% has increased compressive strengths beyond 1% fibre

Table 5 Compressive strength test results of geopolymer composites

Alkaline to binder ratio	Binder proportions	Compressive strength (MPa)											
		Curing period											
		3 days				7 days				28 days			
		Steel (%)											
		0	0.5	1	1.5	0	0.5	1	1.5	0	0.5	1	1.5
0.5	$F_{100}G_0$	11.0	17.2	22.4	21.4	18.0	25.4	30.5	27.8	25.0	35.0	43.6	41.5
	$F_{90}G_{10}$	16.0	21.8	26.0	24.3	22.0	27.9	33.0	29.5	29.0	39.0	44.2	43.0
	$F_{80}G_{20}$	16.0	24.1	28.0	26.3	24.0	30.4	36.3	33.0	32.0	41.0	46.4	44.0
	$F_{70}G_{30}$	21.0	28.0	33.0	28.5	28.0	35.2	39.9	36.1	39.0	45.0	50.8	49.3
	$F_{60}G_{40}$	25.0	31.5	37.0	33.8	34.0	41.5	45.0	41.6	43.0	51.0	55.8	53.0
	$F_{50}G_{50}$	30.0	35.8	40.4	36.1	39.0	46.4	51.8	47.0	50.0	58.0	62.7	60.0
0.6	$F_{100}G_0$	12.0	18.5	23.8	22.6	20.0	27.0	31.0	30.2	31.0	39.6	44.1	42.2
	$F_{90}G_{10}$	16.0	22.4	27.8	25.5	23.0	29.1	34.8	31.5	34.0	40.4	46.8	44.0
	$F_{80}G_{20}$	22.0	27.6	31.4	28.0	28.0	34.0	38.0	35.5	36.0	42.5	48.7	46.5
	$F_{70}G_{30}$	25.0	31.3	36.3	33.1	31.0	38.0	42.8	39.5	43.0	50.2	55.2	53.0
	$F_{60}G_{40}$	29.0	34.6	39.8	36.3	36.0	43.0	48.2	45.0	49.0	56.5	61.0	58.0
	$F_{50}G_{50}$	34.0	40.4	45.0	42.9	41.0	49.0	52.5	50.0	56.0	65.0	69.5	67.0
0.7	$F_{100}G_0$	7.0	13.3	18.6	16.5	16.0	22.5	27.5	25.5	27.0	34.8	40.0	37.5
	$F_{90}G_{10}$	10.0	16.2	21.2	20.0	20.0	25.7	30.5	27.0	29.0	36.1	41.6	39.5
	$F_{80}G_{20}$	12.0	18.6	24.2	22.9	22.0	28.8	33.0	29.5	33.0	39.0	44.4	42.0
	$F_{70}G_{30}$	16.0	23.6	27.8	24.9	25.0	31.2	36.8	33.3	36.0	43.0	49.5	48.0
	$F_{60}G_{40}$	19.0	25.5	30.5	29.5	29.0	34.1	41.0	37.8	40.0	48.0	53.5	51.0

(continued)

Table 5 (continued)

Alkaline to binder ratio	Binder proportions	Compressive strength (MPa)											
		Curing period											
		3 days				7 days				28 days			
		Steel (%)											
		0	0.5	1	1.5	0	0.5	1	1.5	0	0.5	1	1.5
	$F_{50}G_{50}$	23.0	28.9	34.5	32.5	32.0	39.3	44.8	42.3	44.0	54.0	59.0	57.2
0.8	$F_{100}G_0$	5.0	12.5	17.8	15.5	12.0	20.0	25.5	21.5	22.0	31.0	34.6	32.8
	$F_{90}G_{10}$	7.0	14.4	20.0	19.2	16.0	22.5	27.2	23.5	25.0	32.8	37.0	35.0
	$F_{80}G_{20}$	10.0	17.8	21.7	20.5	18.0	26.2	30.8	24.7	29.0	35.5	39.0	37.5
	$F_{70}G_{30}$	13.0	20.0	24.5	23.5	18.5	28.0	32.8	27.5	32.0	37.8	41.5	38.9
	$F_{60}G_{40}$	17.0	22.6	26.8	26.0	22.0	30.0	34.1	32	35.0	39.0	44.3	43.2
	$F_{50}G_{50}$	20.0	26.2	31.5	29.0	27.0	34.5	39.0	37.4	38.0	45.0	49.5	48.2

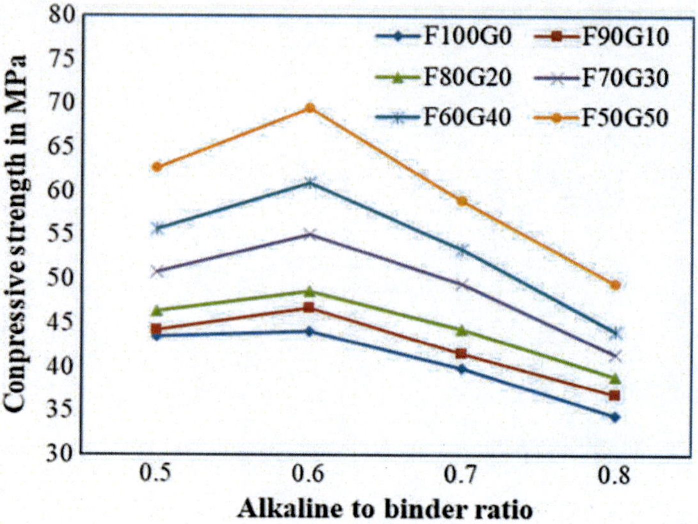

Fig. 6 Variation of compressive strength at 28 days based on binder proportions and alkaline to binder ratio

content and the compressive strength values decreased. This increase in compressive strength up to 1% is due to the action of fibres to the increase in their bond with mortar, thus increases the compressive strength beyond 1%. The workability is reduced due to the higher percentage of fibre content and compaction of geopolymer mortar is severely affected, hence compressive strength decreased. Further, the addition and increase in steel fibres in plain geopolymer mortar increased its compressive strength values. The compressive strengths for 3, 7 and 28 days of 1% volume of fibres were found to be greater than those of 0.5 and 1.5% for all variations of alkaline to binder ratio and all variations of binder proportions (FA & GGBS).

The increase in compressive strength of geopolymer composite is in the range of 5.1–22.6 MPa depending on the steel fibre content, binder compositions and alkaline to binder ratio. The highest values for 28-day compressive strength were 56.0, 69.5, 63.0 and 62.0 MPa for alk/b ratios of 0.5, 0.6, 0.7 and 0.8, respectively, obtained at 1% of steel fibres content with 50 $\times$ 50 binder compositions.

Figure 7 presents the variation of compressive strength for alk/b ratio of 0.5 for the period of 3, 7 and 28 days. The figures show that addition and incorporation of fibres in plain geopolymer mortar for alkaline to binder ratio of 0.5 increased the compressive strength about 13.17–75%. The compressive strength increased by 19.33–75% more than that of plain mortar for 3-day curing, 13.17–69.44% more than that of plain mortar for 7-day curing and 15.38–74.44% more than that of plain mortar for 28-day curing depending on GGBS content and percentage of fibres.

Figure 8 presents the variation of compressive strength for alk/b ratio of 0.6 for the period of 3, 7 and 28 days. The figures show that addition and incorporation of fibres in plain geopolymer mortar for alk/b ratio of 0.6 increase the compressive strength

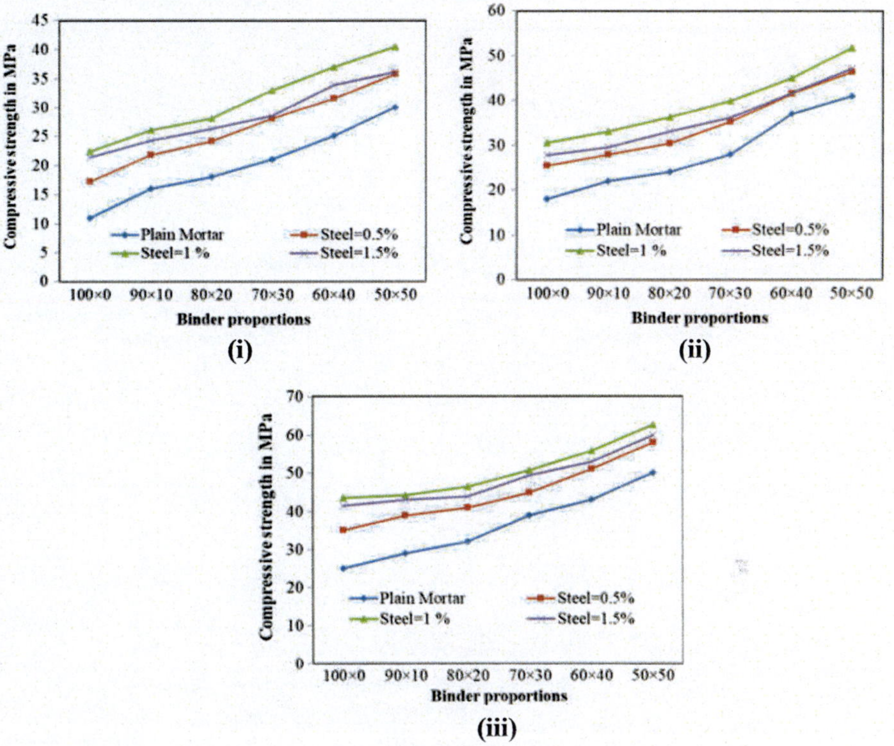

Fig. 7 Compression strength of geopolymer mortar with Alk/B = 0.5 at (i) 3 days, (ii) 7 days and (iii) 28 days

about 15.30–73.75%. The compressive strength increased by 19.82–73.75% more than that of plain mortar for 3-day curing, 19.44–55% more than that of plain mortar for 7-day curing and 15.3–42.25% more than that of plain mortar for 28-day curing depending on GGBS content and percentage of fibres.

Figure 9 presents the variation of compressive strength for alk/b ratio of 0.7 for the period of 3, 7 and 28 days. The figures show that addition and incorporation of fibres in plain geopolymer mortar for alkaline to binder ratio of 0.7 increase the compressive strength about 17.58–75.75%. The compressive strength increased by 25–75.75% more than that of plain mortar for 3-day curing, 17.58–71.87% more than that of plain mortar for 7-day curing and 18.18–48.14% more than that of plain mortar for 28-day curing depending on GGBS content and percentage of fibres.

Figure 10 presents the variation of compressive strength for alk/b ratio of 0.8 for the period of 3, 7 and 28 days respectively. The figures show that additional and incorporation of fibres in plain geopolymer mortar for alk/b ratio of 0.8 increased the compressive strength about 18.42–88.46%. The compressive strength increased between 31 and 88.46% more than that of plain mortar for 3 days curing, between 52 and 83.33% more than that of plain mortar for 7 days curing and between 18.42

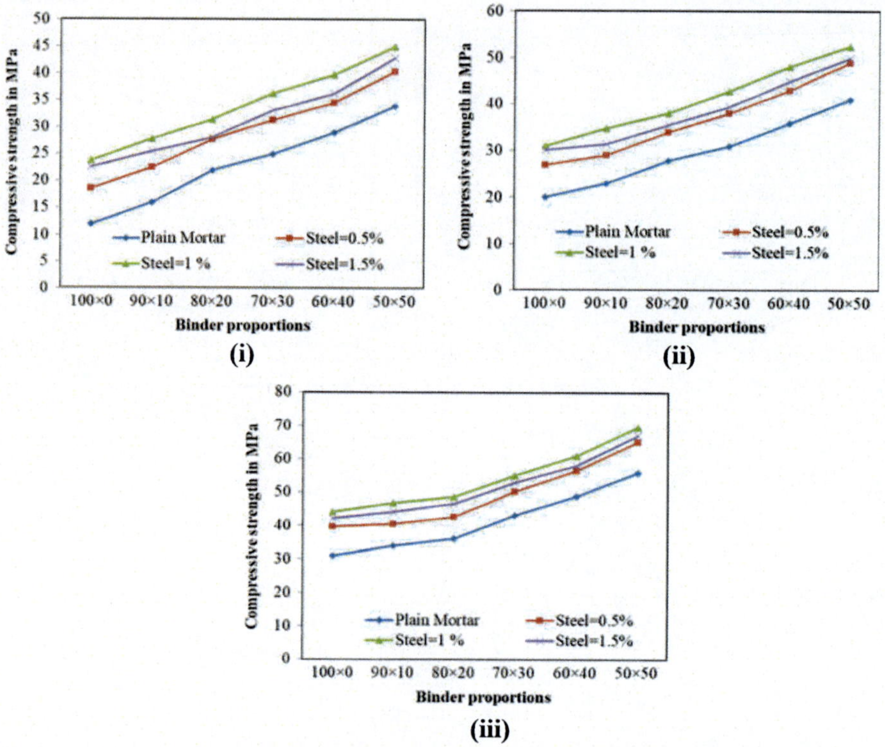

Fig. 8 Compression strength of geopolymer mortar with Alk/B = 0.6 at (i) 3 days, (ii) 7 days and (iii) 28 days

and 67.27% more than that of plain mortar for 28 days curing depending on GGBS content and percentage of fibres.

4.4 Scanning Electron Microscope (SEM)

SEM images of geopolymer composites were analysed in order to assess the effect of steel fibre in geopolymer mortar for microstructural characteristics. SEM images of selected samples of geopolymer mortar reinforced by steel fibres are presented Fig. 11. It can be observed from figure that the steel fibre surface is affected considerably by the geopolymer interfacial properties. Increase in GGBS content leads to enhanced interfacial properties as shown in Fig. 11b, c. These properties have a direct effect on the compressive strength of geopolymer composites. The smooth steel fibre surface is seen on gopolymer sample containing 0% GGBS (Fig. 11a), but for high GGBS content (50%) samples (Fig. 11e), the steel fibre surface is covered with geopolymer matrix.

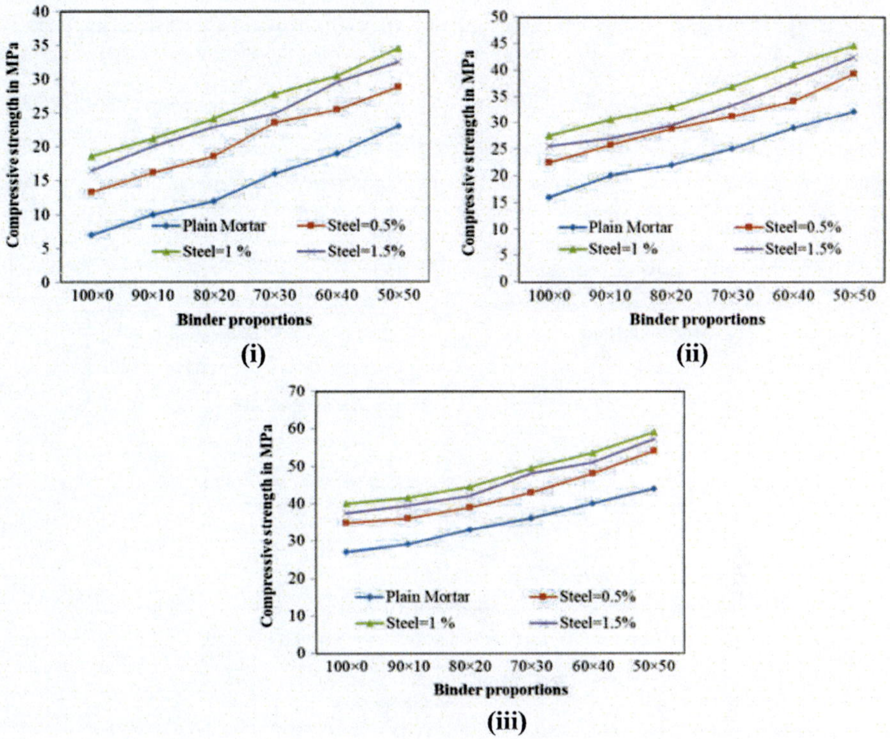

Fig. 9 Compression strength of geopolymer mortar with Alk/B = 0.7 at (i) 3 days, (ii) 7 days and (iii) 28 days

Figure 11d, e also present the SEM images of gopolymer samples containing 40 and 50% GGBS. It can be observed from the figure that a relative steel fibre surface is attached with geopolymer hydration products. This observation is attributed to the relatively good bond between the geopolymer matrix and the fibres [10]. This good bond between the geopolymer matrix and the surface of fibres is responsible for high compressive strength values as discussed in previous section.

5 Conclusion

Effect of steel fibres on flow capacity and compressive strength has been investigated. The microstructure of steel fibre geopolymer mortar specimens was determined using SEM. Based on experimental results obtained from the study, the conclusion can be summarised as follows.

- In all mixes, setting time of geopolymer mortar (both IST and FST) is decreased with incorporation and the increase in GGBS percentage. So to produce the fast

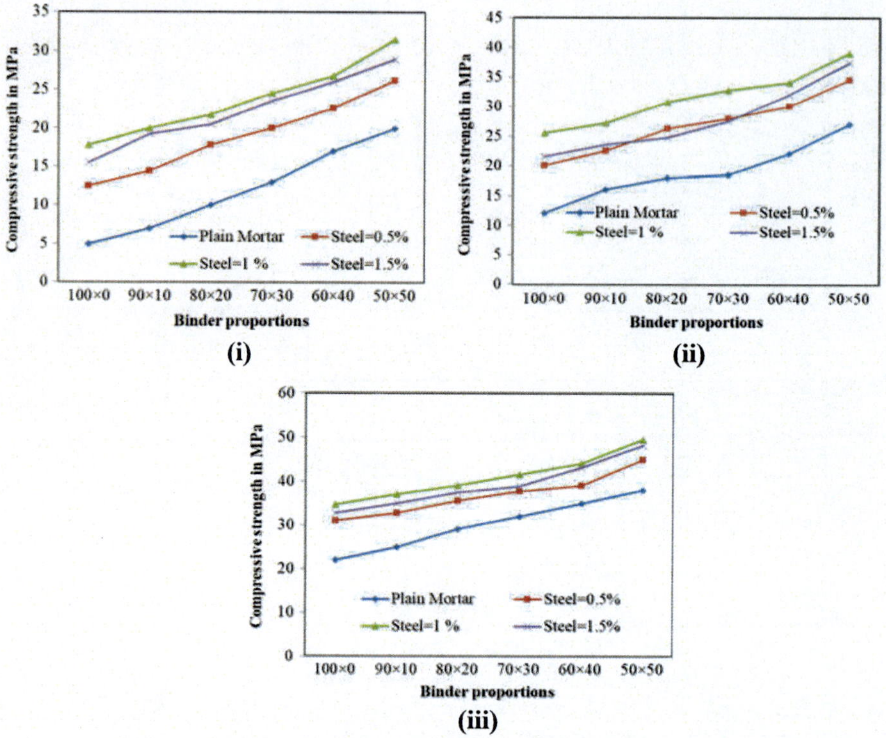

Fig. 10 Compression strength of geopolymer mortar with Alk/B = 0.8 at (i) 3 days, (ii) 7 days and (iii) 28 days

setting geopolymer mortar, partial replacement of FA by GGBS can be a possible solution.

- With the addition and increase in percentage of crimped steel fibres in plain geopolymer, mortar diminishes the flow diameter and flow capacity of mortar mixes in all alkaline to binder ratio. Decrease in flow diameter is between 23.56 and 32.94%.
- Alkaline to binder ratio of 0.6 shows the highest compressive strength values in all variation of steel fibre content and curing period.
- Incorporation and increase in fibre volume increases the compressive strength values of plain geopolymer mortar in the range of 5.1–22.6 MPa depending on the steel fibre content, binder compositions and alkaline to binder ratio.
- The compressive strength of geopolymer composites increases with the increase in GGBS content in the mixes; so to deliver geopolymer mortar with high compressive strength and also fast in setting under ambient curing condition, replacement of FA with GGBS can be a good possible solution.

(a) 0% GGBS ($\mathbf{F_{100}G_0}$)

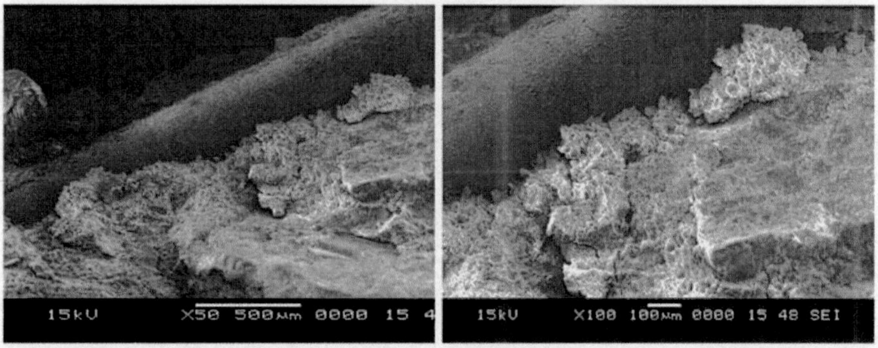

(b) 20% GGBS ($\mathbf{F_{80}G_{20}}$)

(c) 30% GGBS ($\mathbf{F_{70}G_{30}}$)

Fig. 11 SEM images **a** 0% GGBS ($F_{100}G_0$), **b** 20% GGBS ($F_{80}G_{20}$), **c** 30% GGBS ($F_{70}G_{30}$), **d** 40% GGBS ($F_{60}G_{40}$) and **e** 50% GGBS ($F_{50}G_{50}$)

(d) 40% GGBS ($\mathbf{F_{60}G_{40}}$)

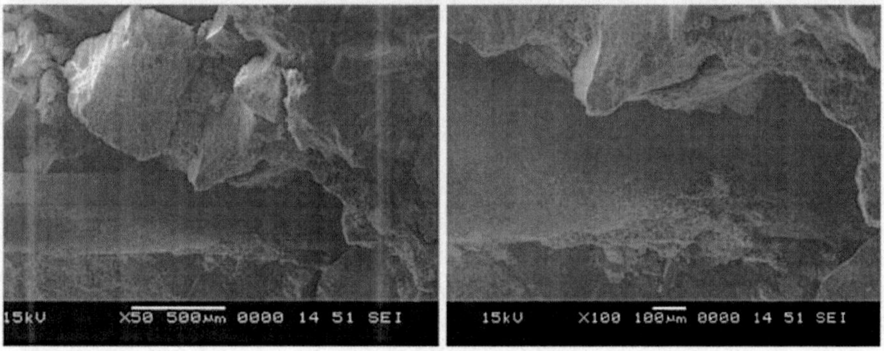

(e) 50% GGBS ($\mathbf{F_{50}G_{50}}$)

Fig. 11 (continued)

- Optimum fibre content showing the maximum strength value in all mixes is 1%. The highest compressive strength values using 1% fibre content, alkaline to binder ratio of 0.6 and 50 × 50 binder compositions is 69.5 MPa at 28-day curing.
- The SEM results indicate that the presence of rough surface of steel fibres and geopolymer hydration products on the surface of steel fibres in the specimens is an evidence of a relatively good bond between the geopolymer matrix and the steel fibre, which increases the compressive strength values.

References

1. Williamson T, Juenger MCG (2016) The role of activating solution concentration on alkali–silica reaction in alkali-activated fly ash concrete. Cem Concr Res 83:124–130
2. Joshi SV, Kadu MS (2012) Role of alkaline activator in development of eco-friendly fly ash based geo polymer concrete. Int J Environ Sci Dev 3(5):417–421
3. Ramesh V, Joy A (2017) Review on the study of fly ash based geopolymer concrete. IJARIIE 3(2):1311–1315. ISSN(O): 2395-4396

4. Wardhonoa A, Lawb DW, Stranob A (2015) The strength of alkali-activated slag/fly ash mortar blends at ambient temperature. In: The 5th international conference of Euro Asia Civil Engineering Forum (EACEF-5). Surabaya 60231, Indonesia, pp 650–656
5. Sayyad AS, Patankar SV (2013) Effect of steel fibres and low calcium fly ash on mechanical and elastic properties of geopolymer concrete composites. Indian J Mater Sci 2013, Article ID 357563:8 pp (Hindawi Publishing Corporation)
6. Rajan BR, Ramujee K (2015) Strength & development of fly ash and GGBS based geopolymer mortar. Int J Recent Adv Eng Technol (IJRAET) 3(1):42–45
7. Lee NK, Lee HK (2013) Setting and mechanical properties of alkali-activated fly ash/slag concrete manufactured at room temperatue. Constr Build Mater 47:1201–1209 (Elsevier)
8. Al-Majidi MH, Lampropoulos A, Cundy A, Meikle S (2014) Development of geopolymer mortar under ambient temperature for in situ applications. University of Brighton, Brighton BN2 4GJ, UK
9. Mohod MV (2015) Performance of polypropylene fibre reinforced concrete. IOSR J Mech Civ Eng (IOSR-JMCE) 12(1) Ver I (Jan–Feb):28–36. e-ISSN: 2278-1684, p-ISSN: 2320-334X
10. Bhalchandra SA, Bhosle AY (2013) Properties of glass fibre reinforced geopolymer concrete. Int J Mod Eng Res (IJMER) 3(4):2007–2010. www.ijmer.com
11. Ramkumar G, Sundarkumar S, Sivakumar A (2015) Development of steel fibre reinforced geopolymer concrete. Int J Adv Res Sci Eng (IJARSE), 4(01):1718–1725
12. Al-Majidi MH, Lampropoulos A, Cundy AB (2017) Steel fibre reinforced geopolymer concrete (SFRGC) with improved microstructure and enhanced fibre-matrix interfacial properties. Constr Build Mater:286307. www.elsevier.com/locate/conbuildmat
13. Natali A, Manzi S, Bignozzi MC (2011) Fiber-reinforced composite materials based on sustainable geopolymer matrix. Nov Procedia Eng 21:1124–1131
14. Porkodi R, Dharmar S, Nagan S (2015) Experimental study on fiber reinforced Self-compacting geopolymer mortar. Int J Adv Res Sci Eng IJARSE 4(01):968–977. http://www.ijarse.com
15. Plizzari GA (2004) Experimental study of fracture behavior of concrete reinforced with steel fibers. Research Report for Officine Maccaferri. Bologna, Italy
16. Ranjbar N, Mehrali M, Behnia A, Pordsari J, Alireza M, Alengaram M, Johnson U, Jumaat, Zamin M (2016) A comprehensive study of the polypropylene fiber reinforced fly ash based geopolymer. PLOS ONE | https://doi.org/10.1371/journal.pone.0147546
17. Guo X, Pan X (2018) Mechanical properties and mechanisms of fiber reinforced fly ash–steel slag based geopolymer mortar. Constr Build Mater 201804:633–641 (Shanghai, China)
18. IS 3812—Part 1 (2013) Pulverized fuel ash—Part 1: for use as pozzolana in cement, cement mortar and concrete. Bureau of Indian Standards, Reaffirmed 2017, New Delhi
19. IS 383 (2016) Specification for coarse and fine aggregates from natural sources for concrete. Bureau of Indian Standards, New Delhi
20. S 8142 (1976) Method of test for determining setting time of concrete by penetration resistance. Bureau of Indian Standards, Reaffirmed 2002, New Delhi
21. IS 4031 (1988) Methods of physical tests for hydraulic cement. Bureau of Indian Standards, Reaffirmed 2019, New Delhi

Durability Properties of Self-compacting Concrete Using Silica Fume

Tribikram Mohanty, Bhargavi Nandan Patra, and Purnachandra Saha

Abstract Vigorous industrialization has led to the voracious generation of waste materials, like silica fume, RHA, ferrochrome ash, fly ash and so on. There were alternative materials that can be used as mineral admixture to cement in concrete. For example, ferrochrome ash which was obtained from Ferro-alloy industries and fly ash which is produced in thermal power plants are a few of them. This research shows the combined effect of fly ash and ferrochrome ash-based concrete in different percentages as partial replacement of cement. This paper is inclusive of results of experiments that have been carried out to get the enhanced mechanical properties of OPC by replacing fly ash with various percentages (10, 20, 30 and 3%) of ferrochrome ash. It was observed from the research that even a small quantity of ferrochrome ash when mixed with 30% of fly ash increases the compressive strength than fly ash alone in concrete. Both ferrochrome ash and fly ash come categorically under industrial waste; hence by using this industrial waste reduces the production of greenhouse gas and helps manage dumping at large. This, at the end, is a method of producing sustainable concrete.

Keywords Ferrochrome ash · Fly ash · Industrial waste · Utilization · Mechanical properties

1 Introduction

Waste garnered by industries like mineral admixture and their by-products might be used in the concrete industry as a partial replacement of cement. The need to meet the increased demand of cement made essential to find out probable substitutions as

T. Mohanty (✉) · B. N. Patra · P. Saha
School of Civil Engineering, KIIT Deemed to be University, Bhubaneswar, India
e-mail: tmohantyfce@kiit.ac.in

B. N. Patra
e-mail: 1654016@kiit.ac.in

P. Saha
e-mail: purnasahafce@kiit.ac.in

B. B. Das et al. (eds.), *Recent Developments in Sustainable Infrastructure*, Lecture Notes in Civil Engineering 75, https://doi.org/10.1007/978-981-15-4577-1_66

production of cement gives rise to CO_2 during $CaCO_3$ calcinations being responsible for about 5% of greenhouse effect. By replacement of cement with certain quantities of waste products such as fly ash, ferrochrome ash, silica fume, red mud, GGBS and so on, CO_2 gas emission can be reduced in concrete industry [1–5]. Deotale et al. [6] observed that by incorporating the RHA, fly ash and steel fibre strength and durability properties can be improved. Acharya and Patra [7] did their research about the mechanical and durability standpoints of concrete with ferrochrome ash (F.ash) as partial replacement of cement. F.ash is a waste material obtained in the industries populating with ferroalloys. Hence, it was observed from the literature review that quality combination of admixtures derived from minerals can improve the different properties of concrete. The objective of this paper is to study the effect of fly ash and ferrochrome ash-based concrete on durability properties of concrete.

2 Materials and Methodology

2.1 Materials

The properties of cement are different and it depends upon their chemical compositions. As per IS:12269-1989, OPC 53 grade cement was used in the research work. Tables 1 and 2 represent the physical and chemical properties of the cement as per IS:269/4831 and IS:12269-1987. Natural coarse and fine aggregates are used in experimental work and their required properties are confirmed according to IS:383-1970. The properties of aggregate are reported in Table 3. BASF (Master gallium sky 8633) is used as super plasticizer for better workability and good strength. Specific gravity of super plasticizer is 1.4. According to IS456:2000 clean water was used

Table 1 Physical properties of 53 grade ordinary Portland cement

Sl. no.	Property	Requirement of IS:12269/1987	Test results
1	Fineness obtained (in m^2/kg)	225 (min)	320
2	Setting time (min) 1. Initial 2. Final	30 (min) 600 (max)	120 185
3	Soundness 1. Le chatelier expansion 2. Autoclave (%)	10 (max) 0.8 (max)	1 0.09
4	Compressive strength (MPa) 1. 3 days 2. 7 days 3. 28 days	27.0 (min) 37.0 (min) 53.0 (min)	48 60 70

Table 2 Chemical properties of 53 grade ordinary Portland cement

Sl. no.	Property	Requirement of IS:12269/1987	Test results
1	Sulphuric anhydride (% by mass)	3 (max)	1.91
2	Loss on ignition (% by mass)	4 (max)	1.04
3	Insoluble residue (% by mass)	3 (max)	0.96
4	Chloride content (% by mass)	0.10 (max)	0.028
5	Lime saturation factor	0.05 (max)	0.90
6	Al_2O_3/Fe_2O_3	0.80 (min)–1.02 (max)	1.36
7	MgO (5% by mass)	6 (by mass)	1.22

Table 3 Physical properties of aggregates

Sl. no.	Property	Fine aggregate	Coarse aggregate
1	Specific gravity	2.65	2.7
2	Fineness modulus	2.47	6.2
3	Water absorption	0.85 (%)	0.4 (%)
4	Free surface moisture	0.90 (%)	–

for mixing and curing of concrete, which was collected from Laboratory of Civil Engineering Department of KIIT, Deemed to be University.

2.2 *Silica Fume*

Silica fume was an amorphous polymer of silicon dioxide and silica, and is also called as micro silica. It is produced from the by-product of silicon and ferrosilicon alloy production (Fig. 1). It forms high performance concrete and is used in Portland cement concrete to get better strength and resistance. It reduces the seeping of concrete to chloride ions, which help in corrosion resistance, especially in chloride-rich environment. The chemical composition is given in Tables 4 and 5.

2.3 *Mix Proportions*

The control mix for M30 is designed without the addition of silica fumes. Silica fumes were incorporated by adding the material in different percentages, ranging from 0 to 30% by weight of cement (SF0, SF5, SF10, SF15, SF20, SF25, SF30). Water/cement ratio is taken as 0.42. Targeted slump is 70 mm. Coarse aggregates

Fig. 1 Silica Fume

Table 4 Physical properties of silica fume

Properties	Unit	Specification
Retention on 45 μm sieve	%	<1.5*
H_2O (When packed)	%	<1.0
Bulk density (*U*)	kg/m^3	200–350
Bulk density (*D*)	kg/m^3	500–700
Specific gravity		3.24

and fine aggregates are taken in the ratio of 40 and 60%, respectively. 1% of super plasticizer is used. According to calculation, desired mix ratio is 1:1.56:0.98.

2.4 *Preparation of Concrete Specimen*

Variable ratios of mix of self-compacted concrete and silica fumes are obtained to carry out the durability properties and compressive strength, and sorptivity on a specimen of size 150 × 150 × 150 (standard BIS specimen). Chloride resistance on standard BIS specimen of diameter 0.15 m and height 0.3 m is used. The curing age for all specimens are 7, 28, 56 and 90 days, respectively, in water at 270 ± 20 °C temp.

2.5 *Mixing Procedure*

To get direct and proper test results of the specimen, we should ensure uniform mixing of concrete. For the batch of concrete that is self-compacted, initially the aggregates classified as coarse are to be weighed for the required amount per mix ratio in a tray named 1; then focusing on the sand, the same procedure is done and then the quantity measured is poured into tray 2 (keep it dry). After measuring the cement, it spreads

Table 5 Chemical composition of silica fume (% by weight)

Constituent	SiO_2	Al_2O_3	CaO	Fe_2O_3	MgO	SO_3	Na_2O	K_2O	P_2O_5	TiO_2
Value (%)	93.61	0.638	0.975	0.319	0.793	0.453	0.31	0.96	0.53	0

homogeneously on surface of sand in tray 2, ensuring proper mixing. This mix is then uniformly spread on the NCA surface in tray 1 and two things are done hence. First—dry mixing, second—mix water with the dry mix to see a uniform colour of concrete. Weigh the concrete now soon in L-shaped box, do a V-funnel test and slump in moulds as per written and known procedure.

For mixing silica fume in self-compacted concrete, the procedure mentioned above is followed to the T except for one step. Before adding sand to coarse aggregate, cement with silica fume is added. Silica fume and cement are mixed, after that sand and cement are mixed with aggregates to achieve various types of concrete with 10–15% by substitute of NCA over volume of normal concrete that is self-compacted.

2.6 Test Procedure

According to IS:7320-1974 and IS:1199-1959, workability of concrete was measured. Compressive strength as per IS:516-1959 was determined. As per ASTM C-1585-11 sorptivity was determined.

3 Testing of Fresh Concrete

To measure the degree of workability for self-compacted concrete slump test, L-box test and V-box test are performed, and silica fume was added with self-compacted concrete (Fig. 2). The factor that has a significant impact on operability is the aspect ratio (l/*d*) of fibres. The ratio of cement in the water used in this study to a slope value of 70 ± 5 mm is very difficult to achieve and is shown in Table 6.

In self compacted concrete, compaction is not required, it is self compacted. After mixing, the mortar is taken out for Slump cone test, L-box test, V-box test to measure the time taken for flow ability.

Fig. 2 L-box, slump cone

Table 6 Slump cone, L-box, V-box test result

Sl. no.	Mix	Slump test (mm)	L-Box test (s)	V-Box test (s)
1	SF0	71	6	6.1
2	SF5	70	5.8	6
3	SF10	69	5.9	5.8
4	SF15	70	5.8	6
5	SF20	71	6	5.9
6	SF25	69	5.9	6.2
7	SF30	69	6.1	5.9

4 Mechanical Properties

The effect of silica fume on mechanical properties such as compressive strength of concrete containing NCA is reported in the following.

4.1 Compressive Strength of Concrete

In concrete industry compressive strength is one of the important aspects for the purpose of structural design. The average strength development by replacement with silica fume in ordinary Portland cement was studied at 7 and 28 days. The variation in compressive strength with different percentage (0, 5, 10, 15, 20, 25, and 30%) of silica fume over OPC is given in Fig. 3.

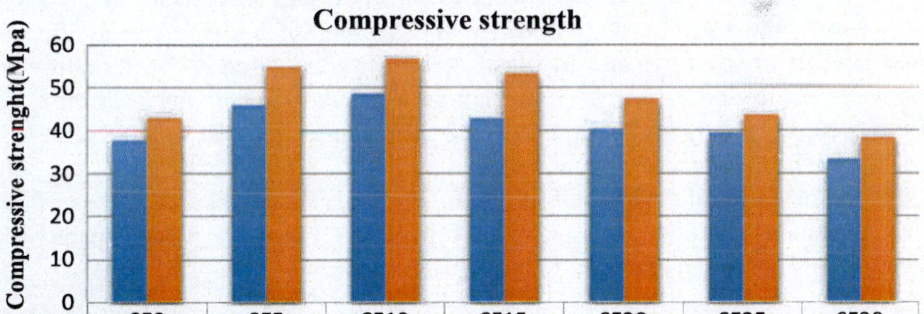

	SF0	SF5	SF10	SF15	SF20	SF25	SF30
7 Days	37.6	45.8	48.4	42.7	40.2	39.4	33.3
28 Days	42.9	54.8	56.7	53.3	47.4	43.6	38.3

Fig. 3 Effect of silica fume on compressive strength of self-compacting concrete

4.2 *Effect of Silica Fume on Compressive Strength*

The suitability of silica fume as replacement of cement was investigated. The compressive strength of 0.15 m cubes containing 0–30% of silica fume was found at 7 and 28 days. The results are presented in Fig. 3. It was observed that due to inclusion of silica fume, the strength gradually increases for all curing ages up to 25% in comparison to normal concrete. The specimen SF10 shows highest compressive strength than that of other specimens. Replacement of SF 0–30% enhanced compressive strength 14.10, 19.65, 17.14, 24.82, 16.86, 10.65 and 15.01% between 7 and 28 days. Replacement of SF 0–25% enhanced compressive strength 27.73, 32.16, 24.24, 10.48 and 1.63% at 28 days as compared to normal concrete. Thus, the results indicated that SF10% replacement may be considered when strength requirement is more than normal.

The increase in strength is due to inclusion of silica fumes. Silica fume is a pozzolanic and non-hazardous industrial by-product material having a very large surface area which is responsible for strength, durability and prolongs the setting time. But the presence of more silica expense of the content of alumina and ferric oxide will make the cement difficult to fuse and form slurry.

5 Durability Properties

5.1 *Sorptivity*

Sorptivity is done in an easier manner that shows the absorption capacity of the specimen to absorb water. It also denotes the void volume and water absorbing capacity. According to the results it can be known that with increase in silica fume addition, the void capacity of specimen also increases. By using silica fume in concrete improves the water absorption capacity. The variation in sorptivity results with different percentage (0, 5, 10, 15, 20, 25, and 30%) of silica fume over OPC is given in Fig. 4. The sorptivity result was determined at the age of 28 days for slightly wet cured specimen of concrete. The result of sorptivity test is indicated as all types of concrete specimens exhibited increase in void space and water absorption capacity with increase in silica fume content.

5.2 *Alkaline Attack Test*

Alkali react with silica (ASA), that is, alkali–silica reaction. It is also known as concrete cancer. Over a period of time the alkaline reacts with the concrete specimen and forms a swelling reaction. As a result, the concrete specimen expands and

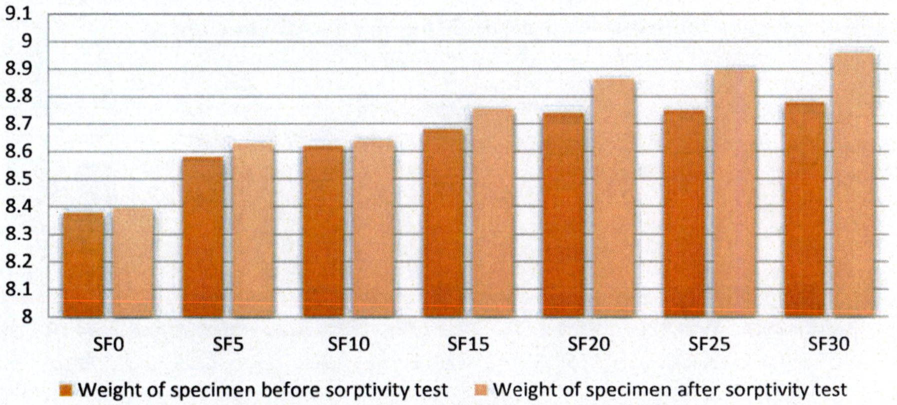

Fig. 4 Sorptivity test results at different percentages with silica fume in kg

Table 7 Weight loss and compressive strength test results after alkaline attack

Specimen name	Initial weight (kg)	Final weight (kg)	Weight loss (kg)	Compressive strength (MPa)
SF0	8.72	8.67	0.5	40.32
SF5	8.78	8.74	0.4	52.44
SF10	8.67	8.56	0.11	53.21
SF15	8.43	8.34	0.9	52.33
SF20	8.82	8.68	0.14	45.65
SF25	8.56	8.45	0.11	40.58
SF30	8.64	8.54	0.10	36.78

forms crack which will result in critical serious structural problem and demolition of structure, and the results are given in Table 7 and Fig. 5.

5.3 *Carbonation Test*

For higher *w*/*c* ratios, the intensity of corrosion is most important. With the presence of $Ca(OH)_2$, silica fume reacts with it and reduces pH of pore solution, thus the process of carbonization was facilitated. Specimen are not affected by carbonation and the result is given in Fig. 6.

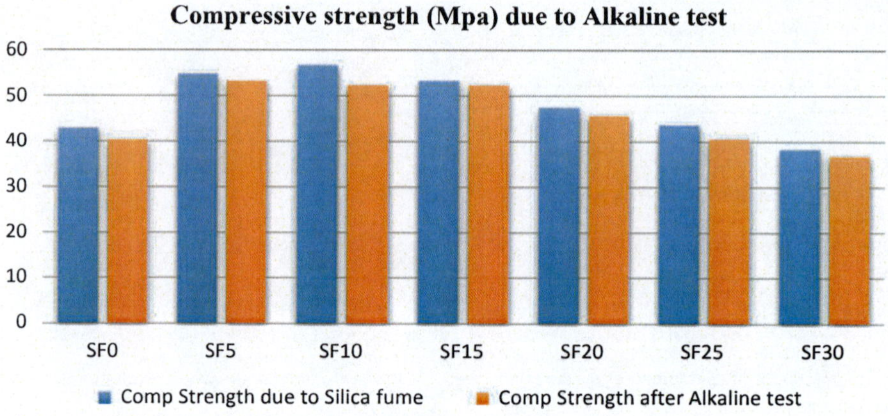

Fig. 5 Compressive strength test results due to alkaline attack

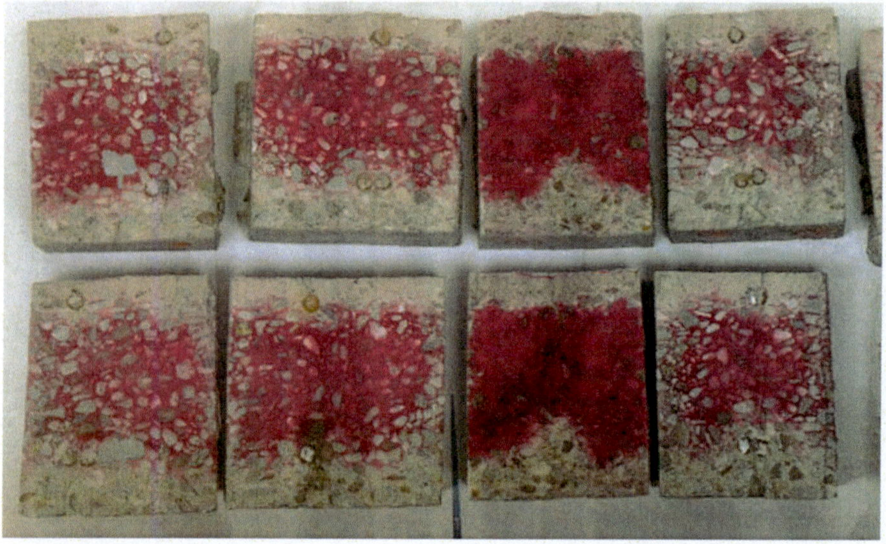

Fig. 6 Cube specimen are not affected by carbonation

5.4 *Chloride Resistance Test*

Chloride penetration resistance of all mixtures with silica smoke content has been increased. It was well known that $Ca(OH)_2$ in wet cement was converted into C–S–H, and reacts with silicate in silica smoke. This is known as pozzolanic reaction. However, it can be considered that chloride penetration resistance for concrete was not only affected by coarse aggregate species but also by the microstructure of wet cement. Weight loss and compressive strength results are given in Table 8 and Fig. 7.

Table 8 Weight loss and compressive strength test results after chloride attack

Specimen name	Initial weight (kg)	Final weight (kg)	Weight loss (kg)	Compressive strength (MPa)
SF0	8.78	8.67	0.11	40.64
SF5	8.56	8.42	0.14	52.24
SF10	8.64	8.56	0.8	53.68
SF15	8.74	8.57	0.17	51.33
SF20	8.68	8.56	0.12	42.65
SF25	8.43	8.34	0.9	39.86
SF30	8.73	8.61	0.12	35.67

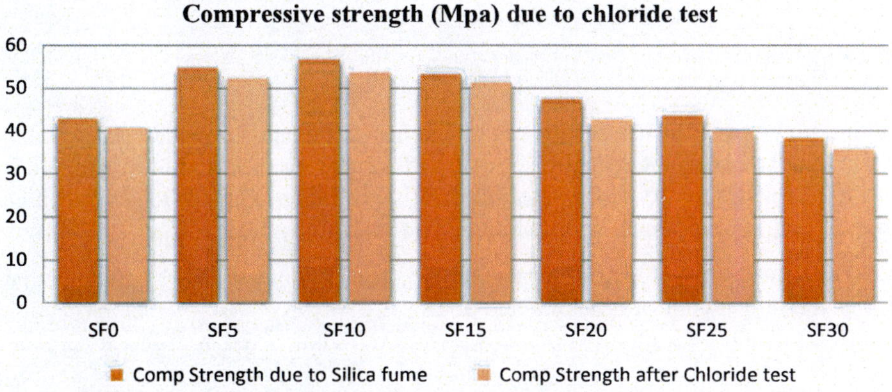

Fig. 7 Compressive strength test results due to chloride attack

Table 9 Compressive strength (MPa) by rebound hammer test

Specimen name	SF0	SF5	SF10	SF15	SF20	SF25	SF30
Compressive strength (MPa)	41.85	53.79	55.45	52.94	46.56	42.75	37.68

5.5 Non-destructive (Rebound Hammer) Test

Rebound hammer test for concrete samples SF0, SF5, SF10, SF15, SF20, SF25, and SF30 is performed at curing age 28 days. The results are shown in Table 9.

6 Standard Deviation

Standard deviation of various concrete mixes at curing age of 7 and 28 days, as shown in table analysis, revealed that the standard deviation of compressive strength at the age of 7 days varies from 0.397 to 1.689 MPa. At the age of 28 days, it varies from 0.382 to 2.024 MPa. The standard deviation from the test results are within the recommended values in Table 8 of IS:456-2000 to estimate the target strength of concrete.

Test	Age (days)	SF0	SF5	SF10	SF15	SF20	SF25	SF30
Compressive strength (MPa)	7	1.689	0.397	0.521	0.911	0.760	0.787	1.261
	28	1.410	1.248	1.584	0.713	2.024	0.382	0.805

7 Conclusions

In this study an effort has been taken to emphasize by using commercial available silica fume to garner the required strength and comparing their durability properties of SCC by addition of various percentages of silica fumes. Conclusions that can be drawn from the experimental observation in current study are:

1. The flow ability value of slump cone, L-box test, V-box test shows optimum, that is 700 mm at control mix, SF5 and SF15.
2. Compressive strength of concrete which is self-compacted at different percentages of silica fume increases as compared to OPC. However, it can be noticed that compressive strength of SF10 shows maximum value.
3. Sorptivity of self-compacted concrete at different percentages of silica fume indicates that with the addition of greater percentage of silica fume, volume of void increases. As a result, water absorbing capacity also increases. SF25 and SF30 show greater void, so water is also absorbed much.
4. Alkaline attack shows the weight loss after immersing in $MgSO_4$ acid of various percentages of silica fume addition. SF20 shows greater weight loss with comparison to others.
5. Carbonation results show that no specimen is affected as it turned purple colour by spraying phenolphthalein indicator on the fractured portion of the cube specimen.
6. Chloride resistance also shows the weight loss after immersing in salt solution of various percentages of silica fume addition. SF15 shows greater weight loss with comparison to others.

References

1. Goodier C (2003) Development of self-compacting concrete. Struct Build 2003(156):405–414
2. Self-consolidating concrete (2017). Wikipedia

3. Jalal MPA, Harandi O, Jfan D (2015) Comparative study on effects of Class F fly ash, nano silica and silica fume on properties of high performance self compacting concrete. Constr Build Mater 94(2015):90–104
4. Dehwah H (2012) Corrosion resistance of self-compacting concrete incorporating quarry dust powder, silica fume and fly ash. Constr Build Mater 2012(37):277–282
5. Ozawa K, Ouchi M (1999) Proceedings of the international workshop on self-compacting concrete. Wikipedia
6. Silica fume (2017). Wikipedia
7. Wongkeo W, Thongsanitgarn P, Ngamjarurojana A, Chaipanich A (2014) Compressive strength and chloride resistance of self-compacting concrete containing high level fly ash and silica fume. Mater Des 64(2014):261–269

The Performance of Geopolymer Concrete Utilizing Wastes as Binder

Sneha Sen and Purnachandra Saha

Abstract Utilizing wastes at present is of utmost importance since it not only leads to the reduction in environmental load and reduction of carbon footprint but also reduction in waste management cost. Geopolymer concrete has been developed as a substitute to conventional cement-concrete where alkali activators react with the binder along with fine and coarse aggregates. The binder for geopolymer concrete can be any wastes which are rich in silica and alumina that will collectively lead to the reduction in the concrete pollution and enhancement of its properties. The objective of this paper is to review the mechanical performance of geopolymer concrete utilizing wastes like fly ash, metakaolin, rice husk ash, palm oil fuel ash, red mud, ferrochrome ash, and so on as binder. It is observed that slag-based geopolymer concrete can be cured in ambient temperature but for the other binder materials, elevated curing temperature is required to attain the target strength. The mechanical properties (like compressive, split tensile, flexural properties) of the geopolymer concrete utilizing fly ash and slags as binder have shown to provide more strength than other binder combinations, and therefore prove to be a better replacement to the conventional cement-concrete.

Keywords Geopolymer concrete · Fly ash · Red mud · Metakaolin · GGBS · Rice husk ash · Palm oil fuel ash

1 Introduction

With the rapid growth of urbanization, vast number of industries are seen to have surfaced which leads to the generation of abundant wastes. This results in landfilling and environmental problems. Thus, the utilization of these wastes is better than disposing them off. In the construction industry, cement has a worldwide demand for being the major component of concrete. But, the manufacture of concrete leads

S. Sen · P. Saha (✉)
School of Civil Engineering, KIIT, Bhubaneswar, Odisha, India
e-mail: purnasahafce@kiit.ac.in; dr.purnasaha@gmail.com

S. Sen
e-mail: 1601068@kiit.ac.in

B. B. Das et al. (eds.), *Recent Developments in Sustainable Infrastructure*, Lecture Notes in Civil Engineering 75, https://doi.org/10.1007/978-981-15-4577-1_67

to huge CO_2 emission which ultimately leads to global warming. Thus, replacing cement with a befitting cementitious material is the need of the hour.

According to past research, the most widely accepted substitute for such a problem is the introduction of geopolymer concrete. It is a mixture of any binder rich in silica or aluminum oxides in the presence of alkali activators (like sodium hydroxide (SH) or sodium silicate (SS)) along with coarse and fine aggregates. Upon comparing ordinary concrete with geopolymer concrete, it was observed that the latter was more pronounced in terms of mechanical and durability properties than the former. The wastes as incorporated in the binder in this review are fly ash (FA), metakaolin (MK), rice husk ash (RHA), ground granulated blast slag (GGBS), palm oil fuel ash (POFA), and red mud (RM)—the substances which are mostly very easily available and rich in silica and/or aluminum oxides.

GGBS or ground granulated blast slag is generated when iron is extracted from its ore during metallurgy process. Inclusion of GGBS on geopolymer concrete brought about a lot of changes to its mechanical properties. It was observed that more the quantity of slag present in the mixture, more was the compressive strength increment. But strength increment was more pronounced with the SS/SH ratio being decreased comparatively. This resulted in the more compact formation of the microstructure of the binder. However, a partial replacement of GGBS with POFA in the binder brought about an even higher compressive strength than the utilization of 100% GGBS. Then again, with the increase in the POFA content in the binder as compared to GGBS led to a reverse nature in the compressive strength development. Inclusion of 50% FA and 50% POFA brought about a higher compressive strength development than 50% GGBS and 50% FA. It was also observed that the increment of Si/Al ratio in the mixture (by changing the composition of alkaline solution) resulted in increment of compressive strength in the geopolymer concrete but this strength was more prominent at higher curing temperatures without adding extra amount of water with the alkali activator. However, a strength reduction was observed with the increased addition of the superplasticizer content in the binder [1–5].

Metakaolin also forms an excellent binder for geopolymer concrete considering that it has a proportionate amount of silica and aluminum oxides. It too leads to the increment of the strength to a substantial amount with either activator concentration or alkali content increment [6–9]. With regard to the chemical composition of the geopolymers containing metakaolin content, it was noted that the addition of the clay resulted in an increase in strength of concrete, and the high concentration of aluminate gel brings about a greater degree of polymerization. Furthermore, a better dissolution of MK particles officiates an accelerated process of formation of silicates and aluminates monomers [10]. Cwirzen et al. [11] reaffirmed the need of a high molarity for the activation of MK by NaOH (exceeding 10 M), to form higher strength geopolymers, substantially with the addition of $Ca(OH)_2$. When compared to RM, MK-based geopolymer showed a better compressive strength increment. But having said this, the only problem with MK-based geopolymer was that it required curing at an elevated temperature for a longer time to gain strength after both compressive as well as splitting tensile tests. Zhang et al. [12] observed that the strength gained in MK-based geopolymer was same as that of ordinary cement concrete at ambient temperatures. But when exposed to higher temperatures, the MK-based geopolymer concrete showed an increment in strength unlike the ordinary cement concrete (where a reduction in strength was observed). This went on increasing until a certain temperature was reached, after which the strength increment in MK-based geopolymer showed a backward trend.

Fly ash is formed in the thermal power plants during coal combustion. Being rich in silica and aluminum oxides, it is commonly used as a binder in the geopolymer concrete. At an ambient temperature condition, the compressive strength was observed to be the highest for FA-based geopolymer mortar [13]. Islam et al. [2] affirmed that at room temperature, an extended pre-curing was essential for the development of strength of the geopolymeric materials with FA as the main binder because elevated temperature curing led to achievement of high strength and shortening of the time of the heat treatment. Furthermore, the increment in NaOH concentration with increase in the compressive strength (as silica and alumina are better filtered in a highly concentrated NaOH solution) was observed. Also, the higher NaOH concentration is more effective in dissolving FA particles and results in a better geopolymerization [14].

Rice husk ash too can be incorporated in the binder material in geopolymer concrete but in addition to GGBS. To prove the above, Kishore et al. [15] found that the strength obtained with 100% FA as the binder was nearly equal to 10% replacement with RHA, thus implying that RHA could also be used as an alternate binder in geopolymer concrete. But, Shaswat et al. [16] found that the inclusion of RHA showed a retarding nature of the compressive strength results when added above 10%.

Hence, incorporating these daily generated wastes will not only lead to reduction in carbon emissions, leaching, air pollution, global warming but also to the reduction of the landfilling problems and so on.

Researchers have previously worked on geopolymer concrete, by utilizing the above-mentioned wastes and have found out their relation with their corresponding individual mechanical properties. Thus, incorporating all the data as obtained previously, this review aims at collaborating the majority of all the results as obtained by them and then recommending the best combination of wastes to produce commercially viable geopolymer concrete, so as to have a sustainable infrastructure.

2 Mechanical Properties of the Wastes

2.1 Compressive Strength

It is defined as the withstanding capacity of any structure or material to loads that tend to reduce its size. It is governed by numerous factors like quality of raw materials, w/c ratio, coarse aggregates to fine aggregates ratio, the age of concrete, compaction, temperature, relative humidity, and curing. The compressive strength of geopolymer concrete decreases with the partial replacements of GGBS with fly ash (FA), metakaolin (MK), silica fume (SF) under ambient curing conditions. It was observed that with the replacement of 60% of GGBS with FA, a comparatively lower decrease in the compressive strength took place than the replacements of GGBS with SF and MK (the latter being the highest) [6, 8, 10, 11].

It was also observed that the fly ash-based metakaolin increases its compressive strength up to a certain temperature and then decreases with further increase in temperature. For pure metakaolin mortars, lower the alkalinity of the activating solution (like sodium hydroxide) used at low temperatures, lower is the strength gained. Strength gain is also influenced by the amount of clay content in the metakaolin-based geopolymer. At ambient temperature, FA-based geopolymer mortar was seen to have the highest compressive strength. At higher temperatures, alkali-activated cementious pastes using class F-fly ash was seen to have higher compressive strength than the one at lower temperature condition. It was also observed that the strength development of the geopolymeric materials was benefited by the extended precuring at room temperature utilizing FA [1, 2, 13].

When 100% palm oil fuel ash (POFA) was incorporated with GGBS, the strength was seen to increase notably. But, with the inclusion of more amounts of POFA than GGBS in the mixture, the compressive strength was seen to decrease. 50% GGBS and 50% POFA led to an observation of obtaining a comparatively higher strength than 50% GGBS and 50% FA. Furthermore, a rise in compressive strength was also observed in a mixture of 50% GGBS, 25% POFA, and 25% FA [2, 13].

Rice husk ash (RHA) too can be incorporated in the binder of a geopolymer concrete but should be used as an addition to GGBS since it has a negative effect on the compressive strength of the geopolymer concrete (a retarding trend of the compressive strength was observed on adding RHA beyond 10%). With the addition of more percent of RHA content, the compressive strength was seen to decrease. Thus, the strength gained was seen to be higher in mortars where grounded RHA was not added [15–17].

The variations in compressive strength due to the combinations of the above-stated wastes in different proportions are shown in Fig. 1.

2.2 *Flexural Strength*

It is the property by which a beam or slab resists its failure in bending. It portrays the maximum stress experienced by the material during its moment of yield. Concrete is known to possess high compression but comparatively low tension. Thus, the direct measurement of flexural strength is associated with a lot of difficulties [18].

On comparing the data from the previous studies, it was observed that at a given concrete age the flexural strength of GPC increased with an increase in the addition of FA content to the binder [19]. It was also observed that in an ambient temperature, the flexural strength of geopolymer concrete increased when GGBS, OPC, or $Ca(OH)_2$

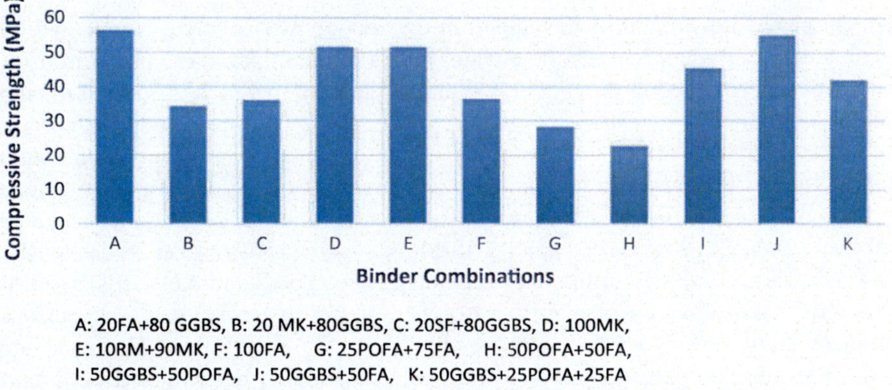

Fig. 1 Graph representing compressive strength versus binder combinations

with FA were incorporated in the binder. In mixtures with 40% alkaline liquid, the flexural strength was observed to have increased upon the addition of GGBS (by 10%), OPC (by 6%), and CH (by 2%). But, the one with 15% GGBFS was unable to achieve a comparatively higher flexural strength than its observed nature. According to Nath et al. [4], for an ambient cured geopolymer, the addition of an extra amount of water along with additives to the binder led to an unpropitious effect on the flexural strength, that is, upon incrementing the amount of additives in the mixture, the flexural strength tended to decrease.

Wastes like RHA can also be incorporated in the binder of the geopolymer concrete given its richness in silicon oxide content. But upon changing the proportions of RHA in addition to GGBS, it was observed that incorporating RHA beyond 10% showed a rather decrease in flexural strength in the geopolymer concrete [16]. Furthermore, it was also observed that, by replacing aggregate with crumb rubber up to 10%, the flexural strength in the geopolymer concrete showed a reduction in flexural strength by up to 20%. However, when the crumb rubber content was increased to 20%, then 30% of the flexure force was then observed to be unchanged but with a reduction by 30% in flexure strength [20].

2.3 *Tensile Strength*

It is another most important property which governs the behavior of the concrete structures. It refers to the capacity of concrete to resist tension and is used to measure the inception and propagation of cracks, anchorage, and shear of the reinforcing steel in concrete. Due to low tension in concrete, the measurement of direct tensile strength is not feasible due to the difficulty in applying the uniaxial tension in the structure. Thus, the split tensile strength test is preferred and is performed, which is basically an indirect method of determining the tensile strength of concrete. The amount at which the tensile strength is developed in the geopolymer concrete is hindered by extra amount of water and alkali activator added to the mixture, the proportions of additives incorporated in the binder, the temperature up to which it is exposed, and so on [1, 5, 20].

Considering the former studies conducted on the same, it was observed that with more slag content in the binder and less SS/SH ratio, the tensile strength developed in the concrete also increased. It was also observed that the geopolymer concrete mixture containing 20% GGBS and 1.5 SS/SH ratio gained 55% higher 28-day tensile strength when compared to the one containing 10% GGBS and 2.5 SS/SH ratio. It was further noted that the tensile strength development in geopolymer concrete was comparatively slower for the mixture with FA as the only binder. However, the rate of tensile strength increment was significantly high when GGBFS (as a part of the binder) was added to the mixture. After 28 days, the mixtures with their contents, GGBS (10%) and GGBFS (20%), achieved a strength which was 25 and 45% (respectively) higher than the mixture with only FA as binder. Comparing the observations thus made from the different tensile strengths of geopolymer mixtures, it was observed that the tensile strength improved with the reduction in SS/SH ratio from 2.5 to 1.5 and also with the addition of extra water in the mixture [1].

3 Discussions

It is observed that geopolymer concrete can be cured at an ambient temperature using GGBS, though it does not undergo pure geopolymerization. MK is one of the best binding elements for the geopolymer concrete because it contains a proportionate amount of silica and aluminum oxides but in concrete with 100% MK as binder had to be cured at an elevated temperature. Up to a certain proportion, the partial replacements of the main binder with the above-stated wastes in geopolymer concrete showed an increase in compressive strength but further increase in the binder quantities showed a reverse nature. To make up for this nature, a partial replacement of the main binder by adding parts of the other waste in small amounts (e.g., 50% GGBS + 25% POFA + 25% FA) showed a better increment in compressive strength than the addition of 100% of a single waste (like MK or FA).

4 Conclusion

In this review, wastes like ground granulated blast slag (GGBS), red mud (RM), fly ash (FA), rice husk ash (RHA), palm oil fuel ash (POFA), and metakaolin (MK) have been incorporated in the binder during the formation of geopolymer concrete.

- The combination was best when 20% FA was used with 80% GGBS in the binder giving the maximum compressive strength as high as 57 MPa.

- An optimum level of compressive strength was gained by incorporating 50% FA with 50% GGBS resulting in 53 MPa.
- Utilization of RHA led to gaining in flexural strength but inclusion of RHA above 10% resulted in a reverse effect.
- Considering the split tensile strength, a rise in strength was observed with the addition of slag content and decrease in sodium hydroxide/sodium silicate (SS/SH) content. However, the tensile strength development was slow where FA was used as the only binder.

References

1. Deb PS, Nath P, Sarker PK (2014) The effects of ground granulated blast-furnace slag blending with fly ash and activator content on the workability and strength properties of geopolymer concrete cured at ambient temperature 62:32–39
2. Islam A, Alengaram UJ, Jumaat MZ, Bashar II (2014) The development of compressive strength of ground granulated blast furnace slag-palm oil fuel ash-fly ash based geopolymer mortar 56:833–841
3. Nath P., Sarker P. K. (2014), "Effect of GGBFS on setting, workability and early strength properties of fly ash geopolymer concrete cured in ambient condition" 66 163–171
4. Nath P, Sarker PK (2017) Flexural strength and elastic modulus of ambient-cured blended low-calcium fly ash geopolymer concrete 130:22–31
5. Das SK (2018) Parametric study of flyash based geopolymer concrete 7(2.31):196–198
6. Duan P, Yan C, Zhou W, Luo W, Shen C (2015) An investigation of the microstructure and durability of a fluidized bed fly ash–metakaolin geopolymer after heat and acid exposure 74:125–137
7. Kuenzel C, Li L, Vandeperre L, Boccaccini AR, Cheeseman CR (2014) Influence of sand on the mechanical properties of metakaolin geopolymers 66:442–446
8. Hadi MNS, Farhan NA, Sheikh MN (2017) Design of geopolymer concrete with GGBFS at ambient curing condition using Taguchi method 140:424–431
9. He J, Zhang J, Yu Y, Zhang G (2012) The strength and microstructure of two geopolymers derived from metakaolin 30:80–91
10. Aboulyat A, Riahi M, Touhami MO, Hannache H, Gomina M, Moussa R (2017) Properties of metakaolin based geopolymer incorporating calcium carbonate. Adv Powder Technol 28:2393–2401
11. Cwirzen A, Provis JL, Penttala V, Cwirzen KH (2014) The effect of limestone on sodium hydroxide-activated metakaolin-based geopolymers 66:53–62
12. Zhang H. Y., Kodur V., Wua B., Yan J., Yuan Z.S. (2018), "Effect of temperature on bond characteristics of geopolymer concrete" 163 277–285
13. Ranjbar N, Mehrali M, Alengaram UJ, Metselaar HSC (2014) Compressive strength and microstructural analysis of fly ash/palm oil fuel ash based geopolymer mortar under elevated temperatures 65:114–121
14. Görhan G, Kürklü G (2014) The influence of the NaOH solution on the properties of the fly ash-based geopolymer mortar cured at different temperatures 58:371–377
15. Kishore GN, Gayathri B (2017) Experimental study on rise husk ash & fly ash based geopolymer concrete using M-sand 225:012273
16. Das SK, Mishra J, Mustakim SM (2018) Rice husk ash as a potential source material for geopolymer concrete: a review 7:81–84
17. Prabu B, Shalini A, Kumar JSK Rice husk ash based geopolymer concrete—a review, 2278–6783

18. Ushaal TG, Anuradha R, Venkatasubramani GS (2015) Flexural behavior of self compacting geopolymer concrete using ggbfs with various replacements of R-sand and M-sand, 1819–6608
19. Xie T, Ozbakkaloglu T (2015) Behavior of low-calcium fly and bottom ash-based geopolymer concrete cured at ambient temperature 41:5945–5958
20. Aly MA, El-Feky MS, Kohail M, Nasr EAR (2019) Performance of geopolymer concrete containing recycled rubber 207:136–144

Effect of Replacement of GGBS and Fly Ash with Cement in Concrete

Rachita Panda and Tanmaya Kumar Sahoo

Abstract Use of ordinary Portland cement as a binding material is well accepted for construction purposes globally. The huge demand for cement in construction leads to a rise in the use of absolute energy from the cement industry and high CO_2 emissions. As a result, the global cement sector is facing increasing difficulties in the conservation of material and energy resources, as well as decreasing its CO_2 emissions. Similarly, many industrial wastes like fly ash, ground granulated blast furnace slag, waste recycled product, cement kiln dust, silica fume, quarry dust, glass waste, rubber waste, red mud are generated nowadays due to growth in industrialization, increasing urbanization and rising standards of living due to technological innovation, which is leading to harmful threats to the environment along with the waste disposal problem. Therefore, in this paper efforts have been made for not only to control environmental pollution but also to reduce the cost of waste management and concrete production by utilizing ground granulated blast furnace slag (GGBS) and fly ash (FA) with partial replacement of cement. Industrial wastes, such as GGBS and FA, are waste by-products from iron industry and thermal power plants, respectively. In the present study, these by-products at different percentages as a partial replacement of cement have been utilized in concrete. The tests were conducted with partial replacement of GGBS at different percentages of 0, 10, 15, 20 and 25% and with FA at different percentage of 0, 5, 7.5, 10, 12.5 and 15% by the dry weight of cement. Tests on workability and compressive strength were carried out on concrete mixes prepared at different percentages of GGBS and FA. The experimental results show a significant improvement in the strength with increase in GGBS and FA content in partially replaced cement-concrete, making it viable for practical uses in the building industry. On the basis of experimental results, it is concluded that with an increase in GGBS and fly ash percentage the compressive strength of concrete increases and the evaluation allows the design of GGBS and FA mixed concrete for medium strength concrete.

R. Panda (✉)
K.I.I.T., Bhubaneswar, India
e-mail: rachitapanda184@gmail.com

T. K. Sahoo
Department of Water Resources, Government of Odisha, Bhubaneswar, India
e-mail: tanmaya162@gmail.com

B. B. Das et al. (eds.), *Recent Developments in Sustainable Infrastructure*, Lecture Notes in Civil Engineering 75, https://doi.org/10.1007/978-981-15-4577-1_68

Keywords Ground granulated blast furnace slag · Fly ash · Concrete · Compressive strength

1 Introduction

Concrete has a significant role worldwide for being used as construction material. The blooming infrastructure and development have led to an increase in construction, hence the demand of cement. In order to meet the requirement of cement, production of cement has been increased which not only enhances the huge requirement of resources for cement industry but also emitting 5–7% CO_2 worldwide [1]. The ultimate strength of ordinary Portland cement-concrete is largely dependent on the formation of C–S–H gel from the hydration process of cement. Hence in recent technologies alternative materials have been selected to get the required strength of concrete replacing cement.

On the other hand, a large amount of industrial wastes has been generated due to the rapid industrialization. These wastes require huge space for dumping or various treatment process to be useful. Hence utilization of these waste products in the preparation of concrete by keeping an eye on the characteristics will reduce the environmental burden, the cost of waste management and the concrete cost of manufacturing.

In the present study, wastes from iron industry and thermal power plant, popularly known as GGBS and FA, respectively, have been utilized in the mixture of concrete. The cementitious behaviour of these wastes is dependent on the specific surface area and the pozzolanic behaviour of its mineral composition. Globally, it is estimated that annual output of fly ash ranges from 0.75 to 1 billion tons [2]. Similarly, GGBS output typically varies from around 300–540 kg for each ton of pig or rough iron created for mineral feed involving 60–65% of iron, whereas during steel production 150–200 kg per ton of slag is produced per ton of liquid steel [3]. As demand for inexpensive energy increases in developing nations, this amount is anticipated to increase in the coming years. Hence bulk utilization of these wastes can be achieved by its civil engineering applications [4].

2 Experimental Study

2.1 Materials Used

2.1.1 Cement

OPC-43 grade cement conforming to **IS: 8112/2013** was collected from Bhubaneswar, Odisha to carry out this experiment. In order to avoid energy consumption of cement industry and environmental protection from excessive CO_2 emission, the use of non-conventional cementitious waste materials have been used in this experiment in the production of concrete.

2.1.2 Fly Ash

Fly ash is a waste generated at thermal power stations constituting about 60–88% of total combustion residues. In this study class F fly ash from NALCO, Talcher, Odisha at Rs. 25/- per bag was collected. As FA is cheaper than cement and has binding property, hence can contribute towards concrete preparation.

2.1.3 Ground Granulated Blast Furnace Slag

GGBS is a consequence of iron industry acquired by heating of the earthy constituents of iron-ore with the limestone and coke at high temperature in the blast furnace about 1500 °C. It was collected from steel industry, Bhubaneswar, Odisha at Rs 200/- per bag. GGBS shows cementitious behaviour, which can be a replacement of cement in concrete.

Portland cement, fly ash and GGBS were the cementitious materials used in this experiment and their physical and chemical characteristics are shown in Table 1.

Table 1 Physical and chemical properties of cement, GGBS and FA

S. no.	Particulars	OPC	GGBS	FA
1	Specific gravity	3.02	2.74	2.43
2	Specific surface area (m^2/kg)	320	558	443
3	Compressive strength (MPa)	40.88	–	–
4	SiO_2 (%)	20.61	33.64	54
	Al_2O_3 (%)	5.028	21.5	26
	CaO (%)	62.61	33.2	18.10
	Fe_2O_3 (%)	3.329	0.2	19.28
	MgO (%)	2.237	9.5	3.30
	SO_3 (%)	2.723	0.66	1.5
	Na_2O (%)	0.328	0.34	–
	K_2O (%)	0.577	0.39	1.34
	P_2O_5 (%)	0.32	–	–
	TiO_2 (%)	0.27	–	1.20
	LOI	1.968	0.56	1.02

Table 2 Physical properties of aggregates

S. no.	Particulars	CA	FA
1	Specific gravity	2.8	2.65
2	Fineness modulus	6.4	2.47
3	Water absorption	0.2 (%)	0.55 (%)

2.1.4 Aggregates

In concrete, aggregates are the major constituents. They are giving the concrete body, reducing shrinkage and increasing strength. The concrete consists of 70–80% of aggregate volume so that any change in the percentage of aggregate will cause insufficient strength. The 10 and 20 mm crushed stones of satisfactory properties were used as coarse aggregates in the present study. The fine aggregates passing through 4.75 mm, conforming to zone II as per the specification of **IS: 383-2016**, were used.

The physical properties of aggregates are given in Table 2.

2.2 *Methods*

2.2.1 Compressive Strength Test of Concrete

The maximum bearing load sustained by the specimen without causing failure of the specimen to the cross-sectional area under gradually applied load is known as the compressive strength. The concrete was prepared by mixing cement, fine aggregate, coarse aggregate homogeneously with required quantity of water. The assembly of concrete was filled the moulds, compacted in three layers, both by hand compaction just as by table vibrator after compaction, giving the specimen a smooth completion and expelling them from the vibrator. The specimens were demoulded following 24 h and moved to restoring tanks where they were allowed to cure for the required number of days. Cubes of size 150 mm have been made according to IS:456 and compression testing machine is used to test the solid blocks. The compression strength was calculated using the formula:

$$\text{Compressive strength} = \frac{\text{load}}{\text{area}} \text{MPa}$$

Cubic blocks of concrete were drawn out of water at each required curing period and retained for surface drying. The blocks have been tested for the desirable quality of strength in a 40T compression testing machine.

2.3 *Concrete Mix Preparation*

A composite mixture of concrete is prepared by adding cement, fine aggregate, coarse aggregate and water. In this study GGBS and FA are mixed with concrete partially replacing cement in various proportions. Mix design of concrete has been done referring IS 10262:2009 for proportioning of all concrete materials for design strength.

Mix design				
W/C ratio	Cement (kg/m^3)	Fine aggregate (kg/m^3)	Coarse aggregate (kg/m^3)	Water (kg/m^3)
0.48	394	790	1043.6	197.16

Cement, fine aggregate and coarse aggregate were taken in the 1:2:2.65 ratio corresponding to concrete grade M-25. All the ingredients of concrete are mixed homogeneously. The preparation of composite concretes were carried out by partially replacing dry weight of cement with 10, 15, 20 and 25 of GGBS and 0, 5, 7.5, 10, 12.5 and 15% of fly ash, respectively. By keeping the percentage of GGBS constant, fly ash percentage has been varied and mix proportions were prepared accordingly. Then compressive strength testing was performed at 7 days and 28 days for each and every sample.

3 Results and Discussion

The development of compressive strength of concrete mixtures was determined at 7 and 28 days of curing with given percentages of GGBS and FA. Workability and strength properties are improved when GGBS is added to concrete. FA provides progressive compressive strength and higher resistance from sulphate attack to the structure. The average of three samples was taken for each proportion of GGBS and FA. The increase in strength completely depends upon the increase in curing period as the addition of water will enhance the chemical reaction of mineral components strengthening the bond. In the reaction process the pozzolanic phenomenon takes place only when the available calcium hydroxide reacts with silica and alumina. The chemical reaction of the Portland cement is conveyed as follows [5]:

$$\text{Cement}\,(C_3S,\ C_2S) + H_2O\,(H) \rightarrow\ C{-}S{-}H{-}\text{gel} + Ca(OH)_2(CH)$$

The pozzolanic reaction is:

$$Ca(OH)_2(CH)\ +\ SiO_2(S)\ +\ H_2O(H) \rightarrow C{-}S{-}H{-}\text{gel}$$

$$Ca(OH)_2(CH) + Al_2O_3(A) + H_2O\ (H) \rightarrow C{-}A{-}H{-}gel$$

Portland cement hydration generates calcium hydroxide, and then pozzolanic reaction occurs by combining silica with calcium hydroxide. The combination of higher GGBS percentage and reduced FA percentage improves concrete strength.

The relation between the compressive strength of concrete at constant GGBS and varying FA content is provided in Figs. 1, 2, 3 and 4.

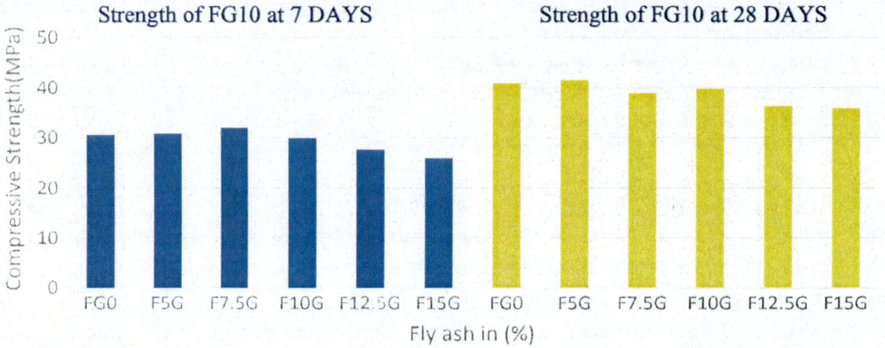

Fig. 1 Compressive strength of samples at 7 days and 28 days having 10% GGBS content at various fly ash content

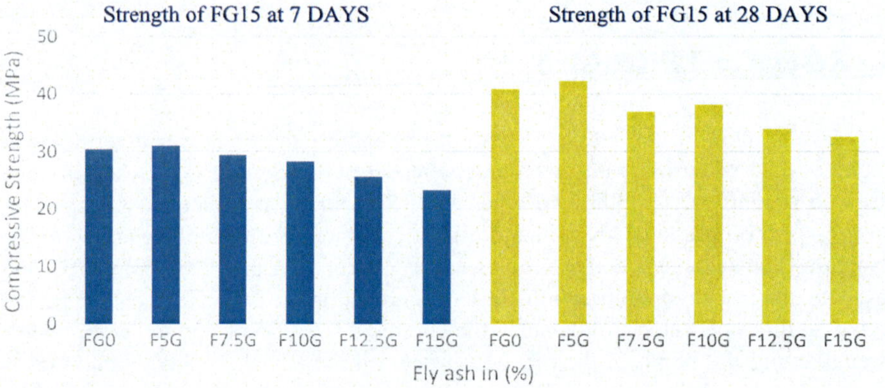

Fig. 2 Compressive strength of samples at 7 days and 28 days having 15% GGBS content at various fly ash content

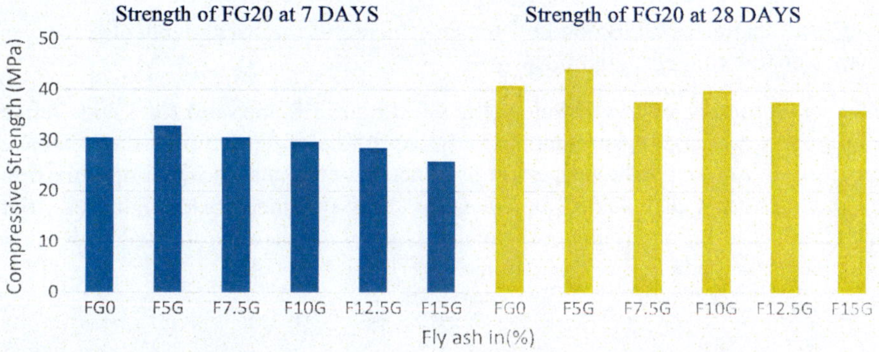

Fig. 3 Compressive strength of samples at 7 days and 28 days having 20% GGBS content at various fly ash content

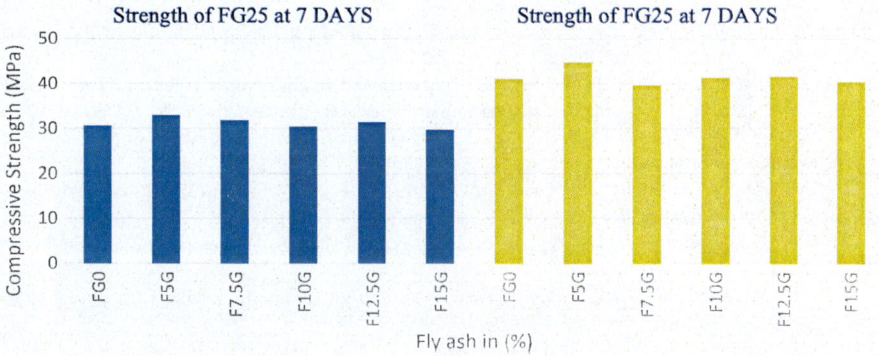

Fig. 4 Compressive strength of samples at 7 days and 28 days having 25% GGBS content at various fly ash content

4 Conclusion

In the present study, the feasibility of using GGBS and FA as a cementitious material with partial replacement of cement was investigated by conducting a series of compressive strength tests at 7 and 28 days with varying percentage of GGBS and FA. The conclusions of the study are as follows:

1. With constant GGBS content and varying FA content, the strength of concrete varies between 30 and 42 MPa, meeting the requirement of medium strength concrete.
2. The higher proportion of GGBS and the lower proportion of FA increases the strength, whereas the lower proportion of GGBS and higher proportion of FA decreases the strength.
3. The maximum strength is obtained at 25% of GGBS and 5% of FA content, which is 42 MPa.

4. The utilization of GGBS and FA not only increases the strength but also reduces the construction cost significantly.

These mix proportions of concrete using GGBS and FA are valid for a wide range of mix proportioning and provide an effective means of determining the influence of industrial wastes on the properties of concrete. They also provide optimization of blending and quality control between the above outcomes. Although these trial mixes are based on a certain set of materials, they can be readily used with other components to produce future outcomes.

References

1. Benhelal E, Zahedi G, Shamsaei E, Bahadori A (2013) Global strategies and potentials to curb CO2 emissions in cement industry. J Prod 51:142–161
2. Dindi A, Quang DV, Vega LF, Nashef E, Abu-Zahra MRM (2019) Applications of fly ash for CO_2 capture, utilization, and storage. J CO_2 Util 29:82–102
3. IBM (2017) Indian minerals yearbook 2017 slag-iron and steel (advance release). Government of India Ministry of Mines, Indian Bureau of Mines 16 Slag-Iron and Steel, New Delhi
4. Jha JN, Choudhary AK, Gill KS, Shukla SK (2014) Behavior of plastic waste fiber-reinforced industrial wastes in pavement applications. Int J Geotech Eng 8(3):277–286
5. Zhou XM, Slater JR, Wavell SE, Oladiran O (2012) Effects of PFA and GGBS on early-ages engineering properties of Portland cement systems. J Adv Concr Technol 10(2):74–85
6. IS: 8112-2013. Specification for 43 grade ordinary Portland cement. Bureau of Indian Standards, New Delhi
7. IS: 10262-2009 and SP 23:1982. Recommended guidelines for concrete Mix. Bureau of Indian Standards, New Delhi

Effect of Silica Fume on Strength Enhancement of Geo-Polymer Mortar in Ambient Curing

Amarendra Kr. Mohapatra, Dillip Kumar Bera, and Ashoke Kumar Rath

Abstract One of the best alternative binder in place of Portland cement is geo-polymer binder. In geo-polymer binder, the main constituent is fly ash which is rich in silicate and alumina. It reacts with sodium hydroxide/silicate solution to produce alumina silicate gel. This gel binds the fine aggregate to form a good paste-like mortar called geo-polymer mortar (GPM). In this paper, the influence of lime and silica fume towards the strength enhancement of GPM mortar has been studied. Most of the author emphasized on hot curing rather than ambient curing for achieving high strength of geo-polymer matrix. This paper also works out to optimize the sodium silicate to hydroxide proportion, molarity of NaOH, various percentage replacement of lime and silica fume partially with fly ash on enhancement of strength of GPM. It was found out that the compressive strength value increased on increase of molarity of NaOH from 6 to 12 M; similarly, for sodium silicate to sodium hydroxide ratio at 2:1 and also at 7.5 and 3% of lime and silica fume replacement, respectively, better strength was shown.

Keywords Geo-polymer mortar · Fly ash · Ambient curing · Lime · Silica fume · Compressive strength

1 Introduction

Normally, ordinary Portland cement is only the versatile binding material used in the making of mortar or concrete. However, it is unlikely to release a great extent of carbon dioxide (CO_2) to the environment at the time of production [1]. Approximately 7% of CO_2 are annually produced during the production of ordinary Portland

A. Kr. Mohapatra
PhD Scholar School of Civil Engineering, KIIT University, Bhubaneswar, India
e-mail: amarwr@gmail.com

D. K. Bera (✉) · A. K. Rath
Faculty School of Civil Engineering, KIIT University, Bhubaneswar, India
e-mail: dberafce@kiit.ac.in

A. K. Rath
e-mail: akrath1947@gmail.com

B. B. Das et al. (eds.), *Recent Developments in Sustainable Infrastructure*, Lecture Notes in Civil Engineering 75, https://doi.org/10.1007/978-981-15-4577-1_69

cement (OPC). Thus CO_2 at about 65% contributes to global warming among all the greenhouse gases [2]. To avoid the above issues, geo-polymer matrix is one of the best, environmental-friendly and similar behavioural alternate binding material like cement. Geo-polymer matrix is an alumino-silicate polymer synthesized from thermal power plant by-product materials like fly ash and sodium hydroxide/silicate solution (alkali-activated solution) [3, 4]. The low calcium fly ash (F-type)-based geo-polymer concrete (GPC) is economical than the concrete and is also helpful to reduce global warming as reported by Madheswaran et al. [5].

Hake and Adam reported in their papers about the use of lime powder in the preparation of fly ash-based GPC to enhance the strength in ambient curing. The compressive strength of GPC in hot curing is higher for a short period of time that of ambient curing. In every construction site the mortar or concrete is cured under ambient environment rather than in hot environment, and hence the study that emphasizes on the performance of GPM under ambient environment curing condition is of special interest [6–8]. In this paper attempt has been taken to optimize the sodium silicate to hydroxide proportion, molarity of NaOH, various percentage replacement of lime and silica fume partially with fly ash to improve the strength properties GPM. Okoyo et al. have added silica fume with fly ash to enhance the properties of GPM [9].

2 Experimental Details

2.1 Materials Used

Fly ash: The fine spherical particle was collected as dry state from National Thermal Power Corporation, Kaniha, Odisha, India and added as a base material to geo-polymer concrete. The chemical compositions and physical properties of fly ash are mentioned in Table 1. The XRD pattern of fly ash is shown in Fig. 1, which shows substantial amorphous phase in its structures with the peaks of crystals of quartz (Q), hematite (H), magnetite (M), and so on. The greatest peak was found at 26° (2θ) with small amount of quartz. 86.82% is passing the average size 45μ of fly ash particles [10]. The low value of loss on ignition (LOI) represents superior pozzolanic activity and lower water demand. SO_3 that is less than 1% shows better volume stability and also durability.

Silica fume: This is an ultrafine particle of spherical in nature, and the average particle size is between 150 and 250 nm, The XRD and SEM images of SF are shown in Figs. 1 and 2b that show fully amorphous in nature.

Lime: This hydrated lime is used as an ingredient for mortar. The calcium hydroxide percentage is more than 88% and was purchased from a local supplier.

Alkali activator solution: Pellets of sodium hydroxide (NaOH) of 97–98% purity and sodium silicate solution (bulk density = 1.5 kg/L, Na_2O = 13.7%, SiO_2 = 29.4%, and water = 55.9% by mass) were collected from local supplier. The NaOH solution

Table 1 Chemical and physical properties of fly ash and silica fume

Materials	Chemical composition (%)										Physical properties	
	CaO	SiO_2	Al_2O_3	Fe_2O_3	SO_3	Na_2O	K_2O	MgO	P_2O_3	LOI (850 °C)	Specific gravity	Blain area (m^2/kg)
Fly ash	1.91	57.06	26.94	5.66	0.04	0.06	1.09	0.72	0.87	2.89	2.24	350
Silica fume	0.49	92.26	0.98	1.97	0.33	–	–	0.96	–	2.1	2.2	21000

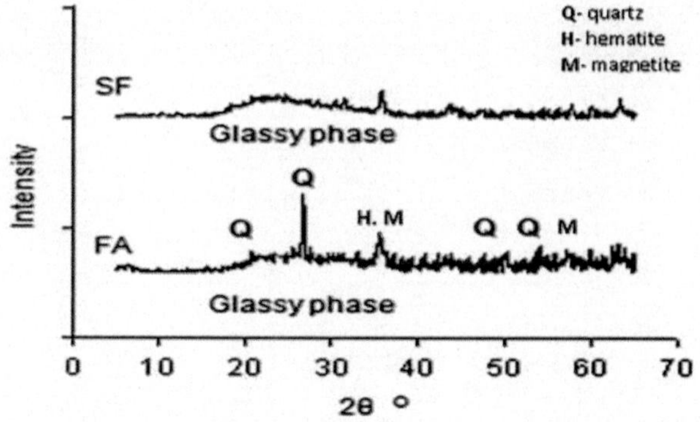

Fig. 1 XRD of fly ash and SF

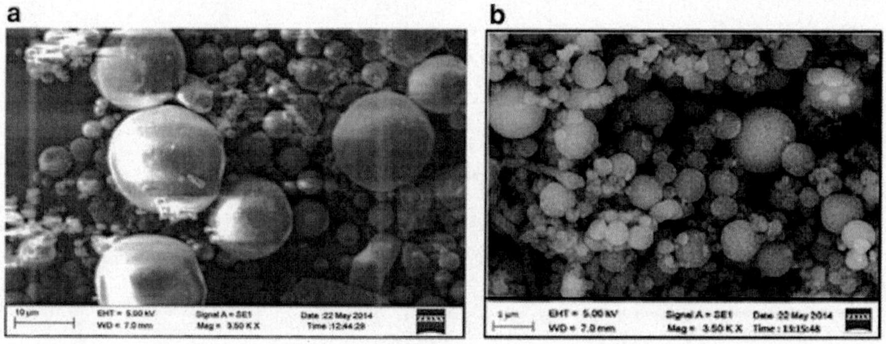

Fig. 2 **a** SEM image of raw fly ash, **b** SEM image of SF

concentration was made by dissolving pellets of NaOH with water as per the molarity requirement [11]. The alkali-activated solution was formed by mixing with different molarity and proportion of NaOH and Na_2SiO_3 solution.

3 Mix Proportion, Casting, Curing and Strength of GP Mortar

3.1 Mix Proportion

In this paper, geo-polymer mortars were made with fly ash to standard sand ratio 1:3 and liquid to FA ratio 0.3. Three series of mixes were designed to optimize: (i) the ratio of sodium silicate to sodium hydroxide (NS:NH) 1, 2 and 3, (ii) percentage dosage of lime 5, 7.5 and 10% by weight of FA, and (iii) silica fume 1, 2, 3 and 4% by partial replacement of FA for different molarity of NaOH (6, 8, 10 and 12 M)

Table 2 Mix proportions of geo-polymer mortar

Series	NaOH concentration (M)	Alkaline solution/FA ratio	NS/NH ratios	Lime (%)	Silica fume (%)
A	6, 8, 10, 12	0.3	1, 2, 3	0	0
B	6, 8, 10, 12	0.3	2	5, 7.5, 10	0
C	6, 8, 10, 12	0.3	2	7.5	1, 2, 3, 4

on compressive strength in ambient curing. The detail of mixes was mentioned in Table 2.

3.2 Casting

The binders employed for the mortar were fly ash, hydrated lime, SF and alkaline activator solution. The alkaline activator solution was got ready one day before the casting. At the beginning, fly ash, lime, silica fume in dry condition were perfectly mixed by taking appropriate quantity of the materials thoroughly. After proper dry mix, the alkali activator solutions of different molarities, that is, 6, 8, 10 and 12 M were added to the dry mix thoroughly to attain a good paste. Mixing was done to obtain a consistent mix within 3–5 min. Then all the mixes were properly compacted in three layers using the standard tamping rod and by vibrating with the help of table vibrator. After proper vibration, the mortar cubes were finally prepared of cube size 7.06 cm for ambient curing.

3.3 Curing

The specimens were demoulded after one day and kept at room temperature at 25 ± 2 °C and 50% relative humidity till the arriving day of testing, that is, 7 and 28 days of compressive strength for different mixes in accordance with the IS: 456-2009. Most of the literatures emphasized on heat curing instead of ambient curing [12] due to the high rate of gain of strength in a small period of time [13–15].

3.4 Compressive Strength (C.S.)

The C.S. was studied after 7 and 28 days of ambient curing. The average strength was calculated out of three test specimens result.

4 Results and Discussion

The various parameters such as NS/NH, molarity, percentages of lime and silica fume at ambient curing mode on compressive strength of fly ash geo-polymer mortar were studied and were determined at the age of 7 and 28 days. Those are given in Tables 3, 4 and 5. In order to identify the various mixes, nomenclature starts with alphabets A, B and C, and are meant for the NS/NH from 1, 2 and 3, respectively, with different molarities from 6 to 12 M. Again in this mix nomenclature, L and S meant for lime and silica fume percentages respectively.

(B12L7.5S4-mix means $NaOH/Na_2SiO_3$ ratio 2, 12 molarity NaOH concentration, 7.5% lime and 4% silica fume)

a. **Na_2SiO_3/NaOH (NS/NH) solutions**

In this paper, the alkaline solution/ash ratio was kept constant at 0.3. The strength varies on increase of NS/NH ratio, shown in Table 3, Figs. 3 and 4. The strength was found maximum at 2:1 [7] ratio than 1 and 3. This may have happened due to the formation of greater amount of soluble silica in sodium silicate solution which retards the geo-polymerization reaction and also due to the difficulty in compaction. Similar results were also mentioned in Thaarrini et al. [16]. Various authors reported alkali solution to fly ash ratio between 0.33 and 0.769, but here 0.3 was kept constant. For the mix samples that have more than 8 M concentration of NaOH, the strength was achieved more than 20 MPa at 1:2 (NS/NH) ratio. As NaOH and silicate solution are quite costly, so less ratio has been taken for present research study, which aims to reduce the cost of materials.

The peak strength of 38.97 MPa was achieved at 12 molarity and NS/NH of ratio 2, that is 8.64% excess than NS/NH ratio 1 and 7.89% excess than NS/NH ratio

Table 3 Compressive strength of GPM with different ratio of NH/NS in MPa

Mix no.	Molarity	7 days	28 days
A6	6 M	11.25	18.08
A8	8 M	14.02	19.43
A10	10 M	17.58	23.54
A12	12 M	22.34	35.87
B6	6 M	**12.25**	**16.94**
B8	8 M	**15.35**	**20.54**
B10	10 M	**19.14**	**25.33**
B12	12 M	**24.56**	**38.97**
C6	6 M	12.27	16.34
C8	8 M	15.14	20.57
C10	10 M	17.87	25.25
C12	12 M	24.32	36.12

Table 4 Compressive strength of GPM at different percentages of lime in MPa

Mix no.	Lime percentage	Molarity	7 days	28 days
B6L5	5% Lime	6 M	12.55	19.01
B8L5		8 M	17.23	26.78
B10L5		10 M	20.25	31.44
B12L5		12 M	24.24	**39.15**
B6L7.5	7.5% Lime	6 M	12.69	19.54
B8L7.5		8 M	19.23	30.12
B10L7.5		10 M	22.54	32.24
B12L7.5		12 M	25.67	**41.25**
B6L10	10% Lime	6 M	12.12	19.21
B8L10		8 M	18.67	28.85
B10L10		10 M	20.87	30.15
B12L10		12 M	25.12	**39.55**

Table 5 Average compressive strength of GPM at different percentages of silica fume in MPa

Mix No.	Silica fume (%)	Molarity	7 days	28 days
B6L7.5S1	1% SF	6 M	13.25	19.89
B8L7.5S1		8 M	19.32	27.12
B10L7.5S1		10 M	21.67	33.24
B12L7.5S1		12 M	26.17	41.87
B6L7.5S2	2% SF	6 M	13.67	20.14
B8L7.5S2		8 M	21.44	28.35
B10L7.5S2		10 M	23.89	34.57
B12L7.5S2		12 M	30.25	42.45
B6L7.5S3	3% SF	6 M	15.33	22.17
B8L7.5S3		8 M	22.26	30.12
B10L7.5S3		10 M	25.24	38.78
B12L7.5S3		12 M	32.77	**44.46**
B6L7.5S4	4% SF	6 M	13.36	20.25
B8L7.5S4		8 M	22.12	28.65
B10L7.5S4		10 M	24.67	35.67
B12L7.5S4		12 M	26.55	41.95

3 after 28 days of ambient curing. Here, maximum C.S. values were obtained at NS/NH-2; hence this was taken as the base for further design mortar mixes.

In the above mix A6, B6, C6 are of NS/NH ratios of 1, 2 and 3, respectively, and 6 represents NaOH molarity.

Fig. 3 Compressive strength of GPM at different ratios of NH/NS in MPa at 7 days

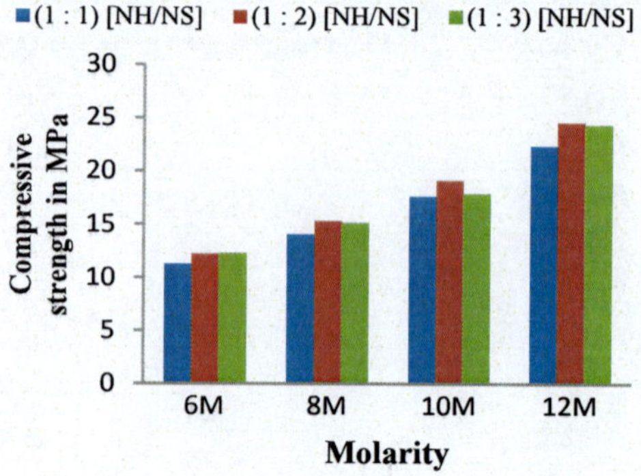

Fig. 4 Compressive strength of GPM at different ratios of NH/NS in MPa at 28 days

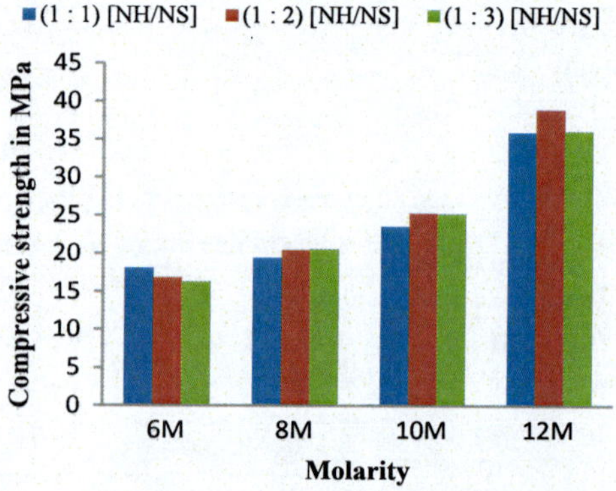

It has also been seen that increase in molarity significantly improves the degree of polymerization and consequently increases the compressive strength.

b. **Percentage of lime**

The C.S. of fly ash geo-polymer mortar at different percentages of lime (5, 7.5, 10%) were determined at the age of 7 and 28 days, and the details are given in Table 4, Figs. 5 and 6.

The peak strength of 41.25 MPa was achieved at 12 molarity and with 7.5% lime, that is 5.85% excess than 0% lime, 5.36% excess than 5% lime and 4.3% excess than 10% lime samples, respectively. At 7.5% lime addition strength was achieved maximum, similarly at 12 molarity. This is due to enhancement of the polymerization process by adding of some percentage of lime. Maximum C.S. values were obtained at 7.5% lime, and hence this was taken as another base for further design mortar mixes.

c. **Percentages of silica fume**

After optimum dosage of lime percentage, that is, 7.5%, again silica fume was added as 1, 2, 3 and 4% by weight of FA for getting better strength than previous mixes and C.S. at 7 and 28 days of ambient curing were mentioned in Table 5, Figs. 7 and

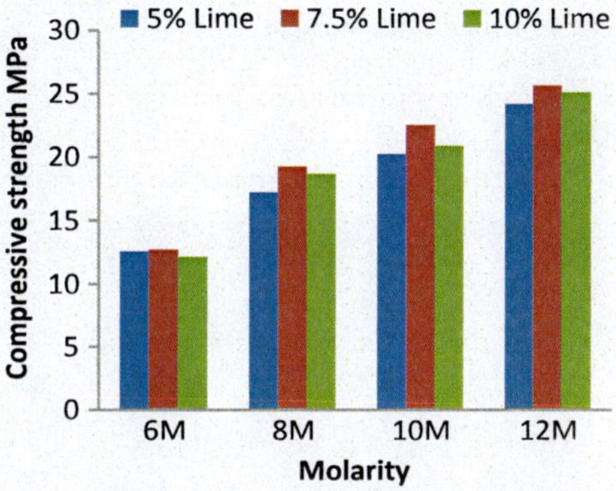

Fig. 5 Compressive strength of GPM at different percentages of lime at 7 days

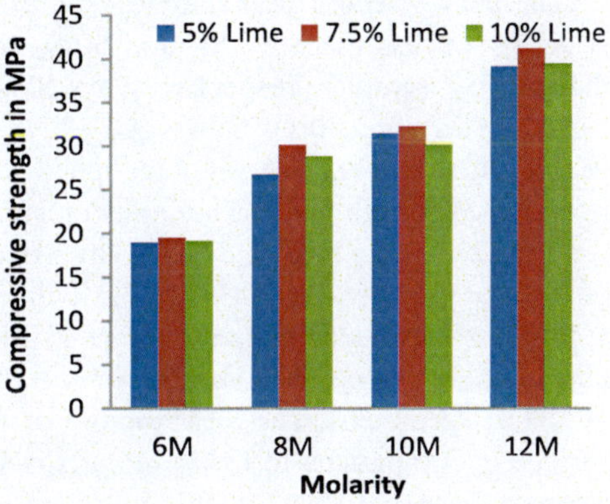

Fig. 6 Compressive strength of GPM at different percentages of lime at 28 days

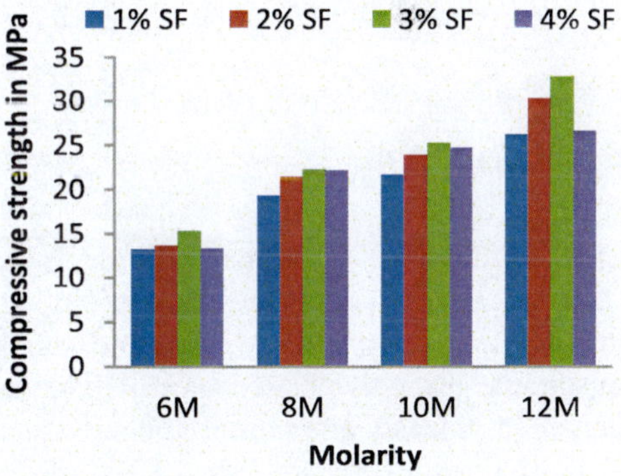

Fig. 7 Compressive strength of GPM with different percentages of silica fume at 7 days

8, for 6 M–12 M NaOH concentration samples. It was observed that at 3% of silica fume substitution samples show highest strength with respect to other substitution percentages of S.F.

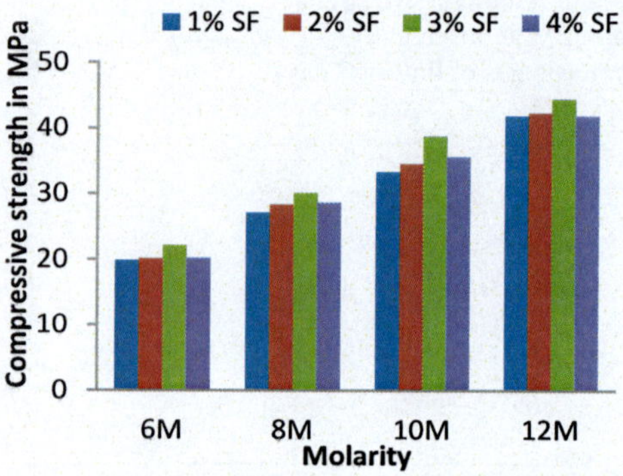

Fig. 8 Compressive strength of GPM with different percentages of silica fume at 28 days

The peak strength was achieved 44.46 MPa in 12 molarity and at 3% SF, that is 7.78% excess than 0% SF and 5.58, 4.5, 5.64% excess with 1, 2 and 4% SF than 3% SF samples, respectively. As SF is a very fine spherical particle that can easily act as pore filling behaviour which makes the structure more compact and denser, but after certain percentage it again conglomerates at particular position that lowers the strength value. The inter-transition zone between geo-polymer paste and sand can be well mixed as detected by SEM picture, mentioned in Fig. 9. The SEM pictures of GPM with addition of 7.5% lime and 3% silica fume substitution sample at 28 days were documented and are given in Fig. 9. Morphology of GPM shows unreacted fly ash percentage is more in 6 M concentration than 12 M concentration of NaOH specimens. The SEM picture of 12 M specimen shows dense and smooth surface in comparison to lower molar specimens. These morphological differences are the main responsible for compressive strength differences. It shows that the overall strength of the geo-polymer mortar depends on the unreacted fly ash particles in the microstructure state [17].

5 Discussion

The morphology of fly ash geo-polymer is helpful to identify the geo-polymer mineral structure, as discussed in the text. Depending on the requirement of crushing strength, geo-polymerization process is flexible to make high strength mortar and concrete. Use of suitable geo-polymer activator in manufacture of building brick is an eco-friendly non-fired process. The mortar appeared as a dense geo-polymer matrix mortar was visualized in the SEM images of 12 M specimen, and subsequently it achieved a high strength of 44.46 MPa [17].

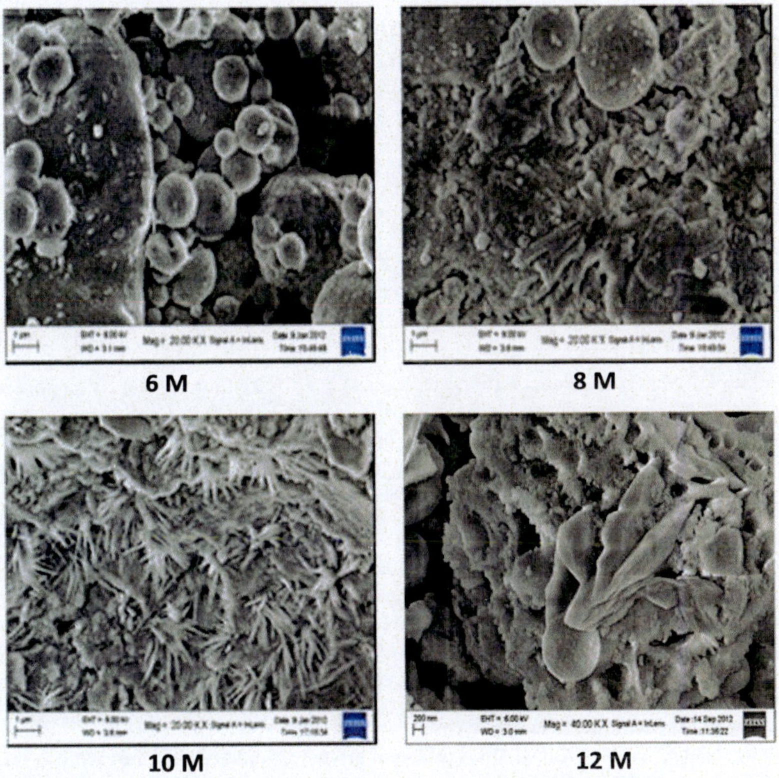

Fig. 9 SEM of mortar sample at 6, 8, 10 and 12 M concentration of NaOH

6 Conclusion

This paper has presented the results of experimental studies of FA-based GPM using lime and SF cured at ambient condition. The following conclusions can be derived as per the results obtained:

- It has been observed that the strength of specimen is directly proportional to the concentration of NaOH solution. The maximum values are obtained at 12 M samples.
- The study has investigated on the proportions of NS/NH ratio and that maximum strength was reported at NS/NH-2 ratio.
- The mortar mixes with lime achieved better compressive strength as compared to the mortar mix without lime. The optimum substitution of lime percentage was found to be 7.5%.
- As on addition of SF the strength was further enhanced with the constant addition of 7.5% lime. The optimum substitution of SF percentage was found to be 3%.

- 12 M specimen morphology was much denser than the other lower molar specimens. The denser gel formation is directly proportional to the enhancement of strength. It also explains that the overall strength of the geo-polymer mortar depends on the unreacted fly ash particles in the microstructure state as seen in SEM pictures.

References

1. Davidovits J (1989) Geo-polymers and geo-polymeric materials. J Therm Anal 35(2):429–441
2. Hardjito D, Wallah SE, Sumajouw DMJ, Rangan BV (2005) Introducing fly ash-based geo-polymer concrete: manufacture and engineering properties. In: 30th Conference on our world in concrete & structures: 23–24, Singapore, pp 271–278
3. Gorhan Gokhan, Kurklu Gokhan (2014) The influence of the NaOH solution on the properties of the fly ash-based geo-polymer mortar cured at different temperature. Compos B 58:371–377
4. Yang KH, Song JK, Ashour AF, Lee ET (2008) Properties of cementless mortars activated by sodium silicate. Constr Build Mater 22:1981–1989
5. Madheswaran CK, Gnanasundar G, Gopala krishnan N (2013) Effect of molarity in geo-polymer concrete. Int J Civil Struct Eng 4(2):106–115
6. Adam AA, Amiri NH, Suarnita IW, Rupang N (2016) The effect of lime addition on the setting time and strength of ambient cured fly ash based geopolymer binder. In: MATEC web of conferences, vol 47, pp 1–5
7. Hake Sandeep L, Damgir RM, Patankar SV (2018) Temperature effect on lime powder-added geopolymer concrete, Hindawi. Adv Civil Eng 1–5
8. Deb PS, Nath P, Sarker PK (2014) The effect of ground granulated blast-furnace slag blending with fly ash and activator content on the workability and strength properties of geo polymer concrete cured at ambient temperature. Mater Design 62:32–39
9. Okoye FN, Durgaprasad J, Singh NB (2016) Effect of silica fume on the mechanical properties of fly ash based-geo-polymer concrete. Ceram Int 42:3000–3006
10. Temuujin J, Van Rissen A, Mackenzie KJD (2010) Preparation and characterization of fly ash based geo-polymer mortars. Constr Build Mater 24:1906–1910
11. Budh CD, Warhade NR (2014) Effect of molarity on compressive strength of geo-polymer mortar. Int J Civil Eng Res 5(1):83–86
12. Shinde BH, Kadam KN (2016) Properties of fly ash based geo-polymer mortar with ambient curing. Int J Eng Res 5:203–206
13. Vijai K, Kumutha R, Vishnuram BG (2010) Effect of types of curing on strength of geopolymer concrete. Int J Phys Sci 5(9):1419–1423
14. Mane S, Jadhav HS (2012) Investigation of geo-polymer mortar and concrete under high temperature. Int J Emer Technol Adv Eng 2(12):384–390
15. Al Bakria AM, Kamarudin H, BinHussain M, Nizar IK, Zarina Y, Rafiza AR (2011) The effect of curing temperature on physical and chemical properties of geo-polymers. Phys Proc 22:286–291
16. Thaarrini J, Venkatasubramani R (2015) Feasibility studies on compressive strength of ground coal ash geo-polymer mortar, period. Polytech Civil Eng 59(3):373–379
17. Radomir Zejak, Irena Nikoli, Dragoljub Ble, Vuk Radmilovi, Velimir Radmilovi (2013) Mechanical and micro structural properties of the fly-ash-based geo-polymer paste and mortar material in technology. Mater Technol 47:535–540

Red Mud as a Controlled Low Strength Material

Bijaya Kumar Das, S. K. Das, and Benu Gopal Mohapatra

Abstract In this paper characterization of red mud as a controlled low strength material (CLSM) is discussed with red mud as a base material and crusher dust or quarry dust, phosphogypsum and fly ash as other mineral admixtures. The highly alkaline red mud is mixed with phosphogypsum (PG) and crusher dust (CD) in different proportions, and fly ash is used to improve the flowability. The material properties of developed CLSM are characterized in both green and hardened state. The bleeding percentage was found to be less than 1% and flow diameter decreased with increase in crusher dust. Higher flow value was observed with phosphogypsum in comparison to with same percentage of crusher dust due to particle size and chemical composition. Higher unconfined compressive strength, California bearing ratio and flowability were observed for CLSM developed with red mud and CD mix in comparison to red mud and PG mixture.

1 Introduction

Industrial waste, which is otherwise hazardous for environment, can be used for production of CLSM in a scientific way, thus incorporating sustainability in infrastructure development [6]. CLSM has the benefit of being used with less on-site labour and equipment, thus making the construction faster even in most restricted area. Having flow consistency of 150–300 mm, it is most workable in trench-filling and back-filling where compaction by mechanical means is nearly impossible. Owing to its self-flowing and self-levelling property it makes the space void-free easing lesser chance for settlement [5], Kaneshiro et al. 2001, [3]. The presence of a high volume

B. K. Das (✉) · B. G. Mohapatra
School of Civil Engineering, KIIT University, Bhubaneswar, India
e-mail: erbijaydas@yahoo.co.in; erbijaydas@yahoo.in

B. G. Mohapatra
e-mail: bmohapatrafce@kiit.ac.in

S. K. Das
Department of Civil Engineering, IIT(ISM), Dhanbad, India
e-mail: saratdas@iitism.ac.in; saratdas@rediffmail.com

B. B. Das et al. (eds.), *Recent Developments in Sustainable Infrastructure*, Lecture Notes in Civil Engineering 75, https://doi.org/10.1007/978-981-15-4577-1_70

of fine particles and porous materials could increase the demand for water and reduce the flowability of CLSM. Therefore, for the selection of suitable materials and design, a proper mix design is important to achieve the desired flowability (>200 mm) without segregation. Re-excavatability is one of the critical properties for back-filling of utility trenches. Generally, CLSM with an unconfined compressive strength of 0.3 MPa or less can be excavated manually [4], NRMCA (2006). No doubt that the processed cost of flowable back-fills is higher than the traditional thermal back-fills (native soils), and it is highly recommended for more frequent usage because of its proven quality and ease in placement that could expedite the construction process and reduce the overall project expenditure [7]. In the present study, bauxite residue, red mud is used as the base material. The red mud is not being used as a natural soil due to its high alkalinity and poses threat to the environment [8].

2 Materials and Their Basic Properties

In this study red mud (RM) was collected from Hindalco, Muri, Jharkhand, India. The fly ash used in the present study was Class F fly ash (FA) and was collected from NALCO. Figures 1 and 2 show the scanning electron micrograph of red mud and fly ash, respectively. The other industrial waste used in the present study is phosphogypsum (PG) which is the by-product formed during the extraction of fertilizer from the phosphate ore with the help of sulfuric acid. The main composition of it is gypsum. Gypsum has a wide applicability in construction industry but phosphogypsum is limited to use because of its radioactivity. Figure 3 shows the PG used for this study.

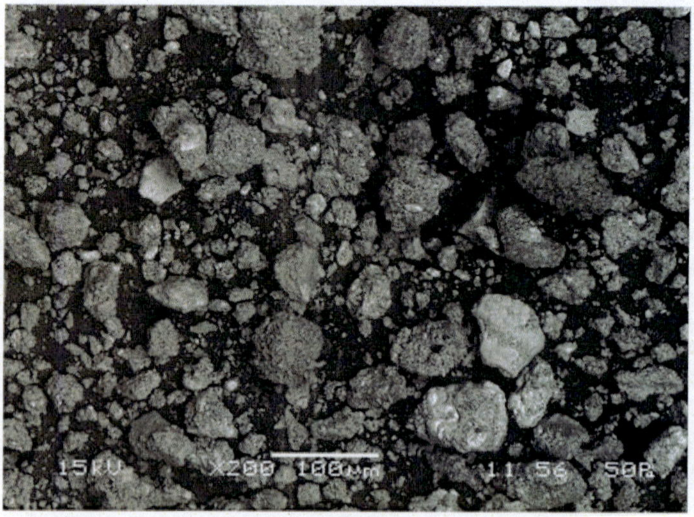

Fig. 1 Scanning electron microscope of red mud

Fig. 2 Scanning electron microscope of NALCO fly ash

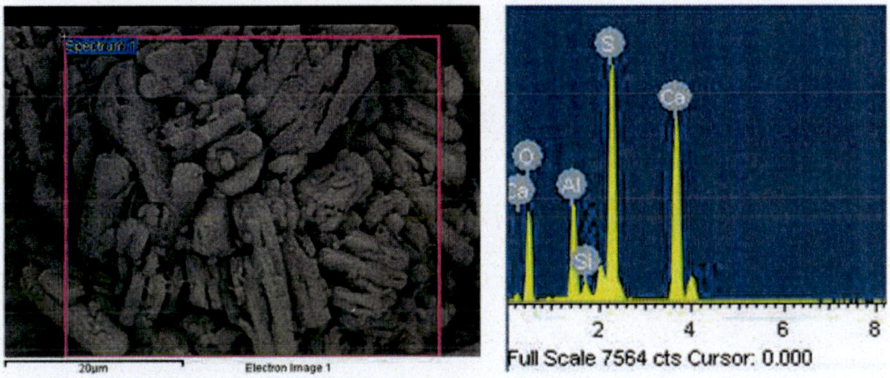

Fig. 3 Scanning electron microscope and EDX spectra of phosphogypsum

The crusher dust (CD) or also called as quarry dust is the by-product of quarrying, mining and the crusher plants where the rocks are crushed into aggregates. These absorb less water and have very good bearing capacity. The CD are non-plastic in nature, have high resistance towards fire and heat and all these properties make it a good construction material (Fig. 4).

The specific gravity of the materials is widely varied with the values as 3.27, 2.74 and 2.24 for red mud, crusher dust and fly ash, respectively.

Fig. 4 Scanning electron micrograph of crusher dust

Table 1 Compaction properties of the industrial waste used in the present study

Material	Light compaction		Heavy compaction	
	OMC (%)	MDD (g/cc)	OMC (%)	MDD (g/cc)
RM	31.5	1.52	27.7	1.59
FA	24.7	1.35	17.9	1.45
CD	13.2	1.76	11.5	1.89

The compaction properties of the basic materials red mud along with that of fly ash and crusher dust are shown in Table 1. As the phosphogypsum is very fine, its compaction properties using proctor compaction may not be representative and not presented here.

3 Methodology

The present study consists of various experiments for the characterization of basic material and characterization of different mixtures as a CLSM material which are listed as follows. The flow consistency, fresh density and bleeding of the fresh CLSM are done as per relevant ASTM standards. The unconfined compressive strength and California bearing ratio tests are done on the hardened sample. The durability of the developed CLSM is studied using slake durability test. Slake durability test is done for testing the durability of the sample against weathering action of water. For this test samples were put into the drums of slake durability and it was rotated in the water at a speed of 20 rpm for 10 min. Then the sample after slaking was collected and kept for oven drying. This procedure was repeated for two cycles and finally the weight retained percentage was calculated.

4 Results and Discussions

In this study various material combination is used to check the applicability of the mix proportion to be used as a CLSM material. There is no guideline regarding the mix design of CLSM and composition with different mix proportion of RM, FA, CD and cement. So the other composition with different mix proportion of RM, FA, PG and cement is considered in the present study. Various laboratory tests, such as flowability, bleeding, UCS, fresh density and CBR, are conducted to find the suitability of the mix as CLSM. These tests were conducted for various compositions with varying water content in order to keep the value of flow in between 150 and 300 mm at two cement contents of 10 and 7%. The effect of crusher dust and phosphogypsum is studied separately.

4.1 Flow Consistency

The flow consistency of the samples at different proportion of the RM, FA and CD is shown in Table 2. Similarly, the flow consistency for different mix of RM, FA and PG is shown in Table 3. It can be observed that for all the mix, the water content is evaluated and the flow value is in between 150 and 300 mm. From the above observation it can also be concluded that the consistency of the material increases with the increase in water content for the same mixture while for different mixture

Table 2 Flow consistency of mix with RM, FA and CD

Mix#	Mix proportion (%)					Flow diameter (mm)
	RM	CD	FA	Cement	Water	
1	70	10	20	10	40	151
2	70	10	20	10	45	230
3	60	20	20	10	40	194
4	60	20	20	10	45	285
5	50	30	20	10	45	218
6	70	10	20	7	40	155
7	70	10	20	7	45	226
8	60	20	20	7	40	193
9	60	20	20	7	45	275
10	50	30	20	7	45	210

Table 3 Flow consistency of mix with RM, FA and PG

Mix#	Mix proportion (%)					Flow diameter (mm)
	RM	PG	FA	Cement	Water	
11	70	10	20	10	45	167
12	70	10	20	10	50	255
13	60	10	30	10	50	214
14	60	10	30	10	55	289
15	50	10	40	10	55	246
16	70	10	20	7	45	171
17	70	10	20	7	50	243
18	60	10	30	7	50	218
19	60	10	30	7	55	291
20	50	10	40	7	55	251

due to increase in finer particle more water content is required to attain the same level of consistency. The above trend can be explained by the theory of the surface area and the plasticity of the RM that the more the surface, more are the absorbed water; and more the content of RM, the more it will hold the water; as a result, the lesser will be available for flow.

4.2 *Density at Different Ages of Mixes*

Since this material is used as fill material or as replacement in back-fill with the compacted material, it becomes very important to find out the density of these materials at different ages, especially when it is freshly prepared. For back-fill material the compaction level plays an important role in the strength, hence density of the material becomes very important parameter. In the present study the density at 7 and 28 days is also investigated along with fresh density and presented in Tables 4 and 5 for the mixes with CD and PG, respectively. The water content plays an important role in affecting the fresh density of the mix. For the same water content, the RM content plays significant role in affecting the density of soil. The hardened density is decreased from the fresh density due to the loss of moisture.

4.3 *Bleeding*

The bleeding of different mixes as shown above is investigated. It was observed that when the bleeding percentage is less than 1.0% both for CD and PG mixes.

Table 4 Density of mix with RM, FA and CD

Mix#	Mix proportion (%)					Density (g/cm^2)		
	RM	CD	FA	Cement	Water	Fresh	7 days	28 days
1	70	10	20	10	40	1.9	1.86	1.8
2	70	10	20	10	45	1.86	1.82	1.77
3	60	20	20	10	40	1.84	1.81	1.76
4	60	20	20	10	45	1.82	1.79	1.74
5	50	30	20	10	45	1.81	1.78	1.73
6	70	10	20	7	40	1.89	1.84	1.79
7	70	10	20	7	45	1.86	1.82	1.77
8	60	20	20	7	40	1.83	1.79	1.73
9	60	20	20	7	45	1.81	1.77	1.72
10	50	30	20	7	45	1.8	1.76	1.72

Table 5 Density of mix with RM, FA and PG

Mix#	Mix proportion (%)					Density (g/cm^2)		
	RM	PG	FA	Cement	Water	Fresh	7 days	28 days
11	70	10	20	10	45	1.89	1.87	1.87
12	70	10	20	10	50	1.87	1.83	1.78
13	60	10	30	10	50	1.85	1.79	1.77
14	60	10	30	10	55	1.82	1.75	1.71
15	50	10	40	10	55	1.78	1.75	1.75
16	70	10	20	7	45	1.81	1.8	1.76
17	70	10	20	7	50	1.79	1.77	1.72
18	60	10	30	7	50	1.65	1.64	1.61
19	60	10	30	7	55	1.57	1.53	1.48
20	50	10	40	7	55	1.54	1.52	1.46

4.4 *Unconfined Compressive Strength (UCS)*

The stress–strain curve of the UCS test of some mixes with CD after 28 days is shown in Fig. 5 and that for the mixes with PG is shown in Fig. 6. It can also be concluded that for the mix proportion there is decrease in the strength value with the increase in water content. It can also be concluded that for different composition for the same water content, the compressive strength of that mix containing red mud in more amount is greater than that with the lower red mud content.

The CBR values of the mixes are found to vary between 37 and 86 with CD and that with PG the CBR values are found to vary from 11 to 100 with different

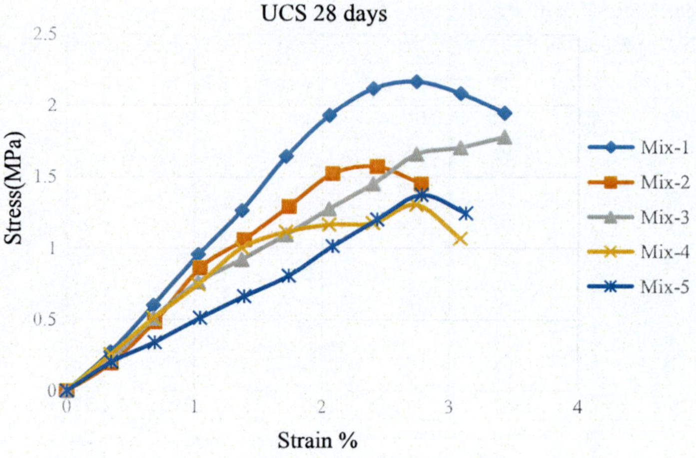

Fig. 5 Stress–strain of mix incorporating CD at 10% cement

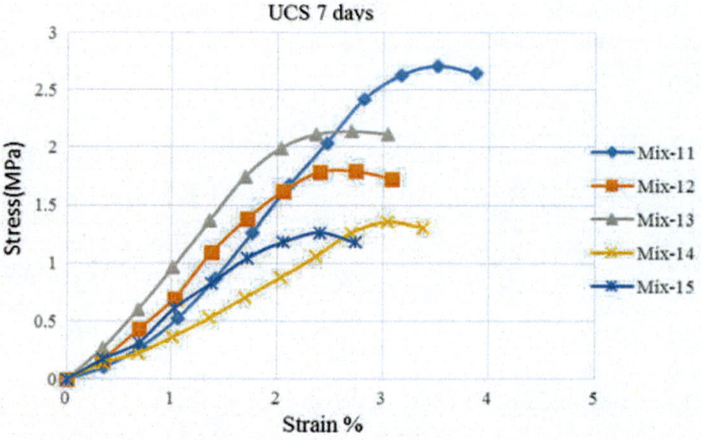

Fig. 6 Stress–strain of mix incorporating PG at 10% cement

mix proportion of PG. Hence, with suitable proportioning of material the developed CLSM can be used as a subbase material.

As the UCS values of some sample are more than 1 MPa, the slake durability tests are conducted on two samples with higher UCS value. Based on Gamble's slake durability classification these materials are low durable. It was also observed that CLSM with CD has higher slake durability index than that with PG.

5 Conclusion

The feasibility of different mix proportions of one made up of RM, FA, PG, cement and water and another made up of RM, FA, CD, cement and water was investigated as a constitute material of CLSM in this study. Experiments were conducted to evaluate the important properties of CLSM with the above material. Based on the experiment conducted the following major conclusions were drawn:

1. All the CLSM mixes developed showed the flowability in the range of 150–300 mm which is considered to be good flowable material. Flowability of the mix containing the varying percentage of CD increases with increase in CD as the specific surface area of RM is higher than that of CD, while the mix containing varying percentage of FA decreases with the increase in FA.
2. The fresh density of the material decreased with the decrease in RM and increase in water content due to the higher density of RM. Bleeding values of all the mixes were below 1%, which is far below the maximum limit of 5% for CLSM.
3. UCS of the prepared mix decreased with decrease in RM content as RM can take more load than other materials used and it was also found that the UCS value for the same mix decreases with increase in water content due to the presence of higher capillary pores. However, all the UCS of 28 days was higher than the minimum value for CLSM.
4. Based on Gamble's slake durability classification, the CLSM materials developed are low durable and CLSM with CD has higher slake durability index than that with PG.

References

1. ASTM D4644 (2016) Standard test method for slake durability of shales and other similar weak rocks. In: ASTM international, West Conshohocken, PA, USA
2. ASTM D4647 (2013) Standard test methods for identification and classification of dispersive clay soils by the pinhole test. In: ASTM international, West Conshohocken, PA
3. Abelleira A, Berke NS, Pickering DG (1998) Corrosion activity of steel in cementitious controlled low-strength materials vs. that in soil. In: Howard AK, Hitch JL (eds) The Design and application of controlled low strength materials (flowable Fill), STM STP 1331, American Society for Testing and Materials
4. ACI 229R-13 (2013) Test methods for controlled low-strength material (CLSM): past, present, and future. In: Howard AK, Hitch JL (eds) The design and application of controlled low-strength materials (Flowable Fill), ASTM STP 1331, American Society for Testing and Materials, 1998, pp 3–10
5. Dockter BA (1998) Comparison of dry scrubber and class c fly ash in controlled low strength materials (CLSM) applications. In: Howard AK, Hitch JL (eds) The design and application of controlled low-strength materials (Flowable Fill), ASTM STP 1331, American Society for Testing and Materials, pp 13–26
6. Folliard KJ, Du L, Trejo D, Halmen C, Sabol S, Leshchinsky D (2008) Development of a recommended practice for use of controlled low-strength material in highway construction. NCHRP Report 597. Transportation Research Board, Washington, D.C

7. Parmar D, Steinmanis J (2017) Underground cables need a proper burial. T&D World Magazine, Geotherm Inc, 1 April 2003. http://tdworld.com/underground-tampd/underground-cables-need-proper-burial. Accessed 23 May 2017
8. Alam Shamsad, Das Bijaya Kumar, Das Sarat Kumar (2018) Dispersion and sedimentation characteristics of red mud. J Hazard Toxic Radioact Waste ASCE 22(4):04018025

Recent Developments in Construction Technology and Management

Genetic Algorithm: An Innovative Technique for Optimizing a Construction Project

Paromik Ray, Dillip Kumar Bera, and Ashoke Kumar Rath

Abstract Time and cost are two basic objectives of any construction project. Optimization of these objectives is the main concern over the last three decades by the construction sectors. Many innovative techniques have been used by the construction companies to optimize the cost and time of a project. Genetic Algorithm (GA) method is one of the most advanced and widely used non-traditional search algorithms based on the mechanics of natural selection and natural genetics. The principle of natural selection is based on the "survival of the fittest" concept coined by Charles Darwin. It is neither an intelligent nor a smart algorithm but it searches for optimal solution in the solution space. The objective is to review GA as an optimizing technique used to generate high-quality solution for optimization process. Reproduction in GA is done by three sophisticated operators—selection, crossover and mutation through which optimal solution is found out only if the condition is true. Hence, GA method is useful optimization process in construction projects. The main advantage present in GA is providing more effective and efficient optimum value in a construction project. Moreover, it also provides optimal trade-off values between project duration and total work done. This concludes that GA can be widely used as an advanced innovative technique for optimization process in future construction project.

Keywords Time and cost · Genetic algorithm · Optimization process · Optimal trade-off · Construction project

---

P. Ray (✉) · D. K. Bera (✉) · A. K. Rath
School of Civil Engineering, KIIT Deemed to Be University, Bhubaneswar, India
e-mail: paromikray@gmail.com

D. K. Bera
e-mail: dkbera.fce@kiit.ac.in

A. K. Rath
e-mail: akrathfce@kiit.ac.in

B. B. Das et al. (eds.), *Recent Developments in Sustainable Infrastructure*, Lecture Notes in Civil Engineering 75, https://doi.org/10.1007/978-981-15-4577-1_71

1 Introduction

According to the driven conditions of today's market, the construction industries are on the rise. All the construction companies including government and private sectors face a number of challenges during the period of construction. The main difficulties that fall during construction time of the project are the scheduled completion of project in the stipulated duration along with completion within the estimated budget of the project. The main parameters that a construction company mainly focuses on are the time and cost of a particular construction project. Also, at the duration of construction, other factors also come into consideration in the form of unpredictable factors, namely, labour performance, economical and social issues, execution errors of contractors, labour strike, design errors, labour strike, climatic changes, etc. These named factors are the majority which is responsible for the delay in the schedule of the project along with the increase in the budget of the project ultimately leading to the excess in the cost liabilities that get added to the budget of the construction project. The problems occurring in the construction site related to these factors are referred to as the Time–Cost Trade-off Problems (TCTO). For overcoming these particular situations that arise in the project, a step is being taken by the project planners and researchers for encountering this problem to process a perfect solution for optimization of TCTO-related problems in a construction project. This resulted in the development of various innovative optimization techniques of TCTO problems in a project site.

Advanced Neural Network (ANN) method, Line of Balance (LOB) method, Discounted Cash Flow (DCF) method, maximum flow–minimal cut theory method, Modified Adaptive Weight Approach (MAWA) method, critical path method and Genetic Algorithm (GA) method are some of the advanced, innovative and widely used techniques or methods which is mainly used for generating the maximum optimized solution of time and cost parameters of a complex and large construction project. The main objective of the review paper is to highlight the Genetic Algorithm (GA) method as the most used, easiest, efficient optimization method or technique in recent times for optimizing TCTO problems. Moreover, a summary or a brief overview of the various innovative techniques along with the recent works done by the researchers in the field of optimization of TCTO problems by using genetic algorithm method. Further, it is concluded that a review is done on the genetic algorithm method being the most efficient optimization technique for solving TCTO problems for future projects and works.

1.1 *Brief Classification of Innovative Techniques*

The table mentioned below (Table 1) highlights the various recent innovative techniques of optimization of time and cost in a project. The methods or techniques are

Table 1 Classification of different optimization techniques

Manual methods	Mathematical technique	Metaheuristic approach	Others
Different optimization methods			
• Maximum Flow Minimal Cut Theory (MFMC) • Least-Cost Scheduling Model • Chance-Constraint Programming Model • Line of Balance Method (LOB)	• Non-Linear Programming (NLP) • Linear Programming (LP) • Integer Programming (IP) • Dynamic Programming (DP) • Inventory Models • Stochastic Models • Simulation Model	• Evolutionary Algorithm (EA) I. Particle Swarm Optimization (PSO) II. Ant Colony Optimization(ACO) III. Modified Shuffled frog Leaping (SFL) • Genetic Algorithm(GA) • Genetic Programming(GP)	• Artificial Neural Network (ANN) • Discounted Cash Flow Method (DCF)

being derived from the previous works of the researchers and scholars who applied the process for optimization.

2 Optimization Process

Optimization is commonly defined as the process of finding the minimum/maximum values with reference to the main derivative or the objective function for satisfying the particular constraints of a function specified within a certain set of function. The main objective or goal of the optimization process is to process the correct or best set of solution of a mathematical solution of a TCTO problem. The main dependent of TCTO (Time–Cost Trade-off) problems is the critical path method of optimization process. Moreover, in a construction project, the duration of the project is the most important parameter for completion in the stipulated time. Also, the increase in scheduled time, increase in cost, unavailability of resources, the presence of risks, and quality and quantity of the materials occurring in the construction project are the main reasons for the requirement of optimization process. The method of Genetic Algorithm (GA) is the most widely, effective and efficient technique for the optimization process. Lastly, many scholars and researchers opted for the method of Genetic Algorithm (GA) as the prime method for optimization of TCTO problems.

3 Genetic Algorithm: An Innovative Technique for Optimization

The method of genetic algorithm belongs to the heuristic group of optimization which is itself an advanced optimization technique of solving TCTO problems in a construction project. This method is a non-traditional search algorithm for solving both single-objective and multi-objective optimizations of TCTO problems depending on the natural algorithm and natural evolution process. This method was first proposed by the scientist named Charles Darwin on the theory of "survival of the fittest". Similarly, the method of Genetic Algorithm (GA) is neither a smart nor an intelligent algorithm which clearly indicates that the method is completely algorithm and mathematical based used for optimization of TCTO problems in a construction project [1]. This method consists of four main operations. At each of the following steps of the genetic algorithm method, these steps select the collected data as the values in a random manner. Further, the process of iteration takes place in generating an optimal solution to a TCTO problem. Moreover, the method of genetic algorithm is also dependent on the process of natural evolution. The optimized solution can be taken as an accurate solution because of the optimization process. The so-called accurate optimized solution obtained by the GA method is completely dependent on the total number of populations and on the fitness level objective function [2]. The method of GA has four main processes, namely, in a sequential order: initialization, selection, crossover and mutation.

Initialization: This process is mainly for the initialization of the collected data values to the fitness value objective function.
Selection: This process is mainly for the selection of the optimized solution values in a random manner. Commonly, this process is referred to as the random selection process.
Crossover: This process is mainly used at the time of interchanging the random values' position for generating the higher and correct optimized solution. This process is generating higher optimized values of a TCTO problem.
Mutation: This process is mainly used for flipping or swapping of more than one single value of the collected data values for generating a highly correct optimized solution of a TCTO problem [3] (Fig. 1).

The above flowchart is representing the main working principle of the optimization process using genetic algorithm method of a TCTO problem in a construction industry. As mentioned above, the flowchart also highlights the four main processes used for optimization process.

4 Working Process of Genetic Algorithm

The Genetic Algorithm (GA) method consists of four main processes, which generates the output of the optimized values of the TCTO problem of a construction project.

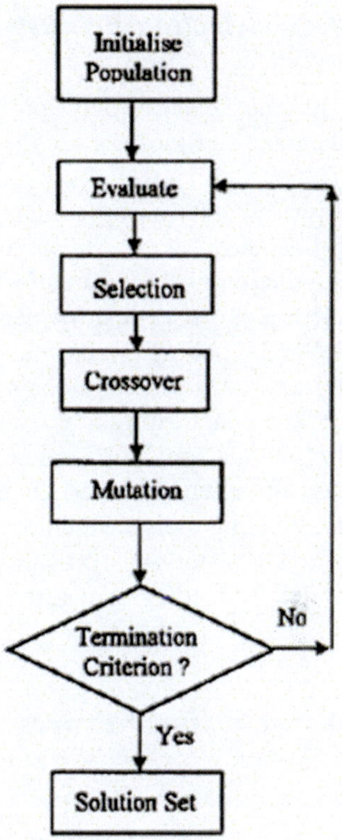

Fig. 1 Flowchart of genetic algorithm method

4.1 Initialization Process

Initially, at the starting of any genetic algorithm optimization process of any mathematical problem, a single variable is being assigned from any previous coded domain or a set. Initially, this easy step of initialization process using genetic algorithm method was found insufficient by many researchers and scholars. Furthermore, it is thought that the process lacked few specific other variables for optimization of TCTO problems [4]. However, more technology and research work lead to the innovation of the optimization technique of the method. Therefore, with the incoming of new technologies and modified parameters in use, it is now very reasonable to quote the initialization process which is one of the important processes for optimization of TCTO problems [5].

4.2 Selection Process

The main selection or the process of improvement of the fitness derivative function of the main chromosomes of the population can be achieved by the process of selection. The selection process is so-called dominating process of genetic algorithm method as it possesses the ability to change the fitness value function of an optimization problem [6]. In this process, the parents are being selected from their group of chromosome population of the data collected. The process of selection of the genetic algorithm method is performed in such a way that the chromosomes having a higher fitness value are having a maximum probability to be selected as the main parents of the fitness value function to be used in the optimization process [6].

The conclusion is that some of the population of the chromosomes of the collected data for optimization problem may contain some of the high-value fitness derivative function than the original derivative function which is the main formula for creating individual chromosomes for the optimization process of a TCTO problem. Ultimately, this process possesses a limitation of a lack of population diversity which leads to a lag in search algorithms in optimization process [7].

4.3 Crossover Process

After the completion of both the processes, initialization and selection processes, the third process is the crossover process. In this process, the derivative function parameters used in the genetic algorithm method create possible two chromosomes in this process. These parameters of the GA method use a random generator function for maintaining the two sets of functions of the domain or a set for the optimization process. This results in the formation of two chromosomes in the optimization process [8]. After the completion of the crossover process, the new set of chromosomes is being added to the set of previous chromosomes set to form a new population.

4.4 Mutation Process

The final process of the genetic algorithm is the mutation process. The main function of the process includes the increasing population diversity of the chromosomes. This process also includes the selection of the inner particles of chromosomes called genes in a random selection manner and then assigning those selected values within the corresponding intervals of pre-defined variables of selected set or domain [9]. That chromosome which follows the method of mutation contains a correct genetic code. The process of mutation enables these created genetic codes that are kept intact and do not get destroyed. Moreover, a non-uniform mutation can also be applied for the optimization process of a TCTO problem which leads to the shortening or narrowing of the chromosomes population created for the optimization process of TCTO problem [10].

5 Use of Genetic Algorithm in Construction Industries

In a construction industry, the use of genetic algorithm for optimization of cost and time or any other parameter will be a valuable reference for the company, as it provides correct optimization values of cost and time, especially in a TCTO problem. Genetic Algorithm (GA) method in a construction industry acts as a useful option for optimizing, controlling, planning and scheduling of all the constructing activities mainly present in the project. It is referred to as a search algorithm containing the tools required for optimization used mainly for assisting the activity makers, planners, or designers in the company in order to achieve and generate a near-optimized solution of a particular TCTO problem. This method theory is extracted from the theory of evolution depending on the theory "survival of the fittest" that is applicable in the construction industry where the construction industry selects the specific and main activities required for construction work [11]. Similarly as mentioned earlier about the mechanisms or processes of GA method, it is the same applicable to construction industries. The method includes the main three processes even in the field of construction industry: (I) The initialization method which is mainly used for generating a set of initial data of total duration of a particular activity along with the total cost budget of that activity; (II) selection process that is mainly used for calculation and selection of the total collected data of the time and cost of the activities along with other parameters including quality, risk factors, material quantity, etc.; and (III) crossover process mainly used for processing and generating an improvement of the selected data containing the time and cost of the activities for a certain period of time [12]. These three main processes are very much applicable for optimization of TCTO problem at a construction industry.

5.1 *Multi-objective Genetic Algorithm*

The optimization of TCTO problem using genetic algorithm method is often referred as multi-objective genetic algorithm method, with single-objective genetic algorithm optimization being referred for single-objective functions especially total time or total budget cost of the project [13]. Moreover, many technologies have evolved over the past decades or in the past generation for the optimization of more than one objective function using GA method. Hence, the concept of multi-objective genetic algorithm came into the world and helpful for solving TCTO problems in a construction industry. With every advanced technology of the GA method along with effective and efficient process of optimization of the TCTO problems, slowly the popularity has been gained and in the coming years, the preference of multi-objective genetic algorithm is higher. Clearly, for multi-objective genetic algorithm function, the optimization of fitness value definite function contains two different parameters containing two different optimum functions for optimization process. The multi-objective GA method does not possess or generate any superior or complex function as an optimizing output but generates a simple variable mathematical function which is very clear as an objective of the TCTO problem for optimization [14]. Thus, multi-objective genetic algorithm method is the most efficient and effective optimization technique for solving TCTO problems.

6 Case-Study: A Live Construction Project

The case study includes a live on-going construction project named "Bose Institute Construction Project" located at the city of Kolkata, West Bengal, India. The construction project includes the construction of G+15-storied educational or an institutional building that is situated in the main town of Salt Lake City of Rajarhat district in the city of Kolkata. The project consists of full-furnished construction of a 15-storied building along with the formation of the basement. The main work of the construction project consists of the main educational building of 15-storied along with two-side building consisting of 5-storied floor each. The main features and the requirement of the construction project are as follows:

Maintaining and ensuring the life-time structural ability of the educational building being constructed for duration of 15 years.

Ensuring proper and efficient construction of 15-storied educational building along with the two buildings of 5-storied each of the construction projects.

Decrease in the negative effects created in the environment and construction of an eco-friendly building structure making positive impact in the environment [15].

As per the construction point of view of the project, there are present around 65–70 activities throughout the project. It is quite difficult and near impossible for optimization of all the activities using genetic algorithm method. It has been advised to select a section or a separate small plan of the construction project for optimization using genetic algorithm method. Thus, for this review paper, a particular section of the construction project has been selected for the optimization using GA method.

6.1 Selection of the Activities

The selected activities for the optimization using GA method consist of the main important 10–12 activities from the part of ground floor of the building to the plinth level of the first floor of the on-going construction project. Other parameters are also included in a tabular form consisting of the duration and the estimated cost of the particular activity selected. The requirement of the cost and duration is much more important for the optimization process using GA method creating a more positive

Table 2 List of activities, duration, cost (I)

Activities	Original duration (Days)	Value of the activity
A. Reinforcement work of column	3	1,09,145
B. Column shuttering work	3	33,750
C. Column concreting work	4	90,813.25
D. Column de-shuttering work	4	7,057.50
E. Staging and shuttering work of beam and slab	3	36,572.60
F. Fabricating, placing and fixing of beam and slab	4	2,44,658

Table 3 List of activities, duration, cost (II)

Activities	Original duration (Days)	Value of the activity
G. Concreting work at ground floor with roof slab	3	1,71,270
H. Curing process	14	2,60,134
I. Removal of shutter staging materials	2	1,48,648
J. Brickwork (main floor along with partition wall)	4	2,60,734

impact and efficiency in the process [16]. The selected activities along with the duration and time are given below (Tables 2 and 3).

6.2 *Deriving the Fitness Function*

This particular construction project consists of the applications of the TCTO problem for optimization process. The derivative function will comprise the main parameters including the total duration and the estimated cost of all the activities [17]. The derivative function includes the optimization of the case study of a TCTO problem:

$$C = \text{dc} + \text{ic} - (\text{id} - q)\text{ie} + \sum_{\forall y} (\text{nd}(y) - \text{pop}(x, y))\text{cs}(y) \tag{1}$$

where C = Total Cost; dc = Direct Cost; ic = indirect Cost = i.e. X id; id = initial duration of the network or duration of the network with normal cost; q = duration of the network after change in duration of the activity; i.e. = indirect expenses per day; nd = normal duration of the activity; pop = matrix of randomly generated durations as per our boundary conditions; cs = cost slope of the activities; x = population index; y = activity whose duration is being changed [15, 18].

The result obtained using the MATLAB 2018 software and a clear comparison of the result obtained is also being reviewed in the paper highlighting the importance of using genetic algorithm method for optimization process. By using the MATLAB 2018 software, an efficient and a clear comparison of the two outputs is obtained with the aid of bar graphs that are being generated from the software. With the help of bar chart generated, it becomes quite easy for processing the optimized solution of the problem [19]. In the bar chart generated, the cost timeline is being.

7 Discussion

The discussion is mainly about the comparison of the two main optimized outputs or results that are obtained by the optimization process using Genetic Algorithm (GA) method. From the above model or the graph shown below, it can be discussed that the optimized value or the output obtained is a satisfactory one. The proposed output or the optimized output obtained for the construction project using Genetic Algorithm (GA) is able to achieve an increase in (10–20)% than the estimated data available for the project. As per the output obtained and by verifying the output with the attached graphs and MATLAB software, the total duration of the optimized result showed an increase in 10%, whereas the total estimated budget of the construction project showed an increase in 20% of the estimated data collected. The optimized output showed a great percent of similarity with the estimated data collected from the construction sector or industry, but it is obvious that the increase in cost and time of the project is beneficial for the completion of project. The difference may arise for the reason of not more than 4–5 activities, which is the main reason for the increase in the estimated cost and total duration of the project.

8 Conclusion

It is concluded that the obtained results are being satisfactory. The objective of the construction involving a TCTO problem was a multi-objective genetic algorithm. Even the single-objective TCTO problem of a construction project also provided correct and satisfactory optimized results of cost and time using genetic algorithm method comprising of using high technology optimized algorithms and other rapidly developed formulas comprising algorithms for processing the optimized result of cost and time of a TCTO problem. The increase in the development of the multi-objective optimization problem using genetic algorithm method is on a major high demand as it also possesses a feature of comparison of two on-going construction projects, thus highlighting the superiority and domination of the genetic algorithm functions on the optimization on TCTO problems. As it is clearly shown in the above case studies, an increase in 10% of the optimized results obtained than the collected data of the construction projects. The estimated cost of the construction project is around 400 crores INR along with the estimated duration of the construction project for 5 years. By optimizing the selected data collected from the companies' file, the total estimated duration has come around in a span of 4 years, but with an increase of the cost of the project as it is known that both the parameters are independent of each other. It can be still debated whether using genetic algorithm method for optimization is the perfect choice in the future or not, but as per reports and concentrating on the comparison of the results obtained, GA method is an effective and efficient method for optimization process of TCTO problem. It is easily used and less laborious method as compared to the various innovative optimization techniques used. Upon the requirements of the planners or the designers of the construction project, the genetic algorithm method can be used as a better option for optimization of TCTO problems.

Obviously, taking into considerations the experience possessed by the researchers and the planners for the optimization process, genetic algorithm might lack the extra edge of being used in a TCTO problem because of the various advanced innovative techniques that is being used in construction projects for optimization process [14]. Moreover, the assurance of the quality of the work is also a major highlight of the construction project, and it is mainly dependent on the work experience of the planners and the designers involved in it. The use of genetic algorithm method for such a particular scenario is completely dependent on the different subjectivity processes of the project along with different advanced characteristics of the operator being used for the project. However, the main intention lies in providing the proper method of quality assurance of the work done along with generation of total innovative quality of work done in the project [13]. It is thus concluded from the comparison of the case studies that GA method is one of the innovative techniques for optimization of TCTO problem of a construction project. The following features are proofs of the advancement of the genetic algorithm.

- Genetic algorithm method works with the process of coding a parameter set, not with a particular parameter.
- Genetic algorithm method searches the main population of the chromosomes rather than a particular population of the chromosomes mainly for global exploration of the project purposes.
- Genetic algorithm method mainly uses the objective function or the derivative function rather than using any complex functions or integers for optimization purpose.
- Genetic algorithm method can use a random choice function for the mechanism for the optimization process acting as a guide for searching the specific algorithm for the improvement and advancement of the derivative or the fitness function.
- Genetic algorithm method also improves the process of easy and simple experimentation for different scenarios of the optimization problem.

The various features act as an aid for the researchers and scholars for using genetic algorithm method for optimization of any type of problem.

References

1. Ashuri B, Tavakolan M (2015) Shuffled frog–leaping model for solving time–cost-resource optimization problems in construction project planning. J Comput Civ Eng:1–11
2. Bhosale V, Shastri SS, Khandare MA (2017) A review of genetic algorithm used for optimizing scheduling of resource constraint construction projects. Int Res J Eng Technol (IRJET) Pub:2869–2872
3. Haque KM, Hasin M, Akhtar A (2012) Genetic Algorithm for project time-cost optimization in fuzzy environment. J Ind Eng Manag:365–381
4. Todowski M, Rajicic D (2016) An initialization procedure in solving optimal power flow by genetic algorithm. IEEE Trans Power Syst:480–487
5. Dommel HW, Tinney WF (2008) Optimal power flow solution. IEEE Trans Power App Syst:1866–1876
6. Michaelwicz Z (2009) Genetic algorithms + data structures = evolution programs. Springer, Berlin, Germany, pp 111–112
7. Walters DC, Sheble GB (2003) Genetic algorithm solution of economic dispatch with valve point loading. IEEE Trans Power Syst:1325–1332
8. Sheble GB, Brittig K (2005) Refined genetic algorithm—economic dispatch example. IEEE Trans Power Syst:117–124
9. Chen PH, Chang HC (2005) Large-scale economic dispatch by genetic algorithm. IEEE Trans Power Syst:1919–1926
10. Bakirtzis AG, Biskas PN, Zoumas CE, Petridis V (2002) Optimal power flow by enhanced genetic algorithm. IEEE Trans Power Syst:229–236
11. Bragadin MA, Kahkonen K (2013) Quality evaluation of construction activities for project control. J Front Constr Eng:17–24
12. Coello CC (2000) An updated survey of GA-based multiobjective optimization techniques. ACM Comput Surv:109–143
13. Bhosale V, Shastri SS, Khandare MA (2016) A review of genetic algorithm used for optimizing scheduling of resource constraint construction projects. Int Res J Eng Technol (IRJET) Pub: 2869–2872
14. Daisy XMZ, Thomas SN, Mohan MK (2004) Applying genetic algorithm techniques for time-cost optimization. Depart Civ Eng:168–176
15. Sorrentino M (2013) Construction time-cost-quality trade- off: a road project case study. J Constr Manage Pub:163–176
16. Ramon SJC (2009) Time, cost and quality in a road building construction. J Constr Eng Manage © ASCE:1271–1274
17. Kandil A, El-Rayes K (2006) MACROS: Multiobjective automated construction resource optimization system. J Manage Eng:126–134
18. El-Rayes K, Kandil A (2005) Time-cost-quality trade-off analysis for highway construction. J Constr Eng Manage © ASCE:477–486
19. Zheng XM, Ng T, Kumarswamy M (2004) Applying a genetic algorithm-based multi-objective approach for time-cost optimization. J Constr Eng Manage:168–176

Exploring the Acceptance of Life Cycle Cost for Residential Projects in India

Abhipsa Guru and Mohibullah

Abstract In the construction project management, demands for sustainable and long-term decision-making methods are increasing as the projects are getting complex. Life cycle costing is the best approach to foresaid conditions for cost management practice. LCC is an economic analyzing method for all costs involved (construction, operation and maintenance) to the construction project over a defined period of time [1]. LCC method assists in finding an economical, effective and purposeful solution for a building project [2]. Best results are obtained by conducting LCC analysis in the earlier time during the feasibility study of the construction project. The questionnaire survey is done to identify all the parameters contributing to life cycle costs. The objective of this study is to explore the level of awareness and acceptance of life cycle cost for residential building in India. The research result reveals critical barriers in acceptance of life cycle cost in India.

Keywords Life cycle cost · Residential building · Operation cost · Maintenance cost

1 Introduction

Cost estimation is done in various construction projects and is dependent on several factors that include the analysis of life cycle approaches, project scope, schedule, location factors and risks. The life cycle costing (LCC) method is used to evaluate the economic performance of a building during the life cycle which takes the following cost into account:

A. Guru (✉) · Mohibullah
Department of Civil Engineering, Construction Engineering and Management, KIIT University, Shikhar Chandi Road, Chandaka Industrial Estate, Bhubaneswar, Patia 751024, Odisha, India
e-mail: guruabhipsa@gmail.com

Mohibullah
e-mail: mohibullah.fce@kiit.ac.in

B. B. Das et al. (eds.), *Recent Developments in Sustainable Infrastructure*, Lecture Notes in Civil Engineering 75, https://doi.org/10.1007/978-981-15-4577-1_72

i. Design and construction cost,
ii. Operation and maintenance cost, and
iii. Disposal cost [3].

Reducing the housing cost and also reducing the life cycle costs that consist of initial investment costs with the operating, maintenance and repair costs will give beneficial results both for the owners and also for the economy of the country. The factors that are taken into account for affecting the initial investments, operating, maintenance and repair cost of the buildings are understood. These factors also affect the costs of the buildings and so preference will be given to good quality and low-cost buildings in the design phase. According to the user's perspective, it will be possible to distinguish the rate at which there is a decrease in operating, maintenance and repair cost depending on the changing characteristics of the building construction. The aim of the study is to distinguish the changes that occur in the initial investment, operating and life cycle costs due to the change made in building envelope properties. At the end, the building's initial investment costs, operating costs and life cycle costs have been distinguished and assessed in line with the changes made in the materials utilized in the building components forming the building envelope along with wall, roof and ground flooring [3]. The goal is also to provide a detailed assessment of Life Cycle Cost and Life Cycle Assessment (LCA) concepts in design of residential house. Comparison is done with the same type of LCA and LCC studies. The future perspective of LCA and LCC studies on buildings is reported. The goal here is also to boost the conception of LCC application in the construction industry.

2 Life Cycle Costing

Economic analysis is done with life cycle costing where all the costs related to design, construction, operation and maintenance of construction project are defined over a given period of time. Life cycle cost includes an initial investment (usually construction cost) and follow-on costs (ordinary payments-energy, utilities, cleaning and maintenance, costs of replacement) and also includes the costs of demolition or disposal [1]. Life cycle costing estimates and assumes all the essential costs throughout the lifetime of building. Life cycle costs take into account construction, maintenance, repair and replacement costs, energy costs and residual value. Life cycle cost also assumes and takes into account all costs at their present value [4]. Future costs (operation, maintenance and demolition) were evaluated using today's prices and an estimate of future inflation. These were again converted into their present value using a suitable discount rate [2].

$$FC = PC * (1 + f)^n \tag{1}$$

where FC = future cost; PC = present cost, f = inflation rate and n = number of years.

Table 1 Life cycle cost of building (LCC) [5]

Cost parameter	Cost item
Investment costs	Design costs, construction costs, land cost, secondary cost related to locating the building, other investment costs, running costs for preliminaries and construction
Operation costs	Maintenance and renovation costs, Power supply costs, water and wastewater costs, waste disposal costs, service fees, insurance, security costs, cleaning costs, administrative fees
End of life costs	Costs of inspection, deconstruction, demolition, disposing of damp material, or any other costs associated with the disposal operations, salvage value

$$DPV = FC/(1+d)^n \tag{2}$$

where DPV = discounted present value; FC = future cost, d = discount rate and n = number of years.

Construction costs include material and labour costs. Maintenance costs consist of replacement over a given time period. The data required for life cycle costing analysis of construction, maintenance and disposal was taken from a standard construction cost guide (Table 1).

Life cycle costing is a tool as well as technique in which comparative assessments are to be made over a defined period of time, considering all relevant economic factors both in terms of initial cost, future operational cost and replacement costs to the end of its life.

3 Research Methodology

The research methodology concentrated on conducting a questionnaire survey. The questionnaire consists of 15 sets of questions. The feedback was collected from various design consultants from India. The questionnaires were given to around 60–70 companies by sending emails or collecting physically in hardcopy at a proposed time. Just after the pilot study was completed, the questionnaire was evolved. It was depicted in a simple checkbox to tick format to enable easy accomplishment. A Likert scale method was adopted for the analysis. The questionnaire is divided into three main parts. The first part is originated from methods of LCC. This part uses a five-point Likert scale. The format of Likert scale was as below:

NEVER HEARD, HEARD, KNOWN, RARELY USED AND OFTEN USED.

3.1 First Component of Questionnaire Survey

(1) Do you use value-oriented method for LCC calculation?
(2) Do you use base case method for LCC calculation?
(3) Do you use net present value method for LCC calculation?
(4) Do you use equivalent annual cost method for LCC calculation?
(5) Do you use approximate LCC method for LCC calculation?

Second part of the questionnaire comprises variables considered in LCC calculation. In this part, participant's utilization of variables in LCC and considered costs in their life cycle costs projections have been calculated utilizing a five-point Likert scale. As a result, the format of Likert scale was presented as below:

CERTAINLY USED, PARTLY USED, UNCERTAIN, PARTLY DISUSED AND CERTAINLY DISUSED.

3.2 Second Component of the Questionnaire Survey

(1) Do you use design cost parameter for LCC calculation?
(2) Do you use the maintenance cost parameter for LCC calculation?
(3) Do you use the energy cost parameter for LCC calculation?
(4) Do you use the service life of building as parameter for LCC calculation?
(5) Do you use the salvage value parameter for LCC calculation?

In the last portion of the questionnaire, questions were selected in order to evaluate participant's awareness and understanding about LCC. The questions were put in five-point Likert scale. Participants' consent and different levels have been assessed in the format given below:

NORMALLY, FREQUENTLY, SOMETIMES, SELDOM and NEVER.

3.3 Final Component of the Questionnaire Survey

(1) Do you give priority to plan and layout of building for LCC calculation?
(2) Do you give priority to material cost for LCC calculation?
(3) Do you give priority to labour cost for LCC calculation?
(4) Do you give priority to minor repair and replacement cost for LCC calculation?
(5) Do you give priority to commissioning cost for LCC calculation?

4 Results

Questions	Never heard	Heard	Known	Rarely used	Often used
1. Do you use value-oriented LCC method for LCC calculation?	38	20	12	6	4
2. Do you use base case method for LCC calculation?	32	25	10	7	6
3. Do you use net present value method for LCC calculation?	10	20	6	9	35
4. Do you use equivalent annual cost method for LCC calculation?	7	2	2	23	46
5. Do you use approximate LCC method for LCC calculation?	38	20	16	2	4
	Certainly used	Partly used	Uncertain	Partly disused	Certainly disused
1. Do you use the design and Construction cost parameter for LCC calculation?	42	20	10	5	3
2. Do you use the maintenance cost parameter for LCC calculation?	4	14	16	20	26

(continued)

(continued)

Questions	Never heard	Heard	Known	Rarely used	Often used
3. Do you use the energy cost parameter for LCC calculation?	32	27	15	2	4
4. Do you use the service life of building as parameter for LCC calculation?	3	10	25	15	27
5. Do you use the salvage value parameter for LCC calculation?	6	10	22	19	23
	Normally	Frequently	Sometimes	Seldom	Never
1. Do you give priority to material cost for LCC calculation?	5	6	10	10	33
2. Do you give priority to labour cost for LCC calculation?	6	7	13	16	38
3. Do you give priority to minor repair and replacement cost of structural components for LCC calculation?	10	7	15	23	25
4.Do you give priority to commissioning cost for LCC calculation?	10	8	15	22	25

(continued)

(continued)

Questions	Never heard	Heard	Known	Rarely used	Often used
5. Do you give priority to material type in design phase for LCC calculation?	9	10	20	23	18

4.1 Analysis of the Methods of LCC

It represents consultant's use and learning level of LCC techniques. For example, rigorous method was graded in general acceptance. Equivalence annual cost approach was also designated in less agreement. From a detailed literature review, order of utilization of LCC evaluation is stated as a number for LCC methods. It is logical that presents worth cost perspective and equivalent cost viewpoint viewed as the most implemented process in literature. It is also understood that the questionnaire solution is primarily used instead of present worth and equivalence cost outlook, value-oriented life cycle cost viewpoint and the base case point of view which are component-based procedures. It illustrates that participating consultants are more concerned with different constituents of product or building than all parameters as a whole. It is noticeable that rigorous and approximate LCC methods are applied to a lesser extent. The responses of all the methods are recorded here.

4.2 Analysis of the Parameters of LCC

Though maintenance cost has been given much priority for building projects, it is viewed that the consultants in India do not give importance to this factor. It is observed that nine criteria are divided into two groups. First is the simple variable which demonstrates cost models largely practiced in LCC assessment. It comprises design cost, construction cost, maintenance cost, operation cost and energy cost [6]. Second is the complex factor which indicates cost model producing all costs. It constitutes life cycle, salvage value and interest rate [1].

4.3 Analysis of Final Part of the Questionnaire

The last part of the questionnaire asked the responder to recognize the importance given to the parameters for LCC calculation. It is observed that the material cost, labour cost, minor repair and replacement cost, material type and commissioning cost all of them are also dependent on the calculation of LCC. They are ranked as per the Likert scale and as per that the priority rating is done so that the importance given to parameters is known. After rating them, it is observed that the consultants give importance to material cost, material type and commissioning cost. The rest of other parameters, labour cost, minor repair and replacement cost, are given less priority for the calculation of 1ife cycle cost according to the ratings and according to observations it is seen. From the responses, it is clear that the consultants have certain knowledge regarding the use of LCC and also they give priority to the various parameters according to the Likert scale range. Thus, the consultants play a very important role in the life cycle cost, and awareness is created regarding the use of life cycle costs.

5 Conclusion

There are various important findings from the research described below:

1. It appears that the component-based process (annual equivalent cost approach and net present value) is the most adopted technique in India. Several consultants do not even ascertain about LCC procedure in advance, although most of them have the understanding of different parameters of LCC analysis [7]. Hence, it could be stated that consultants have the basic notion but in actual during implantation, they usually do not adopt LCC.
2. Maximum consultants follow LCC concept in decision-making and take into account energy cost, design cost and life cycle [8]. It is observed that maintenance cost is ignored in most of the cases. According to feedback received from questionnaire survey, it is established that there is some gap in the knowledge and acceptance of LCC among consultants. It seems that LCC and different systems solution adopted for lowest cost are generally significant for participating consultants but lack of relevant information related to input-cost data and lack of expertise seems to be the major obstacle.
3. Outcomes of the present study state that few consultants perform LCC evaluation in the design process [1]. However, most of the consultants are still with the idea that LCC analysis consumes extra time, and prolonged the entire process of design for which clients may not pay extra costs [4]. Insufficient confidence and awareness of LCC method is the major constraint in its implementation and adoption among the consultants.

References

1. Dwaikat LN, Ali KN Life cycle cost analysis of buildings and life cycle budget development. J Build Eng 303–311
2. Heralova RS (2017) Life cycle costing as an important contribution to feasibility study in construction projects. Procedia Eng 565–570
3. Islam H, Jollands M, Setunge S (2015) Life cycle assessment and life cycle cost implications of residential buildings. J Renew Sustain Energy 129–140
4. Schmidt M, Crawford RH (2017) Developing an integrated framework for assessing life cycle cost of buildings. Procedia Eng 988–995
5. Heralova RS (2014) Life cycle optimization within decision making on alternative designs of public buildings. Procedia Eng 454–463
6. Liu MM, Mi B (2017) Life cycle cost of buildings subjected to earthquakes. J Build Eng 581–589
7. Kneife J (2010) Life cycle cost analysis in buildings. J Build Eng 333–340
8. Noshadravan A, Miller TR, Gregory JG (2017) A life cycle cost analysis of residential buildings including natural hazard risk. J Constr Eng Manag

Genetic Algorithm for Resource Levelling Problem in Construction Projects

Dolasankar Sahu and Mohibullah

Abstract The resource levelling problem (RLP) arises when with sufficient available resources, it is mandatory to have minimum fluctuations in the pattern of resource usage. These fluctuations are very much undesirable because it causes problem in the utilization of labour and monetary concern for the contractor. This kind of problem aims at minimizing total duration or making span of a project-dependent to precedence relations between activities and finite resource availability. A resource levelling approach gives direction to the person doing scheduling to plan activities in a project with smooth resource utilization without exceeding the constraints of resource availability. It is also often useful in minimizing the project duration within limited availability of resources for making project profitable. In RLP, start times of noncritical activities are shifted within their available float times so that the resource usage profiles are levelled best. Existing research has done to analyse various methods and algorithms for the resource levelling problem. In this study, we have discussed the scheduling process of RLP using genetic algorithm technique which is evaluated by a sample problem where resource levelling index (RLI) is improved by 48% to conventional schedule resource profile. Analysis of this study shows that GA could be an effective asset for getting optimized solutions in construction industry.

Keywords Resource levelling · Genetic algorithm · Resource management · Project scheduling

1 Introduction

Project management represents the coordination of all the activities in an efficient way with available resources to achieve effective outcome, completing project on time, on budget and as per predefined quality assured by resource management which is

---

D. Sahu (✉) · Mohibullah
School of Civil Engineering, KIIT Deemed to be University, Bhubaneswar, India
e-mail: dolasankar22@gmail.com

Mohibullah
e-mail: mohibullah.fce@kiit.ac.in

B. B. Das et al. (eds.), *Recent Developments in Sustainable Infrastructure*, Lecture Notes in Civil Engineering 75, https://doi.org/10.1007/978-981-15-4577-1_73

an essential part of project management. Practice of Critical Path Method and Program Evaluation Review Technique approaches has demonstrated to be quite helpful for schedule planning and controlling construction projects. In the project schedule, noncritical activities can change their positions by their floating times available to have better even resource profile [1, 2], while dealing with the resource techniques adopted, Resource Allocation (limited number of resources while keeping project extension to minimum), Resource Smoothening (adjusts activity for a scheduled model so that for resources required to the project do not exceeds predefined duration, attempts to reduce sharp variation in resource histogram linearly) and Resource Levelling (duration can be adjusted based on resource constraints with motive to balance the demand and available supply, attempts to reduce sharp variation in resource histogram) [3–5]. The reasons for levelling the resources are as follows: quantities of construction resources should meet the demand, avoidance of day-to-day resource fluctuation, by maintaining a steady flow of construction resource applications [6]. Start time of noncritical activities shifted according to the float times available for the level of resource utilization.

Studies on resource levelling for scheduling can be categorized into three categories: exact algorithms, heuristics and metaheuristics. Linear programming, branch and bound algorithms come under exact algorithm, which are useful for solutions of small-scale problems (time taking). Thumb rule for heuristic methods can be applied for solving major project problems without any guarantee of finding any optimal solutions. For complex problems, exact algorithm and heuristics may fail to find solutions. Metaheuristics is used for solving complex problems. It consists of an iterative generation process guiding to combine a lower ranked heuristics intelligently mainly with various other concepts used for exploring and achieving spaces related to search and also use different learning methods to structure the data for finding an optimal or near-optimal answer [7].

Earlier, Burgess and killebrew [8] attempted to solve resource levelling problem with objective of minimization of sum of squares of the resource usage to get rectangular shape resource histogram [4, 6, 9–13]. Easa [14] used the objective function sum of absolute deviation between consumptions of resources in consecutive time intervals [15, 9, 16, 17]. Another parameter used is the sum of absolute deviations between a specific time period and average usage of resource [2,14, 16, 18]. Minimum moment algorithm where resource histogram moment minimizes around horizontal axis is proposed by Harris [6] and later packing method is developed by Harris [1]. Several different models are used. Easa [14] presented an integer linear optimization model, a hybrid model. Son and Skibniewski [11] proposed resource levelling combination of local optimizer and simulated annealing. Hegazy [4] used genetic algorithm in minimum moment algorithm. Leu et al. [2] proposed optimization through genetic algorithm and its decision support system application. A neural dynamic model is proposed by Senouci and Adeli [17]. Senouci and Eldin [15] presented a model for genetic algorithm which performs resource-constrained and resource levelling simultaneously. El-Rayes and Jun [19] suggested release and rehire metrics and resource idle days metrics. Doulabi et al. [18] developed a model based on genetic algorithm which allows activity splitting.

The objective of present work is to enhance resource management in the construction project by applying genetic algorithm to level the resource histogram without extending the project duration. To maximize resource utilization efficiency, genetic algorithm is used to reduce unpleasant resource fluctuation which is evaluated by the parameter sum of absolute deviations of daily resource requirements from average daily resource requirements.

2 Resource Levelling

Construction project schedules often cause undesirable resource fluctuations in patterns of utilizations of resources over time that are unsuitable, costly and inefficient to implement on construction. Schedules generated often need to be altered in order to lessen significant fluctuations in resource utilization and make it as smooth as possible over the entire time span [19, 20]. Project scheduling problems are often practitioner's deals during scheduling, while achieving the most effective resource schedule not having duration of the project increased. Project scheduling problems comprise both resource-constrained problems and resource levelling problems, among others. The goal of project scheduling problem where resource is constrained is to lessen the time interval not exceeding the condition of resource availability. And for resource levelling problem target is to attain the most effective resource schedule without increasing the specified duration of project [21].

Use each activity's resource utility index (ratio of the quantity of selected resources to the total quantity of resources required for all competing activities during assigning resource) to maximize each activity's utility and thereafter to optimize the resource allocation and also minimize the project's time span with the expense of time/cost [22].

The purpose for resource levelling is to reduce resource usage fluctuation and smooth the resource histogram and is achieved by shifting noncritical activities within the range of available float time. In project management, slight variance of resource appears a financial load or threat and project baseline schedule determination, which particularize the start time of planned activity while fulfilling both constraints (precedence and deadline) for minimizing the variation in the resource utilizations [23].

3 Genetic Algorithm Concept

Genetic algorithms are search-based techniques inspired by Charles Darwin's most solid theory, that is, evolution theory through natural selection and survival of the fittest principle which is developed by John Holland in 1975. GAs are used for locating optimal solution based on the process of natural selection where stronger individuals stay alive in competing process [2, 4, 15, 24, 25]. It is a search technique for large and complex problems to find optimum or nearly optimum solutions [4,

15]. GAs' advantages from other optimization methods are (a) instead of variables themselves a coding set can be used for variables, and (b) search from a population of solution to improve a single solution [15, 26].

Procedure of genetic algorithm begins with the generation of population of individuals, each of them representing a possible solution. Each individual in population follows evaluation, selection and recombination steps repeatedly called generations. Every individual solution is represented as strings of bits each called chromosomes, and the consisted entities are called genes (each box from string). The entities represent the variable's value and are represented as real number according to the problem or binary value or other user preference [4, 27].

An objective function provides evaluation performance concerning applied parameters to get the solution to the problem. Whether the objective of the problem fulfilled or not is evaluated with the applied constraints through objective function. GA procedure mainly goes through three reproduction operators, i.e. selection, crossover and mutation. Reproduction operator helps to segregate bad solutions and generate better solutions in every iteration. Desirable characteristics of two randomly chosen parent chromosome are combined by their best characteristics to generate an offspring pair during reproduction phase.

Selection procedure helps chromosome to evolve by determining superior to average solutions for recombination to produce new offspring. Objective of selection phase is to give more trials to fittest individual and also to weed out lesser fittest individuals, so that fittest ones get more reproduction opportunities. Roulette wheel and ranking methods are utilized for this process. Better chromosomes are placed in intermediate mating pool where chromosomes get randomly paired. According to John Holland's schema theorem, presumed quantity of a particular partial solution is direct proportionality to ratio of individuals' average performance containing particular partial solution to the whole average population. Accordingly, better schema of above-average individuals has more than average chance for making to next generation [28].

Another reproduction operator is crossover where a fit portion of two parent individuals merges for generating pair of offspring for better solution possibilities. This procedure can be done in many ways, i.e. single-point, two-point and uniform crossover. Crossover rate is the parameter to perform crossover between any pair of strings, and its rate usually considers high to introduce new strings much quickly to population. Crossover rate too high may terminate high-performance strings faster or too low may cause slow exploration rate [25].

In mutation, certain gene position values are randomly changed and are decided by arbitrarily, i.e. bit values are not fixed permanently in strings. According to predefined mutation rate, chromosomes are modified to produce new nonexistent offspring in the population. Mutation rate is considered as low since high rate of mutation may result in random search primitively. The mutation process advantage is to redevelop genetic diversity to get global optima [25, 28] (Fig. 1).

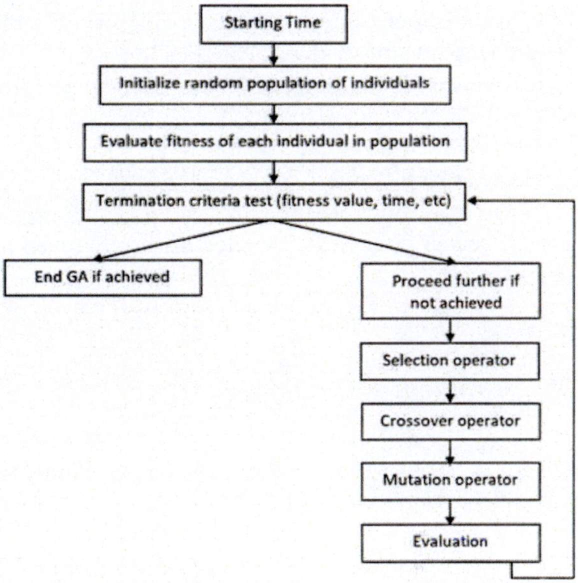

Fig. 1 Genetic algorithm optimization process

3.1 *Genetic Operators*

Genetic/reproduction operators of GA are used to separate inefficient solutions and to generate a new and efficient set of solutions in each iteration. Advantageous features of parents are inherited through offspring during the reproduction by combining the best features of both parents [24]. Three main operators are there in GA: (a) selection, (b) crossover and (c) mutation.

(a) Selection—From population, individuals are selected to recombine and to generate new offspring.
(b) Crossover—A brief part of two fit parent individuals unite to produce two offsprings of different genes. Generally, three types of crossover operations are used for optimization using genetic algorithm method: single-point, two-point and uniform crossover.
(c) Mutation—Introduces modifications in random selection process, thereby presenting a new random individual to get a new offspring.

3.2 *Parameters of Genetic Operators*

Before applying GA, parameters must be set [29]:

Size of Population—Population size to be randomly generated.

Chromosome Length—Linear string of available genes called a chromosome whose length varies with each application.

Crossover Probability—Crossover rate perform between any pairs of two selected Chromosomes/strings. Values vary from 0 to 1.

Mutation Probability—A rate to alter randomly one or more than one genes of a selected string. Values vary from 0 to 1.

GA is a well-directed search system, which uses previous evaluations for guiding the search process for developing an improved performance on function being optimized.

3.3 Problem Formulation

For the proposed problem, the following assumptions, objective and constraints are mentioned (Table 1).

3.3.1 Assumptions

- Activity network and their durations are fixed.
- Daily required resources for each activity are known and are constant throughout activity duration.
- Critical path method is determined, and duration of the project is kept fixed throughout.
- Single-type resource is considered; each activity either gets their required resources or not at all during each day.
- Activities within the network cannot be split.

Table 1 Data for problem

Activity	Predecessor	Duration	Resources
A	–	4	4
B	–	8	2
C	A	4	0
D	B	4	3
E	B	3	6
F	C, D	7	2
G	E	3	4
H	A	11	2
I	F, E	3	5

3.3.2 Objective Function

In this article, resource usage fluctuation is evaluated by 'sum of absolute deviations of daily resource requirements from average daily resource requirements' introduced by Easa in 1989. Motive behind selecting this objective criterion is easy to restore average daily resource usage by suitable plan for resource usage each day that could vary every day.

$$RLI = \sum_{t=1}^{T}\left(\left|\sum_{i=1}^{n} R_{it} - \left(\frac{\sum_{i=1}^{n} R_i D_i}{T}\right)\right|\right) \quad (1)$$

where R_i = daily resource required for activity i

D_i = duration of activity i

R_{it} = sum of resource required for each day t (for all activity on each day)

n = no. of activity.

T = total duration of project.

3.3.3 Constraints

- Resources are allocated for each activity equal to its duration.
- Precedence relationship must not be disrupted in each schedule.
- Earliest start time and latest finish time of each activity must be scheduled.
- Shift range must be within their float time of respective activity.

4 Algorithm Process

4.1 Genetic Algorithm Model

The scheduling procedure to maximize the resource utilization efficiency through resource levelling is organized into three main systems: (i) initialization system for calculation of initial schedule and quantity of float days for each activity in the project, (ii) genetic algorithm system for searching and identifying an optimal schedule set so that it will maximize the efficiency of resource schedule, and (iii) resource levelling system to determine the effect after shifting noncritical activities with their respective available float times. The three systems are described in detail in the following.

4.1.1 Initialization System

This system's objective is to determine initial schedule and float within each activity in the project. The purpose of calculating float times is to determine the upper limit of principal decision variables (quantity of allowable shift days for each activity, respectively). The principal decision variables are maximum shift days (M_{sd}) and quantity of decision variables is the same as quantity of noncritical activities in the whole project. The procedure is as follows:

(a) Input the data for scheduling for each activity which includes daily resource demand, duration and work logic.
(b) Determine early start time (EST_i), early finish time (EFT_i), late start time (LST_i) and late finish time (LFT_i) for each activity 'i' based upon previous step input data.
(c) Determine free float (FF_i) and total float (TF_i) for each activity.

$$FF_i = \text{Min}(\text{successorEST}_i - EFT_i) \tag{2}$$

$$TF_i = LST_i - EST_i \tag{3}$$

where FF_i = free float time of 'i' activity; Min(successorEST_i) = minimum early start time of all successors of activity 'i'; EFT_i = early finish time of 'i' activity; TF_i = total float of 'i' activity; LST_i = late start time of 'i' activity and EST_i = earlier start time of 'i' activity.

(d) For all decision variables (M_{sd}), upper and lower bound must be set. Ensuring that shifting of each noncritical activity must not exceed their respective total float time (TF_i) and not to cause extension to entire project duration this constraint is specified.

$$0 \le M_{sd} \le TF_i \tag{4}$$

In this scheduling model, the principal decision variables are the quantity of shifting days for each noncritical activity (M_{sd}) which are represented by genetic algorithm chromosome.

4.1.2 Genetic Algorithm System

The objective of genetic algorithm system is to explore an optimal schedule that maximizes the efficiency of resource utilization. GAs have been used because of its robustness in finding optimal and near-optimal solutions from large and complex search options. In spite of extensive utilization and capabilities, GAs cannot guarantee for generating an absolute optimal solution in every case. GAs computation procedure is as follows:

(a) Produce an initial solution set (from $k = 1$ to K) for initial population ($Pop_{g=1}$) in the first generation. k consists of randomly produced values for 'maximum shift days variable M_{sd}' (M_1, M_2, M_3, …, M_N) for each solution. Each solution set is designed to produce feasible schedule and resource profile of the project.
(b) Each solution's impact on overall utilization efficiency of resources to be evaluated through resource levelling system that calculates daily resource demand and then resource levelling index (RLI) using Eq. (1) for present levelling system. Evaluated value of the objective function will be used to select and determine the fitness of every solution (k) in the population (Pop_g) for reproduction of new offspring solutions further.
(c) Generate next-generation population ($Pop_{g=g+1}$) solutions on the basis of evaluated fitness values by GA operators, i.e. selection, crossover and mutation. Repeat the steps b and c for a self-decided number of generations (G) to produce near-optimal or optimal solutions or schedules that maximize the efficiency of resource utilization for the project.

4.1.3 Resource Levelling System

In the resource levelling system, effect of noncritical activities shifting within their available float times 'principal decision variables (M_{sd})' is evaluated on overall utilization efficiency of resources, by employing the procedure as follows:

(a) From the initialization system, early start time (EST_i), early finish time (EFT_i), late start time (LST_i) and late finish time (LFT_i) for all activities will be used.
(b) Establish total noncritical activities and determine the impact on overall utilization efficiency of resource by shifting each noncritical activity separately. Evaluation process covers every noncritical activity, only one activity at a time (i^*) beginning from last activity towards start of the project (backward process). Evaluation of noncritical activities selection order is determined by the following primary and tie-breaking incremental set of rules:
    1. Select latest noncritical activity (i^*) with late finish time which has not been shifted in previous shifting cycles.
    2. If rule 1 produces tie between any such activities, then choose the noncritical activity (i^*) which has minimum total floats.
    3. If rule 2 creates a further tie between any such activities, then choose the noncritical activity (i^*) with highest order number (i).
(c) Apply Eq. (5) to evaluate selected shift days number (S_{i*}) for each noncritical activity found in prior step (i^*) on the basis of updated free float (FF_{i*}), maximum shift days (M_{sd*}) and original total float (TF_{i*}). The original total float (TF_{i*}) of each activity (i^*) remains fix but during shifting cycle free float (FF_{i*}) may vary depending on the successor of that activity shifted or not in the earlier cycles. After each shifting cycle (step e), the free float (FF_{i*}) needs to be updated for

identification of selected shift days (S_{i*}) for activity i^* by Eq. (5) to ensure that shift days selected will not make further addition of shifts to its successor.

$$S_{i^*} = \frac{(FF_{n^*} + 1)}{(TF_{n^*} + 1)} \times M_{sd^*} \tag{5}$$

(d) Chosen noncritical activity (i^*) shifts by S_{i*} days and evaluates its rescheduled EST'_{i*} and EFT'_{i*} as follows:

$$EST'_{i*} = EST_{i*} + S_{i*} \tag{6}$$

$$EST'_{i*} = EFT_{i*} + S_{i*} \tag{7}$$

where EST'_{i*} = rescheduled EST of chosen noncritical activity i^*; EFT'_{i*} = rescheduled EFT of the chosen noncritical activity i^*; EST_{i*} = original EST of the chosen noncritical activity i^* prior to shift; and EFT_{i*} = original EFT of chosen noncritical activity i^* prior to shift.

(e) Re-evaluate the FF time of every predecessor of chosen noncritical activity i^* by Eq. (2).

(f) Repeat the process from step b to step e for every activity that is noncritical in the entire project while taking one activity at a moment into consideration.

(g) For rescheduled project, evaluate the daily resource demand.

(h) Calculate the after shifting effect of all noncritical activities according to their respective float times available on overall utilization efficiency of resource by determining the resource levelling index by Eq. (1).

The repetition process of above-mentioned eight steps continues for every solution k in population (Pop_G) of every generation g. Determined value of each solution k is then blended in the optimization function Eq. (1) in GA procedure. The value obtained from optimization function is considered as fitness value for each solution k to regenerate newer offspring solutions in succeeding generations during the search process for optimal solutions. Following a certain generation G, each solution k in the last population Pop_G an optimal or near-optimal solution/schedule will be represented which maximizes the overall efficiency utilization of resources in project.

4.2 *Results and Discussion*

For better performance using GA, parameter adjustment, crossover and mutation rates are absolutely requisite. Here, GA is tested with various crossover scales of 0.2, 0.3, 0.4, 0.5, 0.6, 0.7, 0.8 and 0.9 and mutation scale of 0.2, 0.4 and 0.6, respectively. Two heuristic methods PACK and minimum moment method (MOM) proposed by Harris for resource levelling were evaluated and after 200th-generation GA got the equal optimal value as the PACK beating MOM. Before levelling RLI was 50.36 and

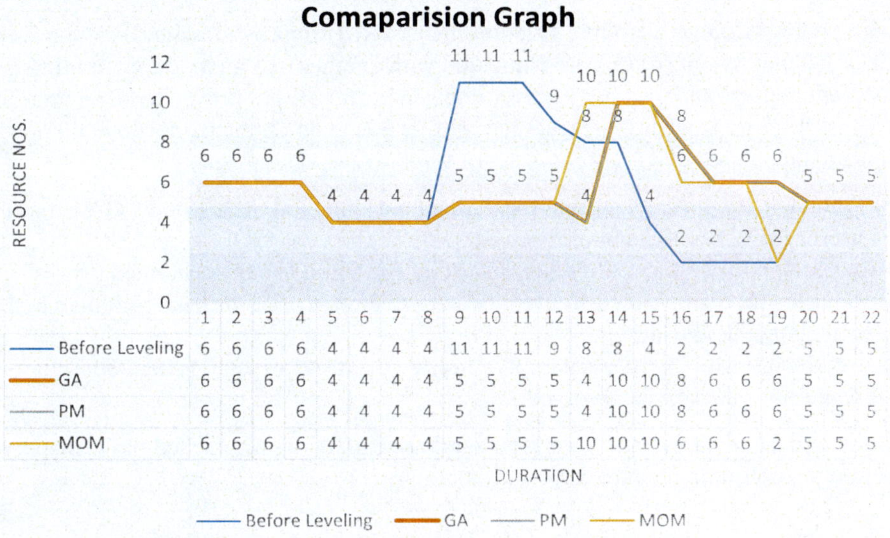

	1	2	3	4	5	6	7	8	9	10	11	12	13	14	15	16	17	18	19	20	21	22
Before Leveling	6	6	6	6	4	4	4	4	11	11	11	9	8	8	4	2	2	2	2	5	5	5
GA	6	6	6	6	4	4	4	4	5	5	5	5	4	10	10	8	6	6	6	5	5	5
PM	6	6	6	6	4	4	4	4	5	5	5	5	4	10	10	8	6	6	6	5	5	5
MOM	6	6	6	6	4	4	4	4	5	5	5	5	10	10	10	6	6	6	2	5	5	5

Fig. 2 Result comparison

after levelling it is 26.36 for both GA and PACK and for MOM approach RLI is 30.36. The smaller the index value, the superior the resource levelling or the greater the resource utilization flow. Between GA's perspective and heuristic approach, the significance difference is that GA approach may suggest many optimal or near-optimal schedules or solutions which may be helpful for different planning considerations for decision-makers but heuristic approach generates single feasible solution.

After applying genetic algorithm to the problem, the maximum resource required is 10 nos. for 2 days and minimum resource required is 4 nos. for 5 days, whereas before levelling maximum and minimum resources were 11 nos. and 2 nos. for 3 days and 4 days, respectively. Through minimum moment method obtained, maximum resource is 10 nos. for 3 days and minimum required is 2 nos. for 1 day and for pack method the same results were obtained as genetic algorithm as shown in Fig. 2. The minimum the resource levelling index, the better the resource profile is levelled. The minimum RLI is achieved through genetic algorithm and pack method, i.e. 26.36.

5 Conclusion

In this present study, an effort has been made to enhance resource management in the construction project by applying genetic algorithm. For maximizing resource utilization efficiency, genetic algorithm is used to reduce unpleasant resource fluctuation and the procedure is carried out in three system processes: initialization, genetic algorithm and resource levelling system. After applying GA to the problem with single resource, resource levelling index (RLI) improved around 48% to initial schedule. Results of GA are compared with another two methods, i.e. minimum moment method and packing method, and genetic algorithm proved to be an important asset for improvising the resource profile. Based on all the works done till now, it is recommended that

- Although results are the same for GA and pack method in this study, GA is more useful for larger and complex projects.
- It is quite difficult to learn when an optimal solution has been attained.
- Longer processing time is required due to random nature of the algorithm which is the main downside.
- Interrelationship with availability profiles of resource and generated activity schedules quality was observed.

GA could be used as a key for resource-based planning and scheduling, thereby deploying optimum resources for a project.

References

1. Harris RB (1990) Packing method for resource levelling pack. J Constr Eng Manag 116(2):331–350
2. Leu SS, Yang CH, Huang JC (2000) Resource levelling in construction by genetic algorithm-based optimization and its decision support system application. Autom Constr 10(1):27–41
3. Project Management Institute (2013) Project management body of knowledge (PMBOK) guide, 5th edn.
4. Hegazy T (1999) Optimization of resource allocation and levelling using genetic algorithms. J Constr Eng Manag 125(3):167–175
5. Moselhi O, Lorterapong P (1993) Least impact algorithm for resource allocation. Can J Civ Eng 20(2):180–188
6. Harris RB (1978) Precedence and arrow networking techniques for construction. Wiley, New York
7. Osman IH, Laporte G (1996) Metaheuristics: a bibliography. Ann Oper Res 63:513–623
8. Burgess AR, Killebrew JB (1962) Variation in activity level on a cyclical arrow diagram. J Ind Eng 13(2):76–83
9. Ahuja HN (1976) Construction performance control by networks, Wiley, New York
10. Bandelloni M, Tucci M, Rinaldi R (1994) Optimal resource leveling using nonserial dynamic programming. Eur J Oper Res 78(2):162–177
11. Son J, Skibniewski MJ (1999) Multiheuristic approach for resource leveling problem in construction engineering: hybrid approach. J Constr Eng Manag 125(1):23–31
12. Hiyassat MAS (2000) Modification of minimum moment approach in resource leveling. J Constr Eng Manag 126(4): 278–284
13. Christodoulou S, Ellinas G, Aslani P (2009) Entropy-based scheduling of resource-constrained construction projects. Autom Constr 18(7):919–928
14. Easa SM (1989) Resource leveling in construction by optimization. J Constr Eng Manag 115(2):302–316
15. Senouci AB, Eldin NN (2004) Use of genetic algorithms in resource scheduling of construction projects. J Constr Eng Manag 130(6):869–877
16. Wagner HM, Giglio RJ, Glaser RG (1964) Preventive maintenance scheduling by mathematical programming. Manag Sci 10(2):316–334

17. Senouci AB, Adeli H (2001) Resource scheduling using dynamics model of Adeli and Park. J Constr Eng Manag 127(1):28–34
18. Doulabi SHH, Seifi A, Shariat SY (2011) An efficient hybrid genetic algorithm for resource leveling via activity splitting. J Constr Eng Manag 137(2):137–146
19. El-Rayes K, Jun DH (2009) Optimizing resource levelling in construction projects. J Constr Eng Manag 135(11):1172–1180
20. Rieck J, Zimmermann J, Gather T (2012) Mixed-integer linear programming for resource levelling problems. Eur J Oper Res 221(1):27–37
21. Ponz-Tienda JL, Yepes V, Pellicer E, Moreno-Flores J (2013) The resource levelling problem with multiple resources using an adaptive genetic algorithm. Autom Constr 29(1):161–172
22. Christodoulou SE, Ellinas G, Michaelidou-Kamenou A (2010) Minimum moment method for resource levelling using entropy maximization. J Constr Eng Manag 136(5):518–527
23. Qiaon J, Li Y (2018) Resource levelling using normalized entropy and relative entropy. Autom Constr 87(1):263–272
24. Al-Tabtabai H, Alex AP (1999) Using genetic algorithms to solve optimization problems in construction. Eng Constr Archit Manag 6(2):121–132
25. Damci A, Polat G (2014) Impacts of different objective functions on resource levelling in construction projects: a case study. J Civ Eng Manag 20(4):537–547
26. Goldberg DE (1989) Genetic algorithms in search, optimization and machine learning. Addison-Wesley, Reading, MA
27. Leu SS, Yang CH (1999) GA-based multicriteria optimal model for construction scheduling. J Constr Eng Manag 125(6):420–427
28. Chan WT, Chua DKH, Kannan G (1996) Construction resource scheduling with genetic algorithms. J Constr Eng Manag 122(2):125–132
29. Mohammed SR, Ameer AAA (2011) Optimization of resource allocation and levelling using genetic algorithms. J Eng 17(4)

Strategies for Construction and Destruction (C&D) Waste Management

Rajarshi Patty, Dillip Kumar Bera, and Ashoke Kumar Rath

Abstract At the present time, waste debris generated from constructions and destructions of buildings, statues, and many more is an important matter of concern for solid waste management, which mainly consists of bricks, cement, stone chips, iron rods, plastics, etc. and also increases rapidly with the growth of construction industry across the world. This waste contributes to a massive amount of dust that is mainly responsible for environmental pollution. Hence, we need well- planned C&D management to recover and reuse the tremendous amount of hazardous substances to make pollution-free environment. This review paper analyzed different C&D management strategies based on the literature review to draw comparisons among these strategies and also to identify the best substitutes from the sustainability perspective. This study observed that 25–35% of the total amount of waste debris that ends up in landfill is the most ordinary and extensively used strategy but it should not be preferred as it has an adverse impact on the environment and the human body. So there are a numerous number of techniques which can be considered for alternate options than landfilling which will have less detrimental effects. From a sustainability point of view, Reduce, Reuse, and Recycling (3R's) are most favored and possibly the best options for C&D waste management which are of recent origin and not yet adopted globally.

Keywords Construction and demolition (C&D) waste · Solid waste · Waste management strategies

---

R. Patty · D. K. Bera (✉) · A. K. Rath
School of Civil Engineering, KIIT Deemed to be University, Bhubaneswar, India
e-mail: dberafce@kiit.ac.in

R. Patty
e-mail: patyrajarshi@gmail.com

A. K. Rath
e-mail: akrath1947@gmail.com

B. B. Das et al. (eds.), *Recent Developments in Sustainable Infrastructure*, Lecture Notes in Civil Engineering 75, https://doi.org/10.1007/978-981-15-4577-1_74

1 Introduction

For the fastest growing country, it should be pointed out that C&D waste management is a crucial matter of concern because waste materials are heavy, bulky, inert, and also a mixture of numerous materials of different characteristics. The term C&D waste associated with solid waste is obtained from activity-related construction, renovation, and demolition including different civil works [1]. The researcher proclaimed that if a small amount of waste materials were reduced, then it would have a dramatic impact on the final profit as well as environment [2]. It is also very hard to choose any particular strategy of C&D waste management. Waste management is a well-planned system of effectively controlling the production, storage, collection, transportation, processing, and disposal or utilization of wastes in an economical manner [3]. 3R's philosophy is widely used in the handling of C&D waste. During the second world war, recycling was taken into account by Germany to reuse demolished concrete, yet most of developing countries like India are not very much conscious of 3R's. Hence, they are still practicing landfilling as the only method of C&D waste handling. Globally, building waste production of 2–3 billion tons per year is estimated, of which 30–40% is concrete [4]. In India, 14.5 MT out of 48 MT of total solid waste generated per annum [5].

The main objective is to study the different sources of C&D waste along with their brief classifications and its hazardous effects and suggest different strategies for safe waste disposal.

Scope of the work is to study the various quantities and constituents generated from the C&D waste. Also, study the C&D waste management strategies of different cities of India and finally suggest the best alternatives to minimize the waste generation.

2 Literature Review

2.1 Background of C&D Waste Management

The main reason for this guideline is to enhance a well-designed environmental protection, sustainable development, and optimum use of resources of C&D management throughout the total project duration. The most traditional disposal mechanism is landfilling, and it has been practicing since the last few decades. Recently, due to exhaustive pressures and insufficient space on landfilling, the 3R's concept must take over as best alternatives in order to reduce waste generation. Recently, many local authorities and government bodies set a target of about 80% recycling of C&D waste. To achieve an environment-friendly C&D waste management, it is recommended to carry out a detailed and accurate research from the very beginning [6].

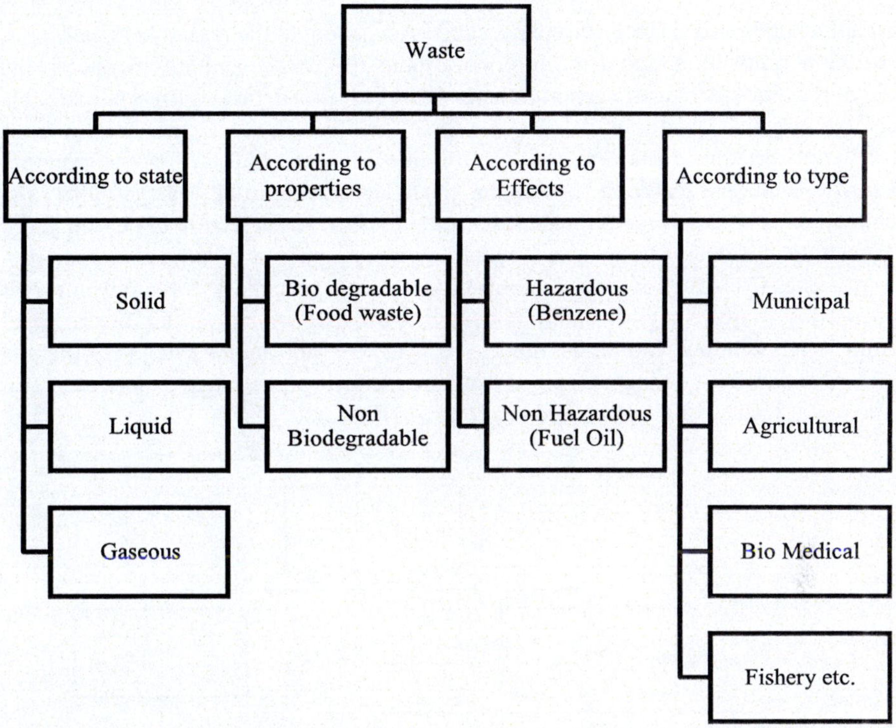

Fig. 1 Classification of wastes

2.2 *Classification of Wastes*

Waste is an inevitable, non-hazardous by-product resulted from human day-to-day activities [7]. Also, this waste is one kind of material/substance which generally depends on the type of materials or regional terminology. The category of waste can be classified according to their state, properties, effects, and type. According to their state, waste is mainly of three types: solid, liquid, and gaseous. Detailed classification of waste is shown below (Fig. 1).

2.3 *C&D Waste Classifications*

In today's market-driven condition, the construction industry is growing enormously all over the world. Similarly, C& D wastes are also increasing simultaneously with the growth of industry, and it creates a major detrimental impact on the environment [8]. This construction waste usually transported to landfill by an open-top roll-off containers like dump trucks or open trailers. According to the researchers, new construction, renovation, and demolition contribute 10%, 40%, and 50%, respectively, out of the total C&D waste generating per year [9]. C&D debris consists of materials such as concrete, wood, brick, metals, asphalt, wallboard, etc. Construction wastes are mainly residues from new construction like used materials during construction, cut-offs, damaged materials, packaging waste, etc. Demolition waste is mostly the collection of all construction materials from building after ejection of certain parts. The volume of demolition waste is much large in number as compared to construction waste. Construction and demolition waste can be mainly classified into two categories: major categories include cement concrete, cement plaster, bricks, steel from RCC, window, and doors, stones, rubbles, etc. and minor categories include electrical fixtures, conduits, GI pipes/Iron pipes/Plastic pipes, glass, panels, etc. [10]. The classification of C&D waste is as follows (Fig. 2).

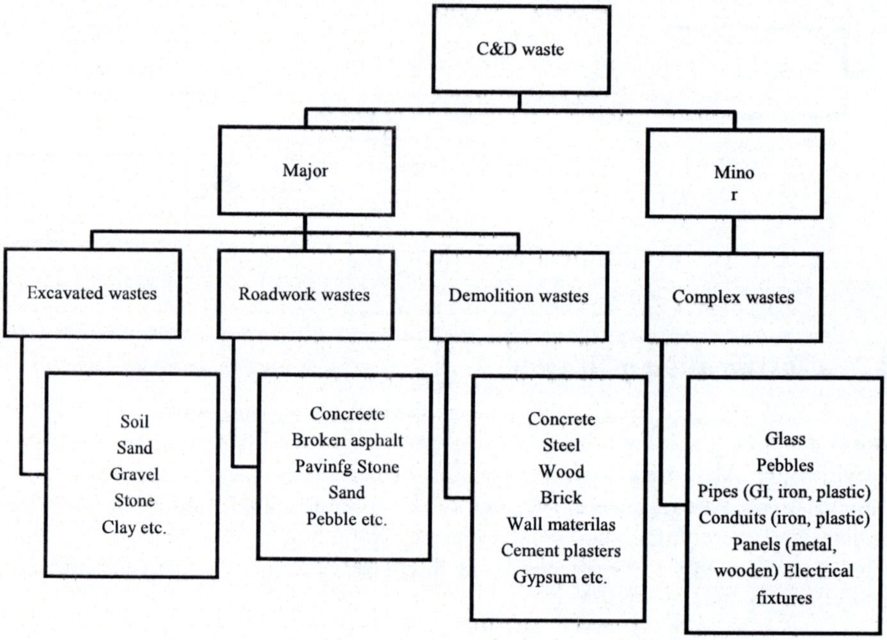

Fig. 2 C&D waste identification from source

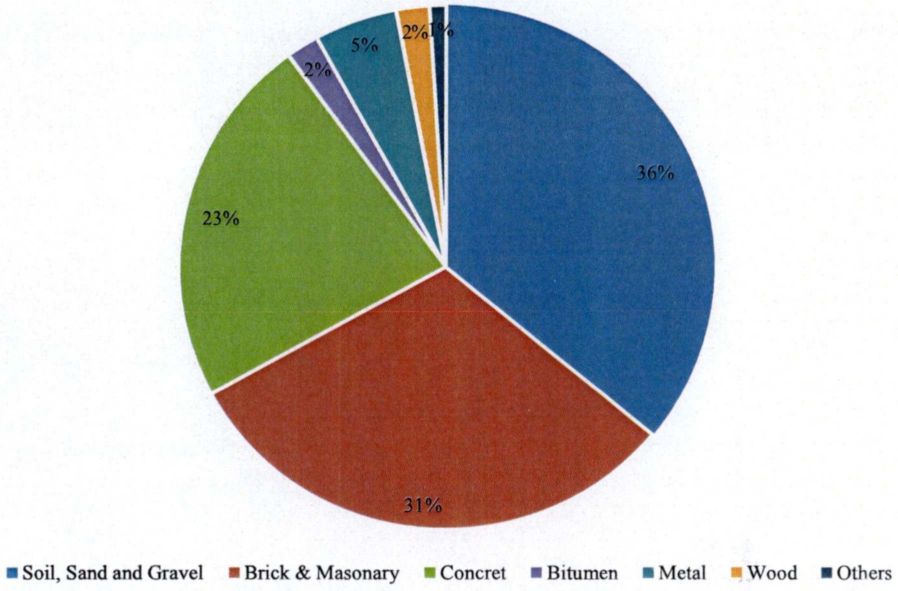

Fig. 3 Percentage distribution of different constituents generated from C&D waste per year in India (2000)

3 C&D Waste Generation Statistics in India

3.1 Composition and Quantity of Various Constituents

Generally, In India, there are two types of construction projects: Small housing project and a large project. Large projects are mainly executed by the contractors on a labor contract basis or turnkey basis, and in this project waste generations range within 3%. In the case of small housing projects, it is mainly carried out by owners on the labor contract basis and waste generation ranges between 5% and 7%. To control the waste generation, strict supervision from contractors or professionals is required. Concrete and masonry construction contribute more than 50% of the waste generated by the construction sector. If recycling is possible of those wastes by converting it to aggregate, then it may offer dual benefits of saving landfill space [6] as mentioned in Fig. 3 and Table 1.

3.2 Waste Generation Statistics of a Few Major Cities in India

In India, most of the major cities are still struggling to adopt a well-planned waste management plant, and the various constituents of waste are not separated prior

Table 1 Quantity and composition of various constitutes generated from C&D waste

Constituents	Quantity (metric tons/Year)	Composition (%)
Soil, sand, and gravel	4.20–5.14	36
Bricks and masonry	3.60–4.40	31
Concrete	2.40–3.67	23
Metals	0.60–0.73	5
Bitumen	0.25–0.30	2
Wood	0.25–0.30	2
Others	0.10–0.15	1

Table 2 C&D waste generation of different cities in India

City	Daily CDW waste generation (tons/day)	Annual CDW generation (million tons/annum)
Mumbai	2,500	0.75
Delhi	4,600	1.38
Bengaluru	875	0.26
Chennai	2,500	0.75
Kolkata	1,600	0.48
Jaipur	200	0.06
Patna	250	0.08
Ahmedabad	700	0.21
Bhopal	50	0.02
Coimbatore	92	0.03

to disposal. Researcher found that 70% of cities in India have lack of awareness and insufficient technology regarding recycling waste. Source said that out of ten major cities in India, Delhi generates the highest amount of C&D waste followed by Chennai, Mumbai, and Kolkata, whereas Bhopal and Coimbatore produce very less amount of waste products as mentioned in Table 2 and Fig. 4.

4 Strategic Plan to Reduce C&D waste

4.1 Sources of C&D Waste Generation in the Construction Industry

Due to the fast growth in the construction sector, it is important to know the source of construction and demolition waste before handling. Basically, there are two types of sources of C&D waste generation present. First one is bulk generators, and another

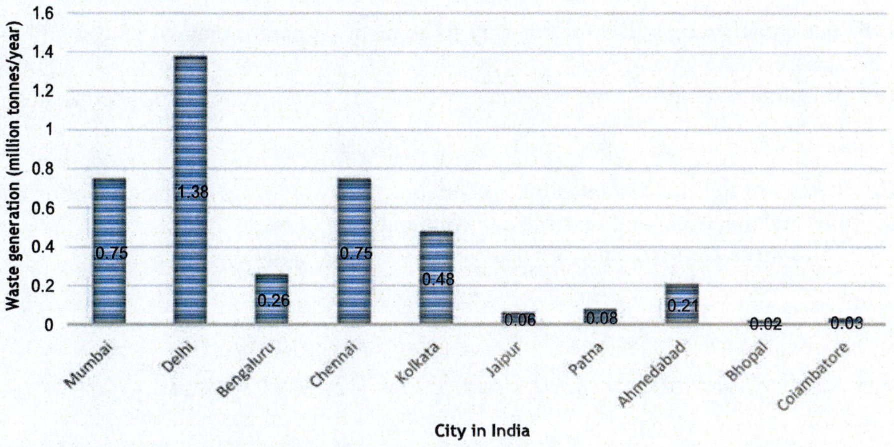

Fig. 4 Waste generated from major Indian cities

one is retail generators or small generators. Bulk generators of waste are considered waste which is resulted from the infrastructure and real estate sector and small/ retail generators mainly result from small commercial enterprise and industrial house buildings [9].

4.2 *Minimization Strategies of C&D Waste*

A well-planned managing and monitoring strategy is highly required in order to minimize waste. To ensure the successful implementation of waste streams, it needs conscientious planning throughout the design, build, and tenancy period, and it can only possible when types, behavior, and production waste are well known [11]. There are several alternatives available for C&D waste management. Construction waste management is dumping in landfill sites. However, space for landfill disposal is decreasing with the increase of waste generation, which may lead to escalating costs as well as harmful effects on the environment [12]. The minimization of waste can be attained through:

(a) Avoiding on-site waste as much as possible.
(b) Mentioning waste specifications and formulations in contract documents.
(c) Developing a plan of site waste management.
(d) Using recycled the content of building materials and products.

This has prompted the need for the best alternatives for waste management. One hierarchical order that is Reduced, Reuse, and Recycle (3R's) is the most favored and possibly the best option for C&D waste management.

The hierarchy of waste management suggests that

(a) The most successful/effective way to achieve an environmental solution is to reduce waste generation.
(b) If further reduction is not possible, then the building materials and products can be reused for a different purpose.
(c) Value of any materials should be recuperated from waste through recycling.
(d) If none of the above suggestions is suitable, then waste is to be disposed of using the best scientific and feasible environmental recourse.

(*Source* Dept. of Env., Transport and Regions, 2000).

4.3 3R's Concept: Present Practice of Waste Disposal

The main fundamental reasons behind 3R's concepts are as follows: One is economical advantages and another one is environmental advantages. The economic advantages comprise minimize the total project cost, reduce the number of waste materials, and enlarged the business sponsorship. The environmental advantages comprise the minimization the environmental pollution which is harmful to the human body [13]. The hierarchy of the 3R concept suggests that:

Reduce: Reduce stands for "minimize/reduce" unnecessary and wasteful products from its original volume. During the design, stage precautions should be taken to identify the waste source in order to minimize the amount of waste generation. The reduction of waste materials can be achieved by design the building with proper standard size and spaces. Also, it is recommended to consult with building professionals to ensure exact calculations of required materials and proper delivery strategies for the construction works so that the wastage can be prevented [14].

Reuse: Reuse means the recovery of C&D waste into the work. Once valuable waste materials have been generated, the ultimate method of managing it is through reuse either on the current site or a nearby site [14].

Recycle: Recycling is a term which is mainly used to describe a series of activities that includes collection and segregation, processing, and remanufacturing. For managing waste products, recycling is the final alternatives and significant commitments for sustainable buildings. It is the process consisting of sorting, cleansing, treating, and reassembling solid waste but it does not include burning. Most importantly, recycling generates usable materials at much lower environmental cost than primary sources [6]. For an effective recycling process, materials that can be reused should be identified early to establish a link with local reuse facilities and contractors on the build process, segregation, collection, and transfer. Recycling preserves our natural resources, energy, and water, and minimizes the emission of greenhouse gas and other pollutants (Figs. 5 and 6).

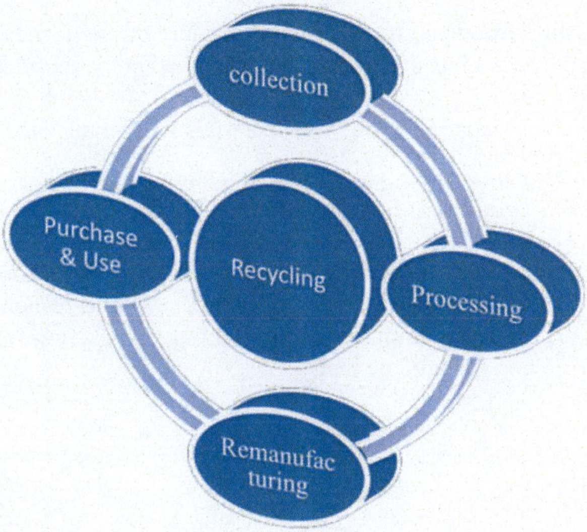

Fig. 5 Recycling method

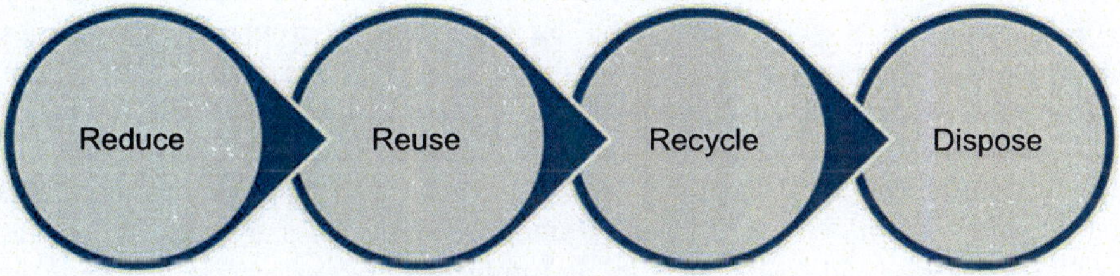

Fig. 6 3R's hierarchy for C&D waste treatment method

4.4 *Construction Waste Management (CWM) Plan*

A CWM plan is the main keystone for successful waste management strategies. It is a comprehensive document which includes all the information needed by all on-site professionals to achieve waste management goals for the project. For CWM plan, contractual clause plays a significant role in ensuring optimal waste production. Site management should follow all the rules and regulations in the contract as proposed earlier so that each and every liability of contractors can be successfully performed. The contractual clause will enable a monthly data collection report from the sub-contractors or supplier to analyze the data of C&D waste generation at the site [6, 13]. A day-to-day report should be prepared to track the progress toward the goal. A comprehensive CWM plan should consist of following points:

(a) Preparing a catalogue of materials which are targeted for 3R's.
(b) Listing out landfill information in detail including tipping fees.

(c) Illustrating the sorting and transporting procedure of the recyclable materials.
(d) Estimating the packaging materials generated whether the supplier can eliminate or recycle packaging.
(e) Recycling information including how materials will be recycled.

A successful CWM plan involves good communication among all the parties: owner, architect, engineer, contractor, and sub-contractor. The Waste Management Plan should define the different responsibilities of different peoples which means of who is responsible for procuring and storing this information, where information will be stored (e.g., on-site, by the architect, by the contractor), who is responsible for using the information to produce operating and financial reports, and how information will be transferred from one party to another.

4.5 Summary

With the growth of the construction industry, the waste product from construction sectors is also increasing rapidly but in India, waste management programs are in its infancy. If C&D waste management is not developed properly and efficiently adopted, then it may create detrimental impacts on the environment which is not good for sustainable movement of the country. Thus, a proper plan of C&D waste management should be made in the design period in order to minimize the volume of C&D waste. Also, C&D minimization and handling should take into consideration in view of insufficient land space; otherwise, this may be an issue regarding landfilling and handling of solid waste. The government should reform the strict rules and policies related to construction waste, and implementation of C&D management should be mandatory for all types of construction. Though landfilling was the most common and widely used waste strategy, recently Reduce, Reuse, and Recycle (3R's) policy is the best minimization strategy and almost all the developed and developing countries are following this waste management strategy. Recycling offers a dual benefit of preserving our natural resources, energy, and water, and minimizes the emission of greenhouse gas and other pollutants.

5 Conclusion

This is the biggest challenge to manage C&D waste in the near future. It has been found that most of the Indian cities are not aware of waste reduction strategies, and still they follow landfilling as a main waste disposal part of the waste management hierarchy. To minimize the waste, a proper plan of waste management must be implemented at the earliest possible stage of the project, and it should be managed accordingly by the nominated project manager. Each party has an equal responsibility for developing, implementing, controlling, and monitoring of C&D waste management plan for the benefit of the nation; also some government rules and regulations should introduce assuming this as a national problem. Waste reduction, its reuse, and finally recycling should be carried out properly so that waste generation can minimize.

References

1. Shen LY, Tam VWY, Tam CM, Drew D (2004) Mapping approach for examining waste management on construction sites. J Constr Eng and Manag 472–441
2. Begum RA, Satari SK, Pereira JJ (2010) Comparison of conventional and industrialized building systems. Am J Environ Sci 383–388
3. Uchegbu SN (1998) Environmental management and protection. Precision Printers and Publishers, Enugu, pp 121–122
4. Nitivattananon V, Borongan G (2007) Construction and demolition waste management: current practices in Asia. In: International conference on sustainable solid waste management, Chennai, India, pp 97–104
5. Pappu A, Saxena M, Asolekar SR (2007) Solid wastes generation in India and their recycling potential in building materials. Indian J Environ Prot 2311–2321
6. Ponnada MR, Kameswari P (2015) Construction and demolition waste management—a review. Int J Adv Sci Technol 19–46
7. Poon CS, Wan Yu AT, Wong SW, Cheung E (2004) Management of construction waste in public housing projects in Hong Kong. J Constr Manag Econ 461–470
8. Gayakwad HP, Sasane NB (2015) Construction and demolition waste management in India. Int Res J Eng Technol 712–715
9. Thomas J, Wilson PM (2013) Construction waste management in India. Am J Eng Res 6–9
10. Shetty RS (2013) Construction and demolition waste—an overview of construction industry in India. Int J Chem Environ Biol Sci 640–642
11. Asaari FAH, Halim HBtA, Isa MH (2004) A study on construction and demolition wastes from buildings in Seberang Perai. In: AWAM international conference on civil engineering, Malaysia, pp 1–5
12. Lingard H, Graham P, Smithers G (2000) Employee perceptions of the solid waste management system operating in a large Australian contracting organisation: implications for company policy implementation. Constr Manag Econ 383
13. Coventry S, Guthrie P (1998) Waste minimisation and recycling in construction. Design manual. CIRIA, London
14. Gray J (2013) Reducing and managing waste. Sustain Build

Comparison Between the Tunnel Form System Formwork and the MIVAN Formwork System in a Multi-unit Building Project

Paromik Ray, Dillip Kumar Bera, and Ashoke Kumar Rath

Abstract Tunnel form system formwork is one type of modern system of formwork that is used for construction purposes. This system has become a prime and permanent construction technique for construction of building in the country along with many foreign countries. This system is an emerging technique for construction purpose due to its efficient construction technology, minimum duration utilized and low estimated budget cost in the construction project. This paper reviews the main comparison of the tunnel form system formwork with the MIVAN formwork technique which is the frequently used technique in the country. Moreover, the working process, advantages, disadvantages along with characteristics and limitations of the formwork system as compared to the behaviour of the structure has also been reviewed. The MIVAN formwork technique also referred to as aluminium formwork provides much faster mode of construction work with easy access all over the country as compared to tunnel form system formwork. This review paper also highlights the failure and hazardous activities of the tunnel form system formwork mentioning with case studies along with the benefits or usage in designing, shuttering process, de-shuttering process and dismantling process. This review paper also highlights the modified and advanced analysis of both the formwork systems and its comparison with the scheduled duration and the estimated cost budget of the construction project. The break-even analysis of both the formwork systems has also been compared in this review paper.

Keywords Tunnel form system formwork · MIVAN technique formwork · Break-even analysis · Cost calculation analysis · Duration calculation analysis

---

P. Ray (✉) · D. K. Bera · A. K. Rath
School of Civil Engineering, KIIT Deemed to be University, Bhubaneswar, India
e-mail: paromikray@gmail.com

D. K. Bera
e-mail: dkbera.fce@kiit.ac.in

A. K. Rath
e-mail: akrathfce@kiit.ac.in

B. B. Das et al. (eds.), *Recent Developments in Sustainable Infrastructure*, Lecture Notes in Civil Engineering 75, https://doi.org/10.1007/978-981-15-4577-1_75

1 Introduction

Tunnel form system formwork is considered as an industrialized building construction technique that mainly allows the casting of the slabs and the walls in a particular single action in a complete rotation. It merges and along with accelerates the speed, the accuracy and the quality of the on-site and the off-site productions and the constructions which are therefore considered as modern methods of construction denoted as MMC. The effect of MMC which is applicable in the reinforced building and the finishing are present sufficiently in a high quality which mainly requires the minimum duration for finishing the work. The walls and the faces of the construction building can be easily and effectively finished by thermal units that are present in sufficient amount as per required.

The major ingredients of these constructed buildings are relatively thin shear walls and slabs as being compared to the conventional reinforced concrete structures that are mainly used. Shear walls of the tunnel form buildings' construction mainly carry the vertical load carrying and the lateral load resisting members due to the main reason for non-attendance of columns and beams. At the time of construction of slabs and walls which is having the identical thickness being casted in a single action operation, the member of cold plates formed at the formation of formwork being reduced at the end of construction operation. Moreover, the concurrent casting of the slabs and the walls formed results in the enormous formation of the on-going construction building. As a result, the formwork system of tunnel system formwork mainly achieves the intensified seismic-type preparation by delaying the main plastic hinge development of the censorious points of the constructed buildings in a wall open area. The system is being created with a well-planned load carriage structural building being used in a vast variance application of the construction. The most successful plan of the building is the section of the monotonous cellular structures such as the hotels, barracks, residential buildings, student lodgings and also the lock-ups. The tunnel form system construction method concrete is poured into a tunnel formwork system of the shear walls and the slab floors at the same duration of the construction. This is a particular operation which is continuous and reoccurring in nature normally which consists of 24 h pattern of each particular floor; the on-going building construction consists of several sub-divisions which can be particularly created with great rapidity and quickly. Generally, this system of formwork the floor plans that are being designed and are made of the same tunnel that is being used or utilized in all the sections of the construction. This particular system of construction technology provides great advantages and proper suitability as compared to MIVAN formwork as used in the foreign countries.

MIVAN formwork as commonly referred to as aluminium system formwork falls in the category of modern formwork systems. It is a common mode of construction in the country as it is quite efficient and quite easily accessible. MIVAN formwork is simple, quick, efficient and high adaptable to the Indian conditions that are quite known for change. This type of formwork has a high standard of work which requires less or minimum finishing and the proper abilities for lasting as main criteria for consideration of the construction project. It is mainly a full-proper engineering system formwork which requires a whole lot of practice for planning and designing as an outperforming system formwork. The process involves the shuttering, de-shuttering as well as dismantling process of columns, beams and slabs comprising of proper steady flow of concrete. This formwork system also gives support to the initial de-shuttering process that mainly attains the main process of air curing along with various curing chemicals that are used in the process of curing in the construction project. MIVAN system is formwork generally very flexible and strong with perfections and quite easily accessible. The elements or the components of the MIVAN system formwork are made of aluminium as it is commonly referred to and also possesses the properties of lightweight in nature. It is mainly used for large span of construction of a building especially columns, beams and slabs of span exceeding 250 mm and more.

2 Tunnel Form System Formwork

This type of formwork system was originated in the foreign and developing countries around 60 years ago from today. Initially, this type of formwork system is used for construction purpose consisting of a number of repetitive cellular structures or concrete structures that are required. Using this particular type of system formwork in construction in the country like India is a big challenging task as it was recently utilized in the country as used frequently in all the developing foreign countries. The increase in admirations of this particular type of formwork system results in maximum production of the building construction structures having regular or monotonous walls or rooms in the on-going project. The walls and slabs are being casted in a monolithic manner that is being used for a different type of formwork both horizontally and vertically that are used for calculating the total duration of the construction project [1]. This particular form of system involves certain parameters of benefits at the time of construction. The benefits include the following.

2.1 *Cost–Benefit*

The tunnel form construction system formwork provides an efficient cost at the time of construction. This particular system is cost-effective, generating remarkable reduction of the estimated cost of the project and high-quality reduction of the cost of the structure without affecting or compromising the design of any structure. This particular process in the formwork system is one of the most favourable processes for the construction of the building project especially the cellular structures with many civil engineers and also many building contractors all over the world have appreciated the efforts of the system of formwork for minimum cost at the time of construction with a potential to carry out more future large construction projects within the estimated budget of the project. Moreover, the smoothness of the surfaces of the walls and the slabs does not specifically require any additional or extra finishing in the form of plaster of paris or any plaster. Early completion of the building project proves to be decisive in the generation of the financial conditions including the rental process or any source of income of the project [2].

2.2 *Building*

The tunnel formwork system is the most modified approach for each operation or program performed. Since the monotonous and rapid continuous working of the formwork system along with its uses in the field of pre-fabricated forms and reinforcing bars or steel cages which simplifies the whole construction process and also delivering useful featureless quantities or parameters along with efficient working process. The use of this specific technology is a common feature to the construction industries or companies, but in the presence of tunnel form system formwork a less dependent on the skilled and efficient workers or labourers is done [3].

2.3 *Quality*

Qualities or quantities of the on-going building project lead to the main increase in the speed of construction with the presence of tunnel formwork system. Exactly, the steel surface present in the formwork generates a high featureless along with high-quality effective finishing techniques in a proper modern decoration with minimum cost of construction estimated budget. Hence, reducing the excess or unrequited portion of construction work of the project ultimately leads to saving or minimizing the extra cost of the estimated budget project [4].

2.4 *Design*

The on-going building construction project with the use of tunnel formwork system consists of a large design plan building which provides an extraordinary design calculations and flexibility conditions by allowing a high level and high accuracy of design parameters by showing a final appearance [4].

2.5 *Safety*

In the tunnel formwork system during the construction work time, at the time of working at a certain range of height in a particular platform or in an edge of the platform, safety is the main factor at the construction project work. With respect to the introduction, the continuous or repetitive nature of the familiarity as well as the functioning and the training acquired by the workers and labourers is very important for the progress of the construction work as well as completion of the project. Hence, the minimum requirements are required in the form of tools, appliances or any specific equipment for the completion of building project in a risk-free and safer manner without any cause of any sort of accidents and finally reducing the cause of accidents [5].

2.6 *Sustainability*

The in situ concreting that takes place at the construction site along with the local availability of the ready-mix concrete supply that is transported from any local area helps in reducing the impacts that are caused by the transportation facilities, thus resulting in fast and quick approach and delivering of the ready-mix concrete and creating a zero wastage or nil wastage that are produced from an overall condition of the construction site and helps in creation of a tidy and neat site for construction project work to progress with minimum cost estimated as per the project, cost savings conditions and proper safety considerations [6]. Moreover, the mass thermal doublets with the presence of exact insulation and radiation along with boiler design of the construction project minimize the heating condition as well as the prices of the boiling condition in the construction project which helps in proper air circulation and reducing the air-conditioning necessity or requirement in the construction project [6].

3 Working Construction Process

The working construction process of the tunnel system formwork mainly is divided into five steps or processes. This working construction process signifies and highlights the process and methods of construction of the tunnel system formwork. The construction steps are as follows.

3.1 STEP 1—Cast Foundation and Starter Walls

The first step of the tunnel formwork construction system includes the main casting of the slab, walls and the floors in the presence of 3- to 5-in. starter walls that are mainly kept for the positioning of the tunnel forms. The starters as per the instructions should be carefully and cautiously assembled or collected along with the vertical arrangements that should also be checked for proper shape and perfect size. The rebars that are fixed in the particular starter walls are present in around 2 feet around the centre of the tunnel formwork system [7].

The starter walls that are mainly used for the setting of tunnels are being stored as well as conformed by large-sized steel angles which are efficiently used. While the conformation process takes place in the top starter inner walls, the steel angles are used by steel hangers that sit on the top of the cross-shaped structure of the building project at a height of every 6 feet. Starter walls are mainly used for building the walls of the bottom that are conformed at the outside wall forms of the constructed building of both the interior and exterior walls as well as the inside portion of the steel angles that are been attached to the outside portion of the wall forms [8].

3.2 STEP 2—Install Tunnel Form

The tunnels that are used in the first step are being placed in a single platform with the aid of a crane that is placed in between the two lines of the wall steel mesh process which are pushed towards a particular point present in the centre of the starter walls for the main alignment as well as the levelled process of the screw jacks in a particular process [8].

After finishing of the above-mentioned process, the forming process of the starter walls is being measured in dimensions of the length and breadth of the starter walls mainly for the alignment and levelling of the screw jacks that are present in the process. Initially, the welded mesh of the starter walls is tied to the rebars by providing the continuous reinforcement for the presence of every floor starting from ground floor to the first floor and moving on to the whole on-going building construction site. The tunnel forms previously formed are being positioned by the crane in the middle portion for the reinforcement rows work, pushing the opposite starter walls of the slab mesh in order to have proper alignment and uniformed with the aid of certain jack drivers present [9].

3.3 STEP 3—Install Electrical Tubes and Opening Block Outs

The electrical wires that are then installed in the previous steps along with the presence of block outs in the various openings of the on-going building construction project such as the doors and the windows are mainly positioned in the openings of the starter walls in the form of pre-cut process of the wire fabric material. The forms of wire fabric materials that are being used in the mentioned process are different in both materials and dimensions of the process being used. The tunnel form system formworks that are being placed in the platforms in the particular process are the main reason for positioning. Spacers are also used in the centre of the steel mash starter walls for the middle of the forms [10].

3.4 STEP 4—Form the Kicker for the Next Level

The starter walls or the kickers of the upcoming floors from the second floor onwards of the construction building project along with the first floors are well casted at the same duration of the construction for better positioning and act as a well support system. The steel angles that are forming an interior as well as exterior starter walls are in the form of cross-shaped pre-casted blocks that are well set between at a certain distance of around 6 feet. The bottom members of each cross-sectional shaped block are well-being centred between the forms of the tunnel system formwork [11].

3.5 STEP 5—Concrete Casting and Curing Process

After the process, the concrete is first allowed to place with the help of an engineer or a distributor along with a formation of the slab, ceiling and the starter walls for each particular floor of the construction building [12], hence enabling a well-positioned monotonous and continuous cycle of the construction project. The whole portion of the construction step is repeated continuously to get the desired output of the construction of tunnel system formwork in the construction project.

4 MIVAN Formwork—Introduction

This is considered as one of the modern formwork techniques for construction purposes. Unlike the previous tunnel system formwork also being used for construction purposes and being a constant use for construction purpose in foreign developing countries, this MIVAN formwork system technique is completely different. This technique of formwork is quite simple, efficient, quick or fast and highly adaptable to any conditions of the construction projects. Since its adaptability to any construction conditions, it is best suited for countries like India because of its easy access throughout the country and less labour of work required for the process of construction. It has a high construction standard of work required for the ability of finishing in the minimum possible duration of the project. The formwork system technique is a fully modern engineered system technique where the whole and skilled practice is required as per planning the fully outstanding details of the project [13].

The following construction process for the MIVAN formwork system technique consists of designing the walls, columns and slabs of the project in one steady flow of the quantity of concrete used for construction. Moreover, the process of de-shuttering columns beams along with shuttering process as well as dismantling process used for the project is done using the MIVAN formwork system. The process of curing especially the process of air curing if required for the project can be done with this system technique formwork with the help of various different curing chemicals used. Lastly, the manufacturing or the construction of this type of construction formwork is made flexible and strong along with the presence of fabrications with perfections and quite easy picking for the construction purpose. Moreover, the whole components or the elements of the MIVAN formwork system technique which is also mentioned below in detail are made out of aluminium and hence can be considered lighter in weight. The MIVAN formwork system technique is also referred to as aluminium formwork system technique as it is being manufactured of aluminium material. The span and length of the usage of the MIVAN formwork are around 250–300 times or sometimes more as calculated in a particular day [14].

4.1 *Components of MIVAN Formwork*

The main and fundamental components required for the MIVAN formwork system technique are the essential components required for the construction of the formwork technique. The dashboard is considered as the essential component. The dashboard which is initially ejected from the aluminium rail section is welded to an aluminium sheet present in the platform. Also, the dashboard is light in weight also called as lightweighted dashboard possessing an outstanding stiffness ratio as compared to the weight ratio, hence providing a clear minimum deviation under the process of concrete loading of the MIVAN formwork. The dashboards are manufactured with respect to the shape and size along with the requirement of construction project as per the designed plan [15].

These dashboards are mainly built of consisting of alloys of aluminium with minimum size of 3–4 mm thickness placed over a layer plate of around 5–6 mm thickness of the back of hard-ribbed dashboard placed in the platforms of MIVAN formwork system techniques. Moreover, these hard-ribbed dashboards are manufactured and processed in the workshops or factories. After the process of manufacturing and generating the dashboards from the workshops, these are flocked or bundled together in a trial erection format for the removal of the measurable site problems regarding the MIVAN formwork system technique [16].

5 Working Construction Process of MIVAN Formwork

The working construction process is very much similar to the traditional tunnel formwork system technique. The steps of the working construction process are being followed below.

5.1 STEP 1—Setting up the Wall Reinforcing Steel

In the first step of construction process of the MIVAN formwork, the wall reinforcing steel is used to form structure to the building of this particular project. This specific reinforcement steel of the formwork helps in giving support to almost 50% of the already designed strength of the structure. Due to these features, the steel mesh is first pre-casted in the factories or the industries before bringing it to the construction site for construction purposes [17].

5.2 STEP 2—Placing the Formwork

In the second step of the construction process, the MIVAN formwork system has already been constructed. The formwork has already been manufactured from the factories and industries. The support of the reinforcement steel is provided only at 50% designed strength of the construction project at the time of setting of concrete. Steel mesh is also provided along with the presence of concrete for providing extra amount of support at the time of construction. Moreover, like the formwork the steel meshes are pre-casted in the factories and industries for proper manufacturing before they are supplied to the construction site for construction purposes [18].

5.3 STEP 3—Pouring of the Concrete

In the third step of the working construction process of MIVAN formwork technique, the formworks are ready-made and are ready to be placed in the position for construction purpose. The pouring of the concrete is the essential process for the working of formwork. Here, freshly made concrete is brought from the industries and are poured inside the formwork by taking the already made shape of the formwork in the position. After the concrete becomes dry and gained weight, the dismantling process of the formwork takes place from the structure. Following the dismantling process, the formwork construction process is completed and already in position ready for its use in construction purpose of the project [19].

6 Break-Even Analysis of the Formwork System

Break-even analysis is a widely used technique in construction or project management for the calculations of product cost or material cost along with the presence of parameters related to finance and accounts. It is used for calculations, solving the problems related to product management and management accounts or finance. This process is basically used for categorizing or classifying the production costs and the material cost into fixed variables which are not liable to change or vary according to the estimated project data. Specifically, the break-even analysis is used for the calculation of total variable cost and the fixed cost of the construction project. By this process of calculation, the level of sales of the total and fixed costs of the construction project makes a profit or a loss is plotted in a line graph to determine and highlight the break-even point as related to the construction project [20].

6.1 Case Study

A case study has been taken, and data has also been collected for the calculation of break-even analysis of the construction project. The data for the calculation has been collected and acquired from an on-going construction site in the city of Mumbai, Maharashtra of the country. The construction project is a residential building that is being constructed in the main part of the city of Mumbai. The data collected has been divided into three portions: Part-A, Part-B and Part-C for easy and less complicated calculations that follow after the calculation of the total estimated project cost and the total fixed variable cost of the particular activities. The calculations are shown and classified on the basis of two portions mainly by using the traditional tunnel system formwork and by using MIVAN aluminium formwork system technique at the time of construction. Specifically, all the parameters are kept constant for calculations preceding the formwork types only changed [21].

As per the data collected, the total construction cost of Part-A of the live project is Rs. 17,14,50,000 with a total number of floors be 23 and total area of each floor be 600 m^2, that of Part-B the total construction project cost is Rs. 18,44,30,000 with a total number of floors be 23 also and the total area of each floor be 650 m^2 and that of Part-C the total construction project is Rs. 16,11,50,000 with a total number of floor be 23 and the total area of each floor be 540 m^2 in whole area of the construction project. The above details are being collected from the on-going construction site to calculate the break-even analysis of the formwork technique system and also comparison among them.

6.2 Cost Calculation

The cost calculation of the selected type of formwork techniques from the collected data of the on-going construction project has been calculated. The estimated cost of the project is an essential and important component for the calculation of break-even analysis of the selected type of formwork in a construction project [22]. By calculation collected from the construction project, it is stated that by using MIVAN formwork system technique, the estimated cost calculates to 16.98 lakhs for Part-A, 18.35 lakhs for Part-B and 15.5 lakhs for Part-C construction of the project along with taking into consideration the equipment cost that is being used for construction purposes as constant with an estimated value of 1.80 lakhs. Similarly, for the conventional system technique formwork, the cost calculation estimates as 10.38 for Part-A, 12.65 lakhs for Part-B and 9 lakhs for Part-C construction of the project along with taking into consideration the equipment cost as constant which is estimated around 1.3 lakhs. Lastly, for the tunnel system technique formwork, the cost calculation is estimated as maximum stated as 50.94 lakhs for Part-A, 55.05 lakhs for Part-B and 46.5 lakhs for Part-C construction of the project by taking the equipment cost as constant estimated as 6.4 lakhs for this particular type of formwork technique. This is the cost calculation collected data of the case study project for determining the break-even analysis of the construction project.

6.3 Duration Calculation

In this case, the duration or the scheduled duration of the on-going live construction project is calculated, as stated by the engineers and researchers that the calculation of scheduled duration of the construction project is an essential and quite difficult task for estimating the break-even analysis of the construction project [23]. The estimated duration is all calculated from the data collected from the live on-going construction project. As per calculations and specifically for calculating the scheduled duration of the construction project, main activities are kept constant for the formwork system technique of all the forms for calculating the break-even analysis of each formwork system technique. Firstly, for the traditional formwork system technique, the calculation estimates as 600 days for Part-A, 600 days for Part-B and also 600 days for Part-C of the construction project. Secondly, for the MIVAN formwork system technique, the calculation estimates as 432 days for Part-A, 432 days for Part-B and also 432 days for Part-C of the construction project. Finally, the tunnel steel system technique generates the minimum duration in days by estimating 87 days in all the three parts of the construction project. In this particular case of calculation of scheduled duration of the construction project, basically up to maximum of 10 common activities have been taken into consideration for the three parts of the construction project. This activity includes starting from the foundation work followed by various sub-activities of the super-structure of the building and ending with the various sub-activities of the finishing work. All these activities are constant in calculating the total scheduled duration of the project for break-even analysis of the construction project. Moreover, a pictorial representation of the calculation of the total duration of the construction project is represented.

Figure 1 concludes that the calculated scheduled duration of the project for conventional system technique formwork is 600 days, MIVAN formwork is 432 days, whereas for tunnel steel formwork system technique is 87 days, the least of all three.

Figure 2 concludes that the calculated scheduled duration of the project for conventional method is estimated maximum among the three system formwork techniques as 54% followed by MIVAN formwork technique as 39% and finally the least is the tunnel system formwork as 8% [24].

Fig. 1 Column chart

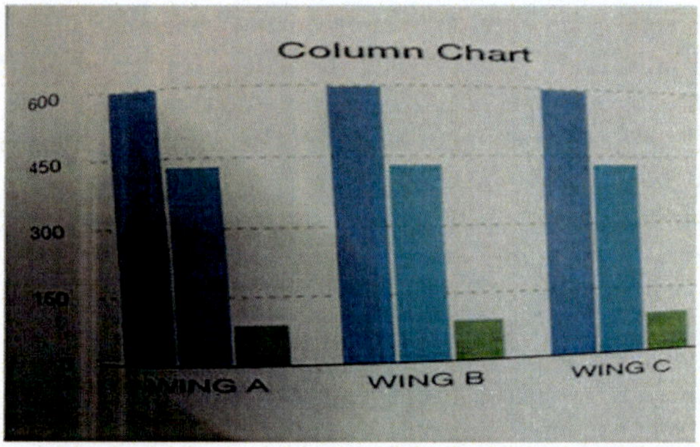

Fig. 2 Pie chart

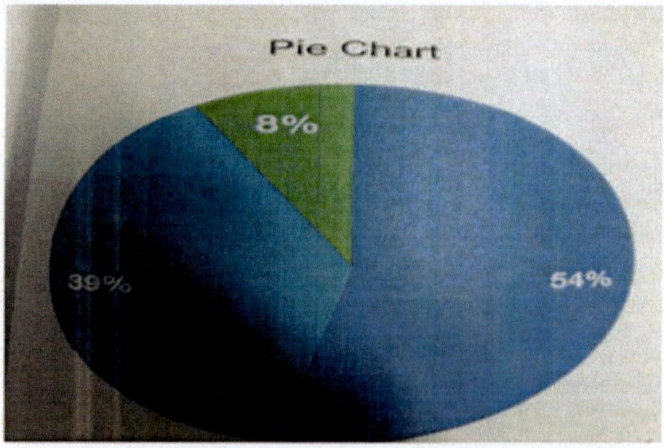

6.4 Calculation and Comparison

The calculation for denoting the break-even analysis of the three techniques of formwork for construction project has been calculated. With the collected data as well as the estimated calculation of the duration and the estimated calculation of the cost budget of the project, the break-even analysis is calculated taking all the parameters and also plotting a line graph for determining the exact break-even analysis of the formwork system technique [25]. The values are being stated individually in separate columns as stated below with the presence of line graph plotted for each particular formwork technique [26].

6.4.1 Tunnel System Formwork—Break-Even Analysis

The calculation for the break-even analysis for the tunnel system formwork technique is estimated at 268. This value is the calculated value of all the parameters that are included in the calculations naming from cost calculation of the formwork (for tunnel system formwork technique), maintenance cost of the construction project which is kept constant including the equipment cost, the scheduled time calculation of the construction project, the profit value of this particular formwork technique at the time of construction work and ending with the number of times the formwork is being used at a particular day or a section of construction work [27]. All these parameter values are included in the calculation to denote the break-even analysis of this particular formwork technique estimated at 268. Finally, a graph is plotted to determine and show pictorially the break-even analysis of this particular formwork [28] (Fig. 3).

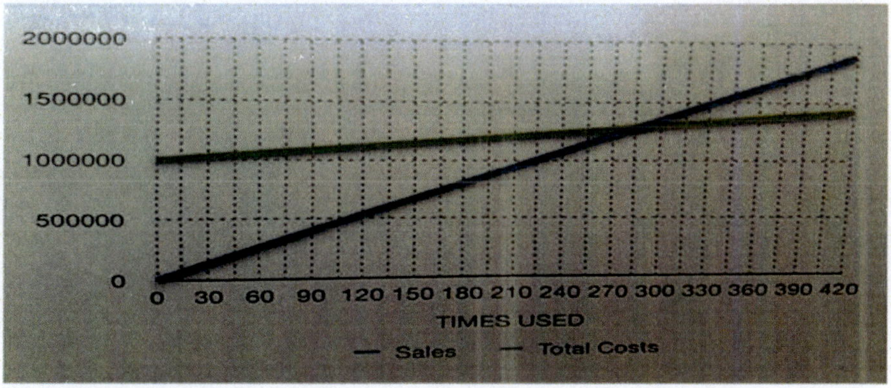

Fig. 3 Graph of tunnel system formwork

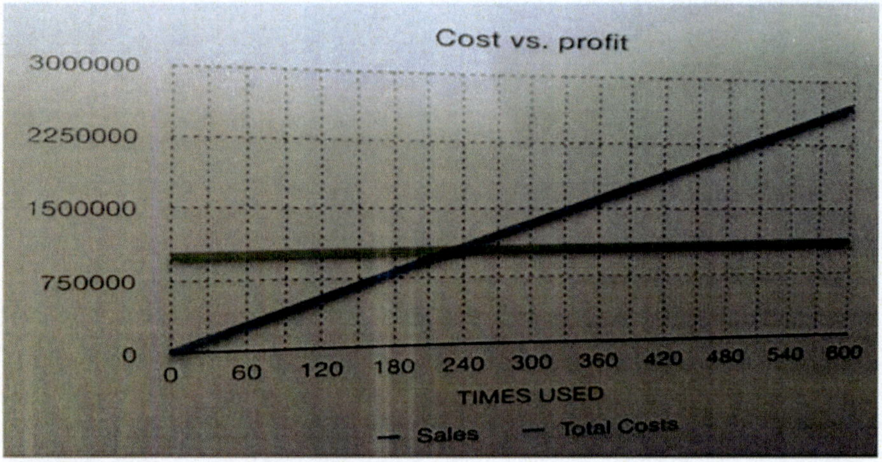

Fig. 4 Graph for MIVAN system formwork

6.4.2 MIVAN System Formwork—Break-Even Analysis

For the MIVAN or aluminium formwork system technique, the calculation for the break-even analysis is estimated at 189. This particular value for this system also includes certain parameter values which start from the values of cost calculation of MIVAN formwork system technique, the maintenance cost of the construction project which is constant and also includes the equipment cost, the scheduled time duration calculation of the formwork system technique, the profit value of this particular formwork system technique used in the working of the construction project and ending with the number of times the formwork is being used in the construction site dependent on both duration and budget of the project [29]. Taking into consideration all these parameter values, the calculated value of break-even analysis is estimated at 189. Finally, a graph is plotted for determining the break-even analysis graphically of the formwork technique system [30] (Fig. 4).

6.4.3 Traditional System Formwork—Break-Even Analysis

Finally, for the traditional system formwork technique, the break-even analysis is also calculated and the estimated value is 260. For this particular formwork, the calculation includes the values of parameters, namely, the cost calculation of this particular formwork, the maintenance cost of the construction project which is obviously constant by also including the equipment cost, the scheduled duration calculation of the construction project, the profit value of the usage of this particular formwork in the construction project and completing with the number of usage of the formwork

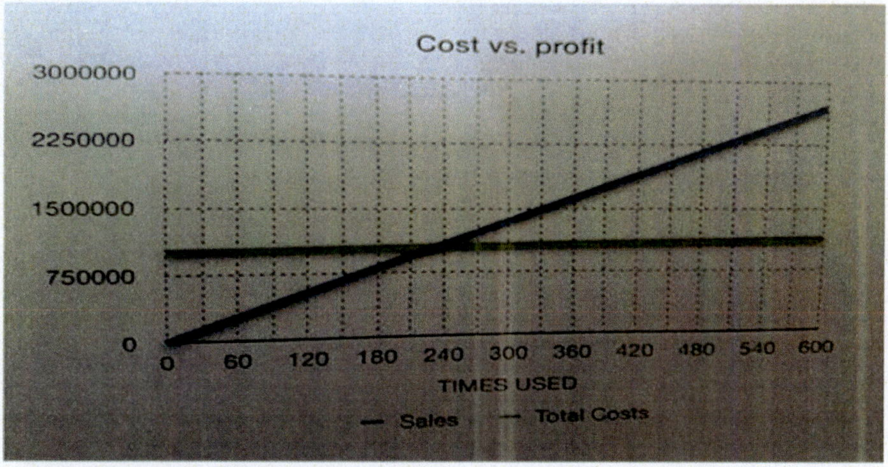

Fig. 5 Graph for traditional system formwork

technique in the construction project. Summarizing all the parameter values, the calculated value of break-even analysis is at 260. In the end, a graph is plotted to clearly determine the break-even analysis of the formwork system technique [31] (Fig. 5).

7 Discussion

The discussion of the review paper is mainly about the comparison between the two techniques of formwork that are used for construction purpose. It also discusses and compares the best use for a construction among both the techniques of formwork in terms of availability, proper usage, flexibility, quick mode of construction and most importantly the break-even analysis by taking into consideration the on-going case study of the construction project. By viewing the proposed output as obtained in the paper, it is clearly advised for the use tunnel formwork system technique for construction due to high break-even analysis value denoting and highlighting the toughness and flexibility that the formwork technique possesses at the time of construction. A survey has been taken by the engineers and the labourers or the workers highlighted the use of steel tunnel formwork system technique for construction purpose in country like India. It has also been stressed to use this technique in the country to minimize the number of accidents that occur almost daily at the time of construction in a site. This particular technique of formwork will definitely provide more safety and security for the skilled workers and labourers working at the project site. Lastly, this particular technique of formwork will improve the quality and mode of construction in the project site and can be used for construction of advanced developed structures in country India and helps in developing and sustainability process of our country. Hence, as discussed the comparison indicates and highlights the use of steel tunnel formwork system technique over aluminium MIVAN formwork system technique for construction work in construction site.

8 Conclusion

The paper on the comparison between the two modern and efficient technique of formwork with related on-going construction project states that the most of the features or the parameters that are being associated with the construction process of both technique formwork systems were originally covered at the time of construction, hence concluding that the tunnel formwork system possesses the ability of furnishing high-quality construction structure at fruitful and credible speed along with generating logical cost. This particular technique will provide a great prospective in application in the country of India with great affordable and designable residential houses or flats due to the increasing population in the country. Therefore, it is also concluded that the quality and the quantity of the construction project along with this formwork technique must be given proper reflections regarding the economical conditions of the country. Moreover, a good quality construction will never be disheartened as un-economical construction project. In this process of comparison and conclusion, it is also observed that the process of time consumption or saving of duration of the project is given to the repairing and adjustment of the poor or weak quality construction work done in the site leading to the delay of the construction work from the estimated duration of the time. It is also concluded with the help of some researchers or the experts that the on-going construction project using the formworks is preferred using lower affordable cost at any range of price.

It is concluded by taking into consideration that safety record of the construction project as safety is of utmost importance at the time of construction. The safety records of both of the formwork system techniques can be improved to a certain extent in future construction works. Many case studies of accidents occurring related to formwork system issues have been taken into serious considerations highlighting the ignorance, rapidity, quickness and carelessness of the workers at the time of construction work. Moreover, the lessons are being learned from these unfortunate incidents in the site regarding safety issues, and it is being studied and researched over and over for minimizing the accidents in the site. In conclusion, it is also quoted "As long the structures are being constructed by human beings with the use of imperfect procedures of construction and materials, maximum possibility of failures occurrence will continue. These failures will likely occur at the time of on-going construction project, thus risking the lives of the labourers and workers in the project site." Hence, the safety of the workers and labourers are of utmost importance.

From the case study taken for the calculation of break-even analysis, it is being concluded that the tunnel formwork system has a break-even point of 268, whereas the MIVAN aluminium traditional formwork system has 189. Hence, concluding that the formwork causing such repetitions or continuity gives a profit to the cost of the formwork. Moreover, the break-even analysis of traditional formwork system is 260.

References

1. Mehta M, Madan M (2013) Building construction. J Civ Eng 1–8
2. Wallace M (2008) Cast walls and slab at the same time with tunnel forms. Aberd J Group 5–8
3. ACI Committee (2001) Guide to formwork for concrete. J Farmington Hill 6–13
4. Reese CD, Eidson JV (2006) Handbook of OSHA construction safety and health. J Health Saf 181–183
5. Feld J, Carper KL (2007) Construction failure. J Health Saf 242–274
6. Kaminetzky D (2001) Design and construction failures. J Health Saf 67–78
7. McKaig T (2002) Building failures Case Stud J Civ Eng 45–56
8. Delatte NJ (2009) Beyond failure. Forensic Case Stud Civ Eng 129–155
9. Joker AS, Mhmdkary M (1999) Tunnel form concrete system. J Build Hous Res Cent 45–54
10. Stinikov V (2018) Ice formwork for high-performance concrete: a model of lean production for prefabricated concrete industry. J Struct Eng 1–8
11. Santilli A, Teixeira S, Puente I (2015) Influence of temperature and concrete reinforcement on vertical formwork design. J Constr Build Mater 188–195
12. Li M, Khelifa M, Khennane A, Ganaoui EM (2019) Structural response of cement-bonded wood composite panels as permanent formwork. J Compos Struct 13–22
13. Lopez DL, Roca P, Liew A, Mele TV, Block P (2019) Tile vaults as integrated formwork for reinforced concrete: construction, experimental testing and a method for the design and analysis of two dimensional structures. J Eng Struct 233–248
14. Spitz N, Coniglio N, Mansori ME, Montagne A, Mezghani S (2018) Quantitative and representative adherence assessment of coated and uncoated concrete-formwork. J Surf Coat Technol 247–256
15. Garg S, Nim KS, Bajpai KK, Misra S (2019) Enhancement in the quality of near surface concrete using some formwork liners. J Constr Build Mater 722–733
16. Huang BT, Li BQ, Xu HS, Li FC (2017) Development of reinforced ultra-high toughness cementitious composite permanent formwork: experimental study and digital image correlation analysis. J Compos Struct 892–903
17. Popescu M, Reiter L, Liew A, Mele TV, Flatt RJ, Block P (2018) Building in concrete with an ultra-lightweight knitted stay-in-place formwork: prototype of a concrete shell bridge. J Struct Eng 322–332
18. Liew A, Sturz YR, Guillame S, Mele TV, Smith RS, Block P (2018) Active control of a rod-net formwork system prototype. J Autom Constr 128–140
19. Lee D, Lim H, Kim T, Cho H, Kang KI (2018) Advanced planning model of formwork layout for productivity improvement in high-rise building construction. J Autom Constr 232–240
20. Goyal R, Mukherjee A, Goyal S (2016) An investigation on bond between FRP stay-in-place formwork and concrete. J Constr Build Mater 741–751
21. Hyun C, Jin C, Shen Z, Kim H (2018) Automated optimization of formwork design through spatial analysis in building information modeling. J Autom Constr 193–205
22. Mansuri D, Chakraborty D, Elzarka H, Deshpande A, Gronseth T (2017) Building information modeling enabled cascading formwork management tool. J Autom Constr 259–272
23. Piechna AK (2017) Comprehensive approach to efficient planning of formwork utilization on the construction site. J Eng Proj Prod Manag 366–372
24. Yang X, Loh P, Legget D (2016) Robotic variable fabric formwork. J Fabr Formwork 3–19
25. Drewniok MP, Cygan G, Golaszewski J (2017) Influence of the rheological properties of SCC on the formwork pressure. In: International scientific conference on sustainable, modern and safe transport, pp 124–129
26. Golafshani EM, Talatahari S (2018) Predicting the climbing rate of slip formwork systems using linear biogeography-based programming. J Appl Soft Comput 263–278
27. Pikus GA, Manzhosov IV (2017) Pressure of fiber reinforced concrete mixtures on vertical formwork panels. In: International conference on industrial engineering, ICIE 2017, pp 836–841

28. Khorsandia N, Valipour H, Bradford M (2018) Deconstructable timber-concrete composite beams with panelised slabs: finite element analysis. J Constr Build Mater 798–811
29. Miller D, Doh JH, Mulvey M (2015) Concrete slab comparison and embodied energy optimisation for alternate design and construction techniques. J Constr Build Mater 329–338
30. Adawi A, Youssef MA, Meshaly ME (2015) Experimental investigation of the composite action between hollowcore slabs with machine-cast finish and concrete topping. J Eng Struct 1–15
31. Boccadoro L, Zweidler S, Steiger R, Frangi A (2017) Calculation model to assess the structural behaviour of LVL-concrete composite members with ductile notched connection. J Eng Struct 106–117

Time Cost Optimization Using Genetic Algorithm of a Construction Project

Paromik Ray, Dillip Kumar Bera, and Ashoke Kumar Rath

Abstract Construction projects often go through delays due to various reasons, which create a dreadful financial influence on the project. For minimizing this scenario, cost and time optimization of a construction project is effectively used. Cost and time optimization method is the most effective and time efficient method with highest achievable performance under specific condition in a construction project. This method is mainly required for cost and time optimization in a construction project. This thesis work also highlights the various innovative techniques that are required for cost and time optimization of the project. Genetic Algorithm (GA) and Particle Swarm Optimization (PSO) methods are considered the advanced innovative techniques which are being used continuously by the construction companies for cost and time optimization. The advance work of Genetic Algorithm(GA) method in the form of GA with Dev-C ++ 4.9.9.2, GA with Line of Balance(LOB), GA with Modified Adaptive Weight Approach (MAWA), GA with Critical Path Method (CPM) along with new methods Linear Programming (LP), Non-Linear Integer Programming Model (NLIP), Discounted Cash Flow Method (DCF), Maximum Flow-Minimal Cut Theory and Artificial Neural Networks Method (ANN) are also included in the various innovative techniques of cost and time optimization process. Furthermore, the method of Genetic Algorithm (GA) which is specified in the thesis work is classified into two parameters where the global parallel GA method provides more effectiveness and efficiency than coarse-grained parallel GA method. Also, it is found through researchers and investigators that the Non-Linear Integer Programming (NLIP) method and Line of Balance (LOB) with GA method both have an efficient and optimum solution for time cost trade-off problem, along with PSO method which is best for Pareto-compromise solution and Direct Cash Flow (DCF) method which optimizes cost and time within the project boundaries. Finally, it is

P. Ray (✉) · D. K. Bera · A. K. Rath
School of Civil Engineering, KIIT Deemed to be University, Bhubaneswar, India
e-mail: paromikray@gmail.com

D. K. Bera
e-mail: dkbera.fce@kiit.ac.in

A. K. Rath
e-mail: akrathfce@kiit.ac.in

B. B. Das et al. (eds.), *Recent Developments in Sustainable Infrastructure*, Lecture Notes in Civil Engineering 75, https://doi.org/10.1007/978-981-15-4577-1_76

observed GA along with its advanced parameters, ANN method and NLP techniques are better for solving time cost trade-off problems.

Keywords Cost and time trade-off · Cost and time optimization · Innovative techniques · Project management

1 Introduction

According to the driven conditions of today's market, the construction industries are on the rise. All the construction companies including government and private sectors face a number of challenges during the period of construction. The main difficulties that fall during construction time of the project are the scheduled completion of project in the stipulated duration along with completion within the estimated budget of the project. The main parameters that a construction company mainly focuses upon are the time and cost of a particular construction project. Also, at the duration of construction other factors also come into consideration in the form of unpredictable factors, namely, labor performance, economical and social issues, execution errors of contractors, labor strike, design errors, labor strike, climatic changes, etc. These named factors are the majority which is responsible for the delay in the schedule of the project along with the increase in the budget of the project ultimately leading to the excess in the cost liabilities that gets added to the budget of the construction project. The problems occurring in the construction site related to these factors are referred to as the Time Cost Trade-off Problems (TCTO). For overcoming these particular situations that arise in the project, a step is being taken by the project planners and researchers for encountering this problem to process a perfect solution for optimization of TCTO-related problems in a construction project. This resulted in the development of various innovative optimization techniques of TCTO problems in a project site.

Advanced Neural Network (ANN) method, Line of Balance (LOB) method, Discounted Cash Flow (DCF) Method, Maximum Flow-Minimal Cut Theory Method, Modified Adaptive Weight Approach (MAWA) Method, Critical Path Method, and Genetic Algorithm (GA) method are some of the advanced, innovative, and widely used techniques or methods which are mainly used for generating the maximum optimized solution of time and cost parameters of a complex and large construction project. The main objective of the review paper is to highlight the Genetic Algorithm Method (GA) as the most used, easiest, and efficient optimization method or technique in recent times for optimizing TCTO problems. Moreover, a summary or a brief overview of the various innovative techniques along with the recent works done by the researchers in the field of optimization of TCTO problems by using genetic algorithm method was given. Further, it is concluded that a review is done on the genetic algorithm method being the most efficient optimization technique for solving TCTO problems for future projects and works.

2 Optimization Process

Optimization is commonly defined as the process of finding the minimum/maximum values with reference to the main derivative or the objective function for satisfying the particular constraints of a function specified within a certain set of function. The main objective or goal of the optimization process is to process the correct or best set of solution of a mathematical solution of a TCTO problem. The main dependent of TCTO (Time Cost Trade-off) problems is the critical path method of optimization process. Moreover, in a construction project the duration of the project is the most important parameter for completion in the stipulated time. Also, the increase in scheduled time, increase in cost, unavailability of resources, presence of risks, and quality quantity of the materials occurring in the construction project are the main reasons for the requirement of optimization process. The method of Genetic Algorithm (GA) is the most widely used, effective, and efficient technique for the optimization process. Lastly, many scholars and researchers have opted the method of Genetic Algorithm (GA) as the prime method for optimization of TCTO problems.

3 Genetic Algorithm: An Innovative Technique for Optimization

The method of genetic algorithm belongs to the heuristic group of optimization which is itself an advanced optimization technique of solving TCTO problems in a construction project. This method is a non-traditional search algorithm for solving both single-objective and multi-objective optimization of TCTO problems depending on the natural algorithm and natural evolution process. This method was first proposed by the scientist named Charles Darwin, namely, on the theory of "survival of the fittest." Similarly, the method of Genetic Algorithm (GA) is neither a smart nor an intelligent algorithm which clearly indicates the method is completely algorithm and mathematical based used for optimization of TCTO problems in a construction project [1]. This method consists of four main operations. At each following steps of the genetic algorithm method, these steps select the collected data as the values in a random manner. Further, the process of iteration takes place in generating an optimal solution of a TCTO problem. Moreover, the method of genetic algorithm is also dependent on the process of natural evolution. The optimized solution can be taken as an accurate solution because of the optimization process. The so-called accurate optimized solution obtained by the GA method is completely dependent on the total number of populations and on the fitness level objective function [2]. The method of GA has four main processes, namely, in a sequential order: initialization, selection, crossover, and mutation.

Initialization: This process is mainly for the initialization of the collected data values to the fitness value objective function.

Selection: This process is mainly for the selection of the optimized solution values in a random manner. Commonly this process is referred as the random selection process.
Crossover: This process is mainly used at the time of interchanging the random values position for generating the higher and correct optimized solution. This process is generating higher optimized values of a TCTO problem.
Mutation: This process is mainly used for flipping or swapping of more than one single value of the collected data values for generating a highly correct optimized solution of a TCTO problem [3] (Fig. 1).

The above flowchart is representing the main working principle of the optimization process using genetic algorithm method of a TCTO problem in a construction industry. As mentioned above, the flowchart also highlights the four main processes used for optimization process.

Fig. 1 Flowchart of genetic algorithm method

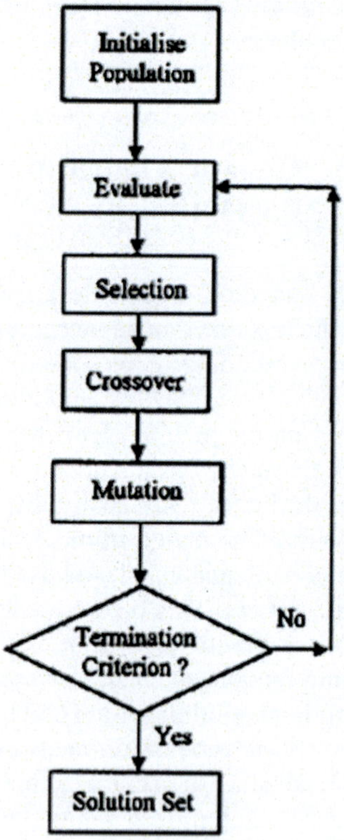

4 Working Process of Genetic Algorithm

The Genetic Algorithm method (GA) consists of four main processes which generate the output of the optimized values of the TCTO problem of a construction project.

4.1 Initialization Process

Initially, at the starting of any genetic algorithm optimization process of any mathematical problem, a single variable is being assigned from any previous coded domain or a set. Initially, this easy step of initialization process using genetic algorithm method was found insufficient by many researchers and scholars. Furthermore, it is thought that the process lacked few specific other variables for optimization of TCTO problems [4]. However, more technology and research work lead to the innovation of the optimization technique of the method. Therefore, with the incoming of new technologies and modified parameters in use, it is now very reasonable to quote the initialization process which is one of the important processes for optimization of TCTO problems [5].

4.2 Selection Process

The main selection or the process of improvement of the fitness derivative function of the main chromosomes of the population can be achieved by the process of selection. The selection process is so-called dominating process of genetic algorithm method as it possesses the ability to change the fitness value function of an optimization problem [6]. In this process, the parents are being selected from there groups of chromosomes population of the data collected. The process of selection of the genetic algorithm method is performed in such a way that the chromosomes having a higher fitness value are having a maximum probability to be selected as the main parents of the fitness value function to be used in the optimization process [6].

Coming to the ultimate conclusion that some of the population of the chromosomes from the collected data of the optimization problem contain some of high value fitness derivative function than the original derivative function which is the main formula for creating individual chromosomes. Ultimately, this process possesses a limitation of a lack of population diversity which leads to a lag in search algorithms in optimization process [7].

4.3 Crossover Process

After the completion of both the process, initialization and selection process, the third process is the crossover process. In this process, the derivative function parameters used in the genetic algorithm method create a possible two chromosomes in this process. These parameters of the GA method use a random generator function for maintaining the two set of functions of the domain or a set for the optimization process. This results in the formation of two chromosomes in the optimization process [8]. After the completion of the crossover process, the new set of chromosomes is being added to the set of previous chromosomes set to form a new population.

4.4 Mutation Process

The final process of the genetic algorithm is the mutation process. The main function of the process includes the increasing the population diversity of the chromosomes. This process also includes the selection of the inner particles of chromosomes called genes in a random selection manner and then assigning those selected values within the corresponding intervals of pre-defined variables of selected set or domain [9]. That chromosome which follows the method of mutation contains a correct genetic code. The process of mutation enables these created genetic codes are kept intact and does not get destroyed. Moreover, a non-uniform mutation can also be applied for the optimization process of a TCTO problem which leads to the shortening or narrowing of the chromosomes population created for the optimization process of TCTO problem [10].

5 Case Study: A Live Construction Project

The case study includes a live ongoing construction project named "Bose Institute Construction Project" located at the city of Kolkata, West Bengal, India. The construction project includes the construction of G + 15 storied educational or an institutional building that is situated in the main town of Salt Lake City of Rajarhat District in the city of Kolkata. The project consists of full-furnished construction of a 15-storied buildings along with the formation of the basement. The main work of the construction project consists of the main educational building of 15 storied along with two-side building consisting of 5-storied floor each. The main features and the requirement of the construction project are as follows:

Maintaining and ensuring the life-time structural ability of the educational building being constructed for duration of 15 years.

Ensuring proper and efficient construction of 15-storied educational building along with the two buildings of 5-storied each of the construction projects.

Decrease in the negative effects created in the environment and construction of an eco-friendly building structure is making positive impact in the environment [11].

As per the construction point of view of the project, there are present around 65–70 activities throughout the project. It is quite difficult and near impossible for optimization of all the activities using genetic algorithm method. It has been advised to select a section or a separate small plan of the construction project for optimization using genetic algorithm method. Thus, for this review paper, a particular section of the construction project has been selected for the optimization using GA method.

5.1 *Selection of the Activities*

The selected activities for the optimization using GA method consist of the main important 10 to 12 activities from the part of ground floor of the building to the plinth level of the first floor of the ongoing construction project. Other parameters are also included in a tabular form consisting of the duration and the estimated cost of the particular activity selected. The requirement of the cost and duration is much more important for the optimization process using GA method creating a more positive impact and efficiency in the process [12]. The selected activities along with the duration and time are given below (Tables 1 and 2).

Table 1 List of activities, duration, cost (I)

Activities	Original duration	Value of the activity
A. Reinforcement work of column	3 days	1,09, 145
B. Column shuttering work	3 days	33,750
C. Column concreting work	4 days	90,813.25
D. Column de-shuttering work	4 days	7,057.50
E. Staging and shuttering work of beam and slab	3 days	36, 572.60
F. Fabricating, placing, and fixing of beam and slab	4 days	2,44,658

Table 2 List of activities, duration, cost (II)

Activities	Original duration	Value of the activity
G. Concreting work at ground floor with roof slab	3 days	1,71, 270
H. Curing process	14 days	2,60, 134
I. Removal of shutter staging materials	2 days	1,48,648
J. Brickwork (main floor along with partition wall)	4 days	2,60,734

5.2 Deriving the Fitness Function

This particular construction project consists of the applications of the TCTO problem for optimization process. The derivative function will comprise the main parameters including the total duration and the estimated cost of all the activities [13]. The derivative function includes the optimization of the case study of a TCTO problem:

$$C = dc + ic - (id - q)ie + \sum_{\forall y} (nd(y) - pop(x, y))cs(y) \tag{1}$$

where C = total cost; dc = direct cost; ic = indirect cost = i.e. X id; id = initial duration of the network or duration of the network with normal cost; q = duration of the network after change in duration of the activity; i.e. = indirect expenses per day; nd = normal duration of the activity; pop = matrix of randomly generated durations as per our boundary conditions; cs = cost slope of the activities; x = population index; and y = activity whose duration is being changed [11] [14].

The result obtained using the MATLAB 2018 software and a clear comparison of the result obtained is also being reviewed in the paper highlighting the importance of using genetic algorithm method for optimization process. By using the MATLAB 2018 software, an efficient and a clear comparison of the two outputs is obtained with the aid of bar graphs that are being generated from the software. With the help of bar chart generated, it becomes quite easy for processing the optimized solution of the problem [15].

5.3 Selection Process

Following the process of derivation or deriving the derivative function as mentioned above, the process of encoding and selection process takes place. In this particular process of selection, the random selection of any selected mentioned activities takes place. With the random selection of any activities, the values of cost and duration at a single selection are taken and are processed or generated in the derivative function for obtaining the optimized output of the TCTO problem by the method of genetic algorithm (GA). Previously, before proceeding with the method of selection process, the process of encoding takes place. This encoding process is not a parameter or process under the category of Genetic Algorithm (GA) method but it leads to the process of optimization of TCTO problem [16]. In this process of encoding, the scheduled time and the estimated duration of the activities as obtained from the construction project are being transformed into coded language in the form of binary digits which include 0 and 1 digit. The binary digits in computing language are termed in 0 and 1 digit. Following the transforming of the numerical digit into coded language, the selection process is utilized for optimization of TCTO problem [17]. For the process of selection, the process of Roulette wheel section selection method is used. This method of selection is efficient, effective, and gives easy access to optimized output for this particular optimized TCTO problem. The method of Roulette wheel section selection consists of top-most part or area of the wheel is being divided into various N (a specific number) portions for obtaining the derivative or the fitness function of the TCTO problem. This wheel is rotated in a particular specific direction which can be clockwise or anti-clockwise depending on the choice of user or person using the method. A fixed pointer is present for denoting the optimized output that is obtained by using the method of Genetic Algorithm (GA). Moreover, for the process of selection using this method, the particular area of the activity is selected where the optimized output will be obtained. The Roulette wheel section selection is determined pictorially using pie-chart graph. A tabular column and a pie-chart graph are given below to determine the optimized output of the TCTO problem using Genetic Algorithm (GA) method [18] (Tables 3 and 4 and Fig. 2).

Table 3 Selection process of the activities (I)

Activity no.	Total duration	Binary coding	Fitness function value	Probability calculation	Cumulative probability	Fitness value
A	3 days	110000	1000	0.051	0.051	0.512
B	3 days	110000	2200	0.011	0.062	1.128
C	4 days	001000	1500	0.076	0.138	0.7692
D	4 days	001000	4000	0.020	0.158	2.0512
E	3 days	110000	1800	0.092	0.250	0.92307

Table 4 Selection process of activities (II)

Activity no.	Total duration	Binary coding	Fitness function value	Probability calculation	Cumulative probability	Fitness value
F	4 days	001000	1100	0.056	0.306	0.5641
G	3 days	110000	2200	0.11	0.416	1.128
H	14 days	0011000	2100	0.10	0.516	1.07692
I	2 days	010000	2100	0.10	0.616	1.07692
J	4 days	001000	1500	0.76	1.00	0.7692
Total = 44 days			**Total = 19500 Avg = 1950**	**Total = 1.00**		

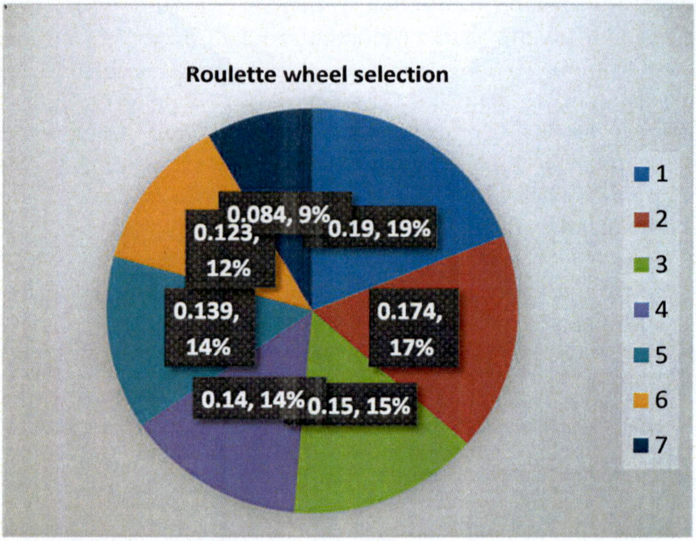

Fig. 2 Pie chart of roulette wheel section

5.4 Crossover Process

With the completion of the two main processes of the Genetic Algorithm (GA) method—initialization and selection, the third process, namely, crossover process is being used for the optimization process of TCTO problem using genetic algorithm method. This process method creates a new population containing chromosomes ad genes for obtaining the optimized output of TCTO problem using Genetic Algorithm (GA) method. A new random function is being used in this process whose main function is to minimize; according to the situation of the project, two sets of function from the derivative function for the optimization process [19] is created. The optimized outputs are generated from the two sets of chromosomes of population created from this crossover process. The new set of output in the form of chromosomes or genes generates or processes the optimized output of the TCTO problems. Moreover, for obtaining the optimized output of the crossover process, the single crossover process is used. The method of crossover can be applied or utilized by different other processes. In other words, there are several classifications for applying the process of crossover in solving TCTO problem using genetic algorithm method. The single crossover process is the most effective, easy-to-use, efficient, and significant output obtaining method for solving the TCTO problem using genetic algorithm method [20]. The detailed process of crossover process used in the solving TCTO-related problem is explained below with related calculations and also with presence of tabular columns (Fig. 3 and Tables 5 and 6).

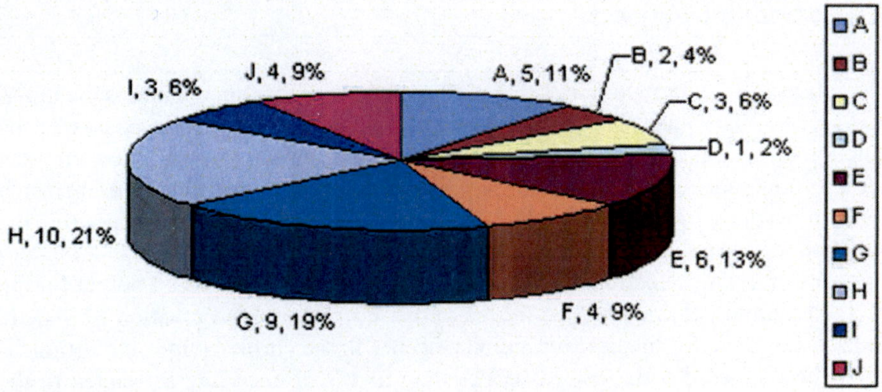

Fig. 3 Pie chart representing crossover process of the activities

Table 5 Crossover process for the activities (I)

Old binary encoding	Random selection	Cross-over random selection	New population	Value of the new population	Function
110000	A	E	110000	3 days	800
110000	B	D	001000	4 days	1000
001000	C	C	001000	4 days	2000
001000	D	B	110000	3 days	3500
110000	E	A	11000	3 days	1500

Table 6 Crossover for the activities (II)

Old binary encoding	Random selection	Cross over random selection	New population	Valne of the new population	Function
001000	F	J	001000	4 days	1000
110000	G	I	010000	2 days	2000
001100	H	H	001100	12 days **(maximum gets mutated)**	1500
010000	I	G	110000	3 days	1500
001000	J	F	001000	4 days	1200
				Total = 42 days	**Total = 16,000**

5.5 Mutation Process

The fourth and final process in the method of genetic algorithm method is the mutation process. The final process of genetic algorithm method mainly possesses the function of adding or generating more or maximum number of population from the chromosomes and genes. The inner or interior micro-genes or particles present inside the chromosomes increase in number as the mutation process takes place. During the mutation process, the microparticles in the form or genes are selected in a random manner or random selection takes place for generating the optimized output for the TCTO problem [21]. Moreover, in this particular process, the creation of genetic codes takes place for the new formation of genes in the chromosomes for optimization. These genetic codes are related to the process of encoding as related to the selection process. These genetic codes are very much strong and intact and are not easily destroyed by any external means or any other processes. The creation of gene microparticles is mainly due to the relation of encoding process to the mutation process [22]. However, the process of mutation can be very long-lasting or time-taking. Hence, for minimizing or shortening the mutation process another process of mutation can be applied which is rarely used. This is the non-uniform mutation process mainly for shortening or narrowing the number of chromosomes for the generation of correct and specific optimized output for solving the TCTO problems. This non-uniform mutation process can be only used at the particular time when the condition of the problem suggests the formation of maximum optimized output or maximum duration being taken for generation of output. This particular mutation process is used once at a time for optimization process [23].

The process of mutation in the method of genetic algorithm is mainly used for obtaining the correct, efficient, and perfect optimized output of the TCTO problem. Also, the process also acts as an aid for maintaining the perfect balance of the chromosomes formed along with the chromosomes already present intact in the optimization process for solving TCTO problem [24]. Moreover, addition of chromosomes along with presence of genes or minor particles inside brings an extra and more specified output to the list of optimized output from the previous processes for the optimization process. Since the process of encoding and mutation is performed parallel to each other, the binary digits are used for generating the optimized output. The process of mutation possesses the feature of conversion of binary digit into its opposite number for solving the TCTO problem using Genetic Algorithm (GA) method. Specifically, the binary conversion involves the conversion of digit 0 to its opposite number 1 and vice versa [25]. Moreover, a mutation probability function is formed at the time of performing the operation of mutation for generating the optimized output for solving the TCTO problem [24]. The process of mutation like other processes also has classifications. The process is classified into four categories. The binary-coded mutation GA process is mostly specified and used for optimization of TCTO problem, since its ability to generate specific, efficient, and to the point optimized output for TCTO problem [26]. Lastly, the graphs and tabular columns are mentioned below for detailed and pictorial representation of the optimized solution (Tables 7 and 8).

Table 7 Mutation process for the activities (I)

Old binary encoding	Random selection	Cross over random selection	New population	Valne of the new population	Function
110000	A	J	010000	2 days	500
110000	B	I	110000	3 days	800
001000	C	H	110000	3 days	1500
001000	D	G	110000	2 days	2500
110000	E	F	010000	2 days	1000

Table 8 Mutation process for the activities (II)

Old binary encoding	Random selection	Cross-over random selection	New population	Value of the new population	Function
001000	F	E	110000	3 days	800
110000	G	D	100000	1 days	1500
001100	H	C	111000	10 days **(maximum gets mutated)**	1200
010000	I	B	010000	2 days	1200
001000	J	A	110000	3 days	1000
				Total = 31 days	**Total = 12,000**

6 Optimized Output

After the completion of the four main processes related to the method of genetic algorithm for solving TCTO problem of a case-study project as well as deriving the derivative function or the fitness value function, the optimized output has to be obtained using various sources with the help of software, coding programs, optimization tools, and binary-coded values [27]. With the help of these particular sources or options, the optimized output is obtained using the method of processes related to GA method for solving TCTO problems. The process of obtaining the optimized output related to software purpose is the most effective, efficient, and correctly specified optimized output for solving TCTO problems. The software used for obtaining the optimized output is by using MATLAB 2018 software [28]. This particular software with the help of certain optimization tools present in the software helps in generation of the optimized output for the TCTO problem. Moreover, many researchers and scholars have opted for utilizing the software MATLAB 2018 for processing the optimized output using GA method [29]. Also, the optimized output can also be obtained with the help of coding programs or using binary coding encoding value process. This process generates the optimized output with the presence of coded programs, conversion of binary digits to encoded values, and also with the help of already generated optimized codes available [30]. These already available optimized coded values or coded programs are only generated for solving problems related to optimization process using genetic algorithm method. This particular method of generation of optimized output using coded programs is less advanced with respect to invention of recent technologies and advanced software available. Moreover, the disadvantage possessed of taking number of man-hours and duration for generation of the optimized output [31]. So, this particular process is less or used minimum in advanced world of technology. Lastly, the advanced and modern technological technique includes the usage of various optimization tools with the aid of software or any other mode of particular optimization software for obtaining the optimized output of a TCTO problem using genetic algorithm method [32]. This particular mode of obtaining the optimized output is less laborious, more advanced, more effective, efficient, and less outcome of huge difference of optimized results. Apart from using MATLAB 2018 software, which is due to recent modern invention has become common mode of optimization tool software for the researchers and scholars. Moreover, MATLAB 2018 software is also being used in the construction companies or live ongoing project for the optimization of TCTO of the construction project [33]. Other optimization tools or software which possess advanced technologies for performing optimization process are CPLEX, IPOPT, Artelys Kniko, MOSEK, TOMLAB, LINDO, etc. [34]. Other than genetic algorithm method, other simple and less-complex techniques are available for optimization process in the mentioned software. Generally, the researchers and scholars prefer using MATLAB 2018 software for performing optimization process to perfection (Figs. 4 and 5).

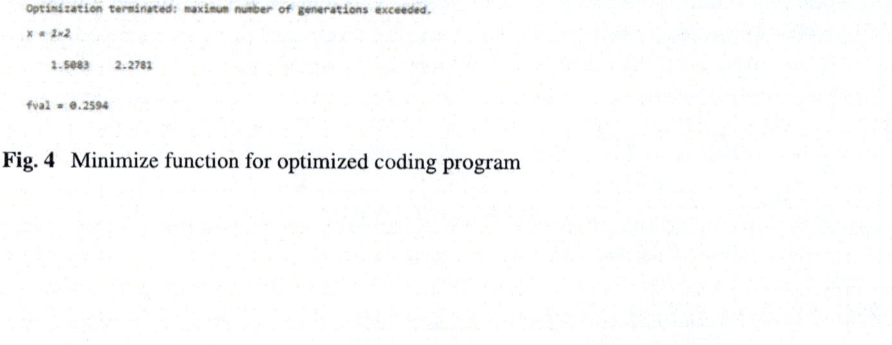

Fig. 4 Minimize function for optimized coding program

Examine the first 10 members of the final population and their corresponding scores. Notice that x(1) is integer-valued for all these population members. The integer ga algorithm generates only integer-feasible populations.

```
disp(population(1:10,:))
```

```
 -5.0000   -0.0000
 -5.0000   -0.0000
 -5.0000    0.0014
 -6.0000    0.0008
-13.0000   -0.0124
-10.0000    0.0011
 -4.0000   -0.0010
       0    0.0072
 -4.0000    0.0010
 -5.0000   -0.0000
```

```
disp(scores(1:10))
```

```
-1.9178
-1.9178
-1.9165
 1.0008
64.0124
25.0011
-1.5126
 2.5072
-1.5126
-1.9178
```

Fig. 5 Optimized output obtained using coding program

7 Discussion

The discussion is mainly about the comparison of the two main optimized outputs or results that are obtained by the optimization process using Genetic Algorithm (GA) method. From the above model or the graph shown below, it can be discussed that the optimized value or the output obtained is a satisfactory one. The proposed output or the optimized output obtained for the construction project using Genetic Algorithm (GA) is able to achieve an increase in (10–20)% than the estimated data available for the project. As per the output obtained and by verifying the output with the attached graphs and MATLAB software, the total duration of the optimized result showed an increase in 10 percent whereas the total estimated budget of the construction project showed an increase in 20 percent of the estimated data collected. The optimized output showed a great percent of similarity with the estimated data collected from the construction sector or industry, but it is obvious that the increase in cost and time of the project is beneficial for the completion of project. The difference may arise of the reason of not more than four to five activities, which is the main reason for the increase in the estimated cost and total duration of the project [34] (Figs. 6 and 7).

8 Conclusion

It is concluded that the obtained results are being satisfactory. The objective of the construction involving a TCTO problem was a multi-objective genetic algorithm. Even, the single-objective TCTO problem of a construction project also provided correct and satisfactory optimized results of cost and time using genetic algorithm

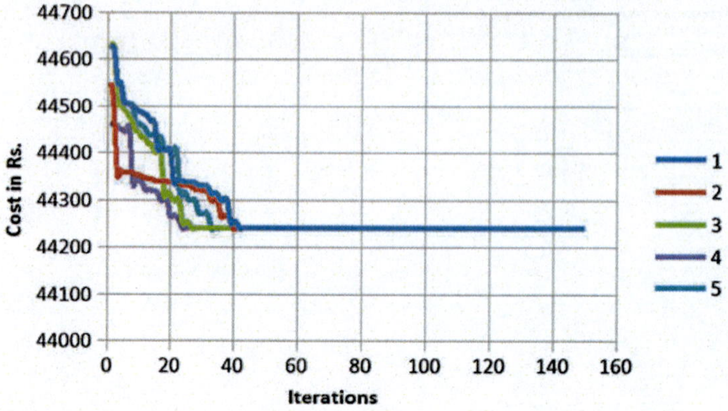

Fig. 6 Cost optimized output in MATLAB 2018 software

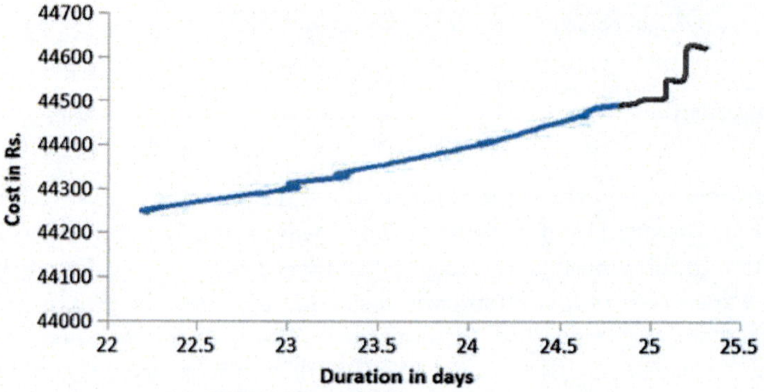

Fig. 7 Time optimized output in MATLAB 2018 software

method comprising of using high technology optimized algorithms and other rapidly developed formulas comprising algorithms for processing the optimized result of cost and time of a TCTO problem. The increase in the development of the multi-objective optimization problem using genetic algorithm method is on a major high demand as it also possesses a feature of comparison of two ongoing construction projects, thus highlighting the superiority and domination of the genetic algorithm functions on the optimization of TCTO problems. As it is clearly shown in the above case studies, an increase in 10 percent of the optimized results is obtained than the collected data of the construction projects. The estimated cost of the construction project is around 400 crores INR along with the estimated duration of the construction project is 5 years. By optimizing the selected data collected from the companies file, the total estimated duration has come around in a span of 4 years, but with an increase of the cost of the project as it is known that both the parameters are independent of each other. It can be still debated whether using genetic algorithm method for optimization is the perfect choice in the future or not, but as per reports and concentrating on the comparison of the results obtained, GA method is an effective and efficient method for optimization process of TCTO problem. It is easily used and less laborious method as compared to the various innovative optimization technique used. Upon the requirements of the planners or the designers of the construction project, the genetic algorithm method can be used as a better option for optimization of TCTO problems [35].

Obviously, taking into considerations the experience possessed by the researchers and the planners for the optimization process, genetic algorithm might lack the extra edge of being used in a TCTO problem because of the various advanced innovative techniques that are being used in construction projects for optimization process [36]. Moreover, the assurance of the quality of the work is also a major highlight of the construction project and it is mainly dependent on the work experience of the planners and the designers involved in it. The use of genetic algorithm method for such particular scenario is completely dependent on the different subjectivity processes of the project along with different advanced characteristics of the operator being used for the project. However, the main intention lies in providing the proper method of quality assurance of the work done along with generation of total innovative quality of work done in the project [37]. It is thus concluded from the comparison of the case studies that GA method is one of the innovative techniques for optimization of TCTO problem of a construction project. The following features are the proof for the advancement of the genetic algorithm:

- Genetic algorithm method works with the process of coding of a parameter set, not with a particular parameter.
- Genetic algorithm method searches the main population of the chromosomes rather than a particular population of the chromosomes mainly for global exploration of the project purposes.
- Genetic algorithm method mainly uses the objective function or the derivative function rather than using any complex functions or integers for optimization purpose.
- Genetic algorithm method can use a random choice function for the mechanism of the optimization process acting as a guide for searching the specific algorithm for the improvement and advancement of the derivative or the fitness function.
- Genetic algorithm method also improves the process of easy and simple experimentation for different scenarios of the optimization problem [38].

The various features act as an aid for the researchers and scholars for using genetic algorithm method for optimization of any type of problem [39].

References

1. Ashuri B, Tavakolan M (2015) Shuffled frog–leaping model for solving time–cost- resource optimization problems in construction project planning. J Comput Civil Eng 1–11

2. Bhosale V, Shastri SS, Khandare MA (2017) A review of genetic algorithm used for optimizing scheduling of resource constraint construction projects. In International Research Journal of Engineering and Technology (IRJET) Publications, (2017), 2869–2872
3. Khan Md, Haque A, Hasina AMd (2012) Genetic algorithm for project time-cost optimization in fuzzy environment. J Ind Eng Manage 365–381
4. Todowski M, Rajicic D (2016) An initialization procedure in solving optimal power flow by genetic algorithm. IEEE Trans Power Syst 480–487
5. Dommel HW, Tinney WF (2008) Optimal power flow solution. IEEE Trans Power App Syst 1866–187
6. Michaelwicz Z (2009) Genetic algorithms + data structures = evolution programs. Springer-Verlag, Berlin, Germany, pp 111–112
7. Walters DC, Sheble GB (2003) Genetic algorithm solution of economic dispatch with valve point loading. IEEE Trans Power Syst 1325–1332
8. Sheble GB, Brittig K (2005) Refined genetic algorithm—economic dispatch example. IEEE Trans Power Syst 117–124
9. Chen PH, Chang HC (2005) Large-scale economic dispatch by genetic algorithm. IEEE Trans Power Syst 1919–192
10. Bakirtzis AG, Biskas PN, Zoumas CE, Petridis V (2002) Optimal power flow by enhanced genetic algorithm. IEEE Trans Power Syst 229–233
11. Sorrentino M (2013) Construction time-cost-quality trade- off: a road project case study. J Constr Manage Publ 163–176
12. Ramon SJC (2009) Time, cost and quality in a road building construction. J Constr Eng Manage 1271–1274
13. Kandil A, El-Rayes K (2006) MACROS: multiobjective automated construction resource optimization system. J Manage Eng 126–134
14. El-Rayes K, Kandil A (2005) Time-cost-quality trade-off analysis for highway construction. J Constr Eng Manage 477–486
15. Zheng XM, Ng T, Kumarswamy M (2004) Applying a genetic algorithm-based multi-objective approach for time-cost optimization. J Constr Eng Manage 168–176
16. Bragadin MA, Kahkonen K (2013) Quality evaluation of construction activities for project control. J Front Constr Eng 17–24
17. Kaiafa S, Chassiakos AP (2015) A genetic algorithm for optimal resource-driven project scheduling. Creat Constr Conf 260–267
18. Alcaraz A, Maroto C (2001) A robust genetic algorithm for resource allocation in project scheduling. Kluwer Academic Publications pp 83–109
19. Yu W, Li B, Ji H, Zhang M, Wang D (2015) Application of multi-objective genetic algorithm to optimize energy efficiency and thermal comfort in building design. J Energy Build 135–143
20. Wang W, Zmeureanu R, Rivard H (2005) Applying multi-objective genetic algorithms in green building design optimization. J Build Environ 1512–1525
21. Dubrow DT, Krarti M (2010) Genetic-algorithm based approach to optimize building envelope design for residential buildings. J Build Environ 1574–1581
22. Agrama AF (2014) Multi-objective genetic optimization for scheduling a multi-storey building. J Automat Constr 119–128
23. Magnier L, Haghighat F (2010) Multiobjective optimization of building design using TRNSYS simulations, genetic algorithm, and Artificial Neural Network. J Build Environ 739–746
24. Li H, Love P(1997) Using improved genetic algorithms to facilitate time-cost optimization. J Constr Eng Manage 233–237
25. Feng CH, Liu L, Burns SA (1997) Using genetic algorithms to solve construction time-cost trade-off problems. J Constr Eng Manage 184–189
26. Shaghagi AR, Markose S (2013) A genetic algorithm (GA) for financial derivatives network construction: A stochastic multi-objective optimization problem. Centre Comput Finance Econ Agents 1–19
27. Sonmez R, Bettemir OH (2012) A hybrid genetic algorithm for the discrete time–cost trade-off problem. J Exp Syst Appl 11428–11434

28. Aminbaksh S, Sonmez R (2016) Discrete particle swarm optimization method for the large-scale discrete time–cost trade-off problem. J Exp Syst Appl 177–185
29. Mongahesmi A, Nikoo MR, Fasaee MAK, Adamowski J (2015)A novel multi criteria decision making model for optimizing time–cost–quality trade-off problems in construction projects. J Exp Syst Appl 3089–3104
30. Saif A, Abbas S, Fayed Z (2015) The PDBO algorithm for discrete time, cost and quality trade—off in software projects with expressing quality by defects. In International Conference on Communication, Management and Information Technology (ICCMIT 2015),pp 930–939
31. Afruzi EN, Najafi AA, Roghanian E, Mazinani M (2014) A multi-objective imperialist competitive algorithm for solving discrete time, cost and quality trade-off problems with mode-identity and resource-constrained situations. J Comput Oper Res 80–96
32. Tavana M, Abtahi AR, Damghani KK (2014) A new multi-objective multi-mode model for solving preemptive time–cost–quality trade-off project scheduling problems. J Exp Syst Appl 1830–1846
33. Eeckhout MVD, Maenhout B, Vanhoucke M (2019) A heuristic procedure to solve the project staffing problem with discrete time/resource trade-offs and personnel scheduling constraints. J Comput Oper Res 144–161
34. Viana A, Sousa JPD (2000) Using metaheuristics in multiobjective resource constrained project scheduling. Eur J Oper Res 359–374
35. Mesejo P, Ibanez O, Cordon O, Cagnoni S (2016) A survey on image segmentation using metaheuristic-based deformable models: state of the art and critical analysis. J Appl Soft Comput 1–29
36. Daisy XMZ, Thomas SN, Mohan MK (2004) Applying genetic algorithm techniques for time-cost optimization. Depart Civil Eng 168–176
37. Bhosale V Shastri SS, Khandare MA (2016) A review of genetic algorithm used for optimizing scheduling of resource constraint construction projects. In International Research Journal of Engineering and Technology (IRJET) Publications, (2016), 2869–2872
38. Shankar LB, Basavarajappa S, Chen CHJ, Kadadevaramath SR (2013) Location and allocation decisions for multi-echelon supply chain network—A multi-objective evolutionary approach. J Exp Syst Appl 551–562
39. Coello CC (2000) An updated survey of GA-based multiobjective optimization techniques. ACM Comput Surve 109–143

Sustainable Infrastructures (Materials & Management)—Planning for Sustainable Infrastructure in Old Town, Bhubaneswar

Sukanya Dasgupta and Nilanjana Roy

Abstract Bhubaneswar, being the first smart city of India has already witnessed the initial application of the smart mobility methods in its infrastructure by the transition of traditional transportation techniques to a sustainable and digitally enhanced system. The paper primarily includes an extensive study of important criteria such as the pedestrian facilities, safety, accessibility and aesthetic appearance [3]. The survey is based on the principles of H. Krambeck's 'Global Walkability index' that was developed for the World Bank. The methodology gives information to several key parameters and identifies the core areas for improvement. The study aims to suggest the walkability values, principles of inclusive and smart mobility and a qualitative analysis of the Old Town Area of Bhuvaneshwar around Bindusagar Lake and other temple complexes including the Lingaraja Temple. With dense and unplanned neighbourhood development, congested roads ad-hock marketplaces, archaic storm water drainage and poor waste management in and around the historic structures, this part of the city remains untouched by recent infrastructure developments. This paper aims at providing sustainable solutions by proposing retrofitting measures to the existing infrastructure. A walkability audit followed up with recommendations has been made, incorporating the latest trends in sustainable infrastructure.

Keywords Traffic management · Walkability · Smart and inclusive mobility · Waste management · Storm water management

S. Dasgupta (✉) · N. Roy
KIIT (DU), Bhubaneswar, Odisha, India
e-mail: sukanya.dasguptafar@kiit.ac.in

N. Roy
e-mail: nilan.roy85@gmail.com

B. B. Das et al. (eds.), *Recent Developments in Sustainable Infrastructure*, Lecture Notes in Civil Engineering 75, https://doi.org/10.1007/978-981-15-4577-1_77

1 Introduction

1.1 Background

Sustainable infrastructure from the perspective of an urban designer or planner is not restricting short-term provision but creating a systematically planned and designed infrastructure that is responsible for its own operational requirements and effects. It has to be sustainable if it is to meet the needs of the future. New facilities and policies are not the only aspects of Sustainable infrastructure design. Based on concepts of Global Sustainable development and Urban sustainability, it also includes adaptation, optimization and efficient reuse of existing infrastructure. This also involves renewal and sometimes redesign .of infrastructure, its detailed future economic projection and analysis, conservation of energy and existing infrastructure, reducing the costs, selection of materials based on its quality, durability and sustainability and reduced and efficient waste management taking into account the Global Climate change and the reversal of negative impacts on the damaged environment [4].

Streets that have been planned for only cars are hazardous for others as it can lead to obesity epidemic due to the lack of space for pedestrians that lowers the daily physical activity in children and adults. On the other hand, 'complete streets' are comparatively more safe and convenient for not just vehicles but also cyclists, pedestrians and people with disabilities [1]. Pedestrianisation and walkability is a significant concept in sustainable urban design approach. Apart from the community and individual health benefits, it also leads to improved social interaction, increase in civic sense and responsibility and also helps in reducing crime. The most important outcome of pedestrianisation is the decrease in vehicular footprint in the communities.

In countries like India, public transport and walking have been a way of life for many. The government has been striving to promote the use of public transit systems. In present times, projects like BRTS/MRTS are a common intervention [2]. However, on the contrary, the increase of private cars and other vehicles on the road has also been witnessed in the recent past. The new flyovers, road widening proposals have created the paradox and have been the catalyst that discourages pedestrian movement [1]. To make matters worse, the footpaths that have been created for pedestrian movements are sometimes used by vehicles compromising pedestrian's safety and convenience.

1.2 Study Area

The city of Bhubaneswar houses 25 km^2 of Old Town area which dates more than a thousand years. In addition to the numerous temples from the Swarna Jaleswar (sixth century A.D.), Vaital and Sirsewara group of temples (700–900A.D.), Brahmeswar group of temples (960–1065 A.D.), Mukteswar group of temples (950–975 A.D.), Rajarani (950–975 A.D.), Lingaraja temple (eleventh century A.D.), Kedareswar group of temples (twelfth century A.D.) there exists many ponds and water tanks in the old town area [4]. Bindusagar Lake situated in the north of Lingaraja Temple is the most famous. The Old Bhubaneswar is located in Zone 18. It is bound by the railway line to the North and West, agricultural fields to the south. The old town stretches along the Tankapani road as a ribbon development with the Rajarani temple complex located along this road. The Old Town Area is composed of congested roads, crowded marketplaces, waterbodies, storm water drains and a larger number of bushes and mounds of solid waste in and around the historic structures. However, the missing essence of the religious precinct is due to the sound of the unmanned moving automobiles. The action area has been illustrated in Fig. 8.

1.2.1 Who's Town Is It Anyway—People or Vehicles?

Otto Königsberger planned Bhubaneswar in 1948, on small neighbourhood unit concept, which would preserve the sense of community. The city is earmarked by distinct landuse pattern, i.e. residential, commercial and institutional, etc. based on a grid-iron road network. However, the old town is based on the 'ashta ayatana' concept. As per the medieval texts, a concept of 'ashta ayatana'—eight sacred precincts each with its temples and prescribed pilgrim routes are ritualistically and symbolically connected to the Lingaraj Temple [4]. Taking reference from the planning concept of the old town which necessarily advocates routes for pilgrims and overlaying it on the current infrastructure as discussed earlier in the paper, the walkability conditions of the pedestrians in this zone needs immediate attention. However, present traffic planning of these precincts does not support easy and efficient pedestrian movement.

2 Literature Review

2.1 Evaluating Walkability

Many subjective factors govern walkability and hence it is rather challenging. Principles of inclusive mobility include (MMDA, 2014).

- A transport system that works for the poor and the vulnerable
- A walkable, bikeable and accessible city
- Moving people, not vehicles
- Mobility with safe and civility
- Planning and communicating better and travel less
- Sharing information to increase connectivity and accessibility
- Making our neighbourhood more accessible
- Changing mindsets and behaviours
- Mobility of all, for all, by all.

Table 1 Audit criteria taking reference of the global walkability index

Component		Indicator
Safety and security	1	Pedestrian victims
	2	Modal conflict
	3	Crossing safety
	4	Crossing exposure
	5	Traffic management
	6	Security
	7	Safety rules and laws regulations
	8	Quality and maintenance of walking path surface
	9	Disability infrastructure
	10	Walking path congestion
	11	Pedestrian amenities
	12	Trees
	13	Environment quality
	14	Connectivity
	15	Parking provision
	16	Scope of expansion/improvement of infrastructure
Policy support	17	Planning for pedestrians
	18	Relevant design guidelines
	19	Heritage conservation rule

Walking audit is a common method for measuring walkability and pedestrian environment. The walkability index that is a commonly used tool for this audit generally has three components: Safety and security ensured for the pedestrians, the convenience available for them and the local governing or statutory support available for walkability in terms of policy or guidelines. In Europe and UK, where pedestrianisation has been a way of urban design principles, pedestrian environment review system (PERS, UK) is an audit tool used in profusion [1].

Table 1 lists out the audit criteria taking reference of the global walkability index with certain additions and alterations which befit the context of the old town. The parameters generated are deliberated to derive the walkability index.

3 Data Analysis

The pedestrians are mainly affected by vehicular dominated transportation system. Accidents in India cause around 80,000 deaths [1]. In cities like Delhi and Mumbai 60–78% are pedestrians, cyclists or street side inhabitants. The old town of Bhubaneswar has also been affected by similar traffic developments. Emphasis

Fig. 1 Lack of pedestrian subways, adequate signalling system and FOB leaves the residents with no choice but to cross the streets using foot. Uncontrolled parking along the sidewalks

should rather be on making the infrastructure safe and sustainable for the pedestrian populace. Introduction of series of CRUT buses, sequence of flyovers, road widening programs in and around the area has been the catalyst in creating this issue.

Major pedestrian activities in the old town are centred on the Bindusagar Lake. The road connecting the temples and the Ravi Talkies node is a hub for social gatherings and also the only path for carrying the idols during Chandan yatra. Vehicular activities in this area can be easily restricted and reoriented. Alternative transit options and complete pedestrianisation might be a possible iteration (Figs. 1, 2, 3, 4 and 5).

The pedestrian count survey (Table 2) at the following junctions reveals that 2 wheelers and 4 wheelers are a major concern for pedestrian movement at this zone.

In certain sections of the Old town in Bhubaneswar, the footpaths are at grade to the road level to provide additional width during the peak traffic hours of the day. However, such a makeshift arrangement has rather increased the rate of accidents and risks for the pedestrians. It is observed that the actual walking distance between 2 destinations has been increased by manifolds due to the mixed use of the land use in this zone. As per the CDP, the zone is residential within heritage zone but the sanctity of the same is not being maintained.

4 Design Considerations Towards a Walkable Old Town

As per urban design principles, there are several iterations that one might consider to make this neighbourhood of the old town more walkable and pedestrian-friendly. However, the pedestrian behaviour pattern analysis and overlaying sustainable infrastructure tools on the pattern will probably be the best designed option. As highlighted earlier, weighing the existing pattern on the scale of the walkability index parameters is discussed below.

Fig. 2 No dedicated vending zones

Fig. 3 No sign of barrier free design

Fig. 4 Lack of dustbins and other street furniture

Fig. 5 Unplanned littering and obstruction caused

4.1 Safety and Security

Providing dedicated sidewalks without any obstruction should be the first mandatory intervention.

- These sidewalks should be equipped with adequate grab bars, designed to correct slope and incorporate all that is required for universal design.

Table 2 Pedestrian count survey

Monitoring sites	7–10 AM			2–5 PM			7–10 PM		
	2 & 3w	LMV	HMV	2 & 3w	LMV	HMV	2 & 3w	LMV	HMV
Ravi talkies square	2375	517	414	2557	581	521	2585	662	572
KedarGouri lane and Bindusagar lane junction	475	87	25	512	95	34	430	100	21

Survey date: 5 April, 19-(LMV—Light motor vehicle, HMV—Heavy motor vehicle)

- Pedestrian crossovers are missing in the entire action area. This crossover should adhere to the given standards and have a synchronised signalling system.
- Rescue or refuge islands must be provided in the zebra crossings for the pedestrians.
- Public awareness programs regarding traffic safety to be organised as is being done by students and other volunteers in other parts of Bhubaneswar.
- Adequate illumination to be ensured in terms of lighting standards the psychological needs of the human safety and security (Fig. 6).
- Fire safety measures, i.e. provision of fire hydrants, breaching inlets and other services to be adequately given. Water clogging is a common issue in this part of the city and the catch basins and other means should be provided.

Fig. 6 Lighting standards for street lighting

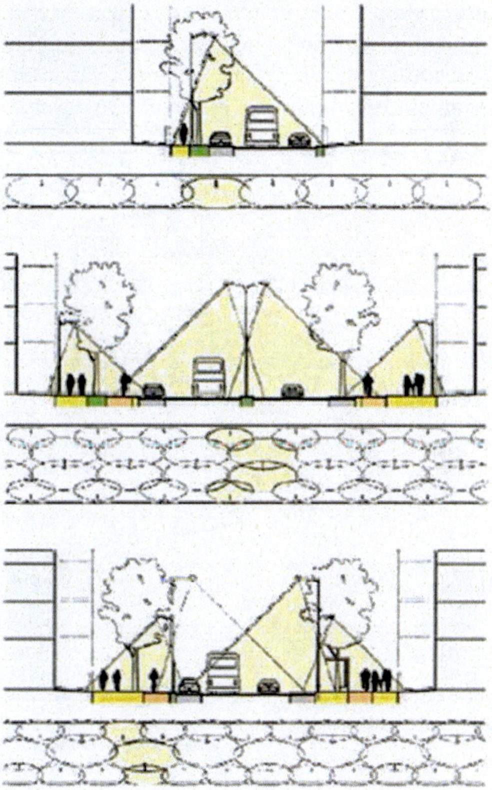

4.2 Convenience and Road Fabric of the Street Front

- Basic street furniture and street hardware need to be installed. Currently, the hoardings and the street advertisements do not have a control to their display mechanism.
- Regulation of street hawkers (giving them unique numbers and designated location) will ease out pedestrian's experience in the vicinity. Few interventions are illustrated vide sketches in Fig. 7.
- Adequate landscape with local tree species to be used for shading purpose. This will also ensure mitigation of the heat island effect.
- Regular maintenance of the provided infrastructure is to be ensured.

Currently, the heritage precinct of this neighbourhood does not find an identity in the shop fronts or the street edges. Measures are already being taken by BDA to ensure that the same gets adhered to. This should be a mandate in this zone. Considering urban design; adherence to architectural and cultural heritage features is imperative in this part of the city.

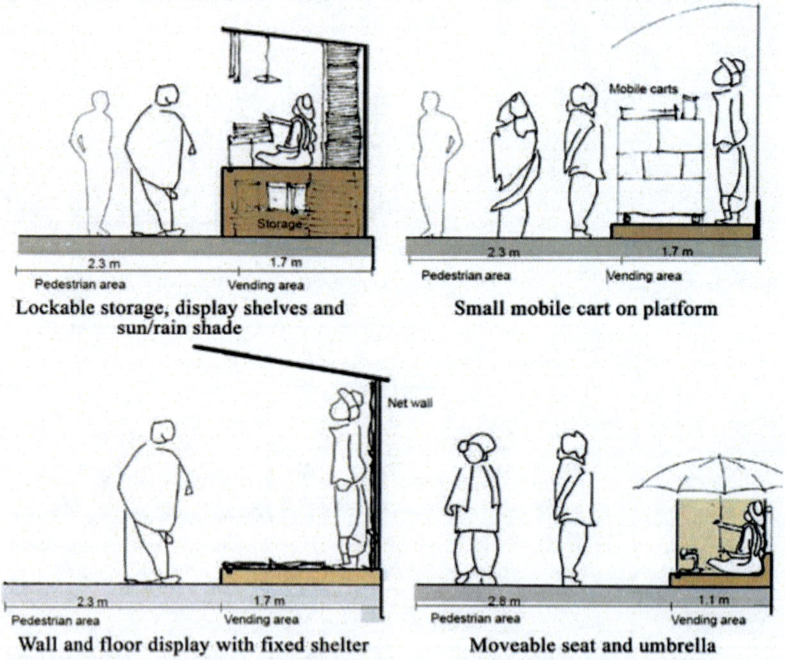

Fig. 7 Few modules of vendor rehabilitation are illustrated above

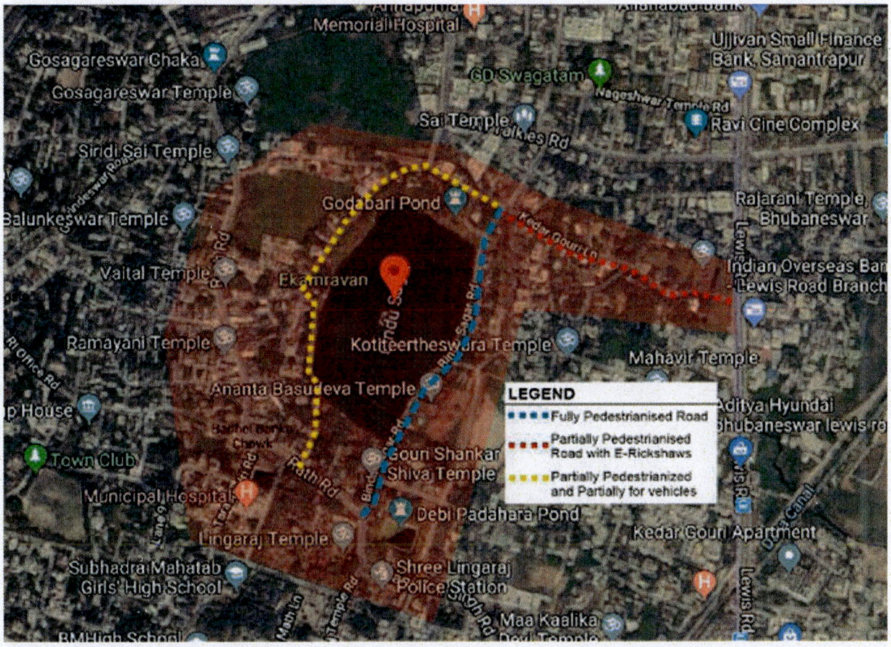

Fig. 8 Action area

4.3 Statutory Support

No urban design or planning intervention is successful without the support of the policymakers. The IRC 103—Guidelines for Inclusive Pedestrian Facilities have been revised by the Transportation Research and Injury Prevention Programme (TRIPP) to upgrade the condition of the pedestrian facilities in India. The success of these policies depends upon its enforcement. These policies should be enforced. Regular audits should be conducted to ensure adherence to the same.

- Breach of the traffic guidelines should have heavy penalties levied on them.
- Free left turns in the pedestrian zones should be stopped and manned.

4.4 Additional Recommendations

- The activities in the densely pedestrian areas are mostly controlled by the public sector. Hence the public and semi-public sector can share a responsibility towards pedestrian movement and can play a major role by providing setbacks for creating 'fully pedestrian lanes'. The Bindusagar lane (Fig. 11) shows no need for any vehicles and is not a major carriageway with the option of using the lane abutting

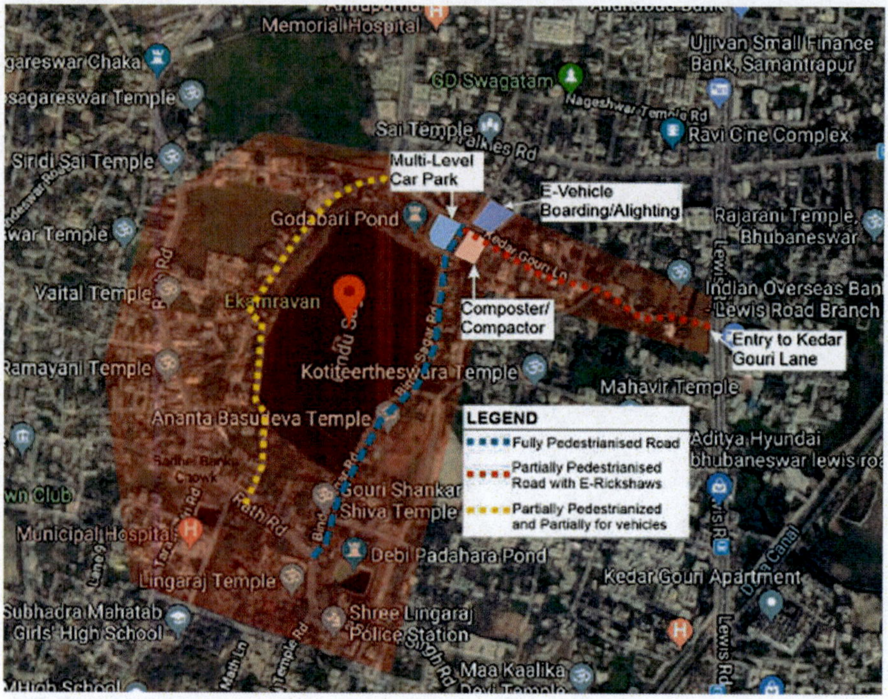

Fig. 9 Proposed recommendations

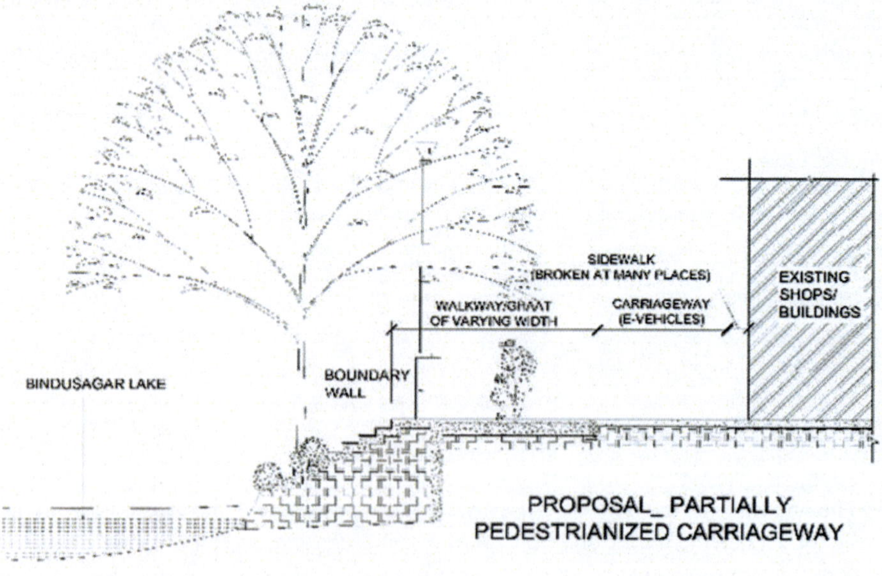

Fig. 10 Proposal—partially pedestrianized carriageway

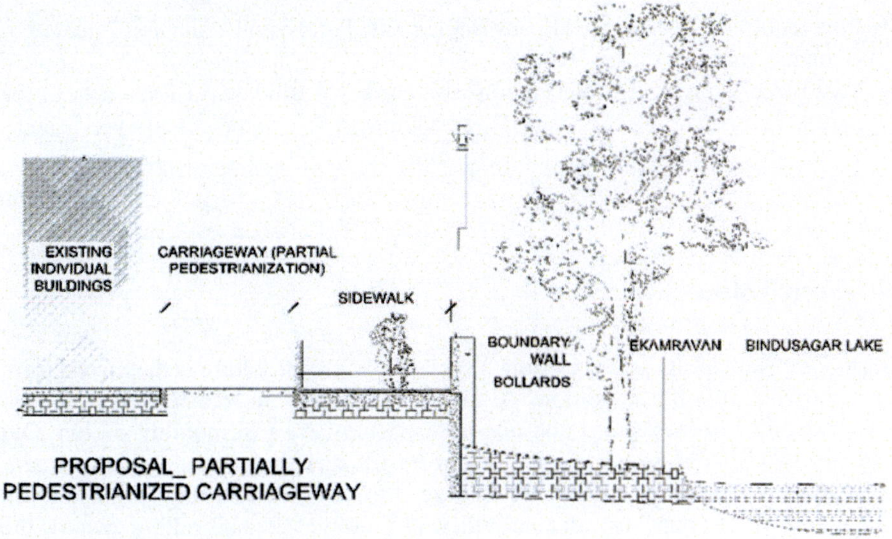

Fig. 11 Proposal—partially pedestrianized carriageway near Ekamravan

the Ekamravan being a transit corridor. Also, the presence of the multiple temples in this zone gives it the feel of a public realm. The residential quarters in the road abutting Bindusagar lane, however, have 2 wheelers and 4 wheelers with the lane being the only access point. A parking block is hence proposed in for accommodating the same considering that this lane is proposed to be completely pedestrian (Fig. 9).

- The lane from Kedar Gouri Park is partially pedestrian. However, this is a main street since it has the St. Xavier's school in close vicinity. It may be proposed that e-rickshaws travel in this zone (Fig. 8). These rickshaws will not only reduce the ingress of polluting vehicles but also maintain the heritage value of this precinct. Necessary arrangements can be made within the existing street section for these sustainable e-rickshaws. (Fig. 10).
- A portion of the road has been kept for e-vehicles which are by far the most sustainable form of transport for the fabric that this precinct demands. These e-vehicles will also act as a mode of transport for the tourists or visitors. Their alighting points and routes have been shown in Fig. 9. These vehicles are not only green but also a barrier-free intervention.
- The lane abutting Ekamravan and Bindusagar Lake has residential units on the other end. Figure 10 illustrates the envisaged proposal of having this road partially pedestrian and partially vehicular. Bollards at every 250 mt will ensure that vehicles do not move on to the pedestrian zone.
- Waste management is another major issue to be catered to in this zone. An OWC (organic waste composter) is proposed in the area besides the parking (Fig. 9).

Interventions like these are already implemented in the unit 1 and unit 2 market of Bhubaneswar city.

- This place is prone to waterlogging and currently Bindusagar Lake acts as the natural rainwater collection tank. However, having rainwater recharge pits in areas of waterlogging will be a sustainable solution.

5 Conclusion

Improving walkability and providing a sustainable infrastructure is the easiest form of improving liveable condition. As discussed earlier, movement towards such infrastructure resolves social and non-partisanship issues in mobility design. Old Bhubaneswar can also be equipped with infrastructure that is modern and sustainable and still maintains its historic lineage. Provision of facilities and amenities like public toilets, benches, adequate sidewalks will get the ball rolling towards the aim. Maintaining and upgrading the infrastructure at regular intervals should be made mandatory. Awareness and academic sessions on walkability must be conducted. The importance of walkability should be further strengthened by making legal reforms aiming to provide equity for pedestrians and their right to walk. For achieving an optimum and desirable urban context, all the developmental organisations need to work in tandem.

Projects for Public Spaces (PPS), an NGO based in New York, has developed philosophy that 'When you plan cities for cars and traffic, you get more cars and traffic. When you plan cities for people, you get more people'. The urban planners and designers of our day cities need to reconceptualize and understand the user in present day context. The urban units for Indian cities should be designed and developed in a manner that it retains the strength and diversity of our sociocultural fabric with the sustainable walkability as an effective contributor [1].

References

1. Barman J, Daftardar C (2010) Planning for sustainable pedestrian infrastructure with upcoming mrts—an appraisal of walkability conditions in lucknow. J Inst Town Plan India J 7(3):64–76
2. Gupta S (2016) Trends and imperatives for inclusive pedestrianisation. Urban Mobility Conference, 2016
3. Malhotra C (2017) Pedestrianisation of commercial Area: A case study of Aminabad, Lucknow. In International Research Journal of Engineering and Technology (IRJET), vol 04 Issue 08
4. Praharaj M (2012) Historic conservation and sustainability: A case of Bindusagar lake, Old Bhubaneswar

Urban Infrastructure and Special Economic Zone (SEZ): Challenges for Corporate Land Alienation

Divyajit Das, Bhubaneswari Bisoyi, Ipseeta Satpathy, and Biswajit Das

Abstract This research paper critically introspects into the area of special economic zones taken up in India by the corporate Moghuls and the challenges it faced from the public during land alienation. Lacunae in transformational enactment of laws have resulted in fighting between people, corporate and governance. The focal study of the paper proves into the case of Urjja Nagar and Dadri during the course of land acquisition by reliance energy. The process of acquisition of alluvial soil in the Ganges and Yamuna Rivers resulted in the protest by farmers. Political stalwarts join the protest against land alienation for SEZ contributing to the development of urban infrastructure. Vehement media report alleged the political power to have cross-crushed the democratic dissent and accused of the corporatisation of democracy, a new normal doctrine. The paper analysed into the concept of public trust doctrine for enabling the urban infrastructure developed without hurdles during the acquisition of land. The paper argues for a concept of compromise and compensation during land alienation to be incorporated in the R&R act. This research paper reviewed the literature available from primary and secondary sources. It is exploratory and descriptive in its approach. The research analyses the factors associated infers into the findings providing a solution to the identified problem. This analysis allowed a permeable amount of land for the villager and fair price for the alienation of land with a share of investment in the company towards the portion of land given to the company as equity.

D. Das
School of Civil Engineering, KIIT University, Bhubaneswar, India
e-mail: divyajitdas10@gmail.com

B. Bisoyi (✉)
Sri Sri University, Cuttack, India
e-mail: bhubaneswari.b@srisriuniversity.edu.in

I. Satpathy · B. Das
School of Management, KIIT University, Bhubaneswar, India
e-mail: ipseeta@ksom.ac.in

B. Das
e-mail: biswajit@ksom.ac.in

B. B. Das et al. (eds.), *Recent Developments in Sustainable Infrastructure*, Lecture Notes in Civil Engineering 75, https://doi.org/10.1007/978-981-15-4577-1_78

Keywords Urban infrastructure · Special economic zone · Corporate land alienation · Property rights · Protest

1 Introduction

With the continually mounting global contest has led towards creation of an ambience where we need to reform robust strategy for elevating our trade exports. The trade policy followed by different countries can be understood by analysing the various macroeconomics policies executed by them. The impact of restrictive policies can lead to impede growth in the developing countries was recognized in 1980s. Various new reforms embarked inorder to facilitate augmentation and amalgamation into the world economy. After pursuing an inward-looking development strategy, India decided to take a historic step of changing tracks in 1991. The liberalization programme was launched in India. Under the influence of Special Economic Zone (SEZ) and foreign trade policy in India growth has been accelerated by providing fiscal incentives, superiority in infrastructure, clearances through single window, and guidelines for reinforcing financial intensification through policy of SEZ. The services provided in this practise area comprises application, registration and track up for SEZ consent, as well as de-registration and management of assets; legal procedure and other credentials. The affiliation accounting firm provides support towards bonding and various types of other services related to accounting and tax aspects [1]. The traditional real estate industry provides various range of services that include title diligence, legal opinions and reports; Joint development agreements, certificates and procedures. The documentation and legal aspect in real estate are done under India's new legislation on Real Estate (Regulation & Development) Act, 2016 [2, 3]. With the new age, the services in the sector of real estates have modified its traditional business and have adapted the technological aspect to provide various services to the customers. The companies in the reals estate provide assistance for registration of the property, arbitration in resolving any dispute and obtaining government clearances when required [4].

2 Necessity of Special Economic Zones

The inspiration for establishing SEZ in India came from China, Shanghai. The success in SEZ gave China, Shanghai and Shenzhen an elevating growth in economy. The known principal aim of SEZ programme in India is to provide a sustainable growth in the export-led by attracting FDI and various private sector parties [5]. Thereby, SEZ can be used as a catalyst in the progression of reforms, fiscal expansion, creation in employability and other development in regional sector. However the dream to accomplish the quantum of flow in FDI is very less for India as compared with China this is because of the perceived meagre position of India in the international platform as reflected by trans-nationality index [6, 7]. The promotion of SEZ is essential for encouraging investors and to deal with the infrastructural deficiencies, complexies in the procedure, bureaucratic disturbances and obstruction elevated by policies such as monetary, trade, labour and taxation [8, 9]. These bottlenecks in the process affect the climate for investment by increase in the production cost and also the transaction cost. The establishment of industrial enclaves SEZs and EPZs have been considered as the strategic tool for escalating the industrial growth in developing countries [10].

3 Concept of Urban Infrastructure Development

Given the predominance of various socio-economic factors poverty is considered to have major influence on reversing the impact on economical growth [11, 12] . Sustainable growth in urban infrastructure development can be progressed when the benefit is broad-based, multi-sectoral and exclusive in nature [13, 14]. Sustainable reduction in poverty can be achieved when the growth is across all sectors, which includes better utilization of land, agricultural productivity and connecting community development through providing economic opportunity and infrastructural development by Special economic zones [15–20]. The prime objectives for establishing of Special Economic Zone are

i. Creation of other financially viable activity;
ii. Increase in the foreign exchange reserve;
iii. Addition in the technological sector and skill;
iv. Growth in the infrastructural facilities;
v. Elevation in the living standard of the people;
vi. Key sector development by promotion of export of goods and services.

4 Urban Infrastructure Project of SEZ Demolish Villages

Urja Nagar coined for the power city of Reliance in the setting of a Special Economic Zone (SEZ) tuned Ujar Nagar (City of Devastation) for the natives. Urban infrastructure growth is not affecting the economic status of the folks in villages. In quoting an article titled ‘corporation state and fascism’ by Vandana Shiva dated 10 August 2006, an US-based social philanthropist named Sandip comment Delhi’s explosive urban sprawl has created demand for grabbing land in Dadri [21, 22].

5 SEZ Resistance

The unique example of SEZ development is considered to have happened in Goa. There, the state government to begin with adopted the policy of SEZ and cancelled it later due to political pressure and by public consultation. In the year 2006, on 5th June the state government of issued an bureaucrat SEZ policy [23–25]. The state government of Goa justified that SEZs are a mean to bring development in terms of both economic and development in industry that would bring large dividend into the state in the long run by generating new employment opportunity. With the pressure from social movements and other political factors, the government of Goa concluded that the current SEZ project is appropriate model to be pursued. On 31 December 2007, the government of Goa became the only government to make an announcement to scap the current SEZ project and denied future possibilities of such projects (The Hindu, January 1, 2008). On the other hand, by April 2015 the Goa state had not withdrawn the 2006 SEZ policy or annulled land allocations earlier made to SEZs. Consequently, those resisting SEZs viewed the announcement made by the chief minister was solely performative and unenforceable devoid of legal pressure [26, 27].

6 Impact of SEZ on Real Estate and Agriculture

Bejhera with Rajendra singh of 'Tarun Bharat Sangh' and Satpal Choudhary of 'Lak Tantrik Party' in May 2006, Dr. Jaipal Singh, one of the fasting protestors said that the villagers came to know of the acquisition only from local newspapers; at the time of unveiling of the power project. No consent was taken and no notice was served to them. When they challenged the injustice and illegality of the acquisition of their land, the authorities told them to either go to court or accept the 'compensation' being awarded. Being aggrieved with the decision of the government and concern for agriculture produce to sustain their livelihood, the farmers turned down the fence and have been ploughing their field to assert their rights [28].

7 GIS and Remote Sensing

Technology can do the estimation of the available landmass very easily and economically. Survey of India in its mapping division can publish aerial photographs of every squire foot of India for bargaining a scale. In deed right now 'google' has done it for North America and some other parts of the world. Similarly each inch of india might have been mapped but not available public like openly. It is similar to the government files and notes, because it might create problems for lots of corporate. National security is a truth ad a pretext, there is some truth in it but are getting sold to the privatization way through the alienation of priceless real estate, even in terms of carbon trading business.

8 Fortification of Entitlements

The policy framed for SEZs suffers from various lacuna of depriving the employment of the community who are reliant on the land and that has led to pervasive turbulence and remonstration measures. Hence, the main objective is to protect the entitlements of the evicted societies. The different types of entitlement that are related to property, liability and inalienability. The rule of property is applied where the value of the entitlement is settled by the seller. The liability rule is applicable in the case where the initial entitlement is violated and in order to protect the entitlement therefore the role of state government is vital. When the state intervenes to forbid the transaction between the buyer and the seller then inalienability rule is applied. The entitlement for the SEZs is protected under the rule of liability. The land acquisition process leads to the contravention of the primary entitlement of the landowners and liability rule is applied for compensating the loss of the entitlement. This process is executed by the state government by protecting the entitlement of the dispossessed [29].

9 Limitations on the Usage of Land

There exist various factors that are applicable for restricting the price depression of agricultural land which includes zoning and land use restrictions. The restriction imposed on these lands is facilitated by government under zoning factor. In addition to these restrictions, there are several restrictions that are imposed by the government for using agricultural land for non-agricultural purpose [30]. The government clearance that has to be obtained for using these agricultural lands for non-agricultural purpose is called Non-Agricultural Use Clearance (NAC). The obligation lies with the party that is interested for using the land for non-agricultural use. Consequently, the farmer cannot take the advantage of the value addition that ascends out of the use of the land in non-agricultural use. It thus bequeaths massive enticements to the buyer at the price that the farmers who primarily own the land and is basically disinterested for giving the land [31]. In the year 2005, Special Economic Zones Act has been amended that illustrates the prior rules and regulations for using the land. Nevertheless, the developers take the benefit of SEZ act by obtaining enormous bands of lands that remain unutilised for years to come.

10 Should Real Estate Be Governed Under ‘Public Trust Doctrine’?

New economy researchers and thinkers view that in a democracy land and resources should be governed on the basis of the public trust doctrine. It should not be done on the ‘eminent domain principle’ [30]. It is observed that during these times of ‘corporate globalization’, the land Acquisition Act can behave as an ‘instrument of fascism’. Therefore there has to be a perfect contemporaneous progressive change of the traditional acts duly amended. Will the LA ac be set aside and enacted afresh matching to the current need is the question. Recent R&R policy of government of Orissa can be referred to on this context. It could be that, let all corporations negotiate directly with farmers and if farmers do not want to sell their farmland, state should not forcefully acquire it and if the state does that, they should get the full market price.

11 Origination of a Market-Based Instrument for Recompense

The implication that arises in the compulsory acquisition of the land under the policy of ‘eminent domain’, there is a necessity to plan a market-based device for compensation in the form of a Special-Purpose Development Corporation (SPDC). In case of SPDC, it acquires incorporated possession of the land and the improvement of project and would deal with the ‘affected people’ a choice among accepting pre-project ‘fair market value’ compensation or pro-rata shares in the SPDC. The compensation for the land is linked with the true value of economic value of the land and subsequently, that land projects are basically linked with social welfare maximizing. As a result, it provides a choice to the landowner that either they can opt for receiving pre-project compensation or SDPC share. Thus, through this mechanism it tries to sustain a market-based mechanism for safeguarding free and impartial return to the property-owners.

12 Conclusion

In order to make the procedure of land acquisition and urban development, it is vital in the part of the state government to make a clarity on the use of the land for only ‘public purpose’. Besides, the state government has to take appropriate steps in order to reduce the transaction cost for the developers and thereby facilitate voluntary acquirement of land. In the present scenario, we have observed competition between different states has exaggerated in order to fascinate various investors for developing SEZs. The rising competition among the investors need to be taken care by the central government to ensure free and rational land acquisition. With the dream of making India a hub for the manufactures, its vital role played the central government to introduce easier rules for the land acquirement procedure in order to interest massive venture from the private players.

References

1. Aggarwal A (2004) Special economic zones: revisiting the policy debate. Econ Polit Wkly 41(43):4533–4536. http://www.jstor.org/stable/4418855
2. Allen DW (1999) Transaction costs. In: Encyclopedia of law and economics, pp 893–926
3. Asif M (2004) land acquisition act: need for an alternative paradigm. Econ Polit Wkly 34(25):1564–1566. http://www.jstor.org/stable/44080906
4. Prabhu C (2009). Special economic zones: an overview. Third Concept 39–44
5. Calabresi G, Melamed A (2009) Property rules, liability rules, and inalienability: one view of the Cathedral. Harv Law Rev 85(6):1089–1128. http://www.jstor.org/stable/1340059
6. Cernea M (2000) Risks, safeguards and reconstruction: a model for population displacement and resettlement. In: Cernea MM, McDowell C (eds) Risks and reconstruction: experiences of resettlers and refugees. The World Bank, Washington, D.C
7. Dalua AK (1993) Environmental impact of large reservoir project on human settlement in Odisha. Ashish Publishing House, New Delhi
8. Dey A (1997) Critique of the Odisha rehabilitation policy. In: Fernandes W, Paranjpye V (eds) Rehabilitation policy and law in India: a right to livelihood. Indian Social Institute, New Delhi
9. Dhar TN (2011) Special economic zones and impacts on land resources and socio-economic implications. Indian J Public Adm LVII(1):15–24
10. Dutta A (2007) Development induced displacement and human rights, economic development project 2007. Legal and institutional concerns, with reference to India
11. Goyal S (1996) Economic perspectives on resettlement and rehabilitation. Econ Polit Wkly xxxi(24)
12. Jeyaraj X (2008) An SEZ with a difference. Soc Action 58:272–282
13. Fernandes W (2004) Rehabilitation policy for the displaced. Econ Polit Wkly 39(12):1191–1193
14. Garada R (2012) Opencast coal mining induced land acquisition: a study on land oustees at Talcher Coalfield, Odisha (India). LAMBERT Academic Publishing AG & Co KG, Germany
15. Ghatak M, Ghosh P (2011) The land acquisition bill: a critique and a proposal. Econ Polit Wkly 65–72
16. Government of India (2004) National policy on resettlement and rehabilitation for project affected families-2003. Published in the Gazette of India, Extraordinary Part-I, Section 1, No-46, dated 17 Feb 2004
17. Kabra NK (2006) Special economic zones: a baneful package. Mainstream 7–13
18. Kothari S (1996)"Whose nation? The displaced as victims of development. Econ Polit Wkly xxxi(24)
19. Ministry of Rural Development (1996) Draft national policy for rehabilitation of persons—displaced as a consequence of acquisition of land. Econ Polit Wkly xxxi(24)
20. Palit C (2004) Short-changing the displaced: national rehabilitation policy. Econ Polit Wkly
21. Pandey B (1998) Depriving the underprivileged for development. Institute for Social and Economic Development, Bhubaneswar, Odisha
22. Parashuraman S (1994) Resettlement and rehabilitation of people displaced by development projects: an evaluation of the Indian experience. TISS, Bombay
23. Pratap K, Sharma S (2007) The prosperous few and the pauperised many: a perspective on special economic zones. Mainstream 17–20

24. Ramanathan U (2011) Land acquisition, eminent domain and the 2011 bill. Econ Polit Wkly xlvI:10–14
25. Robinson WC (2003) Risk and rights: the causes, consequences and challenges of development induced displacement. Brookings Institutions-SAIS project on Internal Displacement, Washington, D.C.
26. Rose CM (1997) The shadow of the Cathedral. Yale Law J 106 (7):2175–2200. http://www.jstor.org/stable/797165
27. Sangvai S (2002) The river and life: people's struggle in the Narmada Valley, 2nd edn. Earth Care, Mumbai
28. Sarkar S (2011) The impossibility of just land acquisition. Econ Polit Wkly xlvi(41):35–38
29. Sharma NK (2014) Special economic zones: socio-economic implications. Econ Polit Wkly 44(20):18–21. http://www.jstor.org/stable/40279007
30. Sinha BK (1996) Draft national policy for Rehabilitation. Objectives and principles. Econ Polit Wkly xxxi(24)
31. Viegas P (1992) The Hirakud Dam oustees: thirty year after. In: Thukral EG (ed) Big dams displaced people rivers of sorrow, rivers of change. Sage, New Delhi

Studies on Integration of Lean Construction and Sustainability in Indian Real Estate Projects

Raja Sekhar Mamillapalli, Ashok Kumar Rath, and Dillip Kumar Bera

Abstract Indian Real estate market has grown largely in recent, which has achieved faster growth among the developing global markets. Quick urbanization and increase in income of the public is the prime reason for the demand and growth of Indian real estate. On the other hand multiple issues like incompetence, productivity declination, waste generation, low sales, etc. are bothering Indian Real estate. To hamper these, lean principles have been introduced in the field of construction which had caused a revolution in the manufacturing process and green building practices to reduce its negative ecological footprint with simultaneous creation of more sustainable environment. The value maximization and waste minimization of a project are achieved through lean construction and its techniques will be integrated with the current sustainable construction techniques to diminish the current wastage without comprising the quality of the construction. When used and combined perfectly, the sustainable parameters and lean principles can be aligned to yield better results and counter support the benefits of both. Lean principles to combine with sustainability parameters in cast-in situ construction in Indian context are a challenging task. Independently both lean construction and sustainability in construction have proven results in Indian real estate projects. However, no studies or framework on integrating the cast-in situ construction of have been attempted either for world across or within Indian context. This paper mainly focuses on the small scale real estate projects. The stakeholders or parties of smaller construction projects argue that the implementation of lean construction in their projects will not yield the results and claim to be the best way that they adopt with their experience, which is not really true on field. To prevent this problem, the adoption of sustainability with respect to the environmental, social, and economic terms will further enhance the building value in the construction market. This paper follows an integration procedure (with reference to PMBOK) of lean and sustainability concepts together are the key to the efficient

R. S. Mamillapalli (✉) · A. K. Rath · D. K. Bera
KIIT Deemed to be University, Bhubaneswar, India
e-mail: nitmtech@gmail.com

A. K. Rath
e-mail: akrathfce@kiit.ac.in

D. K. Bera
e-mail: dberafce@kiit.ac.in

B. B. Das et al. (eds.), *Recent Developments in Sustainable Infrastructure*, Lecture Notes in Civil Engineering 75, https://doi.org/10.1007/978-981-15-4577-1_79

and economic construction in the future. As these require a detailed study of the all the attempts previously made and helps to frame a perfect model which makes it an easier way for the worldwide adoption in the real estate projects.

Keywords Lean construction · Sustainability · Real estate projects · Integration management

1 Introduction

Indian construction industry contributed approximately 7% of GDP on a normal throughout the following decade, in that real home structures alone contribute 11% of construction GDP. Around 55 million people are directly or indirectly employed in Indian construction industry only next to agriculture in terms of in giving employment. Notwithstanding the work age and the monetary significance of the construction part, it strikes the issues, for example, cost invades, squander age, low profitability, and absence of vitality proficiency for its activity, new water prerequisite, venture finish deferral, and absence of expert practices in the business. The construction business assumes a significant position inside the most economies and it impacts nation's GDP through construction productivity to an enormous degree. Development segment impacts different areas firmly and has an immediate effect over different methods of supportability.

Along these lines, more significance ought to be given to distinguish the real difficulties and their answers. Utilizing fitting undertaking the board procedures we can conquer these difficulties and complete a venture inside spending plan and on schedule. Since the most recent 20 years, numerous nations executing the lean and maintainability ideas in their development businesses and getting profited. These ideas were embraced by construction companies to upgrade their profitability and effectiveness of its business activities from top management to on ground site execution personnel.

LC is a generation organization based which is a most ideal approach to manage venture conveyance and considered as a way to deal with plan and oversee work process. LC was born with the objectives of a lean creation structure: work process and lessen squander—to specific frameworks and applies them in a cutting edge venture conveyance process (Lean Construction Institute, 2012). As in numerous enterprises and specialist co-op associations, Lean reasoning has been connected with much achievement and its ideas can be connected to any shape at work, home, or play. To keep pace with the regularly developing multifaceted nature, it needs much improvement to the constructed condition and gaining ground by keeping up a similar productivity increases over different business segments. The worth boost and waste minimization of a venture are accomplished through lean construction. Many accept that lean construction is the route for getting the various components of sustainability. The customary Indian engineering incorporates the idea of maintainability in its very plan. In contrast to selection of lean practices, the administrative offices of the legislature and the proprietor associations are eager to actualize green practices whose essential purpose is to embrace sustainable techniques. It is seen that the responsibility of top administration gives a noteworthy force toward the appropriation of maintainable development rehearses, as it is more because of outer weights than the inward prerequisite. The activities that are taken are for the most part to guarantee the base consistence prerequisite.

Sustainability in manufacturing is adopted to make the process as lean manufacturing with environmental, social, and economic parameters taken into consideration. The contribution of lean to sustainability is basically focus on waste reduction, lean tools will reduce material, and energy wastes during construction and maintenance and also greater operational and product reliability achieved during lean processes will reduce the amount of harmful emissions. However, the other focus is on lean value. For achieving challenging targets, the generic methods were developed in lean product development which can also be used to achieve sustainability targets [1]. It is believed that there exists a reverse action from sustainability to lean construction. The improvements in reliability, discipline, and process efficiency motivated by sustainability concepts are broadly aligned to lean objectives, and thus strengthen the implementation of lean construction.

2 Literature Review

Building materials and processes have a major impact on the sustainable homebuilding and its greatest barriers are higher initial costs would reduce the later operational cost resulting from ill-defined construction processes. Nahmens and Ikuma [2] propose the use of lean construction tool (the SLIK-Safety and Lean Integrated Kaizen) to showcase its effect on triple bottom line of sustainability. Saurav et al. [3] identify five significant lean tools and area of linkages of lean and sustainability on the basis of relative importance index. Researchers say that LC offers many benefits by proper project planning which shortens duration, promotes quality, and sustainability. Bae and Kim [4] listed out some lean tools that showed financial advantages of construction process.

Environmental Management System (EMS) was integrated to LC concepts which is expected to maximize the customer's satisfaction and as well as minimize the waste. EPA [5] found that the implementation of lean principles reduces waste and pollution which covers the environmental and social dimensions.

The sustainable development is the foundation for sustainable construction which has similar concepts particularly in minimizing the waste [6]. Earlier in (1994) defined some key principles in creating a sustainable build environment, Ballard and Howell [7] suggested a few fundamental concepts of lean, Kibert et al. [6] suggested that the implementation of sustainability throughout the organization is more beneficiary rather than only in projects.

By performing different forms of interviews among selected members, Koranda et al. [8] conclude "Commitment and knowledge are the key elements for the desired output and environmental aspect is acquired through LEED principles and economical through Lean concepts for the small construction projects." Dehdasht and Zin [9] propose a hierarchical model which is framed from a volume of studies concentrated on the challenges that may occur during execution, structural and cultural aspect, management and strategy aspect and Last Planner System implementation which also presents a checklist for important barriers in each aspect of sustainability upon lean implementation. A semi-structured format of Delphi interview is conducted for verification of barriers.

Yupeng et al. [10] examine the effects of onsite/offsite prefabricated systems that can be adopted to green building objectives by assessing the interplay between social, economic, and environmental activities and identify prefabricated construction as one such approach.

Besser Freitag [11] developed a theoretical framework by integrating lean construction and sustainability and life cycle of the buildings using a prism method.

Limited literature about integration of lean construction and sustainability and especially in real estate projects. About India, real estate projects with lean and sustainability integration are nowhere data or studies are unavailable. This shows the need and importance of the study for a development of a model for integrating lean and sustainability for real estate projects for better out project delivery with waste minimization and greener development.

Theoretical framework

	LC and sustainability integration w.r.t construction phases	Protection of biodiversity	Energy efficiency	Resource utilization Efficiency	HR relationships	Association with communities	Local partnerships	Productivity Increase	Improvement of Project	Monitoring of Performance
Pre-construction phase	Adopting Lean culture throughout the organization									
	Team arrangement (should be of multi-functional) and Training on lean concepts and techniques.									
	Effective planning of overall construction and continuous assessment of team to follow the lean concepts									
	Training of Staff about Lean Implementation									
	Creating a Master Plan									
	Setting up lean destination									

(continued)

(continued)

	LC and sustainability integration w.r.t construction phases	Protection of biodiversity	Energy efficiency	Resource utilization Efficiency	HR relationships	Association with communities	Local partnerships	Productivity Increase	Improvement of Project	Monitoring of Performance
Construction phase	Warm up the team									
	Upgrading manpower and organizing workstations to modify lean practices									
	Effective observance of 5S program that make to reach construction as "Lean defined"									
	Look ahead planning (Resource planning, JIT procurement and Task make ready checklist)									
	Conduction of Huddle meetings									
	Maintaining Daily progress report									

(continued)

(continued)

	LC and sustainability integration w.r.t construction phases	Protection of biodiversity	Energy efficiency	Resource utilization Efficiency	HR relationships	Association with communities	Local partnerships	Productivity Increase	Improvement of Project	Monitoring of Performance
	Motivating Lean team and workers by giving them some awards									
Post-construction phase	Work Balancing									
	Work Simplification									
	Knowledge Management and Work Standardization									

3 Conclusion

Based on the content analyzed and reviewing the literature on integration of lean and sustainability, it is clearly found some barriers for implementation. They are "Awareness of understanding Lean concepts" and "Top Management commitment." Across the world, the integration of lean with others is implemented easily without many challenges, as most of them are pre-cast constructions. In the context of India, majority of the constructions are cast-in situ it is difficult to implement the integration of lean and sustainability. For that, a theoretical framework is proposed which on implementing in practical conditions may result in better outputs.

4 Extension of Studies

The extension behind this study is to research the requirement for manageability and lean ideas from a point of view of little scale land extend in India. This paper additionally inspects the present condition of utilization of supportability and lean practices on these activities and to demonstrate that such practices would conceivably be mainstreamed. The different focal points of this examination are to build up the connections among manageability and lean ideas to build up a system in which little ventures can coordinate and actualize the two works on accepting that they have a beneficial outcome. The investigations are performed in the areas of little and enormous measured tasks just as green-ensured or non-affirmed ventures. The difficulties which wall structure and development procedure are to be recognized. These can be accomplished through the accompanying:

1. Experimental (case) study
2. Research investigation by Qualitative and Quantitative techniques.

References

1. Lapinski AR, Horman MJ, Riley DR (2006) Lean processes for sustainable project delivery. J Constr Eng Manag 132(10):1083–1091. https://doi.org/10.1061/(asce)0733-9364(2006)132:10(1083)
2. Nahmens I, Ikuma LH (2012) Effects of lean construction on sustainability of modular homebuilding, (June), pp 155–163. https://doi.org/10.1061/(asce)ae.1943-5568.0000054
3. Saurav D et al (2017) Area of linkage between lean construction and sustainability in indian construction industry. Int J Civ Eng Technol (IJCIET) 8(August):623–636
4. Bae J-W, Kim Y-W (2008) Sustainable value on construction projects and lean construction methods. In: Proceedings of IGLC-15, July, Michigan, USA, 3(July), pp 312–321. https://doi.org/10.3992/jgb.3.1.156
5. U.S. Environmental Protection Agency. EPA'S report on the environment (2003 Draft). U.S. Environmental protection agency, Washington, DC, 2008

6. Kibert A et al (2010) Slik operations. J Constr Eng Manag. Elsevier Masson SAS 42(10):26–31. 10.1061/(asce)0733-9364(2006)132:4(384)
7. Ballard G, Howell GA (2004) Competing construction management paradigms, (1992), pp 1–8. https://doi.org/10.1061/40671(2003)39
8. Koranda C et al. (2012) An investigation of the applicability of sustainability and lean concepts to small construction projects, vol 16, pp 699–707. https://doi.org/10.1007/s12205-012-1460-5
9. Dehdasht G, Zin R (2018) Barriers of sustainable lean construction implementation: hierarchical model. Int J Adv Res 6(9):1094–1101. https://doi.org/10.21474/ijar01/7618
10. Luo Y, Riley DR, Horman MJ (2005) Lean principles for pre-fabrication in green design-build (GDB) projects [online]. In: Kenley R (ed) 13th international group for lean construction conference: proceedings. International Group on Lean Construction Sydney, pp 539–548. Availability: ISBN: 1877040347. [cited 13 May 19]
11. Besser Freitag A, Anholon R, Martins de Oliveira V, Vivanco Larrain T (2017) Integration of concepts about lean construction, sustainability and life cycle of buildings: a literature review. Braz J Oper Prod Manag 14(4):486–499

Efficiency Measure of Form Work in Construction: A Comparative Assessment

Sarbesh Mishra

Abstract Architectural planning, engineering, and building are the major areas in the complex, process-driven, field of construction. When it comes to large housing projects in contemporary times, the speed of construction is given paramount importance. Among the emerging technologies in mass housing projects, formwork system happens to be the favourite for two reasons: completing the project in less time and obtaining good concrete finishing without change in the investment. Thanks to the phenomenal rise in building of residential, commercial, and industrial buildings and structures, formwork system is in great demand. Formwork system also owes its great demand due to record building of mega infrastructure projects such as transport infrastructure (R&B), water resources engineering works, sports buildings, stadiums, and so on. Fast-track construction largely depends on formwork system. For example, Mivan, a modern method consisting of an aluminium-based formwork system, is found widely used in the building of housing units and mass residential projects. Suitable to Indian conditions for its low maintenance feature and cost-effectiveness, it is aluminum formwork that can be tailormade to suit various requirements. This paper attempts to compare the technology cost of aluminium-based Mivan formwork with that of conventional construction formwork technology. It works on the hypothesis that Mivan technology is more cost-effective, qualitatively superior, and time-saving as compared to conventional formwork technology.

Keywords Formwork · Formwork system · Mivan · Fast-track construction · Cost-effectiveness · Adaptable · Time-saving · Minimum maintenance

1 Introduction

Distinguishing proof of variables which can diminish the quantity of work hours required to raise, adjust and strip concrete formwork will improve the cost adequacy of development activities. Formwork work costs comprise almost 35% of the all-out expense of work. To keep up work cost at the very least, the board and streamline

S. Mishra (✉)
NICMAR, Hyderabad, India
e-mail: sarbeshmishra@nicmar.ac.in

B. B. Das et al. (eds.), *Recent Developments in Sustainable Infrastructure*, Lecture Notes in Civil Engineering 75, https://doi.org/10.1007/978-981-15-4577-1_80

conditions at the site by concentrating on the components which will improve work efficiency. Redundant structure measurements, legitimate framework choice, proficient booking, and cautious movement coordination can yield huge profitability investment funds. Notwithstanding, deficient material accessibility, adjust, inappropriate sequencing, and disturbances are among recently recognized causes which can add to poor formwork profitability. Improving formwork development proficiency requires recognizable proof and measurement of the elements which can be appeared to affect the formwork group's profitability. Customary formwork for solid development regularly comprised of specially decided requiring talented skilled workers. This sort of formwork frequently had poor security highlights and gave moderate paces of on location development and immense degrees of waste additionally being produced. Formwork framework has huge job in the development procedure, settling on the correct choice by picking the proper formwork framework could prompt reaction to reasonable development. Various frameworks have their very own advantages and picking a right formwork having best backings individual venture necessity non-accessibility gifted and specialists brings about issues of expense and time overruns, substandard development, poor completions spillages, consumption of structures, etc. This normal life expectancy can be enhanced by maintaining a strategic distance from fixes and recovery of structures.

2 Need in Present Indian Construction Market

The earliest formwork frameworks utilized wooden scantlings and timber sprinters as it enables simple shaping and making at site. These wooden scantlings and timber sprinters has started losing their auxiliary and dimensional properties over a period time encompassing the security issues. Significant mishaps occur in strengthened bond concrete (RCC) development in light of mediocre formwork and platform. The speed of development by this framework will outperform speed of a large portion of the other development frameworks. The work in a joint effort with substantial apparatuses (e.g., Tower crane) handles this strategy adequately to accelerate the development, to guarantee quality control and toughness. Appropriation of this framework diminishes, generally speaking, expense of the structure. Our investigation for the most part center around the innovation utilized in Mivan framework, its segments, working cycles, its repetitions. While the presentation of worldwide organizations in India for development exercises has helped in quick development. For building mass lodging works, it is important to have most recent advancements which are fit for fast development and can convey great quality and strong structure in financially savvy way.

3 Scope of the Study

With the gigantic increment deprived for motorization in the development strategies there are number of variables to be considered in touching base at a choice of the most proper strategy and framework or innovation to be utilized for development. In choice among various elective techniques and advancements for the quicker and practical development of structures, it is important to detail a base structure which helps to help the arranging group to pick the most suitable arrangement of formwork according to the particular prerequisite of the undertaking.

4 Characteristics of Good Formwork

1. Good formwork should facilitate easy removal and strip after concreting.
2. Good formwork must ensure sufficient tight joints to prevent the grout of getting lost as grout leakage cause honeycombing of the surface.
3. Good formwork should be sufficient in strength to support the weight of wet concrete, the weight of workers with their equipment, force of vibration, and the force of wind and rain that may result because of seasonal influences.
4. Good formwork should be capable of being re-used. To incorporate this, the formwork must be oil coated to facilitate easy detachment and cleaning should be undertaken immediately after striking.

4.1 Mivan Formwork

The Mivan formwork is comprised of an aluminum compound. While development is in procedure, the formwork should bear, other than its very own weight, the heaviness of wet cement, the live burden because of work, and the effect because of pouring cement and laborers on it. The vibration caused because of vibrators used to smaller the solid ought to likewise be taken consideration off. Hence, the structure of the formwork considering its prerequisites is a fundamental part during the development of the structure. The Mivan formwork ought to have the option to take a live burden including the effect about 370 kg/m^2. It is in any case, common to work with a little factor of well-being in the structure of formwork. The surfaces of formwork ought to be wearing such a way, that after avoidance because of weight of cement and fortification, the surface stays flat, or as wanted by the planner. The sheathing with full live heap of 370 kg/m^2 ought not to avoid more than 0.25 cm and the joists with 200 kg/m^2 of live burden ought not to redirect more than 0.25 cm maintaining the respectability of the details. The measured idea of the Mivan formwork ought to permit simple fixing and evacuation of formwork and the development can continue

Fig. 1 Sample Mivan formwork

rapidly with next to no deviation in dimensional resilience. Moreover, it should be quite flexible and adaptable for any modifications in the layout (Fig. 1).

5 The Advantages of the System

In rapid construction of myriad layouts of architecture, the Mivan formwork is designed for:

(1) Cost-effectiveness.
(2) Design customization to suit project requirements.
(3) Erected using unskilled labor.
(4) Finishing of high quality.
(5) Panel reusability up to 250 times.
(6) Total system forms the complete concrete structure.
(7) Unsurpassed construction speed.

Quality as well as speed must be given due thought alongside cost-effectiveness. Great quality development will never dissuade to activities speed nor should it be uneconomical. Truth be told, tedious fixes and alterations because of low quality work, by and large, postpone the activity and cause extra-budgetary effect on the venture. A few specialists feel that lodging options with low support prerequisites might be favored regardless of whether the underlying expense is high.

6 Limitations

Despite several advantages of Mivan formwork, it also suffers some limitations. Of course, the limitations are insignificant in comparison with its merits and uses. Following are the limitations:

- Due to minuscule sizes, finishing lines are traceable on the concrete surfaces.
- Due to small thickness of components, concealed services are not possible.
- Without strategic planning and uniform elevations, cost-effectiveness cannot be possible.
- As it is castes in RCC, least modification is possible.
- Cost-effectiveness depends on the high volume of work, at least 200 repetitions.
- The formwork necessitates numbers of spacer, wall ties, etc., placeable at 2 feet c/c, which cannot rule out monsoon problems like seepages and leakages.

7 Analysis of Formwork Efficiency

The following four parameters are considered to determine the formwork efficiency, namely:

1. Cost 2. Quality 3. Time 4. Productivity

7.1 Estimation of Cost

Requirement of material is dependent on the design of the construction projects. Labor engagement and the timeline of the project are the consequence of technologies and construction methods. Formwork catalyzes the Speed of the project. We can attribute this as the critical mass of building construction. Uses of latest technologies, the labor estimates, and the time period can be minimized in the project.

Total Project Cost = Materials + Labor + Preliminary Cost + Machineries & Equipment + Waste Material Handling + Safety + Finishes.

Note: Preliminary cost is inclusive of Salaries, Establishment charge, and Security expenses.

In mega building projects, formwork controls the timeline of the construction in view of the effect on floor cycle and facilitates in shortening time to construct the main structure that results in reduction of the total duration of the project. This research compares the cost of total project cost of a residential project in both the types of formwork.

Cost for materials can be computed in two ways:

(a) For material with short life like timber and plywood,

(Cost of material – scrap value)/no of repetitions = cost of material per use

(b) For materials with long life like steel or other metal components

(Cost of material–scrap value) * period of use/material life = cost of material

7.2 *Comparison of Formwork*

This new technology is adopted by construction industry. The construction projects are getting more complex in view of changed human endeavor and the construction industry responded to these unique challenges in terms of time, cost, and qualitative factors. Other than these, there are various factors which help us to choose the type based on the following comparative criteria.

Comparison Between Mivan and Conventional System
In view of the review and correlation from live venture key contributing elements that assume a noteworthy job in choice of formwork are cost, quality, process duration, and number of reiterations. By looking at few parameters through examination and charts from examiner, the accompanying ends can be made.

I. Cost adopted for Mivan formwork is highest among the all formwork systems this is because of the use of aluminum in making of formwork even though the cost of Mivan is high due to higher number of repetitions. The overall cost reduces which makes it favorable in repetitive kind of works.
II. Faster construction can be achieved with Mivan formwork as cycle time is less compared to conventional formwork.
III. As per safety management, Mivan formwork is the best in the industry compared with the others formwork systems.
IV. Changes can be easily accommodated for conventional when compared with Mivan. Because Mivan formwork is already customized and standardized.

8 Case Study: Residential Villa Project in Hyderabad

Our study starts with the case study of the villas located in the Kompally premises; hence, it is ongoing project we can able to get the every information regarding our study, all the information, about the project is given below. The overall project consists of 400+ villas of two types, i.e., east and west facing. Our study is on west facing which include 200 villas; since it is of same kind, so we have preferred Mivan for this project.

Project—"Ashoka-a-la-maison," Kompally
Ashoka-a-la-maison is a reputed builder into premium segment of housings, i.e., villas, situated at Kompally, Hyderabad. The project offers sprawling lawns with serene and beauty of nature having a healthy ratio of constructed to green area in the proportion of 35% and 65%, respectively. The company was formed in 1989 with 28 years of presence has come up with this premium villa segment, named as "Ashoka-a-la-maison."

Key features:

- Large open space
- Even splendid parks
- 80', 60', 40' BT roads
- Underground water supply system
- Decorative street lighting
- 24 * 7 security system
- Sewage treatment plants.

In this project, three different sizes of plot are available, i.e.,

- 250 yards, 300 yards, 400 yards.
- Two facings are available, i.e., East and west.
- 249 yards—40' * 56'-2600sft 3bed rooms
- 300 yards—45' * 60'-3200sft 3bed rooms
- 400 yards—50' * 72'-4200sft 4bed rooms

Source: www.ashokabuilders.com.

9 Conclusion

Formwork system plays a crucial role in successful and timely completion of project. From the above analysis, Mivan formwork is the highly rated formwork system because of its higher number of repetitions in use, with smooth and exact surface finish which brings down overall cost and also provides superior quality types of structures with less time. In case of nonrepetitive and availability of less labor and improper feasibility with site and storage area restrictions, wood formwork is preferred to Mivan formworks. It can be concluded without contradiction that quality and speed be given high priority for cost-effectiveness in large infrastructure or construction.

In India, Mivan technology has great potential in reducing the cost of housing. Mivan formwork helps in timely completion of the project with keeping the attributes intact. This also helps in constructing low cost unit housing and has come up with a formidable alternative to conventional formwork in this ever-growing construction

Fig. 2 Cost comparison

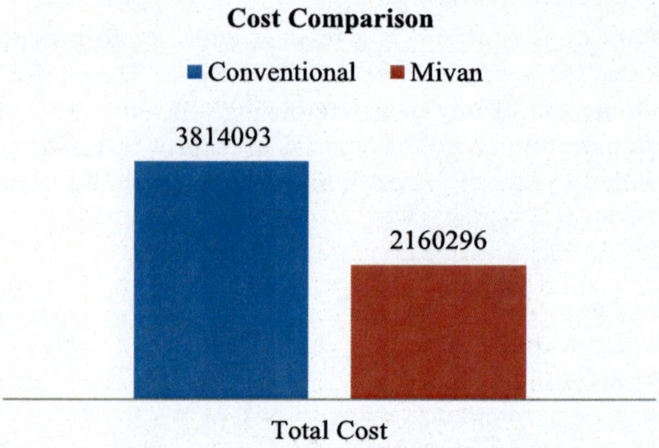

industry. We infer that Mivan form construction is enabling high-quality construction at faster pace and with a reasonable cost. We conclude our study that, by using Mivan formwork, the cost of the construction can be reduced by 43.36% (Ref—Annexure 1).

Annexure—I

See Fig. 2.

Construction Cost Comparison (Annexure—II)

Nature of work	Conventional form work (Rs.)	Mivan form work (Rs.)
Excavation	**30579.46**	**30579.46**
Footings	**300520.21**	**302606.30**
Pcc	29399.63	29399.63
Reinforcement	24116.01	26202.10
Concrete	240142.93	240142.93
Shuttering	6861.64	6861.64
Pedestal	**32168.14**	**32168.14**
RCC	30092.87	30092.87
Shuttering	2075.27	2075.27
Plinth beam	**136578.86**	**136623.85**

(continued)

(continued)

Nature of work	Conventional form work (Rs.)	Mivan form work (Rs.)
Pcc	21066.49	21066.49
Reinforcement	46135.08	46135.08
Concrete	64238.40	64283.40
Shuttering	5138.88	5138.88
Columns	**132391.31**	
Concrete	61461.47	
Reinforcement	44290.46	
Shuttering	26639.38	
GF slab	**519644.50**	**464191.24**
RCC	457390.50	457390.50
Shuttering	62254.00	6800.74
FF slab	**450367.61**	**401240.65**
RCC	394753.10	394753.10
Shuttering	55614.51	6487.55
GF beam	**300713.00**	
RCC	263474.00	
Shuttering	37239.00	
FF beam	**284478.48**	
RCC	245669.85	
Shuttering	38808.63	
GF brick work	**308977.55**	**345136.68**
9″ Wall	218743.55	233950.02
4″ Wall	90234.00	111186.66
FF brick work 9″	**285250.62**	**447750.41**
9″ Wall	285250.62	447750.41
GF plastering	**502507.55**	
Internal	293241.50	
External	209266.05	
FF plastering	**529915.72**	
Internal	312325.45	
External	217590.27	
Total cost of building (INR)	**3814093.00**	**2160296.72**

Bibliography

1. Prathul U, Leeladhar P (2015) Analysis of productivity by Mivan & conventional formwork. Int J Innov Trends Constr 2(4):38–44
2. Sajeet SB, Supreeth SG (2015) Earth quake response of different shapes of Mivan wall tall buildings. Int J Res Eng Technol 4(2):24–32
3. Dhanashri S, Desai DB (2014) Emerging trends in formwork-cost analysis & effectiveness of Mivan formwork over the conventional formwork. J Mech Civ Eng 3(6):27–36
4. Osama M, Khan Jaffer (2010) Analysis of labour productivity. Can J Civ Eng 10(3):286–303
5. Swapnali M, Kumathekar MB (2010) Comparison of the use of traditional and modern formwork systems. Civ Eng Syst Sustain Innov 5(3):43–58
6. Nuzul Azam Haron I (2005) Building cost comparison between conventional and formwork system: a case study of four storey school buildings in Malaysia. Am J Appl Sci 2(4):819–823
7. Gary R, Hanna S (1993) Factors influencing formwork productivity. Can J Civ Eng 2(1):143–158
8. Wijesekara DM (2012) Cost effective and speedy construction for high-rise buildings in Sri Lanka by using aluminium panel system formworks. Int Res J Eng Technol 4(11):1144–1147
9. Standard schedule of rates of Telangana 2018–19 and Rates from CPWD

Recent Developments in Water Resources Engineering and Management

An Approximate Cost Equation of Offshore Wind Turbine Blade

Chandan Kumar Majhi, Satyajeet Nanda, R. C. Pradhan, and Benu Gopal Mohapatra

Abstract Since many years, energy poverty and climate change are the pivotal issues faced by societies. Harnessing offshore wind energy would be a reliable solution to both of these issues. Offshore wind energy is an attractive source of green energy, because of abundant of wind resources in the sea. In compression to the onshore wind turbine, visual and noise impact no more an issue in case offshore wind turbine. Developing offshore wind turbines 15–30 km away from shore experience hostile environmental loads such as ocean currents, storm wave loading, ice loads, and potential ship impact loads significantly contributes the high cost in offshore structure. The turbine blade is an important constituent of the wind turbine. It costs around 22% of the total cost of a wind turbine. In this paper, an approximate cost equation for turbine blade was derived. This cost equation is sensitive to the quantity of various materials used in turbine blade and diameter of the wind turbine. This cost equation can be used to compute the material cost for a wind turbine up to a diameter of 80 m.

Keywords Wind turbine · Turbine blade · Installation · Power output

C. K. Majhi (✉)
M. Tech Student, KIIT University, Bhubaneswar, India
e-mail: Chandankumarmajhi93@gmail.com

S. Nanda
Associate Professor, SCE, KIIT University, Bhubaneswar, India
e-mail: Satyajeet.nandafce@kiit.ac.in

R. C. Pradhan
Assistant Executive Engineer, PWD, Government of Odisha, Bhubaneswar, India
e-mail: Ramapradhan68@gmail.com

B. G. Mohapatra
Dean, SCE, KIIT University, Bhubaneswar, India
e-mail: Dean.Civil@kiit.ac.in

B. B. Das et al. (eds.), *Recent Developments in Sustainable Infrastructure*, Lecture Notes in Civil Engineering 75, https://doi.org/10.1007/978-981-15-4577-1_81

1 Introduction

Widely used coal base energy is no more viable solution for long-term energy requirement as the cost of environmental impact for energy is enormous. From many years, an effort has been made to reduce the adverse environmental impact of coal-based energy, but this has achieved little success. At present, the various national and international agencies are promoting green energy and strongly discourages any further coal-based energy initiative. In recent time, significant advancement has been made to harness energy from wind by using wind turbines. The capacity of the wind turbine now increases to many folds. Large capacity wind turbine comes with a large size blade, which produces a lot of noise and reduces the visibility of that area. Therefore, installing a wind farm nearer to a human habitat is not desirable. A country like India where landmass is being shrunk, hence, is not desirable to install any land base (onshore) wind turbine farm. Alternatively, the wind farm can be installed at 15–30 km away from the shore (offshore), and this is appeared to be very feasible as India consists of very long shoreline. The wind turbine mainly consists of a support structure, rotor blade, tower, and nacelle. The Support Structure and Foundation have to keep the turbine in proper locations against the hostile environmental loading condition.

The initial cost of an offshore wind turbine installation is usually more compared to the onshore wind turbine. However, in the long term, the offshore wind turbine may become more economical as abandon of wind resource and availability of space. The world's wind energy facilities were reached 194,390 MW at the end of 2010, which is about 2% of the total global energy industry. According to GWEC, global wind power will increase by 160% over the next 5 years. As of now 3924 MW in 49 wind farms spread over nine European countries, in which the United Kingdom (UK), Denmark, Netherlands, and Belgium are the top countries in offshore installed capacity (The European Wind Energy Association, 2011). The present trend shows energy industry moving toward dipper water wind and far away from the shore. Countries like the UK and Germany invest in a big way in offshore wind plans and move faster to become a leading country in harnessing green energy resources [1]. In the near future, 10% of electricity demand will be fulfilled through offshore wind turbine [2]. Similarly, in America, many new offshore wind projects are coming in the last few years like the Cape Wind project, Cape Cod, and the Blue-water wind project [3]. According to a study by China Wind Energy Association, Fujian, Guangzhou, Jiangsu, Hainan, Shandong, and Zhejiang have shown great resources for offshore wind energy. It has been estimated that the sea around China posses offshore wind energy generation potential of 43% in shallow water (under 30 m depth) and 34% in deep water (depths beyond 50 m) [4]. An offshore wind turbine is the fastest growing energy initiative that has been taken up by many countries around the world [5–7]). At present, India is more concern in harnessing solar energy, which requires a very large load area and hence not suitable for highly populated countries like India. This research attempted will be made to devolve a cost equation for an offshore wind turbine blade. This cost equation should be sensitive to the diameter of the blade and quantities of various materials used in the manufacturing of wind turbine blade.

2 An Overview of the Cost of Various Components of an Offshore Wind Turbine

Xiaojing et al. [8] estimated that the cost range of electricity generation is between 6 and 12 £ct/ kWh at offshore to that compared with 3–8 £ct/kWh for onshore sites. About 57% of the cost of a wind turbine is contributed to the turbine and support structure. Increase in water depth and coastal distance increases the cost of the wind turbine. The foundation cost for a 1.5 MW wind turbine has increased £317,000 at 7 m depth to £352,000 at 16 m depth. This has increased 11% cost increase. About 60% cost increased as installation moved from shore to 200 km into the sea and 40% cost increased as water depth increases from 10 to 50 m. The turbine cost includes blades, tower, and transformer, which is 33% of the total cost of offshore wind. The support structure like foundation cost is around 24% of the total cost. The grid connection and other material costs about 15% of the total cost. The O&M cost varies according to distance that varies from offshore. The turbine and grid connection cost remains the same with an increase in water depth. The foundation and installation cost will vary with an increase in water depth. Xiaojing [8] reported that operation and maintenance unit cost will increase £100,000 to £300,000 per year per turbine. The various studies suggested that economics of scale can be achieved by a large capacity wind turbine. It is estimated that the cost per kWh can be reduced by 9–17% by increasing the wind turbine capacity. Offshore wind may be expensive today; however, a substantial reduction of cost can be expected over the long term by economies of scale, learning effect, and R&D effort. Primarily, cost reduction in offshore wind depends on R&D, which includes new technique and material.

3 Cost Equation of Wind Turbine Blade

A turbine blade is the most important part of a wind turbine as it observes the wind energy and transmits to the rotor to convert it into the electrical energy. The more is the blade size, the higher will be the energy output. The size of the blade may vary from 10 to 125 m, and the future turbine may come with a turbine blade size of 250 m. Cost of blade influences significantly to the overall project cost as, on an average, the blade cost contributes 22% cost of the total project (Renewable energy technology cost analysis series, 2012). Figure 1 shows the typical cross section of a wind turbine blade. The main material used in the blade is fiberglass, core, resin, adhesive, and root study. In order to derive the cost equation for the blade, first of all in subsequent sections, the relationship between quantities of blade materials and

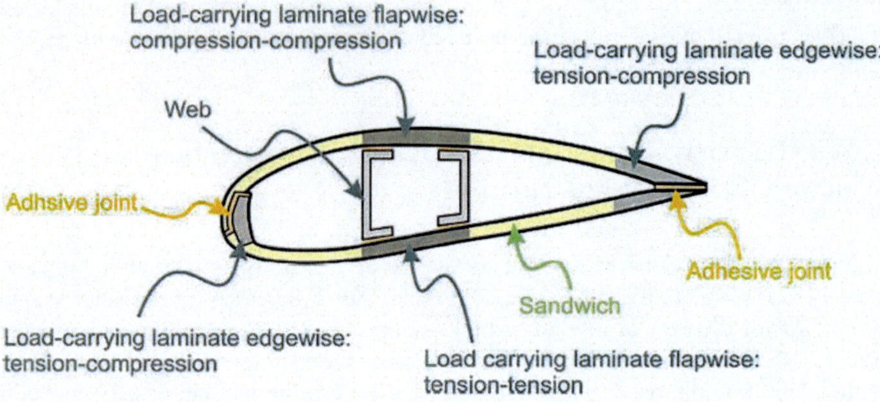

Fig. 1 Cross section of a wind turbine blade [3]

Table 1 Materials used in wind turbine blade (Private communication)

Blade Dimension (m)	30	50	70
Material (kg)			
Fiberglass	2523	11,590	30,587
Core	190	864	2313
Resin	1252	5800	15527
Adhesive	77	189	363
Root studs	66	414	1177

blade diameter will be developed. Thereafter, these relationships shall be multiplied with the unit cost of each material to determine the total material cost.

A database on the material used in the wind turbine blade was prepared. The information on blade materials was collected from various manufacturers through private communication. One such information has been given in Table 1. The major components for the wind turbine blade are fiberglass (around 61%) and resin (around 30%), and percentage of rest of the materials is less than 5. With an increase in blade diameter from 30 to 70 m, the volume of material can increase up to 10 times. The interrelationship between the quantity of each material and the diameter of the wind turbine was derived by applying the method of regression analysis to the database. In the subsequent sections, the result of regression analysis of each material will be discussed.

3.1 Fiberglass

Figure 2 represents the relationship between blade diameter and quantity of fiberglass. Quantity of fiberglass increased exponentially with the diameter of the wind turbine

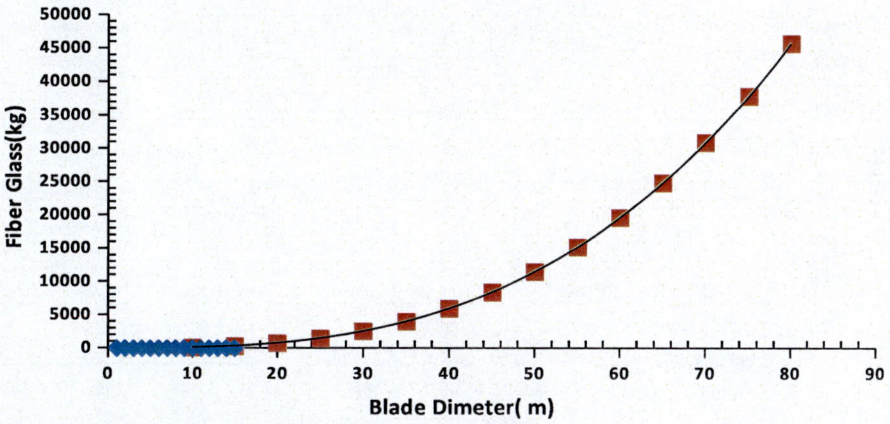

Fig. 2 Relationship between blade diameter and fiberglass

blade. About 45 tones of fiberglass may be required for blade diameter of 80 m. About 18 times of increase in the quantity of fiberglass was found, while the diameter of blade increases from 30 to 80 m. The mathematical expression for the quantity of fiberglass and diameter of the blade can be expressed as

$$Y_{Fg} = 0.1121 * X^{2.95} \quad (1)$$

where Y_{Fg} = quantity of fiberglass in kg, X = blade diameter in m.

3.2 Core

Figure 3 represents the relationship between blade diameter and quantity of corc. Quantity of core increases with the diameter of the blade. Between blade of diameter 30 and 80 m, about 18 times increased in the quantity of core was observed. The interrelationship can be expressed mathematically as

$$Y_{core} = 0.0083 * X^{2.95} \quad (2)$$

where Y_{core} = quantity of Core in kg., X = blade diameter in m.

3.3 Resin

Figure 4 represents the relationship between blade dimension and resin. From Fig. 4, it can be seen that the quantity of resin increased exponentially with the increase in

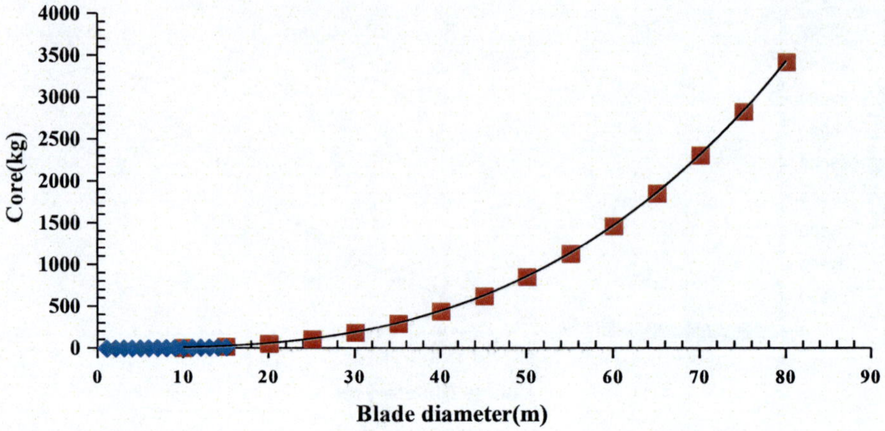

Fig. 3 Relationship between blade diameter and core

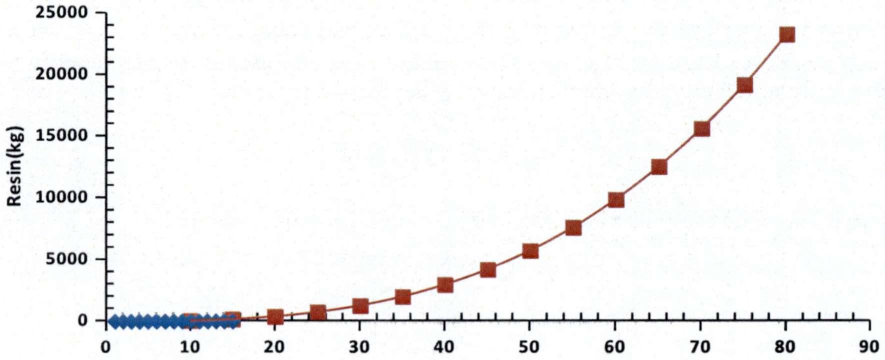

Fig. 4 Relationship between blade diameter and resin

the blade diameter. Forty times increase in resin was observed with the increase in the blade diameter of the wind turbine from 10 to 80 m. The interrelationship between quantity of resin and blade diameter is

$$Y_{resin} = 0.0509 * X^{2.97} \tag{3}$$

where Y_{resin} = quantity of resin in kg and X = Blade diameter in m.

3.4 Adhesive

Figure 5 shows the interrelationship between blade diameter and quantity of adhesive.

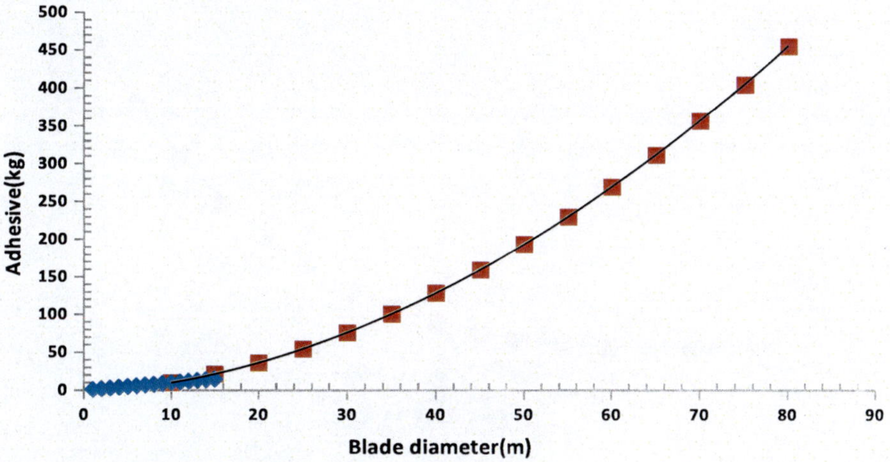

Fig. 5 Relationship between blade diameter and adhesive

From Fig. 5, it can be seen that the quantity of adhesive increased with the increase in the blade diameter. About six times increases in the quantity of adhesive were found, while the diameter of blade changed from 30 to 80 m. Mathematically, the interrelationship can be expressed as

$$Y_{adhesive} = 0.1541 * X^{1.84} \quad (4)$$

where $Y_{adhesive}$ = quantity of adhesive in kg., and X = blade diameter in m.

3.5 *Root Studs*

Figure 6 represents the relationship between the blade diameter and root studs. The interrelationship suggests an increase in root studs with an increase in diameter of the wind turbine blade. More than 28 times increase in root studs was observed when the diameter of the wind turbine blade increase from 30 to 80 m. Mathematically, the interrelationship can be expressed as

$$Y_{rootstud} = 0.0006 * X^{3.41} \quad (5)$$

where YRootstud = quantity of root stud in kg, and X = blade diameter in m.

Equations 1–5 and Figs. 2, 3, 4, 5 and 6 show the relationship between quantities of various materials used in wind turbine blade and blade diameter. Theses relationships were obtained from the regression analysis of a database as discussed above. Table 2 shows more detail of results of the regression analysis.

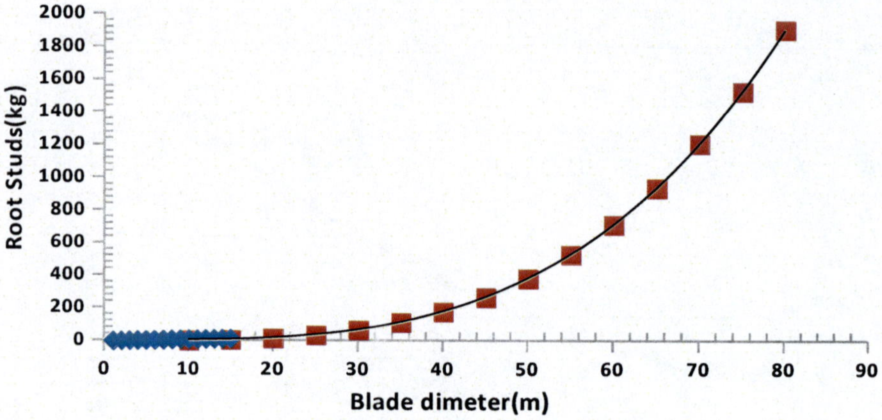

Fig. 6 Relationship between blade diameter and root studs

from the regression analysis. Using Eqs. (1)–(5), the cost of materials used of a given diameter of the blade can be obtained by multiplying the unit cost of the individual material. The cost equation for blade would be as

$$B_C = Y_{Fg} * x_{fg} + Y_{core} * x_{core} + Y_{resin} * x_{resin} + Y_{adhesive} * x_{adhesive} + Y_{rootstuds} * x_{rootstuds} \quad (6)$$

where B_C = total material cost of a wind turbine blade, x_{fg} = price of fiberglass per kg, x_{core} = price of core per kg, x_{resin} = price of resin per kg, $x_{adhesive}$ = price of adhesive per kg, and $x_{rootstuds}$ = price of root stud per kg. Using Eq. 6, change in material cost of a wind turbine blade with the blade diameter is shown in Fig. 7. The material cost increases exponentially with the wind turbine blade diameter. The unit prices of each material used in Fig. 7 are x_{fg} = ₹300/kg, x_{core} = ₹775/kg, x_{resin} = ₹1552/kg, $x_{adhesive}$ = ₹4146/kg, and $x_{rootstuds}$ = ₹350/kg. Equation 6 does not include other costs that may be associated with the wind turbine blade like production cost, transportation cost, profit, etc. Apart from material cost, all other costs are very dynamic in nature and cannot be fixed with the method of regression.

Table 2 Material varies with different blade dimensions

Blade Dimension	10	15	20	25	30	35	40	45	50	55	60	65	70	75	80
Material															
Fiberglass	99.5	328.7	767.7	1482.08	2536.9	3996.4	5924.3	8383.8	11437.7	15148.47	19578.2	24788.7	30841.7	37798.4	45719.8
Core	7.4	24.52	57.3	110.727	189.63	298.3	443.1	627.3	856.187	1134.264	1466.30	1856.96	2310.882	2832.6	3426.93
Resin	47.9	160.1	376.6	731.4	1257.9	1989.5	2959.6	4201.1	5747.1	7630.560	9884.15	12540.7	15632.89	19193.3	23254.5
Adhesive	10.2	21.53	36.4	54.692	76.272	101.33	128.9	159.80	193.670	230.447	270.088	312.549	357.79	405.77	456.477
Root studs	1.56	6.248	16.6	35.781	66.702	112.33	178.21	266.49	381.946	528.937	712.022	935.932	1205.56	1525.9	1902.37

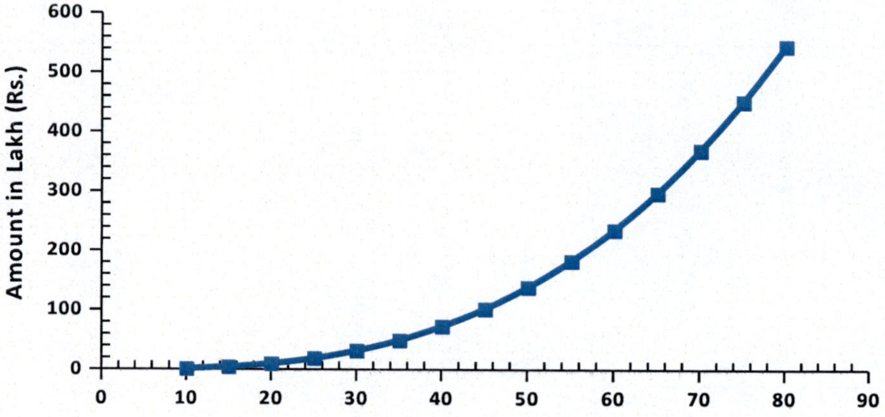

Fig. 7 Cost of the material of wind turbine blade with the diameter

4 Conclusion

Cost equation for wind turbine blade material has been proposed. The cost equation was derived using a regression analysis of a database. The blade material cost varies nonlinearly with the blade dimension. The quantity of blade materials also varies nonlinearly with the blade diameters. The quantity of materials required for a wind turbine blade of 80 m varies in the range of 6–28 times that of a 30 m blade. The material cost of an 80 m wind turbine blade is about 18 times to that of a 30 m wind turbine blade. The proposed cost equation can be used to compute the material cost of an offshore wind turbine.

Bibliography

1. Jones AT, Westwood A (2005) Recent progress in offshore renewable energy. In: Power, San Francisco, USA: IEEE. Lynn and inner dowsing offshore wind farm (2011)
2. McConnel N (2011) Offshore wind power. Alstom power. http://www.centrica.com/files/pdf/centrica_energy/lynn_and_inner_dowsing_newsletter.pdf; http://www.mpi-offshore.com/projects-1/lynn-and-inner-dowsing-wind-farms/
3. Breton SP, Moe G (2009) Status, plans and technologies for offshore wind turbines in Europe and North America. Renew Energy 34(3):646–654
4. Enslow R (2010) China, Norway and offshore wind development: "a win-win relationship?" In: Presented at offshore wind China conference and exhibition. http://www.norway.cn/PageFiles/391359/Azure%20International%20-%20Rachel%20Enslow.pdf, 15–16 March, 2010 [Bergen, Norway]
5. Esteban MD, Diez JJ, López JS, Negro V (2011) Why offshore wind energy? Renew Energy 36(2):444–450

6. Zhang X (2011) Current status and trends of offshore wind energy. Electric Age (3):46–8
7. Kaldellis JK, Zafirakis D (2011) The wind energy revolution: a short review of a long history. Renew Energy 36(7):1887–1901
8. Xiaojing S, Diangui H, Wu G (2012) The current state of offshore wind energy technology development. Energy 41:298–312

A Review: Effect of Turbidity Current on the Reservoir Sedimentation

Bidisha Byabartta, Tanmoy Majumder, Paromita Chakraborty, Jyotiprakash Padhi, and Bitanjaya Das

Abstract Reservoir is frequently vulnerable through damage of capacity owing towards sedimentation. Sedimentation is an inherent procedure in the reservoir. A continuous sediment deposition, which is being transported by the alluvial rivers in the reservoir cannot be fully eradicated rather it can be controlled. Nowadays, reservoir sedimentation is a burning problem, especially in India. For the sustainability of this massive structure (reservoirs) various measures are taken into account. If the sediment inflow is larger than that of the sediment outflow, the effective life of the reservoir is shortened. Turbidity currents generated in the reservoirs play a significant role in the reservoir sedimentation. The term turbidity current is a natural phenomenon which is frequently undecided besides the sediment support appliance (e.g. settling, turbulence, etc.) may be operated simultaneously within a sediment-gravity flow. The turbidity current is the utmost stereotypically a submerged current of typically quickly moving, downhill flow of water produced through the increased density for the high quantities of sediment. It is also responsible for disturbing the vast amounts of sediments in the deepwater. If the turbidity current increases in water, the aforementioned converts thicker besides a smaller amount of flawless for the higher concentration. Because of sedimentation, the yearly loss rate of the storage capacity of reservoir is estimated as 1–2% of the total storage capacity of the reservoir. This paper focuses on an extensive study of the literature and analysed and compared the results of the previous studies on turbidity current of the reservoir sedimentation. It is found that venting of the turbidity current that reaches in the dam is economical as well as beneficial as per the present scenario of the various reservoirs which are facing the sedimentation in it.

Keywords Turbidity current · Reservoir · Sedimentation · Sediment deposition · Sediment transportation

B. Byabartta (✉) · T. Majumder (✉) · P. Chakraborty · J. Padhi · B. Das
Department of Civil Engineering, KIIT University, Bhubaneswar, India
e-mail: bidisha2010byabartta@gmail.com

T. Majumder
e-mail: 1823003@kiit.ac.in

B. B. Das et al. (eds.), *Recent Developments in Sustainable Infrastructure*, Lecture Notes in Civil Engineering 75, https://doi.org/10.1007/978-981-15-4577-1_82

1 Introduction

The sedimentation of the reservoir is dangerous for the development of sustainable water resources. Deposition of sediment in the reservoirs decreases the capacity of storage [1] and positions the hazards of obstructions consumption constructions as well as sediment entrainment in hydropower arrangements [2]. It has been calculated that the loss of storage in reservoir worldwide is in between 0.5 and 1% of total storage capacity of the reservoir. Reservoir sedimentation is one of the major issues threatening reservoir storage capacity. As the reservoir capacity is decreasing gradually due to sedimentation, it is important to look for various measures and by building new reservoir this issue cannot be solved. Turbidity currents are usually defined as sediment-gravity flows in which sediments are suspended by the fluid turbulence. These turbidity currents have a high level of sediment concentrations when occurring during floods and the sediments flow to the most deepest zones of the reservoir along the dam, where sediments get settled while covering the bottom outlet and interfering with the operations of the inlet structures. Due to the density difference caused by the suspended fine particles it results in the flow of turbidity current. These are water flows which are loaded with sediment, the interchange downslope in then the stagnant waters, like the oceans, the lakes and the reservoirs. Their heavy force is increased after postponed matter, which condenses flowing turbid-water heavyweight than the clear water above. We can increase the sustainability of reservoir operation if we can influence the turbidity current in a way that the sentiments do not get deposited in the critical locations.

2 Overview of Turbidity Currents

Turbidity current, the most stereotypically submerged current, is a usually quickly moving, sediment-laden water affecting down the slope. Turbidity current can be instigated by collapsing slope, earthquakes and other geological formations. It involves three parts: the head, the body and the tail. While the body and tail movements of the turbidity current are both outstanding to gravitational forces, the head of the turbidity current is always undertaking by the density changes (the cause of pressure gradient). To explain the flow of appliance of the river-induced turbidity current, the field observations of the turbidity currents in a reproduction lake were approved in the reservoir and its foremost incursion river. Here, Fig. 1 describes the flow mechanism.

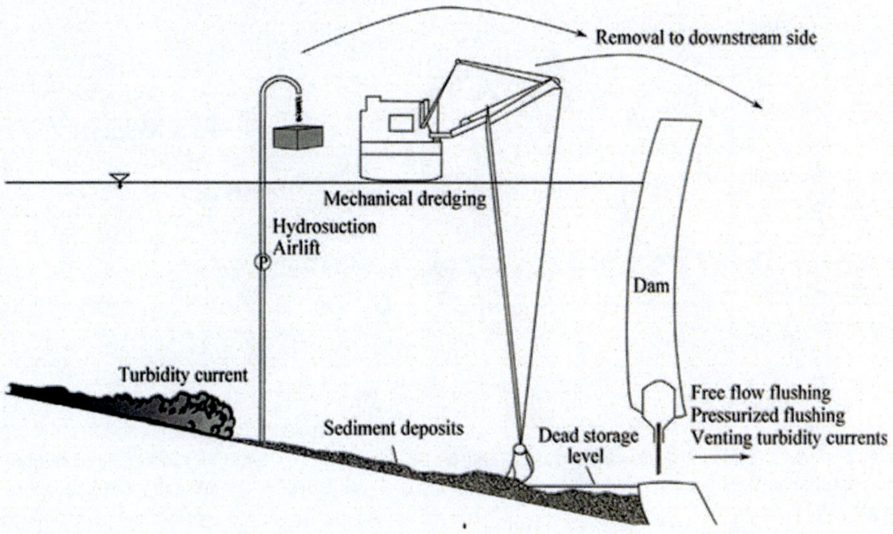

Fig. 1 Turbidity current is roaming along the bed of the reservoir and is vented through bottom outlet

3 Initiation and Evolution of Turbidity Currents

The diversity of beginning mechanisms for the turbidity currents, with practically instantaneous proceedings, for example, seismogenic slumping [3], produces the surge type of currents. This type of current has investigational analogue in the predictable volume of currents created via issue of the lock gate. It might contain an additional or fewer evidently discriminated head, body and tail through an overall reduction together in mean velocity and sediment concentration near the back after the head. The velocity current, transient into a fixed point, will quickly spread its maximum value when the head arrives. After that, the flow diminishes throughout the length of the body and the tail in Fig. 2. The surge type of currents might permit to complete the stages of comparatively steady flow. On behalf of exemplar, the 1929 Grand banks turbidity currents, activated through slope failure subsequent to an earthquake, sustained intended for almost minimum 2–3 h [4]. It was adequately stable towards progress of a huge scale of bedforms in the valley.

Some numerical and analytical representations of the turbidity currents have created an explanation on behalf of the steady flow. But, submission of the steady-flow replicas towards understanding of pledges has not been normally popular, in spite of indication in deposits for the quasi-steady flow [5].

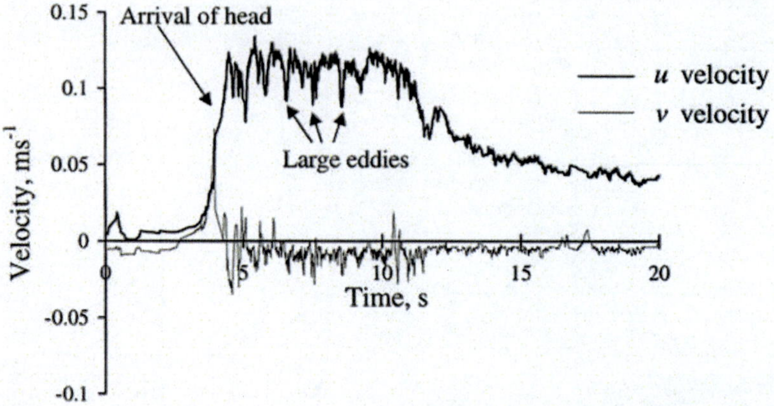

Fig. 2 Time sequences of down stream (u) and vertical velocity (v) restrained in an experimental, lock-exchange brine current, which is by means of the laser Doppler anemometry (from Kneller et al. 1997)

4 Former Lessons on the Turbidity Currents

Previous explanations of lacustrine-turbidity currents were completed by Forel [6] in Lake Geneva, wherever the pedigrees of the underwater canyons in that region of the delta were briefly described. If turbidity currents happened the most conspicuously, noticeably outstanding towards the in height rather stumpy absorptions (a smaller amount of 1 g/l) and suspended-sediment concentrations are essential to mark they appear. Turbidity currents produced in a seawater environment [7] essentially advanced suspended-sediment concentrations (some 35–45 g/l). In the calculation, turbidity currents show an important character in the sediment transport in the reservoir. Sure structures of their present concert are measured. Current major explanations of the turbidity currents in Switzerland were over completed in the Lake Geneva [8]. Simpson (1987) stretches a worthy overall impression of the gravity currents in atmosphere like in laboratory. The actual systematic concentration in the turbidity currents, still, it is happening with the numerous study curriculums on the turbidity currents in the reservoirs. Statistics composed throughout the determining operations on sedimentation and turbidity currents in the Lake Mead, shaped through Hoover Dam, remained shown in U.S. in the section of reclamation then described via Grover and Howard [9] and Howard (1953). Bell [10] has accepted widespread investigational effort happening falls. It has highlighted this reputation of the turbidity currents in sedimentation of reservoirs. In India, in the Krishna–Godavari Basin (the Bay of Bengal, India), there is a depositional model planned for deepwater petroleum reserves and sands (Pliocene). It is based on the investigation of 313 m of predictable cores from the three wells, five depositional facies have been interpreted: (1) sandy slump, sandy debrite, sandy slide and sandy cascading flow; (2)

sandy tidalite; (3) muddy tidalite; (4) muddy slump and debrite and (5) hemipelagite. Slumps and debrites establish up to 99% in one well. Here, the common thing is sand injectites.

5 Trials on Turbidity Currents

Sediment-laden currents: The experiments that were showed in order to regulate the performance of turbidity currents present with non-cohesive silt, which moves down the slope of the bed and it is enveloped by comparable silt. The sediment-laden turbidity currents necessitate to be located clambered by ordinary systems consuming a dimensionless settling velocity. This relation of incurable settling velocity of sediment particles towards certain velocity scale is a measured representative of the currents [11]. Electrostatic forces may affect the fine sediments (for example, clay, chalk and charcoal) which have the properly scaled settling velocity. Then type of tricky is able to remain resolved through consuming the silica flour or glass beads [12]. Upgraded sediment equivalents can be developed by using the grains of sediment density, but there is a resulting decrease in bulk density of current requirements and an upsurge in sediment concentration. Countless investigational studies considerably shorten the system by overlooking the sediment erosion and deposition, and the other solid answers to make the density difference [13].

Froude Scale Modelling: Here is an investigation of the dimensionless governing equations of the turbidity current in order to explore the scale effect. There are three types of flow situations which are measured here: (1) conservative density current; (2) purely depositional turbidity current and (3) mixed depositional or erosional turbidity current. Two commanding dimensionless numbers, the Froude number and the Reynolds number, are seen in the equations. When the metric Froude number comparison is satisfied, then the investigation displays that these results would be an invariant of scale for confirming similar moral Reynolds number then Froude number remains supposed towards vigorously similar.

6 Conclusions

This deliberation finished the environment then standing of high-density dispersal days spinal at smallest towards introduction of a period 'fluxoturbidite' [14], and then appears usual towards remaining pending experimental studies and then theory composed be able to clarify their flexibility method to deposition. On some rate, present theory appears insufficient to clarify the behaviour of certain extremely moveable thick dispersals then influences founded exclusively happening geological interpretations of credits might remain insufficient near solve matters of process.

Constructed proceeding the arithmetical solution, lone container overall assumptions cannot remain drawn, none the less the detailed performance of the turbidity currents be able to get predicted.

References

1. Fan J, Morris GL (1992) Reservoir sedimentation I: delta and density current deposits. J Hydraul Eng 118(3):354–369
2. Schleiss AJ, Franca MJ, Juez C, Cesare GD (2016) Reservoir sedimentation. J Hydraul Res 54(6):594–614
3. Heezen BC, Ewing M (1952) Turbidity currents and Submarine slumps, and the 1929 Grand Banks earthquake. Am J Sci 250:849–873
4. Piper DJW, Aksu AE (1987) The source and origin of the 1929 Grand Banks Turbidity current inferred from sediment budgets. Geo-Mar Lett 7:177–182
5. Kneller BC, Branney MJ (1995) Sustained high-density turbidity currents and the deposition of thick massive sands. J Sedimentol 42(4):607–616
6. Forel FA (1892) Le Leman Monographic Limnologique 1:543
7. Mulder T, Syvitski JPM (1995) Turbidity current generated at river mouths during exceptional discharges to the world oceans. J Geol 103:285–299
8. Lambert AM, Giovanoli F (1988) Records of riverborne turbidity currents and indications of slope failures in the Rhone delta of Lake Geneva. J Limnol Ocean 33(3):458–468
9. Grover NC, Howrad CS (1938) The role of subaqueous debris flow in generating turbidity current. J Sediment Petrol 42:775–793
10. Bell HS (1942) Stratified flow in reservoirs and its use in preventing of silting. United States, Department of Agriculture, Miscellaneous Publication, pp 491–446
11. Middleton GV (1976) Hydraulic interpretation of sand size distributions. J Geol 84:405–426
12. Parker G, Garcia M, Fukushima Y, Yu W (1987) Experiments on turbidity currents over an erodible bed. J Hydraul Res 25(1):123–147
13. Laval A, Cremer M, Beghin P, Ravenne C (1988) Density Surges: Two dimensional experiments. J Sedimentolv 35(1):73–84
14. Carter RM (1975) A discussion and classification of subaqueous mass-transport with particular application to grain-flow, slurry flow and fluxoturbidites. J Earth Sci Rev 11(2):145–177
15. Gould HR, Stewart RH (1953) Continental Sheef Sediments off the west coast of Florida. J Sediment Res 23(2):125

Optimal Use of Land and Water Resources of a River Basin: Case Study

Prabeer Kumar Parhi

Abstract Due to quick rise in population and rapid urbanization more water needs to be diverted for domestic and industrial use thus decreasing the allocation of water for irrigation in the near future. Therefore it is the need of the hour to optimally allocate available land and water resources for irrigated agriculture in a river basin so that maximum benefits from the available resources are received. In view of the above in a given agro-climate region, an optimal cropping pattern can be developed such that the return from the available land resources is maximized with minimum consumption of available water for irrigation. Hence the present study aims at maximizing net return from the cultivable crops in a command area by satisfying land availability, Irrigation water requirement, surface, ground as well as total water availability and minimum and maximum crop area constraints considering the projected population by the year 2051 AD. For the purpose of analysis, the Brahmani Basin in Odisha State has been considered as a case study. For the purpose of optimization Linear Programming (LP) model has been used. The Software package Language for Interactive Discrete Optimizer (LINDO) has been used for solving the LP model. For estimating crop water requirement, the guidelines of Food and Agricultural Organisation (FAO) have been considered. The optimal cropping pattern so developed utilizes 3186.95 MCM water and generates a net profit of Rs 183489.2 lakhs against the existing cropping pattern which uses 4033.21 MCM of water and generates a net profit of Rs 114930 lakhs.

Keywords Cropping pattern · Optimization technique · Linear programming · Brahmani Basin

P. K. Parhi (✉)
Department of Water Engineering and Management, Central University of Jharkhand, Ranchi 835205, India
e-mail: prabeer11@yahoo.co.in

B. B. Das et al. (eds.), *Recent Developments in Sustainable Infrastructure*, Lecture Notes in Civil Engineering 75, https://doi.org/10.1007/978-981-15-4577-1_83

1 Introduction

For optimal use of land and water resources in a river basin, proper policies for optimal use of land and water resources are needed. In a given agro-climate region, cropping pattern should be decided in such a way that it maximizes the return from the available resources while consuming minimum quantity of water. Among the available optimization techniques, the Linear Programming (LP) is found to have maximum acceptance due to its associated ease in solution, capacity to solve large-scale problems and easy availability of software packages. Hence for solving allocation problems LP model is used in the present study. Although large numbers of models [3, 4, 6, 7] have been developed and applied for the optimization of cropping pattern while consuming least amount of water, the present study expands the earlier models and studies its implications and use in Brahmani Basin in Odisha State using LP model.

2 Study Area

Brahmani River is the second largest river in the state of Odisha after Mahanadi River. This river passes through three distinct topographical regions. The upper reach of river runs through hilly catchments of Chhotnagpur Plateau with dense tropical forest, the middle reach through erosion plains and the lower reach through the deltoic plains close to seashore. Within Orissa, there are 21 major tributaries of this river. Rain is the only mode of precipitation in the basin. The average annual rainfall of the basin is 1304.8 mm. The average minimum and maximum annual rainfall of the basin is 969.2 mm and 1574.1 mm, respectively. The basin receives 90% of its total annual rainfall from the southwest monsoon, which visits the basin during June to October. At present the intensity of Kharif and Rabi irrigation is 100% and 62%, respectively, with annual intensity of irrigation 162% [2]. Figure 1 shows the details of Brahmani Basin.

3 Data Used

The Gross Irrigation Requirement (GIR) for different crops during different months of the year has been estimated using FAO Penman Monteith [1] method and average rainfall data for basin. Table 1 shows the GIR (in mm) of different crops in Brahmani Basin. It is also observed that the rainfall received by the area is sufficient to satisfy the water needs of crops during April to September. The net return from different crops (excluding the cost of water) has been calculated and is presented in Table 2.

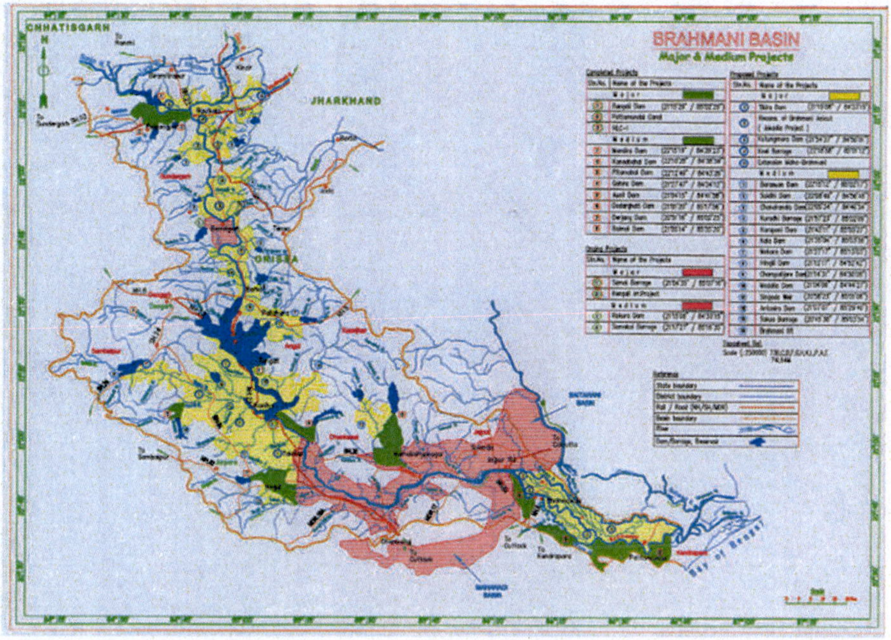

Fig. 1 Details of Brahmani Basin

4 LP Model for Crop Area Allocation

The area allocation model uses a Linear Programming (LP) approach is formulated to allocate the land area among various crops to maximize net return from the crops in the command area. The model is subject to the availability of surface and groundwater, land area limitations under different crops and in different seasons of the year. The model consists of following objective functions and constraints.

4.1 Objective Function

The objective of the present LP model aims at allocation of land area among various crops to maximize the net return from the crops as

Max Z = Max [(Return from Kharif + Return from Rabi) – (Cost of watering from surface source + Cost of watering from ground source)]

Table 1 GIR of different crops in Ramial Command area (in mm)

Crop	Jan	Feb	Mar	Apr	May	Jun	Jul	Aug	Sep	Oct	Nov	Dec
Wheat	129.1	91.2									154.9	129.1
Pulse	36.1										207.2	165.2
Potato	72.3									77.4	63.7	184.2
Vegetables	137.7	44.7									246.1	129.1
Paddy	139.4	235.8	135.9									259.6

Table 2 Net benefit from crops at full irrigation stage

Name of crop	Net benefit Rs/ha
Paddy (R)	18,300
Wheat R)	12,275
Pulses (R)	21,150
Potato (R)	25,400
Vegetable(R)	57,525
Paddy (K)	15,650
Groundnut (K)	13,200
Maize (K)	12,850
Vegetable (K)	48,900

$$\text{Max}\left[\sum_{j=1}^{NCR}\{B_{jk}Y_{jk}-C_{jk}\}A_{jk}+\sum_{j=1}^{NCR}\{B_{jr}Y_{jr}-C_{jr}\}A_{jr}\right] -\left[\sum_{m=1}^{NOM}\{C_S S_m\}+\sum_{m=1}^{NOM}\{C_g G_m\}\right] \quad (1)$$

where NCR = number of feasible crops in any period, B_{jk} = monetary return from jth crop in Rs per quintals during Kharif, Y_{jk} = yield from jth crop in quintals per ha during Kharif, C_{jk} = cost of inputs other then water for jth crop in Rs per ha during Kharif, A_{jk} = Area of jth crop in ha during Kharif, B_{jr}, Y_{jr}, C_{jr} and A_{jr} are corresponding values during Rabi, C_s and C_g are cost of surface and groundwater, respectively (in Rs/ha-m), S_m and G_m are surface and groundwater resources in mth period (ha-m), and NOM = Total number of period in a year (12 months).

4.2 Constraints

The above objective function is optimized with the following constraints.

4.2.1 Land Availability Constraint

The area under various crops during any period of time cannot exceed the cultivable command area (CCA) of the study area. Therefore,

$$\sum_{j=1}^{NCR} A_{jm} \leq A \quad (2)$$

where A_{jm} = area of jth crop (ha) in mth period and A = cultivable command area (ha) of the study area.

4.2.2 Irrigation Water Requirement Constraint

The water requirement for various crops in each month cannot exceed available surface and groundwater resources as

$$\sum_{j=1}^{NCR} A_j I_{jm} \leq S_m + G_m \quad (3)$$

where I_{jm} is the gross irrigation water needs (meter) of jth crop during mth period; S_m and G_m are surface and groundwater resources in mth period (ha-m), respectively.

4.2.3 Surface Water Availability Constraint

The gross annual surface water release cannot exceed total surface water available annually as

$$\sum_{m=1}^{NOM} S_m < \text{Gross availability of surface water} \quad (4)$$

4.2.4 Groundwater Availability Constraint

Total groundwater withdrawal annually cannot exceed annual utilizable groundwater as

$$\sum_{m=1}^{NOM} G_m \leq \text{Annual utilizable groundwater} \quad (5)$$

Assuming monthly groundwater withdrawal cannot exceed 20% annual utilizable groundwater as excessive withdrawal of groundwater may cause undermining. It is written as

$$\sum_{m=1}^{NOM} G_m \leq 20\% \text{ of annual available groundwater} \quad (6)$$

4.2.5 Total Water Availability Constraint

Total water allocated for all purposes cannot exceed the gross availability of water from surface and ground sources as

$$\sum_{m=1}^{NOM} S_m + \sum_{m=1}^{NOM} G_m \text{Total water available} \tag{7}$$

4.2.6 Minimum Crop Area Constraint

In order to satisfy basic food needs (as per present food habit) of the population; the constraint of minimum area allocation for some crops is imposed as

$$A_j \geq A_{\text{min}.j} \tag{8}$$

where $A_{\text{min}.j}$ = Minimum area of *j*th crop (ha).

4.2.7 Maximum Crop Area Constraint

In order to avoid excessive production of certain crops over minimum production requirements, maximum crop area constraint is imposed as

$$A_j \leq A_{\text{max}.j} \tag{9}$$

where $A_{\text{max}.j}$ = Maximum area of *j*th crop (ha).

5 Results and Discussion

Using the software package Language for Interactive Discrete Optimizer (LINDO) the optimal cropping pattern has been determined. It is observed from these results that the optimal cropping pattern so developed utilizes 3186.95 MCM water and generates a net profit of Rs 183489.2 lakhs against the existing cropping pattern which uses 4033.21 MCM of water and generates a net profit of Rs 114930 lakhs. Table 3 shows the comparison of existing crop areas and the crop areas to be allocated for different crops under optimal cropping pattern. It is observed that there is no need of groundwater as total surface water requirement can be met from surface sources only.

Table 3 Areas of crops under different crop plans (proposed and existing)

Crop area (in km^2)	Existing cropping pattern	Optimal cropping pattern
Paddy (K)	5375.96	1102.49
Groundnut (K)	538.72	3966.25
Maize (K)	223.30	1146.25
Vegetables (R)	70	1562.41
Wheat (R)	506.86	846
Pulses(R)	506.85	1692
Oilseed (R)	538.72	0
Paddy (Rabi)	4515.26	2114.57
Sugarcane	77.01	0

6 Conclusions

Following conclusions are derived from the above study:

i. The proposed cropping pattern which maximizes net benefit from crops meeting surface and groundwater availability as well as minimum and maximum crop area constraint gives maximum return to farmers while consuming minimum amount of water and is recommended as the best cropping pattern for the study area.
ii. The proposed crop plan gives 120% more benefit to farmers and consumes 9% less water over the existing plan.
iii. Restriction on over-production of vegetables and potato gives lowest return to farmers, thus necessitating the need of cold storage facility in the locality.
iv. Before introducing the proposed cropping pattern the farmers need to be educated about the benefit and there should be gradual introduction of suggested cropping pattern.

References

1. Allen RG, Pereira LS, Raes D, Smith M (1998) Crop evapo-transpiratio guidelines for computing crop water requirements, irrigation and drain, Paper No 56. FAO, Rome, Italy
2. Basin Planning Report of Brahmani Basin (2002) Department of Water Resources, Government of Odisha, Bhubaneswer
3. Biswas A, Pal BB (2005) Application of fuzzy goal programming technique to land use planning in agricultural system. Int J Manag Sci 33(5):391–398
4. George BA, Reddy BRS, Raghuwanshi NS, Wallender WW (2002) Decision support system for estimating reference evapotranspiration. J Irrig Drain Eng 128(1):1–10

5. Hand Book of Agriculture (2005) Department of Agriculture, Government of Odisha, Bhubaneswar
6. Li J, Inanaga S, Li Z, Eneji AE (2005) Optimizing irrigation scheduling for winter wheat in the North China Plain. Agric Water Manag 76(1):8–23
7. Singh DK, Jaiswal CS, Reddy KS, Singh RM, Bhandarkar DM (2001) Optimal cropping pattern in a canal command area. Agric Water Manag 50(1):1–8

Water Resource Management During Monsoon Months Based on SPI and CZI in Khordha District, India

Jyotiprakash Padhi, Abhilash Mishra, Shubham Choudhary, and Bitanjaya Das

Abstract Droughts took place because of decrease in amount of rainfall occurred over a large time period in a year or season. Management and planning of freshwater, which needs the understanding of historical droughts as well as their impact in the region, interpretation of the drought play a key role. Drought indices can be used for the forecasting as well as monitoring of the drought, as a result of which quantitative information about the characteristics of drought provided to the policy makers, which will be helpful for the planning as well as management of water resources. Therefore, this study was carried out in ten blocks (Balianta, Balipatna, Bhubaneswar, Jatani, Banapur, Begunia, Bolagarh, Chilika, Khordha and Tangi) of Khordha District to find the moderate, severe and extreme drought events using SPI and CZI, threshold limit of rainfall during June, July, August and September based on SPI and water resource management based on drought assessment. Monthly SPI and CZI indices are computed by using the 30 (1988–2017) years' rainfall data. Maximum number of total (moderate + severe + extreme) droughts experienced by Banapur, Balipatna and Khordha, Balipatna and Bolagarh blocks in the months of June, July, August and September, respectively, are based on SPI calculation. Banapur and Balinata, Balipatna, Begunia and Khordha, Balianta and Balipatna and Bolagarh blocks detected with maximum number of total droughts are based on CZI. Rainfall

J. Padhi (✉) · A. Mishra · S. Choudhary · B. Das
KIIT Deemed to be University, Bhubaneswar, Odisha, India
e-mail: jyotiprakash.padhifce@kiit.ac.in

A. Mishra
e-mail: abhilash1601161@gmail.com

S. Choudhary
e-mail: shubhamchoudhary27@gmail.com

B. Das
e-mail: bdasfce@kiit.ac.in

B. B. Das et al. (eds.), *Recent Developments in Sustainable Infrastructure*, Lecture Notes in Civil Engineering 75, https://doi.org/10.1007/978-981-15-4577-1_84

threshold limit was different for different blocks in June, July, August and September due to the spatial and temporal variations of rainfall, and was found to be 89.3, 182.6, 188.4 and 137.4 mm, respectively, in Khordha District. If the rainfall occurred in a particular year below these threshold limits in monsoon months, drought will occur for that month in that year. Therefore, those blocks which were experiencing frequent droughts, water-harvesting structures can be constructed in those blocks. In addition to that, low water requirement crops can be encouraged to grow in those critical blocks.

Keywords Water resource management · Drought · SPI · CZI

1 Introduction

Drought have drawn the attention of environmental, ecological, hydrological, meteorological, geological and agricultural scientists because of its categorization as an environmental disaster [1]. Droughts took place because of decrease in amount of rainfall occurred over a large time period in a year or season [2]. Decrease in the supply of water, failure of crop, deterioration of quality of water, reduction in the generation of power resulted due to the impact of drought on both surface as well as groundwater resources as a result of which also affect a host of economic and social activities [3]. Management and planning of freshwater, which needs the understanding of historical droughts as well as their impact in the region, interpretation of the drought play a key role [1]. The information on the areal extent, duration and severity of drought will be required for planning to tackle the adverse impacts of drought [4]. Drought indices can be used for the forecasting as well as monitoring of the drought as a result of which quantitative information about the characteristics of drought provided to the policy makers, which will be helpful for the management as well as planning of water resources [5]. Therefore, various indices developed by the scientists for monitoring the drought [6–10]. Palmer Drought Severity Index [10], the China-Z index (CZI) [11], Surface Water Supply Index [12], Standardized precipitation index [8] and Effective Drought Index [13] are the various drought indices used for the interpretation of drought. For monitoring of the drought in a particular area, selection of drought indices should be based on the availability of climatic data and also on the ability of the drought index for detecting the spatial as well as temporal variation of a drought event consistently [2]. Sahoo [14] undertaken the study on drought in the KBK Districts by using the drought indices such as SPI, RDI, NDVI and SDI by considering the entire district as a unit. This study was carried out in ten blocks of Khordha District to find the moderate, severe and extreme drought events using SPI and CZI, threshold limit of rainfall during June, July, August and September months based on SPI and water resource management based on drought assessment.

2 Materials and Methodology

For proper management of water resources during the monsoon months based on the analysis of drought, the drought study was undertaken in the Khordha District, Odisha, India, having a geographical area of 2813 km^2 (Fig. 1). The Khordha District is situated between latitude of 19° 55′ N and 20° 25′ N and longitude of 84° 55′ E and 86° 5′ E. The district consists of ten blocks (Balianta, Balipatna, Bhubaneswar, Jatani, Banapur, Begunia, Bolagarh, Chilika, Khordha and Tangi). The normal annual rainfall in the district is 1408 mm with maximum and minimum temperatures of 42.2 and 11.1 °C, respectively. Standardized Precipitation Index (SPI) and China-Z Index (CZI) were estimated by using the past 30 (1988–2017) years' monthly rainfall data for the ten blocks of Khordha District.

2.1 *Standardized Precipitation Index (SPI)*

SPI was proposed by McKee et al. [8] for various time scales in order to quantify the deficit or surplus of precipitation. Dry as well as wet climates can be represented using SPI. Based on SPI value, droughts can be categorized into extreme (SPI ≤ -2.0), severe ($-1.5 >$ SPI > -1.99) and moderate ($-1.0 >$ SPI > -1.49) dry; near normal ($-0.99 <$ SPI < 0.99); moderate wet ($1.0 <$ SPI < 1.49), very wet ($1.5 <$ SPI < 1.99) and extreme wet (SPI ≥ 2.0), according to McKee et al. [8]. In this study, computation of SPI was done for monthly time scale. SPI for the individual month of June, July, August and September was estimated for 30 years on monthly time

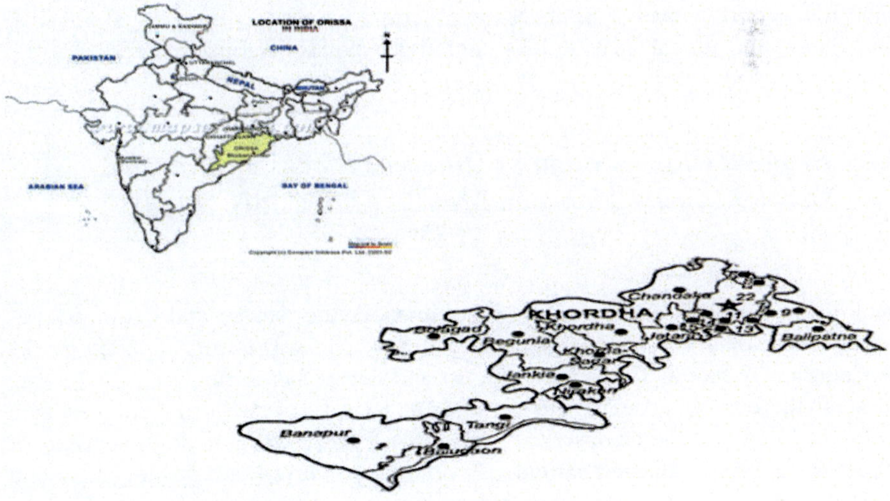

Fig. 1 The study site, Khorda District of Odisha

scale. The value of SPI was computed for these months because the highest amount of rainfall occurred during these months.

2.2 China-Z Index (CZI)

CZI used extensively for monitoring the characteristics of drought conditions across China [5, 11]. The rainfall data follow the Pearson type III distribution and also related to Wilson–Hilferty cube root transformation [15] from chi-square variable to the Z-scale based on the assumption of CZI [16]. In this study, CZI calculated for monthly time scale in June, July, August and September by utilizing the monthly precipitation data for 30 years.

3 Results and Discussion

3.1 Monsoon and Annual Rainfall of Ten Blocks of Khordha District

Maximum, minimum and average monsoon and annual rainfall over the period of 1988–2017 were estimated for ten blocks of Khordha District. Maximum monsoon (2195.0 mm) and annual rainfall (2785.0 mm) observed in Bhubaneswar block, whereas Balipatna block experienced minimum monsoon (401.0 mm) and annual (504.3 mm) rainfall. Highest average monsoon (1211.9 mm) and annual (1621.2 mm) rainfall occurred in case of Bhubaneswar block, whereas lowest average monsoon (844.5 mm) and annual (1125.8 mm) rainfall detected for Balipatna block.

3.2 Drought Classification of Different Blocks of Khordha District During Monsoon Months (June, July, August and September) Based on SPI

In order to find the number of droughts in moderate, severe and extreme categories during monsoon months (June–September), SPI was estimated in ten blocks of Khordha District for June, July, August and September during 1988–2017 and, based on these value, droughts are classified under various categories and presented in Tables 1 and 2. It can be observed from Table 1 that, maximum drought events of moderate (6), severe (4) and extreme (2) droughts occurred in Banapur, Jatani and Balipatna blocks, respectively, for the month of June. Similarly, Banapur, Khordha and Tangi and Bhubaneswar blocks experienced maximum events of moderate (6), severe (3) and extreme (2) droughts, respectively (Table 1) in July. Balipatna, Jatani

Table 1 Moderate, severe and extreme drought events in the months of June and July based on SPI for different blocks of Khordha District

Blocks	Drought (June)			Drought (July)		
	Moderate	Severe	Extreme	Moderate	Severe	Extreme
Balianta	2	2	1	4	0	1
Balipatna	0	2	2	5	2	0
Bhubaneswar	1	3	0	2	0	2
Jatani	1	4	0	3	1	1
Banapur	6	1	0	6	0	0
Begunia	4	0	1	4	1	1
Bolagarh	3	1	0	2	1	1
Chilika	4	2	0	4	1	0
Khordha	4	0	1	4	3	0
Tangi	4	1	1	1	3	0

Table 2 Moderate, severe and extreme drought events in the months of August and September based on SPI for different blocks of Khordha District

Blocks	Drought (August)			Drought (September)		
	Moderate	Severe	Extreme	Moderate	Severe	Extreme
Balianta	2	2	1	1	1	2
Balipatna	4	1	1	1	2	1
Bhubaneswar	2	1	1	4	1	1
Jatani	4	0	0	2	0	4
Banapur	2	3	0	0	1	1
Begunia	3	1	0	1	2	1
Bolagarh	1	3	0	4	3	0
Chilika	3	1	1	3	0	1
Khordha	3	1	0	0	0	2
Tangi	4	1	0	2	0	1

and Tangi blocks experienced maximum four number of moderate drought events (Table 2) in August. Highest number of severe (3) drought events detected in Banapur and Bolagarh blocks where as no extreme drought events occurred in Jatani, Banapur, Begunia, Bolagarh and Khordha blocks in the month of August. Bhubaneswar and Bolagarh, Bolagarh and Jatani blocks experienced maximum number of moderate, severe and extreme drought events in September (Table 2).

Droughts are also classified under different categories based on CZI in the months of June, July, August and September and presented in Tables 3 and 4. Banapur, Jatani and Balipatna blocks experienced maximum events of moderate (7), severe (4) and extreme (2) droughts in the month of June, whereas 6, 3 and 1 moderate, severe and

Table 3 Moderate, severe and extreme drought events in the months of June and July based on CZI for different blocks of Khordha District

Blocks	Drought (June)			Drought (July)		
	Moderate	Severe	Extreme	Moderate	Severe	Extreme
Balianta	4	3	0	3	1	1
Balipatna	0	2	2	5	2	0
Bhubaneswar	2	2	0	3	2	0
Jatani	1	4	0	4	0	1
Banapur	7	0	0	6	0	0
Begunia	3	1	1	5	2	0
Bolagarh	2	0	1	2	2	0
Chilika	3	2	0	4	1	0
Khordha	3	0	1	6	1	0
Tangi	4	1	1	3	3	0

Table 4 Moderate, severe and extreme drought events in the months of August and September based on CZI for different blocks of Khordha District

Blocks	Drought (August)			Drought (September)		
	Moderate	Severe	Extreme	Moderate	Severe	Extreme
Balianta	3	3	0	3	2	0
Balipatna	4	2	0	2	1	1
Bhubaneswar	3	1	0	4	2	0
Jatani	2	2	0	3	0	2
Banapur	2	3	0	0	1	1
Begunia	2	1	0	1	3	0
Bolagarh	1	2	1	6	1	0
Chilika	3	1	1	3	0	1
Khordha	3	1	0	2	1	1
Tangi	4	1	0	2	0	1

extreme drought events occurred in July for Banapur and Khordha blocks, Tangi block and Balianta and Bhubaneswar blocks, respectively (Table 3). It can be observed from Table 4 that, maximum events of moderate (4), severe (3) and extreme (1) droughts detected in Balipatna and Tangi, Balianta and Banapur and Bolagarh and Chilika blocks, respectively, in August based on CZI, whereas Bolagarh, Begunia and Jatani blocks detected with the highest number of drought events of moderate (6), severe (3) and extreme (2) in September.

Total droughts (moderate + severe + extreme) estimated in June, July, August and September for the ten blocks of Khordha District and presented in Figs. 2, 3, 4 and 5. Maximum number of total droughts (7) experienced by Banapur block based

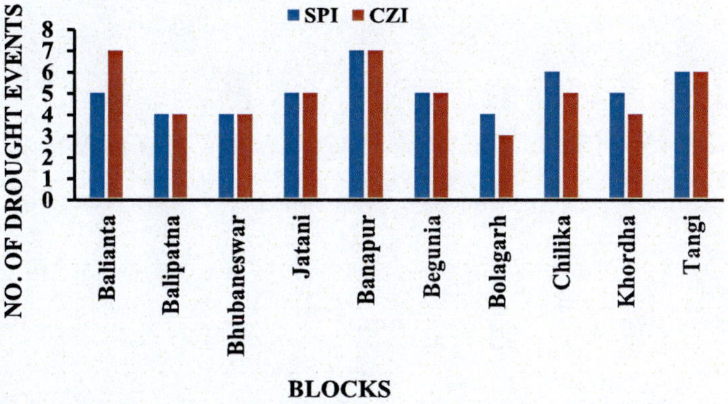

Fig. 2 Total droughts (moderate + severe + extreme) for ten blocks of Khordha District in June based on SPI and CZI

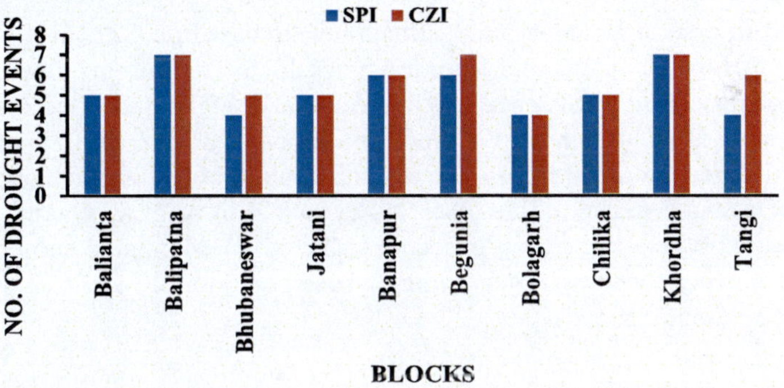

Fig. 3 Total droughts (moderate + severe + extreme) for ten blocks of Khordha District in July based on SPI and CZI

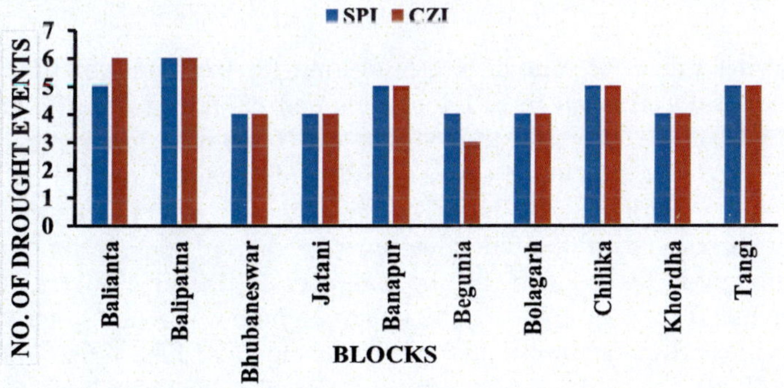

Fig. 4 Total droughts (moderate + severe + extreme) for ten blocks of Khordha District in August based on SPI and CZI

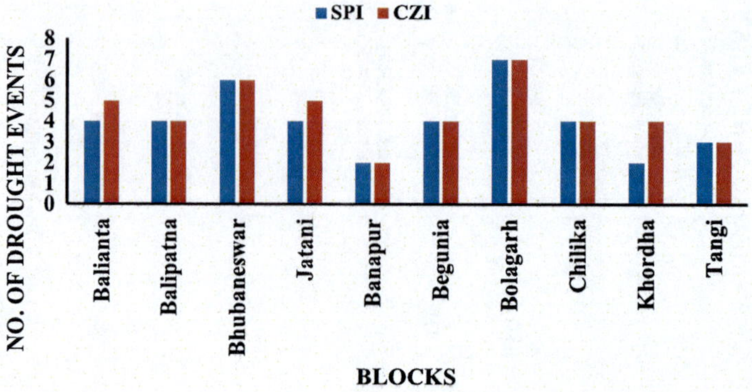

Fig. 5 Total droughts (moderate + severe + extreme) for ten blocks of Khordha District in September based on SPI and CZI

on SPI and in case of Banapur and Balinata blocks based on CZI (Fig. 2). During the month of July, Balipatna and Khordha blocks detected with maximum number of total droughts based on SPI, whereas highest total drought events occurred in case of Balipatna, Begunia and Khordha blocks based on CZI (Fig. 3). Balipatna block can be considered as critical from water resource point of view as maximum number of drought events detected in this block based on SPI in August, whereas Balianta and Balipatna blocks were critical based on CZI (Fig. 4). Maximum number of total droughts (7) experienced by Bolagarh block based on both SPI and CZI during the month of September (Fig. 5).

3.3 Estimation of Threshold Limits of Rainfall in Ten Blocks During Monsoon Months

In order to determine the rainfall threshold limit for the month of June, a linear equation is developed between SPI 1 of June and rainfall of June for ten blocks of Khordha District and presented in Fig. 6 and Table 6. A linear equation (**y = 82.54X + 185.72**) is developed between SPI 1 of June and rainfall of June and presented in Fig. 6 for Balianta block. In this equation, x represents the SPI 1 of June and y corresponds to rainfall value for the month of June. Rainfall threshold limit for drought in June is then calculated by putting the value of x as –1.0 in the linear equation. Threshold limit for the month of June was found to be 103.2 mm in Balianta block. Rainfall threshold limit for the month of June varied from 65.5 to 123.6 mm (Table 5). If the rainfall occurred below this threshold limit, that month of the year can be declared as drought month of the same year. Similarly, threshold limit of rainfall for the month of July is estimated for the ten blocks and presented in Table 5. Threshold limit of rainfall varied from 134.9 to 245.8 mm in the month of July.

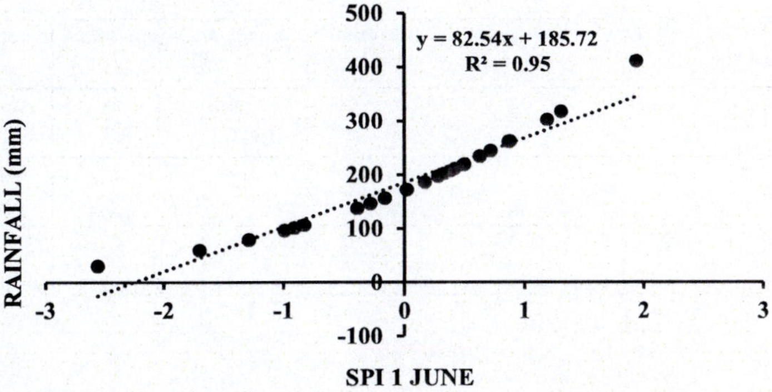

Fig. 6 Relationship between SPI 1 June and rainfall of June for Balianta block

Table 5 Threshold limit of rainfall for the months of June and July in ten blocks of Khordha District

Blocks	June			July		
	Equation	R^2	Threshold value (mm)	Equation	R^2	Threshold value (mm)
Balianta	y = 82.5x + 185.7	0.95	103.2	y = 117.2x + 325.1	0.93	207.9
Balipatna	y = 67.6x + 138.4	0.89	70.8	y = 93.5x + 228.5	0.98	134.9
Bhubaneswar	y = 85.9x + 209.5	0.97	123.6	y = 127.3x + 351.5	0.97	224.3
Jatani	y = 109.8x + 213.2	0.95	103.4	y = 118.2x + 364.0	0.98	245.8
Banapur	y = 99.8x + 175.6	0.96	75.7	y = 140.2x + 306.1	0.98	166.0
Begunia	y = 85.2x + 181.3	0.84	96.1	y = 129.3x + 295.7	0.97	166.4
Bolagarh	y = 79.3x + 146.3	0.98	67.1	y = 133.2x + 288.5	0.96	155.3
Chilika	y = 84.0x + 170.4	0.96	86.3	y = 151.8x + 309.8	0.97	157.9
Khordha	y = 95.7x + 161.2	0.93	65.5	y = 116.3x + 312.0	0.99	195.7
Tangi	y = 71.1x + 172.2	0.96	101.1	y = 117.6x + 290.1	0.96	172.5

Rainfall threshold limits for the months of August and September in the ten blocks of Khordha District are also determined and presented in Table 6. Threshold limit of rainfall varied from 144.2 to 220.9 mm and 93.5 to 155.0 mm in the month of August and September respectively.

Table 6 Threshold limit of rainfall for the months of August and September in ten blocks of Khordha District

Blocks	August			September		
	Equation	R^2	Threshold value (mm)	Equation	R^2	Threshold value (mm)
Balianta	y = 163.8x + 379.7	0.96	215.9	y = 126.9x + 281.9	0.94	155.0
Balipatna	y = 137.8x + 282.1	0.96	144.2	y = 73.6x + 194.8	0.97	121.2
Bhubaneswar	y = 166.8x + 374.1	0.97	207.3	y = 129.5x + 276.5	0.95	147.0
Jatani	y = 166.7x + 375.7	0.98	209.0	y = 92.1x + 247.1	0.93	154.9
Banapur	y = 127.6x + 314.8	0.97	187.1	y = 109.1x + 244.2	0.97	135.1
Begunia	y = 183.0x + 339.6	0.97	156.7	y = 90.6x + 212.0	0.96	121.3
Bolagarh	y = 125.8x + 279.6	0.96	153.9	y = 102.7x + 196.2	0.96	93.5
Chilika	y = 121.0x + 331.4	0.97	210.4	y = 98.9x + 253.1	0.97	154.2
Khordha	y = 170.2x + 349.3	0.97	179.2	y = 108.2x + 248.9	0.93	140.8
Tangi	y = 95.7x + 316.6	0.99	220.9	y = 97.6x + 248.3	0.98	150.7

3.4 *Estimation of Threshold Limits of Rainfall in Khordha During Monsoon Months*

Rainfall threshold limit in June, July, August and September months were calculated by establishing the linear relationship between SPI 1 of June, SPI 1 of July, SPI 1 of August and SPI 1 of September and rainfall of respective months for all the blocks of Khordha District. Rainfall threshold limit was found to be 89.3, 182.6, 188.4 and 137.4 mm for the month of June, July, August and September respectively. If the rainfall in a given year during these months occurred below these threshold limits, drought will occur for that particular month in that year.

3.5 Water Resource Management in Ten Blocks of Khordha District During June, July, August and September Months

Due to spatial and temporal variabilities of rainfall among the ten blocks of Khordha District, the characteristics of drought for these blocks were different based on the interpretation of SPI and CZI value of 1 months' time scale in monsoon months. Four (Banapur, Balipatna, Khordha and Bolagarh) blocks can be considered as critical from water resource management point of view based on SPI value during monsoon months, whereas based on CZI calculation, six (Banapur, Balianta, Balipatna, Begunia, Khordha and Bolagarh) blocks were critical. Adoption of water resource management strategy will be different for different blocks for controlling the drought because of different category of drought experienced by various blocks. Particularly, the blocks detected with maximum number of drought events in the month of July and August were most critical because 65% of the monsoon rainfall occurred in July and August.

4 Conclusion

Detection of drought severity for the ten blocks of Khordha District was evaluated through the computation of drought indices SPI and CZI using monthly precipitation data for identifying the capability of indices during this study. Maximum number of moderate, severe and extreme drought events occurred in 3, 4, 7 and 3 blocks in the month of June, July, August and September respectively based on SPI. Maximum number of total (moderate + severe + extreme) droughts experienced by Banapur, Balipatna and Khordha, Balipatna and Bolagarh blocks in the months of June, July, August and September respectively based on SPI calculation. Banapur and Balinata, Balipatna, Begunia and Khordha, Balianta and Balipatna and Bolagarh blocks detected with maximum number of total droughts based on CZI. Rainfall threshold limit was different for different blocks in June, July, August and September due to the spatial and temporal variation in rainfall. If the rainfall occurred in a particular year below these threshold limits in monsoon months, drought will occur for that month in that year. Particularly, the blocks which were experiencing maximum number of droughts in the month of July and August were most critical because 65% of the monsoon rainfall occurred in July and August. Therefore, those blocks which were experiencing frequent droughts, water-harvesting structures can be constructed in those blocks. In addition to that, low water requirement crops can be encouraged to be grown in those critical blocks. Early warning systems for droughts can be declared by knowing the threshold limit for rainfall so that people can adjust to the drought conditions by preparing themselves by adopting different patterns of cropping for reducing the risk and it will also useful for the people by perceiving the possible occurrences of drought. Interpretation of the drought indices, SPI and CZI led to various useful and practical information for understanding the characteristics of drought better in the study area. The drought study will be useful in planning for the drought preparedness and also the mitigation of drought in a practical and appropriate manner.

References

1. Mishra AK, Singh VP (2010) A review of drought concepts. J Hydrol 391:202–216
2. Morid S, Smakhtin V, Moghaddasi M (2006) Comparison of seven meteorological indices for drought monitoring in Iran. Int J Climatol 26:971–985
3. Riebsame WE, Changnon SA, Karl TR (1991) Drought and Natural Resource Management in the United States: Impacts and Implications of the 1987–1989 Drought. Westview Press, Boulder, CO, p 174
4. Jain VK, Pandey RP, Jain MK, Byun HR (2015) Comparision of drought indices for appraisal of drought characteristics in the Ken River Basin. Weather Clim Extrem 8:1–11
5. Dogan S, Berktay A, Singh VP (2012) Comparison of multi-monthly rainfall-based drought severity indices, with application to semi-arid Konya closed basin, Turkey. J Hydrol 470–471:255–268
6. Gibbs WJ (1987) A drought watch system. World climate programme. WMO/TD No 193, wcp-134, pp 1–22
7. Guttman NB (1998) Comparing the palmer drought index and the standardized precipitation index. J Am Water Resour Assoc 34(1):113–121
8. McKee TB, Doesken NJ, Kleist J (1993) The relationship of drought frequency and duration to time scales. In: Paper presented at 8th conference on applied climatology. American Meteorological Society, Anaheim, CA
9. Meyer SJ, Hubbard KG, Wilhite DA (1993) A crop-specific drought index for corn: I. Model development and validation. Agron J 86:388–395
10. Palmer WC (1965) Meteorologic drought. US Department of Commerce, Weather Bureau, Research Paper No 45, p 58
11. Wu H, Hayes MJ, Weiss A, Hu Q (2001) An evolution of the standardized precipitation index, the China-Z index and the statistical Z-score. Int J Climatol 21:745–758
12. Shafer BA, Dezman LE (1982) Development of a Surface Water Supply Index (SWSI) to assess the severity of drought conditions in Snowpack Runoff Areas. In: Preprints, Western SnowConf., Reno, NV, Colorado State University, pp 164–175
13. Byun HR, Wilhite DA (1999) Objective quantification of drought severity and duration. J Clim 12:2747–2756
14. Sahoo BB (2015) Critical appraisal of different drought indices of drought prediction and their application in KBK Districts of Odisha. Master dissertation, National Institute of Technology Rourkela, Odisha, India
15. Wilson EB, Hilferty MM (1931) The distribution of chi-square. In: Proceedings of the national academy of sciences of the United States of America 17, pp 684–688
16. Kendall MG, Stuart A (1977) The advanced theory of statistics, vol 1, Distribution theory. Charles Griffin Company, London, pp 400–401

Seasonal Variability of Satellite-Derived Aerosol Optical Depth in Smart City, Bhubaneshwar

Sushree Sasmita and Dudam Bharath Kumar

Abstract With the increase in industrialization and urbanization, the environment is being polluted by the addition of harmful products by the anthropogenic activities of human beings. The rapid increase in population accounts for the aerosol concentration that is increasing with time and having a significant impact on climatic conditions. The study focuses on the variation of aerosol optical depth (AOD), Angstrom exponent (AE), concentrated weighted trajectory (CWT) and potential source contribution function (PSCF) in smart city Bhubaneswar in the year 2017. The present study analyses MODIS (Moderate Resolution Imaging Spectroradiometer) data in the year to estimate the AOD level. Results show that the daily mean value for AOD and AE from the satellite data product showed the range falls between 0.16–2.17 and 0.01–1.80, respectively. The corresponding monthly mean for AOD in winter, summer, monsoon and post-monsoon is found to be 0.27 ± 0.16, 0.22 ± 0.05, 0.24 ± 0.1 and 0.18 ± 0.02, respectively. Similarly, the values for AE for the above seasons are obtained as 1.53 ± 0.10, 1.52 ± 0.1, 1.19 ± 1.05 and 1.61 ± 0.09, respectively. The CWT and PSCF are analysed for the seasons of winter (DJF), summer (MAM), monsoon (JJA) and post-monsoon (SON) in the given year.

Keywords Moderate Resolution Imaging Spectroradiometer · Aerosol Optical Depth · Angstrom Exponent · Concentrated Weighted Trajectory · Potential Source Contribution Function

1 Introduction

Large-scale urbanization and industrial development have increased the pollution level with advancement of time. Aerosol pollution has been rising due to rapid industrialization and increase in vehicular pollution. With the technological inventions, the present scenario gives an idea about the global climate changes having a massive impact on living organisms. Population growth also adds to the present situation. The

S. Sasmita · D. B. Kumar (✉)
School of Civil Engineering, Kalinga Institute of Industrial Technology Deemed to be University (KIIT-DU), Bhubaneswar 750124, Odisha, India
e-mail: dudam.kumarfce@kiit.ac.in

B. B. Das et al. (eds.), *Recent Developments in Sustainable Infrastructure*, Lecture Notes in Civil Engineering 75, https://doi.org/10.1007/978-981-15-4577-1_85

changes in seasonal values of aerosol optical depth (AOD) indicate the increase in concentration of aerosols over a particular location. Aerosols are important due to the significant impact on climatic changes as its presence hampers the earth's radiation balance and by scattering or absorbing radiation [1]. Previous studies have been done by including observations from ground, remote sensing and modelling illustrating events like dust storms, rainfall, forest fire and smoke.

Bhubaneswar is one of the recent developing cities for having rapid improvement in infrastructure, technological advancement, green vegetation and trying to attain smart city status.

This study aims at analysing the seasonal changes in MODIS derived AOD and AE in the year 2017 over Bhubaneswar. It also focuses on the analysis of clustering, concentrated weighted trajectory (CWT) and potential source contribution function (PSCF) for the same year. The increase in values of AOD illustrates the increase in mass concentrations of aerosols and variations occur monthly and season wise.

2 Methodology

The effects of aerosols over a location are essential to be evaluated. In order to estimate their impact, we analyse the pollution episodes in all the seasons annually like summer (March), monsoon (JJA), post-monsoon (SON) and winter (DJF) 2017 over smart city Bhubaneswar. Most of these fire events occur in central India, north-eastern states and north-western states in India in summer (MAM) and post-monsoon (SON), respectively. Time series of AOD was obtained from Moderate Resolution Imaging Spectroradiometer (MODIS) satellite of combined land and ocean product collection-6 Version 3.0 (C006 V3.0).

3 Site Description

3.1 Study Area—Bhubaneshwar

The capital of Odisha is Bhubaneswar. Bhubaneswar has its own geographic features that are distinct in nature. The location is 20° 35′ North and 85° 81′ East in Odisha. Geographically, the city has a tropical climate and lies in the tropical zone of India. The three seasons that usually occur in the city are summer, winter, and monsoons. The winter season is moderately cold during the months of November, December and January. During the summer season, the temperature rises very high and heat wave is felt during the months of April and May. It is followed by the monsoons which prevail in the months of June. The pre-monsoon showers occur in the mid week of June. Sometimes the rainfall goes very high resulting in flood situation during monsoon depending on the occurrence of low pressure. The climate of Bhubaneswar

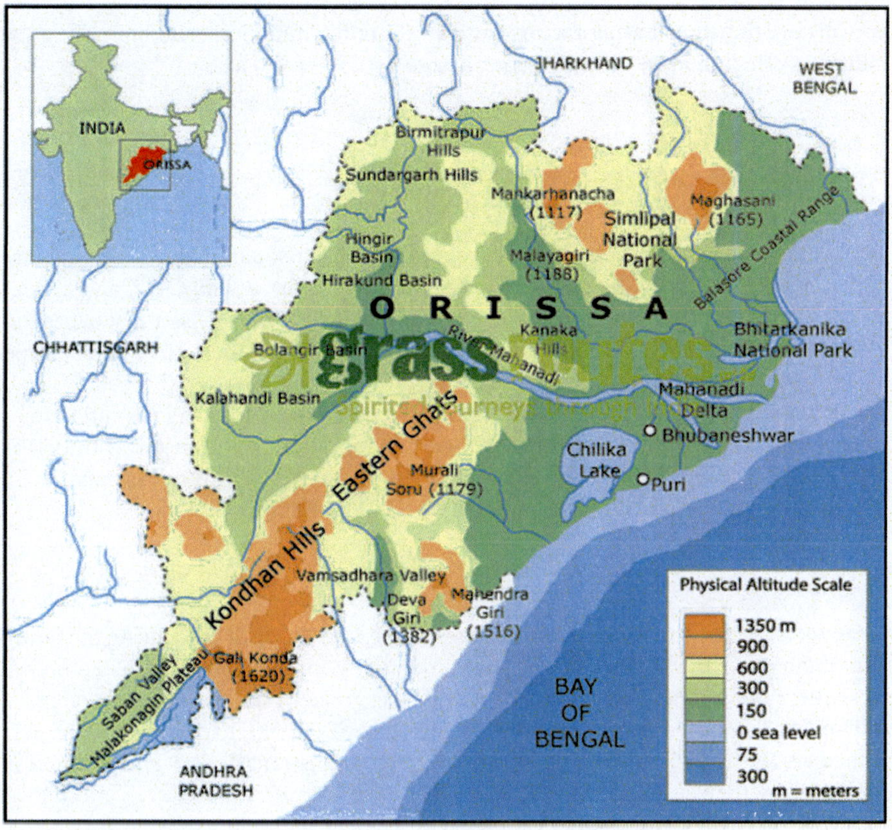

Fig. 1 Map of Odisha

is becoming abnormal due to the following factors like deforestation, depletion of water bodies, massive construction activities, rapid industrialization, immense carbon consumption, global warming and huge expansion of the city in all the directions (Fig. 1).

3.2 Aerosol Optical Properties

Anthropogenic activities and biomass with fossil fuel burning generate aerosols which are present in the environment having an impact on the climate of the area. Aerosol size, interaction among them, optical properties are major contributors for enhancing the AOD value gathered from an area. Aerosols contribute in cloud formation and have an impact resulting in precipitation during the monsoon season. The aerosols are the contributing factors for absorbing the radiations and increasing the scattering effect thereby changing the climate.

3.3 *Source Analysis*

The source analysis was carried out by doing the statistical analysis of air mass back trajectories. Using the obtained data, we analyse the PSCF, CWT and cluster analysis [2]. The source–receptor analysis was carried for the AOD at Bhubaneswar (20.35° N and 85.81° E) during 2017. The analysis shows that the high AOD occurred during winter (DJF) season.

Mathematically, the PSCF is a function of a location as defined by the cell indices i and j, while the number of segments with endpoints that fall in the ij-th cell is denoted by $n_{ij.}$ The total number of end points that fall in the ij-th cell is denoted by n_{ij}. Then the value for PSCF is calculated by the equation:

$$\mathrm{PSCF_{ij}} = \mathrm{m_{ij}}/\mathrm{n_{ij}}$$

where n_{ij} is the total number of air masses falling into the ij-th cell during the study period and m_{ij} is the number of segment trajectory endpoints in the ij-th cell on the days where the source contribution of which was greater than the criterion value.

The CWT analysis determines the significance of potential sources that are performed [2, 3]. The CWT signifies the value of source strength of a grid cell and is calculated by the equation:

$$\mathrm{CWT_{ij}} = \left(\sum_{T=1}^{L} C_t T_{i,j,T}\right) / \sum_{T=1}^{L} T_{i,j,T}$$

where C_T is the AOD refers for the occurrence of back trajectory T, $T_{i,j,,T}$ is the total value of trajectory segment end points in a grid cell (i, j) for back trajectory T divided by the total value of trajectory segments end points for trajectory T, L is the total value of back trajectories over a time period (for each season). C_T and $T_{i,j,T}$ can be determined by counting the number of hourly trajectory segment end points in each grid cell for each trajectory [3].

4 Results and Discussion

Time series of daywise AOD and AE were analysed for the annual data in the year 2017 for the location of Bhubaneswar as shown in Fig. 2. The values of AOD vary from month- and season wise in a gradual progression. The AOD from the satellite data product showed that the winter season (DJF) has the highest value of 0.27 ±

0.16 in comparison to the other season which have a comparatively less value. The summer season (MAM) shows a mean value 0.22 ± 0.05, whereas in the monsoon season (JJA) the mean value is 0.24 ± 0.1 which is less than winter season. There are no observations during the months of July due to rainfall. The post-monsoon (SON) has lowest values of 0.18 ± 0.02 which may is due to rainfall which also occurred during the months of September and October. Among all the months, the AOD is found to be highest in the month of July with value of 0.82 ± 0.13. The lowest values of AOD are obtained in the month of November with value of 0.50 ± 0.21, and no values were found during July owing to rainfall.

The variability of AOD during the seasons may be due to the biomass burning and contribution from transport of aerosols into the atmosphere. It is high during the winter season owing to the fire-counts from open crop burning from the north-eastern side of India. It is confirmed by following the concept of trajectory statistical analysis and analysing the CWT with clustering in trajectory as well (Figs. 3, 4 and 5).

The AE from the satellite data showed almost similar results during all the seasons (Fig. 2). The post-monsoon season (SON) has highest values of 1.61 ± 0.09. Similarly, the winter season (DJF) and summer season (MAM) have values 1.53 ± 0.10 and 1.52 ± 0.1, respectively. The lowest values are obtained in the monsoon season (JJA) is 1.19 ± 1.05 due to rainfall. Similarly, for AE the values are found to

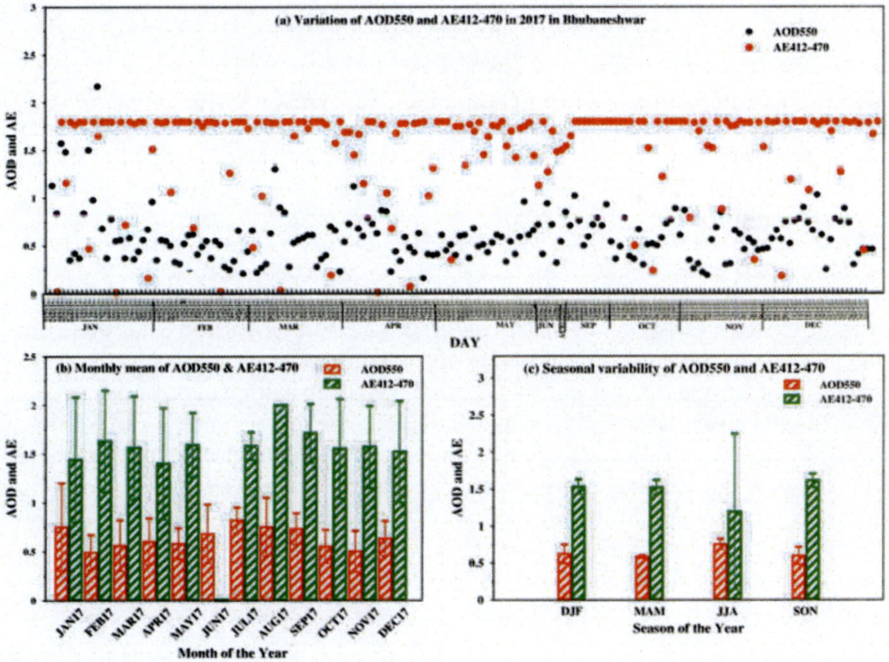

Fig. 2 Plot of AOD and AE showing daily, monthly and seasonal mean data for the year 2017 over Bhubaneswar

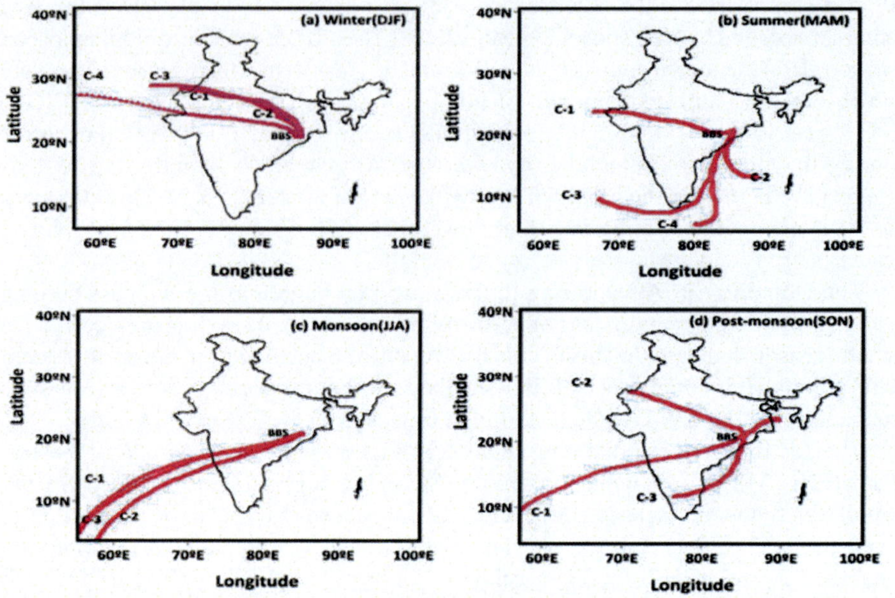

Fig. 3 Figure showing clusters in all the seasons in the year 2017

be highest in the month of August with value to be 2 ± 0 and lowest in the month of June with no values obtained due to rainfall or cloud (Figs. 2, 3, 4 and 5).

5 Conclusion

In summary, we observed that aerosol optical depth from the satellite observations showed the winter season (DJF) has highest value in comparison with other seasons throughout the year. The lowest is observed in monsoon season (JJA) due to rainfall. Trajectory statistical analysis showed high contribution of potential aerosols from regional level (eastern Indo-Gangetic region: November; central eastern India: March) than that of long-range source regions, region though most of the source fields fall over oceanic regions (particularly in March). Through influence of urban emissions, biomass burning events in terms of forest fires in March and those in terms of crop field burning activities in November are likely identified sources through our analysis. The post-monsoon season (SON) has the highest values and the lowest being observed in monsoon (JJA) for AE satellite observations taken. Setting of dispersion model coupled with wind fields may serve the purpose of studying the effects of potential source emissions towards aerosol parameters at the study stations.

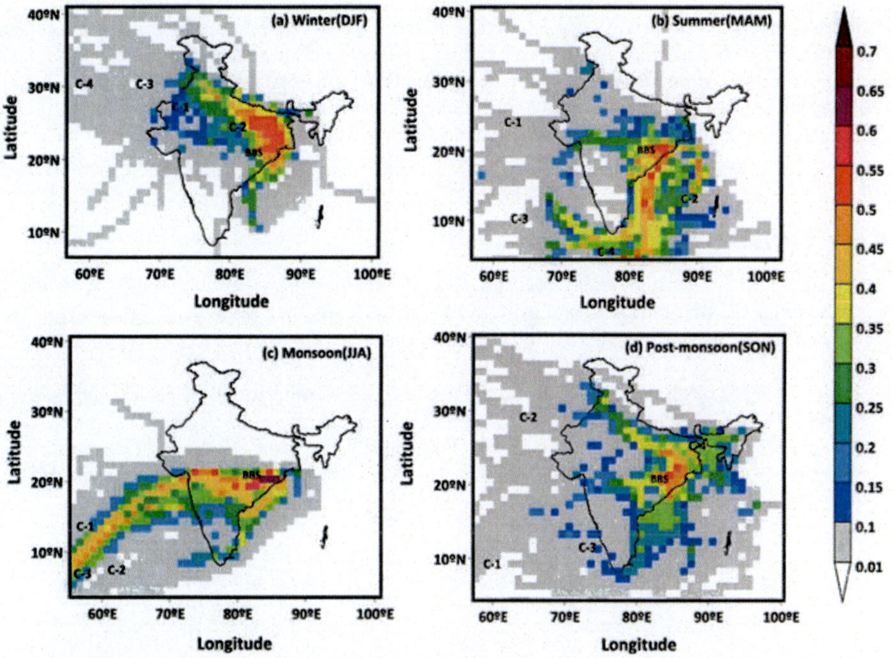

Fig. 4 Plot showing CWT analysis in all the seasons of the year 2017

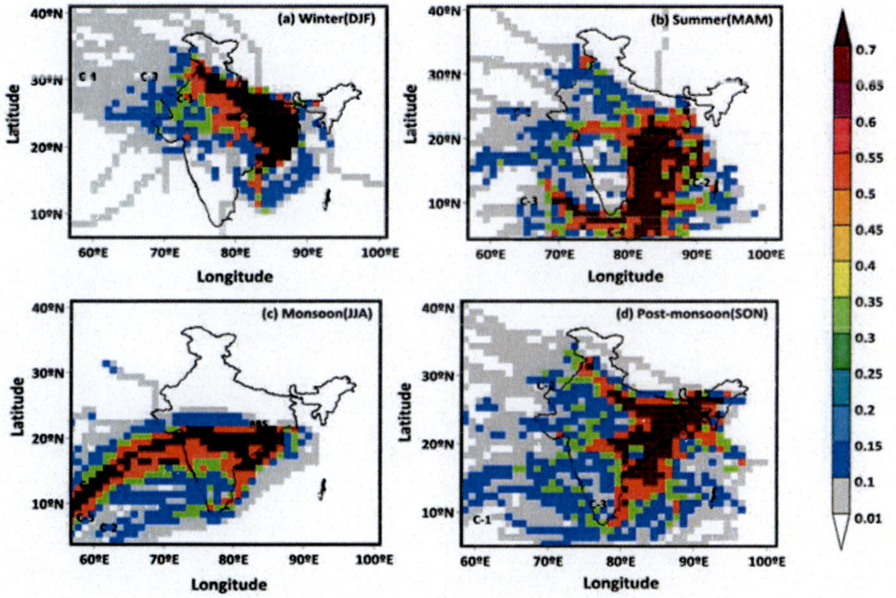

Fig. 5 Plot showing PSCF analysis in all the seasons in 2017

Acknowledgements This work was supported by the Kalinga Institute of Industrial Technology-Deemed to be University (KIIT-DU) (KIIT Bhubaneswar) for its computational facility throughout the project work. The instrument was provided by the Environmental Laboratory in School of Civil Engineering Department, KIIT-Campus-3, Bhubaneswar.

References

1. Khoshsima M, Bidokhti AA, Ahmadi-Givi F (2014) Variations of aerosol optical depth and Angstrom parameters at a suburban location in Iran during 2009–2010. J Earth Syst Sci 123:187–199
2. Stohl A (1998) Computation, accuracy and applications of trajectories—a review and bibliography. Atmos Environ 32(6):947–966
3. Wang YQ, Zhang XY, Arimoto R (2006) The contribution from distant dust sources to the atmospheric particulate matter loadings at XiAn, China during spring. Sci Total Environ 368(2–3):875–883

Modelling Sea Water Intrusion in the Eastern Coast Adjacent to Ersama and Kujanga Blocks of Odisha, India

Rohita Kumar Sethi, Prabhash Kumar Mishra, Deepak Khare, and Khyana Prava Samal

Abstract Coastal belt of the Bay of Bengal has got the human habitation from the time immemorial because of availability of abundant water sources and plain land for cultivation. The requirement of drinking water as well as that for irrigation and industrial use has increased manifold in the region. Surface water is available during monsoon from mid-June to mid-October. Heavy to very heavy rainfall as well as the floods of existing river network keeps the total area water-locked. But after that the total water gets drained into the sea and the area gradually gets dry. People of the area almost depend on ground water source for drinking purpose. About 85% of rural supply schemes depend on ground water. Earlier shallow wells could meet the limited demand. But population growth, urbanisation, increased industrial demand, and the need for high consumptive use pushed for digging deep tube wells/bore wells leading to over exploitation. This has led to the depletion of ground water table as well as potential availability of the aquifers gradually. The Kujanga block will have projected population of 1,22,374 and Ersama block will have 1,20,074 in 2020. Rural area is largely populated. Considering the per capita water demand at 135 l, the overall daily domestic requirement of fresh water is about 33000 m^3. Though the rural demand is much less as compared to that of urban life, it is essential to assess the bare minimum necessity, particularly during dry spells. The people of the area coming under this mix up zone suffer from scarcity in quantity and quality of water as well. There is rampant adaptation of aquaculture, particularly shrimp culture in the area. Shrimp culture ponds along with tidal riverine network enhance saline recharge of the ground water. Further, anthropogenic activities lead to resource depletion, and rising sea water level triggering saline intrusion into the coastal aquifer. So, the present study is attempted to understand the ground water movement in the two

R. K. Sethi (✉) · D. Khare
Department of Water Resources Development and Management, Indian Institute of Technology Roorkee, Roorkee, UA, India
e-mail: rhtemail@gmail.com

P. K. Mishra
National Institute of Hydrology, Roorkee, UA, India

K. P. Samal
KIIT Deemed to be University, Bhubaneswar, Odisha, India

B. B. Das et al. (eds.), *Recent Developments in Sustainable Infrastructure*, Lecture Notes in Civil Engineering 75, https://doi.org/10.1007/978-981-15-4577-1_86

blocks of Ersama and Kujanga, and their impact on human life in short and long run, especially for the coastal zone for sustainable water resources.

Keywords MODFLOW · Ersama · Kujanga · Sea water intrusion · Bay of bengal

1 Introduction

Area coming under the coastal belt of the Bay of Bengal has got the human habitation from the time immemorial. The main cause assigns to availability of abundant water sources and plain land for cultivation as the socio-economic establishments need this aspect worldwide. Gradually the area has turned thickly populated as of now. The requirement of drinking water as well as that for irrigation and industrial use has increased manifold. Huge quantum of surface water is available seasonally, that is during monsoon spell from mid-June to mid-October. Heavy to very heavy rainfall as well as the floods of existing river network keeps the total area water-locked. But after that the total water gets drained into the sea and the area gradually gets dry. People of the area almost depend on ground water source for drinking purpose. About 85% of rural supply schemes depend on ground water. Earlier shallow wells could meet the limited demand. But population growth, urbanisation of some of the area, industrial establishment, and the need of consumptive use in irrigation sector pushed for digging deep tube wells/bore wells leading to over exploitation. This has led to depletion of ground water table as well as potential availability of the aquifers gradually. The rate of ground water recharge cannot meet the depletion rate. Hence it is seen that the water table dips down day-by-day [1–4]. Further during dry spells, that is beyond monsoon, the water table happens to go down the sea level. Hence hydraulic gradient is developed which pushes saline water into the ground water causing a mix up. The people of the area coming under this mix up zone suffer from scarcity in quantity and quality of water as well. There is rampant adaptation of aquaculture, particularly shrimp culture in the area. This leads to spreading of saline cover increased manifold. Shrimp culture ponds along with tidal riverine network enhance saline recharge of the ground water. The anthropogenic activities lead to resource depletion. Rise of sea water level also imparts serious effect over saline intrusion into the coastal aquifer. A little rise of the sea water level can affect extended intrusion [5–8]; IPCC 2007; [9]. So we need to understand the natural processes and their impact on human life in short and long run, especially in the coastal zone for sustainable water resources.

2 Study Area

The study area comprises Ersama and Kujanga blocks of the district Jagatsinghpur in the state of Odisha. It spreads within latitude of 20°03′ to 20°20′ north and longitude of 86°17′ to 86°40′ east. It is bounded by Mahanadi in the north, Balikuda block in the south, Bay of Bengal in the east, and Tirtol block in the west. It is part of the Mahanadi delta. Mahanadi delta is formed by deposition of huge sediment transported from the vast catchment. But tidal incursion and salt water ingress has given rise to wide varieties of sedimentation due to complex hydro-chemical processes going on. A number of rivers flow in the area. The Mahanadi and Paika in the north and Devi River in the south. Many small rivers also drain the area. Most of the rivers are tidal. Besides, the Taladanda canal passes through the area which meets the irrigation and industrial demand of the area.

The area comes under sub-tropical climate zone. Three distinguished seasons exist over here. The dry summer spells from mid-February to mid-June, wet summer from mid-June to mid-October (monsoon season) and then follows winter for the rest part of the year. The annual temperature varies from about 110 °C in winter to about 400 °C in hot summer. The average rainfall of the area is 1500 mm with minimum of 731 mm (2000) and maximum of 1899 mm (1998). The rainfall usually occurs during monsoon with maximum in the month of August. Occasional rains take place during winter due to depression in the Bay of Bengal. Cyclonic storms very often form in the Bay of Bengal and rush into the region causing heavy pouring and experience of high wind damages. These natural calamities occur during months of May and June or November and December. Cultivation and fishing are the main occupation of people. The geomorphic features of the area contain numerous land forms. Of course, major portion of the area is plain. The topographical configuration of the land happens to remain almost submerged in monsoon, but dries up immediately thereafter. Repeated inundation by floods and tidal flushing has added the following features to the landscape. The morphology as well as the quality of the soil has been characterised accordingly.

The overall land form of the deltaic area is flat. Major portion remains plain. The total delta is formed due to fluvial deposition. It is divided into two parts, that is upper and lower deltaic plains. The upper plain remains above the maximum tide level of the sea. It is not influenced by saline contamination of the sea and hence is totally formed by the river alluvium. The lower plain coming under marine influence gets its soil character changed. The southwestern part of the study area belongs to upper plain zone, whereas the rest of the portion comes under lower deltaic plain. The ground water potential available in upper plain area, though quantitatively less, is qualitatively good. But that in lower plain area the reverse is there. Salinity intrusion is the main concern here. Due to mechanical effect of the sea tide, coastal plains are also formed in between the coast and the adjacent inter land. The seaward portion is termed to be young, whereas the landward portions are called older plains. The location map of the study area is shown in Fig. 1.

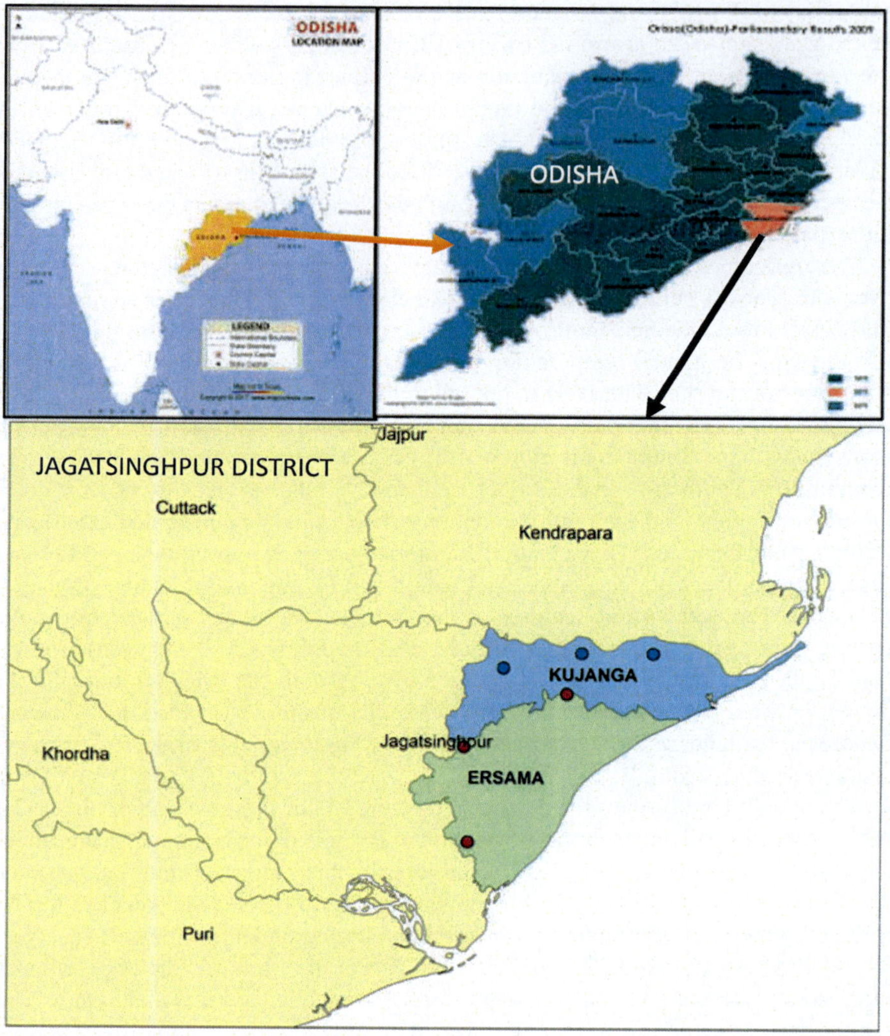

Fig. 1 Location map of the study area

3 Approach and Methodology

3.1 Hydrology and Sea Water Interaction

The delta of the Mahanadi as usual has taken the built-up with gradual sedimentation of alluvium being carried since inception. Still the landscape remains unconsolidated. The thickness of the deposition remains more towards coast. However, it represents a heterogenous pattern due to oscillatory deposition. Impact of versatile pattern in surface inundation and other natural course of environmental events has led to repetitive sequence of sand, gravel, clay, and silt deposition with finer segregation towards coast, which gives out a complex hydrogeological setup to its lithology. The region lying adjacent to the eastern coast is intensively affected by sea water intrusion into the aquifers. The complex aquifer orientation with interplay of fluvial and marine regimes develops a complex groundwater flow model also. The ground water flow towards the sea is obstructed by saline incursion due to tidal and increased shrimp culture inundation recharge. But heterogenous configuration of the aquifer system with advent of the sea water into it retains pockets of fresh water, saline, or brackish water differently at different locations. However, the general study of the relation of fresh and marine water advancement can be done with Ghyben-Herzberg concept.

The study area lies by the coast of the Bay of Bengal. The coast line defines its eastern boundary. So obviously the marine influence propagates into the area both above and below the land surface. Tidal flooding on the impact of lunar forte night pushes into the riverine system of the area regularly. Besides, cyclonic flooding also eventually submerges large part of the area. Over extraction of ground water in the adjacent coastal strip of the area causes decline of the ground water table below the sea water level. Hence, marine water on availing a functional head intrudes into the aquifers of the area (Fig. 2). In the meanwhile, an appreciable patch of the area has already come under this salt water intrusion.

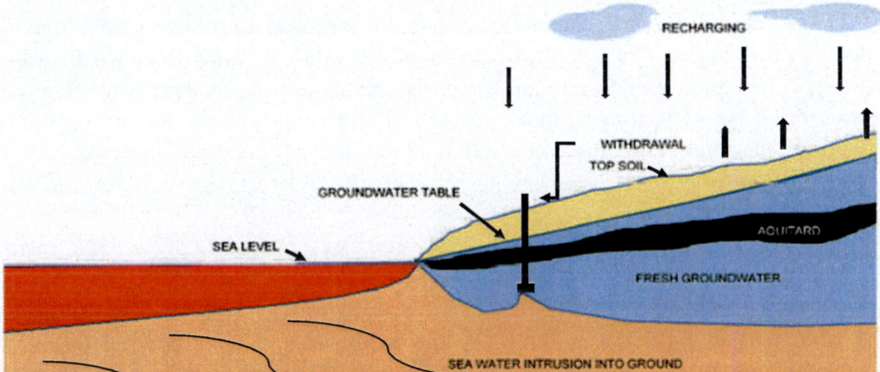

Fig. 2 Saline intrusion into coastal aquifers

3.2 Aquifer Characterisation Through Hydro-Geological Formations

It is seen from the bore-log profiles that number of clay layers encompasses multiple aquifers of different thickness forming a complex system. The topmost layer in Ersama area, as studied from logs of Ersama and Sahadabedi, remains sandy whereas that in Kujanga remains thin clay. The clay thickness is more pronounced in deeper regions and towards coastal belt. The thickness of porous formations of sand and coarse silt vary from few meters to about 100 m. The porous formations are not continuous. Isolated thin patches are also there which attribute to the complexity of the system. Similarly, isolated clay patches are also part of the configuration.

The stratigraphy of the area gives out a total 13 layers of aquifers. But there are three distinct layers of confined aquifers separated by clay layers (aquitards) varying from few centimetres to about 120 m. The topmost confined aquifer is separated from upper phreatic aquifer by a thin clay stratum of some centimetres. The existence of the aquifers is prominent up to 240 m below the ground surface. The minimum thickness of the unconfined aquifer remains in the north-eastern part and that increases south-westerly. The deeper formations constitute sand mixed with gravels. It cannot be ruled out that there are intermittent leakage patches of clay mixed with sand/gravel in between different confined aquifers. This is the reason why spreading of salinity continues to deeper regions.

The bore log profile is given in Fig. 3 and stratigraphic lithology of bore logs in Fig. 4. The lithological stratigraphy of the alluvium system is shown in Fence diagram (Fig. 5). It is depicted that there are three distinct sequences of vertical geological strata formations with discontinuous patches of clay mainly in the deeper aquifers. The alluvium system contains widespread unconfined and confined aquifers.

3.3 Setting of MODFLOW Model

The MODFLOW model was setup for the study area. The structured discretisation package was followed. The study area is encompassed in MODFLOW between 22,14,000 and 22,55,000 m north (South–North) and between 4,28,066 and 4,78,074 m East (West–East). The area is divided into a grid of 50 $\times$ 41 cells with each cell of 1000 $\times$ 1000 m. The grid size is decided basing on data, that is extent of study area and location of observation wells available. The Mahanadi and the Bay of Bengal remain as the hydraulic boundaries of the study area. The cells remaining inside the domain of the modelling were treated to be active. The cells remaining beyond the active zone were termed inactive. The active domain and inactive cells are shown in Fig. 6.

The study on hydro-geological formation of the modelling domain was studied through RockWorks software. The spatial geological stratification shows a very complex configuration of about 13 layers. However, ignoring identity of very thin

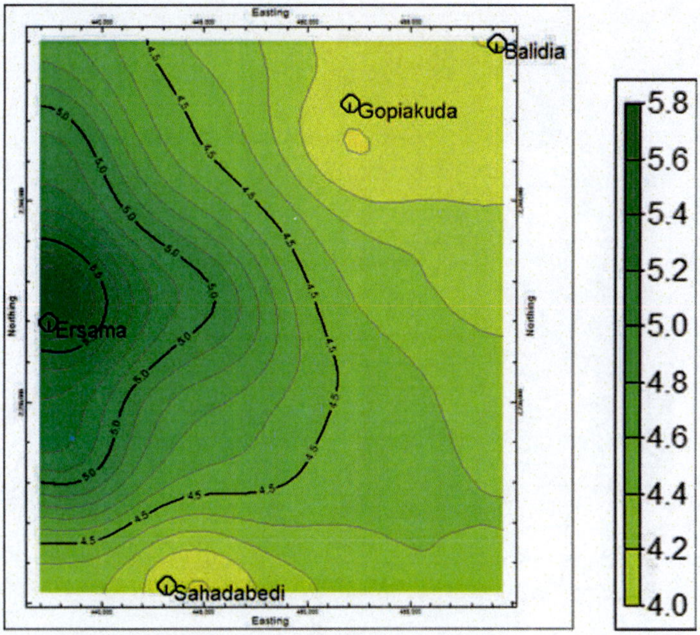

Fig. 3 Bore logs profiles

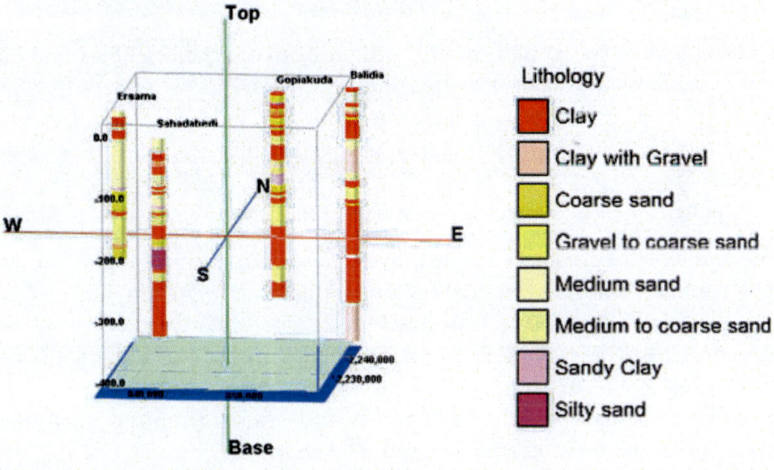

Fig. 4 Stratigraphic lithology of bore logs

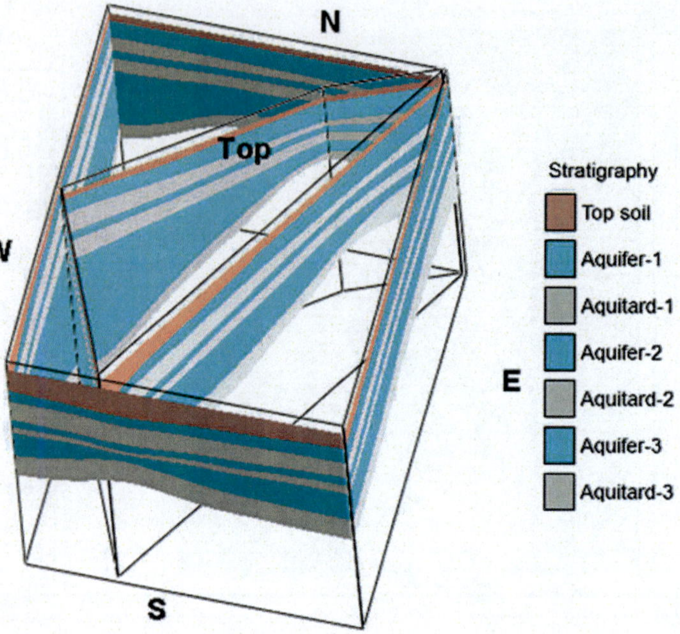

Fig. 5 Stratigraphic panel diagram (effective)

and localised layers, the system portrays three distinct aquifers of variable depth. The stratigraphic works were imported into the MODFLOW domain for vertical discretisation. The vertical discretisation is demonstrated by sectional views. The South–North mid-sectional view is shown in Fig. 7 and that of West–East is viewed in Fig. 8.

The output from vertical discretisation says about the following formations:

(i) Top soil layer of thickness varying up to 76 m representing the phreatic aquifer.
(ii) The confined aquifer separated by a very thin aquitards (Clay layer) from the top unconfined one lies up to 106 m.
(iii) An aquitard of thickness up to 46 m remains as the third distinct layer and spreads up to 112 m deep.
(iv) Then lies a confined aquifer whose thickness varies from few centimetres to about 23 m and remains up to a depth of 135 m.
(v) The second aquitard of thickness about 15 m lies below.
(vi) The third distinct confined aquifer of thickness about 60 m spreads up to 240 m. The vertical discretisation was carried up to a total depth of about 400 m below ground level. The separating aquitards between the unconfined and confined aquifers are not continuous. Hence, the model was setup as a semi-confined one.

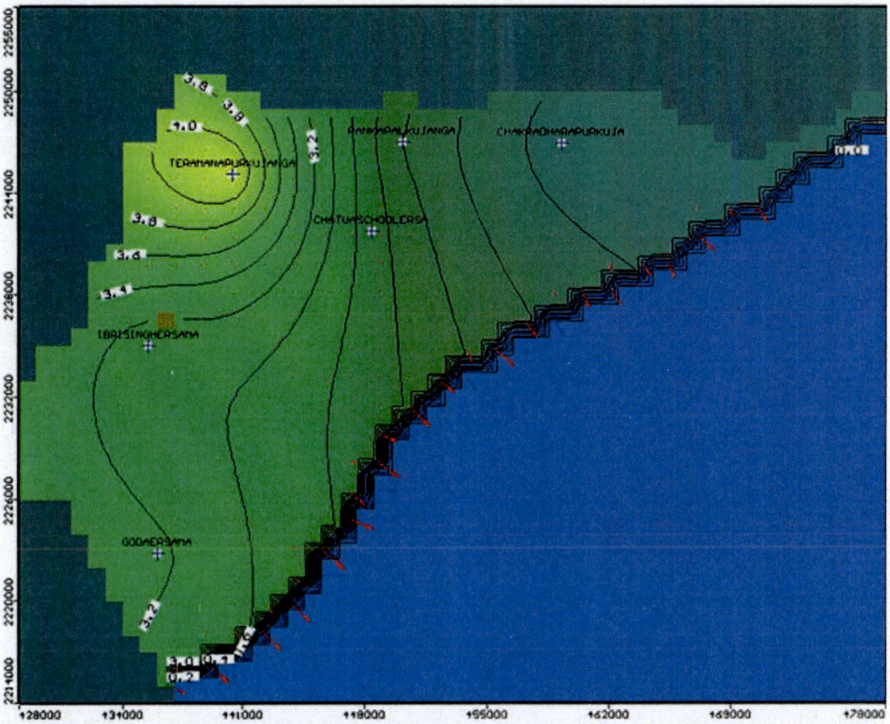

Fig. 6 Active domain and inactive cells

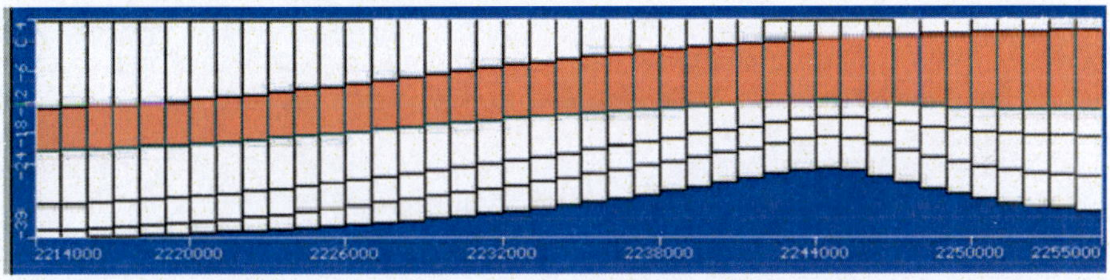

Fig. 7 Model layers at middle row

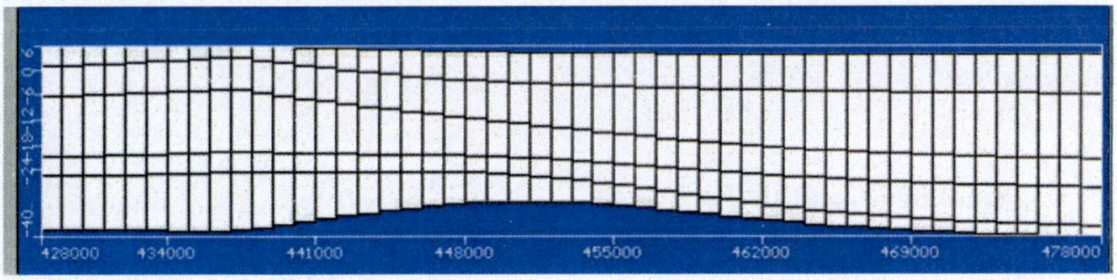

Fig. 8 Model layers at middle column

Table 1 Calibrated parameters

Current parameter values		Previous parameter values	
kx_1	1.000000E-02	kx_1	1.000000E-03
ky_1	1.000000E-02	ky_1	1.000000E-03
kz_1	7.720056E-05	kz_1	1.000000E-04
kx_2	1.518475E-06	kx_2	1.518475E-06
ky_2	1.736110E-06	ky_2	1.736110E-06
kz_2	5.119750E-08	kz_2	5.119750E-08
kx_3	1.763721E-08	kx_3	1.763721E-08
ky_3	4.609676E-07	ky_3	6.944400E-08
kz_3	1.396807E-08	kz_3	1.396807E-08
ss_1	1.000000E-06	ss_1	1.000000E-06
ss_2	2.386075E-05	ss_2	2.386075E-05
ss_3	9.674272E-06	ss_3	9.674272E-06
sy_1	9.738678E-03	sy_1	9.738678E-03
sy_2	5.000000E-03	sy_2	5.000000E-03
sy_3	5.014607E-03	sy_3	1.000000E-02

Calibration period: May-1997 to Dec-2006 (0–3501 days); Validation period: Jan-2007 to Nov-2017 (3532–7489 days)

3.3.1 Data Input

As usual the hydraulic conductivities in three cartesian directions and the specific storage are considered as the dominating parameters. The recharge and discharge data are taken into account to compute these parameters as specific values are not readily available for the study area. The seasonal rise and depletion of the observation wells are considered for the computation. Keeping the available data in mind, assumed values of the parameters are ascertained from the literature (Morris and Johnson 1967). The considered initial guess values of hydraulic conductivities and specific storage are enlisted in Table 1. Subsequent improvements are made by calibration and validation.

3.3.2 Initial Water Level Condition

Flow model is simulated with pre-monsoon water level data of the year 1997 as initial water table condition. Water level values of the cells of grid were generated by SURFER package and then imported to the model domain. The model domain interacted with initial heads varying from 2.07 to 3.57 m above mean sea level.

3.3.3 Boundary Conditions

A groundwater basin is the ideal choice to demarcate the boundary. But it is not always possible to adhere to this choice. So the alternate step is to pick the hydrological features of surface water to assign the boundary. In this study the Mahanadi in the north and the Bay of Bengal in the east are chosen to control the boundary conditions. Other two sides along with these two boundaries are assumed to be ‘No flow boundary’.

3.3.4 Forcing Stress Input

The following stresses were considered in the modelling domain (Fig. 9):

- Ground water levels of observation wells for calibration and validation.
- Extraction of groundwater from both unconfined and confined aquifer.
- Recharging (assumed 20% of rainfall.)
- Temporal river stage variation.

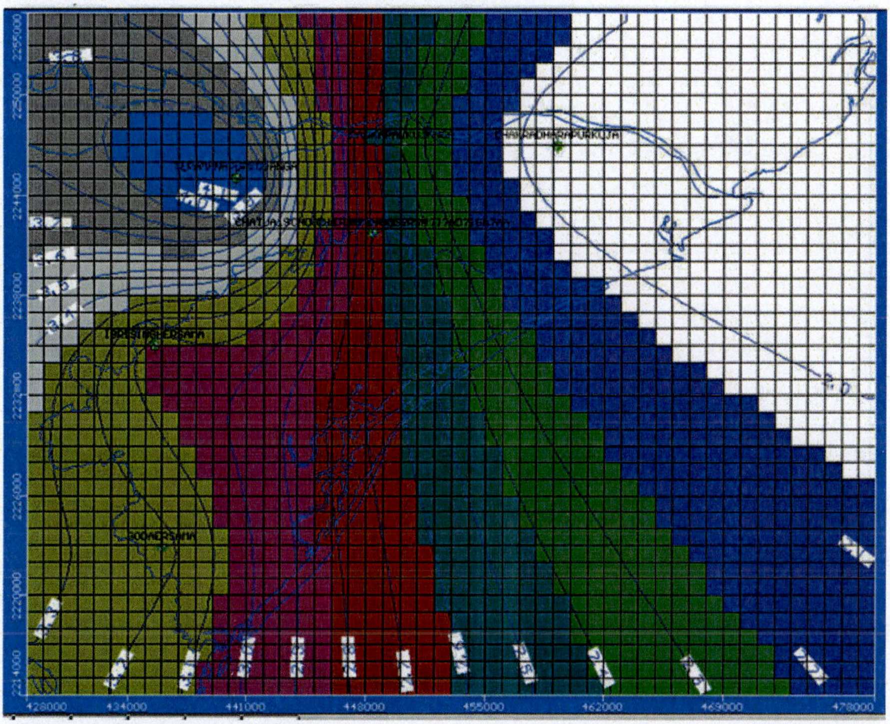

Fig. 9 Applied stress (rainfall)

3.4 Running MODFLOW Model—Calibration and Validation

The following actions were taken to calibrate the parameters for the MODFLOW model.

(i) Assumed parameter values were input into the model. Initial ground water levels put at the top of the topographic elevations considering its steady state. No external stresses were considered on the model domain. The model was allowed to run for sufficient time to execute iterations till steady-state condition is reached.
(ii) Then transient state is considered. All stress inputs are taken into account and improvement of first phase calibrated values of the parameters for a particular set of observed data is affected.
(iii) The response of the model is now compared with another set of observed data to validate it.

The calibration was done over the data for the period from May 1997 to December 2006 and validation over the data for the period from January 2007 to November 2017. Calibration was done for the parameters, that is hydraulic conductivities, K_x, K_y, K_z, storage coefficients, S_x, S_y, S_z, and specific yields S_{yx}, S_{yy}, S_{yz}, as shown in Table 1. Recharge from rainfall, groundwater withdrawal, return flow from the surface, and flows through boundaries were input as stresses (Table 2). Both confined and unconfined aquifers were undertaken for calibration and validation. The rainfall recharge was assumed to be 20% of rainfall due to requirement of specific required data.

Table 2 Recharge (rainfall)

Month	Recharge (mm)
1	2.37
2	3.60
3	3.98
4	3.45
5	11.56
6	39.67
7	59.42
8	61.06
9	43.12
10	40.13
11	8.66
12	1.15
Total	**278.17**

Data for available six observation wells were used to compare response of the simulated model in terms of hydraulic heads. One day was adopted as the time-step size and 30 days (one month) was the stress period length. Input and output figures are treated to be uniform over each time step and stress length. A transient state condition was assigned to the simulation process.

4 Results and Discussion

4.1 The MODFLOW Simulation

The MODFLOW simulation model was developed with calibration of parameters at transient state conditions. The calibration is done on trial and error basis comparing the results with that found through some conventionally accepted evaluation methods such as correlation coefficient (CC), root mean squared values, normalised root mean squared error (NRMSE), residual mean values (RM), absolute residual mean (ARM), standard error of the estimate (SE), and so on. The parameters of the model are said to be calibrated when the response of the model corresponding to these parameters are acceptable with these criteria.

4.2 Model Calibration

The water levels of all the six observation wells for the period from May 1997 to December 2006 were used for calibration of the model. The calibration was done for all the four seasons, that is winter, pre-monsoon, mid-monsoon, and post monsoon. The MODFLOW grid with well positions is shown in Fig. 10. The calibration histogram is given in Fig. 11. The graphs of computed values and observed values are presented in Figs. 12, 13, 14, 15, 16, 17, 18, 19, 20 and 21. The time unit is adopted as a day.

The comparative figures studied over 30 days say that the maximum residual comes to (−)0.791 m at Pankapal. The minimum residual of (−)0.078 m is recorded at Chakradharpur. The residual mean and absolute mean remain (−)0.453 m and 0.453 m, respectively. Standard error of estimate is found to be 0.119 m. The root mean square variation is 0.525 m and normalisation of it comes as 23.127%. The correlation coefficient stands at 0.93.

Similarly, on study of 2496 days for calibration, the maximum residual comes to (−)1.44 m at Teramanapur. The minimum residual of (−)0.203 m is recorded at Chatua. The residual mean and absolute mean remains (−)0.4 and 0.615 m, respectively. Standard error of estimate is found to be 0.275 m. The root mean square variation is 0.733 m and normalisation of it comes as 18.845%. The correlation coefficient stands at 0.966. The comparison breeds a very close correlation.

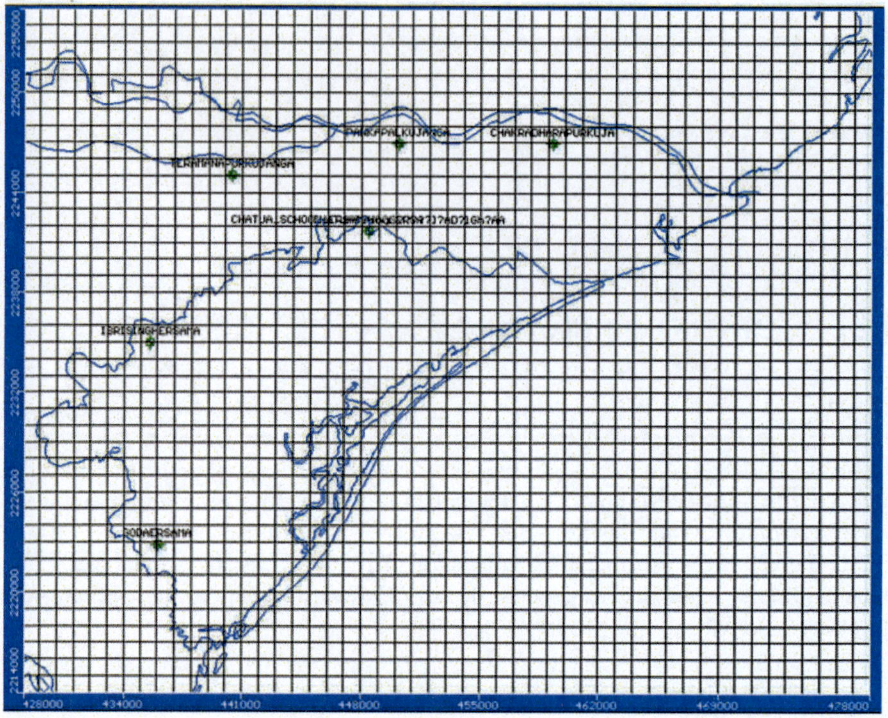

Fig. 10 MODFLOW grid with well positions

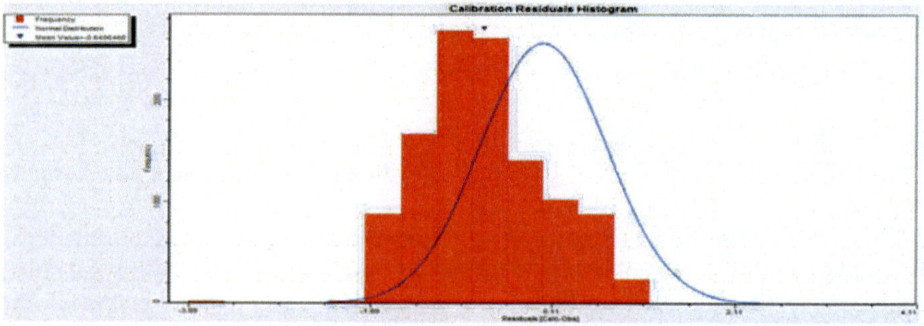

Fig. 11 Calibration residual histogram

4.3 *Model Validation*

During validation for 5087 days the maximum residual comes to 1.7 m at Chakradharpur. The minimum residual of (−)0.045 m is recorded at Chatua. The residual mean and absolute mean remains (−)0.144 m and 0.711 m, respectively. Standard error of estimate is found to be 0.413 m. The root mean square variation is 0.934 m

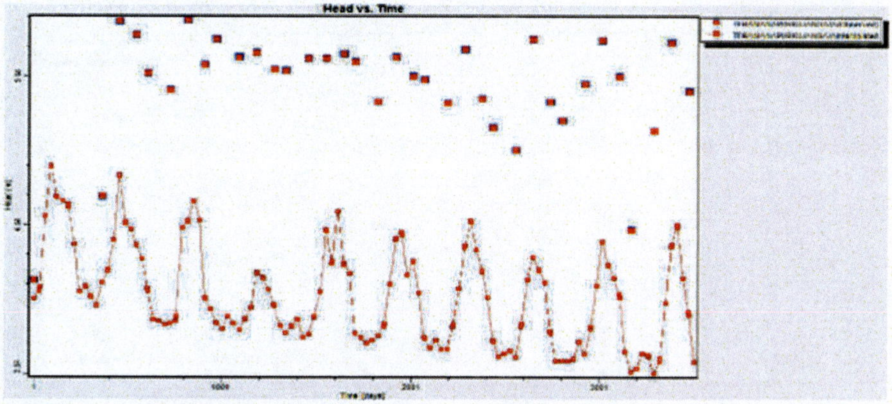

Fig. 12 Head versus time (Teramanapur)

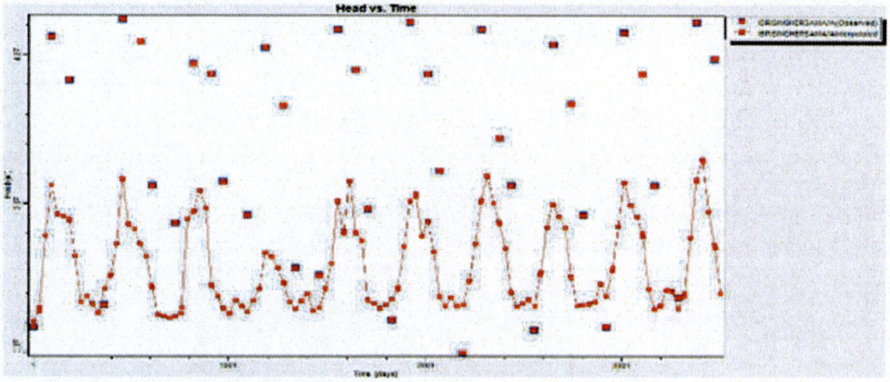

Fig. 13 Head versus time (Ibrisingh)

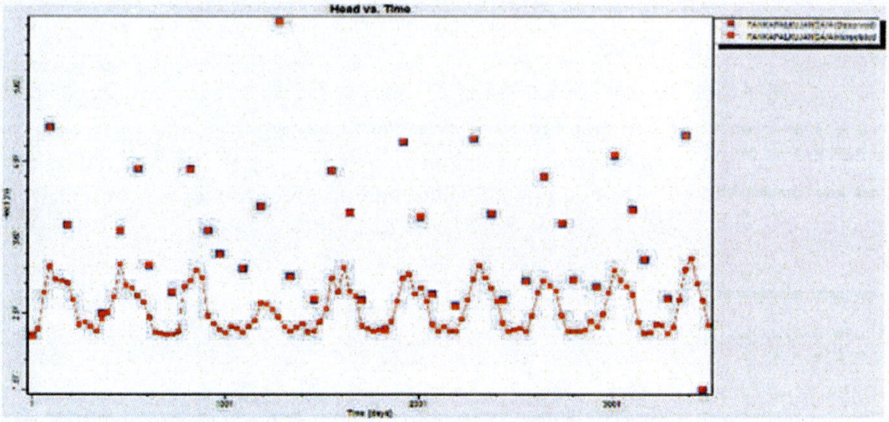

Fig. 14 Head versus time (Pankapal)

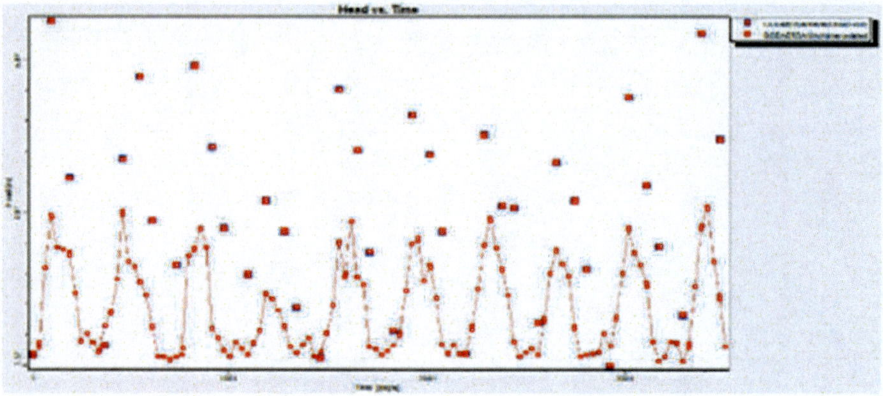

Fig. 15 Head versus time (Goda)

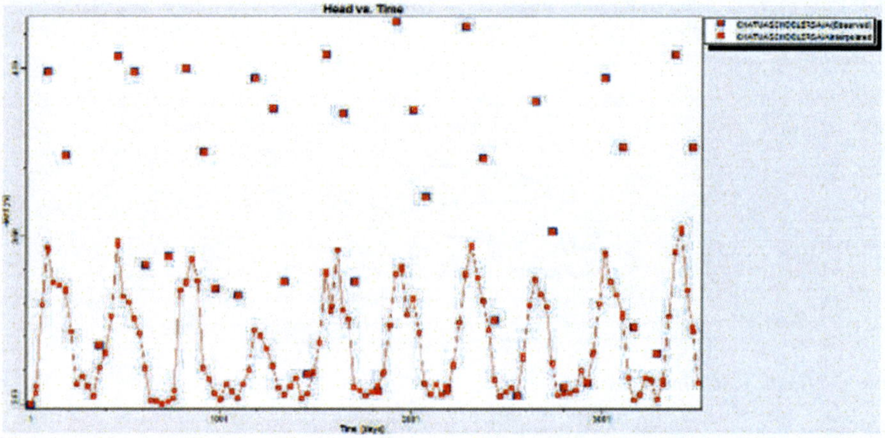

Fig. 16 Head versus time (Chatua)

and normalisation of it comes as 27.164%. The correlation coefficient stands at 0.589. Here it seems that a good correlation is not there which is attributed by the data anomaly. The graphs of computed values and observed values are presented in Figs. 22, 23, 24, 25, 26, 27, 28, 29, 30, 31 and 32.

5 Conclusions

Ersama and Kujanga are two important blocks coming under the coastal belt of the Bay of Bengal. The Kujanga block will have projected population of 1,22,374 and Ersama block will have 1,20,074 in 2020. This will further put pressure in providing

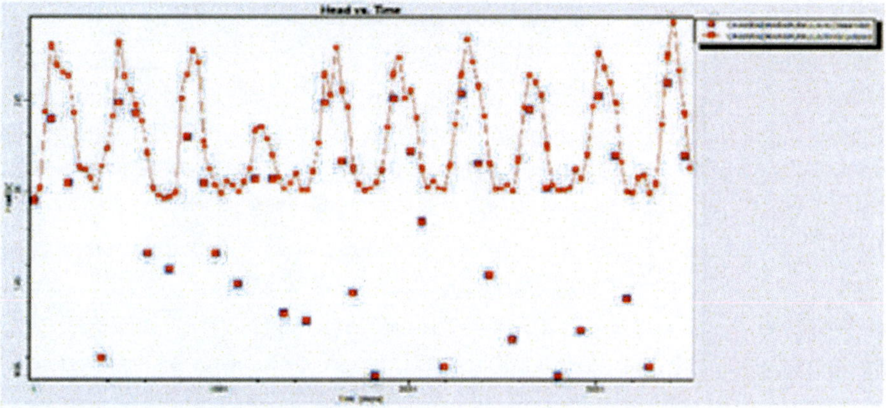

Fig. 17 Head versus time (Chakradharpur)

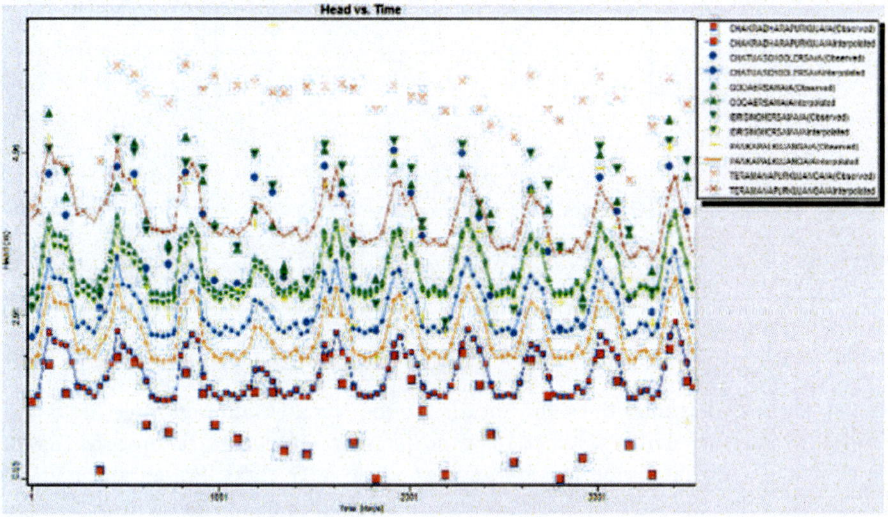

Fig. 18 Head versus time (Series)

drinking water in the region. Surface water is limited in non-monsoon seasons. At the same time, over dependence on ground water for drinking water with limited recharge into the aquifers results in sea water intrusion. The stratigraphy of the area gives out a total 13 layers of aquifers. But there are three distinct layers of confined aquifers separated by clay layers (aquitards) varying from few centimetres to about 120 m. The topmost confined aquifer is separated from upper phreatic aquifer by a thin clay stratum of some centimetres. The existence of the aquifers is prominent up to 240 m below the ground surface. Considering the per capita water demand at 135 litres, the overall daily domestic requirement of fresh water is about 33,000 m^3. The

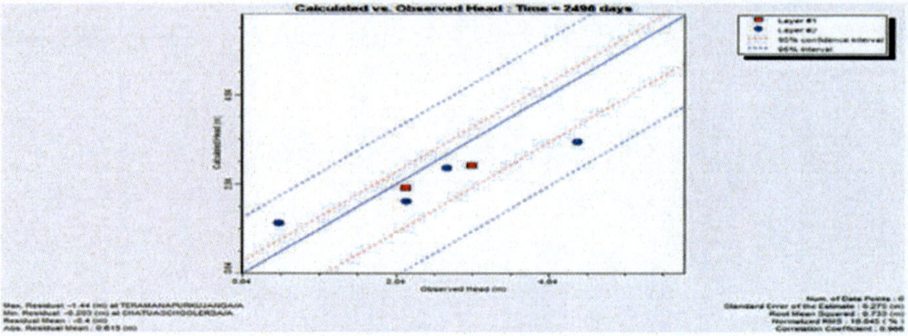

Fig. 19 Calculated versus Observed Head: Time 2496 days

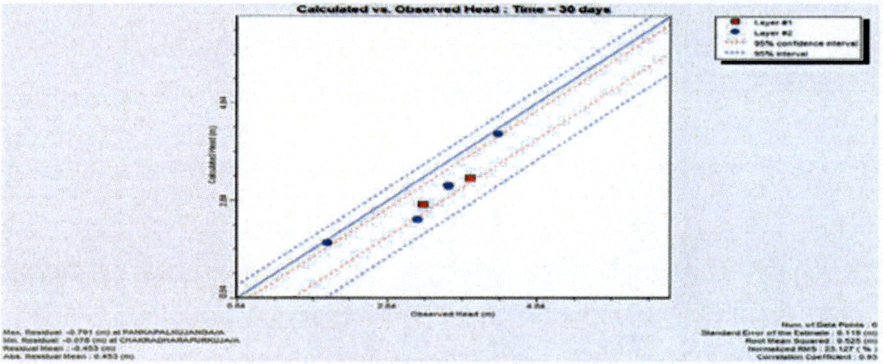

Fig. 20 Calculated versus Observed Head: Time 30 days

annual per head demand of 1089 m^3 needs ground water resource storage of 26,403 HaM. If the 20% rainfall is assumed to be the quantity of recharge, it comes to 18,620 HaM. But as calculated, the feasible recharge is 16,807 HaM. So, an arithmetical deficiency comes out to the tune of around 30%. Further due to saline intrusion the available resource is compromised to a large extent. Further in the study, the ground water movement in the two blocks is simulated using MODFLOW. Boundaries are defined by Mahanadi in the north and the Bay of Bengal in the east. Other two sides are assumed to be ‘No flow boundary’. Ground water levels of observation wells were used for calibration and validation of the model. Extraction of ground water from both unconfined and confined aquifer posed as the stress factor. Recharging was assumed as 20% of rainfall. The model was allowed to run for sufficient time to execute iterations till steady-state condition is reached. The calibration was done over the data for the period from May 1997 to December 2006 and validation over the data for the period from January 2007 to November 2017. In Kujanga block some area comes under irrigation and hence there is dry spell supplementation into ground water. Due to irrigation and bordering of Mahanadi, the ground water in major portion

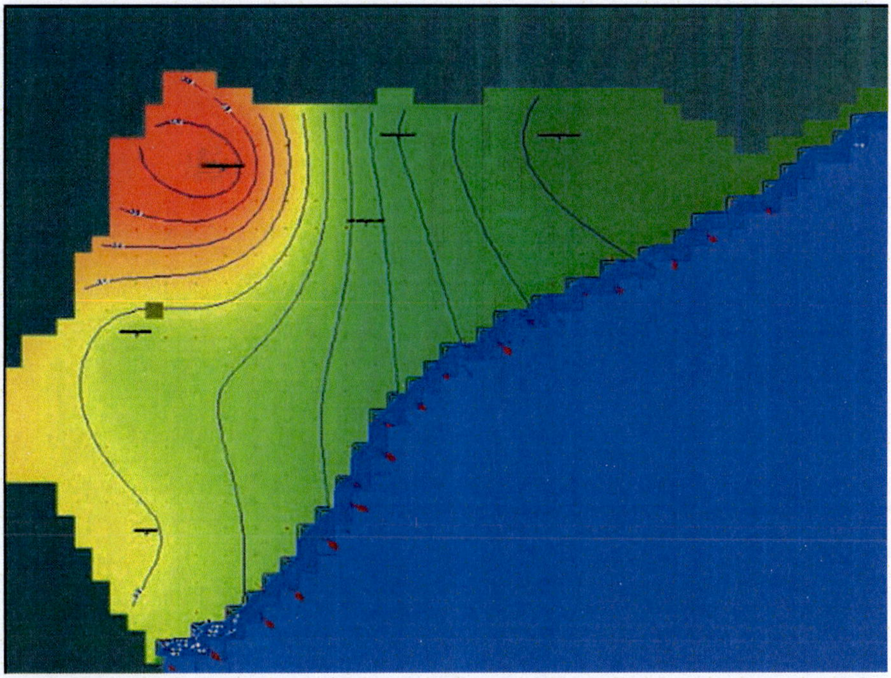

Fig. 21 Water table elevations

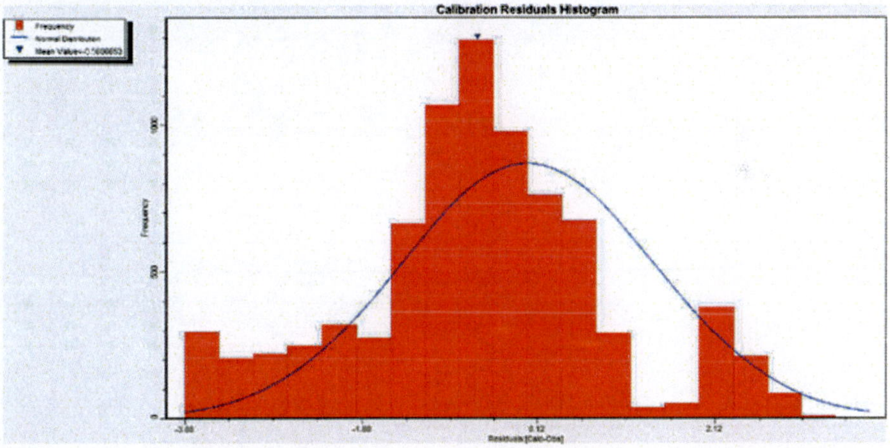

Fig. 22 Validation residual histogram

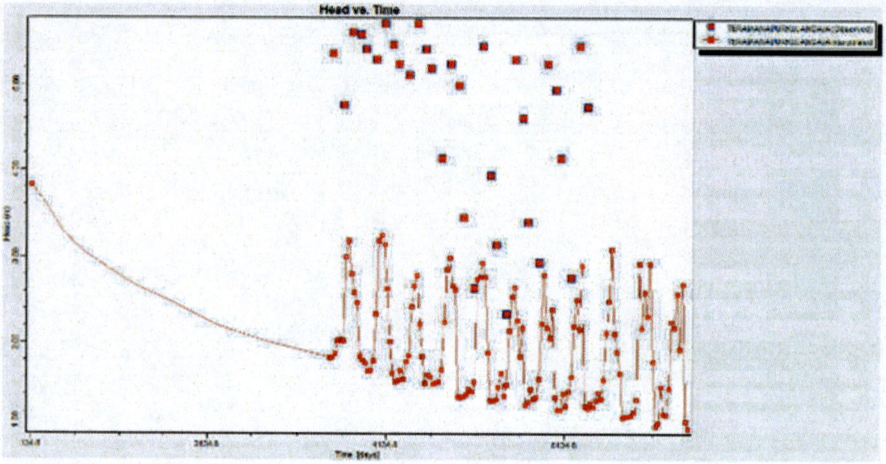

Fig. 23 Head versus time (Teramanapur)

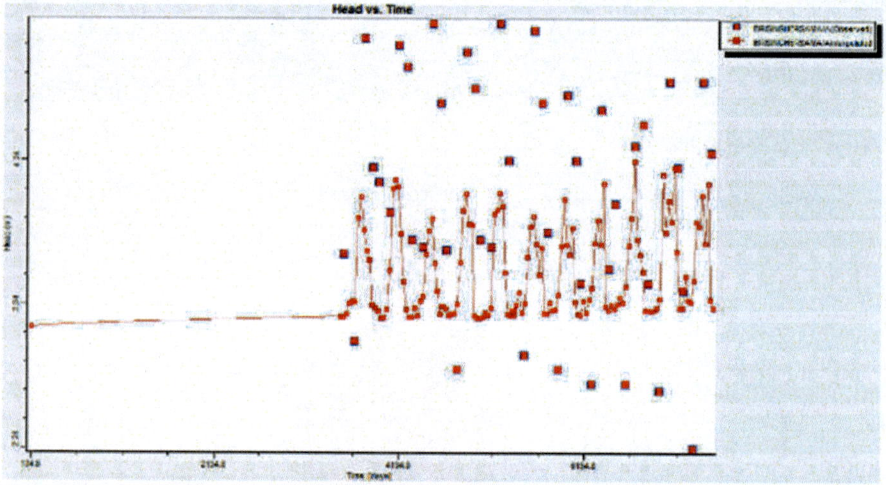

Fig. 24 Head versus time (Ibrisingh)

of this block remains free from saline infestation. But that laxity in Ersama block has pushed it into large-scale sea water intrusion. So, the feasible option to tackle this is to go for artificial recharging. Construction of structural arrangement to check the inflow of sea water during non-monsoon cannot be opted out.

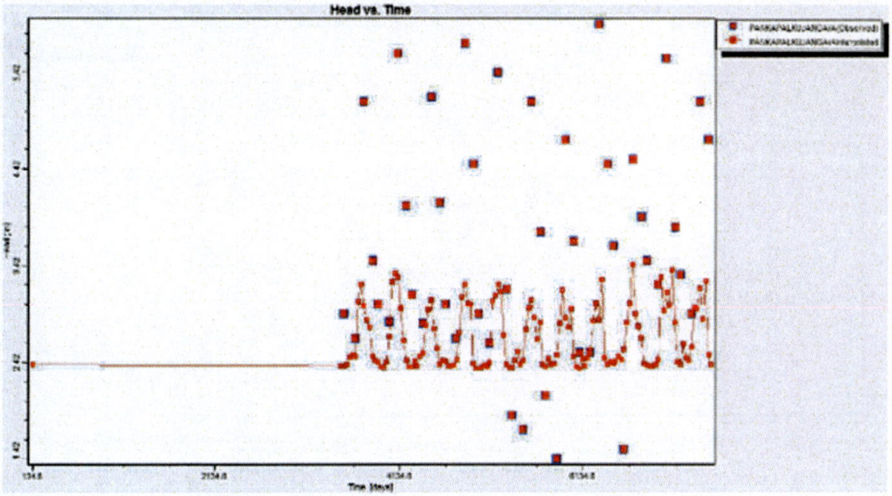

Fig. 25 Head versus time (Pankapal)

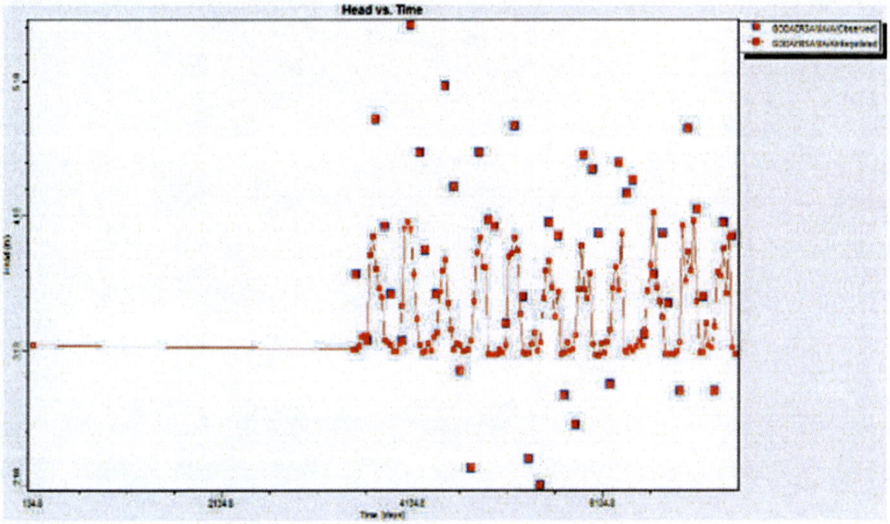

Fig. 26 Head versus time (Goda)

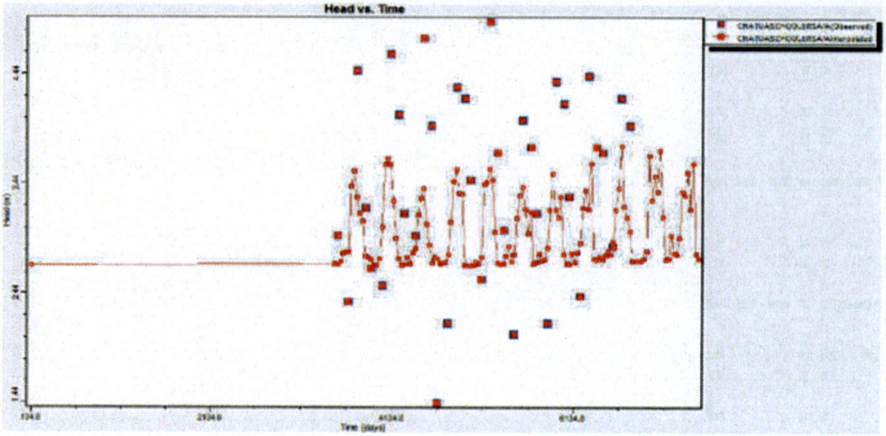

Fig. 27 Head versus time (Chatua)

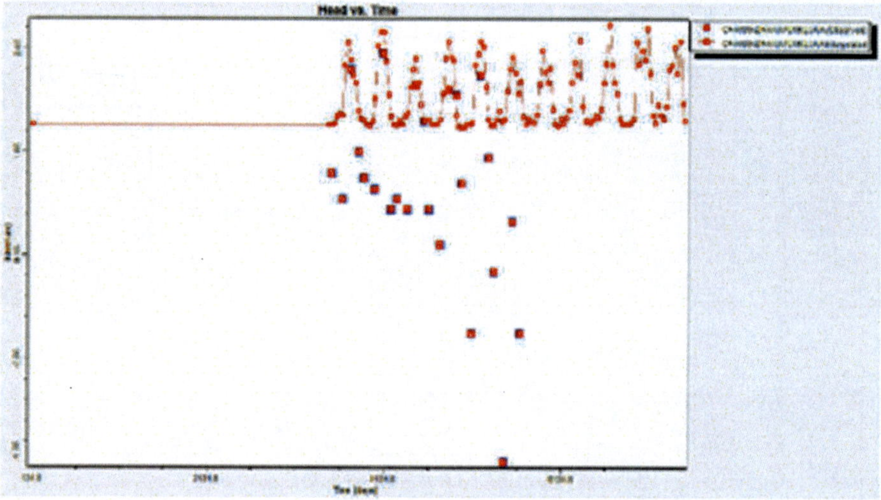

Fig. 28 Head versus time (Chakradharpur)

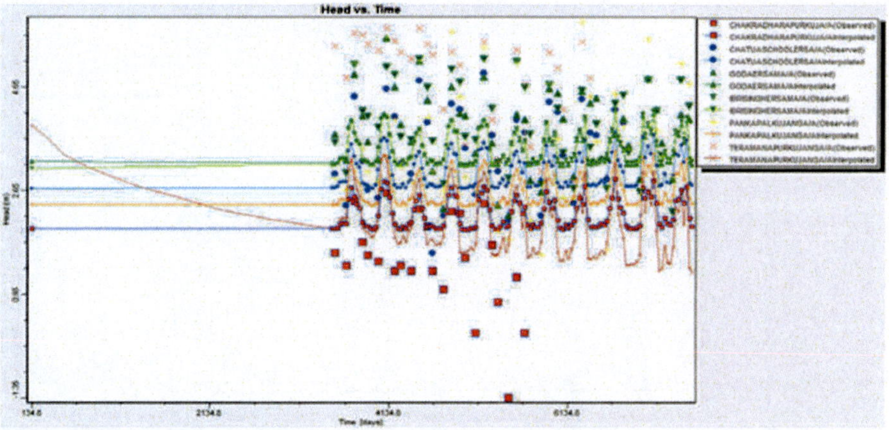

Fig. 29 Head versus time (series)

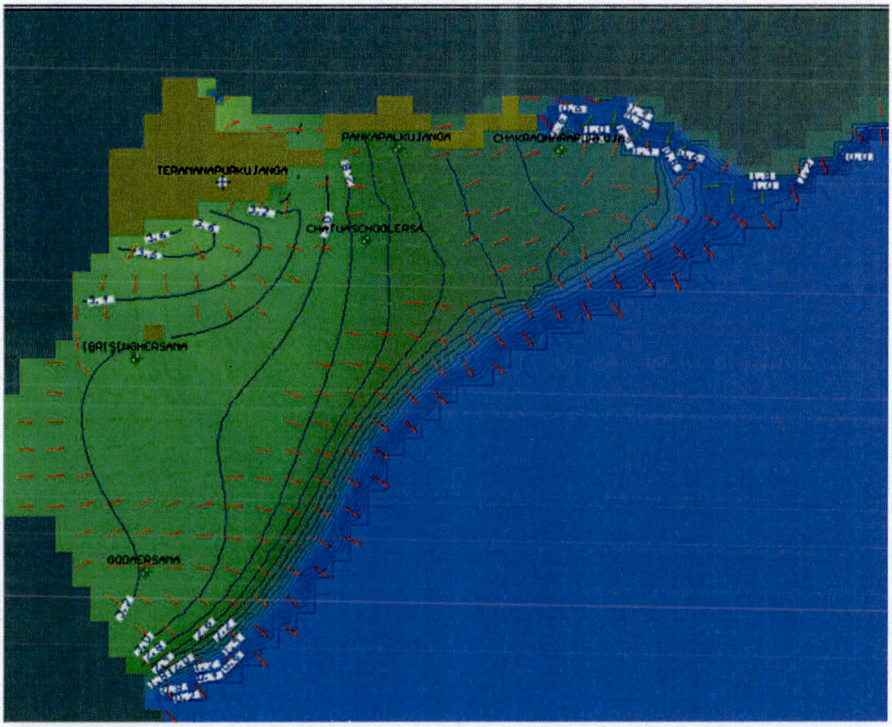

Fig. 30 Head equipotential

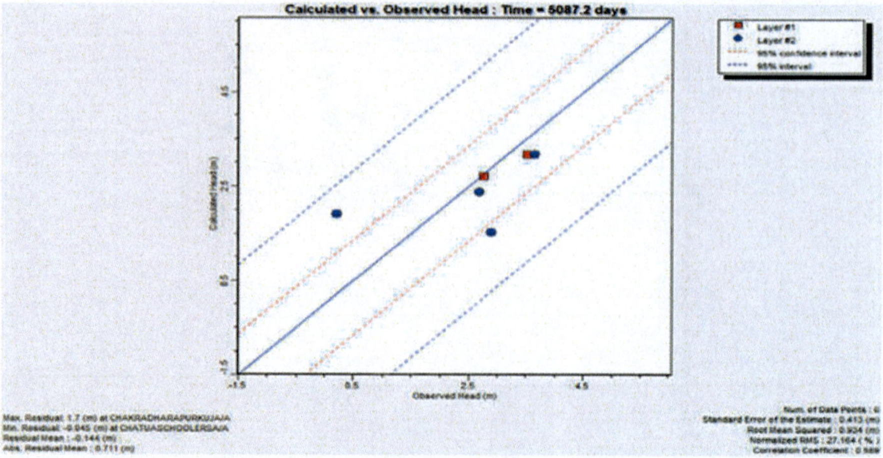

Fig. 31 Calculated versus observed head: time 5087.2 days

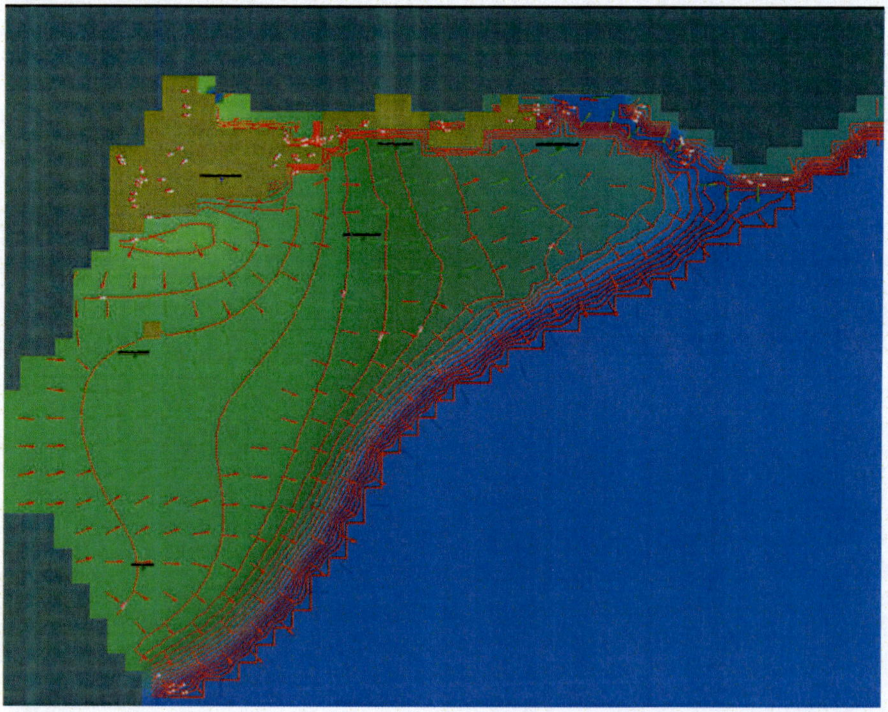

Fig. 32 Water table elevation

References

1. Dhawan V (2017) Water and agriculture in India: background paper for the South Asia expert panel during the Global Forum for Food and Agriculture (GFFA) 2017. OAV German Asia-Pacific Business Association
2. Gopinath S, Srinivasamoorthy K, Saravanan K, Suma CS, Prakash R, Senthilnathan D, Sarma VS (2016) Modeling saline water intrusion in Nagapattinam coastal aquifers, Tamilnadu, India. Model Earth Syst Environ 2(1):2
3. Huisman L, Olsthoorn TN (1983) Artificial groundwater recharge. Pitman Advanced Publishing, Program, Boston
4. Kaledhonkar MJ, Singh OP, Ambast SK, Tyagi NK, Tyagi KC (2003) Artificial groundwater recharge through recharge tubewells: a case study. IE (I) Journal-AG 84:28–32
5. Bear J, Cheng AHD, Sorek S, Ouazar D, Herrera I (eds) (1999) Seawater intrusion in coastal aquifers: concepts, methods and practices, vol 14. Springer Science & Business Media
6. Bobba AG (2002) Numerical modelling of salt-water intrusion due to human activities and sea-level change in the Godavari Delta, India. Hydrol Sci J 47(S1):S67–S80
7. Saxena VK, Mondal NC, Singh VS (2004) Identification of sea-water ingress using strontium and boron in Krishna Delta, India. Curr Sci 586–590
8. Voss CI, Souza WR (1987) Variable density flow and solute transport simulation of regional aquifers containing a narrow freshwater-saltwater transition zone. Water Resour Res 23(10):1851–1866
9. Van Camp M, Mtoni Y, Mjemah IC, Bakundukize C, Walraevens K (2014) Investigating seawater intrusion due to groundwater pumping with schematic model simulations: the example of the Dar es Salaam coastal aquifer in Tanzania. J Afr Earth Sci 96:71–78
10. Parry ML, Canziani OF, Palutikof JP, Van Der Linden PJ, Hanson CE (2007) Contribution of working group II to the fourth assessment report of the intergovernmental panel on climate change. In: Climate change 2007: impacts, adaptation and vulnerability, IPCC, 2007. Cambridge University Press, Cambridge, UK

Groundwater Level Trend Analysis for Sustainable Extraction and Use in Coastal Odisha

Anuradha Panda, Bitanjaya Das, Jyotiprakash Padhi, and Paromita Chakraborty

Abstract Groundwater level indicates groundwater availability, flow and physical characteristics of an aquifer. The availability of groundwater depends on rainfall and recharge pattern. As the use of groundwater is steadily increasing, it becomes imperative to analyze and understand the trend of groundwater recharge and use pattern employing the trend analysis of groundwater level. In this regard, a study of trend analysis of groundwater level has been done for the coastal districts of Odisha, India. The trend analysis is conducted in such coastal districts where the groundwater is affected by salinity problems. The trend analysis is done for seven coastal districts of Odisha: Balasore, Bhadrak, Jagatsinghpur, Kendrapada, Puri, Jajpur and Ganjam. Mann Kendall trend analysis is essential for the exploration of the trend of groundwater level by using the data from 1998 to 2017 for a network of 272 locations categorized in four seasons such as pre-monsoon, mid-monsoon, post-monsoon and winter. Results indicate that for groundwater level there is an increasing trend in 206 locations and decreasing trend in 66 locations during pre-monsoon period; in mid-monsoon period there is an increasing trend in 152 locations and decreasing trend in 120 locations; in post-monsoon period there is an increasing trend in 142 locations and decreasing trend in 130 locations; and in winter there is an increasing trend in 113 locations and decreasing trend in 159 locations. Such analysis can help the planners and managers in decision-making with regard to the extent of groundwater to be extracted and used in different locations and in different seasons. Besides, it will also help in making strategies for rainwater harvesting and augmentation of groundwater recharge activities.

A. Panda · B. Das (✉) · J. Padhi · P. Chakraborty
School of Civil Engineering, KIIT Deemed to be University, Bhubaneswar, India
e-mail: bdasfce@kiit.ac.in

A. Panda
e-mail: anuradha.panda.mama@gmail.com

J. Padhi
e-mail: erjppadhi@gmail.com

P. Chakraborty
e-mail: paromita.chakrabortyfce@kiit.ac.in

B. B. Das et al. (eds.), *Recent Developments in Sustainable Infrastructure*, Lecture Notes in Civil Engineering 75, https://doi.org/10.1007/978-981-15-4577-1_87

Keywords Groundwater level · Mann Kendall trend analysis · Coastal Odisha · Sustainable groundwater use

1 Introduction

Water is an important factor for planning and development activities of any country. The use of water is increasing due to economic and industrial development. India is a tropical country and mainly depends on the monsoon rainfall for its water resources. Groundwater is primarily used for drinking, agriculture and industrial purposes, and its availability depends on rainfall and recharge conditions. During the monsoon period, recharge usually exceeds discharge and the difference goes into storage, hence water level increases. During non-monsoon period, the discharge usually exceeds the recharge and no storage happens, hence water table decreases. The objective of water-level monitoring is to perceive the groundwater level fluctuations and to study the everlasting trend. The trend analysis is important as it will indicate the pattern of seasonal groundwater level and from that we can understand whether extraction and its use are sustainable or not. If the groundwater is not recharged properly, especially in the coastal regions, then to maintain the groundwater level sea water intrusion may take place and the area will suffer from salinity problem. This may lead to problem in drinking water as well as irrigation.

There are various statistical methods which are used for trend analysis and it is different from simple regression method to more advanced non-parametric methods (Mann 1945; Kendall 1955). This is a very popular technique for the analysis of the trends in groundwater levels monitoring. Patie and Singh [6] have identified the pre-monsoon and post-monsoon of groundwater level trend by using Mann-Kendall test and Sen's slope estimator in Karnal district of Haryana. Tabari et al. [9] have investigated groundwater level trend by using Mann-Kendall test and Sen's slope estimator in the Mazandaran province, Iran for the period 1985–2007. Thakur and Thomas [10] studied the application of different methods to identify the trends of groundwater level in few blocks of Sagar district of Madhya Pradesh because this area is affected by severe water scarcity owing to the declining groundwater levels. Lutz et al. [4] had conducted their investigation in Northern Ghana by using this test. Their study shows that the evaluation of groundwater level and the recharge should be done with the limited groundwater level data and climate data. Kumar et al. [3] had evaluated the trend of groundwater level by using this analysis during the period from 1998 to 2012 for 13 locations and four districts of Lucknow. Mann-Kendall test is used to determine the trend of water table fluctuation, spatio-temporal trend analysis and also required for the analysis of the effect of rainfall on groundwater level, according to Abdullahi and Garba [1]. Ribeiro et al. [7] had described the monthly trends of piezometric time series and the magnitude of 23 shallow wells during the period from 1979 to 2008 in Elqui river basin, Chile. Varni et al. [11] had conducted an investigation in Pampa plain, Argentina by following the same analysis.

2 Materials and Methods

2.1 *Study Location*

The state of Odisha in India is situated between the longitudes of 81.27° E and 87.29° E and latitudes of 17.49° N to 22.34° N. The geographical area of the state is 155,707 km^2. Odisha has several rivers and the entire state's geographical area is divided into 11 river basins. The length of coastal Odisha is about 450 km. This coastal region (which extends up to 75 m contour) is a combination of major rivers of Odisha, namely Subarnarekha, Budhabalanga, Baitarani, Bramhani, Mahanadi and Rushikulya. Hence the coastal plain of Odisha is called as "Hexadeltaic region" or "Gift of Six Rivers". Administratively, Odisha is divided into 3 divisions, 58 sub-divisions, 30 districts, 317 tahsils and 314 blocks. The weather in Odisha is highly influenced by sea because of its vast coast line. In the coastal districts, the climate is humid and sticky. The average rainfall of the state is 1489 mm which is different from one place to another. Due to the effect of south-west monsoon, many rivers get flooded during this period. Odisha experiences small rainfall from the retreating monsoon in the month of October–November.

2.2 *Coastal Saline Tract*

Odisha has a vast area of coastal plains with a coast line of 450 km. There is a narrow tract which starts from Chandeneswar in Balasore district (the north-east) and extends up to Brahmagiri in Puri district (south-west). An estimated area of 5,44,667 hectare in coastal alluvial tract is suffering from salinity hazard. The salinity problem is not uniform throughout the coastal tract. The maximum width of this tract is in Mahanadi delta (about 55 km). Puri area is highly affected in salinity problem as in the west of Puri town the aquifers are generally brackish to saline. Parts of seven coastal districts covering 36 blocks suffer from salinity problem. The coastal tract along with the saline tract of Odisha is shown in Fig. 1.

The districts and blocks of Odisha which are suffering from salinity problem are shown in Table 1.

2.3 *Groundwater Status in Coastal Odisha*

As per estimation made by the groundwater estimation committee, the status of replenishable groundwater resources, gross annual draft and average stage of groundwater development of seven coastal districts are shown in Table 2.

Fig. 1 Coastal saline tract (*Source* Directorate of Groundwater Development, Government of Odisha)

Table 1 Salinity affected districts and blocks (*Source* Directorate of Groundwater Development, Government of Odisha)

District	Affected blocks	
	Fully	Partially
Balasore	–	Bahanga, Balasore, Baliapal, Basta, Bhograi, Remuna
Bhadrak	Chandabali	Basudevpur, Tihidi, Dhamnagar
Ganjam	–	Chhatrapur, Chikiti, Ganjam, Khalikote, Rangeilunda
Jajpur	–	Bari, Binjharpur, Dashrathpur
Jagatsinghpur	Erasama	Balikuda, Kujang, Nuagaon
Kendrapada	Mahakalapada, Marshaghai, Rajkanika, Rajnagar	Aul, Derabish, Kendrapada, Pattamundai, Garadapur
Puri	–	Astarang, Brahmagiri, Delang, Gop, Kakatapur, Kansa, Krushna Prasad, Nimapara, Pipili, Puri, Satyabadi
Total 7 districts	6 blocks	36 blocks

2.4 Data Collection

Groundwater level data was collected from Directorate of Groundwater Development, Government of Odisha. The groundwater level data is collected for seven

Table 2 Groundwater status of coastal districts (*Source* Directorate of Groundwater Development, Government of Odisha)

Districts	Annually replenishable groundwater resources (ha m)	Gross annual draft for domestic, industrial and irrigation use (ha m)	Average stage of groundwater development (%)
Balasore	1,10,063	53,471	48.58
Bhadrak	45,409	25,197	55.49
Gajapati	22,670	5,588	24.65
Ganjam	1,14,541	33,408	29.17
Jagatsinghpur	45,029	21,332	47.37
Kendrapada	16,781	8861	52.80
Puri	58,806	10,448	17.77

coastal districts (Balasore, Bhadrak, Jagatsinghpur, Kendrapada, Puri, Jajpur and Ganjam), which include 77 blocks and 272 locations for the time period 1998–2017. The data pertains to four seasons for each location, that is, pre-monsoon, mid-monsoon, post-monsoon and winter. It comes out to be 1088 time series of data sets, each having 20 years duration. The locations of monitoring wells are shown in Fig. 2.

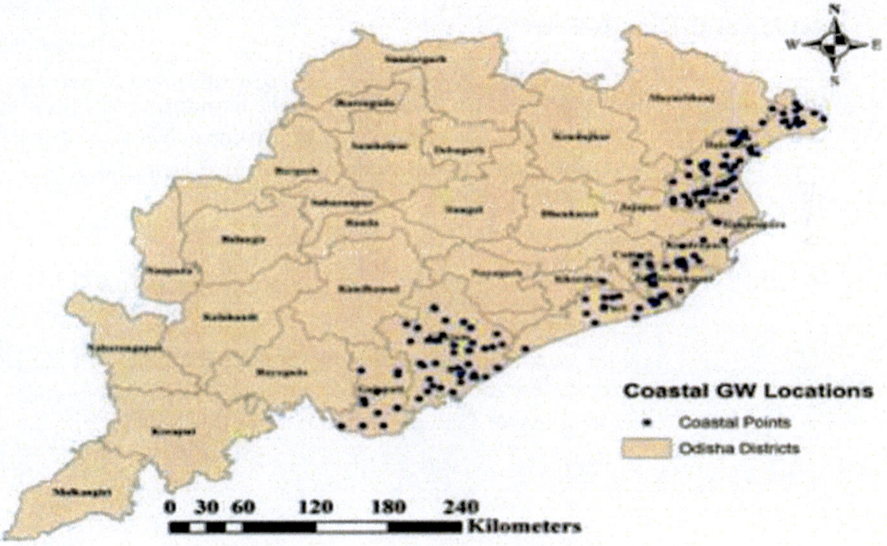

Fig. 2 Map of groundwater monitoring wells of Coastal Odisha

2.5 Methodology

Trend analysis of groundwater level for each location and for each time series was performed by using the Mann-Kendall test. The aim of the Mann-Kendall (MK) test is to evaluate whether there is a monotonic upward trend or downward trend of the variable of interest over time. The monotonic upward trend defines that the variable is consistently increasing, and the monotonic downward trend defines that the variable is consistently decreasing through time but the trend may or may not be linear over time. This test can be used in place of a parametric linear regression analysis which is required to test whether the slope or the magnitude of the estimated linear regression line is other than zero. This analysis requires that the residuals of the fitted regression line would be normally distributed; this is an assumption which is not required by the MK test because this test is a non-parametric test. In this study, different analysis have been adopted for groundwater level analysis such as (i) mean, standard deviation, coefficient of variation of each time series; (ii) groundwater level trend analysis was performed for 272 locations (i.e. for each time series) including pre-monsoon period, mid-monsoon period, post-monsoon period and winter period (total no. of 1088 time series) using Mann-Kendall test; (iii) Z statistics obtained from the trend analysis was analyzed to ascertain increasing or decreasing trend of the time series.

3 Results and Discussion

Analysis was performed for the data sets collected for all the monitoring locations in seven coastal districts of Odisha. Detailed analysis is shown for one coastal district Balasore, and the results of all other districts are compiled and shown also.

3.1 Mean, Standard Deviation and Coefficient of Variation

The graphical representation of mean, standard deviation and coefficient of variation of groundwater level for different monitoring locations of Balasore district (57 locations) and also for different seasons are presented in Figs. 3, 4 and 5.

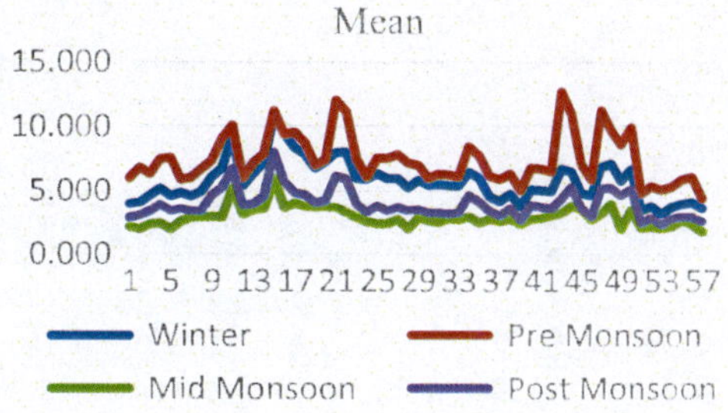

Fig. 3 Mean of groundwater level

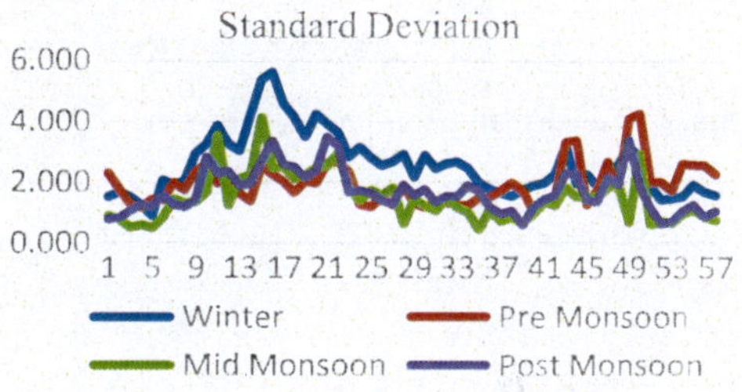

Fig. 4 Standard deviation groundwater level

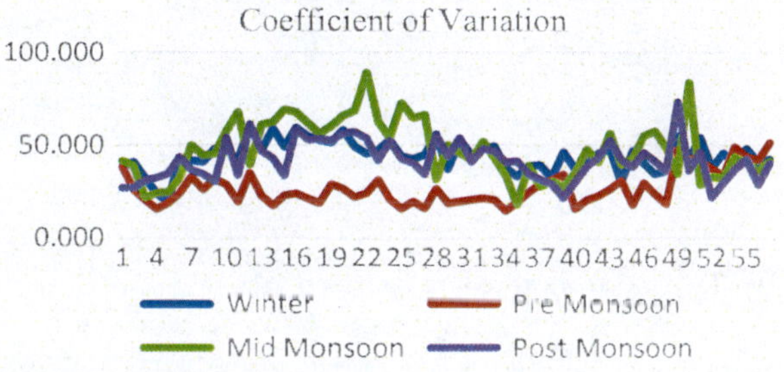

Fig. 5 Coefficient of variation of groundwater level

3.2 *Trend Analysis of Groundwater Level*

Trend analysis of the time series was performed using the Microsoft Excel program MAKESENS [8] which helps in estimating Mann-Kendall test and Sen's slope estimates. The output obtained from using this program for a location in Balasore district for four different seasons are presented in Figs. 6, 7, 8 and 9.

3.3 *"Z" Statistics of Groundwater Level*

Trend analysis was performed for all 57 locations of Balasore district covering four different seasons and the Z statistics obtained from the analysis for different locations and seasons are presented in Figs. 10, 11, 12 and 13. If the Z statistics is negative

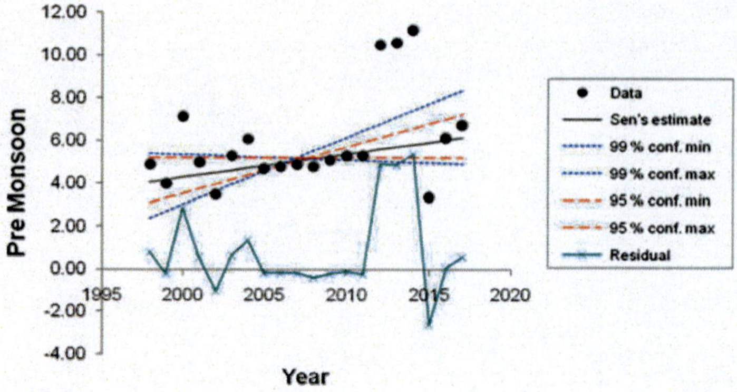

Fig. 6 Trend analysis of pre-monsoon data

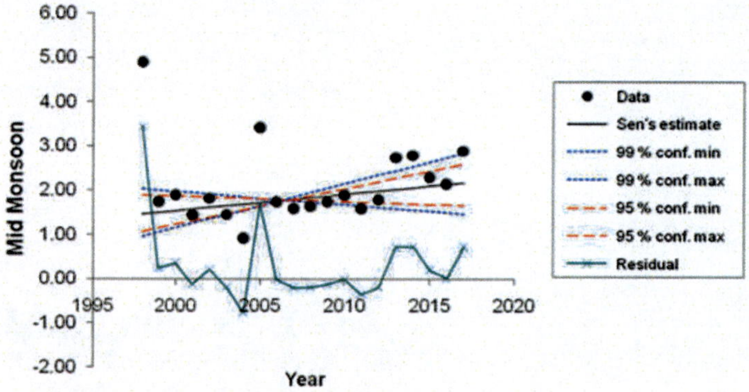

Fig. 7 Trend analysis of mid-monsoon data

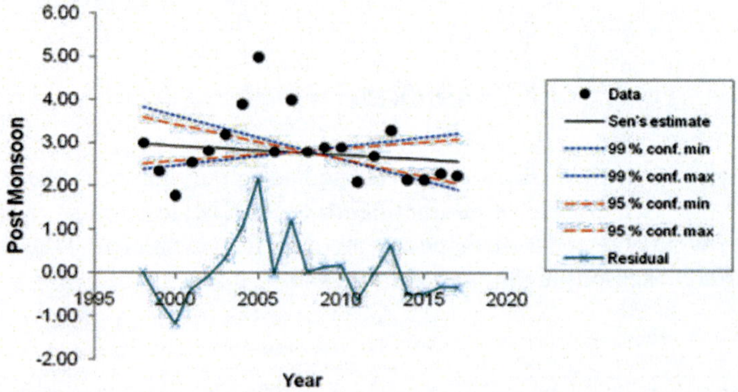

Fig. 8 Trend analysis of post-monsoon data

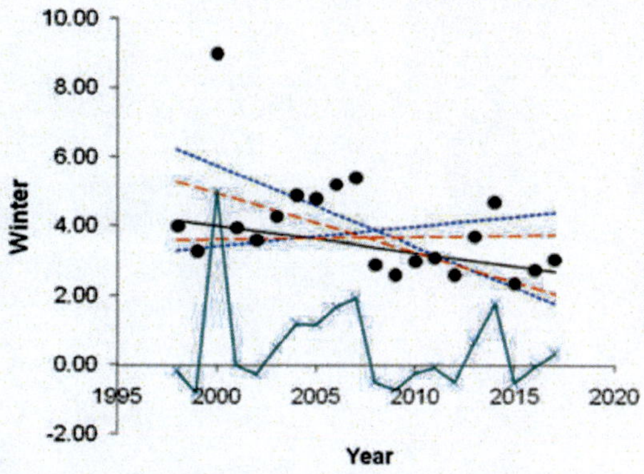

Fig. 9 Trend analysis of winter data

Fig. 10 Z statistics of groundwater levels in Balasore District during Pre Monsoon

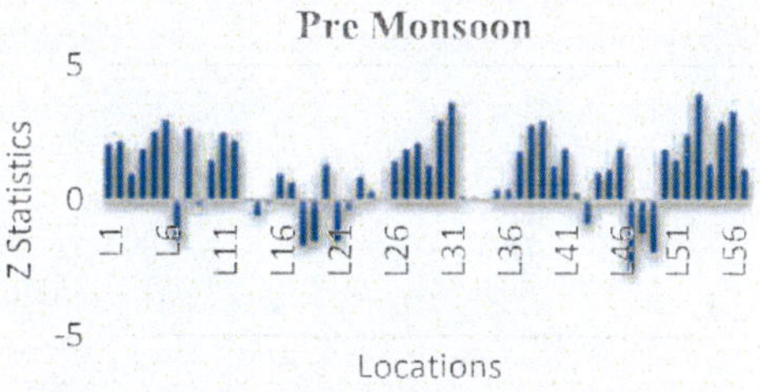

Fig. 11 Z statistics of groundwater levels in Balasore District during Mid Monsoon

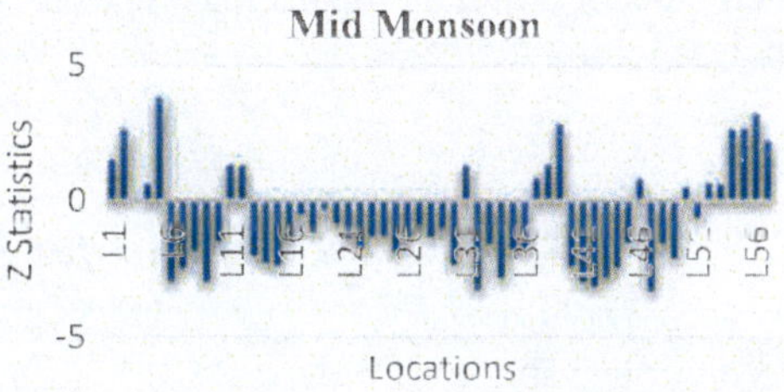

Fig. 12 Z statistics of groundwater levels in Balasore District during Post Monsoon

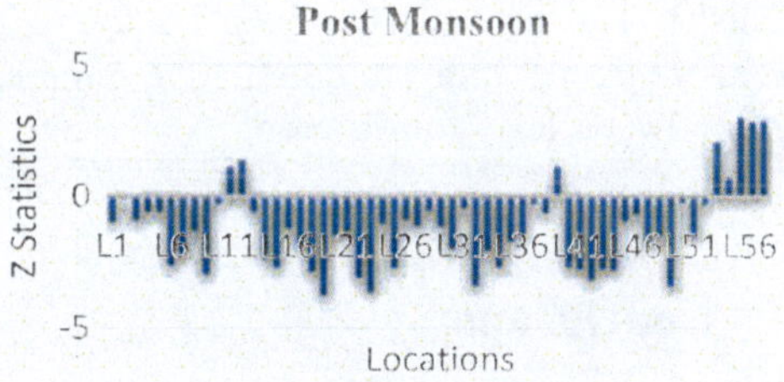

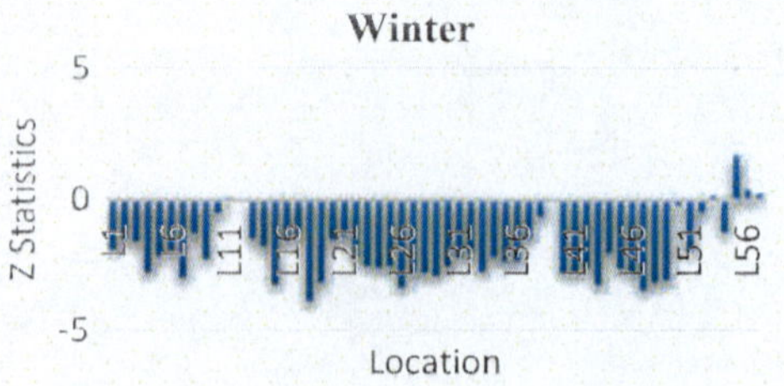

Fig. 13 Z statistics of groundwater levels in Balasore District during Pre Winter

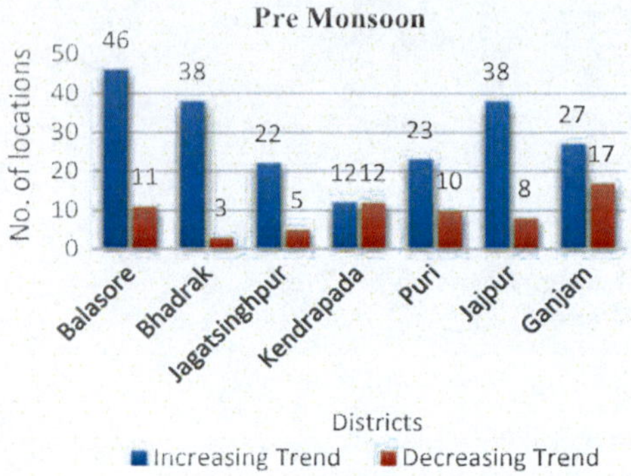

Fig. 14 Increasing and decreasing trend of groundwater levels in coastal districts during Pre Monsoon

then it denotes monotonic decreasing trend and if it is positive it denotes monotonic increasing trend.

3.4 Results of Trend Analysis for All Coastal Districts

As described above for Balasore district, similar trend analysis was performed for all groundwater level data series for different locations in seven districts of coastal Odisha. The results have been compiled and presented in Figs. 14, 15, 16 and 17. The figures indicate cumulative number of increasing and decreasing trend of groundwater level for different seasons and districts of coastal Odisha.

4 Conclusions

The following conclusions can be drawn from the trend analysis of groundwater level of different locations pertaining to different seasons in coastal Odisha.

- In Balasore district, the highest coefficient of variation is found in mid-monsoon period having value 89.05%; in Bhadrak the highest coefficient of variation is found in mid-monsoon period having value 76.815%; in Jagatsinghpur the highest

Fig. 15 Increasing and decreasing trend of groundwater levels in coastal districts during Mid Monsoon

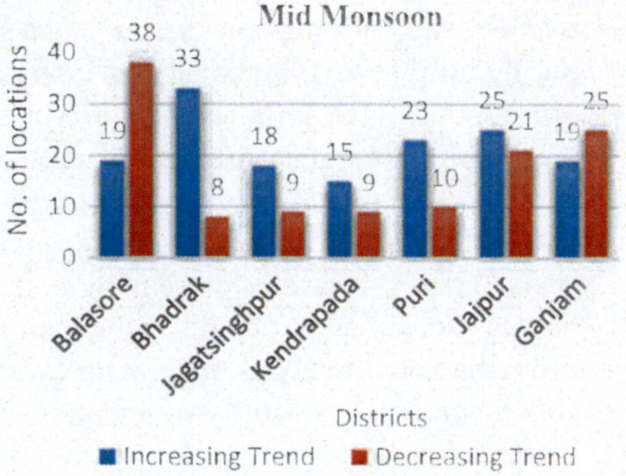

Fig. 16 Increasing and decreasing trend of groundwater levels in coastal districts during Post Monsoon

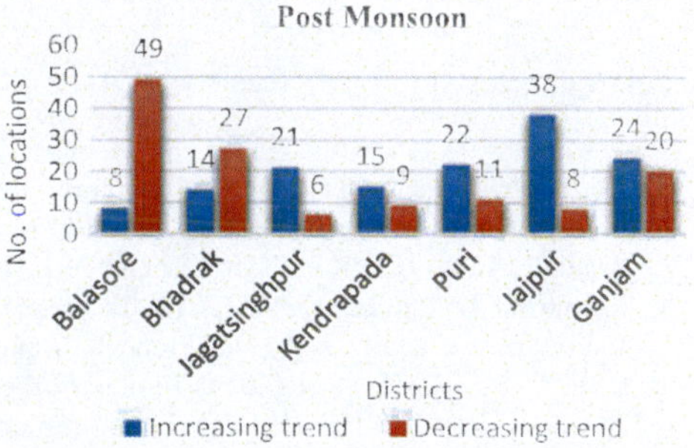

Fig. 17 Increasing and decreasing trend of groundwater levels in coastal districts during Winter

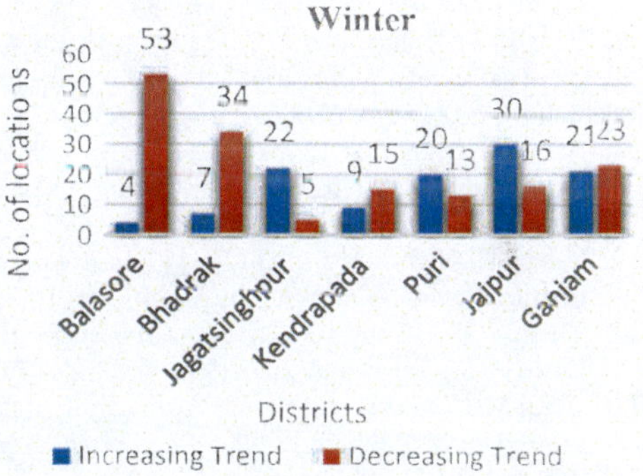

coefficient of variation is found in mid-monsoon period having value 93.582%; in Kendrapada the highest coefficient of variation is found in mid-monsoon period having value 97.165%; in Puri the highest coefficient of variation is found in mid-monsoon period having value 98.84%; in Jajpur the highest coefficient of variation is found in mid-monsoon period having value 96.996%; in Ganjam the highest coefficient of variation is found in mid-monsoon period having value 98.31%. It clearly indicates that the highest coefficient of variation in all coastal district happens in mid-monsoon season which speaks about quite variation of recharging during initial two months of monsoon period.

- For groundwater level, during pre-monsoon period 206 locations are having increasing trend and 66 locations are having decreasing trend. In mid-monsoon period 152 locations are having increasing trend and 120 locations are having decreasing trend. In post-monsoon period 142 locations are having increasing trend and 130 locations are having decreasing trend. In winter 113 locations are having increasing trend and 159 locations are having decreasing trend.
- From the above analysis, it gives an overall idea about the groundwater level trend which can help in identifying locations and season of criticality for groundwater extraction and use.

References

1. Abdullahi MG, Garba I (2016) Effect of rainfall on groundwater level fluctuation in Terengganu, Malaysia. Int J Remote Sens GIS 4:142. https://doi.org/10.4172/jrsg.1000142
2. Kendall MG (1955) Rank Correlation Methods. London: Charles Griffin
3. Kumar P, Chandniha SK et al (2018) Trend analysis of groundwater level using non-parametric test in alluvial aquifers of Uttar Pradesh, India. J Curr World Environ 13(1):44–54
4. Lutz A, Minyila S et al (2015) Fluctuation of groundwater levels and recharge patterns in Northern Ghana. MDPI, Climate https://doi.org/10.3390/cli3010001
5. Mann HB (1945) Non-parametric test against trend. Econometrica 13:163–171
6. Patie GT, Singh DK et al (2015) Time series analysis of groundwater levels and projection of future trend. Int J Geol Soc India 85
7. Ribeiro L, Kretschmer N et al (2014) Evaluating piezometric trends using Mann Kendall test on the alluvial aquifers of the Elqui River basin, Chile. Hydrol Sci J 60:10, 1840–1852. https://doi.org/10.1080/02626667.2014.945936
8. Salmi T, Maatta A (2002) Detecting trends of annual values of atmospheric pollutants by the Mann-Kendall Test and Sen's slope estimates—the excel template application Makesens, Finish Meteorological Institute, FMI-AQ-31
9. Tabari H, Nikbakht J et al (2011) Investigation of groundwater level fluctuations in the north of Iran. Environ Earth Sci. https://doi.org/10.1007/s12665-011-1229-z
10. Thakur GS, Thomas T (2011) Analysis of groundwater levels for detection of trend in Sagar District, Madhya Pradesh. J Geol Soc India 77:303–308
11. Varni M, Comas R et al (2013) Application of the water table fluctuation method to characterize groundwater recharge in the Pampa plain, Argentina. Hydrol Sci J 58:7, 1445–1455. https://doi.org/10.1080/02626667.2013.833663

Recent Developments in Transportation Engineering

Design of Long-Life Pavements for India

Brundaban Beriha and Umesh Chandra Sahoo

Abstract Perpetual or long-life pavements have been proven to be a sustainable pavement solution to meet heavy traffic demand and to address the material scarcity. These are normally designed as full depth asphalt pavements to last more than 50 years without requiring any structural rehabilitation, but need periodic maintenance to address damages which are confined to top few inches of a pavement. However, requirement of huge quantity of bituminous materials makes full depth asphalt pavements very expensive for use in developing countries like India. Studies in countries like UK and China have shown that long-lasting composite pavements with cemented materials and deep strength pavements with thin granular layer sandwiched between asphalt layer and subgrade can also be considered as perpetual pavements. In India, the concept of perpetual pavement is relatively new and introduced only in 2012. Also, the computer program IITPAVE available with these guidelines does not address the design of perpetual pavements in a comprehensive manner. Therefore, a computer program called "PerPave" has been developed under this study to design and compare different perpetual pavement types and to suggest the optimum perpetual pavement composition with minimum cost. It will be helpful for the design engineers in selecting the appropriate perpetual pavement section considering the budget, traffic, and availability of the materials in the locality.

Keywords Long-life pavement · Perpetual pavement · Optimum perpetual pavement composition

1 Introduction

Perpetual pavements or long-life pavements when constructed properly can last more than 50 years without failing structurally. But it needs periodic maintenance to

B. Beriha (✉) · U. C. Sahoo
School of Infrastructure, IIT Bhubaneswar, Bhubaneswar, India
e-mail: bb10@iitbbs.ac.in

U. C. Sahoo
e-mail: ucsahoo@iitbbs.ac.in

B. B. Das et al. (eds.), *Recent Developments in Sustainable Infrastructure*, Lecture Notes in Civil Engineering 75, https://doi.org/10.1007/978-981-15-4577-1_88

address the functional failures arising due to environmental effects which is generally confined to the top layer. These pavements result in high initial construction cost but the life cycle cost is low compared to the conventional pavements considering a period of 50 years [1, 2]. Since 1960, full depth asphalt pavements have been in use as perpetual pavements where all the layers above the subgrade are asphalt layers [6]. Recently, India has included full depth asphalt pavement as perpetual pavement in its design guidelines [5]. However, still it is not gaining popularity as it involves higher initial cost of construction due to the costly bituminous materials.

Countries like UK [7] and China [13] have been successful in building long-life pavements with stabilized material called as long-life composite pavements (UK) or semi-rigid long-life pavements (China). These pavements could be viable alternatives to traditional full depth asphalt pavement in terms of pavement construction and maintenance cost as it would facilitate use of marginal material. Hence, it is necessary for the decision-maker to compare different perpetual pavement alternatives during decision-making process as it can have a huge impact on the overall cost of the project. A user-friendly program called "PerPave" has been developed which facilitates designing and comparing different perpetual pavement alternatives in terms of cost. This paper describes the framework of the program and outlines the different features in it.

2 Types of Perpetual Pavement

Countries like US, UK, Israel, and China have adopted different perpetual pavement types to suite their local conditions. Basically three types of perpetual pavements (Fig. 1) have been in use so far, i.e., full depth asphalt pavement, deep strength pavement, and long-life composite pavement, which are different from each other in terms of materials used in different layers. In full depth asphalt pavement, hot mix asphalt (HMA) is used in all layers above subgrade, whereas deep strength pavement uses a thin granular layer in between thick asphalt layer and subgrade. In case of long-life composite pavements or long-life semi-rigid pavements, thick cemented layer(s) is sandwiched between a thin bituminous layer and subgrade. The

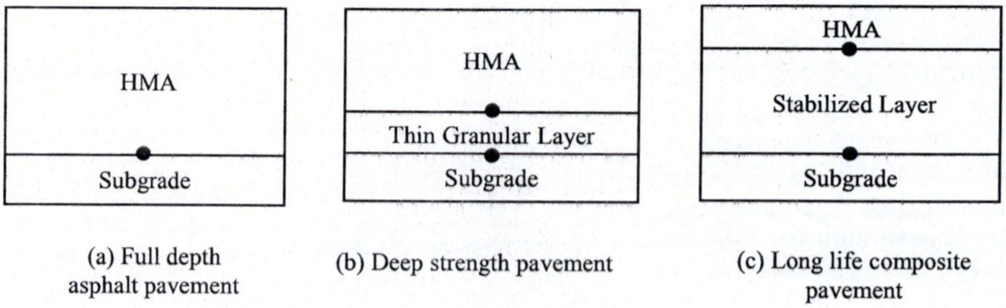

Fig. 1 Types of perpetual pavements with critical locations

common element of all perpetual pavement types is that these do not fail structurally as the major layers experience stresses/strains lower than the endurance limit of the materials composing those layers.

3 Endurance Limit (EL)

The basic design philosophy of perpetual pavement is based on the concept of endurance limit which is the pavement threshold responses at critical locations. When pavement responses due to traffic load at the critical locations are below endurance limit it would have infinite life.

For full depth asphalt pavement and deep strength pavement, these critical responses are horizontal strain at the bottom of asphalt layer and vertical strain at the top of the subgrade which controls the asphalt layer fatigue cracking and subgrade rutting, respectively. Nunn et al. [8] suggested a strain value of 70 and 200 $\mu\varepsilon$ as endurance limit for horizontal strain at the bottom of asphalt layer and vertical strain at the top of subgrade, respectively, which has been adopted in India [5].

In long-life composite pavement, fatigue cracking of stabilized layer is the major distress. The stiff cemented layer in composite pavement significantly reduces the vertical compressive strain above subgrade and also reduces the tensile strain at the bottom of the asphalt layer. Thus, the critical locations for long-life composite pavements are bottom of the asphalt layer, bottom of the cemented layer, and top of the subgrade. For eliminating asphalt fatigue and subgrade rutting, the same endurance limit criteria suggested for full depth asphalt pavement is used for long-life composite pavement. But for stabilized layer fatigue, Parmeggiani [9] proposed strain ratio (flexural strain/breaking strain) value 0.25 as endurance limit for cemented material, whereas Gonzalez et al. [3] suggested a value of 0.26 and 0.33 for cemented material and lean mix concrete, respectively.

4 Design of Perpetual Pavements

The PerRoad computer program developed by Timm and Newcomb [12] is the only program which is dedicatedly used for perpetual pavement design. It combines the mechanistic-empirical design concept and Monte Carlo simulation technique to obtain a probabilistic analysis of pavement responses to different loading configurations, material properties, and layer thicknesses. Using Monte Carlo techniques, random pavement sections are selected considering all the variabilities associated with material property, thickness, and environmental conditions. Similarly, load is also randomly selected from available load spectra. A total of 5000 Monte Carlo cycles is used for the analysis. For each cycle, allowable number of repetition of load until failure and damage is calculated using the following equations:

$$N_{fi} = k_1 (1/\varepsilon_i)^{k_2} \tag{1}$$

$$\text{Damage in individual cycle } (D_i) = 1/N_{fi} \tag{2}$$

where

N_{fi}	Number of cycles until failure,
k_1, k_2	Empirical constants,
ε_i	Pavement strain response at critical location, and
i	Monte Carlo cycle.

Then, total damage is calculated using Miner's hypothesis:

$$\text{Total damage } (D) = \sum_{i=1}^{m} D_i \tag{3}$$

When the ε is below the endurance limit, N_f becomes infinity and damage becomes zero. Then, the average damage per axle is calculated as $D_{avg} = D/m$. Using the damage accumulation rate for a traffic level and expected traffic growth rate, time before the damage reaches the critical point of 0.1 is estimated. If the calculated time exceeds 50 years, pavement is selected as perpetual pavement.

Apart from PerRoad, some agencies uses different pavement analysis programs like Circly (California), Illipave (Illinois), and Kenlayer (Ohio) for design of perpetual pavement [10]. In these programs, trial pavement sections are assumed and the stress and strain outputs are compared with endurance limit. The pavement sections whose responses are under endurance limit are selected as perpetual pavements. However, these programs do not consider the material cost associated with the selected pavement which is an important parameter for the decision-makers in selecting a particular pavement composition for a project, as a different composition may satisfy the criteria of a perpetual pavement with less material cost.

Also, there is no consideration of overloading in the analysis of the existing programs. The effect of overloading on full depth asphalt up to seven times the design axle load for 15–20% of time would not discard the infinite life theory when stress and strain return to endurance limit in normal traffic conditions [11]. But overloading has a significant effect on composite pavements as the cemented layer would experience cracks under heavy loading which will grow further even under normal loading. So considering the effect of overloading a high safety factor should be chosen for design of long-life composite pavement. Hence, a computer program has been developed called 'PerPave' which analyzes different pavement thickness combinations and suggests the optimum perpetual pavement section with minimum construction cost. It also helps in comparing three different perpetual pavement types which will be helpful in decision-making process.

5 Basic Framework of Perpave

The basic framework of the PerPave is shown in Fig. 2 where different pavement trial sections are analyzed with a mechanistic pavement model and their responses are compared with endurance limit criteria. The pavement sections which stratify the endurance limit criteria are considered for cost analysis and the section with minimum cost is chosen as the final design of any pavement. The mechanistic pavement model is based on the solution provided by Huang [4] for linear elastic layer theory.

The entire layout of the PerPave has been divided into six tabs where the first tab contains the basic information about the program. The second tab contains the basic information about the project, e.g., name of the project, length of the project, width of the pavement under the project, etc., which will be used during project cost calculation and report generation. The third, fourth, and fifth tab is used for designing the three perpetual pavement types, i.e., full depth asphalt pavement, deep strength asphalt pavement, and flexible composite pavement, respectively. The final tab summarizes the output from third, fourth, and fifth for comparison.

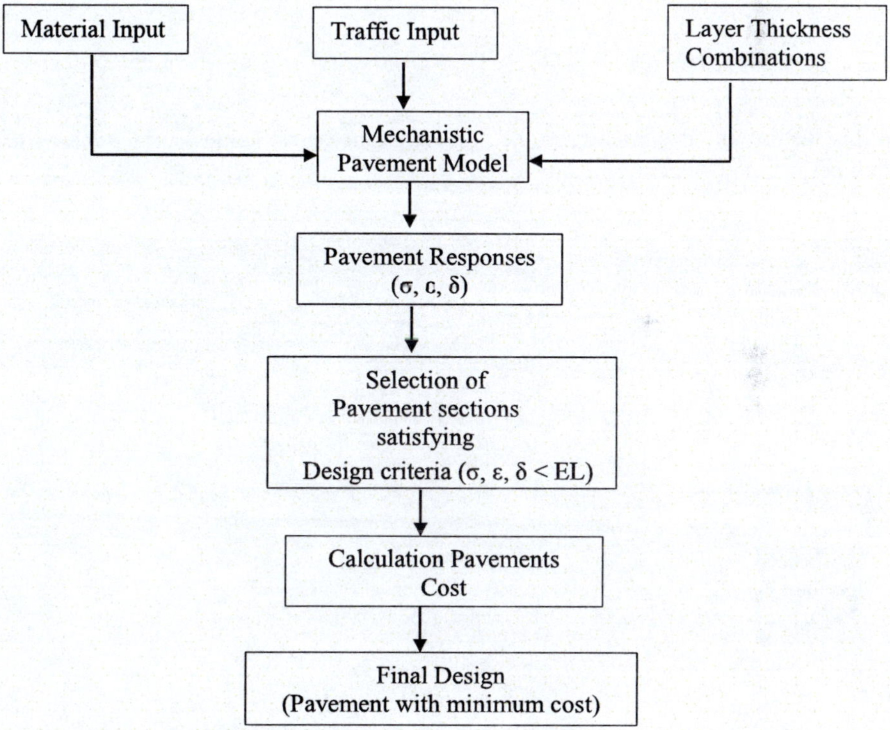

Fig. 2 PerPave design framework for a perpetual pavement type

6 Design Methodology

The basic layout of the major three tabs shown in Fig. 3 used for designing perpetual pavement is similar and consists of different components. This section discusses these components and how these affect the final design.

6.1 Pavement Composition

This section contains information related to the basic composition of the perpetual pavement type and the properties related to each layer, i.e., modulus, Poisson's ratio, thickness range, and material cost. The thickness of each layer is provided in range as the program evaluates all different combinations of thickness to suggest the one with minimum cost. Larger the range of thickness values, larger is the computation time. Hence, logical values should be chosen to minimize the computation time.

6.2 Design Criteria

In this section, the design criteria, i.e., the endurance limit values corresponding to each perpetual pavement type are stored.

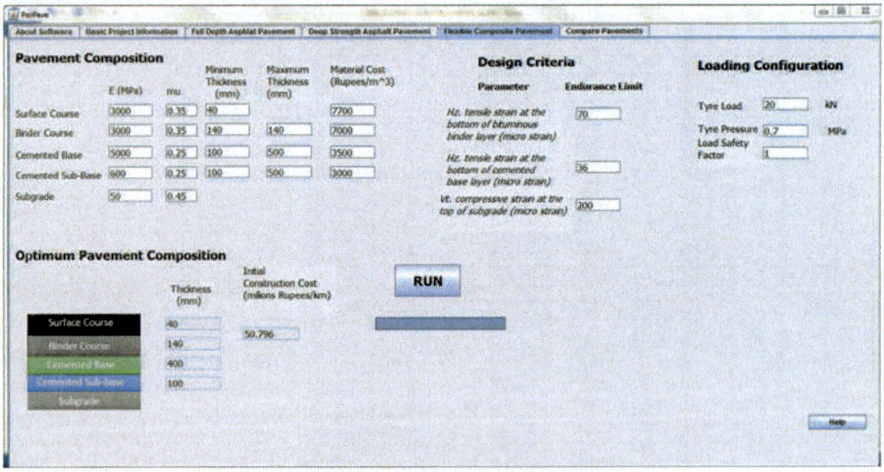

Fig. 3 Layout of PerPave for design of flexible composite pavement

6.3 Loading Configuration

The program does the analysis based on a dual-tire single-axle load as the stresses and strains are high under single-axle load compared to tandem and tridem axle. It allows the user to analyze the pavement with different axle loads and tire pressure. The tire load value corresponds to the single-tire load of the dual-tire configuration. It has an additional feature called "Load Safety Factor" (LSF) which could be used to address the overloading of vehicle. The load safety factor values become "1" in case of no overloading.

6.4 Optimum Pavement Composition

This section shows the layer thickness composition of the perpetual pavement types which yield minimum construction cost. The pavement width, project length, and cost of material are used to calculate the project cost. The project costs of all the perpetual pavement alternatives are compared in the compare pavement tab.

7 Design Example

A case study is taken up to show the output of the PerPave program. Project details are provided in Table 1. The three perpetual pavement types, i.e., full depth asphalt pavement, deep strength pavement, and flexible composite pavement with design inputs as shown in Tables 2, 3, and 4 were analyzed. The modulus and Poisson's ratio value of each layer were taken from IRC 37 (2012) and material cost was collected from the public works department website of Odisha. For flexible composite pavement, the thickness of asphalt layer was fixed to 180 mm as suggested by Nunn [7]. Pavements were analyzed for a standard axle load of 80kN with 0.7 MPa tire pressure. For all perpetual pavement types, LSF was considered 1. The design criteria are given in Table 5.

Table 1 Project details

Project component	Value
Length of the project	10 km
No. of lanes in each direction	2
Lane width	3.5 m
No. of paved shoulder in each direction	1
Shoulder width	1.5 m

Table 2 Full depth asphalt pavement details

Material	Modulus (MPa)	Poisson's ratio	Cost (Rs/m^3)
Surface course	3000	0.35	7700
Binder course	3000	0.35	7000
Asphalt base course	3000	0.35	7000
Subgrade	50	0.45	NA

Table 3 Deep strength pavement details

Material	Modulus (MPa)	Poisson's ratio	Cost (Rs/m^3)
Surface course	3000	0.35	7700
Binder course	3000	0.35	7000
Asphalt base course	3000	0.35	7000
Unbound sub-base	Thickness dependent	0.35	2350
Subgrade	50	0.45	NA

Table 4 Flexible composite pavement details

Material	Modulus (MPa)	Poisson's ratio	Cost (Rs/m^3)
Surface course	3000	0.35	7700
Binder course	3000	0.35	7000
Cemented base course	5000	0.25	3500
Cemented sub-base	600	0.25	2000
Subgrade	50	0.45	NA

Table 5 Design criteria

Pavement type	Critical location	Endurance limit
Full depth asphalt pavement, deep strength pavement and flexible composite pavement	Horizontal tensile strain at the bottom of asphalt base layer	70 $\mu\varepsilon$ [5]
	Vertical strain at the top of the subgrade	200 $\mu\varepsilon$ [5]
Flexible composite pavement	Horizontal tensile strain at the bottom of stabilized base layer	36 $\mu\varepsilon$ [9]

The final design output is shown in Fig. 4. It can be seen from the final output that flexible composite pavement is the costliest alternative with cost of 50.796 million rupees/km followed by deep strength pavement with cost of 50.073 million rupees/km followed by full depth asphalt pavement with cost of 49.266 million rupees/km. For this project case, full depth asphalt pavement was found to be optimum perpetual pavement alternative.

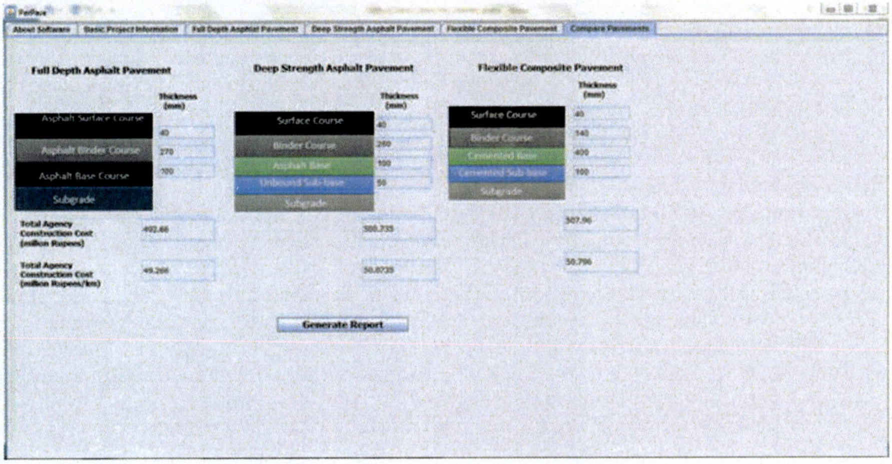

Fig. 4 Comparison of perpetual pavement types for the case study

8 Summary

Perpetual pavements have been proven to be one of the sustainable alternatives to tackle material scarcity along with saving huge maintenance cost. The best perpetual pavement type should be selected for a project to suit the budget and local conditions. The PerPave program would be a handy tool for the pavement designer while selecting a perpetual pavement type for the project. It provides the optimum thickness composition for a perpetual pavement type which not only stratifies the endurance limit design criteria rather yield minimum construction cost. One of the limitations of the current version of PerPave is that it does not consider the maintenance and user delay cost into account while suggesting the best perpetual pavement type. The total life cycle cost should be the deciding factor during selection of a pavement type. Therefore, a life cycle cost calculation part would be added to the PerPave in its later version to make it a complete package for perpetual pavement design.

References

1. Amini AA et al (2012) Life cycle cost comparison of highways with perpetual and conventional pavements. Int J Pavement Eng 13(6):553–568
2. Basu C et al (2013) Perpetual pavement—a boon for the Indian roads. Proc-Soc Behav Sci 104:139–148
3. Gonzalez A, Howard A, de Carteret R (2010) Cost effective structural treatments for rural highways: cemented materials. Austroads Technical Report AP-T168/10, Austroads, Australia
4. Huang YH (2004) Pavement analysis and design
5. Indian Roads Congress (2012) Guidelines for the design of flexible pavements. Indian code of practice, IRC 37

6. Newcomb DE, Buncher M, Huddleston IJ (2001) Concepts of perpetual pavements. Transportation research circular 503: perpetual bituminous pavements
7. Nunn M (2004) Development of a more versatile approach to flexible and flexible composite pavement design. TRL Report 615. Transport Research Laboratory, Crowthorne House
8. Nunn ME et al (1997) Design of long-life flexible pavements for heavy traffic, TRL Report 250. Transport Research Laboratory, Crowthorne, UK
9. Parmeggiani G (2012) Design of long life flexible composite and deep strength asphalt pavements. In: ARRB conference, 25th, Perth, Western Australia, Australia
10. Tarefder RA, Bateman Damien (2010) Design of optimal perpetual pavement structure. J Transp Eng 138(2):157–175
11. Thompson MR, Carpenter SH (2004) Fatigue design principles for long lasting HMA pavements. In: The International Society for Asphalt Pavements' (ISAP) international symposium on long lasting asphalt pavements
12. Timm David H, Newcomb David E (2006) Perpetual pavement design for flexible pavements in the US. Int J Pavement Eng 7(2):111–119
13. Zeng F et al (2016) Design and construction of long-life semirigid pavement: investigation from test sections in Northern and Southern China. No. 16-1913

Comparison of Coconut Root with Other Geotextiles for Transportation Infrastructure

Leonardo Souza and Purnanand P. Savoikar

Abstract To match India's rapid growth, the transportation infrastructure has to keep pace especially in the rural regions. Embankments and hill slope stabilizations comprise a major challenge in this sector. Today the focus of development has shifted to natural products and sustainable infrastructure. In this respect, it is vital to also study a technique that has been used for centuries by the Saraswat civilization that settled in Goa and is today classified as natural geotextiles. This paper compares the Geotextiles used today with this ancient sustainable technique. The benefits of this technique are comparable to those of modern plastics while sustainability is far better.

Keywords Coconut roots · Geotextile · Natural geotextiles · Traditional goan saraswat bunds

1 Introduction

As India is developing at a rapid rate, to support this growth the transportation infrastructure has to keep pace especially in the rural regions that comprise nearly 80% of the total transportation infrastructure in our country. Rural roads once built are rarely repaired unless a catastrophe occurs and hence embankments and hill slope stabilizations comprise a major challenge in this sector. Today the focus of development has shifted from use of plastics to natural products like jute and coir. Bioremediation, biostabilization, bioengineering and sustainable infrastructure are the buzz words driving today's research. In this respect, it is vital to also study a technique—Traditional Goan Saraswat Bunds (TGSBs) that has been used for centuries by the Saraswat civilization that settled in Goa and is today classified as natural geotextiles.

L. Souza (✉) · P. P. Savoikar
Civil Engineering Department, Goa Engineering College, Farmagudi, Goa 403401, India
e-mail: lrdcs@rediffmail.com

P. P. Savoikar
e-mail: psavoikar@gmail.com

B. B. Das et al. (eds.), *Recent Developments in Sustainable Infrastructure*, Lecture Notes in Civil Engineering 75, https://doi.org/10.1007/978-981-15-4577-1_89

1.1 Sustainable Transportation Infrastructure

Sustainability is one of the goals of transportation infrastructure. It promotes co-operation and exchange of information and knowledge about the geotechnical aspects in design, construction, maintenance, monitoring and upgrading of roads, railways and airfields with respect to minimizing carbon footprint and longevity of structures.

1.2 Ancient Saraswat Infrastructure Geotechnics

This paper investigates the feasibility and effectiveness of the traditional practice of using coconut roots for stabilization of earthen embankments used by Ancients Saraswat (Indus) Civilization who fled the disappearance of the Saraswati river settled in Goa [1]. They used a mix of River-Shell-Lime and Coconut-Leaf-Ash to stabilize sandy loamy locally sourced soil for traditional coconut tree lined bunds for transportation, low-land reclamation, rural infrastructure, hill slope stabilization and housing (Fig. 1). This paper seeks to compare this ancient technique with modern techniques and its effectiveness.

2 Literature Review

There are innumerable studies carried out on slope stabilization by different materials. Geotextiles are already a well-established field of research. There is also a vast body of literature developing on natural geotextiles. There is however a dearth of technical literature on coconut tree roots.

Fig. 1 Typical TGSB bunds used for transportation infrastructure in Goa

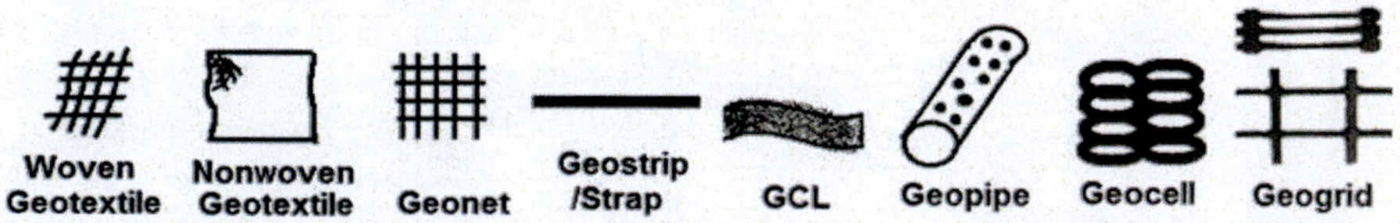

Fig. 2 Forms of geosynthetic reinforcing products

2.1 *Transportation Infrastructure Geosynthetics*

The proceedings of the 1st International Conference on Transportation Geotechnics, [2] deals extensively on geotextiles and other innovations in design and construction techniques used for transportation infrastructure. Palmeira et al. [3], Fay et al. [4] and Mittal and Shukla [5] discussed the various advances in soil reinforcement and their design principles. Geosynthetics is the general term applied to the wide range of products which are now available (Fig. 2).

They are manufactured by many specialized processes using a number of different polymers. The shear strength properties improve tremendously and vertical walls are possible. The deflection of reinforced beds decrease and the subgrade vertical stress increases.

2.2 *Bio-Mediated Soil Improvement*

De Jong et al. [6] and Chou [7] studied the bio-mediated improvement of soil systems, identifying the primary components and interplay between different disciplines, and the ability to alter the in situ soil strength and other properties. The primary properties of soil can realize even a 10-fold change or more include permeability, stiffness, compressibility, shear strength and volumetric behaviour with biomediation due to calcite precipitation.

2.3 *Grassroot Reinforced Soil*

The improvement of soil properties using grassroots has been extensively studied by Gobinath et al. [8], Hengchaovanich [9], Truong [10] and Teerawattanasuk et al. [11]. Vetiver grass has the ability to penetrate through soils mixed with stone or pebbles, through asphalt layer and through hardpan. Vetiver acts as flexible elastic vertical pile roughly 0.5 m diameter and 2–3 m deep (Fig. 3). Their roots tend to travel vertically downwards. Due to this reason, they also don't spread as rapidly as other grasses. They have more utility in vertical anchorage only.

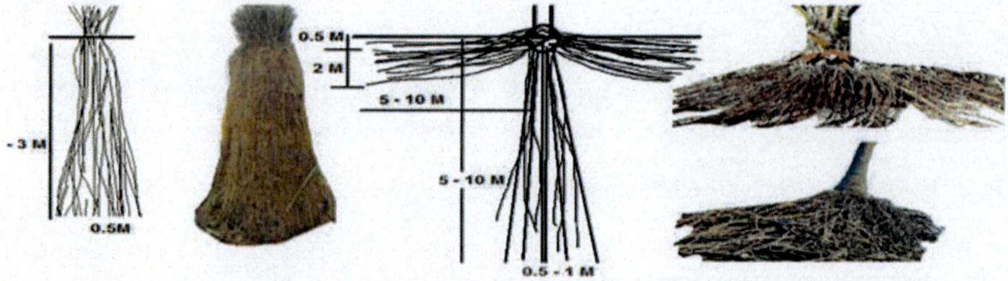

Fig. 3 Vertical roots of vetiver grass and three-dimensional roots of coconut tree

2.4 Coconut Root-Soil Improvement

Souza et al. [12] and Souza and Savoikar [13] researched the ancient Saraswat improvement of soil systems, using a combination of techniques to alter the in situ soil strength and other properties for infrastructure purposes. A combination of paddy straw reinforcement, ash-lime stabilization and coconut tree root natural geonet was successfully used to strengthen and stabilize the embankments and hill slopes for transportation and other infrastructure purposes. Coconut roots have double action, a flexible elastic vertical pile roughly 0.5–1 m diameter and 2–3 m deep and a flexible elastic geomattress roughly 2 m deep and 5–10 m diameter depending on age of the tree and soil conditions (Fig. 3). Coconut roots, unlike other tree roots, are fibrous (not tap root) in nature. They are longer, stronger and tougher than grass, release adhesive secretions in the ground which help in bonding soil particles and increase strength, when they die the hole left is easily filled by migration of sand particles without causing piping effect. They are also more resistant to pests and rodents than normal tree roots.

3 Traditional Goan Saraswat Bunds

Before comparing the ancient and modern techniques it is vital to cast a brief look at the construction practices of TGSBs.

3.1 Coconut Tree Root Bund Construction

The unique feature of the Saraswat soil strengthening strategy was the double row of coconut trees planted on the top of the embankment for soil reinforcement. While normal geosynthetics used provides strength and stability, this ancient technique provided additional scenic beauty and sustainability. This is most vital in rural roads

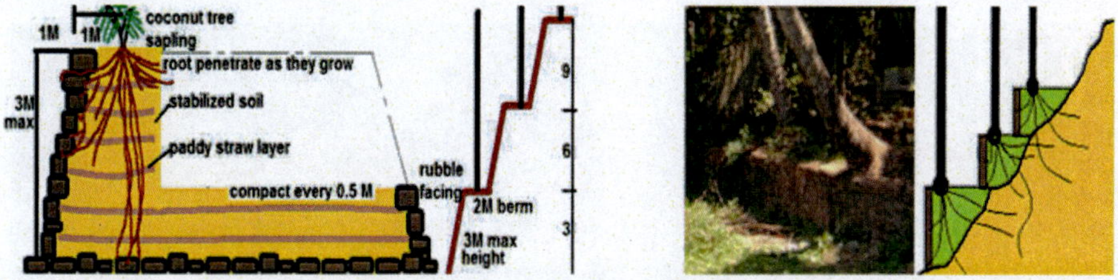

Fig. 4 Embankment and stabilization of hill slopes by TGBS

as repair is rare once construction is completed. They need longer for full functionality than geosynthetics.

3.2 Traditional Constructional Practices

The base was made by dumping rubble, pitching/armour of rubble was laid on both sides first, the sandy soil acted as a filter, a layer of rice paddy straw acted as a separator cum initial geotextile (till roots took over). Layers of 50–30 cm were compacted along with the outer pitching, sourcing of mud was from the nearby riverside, coconut tree was planted on top in two rows in zigzag diamond pattern, replanting schedule for the trees was done every 100 years. Maintaining the line and slope/batter was done using coir string and bamboo poles cut to size (Fig. 4). Roots penetrate subsoil and anchor around stones providing additional reinforcement tying the bund to the subsoil/slope. Thus, the bunds have lasted for the past 8000 years.

3.3 Measurement Techniques

The hand from elbow to fingertip was used as a measure (roughly 50 cm). A side slope of 1 horizontal to 3 vertical was always maintained. Berms are present for every 2.5–3 m.

3.4 Traditional Soil Improvement Techniques

Coconut jaggery molasses was fermented in huge cow dung plastered earthen jars (*bhann*) it acted as an insecticide/fungicide and also improved the workability. Cow dung was mixed with coconut leaf ash and burnt shell lime from locally sourced clams (*khube-tisreo-shinaneo*) added to the fines deficiency and strengthened the soil by pozzolanic effect. There was an antibacterial dung effect, and soil stabilization and

Fig. 5 Compaction of soil using traditional bullock chain and hand tamping

strengthening. Rice straw was used for non-sticky and geotextile effect and also for compaction and initial soil retention.

3.5 Traditional Equipment

Bullock cart transportation, coconut stump rammer, hand implements (for soil, rubble, etc.), 6–8 bullock-chain compacting, etc. was used during the construction of these embankments and slope stabilization structures (Fig. 5).

4 Use in Transportation Infrastructure

Geotextiles are widely used in transport infrastructure. TGSBs were widely used in Goa.

4.1 Uses in Embankments and Slope Stabilization

They are mainly used in reinforcing soils for embankments and hill cuttings. They can be incorporated as part of walls, used behind walls or be a wall themselves as a wraparound facing. TGSBs serve a similar purpose (Fig. 6). They can also be used for hill slope stabilization and to control rockfall and landslides.

In addition to stabilizing embankments and hill slopes both can be used as pavement separators, and support over weak loamy soils. They can also be used to bridge over subsurface cavities and sub-infrastructures like tunnels and sewers. Coconut tree roots though weaker than synthetic geotexiles individually, collectively perform better than standard geotextiles due to their density of root per meter square, superior penetration, greater reach and increased anchorage (Fig. 7). They are better suited

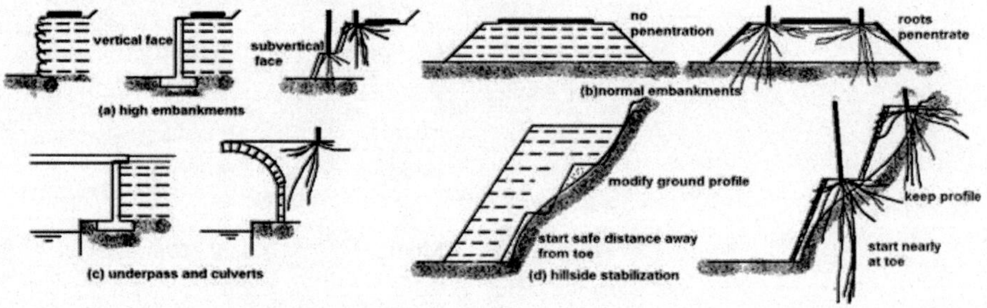

Fig. 6 Reinforced soil walls compared with TGSB in normal soils

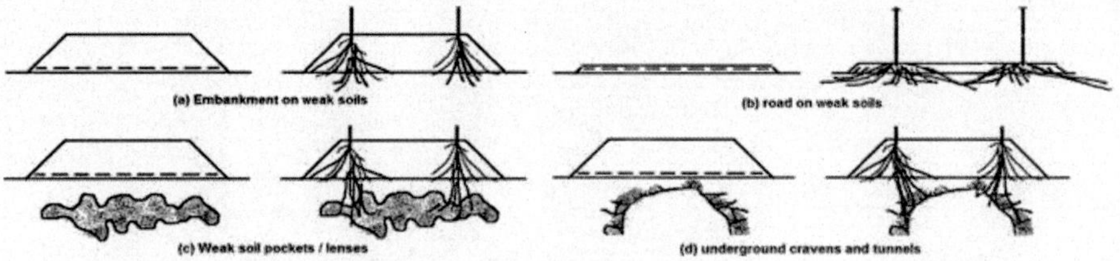

Fig. 7 Reinforced weak loamy subsoils—geosynthetic and coconut root (showing penetration of subsoil and increased anchorage)

Table 1 Spatial requirement of different embankments

	Normal embankment	Panelled walls	TGSB
Side slope	2H: 1 V	0H: 1 V	1H: 3 V
Side angle	20°–30°	90°	70°–75°
Extra space per m height per side	2	0	0.3
% Space per m height per side	200	0	30

for slope stabilization as they penetrate the substrata even rocks binding underlying fractured rock-mass together.

Except in case of panel-faced walls, where the face is vertical there is also 170% savings in the area needed for the construction of TGSB as compared to normal geotextile reinforced embankment (Table 1). This in turn results in equal savings in materials and construction costs.

4.2 *Important Characteristics*

All geotextiles have to be exhaustively tested to determine the optimum textile suitable for a particular situation. Their properties depend on the type of plastic or rubber used and the composite materials present (carbon/glass/steel fibre). Coconut

Table 2 Physical properties

	Geotextiles	TGSB/Coconut roots
Specific gravity	More	1.5
Weight/m	More	0.063 kg
Thickness	Depends on type	5–10 mm
Stiffness	More	8000 kN/m^2
Density	More	20 kN/m^3
Damping factor	0.2	0.3

Table 3 Mechanical properties

	Geotextiles	TGSB/coconut roots
Tenacity	More	Less
Tensile strength	More 500–800 N/mm^2	Less 100–200 N/mm^2
Bursting strength	More	Less
Drapability	Less	More
Compatibility	Less	More
Flexibility	More	Less
Tearing strength	More	Less
Frictional resistance	More	Less due to exo-skin

Table 4 Hydraulic properties

	Geo textiles	TGSB/coconut roots
Porosity	Less	More
Permeability	Less	More
Permittivity	Less	More
Transitivity	More	Less
Turbidity	More	Less
Filtration length	Less	More
Soil retention	Less	More

tree roots also need to be tested for similar characteristics. These characteristics of geotextiles are broadly classified as below (Tables 2, 3, 4, 5 and 6).

4.3 Functions and Uses

Today geotextiles are widely used in transportation infrastructure. They serve various functions. Coconut tree roots in TGSB also serve similar purposes. Table 7

Table 5 Degradation properties

	Geotextiles	TGSB/coconut roots
Biodegradation	None	5–10 years in dead trees
Hydrolytic degradation	Present in heavily acidic and basic soils	Moderate
Photodegradation	At surface reduces up to 1 m	None
Chemical degradation	Present in adverse soils	None
Mechanical degradation	Irreversible	Self-repairable
Attack of rodents/burrowing animals	Present for rubber and plastics	Moderate
Attack of termite	None	In temperate zones

Table 6 Endurance properties

	Geotextiles	TGSB/coconut roots
Elongation	More (2—10%)	Less (1–5%)
Abrasion resistance	Abraded textile is not replaceable	Tree roots develop hard scar tissue to combat damage
Flow	Less	More
Clogging length	Less	More

shows comparison of uses of synthetic and natural roots. Figure 8 shows functions of geotextiles and roots.

4.4 *Functions and Uses*

Today geotextiles are widely used in transportation infrastructure. They serve various functions. Coconut tree roots in TGSB also serve similar purposes. Table 7 shows comparison of uses of synthetic and natural roots. Figure 8 shows functions of geotextiles and roots

Table 7 Comparison of use of synthetic and natural geosynthetics

	Geo textiles	TGSB/coconut roots
Filtration	Underground leachates and water	Underground leachates and water
Drainage	Leachates and water	Leachates and water
Reinforcement	Road and rail embankments	Road embankments
Cushion	Wheel and other loads	Wheel and other loads
Separation	Present	None
Stabilizing	Cuts and hill slopes	Cuts and hill slopes
Soil strengthening	Weak soils	Weak soils
Waterproofing	From rain and for ponding	None
Liquefaction stability	Less	More
Seismic stability	Less	More

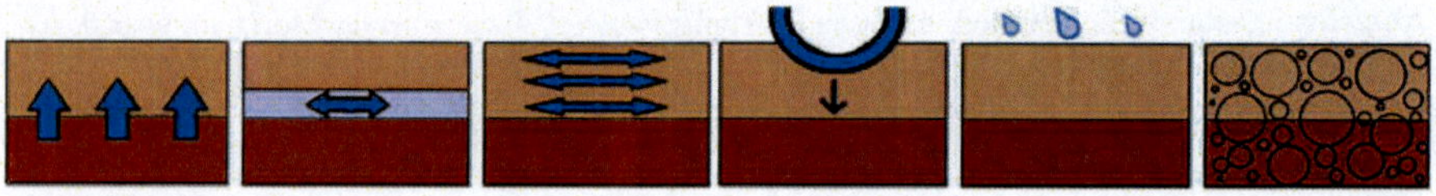

Fig. 8 Functions of geotextiles and roots: filtration, drainage, reinforcement, cushion, waterproofing, separation, liquefaction stability

4.5 *Comparisons of Coconut Tree Root Fibre with Other Fibres*

Coconut tree roots last longer than other plant fibres in the soil due to their high lignin, pectin and cellulose content (Table 8). Coconut tree roots are longer than other individual plant fibres and roots and they perform better. Table 9 compares the strength of coconut roots and other geotextiles.

4.6 *Role of Coconut Tree Roots in Improving Soil Properties*

The mechanical effect of coconut tree root systems in enhancing soil stability is based on three mechanisms (a) Top root mat in surface layers bind soil particle strongly,

Table 8 Comparison of chemical composition of plant fibres and coconut roots

Fibre	Cellulose	Hemi cellulose	Pectin	Lignin	Extractives	Fat as % dry weight
Cotton	91.8	6.3	–	–	1.1	0.7
Flax	71.2	18.5	2.0	2.2	4.3	1.6
Hemp	78.3	5.4	2.5	2.9	–	–
Jute	71.5	13.3	0.2	13.1	1.2	0.6
Coir	36.7	15.2	4.7	32.5	3.1	–
Sisal	73.1	13.3	0.9	11.0	1.3	0.3
Abaca	70.2	21.7	0.6	5.6	1.6	0.2
Coconut roots	37.6	15.4	5.1	32.7	3.0	–

Table 9 Comparison of coconut roots with other geotextiles

Fibres	Strength N/mm^2	Length m	Cross sections mm^2	Intensity/Density per m^2	Expected life
Jute mats	10–120	roll	5–10	1–3 layers	10
Coir mats	30–100	roll	5–10	1–3 layers	5–8
Paddy straw	10–20	1	5	1–3 layers	5
Vertivier grass	20–33	3–5	2	50–100	5–10
Plastic GT	500–800	roll	2–20	1–3 layers	100
Composite GT	1500	roll	2–20	1–3 layers	100
Coconut roots	100–200	5–10	20	50–300	100

increasing in shear strength of soil (c-ϕ), this is aided by the roots of local grasses that grow around the roots (b) The water uptake of roots enhance slope stability, (c) Deep tree root systems act as soil anchor.

4.7 *Mechanism of TGSB Soil Strengthening and Stabilization*

The effect of TGSB in enhancing soil strength and stability is based on three mechanisms: (a) Puzzolanic action of burnt shell lime and coconut leaf ash on soil, (b) the initial effect of paddy straw geofabric and (c) the final impact of coconut root strengthening.

4.8 Additional Advantages Over Geosynthetics

TGSB has several additional advantages vis-à-vis geosynthetics: Aesthetically pleasing enhances scenic beauty, reduces pollution in surrounding areas, handles higher water velocities, fibrous roots compact the soil without displacement, puncture-resistant and regenerates when damaged, encourage other plants and grasses growth, generates income during life, resists storms due to aerodynamic shape, can form a protective side barrier. Additionally, maintenance can be carried out by any semi-trained villager making them ideal for rural infrastructure especially in coastal and hilly regions where coconut trees grow easily.

Coconut tree roots can grow post facto unlike geosynthetics. Roots also get additional strength on strain hardening (scar tissue growth).

5 Conclusions

In the present study, the Saraswat method (TGSB) was compared to geotextiles. From the present study, the following conclusions are made:

1. Geosynthetics give higher short-term strength(max life less than 100 years) but are less sustainable than TGSB (max life more than 1000 s of years).
2. TGSB has equivalent if not better performance than geosynthetics on many functions.
3. Coconut tree root reinforced Transportation infrastructure is best suited especially for rural roads as with minimal repairs they last virtually forever.
4. Most significantly this technique reduces pollution while adding to scenic beauty without compromising on functionality and longevity.

The only disadvantage being that they need longer for full functionality (10 years for full root growth) than geosynthetics.

References

1. Wikipedia (2017) Goud Saraswat Bhramins, https://en.wikipedia.org/wiki/Goud_Saraswat_Bhramins
2. Ellis E, Yu HS, McDowel G, Dawson A, Thom N (2008) Advances in transportation geotechnics. In: Proceedings of the 1st international conference in transportation geotechnics, Nottingham, UK
3. Palmeira EM, Tatsuoka F, Bathurst RJ, Stevenson PE, Zornberg JG (2008) Advances in geosynthetics materials and applications for soil reinforcement and environmental protection works. Electron J Geotech Eng Bouquet 8:1–38
4. Fay L, Akin M, Shi X (2012) Cost-effective and sustainable road slope stabilization and erosion control. Transportation Research Board, Washington, D.C.
5. Mittal A, Shukla S (2015) Geotextile: an overview. In: Proceedings of Indian geotechnical conference IGC2015, Pune

6. De Jong JT, Martinez BC, Mortensen BM, Nelson DC, Waller JT, Weil MH, Ginn TR, Weathers T, Barkouki T, Fujita Y, Redden G, Hunt C, Major D, Tanyu B (2009) Upscaling of bio-mediated soil improvement. In: Hamza M et al (eds) Proceedings of the 17th international conference on soil mechanics and geotechnical engineering
7. Chou CW (2007) Bioimprovement of geotechnical properties of sandy soils. Graduate School of the University of Maryland, College Park, Thesis Master of Science
8. Gobinath R, Ganapathy GP, Akinwumi II (2015) Evaluating the use of lemon grass roots for the reinforcement of a landslideaffected soil from Nilgris district, Tamil Nadu, India. J Mater Environ Sci 6(10):2681–2687
9. Hengchaovanich D (2003) Vetiver system for slope stabilization. APT Consult Co., Ltd, Bangkok, Thailand, diti@samart.co.th
10. Truong P (2013) Review on the application of vetiver system for infrastructure protection. TVNI Veticon Consulting. www.veticon.com.au
11. Teerawattanasuk C, Maneecharoen J, Bergado DT, Voottipruex P, Lam LG (2015) Root strength measurements of vetiver and ruzi grasses. Lowland Technol Int 16(2):71–80
12. Souza L, Naik A, Savoikar P (2016) Stability of traditional 'bunds' - earthen levees—from Goa. In: Proceedings of Indian geotechnical conference, IGC 2016, IIT Madras
13. Souza L, Savoikar P (2017) Ground improvement using coconut leaf ash. In: Proceedings of Indian geotechnical conference, GEONEst 2017
14. DeJong TJ, Mortensenb BM, Martinezb BC, Nelson DC (2010) Bio-mediated soil improvement. Ecol Eng 36:197–210

A Study of Dynamic Traffic Assignment Using Statistical Method for Urban Scenario

K. Sushmitha Singh, Ch. Hardik, and C. Naveen Kumar

Abstract Rapid urban growth leads to a rise in demand for transport and private automobiles in urban regions. The current methods have failed in the current situations to meet the requirements leading to congestion, vehicle pollution and accidents. Vehicular congestion is a major problem in urban road networks. With traffic congestion on roads, the delay to the road users has been increased and the road networks reliability is decreasing. Before decade, four-stage modelling was used as transportation planning tool to evaluate transportation facilities on the system of transportation. But this kind of planning is not sufficient or not applicable to cover the dynamic properties for now existing situations and conditions. Hence, there is a need for new dynamic technique to be applied for the present condition of congestions. The major theme of this dynamic traffic assignment (DTA) is to control the traffic situations in networks through real-time measurement, communication and control. Dynamic traffic assignment was studied for more than four decades and many of the research works are been conducted but there is a lack in dynamic traffic assignment techniques research relating to Indian scenario especially for urban road networks. In this research, the survey is done through a questionnaire which is given to the VNR VJIET College of Engineering students to record their origin, destination, travel time and mode of transport. The origin is taken as JNTUH as all the trips are moving over JNTUH to reach the destination, there are two routes, i.e. one is along with Pragathi Nagar and another is along Nizampet route. The origin, destination, travel times and distance parameters are considered for the study. Travel times and the corresponding volume on the selected routes are taken and linear equations are built and user equilibrium technique is analysed for the obtained equations and the number of trips is calculated.

K. S. Singh · Ch. Hardik · C. N. Kumar (✉)
VNR VJIET, Hyderabad, India
e-mail: Naveenkumar_c@vnrvjiet.in

K. S. Singh
e-mail: Sushmitha.singh787@gmail.com

Ch. Hardik
e-mail: hardikch22@gmail.com

B. B. Das et al. (eds.), *Recent Developments in Sustainable Infrastructure*, Lecture Notes in Civil Engineering 75, https://doi.org/10.1007/978-981-15-4577-1_90

Keywords DTA · Urban road networks · Regression analysis · User equilibrium method

1 Introduction

Traffic congestion is when the vehicles on the road gets slower in their travel speed or stay in standby condition is termed as traffic congestion. This in general terms is also called as traffic jam. Traffic congestion may result in the condition like decrease in capacity levels of the road, due to the various conditions and reasons like occurrence of accidents and road diversions. The congestion is very common scenario in many cities and hence several methods are used to reduce it. Congestion levels are very higher in the many states in India especially like Kolkata, Delhi, Bangalore and Mumbai and it costs approximately to Rs 1.45 crores annually according to Global Consultancy Firm studies. In Hyderabad, the average traffic speed decreased from 27.1 to 18.5 kmph as per the traffic survey due to congestions on the roads. To examine the congestion difficulties and to mitigate them, there is a need for traffic assignment models. Traffic assignment is defined as the selection of route preference between the origin and destination and vice versa. It may be also called a route choice or route assignment.

The main objective of traffic assignment is (i) to estimate zone to zone travel cost (ii) to obtain the reasonable link flow to identify the highly congested links (iii) the available routes between the origin and destination. The inputs that are required are (i) number of routes available between the zones (ii) different routes between the zones (iii) travel time of each and every available route (iv) a decision rule or algorithm to solve the problem statement. The outputs obtained from the results of trip assignment are (i) inter-zonal trips by mode of transport (ii) to predict the resulting flow on each and every link.

Static assignment is the first model or type of traffic assignment which is a deterministic model based on relation of speed and density. Static assignment is to allocate the set of drivers with fixed origin and destination on the network of roads to obtain user equilibrium and system optimal solution. In this, trips are aggregated into time period by zone priority but are not simulated by individual. The advantages of static assignment are route choice capability, integration with travel demand models, computing very fast. The major disadvantages of static assignment are it provides no information regarding the analysis period and aggregate trip tables.

Dynamic traffic assignment is more aggregate, neglecting and great simplifying variations in behaviour and vehicle types. It can handle large networks and maintain route choice. It can simulate each vehicle continuously through time like micromodels. It can be deterministic and well as stochastic. DTA is a new technique used nowadays, it is of practical importance because the models have wide range of applications in the future. DTA is an active study aimed at capturing time-dependent phenomena such as tail spills, flashbacks and time congestions. DTA components of travel will depend on War drop's principles. DTA models can be used to generate forecast of traffic that describe the congestions levels will vary with time. The data required for static and for dynamic is almost same but only the difference is DTA models require path delay operators. There are two major types of DTA models that those deploy rule-based simulations and others which do not depend on simulations but instead depend on equations and inequalities. The simulation-based DTA models are very accurate and provide proper information and position of the vehicles with respect to the time. ITS and ATMIS are integrated with simulation-based modelling. This model helps in visualizing the obtained outputs. The available software for DTA modelling is DYNA SMART P, DYNAMEQ, TRANUS, AIMUSM and others, but this cannot be used because of the commercial proceedings for the simulations. The other type of DTA is based on analytical methods which uses the mathematical equations. The analytical method takes into account the macroscopic and microscopic conduct in terms of optimum conditions, and adhesives with the dynamic Wardrop's principle. The methods available are user equilibrium, system optimum assignment, Frank–Wolfe algorithm, gradient projection method, mathematical programming, variational inequality and, optimal control theory.

This research studies the macroscopic, mesoscopic assignments of the vehicles and an attempt is made to develop a model using user equilibrium, system optimum assignment and Frank–Wolfe algorithm. However, this model mentioned is majorly used to describe the traffic congestions. For this research, the survey is done through questionnaire regarding the origin and destination, their travel times and travel costs.

1.1 Objectives of the Study

i. To develop and validate DTA model for the urban road networks
ii. To estimate the effect of different parameters on the network with respect to time
iii. To calculate the number of trips through user equilibrium equations generated.

2 Literature Study

Few studies were made by researchers regarding the dynamic traffic assignment. Some of them are discussed below.

Azad Abdulhafedh (2017) developed a comparison between three models. The main model offers a harmony DTA to counter the substantial clogged urban system. The model uses Dyna MIT-P re-enactment-based DTA to get arrange execution estimation, for example, time subordinate streams, travel times promotion line length. The second model offers a method to find path marginal cost. DTA model that estimates the impact on travel times and travel routing by mesoscopic simulation on moving bottleneck and network performance.

Duell et al. [1] developed a model and this was the first large scale developed in Australia. A lot of information sources, including a strategic travel model of Sydney, a road network model, family travel surveys, the Sydney GTFS information and traffic volume count data have been compiled and synthesized in Australia. The research team in Sydney wanted to extend the deterministic dynamic traffic assignment model to account for violation of traffic flows.

Fusco et al. [2] involved the study of developing a QDTA model that deals with the time-dependent traffic and its demand and simulating traffic congestion on the urban road networks. This model allows to simulate several hours of traffic on large urban congested roads in very less time using standard computer. The model is capable of producing traffic situations which is compared with the more complex simulations-based DTA models.

Yang and Jayakrishnan [3] The suggested model for dynamic traffic assignments can be easily implemented to assess the strategic and progressive signal coordination scheme for transport plan and operational management as well as for investment projects. The proposed gradient projection algorithm uses the Hessian Matrix in conjunction with step sizer to obtain the best and faster convergence in the model.

Camille (2011) proposed a model named VISTA (visual interactive system for transportation algorithm) and gives the information about the incident presence and availability of alternate routes and also to improve or enhance the incident management and to reroute the traffic from the congested zones.

Zhengu (2008) described the multi-class analytic DTA model. The user class interdependence is modelled. The solution depends on bilevel modelling that is based on an iterative algorithm for optimization.

Shuguang (2003) discussed the average length of queues encountered by cars entering the connection at interval K and shall be calculated on a model based on kinematic wave theory. The tail size is calculated while the link is entered and the link is left.

Liu et al. [7] developed a DTA model with probabilistic time and perceptions, the perceptions considered are under dynamic and stochastic networks. The model considers the risk-taking behaviour of the travellers as stochastic. The solution algorithm shall be the combination of relaxation approach, stochastic network loading and methods of successive average.

Jayakrishnan et al. [8] studied about the modified Greenshield's equation and Bilevel programming. The travel time are not decreasing and convex with the density therefore the link cost function is derived from Greenshield derivations and the results obtained are accurate and can be applied to different networks.

3 Methodology

The program adopted for the research study is shown below (Fig. 1).

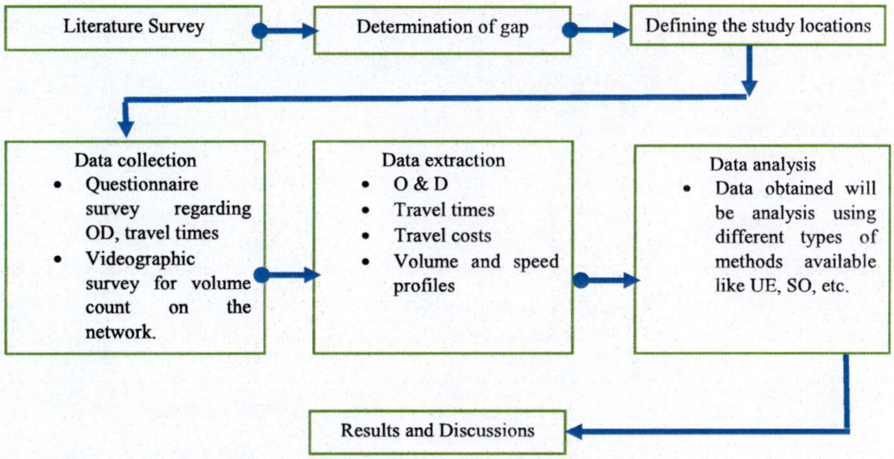

Fig. 1 Schematic approach of methodology of research work

4 Data Collection

Data is collected through the Questionnaire forms from the students of VNR College of Engineering regarding their trip details like origin, destination, travel time, travel mode and route preference. From the obtained data, the trips are categorized as per their localities, each and every trip is analysed from JNTUH to VNR VJIET in spite of any origin. The trips between JNTUH and VNR VJIET are distributed via Nizampet and Pragathi Nagar. The locality maps are below.

The above figure states that to reach VNR VJIET college of engineering, we are provided with two different routes, i.e. one is via Pragathi Nagar and Nizampet route. The above data reflects the travel times for different averaged traffic volumes. The travel times are obtained from questionnaire survey and the volume survey is done through videography to determine the average traffic volume. The actual 5-minute interval volume is estimated from videography survey which can be represented in excel sheet attached separately along with the paper (Figs. 2 and 3, Tables 1 and 2).

4.1 User Equilibrium Assignment

The user balance is based on the first principle of War drop, which defines that the driver does not unilaterally reduce travel costs by turning back to the alternating route. The O-D pair can be described as user balance (UE),

$$F_k(C_k - U) = 0, \tag{1}$$

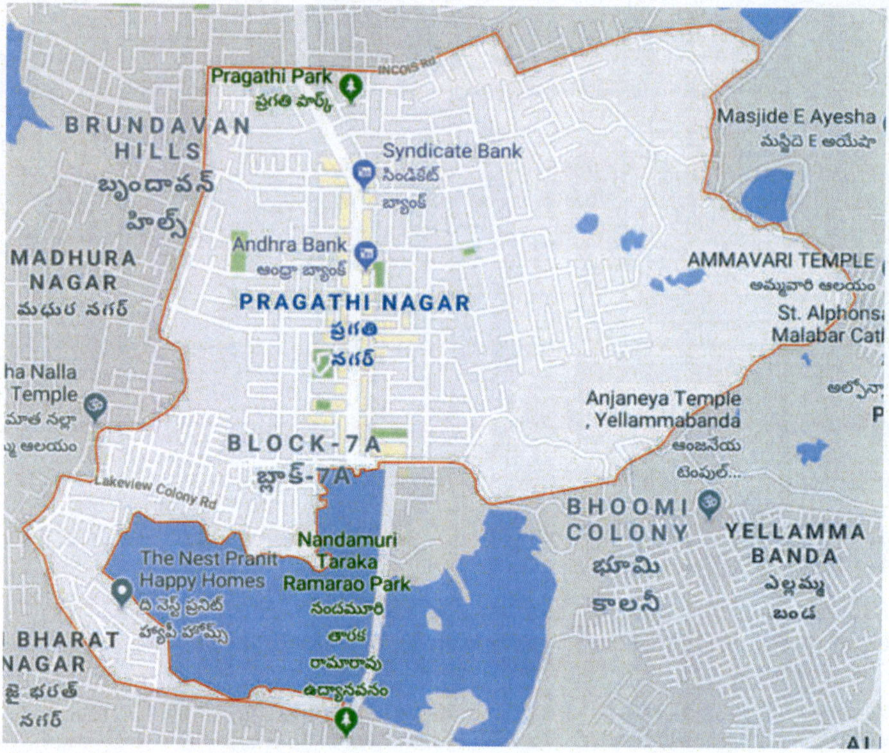

Fig. 2 Pragathi nagar location

$$C_k - U >= 0, \tag{2}$$

where F_k is the flow on path k, C_k is the travel cost on path k, U is the minimum cost, the equation has two states,

(i) If $C_k - U = 0$, from the Eq. 1 $F_k \geq 0$. This means that all used path will have the same travel times.
(ii) If $C_k - U \geq 0$, then from Eq. 1, $F_k = 0$.

This implies the every single utilized way will have travel times more prominent than least-cost ways, where F_k is the stream on the way k, ck is the movement cost on way k and u is the base expense.

4.1.1 Suppositions in User Equilibrium Assignment

(i) The user should have a good knowledge of the travel cost.
(ii) Travel time on a given connection is an element of the stream on that interface as it were.
(iii) Travel time capacities are sure and expanding.

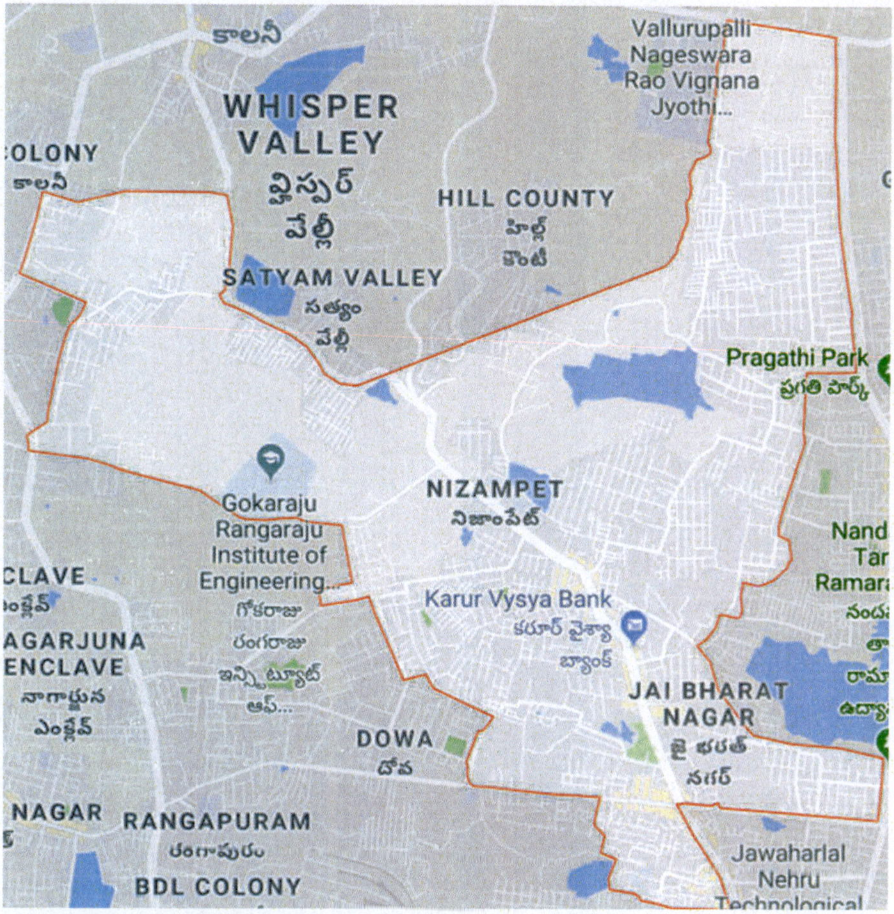

Fig. 3 Nizampet location

The solution for the above balance conditions is given by the arrangement of a proportional nonlinear numerical streamlining program.

$$\text{Minimize, } Z = \sum \int_{0}^{x_a} t_a(x_a)\,\mathrm{d}x \tag{3}$$

The above equations are flow conservation and non-negative limitations. The limitations keep the objective function to a minimum. The equations mentioned will represent the user equilibrium principle. The total can be divided into 2 types, i.e. one which carries the flow and other is which does not carry the flow along the path on which the travel time is greater than or equal to minimum O-D travel times. In the event that the stream design fulfils these conditions no driver can better off

Table 1 Volume data for Pragathi nagar route

Travel times	Averaged volume
10	392
13	574
15	656
17	704
18	704
20	881
23	921
25	1052
30	1224
32	1339
35	1445
40	1544

Table 2 Volume count for Nizampet route

Travel times	Averaged volume
10	786
15	1018
20	1342
25	1718
30	2082
35	2363

by singularly changing courses. All the available different routes may have same or higher travel times. The issue of user balance is convex, because the functions of travel time monotonically increase and the connection between time and flow is autonomous. Therefore, Frank–Wolfe technique can be used further to fix such a convex problem.

4.2 Mathematical Modelling

The statistical analysis like linear regression which explains the relationship between dependent and independent variables has been used to obtain the utility function of the different routes, i.e. Pragathi Nagar and Nizampet route. The obtained volume is considered as independent variable and travel times is considered as dependent variable. The graphs are plotted between volume and travel time, using this graph, a linear equation is obtained through regression analysis for two routes separately and the R^2 value is obtained for both the routes. Hence it is applicable to use the obtained equations. The obtained equations are replicated with user equilibrium equations.

5 Results and Discussions

From the utility equations obtained from linear regression, the further user equilibrium analysis is performed. The R^2 values of the utility equations obtained from the regression analysis are less than 1, hence the equations obtained can be suggested to use for the analysis (Figs. 4 and 5).

$$Y = 0.0153X - 1.0048 \tag{4}$$

$$Y = 0.0258X - 0.6275 \tag{5}$$

The points which make a linear trend line are only taken into consideration for the analysis.

The equations obtained are analysed with the user equilibrium model and the trips are obtained.

$$\text{Min } Z = \int_0^{x_1} 0.0153x_1 - 1.0048 + \int_0^{x_2} 0.0258x_2 - 0.6275 \tag{7}$$

$$.. \tag{8}$$

Subject to
$x_1 + x_2 = 127$. (The total number of trips to reach the destination are 127).

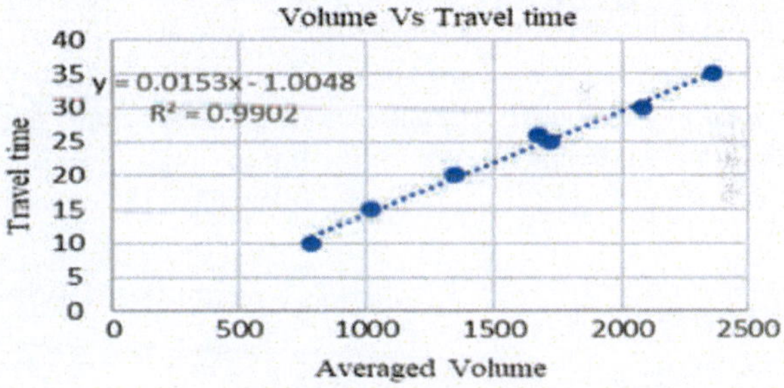

Fig. 4 The plot made between the travel time of Nizampet route and its corresponding volume

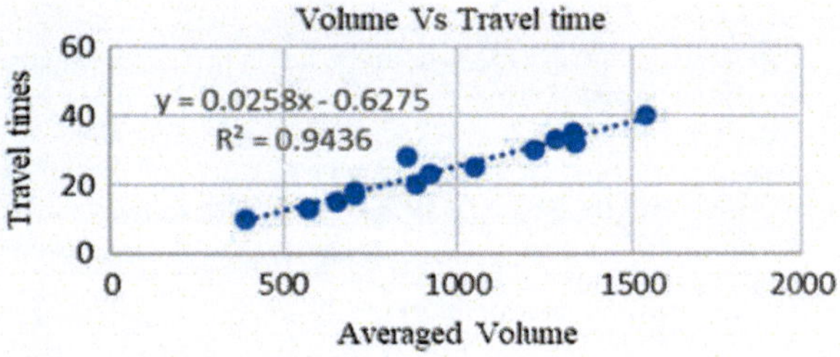

Fig. 5 The plot made between the travel time of Pragathi nagar route with corresponding volume

Substituting $x_2 = 127 - x$, the equation can be formulated as the following equation.

$$\text{Min Z} = 0.0153\frac{x_1^2}{2} - 1.0048x_1 + 0.0258\frac{(127 - x_1)^2}{2} - 0.6275(127 - x_1)^2 \quad (9)$$

Differentiate the above equation and equate to 0 and obtain the values and lead to the solution of $x_1 = 40$ and $x_2 = 87$.

6 Conclusions

The obtained trips from the questionnaire survey are 127 trips from different origins to VNRVJIET among which—trips are made via Pragathi Nagar and—trips are made via Nizampet. But as per the user equilibrium analysis, the obtained results are 40 trips via Pragathi Nagar and 87 trips via Nizampet. The obtained values are quite opposite to that of the real-time trips because the drivers are habituated to the Pragathi Nagar route, as it is the shortest among both the routes even though it has few difficulties. But in reality, by moving to another path, no driver can unilaterally cut travel expenses.

The limitation of this approach is that it assumes that the travel time on the path is the function of the flow on that path but perfectly it is not the scenario because travel time depends on many other factors like number of access points, road conditions, geometric parameter and many more. We assumed that the user has perfect knowledge of the path cost but that might not be real at all the cases as there may be a new user using the road, who does not know about the path cost. Further research can be done by taking few other parameters and further analysis can be done through system optimal and Frank–Wolfe algorithm to obtain the solution.

References

1. Duell M, Amini N, Chand S, Saxena N, Grzybowska H, Waller T (2015) Deploying a dynamic traffic assignment model for the Sydney region. In: Australasian Transport Research Forum (ATRF)
2. Fusco G, Colombaroni C, Gemma A, Sardo SL (2013) A quasi-dynamic traffic assignment model for large congested urban road networks. Int J Math Models Methods Appl Sci 7(4):341–349
3. Yang I, Jayakrishnan R (2012) Gradient projection method for simulation-based dynamic traffic assignment. Transp Res Rec 2284(1):70–80
4. Kamga CN, Mouskos KC, Paaswell RE (2011) A methodology to estimate travel time using dynamic traffic assignment (DTA) under incident conditions. Transp Res Part C Emerg Technol 19(6):1215–1224

5. Wang Z, Huang Z, Luo D (2008) A formulation and solution algorithm for multiple user classes dynamic traffic assignment. In: 2008 7th World congress on intelligent control and automation, pp 3537–3541. IEEE
6. Li S (2003) Dynamic link travel time model in dynamic traffic assignment. In: Proceedings of the 2003 IEEE international conference on intelligent transportation systems, vol 2, pp 1036–1039. IEEE
7. Liu HX, Ban X, Ran B, Mirchandani P (2002) Analytical dynamic traffic assignment model with probabilistic travel times and perceptions. Transp Res Rec 1783(1):125–133
8. Jayakrishnan R, Tsai WK, Chen A (1995) A dynamic traffic assignment model with traffic-flow relationships. Transp Res Part C Emerg Technol 3(1):51–72

A Study on Replacement of Coarse Aggregate with Recycled Concrete Aggregate (RCA) in Road Construction

Bhagyashree Panda, Nazia T. Imran, and Kundan Samal

Abstract Growth in population is influencing the construction activities in the recent times leading to higher demand for various construction materials. This has affected the environment in an adverse way due to the exploitation of depleting natural resources. Aggregate prepared by quarrying rocks and then crushed to various suitable sizes is one such material which is heavily used in construction industry. This procedure leads to serious occupational hazards and also causes disturbance in the ecosystem such as deforestation, air and noise pollution, necessitating the finding of substitute materials for natural aggregates. On contrary, millions of tonnes of concrete are produced every year as wastage due to dismantling of existing structures which are unconsciously dumped or used for filling lower areas. Since concrete acts like solid hardened material as rock, its suitability for use as an aggregate can be explored. In the present study, quarried coarse aggregates, used in both flexible and rigid pavement construction, are replaced with recycled concrete aggregate (RCA). Different strength tests such as toughness and hardness test of RCA along with compressive strength test of cube and tensile strength test of cylinder with 0, 25, 50, 75 and 100% RCA replacement have been performed and the results compared with natural aggregates and with IRC and IS codes. The findings showed that RCA and natural aggregates if mixed in a proportionate ratio can be used in both pavement and building construction.

Keywords Coarse aggregate · Flexible pavement · Recycled concrete aggregate · Rigid pavement · UCS

1 Introduction

Presently hardened concrete waste is increasing all over the world at an alarming rate for many reasons, mostly due to demolition or renovation of existing reinforced concrete or plain cement concrete structures and rigid pavements. Investigations by the Environmental Resources Ltd. (1979) for European Environmental Commission

B. Panda (✉) · N. T. Imran · K. Samal
School of Civil Engineering, KIIT Deemed to Be University, Bhubaneswar, India
e-mail: bhagyashree.pandafce@kiit.ac.in

B. B. Das et al. (eds.), *Recent Developments in Sustainable Infrastructure*, Lecture Notes in Civil Engineering 75, https://doi.org/10.1007/978-981-15-4577-1_91

(EEC) predict that there will be huge increase in construction and demolition concrete waste to 302 million tonnes by the year 2020 from 55 million tonnes in the year 1980, together in the EEC member countries. According to a report in 2008 by Ministry of Environment and Forest (MoEF), with rapid urbanisation and increasing construction activities, India too is generating 0.53 million tonnes/day of construction and demolition waste, with its capital Delhi producing around 5,000 tonnes/day of concrete waste by building, demolition and land-clearing activities (The Hindu, August 29, 2014). Concrete rubble or hardened concrete particles is disposed off to landfills leading to severe adverse environmental impacts as it is not further used in construction and the options of recycling are still meagre. However concrete recycling benefits are now slowly being realised in this age of greater environmental awareness, depleting natural sources of aggregates, more environmental laws, and to keep construction costs down. Moreover the hardened concrete wastes produced are of irregular shapes and sizes which need further processing before being used again in building construction. On contrary, these concrete wastes can be conveniently recycled and used in road construction in place of natural aggregates in various layers of flexible pavement and as coarse aggregates in the construction of rigid pavement, usually termed as recycled concrete aggregate (RCA). The potential use of RCA lies in the fact that aggregates constitutes 75% of concrete and construction as well as demolition wastes consists of 23–35% proportion of concrete followed by 31.5–43% of soil (Municipal Corporation Delhi, 2004 and IL&FS ECOSMART, 2005). RCA can be formed by simply crushing dismantled concrete into smaller sizes, usually in the range of 20–70 mm and can be explored as a substitute of natural coarse aggregates provided the mechanical properties are satisfied. In the present study, RCA is produced by collecting, thoroughly cleaning and crushing waste concrete samples and sieving to get the requisite sizes to meet the requirements so as to replace natural coarse aggregates. Different strength tests such as compressive and tensile strength as well as slump test were conducted on concrete samples formed using varied percentage of coarse aggregate replacement with RCA and compared so as to test the suitability of RCA as aggregate in rigid pavement. Toughness test and hardness test are also conducted using RCA to find its suitability to be used in both flexible and rigid pavement construction (Fig. 1).

Fig. 1 Natural concrete aggregate and RCA

2 Literature Review

Wen et al. [1] evaluated the effects of using recycled aggregate in new concrete pavement and found that replacement of natural coarse aggregate with RCA till 45 percent by volume showed negligible effects on concrete properties but suggested that for new cement concrete pavement construction RCA can be productively replaced with a part of natural coarse aggregates in Washington State. It was also stressed upon the use of high-quality RCA being substituted for natural coarse aggregates. Adnan et al. [2] calculated the performance enhancement of RCA in concrete mix using micronized biomass silica (MBS) which is a pozzolanic material formed from the controlled combustion of rice husk. Results showed that concrete samples at the age of 7, 14, 28 and 90 days containing RCA along with MBS attained higher compressive strength and improved the concrete water permeability. Wen et al. [3] studied the benefits and limitations of using RCA in concrete. It was noticed that RCA decreased the elastic modulus and compressive strength of concrete as compared to natural aggregate whereas increased the drying shrinkage and creep and suggested to incorporate some fly ash in concrete mixture to overcome these limitations. Marinković et al. [4] validated the eco-efficiency of RCA by conducting the environmental assessment of concrete prepared with natural gravel and RCA by analysing the impact of the transport phase and carbon dioxide intake all through the life cycle of concrete structures. Soares et al. [5] evaluated the utilisation of RCA as coarse aggregates with a range of replacement ratios in concrete mixes and then compared with the same compositions of reference concrete mix made up of natural aggregates and superplasticizer. Results verified that concrete mixes containing RCA have an similar mechanical and durability properties as compared to reference concrete. Mukherjee and Barai [6] established that complete substitution of coarse aggregates with RCA and nano-silica showed significant effects on the compressive strength and Interfacial Transition Zone characteristics of concrete mixes prepared using natural aggregates and RCA, with and without nano-silica. Khoshkenari et al. [7] investigated the effect on compressive and splitting tensile strengths of 0–2 mm recycled fine aggregates obtained from normal and high strength concrete. Results showed that fine RCAs from a 30 MPa parent concrete mix, have 3.5 times higher water absorption but density 20% less, hence the density of RCA concrete was about 8–13.5% less than normal aggregate concrete. Pedro et al. [8] experimented on total replacement of coarse natural aggregates by coarse RCA from different sources producing concrete mix with a pre-established mechanical performance of three strength ranges: 15–25, 35–45 and 65–75 MPa. The mixes produced using RCA tried to reproduce the strength of the source concrete. Guerra et al. [9] analysed the effect of various replacement ratios of recycled concrete coarse aggregates on the anchorage strength of ribbed steel bars in concrete by adopting effective w/c ratio of 0.53 and a 125 $\pm$ 10 mm slump keeping other parameters constant. Uygunoğlu et al. [10] investigated the use of recycled aggregate and marble waste in producing of structural cement concrete with different w/c ratios obtained from crushed concrete waste indicated good results in slump-flow, J-ring test, etc. as compared with control series using crushed limestone aggregate.

3 Materials and Methodology

3.1 Materials

Cement exhibit different characteristics depending on the raw materials used, chemical and oxide compositions, additives and fineness resulting in several types of cement being available. The cement used in the experiments is ordinary Portland cement (OPC) of grade 43 in compliance to IS: 8112–1989. The results for physical and chemical properties of the cement were obtained experimentally by conducting suitable tests as per IS 4031:1988 and IS 4032: 1985 are presented in Tables 1 and 2.

Aggregates constitute the bulk of concrete, about 70–80% of volume of concrete, reduce shrinkage, effects economy and have considerable effect on various concrete mix characteristics and its properties. Aggregates were previously thought as inert material chemically but at present some of the aggregate types are known to be chemically active and forms chemical bond at the interface of cement paste and aggregate. Crushed granite of size between 12.5 mm and 10 mm are used for impact and abrasion test of coarse aggregates and 20 mm as the coarse aggregate for strength tests of concrete. The sieve analysis of aggregates is conducted as per the specifications given in IS 2386 (Part 3): 1963. Fine aggregate used in the experiments fulfilled the requisite properties and conforms to the zone as per the specification of IS: 383-1970. The physical properties of fine and coarse aggregates obtained experimentally are specified in Tables 3 and 4, respectively.

Clean potable water for mixing and curing of concrete was obtained as per IS 456: 2000.

Table 1 Physical properties of OPC grade 43

Sl. no	Property	Experimental results	Prerequisite as per IS 8112: 1989
1. 2. 3. 4.	Fineness in m^2/kg • Initial Setting time in minutes • Final Setting time in minutes Soundness • Lechatiler expansion in mm • Autoclave in percentage Compressive strength in Mpa • 72 + 1 h • 168 + 2 h • 672 + 4 h	310 170 235 1.50 0.012 30.26 40.55 53	Minimum 225 Minimum 30 Maximum 600 Maximum 10 Maximum 0.8 Minimum 23.0 Minimum 33.0 Minimum 43.0

Table 2 Chemical properties of OPC grade 43

Sl. No	Properties	Results	Prerequisite as per IS 8112: 1989
1	Lime (percentage by mass)	60.8	6 (maximum)
2	Soluble silica (percentage by mass)	20.6	3 (maximum)
3	Alumina (percentage by mass)	5.3	4 (maximum)
4	Iron oxide (percentage by mass)	4.0	5 (maximum)
5	Magnesia (percentage by mass)	0.8	0.05 (maximum)
6	Sulphuric anhydride (percentage by mass)	1.8	0.66–1.02
7	Loss on ignition (percentage by mass)	3.1	0.66 (minimum)
8	Insoluble reside (percentage by mass)	2.9	
9	Chlorides (percentage by mass)	0.018	
10	Lime saturation factor	0.9	
11	Al_2O_3/Fe_2O_3	1.4	

Table 3 Physical properties of fine and coarse aggregates

Sl. No	Properties	Test results for coarse aggregates	Test results for fine aggregates
1	Specific gravity	2.7	2.65
2	Fineness modulus	6.2	2.47
3	Water absorption	0.4%	0.85%
4	Free surface moisture	–	0.90%

Table 4 Slump value of concrete for different proportions of RCA

Sample	% replacement				
	0%	25%	50%	75%	100%
Cube slump (cm)	25	28	28	28	29
Cylinder slump (cm)	26	27	27	28	29

Recycled concrete aggregates (RCA) used for the study resulted from the processing of inorganic material from dismantled concrete mostly consisting of crushed concrete samples of M25 and M30 grade, disposed off as waste from structural laboratory in School of Civil Engineering, KIIT Deemed to be University which were then washed, cleaned and graded for use as an coarse aggregate replacement in producing various proportions of fresh concrete mix, primarily to check its suitability to be used as road aggregates.

3.2 *Methodology*

Hardness and toughness properties of coarse aggregates used in road construction are closely related and are associated with mechanical degradation, so they are often considered simultaneously. Hence abrasion resistance and impact test were conducted separately on RCA confirming to IS 2386 (Part 4 and Part 5):1963 (Reaffirmed 2002)

3.2.1 Tests Performed on Aggregates

Aggregate Abrasion Test: Although three different tests can be done to characterise aggregate abrasion resistance, Los Angeles abrasion test is accepted and widely used. In this study, 5000 g of RCA and natural aggregate samples are separately tested in LA abrasion testing machine pertaining to Grade B, taking 11 numbers of abrasive charges (4584 ± 25 gm) and revolutions of 500 at a rate of 30 to 34 revolutions/minute. Percentage loss of the material by mass due to combined effect of abrasion, impact and crushing with respect to total weight of material taken was recorded by sieving through 1.70 mm IS sieve.

Aggregate Impact Test: The objective of impact test is to measure the aggregate ability to resist impact loading which is often one of the determining factors in the suitability and service life of a designated material for a particular application. Impact resistance can be quantified by two methods, dynamic tear testing and drop weight testing. In this study, we have used the drop weight testing method to ascertain the impact resistance of RCA and natural aggregates separately by taking oven-dried aggregate sample in IS sieve of 12.60 mm passing and 10 mm retained, tamped 25 times and then subjected to 15 blows by a 13.5 to 14 kg weight hammer, dropping from a free fall height (380 ± 5 mm). The percentage by weight of crushed aggregate passing through 2.36 mm IS sieve to the total sample weight called as aggregate impact value (AIV).

3.2.2 Casting of Concrete Cube and Cylinder

Batching: Aggregates are thoroughly cleaned, after which batching was done for 7 and 28 days cube and 28 days cylinder specimen with 0, 25, 50, 75 and 100% replacement by weight of natural aggregates with RCA for the present study were. For each proportion and type of specimen, three control specimens were tested for compressive strength and splitting tensile strength.

Mixing and Filling up of Concrete in Moulds: Past the batching process, water was added as per the required water-cement ratio and mixing was carried out by shovel until colour uniformity was achieved and a homogeneous concrete mixture was formed. Surplus pouring of water was avoided. The concrete mix was filled up in the moulds in three layers with 25 number of tamping in each layer to fill the voids within the moulds and achieve minimum compaction of concrete.

Compaction: After preparation of the concrete mould, for full compaction of the concrete and removal of any air voids, it was kept in the vibrating table to avoid excess water loss; the mould was then covered with plastic sheet. Past 24 h, the hardened concrete samples were demoulded and then cured till the age of testing, i.e. 7 days and 28 days by submerging in a clean water bath, further which was tested for compressive and tensile strength.

3.2.3 Tests Performed on Concrete

Slump Test: The slump cone test as per IS: 1199: 1959 (Reaffirmed 2004) was carried out for each mix proportion of RCA and natural aggregates in concrete with replacement of RCA by 0, 25, 50, 75 and 100%. It is used to measure the workability of concrete which completely depends upon water-cement ratio.

Compressive strength test: The compressive strength test was conducted according to IS: 4031 (Part 6): 1988 (Reaffirmed 2005) to find out the characteristic compressive strength of the concrete using varying proportions of RCA and natural aggregates. Three specimens of each type of concrete mix were prepared in standard cube mould of 150 mm and were placed in the compression testing machine such that the load was gradually increased at a rate of 140 kg/cm^2/min to the opposite sides of the cube cast. The maximum load at which the specimens fail was recorded. Any unusual feature in the failure type was also noted. The final compressive strength is the average of all the three observations at the curing age of 7 and 28 days.

Tensile strength test: Split cylinder test as per IS 5816: 1999 (Reaffirmed 2004) was conducted to determine the tensile strength of concrete on each concrete sample prepared by using varying proportions of RCA and natural aggregates. The test was conducted after curing of 7 days to find the tensile strength of the cylindrical sample, loaded to failure continuously at a rate of around 14–21 kg/cm^2/min (total load of 9900–14850 kg/minute approximately), in diametrical compression applied along the entire length. The breaking load was noted down.

4 Results and Discussion

Aggregate Abrasion test: For RCA, the abrasion value was found to be 45.34% and for natural aggregates, it was calculated to be 28.9%. Although the abrasion value of RCA was more than natural aggregates but according to IRC, a maximum value of 50% is allowed for bituminous macadam (BM) and water bound macadam (WBM) base courses and 60% for WBM sub-base courses in Indian conditions. A maximum value of 35% is specified for bituminous concrete. Therefore RCA can be used in pavement sub-base and base courses in place of natural coarse aggregates. It can be mixed and used proportionately along with natural coarse aggregates for bituminous concrete and rigid pavements.

Aggregate Impact test: For RCA, the aggregate impact value (AIV) was observed to be 30.263% and for natural aggregates, it was calculated to be 23.065%. It is evident from the above results that there is not much difference in impact value of RCA and natural aggregates both the material have satisfactory impact resistance. According to IRC, for bituminous macadam and WBM base courses, 35 and 40% is the maximum permissible AIV value, respectively. Also for rigid pavements, maximum permissible AIV value of coarse aggregates for wearing surface is 30% and other than wearing surface is 45%. So RCA can be productively used in construction of flexible and rigid pavement in place of natural coarse aggregates.

Slump Value: Slump value was obtained from slump cone test of concrete with varying proportions of RCA and natural aggregates, i.e. for 0, 25, 50, 75 and 100% replacement of RCA is given in the Table 5.

It was noticed that the less concrete slump is obtained for sample having 0% RCA as 25 cm, while that sample having 100% RCA replacement was found to be maximum, i.e. 29 cm. The result showed that concrete slump value made completely with natural coarse aggregates was higher than concrete mix with 100% replacement with RCA as coarse aggregates. This was due to more water absorption by RCA as was observed during the mixing process.

Compressive strength of concrete cube: The table below gives the experimental results of compressive strength of concrete cubes with varying proportions of RCA and natural aggregates for 7 and 28 days curing age, which is the average of 3 observations (Table 6).

From the results, the compressive strength was found to increase slightly with the addition of RCA, giving maximum compressive strength of 24 and 35 Mpa of concrete cubes prepared with 50% replacement of RCA for 7 days and 28 days curing period simultaneously. Also, it was noticed that the compressive strength of concrete with 50% RCA replacement was close to that of the control concrete (0% replacement), i.e. concrete made using natural coarse aggregates completely.

Tensile strength of concrete cylinder: Split cylinder test was performed on each concrete sample to determine the tensile strength of concrete cylinder with varying proportions of RCA and natural aggregates. The table below gives the result for tensile strength of cylinder cured for 28 days.

Table 5 Compressive strength of cubes with different proportions of RCA

Sample	0%	25%	50%	75%	100%
7 days cube strength (Mpa)	22.5	21.7	24	18	21.4
28 days cube strength (Mpa)	33.4	33.6	35	27.1	31

Table 6 Cylinder strength of concrete of different proportions of RCA

Sample	0%	25%	50%	75%	100%
28 days cylinder strength (MPa)	3.83	3.85	3.91	3.67	3.51

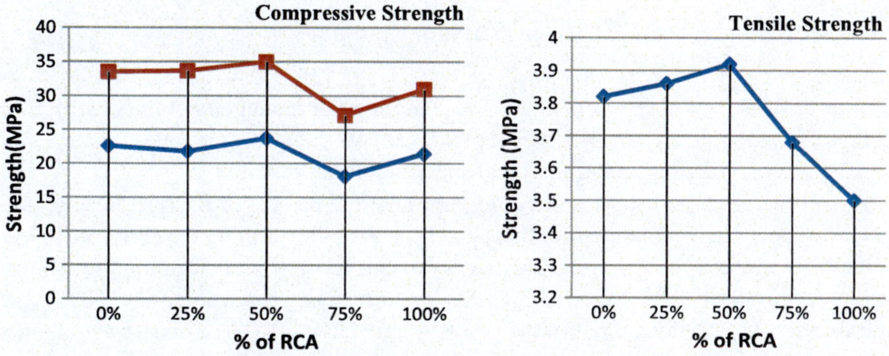

Fig. 2 Graph showing the 7 and 28 days compressive strength of concrete cube and 28 days tensile strength of concrete cylinder with varying % of RCA

From the above results, it was clearly seen that the concrete cylinder tensile strength also showed the same trend as like compressive strength and is maximum to 3.91 Mpa with 50% replacement of RCA. Also the tensile strength of concrete with 50% RCA replacement of RCA was close to that of the control concrete (0% replacement) (Fig. 2).

5 Conclusion

Based on the present study it can be concluded that RCA can be suitably used as a replacement of natural coarse aggregates in concrete for the construction of both building and pavements. The best results were found when the natural coarse aggregates were replaced by 50% of RCA in concrete mix. From the above experimental results, the following conclusions can be made:

- The target compressive strength of concrete can be reached at 20–40% of RCA replacement. For high strength concrete (compressive strength up to 40 MPa) can be achieved by lessening the w/c ratio and adjusting the admixture content.
- The observations indicated that 50% RCA replacement with coarse aggregates gives satisfactory mechanical properties of concrete, namely, compressive strength and tensile strength which can be used to produce high strength concrete mixes for rigid pavement construction.
- Similarly, the hardness and toughness test results also complied that RCA can be effectively used as base and sub-base course material in flexible pavement construction and wearing surface material in concreting in rigid pavement construction in substitution of natural coarse aggregates.
- Also it was observed that with increase in % of RCA replacement with less water content results in low workability due to their high water absorption capacity. So the water content in the concrete mix was continuously monitored.

6 Limitations and Future Work

In future, RCA will play a very crucial role in material management in construction industry making it a cost-effective replacement of natural aggregates, majorly in pavement and building construction because of large quantity of concrete waste. Although research and studies are going on the effective use of RCA, however, due to lack of specifications and guidelines and high water absorption capacity, RCA has some limitations in use as well. Further work can be done to determine the flexure strength, durability and Marshal stability on different proportions of RCA. Also suitable admixtures can be put in the concrete mix having RCA as coarse aggregates to decrease the water absorption and increase workability and thus quality of concrete.

References

1. Wen H, McLean DI, Willoughby K (2015) Evaluation of recycled concrete as aggregates in new concrete pavements. Transp Res Rec J Transp Res Board 2508, 73–78
2. Adnan S.H., Rahman I.A., Loon L.Y (2010) Performance of recycled aggregate concrete containing micronized biomass silica. Int J Sustain Construct Eng Technol 1(2)
3. Wen H, McLean D, Boyle S, Spry T, Mjelde D (2014) Evaluation of recycled concrete as aggregate in new concrete pavement. In: WSDOT Research Project T4120–3, evaluation of recycled concrete, WA-RD 826.1
4. Marinković SB, Malešev M, Ignjatović I (2014) 11 Life cycle assessment (LCA) of concrete made using recycled concrete or natural aggregates. In: Eco-Efficient construction and building materials, pp 239–266
5. Soares D, Brito J, de Ferreira J, Pacheco J (2014) Use of coarse recycled aggregates from precast concrete rejects: Mechanical and durability performance. Construct Build Mater 71:263–272
6. Mukharjee BB, Barai SV (2014) Influence of incorporation of nano-silica and recycled aggregates on compressive strength and microstructure of concrete. Construct Build Mater 71:570–578
7. Khoshkenari AG, Shafigh P, Moghimi M, Mahmud HB (2014) The role of 0–2 mm fine recycled concrete aggregate on the compressive and splitting tensile strengths of recycled concrete aggregate concrete. Mater Des 64:345–354
8. Pedro D, Brito J, de Evangelista L (2014) Influence of the use of recycled concrete aggregates from different sources on structural concrete. Construct Build Mater 71:141–151
9. Guerra M, Ceia F, Brito J, de Júlio E (2014) Anchorage of steel rebars to recycled aggregates concrete. Construct Build Mater 72:113–123
10. Uygunoğlu T, Topçu IB, Çelik AG (2014) Use of waste marble and recycled aggregates in self-compacting concrete for environmental sustainability. J Clean Product 84:691–700